Petroleum Engineering Handbook

Petroleum Engineering Handbook

Larry W. Lake, Editor-in-Chief

I **General Engineering** *John R. Fanchi, Editor*
II **Drilling Engineering** *Robert F. Mitchell, Editor*
III **Facilities and Construction Engineering** *Kenneth E. Arnold, Editor*
IV **Production Operations Engineering** *Joe Dunn Clegg, Editor*
V **Reservoir Engineering and Petrophysics** *Edward D. Holstein, Editor*
VI **Emerging and Peripheral Technologies** *H.R. Warner Jr., Editor*
VII **Indexes and Standards**

Petroleum Engineering Handbook

Larry W. Lake, Editor-in-Chief
U. of Texas at Austin

Volume V - Part B

Reservoir Engineering and Petrophysics

Edward D. Holstein, Editor
Consultant

Society of Petroleum Engineers

Printed in the United States of America

ISBN 978-1-959025-56-6 (print)
ISBN 978-1-55563-335-6 (digital)

Society of Petroleum Engineers
222 Palisades Creek Drive
Richardson, TX 75080-2040 USA

http://store.spe.org/
service@spe.org
1.972.952.9393

Contents

Chapter 9
Oil Reservoir Primary Drive Mechanisms

Mark P. Walsh, Consultant

9.1 Introduction

Muskat defines primary recovery as the production period "beginning with the initial field discovery and continuing until the original energy sources for oil expulsion are no longer alone able to sustain profitable producing rates."[1] Primary recovery is also sometimes referred to as pressure depletion because it necessarily involves the decline of the reservoir pressure. Primary recovery should be distinguished clearly from secondary recovery. Muskat defines secondary recovery as "the injection of (fluids) after the reservoir has reached a state of substantially complete depletion of its initial content of energy available for (fluid) expulsion or where the production rates have approached the limits of profitable operation."[1] One of the most popular secondary-recovery methods is waterflooding. Because primary recovery invariably results in pressure depletion, secondary recovery requires "repressuring" or increasing the reservoir pressure.

Primary recovery includes pressure-maintenance methods. Muskat defines pressure maintenance as "the operation of (fluid) injection into a reservoir during the course of its primary-production history."[1] The main effect of pressure maintenance is to mitigate the reservoir's pressure decline and conserve its energy. The purpose of pressure maintenance is ultimately to improve oil recovery. The most common injected fluids for pressure maintenance are water and separator or residue gas. "Partial" and "complete" pressure maintenance describe the general effectiveness of a given pressure-maintenance operation to retard the rate of pressure decline. Partial pressure maintenance refers to fluid injection while a general state of pressure decline still exists. Full or complete pressure maintenance refers to fluid injection while the reservoir pressure remains essentially constant.

According to Muskat's definition of pressure maintenance, secondary-recovery methods such as waterflooding are not strictly pressure-maintenance operations because they begin after pressure depletion. However, if water injection takes place before the end of pressure depletion, which is not uncommon, it is considered a pressure-maintenance method. If water is injected before the end of primary recovery, the reservoir is classified as an artificial waterdrive. Since Muskat first proposed his definition, others have loosely applied the term pressure maintenance to include any fluid-injection strategy at any stage in the reservoir's production.

Oil reservoirs are classified according to their fluid type. There are three broad oil classes. In order of increasing molecular weight, they are volatile oil, black oil, and heavy oil. Heavy-oil reservoirs are of minor interest during pressure depletion because they typically yield only marginal amounts of oil because of their low dissolved-gas contents and high fluid viscosities. Although heavy-oil reservoirs are not addressed specifically in this chapter, the methods of analysis presented here are equally applicable to them. The distinguishing characteristic between volatile and black oils is the stock-tank-oil content of their equilibrium gases. Equilibrium gases liberated from volatile oils contain appreciable stock-tank or condensable liquids whereas the gases from black oils contain negligible stock-tank liquids. While this distinction leads to only slightly different recovery strategies, it leads to very different methods of analysis and mathematical modeling requirements.

9.2 Volatile- and Black-Oil Fluid Characteristics

The petroleum fluid spectrum is gradational. There is no strict definition of volatile and black oils; there are only general guidelines and characteristics. Despite this lack of precision and the occasional confusion it brings, classification is quite useful and popular.

Molecular weight is a useful yardstick. Black oils typically range from 70 to 150 in molecular weight but may range as high as 190 to 210. In contrast, volatile oils are lower in molecular weight than black oils and typically range from 43 to 70. Oils with molecular weights greater than 210 usually are classified as heavy oils. Fluids with molecular weights of less than 43 are generally gases, which include gas condensates, wet gases, and dry gases. A molecular weight of 43 marks the lower molecular-weight limit of volatile oils.

Black and volatile oils are sometimes subdivided into different fluid types. For instance, volatile oils include near-critical fluids and high-shrinkage oils. Near-critical fluids represent light volatile oils and can include some very rich condensates. High-shrinkage oils represent the high-molecular-weight end of volatile oils and can include some light black oils.

Volatile and black oils are characterized in terms of a number of different properties. **Table 9.1** summarizes their characteristics. This table includes the properties of the full range of petroleum fluids, including gases.

The defining property that distinguishes black and volatile oils is the volatilized-oil content of their equilibrium gases. The volatilized-oil content of a gas represents its condensable liquid portion. Condensable refers to the portion that condenses or "drops out" during pressure reduction and ultimately results as stock-tank liquid. Condensation may take place within the reservoir as the gas passes through the lease separators. Physically, intermediate-hydrocarbon components, typically C_2 through C_7, dominate this fraction. Volatilized oil also is called lease condensate or distillate. Gas condensates and wet gases also contain volatilized oil. Volatilized oil is reported conventionally as part of the crude-oil reserves and production. It should not be confused with and is distinctly different from natural-gas liquids. Natural-gas liquids are derived from the gas-processing plant and are called plant products.

The volatilized-oil content of gases is quantified in terms of their volatilized-oil/gas ratio, typically expressed in units of STB/MMscf or stock-tank m^3 per std m^3 of separator gas. The volatilized-oil/gas ratio of equilibrium gases of black oils is usually less than 1 to 10 STB/MMscf (approximately 0.04 to 0.4 gal/Mscf).The volatilized-oil content of these gases is so low that it usually is ignored. In contrast, the volatilized-oil content of gases from volatile oils is much greater. Their volatilized-oil/gas ratio typically ranges from 10 to 300 STB/MMscf or 0.4 to 8 gal/Mscf.

Several benchmark properties can be correlated with the reservoir fluid's initial molecular weight. **Fig. 9.1** plots the initial formation volume factor (FVF) and initial dissolved gas/oil ratio (GOR) as a function of reservoir-fluid molecular weight for 36 reservoir fluids. The abscissa in Fig. 9.1 spans from a molecular weight of 15 to 180. This range of molecular weights covers the full spectrum of petroleum fluids ranging from dry gases to heavy oils.

Volatile oils exhibit an initial oil FVF in the range of 1.5 to 3.0. Black oils exhibit an initial oil FVF in the range of 1.1 to 1.5. Volatile oils exhibit an initial GOR in the range of

TABLE 9.1—PETROLEUM FLUIDS AND THEIR CHARACTERISTICS

	Oils			Gases	
Characteristic	Heavy Oils and Tars	Black Oils	Volatile Oils	Gas Condensates	Wet and Dry Gases
Initial fluid molecular weight	210+	70 to 210	40 to 70	23 to 40	<23
Stock-tank-oil color	black	brown to light green	greenish to orange	orange to clear	clear
Stock-tank oil-gravity, °API	5 to 15	15 to 45	42 to 55	45 to 60	45+
C_7-plus fraction, mol%	>50	35 to 50	10 to 30	1 to 6	0 to 1
Initial dissolved GOR, scf/STB	0 to 200	200 to 900	900 to 3,500	3,500 to 30,000	30,000+
Initial FVF, B_{oi}, RB/STB	1.0 to 1.1	1.1 to 1.5	1.5 to 3.0	3.0 to 20.0	20.0+
Typical reservoir temperature, °F	90 to 200	100 to 200	150 to 300	150 to 300	150 to 300
Typical saturation pressure, psia	0 to 500	300 to 5,000	3,000 to 7,500	1,500 to 9,000	—
Volatile-oil/gas ratio, STB/MMscf*	0	0 to 10	10 to 200	50 to 300	0 to 50
Maximum vol% liquid during CCE**	100	100	100	0 to 45	0
OOIP, STB/acre-ft (bulk)	1,130 to 1,240	850 to 1,130	400 to 850	60 to 400	0 to 60
OGIP, Mscf/acre-ft (bulk)	0 to 200	200 to 700	300 to 1,000	500 to 2,000	1,000 to 2,200

*At bubblepoint pressure. **Constant composition expansion of reservoir fluid.

900 to 3,500 scf/STB. Black oils exhibit an initial GOR in the range of 200 to 900 scf/STB. These relations establish molecular weight as a credible correlating parameter. McCain[2] has found success in the use of the heptanes-plus content as a correlating parameter.

The inverse of the oil FVF yields a measure of the original oil in place (OOIP) per unit volume of reservoir pore space. Because the oil FVF is greater for volatile oils than black oils, the latter yield greater OOIP per unit volume. Black-oil reservoirs contain 850 to 1130 STB/acre-ft (bulk) while volatile-oil reservoirs contain less, typically 400 to 850 STB/acre-ft.

Although volatile-oil reservoirs contain less oil per unit volume, they typically yield slightly higher oil recoveries than black-oil reservoirs because of their higher dissolved-gas content and lower oil viscosity. Ultimately, volatile-oil reservoirs may yield greater oil reserves than black-oil reservoirs. Light black oils and heavy volatile oils are among the most economically attractive reservoir fluids.

There has been no systematic study to determine the relative percentage of black-oil and volatile-oil reservoirs; however, an examination of the world's 500 largest reservoirs reveals that black-oil reservoirs overwhelmingly dominate the group.[3] One reason there are more black-oil than volatile-oil reservoirs is that the latter are characteristically located at greater depths than the former. As exploration continues to go deeper, more volatile-oil reservoirs can be expected to be discovered.

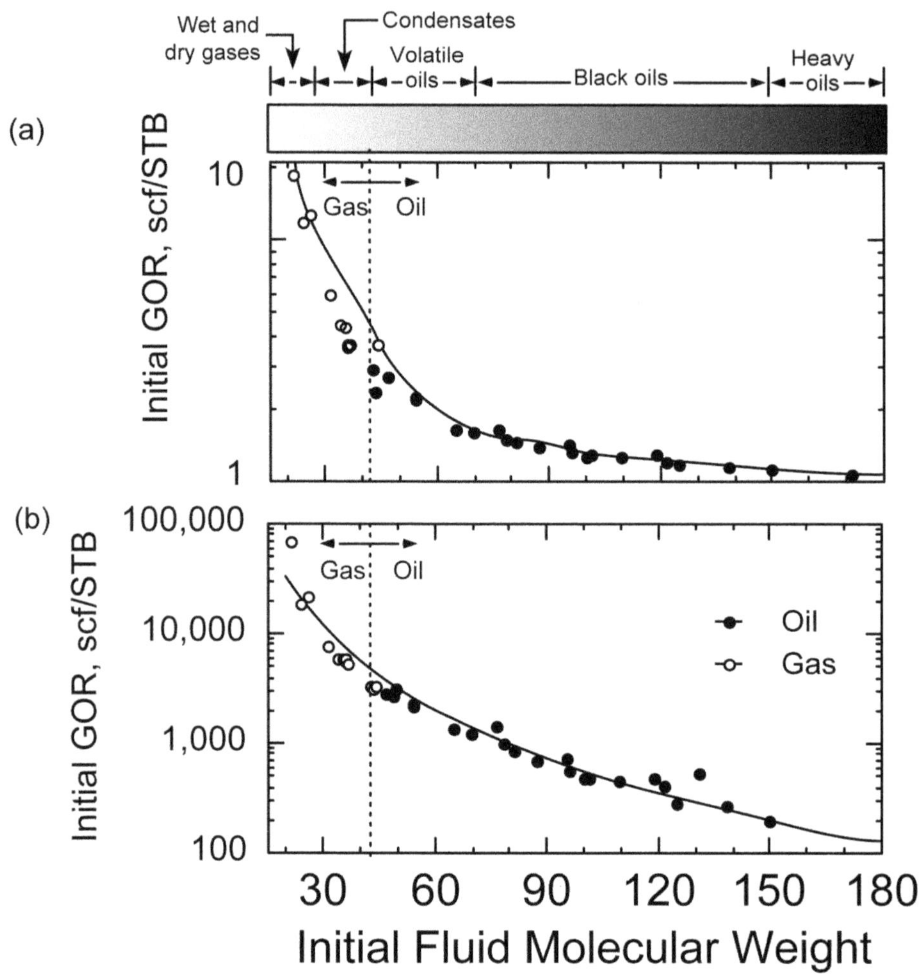

Fig. 9.1—(a) Initial formation volume factor (FVF) and (b) initial dissolved GOR as a function of initial fluid molecular weight.

9.3 Types of Reservoir Energy

The following list outlines the major types of energy available for petroleum production.

- The energy of compression of the water and rock within the reservoir.
- The energy of compression of oil within the reservoir.
- The energy of compression of gas within the reservoir.
- The energy of compression of waters contiguous to and in communication with the petroleum reservoir.
- The gravitational energy that causes the oil and gas to segregate within the reservoir.

Water within the reservoir refers to the water that is originally present within the reservoir at the time of discovery. Oil within the reservoir refers to the oil phase that is originally present at discovery or that may form from the condensation of volatilized oil upon pressure release. Likewise, gas within the reservoir refers to the gas phase that is originally present at discovery or that may form subsequently from the liberation of dissolved gas upon pressure release.

As mechanisms of energy release are provided by the drilling and operation of wells, reservoir pressure declines, fluids expand, flow is induced, and fluids are produced. The net volume of expansion of rock and fluids within the reservoir results in an equal volume of expulsed fluids. The water-bearing reservoirs that adjoin petroleum reservoirs are called aquifers. The expansion of water from the aquifer results in an overflow of water from the aquifer into the petroleum reservoir. The net overflow of water into the petroleum reservoir, in turn, results in an equal volume of fluid expulsion from the petroleum reservoir. Gravity segregation does not directly result in fluid expulsion but causes oil to settle to the bottom and gas to migrate to the top of the reservoir. By producing from only the lower reaches of the reservoir, this process affords a skilled operator a means to recover oil selectively and possibly recover more oil than would otherwise be recovered.

In ranking the types of energy in order of least importance to oil recovery, the energy of the compressed water and rock originally within the reservoir is probably the least important because of the relatively low compressibilities of water and rock. Of equal unimportance is the energy of the compressed oil, although the effects of compressed oil are slightly greater than the effects of compressed water and rock, as evidenced by the slightly greater compressibility of oil (10^{-5} per psi) than water (3×10^{-6} per psi) and rock (6×10^{-6} per psi). Of the energies of the compressed fluids, the effects of compressed gas are undoubtedly the most important because of the greater compressibility of gas. The effects of compressed gas are important even if there is not much free gas initially present, as in the case of an initially undersaturated oil reservoir. In these cases, gas will appear naturally during the course of pressure depletion because of the release of dissolved gas from the oil once the pressure falls below the bubblepoint pressure.

Gravitational forces can be a major factor in oil recovery if the reservoir has sufficient vertical relief and vertical permeability. The effectiveness of gravitational forces will be limited by the rate at which fluids are withdrawn from the reservoir. If the rate of withdrawal is appreciably greater than the rate of fluid segregation, then the effects of gravitational forces will be minimized.

The energy from the compressed waters of aquifers also can be a major factor even though the water has a low compressibility because the size of most aquifers tends to be much larger than the petroleum reservoir. Most oil fields have areas of less than 10 sq mile (6,400 acres), whereas aquifers often have areas of more than 1,000 sq mile.[1]

The energies discussed thus far represent "internal" reservoir energies (i.e., energies originally present within the reservoir and its adjoining geological units at the time of discovery). In addition to these energies, there may be important "external" energies (i.e., energies that originate from outside the reservoir).External energies imply the practice of injecting fluids into the reservoir to augment the reservoir's natural energies. This practice is called pressure maintenance. The two most important injection fluids are compressed water and gas. The resultant action of injected fluids once inside the reservoir is much the same as the fluids originally present. The overall intention of injecting fluids is to add energy to the reservoir to recover more oil or gas than would otherwise be recovered. If gas is injected, it is clear that the intention is to recover more oil than otherwise would be recovered. In addition, the economic attractiveness of this practice relies on the expectation that the additional income derived from the increased oil production will more than offset the additional expenditures and lost or deferred revenues incurred by gas injection. The most common source of gas for gas injection is the gas produced from the reservoir. The chapters on Water Injection and Immiscible Gas Injection in this volume discuss these subjects further.

9.4 Producing Mechanisms

The general performance characteristics of hydrocarbon-producing reservoirs are largely dependent on the types of energy available for moving the hydrocarbon fluids to the wellbore. The predominate energy forms give rise to distinct producing mechanisms. These producing mechanisms are used to help classify petroleum reservoirs.

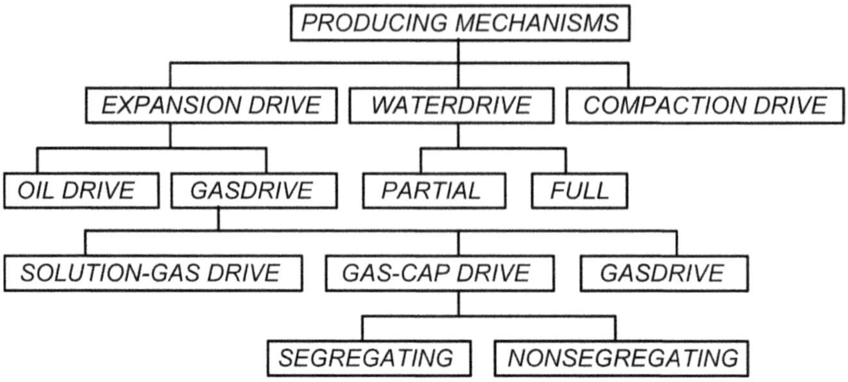

Fig. 9.2—Classification of reservoir-producing mechanisms.

In this section, these producing mechanisms are defined and delineated, although there is not a well-established consensus for some of these definitions. A petroleum reservoir rarely can be characterized throughout its pressure-depletion life by any single producing mechanism. A petroleum reservoir usually is subject to several producing mechanisms over its lifetime; nevertheless, the practice of describing a petroleum reservoir by its predominant producing mechanism is helpful.

Broadly, all commercially productive petroleum reservoirs are divided into either expansion-drive, compaction-drive, or waterdrive reservoirs. An expansion- or compaction-drive reservoir is a predominantly sealed reservoir in which the expansion of fluids and rock originally within the reservoir is responsible for petroleum expulsion from the reservoir. **Fig. 9.2** shows the producing-mechanism system of classification.

In contrast, a waterdrive reservoir is an unsealed petroleum reservoir in communication with water-bearing reservoirs and in which there is appreciable movement of water from the water-bearing reservoir to the petroleum reservoir. If the rate of water intrusion into the reservoir is equal to the volumetric rate of fluid withdrawal from the reservoir, then the reservoir is more descriptively referred to as a complete-waterdrive reservoir. A complete-waterdrive reservoir often experiences, but does not necessarily imply, very little pressure decline. Complete-waterdrive reservoirs may require substantial pressure decline before the water-delivery rate can balance the production rate.

If the rate of water intrusion into the reservoir is substantial but substantially less than the volumetric rate of fluid withdrawal from the reservoir, then the reservoir is referred to as a partial-waterdrive reservoir. In all cases, when a waterdrive is the major producing mechanism, the reservoir pressure will be sensitive to the producing rate. If the reservoir-producing rate is too high relative to the water-influx rate, the waterdrive will lose its effectiveness and the reservoir pressure will decline.

Waterdrives are also classified as edgewater or bottomwater drives, depending on the nature and location of the water encroachment into the reservoir. **Fig. 9.3** shows a schematic of a bottomwater-drive reservoir. Because waterdrive reservoirs experience increasing water content and decreasing hydrocarbon content, they are referred to as nonvolumetric reservoirs. More generally, nonvolumetric reservoirs are reservoirs in which hydrocarbon pore volume (PV) changes during pressure depletion. Conversely, volumetric reservoirs are reservoirs in which hydrocarbon PV does not change during pressure depletion. Because waterdrive reservoirs involve water influx into the reservoir, they also are referred to as water-influx reservoirs.

Pressure depletion causes the internal stress within the reservoir rock to increase. This change produces changes in the grain arrangement and other phenomena that ultimately cause the pore volume of the rock to decrease. The contraction of the reservoir pore volume aids in

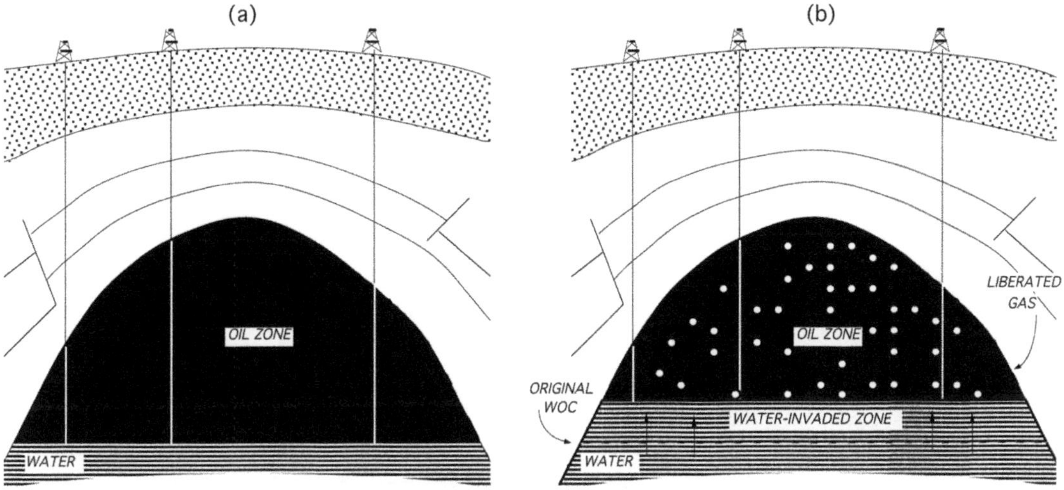

Fig. 9.3—Distribution of water and oil and position of water/oil contact (WOC) in a waterdrive reservoir (a) before production and (b) during depletion.

expelling fluids from the reservoir. The terms "pore-volume contraction" and "rock expansion" are used interchangeably in this chapter to describe this phenomenon, even though very little grain expansion usually takes place. If this phenomenon is a major producing mechanism, the reservoir is a compaction-drive reservoir. Compaction-drive reservoirs are rare because the PV compressibility is usually less than the oil compressibility.

Expansion-drive reservoirs are further classified as oil- or gas-expansion-drive reservoirs depending on whether the oil or gas expansion is the predominant producing mechanism. Dry- and wet-gas reservoirs are gas-expansion-drive reservoirs because they do not contain any free oil at reservoir conditions. More descriptively, a gas-drive reservoir is one in which the expansion of free gas is the predominant producing mechanism. The expanding free gas may originate as initial free gas or as dissolved gas. An oil-drive reservoir, on the other hand, is one in which the expansion of free oil is the predominant producing mechanism.[1] According to these definitions, black-oil and volatile-oil reservoirs are not likely to be oil-drive reservoirs but gas-drive reservoirs because the expansion of gas is ultimately much greater than the expansion of oil. The oil in saturated, black-oil and volatile-oil reservoirs does not expand but contracts during pressure depletion because of the release of dissolved gas. Because the overwhelming majority of expansion-drive reservoirs are gas-drive reservoirs, the term oil-drive reservoir is rarely used. An oil-drive producing mechanism dominates in oil reservoirs only while they are undersaturated.

Gas-drive reservoirs are further subdivided into either solution-gas-drive or gas-cap expansion-drive reservoirs. A gas-cap expansion-drive reservoir is a gas-cap reservoir in which the expanding gas cap is responsible for the majority of the gas expansion. A gas cap is a free-gas zone that overlies an oil zone. The free-gas zone may be pre-existing or may form during the depletion process. Pre-existing gas caps are called primary gas caps. Gas caps that are not originally present but that develop during the depletion process are called secondary or developed gas caps. Secondary gas caps can form from the upward migration of either liberated dissolved gas or from reinjected gas. **Fig. 9.4** shows a schematic of a gas-cap expansion-drive reservoir.

Gas caps are also classified according to their displacement efficiency. At the most favorable extreme, the expanding gas displaces oil in a piston-like manner. At the other limit, the expanding gas displaces oil in a totally diffuse manner. The former are segregation-drive or gravity-drainage gas caps; the latter are nonsegregation-drive gas caps. The boundary between

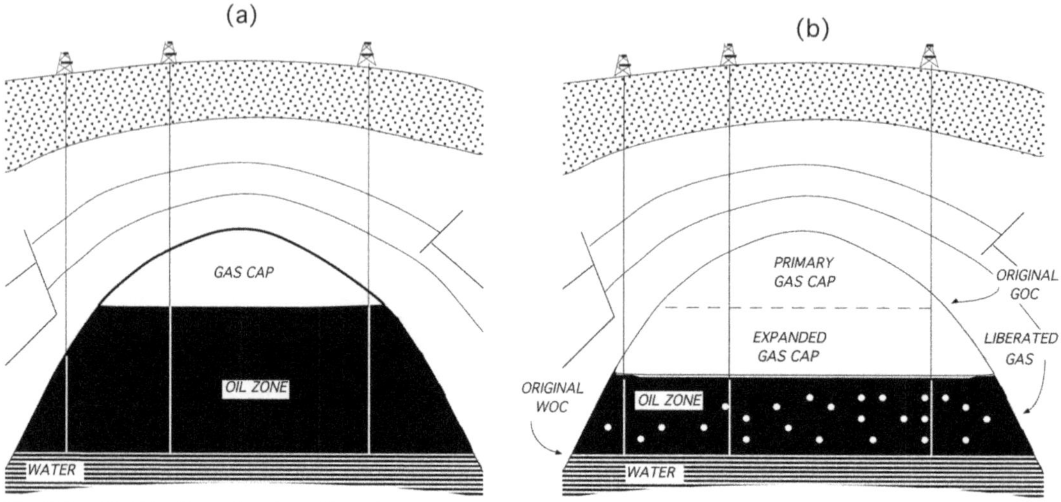

Fig. 9.4—Distribution of water, oil, and gas and position of gas/oil contact (GOC) in a segregating-gas-cap reservoir: (a) before production and (b) during depletion.

the gas-cap zone and oil zone is the gas/oil contact (GOC). Segregation-drive gas caps exhibit a GOC that moves progressively downward during depletion. In contrast, nonsegregation-drive gas caps exhibit a GOC that appears stationary. The gas-cap displacement efficiency depends on the producing rate and vertical permeability. Segregation-drive gas caps tend to have high vertical permeability, while nonsegregation-drive gas caps tend to have low vertical permeability. These two types of gas caps represent limiting cases. In reality, there is a continuum of character between these limits. The exact gas-cap character depends on the actual conditions.

Gas-drive reservoirs that are not gas-cap reservoirs but are dominated by the expansion of solution gas are called solution-gas-drive or dissolved-gas-drive reservoirs. **Fig. 9.5** shows a schematic of a solution-gas-drive reservoir. Gas-drive reservoirs that are neither gas-cap nor solution-gas-drive reservoirs are called gas-drive reservoirs. For example, dry-gas reservoirs are gas-drive reservoirs because they do not qualify as solution-gas-drive or as gas-cap reservoirs. The practice of reinjecting dry gas into and producing wet gas from gas/condensate reservoirs is called gas cycling or cycling.

9.4.1 Recovery Ranges. Table 9.2 lists the approximate primary-recovery range for the different producing mechanisms. The ranges reflect the rank of the reservoir energies. Black-oil reservoirs that exclusively produce by solution-gas-drive mechanism typically recover 10 to 25% of the OOIP by pressure depletion. The American Petroleum Inst. reports an average primary oil recovery of 20.9% for 307 solution-gas-drive reservoirs.[4] In contrast, primary oil recovery from waterdrive, black-oil reservoirs typically ranges from 15 to 50% or higher of the OOIP. Waterdrive, black-oil reservoirs have yielded some of the highest recoveries ever recorded. The primary oil recovery from gas-cap, black-oil reservoirs varies widely depending on whether there is significant gravity drainage. The primary oil recovery from nongravity-drainage, gas-cap, black-oil reservoirs ranges from 15 to 40% of the OOIP. In contrast, the primary oil recovery from gravity-drainage, gas-cap, black-oil reservoirs ranges from 15 to 80% of the OOIP. Primary oil recoveries from gravity-drainage, black-oil reservoirs are among the highest of any black-oil reservoir. Pressure maintenance by gas reinjection is practiced commonly in black-oil reservoirs to improve oil recovery. Black-oil reservoirs subject to gas reinjection without gravity drainage typically recover 15 to 45% of the OOIP. If gas is reinject-

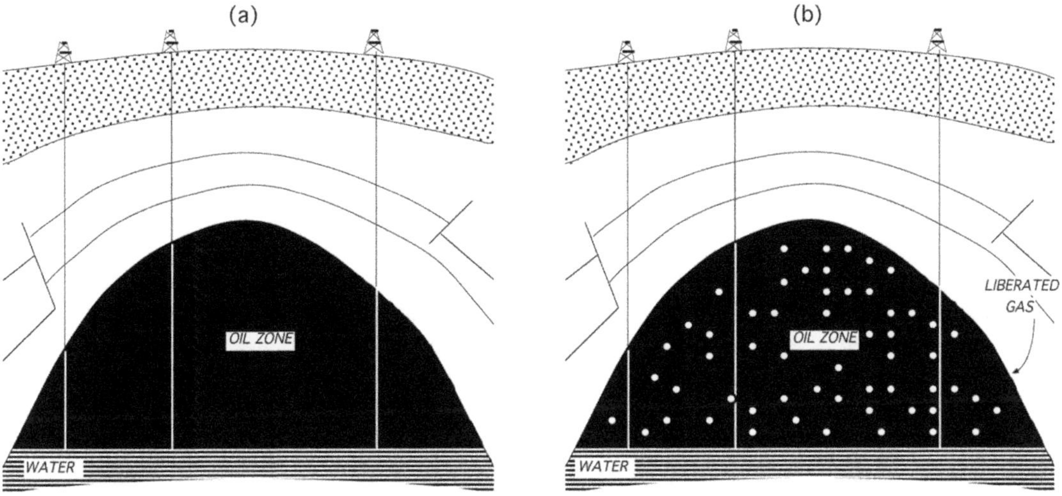

Fig. 9.5—Distribution of water, oil, and gas in a solution-gas-drive reservoir: (a) before production and (b) during depletion.

TABLE 9.2—PRODUCING MECHANISMS AND TYPICAL RECOVERY RANGES

Producing Mechanism	Oil Recovery Range (% OOIP)
Solution-gas drive	10 to 25%
Gas-cap drive without gravity drainage	15 to 40%
Gas-cap drive with gravity drainage	15 to 80%
Gas reinjection with gravity drainage	15 to 80%
Gas reinjection without gravity drainage	15 to 45%
Waterdrive	15 to 60%

ed in a reservoir with active gravity drainage, the primary oil recovery typically ranges from 15 to 80%.

9.5 Fluid Properties

Black and volatile oils, as well as other petroleum fluids, are characterized routinely in terms of their standard pressure/volume/temperature (PVT) parameters: oil FVF (B_o), gas FVF (B_g), dissolved GOR (R_s), and volatilized oil/gas ratio (R_v). These fluid properties, in addition to some others, are prerequisites for a wide variety of reservoir-engineering calculations, including estimating the OOIP and original gas in place (OGIP) and material-balance calculations.

Table 9.3 tabulates and **Fig. 9.6** plots the standard PVT parameters as a function of pressure for a black oil from a west Texas reservoir located at a depth of 6,700 ft with an initial pressure of 3,100 psia and a temperature of 131°F. Only the PVT properties below 2,000 psia are listed. The fluid exhibited a bubblepoint at approximately 1,688 psia and had a molecular weight of 81. **Table 9.4** summarizes its compositional analysis. The fluid has an initial oil FVF of 1.467 RB/STB and dissolved GOR of 838 scf/STB. The equilibrium gas contains negligible volatilized oil. **Fig. 9.7** plots the oil and gas viscosities as a function of pressure.

TABLE 9.3—PVT PARAMETERS FOR WEST TEXAS BLACK OIL

p (psia)	B_o (RB/STB)	B_g (RB/Mscf)	R_s (scf/STB)	μ_o (cp)	μ_g (cp)	Volume Fraction Liquid*
2,000	1.467		838.5	0.3201		1.0000
1,800	1.472		838.5	0.3114		1.0000
1,700	1.475		838.5	0.3071		1.0000
1,640	1.463	1.920	816.1	0.3123	0.0157	0.9780
1,600	1.453	1.977	798.4	0.3169	0.0155	0.9600
1,400	1.408	2.308	713.4	0.3407	0.0140	0.8670
1,200	1.359	2.730	621.0	0.3714	0.0138	0.7540
1,000	1.322	3.328	548.0	0.3973	0.0132	0.6440
800	1.278	4.163	464.0	0.4329	0.0126	0.5130
600	1.237	5.471	383.9	0.4712	0.0121	0.3750
400	1.194	7.786	297.4	0.5189	0.0116	0.2320
200	1.141	13.331	190.9	0.5893	0.0108	0.0970

*From constant composition expansion.

Table 9.5 tabulates and **Fig. 9.8** plots the standard PVT parameters for a volatile oil from a north-central Louisiana reservoir located at a depth of approximately 10,000 ft with an initial pressure of 5,070 psia and a temperature of 246°F.[5,6] The fluid exhibited a bubblepoint at approximately 4,677 psia and had a molecular weight of 47. **Table 9.6** summarizes the initial fluid composition. The fluid has an initial oil FVF of 2.704 RB/STB and dissolved GOR of 2,909 scf/STB. The bubblepoint gas had a volatilized-oil/gas ratio of approximately 120 STB/MMscf. The volatilized-oil/gas ratio decreases with pressure until a pressure of 998 psia is reached. At pressures between 998 and 598 psia, the volatilized-oil/gas ratio increases slightly.

The standard PVT parameters of volatile and black oils are determined experimentally with different laboratory procedures. Black oils are evaluated with a differential-vaporization (DV) experiment;[7,8] in contrast, volatile oils are evaluated with constant volume depletion (CVD).[9,10] Sometimes, however, volatile oils use a specialized DV experiment[11] instead of a CVD experiment. The specialized DV experiment includes a step to measure the volatilized-oil content of equilibrium gases.

The standard PVT parameters for black oils are specified routinely in commercial PVT reports. McCain provides some example PVT reports.[12] The reported PVT parameters, however, may or may not be adjusted for the effects of surface separators. Surface separators maximize the stock-tank liquid yield as fluids pass through them. The oil FVF and dissolved GOR of adjusted properties are characteristically less than unadjusted properties. If the PVT report specifies the adjusted parameters, then no further adjustment is required. If only the raw parameters are specified, then adjustment is needed.

Various empirical methods are used to correct the standard PVT parameters for the effects of separators.[7,13,14] Generally, correction is very important. For example, the unadjusted bubblepoint oil FVF and the dissolved GOR for the example black oil in Table 9.3 are 1.584 RB/STB and 1,007 scf/STB, respectively. On adjustment for separators at 100 psia, the corresponding oil FVF and dissolved GOR are 1.467 RB/STB and 838.5 scf/STB, reflecting increased stock-tank-liquid recovery. Failure to correct the standard PVT parameters for separators can lead to substantial errors in subsequent reservoir-engineering calculations including the volumetric OOIP and OGIP calculations. Volatile oils are even more sensitive to the effects of separators

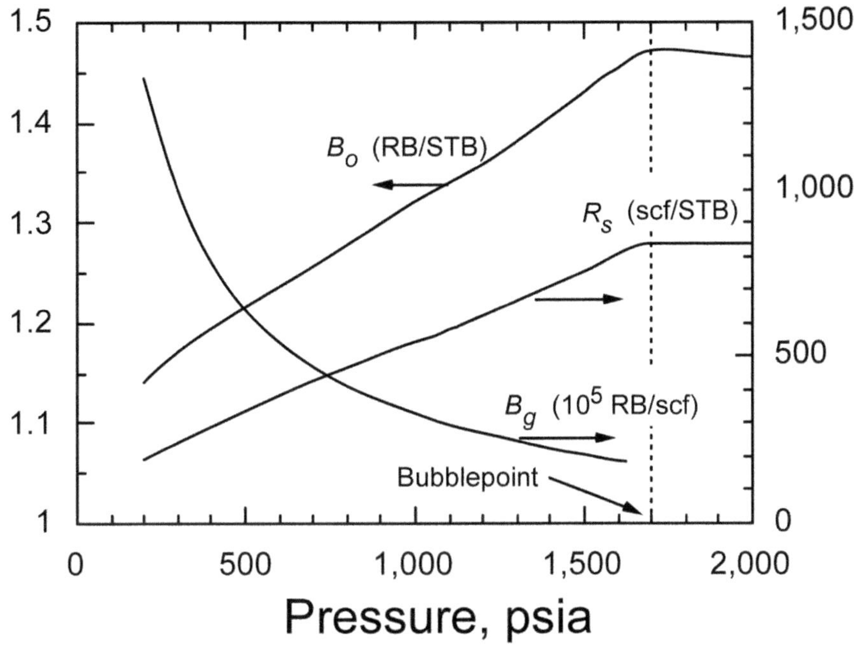

Fig. 9.6—Standard PVT properties as a function of pressure for a west Texas black oil.

than black oils. Volatile oils, however, are subjected to an entirely different laboratory procedure for measurement.

The standard PVT parameters for volatile oils rarely are given in commercial PVT reports. They must be calculated from CVD measurements. In order of increasing complexity, the three methods to calculate standard PVT parameters are the Walsh-Towler algorithm,[15] the Whitson-Torp method,[9] and the equation-of-state (EOS) method.[16,17] The Walsh-Towler algorithm uses recovery data directly from the CVD measurement and computes the corresponding properties. This method is suited for spreadsheet calculation and is fast and simple. The Whitson-Torp method, in contrast, uses equilibrium gas-composition data and computes the properties with Standing's[18] low-pressure *K*-values and a stock-tank-liquid density correlation such as the Alani-Kennedy EOS.[19] This method requires iterative, *K*-value flash calculations. Although this method is more computationally intensive than the Walsh-Towler algorithm, it is more versatile because it allows for arbitrary separator conditions. The EOS method is much more computationally intensive than the other methods. This method tunes a cubic EOS to the attending phase behavior and then uses the EOS to simulate the CVD numerically and estimate the PVT parameters. This method regularly uses commercial software. The methods yield virtually identical results despite their differences.

9.6 Material Balance

The material-balance equation is the simplest expression of the conservation of mass in a reservoir. The equation mathematically defines the different producing mechanisms and effectively relates the reservoir fluid and rock expansion to the subsequent fluid withdrawal. The applicable equation for initially saturated volatile- and black-oil reservoirs is[20–23]

TABLE 9.4—COMPOSITIONAL ANALYSIS OF WEST TEXAS BLACK OIL

Component	Mol%
N_2	0.28
C_1	28.98
CO_2	0.20
C_2	10.34
C_3	12.03
i-C_4	0.56
n-C_4	6.02
i-C_5	1.47
n-C_5	2.93
C_6	3.42
C_{7+}	33.15
Total	100.00
Fluid molecular wt.	81.2

$$\underbrace{G_{fgi}E_g}_{\text{free-gas expansion}} + \underbrace{N_{foi}E_o}_{\text{free-oil expansion}} + \underbrace{WE_w}_{\text{free-water expansion}} + \underbrace{V_{pi}E_f}_{\text{pore-volume contraction}} + \underbrace{W_e}_{\text{water influx}}$$

$$= \underbrace{\left(G_p - G_I\right)\left(\frac{B_g - B_o R_v}{1 - R_v R_s}\right)}_{\text{net gas withdrawal}} + \underbrace{N_p\left(\frac{B_o - B_g R_s}{1 - R_v R_s}\right)}_{\text{oil withdrawal}} + \underbrace{\left(W_p - W_I\right)B_w}_{\text{net water withdrawal}}, \quad \text{(9.1)}$$

where G_{fgi}, N_{foi}, and W are the initial free gas, oil, and water in place, respectively; G_p, N_p, and W_p are the cumulative produced gas, oil, and water, respectively; G_I and W_I are the cumulative injected gas and water, respectively; and E_g, E_o, E_w, and E_f are the gas, oil, water, and rock (formation) expansivities. Most of the equations in this chapter apply to any consistent set of units. A few equations, however, are written assuming English or customary units. Those equations are expressed in SI units in the Appendix.

N_{foi} and G_{fgi} are related to the total OOIP and OGIP, N and G, according to $N = N_{foi} + G_{fgi}R_{vi}$ and $G = G_{fgi} + N_{foi}R_{si}$.

The expansivities are defined as

$$E_o = B_{to} - B_{toi}, \quad \text{(9.2)}$$

$$E_g = B_{tg} - B_{tgi}, \quad \text{(9.3)}$$

$$E_w = B_w - B_{wi}, \quad \text{(9.4)}$$

and $E_f = \dfrac{V_{pi} - V_p(p)}{V_{pi}}$, where B_{to} and B_{tg} are the two-phase FVFs,

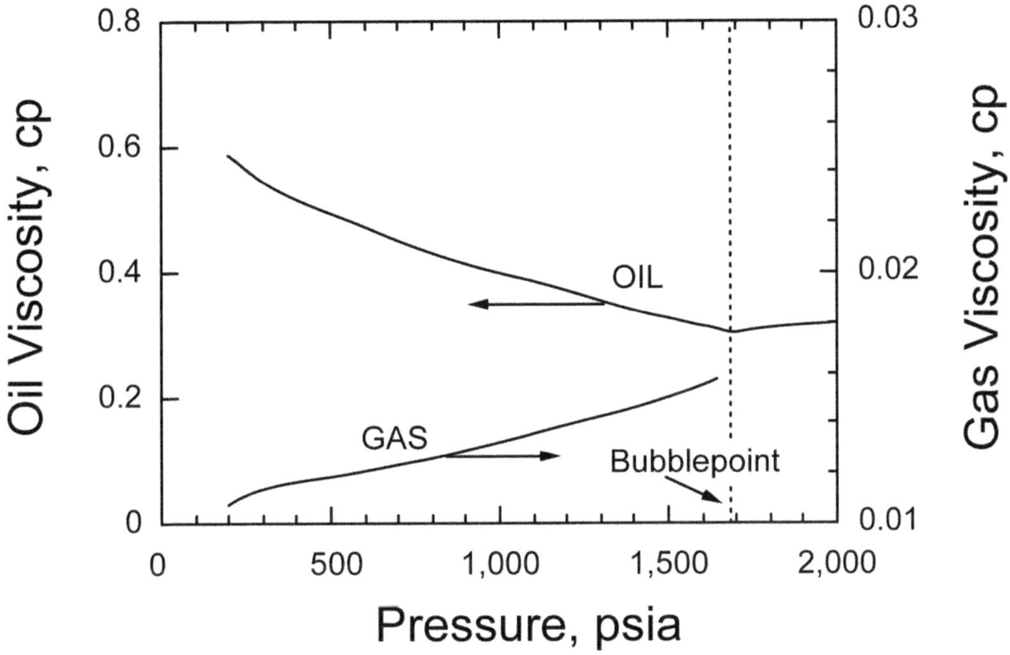

Fig. 9.7—Oil and gas viscosities as a function of pressure for west Texas black oil.

$$B_{to} = \frac{B_o\left(1 - R_{si}R_v\right) + B_g\left(R_{si} - R_s\right)}{1 - R_vR_s}, \qquad (9.5)$$

and

$$B_{tg} = \frac{B_g\left(1 - R_{vi}R_s\right) + B_o\left(R_{vi} - R_v\right)}{1 - R_vR_s}. \qquad (9.6)$$

The rock expansivity is obtained from direct measurement. See Sec. 9.10 for a greater discussion.

Physically, the two-phase FVF is the total hydrocarbon volume per unit volume of oil or gas at standard conditions. The two-phase FVF mimics the observations noted during a constant-composition expansion test. For instance, the two-phase oil FVF is the total hydrocarbon (oil + gas) volume of a saturated oil sample per unit volume of oil at standard conditions. In contrast, the two-phase gas FVF is the total hydrocarbon volume of a saturated gas sample per unit volume of gas at standard conditions. B_{to} and B_{tg} typically are expressed in units of RB/STB and RB/Mscf, respectively. For undersaturated oils, the two-phase oil FVF is equal to the oil FVF; for undersaturated gases, the two-phase gas FVF is equal to the gas FVF.

Eqs. 9.5 and 9.6 account for volatilized oil in the equilibrium gas phase. If volatilized oil is negligible, these equations are simplified. For instance, $B_{to} = B_o + B_g\ (R_{si} - R_s)$ and $B_{tg} = B_g$. These equations apply for black oils. Eq. 9.4 ignores dissolved gas in the aqueous phase.

Eq. 9.1 broadly states that net expansion equals net withdrawal. More specifically, it shows the different forms of expansion and withdrawal. The different forms of expansion such as gas expansion are responsible for the different producing mechanisms.

For the sake of simplicity, Eq. 9.1 is often written in the abbreviated form of

$$G_{fgi}E_{gwf} + N_{foi}E_{owf} + W_e = F, \qquad (9.7)$$

TABLE 9.5—PVT PARAMETERS FOR LOUISIANA VOLATILE OIL

p (psia)	B_o (RB/STB)	B_g (RB/Mscf)	R_s (scf/STB)	R_v (STB/MMscf)	μ_o (cp)	μ_g (cp)	Volume Fraction Liquid*
5,070	2.704		2,909		0.0742		1.000
4,998	2.713		2,909		0.0735		1.000
4,798	2.740		2,909		0.0716		1.000
4,698	2.754		2,909		0.0706		1.000
4,658	2.707	0.830	2,834	116.0	0.0718	0.0375	0.967
4,598	2.631	0.835	2,711	111.0	0.0739	0.0367	0.847
4,398	2.338	0.853	2,247	106.0	0.0847	0.0350	0.747
4,198	2.204	0.874	2,019	94.0	0.0906	0.0327	0.683
3,998	2.093	0.901	1,828	84.0	0.0968	0.0306	0.630
3,798	1.991	0.933	1,651	74.0	0.1028	0.0288	0.584
3,598	1.905	0.970	1,500	66.0	0.1104	0.0271	0.544
3,398	1.828	1.015	1,364	60.0	0.1177	0.0255	0.508
3,198	1.758	1.066	1,237	54.0	0.1242	0.0240	0.471
2,998	1.686	1.125	1,111	49.0	0.1325	0.0227	0.433
2,798	1.632	1.196	1,013	44.0	0.1409	0.0214	0.402
2,598	1.580	1.281	918	39.0	0.1501	0.0203	0.368
2,398	1.534	1.380	833	36.0	0.1598	0.0193	0.336
2,198	1.490	1.498	752	33.0	0.1697	0.0184	0.305
1,998	1.450	1.642	677	30.0	0.1817	0.0175	0.271
1,798	1.413	1.819	608	28.0	0.1940	0.0168	0.239
1,598	1.367	2.035	524	26.0	0.2064	0.0161	0.209
1,398	1.333	2.315	461	25.0	0.2223	0.0155	0.177
1,198	1.305	2.689	406	24.1	0.2438	0.0150	0.146
998	1.272	3.190	344	23.9	0.2629	0.0146	0.117
798	1.239	3.911	283	24.4	0.2882	0.0142	0.089
598	1.205	5.034	212	26.4	0.3193	0.0138	0.063

*From constant composition expansion.

where F = total net fluid withdrawal or production, E_{gwf} = composite gas expansivity, and E_{owf} = composite oil expansivities. F, E_{gwf}, and E_{owf} are defined in

$$F = G_{ps}\left(\frac{B_g - B_o R_v}{1 - R_v R_s}\right) + N_p\left(\frac{B_o - B_g R_s}{1 - R_v R_s}\right) + \left(W_p - W_I\right)B_w, \quad (9.8)$$

$$E_{owf} = E_o + \frac{B_{oi}}{\left(1 - S_{wi}\right)}\left(\frac{S_{wi}E_w}{B_{wi}} + E_f\right), \quad (9.9)$$

$$\text{and } E_{gwf} = E_g + \frac{B_{gi}}{\left(1 - S_{wi}\right)}\left(\frac{S_{wi}E_w}{B_{wi}} + E_f\right). \quad (9.10)$$

The composite expansivities include the connate-water and rock expansivities. Eq. 9.8 includes G_{ps}, which is the cumulative produced sales gas and is defined as $(G_p - G_I)$. F is expressed in reservoir volume units (e.g., RB or res m^3), E_{gwf} is expressed in reservoir volume units per standard unit volume of gas (e.g., RB/scf), and E_{owf} is expressed in reservoir volume units per standard unit volume of oil (e.g., RB/STB).

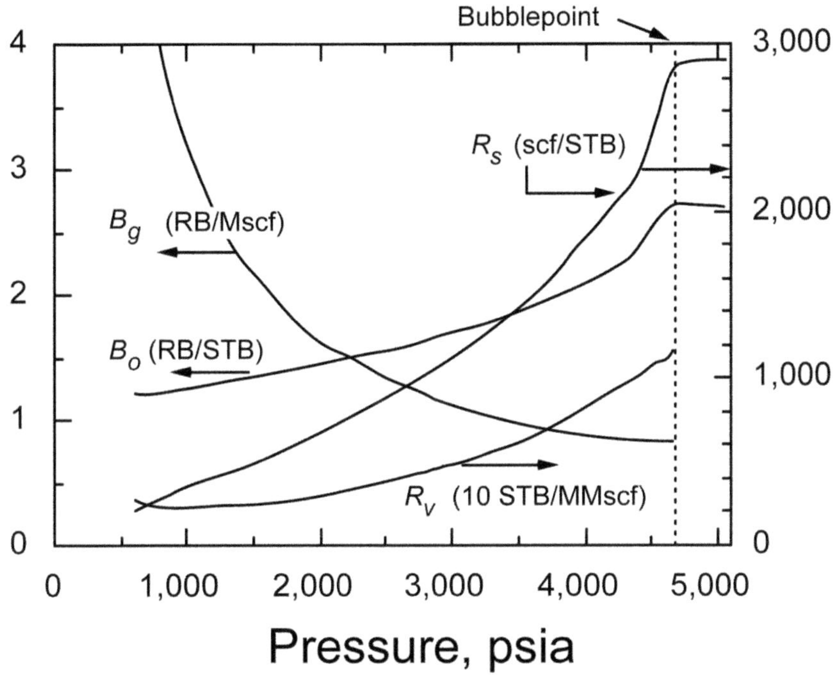

Fig. 9.8—Standard PVT properties as a function of pressure for a Louisiana volatile oil.

TABLE 9.6—COMPOSITIONAL ANALYSIS OF LOUISIANA VOLATILE OIL

Component	Mol%
N_2	1.67
C_1	60.51
CO_2	2.18
C_2	7.52
C_3	4.74
i-C_4	0.00
n-C_4	4.12
i-C_5	0.00
n-C_5	2.97
C_6	1.38
C_{7+}	14.91
Total	100.00
Fluid molecular wt.	46.69

For strictly undersaturated oil reservoirs, no free gas exists (i.e., $G_{fgi} = 0$) and the initial free oil in place is equal to the OOIP (i.e., $N_{foi} = N$) and Eqs. 9.1, 9.7, and 9.8 simplify, respectively, to[20,23,24]

$$\underbrace{NE_o}_{\text{free-oil expansion}} + \underbrace{WE_w}_{\text{free-water expansion}} + \underbrace{V_{pi}E_f}_{\text{pore-volume contraction}} + \underbrace{W_e}_{\text{water influx}}$$

$$= \underbrace{N_p B_o}_{\text{oil withdrawal}} + \underbrace{\left(W_p - W_I\right) B_w}_{\text{net water withdrawal}}, \quad \text{(9.11)}$$

$$N E_{owf} + W_e = F, \quad \text{(9.12)}$$

$$\text{and } F = N_p B_o + \left(W_p - W_I\right) B_w. \quad \text{(9.13)}$$

Eqs. 9.11 through 9.13 ignore gas reinjection.

The material-balance equation also helps explain most oil-recovery strategies. If the material-balance equation is solved for the produced fraction of the original free oil in place, then

$$\frac{N_p}{N_{foi}} = \frac{\dfrac{\left(W_I - W_p\right)B_w + W_e}{N_{foi}} + E_{owf} + \dfrac{G_{fgi}E_{gwf}}{N_{foi}}}{\dfrac{\left(B_o - B_g R_s\right) + \dfrac{G_{ps}}{N_p}\left(B_g - B_o R_v\right)}{\left(1 - R_v R_s\right)}}. \quad \text{(9.14)}$$

Eq. 9.14 succinctly shows that oil recovery increases with water influx (W_e), initial free-gas-cap volume (which is proportional to G_{fgi}), surface water injection (W_I), and surface gas injection (by minimizing gas sales through G_{ps}). It also shows that oil recovery increases by minimizing water production (W_p).

The material-balance equation and its many different forms have many uses including confirming the producing mechanism, estimating the OOIP and OGIP, estimating gas-cap sizes, estimating water-influx volumes, identifying water-influx model parameters, and estimating producing indices.

9.7 Solution-Gas Drives

Oil reservoirs that do not initially contain free gas but develop free gas on pressure depletion are classified as solution-gas drives. The solution-gas-drive mechanism applies once the pressure falls below the bubblepoint. Both black- and volatile-oil reservoirs are amenable to solution-gas drive. Other producing mechanisms may, and often do, augment the solution-gas drive. Solution-gas-drive reservoir performance is used as a benchmark to compare other producing mechanisms.

9.7.1 Stages of Production. Pure solution-gas-drive reservoirs are subject to four stages of idealized production. In chronological order, the four stages are: (1) production while undersaturated; (2) production while saturated but the free gas is immobile; (3) production while saturated and the free gas is mobile and the producing GOR is increasing; and (4) production while saturated and the free gas is mobile and the producing GOR is decreasing. Not all these stages are necessarily realized. For instance, Stage 4 may not be realized if primary recovery is terminated during Stage 3.

The key characteristics of each stage are outlined here.

Stage 1.

- No free gas.
- Producing GOR is equal to initial dissolved GOR.
- Fractional oil and gas recoveries are small and approximately equal.
- Reservoir pressure drops rapidly.

• Duration of stage depends on degree of initial undersaturation. The greater the initial undersaturation, the longer the duration of Stage 1. The stage duration is generally short.

Stage 2.

• Reservoir pressure is less than bubblepoint.
• Free gas appears, but the saturation is small and immobile.
• Producing GOR is slightly less than initial dissolved GOR.
• Rate of pressure decline is mitigated.

Stage 3.

• Free gas becomes mobile.
• Producing GOR increases steadily.
• Fractional gas recovery exceeds fractional oil recovery.
• Longest of all stages; typically consumes 85 to 95% of primary recovery.
• Primary recovery may be terminated during this stage.

Stage 4.

• Reservoir pressure is very low, typically less than 100 to 400 psia.
• Producing GOR decreases.
• Primary recovery often terminated before this stage is realized.

This chronology and these characteristics are an idealization and oversimplification of actual behavior; nevertheless, they are instructive and provide a preliminary basis for understanding scenarios that are more complicated.

9.7.2 Performance. To illustrate solution-gas-drive performance, tank model predictions of a west Texas black-oil reservoir are presented. Though idealized, these simulations, which are from a commercial simulator,[25,26] capture the main features and establish the theory of solution-gas drives.

For the sake of simplicity, the simulations consider the depletion of only a single well in an 80-acre closed area. **Table 9.7** summarizes some of the reservoir and fluid properties. The simulations assume that the PVT parameters in Table 9.3 apply.

Fig. 9.9 shows the performance in terms of pressure, producing GOR, and gas saturation as a function of cumulative produced oil. The four stages of production are depicted. During Stage 1, less than 1% of the OOIP is produced. The pressure decreases from the initial pressure of 2,000 psia to the bubblepoint pressure of 1,688 psia. The producing GOR remains constant and equal to the initial dissolved GOR of 838 scf/STB; no free gas evolves.

During Stage 2, the pressure falls below the bubblepoint; solution gas is liberated; and low, immobile gas saturations form. The cumulative oil recovery reaches approximately 4.5% of the OOIP. The pressure decreases from the bubblepoint to approximately 1,550 psia. The gas saturation increases to approximately 5% PV. The producing GOR actually decreases slightly, but this change is barely noticeable.

During Stage 3, the gas saturation increases to the point at which gas is mobile. Free-gas production begins, and the producing GOR rises steadily. By the end of Stage 3, the cumulative oil recovery is 28% of the OOIP, the pressure has decreased to 200 psia, the gas saturation reaches approximately 35% PV, and the producing GOR reaches approximately 6,700 scf/STB.

During Stage 4, the pressure has reached such a low level that the expansion of gas from reservoir to surface conditions is minimal. Consequently, the producing GOR decreases. By the time the pressure reaches 50 psia, the GOR is only 2,000 scf/STB and the total oil recovery is 32% of the OOIP.

Fig. 9.10 shows reservoir performance as a function of time. This figure plots the pressure, instantaneous producing GOR, cumulative producing GOR, gas saturation, oil rate, gas rate, and fraction of OOIP and OGIP recovered as a function of time. Stage 1 is very short and lasts

TABLE 9.7—RESERVOIR AND FLUID PROPERTIES FOR BLACK-OIL SIMULATIONS

Property	Value
Pattern area, acres	80
Depth, ft	6,700
Number of wells	1
Horizontal permeability, md	5.0
Pay thickness, ft	20
Average porosity,%	31
Connate-water saturation, % PV	20
Temperature, °F	131
Initial pressure, psia	2,000
Bubblepoint pressure, psia	1,688
Initial producing GOR, scf/STB	838.5
Bottomhole pressure, psia	199
Residual-oil saturation to water, % PV	25
Residual-oil saturation to gas, % PV	20
Critical gas saturation, % PV	5
Initial fluid molecular weight	81.2
Initial oil FVF, RB/STB	1.466
Separator pressure, psia	100
Stock-tank-oil density, lbm/ft^3	53.10
Stock-tank-oil molecular weight	151.5
Separator-gas molecular weight	30.7
OOIP, million STB	2.10
OGIP, Bscf	1.76

less than one month. The oil and gas producing rates and pressure decline sharply. The producing rates decline if the bottomhole pressure (BHP) is restricted. The producing rates can remain constant, but only if the minimum BHP is not yet reached.

Stage 2 also is relatively brief, lasting only several months. The reservoir pressure and producing rates also decline sharply but not as quickly as during Stage 1. The decline rate dampens because solution gas is liberated. The producing rates decline if the BHP is restricted. Constant producing rates can be realized only if the minimum BHP is not yet reached.

Stage 3 starts before one year of depletion and continues until the economic limit is reached. In this example, the limit is reached after 13.5 years when the oil-producing rate reaches 20 STB/D. The length of pressure depletion depends strongly on the reservoir permeability and on the prevailing economic conditions. For instance, lower permeabilities will decelerate recovery and protract depletion. The final pressure is 613 psia. This pressure is not low enough to realize Stage 4; therefore, this stage of depletion is not portrayed in Fig. 9.10. The absence of Stage 4 in field cases is not uncommon. The marked increase in the GOR from 838 to 4,506 scf/STB during Stage 3 coincides with marked increase in the gas saturation from 5 to 28.7% PV. At the economic limit, these simulations predict final oil and gas recoveries of 24.2% of the OOIP and 53.1% of the OGIP. Stage 3 clearly dominates the depletion life of a solution-gas-drive reservoir.

The results of this simulation are an oversimplification and idealization of actual performance. Oversimplification stems from the fact that the tank model ignores many important

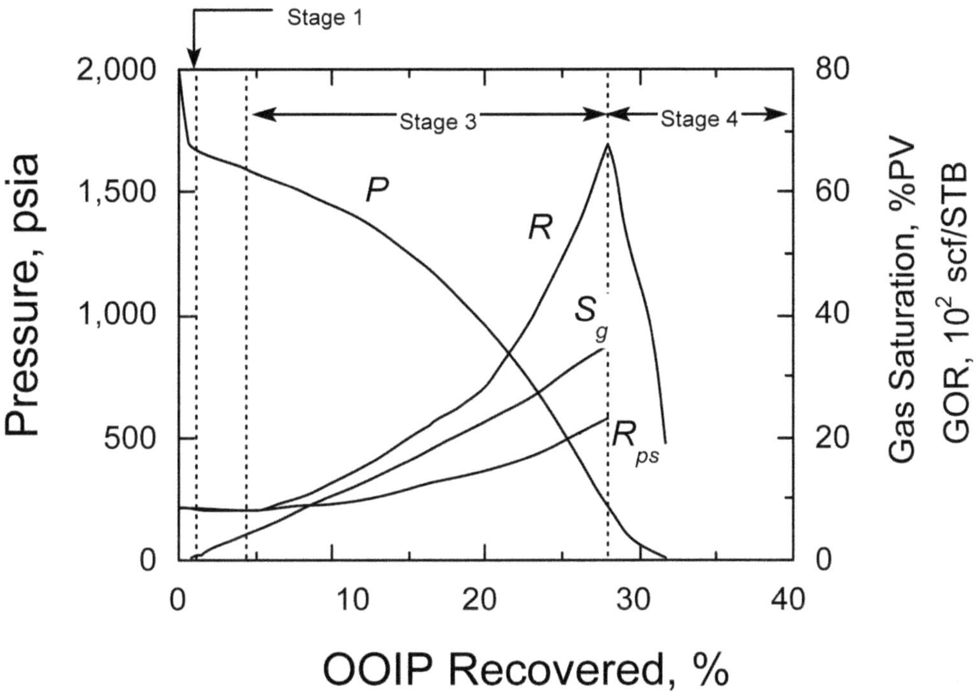

Fig. 9.9—Pressure (*p*), gas saturation (S_g), producing GOR (*R*), and cumulative producing GOR (R_{ps}) as a function of OOIP recovered for a solution-gas-drive, black-oil reservoir.

secondary phenomena. For example, the simulations ignore reservoir heterogeneity, which can be expected to reduce the recoveries by approximately 20 to 50% depending on the degree of heterogeneity. For instance, if a volumetric recovery efficiency of 80% is applied, then the idealized oil recovery of 24.2% corresponds to an adjusted oil recovery of 19.4%. Also, the simulations ignore spatial effects.

Qualitatively, solution-gas-drive, volatile-oil reservoirs act very similarly to their black-oil counterparts. One pronounced quantitative difference, however, is that volatile-oil reservoirs exhibit much greater peak producing GORs. The field example in Sec. 9.7.3 illustrates this difference. This example considers a volatile-oil reservoir that exhibits a peak GOR of approximately 29,000 to 32,000 scf/STB. This GOR is considerably greater than the peak GOR for the example black oil of 6,700 scf/STB. Another difference between volatile- and black-oil reservoirs is that the former often exhibit slightly greater oil recoveries; however, there are numerous exceptions to this trend.

9.7.3 Field Example. Cordell and Ebert[5] report the performance of a volatile-oil reservoir located in north-central Louisiana. **Table 9.8** summarizes some of the pertinent reservoir data. This reservoir produced from the Smackover lime located at an approximate depth of 10,000 ft. The field was discovered in 1953 and was developed with 11 wells on 160-acre spacing. Jacoby and Berry[6] report on the fluid properties of this volatile oil. The standard PVT parameters in Table 9.5, which were developed from laboratory data with the EOS method, are applicable.

Table 9.9 summarizes the reservoir performance in terms of the cumulative oil production, cumulative gas production, and instantaneous producing GOR as a function of pressure. Table 9.9 includes the oil and gas recoveries as a percent of the OOIP and OGIP. These recoveries

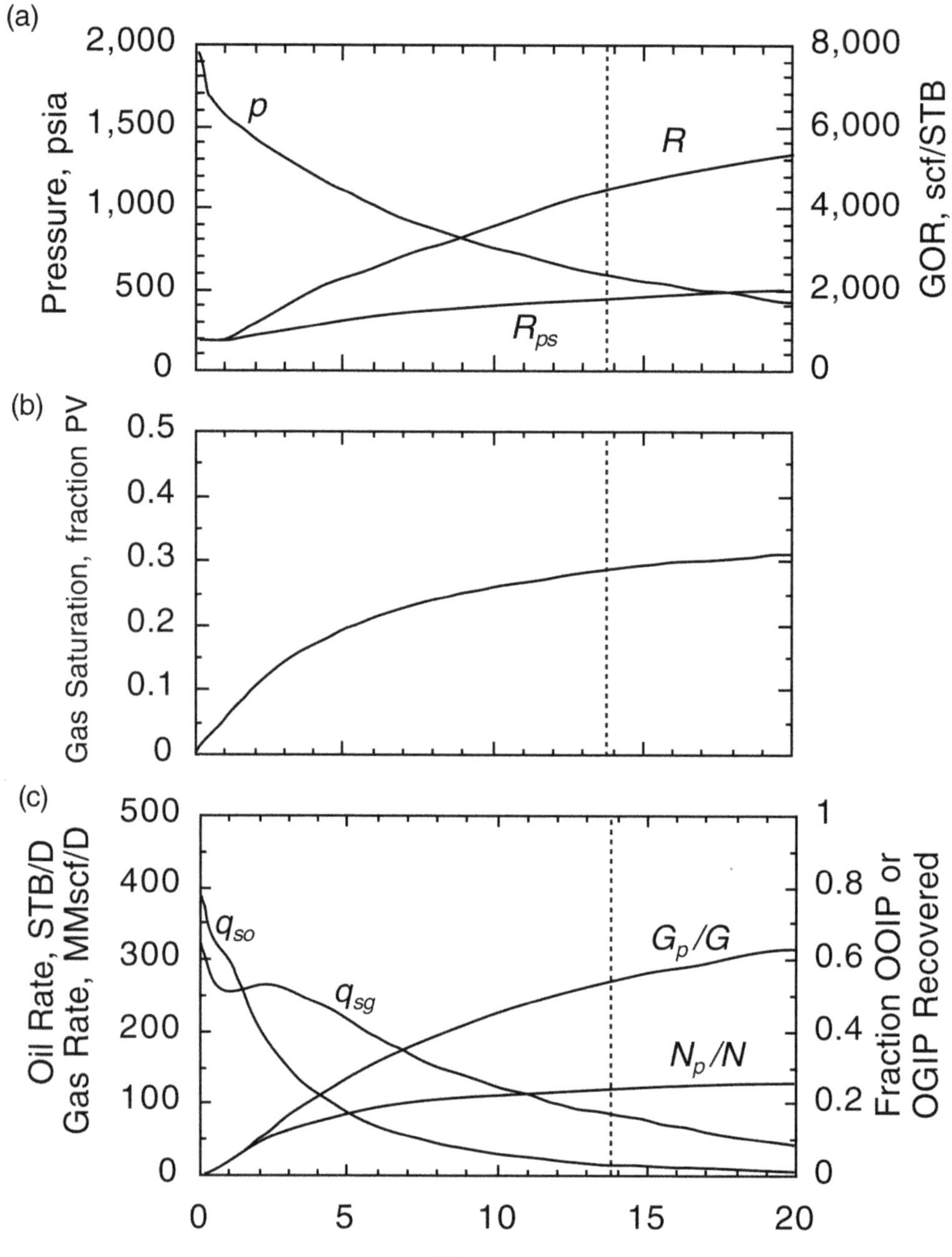

Fig. 9.10—Performance of a solution-gas-drive reservoir: (a) pressure, *p*, instantaneous producing GOR, *R*, and cumulative producing GOR, R_{ps}, histories; (b) gas-saturation history; (c) oil-rate, q_{so}, gas-rate, q_{sg}, oil-recovery, N_p/N, and gas-recovery, G_p/G, histories.

were based on the volumetric OOIP and OGIP estimates of 10.7 million STB and 31.1 Bscf, respectively.

TABLE 9.8—SELECTED RESERVOIR AND FLUID PROPERTIES FOR LOUISIANA VOLATILE-OIL RESERVOIR

Property	Value
Discovery year	1953
Location, state	Louisiana
Formation	Smackover
Lithology	Limestone
Area, acres	80
Depth, ft	10,000
Number of wells	11
Well spacing, acres	160
Horizontal permeability, md	174
Pay thickness, ft	24.4
Average porosity,%	13.6
Connate-water saturation, % PV	28.3
Temperature, °F	246
Initial pressure, psia	5,070
Bubblepoint pressure, psia	4,698
Initial producing GOR, scf/STB	2,909
Initial fluid molecular weight	47
Initial oil FVF, RB/STB	2.703
OOIP, million STB*	10.7
OGIP, Bscf*	31.1

*Volumetric estimate.

Fig. 9.11 shows the reservoir performance in terms of pressure and producing GOR as a function of cumulative oil recovery. This figure qualitatively agrees with the theoretical results for the black oil in Fig. 9.9. A comparison confirms that volatile-oil reservoirs experience much higher producing GORs. For example, this volatile-oil reservoir reaches a peak GOR of 29,000 scf/STB (compared with 6,700 scf/STB for the black oil). This reservoir was produced to a low enough pressure to exhibit Stage 4, which is the period of declining GOR. The GOR began to decline at a reservoir pressure of approximately 800 psia.

9.7.4 Material-Balance Analysis. A material-balance analysis is performed routinely to confirm the suspected producing mechanism and to estimate the OOIP independently. The applicable material-balance equation for a solution-gas-drive reservoir is[20,23,24,27]

$$F - W_e = N E_{owf} \; . \qquad (9.15)$$

Eq. 9.15 is a simplification of Eq. 9.7 and assumes no initial free gas ($G_{fgi} = 0$). Because there is no initial free gas, $N_{foi} = N$. If free gas is present initially, the material-balance methods for gas-cap reservoirs should be applied (see Section 9.8.2). Eq. 9.15 also applies to waterdrives; however, if the following methods are applied to waterdrives, the water-influx history must be reliably known. If the water-influx history is not known, then the methods in Section 9.9.8 regarding waterdrives must be applied. If there is no water influx, then $W_e = 0$.

If a reservoir produces exclusively by solution-gas drive with only supplemental connate-water expansion and pore-volume contraction, then Eq. 9.15 dictates that a plot of F vs. E_{owf} is a straight line, emanates from the origin, and has a slope equal to N. This observation is used to confirm the producing mechanism. If water influx exists and if W_e is known, then an F-vs.-E_{owf} plot is replaced by a ($F - W_e$)-vs.-E_{owf} plot. **Fig. 9.12** shows a ($F - W_e$)-vs.- E_{owf} plot for a volatile-oil reservoir. Once the OOIP is determined, the OGIP is given by $G = R_{si}N$. If an F-vs.-

TABLE 9.9—PRODUCTION HISTORY OF LOUISIANA VOLATILE-OIL RESERVOIR

p (psia)	N_p (million STB)	G_{ps} (Bscf)	R (scf/STB)	N_p (%OOIP)	G_{ps} (%OGIP)
5,070	0.000	0.000	2,909	0.00	0.00
4,998	0.036	0.105	2,909	0.34	0.34
4,798	0.130	0.388	2,909	1.22	1.25
4,698	0.184	0.539	2,909	1.72	1.73
4,658	0.227	0.657	2,834	2.13	2.11
4,598	0.302	0.872	2,749	2.83	2.80
4,398	0.582	1.657	2,914	5.44	5.33
4,198	0.808	2.366	3,256	7.55	7.61
3,998	1.022	3.140	4,026	9.56	10.10
3,798	1.227	4.032	5,087	11.48	12.96
3,598	1.388	4.967	6,494	12.98	15.97
3,398	1.528	5.988	8,068	14.29	19.25
3,198	1.646	7.062	9,541	15.40	22.71
2,998	1.764	8.223	11,290	16.50	26.44
2,798	1.861	9.405	13,093	17.41	30.24
2,598	1.947	10.652	15,248	18.21	34.25
2,398	2.022	11.909	17,340	18.91	38.29
2,198	2.097	13.199	19,738	19.62	42.44
1,998	2.151	14.542	22,595	20.12	46.76
1,798	2.215	15.886	24,433	20.72	51.08
1,598	2.269	17.337	26,574	21.23	55.74
1,398	2.323	18.723	27,913	21.73	60.20
1,198	2.366	20.109	29,090	22.13	64.66
998	2.420	21.539	29,491	22.63	69.26
798	2.463	22.979	29,191	23.04	73.89
598	2.527	24.505	27,534	23.64	78.79

E_{owf} plot is not a straight line, then another producing mechanism, such as a waterdrive or an initial gas cap, exists. The shape of the nonlinearity is important in diagnosing the true producing mechanisms. For instance, if the F-vs.-E_{owf} plot curves upward, this suggests that a waterdrive or an initial gas cap exists. **Fig. 9.13** shows the effect of water influx or an initial gas cap on an F-vs.-E_{owf} plot.

The number of data points in an F-vs.-E_{owf} plot is usually limited by the number of average-reservoir-pressure measurements. Recall that F and E_{owf} are functions of pressure by means of the standard PVT parameters. If two or more data points (other than the origin) exist, then a mathematical criterion must be adopted to determine the "best" line though the data or the "best" estimate of N. If a least-squares criterion is adopted, then the OOIP estimate is[28]

$$N = \frac{n\sum_{j=1}^{n}(F)_j\left(E_{owf}\right)_j - \sum_{j=1}^{n}(F)_j\sum_{j=1}^{n}\left(E_{owf}\right)_j}{n\sum_{j=1}^{n}\left(E_{owf}\right)_j^2 - \sum_{j=1}^{n}\left(E_{owf}\right)_j\sum_{j=1}^{n}\left(E_{owf}\right)_j}, \quad (9.16)$$

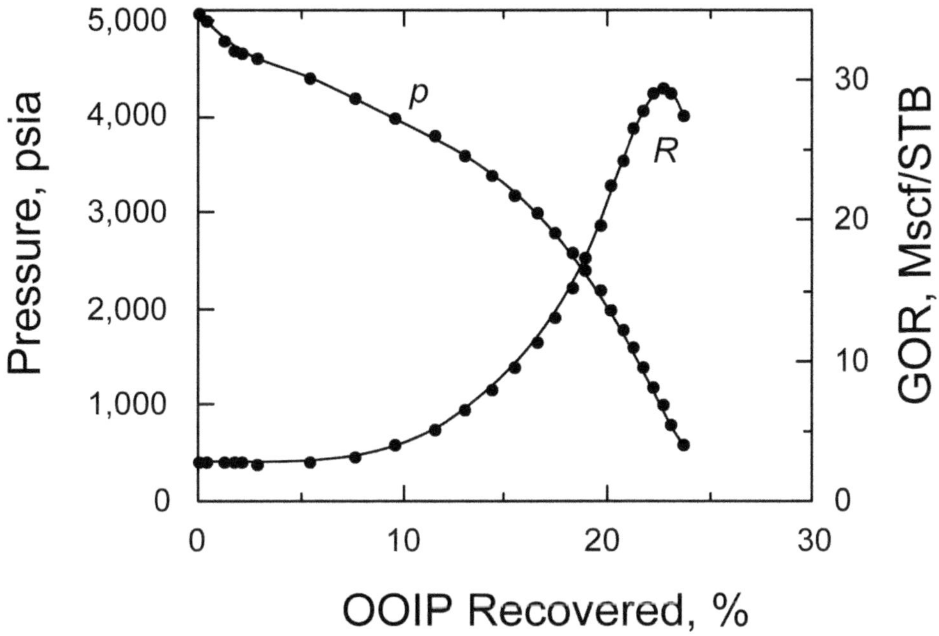

Fig. 9.11—Pressure and producing GOR as a function of OOIP recovered for a Louisiana volatile-oil reservoir.

where subscript j denotes the value at pressure p_j and n is the total number of data points. Eq. 9.16 offers a strictly mathematical means to estimate the OOIP without constructing an F-vs.-E_{owf} plot. In general, however, a plot is recommended because it provides a visual means to assess the scatter of the data. The straightness of the data points is a measure of material balance and confirmation of the solution-gas-drive mechanism.

The composite expansivity E_{owf} implicitly includes and accounts for rock and connate-water expansion. Thus, the methods offered here are applicable equally to reservoirs in which rock and connate-water expansion are important. In practice, rock and connate-water expansion cannot be neglected unless the reservoir is saturated and the pressure is less than approximately 1,500 psia. These phenomena cannot be neglected while the reservoir is undersaturated because their combined effects are not negligible compared to oil expansion. For instance, the relative expansion of oil, rock, and water in an undersaturated west Texas black-oil reservoir was 72, 25, and 3%, respectively. This example also demonstrates that the connate-water expansion is normally insignificant and can be ignored. Not until the pressure falls below the bubblepoint and approximately 1,500 psia will the rock expansion be negligible compared to the net hydrocarbon expansion. If doubt persists as to whether it is safe to ignore rock and connate-water expansion, the safest approach is to include them. To include these phenomena, the rock and connate-water expansivities, E_f and E_w, must be calculated. Sec. 9.10 discusses experimental and empirical methods to estimate E_f. The connate-water expansivity is calculated from Eq. 9.4. This equation ignores dissolved hydrocarbon gases in the water. To include dissolved gases, the water expansivity is calculated from

$$E_w = B_{tw} - B_{twi}, \quad (9.17)$$

where B_{tw} is the two-phase water FVF and is given by

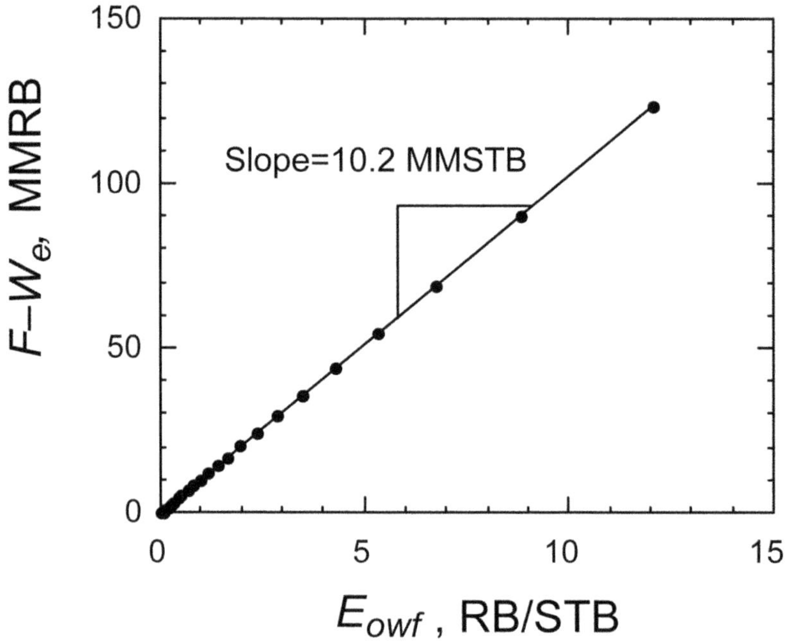

Fig. 9.12—($F - W_e$)-vs.-E_{owf} plot for a volatile-oil reservoir.

$$B_{tw} = B_w + B_g(R_{swi} - R_{sw}), \quad (9.18)$$

where R_{sw} is the dissolved gas/water ratio.[27]

Two common errors occur when applying a material-balance analysis to volatile-oil reservoirs. First, an incorrect set of PVT parameters is used. This occurs if the volatile oil is subjected to a conventional DV test instead of a CVD or a specialized DV experiment that measures volatilized oil. The resulting set of PVT parameters will not reflect the true phase behavior. If this mistake occurs, the volatilized oil/gas ratio, R_v, will be omitted altogether and the resulting values of B_o and R_s will be erroneous and overestimated. Significant errors in these fluid properties will occur if appreciable volatilized oil exists. For example, the volatile oil in Table 9.5 yielded an erroneous initial oil FVF of 3.379 RB/STB and a dissolved GOR of 3,636 scf/STB (errors of approximately 25%) when it was subjected to a standard DV instead of a CVD. The second error commonly occurs if the conventional or black-oil material-balance equation[29,30] is applied instead of the generalized equation in Eq. 9.1. The conventional material balance inherently ignores R_v. Both of these errors will cause the OOIP to be underestimated, which can be quite serious if the volatilized-oil content is appreciable.

Example 9.1: Material-Balance Analysis of a Volatile-Oil Reservoir. Perform a material-balance analysis on the Louisiana volatile-oil reservoir in Sec. 9.7.3. Use the production data in Table 9.9 and the PVT data in Table 9.5 as necessary. Estimate the OOIP (million STB) and confirm the suspected solution-gas-drive producing mechanism if possible. Compare your OOIP estimate to the volumetric estimate of 10.7 million STB reported by Cordell and Ebert.[5]

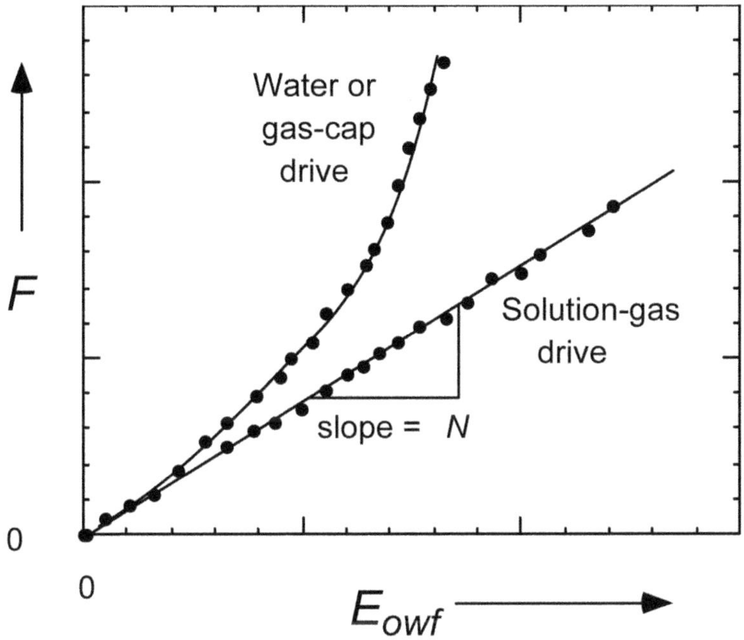

Fig. 9.13—Effect of water influx or an initial gas cap on an F-vs.-E_{owf} plot.

Solution. To confirm the producing mechanism and estimate the OOIP, construct an F-vs.-E_{owf} plot. Because the lower pressure in Table 9.9 is less than 1,500 psia and below the bubblepoint, connate-water expansion and pore-volume contraction can be ignored. Thus, E_{owf} can be replaced by E_o, where $E_o = B_{to} - B_{oi}$ and B_{to} is given by Eq. 9.5. **Table 9.10** tabulates the results of B_{to} and E_o as a function of pressure. For example, at p = 4,398 psia, evaluating Eqs. 9.5 and 9.2 yields

$$B_{to} = \frac{\left(2.338\frac{\text{RB}}{\text{STB}}\right)\left[1-\left(2{,}909\frac{\text{scf}}{\text{STB}}\right)\left(106\times10^{-6}\frac{\text{STB}}{\text{scf}}\right)\right]+\left(0.000853\frac{\text{RB}}{\text{scf}}\right)\left(2{,}909-2{,}247\frac{\text{scf}}{\text{STB}}\right)}{1-\left(106\times10^{-6}\frac{\text{STB}}{\text{scf}}\right)\left(2{,}247\frac{\text{scf}}{\text{STB}}\right)}$$
$$= 2.864\ \frac{\text{RB}}{\text{STB}}$$

and E_o = 2.864 − 2.704 = 0.160 RB/STB.

F is given by Eq. 9.8 when the reservoir is saturated and by Eq. 9.13 when the reservoir is undersaturated. For example, at p = 4,398 psia, evaluating Eq. 9.8 yields

$$F = \left(1.657\times10^{9}\text{scf}\right)\left[\frac{0.000853\frac{\text{RB}}{\text{scf}}-\left(2.338\frac{\text{RB}}{\text{STB}}\right)\left(106\times10^{-6}\frac{\text{STB}}{\text{scf}}\right)}{1-\left(106\times10^{-6}\frac{\text{STB}}{\text{scf}}\right)\left(2{,}247\frac{\text{scf}}{\text{STB}}\right)}\right]$$
$$+\left(0.582\times10^{6}\ \text{STB}\right)\left[\frac{2.338\frac{\text{RB}}{\text{STB}}-\left(0.000853\frac{\text{RB}}{\text{scf}}\right)\left(2{,}247\frac{\text{scf}}{\text{STB}}\right)}{1-\left(106\times10^{-6}\frac{\text{STB}}{\text{scf}}\right)\left(2{,}247\frac{\text{scf}}{\text{STB}}\right)}\right] = 1.638\ \text{million RB}.$$

Table 9.10 tabulates the results at other pressures and the cumulative GOR, $R_{ps} = G_{ps}/N_p$.

TABLE 9.10—MATERIAL-BALANCE CALCULATIONS FOR LOUISIANA VOLATILE OIL

p (psia)	B_{to} (RB/STB)	E_o (RB/STB)	R_{ps} (scf/STB)	F (million RB)
5,070	2.704	0.000	2,906.0	0.000
4,998	2.713	0.010	2,906.0	0.098
4,798	2.740	0.037	2,976.6	0.366
4,698	2.754	0.051	2,925.1	0.510
4,658	2.765	0.061	2,891.9	0.625
4,598	2.785	0.081	2,883.3	0.836
4,398	2.864	0.160	2,847.0	1.638
4,198	2.936	0.233	2,929.7	2.385
3,998	3.019	0.316	3,070.6	3.229
3,798	3.117	0.413	3,286.6	4.238
3,598	3.225	0.522	3,578.6	5.347
3,398	3.351	0.648	3,919.6	6.642
3,198	3.498	0.794	4,291.1	8.124
2,998	3.668	0.965	4,661.5	9.879
2,798	3.863	1.159	5,054.5	11.885
2,598	4.098	1.394	5,471.7	14.287
2,398	4.369	1.666	5,890.0	17.066
2,198	4.695	1.991	6,293.6	20.391
1,998	5.092	2.388	6,761.0	24.471
1,798	5.578	2.875	7,170.6	29.449
1,598	6.202	3.498	7,640.1	35.834
1,398	6.984	4.280	8,060.2	43.843
1,198	8.023	5.319	8,499.8	54.479
998	9.444	6.740	8,901.7	69.044
798	11.501	8.797	9,331.1	90.124
598	14.772	12.068	9,696.9	123.619

Fig. 9.12 shows a plot of F vs. E_o. The slope of this plot is 10.2 million STB, which is an estimate of the OOIP. This estimate agrees closely with the volumetric estimate of 10.7 million STB. The agreement of the volumetric and material-balance OOIP estimates, together with the straightness of the F-vs.-E_o plot, is strong evidence that this reservoir is producing exclusively by a solution-gas-drive mechanism.

If the volatilized oil was ignored and standard PVT parameters based on a conventional DV test were used, the material balance would yield an OOIP estimate of 8.2 million STB, or an error of 23%. Alternatively, if the volatilized oil was ignored and the conventional (black oil) material-balance equation were used instead of the generalized equation defined by Eqs. 9.7 through 9.10, the material balance would yield an OOIP of 9.09 million STB, or an error of 15%.

9.8 Gas-Cap Drives

In some instances, oil reservoirs are discovered with a segregated-gas zone overlying an oil column. The overlying gas zone is referred to as a primary gas cap. In addition to free gas, gas caps usually contain connate water and residual oil. The underlying oil column is sometimes referred to as an oil leg. In other instances, as reservoir pressure declines with production, gas evolves in the reservoir (see Sec. 9.7) and migrates to the top of the structure to add to an existing primary gas cap or to form a gas cap. If properly harnessed, gas caps can enhance oil recovery considerably. The degree with which they improve recovery depends mainly on their size and on the vertical permeability and/or formation dip. Producing wells usually are complet-

TABLE 9.11—GAS-CAP CHARACTERISTICS

Segregating Gas Caps
• Forms expanded gas-cap zone
• Downward-moving GOC
• Segregation mechanisms
• Active segregation—oil-leg gas migrates to gas cap and oil drains downward; and/or
• Passive segregation—gas-cap gas expands and displaces oil at GOC
• High vertical permeability
• High vertical relief
• Rate sensitive
• High oil recovery
Nonsegregating Gas Caps
• No expanded gas-cap zone
• Stationary GOC
• Gas-cap gas diffuses into oil leg
• Low vertical permeability
• Moderate oil recovery

ed only in the oil leg to minimize gas production. Broadly, gas caps are classified as segregating or nonsegregating. **Table 9.11** summarizes the distinguishing characteristics of each.

Segregating gas caps are gas caps that grow and form an enlarged gas-cap zone. Fig. 9.4 shows a schematic of a segregation-drive reservoir. Two different segregation mechanisms are possible: (1) expansion of and frontal displacement by pre-existing gas-cap gas and (2) upward migration of oil-column gas as solution gas is liberated and a free-gas phase forms. The second mechanism involves the simultaneous downward movement of oil to balance the upward flow of gas. This diametric flow pattern is referred to as counterflow.[31] Pirson[32] refers to the first mechanism as passive segregation and the latter mechanism as active segregation. Hall[31] refers to the first mechanism as segregation drive without counterflow and the second mechanism as segregation drive with counterflow. Both mechanisms are time-dependent, and their displacement efficiency depends on the gas/oil density difference, the producing rate, and the vertical permeability.

Both segregation mechanisms yield a progressively descending GOC. The segregation-drive mechanisms can be augmented by crestal gas injection.

If neither of these segregation mechanisms is present, the gas cap is called a nonsegregating gas cap. Nonsegregating gas caps do not form an enlarged gas-cap zone, and their GOC appears stationary. The gas-cap gas expands but the displacement efficiency is so poor that the expanding gas appears to merely diffuse into the oil column. **Fig. 9.14** illustrates the distribution of water, oil, and gas in a nonsegregation-drive gas-cap reservoir.

Broadly, gas caps act to mitigate the pressure decline, extend the life of the reservoir, and ultimately improve the oil recovery. The degree of oil-recovery improvement depends on the size of the gas cap and whether it is a segregation-drive or nonsegregation-drive gas cap.

To understand the mechanics of gas-cap reservoirs, numerical simulation results of segregating and nonsegregating gas caps are presented. Each example uses the fluid-property data in Table 9.3. Each example also uses the reservoir data summarized in Table 9.7, except that the initial pressure is 1,640 psia instead of 2,000 psia and the gas-cap thickness is 10 ft. The gas, oil, and water saturations in the gas cap are 60, 20, and 20%, respectively. The gas cap initial-

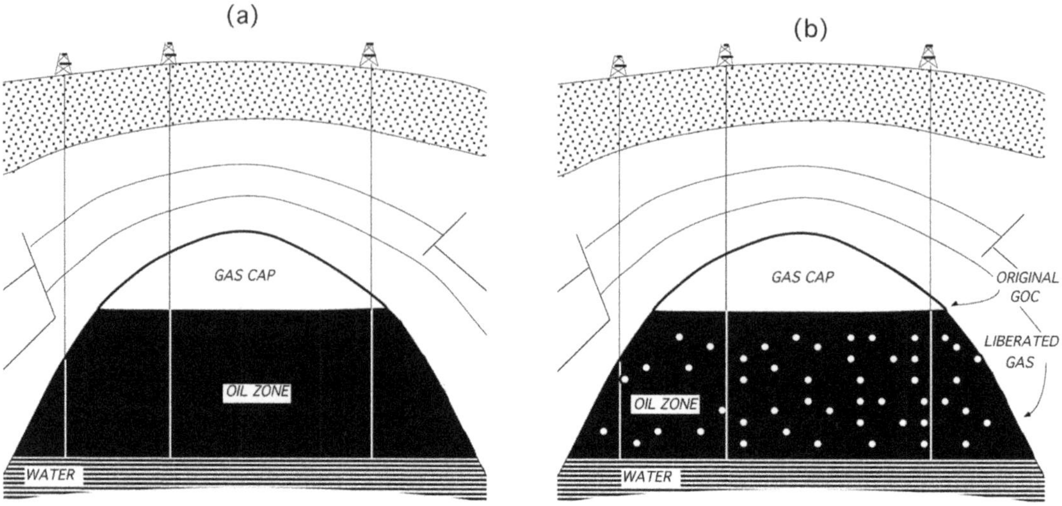

Fig. 9.14—Distribution of water, oil, and gas in a nonsegregating-gas-cap reservoir: (a) at discovery and (b) during depletion.

ly contains 270,000 STB of oil and 816 MMscf of gas; the oil leg initially contains 210 million STB of oil and 1.718 Bscf of gas. The total OOIP = 2.37 million STB, and OGIP = 2.534 Bscf, and m = 0.33. For reference, the segregating- and nonsegregating-gas-cap cases are compared with an identical reservoir without a gas cap (base case).

9.8.1 Performance. ***Nonsegregation-Drive Gas Caps.*** **Fig. 9.15** plots pressure as a function of cumulative oil recovery for a nonsegregation-drive gas-cap reservoir. For comparison, this figure includes the results of the no-gas-cap (base) case. This figure also includes the results of other cases, which are discussed later in this section. All recoveries are reported as a fraction of the original oil-leg OOIP to make direct comparisons valid. The nonsegregation-drive gas-cap case consistently yields higher oil recoveries at a given pressure than the no-gas-cap case, which illustrates the superior recovery performance of gas caps. Viewed another way, the nonsegregation-drive gas-cap case consistently yields a higher pressure at a given oil recovery than the no-gas-cap case, which illustrates the superior pressure-maintenance ability of gas caps.

Fig. 9.16 is a composite figure and shows the performance as a function of time. This figure includes the GOR, gas saturation, oil rate, and oil-recovery histories. The GOR history shows that nonsegregating gas caps eventually yield higher producing GORs than the no-gas-cap reservoir. The higher GOR is caused by higher gas saturation in the oil leg. The higher gas saturation is caused by the gas-cap gas migrating from the gas cap into the oil leg as the pressure declines.

Fig. 9.16 also shows the effect of a nonsegregating gas cap on the oil-rate history. The nonsegregating gas cap consistently yields higher oil rates than without the gas cap. If an economic limit corresponding to a minimum oil rate of 20 STB/D is arbitrarily assumed, then the no-gas-cap case is terminated after 13.8 years while the nonsegregation-drive gas-cap case is terminated after 15.2 years. This comparison shows that the presence of a gas cap extends the primary-recovery life of the reservoir. The curve endpoints denote the time of termination. The no-gas-cap case is slightly different from the black oil case discussed earlier due to the assumed lower original pressure.

Fig. 9.16 includes the fractional oil-recovery history; **Fig. 9.17** shows the gas-recovery history. The curve endpoints denote the time of the economic limit. **Table 9.12** summarizes the

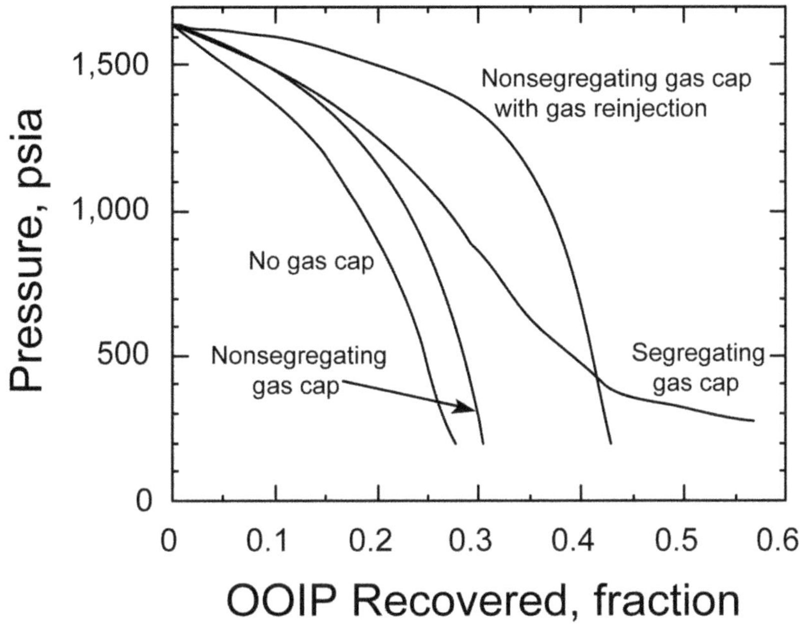

Fig. 9.15—The effect of gas cap and gas reinjection on oil recovery as a function of pressure for a west Texas black-oil reservoir.

conditions at the economic limit. The no-gas-cap and nonsegregation-drive gas-cap cases recover 23.7% and 26.8% of the oil-leg OOIP, respectively. Thus, the nonsegregating gas cap recovers more oil than without the gas cap. The nonsegregating gas cap also is terminated at a higher pressure, producing GOR, gas saturation, and gas rate than without the gas cap. The nonsegregating gas cap recovers 74.9% of the oil-leg OGIP while the no-gas-cap case recovers 52.3% of the oil-leg OGIP. The nonsegregating gas cap recovers more gas because some of the gas-cap gas infiltrates the oil leg and is produced. In conclusion, the presence of a nonsegregating gas cap yields higher ultimate oil and gas recoveries, accelerates recovery, and extends the primary-recovery life of a reservoir.

The effect of a gas cap on oil recovery is related directly to its size relative to the size of the oil leg. The size of the gas cap is described effectively in terms of the dimensionless variable m, which is defined as the ratio of the initial free-gas and free-oil phase volumes (see Eq. 9.23). If all the free gas is located in the gas cap, all the free oil is located in the oil leg, and the oil-leg and gas-cap porosities and connate-water saturations are the same, then m represents the ratio of the gas-cap and oil-leg PV. **Fig. 9.18** shows the effect of m on the final fractional oil recovery for an nonexpanding-gas-cap reservoir. The results in Fig. 9.18 use the same reservoir data as in the previous simulations except different gas-cap sizes are considered. Other reservoir conditions may yield slightly different results. The most noticeable improvement in oil recovery comes as m increases from 0 to 2.0.

The gas-cap size also affects the peak GOR. As the gas cap increases, the peak GOR increases. Fig. 9.18 shows the peak GOR as a function of m for the west Texas reservoir properties. The peak GOR increases with the gas-cap size because more gas-cap gas migrates into the oil column as the gas cap increases. In summary, nonsegregation-drive gas-cap reservoirs tend to yield final fractional oil recoveries in the range of 15 to 40% of the OOIP. Segregation-drive gas-cap reservoirs tend to yield even higher final oil recoveries.

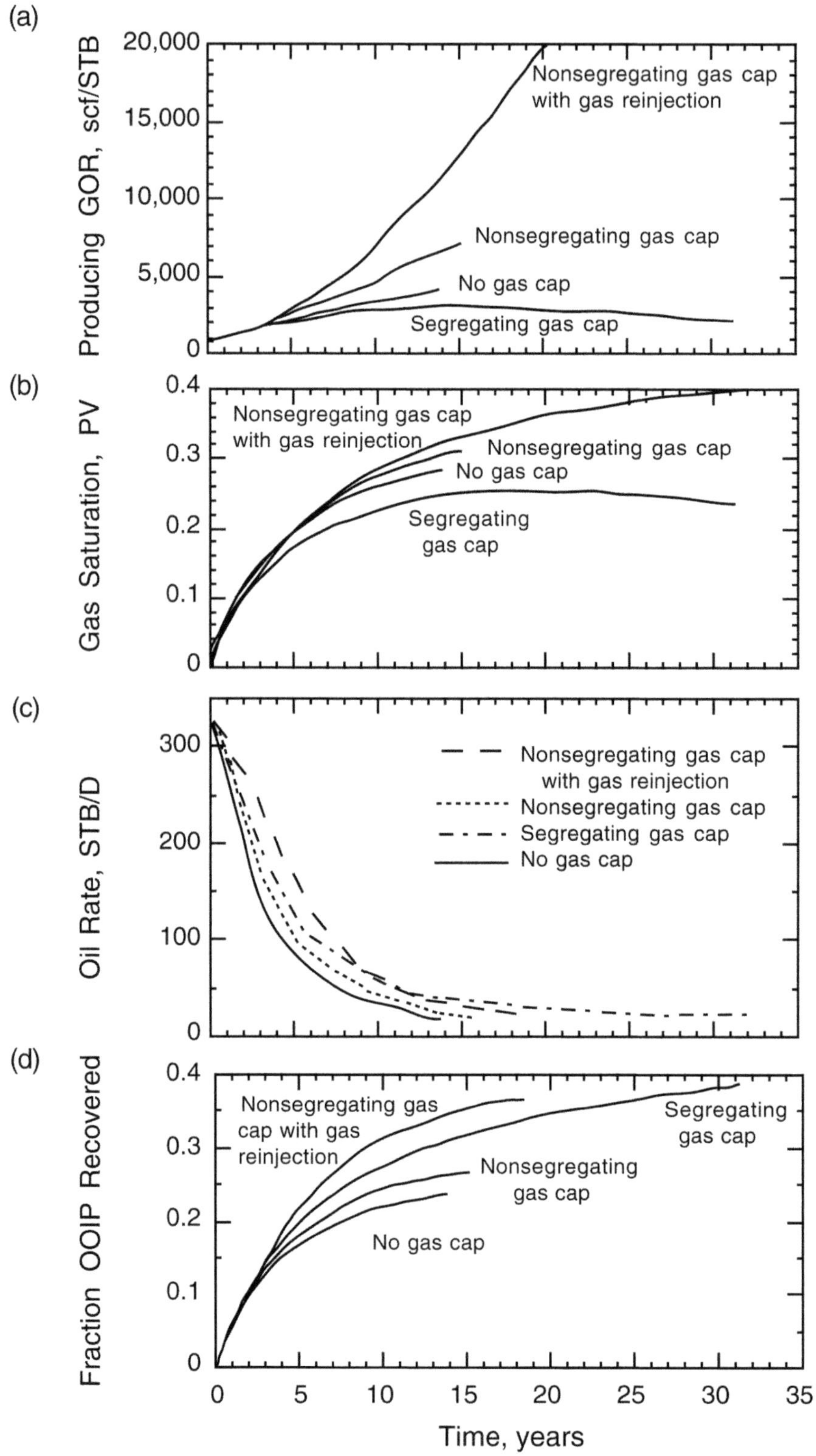

Fig. 9.16—Performance of a gas-cap reservoir: (a) instantaneous producing GOR, (b) gas-saturation, (c) oil-rate, and (d) oil-recovery histories for a west Texas black-oil reservoir.

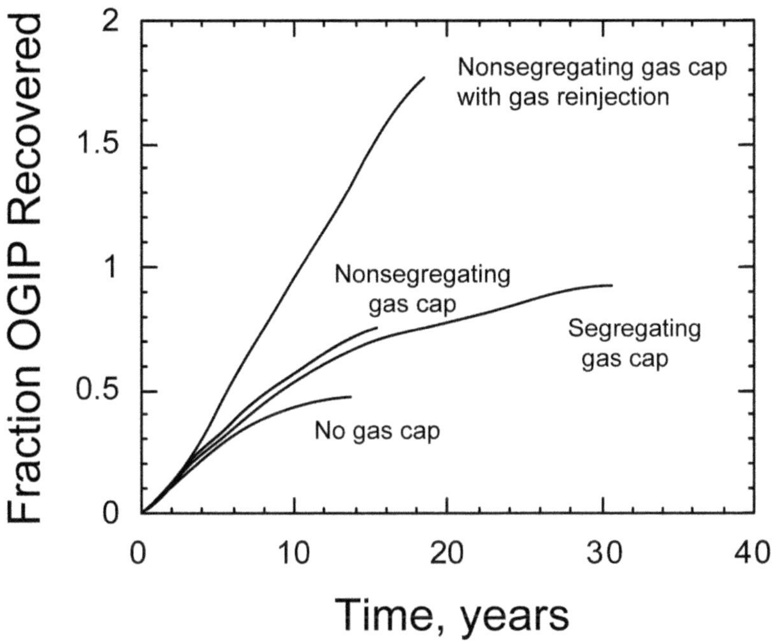

Fig. 9.17—The effect of gas cap and gas reinjection on gas-recovery history for a west Texas black-oil reservoir.

TABLE 9.12—GAS-CAP RESERVOIRS: CONDITIONS AT ECONOMIC LIMIT (20 STB/D)

Case	Time (years)	p (psia)	Recovery Oil (%)	Recovery Gas* (%)	R (scf/STB)	S_g (% PV)	Gas Rate (Mscf/D)
No gas cap (base case)	13.8	605	23.7	52	4,218	28.3	84
Nonsegregating gas cap	15.2	690	26.8	75	7,055	31.2	132
Gas reinjection, nonexpanding gas cap	18.5	993	36.7	177	17,904	35.3	352
Segregating gas cap	31.3	508	38.7	91	2,173	23.5	46

*Expressed as a percent of original oil-leg GIP.

Segregation-Drive Gas Caps. Segregating gas caps are characterized by progressively descending GOCs. The movement of the GOC is caused by active or passive gravity segregation. Active gravity segregation is the simultaneous migration of gas upward and drainage of oil downward. Passive segregation is the natural expansion of the gas-cap gas. Both of these processes involve frontal displacement of oil at the GOC. Frontal displacement helps drive oil to the producing wells. Frontal displacement does not dominate in nonsegregation-drive gas-cap reservoirs. The extent to which gravity segregation occurs depends on the vertical permeability and the rate at which fluids are withdrawn from the reservoir. The greater the vertical permeability and slower the fluid withdrawal, the more pronounced the effects of gravity segregation.

Figs. 9.15 through 9.17 include simulation results of a segregation-drive gas-cap reservoir. These simulations assume properties identical to those of the nonsegregation-drive gas-cap sim-

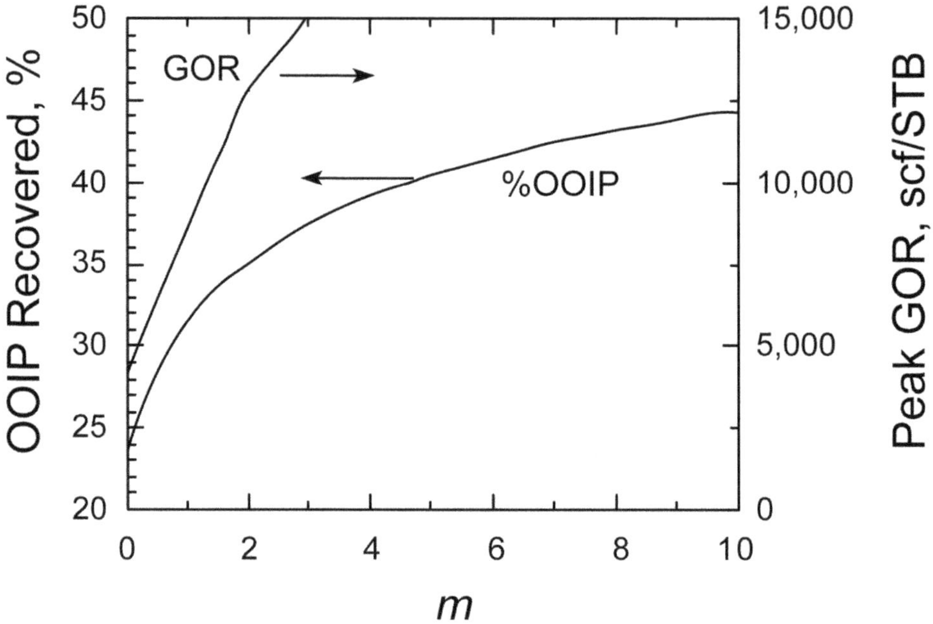

Fig. 9.18—The effect of dimensionless gas-cap size (*m*) on final primary oil recovery and peak producing GOR for a west Texas black-oil reservoir. Recoveries reported as percent of oil-leg OOIP.

ulations except gravity segregation is included. The simulations assume no free-gas production from the gas cap.

Fig. 9.15 shows the pressure as a function of cumulative oil recovery. This figure shows that oil recovery in a segregation-drive gas-cap reservoir at a given pressure is consistently greater than that in a nonsegregation-drive gas-cap or nongas-cap reservoir, especially at low pressures when the effects of gas expansion become pronounced. The oil recovery performance is discussed below.

Fig. 9.16 shows the effect of a segregating gas cap on the GOR history. Only a marginal increase in the GOR is noted; after 15 years, the GOR actually decreases slightly. This type of GOR behavior is characteristic of segregation-drive gas-cap reservoirs.[1,33,34] The segregating gas cap effectively drives and concentrates oil into the shrinking oil leg. The oil leg shrinks as the GOC descends; thus, the segregating gas cap minimizes the gas saturation in the oil leg. The GOR reversal coincides with a reversal in the gas saturation. Fig. 9.16 includes the gas saturation history. The gas saturation steadily increases until it peaks at approximately 0.25 PV; then it decreases. The GOR and gas saturation reversals occur at a moderate to low pressure when the expansion of the gas-cap gas becomes pronounced. The change in the position of the GOC yields a measure of the oil-leg shrinkage. At termination, the GOC has descended approximately 9.3 ft into the original 20-ft oil column.

Fig. 9.16 includes the oil-rate history. The oil rate for the segregating gas cap is consistently higher than for the nonsegregating gas cap or without the gas cap. The oil rate eventually flattens out to between 20 and 50 STB/D and stays within this range for 15 to 31 years. This moderate but steady oil rate explains the superior performance and long life of segregation-drive gas-cap reservoirs. Table 9.12 summarizes and compares the primary-recovery lifetimes of the various cases: the segregating gas cap has a life of 31.3 years; the nonsegregating gas cap has a life of 15.2 years; and the solution-gas drive (base case) has a life of 13.8 years.

Fig. 9.16 includes the cumulative-oil-recovery history. The segregating-gas-cap reservoir recovers 38.7% of the oil-leg OOIP while the nonsegregating-gas-cap and solution-gas-drive reservoirs recover 26.8 and 23.7% of the OOIP, respectively. Such a high recovery level for a segregation-drive reservoir is not uncommon. It is not uncommon for gravity-drainage reservoirs to realize recoveries as high as 60 to 70% of the OOIP; however, they generally require a long time to do so. The curve endpoints in Fig. 9.16 denote the time of the economic limit. The segregating-gas-cap reservoir terminates at a pressure of 508 psia.

Fig. 9.17 shows the gas-recovery history. The segregating-gas-cap reservoir recovers 91.1% of the oil-leg OGIP. This recovery level is considerably greater than the nonsegregating-gas-cap or solution-gas-drive reservoirs (74.9 and 52.3%, respectively). One reason segregating-gas-cap reservoirs tend to yield such high gas recoveries is that they often recover some of the original gas-cap gas, which migrates into the oil leg. In addition, they generally realize lower termination pressures.

The final fractional oil recovery in a segregating-gas-cap reservoir is a strong function of the vertical communication within the reservoir. Vertical communication dictates the extent of segregation. If vertical communication is good, then most of the gas-cap gas will be available for segregation. It will also be available to help drive oil through frontal displacement to the producing wells. If vertical communication is poor, then very little, if any, of the gas-cap gas will segregate. In summary, segregation is controlled principally by three variables: the vertical reservoir permeability, the producing rate, and well spacing. As well spacing and vertical permeability increase and as the producing rate decreases, the effect of gravity segregation increases. For the effects of gravity segregation to be important, however, the well spacing may need to be prohibitively large or the producing rate may need to be prohibitively low. In such reservoirs, the vertical permeability is not high enough to permit much gravity segregation.

The likely role of gravity segregation can be measured in terms of a gravity number, N_g. N_g is defined as the ratio of the time it takes a fluid to move from the drainage radius to the wellbore to the time it takes a fluid to move from the bottom of the reservoir to the top. In oilfield units, the gravity number is

$$N_g = \frac{2.46 \times 10^{-5} k_v \Delta\rho r_e^2}{\mu_o q}, \quad \text{(9.19)}$$

where k_v = vertical permeability, md; $\Delta\rho$ = density difference, lbm/ft^3; r_e = drainage radius, ft; q = producing rate at reservoir conditions, RB/D; and μ_o = oil viscosity, cp.

Gravity segregation is likely pronounced if $N_g > 10$; gravity segregation is likely unimportant if $N_g < 0.10$.For example, if k_v = 10 md, $\Delta\rho$ = 50 lbm/ft^3, r_e = 930 ft, q = 500 RB/D, and μ_o = 1 cp, then N_g = 21.3 and the effects of gravity segregation are likely important. If the vertical permeability is k_v = 0.10 md instead of 10.0 md, then N_g = 0.21 and the effects of gravity segregation are relatively unimportant.

Gas Reinjection. One method for improving oil recovery is to reinject a portion of the produced gas. The reinjected gas helps maintain reservoir pressure. One obvious drawback of gas reinjection is that gas sales revenues are reduced or delayed. The overall intention of gas reinjection is to increase the net profit despite lower gas sales. When there is no sales outlet for produced gas, reinjection can improve oil recovery until a sales outlet is established. Regulations may require reinjection until sales are possible. Inert gases such as nitrogen or carbon dioxide also could be used to supplement or replace natural-gas reinjection.

Figs. 9.15 through 9.17 present simulation results of a gas-reinjection scenario. In this scenario, 70% of the produced wellhead gas is reinjected into the gas cap, and the gas cap is nonsegregating. This means that only 30% of the produced wellhead gas is available for sales. The non-reinjected gas is referred to as sales gas. This term is sometimes a misnomer because not all of the non-reinjected gas is necessarily sold. In practice, some of the sales gas is used routinely as fuel for power or utility requirements.

Fig. 9.15 shows the effect of gas reinjection on pressure as a function of oil recovery. Oil recovery at a given pressure is consistently higher for the gas-reinjection case than for the other cases in Fig. 9.15, except at very low pressures at which the segregating-gas-cap case yields superior performance. Gas reinjection leads to higher oil recoveries because the compressed reinjected gas effectively adds extra energy to the reservoir.

Fig. 9.16 shows the effect of gas reinjection on the GOR history. Gas reinjection leads to very high producing GORs, significantly higher than the other cases. The GOR is higher because the gas saturation is higher. The gas saturation is higher because reinjected gas and initial gas-cap gas migrate into the oil leg during pressure depletion. This occurs because the gas cap is nonsegregating. High producing GORs are a characteristic feature of reservoirs subject to gas reinjection if there is little or no active gravity drainage. High producing GORs mean that large volumes of produced gas will have to be handled and processed at the surface.

Fig. 9.16 includes the effect of gas reinjection on the oil-rate history. This figure shows that the oil rate is higher for the first 8½ years for the gas-reinjection case than for any of the other cases. After 8½ years, the oil rate for the segregating-gas-cap case is slightly greater than the oil rate for the gas-reinjection case. These results demonstrate that gas reinjection is an effective means to arrest the normal oil-rate decline dramatically.

Fig. 9.16 also shows the effect of gas reinjection on the fractional oil-recovery history and that the gas-reinjection case is superior to the other cases. The gas-reinjection case recovers 36.7% of the original oil-leg OOIP at its economic limit of 18½ years. Only the segregating-gas-cap reservoir recovers more oil (38.7%); however, the segregating-gas-cap reservoir requires more time to recover the additional oil.

Fig. 9.17 shows the effect of gas reinjection on the fractional gas-recovery history. The fractional gas recovery is the cumulative produced wellhead gas normalized by the original oil-leg OGIP. The gas-reinjection case recovers 177% of the oil-leg OGIP (see Table 9.12). More than 100% of the oil-leg OGIP is produced because some of the reinjected gas is produced. Because 30% of the produced gas is not reinjected, 0.30 × 177 or 53.1% of the oil-leg OGIP is available for gas sales. This sales-gas recovery is comparable to the case without gas reinjection (52% OGIP).

Reservoirs subject to gravity drainage are especially attractive for gas reinjection. Crestal gas injection into the developing gas cap is the preferred strategy because gravity drainage helps control the movement of the injected gas. Excellent sweep and displacement efficiencies and high oil recoveries can be realized. The Tensleep pool in the Elk Basin field in Wyoming is a good example.[35–37] This pool was projected to recover approximately 64% of the OOIP. See the chapter on Immiscible Gas Injection in Oil Reservoirs in this volume of the *Handbook* for more information on gravity drainage.

9.8.2 Material-Balance Analysis. The purpose of a material-balance analysis includes confirming the producing mechanism and estimating the OOIP, OGIP, and size of the gas cap. The applicable material-balance equation for initially saturated oil reservoirs is[20,22,23]

$$F - We = G_{fgi} E_{gwf} + N_{foi} E_{owf} \; . \quad (9.20)$$

This equation is applicable to all initially saturated reservoirs regardless of the distribution of the initial free gas. For example, this equation is applicable to reservoirs whether the initial free gas is segregated into a gas cap or uniformly dispersed throughout the reservoir. Eq. 9.20 also applies to waterdrives; however, if the following methods are applied to waterdrives, the water-influx history must be reliably known. If the water-influx history is unknown, then the methods Section 9.9.8 must be applied.

The quantities G_{fgi} and N_{foi} are related to N (OOIP) and G (OGIP) by the following equations:

$$G = G_{fgi}\left(1 + \frac{R_{si}B_{gi}}{mB_{oi}}\right) \quad (9.21)$$

$$\text{and } N = N_{foi}\left(1 + \frac{mB_{oi}R_{vi}}{B_{gi}}\right), \quad (9.22)$$

where m is the ratio of the free-gas-phase and free-oil-phase volumes and is defined by

$$m = \frac{V_{fgi}}{V_{foi}} = \frac{G_{fgi}B_{gi}}{N_{foi}B_{oi}} . \quad (9.23)$$

The dimensionless variable m is sometimes called the dimensionless gas-cap volume.

Because G_{fgi} and N_{foi} are independent, they must be determined simultaneously. At least two sets of the independent variables (F, W_e, E_{gwf}, E_{owf}) must be known at two or more pressures (other than the initial pressure) to determine the set (G_{fgi}, N_{foi}). If three or more sets (F, W_e, E_{gwf}, E_{owf}) are known, then multiple sets (G_{fgi}, N_{foi}) can be determined. The optimal set is determined by one of two least-squares solution techniques: iterative or direct methods.

In the iterative method, Eq. 9.20 is expressed as

$$F - W_e = N_{foi}E_t(m), \quad (9.24)$$

where E_t is the total expansivity expressed per unit volume of stock-tank oil and is defined by

$$E_t(m) = E_{owf} + mE_{gwf}\left(\frac{B_{oi}}{B_{gi}}\right) . \quad (9.25)$$

The solution procedure to estimate the OOIP and OGIP involves the following steps:

1. Compute F, E_{gwf}, and E_{owf} for each data point (i.e., average reservoir-pressure measurement).
2. Guess m.
3. Compute $E_t(m)$ with Eq. 9.25.
4. Estimate N_{foi} with a least-squares analysis using Eq. 9.26.

$$N_{foi} = \frac{n\sum_{j=1}^{n}\left(F - W_e\right)_j\left(E_t\right)_j - \sum_{j=1}^{n}\left(F - W_e\right)_j\sum_{j=1}^{n}\left(E_t\right)_j}{n\sum_{j=1}^{n}\left(E_t\right)_j^2 - \sum_{j=1}^{n}\left(E_t\right)_j\sum_{j=1}^{n}\left(E_t\right)_j}, \quad (9.26)$$

where j denotes the data point index and n is the total number of data points.

5. Compute the residual R for each data point with

$$R_j = \left(F - W_e\right)_j - N_{foi}\left(E_t\right)_j . \quad (9.27)$$

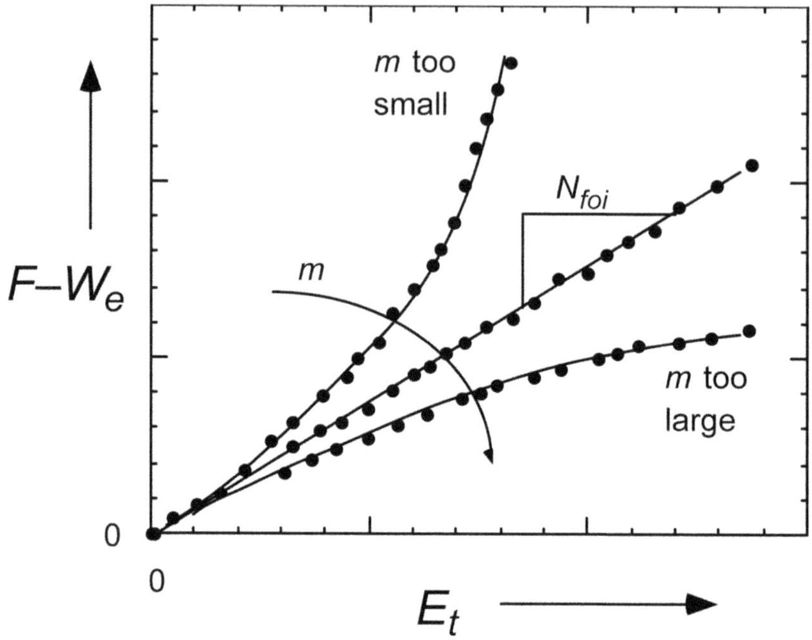

Fig. 9.19—The effect of *m* on a plot of ($F - W_e$) vs. E_t.

6. Compute sum of the squares of residual, R_{ss}, as

$$R_{ss} = \sum_{j=1}^{n} R_j^2 \text{ . } \quad (9.28)$$

7. Return to Step 2 and repeat until R_{ss} is minimized.
8. Compute G, N, and G_{fgi} from Eqs. 9.21 through 9.23.

Minimization algorithms speed solution. This procedure is ideally suited for spreadsheet calculation, especially spreadsheet programs that contain minimization algorithms.

The use of Eq. 9.26 in Step 4 to determine N_{foi} is equivalent to the slope of a $(F - W_e)$-vs.-$E_t(m)$ plot. This graphical solution method can be substituted for Eq. 9.26 in Step 4 if desired. Overall, Steps 2 through 7 are equivalent to the graphical procedure of varying m until the straightest possible $(F - W_e)$-vs.-$E_t(m)$ plot is realized.[38] **Fig. 9.19** shows the qualitative effect of m on the shape of the $(F - W_e)$-vs.-E_t plot. If m is too small, the plot curves upward slightly; if m is too large, the plot curves downward slightly.

Once m is determined, the final $(F - W_e)$-vs.-E_t plot is used to confirm the producing mechanism. The linearity of the plot is a measure of material balance and the applicability of the presumed producing mechanism. If the plot exhibits considerable curvature, then either the presumed mechanism is incorrect or additional producing mechanisms are active. If curvature exists, the shape of the curvature provides insight into the true producing mechanism. For instance, if the plot curves upward, this indicates that net withdrawal exceeds net expansion and that water influx, for example, has been ignored or is possibly underestimated.

As an alternative to the iterative method, Walsh[23,28] presented a direct method. This method is based on least-squares multivariate regression. The least-squares equations are simple but lengthy. The technique is ideally suited for spreadsheet calculation. Walsh's method is especially attractive because it avoids iteration and the complications of attaining and judging convergence.

Havlena and Odeh[38] proposed another solution method in which $(F - W_e)/E_{owf}$ is plotted vs. (E_{gwf}/E_{owf}); the slope of the plot is equal to G_{fgi} and the y-intercept is equal to N_{foi}. This method is popular and attractive because it yields a direct solution. In theory, this method is perfectly acceptable. In practice, however, it has shown to be unreliable because it suffers from hypersensitivity to pressure uncertainty.[28,39] The method has been shown to yield highly erroneous G_{fgi} and N_{foi} estimates in the presence of only small amounts of uncertainty. For instance, Walsh[28] shows that only a 5-psi pressure uncertainty yielded an error of more than 150% in N_{foi} and an error of more than 250% in G_{fgi}. The hypersensitivity is caused by the fact that the divisor (E_{owf}) approaches zero as the pressure approaches the initial pressure. Small errors in E_{owf}, in turn, produce large errors in the quotients $(F - W_e)/E_{owf}$ and (E_{gwf}/E_{owf}). Tehrani[40] calls this problem a "loss in resolving power." Because of this hypersensitivity, this method should be used cautiously.

Walsh[28] tested the direct and iterative methods for their tolerance to uncertainty. He observed sensitivity, but the degree of sensitivity was less than the method of plotting $(F - W_e)/E_{owf}$ vs. (E_{gwf}/E_{owf}). He concluded that material-balance methods for gas-cap reservoirs should be used cautiously.

9.9 Waterdrives

Waterdrive petroleum reservoirs are characteristically bounded by and in communication with aquifers. As pressure decreases during pressure depletion, the compressed waters within the aquifers expand and overflow into the petroleum reservoir. The invading water helps drive the oil to the producing wells, leading to improved oil recoveries. Like gas reinjection and gas-cap expansion, water influx also acts to mitigate the pressure decline. The degree with which water influx improves oil recovery depends on the size of the adjoining aquifer, the degree of communication between the aquifer and petroleum reservoir, and ultimately the amount of water that encroaches into the reservoir. Some of the most prolific oil fields in the world are waterdrive reservoirs. Perhaps the most celebrated example is the East Texas field. The final oil recovery in the East Texas field is projected to be approximately 79%.[41] As this example shows, water influx has the potential to improve oil recovery considerably.

Once a water-influx mechanism has been identified, it is important to monitor the producing wells closely and to minimize water production. Minimizing water production in "edgewater drives" may require systematically shutting in flank wells once the advancing water reaches them. Minimizing water production in "bottomwater drives" may require systematically cementing in lower perforations as the bottom water slowly rises.

An integral part of reservoir surveillance for waterdrives is an active assessment program. The first phase of assessment includes diagnosis, classification, and characterization. The second phase identifies mathematical models that effectively simulate the aquifer, especially its deliverability. This phase includes reliably estimating aquifer model parameters. The third and final phase includes combining aquifer and reservoir models into a common model that can be used to forecast future recovery effectively and to identify optimal depletion strategies. The success of the third phase depends heavily on the success of the preceding two phases.

9.9.1 Waterdrive and Aquifer Classification. Waterdrives are classified in several ways. First, they are classified according to the location of the aquifer relative to the reservoir. If the aquifer areally encircles the reservoir, either partially or wholly, the waterdrive is called a peripheral waterdrive. If the aquifer exclusively feeds one side or flank of the reservoir, the waterdrive is called an edgewater drive. If the aquifer underlays the reservoir and feeds it from beneath, the waterdrive is called a bottomwater drive.

Waterdrives also are classified according to the aquifer's strength and to how well the aquifer delivers recharge water to the reservoir. The aquifer strength also refers to how well the aquifer mitigates the reservoir's normal pressure decline. A strong aquifer refers to one in which the water-influx rate approaches the reservoir's fluid withdrawal rate at reservoir conditions. These reservoirs also are called complete waterdrives and are characterized by minimal pressure decline. Strong aquifers are generally very large in size and highly conductive. A moderate or weak aquifer is one in which the water recharge rate is appreciably less than the reservoir's fluid withdrawal rate. These reservoirs are called partial waterdrives and they are characterized by pressure declines greater than a complete waterdrive but less than a volumetric reservoir. An aquifer's weakness is related directly to its lack in size or conductivity.

9.9.2 Waterdrive Diagnosis. There are several diagnostic indicators to help identify or discount a possible active aquifer.

First, an understanding of the reservoir's geology is important. The entire outer surface of the reservoir must be scrutinized carefully to identify communicating and noncommunicating pathways; communicating pathways represent possible water entry points. Geological maps should be consulted to identify the type of reservoir trap and the trapping surfaces. Trapping surfaces represent impenetrable surfaces and are discounted automatically as possible water-entry points. The remaining outer surfaces need to be evaluated and classified. If no communicating pathways exist, then the reservoir can be confidently discounted as a possible waterdrive; however, if communicating pathways exist, then the reservoir remains a candidate waterdrive.

Second, and perhaps most importantly, the water-cut history of all producing wells should be recorded and regularly monitored. A steady rise in a well's water cut is a good indicator of an active aquifer. Although this is among the best indicators, it is not foolproof. For instance, an increasing water cut might be caused by water coning instead of an active waterdrive. Special precautions need to be exercised to avoid water coning. A rising water/oil contact (WOC) is a good indicator of a bottomwater drive. Special attention should be paid to the location of high-water-cut wells. Their location will help define the position of the reservoir/aquifer boundary in peripheral and edgewater drives.

Third, the change in reservoir pressure also can be a helpful indicator. Strong-waterdrive reservoirs are characterized by a slow or negligible pressure decline. Thus, a slower-than-expected pressure decline can help indicate a waterdrive. Material-balance calculations are important to help identify and confirm a slower-than-expected pressure decline.

The reservoir pressure distribution also can help diagnose an active aquifer. For peripheral-water and edgewater drives, higher pressures tend to exist along the reservoir/aquifer boundary while lower pressures tend to exist at locations that are more distant. A pressure contour map is sometimes helpful to identify pressure disparities.

Fourth, the producing GOR can be a helpful indicator. Strong waterdrives are characterized by small changes in the producing GOR. The small GOR change is directly related to the small pressure decline. Sec. 9.9.3 discusses this and other performance features characteristic of waterdrive reservoirs.

Fifth, a material-balance analysis can help diagnose water influx. Several different types of material-balance analyses such as the McEwen[42] analysis can identify water influx. Sec. 9.9.8 discusses these methods.

9.9.3 Performance. To illustrate the performance of waterdrives, simulation results of an 80-acre segment of a west Texas black-oil reservoir are presented. The segment is assumed to be surrounded by an infinite radial-flow aquifer. The reservoir properties in Table 9.7 apply. The aquifer permeability and porosity are 37 md and 27%, respectively.

Fig. 9.20 shows the effect of water influx on a plot of pressure vs. fractional oil recovery. The initial reservoir pressure is 2,000 psia. Waterdrive and solution-gas-drive performances are

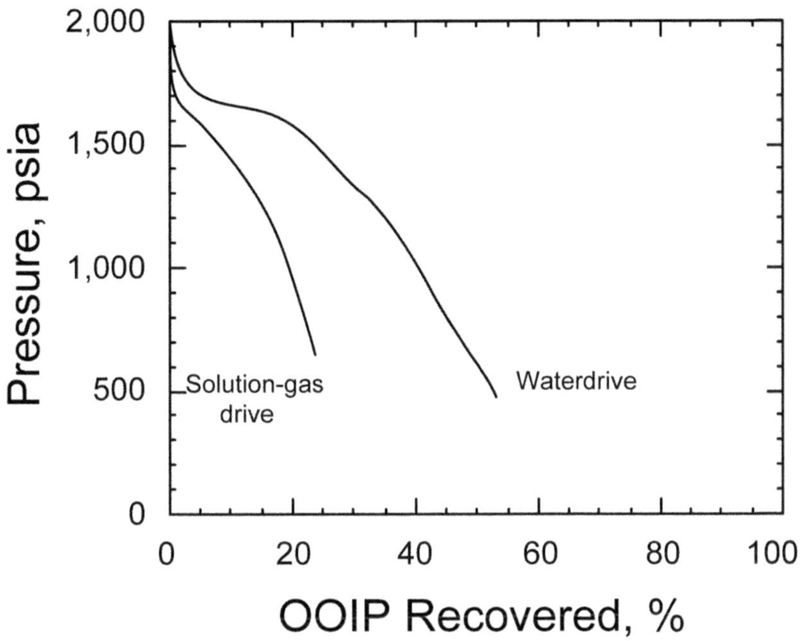

Fig. 9.20—The effect of water influx on pressure vs. percent OOIP recovered at a west Texas black-oil reservoir.

compared. This figure shows that water influx consistently improves the fractional oil recovery at a given pressure. Alternatively, the waterdrive maintains a higher pressure at a given recovery.

Fig. 9.21 shows the reservoir performance with time. This figure contains four separate plots: GOR, saturation, oil-recovery, and pressure histories. Waterdrives typically yield a characteristic GOR history. After a brief increase, the GOR typically levels off. This behavior is explained by the gas-saturation history. The gas saturation increases as soon as the pressure falls below the bubblepoint. After a brief increase, the gas saturation also levels off. The gas saturation levels off because the invading water drives the oil toward the producers. The oil is concentrated in a nonswept region or shrinking oil zone. Without oil displacement, the gas saturation and GOR would rise unabated. The solution-gas drive yields a final GOR of 4,506 scf/STB and gas saturation of 0.287 PV, while the waterdrive yields 1,323 scf/STB and 0.19 PV, respectively. The GOR for the water drive in Fig. 9.21 actually peaks after approximately 10 years and then decreases slightly. A decrease in the GOR reflects good displacement efficiency of oil by water.

As expected, the waterdrive yields a substantially higher recovery. The waterdrive also lengthens the productive life of the reservoir considerably. In this example, the waterdrive recovers 53.2% of the OOIP after 32.6 years, while the solution-gas drive recovers 24.2% of the OOIP after 13.5 years. Both cases assume a terminal oil rate of 20 STB/D. This recovery level indicates a relatively moderate to strong waterdrive. The waterdrive also yields a higher gas recovery (80.5 vs. 53.1%). The water-influx history basically mimics the incremental oil-recovery history. The cumulative encroached water is 58% hydrocarbon pore volume (HCPV) or 0.46 PV. This translates to approximately 1% OOIP incremental recovery for each 0.16 PV (or 2.0% HCPV) of encroached water. The waterdrive in Fig. 9.21 consistently yields higher producing rates than solution-gas drive.

Also as expected, the waterdrive consistently yields a higher pressure at a given time. The waterdrive yields a lower terminal pressure because lower gas saturations are realized at a giv-

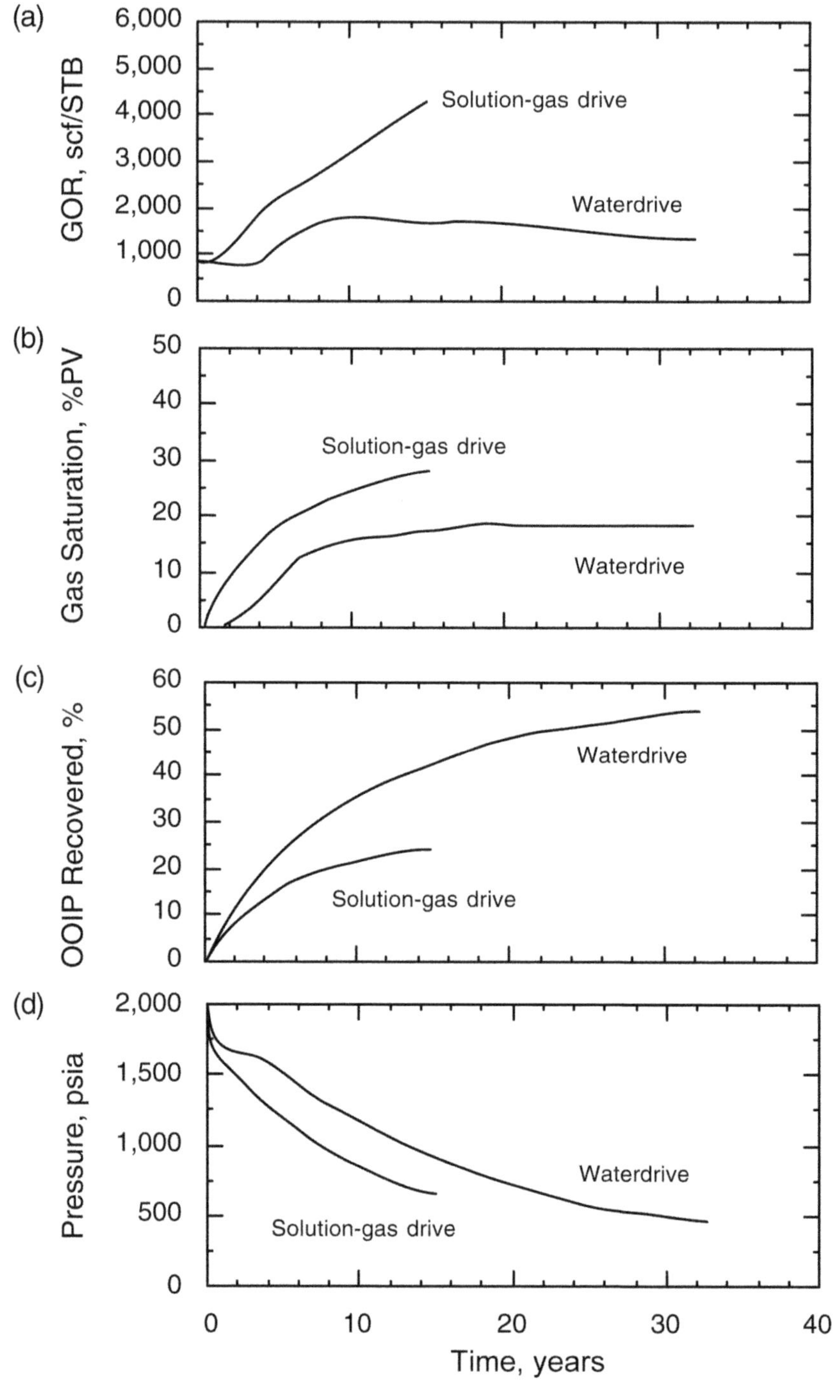

Fig. 9.21—Effect of water influx on (a) GOR, (b) gas-saturation, (c) oil-recovery, and (d) pressure histories for a west Texas black-oil reservoir.

en pressure. The example water and solution-gas drives yield final pressures of 471 and 613 psia, respectively.

The performance trends noted in Figs. 9.20 and 9.21 are not without exception. Waterdrive performance is strongly influenced by the displacement efficiency of oil by water. Figs. 9.20 and 9.21 are representative of moderate-to-good displacement efficiency. If displacement efficiency is poor, lower oil recovery will occur. A less obvious result, however, is that the GOR history will exhibit much different character than already discussed. Instead of rising slightly and leveling off, the GOR acts much like solution-gas drive; namely, the GOR steadily and monotonically increases. This difference occurs because the invading water bypasses substantial oil and fails to drive enough oil toward the producers to arrest the natural increase in the gas saturation. The GOR of a waterdrive can even exceed the GOR of the solution-gas drive if the displacement efficiency is poor enough. Peripheral waterdrives do not tend to be as efficient as bottomwater drives.

In summary, water influx can markedly improve oil recovery in oil reservoirs. The final oil recovery in a waterdrive reservoir depends largely on the net volume of influxed water. The net volume of influxed water is defined as the volume of influxed water less the volume of produced water. As the net volume of influxed water increases, the oil recovery increases. The volume of the influxed water depends mainly on the size of the aquifer and the communication between the aquifer and the reservoir. The maximum possible net volume of influxed water expressed as a fraction of the reservoir PV is

$$(1 - S_{orw} - S_{grw} - S_{wi}), \qquad (9.29)$$

where S_{orw} is the residual oil saturation to water, S_{grw} is the residual gas saturation to water, and S_{wi} is the initial water saturation. As this equation shows, the residual saturations directly affect oil recovery by limiting the net volume of water that can influx into the reservoir. The residual saturations are a direct measure of the displacement efficiency of water. Lower residual oil saturations are preferred over lower residual gas saturations to promote oil production over gas production. The maximum fractional oil recovery for an initially undersaturated black-oil reservoir is

$$E_v(1 - S_{orw} - S_{grw} - S_{wi})/(1 - S_{wi}), \qquad (9.30)$$

where E_v is the volumetric sweep efficiency of the invading water. This equation assumes a complete waterdrive (i.e., no pressure depletion).

9.9.4 Water-Influx Models. Water-influx models are mathematical models that simulate and predict aquifer performance. Most importantly, they predict the cumulative water-influx history. When successfully integrated with a reservoir simulator, the net result is a model that effectively simulates waterdrive performance.

There are several popular aquifer models: the van Everdingen-Hurst (VEH) model,[43] the Carter-Tracy model,[44] the Fetkovich model,[45] the Schilthuis model,[29] and the small- or pot-aquifer model.[46] The first three models are unsteady-state models and are the most realistic. They attempt to simulate the complex pressure changes that gradually occur within the aquifer and between the aquifer and reservoir. As pressure depletion proceeds, the pressure difference between the reservoir and aquifer grows rapidly and then abates as the aquifer and reservoir eventually equilibrate. This pressure interaction causes the water-influx rate to start at zero, grow steadily, reach a maximum, and then dissipate. This particular water-influx-rate history behavior applies to initially saturated oil reservoirs; the behavior for initially undersaturated oil reservoirs is often slightly but distinctly different. The effects of undersaturation on the water-influx performance are discussed in Sec. 9.9.7. The unsteady-state models are far more successful at capturing the real dynamics than other models. In contrast, Schilthuis' steady-state model

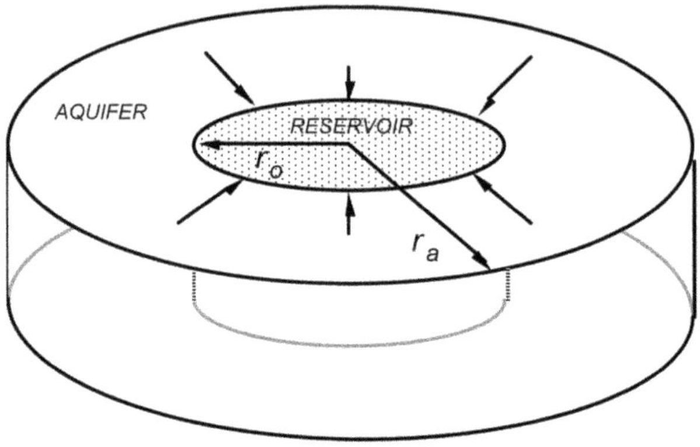

Fig. 9.22—Radial aquifer model.

assumes the aquifer pressure remains constant. The small-aquifer model, however, assumes the aquifer and reservoir pressures are equal.

The VEH model is the most sophisticated of all these models. Its main advantage is its realism. Originally, its main disadvantage was its cumbersome nature. Charts or tables had to be consulted repeatedly to execute a single calculation. To address this limitation, the Carter-Tracy and Fetkovich models were alternatives that were free of tables and charts. These models, however, were only approximations to and simplifications of the VEH model. Since the VEH charts and tables were digitized,[47–49] the need for alternatives has diminished.

Allard and Chen[50] proposed an aquifer model specifically for bottomwater drives. This model included 2D flow. In comparison, the VEH model considered only 1D flow. Simulation practitioners, however, have found that the VEH model is satisfactory in simulating bottomwater drives.[51]

9.9.5 van Everdingen-Hurst (VEH) Model. van Everdingen and Hurst considered two geometries: radial- and linear-flow systems. The radial model assumes that the reservoir is a right cylinder and that the aquifer surrounds the reservoir. **Fig. 9.22** illustrates the radial aquifer model, where r_o = reservoir radius and r_a = aquifer radius. Flow between the aquifer and reservoir is strictly radial. This model is especially effective in simulating peripheral and edgewater drives but also has been successful in simulating bottomwater drives.[50]

In contrast, the linear model assumes the reservoir and aquifer are juxtaposed rectangular parallelepipeds. **Fig. 9.23** shows examples. Flow between the aquifer and reservoir is strictly linear. This model is intended to simulate edgewater and bottomwater drives. The model definition depends on the application. For edgewater drives, the thicknesses of the reservoir and aquifer are identical; the widths of the reservoir and aquifer are also the same, and the aquifer and reservoir lengths are L_a and L_r, respectively (Fig. 9.23a). For bottomwater drives, the width of the reservoir and aquifer are identical; the length (L) of the reservoir and aquifer are also the same; the aquifer depth is L_a, and the reservoir thickness is h (Fig. 9.23b).

van Everdingen and Hurst solved the applicable differential equations analytically to determine the water-influx history for the case of a constant pressure differential at the aquifer/reservoir boundary. This case assumes the reservoir pressure is constant. They called this case the “constant terminal pressure” and reported their results in terms of tables and charts. This solution is not immediately applicable to actual reservoirs because it does not consider a declining reservoir pressure. To address this limitation, van Everdingen and Hurst applied the superposition theorem to a specific reservoir pressure history. This adaptation usually requires

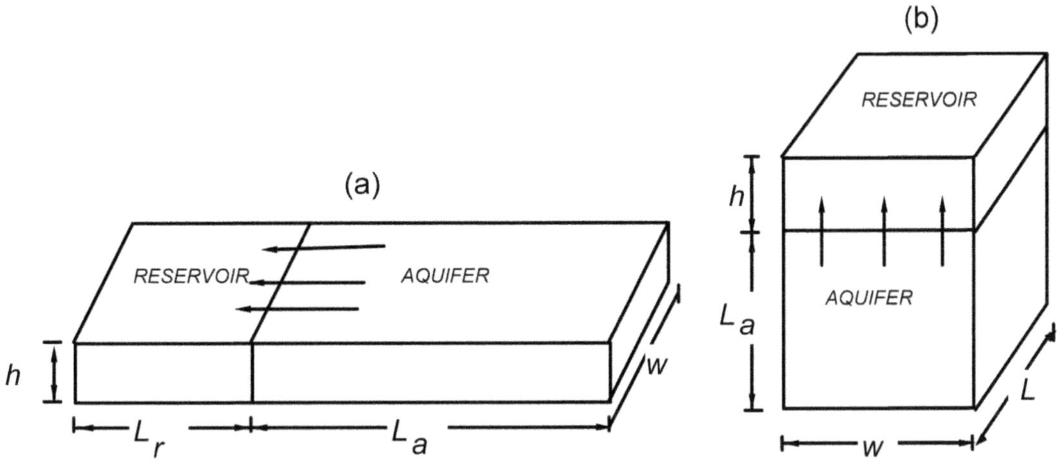

Fig. 9.23—Linear aquifer model for (a) an edgewater drive and (b) a bottomwater drive.

TABLE 9.13—TIME- AND PRESSURE-DOMAIN DISCRETIZATION FOR VEH MODEL

j	t	t_D	$\bar{p}$	$\tilde{p}$	Δp
0	$t_0=0$	t_{D0}	$\bar{p}_0 = p_i$	$\tilde{p}_0 = \bar{p}_0 = p_i$	n/a
1	t_1	t_{D1}	$\bar{p}_1$	$\tilde{p}_1 = (\bar{p}_0 + \bar{p}_1)/2$	$\Delta p_1 = (\tilde{p}_0 - \tilde{p}_1)$
2	t_2	t_{D2}	$\bar{p}_2$	$\tilde{p}_2 = (\bar{p}_1 + \bar{p}_2)/2$	$\Delta p_2 = (\tilde{p}_1 - \tilde{p}_2)$
.	.	.	.	.	.
.	.	.	.	.	.
.	.	.	.	.	.
n	t_n	t_{Dn}	$\bar{p}_n$	$\tilde{p}_n = (\bar{p}_{n-1} + \bar{p}_n)/2$	$\Delta p_n = (\tilde{p}_{n-1} - \tilde{p}_n)$

that the reservoir's pressure history be known. The first step in applying their model is to discretize the time and pressure domains.

Discretization. The time domain is discretized into $(n + 1)$ points $(t_0, t_1, t_2, \ldots, t_n)$, where $t_0 < t_1 < t_2 < \ldots < t_n$ and t_0 corresponds to $t = 0$. The average reservoir pressure domain also is discretized into $(n + 1)$ points $\left(\bar{p}_0, \bar{p}_1, \bar{p}_2, \ldots, \bar{p}_n\right)$, where $\bar{p}_0$ is the initial pressure p_i. The time-averaged pressure between levels j and $j - 1$ is

$$\tilde{p}_j = \left(\frac{\bar{p}_{j-1} + \bar{p}_j}{2}\right) \text{ for } j = 1, 2, \ldots, n \text{ . } \qquad (9.31)$$

The time-averaged pressure at level $j = 0$ is defined as the initial pressure p_i. **Table 9.13** shows discretization of t, $\bar{p}$, and $\tilde{p}$. The time-averaged pressure decrement between levels j and $j - 1$ is

$$\Delta p_j = \tilde{p}_{j-1} - \tilde{p}_j \text{ for } j = 1, 2, \ldots, n \text{ . } \qquad (9.32)$$

No value is defined for $j = 0$. Table 9.13 shows the complete discretization of t, $\bar{p}$, $\tilde{p}$, and Δp.

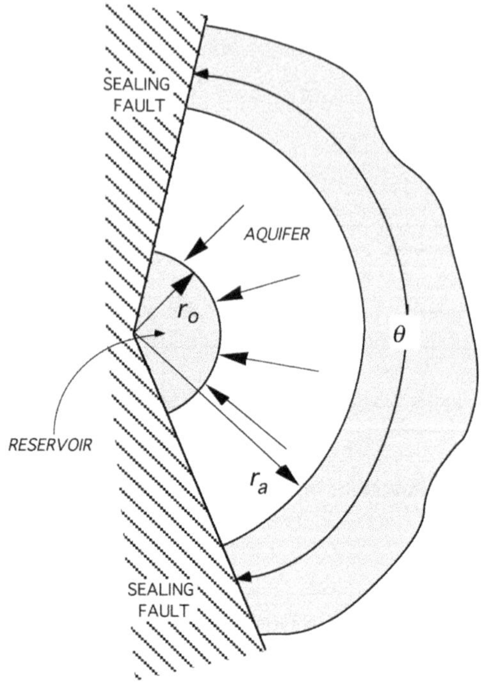

Fig. 9.24—Definition of angle, *θ*, for radial aquifer model.

Cumulative Water Influx. The cumulative water influx at *k*th level is

$$W_{ek}(t) = U \sum_{j=0}^{k-1} \Delta p_{j+1} W_D\left(t_{Dk} - t_{Dj}\right) \text{ for } k = 1, 2, \ldots, n, \qquad (9.33)$$

where U is the aquifer constant and W_D is the dimensionless cumulative water influx. This equation is based on the superposition theorem. The term $W_D(t_{Dk} - t_{Dj})$ is not a product but refers to the evaluation of W_D at a dimensionless time difference of $(t_{Dk} - t_{Dj})$. If we apply Eq. 9.33 for k = 1, 2, and 3, we obtain

$$W_{e1}(t) = U \Delta p_1 W_D\left(t_{D1} - t_{D0}\right),$$

$$W_{e2}(t) = U \Delta p_2\left[W_D\left(t_{D2} - t_{D0}\right) + W_D\left(t_{D2} - t_{D1}\right)\right],$$

$$\text{and } W_{e3}(t) = U \Delta p_3\left[W_D\left(t_{D3} - t_{D0}\right) + W_D\left(t_{D3} - t_{D1}\right) + W_D\left(t_{D3} - t_{D2}\right)\right].$$

The length of the equation grows with the time. The aquifer constant, U, and the dimensionless cumulative water influx, $W_D(t_D)$, depend on whether the radial or linear model is applied.

Radial Model. The radial model is based on the following equations. The effective reservoir radius is a function of the reservoir PV and is

$$r_o = \sqrt{\frac{5.6146 V_{pr}}{\pi \phi_r h f}}, \qquad (9.34)$$

where r_o is expressed in ft, V_{pr} is the reservoir PV expressed in RB, ϕ_r is the reservoir porosity (fraction), and h is the pay thickness in ft. The constant f is $\theta/360$, where θ is the angle that defines the portion of the right cylinder. **Fig. 9.24** illustrates the definition of θ for a radial aquifer model. The dimensionless time is

$$t_D = \frac{2.309 k_a t}{\phi_a \mu_w c_t r_o^2}, \qquad (9.35)$$

where k_a = aquifer permeability (md), μ_w = water viscosity (cp), c_t = total aquifer compressibility (psi^{-1}), ϕ_a = aquifer porosity (fraction), and t is expressed in years. The total aquifer compressibility is the sum of the aquifer and rock compressibilities. The aquifer constant is

$$U = 1.19 h f r_o^2 \phi_a c_t, \qquad (9.36)$$

where U is in units of RB/psi if h is in ft, r_o is in ft, and c_t is in psi^{-1}.

The dimensionless aquifer radius is

$$r_{eD} = \frac{r_a}{r_o}. \qquad (9.37)$$

The dimensionless water influx, W_D, is a function of t_D and r_{eD} and depends on whether the aquifer is infinite acting or finite.

Infinite, Radial Aquifer. The aquifer is infinite acting if r_e approaches infinity or if the pressure disturbance within the aquifer never reaches the aquifer's external boundary. If either of these conditions is met, then W_D is

$$W_D = 2\sqrt{\frac{t_D}{\pi}} + \frac{t_D}{2} - \frac{t_D}{6}\sqrt{\frac{t_D}{\pi}} + \frac{t_D^2}{16} \text{ for } t_D < 1, \qquad (9.38)$$

$$\text{or } W_D = a_7 t_D^7 + a_6 t_D^6 + a_5 t_D^5 + a_4 t_D^4 + a_3 t_D^3 + a_2 t_D^2 + a_1 t_D^1 + a_0 \text{ for } 1 < t_D < 100, \qquad (9.39)$$

where $a_7 = 4.8534 \times 10^{-12}$, $a_6 = -1.8436 \times 10^{-9}$, $a_5 = 2.8354 \times 10^{-7}$, $a_4 = -2.2740 \times 10^{-5}$, $a_3 = 1.0284 \times 10^{-3}$, $a_2 = -2.7455 \times 10^{-2}$, $a_1 = 8.5373 \times 10^{-1}$, $a_0 = 8.1638 \times 10^{-1}$, or

$$W_D = \frac{2 t_D}{\ln t_D} \text{ for } t_D > 100. \qquad (9.40)$$

Marsal[48] presented Eqs. 9.38 and 9.40. Walsh[23,49] presented Eq. 9.39.

Finite, Radial Aquifer. For finite aquifers, Eqs. 9.38 through 9.40 apply if $t_D < t_D^*$, where

$$t_D^* = 0.4\left(r_{eD}^2 - 1\right) \qquad (9.41)$$

$$\text{and } r_{eD} = \frac{r_a}{r_o}. \qquad (9.37)$$

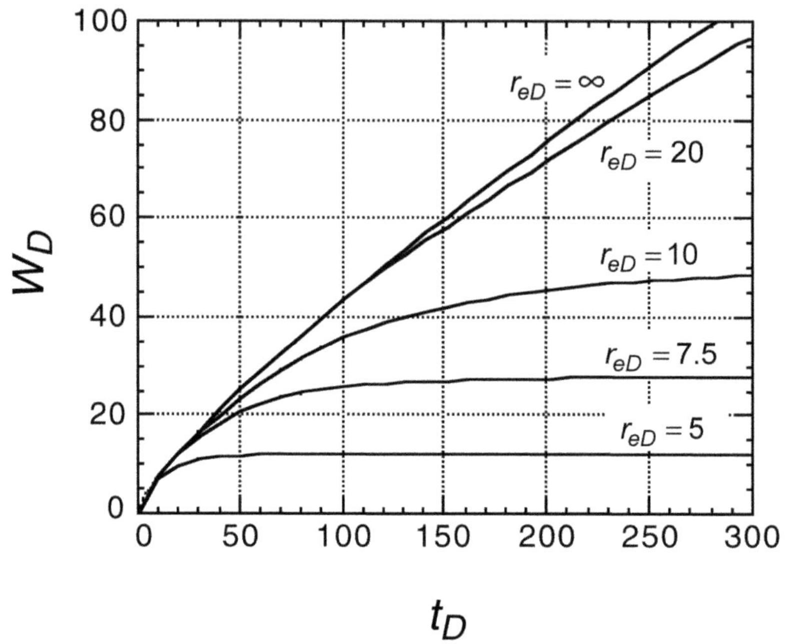

Fig. 9.25—W_D vs. t_D for radial aquifer model.

If $t_D > t_D$*, then

$$W_D = 0.5\left(r_{eD}^2 - 1\right)\left[1 - \exp\left(-\frac{2t_D}{J^*}\right)\right] \text{ for } t_D > t_D{}^*, \quad \text{(9.42)}$$

$$\text{where } J^* = \frac{r_{eD}^4\left(\ln r_{eD}\right)}{\left(r_{eD}^2 - 1\right)} + 0.25\left(1 - 3r_{eD}^2\right). \quad \text{(9.43)}$$

Marsal[48] gave Eqs. 9.41 through 9.43. These equations are effective in approximating the charts and tables by van Everdingen and Hurst. Minor discontinuities exist at some of the equation boundaries. A slightly more accurate but much more lengthy set of equations has been offered by Klins *et al.*[47] **Fig. 9.25** shows W_D as a function of t_D for r_{eD} = 5, 7.5, 10, 20, and ∞. These equations simplify the application of the VEH model enormously.

A finite aquifer can be treated effectively as an infinite aquifer if

$$t_{D\max} \leq 0.4\left(r_{eD}^2 - 1\right) \quad \text{(9.44)}$$

$$\text{or } r_{eD} \geq \sqrt{\frac{t_{D\max}}{0.4} + 1}, \quad \text{(9.45)}$$

where $t_{D\max}$ is the maximum value of t_D. These equations follow from Eq. 9.41. For example, if $t_{D\max}$ is 540 and corresponds to a time of 8 years, then Eq. 9.45 yields ≥ r_{eD} = 38. Therefore, if the aquifer has a dimensionless radius greater than 38, then the aquifer acts indistinguishably from and equivalent to an infinite aquifer at all times less than 8 years.

Linear Aquifer. The aquifer size in the linear model is given in terms of the aquifer/reservoir pore-volume ratio, V_{pa}/V_{pr}.

$$V_{pa} = \left(\frac{V_{pa}}{V_{pr}}\right)V_{pr} \text{ . } \qquad (9.46)$$

The aquifer constant is

$$U = V_{pa}c_t \text{ . } \qquad (9.47)$$

For edgewater drives, the aquifer length is

$$L_a = L_r\left(\frac{V_{pa}}{V_{pr}}\right)\left(\frac{\phi_r}{\phi_a}\right), \qquad (9.48)$$

where L_a and L_r are defined in Fig. 9.23a. For bottomwater drives, the aquifer depth is

$$L_a = h\left(\frac{V_{pa}}{V_{pr}}\right)\left(\frac{\phi_r}{\phi_a}\right), \qquad (9.49)$$

where L_a is defined in Fig. 9.23b. The dimensionless time is

$$t_D = \frac{2.309k_a t}{\phi_a \mu_w c_t L_a^2} \text{ . } \qquad (9.50)$$

Eqs. 9.50 and 9.35 use the same units except L_a is given in ft. One difference between the linear and radial models is that t_D is a function of the aquifer size for the linear model, whereas t_D is independent of the aquifer size for the radial model. This difference forces a recalculation of t_D in the linear model if the aquifer size is changed. The dimensionless cumulative water influx is

$$W_D = 2\sqrt{\frac{t_D}{\pi}} \text{ for } t_D < 0.47 \qquad (9.51)$$

$$\text{and } W_D = 1 - 0.065807t_D^{-1.6512} \text{ for } t_D < 0.47. \qquad (9.52)$$

Eq. 9.51 is by Marsal,[48] and Eq. 9.52 is by Walsh.[23,49] **Fig. 9.26** shows W_D as a function of t_D. The aquifer can be treated as infinite if the aquifer length is greater than the critical length.

$$L_{ac} = \sqrt{\frac{2.309k_a t_{max}}{0.5\phi_a \mu_w c_t}}, \qquad (9.53)$$

where t_{max} is the maximum time expressed in years and L_{ac} is in units of ft. Eqs. 9.50 and 9.53 use the same units. Alternatively, the aquifer is infinite-acting if $t_D \leq 0.50$. If infinite-acting and an edgewater drive, W_e can be evaluated directly without computing W_D and is

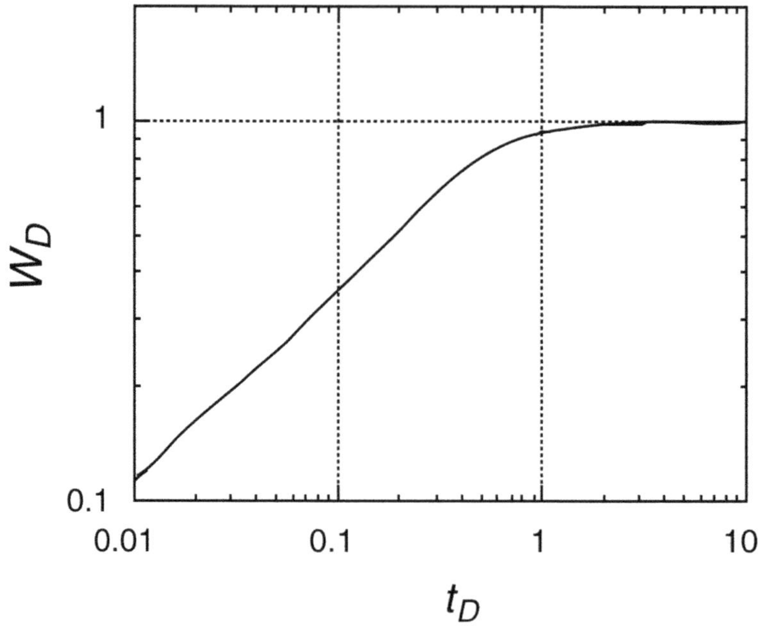

Fig. 9.26—W_D vs. t_D for linear aquifer model.

$$W_e = 0.3054 hw\Delta p\sqrt{\frac{\phi k_a c_t t}{\mu w}}, \quad (9.54)$$

where the units in Eq. 9.35 apply, and W_e is in units of RB and h and w are in units of ft.

Calculation Procedure. The calculation requires historical average reservoir-pressure data. Given this data, the following procedure is used to compute the water-influx history. This method applies for radial and linear models, except where noted:

1. Discretize the time and average reservoir-pressure domains and define t_j and $\bar{p}_j$ for (j = 0, 1, ..., n) according to Table 9.13.
2. Compute the time-averaged reservoir pressure $\tilde{p}_j$ for (j = 1, 2, ..., n) with Eq. 9.31. Note that $\tilde{p}_0 = p_i$.
3. Compute the time-averaged incremental pressure differential Δp_j for (j= 1, 2, ..., n) with Eq. 9.32.
4. Compute t_{Dj} for (j = 0, 1, ..., n) with Eq. 9.35 for radial aquifers or with Eq. 9.50 for linear aquifers.
5. Steps 5 through 9 create a computational loop that is repeated n times. The loop index is k, where k = 1, ..., n. For the kth time level, compute $(t_{Dk} - t_{Dj})$ for (j = 0, ..., k – 1).
6. For the kth time level, compute $W_D(t_{Dk} - t_{Dj})$ for (j = 0, ..., k – 1).
7. For the kth time level, compute $\Delta p_{j+1} W_D(t_{Dk} - t_{Dj})$ for (j = 0, ..., k – 1).
8. For the kth time level, compute W_{ek} with Eq. 9.33.
9. Increment the time from level k to k + 1, and return to Step 5 until $k > n$.

This procedure is highly repetitive and well suited for spreadsheet calculation. Example 9.2 illustrates the procedure.

9.9.6 Determining Water-Influx Model Parameters. The minimum parameters that need to be specified in the radial model are the aquifer constant, U, the time constant, k_t, and the dimensionless aquifer radius, r_{eD}. The time constant combines a number of constants, is the proportionality constant between the dimensionless and real time, and is defined by

$$t_D = k_t t \;. \quad \text{(9.55)}$$

Physically, the time constant represents the aquifer conductivity. In summary, U and k_t are defined as

$$U = 1.19 h f r_o^2 \phi_a c_t = \frac{2V_{pr}\phi_a c_t}{\phi_r}, \quad \text{(9.56)}$$

$$k_t = \frac{2.309 k_a}{\phi_a \mu_w c_t r_o^2}, \quad \text{(9.57)}$$

$$\text{and } r_{eD} = r_e / r_o \;. \quad \text{(9.37)}$$

Eq. 9.56 assumes the same units as Eq. 9.36, and V_{pr} is given in res bbl. Eq. 9.57 assumes the same units as Eq. 9.35.

The minimum parameters that need to be specified in the linear model are the aquifer constant, time constant, and aquifer/reservoir PV ratio (V_{pa}/V_{pr}). The aquifer constant and time constant are

$$U = V_{pa} c_t . \quad \text{(9.58)}$$

$$\text{and } k_t = \frac{2.309 k_a}{\phi_a \mu_w c_t L_a^2} \;. \quad \text{(9.59)}$$

Eq. 9.59 assumes the same units as Eq. 9.35.

There are three common methods to estimate model parameters: direct measurement, history matching, and material balance. The first two methods are described in the following sections. The material-balance method, through the McEwen method,[42] is described in Section 9.9.8.

Direct Measurement. This method estimates model parameters from direct measurement of the independent constants. Though ideally preferred, this method is rarely possible because of the uncertainty of some of the constants.

For the radial model, the model parameters (U, k_t, and r_{eD}) are a function of the following constants: r_e, r_o, k_a, h, f, ϕ_a, c_t, and μ_w. These constants follow from inspection of Eqs. 9.36, 9.37, and 9.57. The uncertainty among these constants varies. Of these constants, r_e, r_o, and k_a are perhaps the most uncertain. Qualitatively, these constants are related to the aquifer size, reservoir size, and aquifer conductivity (i.e., V_{pa}, V_{pr}, and k_t).

For the linear model, the model parameters (U, k_t, and V_{pa}/V_{pr}) are a function of the following constants: V_{pa}, V_{pr}, k_a, ϕ_a, c_t, μ_w, and L_a. This list follows from inspection of Eqs. 9.47 and 9.59. Of these constants, V_{pa}, V_{pr}, L_a, and k_a are the most uncertain. If h is approximately known, then V_{pa} and L_a are not independent but related through Eqs. 9.48 or 9.49. Thus, V_{pa}, V_{pr}, and k_a are the most uncertain independent constants. Qualitatively, these constants are related to the aquifer size, reservoir size, and aquifer conductivity (i.e., V_{pa}, V_{pr}, and k_t, respectively). Note the similarity between the radial and linear models.

TABLE 9.14—RESERVOIR PERFORMANCE DATA FOR EXAMPLE 9.2

j	Time (years)	p (psia)	Cumulative Water Influx (million RB)
1	0.000	1,640	0.000
2	0.368	1,600	0.062
3	2.439	1,400	1.272
4	4.957	1,200	4.156
5	7.732	1,000	8.455
6	11.926	800	15.494
7	18.126	600	26.014
8	30.044	400	42.243

In summary, because of the uncertainty of the aquifer size and conductivity and reservoir size, it is difficult to estimate reliably the water-influx model parameters. Nevertheless, every attempt should be made to estimate the median, variance, and range of each constant and the model parameters. This information is helpful in the history-matching method.

History Matching. If the water-influx history can be estimated, then model parameters can be estimated from history matching. When history matching is used, only the most uncertain constants should be treated as adjustable parameters: preferably only r_e, r_o, and k_a for the radial model or V_{pa}, V_{pr}, and k_a for the linear model. Unless only one adjustable parameter exists, history matching is usually complicated by nonuniqueness.[40,42,52–54] Nonuniqueness, however, can be minimized by limiting the range of parameter adjustment to realistic ranges. Example 9.2 illustrates the history-matching procedure.

Example 9.2: History Matching Water Influx. **Table 9.14** summarizes the cumulative water influx and average reservoir pressure as a function of time for an initially saturated, black-oil reservoir. Areally, the reservoir is approximately semicircular, bounded on one side by a sealing fault and the other side by an aquifer. **Fig. 9.27a** shows a schematic representation of the reservoir. Assume the reservoir and aquifer properties in **Table 9.15** apply.

Assume a radial-flow aquifer with $f = 0.50$. Find the optimal aquifer size (V_{pa} and r_{eD}) that best simulates the water-influx performance. Plot and compare the actual and predicted water-influx histories.

Find the optimal aquifer size (V_{pa}/V_{pr} and L_a) that best matches the water-influx performance assuming a linear-flow aquifer. Assume the reservoir width is $w = 2r_o$ and length is $L_r = \pi r_o/4$, where r_o is given by Eq. 9.34. Fig. 9.27b schematically shows the areal interpretation. Plot and compare the actual and predicted water-influx histories. Which model (linear or radial) best matches the data?

Solution. Compute the effective reservoir radius from

$$r_o = \sqrt{\frac{5.6146 V_{pr}}{\pi \phi_r h f}} = \sqrt{\frac{(5.6146 \text{ ft}^3/\text{bbl})(385 \times 10^6 \text{ RB})}{\pi(0.31)(20 \text{ ft})(0.5)}} = 14{,}892 \text{ ft}, \quad (9.34)$$

where $f = 0.50$. The total compressibility is the sum of the rock and water compressibilities or $c_t = 5.88 \times 10^{-6}$ psi^{-1}. The time constant, k_t, is given by Eq. 9.57 and is $k_t = 0.8682$ years^{-1}. U is given by Eq. 9.56 and is 3,955 RB/psi. **Table 9.16** tabulates t_D, $\tilde{p}$, and Δp.

The solution procedure varies r_{eD} until the best match is obtained. **Tables 9.17 through 9.19** give the details of the calculation for $r_{eD} = 5.0$. Table 9.17 tabulates ($t_{Dk} - t_{Dj}$) for k and j

TABLE 9.15—RESERVOIR AND FLUID PROPERTIES FOR EXAMPLE 9.2

Property	Value
Reservoir pore volume, million RB	385
Reservoir porosity, %	31
Reservoir permeability, md	5
Pay thickness, ft	20
f	0.50
Initial water saturation, %	20
Temperature, °F	131
Water viscosity, cp	0.48
Water compressibility, 1/psi	2.88
Rock compressibility, 10^{-6} PV/PV/psi	3.0
Aquifer porosity, %	27
Aquifer permeability, md	64.2
Aquifer thickness, ft	20
Water salinity, ppm	20,000

= 0, 1, ..., 7. **Table 9.18** tabulates $W_D(t_{Dk} - t_{Dj})$ for k and j = 0, 1, ..., 7. Table 9.19 tabulates $\Delta p_{j+1}W_D(t_{Dk} - t_{Dj})$ for k and j = 0, 1, ..., 7. Table 9.16 tabulates $\Sigma\Delta p_{j+1}W_D(t_{Dk} - t_{Dj})$ for k = 1, 2, ..., 7 for r_{eD} = 5 and W_{ej} for j = 0, 1, ..., 7 for r_{eD} = 5. **Fig. 9.28** plots W_e vs. t and compares model and actual results; excellent agreement is noted. These calculations were repeated for other values of r_{eD} but r_{eD} = 5.0 was found to give the best fit of the actual and model water-influx histories. Notice how well the radial model matches the history. A dimensionless aquifer radius of r_{eD} = 5.0 corresponds to V_{pa}/V_{pr} = 20.8. This means that the aquifer is 20.8 times larger than the reservoir.

To determine whether this aquifer can be treated as infinite acting, we evaluate Eq. 9.45 with $t_{D\max}$ = 26.08. This calculation yields r_{eD} = 8.1. Because this value of r_{eD} is greater than the history-matched value of r_{eD} = 5.0, this aquifer cannot be treated as infinite.

For the linear aquifer model, the geometry dictates that L_r = 11,696 ft if $L_r = \pi r_o/4 = \pi$ (14,892)/4.With the same trial-and-error procedure as used for the radial aquifer, the linear aquifer yields V_{pa}/V_{pr} = 12 for the best match between the actual and predicted water-influx data. This value of V_{pa}/V_{pr} yields L_a = 161,145 ft, U = 27,271 RB/psi, and k_t = 0.0074 years^{-1}. Fig. 9.28 compares the predicted and actual data and shows that the match is poor. This comparison reveals that the linear model is not preferable to simulate water influx for this reservoir.

9.9.7 Aquifer Performance. The aquifer performance is described in terms of the delivery rate, average aquifer pressure, and cumulative water-influx volume as a function of time. The aquifer pressure characteristically lags behind the reservoir pressure and is estimated by

$$\bar{p}_a = p_i - \frac{W_e(t)}{c_t V_{pa}} \quad \text{...} \quad (9.60)$$

The aquifer delivery rate is $q_w = \partial W_e(t)/\partial t$, which is determined from the slope of the W_e vs. t curve.

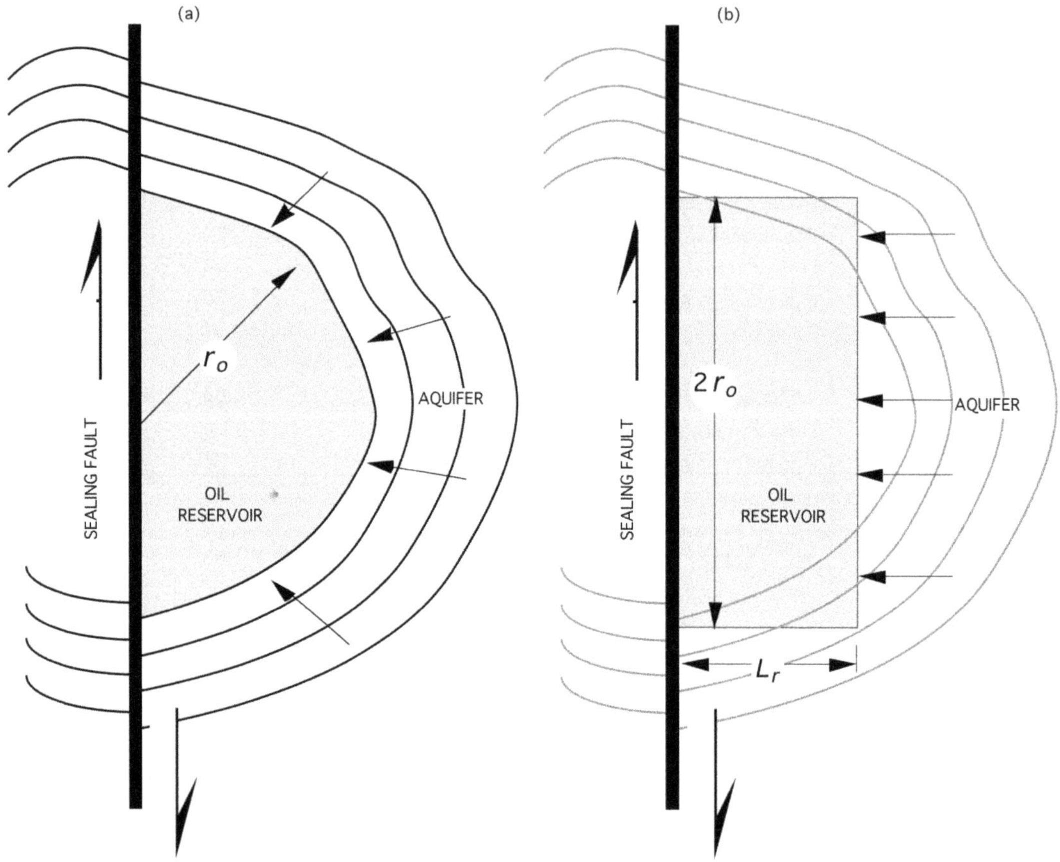

Fig. 9.27—(a) Areal schematic of a reservoir partially surrounded by a radial aquifer; (b) interpretation of reservoir geometry when linear aquifer model is applied.

Fig. 9.29 shows the aquifer pressure and delivery-rate history for the data in Example 9.2. This figure includes the reservoir pressure history for comparison. The qualitative results in Fig. 9.29 are representative of many aquifers. The water-delivery rate is initially zero and increases rapidly. It peaks after approximately 12 to 14 years and then slowly decreases. The aquifer and reservoir pressures start at equivalent values. The reservoir pressure declines more quickly than the aquifer pressure. The pressure differential between the aquifer and reservoir grows and is approximately 250, 350, and 500 psia, respectively, after 2, 5, and 10 years. The pressure differential peaks after 12 to 14 years and then begins to dissipate. The pressure differential and delivery rate decline together.

The aquifer performance noted in Fig. 9.29 is not without exception. The qualitative performance in Fig. 9.29 is characteristic of an initially saturated reservoir. Aquifers feeding initially undersaturated reservoirs may behave quite differently. The difference stems from the difference in the reservoir pressure histories. The reservoir pressure in initially undersaturated oil reservoirs initially declines much more quickly than in initially saturated reservoirs. Consequently, initially undersaturated reservoirs create a substantial pressure differential between the reservoir and aquifer much sooner than initially saturated reservoirs. Of course, this distinction depends on the degree of undersaturation. If the reservoir is significantly undersaturated, a large pressure differential between the reservoir and aquifer is quickly established. This large pressure differential, in turn, promotes water influx; consequently, the water-influx rate increas-

TABLE 9.16—PREDICTED WATER-INFLUX HISTORY FOR RADIAL AQUIFER IF r_{eD}=5.0

j	t (years)	$\bar{p}$ (psia)	t_D	$\bar{p}$ (psia)	Δp (psi)	$\Sigma\Delta pW_D$ (psi)	W_e (million RB)
0	0	1,640	0	1,640		0	0
1	0.368	1,600	0.32	1,620	20	16	0.062
2	2.439	1,400	2.12	1,500	120	322	1.275
3	4.957	1,200	4.30	1,300	200	1,054	4.170
4	7.732	1,000	6.71	1,100	200	2,147	8.491
5	11.926	800	10.35	900	200	3,939	15.580
6	18.126	600	15.74	700	200	6,620	26.182
7	30.044	400	26.08	500	200	10,735	42.460

TABLE 9.17—SUMMARY OF ($t_{Dk}-t_{Dj}$) CALCULATIONS FOR RADIAL AQUIFER

		$t_{Dk}-t_{Dj}$						
k	t_{Dk}	j=0	j=1	j=2	j=3	j=4	j=5	j=6
0	0.00	0.00	0.00	0.00	0.00	0.00	0.00	0.00
1	0.32	0.32	0.00	0.00	0.00	0.00	0.00	0.00
2	2.12	2.12	1.80	0.00	0.00	0.00	0.00	0.00
3	4.30	4.30	3.98	2.19	0.00	0.00	0.00	0.00
4	6.71	6.71	6.39	4.60	2.41	0.00	0.00	0.00
5	10.35	10.35	10.03	8.24	6.05	3.64	0.00	0.00
6	15.74	15.74	15.42	13.62	11.43	9.02	5.38	0.00
7	26.08	26.08	25.76	23.97	21.78	19.37	15.73	10.35

es more rapidly in initially undersaturated reservoirs than initially saturated reservoirs. Once the bubblepoint is reached, the pressure differential between the aquifer and reservoir may decline temporarily. Later, the pressure differential may increase, reminiscent of an initially saturated reservoir, as in Fig. 9.29. The net effect is that water recharge rate may oscillate in an initially undersaturated oil reservoir.

9.9.8 Material-Balance Analysis. The objectives of a material-balance analysis include confirming the producing mechanism, estimating the OOIP, estimating the water-influx history, and estimating the water-influx model parameters. The water-influx model parameters are needed to forecast future water influx and oil recovery.

The analysis depends on the known and unknown constants and variables. Three scenarios are considered: the water-influx history is known but the OOIP is unknown, the water-influx history is unknown but the OOIP is known, and both the water-influx history and the OOIP are unknown.

Water Influx Known, OOIP Unknown. If the water-influx history is known and the OOIP unknown, the material-balance methods in the previous sections are directly applicable. For instance, if the reservoir is initially undersaturated, then an (F – W_e)-vs.-E_{owf} plot can be used

TABLE 9.18—SUMMARY OF $W_D(t_{Dk}-t_{Dj})$ CALCULATIONS FOR RADIAL AQUIFER

		$W_D(t_{Dk}-t_{Dj})$						
k	t_{Dk}	j=0	j=1	j=2	j=3	j=4	j=5	j=6
0	0.00	0.00	0.00	0.00	0.00	0.00	0.00	0.00
1	0.32	0.79	0.00	0.00	0.00	0.00	0.00	0.00
2	2.12	2.51	2.27	0.00	0.00	0.00	0.00	0.00
3	4.30	4.06	3.84	2.56	0.00	0.00	0.00	0.00
4	6.71	5.24	5.39	4.25	2.73	0.00	0.00	0.00
5	10.35	7.05	6.91	6.06	5.18	3.61	0.00	0.00
6	15.74	8.87	8.78	8.25	7.48	6.45	4.76	0.00
7	26.08	10.71	10.67	10.45	10.13	9.71	8.87	7.04

TABLE 9.19—SUMMARY OF $\Delta p_{j+1} W_D(t_{Dk}-t_{Dj})$ CALCULATIONS FOR RADIAL AQUIFER

		$\Delta p_{j+1} W_D(t_{Dk}-t_{Dj})$						
k	t_{Dk}	j=0	j=1	j=2	j=3	j=4	j=5	j=6
0	0.00	0.00	0.00	0.00	0.00	0.00	0.00	0.00
1	0.32	15.70	0.00	0.00	0.00	0.00	0.00	0.00
2	2.12	50.20	272.20	0.00	0.00	0.00	0.00	0.00
3	4.30	81.10	461.00	512.30	0.00	0.00	0.00	0.00
4	6.71	104.70	646.40	850.10	545.60	0.00	0.00	0.00
5	10.35	140.90	829.00	1,212.50	1,035.30	721.30	0.00	0.00
6	15.74	177.40	1,054.20	1,650.20	1,496.30	1,289.70	951.80	0.00
7	26.08	214.10	1,280.60	2,090.20	2,026.60	1,941.30	1,773.90	1,408.40

to confirm the mechanism and estimate the OOIP. The water-influx model parameters can be determined by history matching.

Water Influx Unknown, OOIP Known. If the OOIP is known and the water-influx history is unknown, then the material-balance equation can be used to estimate the water-influx history. Solving Eq. 9.7 for W_e yields

$$W_e = G_{fgi} E_{gwf} + N_{foi} E_{owf} - F \; . \quad (9.61)$$

If the OOIP (N) and OGIP (G) are known, G_{fgi} and N_{foi} are computed from Eqs. 9.21 and 9.22. Eq. 9.61 is applied for each historical average pressure measurement to compute the cumulative water influx. Example 9.3 illustrates this method. Once the water-influx history is estimated, the aquifer parameters can be estimated from history matching.

Water Influx Unknown, OOIP Unknown. This case simultaneously determines the OOIP, water-influx history, and water-influx model parameters. This is a challenging problem. Woods and Muskat[52] were among the first to study this problem and they noted that the solution was complicated by nonuniqueness. Others, too, have noted nonuniqueness.[40,42,53,54] Despite these complications, certain techniques have proved useful and some approaches are better than others are. The solution method is based on the work of McEwen[42] and depends on whether the radial or linear version of the VEH model is applied.

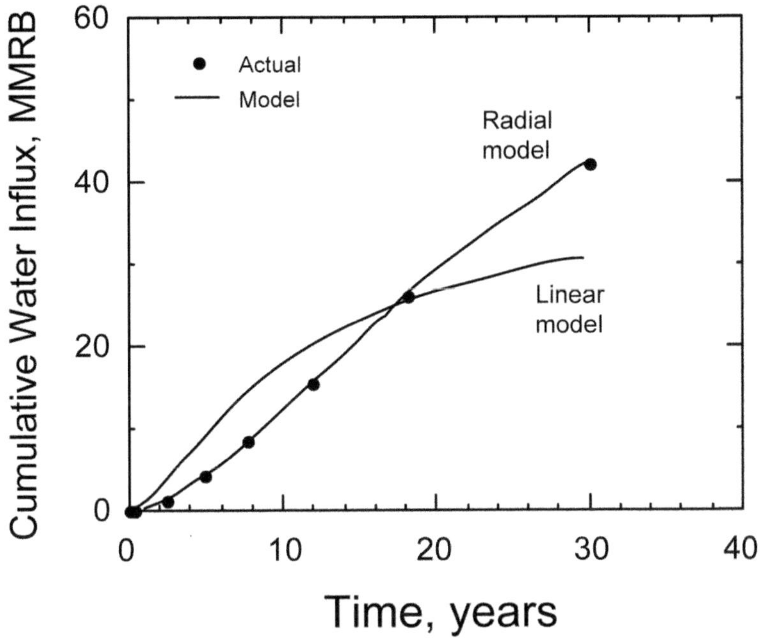

Fig. 9.28—Comparison of actual and predicted water-influx histories for Example 9.2.

Radial Aquifer. This method simultaneously determines the OOIP, water-influx history, and model parameters r_{eD} and k_t. The aquifer constant U is then subsequently determined. The water influx is

$$W_e = U \Sigma \Delta p W_D \,. \qquad (9.62)$$

This equation is an abbreviation of Eq. 9.33. The summation $\Sigma \Delta p W_D$ is a function of only r_{eD} and k_t. McEwen noted that U is related to N_{foi} and G_{fgi} through

$$U = \frac{2\left(N_{foi}B_{oi} + G_{fgi}B_{gi}\right)c_t}{\left(1 - S_{wi}\right)}\left(\frac{\phi_a}{\phi_r}\right). \qquad (9.63)$$

Substituting this equation into Eq. 9.62 and substituting this result into Eq. 9.7 yields

$$F = N_{foi}E_{ow} + G_{fgi}E_{gw}, \qquad (9.64)$$

where

$$E_{ow} = E_{owf} + \frac{2B_{oi}c_t\Sigma \Delta p W_D\left(k_t, r_{eD}\right)}{\left(1 - S_{wi}\right)}\left(\frac{\phi_a}{\phi_r}\right) \qquad (9.65)$$

and $$E_{gw} = E_{gwf} + \frac{2B_{gi}c_t\Sigma \Delta p W_D\left(k_t, r_{eD}\right)}{\left(1 - S_{wi}\right)}\left(\frac{\phi_a}{\phi_r}\right). \qquad (9.66)$$

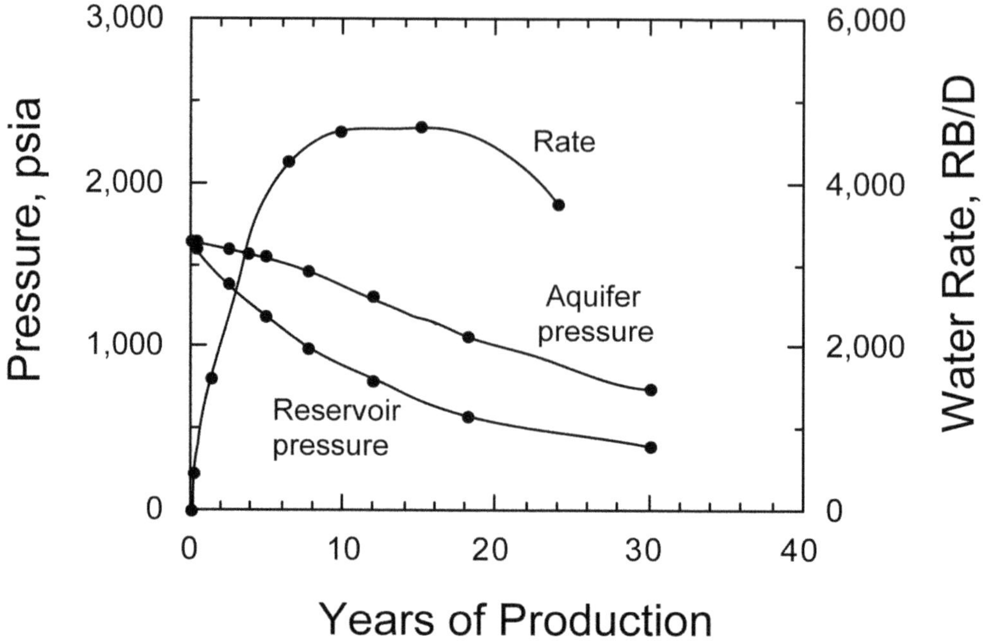

Fig. 9.29—Predicted aquifer pressure and water-influx rate histories for Example 9.2.

For the case of an initially undersaturated oil reservoir,

$$F = NE_{ow} \; . \qquad (9.67)$$

This equation shows that a plot of F vs. E_{ow} is a straight line, emanates from the origin, and has a slope equal to N, the OOIP. This observation provides a means to confirm the producing mechanism.

E_{ow}, however, is a function of k_t and r_{eD}, and these parameters are unknown *a priori*. Thus, the problem reduces to one of finding the optimal k_t and r_{eD} that minimizes the material-balance error. Graphically, this is equivalent to varying k_t and r_{eD} until the straightest possible line is realized. The slope of the line equals the OOIP, and the OGIP is the product NR_{si} if the reservoir is initially undersaturated. Mathematically, the material-balance error is minimized when the sum of the squares of the residual is minimized. The residual for point i is

$$R_i = (F)_i - N\left(E_{ow}\right)_i \; . \qquad (9.68)$$

The sum of the squares of the residual is

$$R_{ss} = \sum_{i=1}^{n} \left(R_i\right)^2 , \qquad (9.69)$$

where n is the total number of data points.

In summary, the McEwen method to simultaneously estimate N, r_{eD}, and k_t is as follows:

1. Estimate a limited range of realistic values for k_t if possible.
2. Compute F for each data point with Eq. 9.8 if saturated or Eq. 9.13 if undersaturated.
3. Guess k_t and r_{eD}.

4. Compute E_{ow} for each data point with Eq. 9.65.
5. Compute N with least-squares linear regression or graphically from the slope of an F-vs.-E_{ow} plot.
6. Compute R_i for each data point with Eq. 9.68.
7. Compute R_{ss} with Eq. 9.69.
8. If the R_{ss} is minimized, then go to Step 9; otherwise, return to Step 3.
9. Compute the aquifer constant with Eq. 9.63.
10. Compute the water influx for each data point with Eq. 9.62.

If least-squares linear regression is used to compute N in Step 5, an equation analogous to Eq. 9.16 is used (where E_{ow} is substituted for E_{owf}). This solution method is iterative because the material-balance error must be minimized. This calculation is carried out with a trial-and-error method or a minimization algorithm. Least-squares linear regression and minimization algorithms have become standard features in commercial spreadsheets.

McEwen's method also can be applied to initially saturated reservoirs; however, the solution procedure must be expanded and modified slightly. More specifically, the solution procedure is the same as for initially undersaturated reservoirs except Steps 5 and 6. Step 5 must be modified to include the simultaneous calculation of N_{foi} and G_{fgi} by multivariate, least-squares (planar) regression.[23,28] Step 6 must be modified, and the residual for point i is computed with $R_i = (F)_i - G_{fgi}(E_{gw})_i - N_{foi}(E_{ow})_i$.

Linear Aquifer. McEwen's method for linear aquifers is very similar to the radial model. The method simultaneously determines the OOIP, water-influx history, and the model parameters $[(V_{pa}/V_{pr})$ and $k_t]$.

The aquifer constant U is related to V_{pa} through

$$U = \left(\frac{V_{pa}}{V_{pr}}\right)\frac{\left(N_{foi}B_{oi} + G_{fgi}B_{gi}\right)c_t}{\left(1 - S_{wi}\right)} \quad \text{(9.70)}$$

Substituting this equation into Eq. 9.62 and substituting the result into Eq. 9.7 gives Eq. 9.64, where

$$E_{gw} = E_{gwf} + \left(\frac{V_{pa}}{V_{pr}}\right)\frac{B_{gi}c_t\Sigma\Delta pW_D\left(k_t\right)}{\left(1 - S_{wi}\right)}, \quad \text{(9.71)}$$

$$E_{ow} = E_{owf} + \left(\frac{V_{pa}}{V_{pr}}\right)\frac{B_{oi}c_t\Sigma\Delta pW_D\left(k_t\right)}{\left(1 - S_{wi}\right)}, \quad \text{(9.72)}$$

and $\Sigma\Delta pW_D$ is a function of only k_t and not (V_{pa}/V_{pr}). Eqs. 9.71 and 9.72 are analogous to Eqs. 9.65 and 9.66 in the radial model.

The solution procedure for the linear model is identical to that of the radial model except that k_t and V_{pa}/V_{pr} are optimized to minimize the material-balance error. Once V_{pa}/V_{pr} is determined, the aquifer constant U is determined from Eq. 9.70 and W_e is determined from Eq. 9.62. Example 9.4 illustrates an example of the McEwen method.

Numerous alternative material-balance methods have been proposed to analyze waterdrive reservoirs. Some are very popular and widely used. While most are theoretically valid, most are also unreliable. van Everdingen *et al.*,[55] for instance, proposed plotting F/E_o vs. $(\Sigma\Delta pW_D)/E_o$. The slope of this plot equals U and the y-intercept equals N. van Everdingen *et al.* proposed varying k_t until the straightest possible line was obtained. Later, Havlena and Odeh[38] popularized this method and modified it to include the aquifer size (r_{eD}) as an additional unknown and determinable parameter. Dake[30] also advocated this method. Chierici *et al.*[53] proposed a variation of this method with a $F/(\Sigma\Delta pW_D)$ vs. $E_o/(\Sigma\Delta pW_D)$ plot. McEwen[42] studied the method of van Everdingen *et al.* and noted hypersensivity to pressure uncertainty. He observed unacceptably large errors and deemed the method unreliable. Later, Tehrani[40] presented a systematic analysis of these methods and confirmed McEwen's conclusions. Wang and Hwan[39] confirmed Tehrani's findings. Sills[54] presents a review and comparison of the McEwen, Havlena-Odeh, and van Everdingen-Timmerman-McMahon methods.

To help diagnose waterdrives, Campbell[56] proposed plotting F/E_{owf} vs. N_p for initially undersaturated oil reservoirs. This method is analogous to Cole's[57] popular method of plotting F/E_{gwf} vs. G_p for gas reservoirs. In theory, an active waterdrive is indicated if the plot varies appreciably from a horizontal line. The degree of curvature is a qualitative measure of the waterdrive strength. The curve emanates from a y-intercept equal to the OOIP. The shape of the curve mimics and is related to the attending water recharge rate history. Cole and Campbell plots are attractive because of their simplicity and are widely reported. Unfortunately, in practice, they are not always reliable because of hypersensitivity caused by uncertainty. The origin of the hypersensitivity is analogous to the problems noted by McEwen,[42] Tehrani,[40] Wang and Hwan,[39] and Walsh[28] for other types of material-balance plots. The quotient F/E_{owf} approaches infinity initially because the E_{owf} approaches zero. A systematic reservoir pressure error of only 1 to 2%, for instance, can lead to erroneous conclusions regarding water-influx diagnosis. These facts complicate the interpretation. For these reasons, Campbell plots should be used cautiously to diagnose water influx and used very cautiously to estimate the OOIP. If they are used, the early-time data should be weighted minimally. The reliability of these plots increases with pressure depletion; however, water influx mitigates pressure depletion and delays reliability. Unfortunately, water-influx diagnoses are sought as early as possible, which further complicates and compromises the use of these plots.

Example 9.3: Estimating Water Influx With Material Balance. **Table 9.20** summarizes the cumulative oil and gas production as a function of time and average reservoir pressure for a black-oil reservoir. The discovery (initial) pressure is 1,640 psia, and production data are tabulated through a pressure of 800 psia.

Volumetric measurements estimate an OOIP of 210,420 thousand STB with no initial free gas. Estimate the cumulative water influx (RB) at each pressure in Table 9.20 with material balance. Assume that the standard PVT parameters in Table 9.3 apply.

Solution. Eq. 9.61 gives the cumulative water influx. Because there is no initial free gas, $G_{fgi} = 0$, $N_{foi} = N$, and Eq. 9.61 simplifies to

$$W_e = NE_o - F, \quad (9.73)$$

where rock and water expansion are ignored and F and E_o are given by Eqs. 9.8 and 9.2, respectively. E_o is a function of B_{to}, which is given by Eq. 9.5.

Table 9.21 tabulates the results. **Fig. 9.30** plots the water-influx history.

Example 9.4: Determining Water-Influx Parameters and OOIP. van Everdingen *et al.*[55] studied water influx in an initially undersaturated oil reservoir located in the Wilcox formation at a depth of 8,100 ft subsea. The accumulation covered approximately 1,830 acres. The maximum gross and net thicknesses were 37 and 26 ft, respectively. The reservoir fluid exhibited an initial oil FVF of 1.538 RB/STB and a GOR of 900 scf/STB. **Table 9.22** reports the reservoir and aquifer properties.

TABLE 9.20—RESERVOIR PERFORMANCE DATA FOR EXAMPLE 9.3

Time (years)	p (psia)	Cumulative Oil Recovery (million STB)	Cumulative Gas Recovery (Bscf)
0.000	1,640	0.000	0.000
0.368	1,600	3.619	2.921
2.439	1,400	19.938	19.208
4.957	1,200	32.262	38.535
7.732	1,000	40.399	56.131
11.926	800	47.873	75.978

TABLE 9.21—SUMMARY OF WATER-INFLUX CALCULATIONS FOR EXAMPLE 9.3

p (psia)	B_{to} (RB/STB)	E_o (RB/STB)	F (million RB)	Cumulative Water Influx (million RB)	Time (years)
1,640	1.4630	0.0000	0.00	0.00	0.000
1,600	1.4880	0.0250	5.32	0.062	0.368
1,400	1.6450	0.1820	39.57	1.270	2.439
1,200	1.8916	0.4286	94.34	4.150	4.957
1,000	2.2142	0.7512	166.53	8.450	7.732
800	2.7438	1.2808	285.00	15.500	11.926

van Everdingen *et al.*[55] estimated the OOIP at 24 to 25 million STB from volumetric measurement. From the oil, gas, and water production data and standard PVT properties, they computed F and E_o as a function of pressure. Production took place from 1942 through 1950. Approximately 6.965 million STB of oil and 7.1 Bscf of gas were produced during this period. **Table 9.23** gives the average reservoir pressure, total fluid withdrawal, and oil-phase expansivity as a function of time.

Use McEwen's method to find the optimal dimensionless aquifer radius, r_{eD}, and the aquifer time constant, k_t (years^{-1}). Plot F vs. E_{ow}. Also, compute the aquifer constant, U (RB/psi). Estimate the OOIP (million STB). Estimate the water delivery rate (RB/D) and average aquifer pressure (psia), and plot the histories.

Solution. Though the time constant, k_t, is unknown, first estimate a realistic range of values based on reservoir and aquifer properties. The time constant is given by Eq. 9.57. The total compressibility is the sum of the rock and water compressibility or 6.8×10^{-6} psi^{-1}. The aquifer porosity is 20.9%. The water viscosity is 0.25 cp. The aquifer permeability is 275 md. The only unspecified quantity on the right side of Eq. 9.57 is the effective reservoir radius. Although this quantity is unknown because the size of the reservoir is uncertain, it can be estimated from

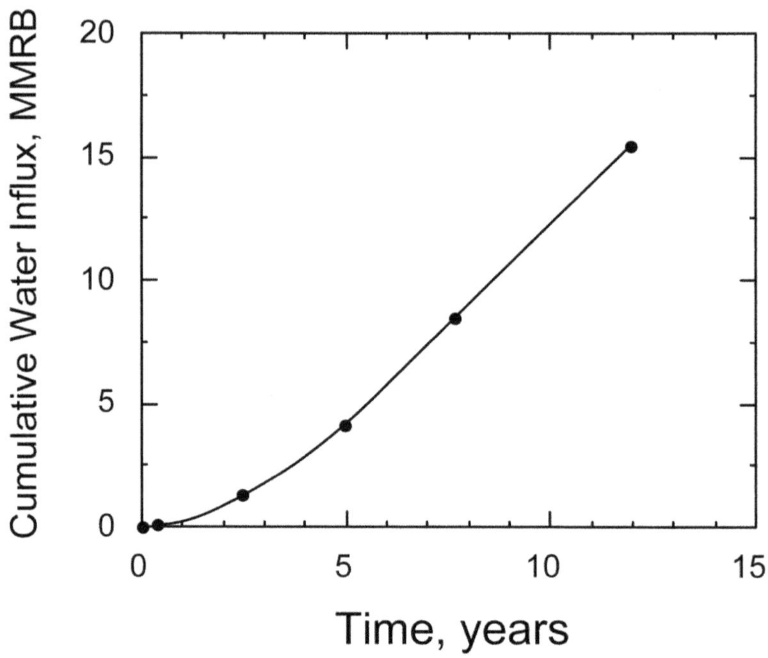

Fig. 9.30—Computed water-influx history for Example 9.3.

$$r_o = \sqrt{\frac{5.6146V_{pr}}{\pi \phi_r h f}} \quad \text{...} \quad (9.34)$$

To use this equation, estimate the reservoir PV, which is given by

$$V_{pr} = \frac{N_{foi}B_{oi} + G_{fgi}B_{gi}}{\left(1 - S_{wi}\right)} \quad \text{...} \quad (9.74)$$

Evaluating this equation on the basis of N_{foi} = 25 million STB yields V_{pr} = 45.23 million RB. Evaluating Eq. 9.34 for f = 1, h = 26 ft, and ϕ_r = 19.9% yields r_o = 3,951 ft. Evaluating Eq. 9.57 yields k_t = 116 years^{-1}. Thus, a liberal range for k_t is 10 to 1,000 years^{-1}.

Next, compute E_{ow} from Eq. 9.65. This equation requires computing $\Sigma\Delta pW_D$, which is a function of r_{eD} and k_t. The overall solution procedure contains the following steps:

1. Assume values of r_{eD} and k_t.
2. Compute t_D for each historical data point with Eq. 9.35.
3. Compute $\Sigma\Delta pW_D$ for each historical data point with the VEH model.
4. Compute E_{ow} for each historical data point with Eq. 9.65.
5. Plot F vs. E_{ow}.
6. Determine the OOIP from the slope.
7. Compute the residual, R_i, for each data point with Eq. 9.68 and compute R_{ss} with Eq. 9.69.
8. Change r_{eD} and k_t and return to Step 1 until a minimum R_{ss} is obtained.

Table 9.24 summarizes the results for the case of r_{eD} = 20 and k_t = 17 years^{-1}. The step-by-step calculations to compute $\Sigma\Delta pW_D$ are omitted for the sake of brevity. Example 9.2 illustrates these calculations.

TABLE 9.22—RESERVOIR AND AQUIFER PROPERTIES FOR WILCOX FORMATION BLACK OIL

Property	Value
Reservoir Properties	
Porosity, %	19.90
Permeability, md	236
Initial water saturation, %	15
Temperature, °F	240
Initial pressure, psia	3,793
Bubblepoint pressure, psia	3,690
Aquifer Properties	
Porosity, %	20.90
Permeability, md	275
Water compressibility, 10^{-6} PV/PV/psi	3.3
Rock compressibility, 10^{-6} PV/PV/psi	3.5
Water viscosity, cp	0.25
Water salinity, ppm	20,000

Fig. 9.31 plots F vs. E_{ow} for r_{eD} = 20 and k_t = 17 years^{-1}. The slope of this plot is 18.7 million STB, which is an estimate of the OOIP. The material-balance error is given by the R_{ss}, which is 0.015 million res bbl.

The values of r_{eD} and k_t were varied until a minimum R_{ss} was realized. **Table 9.25** summarizes the results for the following values of r_{eD}: 10, 15, 20, and 30. No values greater than r_{eD} = 30 were considered because they yielded identical results and were equivalent to an infinite-acting aquifer.

The results in Table 9.25 are summarized graphically in **Fig. 9.32**. This figure plots the standard error vs. the time constant for r_{eD} = 10, 20, and 30. The standard error is defined as the square root of R_{ss}/n, where n is the total number of data points. These results show that a minimum error is realized for the following range of properties: r_{eD} = 20 to 30 and k_t = 16 to 17 years^{-1}. **Table 9.26** summarizes the range of constants yielded by this range of values.

The material-balance OOIP estimate (18.7 to 20.3 million STB) approximately agrees with the volumetric estimate (24 to 25 million STB). Technically, r_{eD} = 20 and k_t = 17 years^{-1} yields a minimum error; however, comparable errors are realized for the stated range of r_{eD} and k_t. The boundary between infinite- and noninfinite-acting aquifers occurs at r_{eD} = 20.1. Thus, an aquifer with r_{eD} = 20 is not quite but very nearly an infinite-acting aquifer. These observations help explain why values of r_{eD} in the range 20 to 30 yield virtually identical errors and similar results. On the basis of simulating reservoir performance through 9 years, selecting any value of r_{eD} greater than 20 is practically acceptable. A value of r_{eD} = 30, however, yields an OOIP estimate (20.2 million STB) slightly closer to the volumetric OOIP estimate than the OOIP estimate (18.8 million STB) for r_{eD} = 20.On this basis, r_{eD} = 30 may be preferable. Without additional historical data and history matching, a more conclusive determination of r_{eD} beyond the stated range is not possible.

Fig. 9.31 includes an F-vs.-E_{ow} plot for r_{eD} = 10 and k_t = 11 years^{-1}. The F vs. E_{ow} plots for r_{eD} = 20 and r_{eD} = 30 are virtually indistinguishable. The plot for r_{eD} = 10 exhibits appreciable curvature; in contrast, the plot for r_{eD} = 20 is linear. The degree of curvature is a measure of the lack of material balance. The upward curvature of the former plot indicates that water influx is underpredicted.

Fig. 9.33 shows the predicted water-influx history. This figure assumes r_{eD} = 20. A cumulative water-influx volume of 14.83 million res bbl is predicted after 9 years. This amount of

TABLE 9.23—RESERVOIR PERFORMANCE DATA FOR EXAMPLE 9.4

j	t (quarters of years)	$\bar{p}$ (psia)	Total Fluid Withdrawal F (million RB)	Oil-Phase Expansivity E_o (RB/STB)
1	0	3,800	0.000	0.00000
2	2	3,768	0.075	0.00075
3	4	3,699	0.266	0.00267
4	6	3,613	0.831	0.01296
5	8	3,511	1.690	0.02725
6	10	3,444	2.600	0.03731
7	12	3,375	3.656	0.04827
8	14	3,309	4.770	0.05936
9	16	3,277	5.813	0.06497
10	18	3,263	6.790	0.06747
11	20	3,230	7.914	0.07348
12	22	3,174	9.138	0.08409
13	24	3,116	10.499	0.09564
14	26	3,057	11.949	0.10801
15	28	3,015	13.455	0.11721
16	30	3,040	14.533	0.11169
17	32	3,062	15.497	0.10693
18	34	3,060	16.547	0.10736
19	36	3,060	17.576	0.10736

water influx equates to 46.0% of the (reservoir) HCPV (assuming an OOIP of 18.7 million STB and a HCPV of 33.83 million res bbl). This substantial amount of water influx, together with the relatively small pressure decline (from 3,800 to 3,060 psia), suggests a moderate to strong waterdrive.

The pressure data in Table 9.23 show that the reservoir pressure actually increased after 7 years. This example illustrates another consequence of the VEH model; namely, the model can treat pressure increases as well as pressure decreases. The operators offered no explanation regarding why the reservoir pressure began to increase after seven years. It was noted that the pressure remained approximately constant and increased during periods of marked GOR decline. It was suspected that a secondary gas cap formed as the reservoir pressure declined below the bubblepoint. Upstructure wells reportedly began producing free gas as the gas cap formed and grew. The GOR grew as the free-gas production grew. The GOR declined when production from the high-GOR wells was curtailed. Though this scenario explains the GOR behavior and some of the pressure behavior, it does not explain why the reservoir pressure actually increased slightly during the last two years of production.

9.10 Compaction Drives

If pore-volume contraction contributes prominently to overall expansion while the reservoir is saturated, then the reservoir is classified as a compaction drive. Compaction-drive oil reservoirs are supplemented by solution-gas drive if the reservoir falls below the bubblepoint; they may or may not be supplemented by a water or gas-cap drive.

Compaction drives characteristically exhibit elevated rock compressibilities, often 10 to 50 times greater than normal. Rock compressibility is called PV, or pore, compressibility and is expressed in units of PV change per unit PV per unit pressure change. Rock compressibility is a function of pressure. Normal compressibilities range from 3 to 8 $\times$ 10^{-6} psi^{-1} at pressures

TABLE 9.24—PREDICTED WATER-INFLUX HISTORY FOR RADIAL AQUIFER IF r_{eD}=20 and k_t=17 years^{-1}

j	t (years)	$\bar{p}$ (psia)	t_D	$\tilde{p}$ (psi)	Δp (psi)	$\Sigma\Delta pW_D$ (psi)	E_{ow} (RB/STB)	$W_e(t)$ (million RB)	Average Water Rate (RB/D)	Time at Rate (years)
0	0.0	3,800	0	3,784		0	0.000	0.000	0	0
1	0.5	3,768	9	3,734	16.0	106	0.003	0.049	589	0.25
2	1.0	3,699	18	3,656	50.5	509	0.016	0.233	2,241	0.75
3	1.5	3,613	27	3,562	77.5	1,300	0.047	0.596	4,401	1.25
4	2.0	3,511	36	3,478	94.0	2,507	0.092	1.151	6,718	1.75
5	2.5	3,444	45	3,410	84.5	4,009	0.141	1.840	8,357	2.25
6	3.0	3,375	54	3,342	68.0	5,698	0.196	2.616	9,398	2.75
7	3.5	3,309	63	3,293	67.5	7,599	0.256	3.488	10,572	3.25
8	4.0	3,277	72	3,270	49.0	9,586	0.313	4.400	11,057	3.75
9	4.5	3,263	81	3,247	23.0	11,529	0.366	5.292	10,808	4.25
10	5.0	3,230	90	3,202	23.5	13,498	0.423	6.195	10,955	4.75
11	5.5	3,174	99	3,145	44.5	15,643	0.489	7.180	11,937	5.25
12	6.0	3,116	108	3,087	57.0	18,007	0.562	8.265	13,152	5.75
13	6.5	3,057	117	3,036	58.5	20,569	0.640	9.441	14,252	6.25
14	7.0	3,015	126	3,028	50.5	23,262	0.719	10.677	14,985	6.75
15	7.5	3,040	135	3,051	8.5	25,821	0.780	11.852	14,235	7.25
16	8.0	3,062	144	3,061	-23.5	28,120	0.835	12.907	12,791	7.75
17	8.5	3,060	153	3,060	-10.0	30,282	0.891	13.899	12,029	8.25
18	9.0	3,060	162	900	1.0	32,315	0.944	14.832	11,309	8.75

greater than approximately 1,000 psia. In contrast, elevated rock compressibilities can reach as high as 150 × 10^{-6} psi^{-1} or higher at comparable pressures.[58]

In general, compaction-drive reservoirs are rare; however, strong compaction drives do exist. The Ekofisk field in the Norwegian sector of the North Sea, with reserves in excess of 1.7 billion bbl, lies at a depth of 9,300 ft below sea level in 235 ft of water. The reservoir is a chalk formation that exhibits porosities in the range of 25 to 48%.[59] The operators reported rock compressibilities as high as 50 to 100 × 10^{-6} psi^{-1}.[60] Extreme compressibilities such as these can account for 70 to 80% of the expansion above the bubblepoint and 20 to 50% or more of the expansion below the bubblepoint.

9.10.1 Performance. Compaction-drive oil reservoirs act like their noncompaction counterparts except that they exhibit enhanced recoveries. For instance, a solution-gas-drive, compaction-drive reservoir will act qualitatively like a normal solution-gas-drive reservoir except the oil recovery will be greater. The enhanced recoveries are a direct consequence of the extra rock expansion that compaction-drive reservoirs naturally possess.

The excessive compaction noted in compaction-drive reservoirs has contributed to some production problems. For example, the compaction has been linked to a decline in reservoir permeability, fracture closure, and subsidence.[58–60] In most cases, however, these problems have been manageable, and the net result of compaction has been very favorable.

9.10.2 Material-Balance Analysis. The material-balance methods discussed in the previous sections are equally applicable to compaction drives. The only difference is that rock expansion cannot be ignored. Including rock expansion requires evaluating the rock expansivity, E_f.

The most accurate and reliable method is to measure E_f as a function of pressure. This method is strongly recommended if a compaction-drive mechanism is suspected because of the sensitivity of the analysis to E_f. **Table 9.27** summarizes the experimental results for a high-pressure Gulf Coast gas reservoir.[61] The initial reservoir and hydrostatic pressure was 9,800

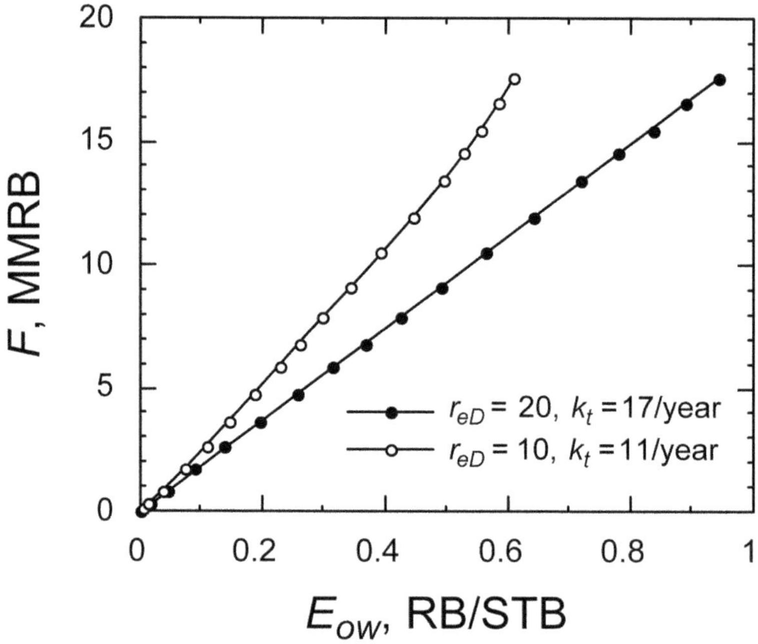

Fig. 9.31—F-vs.-E_{ow} plot for Example 9.4.

TABLE 9.25—MATERIAL-BALANCE ERROR AS A FUNCTION OF r_{eD} and k_t

k_t (year^{-1})	r_{eD}=10 Standard Error (million RB)	N (million STB)	k_t (year^{-1})	r_{eD} =15 Standard Error (million RB)	N (million STB)	k_t (year^{-1})	r_{eD} =20 Standard Error (million RB)	N (million STB)	k_t (year^{-1})	r_{eD} =30 Standard Error (million RB)	N (million STB)
1	1.220	76.7	1	1.220	76.7	1	1.220	76.7	1	1.220	76.7
5	0.591	41.3	3	0.729	51.9	5	0.485	40.7	10	0.189	27.4
10	0.403	29.0	10	0.229	27.5	10	0.189	27.4	15	0.046	21.1
11	0.397	27.7	20	0.115	18.2	15	0.046	21.1	16	0.035	20.2
20	0.579	20.8	23	0.102	16.7	17	0.029	18.7	20	0.062	17.3
50	1.295	15.5	30	0.133	14.2	20	0.040	17.5	50	0.186	9.0
100	1.860	14.0	100	0.900	8.1	30	0.064	13.4	100	0.242	5.4
1,000	2.436	13.1	300	1.897	6.6	50	0.070	9.5	1000	1.478	1.8
			1000	2.395	6.2	100	0.256	6.3			1.8
						500	1.740	3.9			

psia. The rock expansivity ranged between 0 and 8.07%. The porosity decreased from 16.7 to 15.5% over the course of the test. This particular sample exhibited higher than normal expansion.

The rock expansivity and compressibility are related through

$$E_f(p) = \int_{p}^{p_i} c_f \mathrm{d}p, \qquad (9.75)$$

where c_f is the rock compressibility. This equation assumes that the fractional change in PV is small. Physically, the rock expansivity represents the fractional change in PV while, in con-

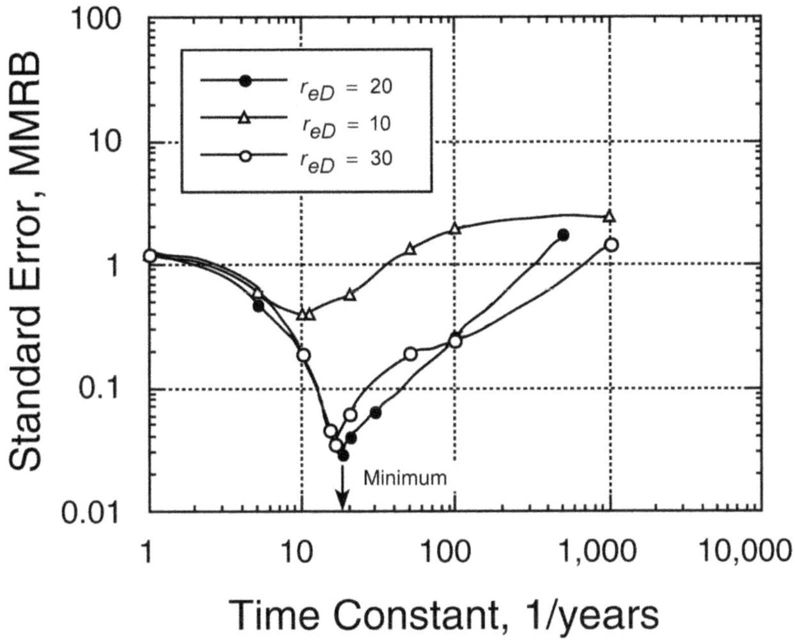

Fig. 9.32—Material-balance error vs. time constant for several values of r_{eD} for Example 9.4.

TABLE 9.26—RANGE OF CONSTANTS

OOIP, million STB	18.7 to 20.2
OGIP, Bscf	16.8 to 18.2
U, RB/psi	459 to 477
Standard error, million RB	0.029 to 0.035

trast, the rock compressibility represents the rate of change in fractional PV with pressure. While the former is more pertinent to material-balance calculation, experimental data are often reported in terms of the latter. Table 9.27 includes rock-compressibility measurements. If c_f is known as a function of pressure, then the integral on the right side of Eq. 9.75 can be evaluated numerically to determine $E_f(p)$. If c_f is relatively independent of pressure, then Eq. 9.75 can be simplified to

$$E_f(p) = c_f(p_i - p) \; . \qquad (9.76)$$

This method of estimating E_f is not usually preferable because c_f is rarely constant. **Fig. 9.34** illustrates a case and plots the rock compressibility as a function of pressure from the data in Table 9.27. Several features are worth noting, and many of these features are characteristic of compaction drives. First, the rock compressibility ranges between 4 to 21 × 10^{-6} psi^{-1}, which is a greater-than-normal range. Second, the rock compressibility clearly is not independent of pressure. Third, the compressibility declines sharply as the pressure first declines below the initial pressure. This phenomenon is largely attributed to grain rearrangement. Fourth, the rock com-

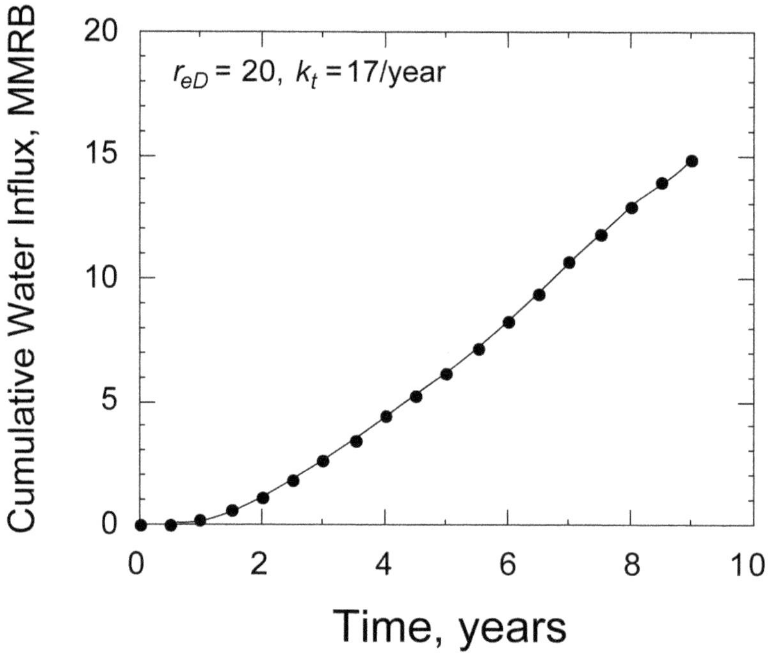

Fig. 9.33—Predicted water-influx history for Example 9.4.

TABLE 9.27—EXPERIMENTAL ROCK-COMPRESSIBILITY RESULTS FOR A HIGH-PRESSURE, GULF COAST RESERVOIR

Overburden Pressure (psia)	PV (cm^3)	Bulk Volume (cm^3)	Porosity (%)	Pressure (psia)	Rock Expansivity	Rock Compressibility (10^{-6} PV/PV/psi)
200	3.420	20.530	16.66	9,800	0.0000	15.1
1,000	3.379	20.498	16.48	9,000	0.0120	12.5
2,000	3.337	20.447	16.32	8,000	0.0243	10.2
3,000	3.303	20.413	16.18	7,000	0.0342	8.2
4,000	3.276	20.382	16.07	6,000	0.0421	5.8
5,000	3.257	20.367	15.99	5,000	0.0477	4.3
6,000	3.243	20.353	15.93	4,000	0.0518	4.0
7,000	3.230	20.340	15.88	3,000	0.0556	5.3
8,000	3.213	20.323	15.81	2,000	0.0605	11.3
9,000	3.177	20.329	15.63	1,000	0.0711	20.9
9,500	3.144	20.254	15.52	500	0.0807	

pressibility increases at pressures below 4,000 psia. This phenomenon is attributed to pore collapse.

Once $E_f(p)$ is estimated, the material-balance methods in the previous sections can be applied to estimate the OOIP and confirm the producing mechanism.

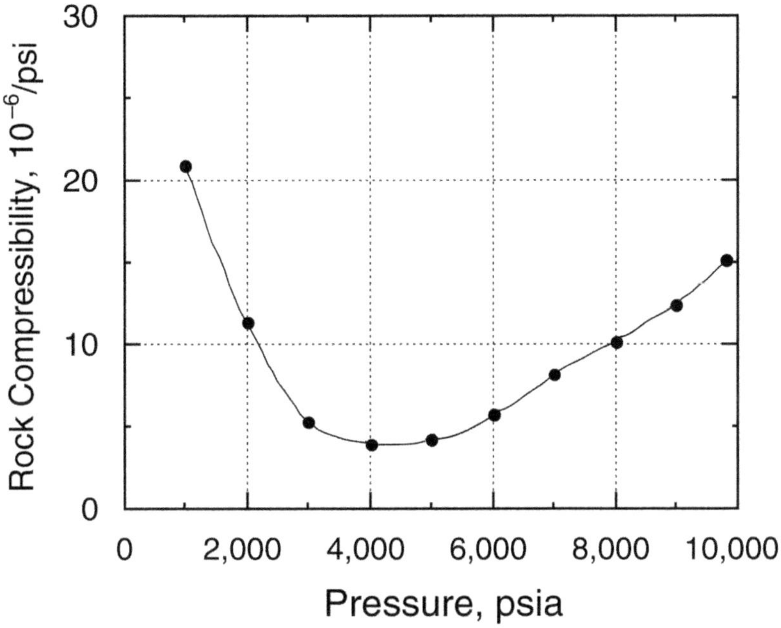

Fig. 9.34—Rock compressibility as a function of pressure for a Gulf Coast reservoir.

9.11 Water and Gas Coning

Coning is a production problem in which gas-cap gas or bottomwater infiltrates the perforation zone in the near-wellbore area and reduces oil production. Gas coning is distinctly different from, and should not be confused with, free-gas production caused by a naturally expanding gas cap. Likewise, water coning should not be confused with water production caused by a rising WOC from water influx. Coning is a rate-sensitive phenomenon generally associated with high producing rates. Strictly a near-wellbore phenomenon, it only develops once the pressure forces drawing fluids toward the wellbore overcome the natural buoyancy forces that segregate gas and water from oil.

Under ideal conditions in which no coning exists, flow is principally horizontal and mainly oil is produced. **Fig. 9.35** illustrates a producing well with no coning. When coning exists, however, the overlying gas is drawn downward or bottomwater is drawn upward and into the oil column. Coning trades oil production for gas or water production. **Fig. 9.36** illustrates a producing well subject to gas and water coning.

Two strategies commonly are used to minimize coning. One approach is partial perforation or penetration. In this approach, only a limited portion of the pay thickness is perforated. If gas coning is anticipated, the pay thickness near the GOC is not perforated. If water coning is anticipated, the pay thickness near the WOC is not perforated. In instances in which severe coning is expected, only a small portion of the pay thickness may be perforated. The variables in **Fig. 9.37** define the length of the perforation interval, b, and its position within the pay thickness, h. The distance L_g is the distance between the top of the pay and the uppermost perforation, and the distance L_w is the distance between the bottom of the pay and the lowest perforation. The quotient b/h is the partial perforation fraction. Although this strategy will reduce and can eliminate coning problems, it suffers an obvious drawback; namely, it temporarily reduces oil production in the hope of eventually avoiding coning.

A second remedial strategy is based on the observation that there is a critical producing rate below which the cone stabilizes and will not reach the perforations. This critical rate is a

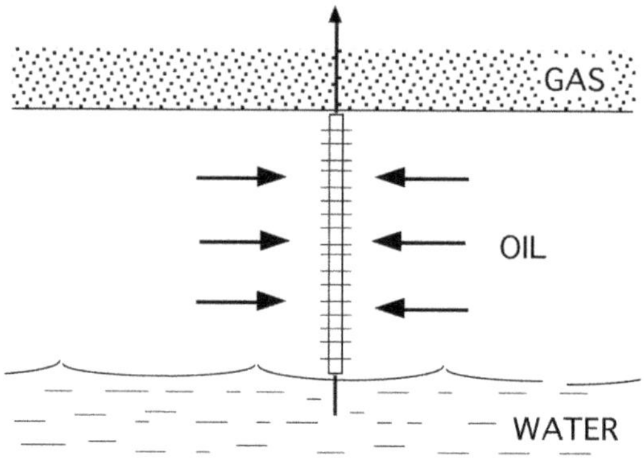

Fig. 9.35—A producing well with no coning.

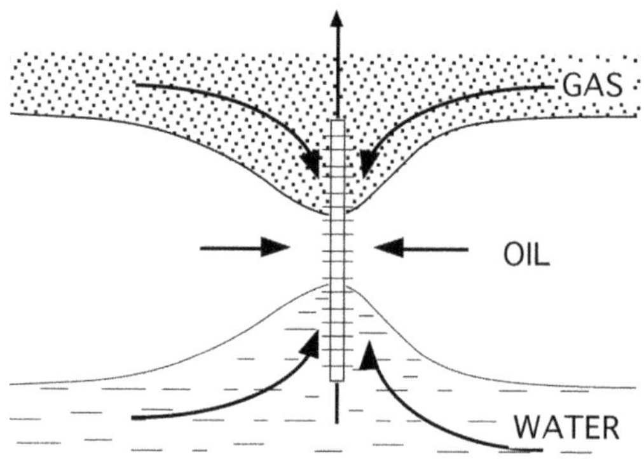

Fig. 9.36—A producing well subject to gas and water coning.

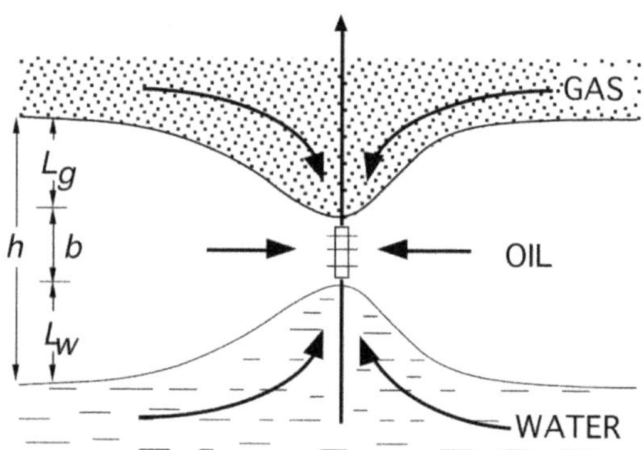

Fig. 9.37—Definition of variables for a partially perforated producing well.

function of the perforation length. As the perforation length increases, the critical producing rate decreases. Often, the critical producing rate is much less than the possible producing rate. This difference creates an operational decision: produce at a rate greater than the critical and eventually risk coning, or produce at a rate less than the critical and temporarily sacrifice oil production. If the critical rate is less than the minimum economic rate, then the operator has no choice but to produce above the critical rate or abandon the well.

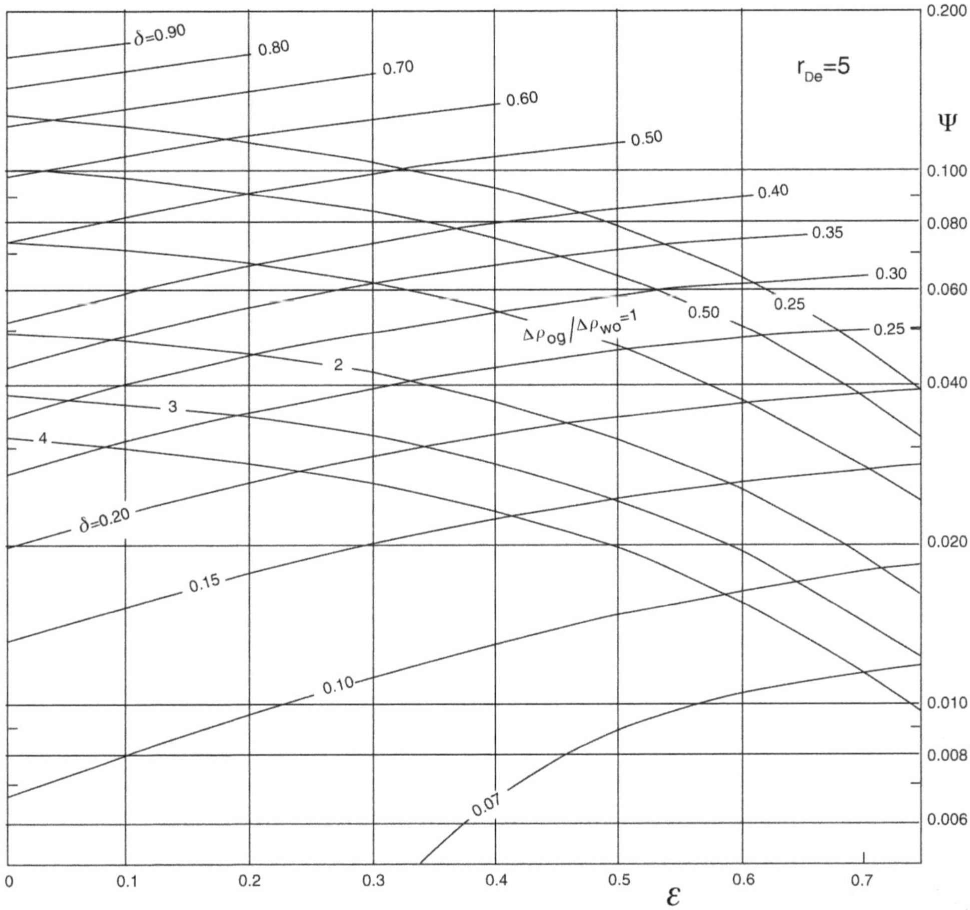

Fig. 9.38—ψ as a function of ε, δ, and $\Delta\rho_{og}/\Delta\rho_{wo}$ for r_{De}= 5.[66]

To combat coning, a hybrid strategy is often used whereby a combination of partial perforation and a reduced producing rate is used. One especially unattractive consequence of gas coning is that it prematurely depletes the gas-cap gas and diminishes the gas-cap producing mechanism. Fortunately, gas coning is not as problematic as water coning because the density difference between oil and gas is greater than the difference between water and oil. This density difference through gravity segregation helps mitigate coning.

To develop an effective remedial strategy against coning, certain theoretical aspects regarding coning must be understood. Mathematically, coning is a challenging problem because of its complexity. To develop tractable analytical solutions, tenuous assumptions must be invoked. These assumptions limit the practical applicability of these solutions. The most reliable way to study coning is with a specially designed finite-difference simulator.[61,62,63] Nevertheless, certain analytical solutions and empirical correlations can be helpful and serve as a preliminary guide.

Muskat and Wyckoff [64] and Chaney *et al.*[65] were among the first to contribute substantively to this problem. Since their efforts, several other authors have contributed to the body of literature.[66–73] Many of these works have led to similar correlations. Wheatley[72] presented a comparison of some popular correlations. As a representative sample, the correlations of Schols[71] and Chierici *et al.*[66] are presented here. Both works apply to both water and gas coning. Both efforts also use the following equation to compute the critical producing rate:

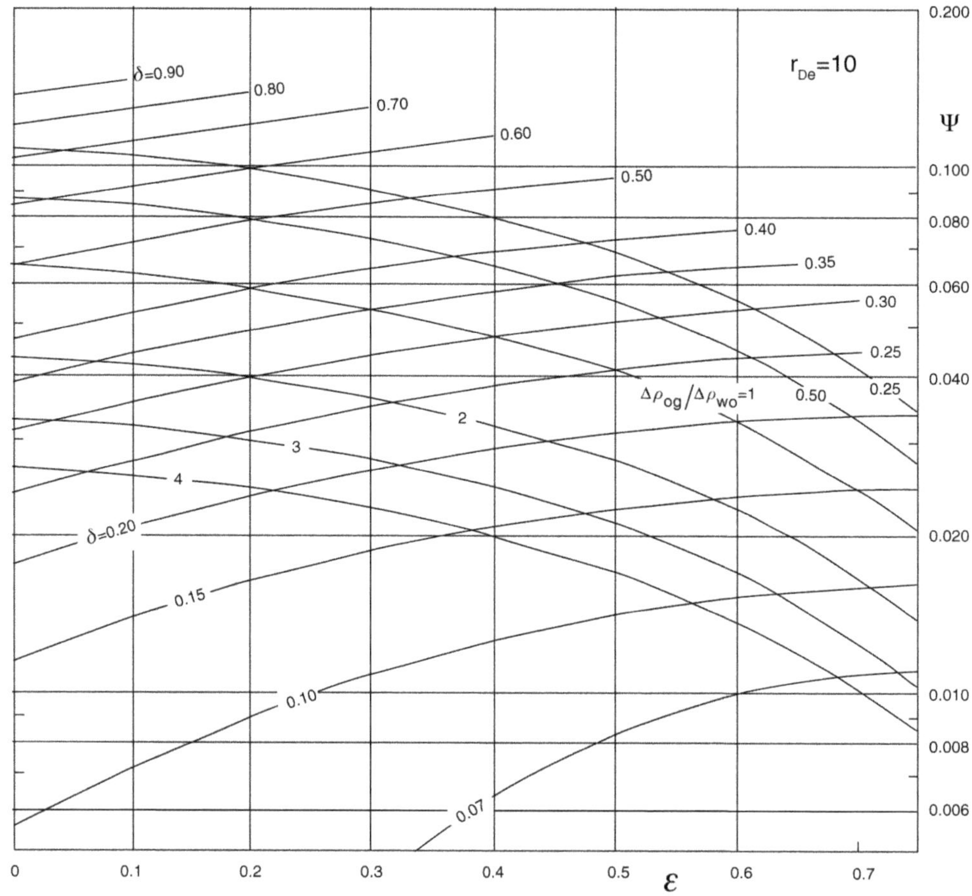

Fig. 9.39—ψ as a function of ε, δ, and $\Delta\rho_{og}/\Delta\rho_{wo}$ for r_{De} = 10.[66]

$$q_c = \frac{0.003073 k_o h^2 \Delta\rho}{\mu_o B_o} q_{Dc}, \qquad (9.77)$$

where $\Delta\rho$ = density difference (g/cm³), B_o = average oil FVF, μ_o = average oil viscosity (cp), k_o = oil permeability (md), q_{Dc} = dimensionless critical producing rate, h = pay thickness (ft), and q_c is given in STB/D. The oil permeability, k_o, is the product of the horizontal permeability and the oil relative permeability. The dimensionless critical rate, q_{Dc}, is specified by correlation.

9.11.1 Schols' Correlation. Schols' correlation[71] is based on a numerical simulation study. The dimensionless critical producing rate is

$$q_{Dc} = \frac{1}{2\pi}\left[0.432 + \frac{\pi}{\ln\left(\frac{r_e}{r_w}\right)}\right]\left[1 - \left(\frac{b}{h}\right)^2\right]\left(\frac{r_e}{h}\right)^{-0.14}, \qquad (9.78)$$

where b is the length of the perforation interval and r_e is the drainage radius.

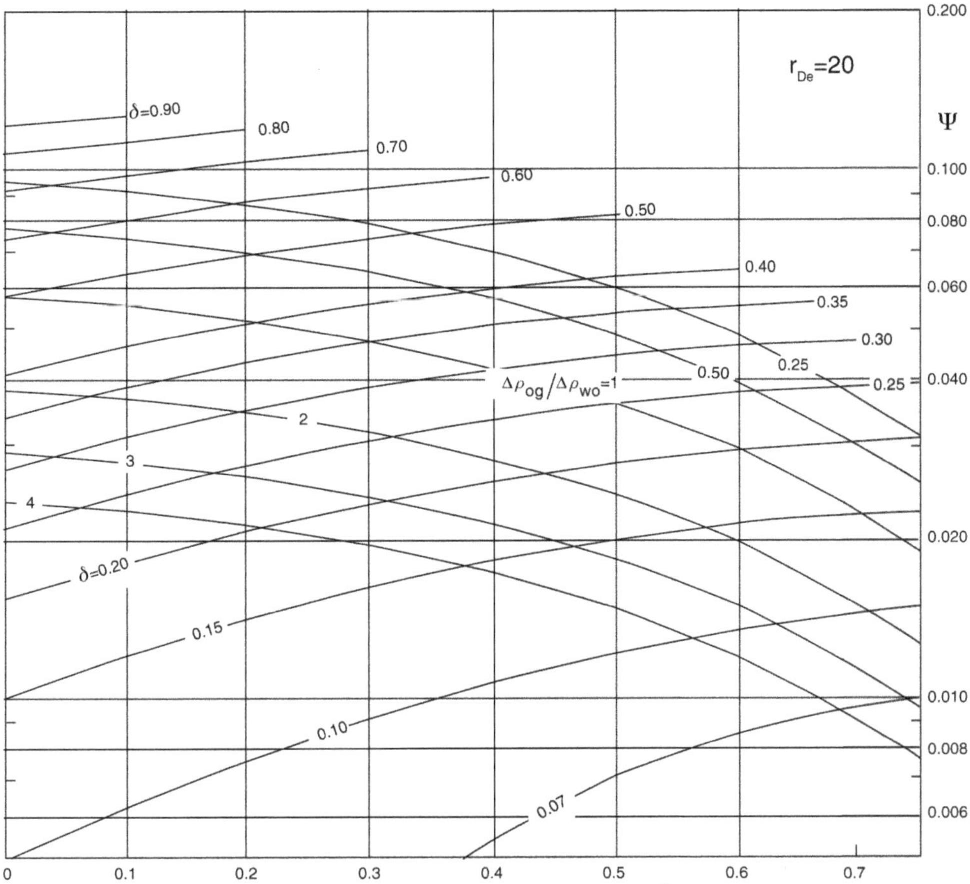

Fig. 9.40—ψ as a function of ε, δ, and $\Delta\rho_{og}/\Delta\rho_{wo}$ for r_{De} = 20.[66]

This correlation applies to both water and gas coning; however, it directly applies only to cases in which water or gas coning exist separately. In other words, it does not directly apply to cases in which water and gas coning act simultaneously. The correlation can be used to predict the critical rate for a pre-existing completion or to predict the optimum perforation length for a future completion. In this latter application, the optimum perforation length is defined as the length at which the critical and theoretical producing rates are equal. The theoretical producing rate of a partially penetrating well is computed in Section 9.11.3. Example 9.5 illustrates the former application while Example 9.6 illustrates the latter application.

9.11.2 Chierici *et al.* Correlation. The correlation by Chierici *et al.*[66] was based on a potentiometric study. This was one of the most sophisticated correlations. It allows the vertical permeability to differ from the horizontal permeability. This can be an important factor because coning vanishes as the vertical permeability approaches zero. This correlation also treats the problem of simultaneous gas/water coning. This is important in situations in which a gas cap and bottomwater coexist.

The correlation of Chierici *et al.* was specified in terms of a series of charts. The charts used the following nomenclature. The dimensionless critical rate is denoted as ψ (previously defined as q_{Dc}) and the following dimensionless variables are defined as

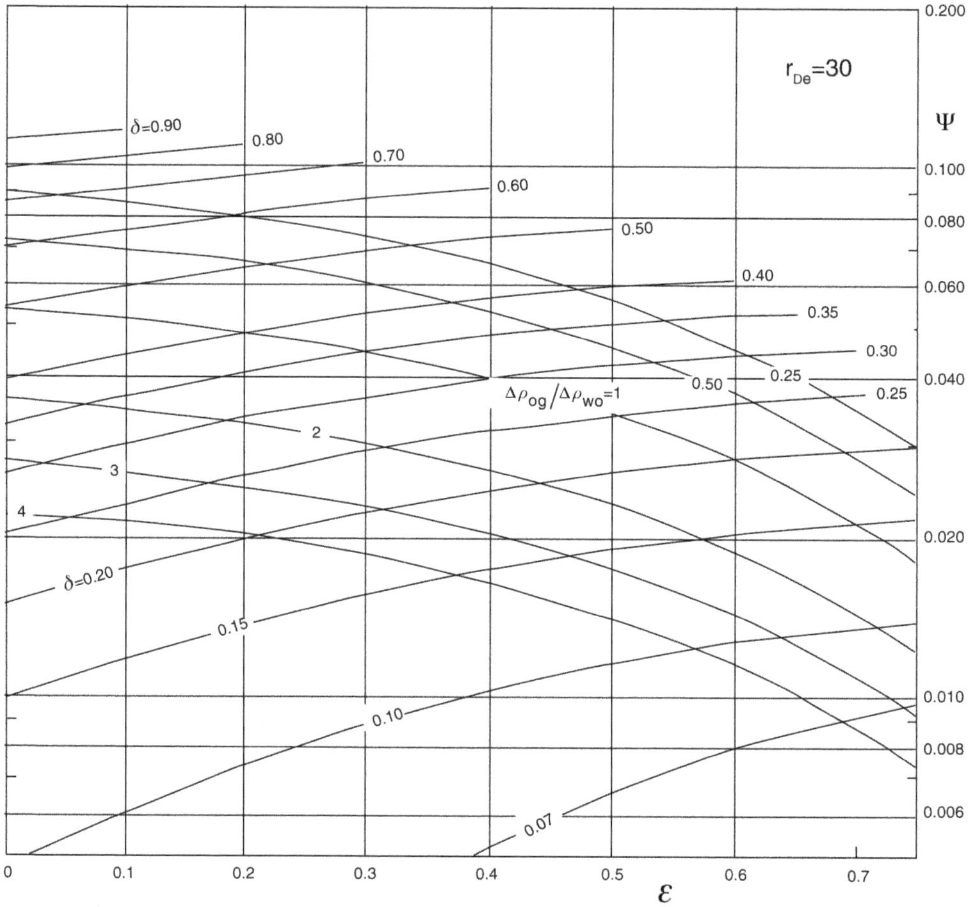

Fig. 9.41—ψ as a function of ε, δ, and $\Delta\rho_{og}/\Delta\rho_{wo}$ for r_{De} = 30.[66]

$$\delta_g = L_g / h, \quad (9.79)$$

$$\delta_w = L_w / h, \quad (9.80)$$

$$\varepsilon = b / h, \quad (9.81)$$

$$\text{and } r_{De} = (r_e / b)(k_v / k_H)^{1/2} . \quad (9.82)$$

The dimensionless critical rate, ψ, is a function of r_{De}, ε, and δ. **Figs. 9.38 through 9.44** show the charts. Each chart corresponds to a different value of r_{De}. Specific charts exist for r_{De} = 5, 10, 20, 30, 40, 60, and 80.

The charts are used differently depending on whether they are used to compute the critical rates or optimize the perforation length. The problem of optimizing the perforation length for simultaneous gas and water coning is more complicated than separate water or gas coning. The charts simplify the problem; however, the solution procedure depends on the application.

Calculating Critical Rates. For a pre-existing perforation length, b, the critical rates are computed with the following procedure:

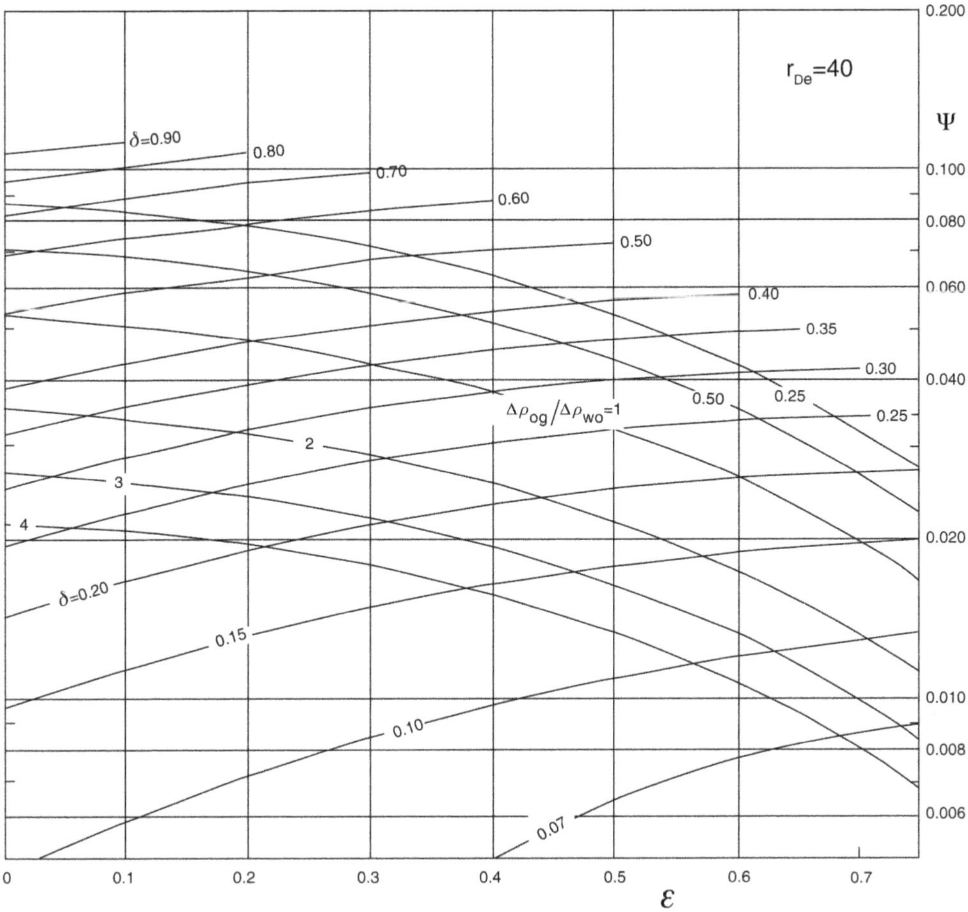

Fig. 9.42—ψ as a function of ε, δ, and $\Delta\rho_{og}/\Delta\rho_{wo}$ for r_{De} = 40.[66]

1. Compute δ_g, δ_w, ε, and r_{De} with Eqs. 9.79 through 9.82.

2. Locate the correct chart or charts, depending on the value of r_{De}. Interpolation between two charts may be required.

3. Compute ψ_g from the charts based on r_{De}, ε, and δ_g; then, compute ψ_w from the charts based on r_{De}, ε, and δ_w.

4. Compute the critical rates to avoid water and gas coning with Eq. 9.77.

This procedure ignores the curves labeled $\Delta\rho_{og}/\Delta\rho_{wo}$. The calculation procedure is simplified if only bottomwater or a gas cap exists. If no bottomwater exists, then the calculation of ψ_w and δ_w can be ignored. Conversely, if no gas cap exists, then the calculation of ψ_g and δ_g can be ignored. Example 9.5 illustrates an application.

Calculating Optimum Perforation Length. For a bottomwater, gas-cap reservoir, the procedure to calculate the optimum perforation (length and position) uses the curves labeled $\Delta\rho_{og}/\Delta\rho_{wo}$ and is as follows:

1. Compute r_{De} with Eq. 9.82. Also, compute $\Delta\rho_{og}/\Delta\rho_{wo}$.

2. Assume a value of ε.

3. Compute ψ and δ with the charts. These values correspond to ψ_g and δ_g.

4. Compute $L_g = h\delta_g$ and $L_w = h(1 - \varepsilon) - L_g$.

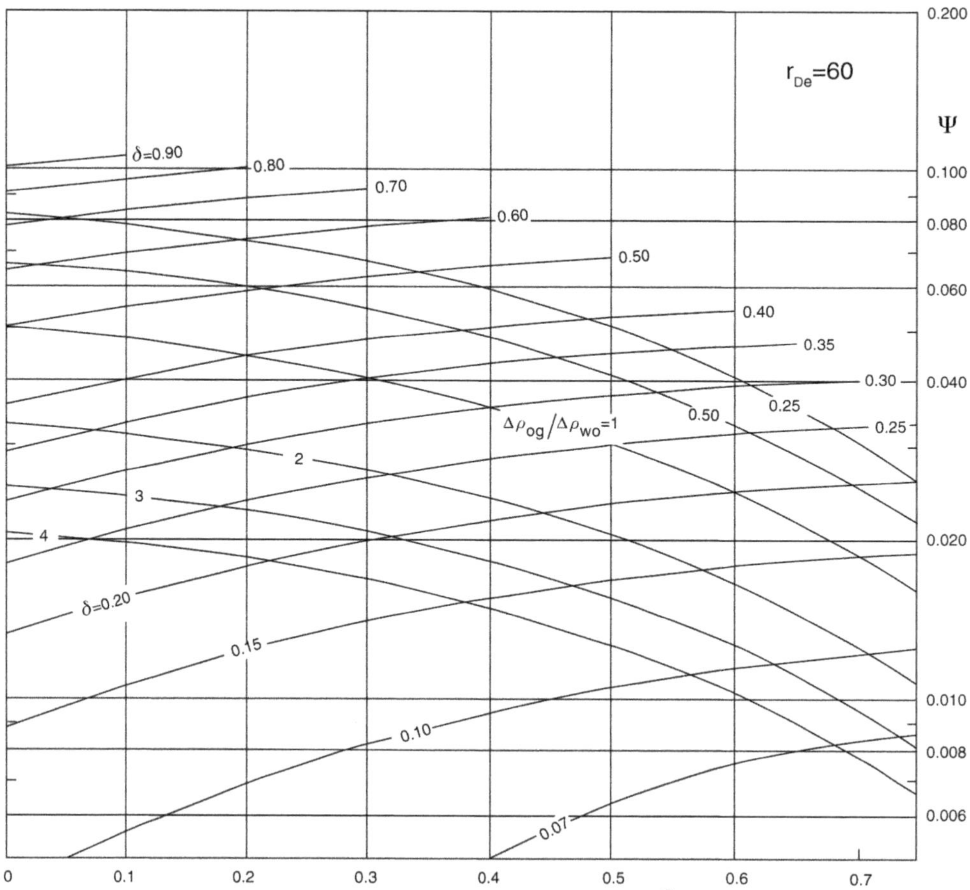

Fig. 9.43—ψ as a function of ε, δ, and $\Delta\rho_{og}/\Delta\rho_{wo}$ for r_{De} = 60.[66]

5. Compute the critical rate from Eq. 9.77. Only one critical rate is needed because the procedure assumes equal critical rates for gas and water coning.

6. Return to Step 2. Assume a new value of ε, and repeat the calculation until an adequate range of ε is covered.

7. Compute the theoretical producing rate of a partially penetrating well as a function of ε with Eq. 9.83. (See Section 9.11.3 below).

8. Plot the critical and theoretical producing rates as a function of ε. The value of ε at which the critical and producing rates intersect yields the optimal perforation interval.

This procedure is simplified if only a gas cap or bottomwater exists. The same procedure applies except $\Delta\rho_{og}/\Delta\rho_{wo}$ is not calculated and $\delta_g = 1 - \varepsilon$ (if there is no bottomwater and only gas coning is a problem) or $\delta_w = 1 - \varepsilon$ (if there is no gas cap and only water coning is a problem). ψ is computed for a range of ε until an optimal value of ε is identified. Example 9.6 illustrates this procedure for a gas-cap reservoir. Example 9.7 illustrates an application for a gas-cap, bottomwater reservoir.

9.11.3 Partially Penetrating Wells. The theoretical producing rate of a partially penetrating well is needed to compute the optimum perforation length. Partially penetrating wells are wells that do not fully penetrate or are not fully perforated throughout the pay thickness. If vertical

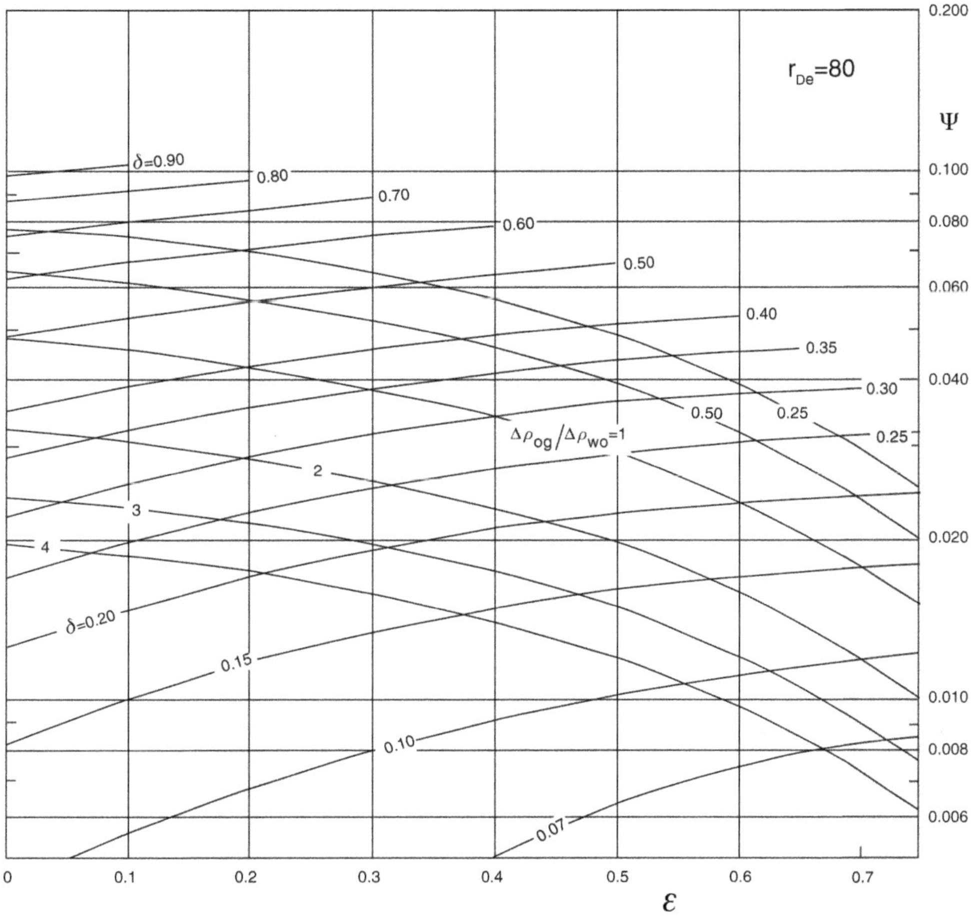

Fig. 9.44—ψ as a function of ε, δ, and $\Delta\rho_{og}/\Delta\rho_{wo}$ for r_{De} = 80.[66]

permeability exists, these wells will produce fluids from above and below the perforations. **Fig. 9.45** illustrates fluid delivery into a partially penetrating producing well. Under these circumstances, fluid flow is obviously not strictly horizontal. The producing rate in partially penetrating wells with nonzero vertical permeability is greater than the rate with no vertical permeability.

Partially penetrating wells commonly are used to minimize coning. The critical rate gives a producing rate below which no coning will occur. Often, however, the critical flow rate is much less than maximum possible flow rate. To judge the difference, estimates for the flow rate of a partially penetrating well are needed. Several authors have offered analytical expressions to estimate the flow rate.[1,74,75] Most efforts yield estimates within a few percent of one another. The Kozeny expression,[75] for example, is

$$q = \frac{0.00708 k_H b \left(p_e - p_w\right)}{\mu_o B_o \left[\ln\left(\frac{r_e}{r_w}\right) + s \right]} \left[1 + 7\sqrt{\frac{r_w}{2b}\sqrt{\frac{k_v}{k_H}}} \cos\left(\frac{\pi b}{2h}\right) \right], \quad (9.83)$$

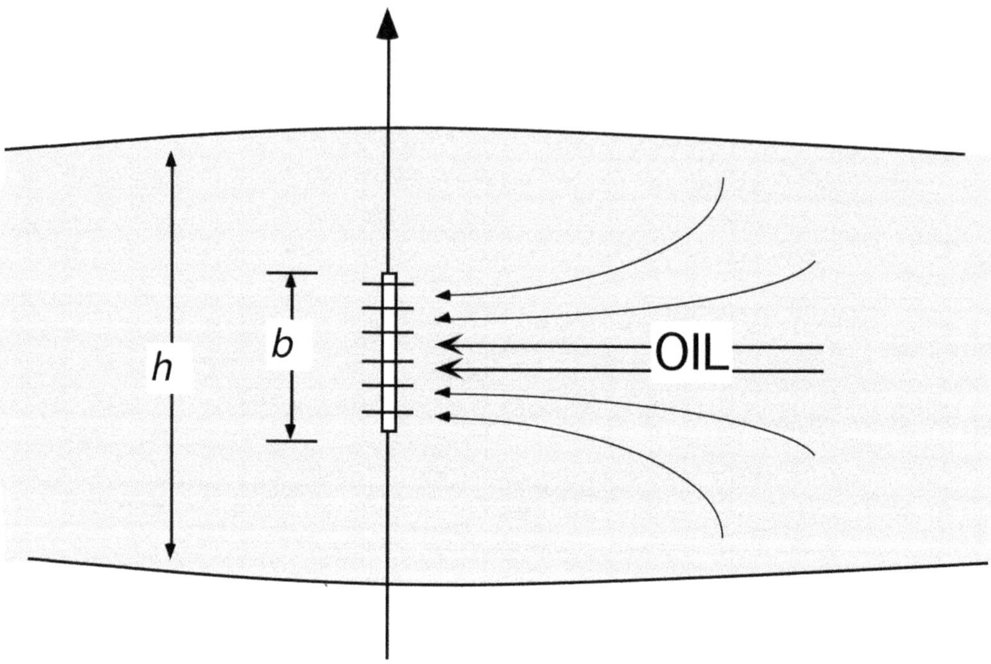

Fig. 9.45—Fluid delivery into a partially penetrating producing well.

where k_H = horizontal permeability, k_v = vertical permeability, p_e = pressure at drainage radius, p_w = wellbore pressure, s = skin factor, q = oil producing rate, and the argument of the cosine assumes radians. This equation assumes units of md, ft, psi, cp, and STB/D. This equation also assumes steady-state flow in a circular drainage area, where r_e and r_w are the drainage and wellbore radii, respectively. Eq. 9.83 gives the Kozeny equation in SI units.

9.11.4 Variables Affecting Coning. The ratio of q_c/q is a measure of the tendency not to cone. As q_c increases or q decreases, the likeliness to avoid increases. According to Eqs. 9.77 and 9.83, the ratio q_c/q is proportional to

$$\frac{q_c}{q} \propto \frac{h^2 \Delta\rho \left[\ln\left(\frac{r_e}{r_w}\right) + s \right]}{b} \quad \text{(9.84)}$$

This expression shows that the likeliness to control coning increases as the penetration interval b decreases. Eq. 9.84 also shows that the likeliness to control coning increases as the pay thickness increases, density difference increases, well spacing increases, and perforation length decreases. Horizontal permeability does not affect the likelihood of success. This expression also suggests that controlling coning in a thin reservoir may be difficult.

9.11.5 Additional Measures To Control Coning. Other techniques have been applied to control coning. These include placing an artificial barrier above or below the pay to suppress vertical flow,[76] injecting oil to control gas coning,[77] or the use of horizontal wells. Barriers composed of cement and high-molecular-weight polymers have been tried. Another, although expensive, technique is to drill additional wells and produce them at the critical rate.

Example 9.5: Computing Critical Rate To Prevent Coning. Compute the critical rate (STB/D) for a well in a gas-cap reservoir with the following characteristics: k_H = 60 md, k_v = 39 md, h = 150 ft, r_e = 933 ft (80-acre well spacing), r_w = 0.5 ft, B_o = 1.25 RB/STB, μ_o = 1.11 cp, ρ_o = 0.714 g/cm^3, and ρ_g = 0.098 g/cm^3. The well is completed in only the lower 60 ft of the pay thickness.

Solution. Schols' Correlation. With Eq. 9.78, compute q_{Dc}. This yields

$$q_{Dc} = \frac{1}{2\pi}\left(0.432 + \frac{\pi}{\ln \frac{933}{0.5}}\right)\left[1 - \left(\frac{60}{150}\right)^2\right]\left(\frac{933}{150}\right)^{-0.14} = 0.0879.$$

The critical rate is then computed with Eq. 9.77, which yields

$$q_c = \frac{0.003073(60\ \text{md})(150\ \text{ft})^2(0.714 - 0.098\ \text{g/cm}^3)}{(1.11\ \text{cp})\,(1.25\ \text{RB/STB})}(0.0879) = 161.9\ \text{STB/D}.$$

Chierici *et al.* Correlation. First, evaluate L_g = 90 ft, δ_g = 90/150 = 0.60, ε = 60/150 = 0.40, and r_{De} = 5.01. With the chart for r_{De} = 5, obtain ψ = 0.120. Evaluating Eq. 9.77 yields

$$q_c = \frac{0.003073(60\ \text{md})(150\ \text{ft})^2(0.714 - 0.098\ \text{g/cm}^3)}{(1.11\ \text{cp})\,(1.25\ \text{RB/STB})}(0.120) = 221\ \text{STB/D}.$$

Example 9.6: Computing Optimum Perforation Length To Prevent Coning. For an uncompleted well in the gas-cap reservoir in Example 9.5, compute the optimum perforation length at a reservoir pressure of 1,800 psia. Assume a wellbore pressure of 1,500 psia and a skin factor of 10.

Solution. Schols' Correlation. Compute the dimensionless critical rate and critical rate as a function of b/h. **Table 9.28** summarizes the results for b/h = 0 to 1.

Next, estimate the theoretical producing rate for a partially penetrating well. With Eq. 9.83 for b = 60 ft yields

$$q = \frac{0.00708(60\ \text{md})(60\ \text{ft})(1{,}800 - 1{,}500\ \text{psi})}{(1.11\ \text{cp})(1.25\ \text{RB/STB})\left(\ln \frac{933}{0.5} + 10\right)}\left(1 + 7\sqrt{\frac{(0.5\ \text{ft})}{2(60\ \text{ft})}}\sqrt{\frac{39}{60}}\cos\left(\frac{\pi(60\ \text{ft})}{2(150\ \text{ft})}\right)\right) = 430\ \text{STB/D}.$$

Table 9.28 summarizes the results at other values of b. The rate varies from 0 to 808 STB/D, depending on the length of the perforation interval. The optimum perforation length corresponds to the value of b at which the critical and theoretical producing rates are equal. **Fig. 9.46** shows a plot of q and q_c vs. b/h. The curves intersect at approximately b/h = 0.15. This corresponds to b = 22.5 ft, L_g = 127.5 ft, and q = 187 STB/D. In conclusion, only the lower 22.5 ft of the pay thickness should be perforated to avoid gas coning.

Chierici *et al.* Correlation. First, compute δ_g = 1–b/h for a range of b/h. Next, compute r_{De}. From Example 9.5, r_{De} = 5. With the appropriate chart, compute ψ for the range of b/h. The charts use b/h as ε. Then compute the critical rate with Eq. 9.77. **Table 9.29** summarizes the results. Fig. 9.46 plots q_c vs. b/h.

TABLE 9.28—SUMMARY OF CALCULATIONS WITH THE SCHOLS[71] CORRELATION

b/h	q_{Dc}	q_c (STB/D)	Oil-Producing Rate (STB/D)
0	0.1046	192.7	0.0
0.1	0.1036	190.8	145.7
0.2	0.1004	185.0	250.1
0.3	0.0952	175.4	344.0
0.4	0.0879	161.9	429.8
0.5	0.0785	144.5	508.2
0.6	0.0670	123.3	579.9
0.7	0.0534	98.3	645.1
0.8	0.0377	69.4	704.5
0.9	0.0199	36.6	758.8
1	0.0000	0.0	808.9

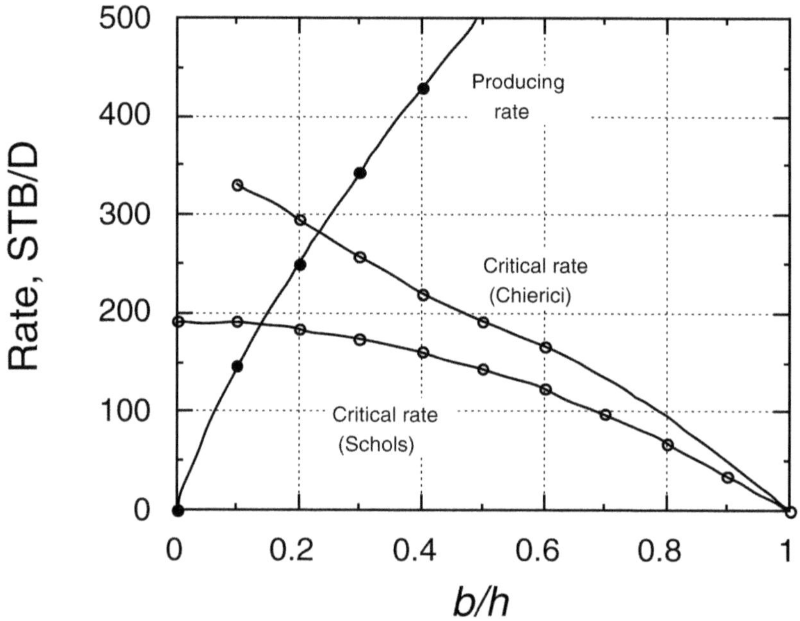

Fig. 9.46—Producing and critical rates as a function of *b/h*.

The optimal perforation length corresponds to the value of b/h at which the critical and producing rates are equal. This approximately occurs for b/h = 0.25. This corresponds to b = 37.5 ft, L_g = 112.5 ft, and q = 275 STB/D. In conclusion, only the lower 37.5 ft of the pay should be perforated. The method of Chierici *et al.* yields a wider perforation interval than the method of Schols (37.5 vs. 22.5 ft). The Chierici *et al.* method is consistently more liberal than Schols' method.

Example 9.7: Optimum Perforation Length To Prevent Coning in a Bottom-Water, Gas-Cap Reservoir. Assume the gas-cap reservoir in Examples 9.5 and 9.6 is

TABLE 9.29—SUMMARY OF CALCULATIONS WITH THE CHIERICI et al.[66] CORRELATION

b/h	δ_g	$\psi = q_{Dc}$	q_c (STB/D)	Oil-Producing Rate (STB/D)
0	1			0.0
0.1	0.9	0.180	331.5	145.7
0.2	0.8	0.160	294.7	250.1
0.3	0.7	0.140	257.9	344.0
0.4	0.6	0.120	221.0	429.8
0.5	0.5	0.105	193.1	508.2
0.6	0.4	0.090	165.8	579.9

TABLE 9.30—RESULTS FOR EXAMPLE 9.7

b/h	δ_g	$\psi = q_{Dc}$	q_c (STB/D)	Oil-Producing Rate (STB/D)
0				0.0
0.05	0.40	0.059	108.7	86.7
0.10	0.37	0.058	106.8	145.7
0.20	0.33	0.055	101.3	250.1
0.30	0.28	0.048	88.4	344.0
0.40	0.24	0.040	73.7	429.8

underlain by water. The water density is 1.092 g/cm^3. Compute the optimum perforation length and position of the perforation interval with the Chierici *et al.* correlation.

Solution. First, compute $\Delta\rho_{og}/\Delta\rho_{wo}$. This yields (0.741 – 0.098)/(1.092 – 0.741) = 1.83. Next, determine δ_g and ψ from the charts for a range of ε. If ε = 0.40, for example, use the chart corresponding to r_{De} = 5 to determine that δ_g = 0.24 and ψ = 0.040. **Table 9.30** summarizes the results for a range of ε values from 0.05 to 0.40. Next, compute the critical rate for each value of ψ with Eq. 9.77. Finally, compute the theoretical producing rate with Eq. 9.83.

An examination of Table 9.30 shows that the critical and producing rates are equal at $\varepsilon = b/h$ = 0.075. This value of b/h corresponds to L_g = (0.385) (150) = 57.8 ft, b = (0.075) (150) = 11.3 ft, and L_w = 150 – 57.8 – 11.3 = 81 ft. In conclusion, the well should be perforated with an 11.3-ft interval located 57.8 ft below the GOC and 81 ft above the WOC.

Nomenclature

a = constant
b = perforation length, L, ft
B_g = gas FVF, RB/scf
B_o = oil FVF, RB/STB
B_{tg} = two-phase gas FVF, RB/scf
B_{to} = two-phase oil FVF, RB/STB
B_{tw} = two-phase water/gas FVF, RB/STB
B_w = water FVF, RB/STB
c_f = rock compressibility, Lt^2/m, 1/psi

c_t = total aquifer compressibility, Lt²/m, 1/psi
E_f = rock (formation) expansivity
E_g = gas expansivity, RB/scf
E_{gw} = expansivity for McEwen method, RB/scf
E_{gwf} = composite gas/water/rock FVF, RB/scf
E_o = oil expansivity, RB/STB
E_{ow} = expansivity for McEwen method, RB/STB
E_{owf} = composite oil/water/rock FVF, RB/STB
E_t = total expansivity, RB/STB
E_v = volumetric sweep efficiency, fraction
E_w = water expansivity, RB/STB
f = reservoir angle
F = total fluid withdrawal, L³, RB
G = total original gas in place, L³, scf
G_{fgi} = initial free gas in place, L³, scf
G_i = cumulative gas injected, L³, scf
G_p = cumulative produced gas, L³, scf
G_{ps} = cumulative produced sales gas, L³, scf
h = pay thickness, L, ft
J^* = constant
k = permeability, L², md
k_a = aquifer permeability, L², md
k_H = horizontal permeability, L², md
k_t = time constant, 1/t, 1/years
k_v = vertical permeability, L², md
K = gas/oil equilibrium ratio or constant
L_a = aquifer length, L, ft
L_g = perforation gap between uppermost perforation and GOC, L, ft
L_r = reservoir length, L, ft
L_w = perforation gap between lowermost perforation and WOC, ft, L
m = initial free-gas-phase to free-oil-phase volume ratio (gas-cap size)
n = total number of data points
N = total original oil in place, L³, STB
N_{foi} = initial free oil in place, L³, STB
N_g = dimensionless gravity number
N_p = cumulative produced oil, L³, STB
p = pressure, m/Lt², psi
$\tilde{p}$ = time-averaged pressure, m/Lt², psi
$\bar{p}$ = average reservoir pressure, m/Lt², psi
$\bar{p}_a$ = average aquifer pressure, m/Lt², psi
p_e = pressure at drainage radius, m/Lt², psi
p_w = wellbore pressure, m/Lt², psi
q = producing rate at reservoir conditions (RB/D) or surface conditions (STB/D), L³/t
q_c = critical coning rate, STB/D, L³/t
q_{Dc} = dimensionless critical coning rate

q_{sg} = surface producing gas rate, L^3/t, scf/D
q_{so} = surface producing oil rate, L^3/t, STB/D
q_w = aquifer delivery rate, L^3/t, RB/D
r_a = aquifer radius, L, ft
r_{De} = dimensionless radius including vertical permeability
r_e = reservoir drainage radius
r_{eD} = ratio of reservoir to aquifer radii
r_o = effective reservoir radius, L, ft
r_w = wellbore radius, L, ft
R = instantaneous producing GOR, scf/STB
R_j = residual for point j, L^3, RB
R_p = cumulative GOR, scf/STB
R_{ps} = cumulative sales GOR, scf/STB
R_s = dissolved GOR, scf/STB
R_{ss} = sum of squares of the residual, L^6, RB^2
R_{sw} = dissolved-gas/water ratio, scf/STB
R_v = volatilized-oil/gas ratio, STB/MMscf
s = skin factor
S_g = gas saturation, fraction
S_{grw} = residual gas saturation to water flow, fraction
S_{orw} = residual oil saturation to water flow, fraction
S_{wi} = initial water saturation, fraction
t = time, t, years
t_{max} = maximum time, t, years
t_D = dimensionless time
t_{D*} = dimensionless time boundary between infinite-acting and finite aquifers
t_{Dmax} = maximum dimensionless time
U = aquifer constant, L^4t^2/m, RB/psi
V_{fgi} = initial volume of free gas, L^3, RB
V_{foi} = initial volume of free oil, L^3, RB
V_p = reservoir PV, L^3, RB
V_{pa} = aquifer PV, L^3, RB
V_{pi} = initial reservoir PV, L^3, RB
V_{pr} = reservoir PV, L^3, RB
w = reservoir width, L, ft
W = initial water in place, L^3, STB
W_D = dimensionless cumulative water influx
W_e = cumulative water influx, L^3, RB
W_I = cumulative injected water, L^3, STB
W_p = cumulative produced water, L^3, STB
δ_g = dimensionless perforation gap below GOC
δ_w = dimensionless perforation gap above WOC
Δp = difference of time-averaged pressure, m/Lt^2, psi
$\Delta\rho$ = density difference, m/L^3, lbm/ft^3 and g/cm^3
$\Delta\rho_{og}$ = oil/gas density difference, m/L^3, g/cm^3
$\Delta\rho_{wo}$ = water/oil density difference, m/L^3, g/cm^3

ε = b/h
ϕ_a = aquifer porosity, fraction
ϕ_r = reservoir porosity, fraction
θ = angle
μ_g = gas viscosity, m/Lt, cp
μ_o = oil viscosity, m/Lt, cp
μ_w = water viscosity, m/Lt, cp
ρ_g = gas density, m/L^3, g/cm^3
ρ_o = oil density, m/L^3, g/cm^3
ρ_w = water density, m/L^3, g/cm^3
ψ = dimensionless critical rate
ψ_g = dimensionless critical gas coning rate
ψ_w = dimensionless critical water coning rate

Subscripts

i = initial condition
j = index
k = index

References

1. Muskat, M.: *Physical Principles of Oil Production,* McGraw-Hill Book Co. Inc., New York City (1949).
2. McCain, W.D. Jr.: "Heavy Components Control Reservoir Fluid Behavior," *JPT* (September 1994) 746.
3. Carmalt, S.W. and St. John, B.: "Giant Oil and Gas Fields," *Future of Petroleum Provinces of the World,* M.T. Halbouty (ed.), American Assn. of Petroleum Geologists (1984).
4. *Bull. D-14, Statistical Analysis of Crude Oil Recovery and Recovery Efficiency,* second edition, API, Dallas (1984).
5. Cordell, J.C. and Ebert, C.K.: "A Case History—Comparison of Predicted and Actual Performance of a Reservoir Producing Volatile Crude Oil," *JPT* (November 1965) 1291.
6. Jacoby, R.H. and Berry, V.J. Jr.: "A Method for Predicting Depletion Performance of a Reservoir Producing Volatile Crude Oil," *Trans.*, AIME (1957) **210,** 27.
7. Amyx, J.W., Bass, D.M., and Whiting, R.L.: *Petroleum Reservoir Engineering—Physical Properties*, McGraw-Hill Book Co. Inc., New York City (1960).
8. Dodson, C.R., Goodwill, D, and Mayer, E.H.: "Application of Laboratory PVT Data to Reservoir Engineering Problems," *Trans.*, AIME (1953) **198,** 287.
9. Whitson, C.H. and Torp, S.B.: "Evaluating Constant-Volume Depletion Data," *JPT* (March 1983) 610.
10. Ahmed, T.: *Hydrocarbon Phase Behavior,* Gulf Publishing Co., Houston (1989).
11. Reudelhuber, F.O. and Hinds, R.F.: "A Compositional Material Balance Method for Prediction of Recovery from Volatile Oil Depletion Drive Reservoirs," *Trans.*, AIME (1957) **210,** 19.
12. McCain, W.D.: *The Properties of Petroleum Fluids,* PennWell Publishing Co., Tulsa (1990).
13. Poettmann, F.H. and Thompson, R.S.: "Discussion of Engineering Applications of Phase Behavior of Crude Oil and Condensate Systems," *JPT* (November 1986) 1263.
14. Moses, P.L.: "Engineering Applications of Phase Behavior of Crude Oil and Condensate Systems," *JPT* (July 1986) 715.
15. Walsh, M.P. and Towler, B.F.: "Method Computes PVT Properties for Gas Condensates," *Oil & Gas J.* (31 July 1995) 83.
16. Coats, K.H. and Smart, G.T.: "Application of Regression-Based EOS PVT Program to Laboratory Data," *SPERE* (May 1986) 277.

17. Cook, R.E., Jacoby, R.H., and Ramesh, A.B.: "A Beta-Type Reservoir Simulator for Approximating Compositional Effects During Gas Injection," *SPEJ* (October 1974) 471.
18. Standing, M.B.: "A Set of Equations for Computing Equilibrium Ratios of a Crude Oil/Natural Gas System at Pressures Below 1,000 psia," *JPT* (September 1979) 1193.
19. Alani, G.H. and Kennedy, H.T.: "Volumes of Liquid Hydrocarbons at High Temperatures and Pressures," *Trans.*, AIME (1960) **219,** 288.
20. Walsh, M.P.: "A Generalized Approach to Reservoir Material Balance Calculations," *J. Cdn. Pet. Tech.* (January 1995) 55–63.
21. Walsh, M.P.: "New, Improved Equation Solves for Volatile Oil and Condensate Reserves," *Oil & Gas J.* (22 August 1994) 72.
22. Walsh, M.P., Ansah, J., and Raghavan, R.: "The New, Generalized Material Balance as an Equation of a Straight Line: Part 2—Applications to Saturated and Non-Volumetric Reservoirs," paper SPE 27728 presented at the 1994 SPE Permian Basin Oil and Gas Recovery Conference, Midland, Texas, 16–18 March.
23. Walsh, M.P. and Lake, L.W.: *A Generalized Approach to Primary Hydrocarbon Recovery,* Elsevier, Amsterdam (2003).
24. Walsh, M.P., Ansah, J., and Raghavan, R.: "The New, Generalized Material Balance as an Equation of a Straight Line: Part 1—Applications to Undersaturated, Volumetric Reservoirs," paper SPE 27684 presented at the 1994 SPE Permian Basin Oil and Gas Recovery Conference, Midland, Texas, 16–18 March.
25. Society of Petroleum Engineers: *QUICKSIM—A Modified Black-Oil Tank Model, User's Guide,* SPE Software Catalog, Richardson, Texas (2001).
26. Walsh, M.P.: "QUICKSIM—A Modified Black-Oil Tank Model," Version 1.6, Petroleum Recovery Research Inst., Austin, Texas (2000).
27. Walsh, M.P.: "Discussion of Application of Material Balance for High-Pressure Gas Reservoirs," *SPEJ* (December 1998) 402.
28. Walsh, M.P.: "Effect of Pressure Uncertainty on Material-Balance Plots," paper SPE 56691 presented at the 1999 SPE Annual Technical Conference and Exhibition, Houston, 3–6 October.
29. Schilthuis, R.J.: "Active Oil and Reservoir Energy," *Trans.*, AIME (1936) **118,** 33.
30. Dake, L.P.: *Fundamentals of Reservoir Engineering,* Elsevier Scientific Publishing Co., Amsterdam (1978).
31. Hall, H.N.: "Analysis of Gravity Drainage," *JPT* (September 1961) 927; *Trans.,* AIME, **222.**
32. Pirson, S.J.: *Oil Reservoir Engineering,* McGraw-Hill Book Co. Inc., New York City (1958).
33. Katz, D.L.: "Possibilities of Secondary Recovery for the Oklahoma City Wilcox Sand," *Trans.,* AIME (1942) **146,** 28.
34. Hill, H.B. and Guthrie, R.K.: "Engineering Study of the Rodessa Oil Field in Louisiana, Texas, and Arkansas," Report Investigation 3715, U.S. Bureau of Mines, Washington, DC (August 1943) 87.
35. Garthwaite, D.L. and Krebill, F.K.: "Supplement 1962: Pressure Maintenance by Inert Gas Injection in the High Relief Elk Basin Field," *Field Case Histories, Oil Reservoirs,* Reprint Series, SPE, Richardson, Texas (1962) **4.**
36. Garthwaite, D.L.: "Supplement, 1975: Pressure Maintenance by Inert Gas Injection in the High Relief Elk Basin Field," *Field Case Histories, Oil and Gas Reservoirs,* Reprint Series, SPE, Richardson, Texas (1975) **4a.**
37. Stewart, F.M., Garthwaite, D.L., and Krebill, F.K.: "Pressure Maintenance by Inert Gas Injection in the High Relief Elk Basin Field," *Trans.,* AIME (1955) **204,** 49.
38. Havlena, D. and Odeh, A.S.: "The Material Balance as an Equation of a Straight Line," *JPT* (August 1963) 896; *Trans.,* AIME, **228.**
39. Wang, B. and Hwan, R.R.: "Influence of Reservoir Drive Mechanism on Uncertainties of Material Balance Calculations," paper SPE 38918 presented at the 1997 SPE Annual Technical Conference and Exhibition, San Antonio, Texas, 5–8 October.
40. Tehrani, D.H.: "An Analysis of a Volumetric Balance Equation for Calculation of Oil In Place and Water Influx," *JPT* (September 1985) 1664.
41. Roadifer, R.E.: "Size Distributions of World's Largest Known Oil, Tar Accumulations," *Oil & Gas J.* (24 February 1986) 93.

42. McEwen, C.R.: "Material Balance Calculations with Water Influx in the Presence of Uncertainty in Pressures," *SPEJ* (June 1962) 120.
43. van Everdingen, A.F. and Hurst, W.: "The Application of the Laplace Transformation to Flow Problems in Reservoirs," *Trans.,* AIME (1949) **186,** 305.
44. Carter, R.D. and Tracy, G.W.: "An Improved Method for Calculating Water Influx," *Trans.,* AIME (1960) **219,** 415.
45. Fetkovich, M.J.: "A Simplified Approach to Water Influx Calculations—Finite Aquifer Systems," *JPT* (July 1971) 814.
46. Coats, K.H.: "Mathematical Methods for Reservoir Simulation," presented by the College of Engineering, U. of Texas at Austin, 8–12 June 1970.
47. Klins, M.A., Bouchard, A.J., and Cable, C.L.: "A Polynomial Approach to the van Everdingen-Hurst Dimensionless Variables for Water Encroachment," *SPERE* (February 1988) 320.
48. Marsal, D.: "Topics of Reservoir Engineering," Course Notes, Delft U. of Technology (1982).
49. Walsh, M.P.: *A Generalized Approach to Petroleum Reservoir Engineering,* Petroleum Recovery Research Inst., Austin, Texas (1996).
50. Allard, D.R. and Chen, S.M.: "Calculation of Water Influx for Bottomwater Drive Reservoirs," *SPERE* (May 1988) 369.
51. Field, M.B., Givens, J.W., and Paxman, D.S.: "Kaybob South—Reservoir Simulation of a Gas Cycling Project with Bottom Water Drive," *JPT* (April 1970) 481.
52. Woods, R.W. and Muskat, M.M.: "An Analysis of Material-Balance Calculations," *Trans.,* AIME (1945) **160,** 124.
53. Chierici, G.L., Pizzi, G, and Ciucci, G.M.: "Water Drive Gas Reservoirs: Uncertainty in Reserves Evaluation From Past History," *JPT* (February 1967) 237.
54. Sills, S.R.: "Improved Material-Balance Regression Analysis for Waterdrive Oil and Gas Reservoirs," *SPERE* (May 1996) 127.
55. van Everdingen, A.F., Timmerman, E.H., and McMahon, J.J.: "Application of the Material Balance Equation to a Partial Water-Drive Reservoir," *Trans.,* AIME (1953) **198,** 51.
56. Campbell, R.A. and Campbell, J.M.: *Mineral Property Economics, Vol. 3: Petroleum Property Evaluation,* Campbell Petroleum Series, Norman, Oklahoma (1978).
57. Cole, F.W.: *Reservoir Engineering Manual,* Gulf Publishing Co., Houston, (1969).
58. Cook, C.C. and Jewell, S.: "Simulation of a North Sea Field Experiencing Significant Compaction Drive," *SPERE* (February 1996) 48.
59. Sulak, R.M.: "Ekofisk Field: The First 20 Years," *JPT* (October 1991) 1265.
60. Sulak, R.M., Thomas, L.K., and Boade, R.R.: "3D Reservoir Simulation of Ekofisk Compaction Drive," *JPT* (October 1991) 1272.
61. Fetkovich, M.J., Reese, D.E., and Whitson, C.H.: "Application of a General Material Balance for High-Pressure Gas Reservoirs," *SPEJ* (March 1998) 3.
62. Letkeman, J.P. and Ridings, R.L.: "A Numerical Coning Model," *SPEJ* (December 1970) 418.
63. MacDonald, R.C. and Coats, K.H.: "Methods for Numerical Simulation of Water and Gas Coning," *SPEJ* (December 1970) 425.
64. Muskat, M. and Wyckoff, R.D.: "An Approximate Theory of Water-Coning in Oil Production," *Trans.,* AIME (1935) **114,** 144.
65. Chaney, P.E. *et al.*: "How to Perforate Your Well to Prevent Oil and Gas Coning," *Oil & Gas J.* (7 May 1956) 108.
66. Chierici, G.L., Ciucci, G.M., and Pizzi, G.: "A Systematic Study of Gas and Water Coning By Potentiometric Models," *JPT* (August 1964) 923.
67. Bournazel, C. and Jeanson, B.: "Fast Water-Coning Evaluation Method," paper SPE 3628 presented at the 1971 SPE Annual Meeting, New Orleans, 3–6 October.
68. Hoyland, L.A., Papatzacos, P., and Skjaeveland, S.M.: "Critical Rate for Water Coning: Correlation and Analytical Solution," *SPERE* (November 1989) 495.
69. Kuo, M.C.T. and DesBrisay, C.L.: "A Simplified Method for Water Coning Predictions," paper SPE 12067 presented at the 1983 SPE Annual Technical Conference and Exhibition, San Francisco, 5–8 October.
70. Meyer, H.I. and Garder, A.O.: "Mechanics of Two Immiscible Fluids in Porous Media," *J. of Applied Physics* (November 1954) 1400.

71. Schols, R.S.: "An Empirical Formula for the Critical Oil Rate," *Erdöl Erdgas,* Zeitschrift (January 1972) 6–11.
72. Sobocinski, D.P. and Cornelius, A.J.: "A Correlation for Predicting Water Coning Time," *JPT* (May 1965) 594.
73. Wheatley, M.J.: "An Approximate Theory of Oil/Water Coning," paper SPE 14210 presented at the 1985 SPE Annual Technical Conference and Exhibition, Las Vegas, Nevada, 22–25 September.
74. Saidikowski, R.M.: "Numerical Simulations of the Combined Effects of Wellbore Damage and Partial Penetration," paper SPE 8204 presented at the 1979 SPE Annual Technical Conference and Exhibition, Las Vegas, Nevada, 23–26 September.
75. Kozeny, J.: *Wasserkraft und Wasserwirtschaft* (1933) **28,** 101.
76. Howard, G.C. and Fast, C.R.: "Squeeze Cementing Operations," *Trans.,* AIME (1950) **189,** 53.
77. Richardson, J.G. and Blackwell, R.J.: "Use of Simple Mathematical Models for Predicting Reservoir Behavior," *JPT* (September 1971) 1145.

General References

Amyx, J.W., Bass, D.M., and Whiting, R.L.: *Petroleum Reservoir Engineering—Physical Properties,* McGraw-Hill Book Co. Inc., New York City (1960).

Archer, J.S. and Wall, P.G.: *Petroleum Engineering: Principles and Practice,* Graham and Trotman, London (1992).

Bradley, H.B.: *Petroleum Engineering Handbook,* SPE, Richardson, Texas (1987).

Calhoun, J.C. Jr.: *Fundamentals of Reservoir Engineering,* U. of Oklahoma Press, Norman, Oklahoma (1951).

Caudle, B.H.: *Fundamentals of Reservoir Engineering—Parts 1 and 2,* SPE, Richardson, Texas (1967).

Clark, N.J.: *Elements of Petroleum Reservoirs,* Society of Petroleum Engineers of AIME, Dallas (1960).

Craft, B.C. and Hawkins, M.F. Jr.: *Applied Petroleum Reservoir Engineering,* Prentice-Hall Inc., Upper Saddle River, New Jersey (1959).

Craft, B.C., Hawkins, M.F. Jr., and Terry, R.E.: *Applied Petroleum Reservoir Engineering,* second edition, Prentice-Hall Inc., Upper Saddle River, New Jersey (1991).

Dake, L.P.: *Fundamentals of Reservoir Engineering,* Elsevier Scientific Publishing Co., New York City (1978).

Guerrero, E.T.: *Practical Reservoir Engineering,* Petroleum Publishing Co., Tulsa (1951).

Hagoort, J.: *Fundamentals of Gas Reservoir Engineering,* Elsevier, New York City (1988).

Katz, D.L. *et al.: Handbook of Natural Gas Engineering,* McGraw-Hill Book Co. Inc., New York City (1959).

Koederitz, L.F., Harvey, A.H., and Honarpour, M.: *Introduction to Petroleum Reservoir Analysis,* Gulf Publishing Co., Houston (1989).

Muskat, M.: *Physical Principles of Oil Production,* McGraw-Hill Book Co. Inc., New York City (1949) National Human Resources Development Corp., reprinted (1981).

Pirson, S. J.: *Oil Reservoir Engineering,* McGraw-Hill Book Co. Inc., New York City (1958).

Smith, C.R., Tracy, G.W., and Farrar, R.L.: *Applied Reservoir Engineering—Vols. 1 and 2,* Oil and Gas Consultants Intl., Tulsa (1992).

Standing, M.B.: *Volumetric and Phase Behavior of Oil Field Hydrocarbon Systems,* SPE, Dallas (1951).

Timmerman, E.H.: *Practical Reservoir Engineering—Parts 1 and 2,* PennWell Books, Tulsa (1982).

Walsh, M.P. and Lake, L.W.: *A Generalized Approach to Primary Hydrocarbon Recovery,* Elsevier, Amsterdam (2003).

Appendix—Key Equations in SI Units

$$N_g = \frac{2.63 \times 10^{-3} k_v \Delta \rho r_e^2}{\mu q}, \quad (9.19)$$

$$t_D = \frac{0.03110 k_a t}{\phi_a \mu_w c_t r_o^2}, \quad (9.35)$$

$$U = 4.89 \times 10^{-3} h f r_o^2 \phi_a c_t, \quad (9.36)$$

$$t_D = \frac{0.03110 k_a t}{\phi_a \mu_w c_t L_a^2}, \quad (9.50)$$

$$L_{ac} = \sqrt{\frac{0.03110 k_a t_{max}}{2 \phi_a \mu_w c_t}}, \quad (9.53)$$

$$q_c = \frac{5.256 \times 10^{-3} k_o h^2 \Delta \rho}{\mu_o B_o} q_{Dc}, \quad (9.77)$$

and $$q = \frac{5.35 \times 10^{-4} k_H b (p_e - p_w)}{\mu_o B_o \left(\ln \frac{r_e}{r_w} + s \right)} \left[1 + 7 \sqrt{\frac{r_w}{2b}} \sqrt{\frac{k_v}{k_H}} \cos \left(\frac{\pi b}{2h} \right) \right], \quad (9.83)$$

where the following units apply: k is given md; $\Delta \rho$ is in g/cm^3; b, h, r, and L_a are in m; q is in std m^3/day; μ is in mPa-sec or cp; t is in years; c is in 1/kPa; U is in m^3/kPa; and k_t is in 1/years.

SI Metric Conversion Factors

acre ×	4.046 856	E + 03	= m^2
acre-ft ×	1.233 489	E + 03	= m^3
bbl ×	1.589 873	E − 01	= m^3
cp ×	1.0*	E − 03	= Pa·s
ft ×	3.048*	E − 01	= m
°F	(°F − 32)/1.8		= °C
gal ×	3.785 412	E − 03	= m^3
lbm ×	4.535 924	E − 01	= kg
psi ×	6.894 757	E + 00	= kPa
sq mile ×	2.589 988	E + 00	= km^2
std ft^3 ×	2.831 685	E − 02	= m^3

*Conversion factor is exact.

Chapter 10
Gas Reservoirs

Mark A. Miller, Consultant and **E.D. Holstein,** Consultant

10.1 Introduction

This chapter addresses the flow characteristics and depletion strategies for gas reservoirs. The focus will be primarily on nonassociated accumulations, but much of the fluid behavior, flow regimes, and recovery aspects are also applicable to gas caps associated with oil columns.

In this chapter, gas reservoirs have been divided into three groups; dry gas, wet gas, and retrograde-condensate gas. A dry-gas reservoir is defined as producing a single composition of gas that is constant in the reservoir, wellbore, and lease-separation equipment throughout the life of a field. Some liquids may be recovered by processing in a gas plant. A wet-gas reservoir is defined as producing a single gas composition to the producing well perforations throughout its life. Condensate will form either while flowing to the surface or in lease-separation equipment. A retrograde-condensate gas reservoir initially contains a single-phase fluid, which changes to two phases (condensate and gas) in the reservoir when the reservoir pressure decreases. Additional condensate forms with changes in pressure and temperature in the tubing and during lease separation. From a reservoir standpoint, dry and wet gas can be treated similarly in terms of producing characteristics, pressure behavior, and recovery potential. Wellbore hydraulics may be different. Studies of retrograde-condensate gas reservoirs must consider changes in condensate yield as reservoir pressure declines, the potential for decreased well deliverability as liquid saturations increase near the wellbore, and the effects of two-phase flow on wellbore hydraulics.

A comprehensive discussion of gas and condensate properties and phase behavior can be found in several chapters of the General Engineering section of this *Handbook.* Aspects of predicting wellbore hydraulics are covered in the Production Operations Engineering section of this *Handbook.* Lease equipment for processing gas and pipelining considerations are covered in several chapters of the Facilities Engineering section of this *Handbook.* The reader may want to refer to these chapters to understand some of the nomenclature and concepts referred to in the present chapter.

10.2 Properties of Natural Gases

Natural petroleum gases contain varying amounts of different (primarily alkane) hydrocarbon compounds and one or more inorganic compounds, such as hydrogen sulfide, carbon dioxide,

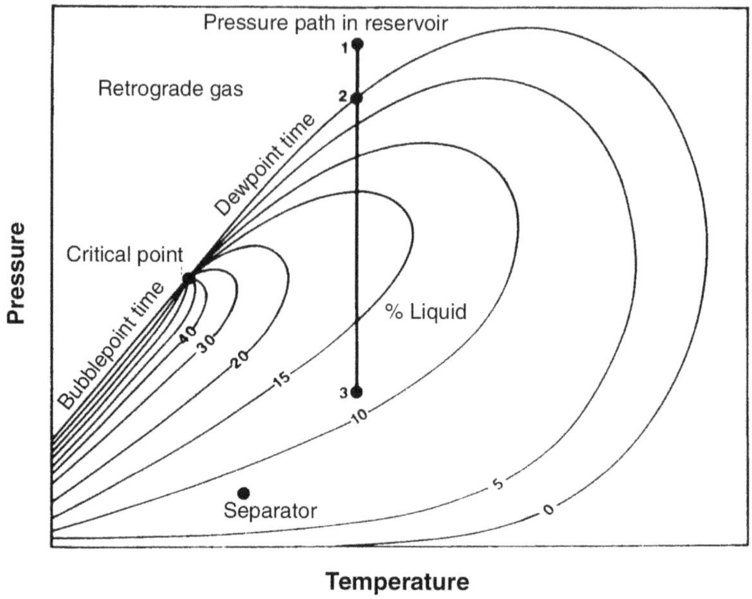

Fig. 10.1—Phase diagram of a retrograde-condensate gas.[1] (Source: *The Properties of Petroleum Fluids,* second edition, by William D. McCain Jr. Copyright PennWell Books, 1990.)

nitrogen (N_2), and water. Characterizing, measuring, and correlating the physical properties of natural gases must take into account this variety of constituents.

10.2.1 Phase Behavior of Natural-Gas Reservoirs. A widely accepted system for categorizing petroleum reservoir fluids is based on five classes: low-shrinkage (crude) oils, high-shrinkage (volatile) oils, retrograde-condensate gases, wet gases, and dry gases. Typical phase diagrams for the gas categories are shown in **Figs. 10.1 through 10.3**.

A retrograde-condensate fluid has a phase envelope such that reservoir temperature lies between the critical temperature and the cricondentherm (Fig. 10.1). As a result, a liquid phase will form in the reservoir as pressure declines, and the amount and gravity of produced liquids will change with time. Condensate liquids are generally "water white" or light in color (brown, orange, or greenish), with gravities typically between 40 and 60°API. Producing-liquid yields can be as high as 300 STB/MMscf. McCain[1] suggests that when yields are below approximately 20 STB/MMscf, even though phase-behavior considerations may show retrograde behavior, the amount of liquid dropout in the reservoir is insignificant. The primary difficulties in producing condensate reservoirs are as follows: (a) liquid deposition near the wellbore causes a decrease in gas deliverability that can approach 100% in a reservoir with less than 50-md permeability, and (b) a large amount of the most valuable hydrocarbon components is left in the reservoir rather than produced.

In a wet-gas reservoir, temperature is higher than the cricondentherm (Fig. 10.2). Therefore, a liquid phase never forms in the reservoir. Considerable liquid can still form (condense) at surface conditions or even in the wellbore. The term "condensate" is often applied to any light hydrocarbon liquid produced from a gas well. However, the term "condensate reservoir" should be applied only to situations in which condensate is actually formed in the reservoir because of retrograde behavior. Wet-gas reservoirs can always be treated as containing single-phase gas in the reservoir, while retrograde-condensate reservoirs may not. Wet-gas reservoirs generally produce liquids with gravities similar to those for retrograde condensates, but with yields less than approximately 20 STB/MMscf.[1]

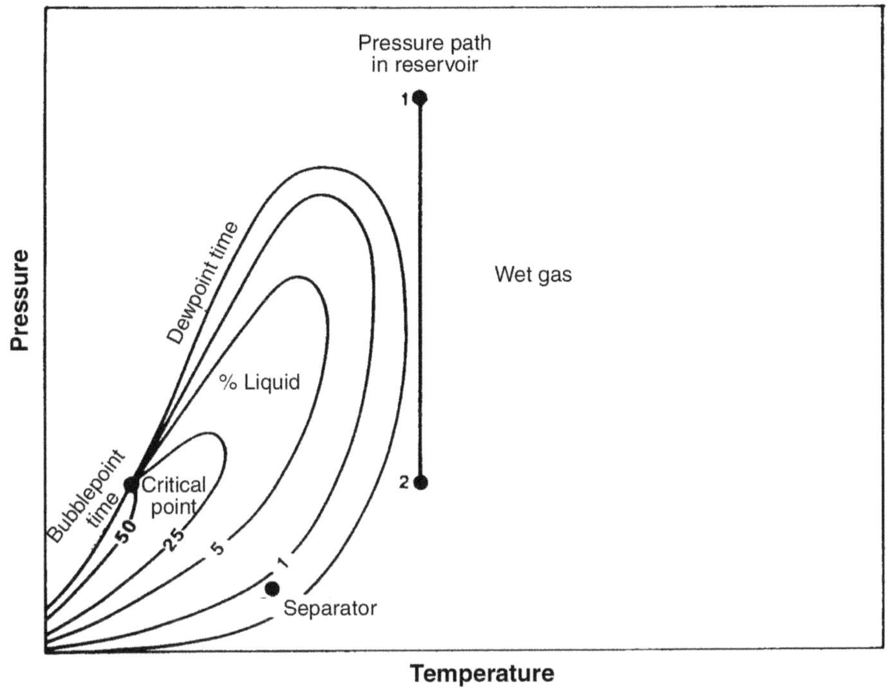

Fig. 10.2—Phase diagram of a wet gas.[1] (Source: *The Properties of Petroleum Fluids,* second edition, by William D. McCain Jr. Copyright PennWell Books, 1990.)

In a dry-gas reservoir, the cricondentherm is much lower than the reservoir temperature (Fig. 10.3), resulting in little or no liquid production at the surface. A somewhat arbitrary cut-off liquid yield of 10 STB/MMscf is sometimes used to distinguish dry-gas reservoirs.

While the difference between retrograde-condensate and wet gases is notable, there is much less distinction between wet and dry gases. For both wet and dry gases, reservoir engineering calculations are based on a single-phase reservoir gas. The only issue is whether there is a sufficient volume of produced liquid to be considered in such calculations as material balance or wellbore hydraulics. Retrograde systems require more-complex calculations using equations of state (EOSs) and other advanced engineering methods.

10.2.2 Pressure/Volume/Temperature (PVT) Behavior. The chapters on Fluid Sampling, Gas Properties, and Thermodynamics and Phase Behavior in the General Engineering section of this *Handbook* contain a detailed explanation of PVT and thermodynamic relationships for ideal and real gases. Some fundamental relations are repeated here as background for the remainder of this chapter.

The basis of gas PVT behavior is the ideal gas law, and by extension the real gas law:

$$pV_m = zRT \quad \text{(10.1)}$$

The universal gas constant in practical units is

$$R = 10.7315 \frac{\text{psia} \cdot \text{ft}^3}{\text{lb mol} \cdot {}^\circ\text{R}} = 1.98586 \frac{\text{btu}}{\text{lb mol} \cdot {}^\circ\text{R}}$$

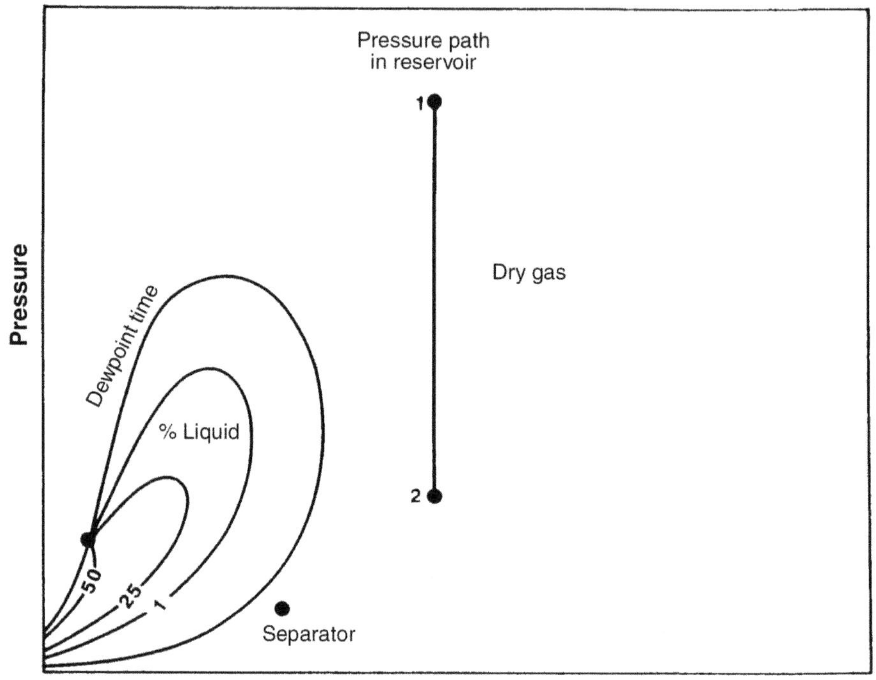

Fig. 10.3—Phase diagram of a dry gas.[1] (Source: *The Properties of Petroleum Fluids,* second edition, by William D. McCain Jr. Copyright PennWell Books, 1990.)

$$= 8.31441 \frac{\text{kPa} \times \text{m}^3}{\text{kmol} \times \text{K}} = 8.31441 \frac{\text{kJ}}{\text{kmol} \cdot {}^\circ\text{R}} \quad \text{(10.2)}$$

For economic reasons, most (but not all) pressure gauges read zero pressure when pressure is equal to the ambient atmospheric pressure. Therefore, atmospheric pressure must be added to gauge pressures to convert them to an absolute basis. For most engineering purposes, atmospheric pressure is usually taken to be 14.7 psia (101 kPa). For precise scientific and engineering applications, actual atmospheric pressure (i.e., barometric pressure, which varies with both location and time) should be used. Standard temperature and pressure are set by different governmental agencies and should be determined for a specific field or reservoir to be sure that reserves and produced quantities are reported with the correct bases. The SPE standard temperature is 59°F (15°C), and the standard pressure is 14.696 psia (101.325 kPa).

Application of Eq. 10.1, in a practical sense, must consider how to determine the various factors for petroleum gases that are mixtures of several components. Such determinations would include apparent molecular weight and z (using pseudoreduced pressure and temperature and corrections for nonhydrocarbon components).

10.2.3 Gas Density and Formation Volume Factor. The density of a gas can be calculated from the real-gas law once a z factor has been determined. For pressure in psia and temperature in °R, density in lbm/ft³ is given by

$$\rho_g = 2.6988 \frac{p\gamma_g}{zT} \quad \text{(10.3)}$$

For pressure in kPa, temperature in K, and density in kg/m^3,

$$\rho_g = 3.4834 \frac{p\gamma_g}{zT} \text{ . } \quad (10.4)$$

The gas formation volume factor is defined as the volume occupied by a gas at reservoir conditions divided by the volume at standard conditions:

$$B_g = \frac{p_{sc}}{T_{sc}} \frac{zT}{p}$$

$$= 28.279 \frac{zT}{p} \text{ft}^3 / \text{Mscf} = 5.0368 \frac{zT}{p} \text{bbl} / \text{Mscf} \text{ . } \quad (10.5)$$

The second and third lines of Eq. 10.5 give B_g using standard pressure of 14.696 psia and standard temperature of 60°F.

In SI units using SPE standard pressure and temperature,

$$B_g = 2.8438 \frac{zT}{p} \text{m}^3 / \text{std m}^3 \text{ . } \quad (10.6)$$

10.2.4 Viscosity. Most gas viscosities range from 0.01 to 0.03 cp, making them difficult to measure accurately. Accurate determination of gas viscosities has low economic value. Instead, values are normally determined from one of two correlations.

The first one in common use today, from Lee *et al.*,[2] is given in equation form as

$$\mu_g = K \times 10^{-4} \exp\left(X\rho_g^Y\right), \quad (10.7)$$

where $K = \frac{\left(9.4 + 0.02M_g\right)T^{1.5}}{209 + 19M_g + T}$, $X = 3.5 + \frac{986}{T} + 0.01M_g$, $Y = 2.4 - 0.2X$, and $\rho_g = \frac{1}{669.95} \frac{pM_g}{zT} = \frac{1}{23.132} \frac{p\gamma_g}{zT}$.

The gas density in Eq. 10.7 is in g/cm^3 when p and T are in oilfield units (psia, °R). The equivalent formula for SI units (pressure and temperature in kPa and K, respectively) is

$$\rho_g = \frac{1}{8314.4} \frac{pM_g}{zT} = \frac{1}{287.08} \frac{p\gamma_g}{zT} \text{ . } \quad (10.8)$$

Fig. 10.4 shows gas viscosities generated from this correlation for a 0.80-gravity natural gas.

Another common correlation[3] entails a two-step graphical process and is cumbersome for computer applications. Because gas viscosities are seldom needed with great accuracy, the Lee *et al.*[2] correlation is most applicable for modern reservoir-engineering practice.

10.2.5 Determining Reservoir-Fluid Properties. Condensation of liquids from wet-gas and retrograde-condensate fluids in the production system means that gas produced from separation equipment may be significantly different from the gas that flows into the wellbore from the reservoir. In general, separator gas will be lower in gravity and will have fewer high-molecular-weight hydrocarbons present in the mixture.

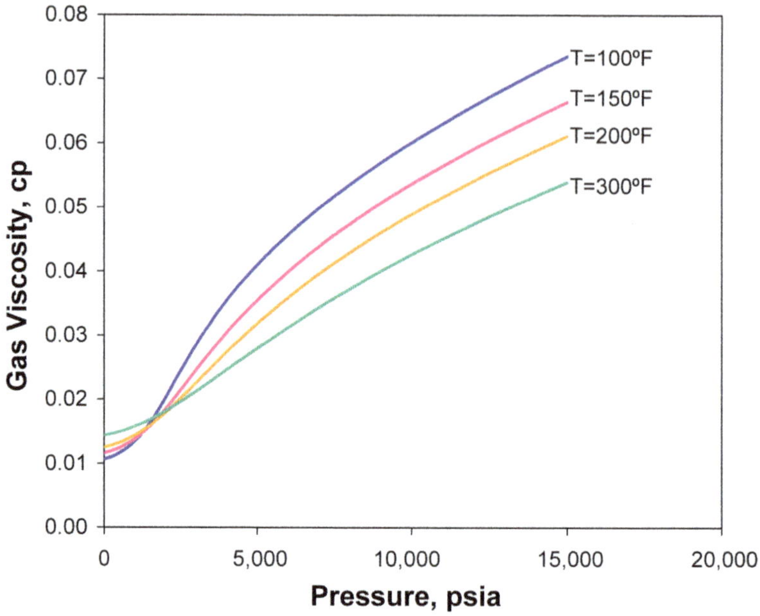

Fig. 10.4—Viscosity of 0.80-gravity natural gas vs. pressure and temperature according to the Lee *et al.*[2] correlation.

For proper laboratory measurements, a bottomhole sample should be collected. An alternative is a reconstituted sample that is created by mixing the separator-produced gas and liquid in proportion to their relative production rates. When the compositions of liquid and gaseous streams have both been measured, the composition of the mixture can be determined from

$$z_j = x_j \tilde{n}_o + y_j \tilde{n}_g \, . \quad (10.9)$$

Note that

$$\sum_{j=1}^{n_c} x_j = \sum_{j=1}^{n_c} y_j = \sum_{j=1}^{n_c} z_j = 1 . \quad (10.10)$$

$$\tilde{n}_o + \tilde{n}_g = 1 . \quad (10.11)$$

Relative molar amounts can be determined by converting measured produced volumes either to moles or to equivalent standard cubic feet. For a gas phase, conversion of a produced volume referenced to standard conditions to moles is

$$n_g = \frac{p_{sc} V_g^{sc}}{R T_{sc}} = \frac{V_g^{sc}}{V_{msc}} \, . \quad (10.12)$$

For a liquid, conversion of a volume to moles is

$$n_o = V_o \frac{\rho_o}{M_o} \quad \text{(10.13)}$$

If liquid volume is measured at standard conditions, density can be calculated from specific gravity or API gravity. If liquid molecular weight is not measured, it can be approximated with the Gold *et al.*[4] correlation:

$$M_o = \frac{5{,}954}{°\text{API} - 8.8} = \frac{42.43\gamma_o}{1.008 - \gamma_o} \quad \text{(10.14)}$$

An alternative to converting measured volumes to moles is to convert all measured volumes to equivalent standard volumes because standard volume is directly proportional to moles (through the value of the standard molar volume). This procedure has the advantage that measured gas volumes need not be converted (this is necessary only for liquid volumes).

Liquid volumes are converted to equivalent gas standard volumes using a parameter called the gas equivalent of oil. This parameter represents the effective standard volume occupied by hydrocarbons that are liquid at surface conditions but are in the gas phase at downhole conditions. This parameter is calculated by

$$GE_o = R\rho_{\text{H}_2\text{O}}^{sc} \frac{T_{sc}}{p_{sc}} \frac{\gamma_o}{M_o} \quad \text{(10.15)}$$

GE_o is calculated with oilfield units in Mscf/STB (the second term below uses standard conditions of 14.696 psia and 60°F) by

$$GE_o = 3.7615 \frac{T_{sc}}{p_{sc}} \frac{\gamma_o}{M_o} = 133.01 \frac{\gamma_o}{M_o} \quad \text{(10.16)}$$

In SI units, GE_o in std m^3/std m^3 is (using standard conditions of 101.325 kPa and 15°C)

$$GE_o = 8{,}314.4 \frac{T_{sc}}{p_{sc}} \frac{\gamma_o}{M_o} = 23{,}645 \frac{\gamma_o}{M_o} \quad \text{(10.17)}$$

Liquid-production volumes are multiplied by GE_o to determine the equivalent standard volume as gas in the reservoir; that is,

$$V_{o,\text{ eq}}^{sc} = GE_o V_o^{sc}, \quad \text{(10.18)}$$

where V_o^{sc} is the actual oil volume measured at stock-tank conditions, and $V_{o,\text{ eq}}^{sc}$ is the gas-equivalent volume of the oil.

These equations also can be used to determine the equivalent gas production of pure hydrocarbons separated from gas in a processing plant. Values of specific gravity and molecular weight for pure components can be found from standard sources such as the Gas Processors Suppliers Assn. (GPSA) Handbook.[5] In addition, if liquid production is measured at separator, rather than stock-tank, conditions, Eq. 10.16 or Eq. 10.17 can be used with separator temperature and pressure rather than standard temperature and pressure.

Relative volumes of gas and liquid phases can then be calculated as

$$\tilde{n}_g = \frac{n_g}{n_g + n_o} = \frac{V_g^{sc}}{V_g^{sc} + V_{o,\,eq}^{sc}}$$

$$\tilde{n}_o = 1 - \tilde{n}_o = \frac{n_o}{n_g + n_o} = \frac{V_{o,\,eq}^{sc}}{V_g^{sc} + V_{o,\,eq}^{sc}} \quad \text{.......... (10.19)}$$

When wellstream composition is unavailable, correlations must be used to determine gas properties, requiring the calculation of the wellstream (mixture) gravity on the basis of the gravity of the separator gas (often called dry gas) and the specific gravity of the produced liquid (condensate or oil):

$$\gamma_{wg} = \frac{\gamma_g + \frac{V_{msc}\,\rho_{H_2O}^{sc}}{M_{air}}\gamma_o Y}{1 + V_{msc}\,\rho_{H_2O}^{sc}\frac{\gamma_o}{M_o}Y}, \quad \text{.......... (10.20)}$$

where the subscript *g* refers to the separator-gas gravity, *wg* refers to the wellstream gas, and *o* refers to the produced condensate (oil). *Y* is the produced condensate yield. Gold's correlation can be used to estimate condensate molecular weight.

In oilfield units with yield in STB/MMscf,

$$\gamma_{wg} = \frac{\gamma_g + \frac{1}{7707.2}\frac{T_{sc}}{p_{sc}}\gamma_o Y}{1 + \frac{1}{266.11}\frac{T_{sc}}{p_{sc}}\frac{\gamma_o Y}{M_o}} = \frac{\gamma_g + \frac{\gamma_o Y}{217.95}}{1 + \frac{\gamma_o Y}{7.5254 M_o}} \quad \text{.......... (10.21)}$$

In SI units with yield in std m³/std m³,

$$\gamma_{wg} = \frac{\gamma_g + 286.80\frac{T_{sc}}{p_{sc}}\gamma_o Y}{1 + 8{,}306.4\frac{T_{sc}}{p_{sc}}\frac{\gamma_o Y}{M_o}} = \frac{\gamma_g + 815.61\gamma_o Y}{1 + 23{,}622\frac{\gamma_o Y}{M_o}} \quad \text{.......... (10.22)}$$

10.2.6 Measuring Retrograde Behavior. When a liquid phase begins to form in the reservoir, the produced stream is no longer representative of the reservoir-fluid composition, but rather only the composition of the fluids entering the wellbore. Situations in which the liquid content of gases is high require the use of advanced laboratory tests and/or equation-of-state modeling to measure and predict these multiphase effects.

Laboratory measurements of the PVT behavior of condensate systems are similar to tests used for black oils; however, the primary interest becomes the measurement of relatively small amounts of condensed liquid. In general, systems with producing gas/oil ratios of 15,000 scf/STB (67 STB/scf) have a liquid dropout of approximately 4 to 6% by volume, while reservoirs with ratios around 50,000 scf/STB ordinarily have liquid dropouts of less than 1% by volume.[1]

Two types of tests are generally run on retrograde fluids: constant-composition expansion (CCE) and constant-volume depletion. For examples refer to **Tables 10.1 through 10.3**.

Table 10.1 gives the compositions of separator gas and liquid streams and other data used in making a recombined sample for analysis of reservoir-fluid composition.

TABLE 10.1—EXAMPLE FIELD AND ANALYSIS DATA FOR RETROGRADE-FLUID STUDY[6]

Field data and analyses for condensate well No. 7 (courtesy of Core Lab Inc.)
Original reservoir pressure at 11,300 ft, 7,000 psig
Reservoir temperature, 256°F (398 K)
At sampling bottomhole pressure 1,505 psig (10.4 MPa), tubing rd. 780 psig (5.5 MPa)
Separator liquid, 185 B/D (46.5°API)
Primary separator gas, 869 Mcf/day (gravity 0.699)
Gas/liquid ratio, 4,697 scf/bbl, 212.9 bbl/MMcf

Component	Separator liquid, mol%	Separator Gas Mol%	Separator Gas Gallons per thousand ft^3, GPM	Wellstream Mol%	Wellstream GPM
Hydrogen sulfide	0.00	0.00		0.00	
Carbon dioxide	0.00	0.01		0.01	
Nitrogen	0.01	0.13		0.11	
Methane	10.76	83.01		68.93	
Ethane	6.17	9.23	2.454	8.63	2.295
Propane	–8.81	4.50	1.231	5.34	1.461
iso-butane	2.85	0.74	0.241	1.15	0.374
n-Butane	7.02	1.20	0.376	2.33	0.730
iso-pentane	3.47	0.31	0.113	0.93	0.338
n-Pentane	3.31	0.25	0.090	0.85	0.36
Hexanes	8.03	0.21	0.085	1.73	0.702
Heptanes plus	49.57	0.41	0.185	9.99	6.006
	100.00	100.00	4.775	100.00	12.212
Properties of Heptane Plus					
API gravity at 60°F	39.0				
Density, g/cm^3 at 60°F	0.8293				
Molecular weight	160	103		158	

Calculated separator-gas gravity = 0.699
Calculated gross heating value for separator gas = 1,230 Btu/ft^3 of dry gas at 14.65 psia and 60°F
Primary separator gas collected at 440 psig and 87°F
Primary separator liquid collected at 440 psig and 87°F
Primary-separator-gas/separator-liquid ratio = 3,944 scf/bbl at 60°F
Primary-separator-liquid/stock-tank-liquid ratio = 1.191 bbl/bbl at 60°F
Primary-separator-gas/wellstream ratio = 805.19 Mcf/MMcf
Stock-tank-liquid/wellstream ratio = 171.4 bbl/MMcf

From Katz, D.L.: *Handbook of Natural Gas Engineering;* © The McGraw-Hill Companies Inc. (1990)

A CCE using a visual cell furnishes the dewpoint of the reservoir fluid at reservoir temperature and the total volume of the reservoir fluid as a function of pressure. The volume of liquid formed at pressures below the dewpoint can also be measured. Table 10.2 shows the results of such a test. The term "relative volume" refers to the volume of gas plus liquid compared to the dewpoint volume. Retrograde-liquid volume is given as a percent of pore space,

TABLE 10.2—EXAMPLE VOLUMETRIC BEHAVIOR OF CONDENSATE WELL FLUID[6]

Volumetric and retrograde-liquid measurements on condensate well fluid, well No. 7 (courtesy of Core Lab Inc.)

Pressure, psig	Relative Volume	Retrograde-Liquid Volume, percent of pore space
7,500	0.9341	
7,000 (reservoir pressure)	0.9523*	
6,500	0.9727	
6,300	0.9834	
6,200	0.9891	
6,100	0.9942	
6,010 (dewpoint pressure)	1.0000⁺	0.0
5,950	1.0034	Trace
5,900	1.0076	0.1
5,800	1.0138	0.2
5,600	1.0267	0.5
5,300	1.0481	2.0
5,000 (first depletion level)	1.0749	7.8
4,500	1.1268	—
4,000	1.2024	21.3
3,500	1.3096	—
3,000	1.4689	25.0
2,500	1.7169	—
2,100	2.0191	24.4
1,860	2.2747	—
1,683	2.5150	—
1,460	2.9087	—
1,290	3.3173	—
1,200	—	22.5
1,160	3.7153	—
1,050	4.1342	—
700	—	21.0
0	—	17.6

*Gas expansion factor = 1.545 Mcf/bbl

⁺Gas expansion factor = 1.471 Mcf/bbl

From Katz, D.L.: *Handbook of Natural Gas Engineering;* © 1990 The McGraw-Hill Companies Inc. (1990)

which essentially shows how the average condensate saturation changes with average reservoir pressure. **Fig. 10.5** is a graphical representation of the relative condensate volume.

Visual cells also can be used to simulate pressure depletion. The validity of these tests is based on the assumption that the retrograde liquid that condenses in the reservoir will not be mobile. This assumption is valid except for very rich gas/condensate reservoirs. If significant retrograde liquid becomes mobile and migrates to producing wells, gas/liquid relative permeability data should be measured and used to adjust the predicted recovery.

Table 10.3 is an example of a visual-cell depletion study on the same retrograde gas for which properties are shown in Tables 10.1 and 10.2.

TABLE 10.3—DEPLETION STUDY AT 256°F

	Reservoir Pressure (psig)							
	6,010	5,000	4,000	3,000	2,100	1,200	700	700*
Component	Composition of Produced Well Stream (mol%)							
Inerts	0.12	0.13	0.13	0.14	0.14	0.13	0.12	0.01
C_1	68.93	70.69	73.60	76.60	77.77	77.04	75.13	11.95
C_2	8.63	8.67	8.72	8.82	8.96	9.37	9.82	4.10
C_3	5.34	5.26	5.20	5.16	5.16	5.44	5.90	4.80
i-C_4	1.15	1.10	1.05	1.01	1.01	1.10	1.26	1.57
n-C_4	2.33	2.21	2.09	1.99	1.98	2.15	2.45	3.75
i-C_5	0.93	0.86	0.78	0.73	0.72	0.77	0.87	2.15
n-C_5	0.85	0.76	0.70	0.65	0.63	0.68	0.78	2.15
C_6	1.73	1.48	1.25	10.80	1.01	1.07	1.25	6.50
C_7+	9.99	8.84	6.48	3.82	2.62	2.25	2.42	63.02
MW C_7+	158	146	134	123	115	110	109	174
Density C_7+	0.827	0.817	0.805	0.794	0.784	0.779	0.778	0.837
Z								
Equil Gas	1.140	1.015	0.897	0.853	0.865	0.902	0.938	
Two-Phase	1.140	1.016	0.921	0.851	0.799	0.722	0.612	
G_p % of G	0	6.6	17.5	32.9	49.9	68.1	77.9	

* Equilibrium liquid phase

The depletion study begins by expanding the reservoir fluid in the cell until the first depletion pressure is reached (5,000 psig in this example). The fluid in the cell is brought to equilibrium, and the volume of retrograde liquid is observed. Gas is removed from the top of the cell while a constant pressure is maintained until the hydrocarbon volume of the cell is the same as when the test began. The gas volume removed is measured at the depletion pressure and reservoir temperature, analyzed for composition, and measured at atmospheric pressure and temperature.

The ideal-gas law can be used to calculate the "ideal volume" at the depletion pressure and reservoir temperature of the gas withdrawn from the cell. Dividing the ideal volume by the actual volume yields the deviation factor, z, for the produced gas. This is listed in **Table 10.4** under z for the equilibrium gas. The actual volume of gas remaining in the cell at this point is the gas originally in the cell at the dewpoint pressure minus the gas produced at the first depletion level. Dividing the actual volume remaining in the cell into the calculated ideal volume remaining in the cell at this first depletion pressure yields the two-phase deviation factor shown. The two-phase z factor is an equivalent z factor that includes the total volume of gas plus liquid:

$$z_{2\phi} = \frac{p\left(V_g + V_l\right)}{\left(n_g + n_l\right)RT} \quad \text{.......... (10.23)}$$

The two-phase z factor is the correct value to apply to such things as p/z analysis of retrograde-condensate reservoirs.

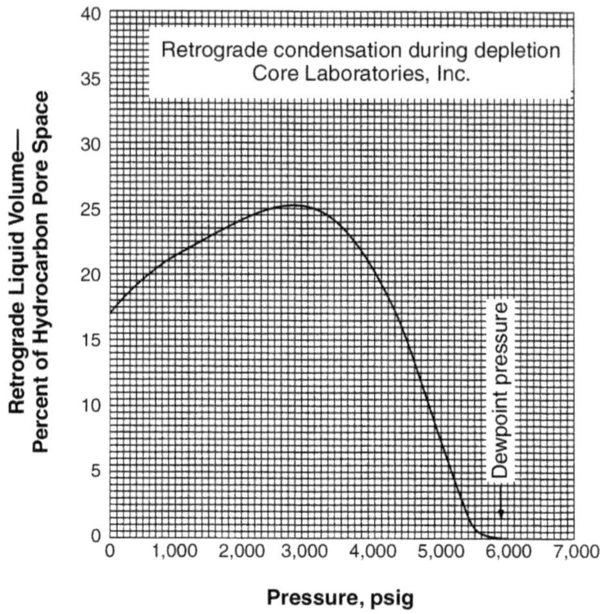

Fig. 10.5—Example retrograde-liquid volume vs. pressure.[6]

A series of expansions and constant-pressure displacements is repeated at each depletion pressure until an arbitrary abandonment pressure is reached. The abandonment pressure is considered arbitrary because no engineering or economic calculations have been made to determine this pressure for the purpose of the reservoir-fluid study.

At the final depletion pressure, the compositions of both the produced well stream and the retrograde liquid are measured. These data are included as a control composition in the event that the study is used for compositional material-balance purposes.

The composition data can be used with equilibrium constants (determined by either laboratory measurements or general correlations) to determine recovery at the various stages of pressure depletion represented by the laboratory measurements. In this case, initial condensate content was 181.74 STB/Mrcf (213 STB/Mscf), and the amount recovered from the dewpoint to 700 psig was 51.91 STB/Mrcf. The gas formation volume factor was determined to be 0.6472 RB/Mscf at initial conditions and 0.6798 RB/Mscf at the dewpoint. If a hydrocarbon pore space of 500 × 10^6 ft^3 were determined from volumetric calculations, then from these data and those presented in Table 10.3, recoveries by pressure depletion would be

$$G = (500 \text{ MMcf})(178.1)/(0.6472) = 137 \text{ Bscf}$$

$$G_p = G\left(\frac{B_{gdp} - B_{gi}}{B_{gdp}}\right) + G\left(\frac{B_{gi}}{B_{gdp}}\right)(0.779) = 108 \text{ Bscf} (79\ \%\ \text{OGIP})$$

$$N_p = G\left(\frac{B_{gdp} - B_{gi}}{B_{gdp}}\right)(181.74)(1{,}000) + G\left(\frac{B_{gi}}{B_{gdp}}\right)(51.91)(1{,}000)$$
$$= 8 \text{ MMSTB} (33\ \%\ \text{OOIP}) \text{ . } \qquad (10.24)$$

These calculations indicate the large amount of liquid remaining in the reservoir at depletion even with excellent drainage to the wells. Further reductions in recovery would be expected because of areas of the reservoir inadequately drained with existing wells.

TABLE 10.4—DEFINITIONS OF A "GENERIC" POTENTIAL DIFFERENCE $\Delta\psi$ FOR DIFFERENT FLUIDS

	$\Delta\psi$	Oilfield Units	SI Units
Liquids	Δp	psia	kPa
Real gases	$\Delta m(p)$	$psia^2/cp$	GPa/s
High-pressure gases (approximate)	Δp	psia	kPa
Low-pressure gases (approximate)	Δp^2	$psia^2$	kPa^2

To deal with such phase-behavior effects in more than an empirical manner requires the use of PVT simulators. These simulators are based on EOSs that describe the phase volumes and compositions of liquid and gaseous phases as functions of pressure and temperature. Because hydrocarbon molecules interact with each other in solution, the coefficients in an EOS are not always adequately known. PVT tests such as those described in this section, along with the known composition of the original fluid, can be used to "tune" an appropriate EOS to achieve results that nearly match the measurements. Once this tuning process is complete, those coefficients can then be used to make predictions under differing operating conditions with some degree of reliability.

10.2.7 EOSs. When the effects of complex phase behavior on phase compositions and physical properties cannot be calculated accurately with simple approaches, it is often desirable to use an EOS. An EOS approach is often necessary when dealing with volatile oils and retrograde-condensate gases.

EOSs provide a numerical method for calculating both composition and relative amount for each phase present in the system. In reservoir simulation, EOS calculations are typically restricted to two hydrocarbon phases: a liquid (oleic) phase and a gaseous phase. However, there are situations in which an aqueous phase is included in the EOS calculations, or even in which a third hydrocarbon-containing phase may be present (e.g., in CO_2 flooding). These are generally done in more advanced compositional simulators.

The two most common EOSs used in petroleum engineering applications are the Peng-Robinson and Soave-Redlich-Kwong equations, which historically were derived from van der Waals' equation. These three equations are called "cubic" because they result in a cubic representation for the molar volume. The basic equations are as follows:

Ideal Gas

$$pV_m = RT \quad (10.25)$$

van der Waals

$$\left(p + \frac{a_c}{V_m^2}\right)\left(V_m - b\right) = RT \quad (10.26)$$

Soave-Redlich-Kwong

$$\left[p + \frac{a_c\alpha(T)}{V_m\left(V_m + b\right)}\right]\left(V_m - b\right) = RT \quad (10.27)$$

Peng-Robinson

$$\left[p+\frac{a_c\alpha(T)}{V_m(V_m+b)+b(V_m-b)}\right](V_m-b)=RT\ . \qquad (10.28)$$

The parameters a_c, $\alpha(T)$, and b are determined empirically from experimental data (for pure components, the data are critical temperature and pressure and a specified point on the vapor-pressure curve), $\alpha(T)$ being a function of temperature and having a value of 1 at the critical temperature. Note that the parameters have different values depending on the equation.

The reader is referred to texts such as those by Ahmed,[7] Pedersen *et al.*,[8] McCain,[1] and Whitson and Brule.[9]

10.3 Petrophysical Properties

Petrophysical properties required for typical reservoir engineering purposes include porosity, pore-volume compressibility, permeability, relative-permeability-vs.-saturation curves, capillary-pressure-vs.-saturation curves, and liquid saturations. Additional data are sometimes required as well, but typically not for natural-gas reservoirs.

An extensive discussion of the methods for measuring, analyzing, and interpreting petrophysical data on these and other petrophysical properties can be found in several chapters of the General Engineering section of this *Handbook* and in subchapters of this section.

Two petrophysical properties of interest in gas engineering work are the Klinkenberg effect and non-Darcy flow.

10.3.1 Klinkenberg Effect. Low-pressure (i.e., laboratory) measurements give rise to what is termed the Klinkenberg[10] or "slippage" effect because the mean free path of gas molecules is approximately the same size as the pores in a reservoir rock, meaning that gas molecules are so far apart that the gas does not behave as a continuum fluid, resulting in erroneously high apparent permeability. At low pressures, measured gas permeabilities can be empirically related to effective liquid (or high-pressure gas) permeabilities by

$$k_g=k_l\left(1+\frac{b}{\bar{p}}\right)\ . \qquad (10.29)$$

The effective liquid permeability can be determined in the laboratory by measuring gas permeabilities at different average core pressures. A plot of k_g vs. $1/\bar{p}$ yields an intercept equal to k_l (**Fig. 10.6**) of 24 md compared with 48 md at low pressure. The Klinkenberg effect is unimportant at reservoir pressures.

10.3.2 Non-Darcy Flow. At high fluid velocities, Darcy's law may not always be accurate. An additional energy loss is often apparent above that predicted from the laminar-flow relationship suggested by Darcy's law. This effect has sometimes been called *turbulence* or *inertial turbulence* based on analogies with pipe flow. The effect, however, is probably caused by multiple factors, including pore-scale as well as reservoir-scale phenomena. Because of the lack of understanding of the fundamental nature of such phenomena, it is usually simply referred to as non-Darcy flow.

The most common expression of the non-Darcy effect is through the Forchheimer[11] equation:

$$-\nabla p=\frac{\mu}{k}\vec{u}+\beta\rho\,|\,u\,|\,\vec{u}\ , \qquad (10.30)$$

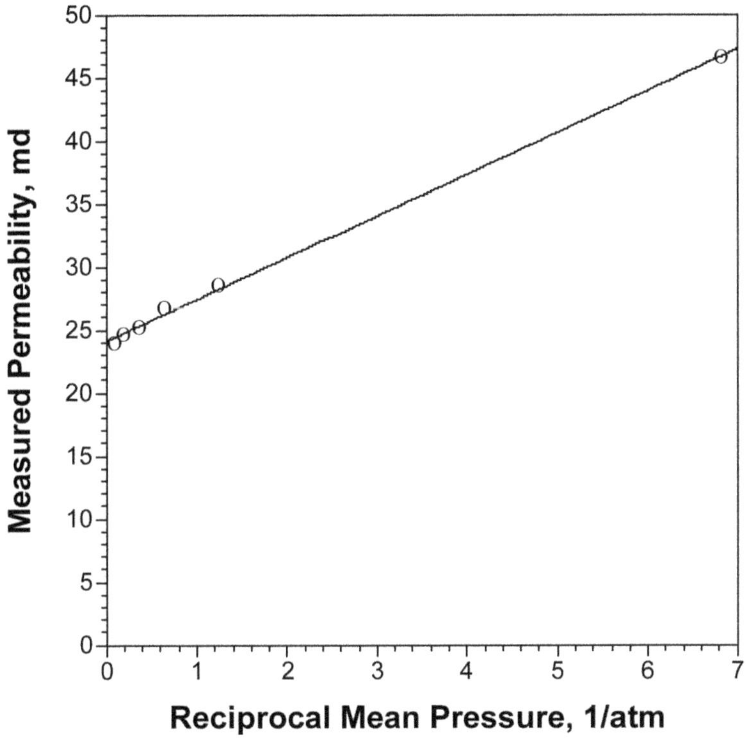

Fig. 10.6—Example of Klinkenberg[10] permeability correction.

where β is called the non-Darcy velocity coefficient, having units of L^{-1}, and u is the volumetric flux (q/A) through the rock.

Note that Eq. 10.30 introduces a velocity-squared term into Darcy's law. This effect shows up as a flow-rate-squared term in flow relationships (e.g., well-deliverability equations) that involve Darcy's law.

Although non-Darcy flow can occur at all points in a reservoir, in practice it is only significant in the near-well region, where gas velocities are highest owing to radial-flow effects and expansion of gas volume at low pressure. For this reason, non-Darcy flow is incorporated primarily as a flow-rate-dependent skin factor and is seldom, if ever, incorporated to calculate flow away from the wellbore. The magnitude of the non-Darcy effect must generally be measured empirically at reservoir conditions using well tests.

10.4 Well Performance

This section deals with methods for analyzing and predicting the performance of producing natural-gas wells. Steady-state-, pseudosteady-state-, and transient-flow concepts are developed, resulting in a variety of specific techniques and empirical relationships for both testing wells and predicting their future performance under different operating conditions. The information included here is a condensation of the concepts and equations developed in detail in the chapter on fluid flow and well analysis in this section as they apply to gas reservoirs.

10.4.1 Basic Equations. The basis for all well-performance relationships is Darcy's law, which in its fundamental differential form applies to any fluid—gas or liquid. However, different forms of Darcy's law arise for different fluids when flow rates are measured at standard conditions. The different forms of the equations are based on appropriate equations of state (i.e., density as a function of pressure) for a particular fluid. In the resulting equations, presented next, flow rate is taken as being positive in the direction opposite to the pressure gradient, thus dropping the minus sign from Darcy's law. When multiple-line equations are presented, the first will be in fundamental units, the second in oilfield units, and the third in SI units.

Four different fluid representations are considered: (1) liquid (small and constant compressibility), (2) real gas, (3) approximate high-pressure gas, and (4) approximate low-pressure gas. Because this chapter is devoted to well performance, radial forms of the flow equations are presented. Linear and spherical forms follow a similar development.

Steady-State Radial Horizontal Liquid Flow.

$$q_o = \frac{2\pi k h}{\ln\left(r_2/r_1\right)} \int_{p_1}^{p_2} \frac{d\tilde{p}}{B_o(\tilde{p})\mu_o(\tilde{p})} \quad \text{.......... (10.31)}$$

Steady-State Radial Horizontal Gas Flow.

$$q_g = \frac{2\pi k h T_{sc}}{p_{sc} T \ln\left(r_2/r_1\right)} \int_{p_1}^{p_2} \frac{\tilde{p}\, d\tilde{p}}{\mu_g(\tilde{p}) z(\tilde{p})} \quad \text{.......... (10.32)}$$

For liquids, the product of $B_o\mu_o$ is approximately constant over a fairly wide pressure range so that for practical purposes, Eq. 10.31 can be written as

$$q_o \approx \frac{2\pi k h \Delta p}{\bar{B}_o \bar{\mu}_o \ln\left(r_2/r_1\right)}, \quad \text{.......... (10.33)}$$

where $\Delta p = p_2 - p_1$ and $\bar{B}_o\bar{\mu}_o$ is evaluated at some average pressure between p_1 and p_2. The exact pressure at which the oil formation volume factor and viscosity are evaluated is not critical because the product of $\bar{B}_o\bar{\mu}_o$ is approximately constant.

Because this approximation is not generally valid for gases, the steady-state radial gas-flow equation is written as

$$q_g = \frac{\pi k h T_{sc} \Delta m}{p_{sc} T \ln\left(r_2/r_1\right)}, \quad \text{.......... (10.34)}$$

where the real-gas potential Δm[12] is defined by

$$\Delta m = m(p_2) - m(p_1) = 2\int_{p_1}^{p_2} \frac{\tilde{p}\, d\tilde{p}}{\mu_g(\tilde{p}) z(\tilde{p})} \quad \text{.......... (10.35)}$$

Fig. 10.7 shows a typical plot of $p/\mu z$ vs. p. Twice the area under the curve between any two pressures represents the real-gas potential difference. Note that at high pressures, $p/\mu z$ is approximately constant. Also, although it is not readily apparent from the plot, at low pressures the product μz is approximately constant.

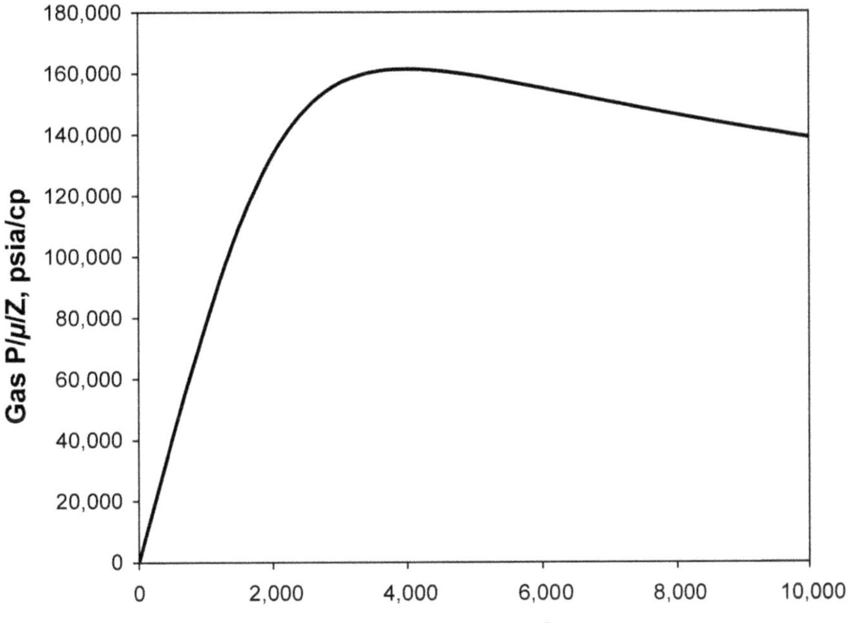

Fig. 10.7—Typical plot of *p/μz* vs. *p*.

Values for Δm are usually determined by numerically integrating Eq. 10.35 using gas viscosity and z-factor data from measurements or correlations. Typical results are shown in **Fig. 10.8**.

Note that for ideal gases, $z = 1$ and does not vary with pressure, resulting in the identity $\Delta m = \frac{p_1^2 - p_2^2}{\mu_g}$. The steady-state radial-flow equation for an ideal gas is thus

$$q_g = \frac{\pi k h T_{sc} \Delta p^2}{\mu_g p_{sc} T \ln\left(r_2 / r_1\right)}, \quad \text{(10.36)}$$

where $\Delta p^2 = p_2^2 - p_1^2$.

To avoid numerical evaluation of Δm, which can be time-consuming if done by hand, it is sometimes useful to have two other approximate real-gas forms of each of the flow equations. These approximate forms are derived by noting that for most natural gases at low pressures (i.e., less than approximately 2,000 psia, or 14 MPa), the product $\mu_g z$ is approximately constant. Under these conditions,

$$\Delta m = 2\int_{p_1}^{p_2} \frac{\tilde{p}\,\mathrm{d}\tilde{p}}{\mu_g\left(\tilde{p}\right) z\left(\tilde{p}\right)} \mathrm{d}p \approx \frac{\Delta p^2}{\bar{\mu}_g \bar{z}} \;. \quad \text{(10.37)}$$

Both μ_g and z should be evaluated at some average pressure between the two pressures. The specific value of average pressure used is not very significant because the product $\mu_g z$ is relatively constant, as demonstrated in **Fig. 10.9**.

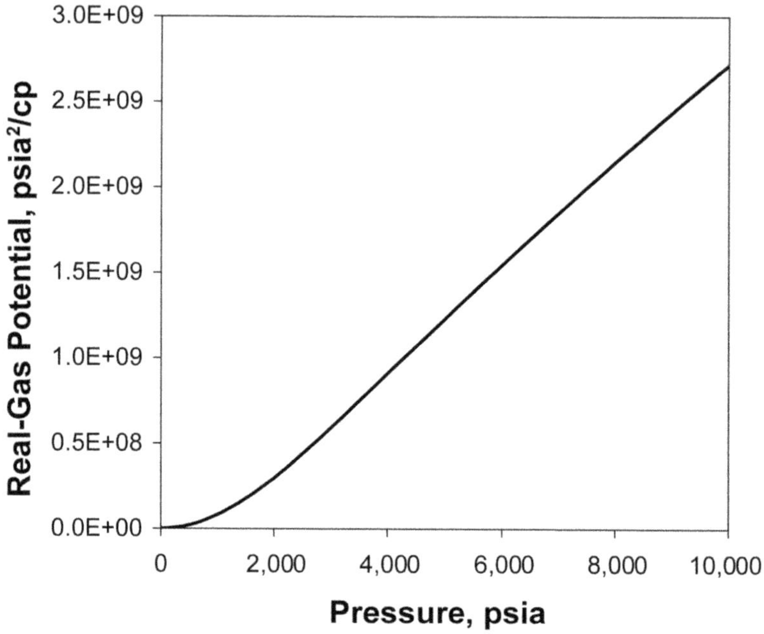

Fig. 10.8—Example graph of real-gas potential vs. pressure.

At high pressures (i.e., greater than approximately 2,000 psia, or 14 MPa), for most natural gases the product of $\mu_g\beta_g$ is relatively constant (see **Fig. 10.10** for an example).

This means that

$$\Delta m = 2\int_{p_1}^{p_2} \frac{\tilde{p}d\tilde{p}}{\mu_g(\tilde{p})z(\tilde{p})} = 2\frac{p_{sc}T}{T_{sc}}\int_{p_1}^{p_2} \frac{\tilde{p}d\tilde{p}}{\mu_g(\tilde{p})B_g(\tilde{p})} \approx 2\frac{p_{sc}T}{T_{sc}}\frac{\Delta p}{\bar{\mu}_g\bar{B}_g} \text{ . } \qquad (10.38)$$

Note that the conversion between the real-gas potential in oilfield vs. SI units is

$$\left[\Delta m(p) - \frac{\text{GPa}}{\text{s}}\right] = 47.538\left[\Delta m(p) - \frac{\text{psia}^2}{\text{cp}}\right] \text{ . } \qquad (10.39)$$

So that one simplified set of equations can be used throughout the remainder of the chapter, some additional parameters will be defined. First, a "generic" potential difference, $\Delta\psi$, can be expressed for each of the fluid cases according to Table 10.4.

A general radial-flow equation can then be expressed for all cases as

$$q_{sc} = \frac{kh}{\beta \ln\left(r_2/r_1\right)}\Delta\psi, \qquad (10.40)$$

where β is given by the following expressions, which include the unit conversions necessary to apply Eq. 10.40. The first line of each equation is in fundamental units, the second in oilfield units, and the third in SI units.

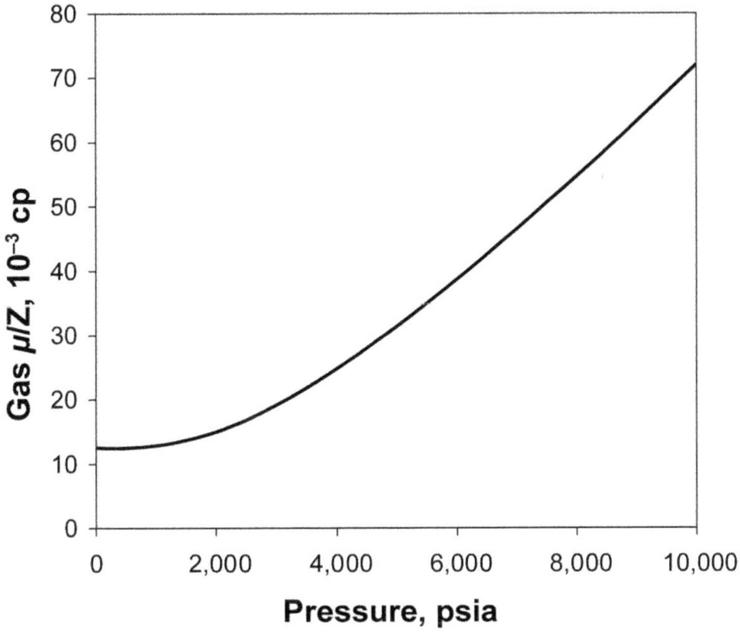

Fig. 10.9—Example showing that $\mu_g z$ is approximately constant at < 2,000 psia.

Liquids (β units are psi-md-ft-D/STB, kPa-m^2-m-d/std m^3).

$$\beta = \frac{\bar{B}\bar{\mu}}{2\pi}$$
$$= 141.21\bar{B}\bar{\mu} = 1.8421\bar{B}\bar{\mu} \,. \quad (10.41)$$

Real Gases (β units are psi^2-md-ft-D/Mscf/cp, kPa^2·m^2·m·d/std m^3/mPa·s).

$$\beta = \frac{p_{sc}T}{\pi T_{sc}}$$
$$= 50{,}300\frac{p_{sc}T}{T_{sc}} = 1422.4T$$
$$= 3.6841\frac{p_{sc}T}{T_{sc}} = 1.2955T \,. \quad (10.42)$$

High-Pressure Gases (approximate) (β units are psi-md-ft·D/Mscf, kPa·m^2·m·d/std m^3).

$$\beta = \frac{\bar{B}_g\bar{\mu}_g}{2\pi}$$
$$= 25.150\bar{B}_g\bar{\mu}_g = 1.8421\bar{B}_g\bar{\mu}_g \,. \quad (10.43)$$

Low-Pressure Gases (approximate) (β units are psia^2-md-ft-D/Mscf, kPa^2·m^2·m·d/std m^3).

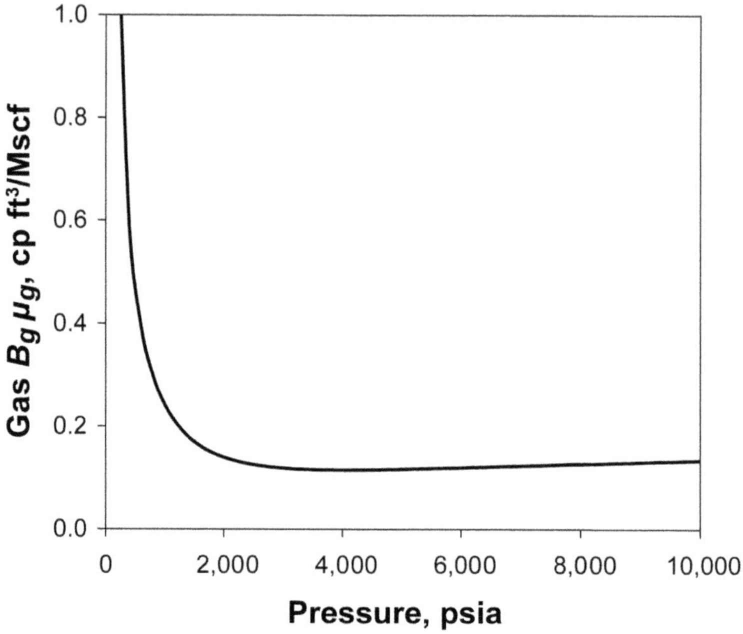

Fig. 10.10—Example graph showing that the product $\mu_g B_g$ is approximately constant for gases at pressures greater than approximately 2,000 psia.

$$\beta = \frac{p_{sc} T \bar{\mu}_g \bar{z}}{\pi T_{sc}}$$

$$= 50{,}300 \frac{p_{sc} T \bar{\mu}_g \bar{z}}{T_{sc}} = 1422.4 T \bar{\mu}_g \bar{z}$$

$$= 3.6841 \frac{p_{sc} T \bar{\mu}_g \bar{z}}{T_{sc}} = 1.2955 T \bar{\mu}_g \bar{z} \; . \qquad (10.44)$$

To concentrate on the specifics of well flow, in the remainder of the chapter the subscript *e* will refer to an external drainage radius, and the subscript *wf* will refer to pressure at the inlet sandface of a flowing well.

10.4.2 Skin. Sometimes wells experience near-wellbore phenomena (e.g., fractures and mud-filtrate damage) that cause production to be different from that calculated by Darcy's law. These near-wellbore effects are often very complex. Their total effect is normally characterized with the use of a skin factor, *S,* which appears in the steady-state radial-flow equation as

$$q_{sc} = \frac{k h \Delta \psi}{\beta \left[\ln \left(r_e / r_w \right) + S \right]} \; . \qquad (10.45)$$

Skin is a dimensionless parameter treated mathematically as an infinitely thin damaged or stimulated zone, regardless of the actual dimensions of the altered zone. Positive values indicate well damage (decreased productivity). Negative values indicate well stimulation (increased productivity). **Fig. 10.11** shows a typical pressure profile for a well with a positive skin. Wells with formation damage, partially penetrating wells, and wells with significant pressure drops in

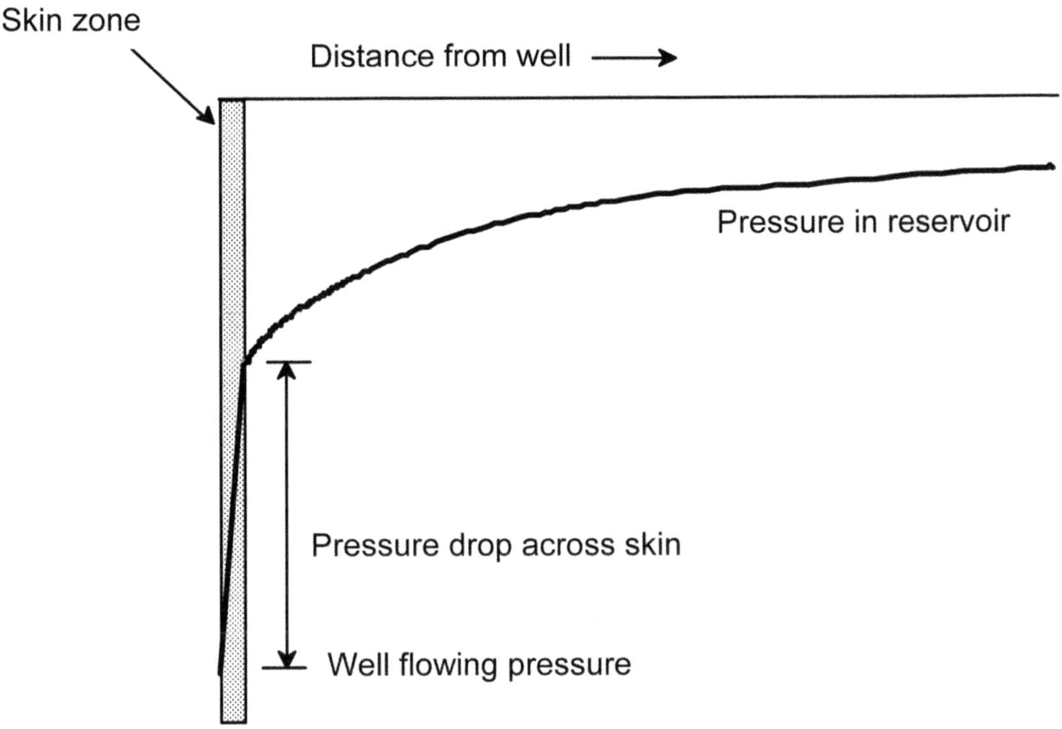

Fig. 10.11—Pressure distribution in a reservoir with a positive skin.

their completions have positive skins. Hydraulically fractured wells have negative skins. In general, skin must be determined empirically, usually from pressure-transient tests. Further discussion of the physical meaning of skin is given in the section on pressure-transient analysis.

10.4.3 Non-Darcy Flow. In gas wells, there may be a significant non-Darcy component of flow that results in an additional potential difference that depends on the square of the flow rate. The non-Darcy effect appears in well-deliverability equations as a flow-rate-dependent skin,

$$S' = S + Dq_g, \quad \text{(10.46)}$$

where S is the fixed or "mechanical" skin, and S' represents the total apparent skin including non-Darcy effects. D is called the non-Darcy-flow coefficient, having units of D/Mscf or d/std m^3. Although the non-Darcy coefficient may be calculated from laboratory measurements of the non-Darcy coefficient, it is typically determined in the field from well tests.

10.4.4 Transient Flow. At early times after a well has been put on production and at early times after a well has been shut in, flow occurs in a transient mode, making the steady-state forms of Darcy's law inappropriate. To mathematically represent transient flow, the relationship between density and pressure and material-balance (continuity) relationships must also be considered. When combined with Darcy's law, the result is the diffusivity equation, which in radial coordinates is

$$\frac{1}{r}\frac{\partial}{\partial r}\left(r\frac{\partial \psi}{\partial r}\right) = \frac{\phi \mu c_t}{k}\frac{\partial \psi}{\partial t} . \quad \text{(10.47)}$$

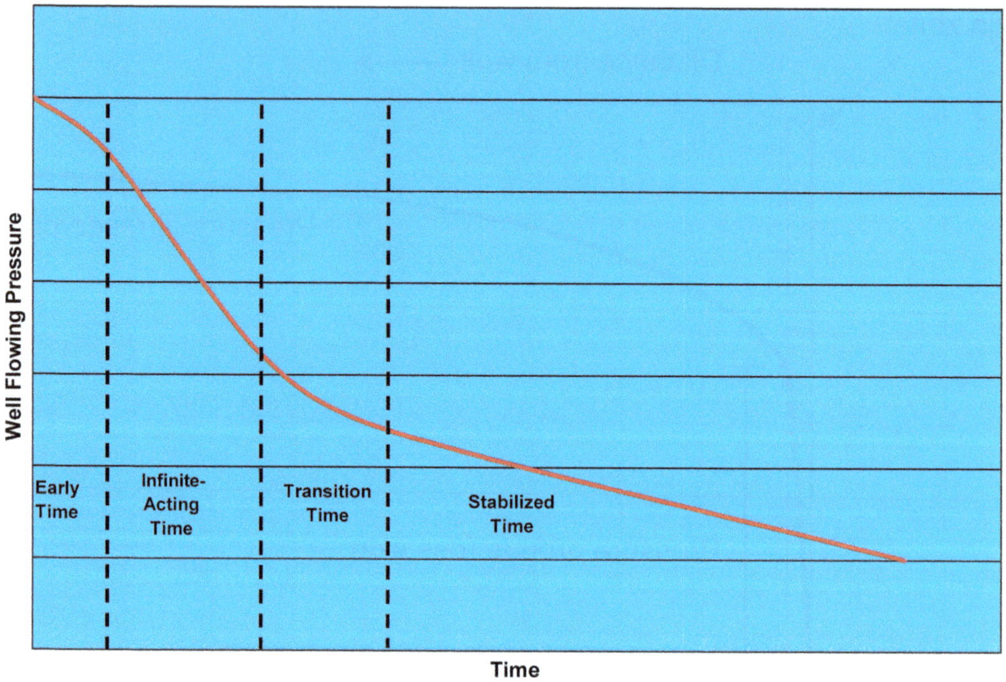

Fig. 10.12—Pressure/time history for a well produced at constant rate. Stabilized time occurs after all drainage-volume boundaries are fully "felt" and the well enters either steady-state or pseudosteady-state flow.

Note that the generic potential ψ is used in Eq. 10.47. This will be discussed further with regard to transient solutions for different fluid systems.

It is useful to distinguish between four different time periods when dealing with solutions to Eq. 10.47: (1) early time, (2) infinite-acting time, (3) transition time, and (4) stabilized time (**Fig. 10.12**).

Early time is dominated by wellbore, rather than reservoir, effects. During this time, little can be determined about the reservoir. This time may last from a few minutes to a few days.

During infinite-acting time, however, well response is the same as a well being produced from an infinite reservoir. Most pressure-transient tests analyze data during this time period. Because all reservoirs are finite, however, this time must end. It does so when the well response is affected by a part of the outer boundary of the well's drainage volume.

Steady-state flow is characterized by pressures being constant with time, requiring that the outer boundary of the system be maintained at constant pressure and the well be kept at either constant pressure or constant rate. This flow regime applies to certain water-influx situations or fluid-injection projects.

Pseudosteady-state flow occurs at late time in closed systems with a well produced at constant rate. Although pressures still change with time in pseudosteady state, all pressures everywhere in the reservoir decline at the same rate. This means that the pressure profile reduces uniformly throughout the reservoir.

Transition time occurs between infinite-acting and late time. During the transition period, the outer drainage boundaries are being felt in succession, causing the shift from infinite-acting to late time to occur over some length of time. In regularly shaped drainage areas (e.g., circles and squares), transition time may not exist. In irregularly shaped drainage areas, particularly with a well placed off-center, transition time can be quite long.

Infinite-Acting Flow. The time of primary interest in pressure-transient testing is the infinite-acting period, the mathematical solution for which comes from the diffusivity equation, expressed with the following dimensionless variables.

$$r_D = \frac{r}{r_w} \; . \qquad (10.48)$$

$$t_D = \frac{kt}{\phi \mu c_t r_w^2}$$

$$= \frac{1}{3{,}792.5} \frac{kt}{\phi \mu c_t r_w^2} = \frac{1}{277.78} \frac{kt}{\phi \mu c_t r_w^2} \; . \qquad (10.49)$$

$$\psi_D = \frac{kh\Delta\psi}{q_{sc}\beta}$$

$$\Delta\psi = \psi_i - \psi(r, t) \; . \qquad (10.50)$$

The subscript i refers to the initial conditions of the well's drainage volume. Time is in hours for both dimensionless-time equations.

At sufficiently large values of t_D / r_D^2, which are typical for most reservoir conditions, the solution can be mathematically approximated by

$$\psi_D(t_D, r_D) = \frac{1}{2}\left[\ln\left(\frac{t_D}{r_D^2}\right) + 0.80908 \right] = 1.1513 \log\left(\frac{2.2458 t_D}{r_D^2}\right) . \qquad (10.51)$$

This approximation is good to within 2% accuracy for $t_D / r_D^2 > 5$, within 1% accuracy for $t_D / r_D^2 > 8.5$, and essentially identical to $t_D / r_D^2 > 100$.[13]

Noting that the dimensionless potential at the well differs from the dimensionless potential at the wellbore radius ($r_D = 1$) by the skin factor, the basic radial well flow equation used for most well-testing purposes is

$$\psi_{Dw} = 1.1513 \log\left(2.2458 t_D\right) + S' . \qquad (10.52)$$

A graph of ψ_{wf} vs. the logarithm of time is a straight line. This slope is the basis for much pressure-drawdown and -buildup testing.

When predicting well performance, it is important to recognize when data are being collected during transient flow and to take into account the continuing decline in well deliverability until a steady-state or pseudosteady-state condition is established.

Pseudosteady State. In a closed drainage volume, once all the outer boundaries have been fully felt, a constant-rate well will experience pseudosteady-state flow. Because all pressures in the reservoir decline at the same rate during pseudosteady-state flow, the difference between reservoir pressures and the well pressure remains constant, even though both individually are changing with time. Because the resulting equation does not explicitly show a time dependence, the term *pseudosteady state* is used. Some authors also refer to this time period as "semisteady state."

Because pressure differences remain constant during pseudosteady-state flow, the following equations can be written to represent well performance during this period.[13]

$$\psi_{Dw} - \bar{\psi}_D = \frac{1}{2} \ln\left(\frac{2.2458A}{C_A r_w^2}\right) + S' = \ln\left(\sqrt{\frac{2.2458A}{C_A r_w^2}}\right) + S', \quad (10.53)$$

where C_A is the Dietz shape factor, dimensionless.

Because $\psi_{Dw} - \bar{\psi}_D = \frac{kh}{q_{sc}\beta}(\bar{\psi} - \psi_{wf})$, Eq. 10.53 can also be written in a more usable engineering form as

$$q_{sc} = \frac{kh}{\beta} \frac{\Delta\psi}{\ln\left(\sqrt{\frac{2.2458A}{C_A r_w^2}}\right) + S'}, \quad (10.54)$$

where $\Delta\psi = \bar{\psi} - \psi_{wf}$.

The shape factor can be determined empirically from mathematical solutions to the diffusivity equation in closed systems. **Tables 10.5 and 10.6** provide a list of shape factors for different drainage shapes and well placements.

For a well in a closed circle, the pseudosteady-state equation can also be written as follows, which is equivalent to having a shape factor of 31.62.

$$\psi_{Dw} - \bar{\psi}_D = \ln\left(\frac{r_e}{r_w}\right) - \frac{3}{4} + S' = \ln\left(\frac{0.472 r_e}{r_w}\right) + S'. \quad (10.55)$$

This equation also works for a well in the center of a closed square or other regular shape by calculating an equivalent-radius circle:

$$r_e = \sqrt{\frac{A}{\pi}}. \quad (10.56)$$

Transient Drainage Radius. To simplify the flow equations, it is sometimes useful to use what has been called a transient drainage radius, defined by

$$\ln\left(\frac{r_d}{r_w}\right) = \psi_D(r_D = 1) - \bar{\psi}_D. \quad (10.57)$$

This drainage radius is defined so that it represents the radius out to which there is a significant pressure drop.

During infinite-acting time, $\bar{\psi}_D \approx 0$ because there has been little withdrawal from the reservoir, which means that

$$\ln\left(\frac{r_d}{r_w}\right) \approx \frac{1}{2}\left(\ln t_D + 0.80908\right), \quad (10.58)$$

or, equivalently,

$$r_d = 1.4986\sqrt{t_D r_w^2}. \quad (10.59)$$

TABLE 10.5— PSEUDOSTEADY-STATE SHAPE FACTORS FOR VARIOUS RESERVOIR SHAPES[13]

IN BOUNDED RESERVOIRS	C_A	$\ln C_A$	$\frac{1}{2}\ln\left(\frac{2.2458}{C_A}\right)$	EXACT FOR $t_{DA}>$	LESS THAN 1% ERROR FOR $t_{DA}>$	USE INFINITE SYSTEM SOLUTION WITH LESS THAN 1% ERROR FOR $t_{DA}>$
	31.62	3.4538	-1.3224	0.1	0.06	0.10
	31.6	3.4532	-1.3220	0.1	0.06	0.10
	27.6	3.3178	-1.2544	0.2	0.07	0.09
	27.1	3.2995	-1.2452	0.2	0.07	0.09
	21.9	3.0865	-1.1387	0.4	0.12	0.08
	0.098	-2.3227	+1.5659	0.9	0.60	0.015
	30.8828	3.4302	-1.3106	0.1	0.05	0.09
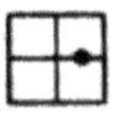	12.9851	2.5638	-0.8774	0.7	0.25	0.03
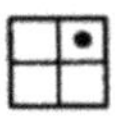	4.5132	1.5070	-0.3490	0.6	0.30	0.025
	3.3351	1.2045	-0.1977	0.7	0.25	0.01
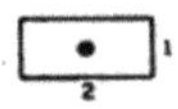	21.8369	3.0836	-1.1373	0.3	0.15	0.025
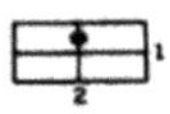	10.8374	2.3830	-0.7870	0.4	0.15	0.025
	4.5141	1.5072	-0.3491	1.5	0.50	0.06
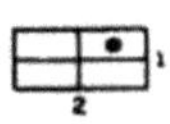	2.0769	0.7309	+0.0391	1.7	0.50	0.02
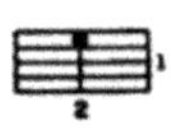	3.1573	1.1497	-0.1703	0.4	0.15	0.005

From Katz, D.L.: *Handbook of Natural Gas Engineering;* © The McGraw-Hill Companies Inc. (1990)

Note that the product $t_D r_w^2$ is simply dimensionless time without the wellbore radius in the denominator.

During pseudosteady-state time,

$$r_d = 1.4986\sqrt{\frac{A}{C_A}} \; . \qquad (10.60)$$

TABLE 10.6—MORE PSEUDOSTEADY-STATE SHAPE FACTORS FOR VARIOUS RESERVOIR SHAPES[13]

	C_A	ℓnC_A	$\frac{1}{2}\ell n\left(\frac{2.2458}{C_A}\right)$	EXACT FOR $t_{DA}>$	LESS THAN 1% ERROR FOR $t_{DA}>$	USE INFINITE SYSTEM SOLUTION WITH LESS THAN 1% ERROR FOR $t_{DA}>$
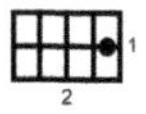	0.5813	-0.5425	+0.6758	2.0	0.60	0.02
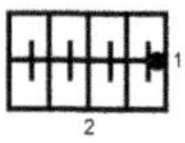	0.1109	-2.1991	+1.5041	3.0	0.60	0.005
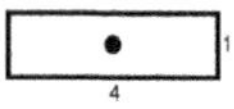	5.3790	1.6825	-0.4367	0.8	0.30	0.01
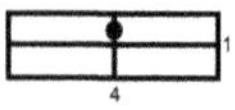	2.6896	0.9894	-0.0902	0.8	0.30	0.01
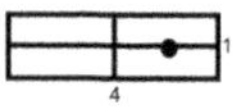	0.2318	-1.4619	+1.1355	4.0	2.00	0.03
	0.1155	-2.1585	+1.4838	4.0	2.00	0.01
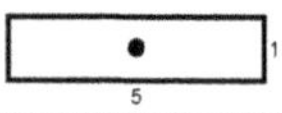	2.3606	0.8589	-0.0249	1.0	0.40	0.025
IN VERTICALLY FRACTURED RESERVOIRS						
	USE $(x_e/s_f)^2$ IN PLACE OF A/r_w^2 FOR FRACTURED SYSTEMS					
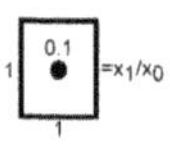	2.6541	0.9761	-0.0835	0.175	0.08	CANNOT USE
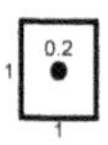	2.0348	0.7104	+0.0493	0.175	0.09	CANNOT USE
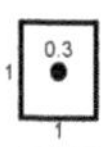	1.9986	0.6924	+0.0583	0.175	0.09	CANNOT USE
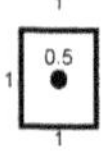	1.6620	0.5080	+0.1505	0.175	0.09	CANNOT USE
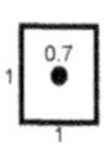	1.3127	0.2721	+0.2685	0.175	0.09	CANNOT USE
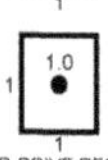	0.7887	-0.2374	+0.5232	0.175	0.09	CANNOT USE
IN WATER-DRIVE RESERVOIRS						
	19.1	2.95	-1.07	–	–	–
IN WATERS OF UNKNOWN PRODUCTION CHARACTER						
	25.0	3.22	-1.20	–	–	–

From Katz, D.L.: *Handbook of Natural Gas Engineering;*

For a circular or equivalent square drainage area, it can be shown that

$$r_d = 0.472 r_e \ . \qquad (10.61)$$

The deliverability equation for wells can then be written in the following simplified form for all times:

$$q_{sc} = \frac{kh}{\beta} \frac{\bar{\psi} - \psi_{wf}}{\ln\left(\frac{r_d}{r_w}\right) + S'} . \quad (10.62)$$

Note that pressures decline in proportion to the logarithm of time during infinite-acting time, while in pseudosteady state, pressures decline approximately in direct proportion to time.

Another way to determine the onset of pseudosteady state is through knowledge of the drainage shape and reservoir parameters. Tables 10.5 and 10.6 give times for the end of the infinite-acting period and for the beginning of time when the pseudosteady-state equation can be used. These tables use a dimensionless time based on drainage area,

$$t_{DA} = \frac{kt}{\phi \mu c_t A} = t_D \frac{r_w^2}{A}$$

$$= \frac{1}{3{,}792.5} \frac{kt}{\phi \mu c_t A} = \frac{1}{277.78} \frac{kt}{\phi \mu c_t A} . \quad (10.63)$$

10.4.5 Estimating Drainage Shapes. With little effort, it is possible to make a reasonable approximation of well drainage volumes and shapes. The process is based on the following assumptions:

1. The volume drained by an individual well is proportional to its flow rate.
2. Distance to a "no-flow" boundary between pairs of wells is proportional to each competing well's flow rate.

The following technique can then be used to estimate the drainage area at a given time. Assign a flow rate to each well based on an average production over some reasonable time period.

1. Using Assumption 2 above, assign no-flow points between pairs of wells.
2. Sketch no-flow lines by connecting up no-flow points.
3. Adjust the lines near reservoir boundaries to ensure that Assumption 1 is satisfied. Make adjustments for variations in thickness and known geologic features (e.g., faults).

Although this process is fairly rough, it can give a reasonable estimate of the drainage areas of each well in a reservoir.

10.4.6 Radius of Investigation. Tables 10.5 and 10.6 list times for the end of the infinite-acting-flow period and the beginning of the pseudosteady-state-flow period. These times can be used to define a "radius of investigation." The physical meaning of the radius of investigation is that it represents the minimum radius at which a boundary could exist, but which has not yet been "felt" at a given time (usually meaning the time at the end of a drawdown test that has not yet been affected by a reservoir boundary).

Thus, if t_{DA}^* is the area-based dimensionless time that defines the end of the infinite-acting period for a circular drainage area,

$$r_{\text{inv}} = \frac{1}{\sqrt{\pi t_{DA}^*}} \sqrt{t_D r_w^2} . \quad (10.64)$$

Depending on the level of precision that would define the end of the infinite-acting period, the radius of investigation would have different values. If a 1% criterion is used, t_{DA}^* is equal to 0.06, and

$$r_{inv} = 2.3\sqrt{t_D r_w^2}\ . \qquad (10.65)$$

Another common cutoff criterion is a t_{DA}^* of 0.1, after which the pseudosteady-state solution is listed by Earlougher[13] as being "exact" (probably to within the number of significant digits of one's computer). With this criterion,

$$r_{inv} = 1.8\sqrt{t_D r_w^2}\ . \qquad (10.66)$$

For practical purposes, it is reasonable to use what many authors recommend for the radius of investigation,

$$r_{inv} = 2\sqrt{t_D r_w^2}\ . \qquad (10.67)$$

10.4.7 Pressure-Transient Testing of Gas Wells. Refer to the chapter on well testing in this section of the *Handbook* for detailed discussions of interpreting pressure-transient tests of gas wells.

10.4.8 Deliverability Testing. Gas wells have historically been tested at a series of bottom-hole pressures and rates to develop stabilized-deliverability relationships. One of the reasons for this is the importance of the non-Darcy flow contribution to well performance. A single-rate test cannot address this contribution. This section describes the types of tests typically run, along with their appropriate analysis techniques.

Stabilized-Deliverability Test. A stabilized deliverability, sometimes called a four-point or backpressure test, is conducted by producing a well at four rates (**Fig. 10.13**). In this test, each rate is run long enough to reach stabilized conditions. To be theoretically valid, "stabilized" should mean to pseudosteady state, although in practice tests are sometimes run only until little variation in well flowing pressure is observed.

The most widely used method of presenting such data was first suggested by Rawlins and Schellhardt.[15] This method is not based on the pseudosteady-state-flow equations but is based on an empirical observation that Δp^2 vs. q_g plotted on a log-log graph typically lies on a straight line (**Fig. 10.14**). The equation for a straight line on a log-log graph is

$$q_g = C\left(\bar{p}^2 - p_{wf}^2\right)^n = C\left(\Delta p^2\right)^n\ . \qquad (10.68)$$

The slope of the straight line on a log-log graph is $1/n$. The easiest way to determine slopes on a log-log graph is to recall that differences in the run and rise of the line must be taken in terms of logarithms, so that if $\left(q_{g1}, \Delta p_1^2\right)$ and $\left(q_{g2}, \Delta p_2^2\right)$ are two points on the straight line,

$$\text{Slope} = \frac{\log \Delta p_2^2 - \log \Delta p_1^2}{\log q_{g2} - \log q_{g1}} = \frac{\log\left(\Delta p_2^2 / \Delta p_1^2\right)}{\log\left(q_{g2}/q_{g1}\right)}\ . \qquad (10.69)$$

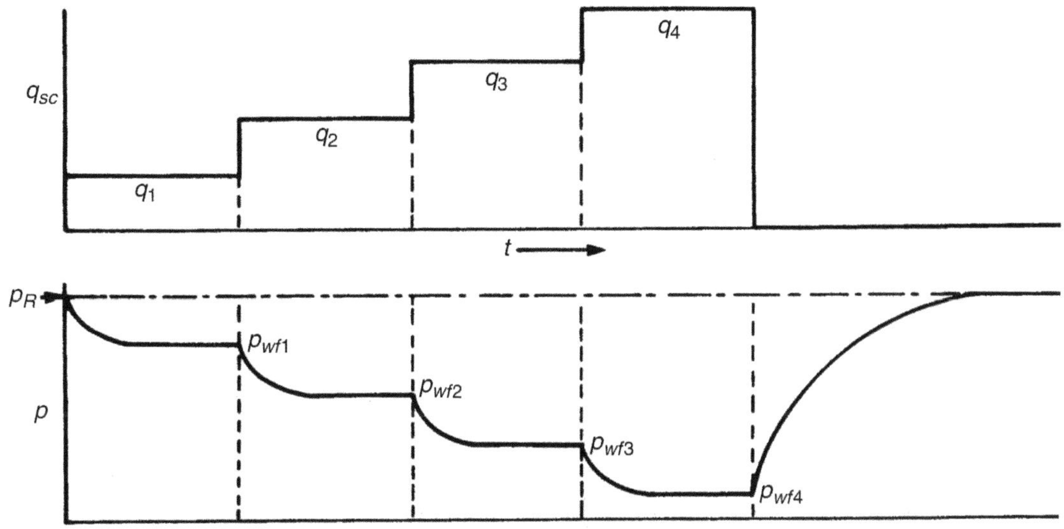

Fig. 10.13—Rate vs. time history of a stabilized deliverability test.[14]

A value of $n = 1$ corresponds to Darcy (laminar) flow, while $n = 0.5$ (slope = 2) corresponds to completely turbulent flow.

A parameter usually presented with this data is the absolute open flow (AOF) potential. The AOF may be found by extrapolating the log-log plot and reading the flow rate at a value of $\Delta p^2 = \bar{p}^2$ (i.e., $p_{wf} = 0$). The value of C is calculated from any point on the straight line but is perhaps easiest to calculate as

$$C = \frac{\text{AOF}}{\bar{p}^{2n}} \quad \text{(10.70)}$$

An alternative is to write the deliverability equation in terms of AOF instead of the parameter C:

$$q_g = \text{AOF}\left(1 - \frac{p_{wf}^2}{\bar{p}^2}\right)^n \quad \text{(10.71)}$$

The AOF is sometimes used as a measure of a well's potential flow capacity for regulatory and other purposes.

A more accurate way to analyze the stabilized deliverability test is to use the following equation, which is simply an alternative algebraic form of Eq. 10.62.

$$\Delta\psi = \bar{\psi} - \psi_{wf} = aq_g + bq_g^2,$$

$$a = \frac{\beta}{kh}\left[\ln\left(\frac{r_d}{r_w}\right) + S\right],$$

$$b = \frac{\beta}{kh}(D) \quad \text{(10.72)}$$

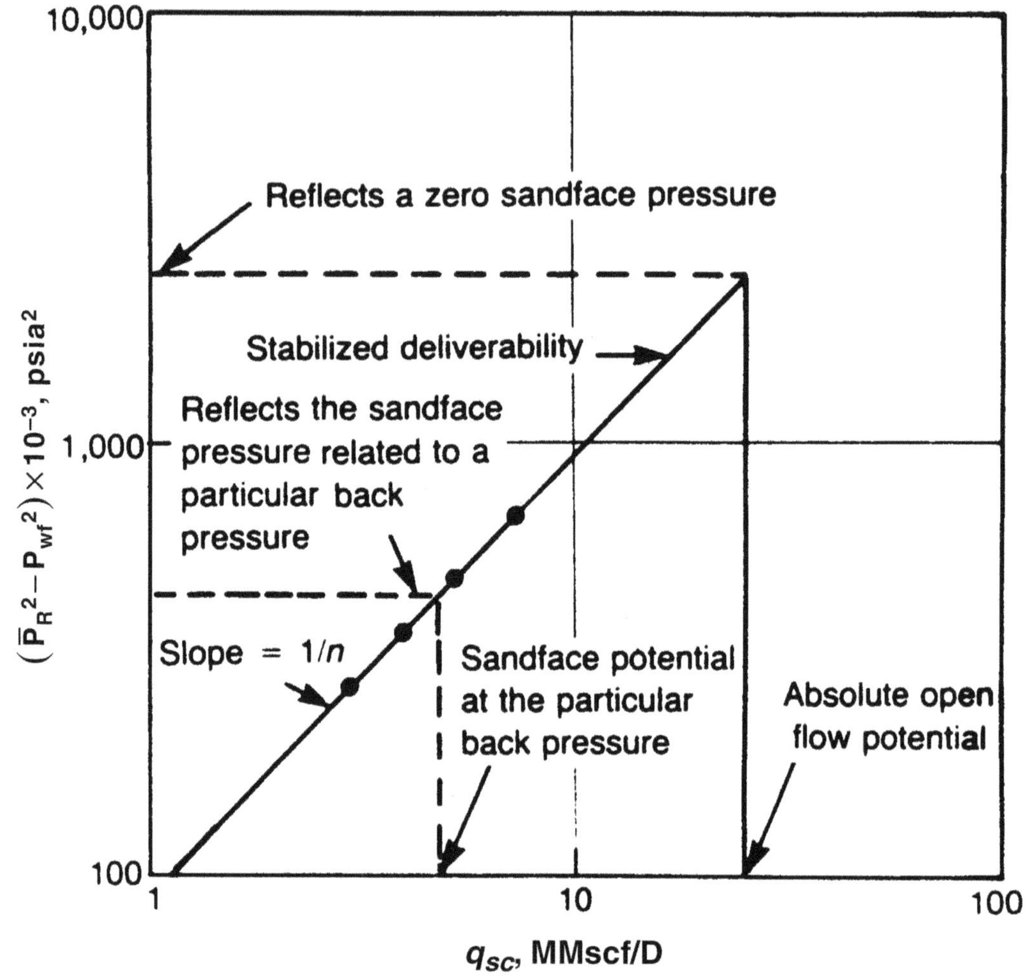

Fig. 10.14—Typical gas-well stabilized-deliverability plot.[14]

A Cartesian plot of $\Delta\psi/q_g$ vs. q_g will yield a slope of b and an intercept of a (**Fig. 10.15**). The advantage of this approach is that a and b have physical meaning based on reservoir parameters and thus can be compared to what is known about well and reservoir properties. C and n, on the other hand, have no such physical meanings. One disadvantage of this approach is that it is often more difficult to find a good straight line. However, this is a vagary of using Cartesian rather than log-log plots and does not represent any actual degradation of the data.

Once a and b are determined from the graph, the AOF using this approach is calculated by

$$\text{AOF} = \frac{\overline{\psi}}{a} \frac{2}{1 + \sqrt{1 + 4b\overline{\psi}/a^2}} \quad \text{(10.73)}$$

The second factor in Eq. 10.73 represents the reduction in the AOF caused by non-Darcy effects. The AOF determined from the a/b approach is usually lower than that calculated from the C/n approach.

Isochronal Tests. Another type of test often run on gas wells is the isochronal test. The difference between an isochronal test and a stabilized-deliverability test is that the flow periods

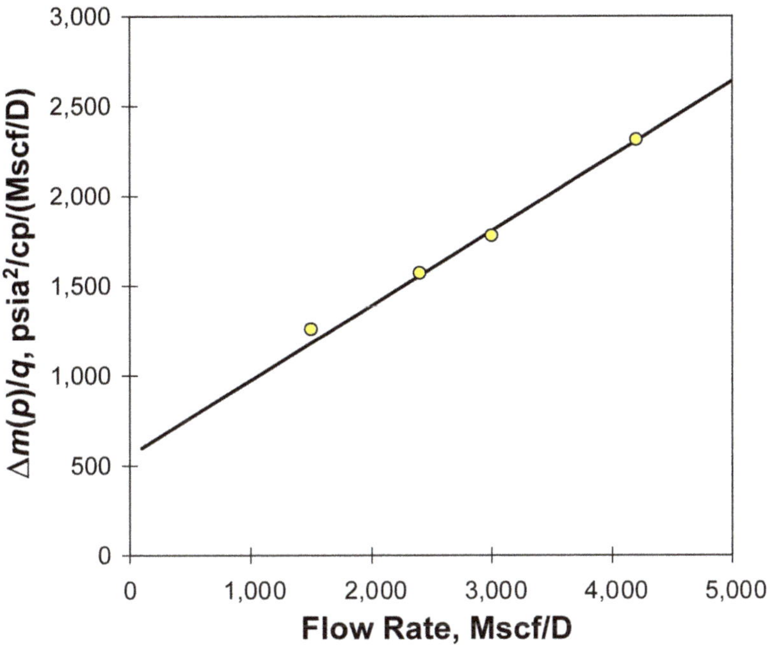

Fig. 10.15—Theoretical gas-well stabilized-deliverability plot.

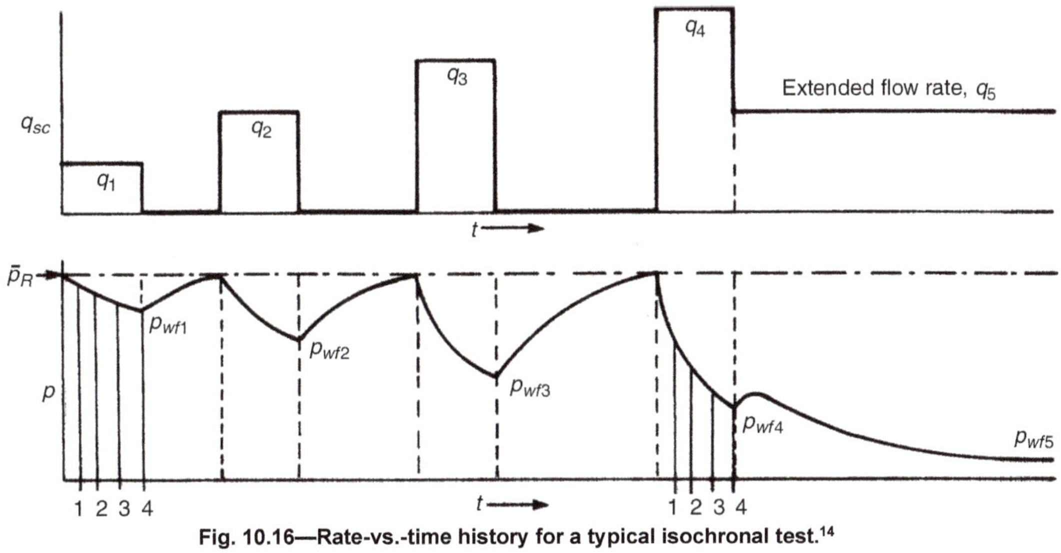

Fig. 10.16—Rate-vs.-time history for a typical isochronal test.[14]

are not run long enough to reach stabilized flow. This is done to shorten testing time and to conserve gas, particularly where no pipeline is available. **Fig. 10.16** shows a rate-vs.-time history for a typical isochronal test. Note that although the shut-in times are sufficiently long to approach initial reservoir pressure, the producing times are not long enough to reach pseudosteady state.

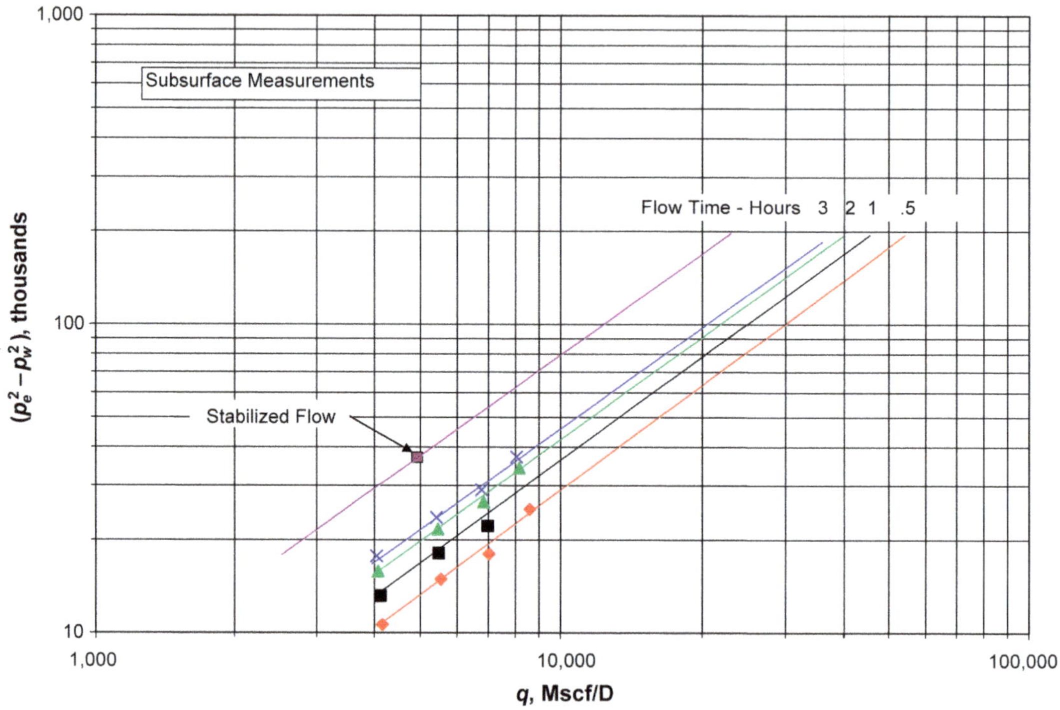

Fig. 10.17—Results of a typical isochronal test.

Lines are drawn through a Δp^2-vs.-q_g (**Fig. 10.17**) or $\Delta\psi/q_{sc}$-vs.-q_{sc} plot (**Fig. 10.18**) at common producing times; that is, during each flowing period, pressures are read at fixed times since the initiation of flow.

For a given producing time, a line through the data should have a constant slope. This slope should be the same no matter what the producing time because the transient drainage radius does not depend on flow rate. Thus, from each flow period, data are plotted for identical producing times (Fig. 10.17 or Fig. 10.18).

Finally, an extended flow period (to stabilization) is run at one rate. A line is passed through this single stabilized point, with the same slope as the other isochronal lines. If properly conducted, this test has been shown to give results comparable to a stabilized-deliverability test.

A modified isochronal test is also sometimes run in which the shut-in times are also shortened (**Fig. 10.19**). This type of test also works well if the value of pressure at the end of the last shut-in period is used in place of the average reservoir pressure.

10.4.9 Using Gas-Well Deliverability Relationships. Single-well deliverability equations can be used for a variety of purposes, including the following:

- Prediction of flow-rate changes caused by changing reservoir pressure (i.e., during reservoir depletion over time).
- Prediction of flow-rate changes caused by changing well flowing pressure resulting from production-equipment changes (e.g., compression).
- Prediction of bottomhole-flowing-pressure changes caused by changing well rates.

In general, the most theoretically valid deliverability equation should be used:

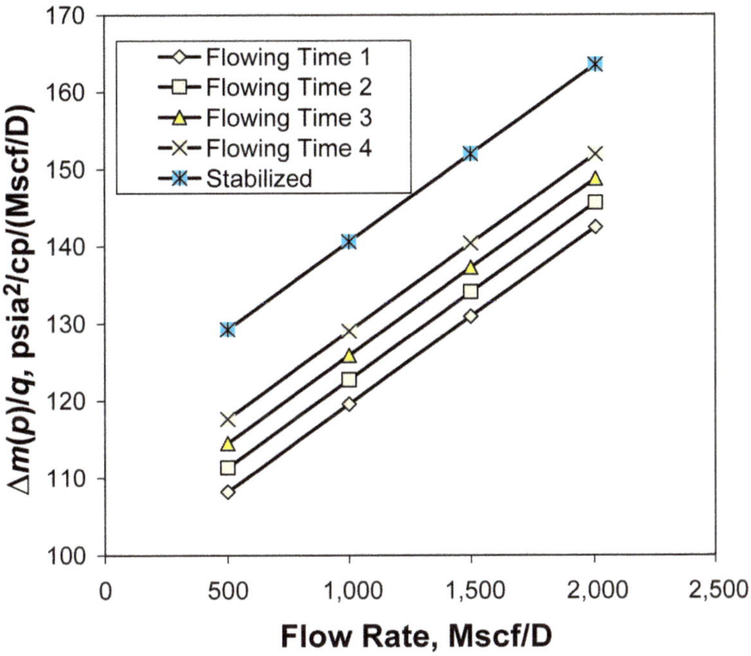

Fig. 10.18—Modified isochronal test analyzed by *a*/*b* analysis.

$$\bar{\psi} - \psi_{wf} = aq_g + bq_g^2 \text{ . } \qquad (10.74)$$

The use of Eq. 10.74 to solve for well flowing pressure is straightforward, except for the requirement to convert the generic potential ψ to actual pressure when using the real-gas potential $m(p)$. The conversion of $m(p)$ to p can be done either graphically or numerically with a computational algorithm (preferred).

Eq. 10.74 is a quadratic equation in flow rate, so when flow rate is being calculated from known pressures, the following can be used.

$$q_g = \frac{\left(\bar{\psi} - \psi_{wf}\right)}{a} \frac{2}{1 + \sqrt{1 + 4b\left(\bar{\psi} - \psi_{wf}\right)/a^2}} \text{ . } \qquad (10.75)$$

In some circumstances, it may be desirable to use either the C/n equation,

$$q_g = C\left(\bar{p}^2 - p_{wf}^2\right)^n = C\left(\Delta p^2\right)^n, \qquad (10.76)$$

or a productivity index (PI) equation,

$$q_g = PI\left(\bar{\psi} - \psi_{wf}\right), \qquad (10.77)$$

for deliverability calculations. Both can be used in a similar manner, as described above.

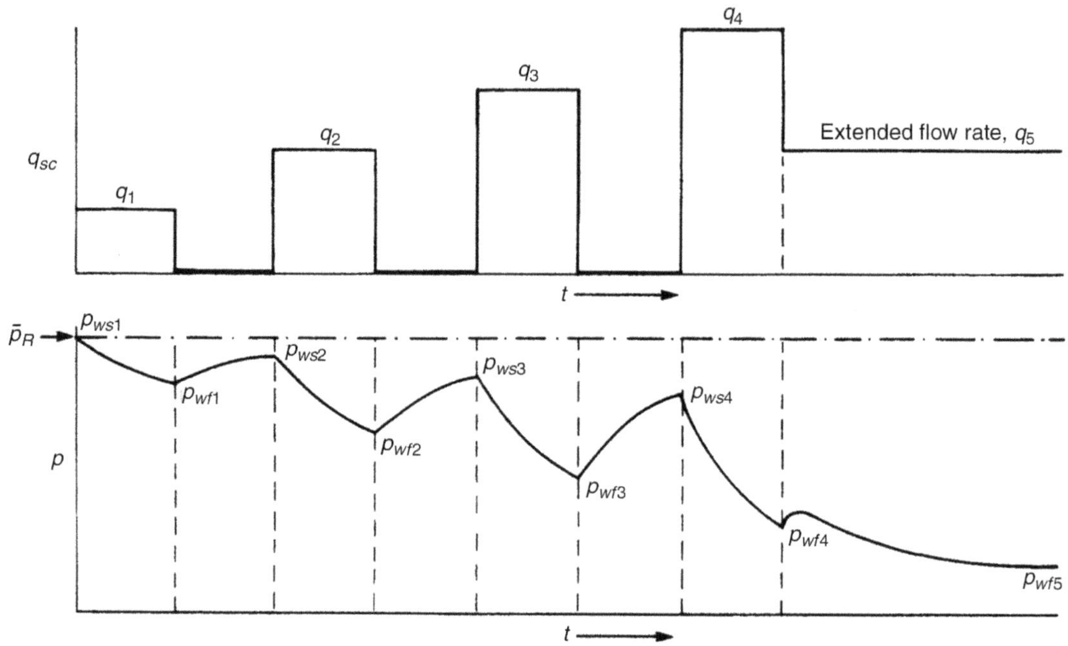

Fig. 10.19—Rate vs. time history for a typical modified isochronal test.[14]

10.4.10 Effects of Skin. There are several ways to look at the physical effects attributable to skin. The first is in terms of a flow efficiency (FE), which is defined as the well productivity with skin compared to the no-skin case:

$$\text{FE} = \frac{\ln\left(r_d / r_w\right)}{\ln\left(r_d / r_w\right) + S'} \ . \qquad (10.78)$$

The effective potential drop caused by skin also can be calculated as

$$\Delta\psi_{\text{skin}} = \frac{S'm}{1.1513} \ . \qquad (10.79)$$

Some prefer to consider skin in terms of an effective wellbore radius,

$$r_w' = r_w e^{-S} \ . \qquad (10.80)$$

It should be remembered that the skin in these equations is rate-dependent because of the non-Darcy effect. To determine the two different components of skin, mechanical-vs.-non-Darcy pressure-buildup or -drawdown tests must be run at more than one rate. If multirate transient tests are run, a simple plot of S' vs. q_g will yield a slope of D and an intercept of S.

10.5 Volumes and Recovery

This section discusses various aspects of gas reservoir performance, primarily to determine initial gas in place and how much is recoverable. The equations developed in this section also will be used to form the basis of forecasting future production rates by capturing the relationship between cumulative fluid production and average reservoir pressure.

10.5.1 Gas in Place. ***Volumetric Determination.*** Original gas in place (OGIP) can be estimated volumetrically with geological and petrophysical data:

$$G = \frac{Ah\phi S_{gi}}{B_{gi}} \text{ . } \quad (10.81)$$

In oilfield units with gas measured in Bscf and B_{gi} in ft³/Mscf,

$$G = \frac{1}{22.957}\frac{Ah\phi S_{gi}}{B_{gi}}$$

$$= \frac{1}{22{,}957}\frac{T_{sc}}{p_{sc}}\frac{Ah\phi S_{gi}(p/z)_i}{T} = \frac{1}{649.21}\frac{Ah\phi S_{gi}(p/z)_i}{T} \text{ . } \quad (10.82)$$

In SI units with gas measured in std m³ and A in m²,

$$G = \frac{Ah\phi S_{gi}}{B_{gi}}$$

$$= \frac{T_{sc}}{p_{sc}}\frac{Ah\phi S_{gi}(p/z)_i}{T} = 2.8438\frac{Ah\phi S_{gi}(p/z)_i}{T} \text{ . } \quad (10.83)$$

Material-Balance Determination of OGIP. Material-balance equations provide a relationship between original fluids in place, cumulative fluid production, and average reservoir pressure. For many gas reservoirs, a simple material-balance equation can be derived on the basis of the following assumptions:

- The gas-filled pore volume is constant.
- Gas dissolved in water or liberated from the rock is negligible.
- Reservoir temperature is uniform and constant.

With these assumptions, the real-gas law can be used to derive

$$V_g = \frac{RT}{V_{msc}}\frac{z_i}{p_i}G = \frac{RT}{V_{msc}}\frac{z}{p}\left(G - G_p\right) \text{ . } \quad (10.84)$$

This equation can be rearranged to get the usual volumetric gas material-balance equation,

$$\frac{G_p}{G} = 1 - \frac{p/z}{p_i/z_i} \text{ . } \quad (10.85)$$

This equation is the basis for the p/z-vs.-G_p graph used to analyze gas reservoirs.

10.5.2 Determining Average Reservoir Pressure. Reservoir engineers have often used pressure contour maps or some approximate methods to determine field average reservoir pressure for p/z analysis. Usually, however, individual well pressures are based on extrapolation of pressure buildup tests or from long shut-in periods. In either case, the average pressure measured does not represent a point value, but rather is the average value within the well's effective drainage volume (see Sec. 10.4.5).

By combining the assumptions used to assign drainage shapes and considerations of the gas law, the following procedure could be used for developing an average reservoir pressure at any point in time.

1. Be certain to determine average reservoir pressure accurately. Sometimes, shut-in times are inadequate to achieve complete buildup. When this occurs, one way to approximate reservoir pressure from a long shut-in is to use the Matthew-Brons-Hazebroek method[16] estimating the semilog straight-line slope from reservoir properties rather than a buildup test. Of course, buildup tests are the preferable way to determine average reservoir pressure when economically feasible. An alternative way to ramp up an incomplete buildup is to run a buildup test early in the life of a well, noting the time to complete buildup and the percentage buildup at shorter times. Then, these percentages can be used in subsequent shut-in tests of shorter times than those required for full buildup.

2. For each well, make a graph of p/z vs. G_p. In general, these graphs will not necessarily yield a straight line. If the well's drainage volume is changing with time, these will be curves. Either way, pass a smooth curve (not necessarily straight) through the data points.

3. To estimate the average p/z for a given well's drainage volume at a given point in time, first determine the cumulative gas produced for that well at the desired time, then use the value of G_p to get a value of well p/z from the graphs created in Step 2.

4. Estimate the average production rate for each well at the desired time. This should be some reasonable average "eye-balled" from production curves, and not necessarily a specific daily rate.

5. Determine the reservoir average p/z as the average of the individual-well values (Step 2), weighted by their production rate:

$$(p/z)_{\mathrm{res}} = \frac{\sum_{j=1}^{n_w} (p_j/z_j) q_j}{\sum_{j=1}^{n_w} q_j}, \qquad (10.86)$$

where n_w is the number of active producing wells. This procedure works reasonably well and is straightforward.

Accurate determination of average reservoir pressure is particularly difficult in tight gas sands. Shut-in pressures may not be near average reservoir pressure for several months or years, obviously too long to be of any value. In addition, low-permeability reservoirs can have significant pressure differences across the field because certain areas can be drained more effectively than others.

Poston and Berg[17] discuss methods for adjusting p/z plots for the lack of sufficient buildup time in determining average reservoir pressures. Although these methods have some validity, they also are prone to large errors because of data uncertainties. A recommended practice is, where feasible, to perform advanced pressure-transient-analysis methods on pressure-buildup tests to provide the means to extrapolate to expected values of average reservoir pressure. Such methods rely on the extrapolation of buildup pressures followed by a correction that incorporates drainage shape and volume. The problem with these techniques is that the correction for drainage shape and volume can be very significant (because of low reservoir permeabilities), and these techniques are highly uncertain, given the extent of heterogeneities and compartmentalization in typical tight gas reservoirs. Calibration of pressure-buildup analyses against actual well responses and reservoir simulation history-match studies can be helpful.

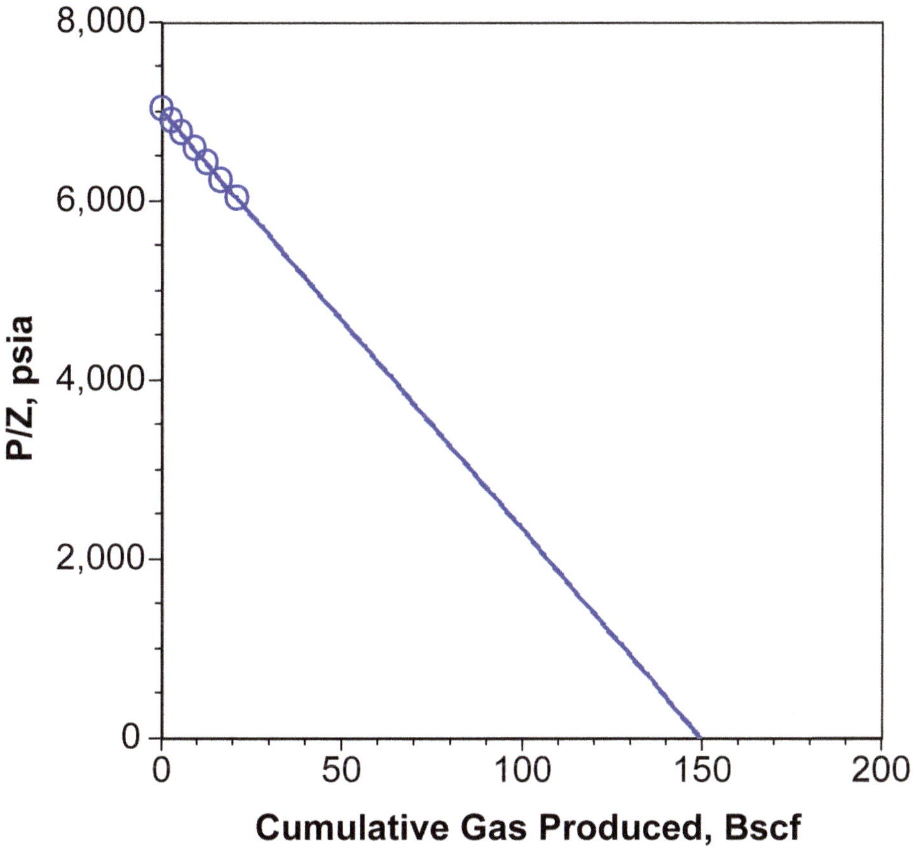

Fig. 10.20—Example *p/z*-vs.-G_p curve for a volumetric gas reservoir.

Another problem in tight reservoirs is the variation in average reservoir pressure across the field, both because of reservoir compartmentalization and because of low permeabilities. In higher-permeability gas reservoirs, this problem is generally not so severe, meaning that the average reservoir pressure needs to be measured only in a few wells to generate accurate *p/z* analyses. When variations in average reservoir pressure are large, however, methods need to be used to account for differences across the field.

Although reservoir simulation is one possibility to deal with this problem, it is also possible to use a "compartmentalized-reservoir model"[18] to incorporate these effects. This model treats the reservoir as a set of communicating "tanks." The technique is basically a history-matching process that uses compartment volumes and compartment-to-compartment transmissibilities as tuning parameters. This method is important and has technical value, although it does not address the problem of required long shut-in times. The method has been applied to several reservoirs with very good results; hence, its use is recommended.

10.5.3 Volumetric Reservoirs. In volumetric dry- and wet-gas reservoirs, *p/z* vs. cumulative gas production will be a straight line intercepting the gas-production axis at the OGIP. An example is given in **Fig. 10.20**. The intercept (G_p = 0) on the *p/z* axis is p_i/z_i, and the intercept on the G_p axis (p/z = 0) is G. This graph provides a convenient method of using average-reservoir-pressure data to estimate OGIP and recoverable reserves once an abandonment *p/z* is established. When these plots are applicable, results for OGIP are generally considered very accurate after approximately 10% of gas reserves have been produced (sometimes a bit earlier).

When only a small amount of early data is available, the OGIP can be determined from any point (G_p, p/z) by

$$G = \frac{G_p}{1 - \frac{p/z}{p_i/z_i}} . \qquad (10.87)$$

When placing the straight line through p/z data, it is usually prudent to consider the first point (i.e., at field discovery pressure) as more accurate than others. Because there has been little field depletion and there has been sufficient time for pressures to return to stabilized conditions, discovery-pressure measurements are generally reliable. Regression approaches to placing the p/z-vs.-G_p line should take this into consideration.

Volumetric estimates based on cores, well logs, fluid analyses, and geological estimates of reservoir size provide a "rock-based" estimate of gas in place, while material-balance relationships provide a "fluids-based" or "pressure-based" estimate. These two types of estimates are essentially independent (except for the use of consistent values of B_{gi}). Thus, when the two estimates are comparable, there is greater certainty in the OGIP estimate. It is now a common practice to develop geologic and simulation models of a reservoir to determine reserves and depletion strategies and to evaluate alternative development scenarios. Acquisition of p/z data can provide another measure of the volume being drained by a well or set of wells in what is thought to be a common reservoir. Differences between models and p/z data can be a valuable tool in managing a reservoir and detecting opportunities for additional development or deferral of expenditures that become unnecessary because of a change in the size of a resource.

Well-deliverability forecasts can be used to predict the economic limit of production for a field (income = costs) and the resulting p_a/z_a. Recoverable reserves then become

$$G_{pa} = G\left(1 - \frac{p_a/z_a}{p_i/z_i}\right) . \qquad (10.88)$$

10.5.4 Highly Compressive Reservoirs. For some high-pressure gas reservoirs (e.g., geopressured or abnormally pressured reservoirs), the combined rock and water compressibility can result in a nonlinear p/z plot (**Fig. 10.21**). Ignoring this effect can lead to large overestimates of the OGIP. Local knowledge is the best source of information about whether these effects should be considered. Such performances usually should be suspected for geopressured reservoirs.

If the pore volume and water can be considered to have constant compressibility, then the change of gas-filled pore volume with pressure is

$$\frac{V_g}{V_{gi}} = \frac{e^{c_f \Delta p} - S_{wi} e^{-c_w \Delta p}}{1 - S_{wi}} . \qquad (10.89)$$

Using the first two terms in a Taylor series expansion for the exponential function,

$$\frac{V_g}{V_{gi}} \approx 1 - c_e \Delta p, \qquad (10.90)$$

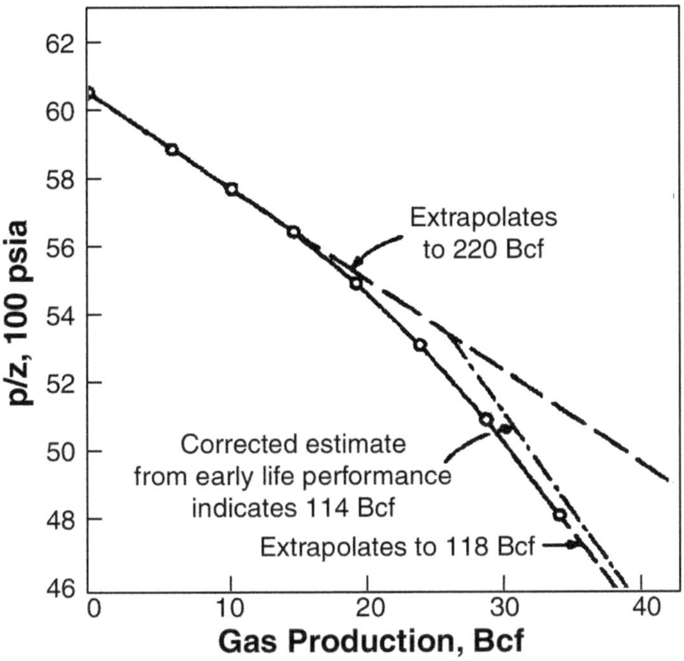

Fig. 10.21—*p/z* vs. cumulative production, North Ossum field, Lafayette Parish, Louisiana, NS2B reservoir.[19]

$$\text{where } c_e = \frac{c_f + S_{wi}c_w}{1 - S_{wi}} \text{ . } \qquad (10.91)$$

The material-balance equation for a compressive reservoir then becomes

$$\frac{G_p}{G} = 1 - \left(1 - c_e \Delta p\right) \frac{p/z}{p_i/z_i} \text{ . } \qquad (10.92)$$

Eq. 10.92 suggests that if the effective reservoir compressibility, c_e, can be estimated, then the *p/z* plot for such reservoirs can be linearized by multiplying the *p/z* values by $1 - c_e \Delta p$. However, there is typically very little knowledge of the effective system compressibility, meaning that this relationship is of limited practical use for reservoir engineering purposes. In addition, the effective system compressibility may even change with time, typically becoming smaller as reservoir pressure declines and the reservoir rocks compact.

There is sometimes a change in the slope of the *p/z* plot when an abnormally pressured gas reservoir reaches normal pressure, as shown in Fig. 10.21. Approaches suggested for analyzing geopressured gas reservoirs include methods to account for some unusually high apparent values of the effective pore-volume compressibility. Soft-sediment compaction, shale dewatering, and limited aquifer influx are among the physical effects proposed by various authors. For further information on this topic, the reader is referred to papers by Hammerlindl,[19] Roach,[20] Prasad and Rogers,[21] Bernard,[22] Fetkovich *et al.*,[23] Ambastha,[24] Yale *et al.*,[25] El Sharkawy,[26] and Gan and Blasingame.[27] Poston and Berg[17] also provide an evaluation of different methods of accounting for pressure support experienced in geopressured reservoirs.

Many overpressured reservoirs, however, do not demonstrate the change in slope, as illustrated by **Fig. 10.22**. These data are from four wells in a common reservoir with an initial

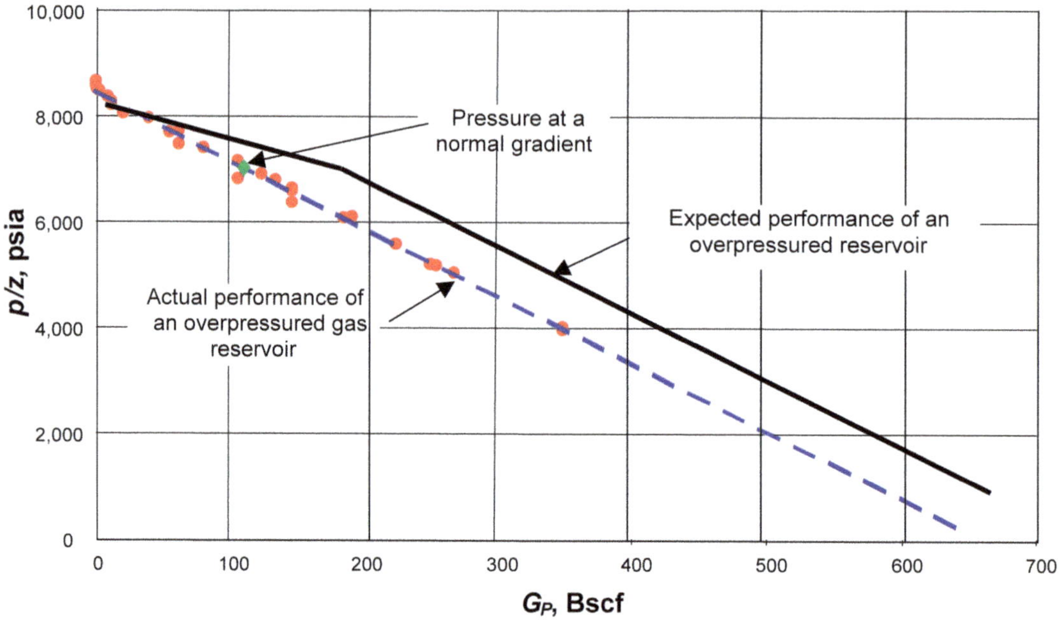

Fig. 10.22—Actual vs. expected *p/z* curves for an overpressured gas reservoir.

pressure gradient >0.65 psi/ft. As indicated, there was no change in slope when the reservoir pressure reached a normal gradient. The reservoir consists of a competent sandstone that may have a low effective compressibility.

Considering the previous discussions, using early *p/z* has many uncertainties. Best practices would suggest that early-time analyses use ranges for effective pore-volume compressibility based (where possible) on analogous or similar regionally located reservoirs to reduce the high uncertainty in early data and its potential dramatic effect on estimates of OGIP.

Abandonment conditions for highly compressive reservoirs are determined in the same manner as those for volumetric reservoirs.

10.5.5 Waterdrive Reservoirs. Fig. 10.23 shows a typical *p/z* plot for a gas reservoir with an active waterdrive. Note that for a given value of cumulative gas production, pressures are higher than for a volumetric reservoir.

The material-balance equation for a waterdrive reservoir is

$$\frac{G_p}{G} = 1 - \left[1 - \frac{(W_e - W_p B_w)}{G B_{gi}}\right] \frac{p/z}{p_i/z_i} \quad \text{(10.93)}$$

If it can be assumed that volumetric estimates of G are accurate, then Eq. 10.93 can be rearranged to calculate a water-influx history for comparison against contact mapping.

$$W_e = W_p B_w + G B_{gi}\left[1 - \left(1 - \frac{G_p}{G}\right)\frac{p_i/z_i}{p/z}\right] \quad \text{(10.94)}$$

This can be used with different aquifer models to determine how to predict future water influx. This will be discussed further in the next section.

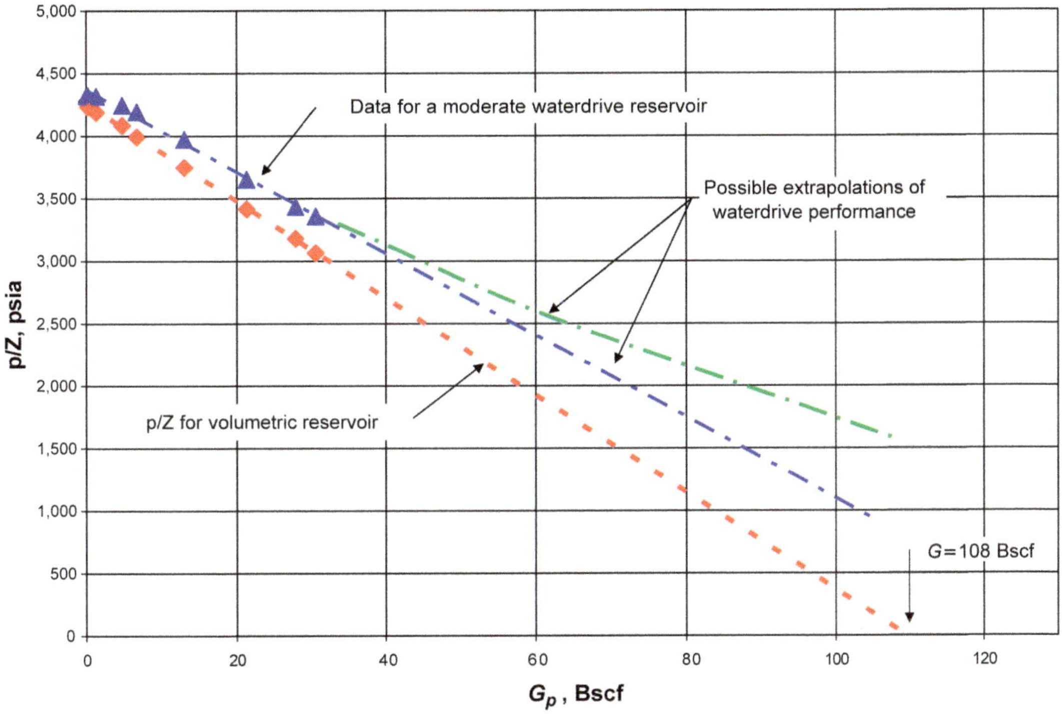

Fig. 10.23—Volumetric-vs.-waterdrive *p/z* trends.

Using Eq. 10.93 to estimate G requires information about the cumulative water influx. Estimates may be obtained by mapping water movement using watered-out wells, logging surveys, or data from infill wells.

By writing Eq. 10.93 in terms of a volumetric sweep efficiency, E_v, and the residual gas saturation to water displacement, S_{gr}, the material-balance calculation can then be written as

$$\frac{G_p}{G} = 1 - \left[1 - E_v\left(1 - \frac{S_{gr}}{S_{gi}}\right)\right]\frac{p/z}{p_i/z_i} \quad \text{.......................................} \quad (10.95)$$

If E_v is taken to be the estimated volumetric sweep efficiency at abandonment, then this equation represents all possible abandonment conditions regardless of the rate of water influx. An abandonment line can then be drawn on the *p/z* plot, the bottom point of which is at $p/z = 0$, $G_p = G$, and the top point of which is at $p/z = p_i/z_i$. A straight line connecting these two points is the locus of all possible abandonment points. The intersection of this abandonment line and the actual p/z-vs.-G_p line, as illustrated in **Fig. 10.24**, gives an estimate of ultimate recovery.

Many factors affect gas-recovery potential when a waterdrive is active. The main considerations are how much of the reservoir will be invaded by water, what the pressure decline will be, and what the trapped-gas saturation will be behind the waterfront. Simple 3D simulations can be used to study the strength of an aquifer based on its size relevant to the gas volume and the variation in rock quality. A good reservoir description is the key to a successful prediction. How the reservoir is affected is also a function of withdrawal rates. Strong aquifers can sustain reservoir pressure and result in low recoveries. Gas saturations trapped behind the invading water are typically about one-third of the initial hydrocarbon saturation. If volumetric sweep is

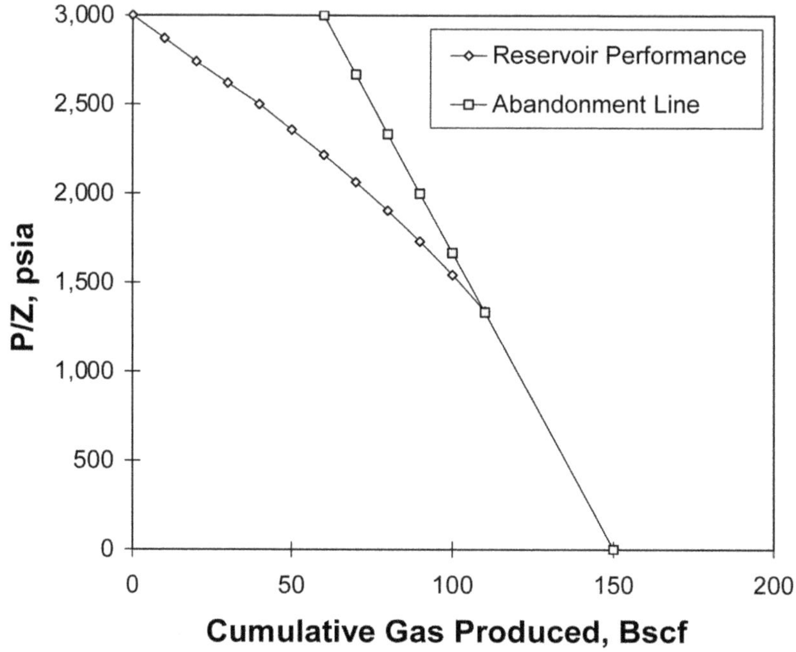

Fig. 10.24—Typical *p/z* plot for a waterdrive gas reservoir.

high and there is little pressure depletion ($B_{ga} \approx B_{gi}$), then a reservoir with S_{wi} equal to 25% of pore volume would experience a recovery of

$$E_{Rg} = \left(\frac{1 - S_{wi} - S_{gr}}{1 - S_{wi}}\right)\frac{B_{gi}}{B_{ga}}E_v = \left(\frac{1 - 0.25 - 0.25}{1 - 0.25}\right)0.9 = 0.6. \qquad (10.96)$$

Recoveries as low as 50% of OGIP can occur in adverse circumstances, but more often recoveries exceed 70% of OGIP owing to partial pressure depletion. There are documented cases[28,29] in which a significant increase in offtake rate has resulted in pressure depletion of a waterdrive reservoir when withdrawal rates were sufficient to outrun invading water. Recovery from reservoirs that exceed 1 Tcf OGIP with permeabilities greater than 250 md will normally be unaffected by an aquifer of any size if field depletion occurs over a 20-year period or less. Again, a simple simulation model will confirm how potentially effective an aquifer may be.

The effect of a weak to moderate waterdrive is often difficult to detect with a simple *p/z* plot. Often, a straight-line plot will occur (Fig. 10.23) and will lead to incorrect estimates. Cole[30] has suggested an improved method. If the expansibility of water is small compared to gas expansibility, then the material balance can be arranged as

$$\frac{G_p B_g}{B_g - B_{gi}} = G + \frac{W_e - W_p B_w}{B_g - B_{gi}} \; . \qquad (10.97)$$

Cole's methodology is to plot the left side of the equation against G_p. The shape of the resulting plot will vary depending on the existence and strength of a waterdrive, as illustrated by **Fig. 10.25**.

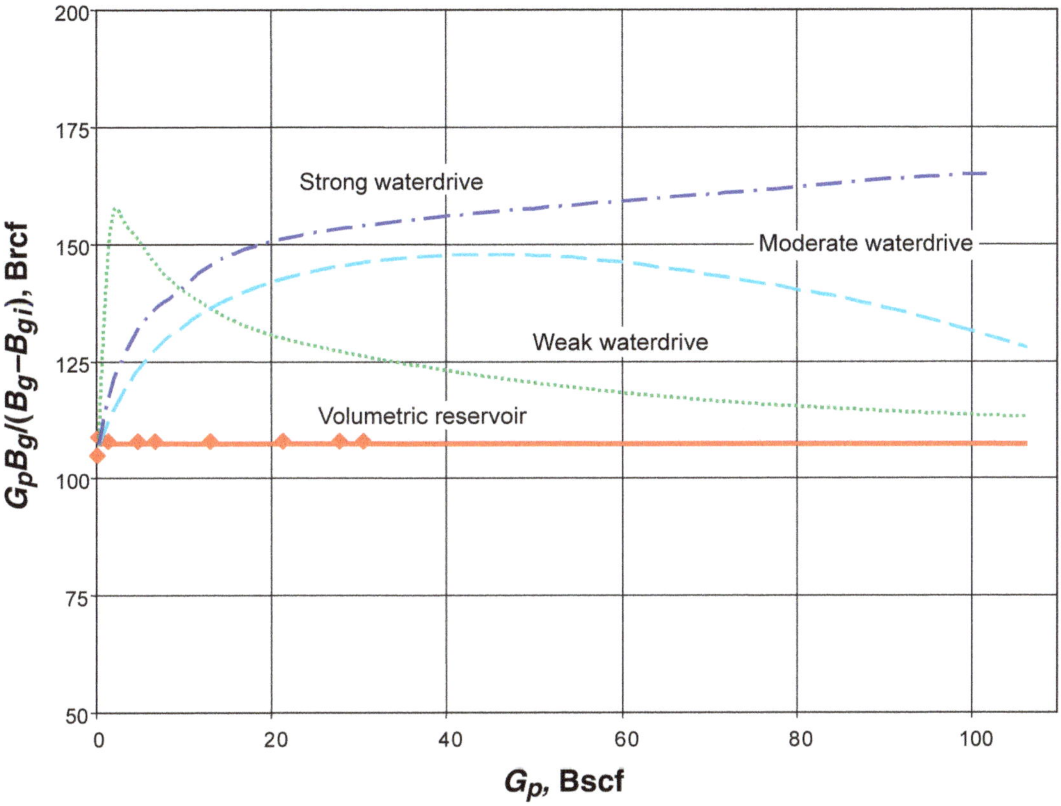

Fig. 10.25—Plot for volumetric and waterdrive gas reservoirs (after Cole[30]).

Data from a volumetric reservoir will plot as a horizontal line. A weak waterdrive yields an early increase in ordinate values followed by a negative slope. The initial increase may not be detected because many pressure measurements are needed very early in the producing life of the reservoir. Moderate to strong waterdrives give overstated OGIP values. This plot is very sensitive to the effects of water influx and is a good qualitative tool. Back extrapolating the plot to OGIP has been suggested. In practice, the slope usually changes with each pressure measurement, and extrapolation is difficult to impossible.

Pletcher[31] has suggested a further modification to the Cole plot to account for rock and water compressibility. In doing so, Eq. 10.97 becomes

$$\frac{F}{E_t} = G + \frac{W_e}{E_t}, \quad \text{(10.98)}$$

where $F = G\left(E_g + E_{fw}\right) + W_e$,

$$E_g = B_g - B_{gi},$$

$$E_{fw} = \frac{\left(B_{gi}\right)\left(S_{wi}c_w + c_f\right)\left(p_i - p\right)}{\left(1 - S_{wi}\right)},$$

$$\text{and } E_t = E_g + E_{fw}\,. \quad \text{(10.99)}$$

The shapes of the resulting plots are the same as those in Fig. 10.25, but they do avoid the negative slope of a Cole plot that results from an abnormally pressured reservoir with no water influx.

10.5.6 Retrograde-Condensate Reservoirs. Depletion behavior of retrograde-condensate reservoirs can be handled through the p/z analyses discussed previously, with the caveat that the z factor must be the two-phase z factor (see Sec. 10.2). Two-phase z factors either may be obtained from laboratory tests or predicted from composition with an EOS. In wet-gas and retrograde-condensate reservoirs, cumulative gas produced must include both gas and liquid (as equivalent gas) production. This is particularly important for high-liquid-yield gases.

In the calculation of future reserves for planning purposes, it is usually necessary to break out gas and liquid reserves separately, perhaps even by individual gas component. For wet-gas reservoirs, liquid yields from a particular gas can be expected to remain constant with time, so long as the gas is processed in the same manner. Changes in separator conditions and/or gas-processing facilities could result in changing liquid yields, however.

Retrograde-condensate reservoirs, on the other hand, will produce at a variable yield as the reservoir pressure declines. Determination of the expected yields can be based on laboratory tests and/or EOS calculations. The PVT test presented in Sec. 10.2 shows how yields of the various gas components can vary over time.

10.5.7 Pressure Maintenance and Cycling Operations. Pressure maintenance of a retrograde-condensate gas reservoir can exist by virtue of an active waterdrive, water-injection operations, gas-injection operations, or combinations of all of these. Certain reservoirs may contain fluids near their critical points and are thereby candidates for special recovery methods, such as the injection of specially tailored gas compositions to provide miscibility and phase-change processes that could improve recovery efficiency. These usually are not regarded as gas/condensate cases. All these improved-recovery methods are best studied with simple-to-complex computer models. Simple models can be used initially to screen prospects, and then more-detailed studies including compositional considerations can be conducted.

Waterdrive and Water-Injection Pressure Maintenance. Recovery from retrograde-condensate gas reservoirs with a waterdrive or water injection is subject to the same considerations as described in the chapter on Waterflooding (in the Reservoir Engineering and Petrophysics section of this *Handbook*) for water injection into oil reservoirs. To make a recovery assessment, the first requirement is a good description of the rock and fluid characteristics of the reservoir and the aquifer. Variations in the permeability of various strata, mobility ratios, and gravity-stable advance of the water front will affect the volumetric sweep. S_{gr} should be approximately the same as described earlier for dry- and wet-gas displacements by water. The favorable mobility ratio can result in a high volumetric sweep. There is strong evidence, however, that displacement efficiency by water is not high. While Buckley *et al.*[32] indicated that the displacement efficiency of water displacement of gas can be as high as 80 to 85%, experiments and field observations by Geffen *et al.*[33] indicate that it may be as low as 50%. All things considered, the recovery of gas condensate in the vapor phase by water injection is likely to be appreciably lower than by cycling, and any consideration of water injection for gas/condensate recovery should be accompanied by detailed experimental work on cores from the specific reservoir involved. This will help to determine whether water can, in fact, accomplish a high-enough displacement efficiency to justify its use.

Premature water breakthrough can, and often does, result in "load up" and loss of the ability of a well to flow. It is difficult to obtain economical flow rates by artificial lift. This loss of productivity may result in premature abandonment of the project. The problems would be particularly serious for deeper reservoirs in which the cost of removing water would be a significant factor. Yuster[34] discusses possible remedial methods for drowned gas wells. Bennett and Auvenshine[35] discuss dewatering gas wells. Dunning and Eakin[36] describe an inexpensive method to remove water from drowned gas wells with foaming agents.

Generally, the use of water injection for maintaining pressure in a gas/condensate reservoir will be unattractive where a wide range of permeabilities exists in a layered reservoir and selective breakthrough of large volumes occurs early in the life of the reservoir.

Dry-Gas Injection. Comparative economics determine whether a gas/condensate reservoir should be produced by pressure depletion or by pressure maintenance (i.e., does the additional condensate recovery justify the cost of compressing, injecting, and processing the injected gas?). Delayed gas sales also may be a factor. The objective of using dry-gas injection is to maintain the reservoir pressure usually above or near the dewpoint to minimize the amount of retrograde condensation. Dry field gases are miscible with nearly all reservoir gas/condensate systems; methane normally is the primary constituent of dry field gas. Dry-gas cycling of gas/condensate reservoirs is a special case of miscible-phase displacement of hydrocarbon fluids for improving recovery. Experimentation has shown that the displacement of one miscible fluid by another that is miscible is highly efficient on a microscopic scale; usually, the efficiency is considered 100% or very nearly so. Cycling does result in liquid recoveries at economical rates while avoiding waste of the produced gas when a market for that gas is not available.

Inert-Gas Injection. The use of inert gas to replace voidage during cycling of gas/condensate reservoirs can be an economical alternative to dry natural gas. One of the first successful inert-gas-injection projects was in 1949 at Elk Basin, Wyoming,[37] where stack gas from steam boilers was used for injection. In 1959, the first successful use of internal-combustion-engine exhaust was seen in a Louisiana oil field.[38] The first use of pure cryogenically produced N_2 to prevent the retrograde loss of liquids was in the Wilcox 5 sand in the Fordoche field located in Pointe Coupee Parish, Louisiana.[39] In the Fordoche field, the N_2 amounted to approximately 30% of the natural-gas/N_2 mixture injected.

Studies by Moses and Wilson[40] confirmed that the mixing of N_2 with a gas/condensate fluid elevated the dewpoint pressure. Moses and Wilson also presented data to show that the mixing of a lean gas with a rich-gas condensate would result in a fluid with a higher dewpoint pressure. The increase in dewpoint pressure was greater with N_2 than with the lean gas. In the same study, results are presented from slimtube displacement tests of the same gas/condensate fluid both by pure nitrogen and by a lean gas. In both displacements, more than 98% recovery of reservoir liquid was achieved. These results also were observed by Peterson[41] using gas-cap gas material from the Painter field located in southwest Wyoming. The authors concluded that the observed results were obtained because of multiple-contact miscibility.

Cryogenic-produced N_2 possesses many desirable physical properties.[42] Those that make nitrogen most useful for a cycling fluid are that it is totally inert (noncorrosive) and that it has a higher compressibility factor than lean gas (requires less volume). The latter advantage is partially offset by increased compression requirements when compared with lean gas.

The use of inert gas as a cycling fluid offers both advantages and disadvantages. The major advantages are early sale of residue gas and liquids and a higher recovery of total hydrocarbons because the reservoir contains large volumes of inert gas rather than hydrocarbons at abandonment. Disadvantages are production problems and increased operating costs caused by corrosion if combustion or flue gas is used, possible additional capital investments and operating costs to remove inert gas from the sales gas (a condition aggravated by early breakthrough of inerts), and potential costs to pretreat before compression and/or to fund reinjection facilities.

10.6 Forecasting

This section explores the fundamental relationships underlying gas-reservoir performance and presents some simple techniques for forecasting production rate vs. time.

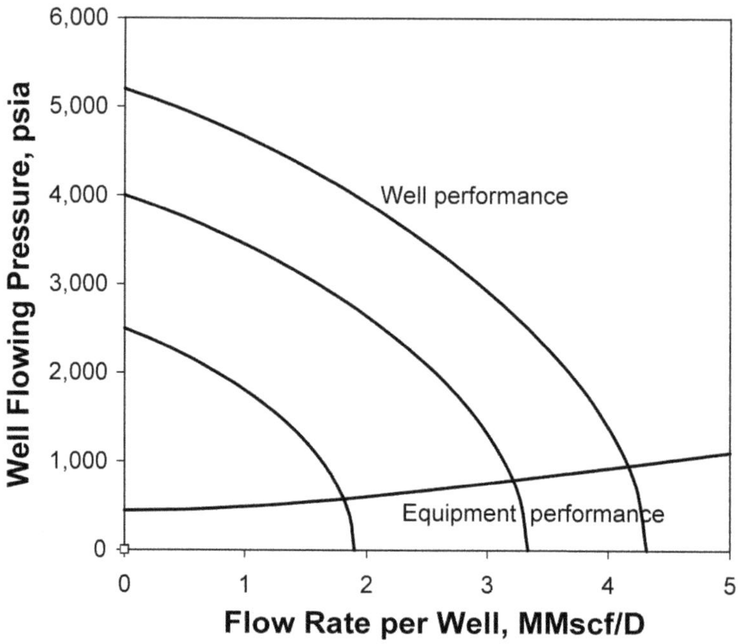

Fig. 10.26—Example well- and equipment-performance curves.

10.6.1 System Performance. One way to envision the different factors affecting the performance of a gas reservoir is to define the production "system" with three components: (1) well deliverability (developed in Sec. 10.4), (2) wellbore hydraulics, and (3) production-equipment constraints. Rate-vs.-time behavior is governed by the combined effect of these three parts, which in turn have performance characteristics that vary with pressure and production rate. Wellbore and system constraints include

- Tubing and choke sizes.
- Amount of entrained liquids (condensate and water).
- Accumulation of sand or debris in the wellbore.
- Flowline pressure drops.
- Compression ratios across compressors.
- Pressure losses in separators.

If these relationships are plotted on the same presentation, the resulting graph will look like **Fig. 10.26**.

At low flow rates, the equipment-performance curve is nearly horizontal, reflecting the small flowing frictional pressure drops in the system. If there is liquid holdup in the production tubing, multiphase-flow calculations can show the curve bending upward at low production rates.

The curves represent maximum performance with existing completion, well, and equipment configurations. At any time, of course, it is possible to operate a well below its maximum performance characteristics by adjusting such things as a choke size at the wellhead. Because this is effectively a "zero-expense" change, it represents how a system is "tuned" to operate at some predetermined rate or other operating condition. Perforating additional producing intervals, using stimulation treatments, lowering separator pressure, or installing compressors, larger flowlines, and/or production tubing can change the performance curves.

The concave downward lines (flow rate increasing with decreasing well pressure) are the well-performance curves for different average reservoir pressures. Note that the value of well flowing pressure at zero flow rate is the average reservoir pressure and that the value of flow

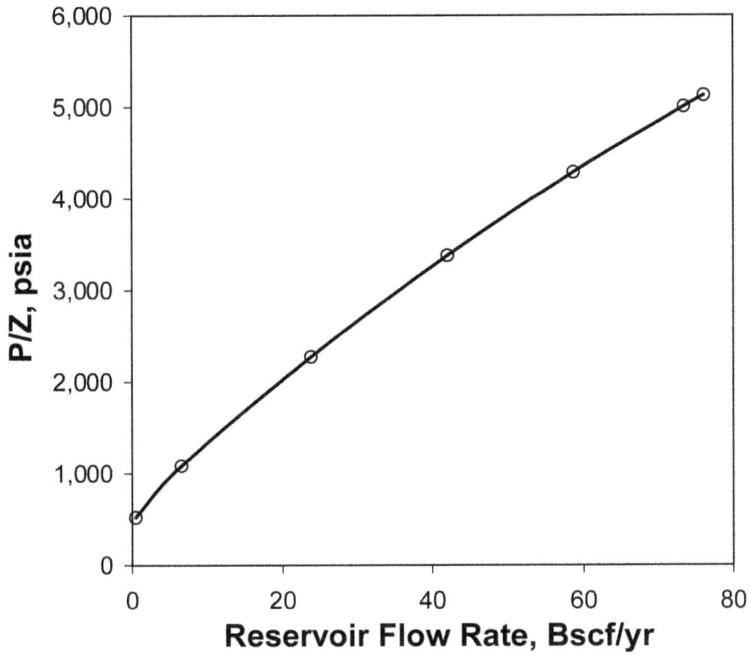

Fig. 10.27—Example reservoir-deliverability curve.

rate at zero well flowing pressure is the AOF. As reservoir pressure declines, of course, the well-performance curves move down and to the left. These curves may be altered by operational changes that affect well deliverability. Such processes as hydraulic fracturing, acidizing, or reperforating will increase well productivity and move up the AOF point.

The intersection of the equipment- and well-performance curves represents the operating point of the well at a given value of average reservoir pressure. Note that if the equipment-performance curve represents maximum (full-open choke) performance, then the intersection point also represents maximum rate performance.

10.6.2 Reservoir Deliverability. The intersection points of the well- and equipment-performance curves can be used to construct a relationship between average reservoir *p/z* and maximum well deliverability such as that shown in **Fig. 10.27**. This curve represents the rate that the combined well and equipment design is capable of delivering at any particular average reservoir pressure. Note that this curve can be used to determine the abandonment *p/z* by knowing the economic limit (i.e., the minimum economic rate).

For the purposes of forecasting total reservoir performance, it is necessary to develop a graph such as Fig. 10.27 for the reservoir and not just for an individual well. One way to do this is to construct individual graphs for each well and then to add up the flow rates for each well at a given value of average reservoir *p/z*. Alternatively, one could construct the graph for an average well in the reservoir and then simply multiply by the total number of wells.

Another method would be to first determine reservoir deliverability constants *a* and *b* (that is, *a* and *b* should represent the following deliverability equation for the reservoir):

$$\overline{\psi} - \psi_{wf} = aq_R + bq_R^2, \qquad (10.100)$$

where q_R is the total reservoir production rate.

There are many ways to develop this sort of relationship, including an analysis of reservoir performance in the same manner as well performance with techniques explained in Sec. 10.4. A simple way would be to obtain $\bar{a}$ and $\bar{b}$ for the individual wells in the reservoir, and then, by assuming that the average well produces at the total reservoir rate divided by the number of wells, n_w,

$$\bar{\psi} - \psi_{wf} = \frac{\bar{a}}{n_w} q_R + \frac{\bar{b}}{n_w^2} q_R^2 . \tag{10.101}$$

This equation could then be used with an average well-equipment-performance curve.

10.6.3 Forecasting Methods. It was seen in Sec. 10.5 that through material-balance relationships, it is possible to evaluate expected reservoir pressures knowing cumulative production. The material-balance equation combined with the reservoir-deliverability relationship thus leads to a set of two relationships that must hold simultaneously:

$$q = \text{Function}\left(\bar{p}\right) \tag{10.102}$$

$$\text{and } \bar{p} = \text{Function}\left(G_p\right) = \text{Function}\left[\int_0^t q(t)\mathrm{d}t\right] . \tag{10.103}$$

For simple gas-reservoir situations, it is possible to solve these two relationships either graphically or computationally.

Consider, for example, a typical gas-reservoir production scenario in which the reservoir is first put on production at some fixed reservoir rate for an extended period of time. This fixed rate may result from sales-contract considerations or limitations on processing equipment and pipelines. Let q_c denote the fixed initial reservoir production rate and t_c denote the time of the fixed-rate period.

It is possible to use the reservoir-deliverability and p/z material-balance curves to determine the length of the constant-rate period given the rate or vice versa. First, consider how to determine the time.

For a given rate, the reservoir-deliverability curve can be used to determine the lowest reservoir p/z that will deliver this rate. Because the reservoir-deliverability curve represents the maximum reservoir deliverability before the end of the constant-rate period, the reservoir will be produced at less-than-maximum rates. The value of p/z at the end of the constant-rate period can be entered into the p/z material-balance plot to determine G_{pc} at the end of the constant-rate period. The constant-rate time period is then

$$t_c = \frac{G_{pc}}{q_c} . \tag{10.104}$$

For instance, refer to **Fig. 10.28**, where reservoir deliverability and G_p vs. p/z are plotted. If rates from the field were limited to 40 Bscf/yr by contract, equipment, or pipeline limitations, then the lowest p/z that will support this rate is equivalent to a cumulative production of 400 Bscf. The period of constant production would be 10 years.

Similarly, other rates can be selected and equivalent periods of production calculated to develop the inset curve in Fig. 10.28.

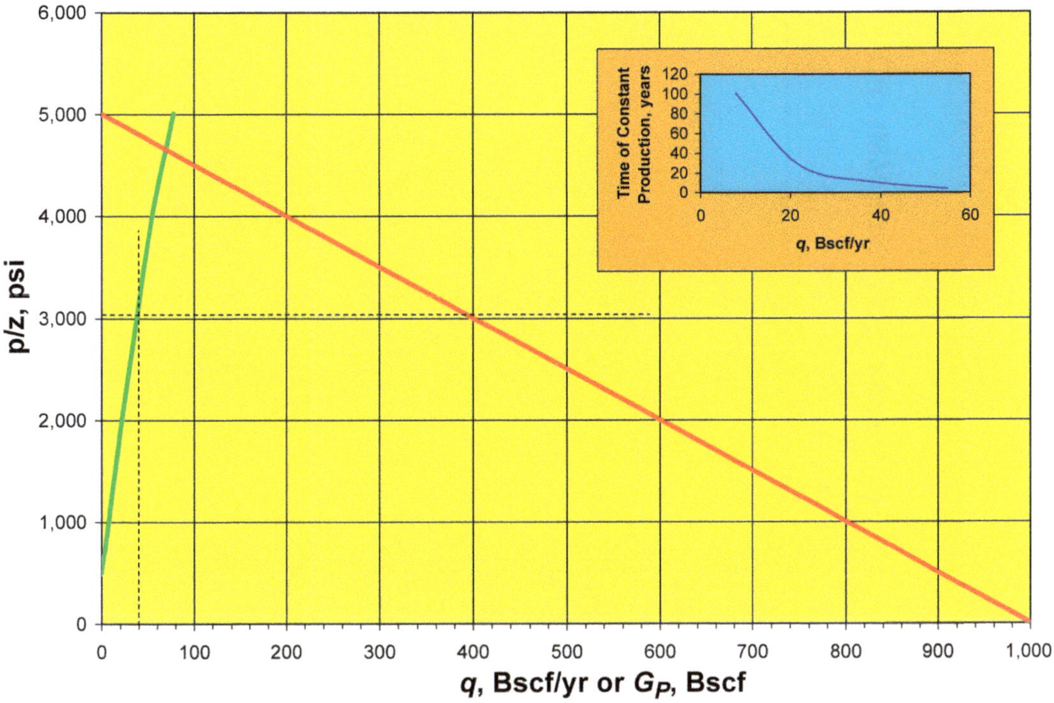

Fig. 10.28—Example reservoir-deliverability and material-balance curves.

These procedures should be based on average sustainable rates that account for down time and other factors that reduce the composite well deliverability. Care also should be taken to account for the normal "ramp-up" of production that occurs during development drilling.

At the end of the constant-rate period, the reservoir will, barring well or equipment changes, go on decline. At this point, the reservoir will produce at its maximum rate according to the reservoir-deliverability curve. A simple procedure can be used to forecast this period as well by discretizing future production into increments.

Consider that a prediction will be made by discretizing the future cumulative gas produced in an increment ΔG_p. Over the time period that this amount of gas is to be produced, there will be some average flow rate $\bar{q}$. The time to produce the incremental gas production can be approximated by

$$\Delta t = \frac{\Delta G_p}{\bar{q}} \ . \quad (10.105)$$

Assume that q_j is the flow rate at the end of the time increment, and q_{j-1} is the flow rate at the beginning of the time increment. Approximating the average flow rate during flow period j as

$$\bar{q}_j = \frac{q_j + q_{j-1}}{2}, \quad (10.106)$$

$$\text{then } \Delta t_j = \frac{\Delta G_{pj}}{\bar{q}_j} = \frac{2\Delta G_{pj}}{q_j + q_{j-1}} \ . \quad (10.107)$$

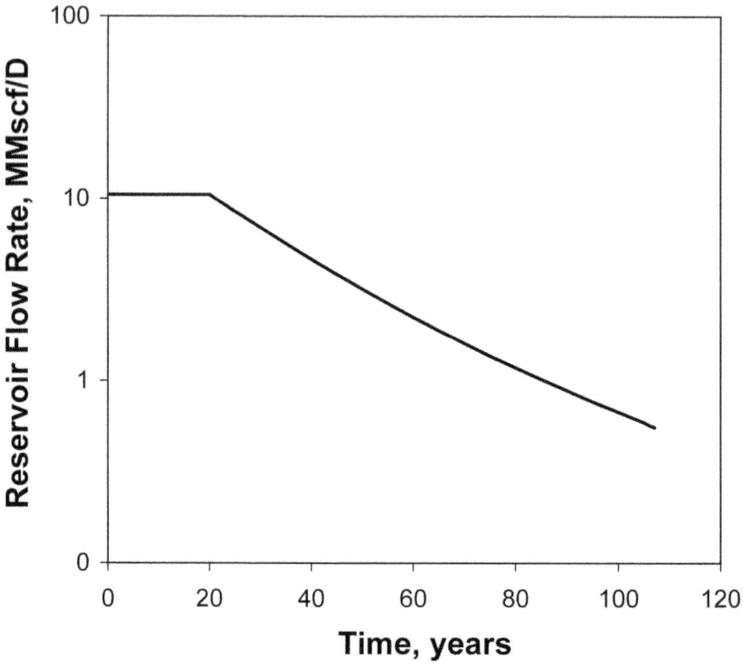

Fig. 10.29—Example production rate-vs.-time curve for a gas reservoir.

Total time since the beginning of the decline period is determined by

$$t_j = t_{j-1} + \Delta t_j \; . \qquad (10.108)$$

A plot of q_j (not $\bar{q}_j$) vs. t_j is the desired forecast. Improved accuracy may be achieved by using smaller values of ΔG_p. An example decline graph can be seen in **Fig. 10.29**.

The above procedures can be used to evaluate different reservoir-development scenarios. For example, infill drilling or adding compressors would improve the reservoir-deliverability relationship (and thus change the forecast) by sustaining or increasing early-time production but causing the reservoir to have a steeper decline. Whether this would be desirable would depend on economic considerations.

These procedures also suggest ways that past reservoir performance may be evaluated. For example, a historical reservoir-deliverability curve can be generated by plotting reservoir p/z vs. q_R during the decline period. This curve could then be adjusted for changes in operating equipment to determine future performance characteristics.

10.6.4 Water Influx. If a gas reservoir is under waterdrive conditions, there is an additional requirement to forecast the amount of water influx to be expected. The difficulty in performing water-influx calculations for most reservoir situations is in knowing the performance characteristics of the aquifer. As mentioned before, computer models ranging from simple to very complex are now available for general use and represent the best method of predicting the amount and timing of water influx. The models also allow investigation of the effects of varying aquifer size and rock characteristics. Matching of performance data will significantly improve the reliability of model projections.

10.6.5 Retrograde-Condensate Reservoirs. The options discussed before for depleting retrograde-condensate reservoirs are:

- Produce by depletion.
- Recycle processed produced gas.
- Recycle processed produced gas plus other makeup gas.
- Inject N_2 or other inert gas.
- Waterdrive or injected water.

With straight depletion, a considerable amount of the heaviest and most valuable hydrocarbons will be left in the reservoir. Also, reservoir fluids passing through the low-pressure region around the wellbore experience retrograde condensation, resulting in a large liquid saturation buildup and a significant decrease in gas permeability. This is important from at least two standpoints. First, the composition history of produced fluids early in the life of the reservoir may diverge from predictions that assume uniform pressure in the reservoir at any instant of time. Second, and more importantly, well deliverability will be decreased significantly, affecting both timing and ultimate recovery. Wells can even cease to produce in reservoirs with permeabilities less than 10 md. The effects of liquid buildup on well deliverability need special consideration in low-permeability environments.

Recycling of gas aids recovery from condensate reservoirs in two ways. First, reservoir pressure is maintained, if not above the dewpoint, at least at pressures that minimize liquid deposition. If makeup gas is injected, it is, of course, possible to never go below the dewpoint pressure. The economics of recycling produced gas must weigh the additional recovery benefits against the additional handling and injection costs, the deferred revenue of recovered injected gas, and the lost revenue of injected gas that will remain in the reservoir. These factors make most hydrocarbon gas-injection schemes uneconomical if a gas sale is possible. But where gas sales are to be delayed for many years (e.g., Prudhoe Bay), the resulting economics support a cycling program.

If, however, reservoir pressure has dipped below the dewpoint pressure, injected lean gas can also serve to revaporize a significant part of the deposited condensate because of phase-behavior effects. There has been some interest in the use of N_2 as an injected gas for condensate reservoirs. There is evidence that phase behavior is nearly equivalent to methane injection, and costs may be somewhat lower.

These effects, however, are complex. Injected gas may sweep inefficiently between injection and production wells, causing poor economics. Because of the complexities of this process, cycling operations should not be undertaken without the aid of reservoir-simulation studies. Compositional reservoir-simulation models are available that can easily handle the thermodynamics and fluid-flow characteristics of a recycling project. Characterization of phase behavior is generally done through an EOS, which can be tuned to laboratory PVT tests.

Gas Requirements in Cycling Operations. Miller and Lents[43] expected to cycle the equivalent of approximately 115% of the gas in place to recover some 85% of the wet-gas reserves of the Cotton Valley Bodcaw reservoir. Brinkley[44] indicated cycling-gas volumes of as much as 130% of original wet gas in place for various reservoirs. The requirements for a given reservoir will be driven mostly by economic considerations. The makeup gas needed for constant-pressure cycling is mainly the volume required to replace shrinkage by liquid recovery and the amount consumed for various fuel needs. For some composition, temperature, and pressure ranges, the removal of high-molecular-weight constituents from the produced wet gas may result in a higher compressibility factor for the injected dry gas; hence, the greater volume per mole injected may require little or no makeup gas for constant-pressure cycling.

The amount of gas not available for injection because of consumption for operating needs should be taken into account when determining makeup-gas requirements if pressure is to be maintained. The amount of fuel for compression and treatment plants depends mainly on the total amount of gas to be returned to the reservoir and the discharge pressure for the plant. Discharge pressure, in turn, depends on the total rate of injection demanded, the number of injection wells, and their intake capacities throughout the life of the operation. Other factors affecting the amount of gas required for overall operations are type of plant, type of liquid-recovery system used, and auxiliary field requirements (such as for drilling, completion, and well testing; camp fuel and power for maintenance shops, general service facilities, and employee housing; and other factors that vary from one case to another).

Moore[45] reports that fuel consumption for the compression plant alone varies from 7 to 12 ft^3/bhp-hr; this is probably for gases with heat values of approximately 1,000 Btu/scf.

If the fuel consumption is 8 ft^3/bhp-hr and the compression ratio is 15 (compressing from, say, 461 to 7,000 psia), fuel requirements would be 34.4 Mscf/MMscf injected. For an example reservoir originally containing 131 Bscf of wet gas, which might be cycled the equivalent of 1.25 times, the approximate compressor fuel consumption would be 5.6 Bscf, or approximately 3% of the gas handled through the plant.

Treatment-plant fuel and other plant needs added to compressor fuel bring the range of consumption inside the plant fence to 3 to 7% of the gas handled by a cycling plant. In addition to these needs and others mentioned earlier, possible gas losses that can occur in a cycling operation are: gas used in "blowing down" wells, should this be necessary for cleaning or treating purposes; small gas leaks at compressor plants and in field lines; and gas leaks resulting from imperfect seals or corrosion in well tubing, casing, and cement jobs. Remedial workover operations should be planned immediately when there is evidence of appreciable loss of gas between the compression plant and the reservoir sandface or between the outflow-well sandface and the plant intake.

10.7 Summary

The study of gas reservoirs must consider the unique nature of gaseous fluids and their behavior in the reservoir and during production. This chapter has provided methods for characterizing gas reservoirs and their contents, including deliverability calculations, OGIP determination, and recovery potential. Differences in phase behavior between dry, wet, and retrograde-condensate gases result in different depletion schemes that may need additional altering if a strong waterdrive exists. Recycling of produced hydrocarbon gases or injection of nonhydrocarbon fluids may be justified in retrograde-condensate gas reservoirs, particularly if gas sales will be delayed for several years.

Nomenclature

a = empirical constant
A = drainage area, reservoir area, L^2
AOF = absolute open flow potential, std L^3/t
b = empirical constant
B = formation volume factor, L^3/std L^3
B_{gi} = initial gas formation volume factor, L^3/std L^3
c = compressibility, Lt^2/m
c_f = pore-volume compressibility, Lt^2/m
c_w = water compressibility, Lt^2/m
C = constant in gas-deliverability equation
C_A = Dietz shape factor, dimensionless
D = non-Darcy-flow coefficient, t/std L^3
E_{fw} = cumulative formation and water expansion, L^3
E_g = cumulative gas expansion, L^3
E_R = recovery efficiency, fraction
E_t = total cumulative expansion, L^3
E_v = volumetric sweep efficiency, fraction

F = cumulative reservoir voidage, L^3
G = original gas in place, std L^3
GE = gas equivalent, std L^3/std L^3
G_{pc} = cumulative gas production during a period of constant rate, std L^3
h = average reservoir thickness, L
k_g = measured gas permeability, L^2
k_l = effective liquid permeability, L^2
K = parameter in Lee *et al.*[2] viscosity correlation
m = real-gas potential, m/Lt^2
M = molecular weight
n = number of moles of gas or exponent in gas-deliverability equation
n_c = total number of components in gas mixture
n_w = number of wells
$\tilde{n}_g$ = relative number of total moles in gaseous phase, fraction
$\tilde{n}_o$ = relative number of total moles in oil phase, fraction
N_p = cumulative condensate production, std L^3
p = pressure, m/Lt^2
$\bar{p}$ = average pressure, m/Lt^2
$\tilde{p}$ = variable of integration in real-gas potential equation, m/Lt^2
PI = productivity index, std $L^3/t/m/Lt^2$
q = production rate, std L^3/t
q_c = production rate during period of constant rate, std L^3/t
q_R = total reservoir gas production rate, std L^3/t
r_1 = radial distance at which pressure p_1 is measured, L
r_2 = radial distance at which pressure p_2 is measured, L
R = universal gas constant, mL^2/nt^2T
S = mechanical skin, dimensionless
S' = total skin, dimensionless
S_{gi} = initial average gas saturation, fraction
S_{wi} = initial water saturation, fraction
t = time, t
t_c = time of constant-rate production, t
T = temperature, T
u = volumetric flux (q/A), $L^3/t/L^2$
V = volume, L^3
V_m = molar volume, L^3/n
W_e = cumulative water influx, L^3
W_p = cumulative water produced, std L^3
x_j = mole fraction of component j in liquid phase
X = parameter in Lee *et al.*[2] viscosity correlation
y_j = mole fraction of component j in gaseous phase
Y = produced condensate yield, std L^3/std L^3
z = gas deviation factor, dimensionless
z_j = mole fraction of component j in mixture
α = cubic equation-of-state parameter
α_c = empirical constant
β = defined in Eq. 10.41

ρ = density, m/L^3
ϕ = porosity, fraction
γ = specific gravity (air = 1.0 for gas)
μ = viscosity, cp
ψ = generic potential
$\Delta\psi$ = generic-potential difference

Subscripts

2ϕ = two-phase
a = abandonment
D = dimensionless
e = at the drainage radius
g = gas
i = initial
inv = investigation
l = liquid
m = molar
o = oil
p = produced
r = residual
sc = standard conditions
t = total
v = vertical
w = well or wellbore
wf = wellbore face
wg = wellstream gas

References

1. McCain, W.D. Jr.: *The Properties of Petroleum Fluids*, PennWell, Tulsa (1990).
2. Lee, A.L., Gonzalez, M.H., and Eakin, B.E.: "The Viscosity of Natural Gases," *JPT* (August 1966) 997; *Trans.,* AIME, **237.**
3. Carr, N.L., Kobayashi, R., and Burrows, D.B.: "Viscosity of Hydrocarbon Gases Under Pressure," *Trans.*, AIME (1954) **201,** 264.
4. Gold, D.K., McCain, W.D. Jr., and Jennings, J.W.: "An Improved Method for the Determination of the Reservoir-Gas Specific Gravity for Retrograde Gases," *JPT* (July 1989) 747; *Trans.,* AIME, **287.**
5. *Engineering Data Book*, Gas Processors Suppliers Assn., Tulsa (1987).
6. Katz, D.L. and Lee, R.L.: *Natural Gas Engineering—Production and Storage*, McGraw-Hill, New York City (1990).
7. Ahmed, T.: *Hydrocarbon Phase Behavior*, Gulf Publishing Co., Houston (1989).
8. Pedersen, K.S., Fredenslund, A., and Thomassen, P.: *Properties of Oils and Natural Gases*, Gulf Publishing Co., Houston (1989).
9. Whitson, C.H. and Brule, M.R.: *Phase Behavior,* Monograph Series, SPE, Richardson, Texas (2000) **20.**
10. Klinkenberg, L.J.: "The Permeability of Porous Media to Liquids and Gases," *Drill. & Prod. Prac.,* API (1941) 200.
11. Forchheimer, P.: "Wasserbewegung durch Boden," *Zeitz ver deutsch Ing.* (1901) **45,** 1731.
12. Al-Hussainy, R., Ramey, H.J. Jr., and Crawford, P.B.: "The Flow of Real Gases Through Porous Media," *JPT* (May 1966) 637.
13. Earlougher, R.C. Jr.: *Advances in Well Test Analysis,* Monograph Series, SPE, Dallas (1977) **5.**
14. Beggs, H.D.: *Gas Production Operations*, OGCI Publications, Tulsa (1984).

15. Rawlins, E.L. and Schellhardt, M.A.: *Backpressure Data on Natural Gas Wells and Their Application to Production Practices,* U.S. Bureau of Mines Monograph 7 (1936).
16. Matthews, C.S., Brons, F., and Hazebroek, P.: "A Method for Determination of Average Reservoir Pressure in a Bounded Reservoir," *Trans.,* AIME (1954) **201,** 182.
17. Poston, S.W. and Berg, R.R.: *Overpressured Gas Reservoirs,* SPE, Richardson, Texas (1997).
18. Payne, D.A.: "Material-Balance Calculations in Tight-Gas Reservoirs: The Pitfalls of *p/z* Plots and a More Accurate Technique," *SPERE* (November 1996) 260.
19. Hammerlindl, D.J.: "Predicting Gas Reserves in Abnormally Pressured Reservoirs," paper SPE 3479 presented at the 1971 SPE Annual Meeting, New Orleans, 3–6 October.
20. Roach, R.H.: "Analyzing Geopressured Reservoirs—A Material Balance Technique," paper SPE 9968 available from SPE, Richardson, Texas (1981).
21. Prasad, R.K. and Rogers, L.A.: "Superpressured Gas Reservoirs: Case Studies and a Generalized Tank Model," paper SPE 16861 presented at the 1987 SPE Annual Technical Conference and Exhibition, Dallas, 27–30 September.
22. Bernard, W.J.: "Gulf Coast Geopressured Gas Reservoirs: Drive Mechanism and Performance Prediction," paper SPE 14362 presented at the 1985 SPE Annual Technical Conference and Exhibition, Las Vegas, Nevada, 22–25 September.
23. Fetkovich, M.J., Reese, D.E., and Whitson, C.H.: "Application of a General Material Balance for High-Pressure Gas Reservoirs," *Gas Reservoir Engineering,* Reprint Series, SPE, Richardson, Texas (1999) **52,** 48.
24. Ambastha, A.K.: "Evaluation of Material Balance Analysis Methods for Volumetric, Abnormally-Pressured Gas Reservoirs," *J. Cdn. Pet. Tech.* (October 1993) **32,** 8, 19.
25. Yale, D.P. *et al.:* "Application of Variable Formation Compressibility for Improved Reservoir Analysis," paper SPE 26647 presented at the 1993 SPE Annual Technical Conference and Exhibition, Houston, 3–6 October.
26. El Sharkawy, A.M.: "Analytical and Numerical Solutions for Estimating the Gas In-Place for Abnormal Pressure Reservoirs," paper SPE 29934 presented at the 1995 SPE International Meeting on Petroleum Engineering, Beijing, 14–17 November.
27. Gan, R.G. and Blasingame, T.A.: "A Semianalytical *p/z* Technique for the Analysis of Reservoir Performance From Abnormally Pressured Gas Reservoirs," paper SPE 71514 presented at the 2001 SPE Annual Technical Conference and Exhibition, New Orleans, 30 September–3 October.
28. Brinkman, F.P.: "Increased Gas Recovery From a Moderate Water Drive Reservoir," *JPT* (December 1981) 2475.
29. Moltz, A.K.: "Modeling a Repressured Waterdrive Gas Reservoir," paper SPE 22936 presented at the 1991 SPE Annual Technical Conference and Exhibition, Dallas, 6–9 October.
30. Cole, F.W., *Reservoir Engineering Manual,* Gulf Publishing Co., Houston (1969) 285.
31. Pletcher, J.L.: "Improvements to Reservoir Material-Balance Methods," *SPEREE* (February 2002) 49.
32. *Petroleum Conservation,* S.E. Buckley *et al.* (eds.), AIME, New York City (1951).
33. Geffen, T.M. *et al.:* "Efficiency of Gas Displacement From Porous Media by Liquid Flooding," *Trans.,* AIME (1952) **195,** 29.
34. Yuster, S.T.: "The Rehabilitation of Drowned Gas Wells," *Drill. & Prod. Prac.,* API (1946) 209–16.
35. Bennett , E.N. and Auvenshine, W.L.: "Dewatering of Gas Wells," *Drill. & Prod. Prac.,* API (1956) 224–30.
36. Dunning, H.N. and Eakin, J.L.: "Foaming Agents Are Low-Cost Treatments for Tired Gassers," *Oil and Gas J.* (2 February 1959) **57,** No. 6, 108.
37. Bates, G.O., Kilmer, J.W., and Shirley, H.T.: "Eight Years of Experience with Inert Gas Equipment," paper S7-PET-34 presented at the 1957 ASME Petroleum Mechanical Engineering Conference, September.
38. Barstow, W.F.: "Fourteen Years of Progress in Catalytic Treating of Exhaust Gas," paper SPE 4574 presented at the 1973 SPE Annual Meeting, Las Vegas, Nevada, 30 September–3 October.
39. Eckles, W.W. and Holden, W.W.: "Unique Enhanced Oil and Gas Recovery Project for Very High Pressure Wilcox Sands Uses Cryogenic Nitrogen and Methane Mixture," paper SPE 9415 presented at the 1980 SPE Annual Technical Conference and Exhibition, Dallas, 21–24 September.

40. Moses, P.L. and Wilson, K.: "Phase Equilibrium Considerations in Utilizing Nitrogen for Improved Recovery From Retrograde Condensate Reservoirs," paper SPE 7493 presented at the 1978 SPE Annual Technical Conference and Exhibition, Houston, 1–3 October.
41. Peterson, A.V.: "Optimal Recovery Experiments with N_2 and CO_2," *Pet. Eng. Intl.* (November 1978) 40.
42. *Physical Properties of Nitrogen for Use in Petroleum Reservoirs, Bull.*, Air Products and Chemical Inc., Allentown, Pennsylvania (1977).
43. Miller, M.G. and Lents, M.R.: "Performance of Bodcaw Reservoir, Cotton Valley Field Cycling Project, New Methods of Predicting Gas-Condensate Reservoir Performance Under Cycling Operations Compared to Field Data," *Drill. & Prod. Prac.*, API (1946) 128–49.
44. Brinkley, T.W.: "Calculation of Rate and Ultimate Recovery from Gas Condensate Reservoirs," paper lO28-G presented at the 1958 SPE Petroleum Conference on Production and Reservoir Engineering, Tulsa, 20–21 March.
45. *Proc.*, Ninth Oil Recovery Conference Symposium on Natural Gas in Texas, College Station, Texas (1956).

SI Metric Conversion Factors

°API		141.5/(131.5 + °API)		= g/cm³
bbl	×	1.589 873	E − 01	= m³
Btu	×	1.055 056	E + 00	= kJ
cp	×	1.0*	E − 03	= Pa·s
ft	×	3.048*	E − 01	= m
ft³	×	2.831 685	E − 02	= m³
°F		(°F + 459.67)/1.8		= K
in.³	×	1.638 706	E + 01	= cm³
lbm	×	4.535 924	E − 01	= kg
psi	×	6.894 757	E + 00	= kPa
psi²	×	4.753 8	E + 01	= kPa²

*Conversion factor is exact.

Chapter 11
Waterflooding

H.R. (Hal) Warner Jr., SPE, Warner Consulting Services

11.1 Introduction

This chapter concerns the use of water injection to increase the production from oil reservoirs, and the technologies that have been developed over the past 50+ years to evaluate, design, operate, and monitor such projects. Use of water to increase oil production is known as "secondary recovery" and typically follows "primary production," which uses the reservoir's natural energy (fluid and rock expansion, solution-gas drive, gravity drainage, and aquifer influx) to produce oil.

The principal reason for waterflooding an oil reservoir is to increase the oil-production rate and, ultimately, the oil recovery. This is accomplished by "voidage replacement"—injection of water to increase the reservoir pressure to its initial level and maintain it near that pressure. The water displaces oil from the pore spaces, but the efficiency of such displacement depends on many factors (e.g., oil viscosity and rock characteristics). In oil fields such as Wilmington (California, U.S.A.) and Ekofisk (North Sea), voidage replacement also has been used to mitigate additional surface subsidence. In these cases, the high porosity of the unconsolidated sandstones of the Wilmington oil field's reservoirs and of the soft chalk reservoir rock in the Ekofisk oil field had compacted significantly when the reservoir pressure was drawn down during primary production.

This chapter discusses various aspects of waterflooding briefly, and it mainly covers the fundamental considerations involved when engineering, designing, operating, and monitoring a waterflood. Waterflooding technology is a large and well-covered topic in the literature, and the ability of this chapter to cover all aspects of these technologies is limited. Over the past 40 years, SPE has published three significant and in-depth books written by Craig,[1] Willhite,[2] and Rose *et al.*[3] that address waterflooding technology.

SPE Reprint Series Vol. 2a[4] contains the most significant SPE technical papers that describe waterflooding technology as of 1973. A recent search of the SPE eLibrary using keyword "waterflood" identified more than 5,500 SPE technical papers, including 650+ whose titles contain "waterflood."

11.1.1 Historic Context. In the early days of the oil industry, saline water or brine frequently was produced from a well along with oil, and as the oil-production rate declined, the water-production rate often would increase. This water typically was disposed of by dumping it into nearby streams or rivers. In the 1920s, the practice began of reinjecting the produced water into porous and permeable subsurface formations, including the reservoir interval from which the oil and water originally had come. By the 1930s, reinjection of produced water had become a common oilfield practice.

Reinjection of water was first done systematically in the Bradford oil field of Pennsylvania, U.S.A.[5] There, the initial "circle-flood" approach was replaced by a "line flood," in which two rows of producing wells were staggered on both sides of an equally spaced row of water-injection wells. In the 1920s, besides the line flood, a "five-spot" well layout was used (so named because its pattern is like that of the five spots on a die).

Much of waterflooding's technology and common practice developed in the U.S. between 1940 and 1970. By the mid-1940s, the onshore U.S. oil industry was maturing and primary production from many of its reservoirs had declined significantly, whereas most reservoirs elsewhere in the world were in the early stages of primary production. Also, in the U.S., thousands of wells had been drilled that were closely spaced, so that the effects of water injection were more obvious and so were more quickly understandable.

In addition to the need to dispose of saline water that was produced along with the oil, several other factors made waterflooding a logical and economical method for increasing recovery from oil fields. Very early on, it was recognized that in most reservoirs, only a small percentage of the original oil in place (OOIP) was being recovered during the primary-production period because of depletion of the reservoirs' natural energy. Additional recovery methods were needed to produce the large quantity of oil that remained. Water injection's early success in lengthening the oil-production period by years made waterflooding the natural step after primary production to recover additional oil from reservoirs whose oil-production rate had declined to very low levels.

Other key factors that drove waterflooding's development and increasing use were that water is inexpensive; that water generally is readily available in large quantities from nearby streams, rivers, or oceans, or from wells drilled into shallower or deeper subsurface aquifers; and that water injection effectively made production wells that were near the water-injection wells flow or be pumped at higher rates because of the increased reservoir pressure. Concurrently, the scientific reasons behind waterflooding's success were identified (i.e., that water has viscosity, density, and wetting properties, compared to oil, that affect how efficiently it will displace various oils from reservoir rock).

By the 1970s, most onshore oil fields in the U.S., USSR, and China, for which waterflooding was the logical recovery process, were being produced by use of this technology in various well-pattern arrangements. Some U.S. offshore oil fields and oil fields elsewhere in the world were receiving water injection as deemed appropriate by their owners and operators. Since then, many large-scale water-injection projects have been applied to oil reservoirs in locations ranging from far offshore in the North Sea to the Arctic regions to desert areas.

11.1.2 Chapter Topics. This chapter presents the subject of waterflooding in nine sections and follows the logic sequence used by Willhite[2]:

- Microscopic efficiency of immiscible displacement.
- Macroscopic displacement efficiency of a linear waterflood.
- Reservoir-geology considerations in the design and operation of waterfloods.
- Immiscible displacement in two dimensions—areal.
- Vertical displacement in linear and areal models.
- Waterflood design.
- Waterflood monitoring.
- Field case studies: waterflood examples.
- Summary and conclusions.

The sections begin with a discussion of unit-displacement efficiency, which is how water displaces oil from a porous and permeable reservoir rock on a microscopic scale. This is the level of analysis that is applied when water-/oil-flow measurements are made on small core-plug samples in a laboratory. Calculations for determining how well waterflooding will work on a reservoir scale must include the effects of geology, gravity, and geometry (vertical, areal, and well-spacing/-pattern arrangement). The formula for overall waterflood oil-recovery efficiency E_R might be simply stated as the product of three independent terms:

$$E_R = E_D E_I E_A, \quad (11.1)$$

where E_D = the unit-displacement efficiency, E_I = the vertical-displacement efficiency, and E_A = the areal-displacement efficiency. Of course, assuming independence of these three factors is not valid for real oil reservoirs.

Also, a waterflood is a dynamic process that lasts for several decades; hence, after the project has been initiated, there are opportunities to modify the original waterflood design and operating guidelines on the basis of analysis of the actual field production data. This is why real-time monitoring of waterflood performance is required, both at the injection wells and at the production wells.

Throughout this chapter, keep in mind that the most important aspect of evaluating a field waterflooding project is understanding the reservoir rocks. This understanding begins with knowing the depositional environment at the pore and reservoir levels and possibly also several levels in between. Second, the diagenetic history of the reservoir rocks must be ascertained. Then, the structure and faulting of the reservoir must be determined to understand the interconnectivities among the various parts of the reservoir, particularly the injector/producer connectivity. Finally, the water/oil/rock characteristics need to be understood because they control wettability, residual oil saturation to waterflooding, and the oil relative permeability at higher water saturations. Because of these needs, there always should be a developmental geologist on the waterflood-evaluation team.

This chapter does not include a separate section on oil properties and their impact on waterflooding recovery efficiency, but oil properties are important to technical and economic success and are discussed as appropriate throughout this chapter. The key oil properties are viscosity and density at reservoir conditions. In a porous medium, the mobility of a fluid is defined as its endpoint relative permeability divided by its viscosity; hence, a fluid with a low viscosity ($\leq$ 1 cp) has a high mobility unless its relative permeability is very low. Similarly, a low-API crude oil ($\leq$ 20°API) has a high viscosity and a very low mobility unless it is heated to high temperatures. Because water's viscosity at reservoir temperatures generally is much lower than or, at best, equal to that of the reservoir oil, the water-/oil-viscosity ratio is generally much greater than 1:1. As discussed in some detail later in this chapter, the water-/oil-mobility ratio is a key parameter in determining the efficiency of the water/oil displacement process, with the recovery efficiency increasing as the water-/oil-mobility ratio decreases.

11.1.3 Topics Covered Elsewhere in This *Handbook*. Topics that concern typical water-injection operations but that are not discussed in this chapter are (1) aquifer or bottomwater drives with water injected into underlying aquifer intervals; (2) use of numerical reservoir simulators to analyze waterflood performance; and (3) enhanced-oil-recovery (EOR) methods that involve either continuous or alternating water injection. Discussions of these and other aspects of waterflooding technologies can be found elsewhere in this *Handbook,* including in the numerical-reservoir-simulation chapter in the Reservoir Engineering and Petrophysics volume, in the General Engineering volume chapters about crude-oil properties and water properties, in the Production Operations Engineering volume chapters about handling of oil and water production

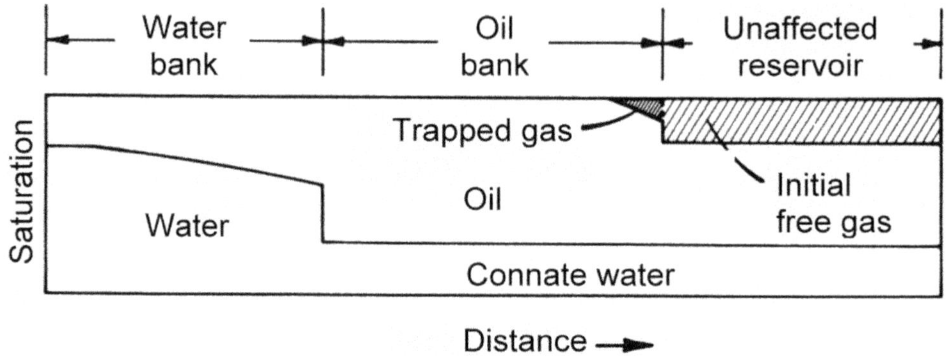

Fig. 11.1—Saturation profile during a waterflood.[1]

during waterflooding operations, and in the Facilities and Construction Engineering volume chapters about surface facilities that are required for waterflood operations.

11.1.4 Limitations of Waterflood Technology. Waterflooding can increase the volume of oil recovered from a reservoir; however, it is not always the best technology to use and it can have complicating factors. When evaluating how best to produce a particular oil reservoir, a petroleum engineer should include waterflooding in the options that are analyzed, both technically and economically. Those evaluations should include such potentially complicating factors as compatibility of the planned injected water with the reservoir's connate water; interaction of the injected water with the reservoir rock (clay sensitivities, rock dissolution, or generally weakening the rock framework); injection-water treatment to remove oxygen, bacteria, and undesirable chemicals; and the challenges involved in separating and handling the produced water that has trace oil content, naturally occurring radioactive materials (NORMs), and various scale-forming minerals.

11.2 Microscopic Efficiency of Immiscible Displacement

This section discusses the conceptual aspects of the displacement of oil by water in reservoir rocks. **Fig. 11.1** is a schematic diagram of the water/oil displacement process. At the pore level (i.e., where the water and oil phases interact immiscibly when moving from one set of pores to the next), wettability and pore geometry are the two key considerations. The interplay between wettability and pore geometry in a reservoir rock is what is represented by the laboratory-determined capillary pressure curves and water/oil relative permeability curves that engineers use when making OOIP and fluid-flow calculations. The sections below discuss these basic concepts and their implications for initial water- and oil-saturation distribution, for relative permeability, and for how initial gas saturation will affect water/oil flow behavior.

11.2.1 Wettability. Wettability is defined in terms of the interaction of two immiscible phases, such as oil and water, and a solid surface, such as that of the pores of a reservoir rock. For understanding wettability concepts and for simple laboratory determinations, the solid surface is taken as a smooth flat surface. **Fig. 11.2** illustrates two styles of wettability: water-wet and oil-wet. Eq. 11.2 describes the force relationship that is in balance for the drop of water that is on the solid surface and is surrounded by oil. The interfacial tension (IFT) between the oil and water phases varies depending on the compositions of the phases but generally is relatively high, in the 10- to 30-dyne/cm range. The contact angle θ is used to define which fluid phase is more wetting—for low contact angles, the water phase is more wetting, whereas for high contact angles, the oil phase is more wetting.

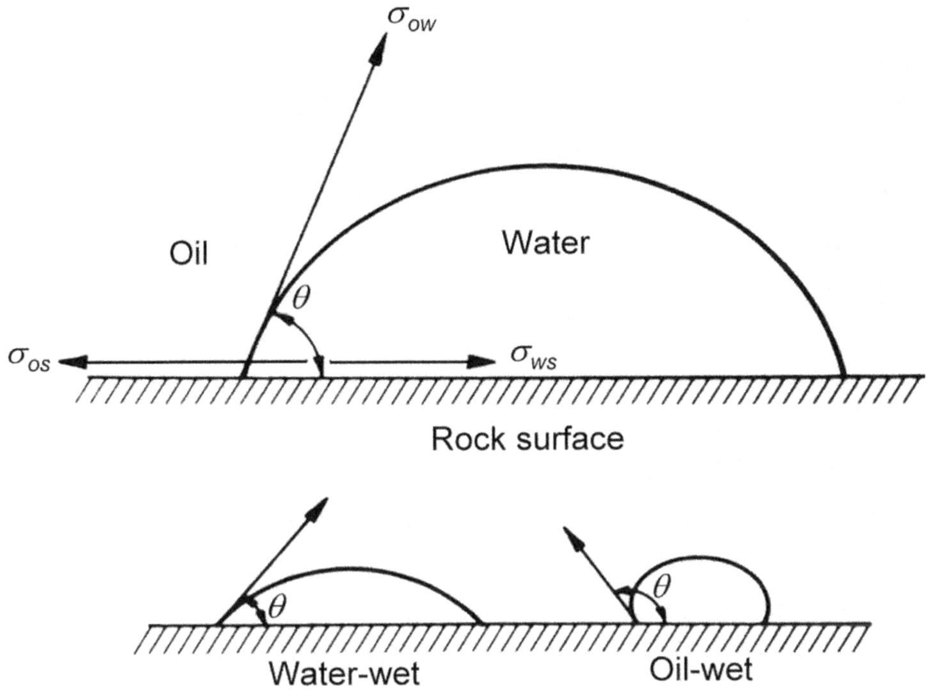

Fig. 11.2—Wettability of oil/water/solid system.[2]

$$\sigma_{os} - \sigma_{ws} = \sigma_{ow} \cos \theta, \quad (11.2)$$

where σ_{os} = the IFT between the oil and solid phases, σ_{ws} = the IFT between the water and solid phases, and σ_{ow} = the IFT between the oil and water phases.

The particular contact angle depends on many variables, including the composition of the crude oil and the amount of gas in solution; the salinity and pH of the connate brine; the mineralogy of the rock surfaces; and the salinity and pH of the injected water that is used for waterflooding. The concentration of surface-active components (e.g., asphaltenes) that are in the crude oil and that can adsorb on the rock surfaces affects wettability.

Reservoir rocks typically are described as being water-wet, oil-wet, or intermediate-wet. A water-wet rock surface is one that has a strong preference to be coated, or "wetted," by the water phase, so that there will be a continuous water phase on the rock surfaces. Oil-wet rocks prefer to be coated with oil instead of water. Strongly oil-wet rocks have been created for laboratory studies but, as discussed below, are unlikely to exist in real reservoirs. Intermediate-wet reservoir rocks have been found in several oil reservoirs. The term "dalmatian wetting" describes reservoir rocks that have both oil-wet and water-wet surfaces. **Fig. 11.3** illustrates two styles of intermediate-wetting.

Two types of laboratory measurements commonly are used to estimate wettability. First, the crude-oil/brine IFT values can be measured on smooth rock surfaces of various mineralogies. Second, Amott tests can be run on the reservoir rock to determine the extents to which it imbibes oil and brine. When running the Amott tests, it is critical to initialize the core plugs as close to original reservoir conditions as possible either by using well-preserved core samples or by aging the core plugs in the presence of reservoir crude oil. High-quality water/oil capillary-pressure (P_c) and water/oil relative permeability (k_{rwo}) data, both of which are strongly affected

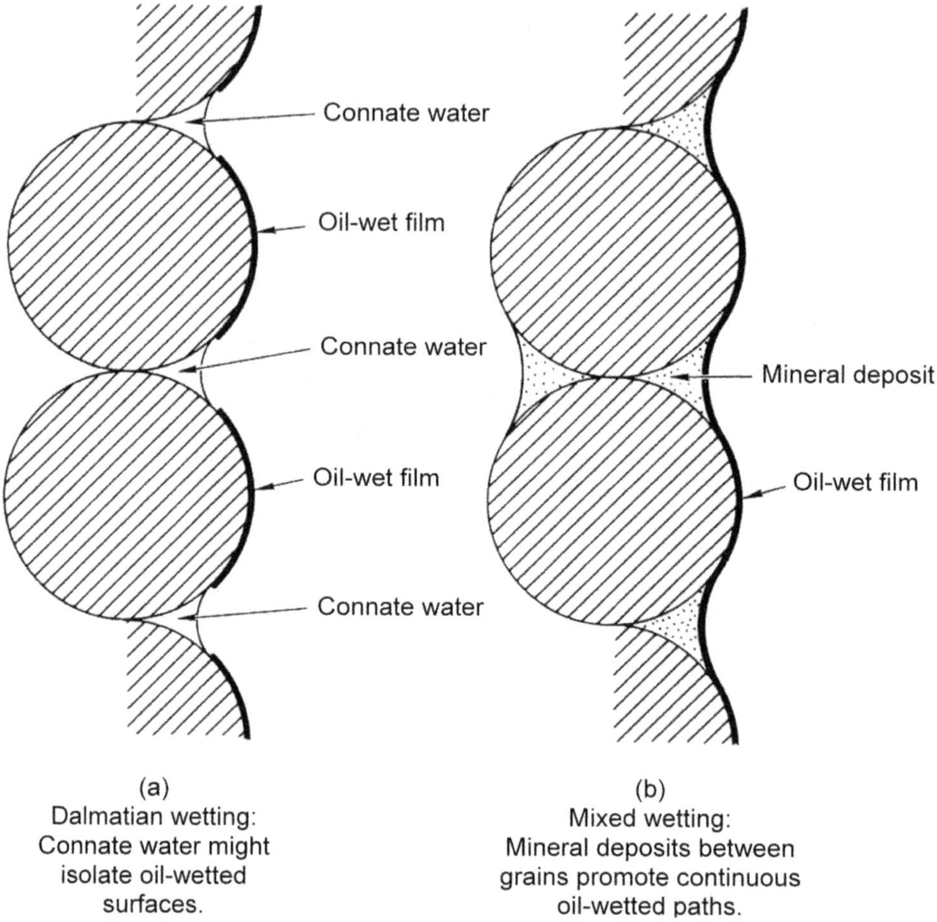

Fig. 11.3—Relationship of mineralogy to wetting conditions: (a) dalmatian wetting and (b) mixed wetting.[2]

by rock wettability, are needed as input to waterflood calculations, whether using simple engineering methods or complex numerical reservoir simulators.

11.2.2 Pore Geometry. The pore geometry for any reservoir rock is the result of its depositional and diagenetic history. The depositional environment determines a rock's grain size and sorting. Post-depositional diagenetic changes caused by various types of cementation, leaching, and clay alteration will impact a rock's pore characteristics whether the rock is primarily silica or carbonate. A chapter in the General Engineering volume of this *Handbook* discusses many of these factors and how they affect the single-phase permeability of a reservoir rock.

Figs. 11.4 and 11.5 show photomicrographs and k_{rwo} curves for a sandstone with large, well-connected pores and for one with small, well-connected pores, respectively. These illustrate just one of many possible differences in pore geometry. Pore distributions in carbonate rocks often are more complicated because of vug networks and fractures. Also, there are many scales of pore-geometry heterogeneities; a core plug has one scale of pore-size variation, but other important variations are found at each higher scale.

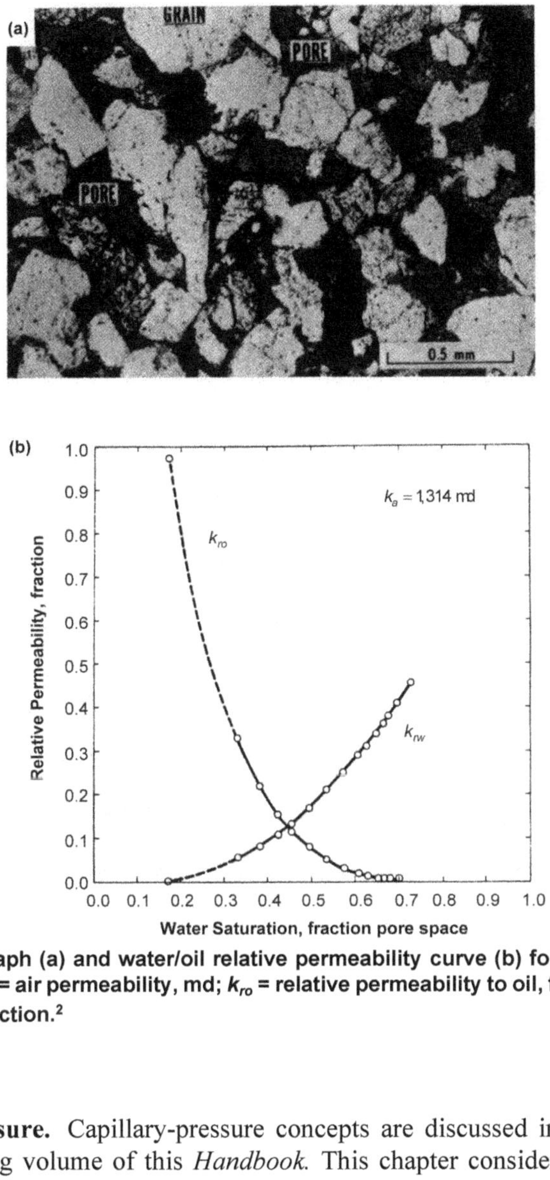

Fig. 11.4—Photomicrograph (a) and water/oil relative permeability curve (b) for a sandstone with large, well-connected pores. k_a = air permeability, md; k_{ro} = relative permeability to oil, fraction; and k_{rw} = relative permeability to water, fraction.[2]

11.2.3 Capillary Pressure. Capillary-pressure concepts are discussed in detail in a chapter of the General Engineering volume of this *Handbook.* This chapter considers the characteristics of and differences between the drainage and imbibition capillary-pressure/water-saturation (P_c/S_w) curves. Capillary pressure affects waterflood performance and engineering calculations because the extent to which the water/oil flood front is vertically and horizontally "smeared out" during the waterflood is controlled by the P_c/S_w imbibition curve.

Reservoir rocks are considered to be water-wet initially because all reservoir rocks were deposited in water-filled environments or were immersed in water soon after deposition, when their overlying sediments were deposited. The drainage P_c/S_w curve describes the drainage process, or the P_c/S_w relationship while the nonwetting-fluid phase (oil) displaces the wetting-fluid phase (brine) from various parts of the pore system, thus decreasing the wetting-phase saturation. If during the displacement process the process is reversed and the wetting-phase saturation increases, it is known as imbibition (the imbibing of the wetting phase).

Fig. 11.6 shows the drainage and imbibition P_c/S_w characteristics of a strongly water-wet rock. The minimum wetting or water saturation from the drainage process is termed the con-

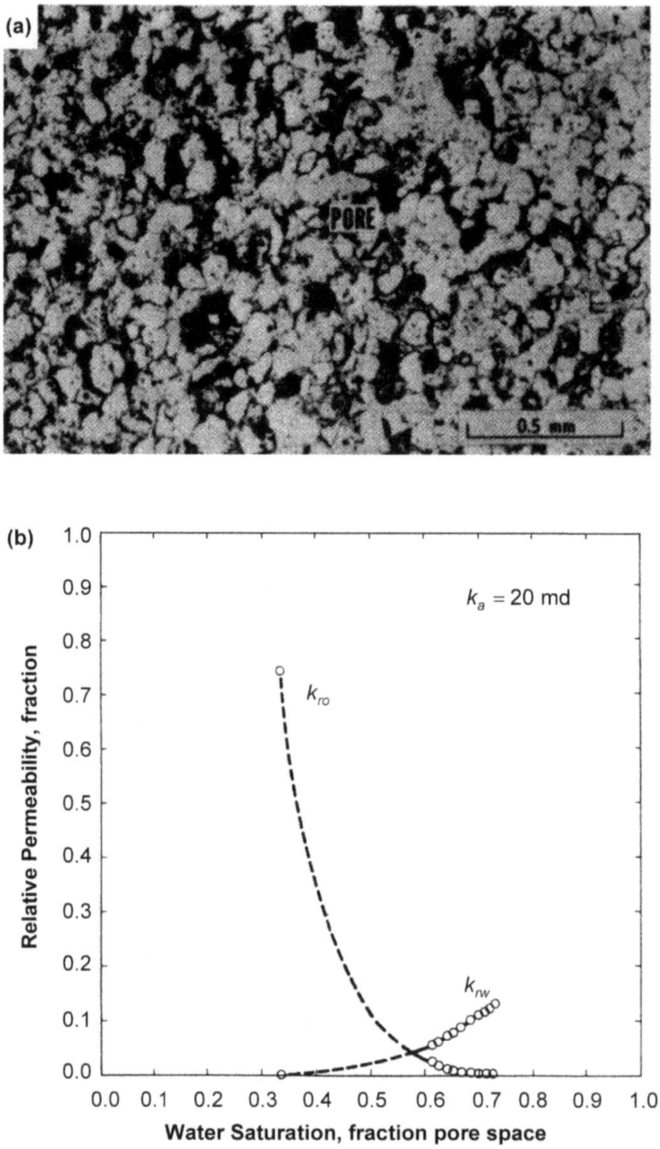

Fig. 11.5—Photomicrograph (a) and water/oil relative permeability curve (b) for a sandstone with small, well-connected pores.[2]

nate (or irreducible) water saturation. The maximum water saturation from the imbibition process defines the minimum nonwetting-phase saturation, or (for waterflooding considerations) the residual oil saturation to waterflooding S_{orw}. **Figs. 11.7 and 11.8** show the drainage and imbibition P_c/S_w curves from the laboratory tests of an oil-wet rock and a rock with intermediate wettability, respectively. Fig. 11.8 includes both the spontaneous (number 2 on curve) and the forced (number 3 on curve) portions of the imbibition curve. Spontaneous imbibition occurs without any pressure being applied to the test apparatus, whereas obtaining the forced imbibition portion of the curve requires an external pressure to be applied. Note that $P_c = 0$ does not define the S_{orw}.

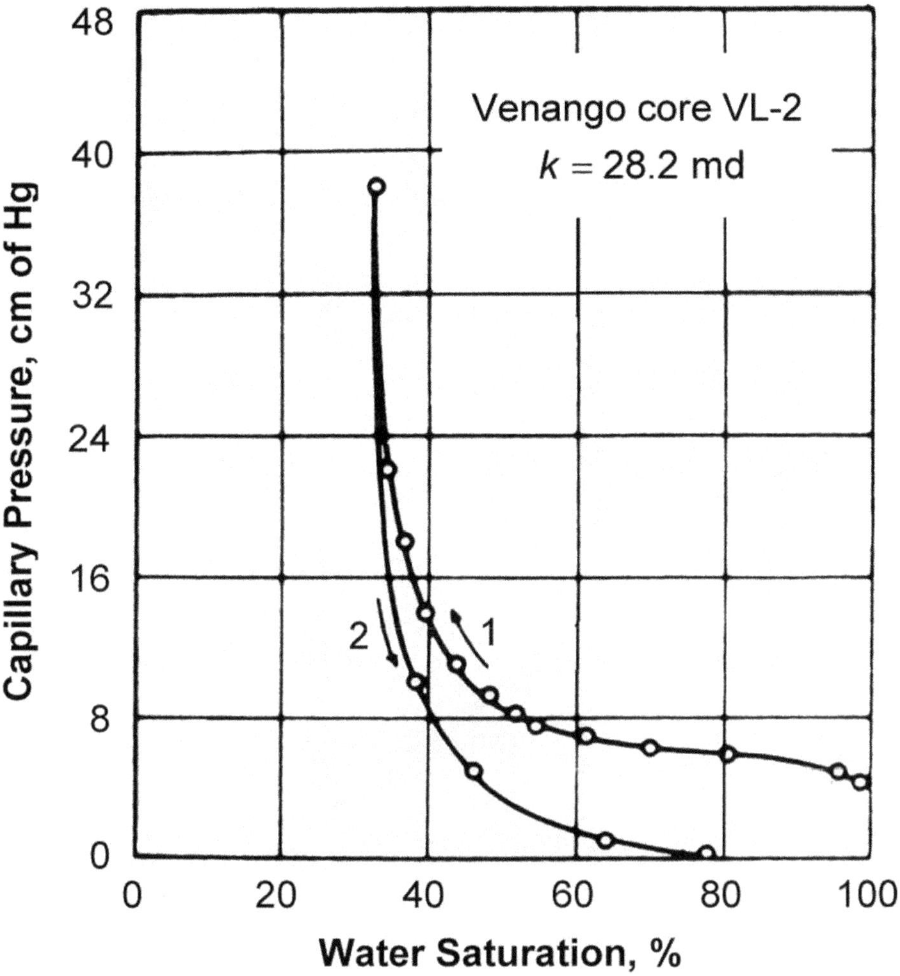

Fig. 11.6—Capillary pressure characteristics for a strongly water-wet rock. Curve 1 represents drainage, and Curve 2 represents imbibition.[2]

11.2.4 Initial Water-/Oil-Saturation Distribution. An oil field's initial water-/oil-saturation distribution depends on its hydrocarbon history and has a significant effect on its waterflooding potential. The pore system in a reservoir rock contains a very large number of pore bodies whose filling by oil is controlled by the diameters of the pore throats that link them.

During the oil-filling process, the oil first enters through the largest pore throats, and all other parts of the pore system remain filled with connate brine. As more oil enters the reservoir trap, the oil column lengthens downward. Just above the oil/water contact, only the pores that are accessible from the largest pore throats fill with oil. At the top of the oil column, where the capillary pressure is greatest, not only the largest pores are oil-filled, but also some that have smaller pore throats. The very fine pore spaces remain filled with connate brine.

This process continues until the oil column reaches its maximum length. This whole process is the drainage cycle of the P_c/S_w curves. At this point in the process, oil is filling the largest pores and water is filling the smallest pores; however, the P_c/S_w drainage curve governs the percentage of each. Connate brine will remain as films on the surfaces of the largest

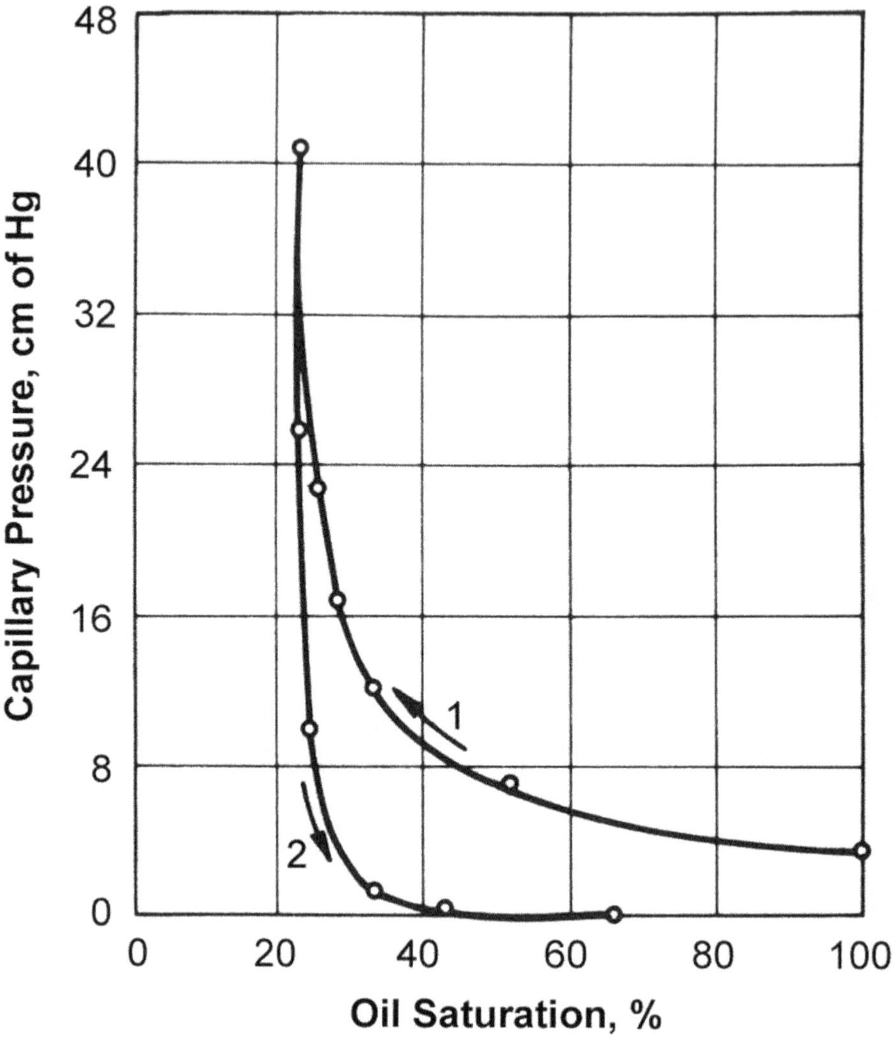

Fig. 11.7—Water/oil capillary pressure characteristics for Tensleep Sandstone oil-wet rock. Curve 1 represents drainage, and Curve 2 represents imbibition.[2]

pores, but surface-active components of the crude oil might adsorb on some of the pore surfaces, rendering them oil-wet. Hence, the overall system can have mixed-wet characteristics.

There are oil fields that, although initially filled through a drainage process, when discovered were on the imbibition cycle because of a complicated hydrocarbon or structural history. Portions of several west Texas San Andres carbonate reservoirs and the Prudhoe Bay field of Alaska are examples of such oil fields.

This original water-/oil-saturation distribution is important to understand for waterflooding because it controls the efficiency of the waterflood in portions of the reservoir. It also relates directly to the residual oil saturation that can be achieved at the end of a waterflood.

11.2.5 Relative Permeability. Relative permeability (k_r) concepts are discussed in detail in a chapter of the General Engineering volume of this *Handbook.* For the purposes of this chapter, their important aspect is the characteristics of imbibition oil/water k_r curves because these gov-

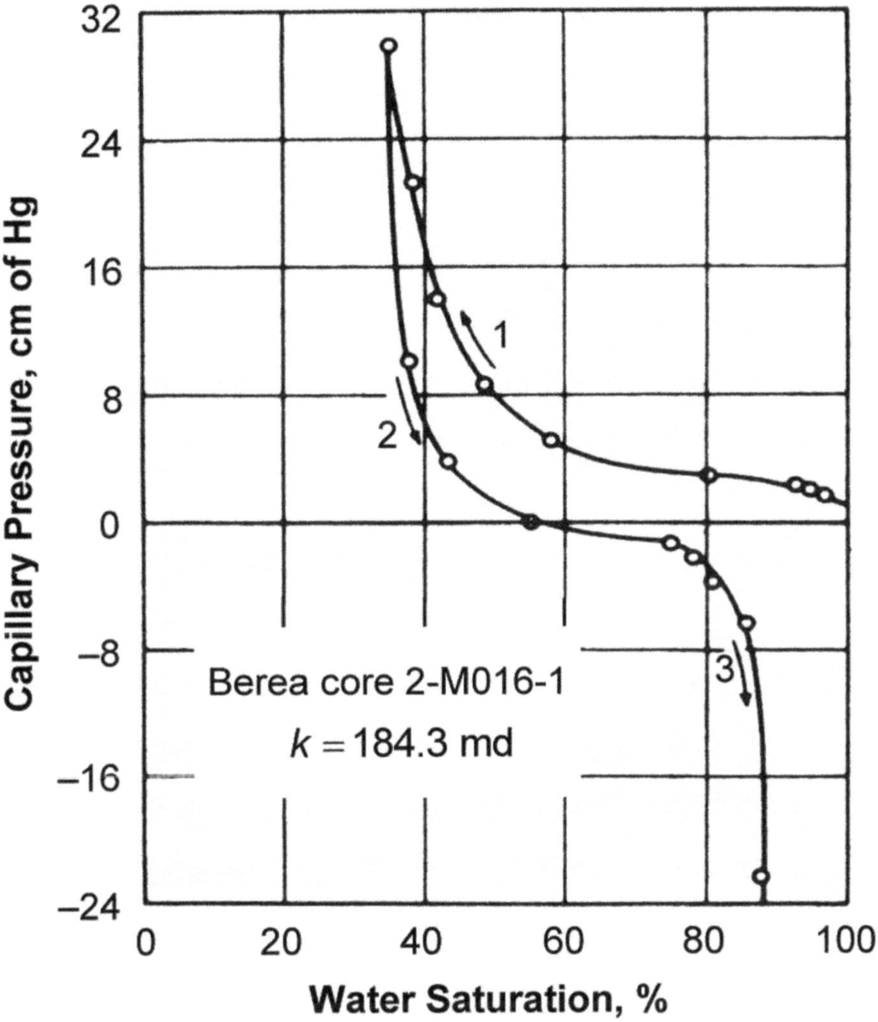

Fig. 11.8—Water/oil capillary-pressure characteristics for intermediate wettability. Curve 1 represents drainage, Curve 2 represents spontaneous imbibition, and Curve 3 represents forced imbibition.[2]

ern the nature and efficiency of the waterflood displacement and how much of the OOIP will be recovered before the waterflood economic limit is reached.

The shapes of the imbibition water/oil k_r curves depend on pore geometry and wettability. As noted earlier, Figs. 11.4 and 11.5 show the differences between these curves for a sandstone with large, well-connected pores and one with small, well-connected pores. The k_{rw} is greatly reduced for the sandstone with small pores at all saturation levels. **Fig. 11.9** shows the effect of wettability, as measured by the U.S. Bureau of Mines (USBM) Amott wettability index, on the water/oil k_r curves. As is expected for a change from water-wet to oil-wet in such laboratory tests, the water k_r curve rises with increasing oil-wetness and the oil k_r curve decreases.

Most importantly, laboratory-determined water/oil k_r data should be obtained at the best approximation of reservoir conditions. Salathiel describes the importance of this to actual field oil/water displacement.[6] **Fig. 11.10** shows the results of Salathiel's laboratory experiments that relate to the East Texas oil field. These curves show that the oil relative permeability for water-wet conditions is significantly different than for mixed-wet conditions. In water-wet conditions,

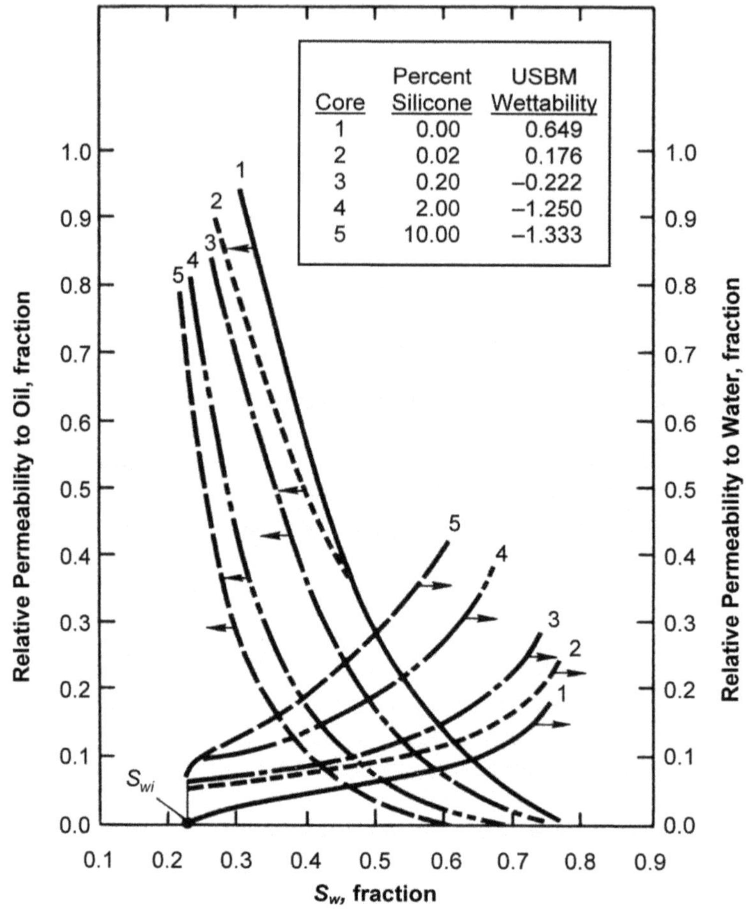

Fig. 11.9—Oil and water relative permeabilities for Squirrel-sandstone cores for water-wet and oil-wet conditions.[2]

the oil phase becomes discontinuous and loses its mobility quickly. In mixed-wet conditions, the oil maintains phase continuity by means of the oil-wetted rock surfaces and slowly drains to significantly lower oil saturations. The comparison of the laboratory results to the actual field production data and the residual-oil-saturation-pressure core data showed that the reservoir had mixed-wettability, yielding S_{orw} values of < 10% PV in many portions of the reservoir.

11.2.6 Residual Oil Saturation. For waterflooding, the two most important numbers for a reservoir rock are the connate-water saturation S_{wc} and the S_{orw}. The S_{wc} determines how much oil initially is in each unit volume of rock when the reservoir is discovered. The S_{orw} is how much of the OOIP will remain in rock that will be well swept by injected-water volumes. Assuming that the oil-formation-volume factor is the same at the beginning and the end of the waterflood, the equation for the unit-displacement efficiency is:

$$E_D = 1 - \frac{S_{orw}}{S_{oi}}, \quad \text{(11.3)}$$

where S_{oi} = initial oil saturation $(1 - S_{wc})$.

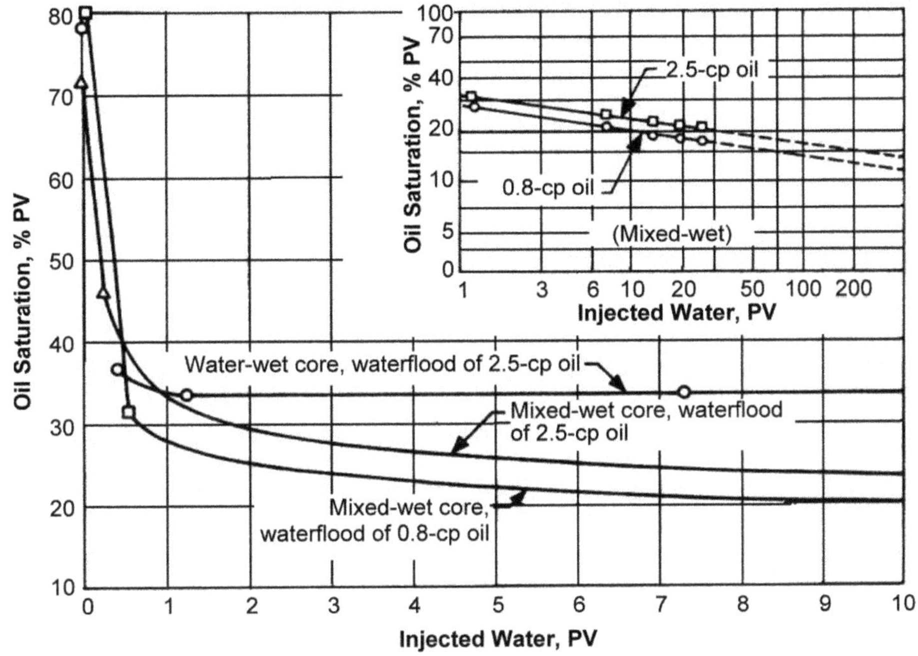

Fig. 11.10—Comparison of waterflood behavior for mixed-wet and water-wet cores. Insert shows extension of mixed-wet-core flooding data.[2] PV = pore volume.

TABLE 11.1—RESIDUAL OIL SATURATION AFTER 25 PV OF WATERFLOODING[6]

Rock Sample	Permeability, md	Porosity, % BV	S_w at Time of "Contact," % PV	S_o After 25 PV of Waterflooding: Water-Wet	Mixed-Wet
Boise (sandstone)	1,094	29.3	13.5	33.5	20.5
Upper Austin (sandstone)	596	28.0	20.0	30.0	22.9
Woodbine Outcrop (sandstone)	690	33.0	17.0	27.3	30.7
Upper Noodle (limestone)	620	21.2	18.9	40.5	28.1
Lissie (sandstone)	536	21.9	7.2	42.5	29.1

The S_{orw} is the endpoint of the water/oil imbibition k_{ro} curve, which was discussed above; however, for simple waterflood calculations this value is the most critical one. **Table 11.1** compares Salathiel's S_{orw} results for the water-wet conditions to those for the mixed-wet conditions.[6] The S_{orw} for the mixed-wet samples generally was 10% PV lower than for the water-wet samples. In the water-wet conditions, more of the oil phase gets "snapped off" and therefore trapped and immobilized as isolated oil globules by the increasing water saturation. Jerauld and Rathmell[7] found similar results for the Prudhoe Bay field.

S_{orw} can be measured several ways. It can be determined as part of all relative permeability laboratory studies. Historically, short core-plug "floodpot" tests have been run in the laboratory, and only the rock sample's porosity, absolute air permeability, S_{wc}, S_{orw}, and permeability at the two endpoint saturations have been reported. It is important to ensure that these laboratory tests are conducted long enough for the displacement to be taken to its true endpoint. They can be performed either as displacement tests or by using a centrifuge to measure these data. Displacement tests historically have been used, but because of improvements in centrifuge technology, the centrifuge approach is becoming more common. Usually, floodpot-test times are inadequate to reach a true S_{orw}. Imbibition capillary pressure measurements obtain more-reliable values for water-wet porous media.

Generally, S_{orw} is inversely related to S_{wi}. This can be understood in terms of the pore spaces that become filled with water and oil. While the S_{wi} decreases (or the S_{oi} increases), the oil phase occupies more of the pores and fills more of the smaller pore spaces. When water displaces the oil, the advancing waterfront traps more of the oil, especially if the rock is water-wet.

The performance of a waterflood depends on the impact of viscous and capillary forces on S_{orw} and k_r. At reservoir flow rates, the viscous forces do not vary enough to make a significant difference in k_r and S_{orw}; however, under laboratory conditions, viscous and capillary forces are major considerations because short core-plug displacement tests actually measure pressure drops and fluid-production volumes as a function of time that include large capillary end effects. These data must be entered into interpretative calculations to derive the water/oil P_c/S_w and k_r curves that are used later in field waterflood calculations. The laboratory personnel must choose what length of core plugs to test, what flow rates and pressure drops to apply, whether to make the measurements at steady-state or unsteady-state conditions, and how to interpret these data.

11.2.7 Initial Gas Saturation S_{gi}. In many oil reservoirs, a free-gas saturation formed during the early production period because the waterflood was not initiated before the reservoir pressure had dropped through the oil bubblepoint pressure. For many years, the effect of this gas saturation on S_{orw} has been a subject of considerable technical interest. **Fig. 11.11** summarizes the experimental results of several investigators and shows the impact of S_{gt} on S_{orw} for water-wet rocks. The S_{orw} decreased as S_{gt} increased. Because gas is the most nonwetting of the fluid phases, the residual gas phase occupies the center of the pore bodies and hence can reduce the volume of oil that is trapped.

11.2.8 Other Considerations. Historically, most laboratory tests have been run at surface temperature and pressure conditions using dead crude oils and constant brine salinity when measuring water/oil P_c/S_w and k_r data. Over the past decade, U.S. researchers at the U. of Wyoming and the U. of Texas have published papers concerning studies of the effect of temperature, salinity, and oil composition on wettability and waterflood oil recovery.[8,9] Those studies show that oil recovery increases with higher temperature, and generally also with variation in salinity.

11.2.9 Mobility Ratio. The mobility of a phase (Eq. 11.4) is defined as its relative permeability divided by its viscosity. Hence, mobility combines a rock property (relative permeability) with a fluid property (fluid viscosity). The water/oil relative permeability is assumed to depend only on the saturations of the two fluid phases.

$$\lambda_i = \left(\frac{k_i}{\mu_i}\right), \qquad (11.4)$$

where λ_i = mobility of fluid phase i, k_i = relative permeability of fluid phase i, and μ_i = viscosity of fluid phase i. Mobility relates to the amount of resistance to flow through a reservoir rock that a fluid has at a given saturation of that fluid. Because viscosity is in the denominator

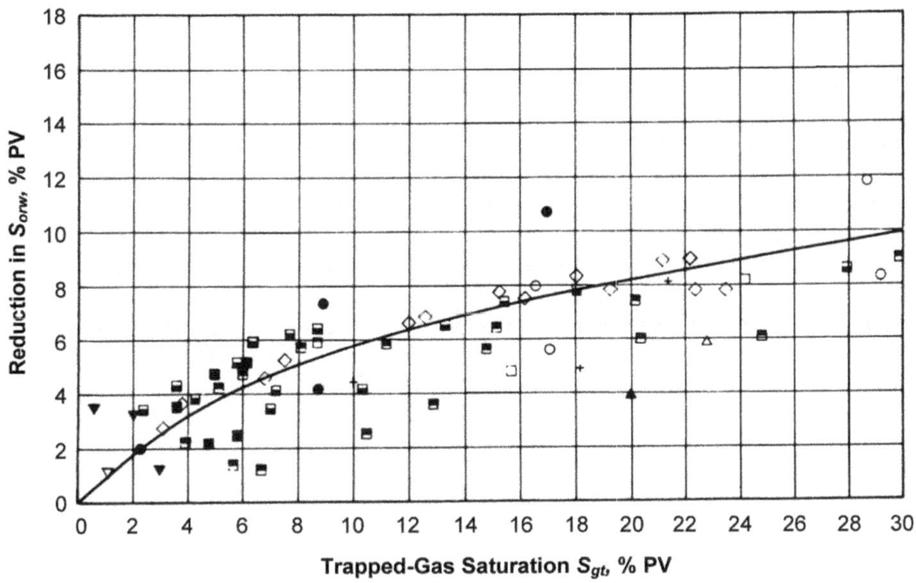

Fig. 11.11—Effect of trapped-gas saturation on waterflood oil recovery for preferentially water-wet rocks.[2]

of this equation, low-viscosity fluids generally have high mobility and high-viscosity fluids generally have low mobility.

The mobility ratio M generally is defined as the mobility of the displacing phase (for waterflooding, water) divided by the mobility of the displaced phase (oil). Eqs. 11.5 and 11.6 present two forms of the mobility-ratio equation:

$$M = \frac{k_{rw}}{\mu_w} \frac{\mu_o}{k_{ro}}, \quad \text{(11.5)}$$

where μ_w = viscosity of water, cp; μ_o = viscosity of oil, cp; k_{rw} = relative permeability to water; and k_{ro} = relative permeability to oil.

The mobility ratio also can be expressed as the product of the two fluids' relative permeability and viscosity ratios.

$$M = \frac{k_{rw}}{k_{ro}} \frac{\mu_o}{\mu_w}. \quad \text{(11.6)}$$

Mobility ratios are considered to be either "favorable" or "unfavorable." A favorable mobility ratio is a low value (≤ 1); this means that the displaced phase (oil) has a higher mobility than does the displacing phase (water). An unfavorable mobility ratio (> 1) is the other way around. In practical terms, a favorable mobility ratio means that the displaced oil phase can move more quickly through the reservoir rock than can the displacing water phase.

For simple waterflooding calculations, the mobility ratio is calculated at the endpoint relative permeability values for the two phases. Hence, the equation to be used for the waterflood mobility ratio is:

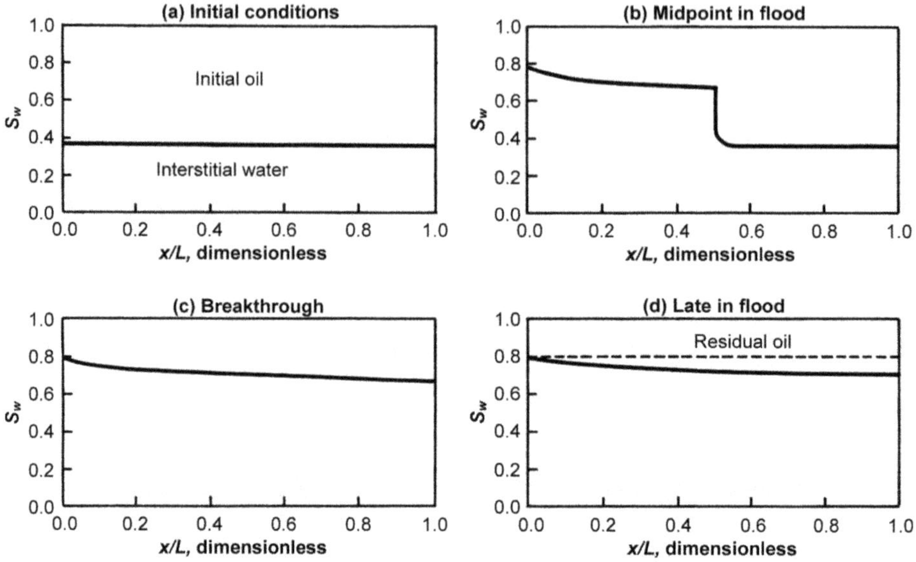

Fig. 11.12—Saturation distribution during different stages of a waterflood.[2] *L* = length, ft; *x* = *x*-direction length, ft; and *x/L* is dimensionless and varies from 0 to 1.

$$M = \frac{k_{rwe}}{k_{roe}} \frac{\mu_o}{\mu_w}, \quad (11.7)$$

where k_{rwe} = relative permeability to water at the endpoint (S_{orw}) and k_{roe} = relative permeability to oil at the endpoint (S_{wi}). This mobility ratio assumes a plug-like displacement between the oil phase at connate-water saturation before the flood front and the water phase at residual oil saturation behind the flood front.

Because, in most reservoir situations, water's viscosity is lower than oil's, the viscosity ratio is unfavorable for water to displace oil efficiently; however, as Figs. 11.4, 11.5, and 11.9 show, the relative permeability of water at residual oil saturation is lower by a factor of two to eight than that of oil at connate-water saturation. Hence, for many reservoirs, the mobility ratio is close to unity (favorable) if the oil viscosity is greater than the water viscosity at reservoir conditions only by a factor of five.

11.3 Macroscopic Displacement Efficiency of a Linear Waterflood

This section discusses the mathematical aspects of water/oil displacement for homogeneous linear systems. The presentation here is brief and does not include the intermediate steps of the mathematical derivation of the key equations. The details of these mathematical derivations are available in Willhite.[2]

The displacement of oil by water from a porous and permeable rock is an unsteady-state process because the saturations change with time and distance from the injection point (see schematic diagram of Fig. 11.1). These saturation changes cause the relative permeability values and pressures to change as a function of time at each position in the rock. **Fig. 11.12** illustrates the various stages of an oil/water displacement process in a homogeneous linear system.

The mathematical derivation of fluid-flow equations for porous media begins with the simple concept of a material-balance calculation: accumulation equals fluid in minus fluid out. This equation is written for the whole system and for each of the phases: water, oil, and gas. Eqs. 11.8 and 11.9 are the equations for the conservation of mass for a water/oil homogeneous linear system:

$$-\frac{\partial}{\partial x}(\rho_o u_{ox}) = \frac{\partial}{\partial t}(\rho_o S_o \phi) \qquad (11.8)$$

and

$$-\frac{\partial}{\partial x}(\rho_w u_{wx}) = \frac{\partial}{\partial t}(\rho_w S_w \phi), \qquad (11.9)$$

where x = position in x-coordinate system, ft; ρ_o = oil density, lbm/ft^3 or g/cm^3; u_{ox} = oil velocity in the x direction, ft/day; t = time, days; S_o = oil saturation, fraction PV; ϕ = porosity, fraction BV; ρ_w = water density, lbm/ft^3 or g/cm^3; u_{wx} = water velocity in the x direction, ft/day; and S_w = water saturation, fraction.

Assuming that the oil and water are incompressible and that the porosity is constant, these equations become:

$$-\frac{\partial q_o}{\partial x} = A\phi \frac{\partial S_o}{\partial t} \qquad (11.10)$$

and

$$-\frac{\partial q_w}{\partial x} = A\phi \frac{\partial S_w}{\partial t}, \qquad (11.11)$$

where q_o = oil-production rate, B/D; A = cross-sectional area available for flow, ft^2; and q_w = water-production rate, B/D.

Next, the equations for fractional flow of oil and water are incorporated into these equations. The three fractional-flow equations are:

$$f_o = \frac{q_o}{q_t} = \frac{q_o}{q_w + q_o}, \qquad (11.12)$$

$$f_w = \frac{q_w}{q_t} = \frac{q_w}{q_w + q_o}, \qquad (11.13)$$

and

$$f_o + f_w = 1.0, \qquad (11.14)$$

where f_o = fractional flow of oil; q_t = the total production rate, B/D; and f_w = fractional flow of water.

Substituting Eq. 11.13 into Eq. 11.11 yields:

$$-\frac{\partial f_w}{\partial x} = \frac{\phi A}{q_t}\frac{\partial S_w}{\partial t}. \qquad (11.15)$$

11.3.1 Buckley-Leverett Solution. Further mathematical manipulation of these equations obtains the Buckley-Leverett equation (Eq. 11.16), or frontal-advance equation. To derive this equation, it is assumed that the fractional flow of water is a function only of the water saturation and that there is no mass transfer between the oil and water phases.

$$\left(\frac{dx}{dt}\right)_{S_w} = \frac{q_t}{\phi A}\left(\frac{\partial f_w}{\partial S_w}\right)_t . \qquad (11.16)$$

This equation shows that in a linear displacement of water displacing oil, each water saturation moves through the rock at a velocity that is computed from the derivative of the fractional flow with respect to water saturation.

The general form of the fractional-flow equation for water is:

$$f_w = \frac{1}{1+\left(\frac{k_o}{k_w}\right)\left(\frac{\mu_w}{\mu_o}\right)} + \frac{\frac{k_o A}{\mu_o q_t}\left[\frac{\partial P_c}{\partial x} + (\rho_o - \rho_w) g \sin\alpha\right]}{1+\left(\frac{k_o}{k_w}\right)\left(\frac{\mu_w}{\mu_o}\right)} , \qquad (11.17)$$

where k_o = permeability to oil, darcies; g = gravity constant; α = reservoir dip angle, degrees; and k_w = permeability to water, darcies. This equation includes terms for capillary pressure variation (as a function of saturation) in the linear direction and for the linear system possibly dipping at angle α.

Assuming that the gradient in P_c with position is very small and that the linear system is horizontal reduces Eq. 11.17 to:

$$f_w = \frac{1}{1+\left(\frac{k_o}{k_w}\right)\left(\frac{\mu_w}{\mu_o}\right)} . \qquad (11.18)$$

Fig. 11.13 presents a typical fractional-flow curve that would be calculated from Eq. 11.18. This figure also shows a tangent to the fractional-flow curve that originates at the initial water saturation. The tangent point defines the "breakthrough" or "flood-front" saturation S_{wf}. This saturation is equivalent to the saturation that Buckley and Leverett obtained through intuitive arguments.[10] It subsequently was recognized that this tangent intersects the fractional-flow curve at the saturation that is common to the stabilized and the nonstabilized zones.[11]

The frontal-advance equation (Eq. 11.16) cannot predict the saturation profile between the connate-water saturation and the breakthrough saturation. An approximation that was developed from the Buckley-Leverett solution considers the saturation change to be a step increase ("shock") from the connate-water saturation S_{wc} to the flood-front saturation S_{wf}. **Fig. 11.14** shows this saturation profile. The shock occurs because all saturations that are less than S_{wf} travel at the velocity of the flood front. Saturations that are greater than S_{wf} travel at velocities that are determined from Eq. 11.16 by calculating the derivative of the fractional-flow curve at each S_w value.

That the Buckley-Leverett solution is reasonable has been experimentally verified. **Fig. 11.15** compares experimental results with calculated values for two oils that have nearly a hundred-fold difference in viscosity.

Fig. 11.16 shows the viscosity ratio's effect on the water fractional-flow behavior. The viscosity ratio is a key parameter; the efficiency of the linear displacement process of water displacing oil changes and is substantially different when the oil's and the water's viscosity is the same compared to when the oil's viscosity is much higher than the water's.

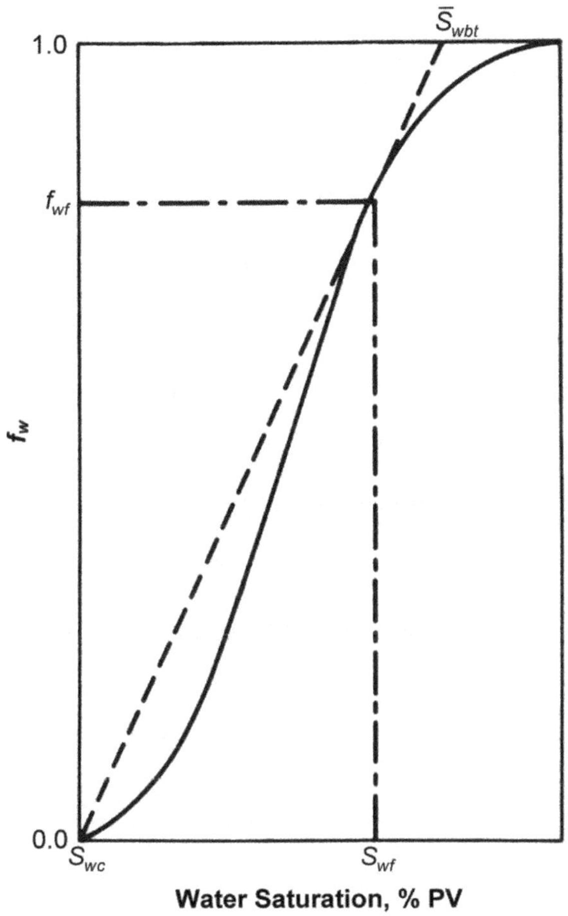

Fig. 11.13—Determination of flood-front saturation.[2] f_{wf} = fractional flow of water at the flood front, $\bar{S}_{wbt}$ = average water saturation behind the flood front at water breakthrough, and S_{wf} = water saturation at the flood front.

11.4 Reservoir-Geology Considerations in the Design and Operation of Waterfloods

This section briefly discusses the geologic considerations in assessing waterflood performance and then describes the areal and vertical aspects of waterflood performance and analysis. (See Chap. 1 in this volume of the *Handbook* for a review of the geologic factors required for detailed reservoir engineering of oil and gas fields.) This section considers the macroscopic aspects of the geology that affects waterfloods; the microscopic aspects were discussed earlier in Sec. 11.2.

All oil reservoirs are heterogeneous rock formations. The primary geological consideration in waterflooding evaluation is to determine the nature and degree of heterogeneities that exist in a particular oil field. Reservoir heterogeneities can take many forms, including

• Shale, anhydrite, or other impermeable layers that partly or completely separate the porous and permeable reservoir layers.

• Interbedded hydrocarbon-bearing layers that have significantly different rock qualities—sandstones or carbonates.

• Varying continuity, interconnection, and areal extent of porous and permeable layers throughout the reservoir.

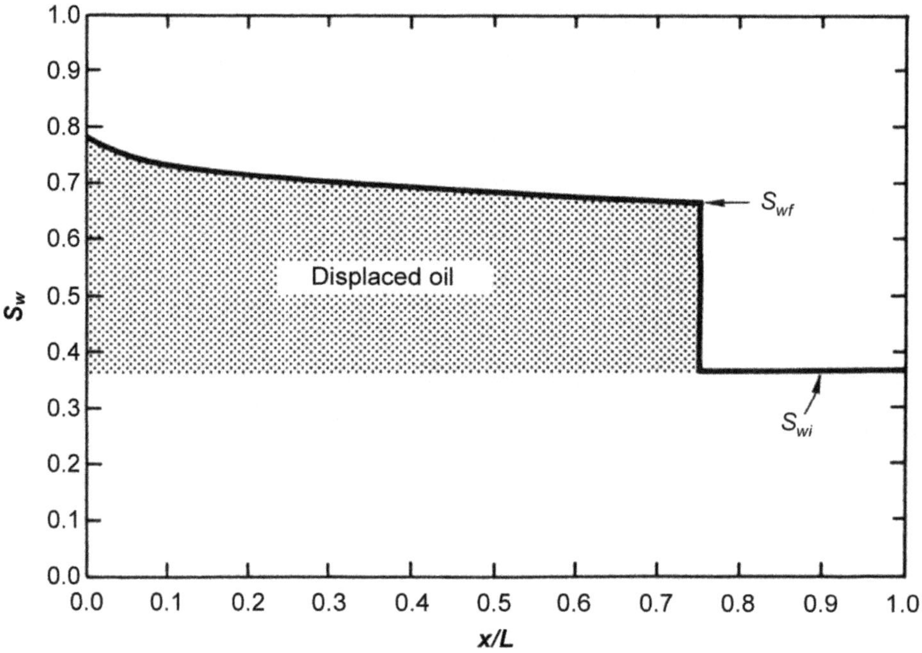

Fig. 11.14—Saturation profile computed from Buckley-Leverett solution.[2]

• Directional permeability trends that are caused by the depositional environment or by diagenetic changes.

• Fracture trends that developed because of regional tectonic stresses on the rock and the effects of burial and uplift on the particular rock layer.

• Fault trends that affect the connection of one part of an oil reservoir to adjacent areas, either because they are flow barriers or because they are open conduits that allow unlimited flow along the fault plane.

The structure of the reservoir and how it affects waterflood performance is another geological consideration. Structure creates dipping beds that dip at various angles. The interplay between the bed angle, gravity, and the oil/brine density difference at reservoir conditions significantly affects the relative vertical and horizontal flow behaviors. Structural considerations also can include whether the oil column has an underlying aquifer or an overlying gas cap, either of which can significantly affect the likelihood of successfully waterflooding the oil column.

Geologists and geophysicists must assess such geological and structural aspects of a reservoir. Geologists use cores and routine-core-analysis data to develop an understanding of the depositional environment and post-depositional diagenesis and to characterize the reservoir's internal architecture. Using seismic data, geophysicists can discern the major faults, as well as trends in rock quality, since cores and well logs are essentially pin pricks into the overall reservoir.

The technical team that is evaluating and monitoring waterflood performance should include a geologist and a geophysicist. Including a geostatistician on the technical team, as well, will help to ensure that the geoscientists' reservoir description is properly translated into engineering calculations, whether those are simpler calculations or are detailed numerical reservoir simulations.

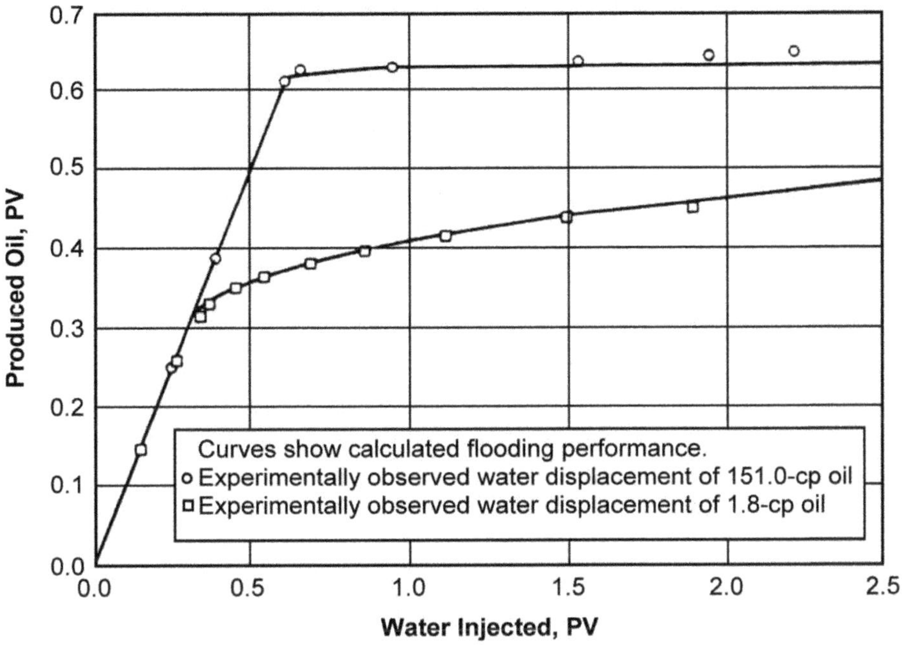

Fig. 11.15—Comparison of waterflooding response with results calculated from frontal-advance equation.[2]

For a waterflood, the reservoir description must be developed on the scale that is required for the quantitative evaluation (i.e., it must be "fit-for-purpose"). A variety of approaches (e.g., object- and pixel-based techniques) can be used.[12] The "flow unit" is a concept that frequently is used by geologists and that would be useful to engineers. "A flow unit is a volume of the total reservoir rock within which geological and petrophysical properties that affect fluid flow are internally consistent and predictably different from properties of other rock volumes (e.g., flow units)."[13]

The process of evaluating a reservoir's geology begins when the reservoir is discovered and is placed on primary production. After a waterflood has been initiated, the production- and injection-well data provide additional insight into the internal characteristics of the rock volume that is being flooded. In fact, the waterflood production-well data (the water and oil rates as a function of time) are critical because they are the first data that relate directly to the interwell connectivity within the reservoir and that validate or cause modification of the geoscientists' concepts of the various levels of reservoir heterogeneities.

During a waterflood, tracers can be injected to track which injector/producer pairs are well connected and which are poorly connected. (See the chapter on tracers in this volume of the *Handbook.*) Other monitoring techniques include the use of specially drilled observation wells and 4D-seismic interpretations to track the directionality and shape of the higher-pressure water-swept reservoir areas that are centered on the injection wells.

11.5 Immiscible Displacement in Two Dimensions—Areal

Historically, when computer capabilities were limited, the 3D aspects of a waterflood evaluation were simplified so that the technical problem could be treated as either a 2D-areal problem or a 2D-vertical problem. This section reviews the methods for treating the waterflood analysis as a 2D-areal technical problem. To simplify 3D to 2D areal, either the reservoir must be assumed to be vertically a thin and homogeneous rock interval (hence having no gravity

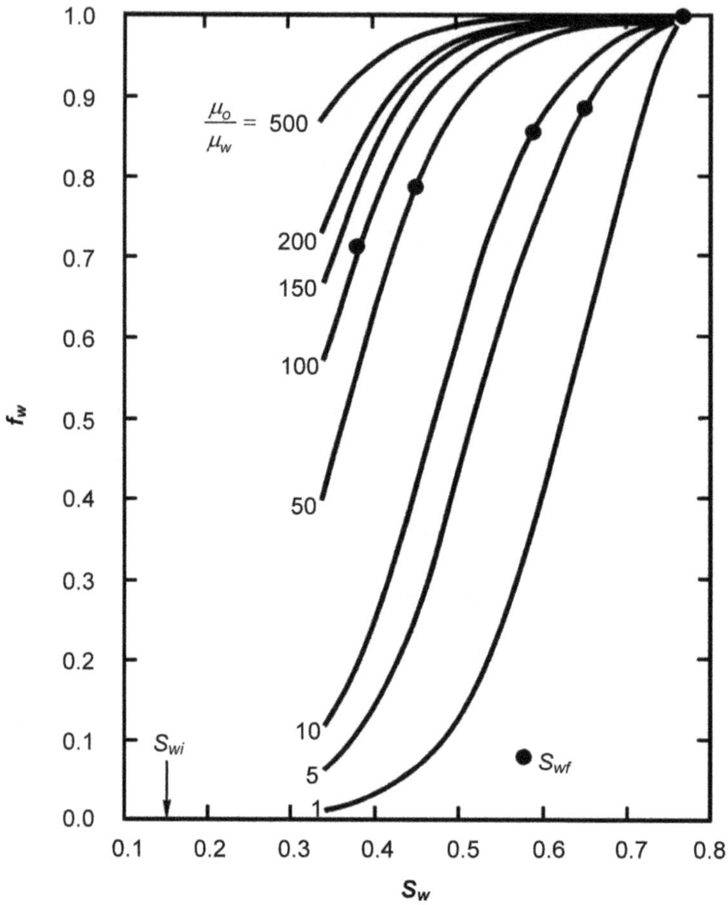

Fig. 11.16—Effect of viscosity ratio on fractional flow.[2]

considerations) or one of the published techniques to handle the vertical heterogeneity and expected gravity effects within the context of a 2D-areal calculation must be used.

The primary areal considerations for a waterflood involve the choices of the pattern style (see **Fig. 11.17**) and the well spacing. Maximizing the ultimate oil recovery and economic return from waterflooding requires making many pattern- and spacing-related decisions when secondary recovery is evaluated. This has been particularly true for onshore oil fields in the U.S. in which a significant number of wells were drilled for primary production. Many SPE technical papers have been published addressing these issues. (Sec. 11.9.4 West Texas Carbonate Waterfloods discusses these topics further and references several relevant papers.) For offshore oil fields where the maximum number of wells is limited, the optimal waterflood injection-well/production-well layout is best determined by the use of 3D numerical reservoir simulation.

The discussion below describes various factors that affect waterflood performance and discusses some of the 2D-areal calculation methodologies that have been developed. This discussion does not cover the use of modern numerical reservoir-simulation models in a 2D mode for analyzing a reservoir's waterflood performance, either for history matching or for future-performance projections. See Chap. 17 in this volume of the *Handbook* for a review of numerical reservoir simulation.

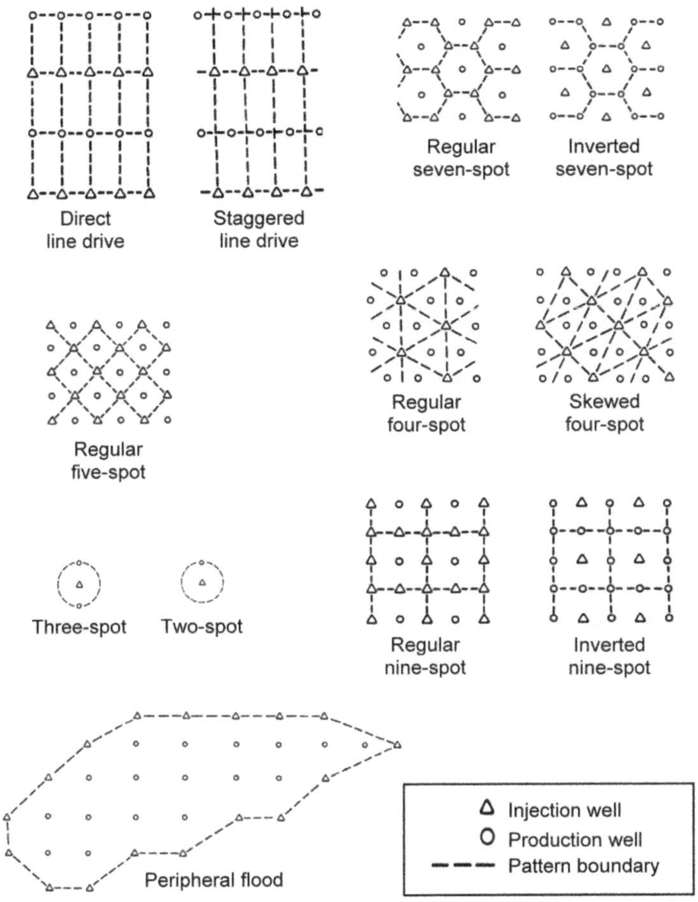

Fig. 11.17—Common waterflood-pattern configurations.[3]

11.5.1 Waterflood Displacement in a Five-Spot Pattern. Because it has been studied extensively, taking a look at the five-spot pattern (Fig. 11.17) provides an effective review of the areal aspects of waterflooding. **Fig. 11.18** plots the waterflood performance of a five-spot experimental homogeneous sandpack for range of water/oil viscosity ratios from 0.83 (slightly favorable) to 754.0 (very unfavorable). **Fig. 11.19** shows the X-ray shadowgraphs of a waterflood of two five-spot homogeneous sandpacks, one whose mobility ratio is favorable (M = 0.40) and one whose mobility ratio is unfavorable (M = 1.43). **Fig. 11.20** plots the breakthrough areal-sweep efficiency as a function of M for a five-spot-pattern flood.

Figs. 11.18 through 11.20 show that the water/oil viscosity ratio is critical to the efficiency of the waterflood displacement. When the oil is more viscous than the water, the water areally displaces the oil less efficiently. When the oil is less viscous than the water, the water areally displaces the oil very efficiently.

11.5.2 Streamtube Waterflood Calculations. In the 1960s, Higgins and Leighton published a technique for analyzing waterfloods using what has been termed streamtubes, or stream channels, and using the concept of streamlines.[14,15] Streamlines are the paths that fluid particles follow when they move from the injector to the producer; a streamtube is the region between two streamlines. The Higgins and Leighton technique approximates the displacement problem by use of a set of streamtubes in which fluid flows from the injection well to the production

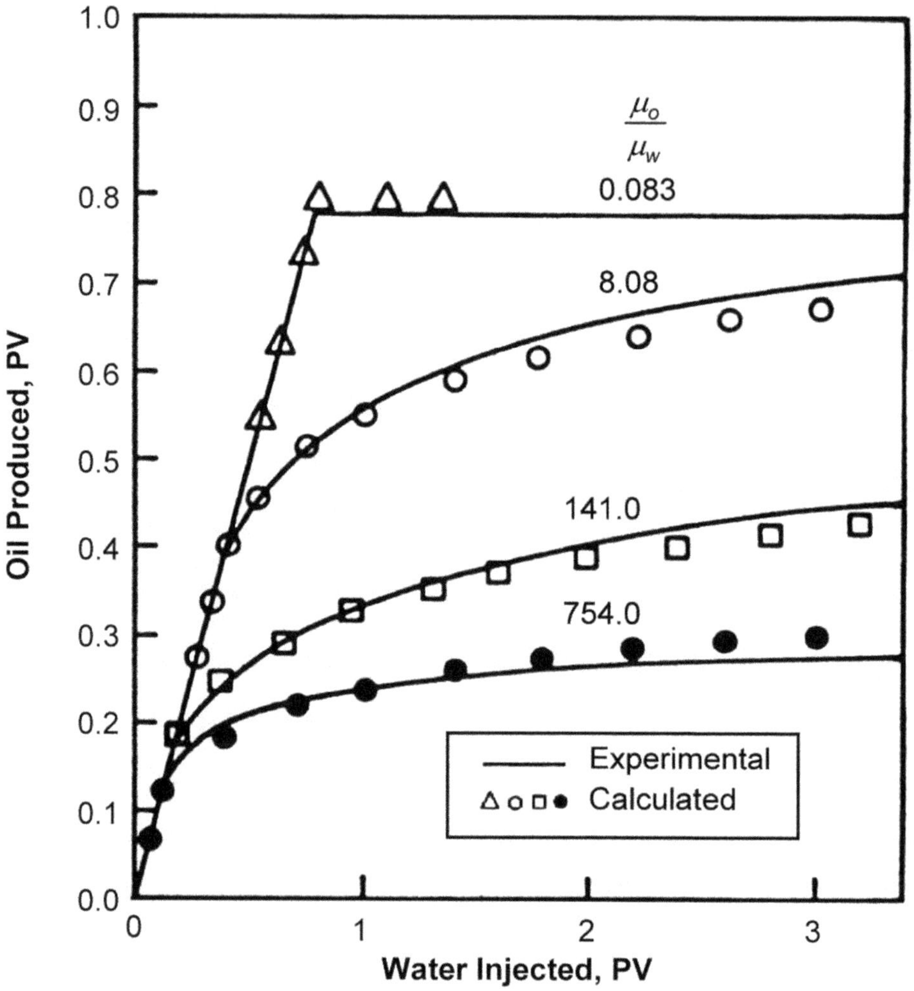

Fig. 11.18—Oil recovery from waterflooding five-spot pattern models.[2]

well and in which no fluid flows perpendicular to the streamlines (**Fig. 11.21**). Streamlines for the steady flow of a single phase (unit mobility ratio) have been determined for regular displacement patterns. Streamlines for arbitrary arrangements of injection and production wells can be determined by superposition of numerical solutions (see **Fig. 11.22**).[16]

Streamtube techniques originally were developed for areal waterflood analysis. These techniques have continued to be popular with advancements in the mathematical techniques and computing power. (Also see the discussion about streamtube techniques in the Simulation chapter in this volume of the *Handbook*). The more advanced streamtube models can handle stratified-layer systems. Many universities and oil companies have developed sophisticated streamtube models and applied them to waterflood calculations. Such models also are available from commercial software providers to the oil industry.[17]

Any 2D approach to waterflood analysis is an approximation. How accurate an approximation it is will depend on how vertically stratified the reservoir is and how significant the gravity considerations are in the real reservoir compared to what the numerical modeling assumes about them. For many situations, 2D-streamtube models have been used successfully to model historic waterflood performance and to project future waterflood performance.[18]

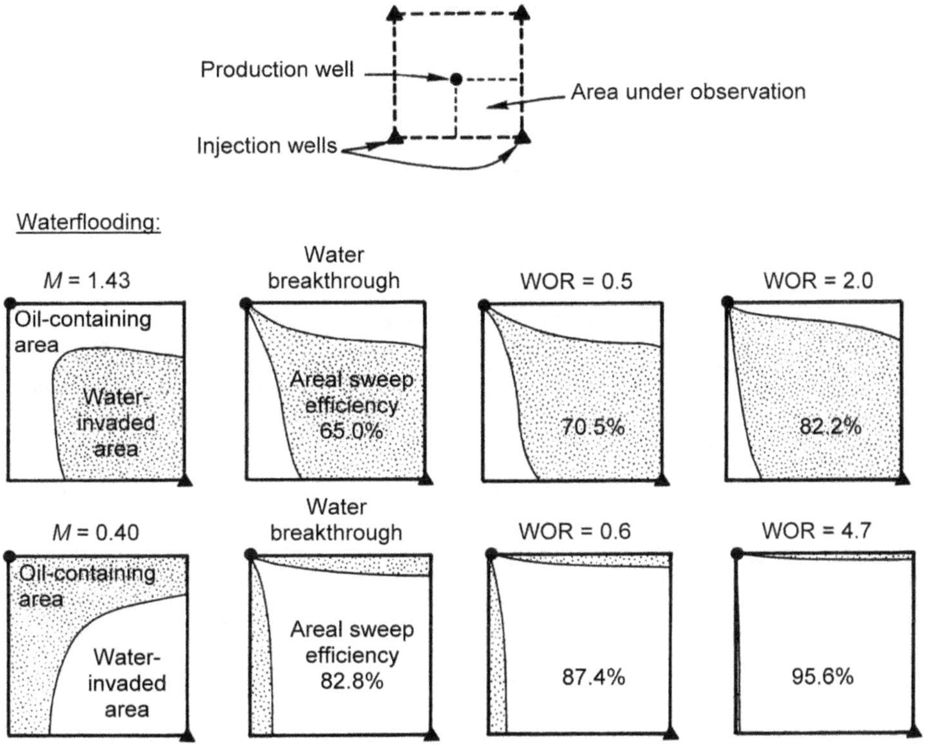

Fig. 11.19—X-ray shadowgraphs of flood progress in experimental scaled five-spot pattern models.[2] (WOR = instantaneous producing water/oil ratio.)

11.5.3 Waterflood-Pattern and Well-Spacing Considerations. As Fig. 11.17 shows, a variety of geometric injector/producer pattern layouts can be used when waterflooding an oil reservoir. These geometric layouts are designed to produce an efficient waterflood for the whole of the reservoir, assuming the rock is homogeneous. The producer/injector ratio typically is chosen on the basis of the expected injection rates for the water injectors and the total fluid-production rates for the production wells. The goal is to have voidage replacement, with the injected volume equal to the produced volume. Injection and produced volumes will depend on the transmissibility of the system, the maximum pressure that can be applied at the injector, and the minimum pressure that can be achieved at the producer. It has been customary to limit injection pressures to lower than what would cause fracturing. As will be discussed later, this guideline can be relaxed in some circumstances.

Other considerations with the injector/producer-pattern layout are the anisotropic permeability in the reservoir and the orientation of any natural-fracture systems that are within the reservoir interval. **Fig. 11.23** shows the correct and incorrect orientation of lines of injectors and producers in this respect. The goal is to make these natural aspects of the reservoir improve the areal sweepout as the water displaces the oil, rather than allowing them to affect it negatively.

The choice of waterflood well spacing is primarily a function of four considerations: incremental oil recovery; increased oil-production rate; reduced operating costs (producing more of the oil at lower water cuts); and costs to drill additional wells, including additional platform space at offshore locations. The initial well spacing typically is chosen on the basis of engineering studies, assuming that there will be a logical phased-development plan for the oil field,

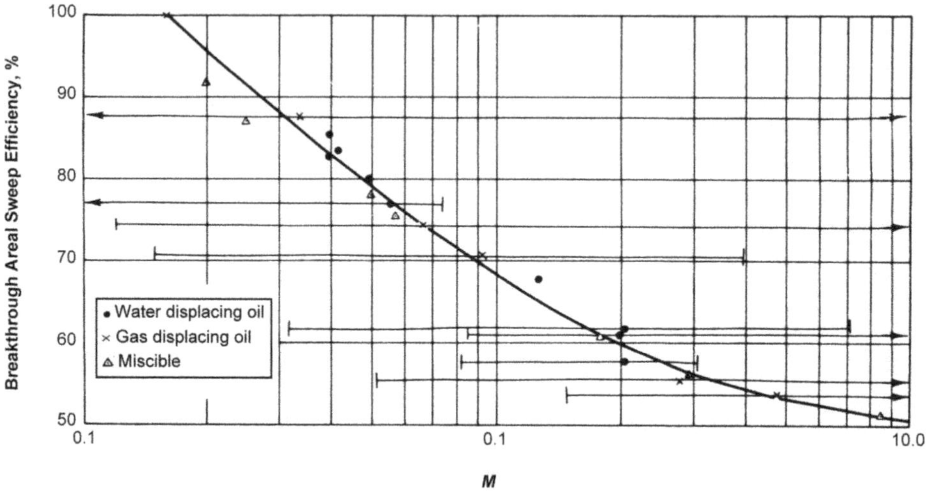

Fig. 11.20—Correlation of areal sweep efficiency at breakthrough with mobility ratio for miscible and immiscible displacement in five-spot-pattern floods.[2] On this figure, the mobility *M* has been calculated using the average water saturation in the water-contacted portion of the reservoir. The horizontal lines indicate what the range of mobility ratios *M* would be if it were calculated using the water mobility at floodout conditions (the right extremity of the horizontal bar) or using the water mobility at the flood front (the left extremity of the bar), with the arrows indicating that the values were off the scale of this plot.

including possibilities for various levels of infill drilling and pattern realignments. Low-permeability reservoirs typically have closer well spacings than do high-permeability ones because of their reduced ability to move fluids rapidly between widely spaced injectors and producers. The initial well spacing usually is somewhat wider than what is likely to be used toward the end of field life. The number of injection and production wells also is tied to the capacity of the injection facilities and production-fluid-separation facilities.

The incremental oil recovery can be a function of the well spacing and depends on the continuity of the porous and permeable reservoir-rock layers. If the reservoir layers are highly continuous, then reducing the well spacing will not have much effect on the ultimate oil recovery by waterflooding; however, for layers that are discontinuous over the reservoir, closer well spacing will provide more reservoir continuity between injector/producer pairs. This aspect of waterflooding has been studied in depth for a variety of U.S. oil fields, mainly west Texas carbonate reservoirs, in which reservoir-layer continuity is a major issue. Many papers have been published on this topic.

For example, Barber *et al.*[19] reviewed nine sandstone and carbonate reservoirs' responses to infill drilling. **Figs. 11.24 and 11.25** show aspects of reservoir discontinuity in west Texas Clearfork and San Andres carbonate reservoirs. Fig. 11.25 shows that the San Andres reservoirs have significantly greater reservoir continuity at a 10-acre well spacing than at 20- and 40-acre well spacings. For these reservoirs, "[c]ontinuity calculations made after infill drilling indicated the pay zones to be more discontinuous than when calculations were made before infill drilling."[19] From the experience in those nine fields, the authors concluded that "the ultimate well density in any given field can be determined only after several years of field performance provide sufficient information on reservoir continuity and recovery efficiencies."[19]

Another paper on this topic is by researchers at Texas A&M U.,[20] who spent several years analyzing the impact of well-spacing reduction on west Texas Clearfork and San Andres waterflood performance from various geographic areas of the Permian Basin. The authors found infill drilling to be more effective for the San Andres units than for the Clearfork units. In the San Andres units, they determined the infill drilling to be more effective for the units in the

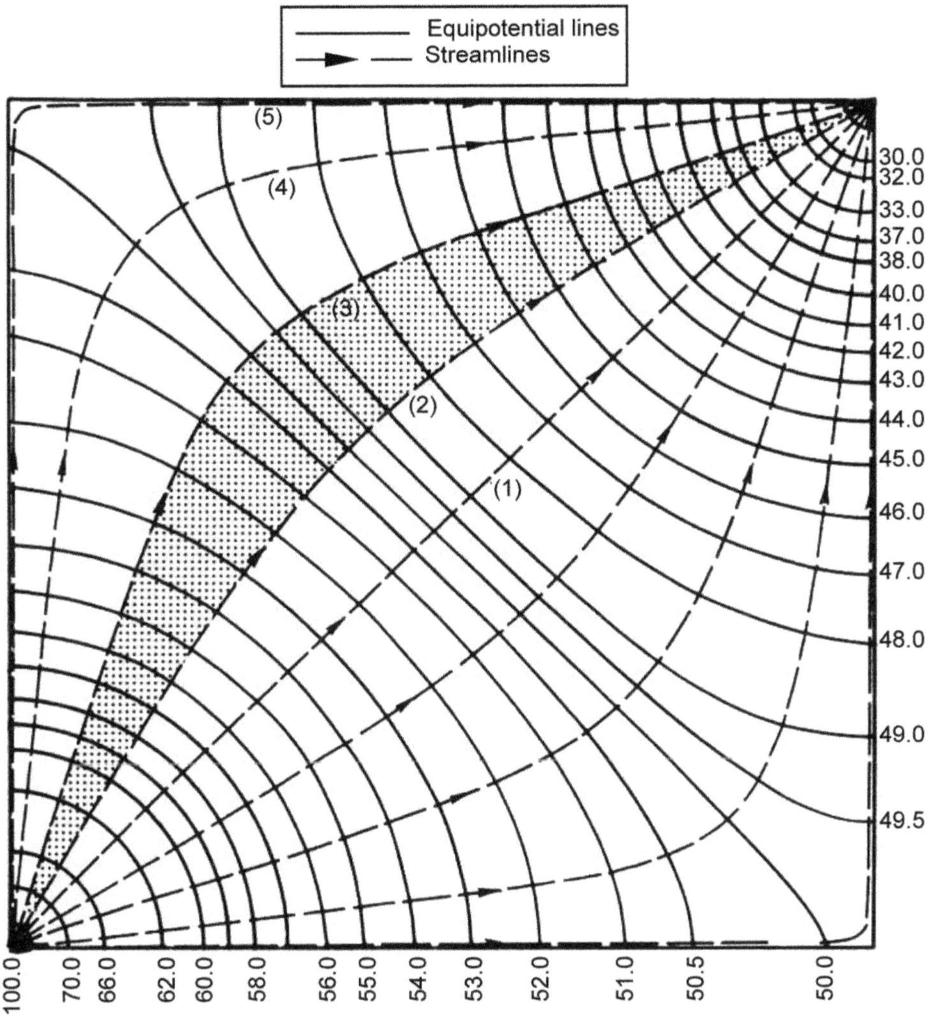

Fig. 11.21—Streamlines and equipotential lines for single-phase flow in a quadrant of a five-spot pattern.[2] For illustrative purpose, for this homogeneous quarter-of-a-five-spot pattern and single-phase flow, a set of streamlines has been shown as (1) through (5); the numbers along the axes are lines of equal potential scaled from 100 at the injection point to zero at the production point. By definition, streamlines are perpendicular to lines of equal potential.

Northern Shelf area than for those in the Central Basin Platform area. In those analyses, they calculated a 2 to 5% additional recovery of OOIP for a 9-acre well spacing, compared to a 22-acre well spacing.[20]

As noted above, many more papers have been published on this topic, but these two papers identify the key well-spacing issues. As both studies conclude, one does not know *a priori* what the optimal well spacing will be. The technical team must analyze the available data before starting a waterflood, then continue to evaluate the production and injection data to determine the waterflood's efficiency and the extent to which infill-well locations should be drilled or patterns realigned.

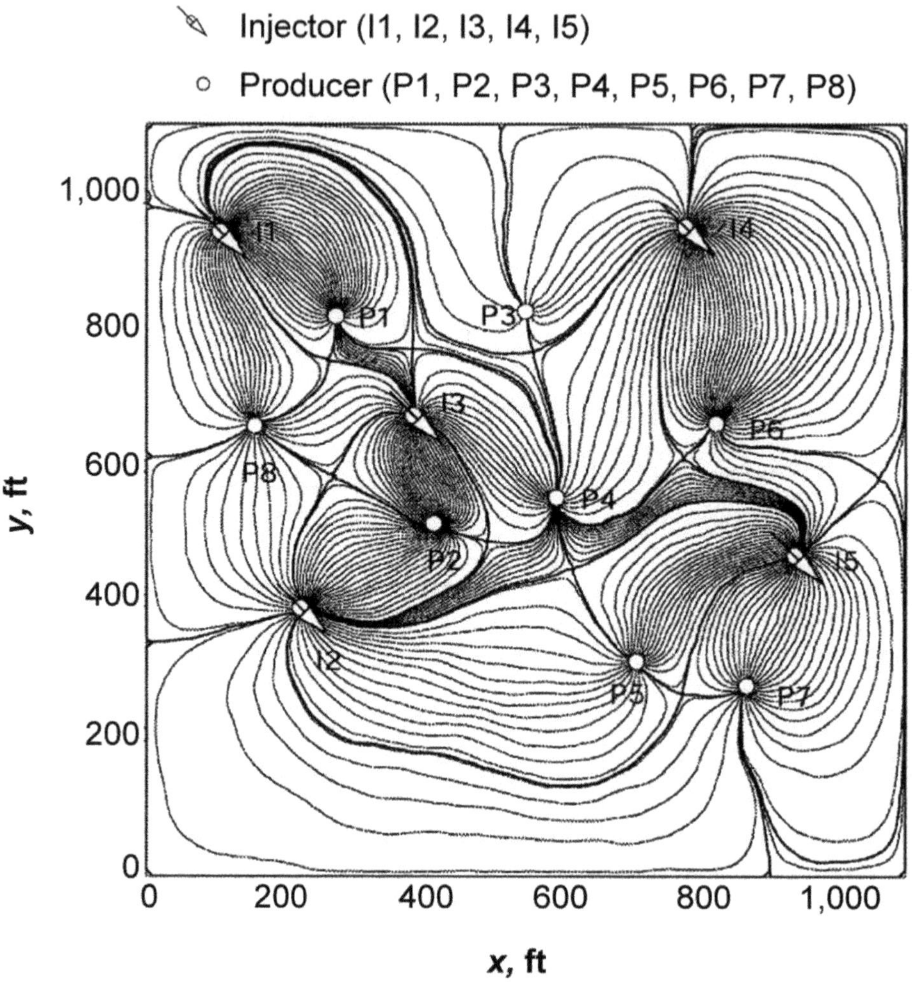

Fig. 11.22—Geometries of streamtubes of a heterogeneous multiwell case (113 × 113 grid, heterogeneous, M = 10).[16]

11.5.4 Horizontal Wells, Multilateral Wells, and Fracture Orientation. Until the 1980s, all waterflood analysis assumed that the wellbores would penetrate the reservoir vertically and that water should not be injected above the formation parting pressure. Accordingly, mathematical formulations assumed areally that injection and production wells were point sources and sinks. This style of analysis is evident in the streamline patterns in Fig. 11.22.

The advent of horizontal-well and multilateral-well technologies has brought with it the ability to create line injection-well sources and line production-well sinks, which has changed how a waterflood pattern might be developed. Theoretical calculations show that parallel lines of horizontal injectors and horizontal producers will increase oil recovery and in the limit are the perfect line-drive pattern arrangement.

When fracturing injection wells, a concern has been that a fracture might extend from the oil-reservoir intervals into adjacent porous and permeable layers, into which considerable injection brine could be lost (i.e., thief zones); however, given that the principal orientation of any hydraulic fracture is known and that the fractures can be restricted to the oil-reservoir interval, hydraulic fractures can improve the areal sweepout during waterflooding in much the same

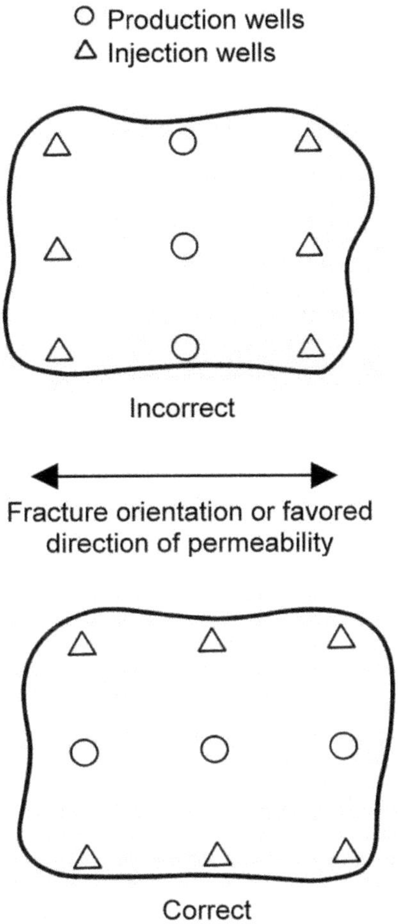

Fig. 11.23—Correct and incorrect pattern alignment with anisotropic permeability, or an oriented fracture system.[3]

way that horizontal wells can. This is true of fractures from the injection wells and the production wells, given that an appropriate pattern style is used. On the other hand, if they are oriented so that the fracture tips are significantly closer to each other than the vertical wellbores are, hydraulic fractures from the injectors and producers will yield poorer areal sweepout of the reservoir during waterflooding than would unfractured vertical wells.

An excellent example of a situation in which fracturing was needed is the waterflooding of diatomite reservoirs in California, U.S.A. These reservoirs have high porosity, but permeabilities of 0.1 md or less. Conventional injection methods yielded uneconomic rates; however, positioning the injectors and producers normal to the induced-fracture orientation established reasonable rates without significant loss of injected water to adjacent porous and permeable intervals and without premature water breakthrough.[21,22]

11.6 Vertical Displacement in Linear and Areal Models

The previous section's discussion of waterflood displacement behavior assumed that the vertical saturation distribution was homogeneous at all areal locations. This section discusses the impact of vertical variations in permeability and the effect of gravity on simple 2D reservoir situations in which the areal effects are ignored. Gravity effects always are present because for

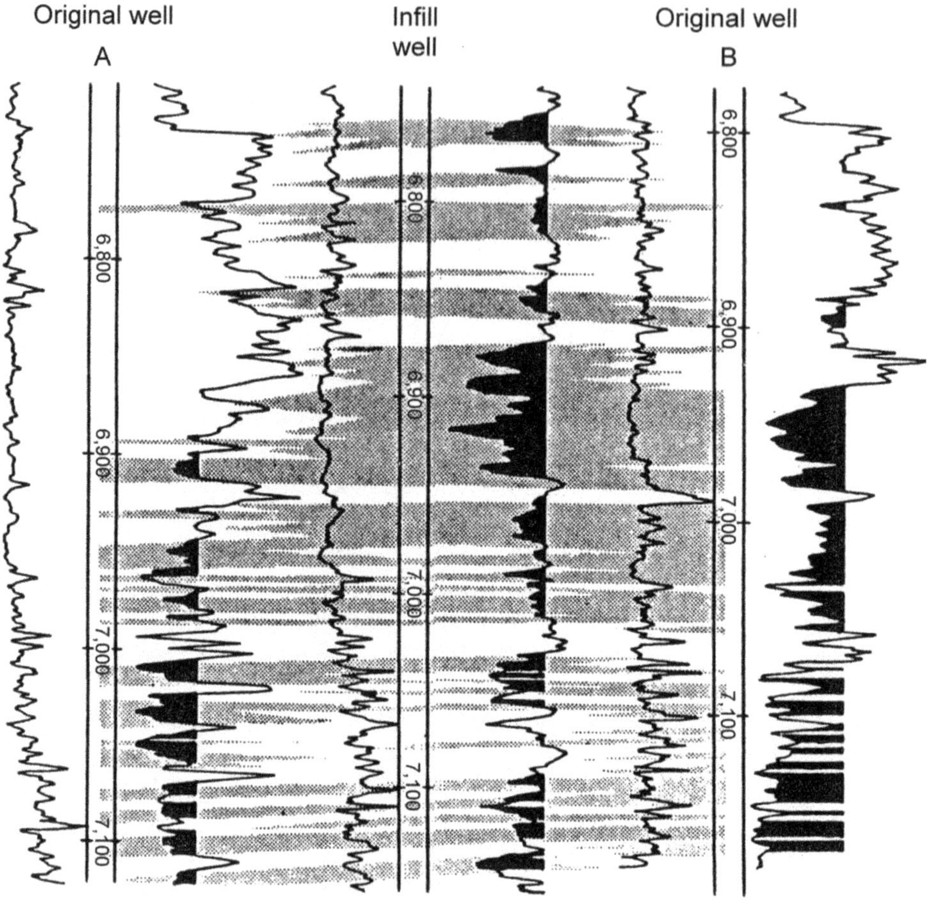

Fig. 11.24—Type cross section for Fullerton Clearfork reservoir.[19]

any potential waterflood project, oil always is less dense than water, even more so after the gas is included that is dissolved in the oil at reservoir conditions.

Three particular situations are discussed here:

- Stratified systems with noncommunicating layers for various mobility ratios.
- Homogeneous systems with gravity (including dipping beds).
- Stratified systems with communicating layers and assumed vertical fluid equilibrium.

The discussion below does not include the P_c effects on vertical saturation distributions. Through countercurrent imbibition, P_c effects help to counteract nonequilibrium water/oil saturation distributions. The mathematics of including P_c effects makes the problems too complicated for inclusion in this chapter. Standard numerical reservoir simulators—which are needed for a complete analysis of real reservoir situations—account for the effects of countercurrent imbibition caused by P_c effects, as well as for the water/oil density and viscosity differences that lead to injection-water gravity underrunning and the layer-by-layer permeability variations, with or without communicating layers.

Before presenting some of the technical-literature techniques for studying the vertical displacement characteristics of water/oil displacement, one must first define some measure of the vertical permeability variations. Dykstra and Parsons[23] developed a method that is based on routine-core-analysis data. In that approach, the routine-core-analysis permeability data for the pay intervals are arranged in descending order, and the percent of the total number of values

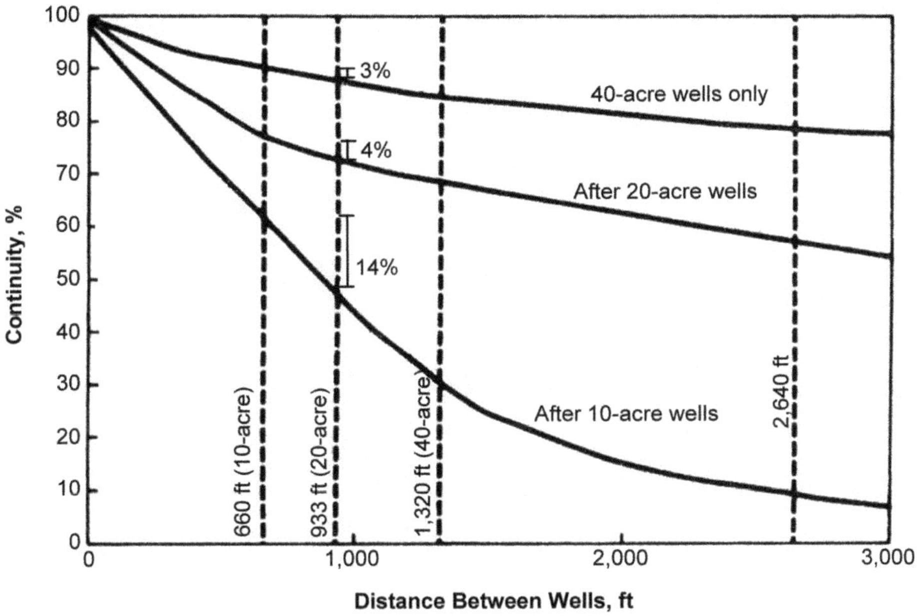

Fig. 11.25—Continuity progression for Means San Andres unit.[19] The lines on this plot show the estimated percent continuity before drilling infill wells. After each set of infill wells was drilled, the actual reservoir continuity was found to be less than predicted. This is why the "After 10-acre wells" line is the lowest of the three lines. The 3, 4, and 14% vertical bars indicate the expected increase in percent continuity estimated for a 10-acre well spacing from the 40-acre well-spacing data (3%) and from the 20-acre well-spacing data (4%), compared to what actually was found after the 10-acre infill wells were drilled (14%).

that exceeds each entry is calculated. The values then are presented as a log-probability plot (see **Fig. 11.26**). A reasonably straight line is drawn through the data, with the points in the 10-to-90% range being more heavily weighted. This straight line is a measure of the dispersion and the heterogeneity of the reservoir rock. What has come to be known as the "Dykstra-Parsons coefficient of permeability variation" V is defined as:

$$V = \frac{k_{50} - k_{84.1}}{k_{50}}, \quad (11.19)$$

where k_{50} = median permeability value, md, and $k_{84.1}$ = permeability at 84.1% probability (one standard deviation), md.

Fig. 11.27 shows the relationship of the Dykstra-Parsons V values to varying degrees of rock heterogeneity. Note that in Fig. 11.27, the V for most reservoirs ranges from 0.5 to 0.9.

11.6.1 Stratified Systems With Noncommunicating Layers. Over the years, several waterflood prediction methods have been proposed and published that account for the vertical variations in rock properties, particularly permeability. These simple methods assumed that every rock layer acts independently of all other rock layers (even at 1-ft increments in the reservoir) and that each rock layer is continuous from the injection well to the production well. These early methods were developed when the ability to make detailed, complicated engineering calculations was limited. They focused on how to account for (1) the effect of the vertical permeability variation with minimal consideration of the mobility ratio and (2) the effect of

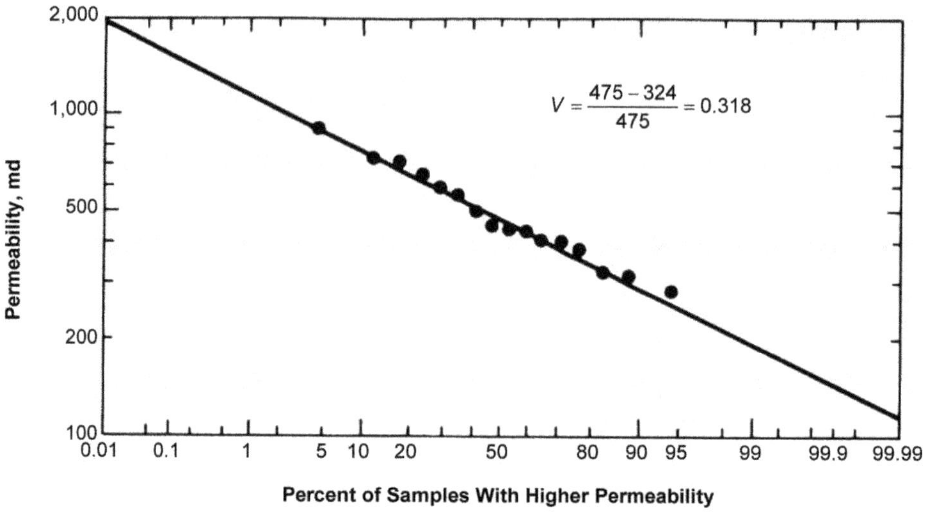

Fig. 11.26—Plot of permeability data on log-normal paper.[2]

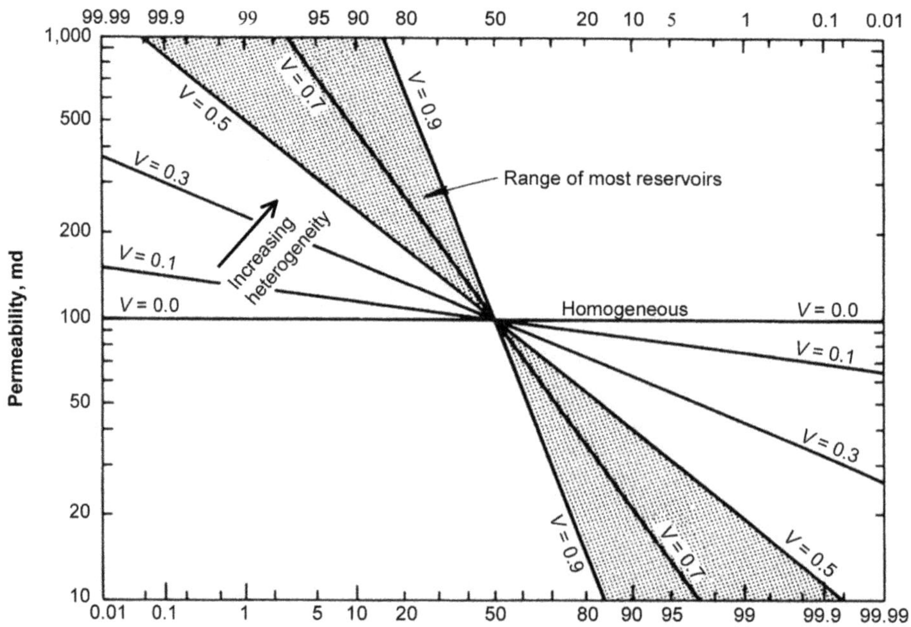

Fig. 11.27—Characterization of reservoir heterogeneity by permeability variation *V*.[2]

vertical permeability variation and mobility ratio, assuming constant pressure at the injection and production wells.

Stiles[24] developed one of the earliest methods, for which only the permeability-thickness (*kh*) distribution of the vertical reservoir interval and the mobility ratio at endpoint conditions need to be known (see Eq. 11.6). The water/oil ratio (WOR) F_{wo} after water breakthrough as a function of the fraction of the total flow capacity *C* represented by layers having water breakthrough is defined as:

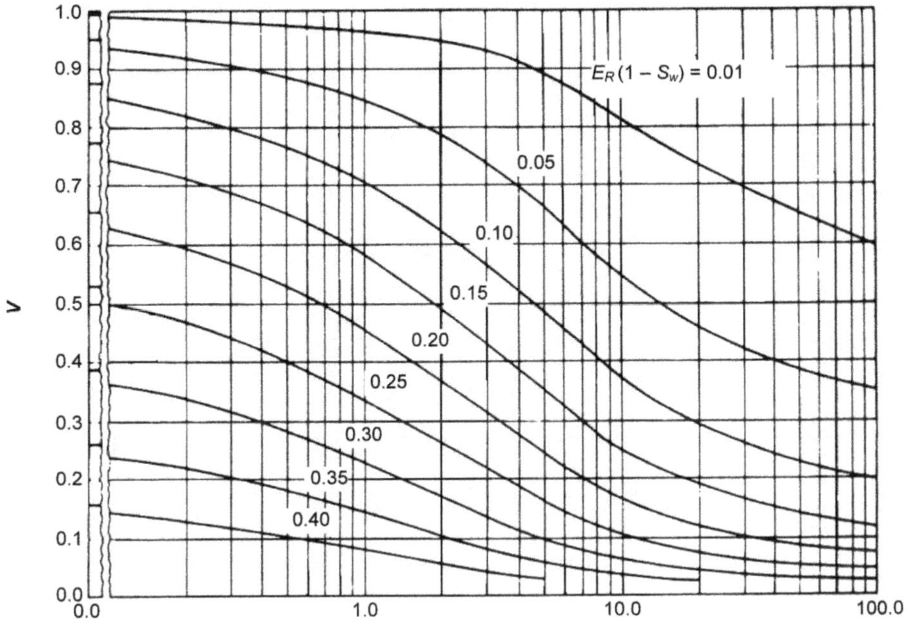

Fig. 11.28—Permeability variation plotted against mobility ratio, showing lines of a constant E_R (1 – S_w) for a producing WOR of 1.[1]

$$F_{wo} = \frac{C}{1-C}\frac{k_{rw}}{\mu_w}\frac{\mu_o}{k_{ro}}B_o, \quad (11.20)$$

where B_o = the oil formation-volume factor, RB/STB.

A more sophisticated method that is widely used is that of Dykstra and Parsons.[23] Their method is based on calculations for linear layered models and assumes no crossflow and the use of the results of more than 200 floodpot tests that were performed on more than 40 California oil-reservoir core samples. This Dykstra-and-Parsons method takes into account initial fluid saturations, mobility ratios, producing WORs, and fractional oil recoveries. The permeability variation was taken into account by use of V, as defined in Eq. 11.19 above. **Figs. 11.28 through 11.31** plot the results of the Dykstra-and-Parsons technique as V vs. M for four WOR levels (1, 5, 25, and 100).

11.6.2 Homogeneous Reservoirs Subject to Gravity Effects. In essentially all reservoirs, even those with close well spacings, the horizontal distance between an injector well and a producer well is very long relative to the vertical thickness of the reservoir pay interval. This means that gravity plays a major role in the water/oil-displacement process, given that the fluids can move vertically within the pay interval. For conceptual and calculation purposes, the limiting case is to assume that gravity forces dominate the water/oil-displacement process, that gravity segregation of the oil and water is complete, and that the system is in "vertical equilibrium." This means that vertically the gravity and capillary forces are in balance and that the vertical saturation distribution is governed by the P_c/S_w function.

The first and simplest homogeneous reservoir situation described here is a reservoir whose permeability is constant throughout the pay interval. Craig[1] studied a set of scaled laboratory vertical models experimentally and developed a correlation between the sweep efficiency at breakthrough and the values of the scaling parameter:

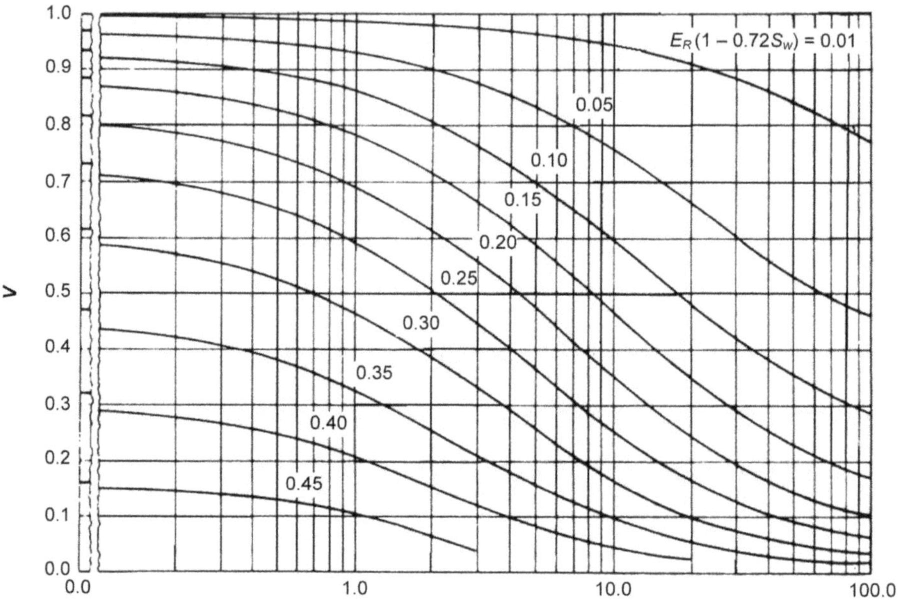

Fig. 11.29—Permeability variation plotted against mobility ratio, showing lines of a constant E_R $(1 - 0.72S_w)$ for a producing WOR of 5.[1]

$$\frac{(\Delta p)_h}{(\Delta p)_V} = \frac{u_t \mu_o L}{k_x g \Delta \rho h}, \qquad (11.21)$$

where $(\Delta p)_h$ = pressure difference in the horizontal direction, psi; $(\Delta p)_V$ = pressure difference in the vertical direction, psi; u_t = horizontal Darcy velocity, ft/D; k_x = permeability in the x direction, darcies; and $\Delta\rho$ = water/oil density difference, lbm/ft^3.

As **Fig. 11.32** shows, the sweep efficiency as it is related to the scaling parameter is a strong function of mobility ratio. **Fig. 11.33** compares the fractional flow of water for a homogeneous system with vertical equilibrium to the fractional flow of water calculated from the original laboratory water/oil relative permeability curves. The effect of the water moving along the base of the reservoir interval because of the gravity effects—but with the P_c/S_w curve controlling the vertical distribution of the water and oil saturations—is that the water breaks through earlier and the WOR rises more slowly.

Another reservoir situation that involves gravity effects is a homogeneous reservoir with dipping beds. If the rate of water injection in a waterflood is too low for vertical equilibrium to occur, there will be gravity-stabilized flow between the water and the oil. Dietz[25] has derived a relationship to predict the critical velocity q_c required to propagate a stable interface through a linear system in which gravity forces dominate, but in which pistonlike displacement occurs and P_c effects are neglected:

$$q_c = \frac{\left(4.9 \times 10^{-1}\right) k k_{rw} A \left(\rho_w - \rho_o\right) \sin \alpha}{\mu_w (M - 1)}, \qquad (11.22)$$

where ρ_o = oil density, lbm/ft^3; ρ_w = water density, lbm/ft^3; and α = dip angle, degree.

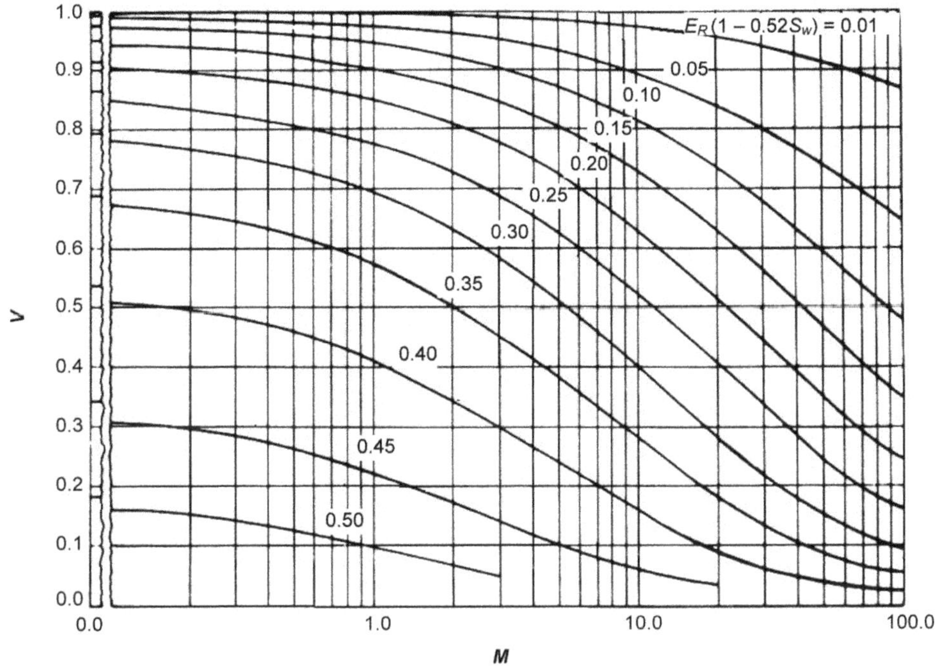

Fig. 11.30—Permeability variation plotted against mobility ratio, showing lines of a constant E_R (1 – 0.52S_w) for a producing WOR of 25.[1]

When the oil/water interface is stable, the velocities of oil and water are equal at every point in the interface. The interface is linear and will move at a constant velocity through the system as long as $q < q_c$. The stable linear interface will not necessarily be flat; however, it will be stable with a slope β, as defined by Eq. 11.23:

$$\frac{dy}{dx} = -\tan\beta = \frac{(M-1-G)}{G}\tan\alpha, \quad\text{(11.23)}$$

where y = the position in y-coordinate system, ft, and G is dimensionless and defined by Eq. 11.24:

$$G = \frac{\left(4.9\times10^{-1}\right)kk_{rw}A\left(\rho_w-\rho_o\right)\sin\alpha}{\mu_w q_t}. \quad\text{(11.24)}$$

Figs. 11.34a and 11.34b depict gravity-stable situations for two different mobility ratios.[26] Fig. 11.34c depicts the unstable situation for an unfavorable mobility ratio where the displacement rate is too high for the water and oil to maintain vertical equilibrium.

11.6.3 Stratified Systems With Communicating Layers and Assumed Vertical Equilibrium. One of the systems that have been analyzed with simple calculations is that of water/oil displacement with vertical permeability variations and gravity effects, but with capillary pressure neglected. Dake[26] explores this in his reservoir-engineering textbook. Dake's illustrative example assumes a three-layer system. He assumes the permeability variation to be highest to lowest from top to bottom, and then compares those results with results from assuming the reverse,

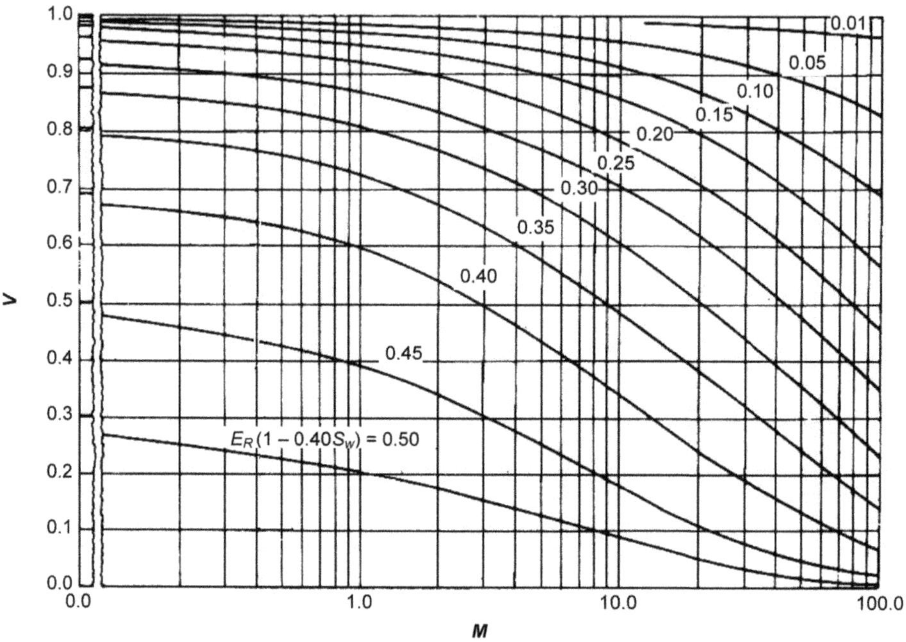

Fig. 11.31—Permeability variation plotted against mobility ratio, showing lines of a constant E_R $(1 - 0.40S_w)$ for a producing WOR of 100.[1]

the layer with highest permeability variation then on the bottom. **Fig. 11.35** lists the properties of the three layers, and **Fig. 11.36** presents the averaged relative permeability curves. **Fig. 11.37** shows the pseudocapillary pressure (a) and fractional-flow curves (b). Note that Fig. 11.37b includes smoothed and unsmoothed versions of the fractional-flow curve. The smoothed version is the curve that would be used for a Welge-type fractional-flow calculation.

Dake's example shows that in waterflooding where gravity effects are significant, having the high-permeability layers at the top of the reservoir interval allows a much more efficient oil displacement than when the high-permeability layers are at the bottom of the reservoir interval. This is because gravity causes the water to slump, and when the lower-permeability layers are the base, the water must move more slowly than the oil from the injector to the producer.

So far, the chapter discussions have highlighted the development of techniques for understanding and analyzing key aspects of oil/water displacement. These techniques predate modern computers; hence, they were developed to simplify the real reservoir problem sufficiently to allow various engineering calculations to be made.

Of course, the availability of modern computers and advanced numerical-reservoir-simulation software has rendered many of these simplifying assumptions unnecessary when quantifying waterflood-type water/oil displacements in real reservoirs. Nevertheless, these historic techniques have been discussed here to provide an understanding of the dynamics of the water/oil-displacement process and the primary variables that influence the recovery efficiency, as background to the discussion of waterflood design in the next section.

11.7 Waterflood Design

The design of a waterflood has many phases. First, simple engineering evaluation techniques are used to determine whether the reservoir meets the minimum technical and economic criteria for a successful waterflood. If so, then more-detailed technical calculations are made. These include the full range of engineering and geoscience studies. The geologists must develop as

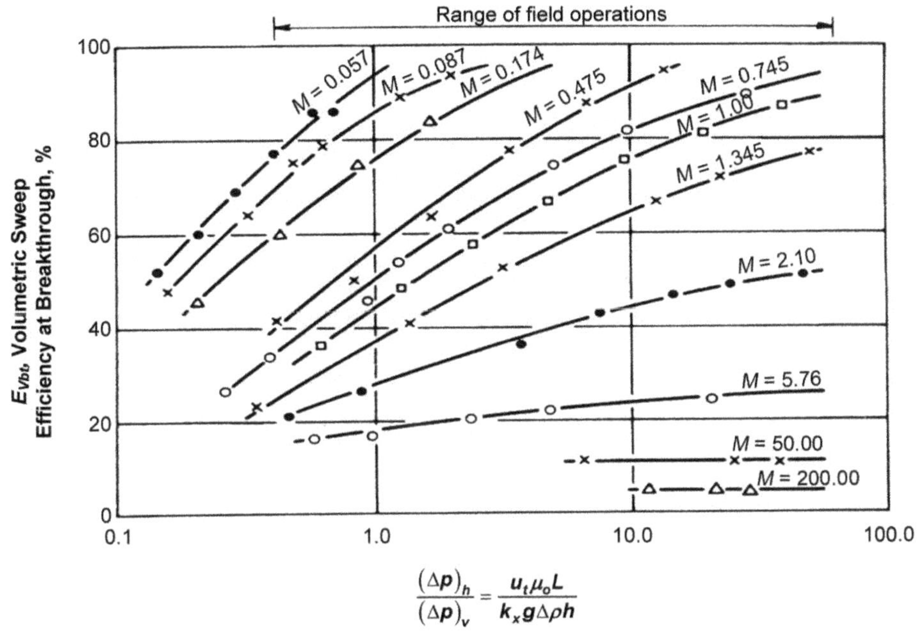

Fig. 11.32—Correlation of scaled-model experiments to determine the effect of gravity segregation on volumetric sweep efficiency in uniform linear systems.[2]

complete an understanding as possible of the internal character of the pay intervals and of the continuity of nonpay intervals. This preflood understanding often is limited because the injector/producer wells' connectivity has not been determined quantitatively. Interference testing can provide insight into connectivity when its cost is justifiable. Data gathered from smart wells can be particularly helpful in determining connectivity in high-cost environments where there is a limited number of wellbores. Analogs also can prove useful. Otherwise, little definitive data will be available until after there has been significant fluid movement from the injectors toward the producers.

The engineer will make a number of reservoir calculations to determine the well spacing and pattern style that will be used in a particular flood. These choices are based on the available understanding of the reservoir geology, the proposed design of surface facilities (particularly water-injection volumes), and any potential limits on the numbers of injectors and producers. Such factors are interrelated in terms of capital and operating costs and oil-, water-, and gas-producing rates to define the overall economics of the project. In making these preliminary calculations, facility capacities need to be flexible because as the waterflood progresses, there almost certainly will be modifications to the original designs and operating plans.

Fig. 11.17 shows a variety of injector/producer pattern layouts that can be considered. In reality, the existing wellbore locations might limit the pattern layout to a nonsymmetrical arrangement like that shown in **Fig. 11.38**. Also, as shown in Fig. 11.23, the orientation of the rows of producers and injectors must take into account any permeability anisotropy and natural-fracture orientation. At offshore locations, the number of well slots on the drilling platforms limits the number of producers and injectors and their layout.

In this section, a number of waterflood design considerations will be discussed briefly. (Rose *et al.*[3] is entirely devoted to this topic.) The design aspects discussed below include

- Injection-water sensitivity studies.
- Injection wells, injectivity, and allocation approaches, including well fracturing.

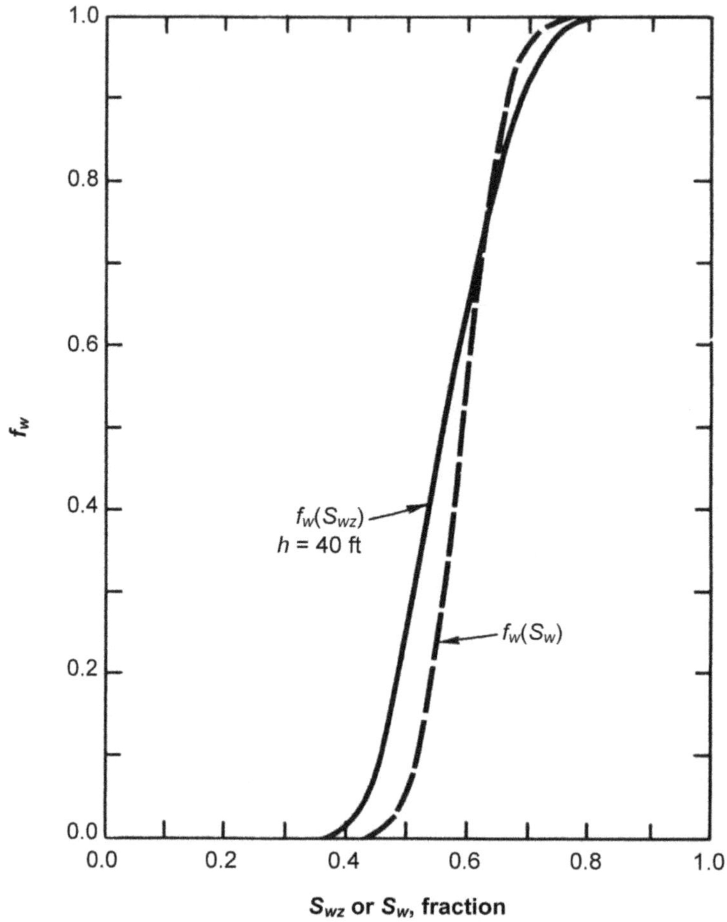

Fig. 11.33—Comparison of the fractional-flow curve for vertical equilibrium with the fractional-flow curve obtained from fluid and rock properties.[2] (h = reservoir thickness, ft, and S_{wz} = thickness-averaged water saturation for vertical equilibrium case with h = 40 ft, fraction PV.)

- Pilot waterflooding.
- Production wells.
- Surface facilities for injection water.
- Surface facilities for produced fluids.

11.7.1 Injection-Water-Sensitivity Studies. The factors to which injection-water-sensitivity studies relate are water-source and -volume options, source-water/connate-water compatibility, and source-water/reservoir-rock interactions. After the preliminary reservoir evaluation indicates that waterflooding is likely to be economically justified and that it will increase significantly the volume of oil recovered, the next consideration is to find an acceptable source from which to obtain enough water for the proposed waterflood project. **Fig. 11.39** schematically shows the variety of natural sources for such water. Onshore locations typically obtain injection water from subsurface aquifer intervals or nearby streams or rivers. Nearshore and offshore waterflood projects typically use seawater.

Source-water/connate-water compatibility mainly concerns whether mixing the two waters causes any precipitation of insoluble carbonate or sulfate compounds that might impair reservoir permeability. Although permeability impairment typically is not a major consideration,

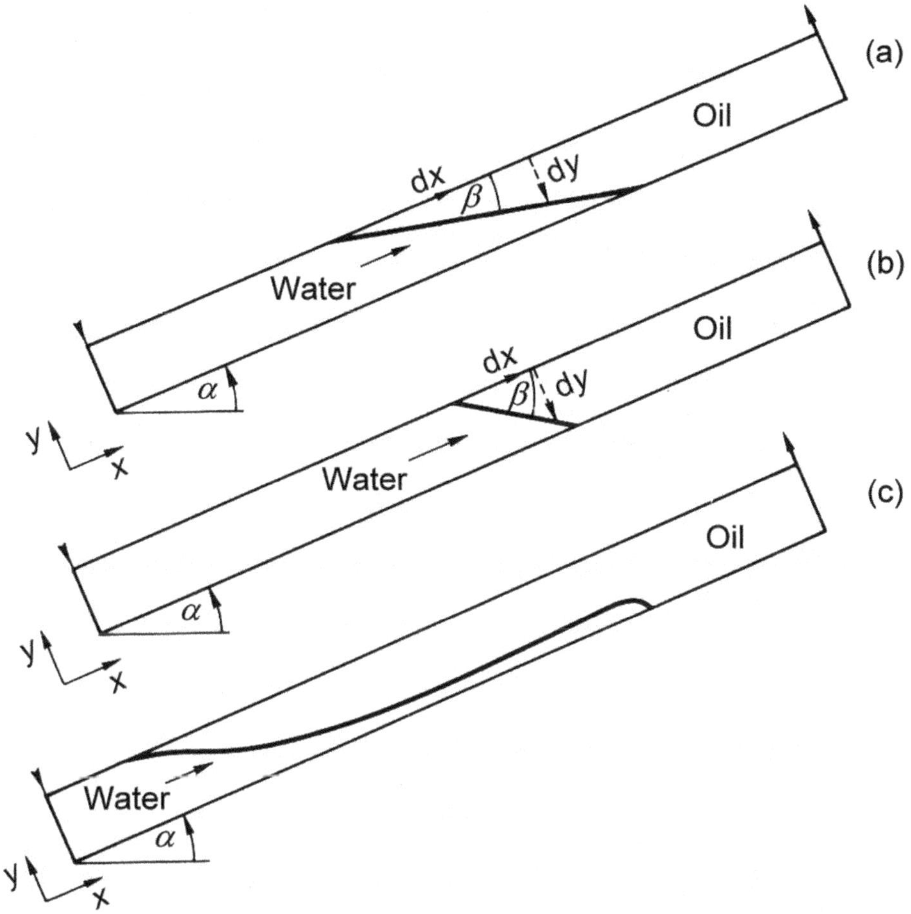

Fig. 11.34—Stable [(a) and (b)] and unstable (c) displacement in gravity-segregated displacement.[2] For (a), $G > M - 1$, $M > 1$, and $\beta > \alpha$. For (b), $G > M - 1$, $M < 1$, and $\beta > \alpha$. For (c), $G < M - 1$.

precipitation and scale buildup in pumps and other surface water-handling equipment can cause costly downtime and repairs.

Potential sensitivity of the reservoir pay intervals to the injection water is a major consideration. For sandstone reservoirs that contain various types of clay, the key consideration is whether there exists clay sensitivity to the difference between the connate-water salinity and the injection-water salinity, particularly for freshwater injection-water sources. Such sensitivity can occur either as clay swelling or as mobilization of clay fines, both of which can reduce reservoir permeability significantly. For high-porosity chalk reservoirs, the injection-water/reservoir-rock interaction might weaken the rock framework and cause pore collapse and surface subsidence.[27]

Another aspect of injection-water sensitivity is the amount and size of suspended particulate being carried by the injection water. This is a concern mainly when using surface water sources for the injection water. An example of where this is a significant consideration is the Kuparuk oil field on the North Slope of Alaska, U.S.A., where nearshore ocean water is the waterflood injection water. There, the spring runoff down the rivers from the Brooks Mountains can cause the nearshore ocean water to contain unacceptable amounts of solid particulate for several weeks of the year. Similar problems occur in the Gulf of Mexico in fields near the

Layer 3	$h = 10$ ft, $k = 200$ md, $\phi = 0.20$, $k_{rw} = 0.4$, and $k_{ro} = 0.8$
Layer 2	$h = 20$ ft, $k = 100$ md, $\phi = 0.17$, $k_{rw} = 0.3$, and $k_{ro} = 0.6$
Layer 1	$h = 10$ ft, $k = 50$ md, $\phi = 0.15$, $k_{rw} = 0.25$, and $k_{ro} = 0.5$

Fig. 11.35—Individual layer properties for a three-layer example (after Ref. 26). (ϕ = porosity, fraction BV.)

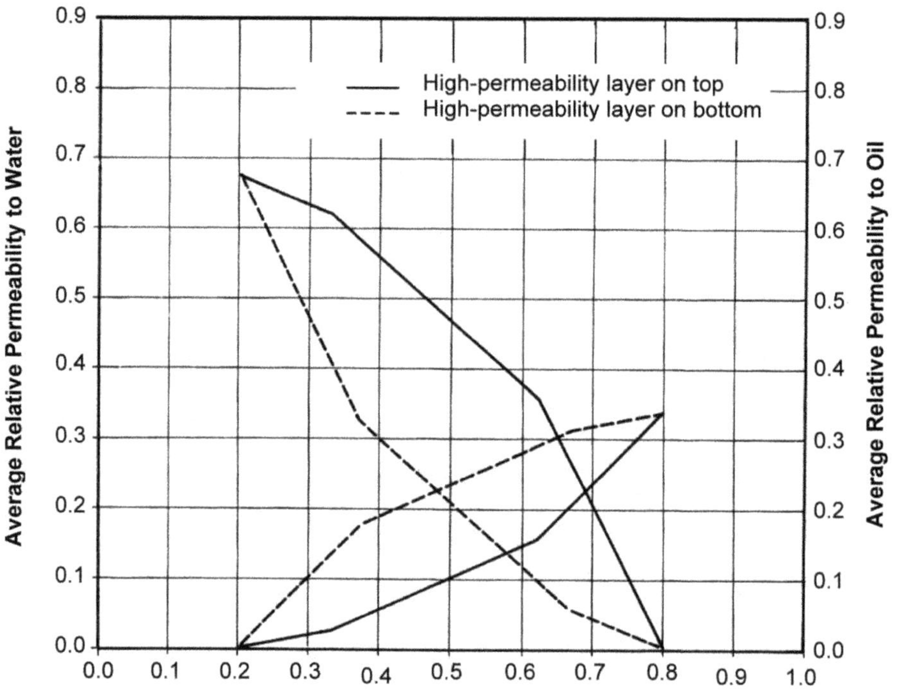

Fig. 11.36—Averaged relative permeability curves for a three-layer example (after Ref. 26).

mouth of the Mississippi River. Also in the Gulf of Mexico, water that is drawn from too near the surface often contains organic matter that can reduce injectivity.

11.7.2 Injection Wells, Injectivity, and Allocation Approaches, Including Well Fracturing. Several aspects of the design and operation of water-injection wells are critical to their success. The first is that these wells must have sufficient injectivity to flow the desired volume of water into the reservoir each day. The expected injectivity can be calculated on the basis of routine core analysis, special core analysis and/or log data, and the existing production wells' productivity; however, well injectivity often is not known until water actually is injected into the reservoir interval. This is because the near-wellbore "skin" (a rock volume of reduced permeability around the wellbore) is not known until an actual well test is conducted. Injection wells

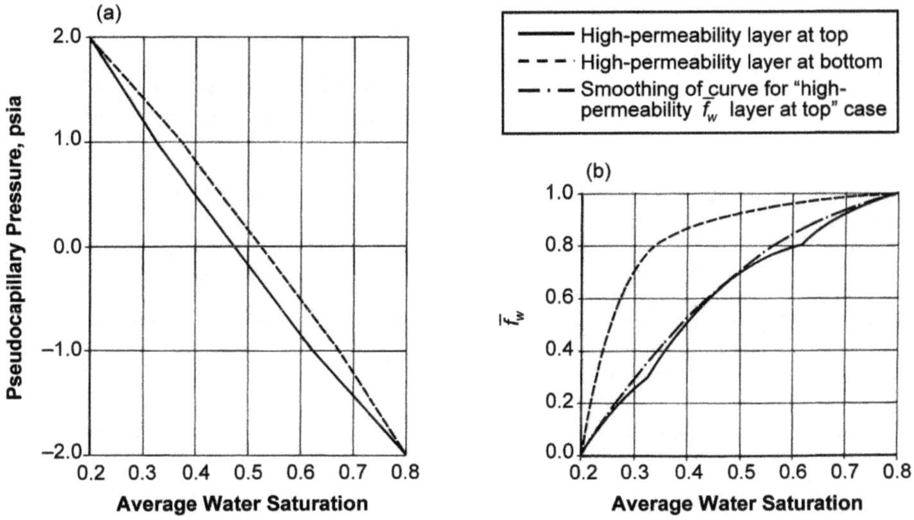

Fig. 11.37—Pseudocapillary pressures (a) and fractional-flow curves (b) for a three-layer example (after Ref. 26).

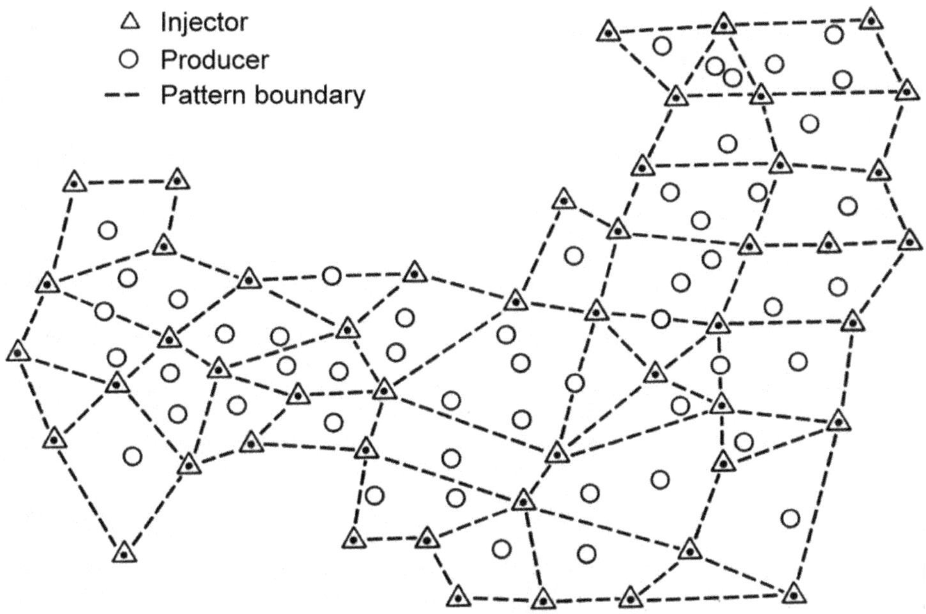

Fig. 11.38—Irregular five-spot pattern layout.[3]

can be fractured to eliminate positive skin in the near-wellbore region; however, fracturing must be done carefully to avoid fracturing out of the reservoir interval and into adjacent porous and permeable intervals into which injection water can be lost.

An aspect of well injectivity that has been studied during the last 20 years is the change in rock stresses that is caused by the cooling effect of the injection water on the near-wellbore region around injectors. This happens particularly in Arctic and offshore waterflood operations, where the injection-water temperatures can be considerably below the reservoir temperature

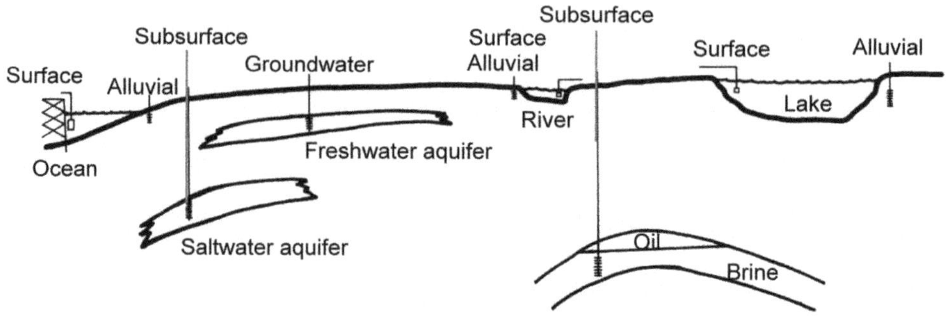

Fig. 11.39—Possible injection-water sources.[3]

(i.e., more than 100°F difference). Perkins and Gonzalez[28] have studied this phenomenon and found that the cooling effect reduces the earth stresses by several hundred psi. Hence, in the reservoir, a small area around water injectors' wellbores will fracture more easily, giving that area enhanced permeability (or negative skin).[28] For the Prudhoe Bay field on the North Slope of Alaska, U.S.A., the fracture gradient was reduced to as low as 0.50 psi/ft from the original fracture gradient of 0.60 to 0.70 psi/ft.[29]

Another critical aspect of water-injection-well design and operation is the allocation of water to zones being waterflooded. Having the ability to allocate injection water as desired to the various waterflooded intervals is important for waterflood success because the overall waterflood is controlled primarily at the injection wells, not at the production wells. This is not an issue if there is only one reservoir interval, but in many oil fields, there are multiple reservoir intervals being waterflooded at the same time. If possible, the injection-well bottomhole-tubing, packer, and perforation configuration should be designed to allow control of the relative volumes of water that are injected into the various intervals being waterflooded. This can be accomplished if each injection well is perforated in only one reservoir interval, but one reservoir interval per injector is unlikely to be cost-effective compared to the alternative of fewer wells with more-complicated arrangements of chokes, tubing strings, and packers, particularly if there are multiple pay intervals stacked on top of each other.

Optimum completion design is site-specific and must be based on mechanical and reservoir characteristics for the project at hand.

11.7.3 Pilot Waterflooding. Pilot waterfloods seldom are used today because of the wealth of experience in waterflooding; however, in many situations, they have been conducted to provide more quantitative data on the potential for successful waterflooding on a fieldwide scale.[30] Such pilot waterfloods definitely provide useful data concerning water injectivity, tendencies for early water breakthrough, and additional recovery potential. Determining recovery potential requires a pilot waterflood that is designed to represent what will happen in a full-scale application. Too often, one-pattern pilot waterfloods have been conducted that do not represent the confined injection/production relationship that is needed. Also, if the pilot waterflood is conducted on a well spacing that is considerably smaller than that used for the full-field waterflood (so that injector/producer connectivity data can be obtained sooner), the information it provides might be misleading about the injector/producer connectivity on the larger well spacing of the full-field waterflood. Thus, definitive objectives of a pilot waterflood should be established, and the pilot project should be designed and operated accordingly.

11.7.4 Production Wells. In many cases, the water-injection wells are drilled as new wells; however, the production wells typically are those that already are producing from the oil field. For waterflooding, producers should be completed in the same intervals in which the injection wells are completed. If the production wells are completed in several reservoir intervals, it is best to have sufficient length between the perforated reservoir intervals to allow workover operations to shut off those intervals that are producing much water and little oil by either cement-squeeze operations or by setting a packer in the production tubing.

11.7.5 Surface Facilities for Injection Water. Maintaining high water quality is important for sustaining injectivity, reducing corrosion-related costs, and minimizing equipment plugging. The American Petroleum Institute (API) has published recommendations for analysis of oilfield waters[31] and for biological analysis of injection waters.[32] The industry also has adopted standardized procedures for membrane/filterability tests.[33]

The water-injection surface facilities prepare the water chemically for injection and pressure the water to the desired wellhead injection pressure. Depending on the source of the injection water, the water might need treatment to remove oxygen, prevent scale and corrosion, and chelate the iron. It also might need microbiological treatment and to be filtered to remove particulates.[34,35] What injection-water preparation techniques are used will vary from one waterflood project to the next. (See chapters in the Facilities and Construction Engineering volume for extensive information on preparing water for injection.) This section specifically discusses surface and produced waters, but the techniques that are covered here also are applicable to water that is produced from aquifers.

One major consideration in injection-water treatment is to prevent the reservoir from being "inoculated" with sulfate-reducing bacteria that can cause a reservoir to develop an in-situ H_2S concentration during the waterflood. This particularly is a problem when using ocean water, which contains both the sulfate-reducing bacteria and the sulfate ions that are their food supply. Once the sulfate-reducing bacteria have been introduced into a reservoir, they are essentially impossible to kill; however, they can be controlled with the injection of bactericides such as formaldehyde.[36,37]

Pressuring water to the desired injection pressure is the final step before it is piped to the injection wells. Chapters in the Facilities and Construction Engineering volume of this *Handbook* cover the types of equipment required for increasing the pressure of the injected water. The wellhead injection pressure is calculated by subtracting the weight of the injection-water column from the desired bottomhole pressure, and then adding friction-flow pressure losses down the wellbore.

In a few reservoir situations, "dumpflooding" has been practiced. This is where a water-bearing formation above or below the oil reservoir is perforated, as is the oil-reservoir interval in those same wellbores. Water then is allowed to flow directly from the water-bearing formation into the oil-bearing formation, without ever bringing that water to the surface for any treating or pumping. This is a very simple approach to waterflooding, but generally it has been unsuccessful because the rate of water injection is uncontrolled and limited to the pressure difference between the two formations, which decreases with time as the water-bearing interval is depleted, particularly near the wellbore, and as the oil reservoir interval near the wellbore pressures up.

11.7.6 Surface Facilities for Produced Fluids. The facilities for handling produced fluids for a waterflood must be designed with considerable flexibility. These facilities must handle a wide range of gas-, oil-, and water-production rates over the course of the waterflood, typically a period of several decades. Several chapters in the Facilities and Construction Engineering volume of this *Handbook* cover these issues in detail.

Initially, the production wells are likely to handle only oil and gas, without water production. When water breakthrough occurs, the water volumes will increase and, over time, water will become the great majority of the produced fluids. Accordingly, a variety of water issues must be considered. First is whether the produced fluids can be separated easily or must be treated with heat and/or chemicals in the surface equipment to achieve the desired level of separation. Second is whether the precipitation of scale in the producing wells or the production surface facilities is causing complications. Regarding scaling tendencies and because of increasing environmental concerns, the handling of NORMs has become an issue with respect to produced-water discharges.[38]

Over the duration of a waterflood and as produced-water volumes increase, there is likely to be the need and desire to reinject the produced water. In this situation, the produced water must be treated so that its oil and particulate content is sufficiently small that, when the water is reinjected, these very small oil droplets will not reduce the injectivity of the water injectors.[39] Oil fields in the North Sea and on Alaska's North Slope have had to reinject large volumes of produced water.[40,41] Regarding injectivity losses, experimental coreflood data tend to be more pessimistic than is actual injector performance in the field.[37] In all cases, to reinject produced water successfully, that water must be treated to meet specifications determined to minimize those injectivity losses.[41]

11.8 Waterflood Monitoring

Because a waterflood project spans several decades, it is monitored continuously and routinely by engineers who are responsible for its operations, as well as periodically using more-detailed and specialized technical studies (e.g., full-field numerical-reservoir-simulation studies). There are many opportunities to modify and improve the waterflood as data are acquired and analyzed.

The basics of a waterflood analysis center on material-balance concepts. Applying material-balance concepts means that initially there is "reservoir fill-up" if the reservoir previously had some years of primary production. During this period, the reservoir is repressured to its original reservoir pressure because the injected-water volumes will be substantially greater than the produced-fluid volumes. Thereafter, the waterflood will be operated as a voidage-replacement process.

The earliest waterflood-monitoring techniques were developed soon after the first field applications of waterflooding; they were based on simple plots, maps, and calculations. Among these were the plots published by Dyes *et al.*[42] that estimated the water breakthrough and post-breakthrough behavior of various waterflood pattern configurations. **Figs. 11.40 and 11.41** are examples of these plots for a five-spot pattern and a direct-line-drive pattern, respectively, and can be used to make "first-estimate" waterflood calculations.

In this section, the aspects of waterflood monitoring that are discussed are

- Data acquisition: routine data gathering.
- Special-data acquisition: infill and observation wells and 4D-seismic data.
- Simple waterflood-analysis techniques: *X*-plot, log(WOR) vs. cumulative-oil-production plots, and decline curves.
- Sophisticated waterflood-monitoring techniques.

11.8.1 Data Acquisition—Routine-Data Gathering. Waterflood monitoring begins with the acquisition of the routine data that are necessary for engineering calculations. The routine data include well-by-well daily oil-, gas-, and water-production rates; well-by-well water-injection rates; injection wellhead pressures; and production-well pressure data. Often, the well-by-well daily rate data are back-calculated from the gathering center's total produced volumes of these three phases, and then are allocated back to the individual wells on the basis of periodic individual well tests.

Next, the production- and injection-well data are allocated to the individual reservoir intervals, if multiple reservoir intervals are commingled. This well-by-well data allocation requires that spinner surveys (or their equivalent) be run periodically in the individual wells to determine how much of the fluid is coming from each of the perforated intervals. The spinner surveys should be run both with the well flowing and with the well shut in. Data from these

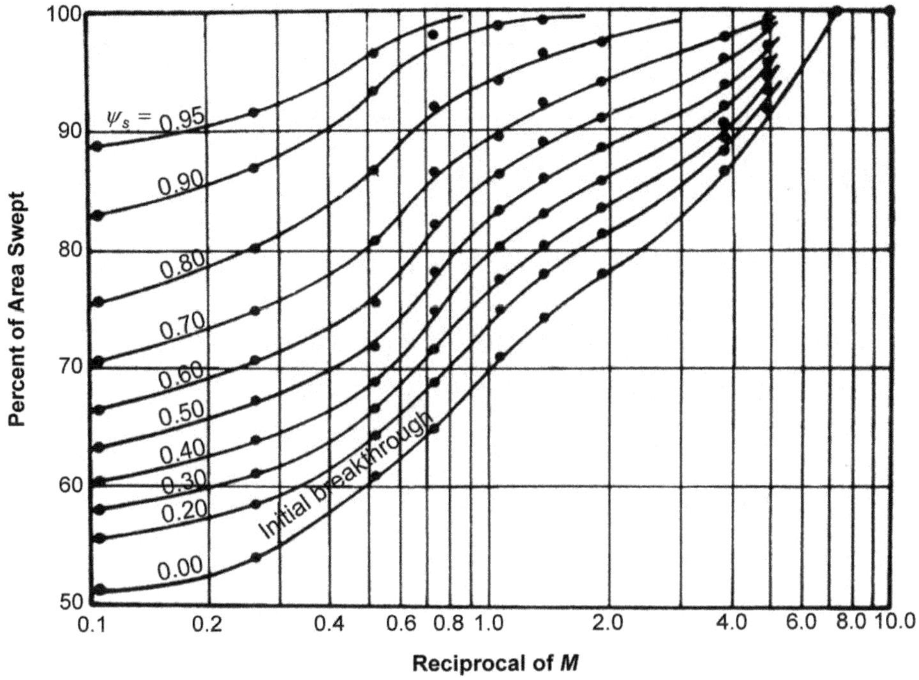

Fig. 11.40—Effect of mobility ratio on oil production for five-spot pattern.[1] (ψ_s = fraction of total flow coming from the swept portion of the pattern.)

surveys, along with pressure-buildup and -falloff data, help in the estimation of the reservoir pressure in the various reservoir intervals. Also, a variety of production logs should be run to estimate changes in fluid saturations in the near-wellbore region as the waterflood progresses.

The field engineers can use all these data in the types of calculations that are described below.

11.8.2 Special-Data Acquisition: Infill and Observation Wells and 4D-Seismic Data. During the waterflood, there are likely to be opportunities to gather additional data away from the injection and production wellbores. These data can take several forms.

Infill wells are likely to be drilled in locations where oil should not have been displaced by the injected water. Consequently, these locations are not good locations for determining how effectively the various portions of the reservoir intervals are being swept by injected water. After drilling, the hydrocarbon distribution actually present at an infill-well location is evaluated using openhole or cased-hole logging. Where formation waters have a sufficiently high salinity, pulsed-neutron thermal-decay time and resistivity logging may be used to evaluate the residual oil saturations. In lower-water-salinity conditions, carbon/oxygen and resistivity logging are used. Using these techniques, the locations of fully flooded, partly flooded, and unflooded reservoir intervals can be determined in new wells and in existing producing wells. Multiple logging runs over time in a producing well allow the monitoring and management of a waterflood. If wells are drilled later during the waterflood, then the residual-oil-saturation distribution can be obtained by use of special coring procedures or special tracer tests, as described elsewhere in this volume of the *Handbook*.

Special observation wells sometimes are drilled at a location in the oil reservoir where the water/oil flood front should be detectable as it passes. Historically, most of these wells were cored but had steel casing, such that standard logging methods could not be used; however, before the development of through-steel-casing resistivity logging, fiberglass casing together

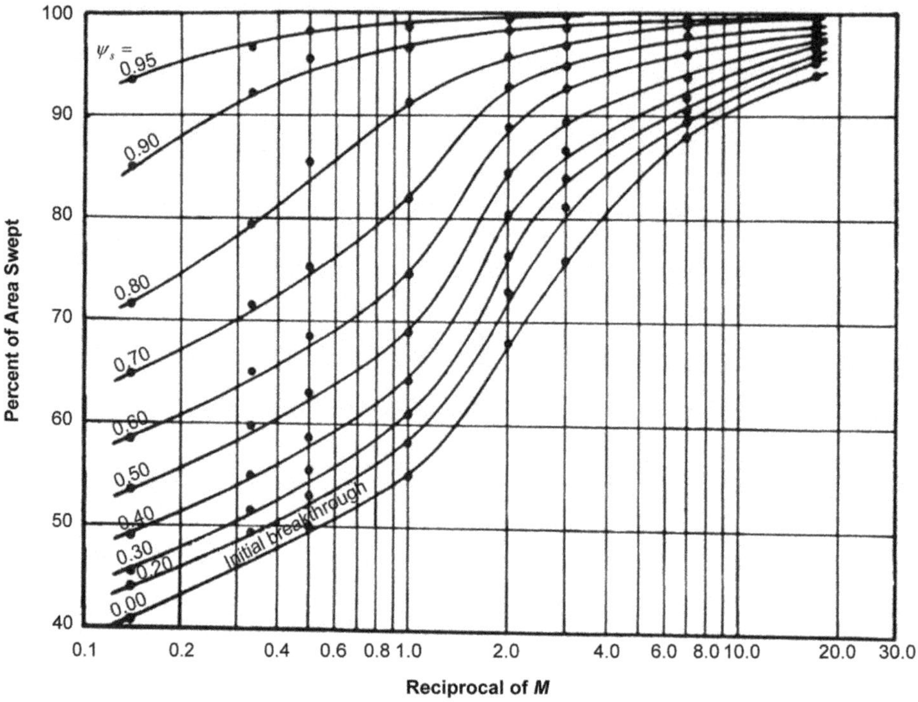

Fig. 11.41—Effect of mobility ratio on oil production for the direct-line-drive (square pattern). In this case, the distance between adjacent rows of wells is equal to the distance between like wells in a row.

with induction-resistivity logging occasionally were used to observe the water/oil displacement process over time.

Recently, the 4D-seismic technique has been developed to determine in what directions the water is moving from the injection wells.[43,44] The 4D-seismic technique compares 3D-seismic data that were obtained before the start of waterflooding to a second or third 3D survey that was conducted some years later. This allows an areal visualization of where there are high-pressure areas caused by water injection and where there are low-pressure areas caused by production (typically from the presence of some free-gas saturation near the production wellbores). Also, the 4D picture might show which portions of the reservoir pay intervals are well connected and which are not.

11.8.3 Simple Waterflood-Analysis Techniques: *X*-Plot, Log(WOR) vs. Cumulative-Oil-Production Plots, and Decline Curves. The initial engineering analysis of waterflood performance is done to ensure that the water is being injected into the wells at the desired well-by-well rates and, if several oil-reservoir intervals are perforated in each injector, that each of those reservoir intervals is taking its appropriate share of the injected water from each of the injection wells. This may appear trivial, but many waterfloods have had significant problems in this regard. Before the waterflood begins, the engineers must estimate how much water should be flowing into each injector. Those estimates are based on fill-up and voidage-replacement calculations for that area of the oil reservoir and for the reservoir as a whole.

A "bubble map" can be used for visualizing the advance and relative volumes of the injected water. **Fig. 11.42** shows the injection wells' bubble map for one of the staggered line-drive pattern elements of one of the oil-reservoir intervals of the Long Beach Unit (LBU) area of the Wilmington oil field in California, U.S.A.[45] Bubble maps are created for each injector by dividing the volume of water injected by the movable oil per vertical reservoir interval [thickness ×

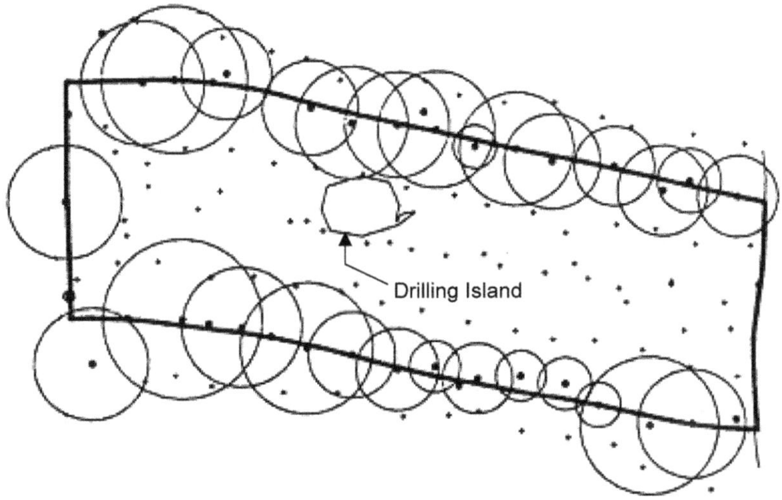

Fig. 11.42—Layer injection-bubble map for Voidage Block 3 FO sand interval—LBU of the Wilmington oil field in California, U.S.A.[45]

porosity × $(1 - S_{wc} - S_{orw})$] to calculate the area that should have been swept by that volume of injected water; then presenting these areas on a map as circles of various sizes centered on each of the injection-well locations.

Another aspect of waterflood monitoring is to track the performance of the production wells. As noted earlier, some oil reservoirs will have had a considerable period of primary production during which the reservoir pressure was drawn down below the bubblepoint pressure and a gas saturation developed. These reservoirs first will "fill up," which means that they will go through a period during which the injection water displaces mobile gas and increases the reservoir pressure to force free gas back into solution in the oil. During the early part of the fill-up period, the production wells will see minimal response because of water injection. When the reservoir pressure has returned to its original value, the water injection becomes a voidage-replacement operation. During the voidage-replacement period, the engineer must make sure (1) that the volume of injected water equals the reservoir voidage of oil, gas, and water production, and (2) that the various areas or patterns of the reservoir are being balanced to maximize the oil recovery and to minimize water-handling operating costs.

A number of graphs of the production and injection data can be prepared to help analyze the waterflood performance. For example, a conformance plot is a plot of cumulative oil recovery (or oil-recovery efficiency) vs. net displaceable hydrocarbon PV injected. A waterflood-performance envelope is defined by drawing an obtuse triangle that is bounded by recovery at the start of waterflooding, the maximum oil recovery at 100% of the x-axis, and a third point, F_{pvg}, which is related to the net injection required to displace the existing gas saturation at the start of the waterflood and is defined as:

$$F_{pvg} = S_g / (1 - S_{wi} - S_{orw}), \quad (11.25)$$

where S_g = gas saturation, fraction.

Because actual performance cannot fall outside the performance envelope, this plot is a check on fluid allocation for a pattern. This plot also can indicate when injection water is being lost to thief zones. When slope changes are noted in this plot, the possible causes should be investigated. **Fig. 11.43** is a conformance plot for the Kuparuk River oil field on Alaska's North Slope.[46]

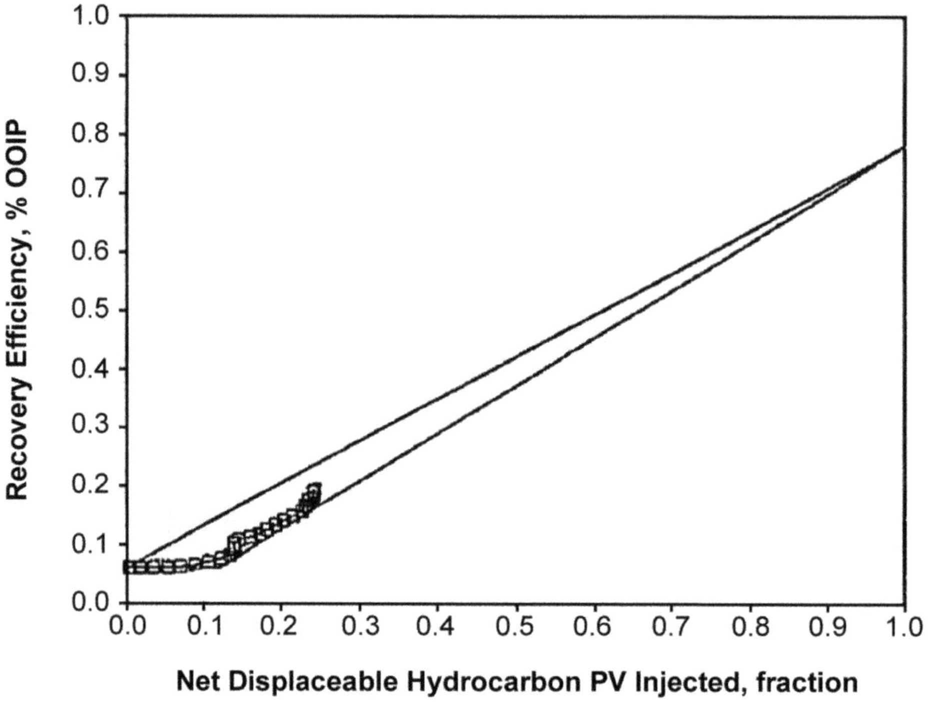

Fig. 11.43—Example of a conformance plot.[46]

Other plots that typically are made after water breakthrough are the *X*-plot and its close cousin, the log(WOR) vs. cumulative-oil-production plot. The *X*-plot technique, developed by Ershaghi and coworkers, is based on the leaky-piston-displacement concepts of Buckley and Leverett[47] and assumes that the plot of log (k_{rw}/k_{ro}) vs. S_w is a straight line.[48,49] The *X*-plot is a graph of recovery, E_R vs. X, where $X = \ln\left(\frac{1}{f_w} - 1\right) - \frac{1}{f_w}$. This plot's underlying assumption suggests that its usefulness theoretically is limited to higher water cuts.

The "cut-cum" plot, or plot of log(WOR) vs. cumulative oil production, easily can be made using production data for each well and for reservoirs as a whole. These plots generally are useful predictors of future waterflood performance because there is a considerable period of straight-line behavior on these plots for many wells and reservoirs when the waterflood is fully developed without major variations in field operations. Analysis of log(WOR) vs. cumulative-oil-production plots for waterflood analysis has been conducted using numerical reservoir simulation. These simulations show that the linear trends on this type of plot are found even at low WOR values and also are found for a variety of reservoir layering, flood configurations, and operational changes.[50]

Decline curves also have been used for waterflood analysis. Where waterfloods were initiated in depleted sandstone reservoirs, empirical correlations were developed to estimate the likely oil-rate increases during fill-up and while the oil bank moves toward the production wells, and then to estimate the oil-rate declines as the WOR increased later in the life of the waterflood.[51] **Fig. 11.44** shows the production response for this situation. For the oil-rate-increase and -decline periods, the plot has exponential oil-rate-vs.-time characteristics.

Hall plots and Hearn plots often are used to monitor injection wells. Use of these plots helps to maximize water-injection rates, which accelerates oil production from offsetting producers. On a Hall plot, the bottomhole injection pressure is plotted vs. cumulative water injection

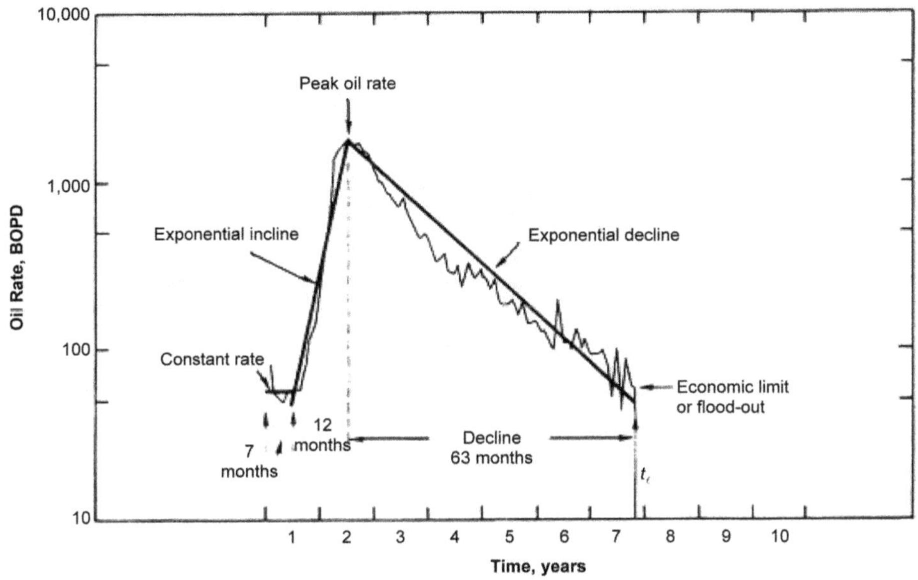

Fig. 11.44—Production response of a fully developed waterflood with water input to all injection wells at the beginning of the flood.[2]

to monitor reservoir fill-up and average-reservoir-pressure increase. On a Hearn plot, the inverse injectivity index is plotted vs. cumulative water injection. Both types of plots also can be used to determine whether the water-injection rates are being kept below those allowed by the fracture-parting pressure.[52]

A combination of these simple plot and calculation techniques typically is used for waterflood analysis. By plotting both oil rate vs. time and log(WOR) vs. cumulative oil, the engineer can better understand how well the waterflood is performing compared to original estimates and can determine what changes, if any, are needed.

Keep in mind that most of these techniques are premised on continued current operations. If there are significant changes in the allocation of injection water or if there are well workovers or pattern realignments, then the trends on these plots might not remain straight lines from which future waterflood performance is easy to predict.

11.8.4 Sophisticated Waterflood-Monitoring Techniques. The modern numerical reservoir simulator is the best tool for performing waterflood analysis, including history matching of past performance and projection of future performance for continued current operations or for various operational and well changes. However, this tool generally is used with an updated history match only every 5 to 10 years.

Between major studies, the field engineers typically use simpler surveillance methods that have been upgraded by the availability of the notebook computers that have sizeable hard disks and rapid computing capabilities. New software packages have been developed that analyze trends and can handle large amounts of electronically acquired data.

One example of the use of this approach is the set of techniques used for the LBU area of the Wilmington oil field, which has 1,200 wells, multiple oil reservoirs, and 27 years of waterflood history.[45] **Fig. 11.45** shows the logic used for the surveillance calculations. As Fig. 11.45 shows, a large waterflood involves massive amounts of data that must be handled logically and consistently so that the engineers can obtain useful results.

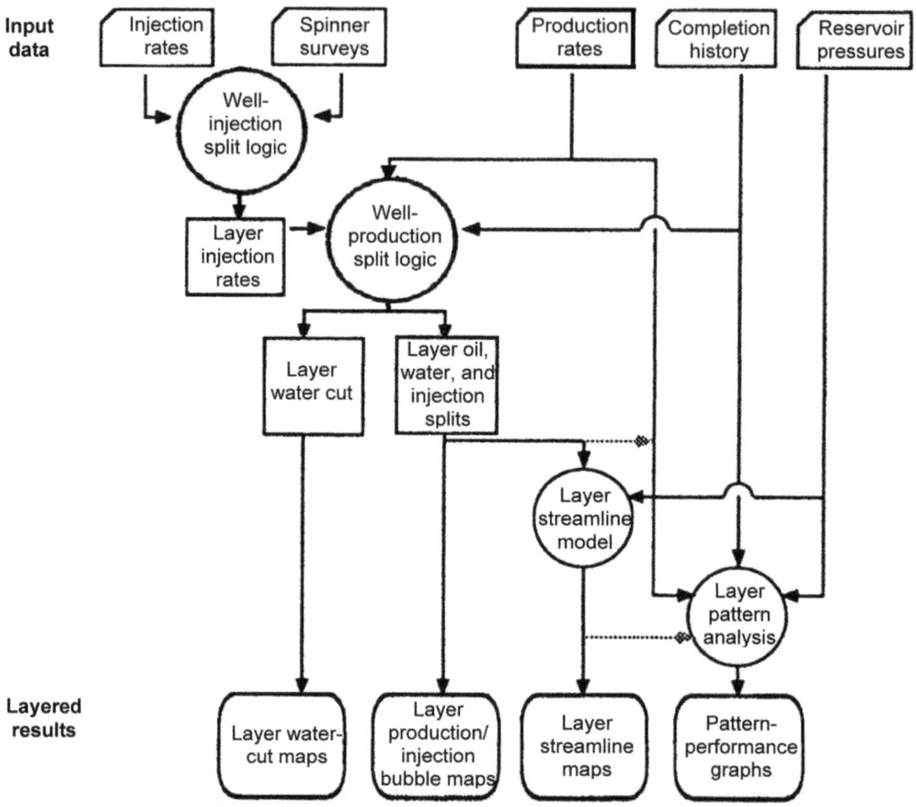

Fig. 11.45—Surveillance-overview logic for a layered waterflood.[45]

The routine waterflood-analysis calculations that are used for the Kuparuk River oil field[46,53,54] are another example of this approach. The Kuparuk River oil field covers more than 200 square miles and contains 600 patterns, with two separate reservoir intervals. Again, a massive amount of well data has been gathered over more than a decade of waterflooding and primary production. All of the simple calculation procedures that were discussed in earlier sections of this chapter [decline curves, log(WOR) vs. cumulative-oil-production plots, and *X*-plots] are used to evaluate well-by-well and pattern performance. **Figs. 11.46 and 11.47** show relationships among the various calculation and database modules within the material-balance shell and the object relationships.[54]

These two examples are of very large oil fields. Oil fields with fewer injection and production wells will require less-extensive waterflood-monitoring calculations; however, all the types of material-balance calculations will be required to determine whether the waterflood is performing as expected and whether significant operational changes are needed.

Also keep in mind that the usefulness of all these calculations depends greatly on the quality of the input data. The original field data must be reviewed for completeness and accuracy before entering them to these types of calculations. Where gaps in the data or clearly erroneous numbers are found, the engineers must judge how to edit the data or adjust calculations for such data problems. Where total field data has been allocated to the individual wells, the allocation procedures should be checked, including whether they have changed over time and thus caused changes in the slopes of some of plotted data—changes that are not real, but that are artifacts of the data-allocation procedures.

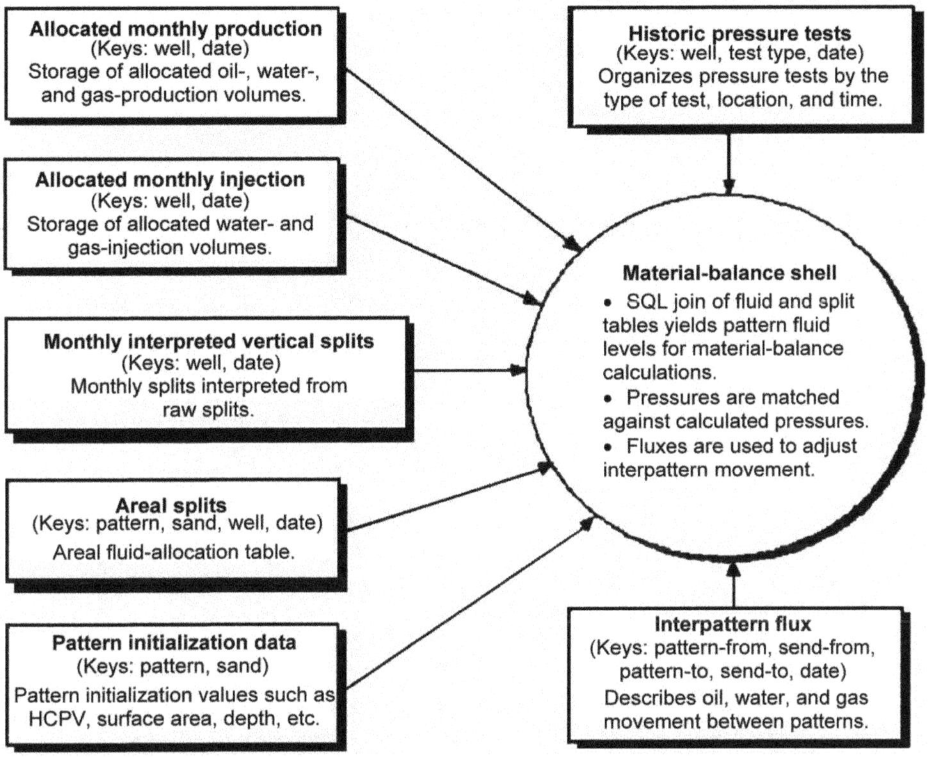

Fig. 11.46—Relationship of relational database management system (RDBMS) tables to material-balance shell logic.[54] (HCPV = hydrocarbon pore volume, and SQL = structured query language.)

11.9 Waterflood Field-Case-Studies Examples

Several oil fields have been discussed in this chapter as examples of the use of various waterflood concepts and calculation procedures. The SPE technical literature contains many papers about major fields that have been waterflooded for a decade or more. In this section, a few of these oil fields are described briefly to highlight these waterflooding projects from their inception to the present, including how they were modified as they proceeded and as detailed engineering analyses were completed, and what complications were encountered. As noted earlier, many technical papers are available on these topics, but here only one to three publications are referenced for each oil field.

The waterfloods in these examples involve oils that range from light (30 to 40°API, with reservoir viscosities of 0.2 to 0.4 cp) to heavy (15 to 20°API, with reservoir viscosities of 30 to 70 cp). In all cases, the waterfloods have maintained reservoir pressure and increased oil recovery very successfully. A few other waterfloods also will be discussed briefly to highlight additional waterflooding considerations. The fields and waterfloods that are discussed here are Ekofisk (North Sea, offshore Norway), Wilmington (southern California, U.S.A.), Kuparuk River (Alaska North Slope, U.S.A.), west Texas carbonate waterfloods (U.S.A.), and Kirkuk (Iraq).

Three of these fields are various types of carbonate, and two are sandstone. Geographically, one is in the North Sea; one in the Long Beach, California, harbor area; one in the Arctic on Alaska's North Slope; and two are in onshore oilfield areas. In some of these oil fields, the waterflood has been followed by EOR that uses miscible- or immiscible-gas injection in which some water injection has been continued to provide mobility control.

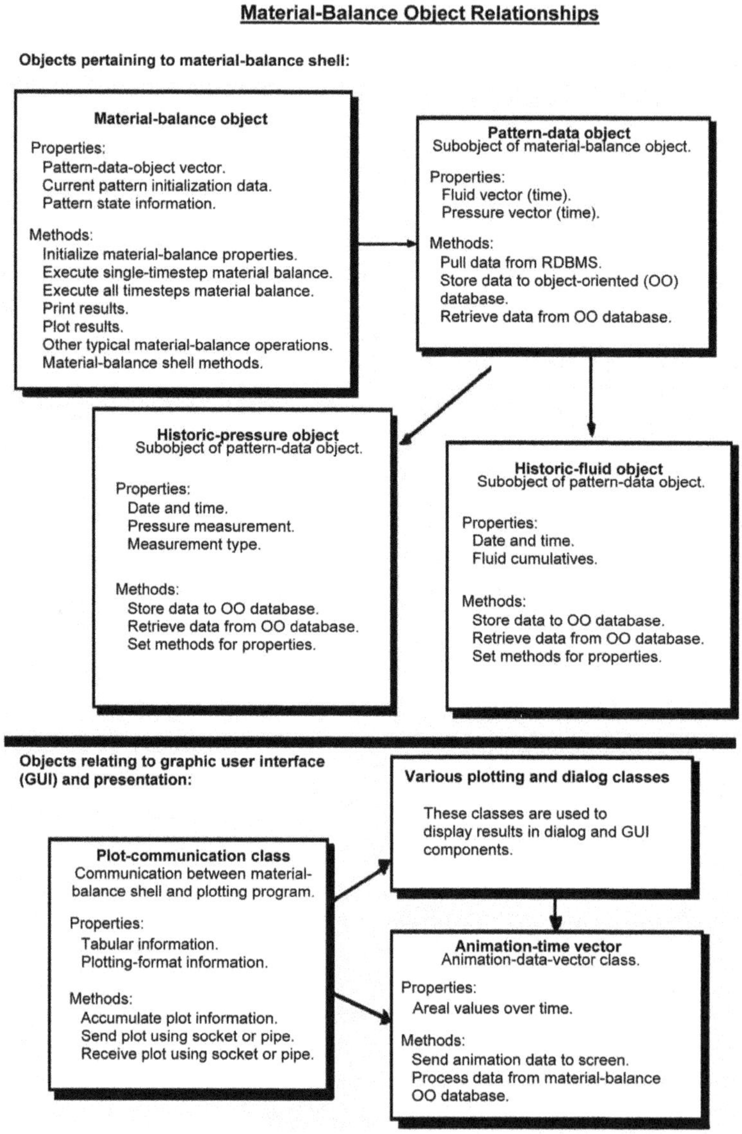

Fig. 11.47—Material-balance object relationships.[54]

The reservoir-management chapter in this volume of the *Handbook* also describes the performance of several fields that are being waterflooded. That discussion's emphasis is on reservoir management; however, it does include some information on recovery performance and depletion strategies.

11.9.1 Ekofisk (North Sea).[55–57] The Ekofisk oil field is in the North Sea, south of Norway. It is a large, carbonate reservoir [(6.4 billion bbl stock tank original oil in place (STOOIP)] that has two zones, Ekofisk and Tor, that are high-porosity, fractured chalks with matrix permeabilities of approximately 1 md and effective permeabilities that range from 1 to 50 md. Discovered in 1969, the Ekofisk field was found at very high pressure [7,120 psia at 10,400 ft true vertical depth subsea (TVDSS)] but with an initial bubblepoint pressure that was 1,600 psi below initial reservoir pressure. Ekofisk's oil is 38°API, has a viscosity of approximately 0.25

cp, and has a solution gas/oil ratio (GOR) of more than 1,500 scf/STB. Primary production began in June 1971 and peaked in 1976 at 350,000 BOPD from 30 production wells (with 8 gas reinjection wells). **Fig. 11.48** shows the structure of the Ekofisk field and its injection- and production-well locations as of early 2003. Laboratory-test results indicated that waterflooding by water imbibition into the low-permeability chalk was favorable for the Tor formation, but the laboratory results for the Ekofisk formation were variable.[53] A waterflood pilot of the Tor formation was initiated in 1981, and favorable results were obtained by 1983. A 30-slot water-injection platform with an injection capacity of 375,000 BWPD started up in 1987. A second waterflood pilot was run in the lower Ekofisk formation from 1985 to 1987, and its results were positive.[30] Further laboratory studies of the upper Ekofisk formation were somewhat negative, but a water-injection test into this zone and a coring of a sidetrack well 6 months later indicated that the upper Ekofisk formation also could be waterflooded successfully. By the early 1990s, all three intervals were receiving injected water.

Surface subsidence was a major issue at Ekofisk. By 1984, the seabed had subsided approximately 10 ft, prompting a major project to jack up the offshore platforms. A major field study in 1992 concluded that using water-injection pressure maintenance to arrest the reservoir pressure decline could minimize future seabed subsidence. Voidage replacement was achieved in 1993. Additional laboratory studies found that water injection had induced shear in the chalk. Shear failure and water-weakening of the rock matrix causes additional deformation of the chalk, even under conditions of constant or decreasing stress levels. Despite the use of voidage-replacement waterflooding, seabed subsidence continued until 1998, when the subsidence rates slowed dramatically because the water-weakening effect was expended and the reservoir pressure had increased.

In 1997, production began from the 50-slot production platform 2/4 X, and in 1998 full processing of the Ekofisk fluids was handled by the 2/4 J processing platform. These new "Ekofisk II" facilities replaced the aging original facilities and were designed to increase operational efficiency and to allow safe and economical production until at least 2028, the end of the current license period. The current best estimate of the ultimate recovery factor from the start of production through waterflooding is 44% of OOIP.

During the past 20 years of waterflooding, many operational changes have been made at Ekofisk. The changes were a logical progression that was based on laboratory studies, field pilot tests, and engineering analyses of field production and pressure data. All of this has led to a very successful waterflood project, but with a few unexpected complications that the engineers had to handle in the course of the waterflood project.

11.9.2 Wilmington Oil Field (California). The LBU area of the Wilmington oil field (southern California, U.S.A.) is mainly under the Long Beach harbor and contains more than 3 billion bbl of OOIP.[45,58,59] This oil field is a large anticline that is crosscut by several faults with displacements of 50 to 450 ft. It consists of seven zones between 2,500 and 7,000 ft TVDSS, the upper six of which are turbidite deposits of unconsolidated to poorly consolidated sandstone (1 to 1,000 md and 20 to 30% BV porosity) interbedded with shales. The gross thickness of 3,300 ft contains 900 ft of sandstone. In the largest zone, the Ranger (> 2 billion bbl of OOIP), the oils range from 14 to 21°API gravity and from 20- to 70-cp viscosity; hence, any waterflood would operate under a very unfavorable mobility ratio, and early water breakthrough would be expected.

From its discovery in 1936 to the 1950s, most of the onshore portion of this oil field (the non-LBU area of the Wilmington oil field) was produced using the pressure-depletion oil-recovery mechanism. Because of this, there was significant surface subsidence—up to 29 ft, with some areas dropping from several feet above sea level to below sea level (but protected by dikes). Development of the LBU area was delayed until an agreement with the government

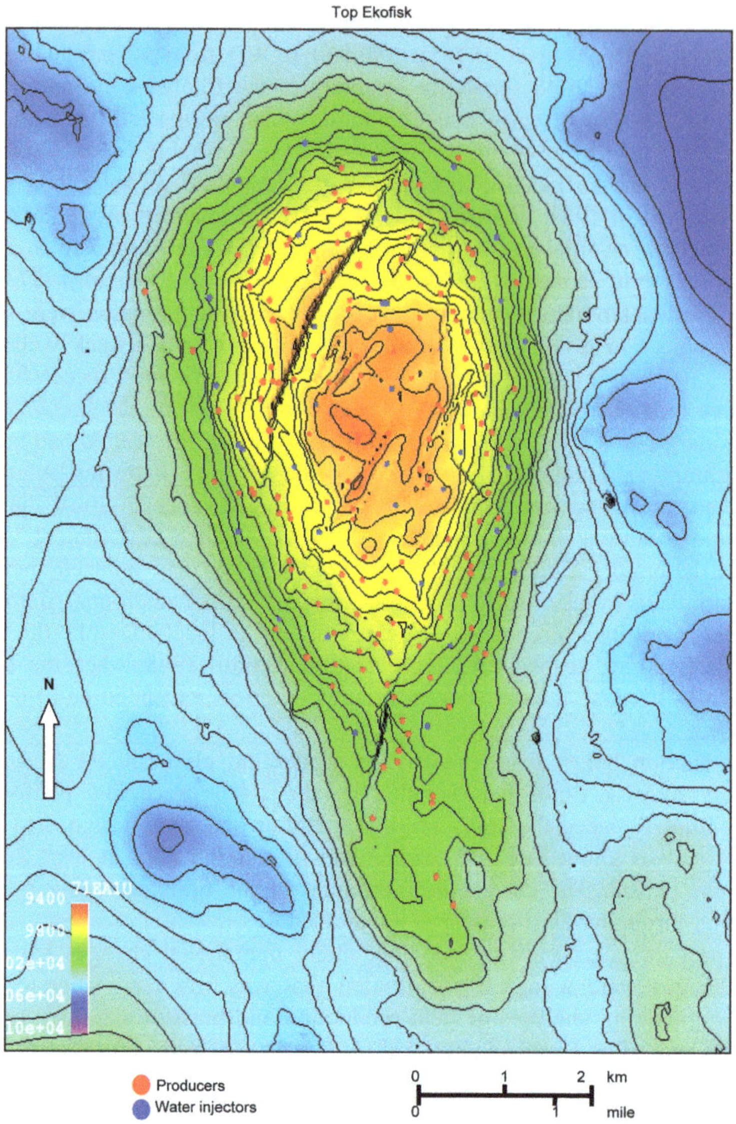

Fig. 11.48—Areal map of Ekofisk field with injection- and production-well locations (image provided courtesy of ConocoPhillips).

was reached that required voidage-replacement waterflooding to be implemented from its beginning to prevent further subsidence. LBU was developed in the mid-1960s, with the directional drilling of more than 1,000 wells from four artificial islands and from the nearby pier area. The early completions consisted of gravel-packed slotted liners that were up to 1,000 ft long in the injectors and the producers.

The LBU Ranger-zone waterflood was a 3:1 staggered line drive on a 10-acre well spacing (see **Fig. 11.49**); peripheral waterflooding has been used in the other zones. Oil production peaked in 1969 at 150,000 BOPD. The oil production has declined slowly since then, and the water production rate has increased steadily over the years. Water has been injected at rates of up to 1 million BWPD. Current oil production is approximately 32,000 BOPD at an average water cut of approximately 96%. To date, total production is > 940 million bbl of oil.

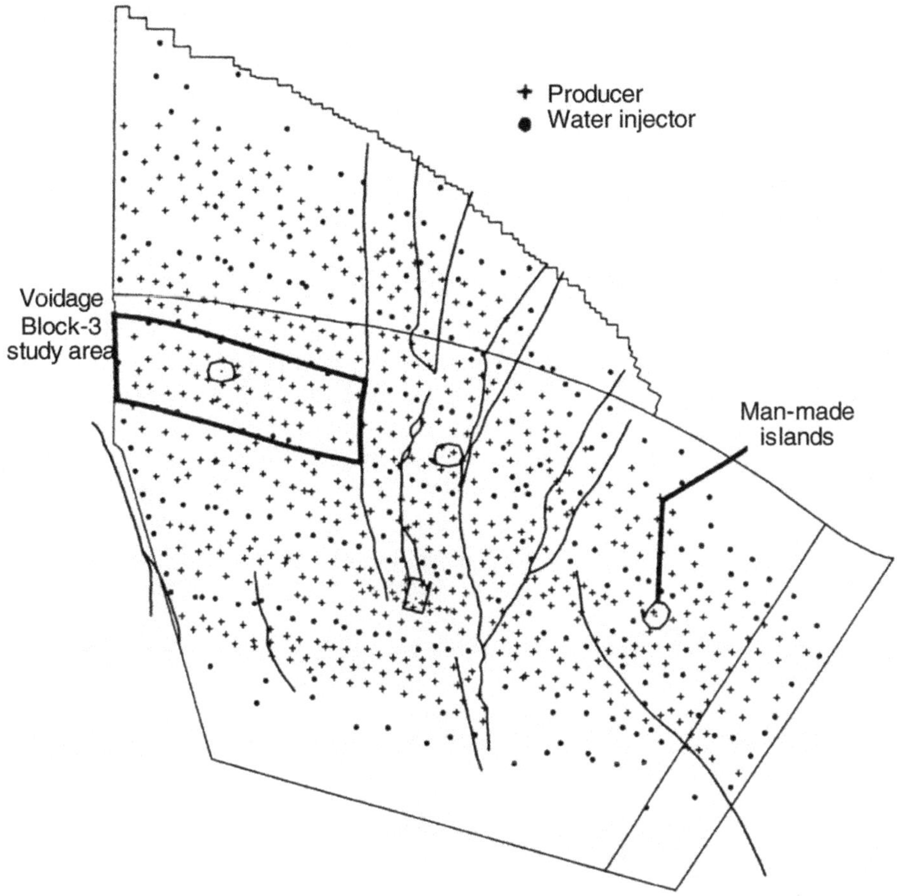

Fig. 11.49—Areal map of injection and production wells in the Ranger-zone waterflood, LBU, Wilmington oil field.[45]

As years of oil-production and water-production and -injection data were acquired, the field engineers determined that some of the initial well-completion practices needed to be changed. The early completion techniques caused thief zones to develop in the higher-permeability sands, so that the deeper sands could not be pumped off. Between 1979 and 1986, by the drilling of 450 new wells into the lower portions of the Ranger and Terminal zones, this massive interval was redeveloped as two or three separate intervals.[58] This subzoning added 160 million bbl of oil reserves and increased the field rate by 30,000 BOPD.

In 1991, an optimized waterflood program was undertaken to reduce the volume of produced water, recomplete wells in sands that were not being well swept by the waterflood, and drill new wells in selected locations to improve the performance of the waterflood.[59] Pattern surveillance was used to quantify areas with low water throughputs and to guide the selection of new well locations. To aid the optimization studies, detailed studies of the geology of the field were undertaken and some 3D-seismic data were acquired and interpreted. The 3D-seismic data significantly changed the understanding of orientation of some of the faulting patterns within the field area; the previous fault patterns had been developed from the data gathered during drilling of 1,200 directional wells and from interpretation of early 2D-seismic data. Also in thicker sands, more than 50 horizontal wells have been drilled to capture bypassed "attic"

TABLE 11.2—ESTIMATED ULTIMATE-RECOVERY FACTORS FOR THE LBU PORTION OF THE WILMINGTON OIL FIELD

Zone	OOIP, million bbl oil	Recovery Factor, % OOIP
Ranger West	1,335	38
Ranger East	875	29
Terminal (upper and lower)	415	35

oil and banked oil along faults. This optimized waterflood program has added 125 million bbl of oil reserves and has the potential to add another 90 million bbl of reserves.

Over the past 30 years, the field engineers have monitored and modified the original waterflood design using the full variety of waterflood-analysis techniques, including bubble maps, log (WOR) vs. cumulative-oil–production plots, X-plots, streamtube models, and numerical reservoir simulation for selected intervals within the Ranger zone over most of the LBU area.

One last point: Because of its very unfavorable mobility ratio, the LBU waterflood has undergone several decades of water injection and production with 80 to 97% water cuts (5 to 25 bbl water produced per bbl oil). The injected water has stripped considerable amounts of the dissolved solution gas from the oil in the reservoir sands. This is observed in the producing GORs, which—if all the gas were assumed to be from the oil—would indicate that free gas is being produced along with the oil and water. Instead, what is happening is that each barrel of "dead" injected water that cycles through the reservoir sands is extracting 5 to 10 scf of gas per STB of reservoir oil. This causes the reservoir oil to contain less solution gas and to increase in viscosity as the waterflood progresses.

Estimated ultimate-recovery factors for the LBU as of 2005 are shown in **Table 11.2**.

Overall, the LBU area of the Wilmington oil field has been a very successful waterflood of a lower-API-gravity, more-viscous oil. For much of the waterflood period, the water cuts have been high, 80 to 97%. The LBU waterflood has successfully prevented further surface subsidence.

11.9.3 Kuparuk River (Alaska North Slope). The Kuparuk River oil field is west of the supergiant Prudhoe Bay oil field on Alaska's North Slope and was discovered in 1969.[12,46,50,54] It has approximately 5.9 billion bbl of STOOIP and covers more than 200 sq. miles (see **Fig. 11.50**). The sandstone reservoir consists of two zones [A (62% of STOOIP) and C (38% of STOOIP)] that are separated by impermeable shales and siltstones. Sales oil is approximately 24°API with a viscosity at reservoir conditions of approximately 2.5 cp. The reservoir oil was approximately 300 to 500 psi undersaturated at the original reservoir pressure of approximately 3,300 psia. The reservoir is broken into segments by several north-to-south faults (density of approximately three faults per mile) that have sufficient throw to totally offset adjacent portions of the reservoir. The two major stratigraphic flow units exhibit considerably different properties, with the lower A zone having lower permeability, the C zone having considerably higher permeability, and the difference between them being approximately an order of magnitude.

The oil field was developed in stages, starting with the initially discovered eastern portion. Initial development was on 160-acre well spacing, and production started in 1981. The waterflood began in 1983. Expansion of the water-handling and -injection facilities led to full-field waterflooding in 1985. To date, more than 600 patterns have been developed with approximately 850 wells from 42 drillsites. The pattern style—a 320-acre, nominally 1:1 east/west line drive—takes into consideration the fault alignment. Annual average production peaked at more than 320,000 BOPD in 1992. Water production began in 1983, and the WOR slowly but steadily increased over the years to a value > 1 by 1990. The two intervals have been flooded at differ-

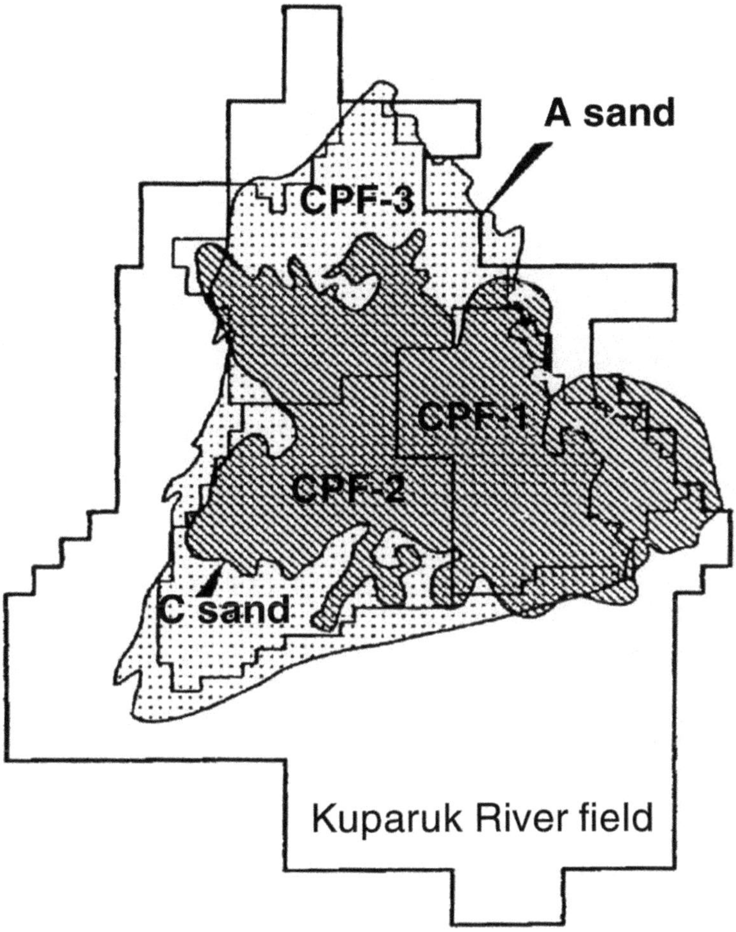

Fig. 11.50—Areal map of reservoir sands, Kuparuk River oil field.[53] (CPF = central production facility. On this map, CPF 1, CPF 2, and CPF 3 refer to the areas of the Kuparuk River field that are producing into these three sets of facilities.)

ent rates because of their different reservoir properties. Practically all of the wells have been hydraulically fractured to enhance well productivity and injectivity.

The most significant aspects of the Kuparuk River waterflood are the field pattern development taking into account the reservoir-fault alignment, the allocation of water injection into the dually completed water-injection wells, and the souring of what was originally an H_2S-free oil reservoir. Produced water is treated and reinjected with make-up seawater to balance pattern withdrawals. During the waterflood, the reservoir became inoculated with sulfate-reducing bacteria.[33,34] Because of the sulfate content of the seawater, these bacteria flourished and multiplied at reservoir conditions, so that the produced gas began to contain H_2S.

Estimated ultimate-recovery factors for the Kuparuk River oil field are given in **Table 11.3**.

The Kuparuk River waterflood has been very successful because of the field engineers' constant monitoring and active intervention over the past two decades. Over the years, EOR by immiscible water-alternating-gas (IWAG) and miscible water-alternating-gas (MWAG) injection has been used in various areas of the oil field to gain additional oil recovery (10% for Zone C

TABLE 11.3—ESTIMATED ULTIMATE-RECOVERY FACTORS FOR THE KUPARUK RIVER OIL FIELD

Zone	OOIP, million bbl oil	Recovery Factor, % OOIP
C	3,700	40
A	2,200	41

and 6% for Zone A). Large-scale MWAG injection started in 1996 and has expanded to include more than 50% of the patterns.

11.9.4 West Texas Carbonate Waterfloods. San Andres and Clearfork are two carbonate reservoir intervals that are present over a considerable area of the Permian Basin in west Texas. These reservoirs (e.g., Wasson, Slaughter, Seminole) contain several billion bbl of approximately 30°API oil. They are very-layered, heterogeneous carbonates and dolomites that have large variation in permeability from layer to layer. Interestingly, because of the complex hydrocarbon-accumulation history of this basin, much of this area has an underlying interval that contains residual oil saturation.

Most of these reservoirs were discovered in the late 1930s and the 1940s. Even where both the San Andres and the Clearfork were found to be oil-bearing, surface-lease and unitization considerations required the reservoirs to be developed by drilling separate sets of wells. These reservoirs had approximately two decades of primary production.

From the 1960s through the 1980s, almost all of these reservoirs underwent waterflooding. During this period, many SPE technical papers addressed aspects of these waterfloods. Of particular interest was the continuity, or discontinuity, of the pay intervals and the extent to which infill drilling from 40-acre well spacing to 20 acres, and possibly to 10 acres, could be economically justified by increased oil recovery.[19,20,60] Another issue was what pattern style was the best to use.[61] The recovery factors for the sum of primary production and waterflooding ranges from 25 to 60% of OOIP; the variation is caused mainly by geological factors in these different geographical areas and by differences between the San Andres and Clearfork reservoirs.[20]

Pressure coring and sponge coring are technologies that were developed specifically for these waterfloods to determine the residual oil saturation to waterflooding in these carbonate reservoirs.[62] These techniques were used to determine the variation in oil saturation from well-swept higher-permeability layers to poorly swept lower-permeability layers.

The west Texas carbonate waterfloods have proved quite successful and have recovered significant additional oil. Also, the San Andres reservoirs' waterfloods have provided much of the data and related technical studies that were needed to justify implementation of EOR by using CO_2 injection. During the tertiary EOR projects, water injection in the WAG mode has continued to control the mobility of the low-viscosity CO_2.

11.9.5 Kirkuk (Iraq). There is only one SPE technical paper about the supergiant reservoir, the Kirkuk oil field of Iraq. This field is given as an example here because its reservoir's geology is quite different from the other reservoirs discussed above.[63]

The primary pay interval for the Kirkuk field is the 1,200-ft-thick Main Limestone. This interval consists of a series of extensively fractured limestones, some porcelaneous and some dolomitized. These limestones were deposited in a variety of environments—back-reef/lagoonal, fore-reef, and basinal—and have a wide range of porosity and permeability properties. The oil is contained both in an extensive, extremely permeable but low-capacity fracture system and in a low-permeability but high-capacity, matrix-pore system. Also, the reservoir is underlain by a fieldwide aquifer. The oil gravity is approximately 36°API and was approximately 500 psi undersaturated at the original reservoir pressure of 1,100 psia.

Kirkuk began production in 1934, and 2 billion bbl of oil were produced before water injection was implemented in 1961. From 1961 to 1971, 3.2 billion bbl of oil were produced under pressure maintenance by waterdrive using river water. The 1971 production rate was approximately 1.1 million BOPD. Since then, the field has continued to produce large volumes of oil by voidage-replacement water injection; however, few production details for recent years appear in the technical literature.

The interesting technical aspects of this type of reservoir are the determination of the ultimate oil recovery from the matrix and the time scale of matrix oil recovery. Laboratory experiments can be run using matrix rock samples to determine the water/oil imbibition behavior; however, what matters is the actual reservoir's matrix/fracture interaction because the fracture density varies considerably. The early water injection showed that within the fracture network there was rapid communication over a distance of more than 20 miles. Water injection initially was peripheral; however, because of low injectivity caused by lack of downdip fracturing, injection was shifted to seven injection wells in the saddle area between the two principal domes of this oil field, one of which had an injection capacity of more than 400,000 BWPD.

A 90-day temporary production stoppage in 1967 allowed unique field data to be acquired regarding the matrix/fracture interaction because of the observed changes in the oil/water contact (OWC). It was observed that the OWCs fell in the areas where they were the highest and rose in the areas where they were the lowest. These OWC changes were the result of the countercurrent imbibition process between the fracture network and matrix pore system. From these data, the time-delay function could be calculated on the basis of observed field data. Depending on the assumptions, the half-life was estimated to be 3 to 5 years and the ultimate recovery was estimated at 30 to 45% of the OOIP.

11.10 Summary and Conclusions

This chapter has described the technical aspects of waterflooding, but only briefly compared to the vast amount of SPE technical literature on this subject. It also presented several field case studies for a few very large oil fields. Note, though, that reservoirs of all sizes have been waterflooded: very large fields having hundreds to 1,000+ wells; medium-sized fields; and small fields, for which the total waterflood is only a few patterns.

The conclusions concerning waterflooding are:

- Waterflooding is the most commonly used secondary-oil-recovery method. This is because water is inexpensive and readily available in large volumes and because water is very effective at substantially increasing oil recovery.
- The level of effectiveness of a waterflood depends on the mobility ratio between the oil and water, and the geology of the oil reservoir. Waterflooding is effective because almost all reservoir rocks are either water-wet or mixed-wet. The depositional and diagenetic characteristics of a reservoir control major aspects of the water/oil displacement process. These characteristics can either enhance waterflood performance or have detrimental effects on the WOR as a function of time. Often, the details of a reservoir's internal geology are not known until production wells start producing injected water.
- Gravity effects—i.e., the interplay between the gravity/density effects and the geologic layering of a reservoir—are important in waterfloods because at reservoir conditions, oil always is less dense than connate brine or injected water. This interplay can either help or hurt waterflood performance relative to a homogeneous system.
- Waterflooding is a process that typically takes several decades to complete. Hence, continuous, routine field production and pressure data must be taken for monitoring and analyzing waterflood performance. Occasionally, more-expensive, special-data acquisition programs (i.e., 3D- or 4D-seismic data) are run to assist the evaluation process. A variety of engineering tools

have been developed to analyze waterflood performance, ranging from simple plots of field production data to full-field numerical-reservoir-simulation models.

• Waterfloods are dynamic processes the performance of which, as production wells respond to the injection of water, can be improved by modification of operations by the technical team. Such modifications include changing the allocation of injection water among the injection wells and the waterflooded intervals, drilling additional wells at infill locations, and/or modifying the pattern style.

• Waterflooding has been used successfully in oil fields of all sizes and all over the world, in offshore and onshore oil fields.

Nomenclature

A = cross-sectional area available for flow, ft^2
B_o = oil formation-volume factor, RB/STB
C = total flow capacity
E_A = areal displacement efficiency
E_D = unit-displacement efficiency
E_I = vertical displacement efficiency
E_R = waterflood oil-recovery efficiency
E_V = volumetric sweep efficiency
E_{Vbt} = volumetric sweep efficiency at breakthrough
f_o = fractional flow of oil
f_w = fractional flow of water
$\bar{f}_w$ = average fractional flow of water
f_{wf} = fractional flow of water at flood front
F_{wo} = water/oil ratio
F_{pvg} = fraction of displaceable pore volume that is gas saturated
g = gravity constant
G = value as defined by Eq. 11.24, dimensionless
h = reservoir thickness, ft
i = fluid phase i
k = absolute permeability, darcies
k_{50} = median permeability value, md
$k_{84.1}$ = permeability at 1 standard deviation above mean value, md
k_a = air permeability
k_i = permeability of fluid phase i, darcies
k_o = permeability to oil, darcies
k_r = relative permeability
k_{ro} = relative permeability to oil, fraction
k_{roe} = relative permeability to oil at endpoint (S_{wi})
k_{rw} = relative permeability to water, fraction
k_{rwe} = relative permeability to water at endpoint (S_{orw})
k_{rwo} = water/oil relative permeability
k_w = permeability to water, darcies
k_x = permeability in the x direction, darcies
L = length, ft
M = mobility ratio
P_c = capillary pressure, psia
q_c = critical rate, B/D
q_o = oil-production rate, B/D
q_t = total production rate, B/D

q_w = water-production rate, B/D
S_{gi} = initial gas saturation, fraction PV
S_{gt} = trapped gas saturation, fraction PV
S_o = oil saturation, fraction PV
S_{oi} = initial oil saturation or $(1 - S_{wc})$, fraction PV
S_{orw} = residual oil saturation to waterflooding, fraction PV
S_w = water saturation, fraction PV
$\bar{S}_{wbt}$ = averaged water saturation behind the flood front at breakthrough, fraction PV
S_{wc} = connate-water saturation, fraction PV
S_{wf} = breakthrough or flood-front water saturation, fraction PV
S_{wi} = initial water saturation, fraction PV
S_{wz} = thickness-averaged water saturation for vertical equilibrium, fraction PV
t = time, days
u_{ox} = oil Darcy velocity in the x direction, ft/day
u_t = total Darcy velocity, ft/day
u_{wx} = water Darcy velocity in the x direction, ft/day
V = Dykstra-Parsons coefficient of permeability variation
x = position in x-coordinate system, ft
X = the X-plot value X, which is $\ln\left(\frac{1}{f_w} - 1\right) - \frac{1}{f_w}$
y = position in y-coordinate system, ft
α = reservoir dip angle, degrees
β = slope, degrees
$(\Delta p)_h$ = pressure difference in horizontal direction, psi
$(\Delta p)_V$ = pressure difference in vertical direction, psi
$\Delta\rho$ = water/oil density difference, lbm/ft³ or g/cm³
θ = contact angle, degrees
λ_i = mobility of fluid phase i, darcies/cp
μ_i = viscosity of fluid phase i, cp
μ_o = oil viscosity, cp
μ_w = water viscosity, cp
ρ_o = oil density, lbm/ft³ or g/cm³
ρ_w = water density, lbm/ft³ or g/cm³
σ_{os} = interfacial tension between the oil phase and solid phase, dyne/cm
σ_{ws} = interfacial tension between the water phase and solid phase, dyne/cm
σ_{ow} = interfacial tension between the oil phase and water phase, dyne/cm
ϕ = porosity, fraction BV
ψ_s = fraction of total flow coming from the swept portion of the pattern

Acknowledgments

SPE members have published numerous papers concerning waterflood technology over the past 60 years. SPE has several major publications about waterflood technology, including two monographs,[1,3] one textbook,[2] and one *Reprint Series* volume[4]. I have relied on these major publications and on a number of these technical papers in preparing this waterflood chapter for SPE's *Petroleum Engineering Handbook*. I have particularly relied on Willhite's *Waterflooding* textbook.[2] From the outset, I recognized that much of the summary of waterflood technologies presented here could only paraphrase portions of Willhite's textbook. Willhite describes the waterflood technical calculations in much more detail than is appropriate for this *Handbook* chapter. Also, I have used a number of the same figures as Willhite did; for the reader's convenience, I have referenced them to his textbook and not to the many SPE technical papers of the past 50 years in which they originally were published. **Appendix A** cross-references to Willhite[2] all the tables, figures, and equations from it that are used in this chapter.

References

1. Craig, F.F. Jr.: *Reservoir Engineering Aspects of Waterflooding,* Monograph Series, SPE, Richardson, Texas (1980) **3.**
2. Willhite, G.P.: *Waterflooding,* Textbook Series, SPE, Richardson, Texas (1986) **3.**
3. Rose, S.C., Buckwalter, J.F., and Woodhall, R.J.: *Design Engineering Aspects of Waterflooding,* Monograph Series, SPE, Richardson, Texas (1989) **11.**
4. *Waterflooding,* SPE Reprints Series, SPE, Richardson, Texas (1973) **2a.**
5. Fettke, C.R.: "The Bradford oil field, Pennsylvania and New York," Mineral Resources Report M21, Pennsylvania Geological Survey, Harrisburg, Pennsylvania (1938) 298–301.
6. Salathiel, R.A.: "Oil Recovery by Surface Film Drainage in Mixed-Wettability Rocks," *JPT* (October 1973) 1216; *Trans.,* AIME, **255.**
7. Jerauld, G.R. and Rathmell, J.J.: "Wettability and Relative Permeability of Prudhoe Bay: A Case Study in Mixed-Wet Reservoirs," *SPERE* (February 1997) 58.
8. Sharma, M.M. and Filoco, P.R.: "Effect of Brine Salinity and Crude-Oil Properties on Oil Recovery and Residual Saturations," *SPEJ* (September 2000) 293.
9. Zhou, X., Morrow, N.R., and Shouxiang, M.: "Interrelationship of Wettability, Initial Water Saturation, Aging Time, and Oil Recovery by Spontaneous Imbibition and Waterflooding," *SPEJ* (June 2000) 199.
10. Buckley, S.E. and Leverett, M.C.: "Mechanism of Fluid Displacement in Sands," *Trans.,* AIME (1942) **146,** 107.
11. Terwilliger, P.L. *et al.:* "An Experimental and Theoretical Investigation of Gravity Drainage Performance," *Trans.,* AIME (1951) **192**, 285.
12. Scheihing, M.H., Thompson, R.D., and Seifert, D.: "Multiscale Reservoir Description Models for Performance Prediction in the Kuparuk River Field, North Slope of Alaska," paper SPE 76753 presented at 2002 SPE Western Regional/AAPG Pacific Section Joint Meeting, Anchorage, 20–22 May.
13. Ebanks, W.J., Scheihing, M.H., and Atkinson, C.D.: "Flow Units for Reservoir Characterization," *AAPG Methods in Exploration, No. 10,* D. Morton-Thompson and A.M. Woods (eds.), AAPG, Tulsa (1993) 282–285.
14. Higgins, R.V. and Leighton, A.J.: "Computer Prediction of Water Drive of Oil and Gas Mixtures Through Irregularly Bonded Porous Media Three-Phase Flow," *JPT* (September 1962) 1048; *Trans,* AIME, **225.**
15. Higgins, R.V. and Leighton, A.J.: "Matching Calculated With Actual Waterflood Performance by Estimating Some Reservoir Properties," *JPT* (May 1974) 501.
16. Baek, M. and Hewett, T.A.: "A Hybrid Streamtube Simulator Using a Semianalytical Method," paper SPE 63151 presented at the 2000 SPE Annual Technical Conference and Exhibition, Dallas, 1–4 October.
17. Emanuel, A.S. and Milliken, W.J.: "The Application of Streamtube Techniques to Full-Field Waterflood Simulation," *SPERE* (August 1997) 211.
18. Grinestaff, G.H. and Caffrey, D.J.: "Waterflood Management: A Case Study of the Northwest Fault Block Area of Prudhoe Bay, Alaska, Using Streamline Simulation and Traditional Waterflood Analysis," paper SPE 63152 presented at the 2000 SPE Annual Technical Conference and Exhibition, Dallas, 1–4 October.
19. Barber, A.H. *et al.:* "Infill Drilling to Increase Reserves—Actual Experience in Nine Fields in Texas, Oklahoma, and Illinois," *JPT* (August 1983) 1530.
20. Lu, G.F., Brimhall, R.M., and Wu, C.H.: "Geographical Distribution and Forecast Models of Infill Drilling Oil Recovery for Permian Basin Carbonate Reservoirs," paper SPE 26503 presented at the 1993 SPE Annual Technical Conference and Exhibition, Houston, 3–6 October.

21. Fast, R.E., Murer, A.S., and Zambrano, L.G.: "Lost Hills Diatomite Simulation Study: Predicting Waterflood Performance in a Low-Permeability, Compacting Reservoir," paper SPE 26627 presented at the 1993 SPE Annual Technical Conference and Exhibition, Houston, 3–6 October.
22. Wright, C.A. and Conant, R.A.: "Hydraulic Fracture Reorientation in Primary and Secondary Recovery from Low-Permeability Reservoirs," paper SPE 30484 presented at the 1995 SPE Annual Technical Conference and Exhibition, Dallas, 22–25 October.
23. Dykstra, H. and Parsons, R.L.: "The Prediction of Oil Recovery by Water Flood," *Secondary Recovery of Oil in the United States, Principles and Practice,* second edition, API, Washington, DC (1950) 160–174.
24. Stiles, W.E.: "Use of Permeability Distribution in Water Flood Calculations," *Trans.,* AIME (1949) **186,** 9.
25. Dietz, D.N.: "A Theoretical Approach to the Problem of Encroaching and By-Passing Edge Water," *Proc.,* Akad. Wetenschappen, Amsterdam, (1953) **56** (1), series B, 83.
26. Dake, L.P.: *Fundamentals of Reservoir Engineering,* Elsevier Scientific Publishing Co., New York City (1978) 399–413.
27. Teufel, L.W. and Rhett, D.W.: "Failure of Chalk During Waterflooding of the Ekofisk Field," paper SPE 24911 presented at the 1992 SPE Annual Technical Conference and Exhibition, Washington, DC, 2–4 October.
28. Perkins, T.K. and Gonzalez, J.A.: "The Effect of Thermoelastic Stresses on Injection Well Fracturing," *SPEJ* (February 1985) 78.
29. Garon, A.M., Lin, C.Y., and Dunayevsky, V.A.: "Simulation of Thermally Induced Waterflood Fracturing in Prudhoe Bay," paper SPE 17417 presented at the 1988 SPE California Regional Meeting, Long Beach, California, 23–25 March.
30. Sylte, J.E., Hallenbeck, L.D., and Thomas, L.K.: "Ekofisk Formation Pilot Waterflood," paper SPE 18276 presented at the 1988 SPE Annual Technical Conference and Exhibition, Houston, 2–5 October.
31. *RP 45, Recommended Practice for Analysis of Oilfield Waters,* second edition, API, Washington, DC (1968, reissued 1981, third edition 1 August 1998).
32. *RP 38, Recommended Practice for Biological Analysis of Subsurface Injection Waters,* third edition, API, Washington, DC (1975, reissued 1982).
33. *TM0173-2005, Methods for Determining Quality of Subsurface Injection Water Using Membrane Filters,* NACE, Houston (1973, reaffirmed 2005).
34. Mitchell, R.W.: "The Forties Field Sea-Water Injection System," *JPT* (June 1978) 877.
35. Hamouda, A.A.: "Water Injection Quality in Ekofisk—UV Sterilization and Monitoring Techniques," paper SPE 21048 presented at the 1991 SPE International Symposium on Oilfield Chemistry, Anaheim, California, 20–22 February.
36. Kriel, B.G. *et al.:* "The Efficacy of Formaldehyde for the Control of Biogenic Sulfide Production in Porous Media," paper SPE 25196 presented at the1993 SPE International Symposium on Oilfield Chemistry, New Orleans, 2–5 March.
37. Frazer, L.C. and Bolling, J.D.: "Hydrogen Sulfide Forecasting Techniques for the Kuparuk River Field," paper SPE 22105 presented at the 1991 International Arctic Technology Conference, Anchorage, 29–31 May.
38. Hart, A.D., Graham, B.D., and Gettleson, D.A.: "NORM Associated with Produced Water Discharges," paper SPE 29727 presented at the 1995 SPE/EPA Exploration and Production Environmental Conference, Houston, 27–29 March.
39. Hjelmas, T.A. *et al.:* "Produced Water Reinjection: Experiences from Performance Measurements on Ula in the North Sea," paper SPE 35874 presented at the 1996 SPE Conference on Health, Safety, and Environment in Oil and Gas Exploration and Production, New Orleans, 9–12 June.
40. Hsi, C.D. *et al.:* "Formation Injectivity Damage Due to Produced Water Reinjection," paper SPE 27395 presented at the 1994 SPE Formation Damage Control Symposium, Lafayette, Louisiana, 7–10 February.
41. Martins, J.P. *et al.:* "Produced-Water Reinjection and Fracturing in Prudhoe Bay," *SPERE* (August 1995) 176.
42. Dyes, A.B., Caudle, B.H., and Erickson, R.A.: "Oil Production after Breakthrough—as Influenced by Mobility Ratio," *Trans.,* AIME (1954) **201,** 201.

43. O'Donovan, A.R., Smith S.G., and Kristiansen, P.: "Foinaven 4D Seismic—Dynamic Reservoir Parameters and Reservoir Management," paper SPE 63294 presented at the 2000 SPE Annual Technical Conference and Exhibition, Dallas, 1–4 October.
44. Staples, R. *et al.:* "Integrating 4D seismic to optimize production," paper SPE 78346 presented at the 2002 SPE European Petroleum Conference, Aberdeen, 29–31 October.
45. Woodling, G.S. *et al.:* "Layered Waterflood Surveillance in a Mature Field: The Long Beach Unit," paper SPE 26082 presented at the 1993 SPE Western Regional Meeting, Anchorage, 26–28 May.
46. Chapman, L.R. and Thompson, R.R.: "Waterflood Surveillance in the Kuparuk River Unit With Computerized Pattern Analysis," *JPT* (March 1989) 277.
47. Buckley, S.E. and Leverett, M.C.: "Mechanisms of Fluid Displacement in Sands," *Trans.,* AIME (1942) **146,** 107–116.
48. Ershaghi, I. and Omorigie, O.: "A Method for Extrapolation of Cut vs. Recovery Curves," *JPT* (February 1978) 203.
49. Ershaghi, I. and Abdassah, D.: "A Prediction Technique for Immiscible Processes Using Field Performance Data (includes associated papers SPE 13392, 13793, 15146 and 19506)," *JPT* (April 1984) 664.
50. Lo, K.K., Warner, H.R. Jr., and Johnson, J.B.: "A Study of the Post-Breakthrough Characteristics of Waterfloods," paper SPE 20064 presented at the 1990 SPE California Regional Meeting, Ventura, California, 4–6 April.
51. Bush, J.L. and Helander, D.P.: "Empirical Prediction of Recovery Rate in Waterflooding Depleted Sands," *JPT* (September 1968) 933.
52. Jarrell, P.M. and Stein, M.H.: "Maximizing Injection Rates in Wells Recently Converted to Injection using Hearn and Hall Plots," paper SPE 21724 presented at the 1991 SPE Production Operations Symposium, Oklahoma City, Oklahoma, 7–9 April.
53. Currier, J.H. and Sindelar, S.T.: "Performance Analysis in an Immature Waterflood: The Kuparuk River Field," paper SPE 20775 presented at the 1990 SPE Annual Technical Conference and Exhibition, New Orleans, 23–26 September.
54. Hedges, P.L. and Scherer, P.W.: "Group Oriented Software Solution for Pattern Material Balance of an Areally Extensive Field, Kuparuk River Field, Alaska," paper SPE 35987 presented at the 1996 SPE Petroleum Computer Conference, Dallas, 2–5 June.
55. Torsaeter, O.: "An Experimental Study of Water Imbibition in Chalk From the Ekofisk Field," paper SPE 12688 presented at the 1984 SPE Enhanced Oil Recovery Symposium, Tulsa, 15–18 April.
56. Hermansen, H. *et al.:* "Twenty Five Years of Ekofisk Reservoir Management," paper SPE 38927 presented at the 1997 SPE Annual Technical Conference and Exhibition, San Antonio, Texas, 5–8 October.
57. Agarwal, B. *et al.:* "Reservoir Characterization of Ekofisk Field: A Giant, Fractured Chalk Reservoir in the Norwegian North Sea—History Match," *SPEREE* (December 2000) 534.
58. James, D.M., Mayer, E.H., and Scranton, J.M.: "Use of Economic and Reservoir Models in Planning the Ranger Zone Flood, Long Beach Unit, Wilmington Field," paper SPE 1855 presented at the 1967 SPE Annual Meeting, New Orleans, 1–4 October.
59. Robertson, J.A., Blesener, J.A., and Soo Hoo, S.: "Subzone Redevelopment of the Long Beach Unit, Wilmington Oil Field: A Case Study," *JPT* (October 1987) 1229.
60. Maguson, W.L. and Knowles, J.D.: "Denver Unit 10-Acre Infill Pilot Test and Residual Oil Testing," paper SPE 6385 presented at the 1977 SPE Permian Basin Oil and Gas Recovery Conference, Midland, Texas, 10–11 March.
61. Christman, P.G.: "Modeling the Effects of Infill Drilling and Pattern Modification in Discontinuous Reservoirs," *SPERE* (February 1995) 4.
62. Bilhartz, H.L. Jr. and Charlson, G.S.: "Coring for In-Situ Saturations in the Willard Unit CO_2 Flood Mini-Test," paper SPE 7050 presented at the 1978 SPE Symposium on Improved Methods of Oil Recovery, Tulsa, 16–17 April.
63. Al-Naqib, F.M. *et al.:* "Water Drive Performance of the Fractured Kirkuk Field of Northern Iraq," paper SPE 3437 presented at the 1971 SPE Annual Meeting, New Orleans, 3–6 October.

TABLE 11A.1—CHAPTER TABLES, FIGURES, AND EQUATIONS CROSS-REFERENCED TO WILLHITE[2]

This chapter's:	Is or is modified from Willhite's:	This chapter's:	Is or is modified from Willhite's:
Table 11.1	Table 2.5	Fig. 11.33	Fig. 5.15
Fig. 11.2	Fig. 2.5	Fig. 11.34	Fig. 5.19
Fig. 11.3	Fig. 2.9	Fig. 11.44	Fig. 6.24
Fig. 11.4	Fig. 2.33	Eq. 11.2	Eq. 2.1
Fig. 11.5	Fig. 2.34	Eq. 11.3	Eq. 2.17
Fig. 11.6	Fig. 2.19	Eq. 11.4	Eq. 3.205
Fig. 11.7	Fig. 2.23	Eq. 11.7	Eq. 3.207
Fig. 11.8	Fig. 2.24	Eq. 11.8	Eq. 3.46
Fig. 11.9	Fig. 2.37	Eq. 11.9	Eq. 3.47
Fig. 11.10	Fig. 2.52	Eq. 11.10	Eq. 3.5
Fig. 11.11	Fig. 2.58	Eq. 11.11	Eq. 3.51
Fig. 11.12	Fig. 3.6	Eq. 11.12	Eq. 3.54
Fig. 11.13	Fig. 3.9	Eq. 11.13	Eq. 3.55
Fig. 11.14	Fig. 3.10	Eq. 11.14	Eq. 3.56
Fig. 11.15	Fig. 3.38	Eq. 11.15	Eq. 3.57
Fig. 11.16	Fig. 3.39	Eq. 11.16	Eq. 3.63
Fig. 11.18	Fig. 4.5	Eq. 11.17	Eq. 3.73
Fig. 11.19	Fig. 4.6	Eq. 11.18	Eq. 3.79
Fig. 11.20	Fig. 4.8	Eq. 11.19	Eq. 5.178
Fig. 11.21	Fig. 4.11	Eq. 11.21	Eq. 5.136
Fig. 11.26	Fig. 5.44	Eq. 11.22	Eq. 5.75
Fig. 11.27	Fig. 5.45	Eq. 11.23	Eq. 5.76
Fig. 11.32	Fig. 5.25	Eq. 11.24	Eq. 5.72

Appendix A—Cross-Referencing to Willhite[2]

Table 11A.1 cross-references to Willhite[2] all the tables, figures, and equations from it that are used in this chapter.

SI Metric Conversion Factors

acre	×	4.046 873	E + 03	= m^2
°API		141.5/(131.5 + °API)		= g/cm^3
bbl	×	1.589 873	E − 01	= m^3
cp	×	1.0*	E − 03	= Pa·s
dyne	×	1.0*	E − 02	= mN
dyne/cm	×	1.0*	E − 00	= mN/m
°F		(°F − 32)/1.8		= °C
ft	×	3.048*	E − 01	= m
ft/D	×	3.527 778	E − 03	= mm/s
ft^2	×	9.290 304	E − 02	= m^2
ft^3	×	2.831 685	E − 02	= m^3
in.	×	2.54*	E − 00	= cm
$in.^2$	×	6.451 6*	E + 00	= cm^2

$in.^3$	×	1.638 706	E + 01	= cm^3
lbm	×	4.535 924	E − 01	= kg
lbm/ft^3	×	1.601 846	E + 01	= kg/m^3
mile	×	1.609 344*	E − 00	= km
psi	×	6.894 757	E − 00	= kPa
psi/ft	×	2.262 059	E + 01	= kPa/m
scf/STB	×	1.801 175	E − 01	= std m^3/m^3
sq mile	×	2.589 988	E + 00	= km^2

*Conversion factor is exact.

Chapter 12
Immiscible Gas Injection in Oil Reservoirs

H.R. (Hal) Warner Jr., Warner Consulting Services and
E.D. Holstein, Consultant

12.1 Introduction

This chapter concerns gas injection into oil reservoirs to increase oil recovery by immiscible displacement. The use of gas, either of a designed composition or at high-enough pressure, to result in the miscible displacement of oil is not discussed here; for a discussion of that topic, see the chapter on miscible flooding in this section of the *Handbook*. A variety of gases can and have been used for immiscible gas displacement, with lean hydrocarbon gas used for most applications to date. Historically, immiscible gas injection was first used for reservoir pressure maintenance. The first such projects were initiated in the 1930s and used lean hydrocarbon gas (e.g., Oklahoma City field and Cunningham pool in the United States[1] and Bahrain field in Bahrain[2,3]). Over the decades, a considerable number of immiscible gas injection projects have been undertaken, some with excellent results and others with poor performance. Reasons for this range of performance are discussed in this chapter. At the end of this chapter, a variety of case studies are presented that briefly describe several of the successful immiscible gas injection projects.

Gas injection projects are undertaken when and where there is a readily available supply of gas. This gas supply typically comes from produced solution gas or gas-cap gas, gas produced from a deeper gas-filled reservoir, or gas from a relatively close gas field. Such projects take a variety of forms, including the following:

- Reinjection of produced gas into existing gas caps overlying producing oil columns.
- Injection into oil reservoirs of separated produced gas for pressure maintenance, for gas storage, or as required by government regulations.
- Gas injection to prevent migration of oil into a gas cap because of a natural waterdrive, downdip water injection, or both.
- Gas injection to increase recoveries from reservoirs containing volatile, high-shrinkage oils and into gas-cap reservoirs containing retrograde gas condensate.
- Gas injection into very undersaturated oil reservoirs for the purpose of swelling the oil and hence increasing oil recovery.

The primary physical mechanisms that occur as a result of gas injection are (1) partial or complete maintenance of reservoir pressure, (2) displacement of oil by gas both horizontally and vertically, (3) vaporization of the liquid hydrocarbon components from the oil column and possibly from the gas cap if retrograde condensation has occurred or if the original gas cap contains a relict oil saturation, and (4) swelling of the oil if the oil at original reservoir conditions was very undersaturated with gas. Gas injection is particularly effective in high-relief reservoirs where the process is called "gravity drainage" because the vertical/gravity aspects increase the efficiency of the process and enhance recovery of updip oil residing above the uppermost oil-zone perforations.

The decision to apply immiscible gas injection is based on a combination of technical and economic factors. Deferral of gas sales is a significant economic deterrent for many potential gas injection projects if an outlet for immediate gas sales is available. Nevertheless, a variety of opportunities still exist. First are those reservoirs with characteristics and conditions particularly conducive to gas/oil gravity drainage and where attendant high oil recoveries are possible. Second are those reservoirs where decreased depletion time resulting from lower reservoir oil viscosity and gas saturation in the vicinity of producing wells is more attractive economically than alternative recovery methods that have higher ultimate recovery potential but at higher costs. And third are reservoirs where recovery considerations are augmented by gas storage considerations and hence gas sales may be delayed for several years.

Nonhydrocarbon gases such as CO_2 and nitrogen can and have been used.[4] In general, calculation techniques developed for hydrocarbon-gas injection and displacement can be used for the design and application of nonhydrocarbon, immiscible gas projects. Valuing the use of such gases must include any additional costs related to these gases, such as corrosion control, separating the nonhydrocarbon components to meet gas marketing specifications, and using the produced gas as fuel in field operations.

The sections in this chapter are presented in the following order:

- Microscopic and Macroscopic Displacement Efficiency of Immiscible Gas Displacement
- Gas/Oil Compositional Effects During Immiscible Gas Displacement
- Reservoir Geology Considerations Regarding Immiscible Gas Displacement
- General Immiscible Gas/Oil Displacement Techniques
- Vertical or "Gravity Drainage" Gas Displacement
- Calculation Methods for Immiscible Gas Displacement
- Immiscible Gasflood Monitoring
- Field Case Studies: Immiscible Gas Injection Examples
- Miscellaneous Topics
- Summary and Conclusions

The purposes of this chapter include listing the physical criteria that separate the successful gas injection operations from the unsuccessful ones, describing the reservoir and process variables that must be defined and quantified, and demonstrating some of the simple techniques available for predicting and evaluating field performance. Some of these calculations can be performed with spreadsheets or, more tediously, with hand-held calculators. Modern numerical reservoir simulators are commonly used to calculate the projected performance of applying immiscible gas injection to a particular reservoir. For reservoirs with several years of immiscible gas injection, these same simulators can be used to history match past performance and to project future performance under various scenarios (e.g., continuing current operations, evaluating various new producing wells options, or comparing surface facility operational alternatives). See the chapter on reservoir simulation in this section of the *Handbook*.

Specifically not included in this chapter is any discussion of the factors to consider in implementing a gas injection project, such as gas compression needs, gas distribution systems, wellbore configurations, and vessel selection and sizing for handling produced fluids. These subjects are covered in various chapters in the Production Operations Engineering and Facilities and Construction Engineering sections of the *Handbook*.

TABLE 12.1—COMPARISON OF MEASURED AND COMPUTED SURFACE TENSIONS OF VARIOUS RESERVOIR FLUID SYSTEMS[8]

		Surface Tension, dynes/cm	
Reservoir Fluid	Pressure, psia	Measured	Computed
Asphalt-free oil system at 73°F	2,275	2.7	2.6
Asphalt-free oil and rich gas at 73°F	2,275	0.55	0.52
Asphalt-free oil and lean gas at 73°F	2,275	4.1	3.8
Equilibrium liquid and gas phases of	2,150	5.5	4.7
Oil A at 130°F	1,650	6.7	7.4
	1,150	10.1	10.3
	185	19.5	18.7
Equilibrium liquid and gas phases of	2,815	1.3	1.2
Oil C at 180°F	3,315	2.3	2.1
	2,815	3.3	3.1
	2,315	4.6	4.8
Equilibrium liquid and gas phases of	2,010	6.0	5.6
Oil D at 170°F	1,610	8.5	8.1
	1,110	10.3	11.5

12.2 Microscopic and Macroscopic Displacement Efficiency of Immiscible Gas Displacement

The conceptual aspects of the displacement of oil by gas in reservoir rocks are discussed in this section. There are three aspects to this displacement: gas and oil viscosities, gas/oil capillary pressure (P_c) and relative permeability (k_r) data, and the compositional interaction, or component mass transfer, between the oil and gas phases. The first two topics are discussed in this section; the third is discussed in the next section.

12.2.1 Gas/Oil Viscosity and Density Contrast. One must first understand the viscosity and density differences between gas and oil to appreciate why the gas/oil displacement process can be very inefficient. Gases at reservoir conditions have viscosities of ≈0.02 cp, whereas oil viscosities generally range from 0.5 cp to tens of centipoises. Gases at reservoir conditions have densities generally one-third or less than that of oil. Thus, gas is generally one to two orders of magnitude less viscous than the oil it is trying to displace. Regarding the fluid density difference, gas is always considerably "lighter" than the oil; hence, gas, when flowing, will segregate by gravity to the top of the reservoir or zone and oil will "sink" simultaneously to the bottom of the reservoir or zone.

Another gas/oil property that must be known for calculations at reservoir conditions is the interfacial tension (IFT) between the oil and gas fluid pair. This value is needed at reservoir conditions for the conversion of gas/oil capillary pressure data from surface to reservoir conditions. A number of technical papers discuss the calculation of IFT from compositional information about the oil and gas phases.[5–9] **Table 12.1** from Firoozabadi *et al.*[8] shows several reservoir oil-gas fluid-pair IFT values (measured and calculated) as a function of temperature and pressure. As the pressure increases, the IFT values decrease, although not low enough for miscible displacement to occur. Although not illustrated in the table, it should be noted that the IFT between nitrogen and oil is higher than that between a lean natural gas and the same oil.

12.2.2 Gas/Oil Capillary Pressure and Relative Permeability. The gas/oil capillary pressure and relative permeability data are typically measured by commercial laboratories using routine

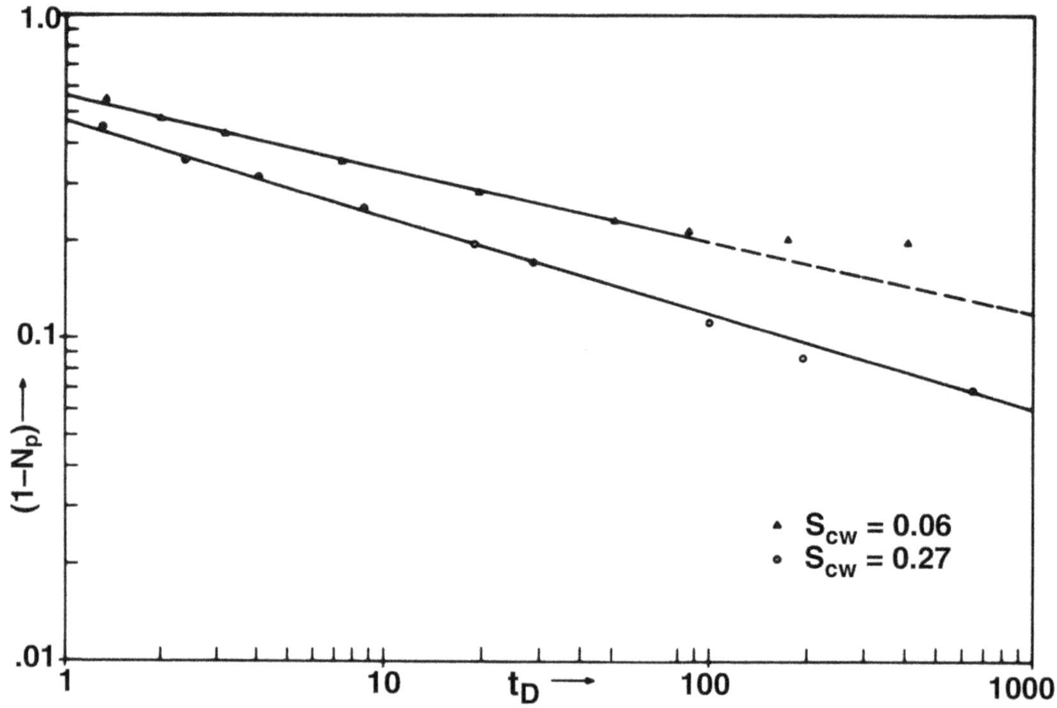

Fig. 12.1—Gas/oil displacement results for Berea cores; oil production as a function of time. This figure shows that long drainage times are required for displacement of oil to low saturation values.[10]

special core analysis procedures. Gas-oil capillary pressure data can be measured with either porous-plate or centrifuge equipment. One approach for obtaining gas/oil relative permeability data is the viscous displacement method in which gas displaces oil. A second method is the centrifuge method, which is generally used to obtain capillary pressure and relative permeability information simultaneously.

In all cases, gas is the nonwetting phase in this displacement; hence, it will preferentially flow through the largest pores first. However, what is very important in the determination of the oil relative permeability is the distribution of the oil phase in the core sample because in real reservoirs connate water occupies the smallest pores. As shown by Hagoort,[10] initial water saturation has a significant effect on oil relative permeability during the gas/oil displacement (centrifuge experiments). The water phase will occupy a greater percentage of the smaller pore spaces as the connate water saturation increases. As a result, the pore structure appears more streamlined to the oil and gas phases. The oil relative permeability at higher connate water saturations is considerably higher (see **Fig. 12.1** and the discussion of capillary pressure and relative permeability concepts in the chapter on those topics in the General Engineering Section of the *Handbook*).

The other key aspect of the oil relative permeability (k_{ro}) is the determination of its value as the oil saturation decreases. Because oil relative permeability becomes quite low but nonzero, the time to reach equilibrium in laboratory core plug measurements can be very long. Fig. 12.1 presents experimental results for cumulative oil recovery as a function of drainage time and shows that the oil continues to flow but more and more slowly (linearly as a function of the logarithm of t_D); Hagoort[10] found similar behavior for the four different rock types he tested.

If the gas/oil relative permeability data were measured with the viscous displacement technique (the extended Welge technique as described by Johnston *et al.*),[11] extra care is needed in applying these data. First, the displacement of oil by gas is at an unfavorable mobility ratio (see discussion below) that makes the process unstable. Second, a displacement is adversely affected by capillary end effects that, for the gas/oil system, cannot be overcome by high gas throughput rates. At low oil saturations, the region of most interest, the capillary end effect is the greatest.[10]

Finally, one method developed to affect the gas/oil relative permeability and to reduce gas mobility is to inject water alternately with gas (WAG). This procedure was proposed by Caudle and Dyes.[12] Although the method was proposed for use in miscible gasfloods, the concept applies equally to immiscible gas displacements. This technique has been used in many west Texas CO_2 miscible gas projects, in the Prudhoe Bay miscible flood,[13] and in the Kuparuk immiscible and miscible gas injection processes.[14,15] The three-phase gas, oil, and water relative permeabilities are calculated in numerical reservoir simulators with algorithms developed over the past several decades.[16]

12.2.3 Mobility Ratio. The mobility of a fluid (Eq. 12.1) is defined as its relative permeability divided by its viscosity. Mobility combines a rock property, permeability, with a fluid property, fluid viscosity. Gas-oil relative permeabilities are assumed to be dependent on the saturations of the two fluid phases and independent of fluid viscosity:

$$M_i = \left(k k_{ri} / \mu_i\right) . \qquad (12.1)$$

A fluid's mobility relates to its flow resistance in a reservoir rock at a given saturation of that fluid. Because viscosity is in the denominator of this definition, gases, which are very-low-viscosity fluids, have very high mobility.

Mobility ratio is generally defined as the mobility of the displacing phase (in the gas/oil case, gas) divided by the mobility of the displaced phase, which is oil. Eq. 12.2 presents two forms of the mobility ratio equation:

$$M = \frac{k_{rg}}{\mu_g} \bigg/ \frac{k_{ro}}{\mu_o} = \frac{k_{rg}\mu_o}{k_{ro}\mu_g} . \qquad (12.2)$$

Eq. 12.2 can also be written in more familiar engineering terms as the ratio of the two fluids' relative permeability values multiplied by the ratio of the two fluids' viscosities.

$$M = \left(\frac{k_{rg}}{k_{ro}}\right)\left(\frac{\mu_o}{\mu_g}\right) . \qquad (12.3)$$

For simple calculations, the mobility ratio is calculated at the endpoint relative permeability values for the two phases. Hence, the equation that practical engineers use for the gas/oil mobility ratio is

$$M = \frac{k_{g@Sorg}\ \mu_o}{k_{o@Swi}\ \mu_g} . \qquad (12.4)$$

All displacements of oil by gas are at "unfavorable" mobility ratios, with typical values of 10 to 100 or more.

12.2.4 Gas/Oil Linear Displacement Efficiency. The equations that characterize the mechanics of oil displacement by an immiscible fluid were developed by Buckley and Leverett[17] using relative permeability concepts and Darcy's law describing steady-state fluid flow through porous media. The resulting fractional flow equation describes quantitatively the fraction of displacing fluid flowing in terms of the physical characteristics of a unit element of porous media. Assumptions inherent in their work are steady-state flow, constant pressure, no compositional effects, no production of fluids behind the gas front, no capillary effects, movement of advancing gas parallel to the bedding plane, immobile water saturation, and uniform cross-sectional flow (no gravity segregation of fluids within the element). Subsequent work by Welge[18] made solving the displacement equations easier.

The Welge equation for the fractional flow of gas at any gas saturation (S_g) is calculated as follows:

$$f_g = \frac{1 + \left(0.044 k k_{ro} \Delta\rho A \sin\alpha \Big/ q_t \mu_o \right)}{1 + 1/M}, \quad (12.5)$$

where

A = area of cross section normal to the bedding plane, ft^2,
f_g = fraction of flowing stream that is gas,
k = permeability, darcies,
k_{ro} = relative permeability to oil, fraction,
k_{rg} = relative permeability to gas, fraction,
M = mobility ratio, $\frac{k_{rg}\mu_o}{k_{ro}\mu_g}$,
q_T = total flow rate through area A, res ft^3/D,
α = angle of dip, positive downdip, degrees,
$\Delta\rho$ = density difference, $\rho_g - \rho_o$, lbm/ft^3,
μ_o = viscosity of oil, cp, and
μ_g = viscosity of gas, cp.

When gravity is negligible, this equation becomes the more familiar Buckley-Leverett equation:

$$f_g = \frac{1}{1 + 1/M}. \quad (12.6)$$

Fig. 12.2 is a typical plot resulting from these calculations. The importance of the gravity term is indicated.

To relate the fraction of gas flowing to time, Buckley and Leverett developed the following material-balance equation:

$$L = \frac{q_T t}{\phi A}\left(\frac{\mathrm{d}f_g}{\mathrm{d}S_g}\right), \quad (12.7)$$

where

L = length, ft,
S_g = gas saturation, fraction,
t = time, days, and
ϕ = porosity, fraction.

The value of the derivative $\mathrm{d}f_g/\mathrm{d}S_g$ may be obtained for any value of gas saturation by determining slopes at various points on the f_g vs. S_g curve. These slopes can be determined manually or, more precisely, using the method presented by Kern[19] for computer spreadsheets.

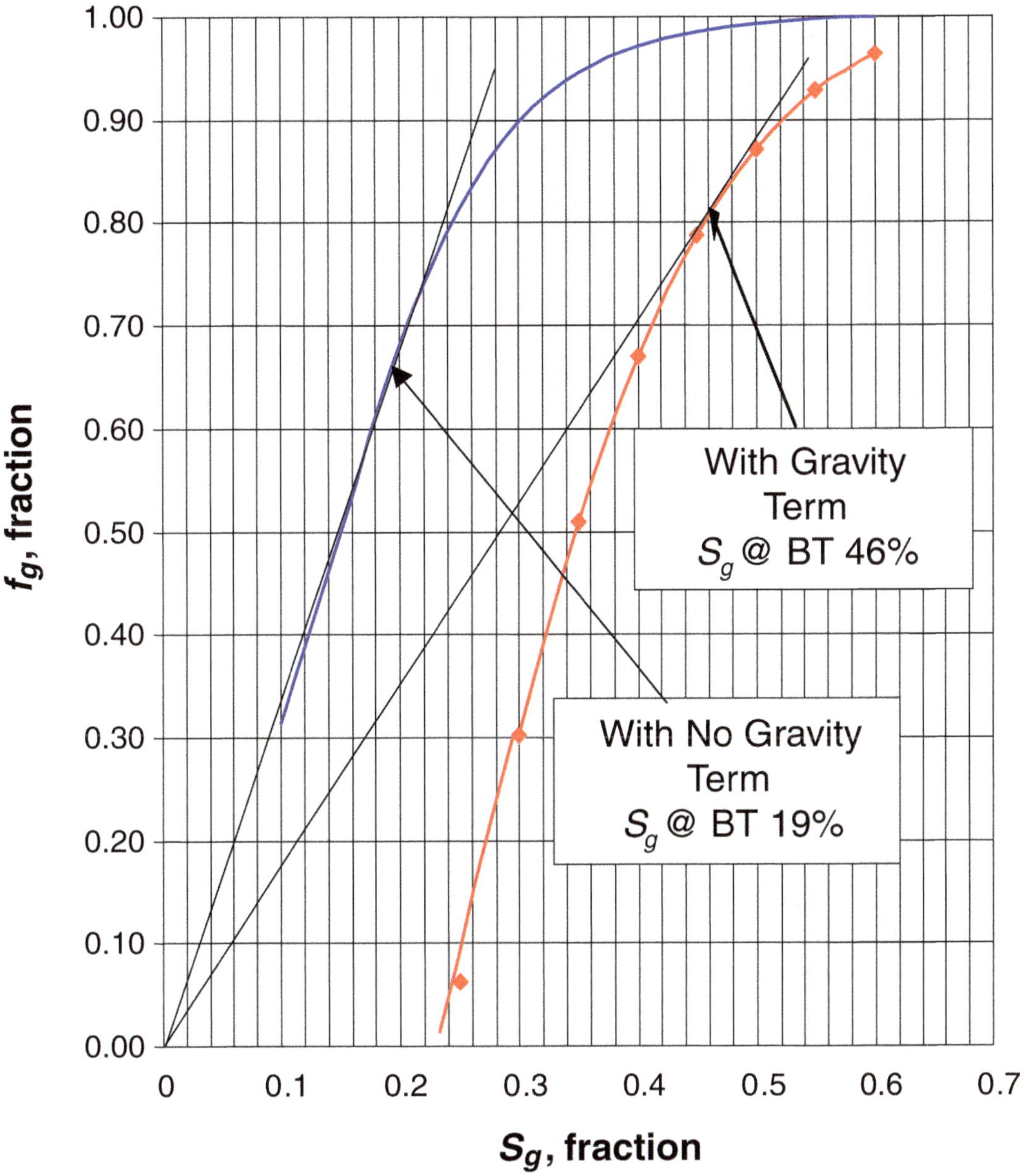

Fig. 12.2—Buckley-Leverett fractional gas flow plot (based on data from the Hawkins field).

Fig. 12.3 illustrates calculated gas-saturation distributions derived from the no-gravity and with-gravity fractional flow curves shown in Fig. 12.2. The area beneath each curve represents the gas-invaded zone. The saturation profile calculation results in lengths that first increase as saturation decreases and then decrease at lower saturations. While correct from a material-balance standpoint, it has been customary to square off the leading edge of the curve at the breakthrough saturation to account for capillary pressure that was neglected in the original derivation of the equation.

The gas/oil displacement efficiency, the percent of the oil volume that has been recovered, can be calculated for any period of gas injection by integrating the volume of the gas-invaded

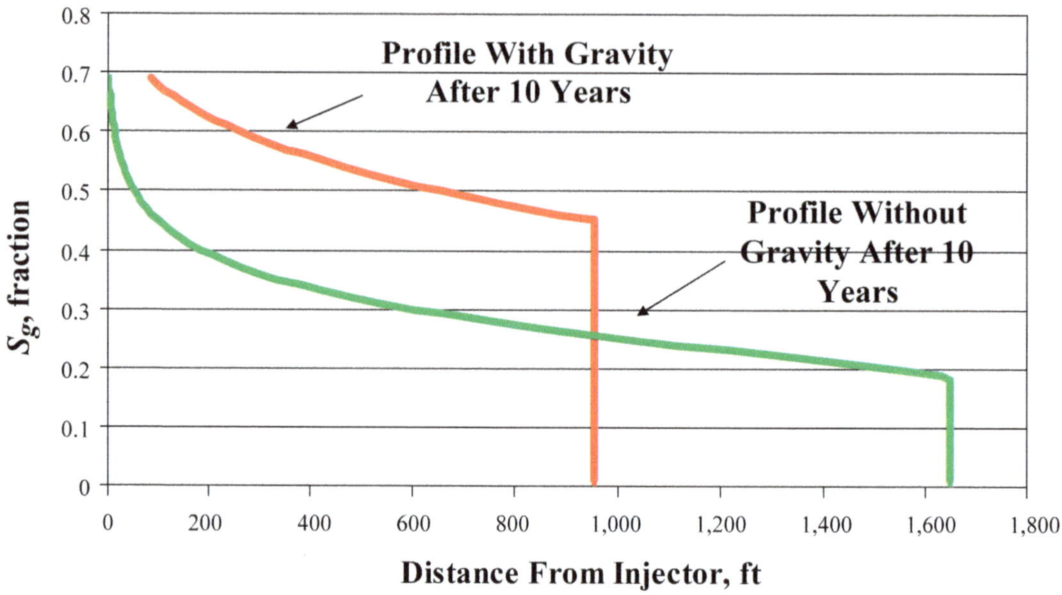

Fig. 12.3—Typical Buckley-Leverett saturation profiles.

zone as a function of gas saturation (S_g). Hence, the fractional flow curves (Fig. 12.2) are used to generate saturation profiles (Fig. 12.3) that lead to values for the gas/oil displacement efficiency. In the next section, several of the factors affecting this efficiency are discussed.

12.2.5 Factors Affecting Gas/Oil Displacement Efficiency. The fractional-flow and material-balance equations discussed above are important for understanding the effects on the efficiency of the gas/oil displacement process of (1) initial saturation conditions, (2) fluid viscosity ratios, (3) relative permeability ratios, (4) formation dip, (5) capillary pressure, and (6) factors of permeability, density difference, rate of injection, and cross section open to flow.

Initial Saturation Conditions. If gas injection is initiated after reservoir pressure has declined below the bubblepoint, the gas saturation will decrease the amount of displaceable oil. If the free gas saturation exceeds the breakthrough saturation, no oil bank will be formed. Instead, oil production will be accompanied by immediate and continually increasing gas production. Laboratory investigations and mathematical analyses have demonstrated this influence of gas saturation on gas displacement performance.[20]

S_{wi} has been shown to have no influence on displacement efficiency at gas breakthrough, but it directly affects the displaceable oil volume.[21] If S_{wi} is mobile, the displacement equations are not directly applicable because they were developed for two-phase flow. Approximations of gas displacement performance can usually be made when three phases are mobile by treating the water and oil phases as a single liquid phase. Displacement calculations can then be made with k_{rg} and k_{ro} data determined from core samples containing an immobile water saturation. Oil recovery can be differentiated from total liquid recovery on the basis of material balance calculations incorporating an estimated minimum interstitial water saturation.

Fluid Viscosities. The effect of oil viscosity on fractional flow is illustrated in **Fig. 12.4**. In this plot, the S_g at breakthrough increases from 12 to 38% with a 10-fold decrease in oil viscosity.

Relative Permeability Ratios. The concepts of relative permeability can be applied equally well to complete or partial pressure-maintenance operations. Relative permeability, a character-

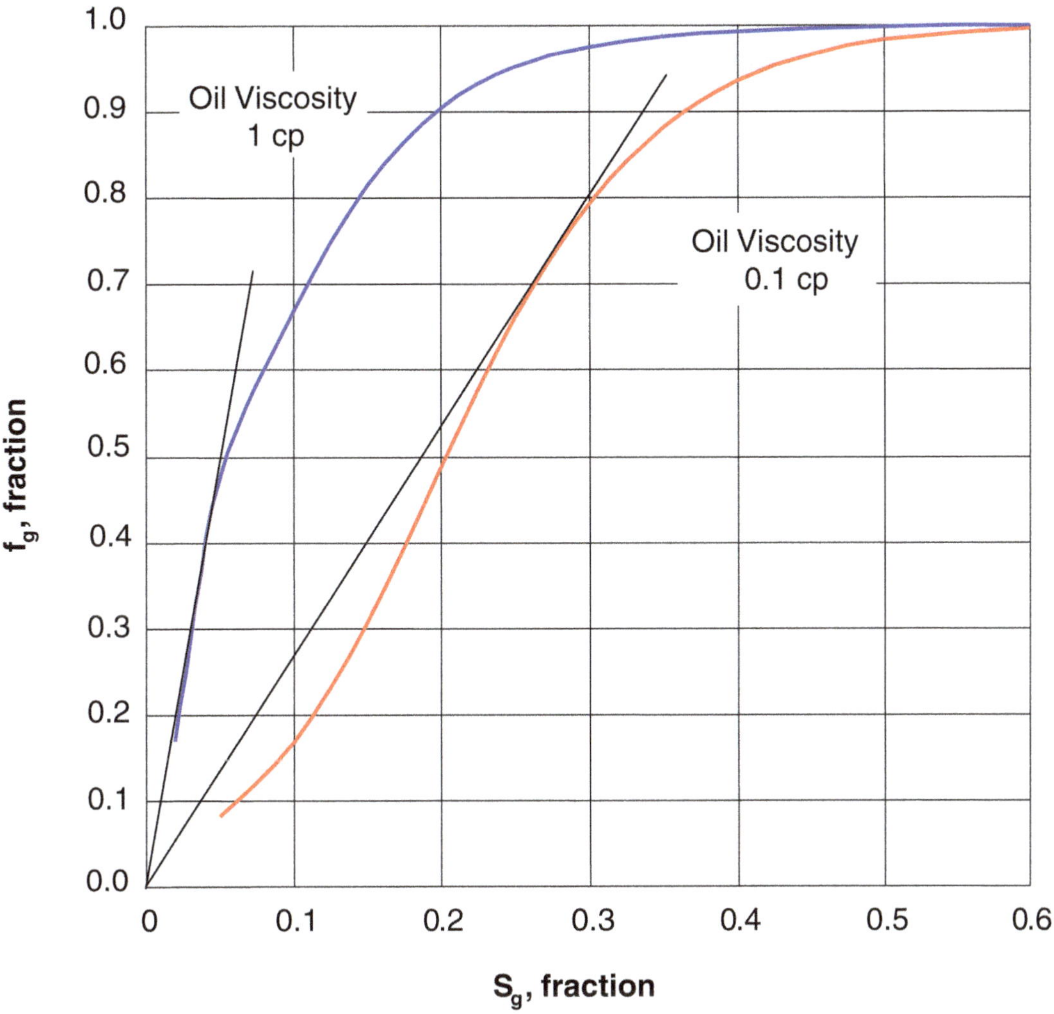

Fig. 12.4—Effect of viscosity on gas/oil fractional flow.

istic of the reservoir rock, is a function of fluid saturation conditions. It is important that calculations be based on dependable data obtained by laboratory analyses at reservoir conditions using representative core samples. If possible, the laboratory-determined data should be supplemented by relative permeabilities calculated from field performance data.

Formation Dip. When formation dip aids gravity, as illustrated in Fig. 12.2, fractional flow behavior is significantly improved if permeability is high enough and withdrawal rates do not exceed gravity-stable conditions. Gravity drainage is discussed later in this chapter.

Capillary Pressure. Capillary forces are opposite gravity drainage forces and directionally decrease displacement efficiency. However, capillary forces can often be ignored as insignificant for projects with rates of displacement normally used. Only at extremely low rates of displacement, where viscous forces become negligible, is the saturation distribution controlled to a significant extent by the balance between capillary and gravitational forces. Another place where capillary forces are considered important is many of the large carbonate reservoirs of the Middle East where the matrix-blocks/fracture-system interaction can significantly affect overall reservoir performance.

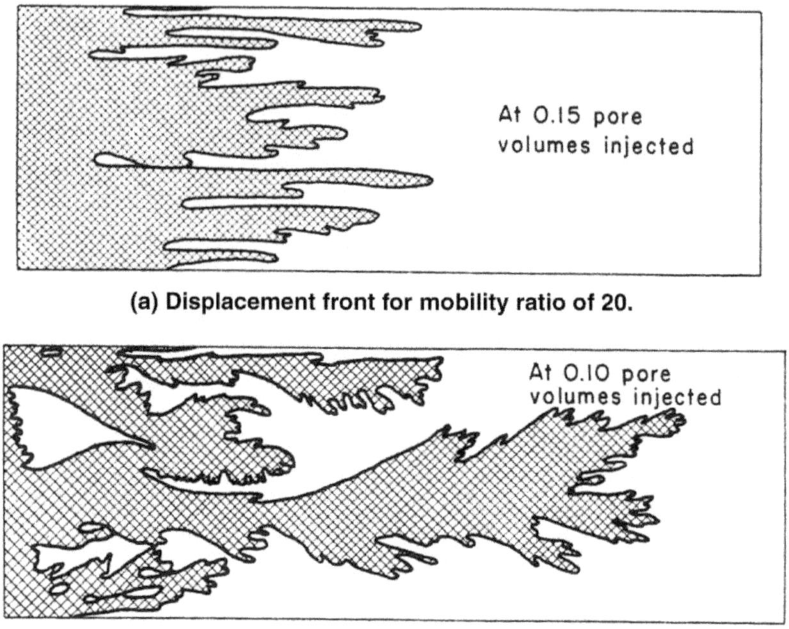

(a) Displacement front for mobility ratio of 20.

(b) Displacement front for mobility ratio of 383.

Fig. 12.5—Cross-sectional view of gas/oil displacement front [at 0.15 pore volumes injected (PVI)] for mobility ratios of (a) 20 and (b) 383 at 0.10 and 0.15 PVI, respectively.[22]

Other Factors. Higher permeability, greater density difference between oil and gas, and a lower displacement rate all improve the displacement efficiency.

12.2.6 Unfavorable Mobility Ratio Causes Viscous Flow Instabilities. Displacements that take place at very unfavorable mobility ratios are unstable, and viscous fingering occurs. This is the situation for essentially all gas/oil displacements, especially if the displacement is occurring horizontally. The impact of such instabilities is illustrated in **Figs. 12.5 and 12.6.**[22,23] Both figures were drawn from technical literature concerning miscible displacement laboratory experiments using homogeneous sandpacks, but the observed effects would be the same for immiscible gas displacing oil at very unfavorable mobility ratios. Fig. 12.5 shows, in cross-sectional view, the nature of viscous fingering for two highly unfavorable mobility ratios (the two fluids have equal densities). The flood front in both cases is very unstable. Fig. 12.6 shows, in areal view, the effect of mobility ratio on the displacement process in a quarter of a five-spot pattern for mobility ratios from 0.151 to 71.5. For mobility ratios from 4.58 to 71.5 (cases D through F), the flood front is very unstable, and breakthrough occurs via narrow fingers of the injected fluid; these cases show how the process of gas displacing oil would occur.

In both of these illustrations, the cause of the viscous fingering was a slight perturbation in the flow field that grew into the viscous finger once the perturbation occurred. In real reservoir situations, there are two physical aspects that enhance the viscous-fingering phenomenon. First, real reservoirs are very heterogeneous, so a variety of styles of permeability heterogeneities can initiate viscous fingering. Second, in a cross-section immiscible gas/oil displacement process, gas is always less dense than oil. Hence, there is a gas/oil density difference, and the force of gravity causes the gas to override the oil and initiate a viscous finger of the high-mobility gas phase along the top of the reservoir interval.

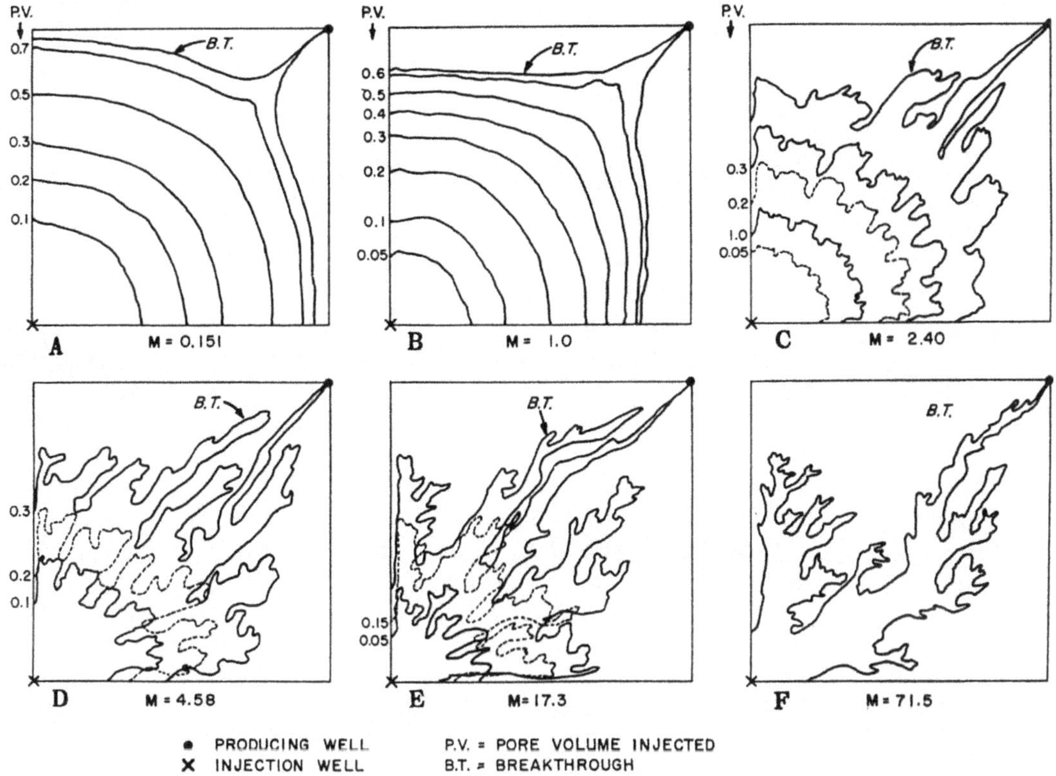

Fig. 12.6—Gas/oil displacement fronts for various mobility ratios (0.151 to 71.5) and PVI until breakthrough, quarter of a five-spot pattern.[23]

If the gas/oil displacement is occurring vertically with gas generally displacing oil downward, gravity will work to stabilize the flood front between the gas and oil, although if the rate is too high, instabilities in the form of gas cones or tongues can occur.

12.3 Gas/Oil Compositional Effects During Immiscible Gas Displacement

This section briefly discusses the various mass transfer compositional aspects of the immiscible gas/oil displacement process. The implications of these compositional effects are very dependent on the oil composition, the composition of the injected gas, and the surface facilities and pipelines available in a particular field situation. The injected gas/oil composition interactions can be categorized as either swelling effects (gas dissolving into the oil phase) or stripping effects (various components from the oil transferring to the gas phase).

12.3.1 Swelling Compositional Effects. The most obvious compositional effect in the immiscible gas/oil displacement process is that, if the oil is not saturated with gas at the reservoir pressure or if the reservoir pressure is increased as a result of the gas injection, the volume of gas dissolved in the oil will increase until the oil is saturated at that pressure. At the same time and because of the increased volume of gas in solution in the oil, the oil formation volume factor (FVF) will increase. This phenomenon, commonly called swelling, can increase the efficiency of the gas/oil displacement process.

The significance of the swelling effect is dependent on the oil reservoir situation. For an oil reservoir in which there is a gas cap, the underlying oil column will already be fully or nearly saturated with gas at the reservoir pressure. Hence, there will be very little impact on the gas/

oil displacement process as a result of the interaction between the reservoir oil and injected gas. However, for an oil reservoir in which there is no gas cap and where the oil bubblepoint pressure is very low compared with the original reservoir pressure, the swelling effect can be a very significant part of the gas/oil displacement process. The Swanson River field (Cook Inlet, Alaska) is an example of this latter situation (original reservoir pressure, 5,580 psi; oil bubblepoint pressure, 1,350 psi).[24] The change in oil FVF from a bubblepoint of 1,350 psi to being saturated at 5,580 psi is from 1.21 to ≈1.80 RB/STB. The application of immiscible gas injection to the Swanson River field is discussed in the Case Studies section of this chapter.

In some of the simple calculation techniques discussed below, the swelling effect is included. In a "black oil" type of numerical reservoir simulator, the swelling effect is taken into account because, although there are only two hydrocarbon components (gas and oil) for the two hydrocarbon phases, the swelling effect is incorporated by means of the entered table of oil pressure/volume/temperature (PVT) properties (e.g., FVF, gas in solution, oil viscosity) as a function of pressure. In other words, in this type of reservoir simulator, the gas hydrocarbon component exists either as a free-gas phase or as gas dissolved in the oil; the oil hydrocarbon component exists only as part of the liquid phase. A compositional numerical reservoir simulator will automatically take the swelling phenomenon into account in its equation-of-state phase-behavior calculations. See the chapter on numerical simulation in this section of the *Handbook* for detailed discussion of the various types of numerical reservoir simulators and their applications.

12.3.2 Stripping Compositional Effects. The other key compositional aspect of the immiscible gas/oil displacement process is the vaporizing (or stripping) by the lean injected gas of some hydrocarbon components of the oil, particularly the intermediate hydrocarbon components (C_3 through C_8). In most cases, the injected gas is very lean natural gas that is the residue gas from a nearby gas processing plant and composed primarily of methane. At such gas processing plants, the propane and heavier hydrocarbon components typically have been condensed from the entering produced gas; in some cases, ethane is also extracted from the gas. Such a lean injected gas will, when in contact with the oil at reservoir conditions, vaporize various hydrocarbon components from the oil until the gas and oil phases have reached compositional equilibrium.

In immiscible gas/oil displacements using nitrogen (N_2), carbon dioxide (CO_2), or some combination of these gases (such as flue gas, 88% N_2, 12% CO_2), these nonhydrocarbon injected gases can also vaporize various hydrocarbon components until gas/oil equilibrium is reached at reservoir conditions. This phenomenon has been observed in the Hawkins nitrogen injection project. Nitrogen is not as efficient as methane at stripping hydrocarbon components from the oil. Carbon dioxide, because its phase behavior is much like that of propane, can vaporize a considerable amount of hydrocarbon components from the oil at reservoir conditions.

The significance of the stripping effect depends on the oil composition. Immiscible gas/oil injection projects have been applied to reservoirs with oil gravities from 24 to 43°API or more.[14,25,26] In all cases, the stripping effect increases the recovery of hydrocarbons from the oil reservoir, but lighter oils have a much greater percentage of their components vaporized by cycling gas through the reservoir and operating at higher gas/oil ratios (GORs).

In some of the simple calculation techniques discussed below, aspects of the stripping of oil by lean injected gas can be approximated. Black-oil numerical reservoir simulators cannot handle the vaporization of hydrocarbon components from the oil into the gas phase. A compositional numerical reservoir simulator must be used to quantify this effect for a particular oil/injected-gas reservoir situation. These calculations are based on the use of an equation-of-state fluid characterization that is "tuned" to PVT laboratory data for the particular oil and potential injected-gas compositions. The compositional simulator also can quantify the effects of various surface facility configurations, including associated gas plants.

Calculation Methods. These compositional effects are too complicated to be quantified with hand-calculation methods. The black-oil numerical reservoir simulator can be used in limited ways to make these calculations. The black-oil model can reasonably handle the swelling effect, but it cannot handle the stripping effect at all.

The compositional effects of immiscible gas injection are best calculated using a compositional numerical reservoir simulator. To use such a model most accurately, considerable gas/oil PVT data need to be taken with the reservoir oil mixed with a range of gas compositions. These measurements should include a variety of special measurements, such as swelling experiments and stripping experiments, that can be used to develop a more accurate equation-of-state fluid characterization. Then the reservoir model can be used to quantify the performance of possible immiscible gas injection projects and associated surface facilities that might be built.

Surface Facility Considerations. In most applications of immiscible gas injection, during the early years of the projects, hydrocarbons are produced at about the original solution GOR. During this period, the impact of surface-facility design on the volume of produced hydrocarbon liquids is relatively limited. Later in the life of the projects when the producing GOR rises, the surface facilities are far more important in terms of both the volume of gas that can be handled and the extent to which this gas is processed. During this late period, the rate of oil production will be limited by the surface facilities' gas-handling capacity.[27]

The gas recovered from the typical series of oilfield separators will contain a large amount of ethane through butane components and decreasing but significant amounts of the various heavier hydrocarbon components. Early generations of gas-processing plants were capable of separating out components that would condense down to temperatures of -10 to -20°F. Modern gas-processing plants operate at temperatures of -40°F or considerably lower; at these low temperatures, essentially all the ethane and heavier hydrocarbons are recovered, leaving a residue gas consisting primarily of methane. See the chapters in the Facilities and Construction Engineering section of the *Handbook* for additional discussion of oilfield surface separators and gas-processing plants.

Another related factor is the type of pipeline networks available to transport the hydrocarbon products to market. In some geographical areas, only those hydrocarbon components that can be stabilized in a crude oil stream can be transported and sold because there are only crude oil pipelines in that area. In much of the United States and Canada, a variety of crude oil and natural gas liquid (NGL) pipelines have been built so that the lighter, liquefied hydrocarbon components can also be marketed, particularly to the petrochemical industry of the U.S. Gulf Coast. In other parts of the world, as the number of liquefied petroleum gas tankers has increased, worldwide markets for the lighter hydrocarbons have developed. As a result, more oil fields have had large-scale gas-processing projects built to recover and market propane and heavier hydrocarbons from the produced gas streams.

12.4 Reservoir Geology Considerations Regarding Immiscible Gas Displacement

Many aspects of reservoir geology interplay with the immiscible gas/oil displacement process to determine overall recovery efficiency. Because there is always a considerable density difference between gas and oil, the extent to which vertical segregation of the fluids occurs and can be taken advantage of or controlled is critical to the success of gas displacing oil. See the reservoir geology chapter in this section of the *Handbook* for considerably more discussion of this topic.

12.4.1 General Geological Considerations. As with any oil recovery process involving the injection of one fluid to displace oil in the reservoir, the internal geometries of the reservoir interval have a controlling effect on how efficiently the injected fluid displaces the oil from the

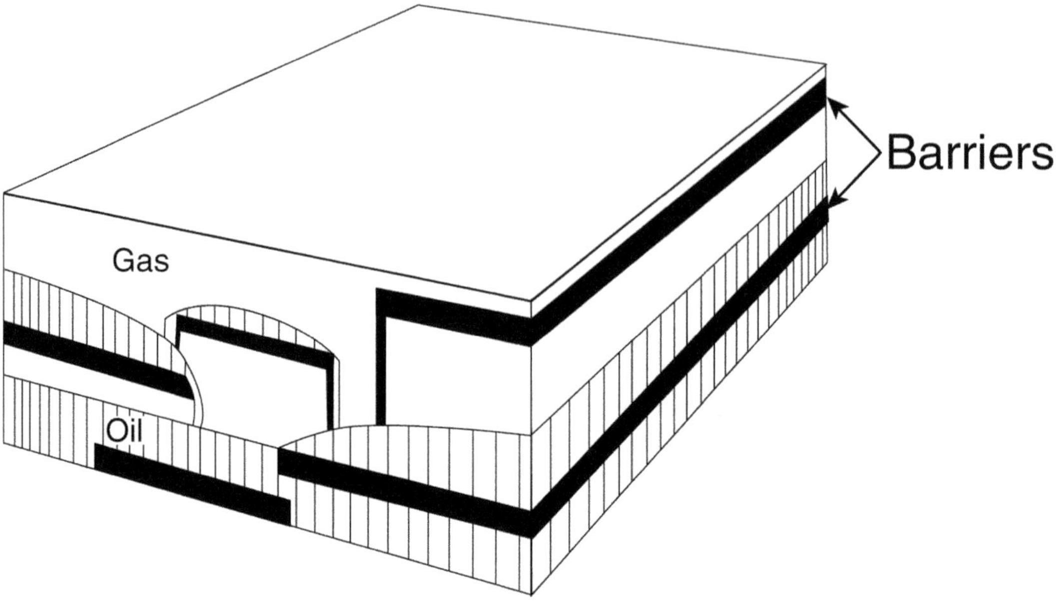

Fig. 12.7—Schematic of oil draining off small vertical flow barriers.[28]

whole of the reservoir. For the immiscible gas/oil displacement process, the key factors are stratigraphy and structure.

Stratigraphy. The stratigraphy of a reservoir is determined primarily by its depositional environment. First and foremost is how layered the reservoir is in terms of both how heterogeneous the various sand intervals are and the scale at which shales or other barriers to vertical flow are interbedded with the sands. Another very important aspect is how continuous the shale intervals are. With fully continuous shales, a reservoir interval should be divided into compartments that will not interact with each other. Unless the reservoir has a steep dip, such shales will negatively affect the gas/oil displacement process.

Less continuous shales can result in better distribution of injected gas without a strong negative impact on the gas/oil gravity drainage process (see **Fig. 12.7** for an illustrative cross-sectional view of sands interbedded with discontinuous shales). Richardson *et al.*[28] analyzed the effects of such limited-size shales. First, they determined geologic factors that control shale dimensions and continuity for sandstone deposits. There is a wide range of shales whose dimensions depend on the depositional environment, with marine shales being the most extensive and flood-plain and interdistributary shales being of smaller areal dimensions (1,300 to 5,250 ft wide by 5,250 to 10,500 ft long). Next, simple 2D calculations of oil drainage off small shales were made, assuming that the various beds were horizontal. They concluded that, "The time required for oil drainage from a barrier is proportional to its width squared and viscosity, and inversely proportional to the horizontal permeability and density difference. Lateral drainage off small barriers can be rapid, and recoveries may be reduced only slightly."[28]

Structure. The structural aspects of a particular reservoir consist of several parts: the closure or vertical thickness of the hydrocarbon column, dip angle of the beds, and size and relative thickness of the gas cap compared with the oil column. Thick reservoirs (> 600 ft of oil column) are the best for application of the immiscible gas/oil drainage process with gas injection at the crest of the structure and oil production from as far downdip as possible. Dip angle is important to the efficiency of the displacement process because a higher dip angle generally means that the effective vertical permeability is increased.

The relative size of the oil column compared with the gas cap affects the performance of a particular reservoir. The gas/oil gravity drainage process has been applied to reservoirs that have, relative to the size of the oil column, very small gas caps[26] and to some with very large gas caps.[29] Success has been achieved over the full range of ratios of gas cap to oil column size. The advantage of having a large initial gas cap is that the reservoir pressure drops very slowly as the oil is produced compared with a situation with a relatively small gas cap in which the reservoir pressure falls quite rapidly until the secondary gas cap grows sufficiently.

Other Geological Factors. Within the reservoir sandstone layers, the nature of the sand layering can strongly affect the efficiency of the gas/oil displacement. In those depositional environments in which the highest-permeability sands are on the bottom of the reservoir interval, the gas/oil displacement process will be far more efficient, especially compared with the situation in which the depositional environment results in the highest permeability toward the top of the reservoir interval. The reason is that, in the first situation, the gravity override of the gas is slowed by the vertical distribution of permeability, but in the latter situation, the gas gravity override is enhanced.

Even if the reservoir were totally homogeneous, a horizontal gas/oil displacement process would not be very efficient because the gas will strongly override the oil and, because of its high mobility, will rapidly travel from the injection wells to the production wells. For reservoirs with many “random” heterogeneities, the gas/oil displacement process will be aided because heterogeneities inhibit growth of low-viscosity fingers by forcing them to travel a more circuitous path between the injector and producer.

Vertical Permeability at Various Scales. The challenge in making calculations for the immiscible gas/oil displacement process at the reservoir scale is to quantify properly the vertical permeability of the reservoir as a whole. There are several scales at which vertical permeability affects the gas/oil displacement process. The engineer has quantitative data on the vertical permeability from routine core analysis performed foot by foot on small core plugs from the reservoir interval. The next larger scale concerns the areal extent of any impermeable layers observed in the cores, be they 1 in., 6 in., or several feet thick. Geologists typically estimate the areal dimensions of these impermeable layers from their training, experience, and studies of outcrops from similar depositional environments.

Despite all the technical work performed before this process is applied to a particular reservoir, the actual effective vertical permeability and its distribution will not be fully known until some years later. The vertical permeability can be quantified by observing reservoir performance. Typically, gravity-drainage immiscible gas/oil displacements are undertaken with the assumption of good vertical permeability so that if actual reservoir performance matches the projections, then the vertical permeability is as high as previously assumed. However, if the reservoir performance is poorer than expected, a likely cause is lower vertical permeability and/or heterogeneities in the vertical permeability distribution over the reservoir.

Carbonate Reservoirs. The geologic discussion above primarily concerns sandstone reservoirs, although many of the general concepts also apply to carbonate reservoirs. Because diagenetic changes often alter the original framework of a carbonate reservoir far more than what occurs in sandstone reservoirs and because some types of carbonate reservoirs do not have sandstone equivalents, this section briefly discusses some differences in carbonate reservoirs.

One type of carbonate deposits that results in reservoirs with thick vertical dimensions, especially compared with their areal dimensions, is the carbonate reef deposit. The style of this deposit with the greatest vertical-to-horizontal aspect is called a pinnacle reef. Reef deposits typically contain large vugs. The key question is, How interconnected are these vugs? Diagenetic processes can isolate these vugs or can provide various types of pore-to-pore interconnections. For example, in New Mexico, the Abo reef trend developed on the northern margin of the Delaware basin. The original reef framework of hydrocorals, sponges, and algae has been totally dolomitized to create a pore system consisting only of vugs, fractures, and fissures.[26]

In carbonate reservoirs, the diagenetic process includes both chemical alteration, such as dolomitization, and cementing and leaching processes. Cementation with calcite, anhydrite, or other insoluble chemicals can have a significant negative impact on the reservoir's pore system. Leaching has the opposite effect and generally enhances the reservoir quality, although leaching may increase the range of heterogeneities and lead to some superconductive flow paths in portions of a reservoir. As carbonate rocks become more brittle because of chemical alteration, fracturing commonly occurs. The geologist and petrographer must examine the cores in great detail to determine the number and sequence of cementation, leaching, and fracturing events that have altered a particular rock interval over geologic time.

Middle East Carbonate Matrix/Fracture-System Reservoirs. A particular style of reservoir in which a considerable number of immiscible gas/oil gravity drainage projects have been applied is the Middle East carbonate matrix-block/fracture-system reservoirs. Most of these reservoirs are very large folded anticlinal structures with dimensions of tens of miles long by several miles wide and with hydrocarbon columns hundreds of feet to several thousand feet thick.

In these carbonate reservoirs, the matrix is high porosity but low permeability (generally ≤ 1 md), and the fracture system created matrix blocks with dimensions ranging from a few feet to > 10 ft.[30,31] The fractures can be up to a tenth of an inch wide, so the effective interwell permeabilities are very high.

The geologic complications of these matrix-block/fracture-system reservoirs concern the way that the matrix and fractures are interconnected and fluid is transferred between these two portions of the pore system. This combines (1) the interaction of the fractures with the matrix along the faces of vertical fractures, (2) the interaction of one matrix block with its neighboring matrix blocks if capillary continuity exists along such surfaces (see **Fig. 12.8** for the schematic oil saturation profiles for cases without and with capillary continuity), and (3) possible fluid transfer along matrix/fracture surfaces if portions of the fracture system are inclined planes and neither vertical nor horizontal. Also, the presence of cementation along some of the fracture surfaces is very important to fluid transfer because the fractures will rapidly transport fluid, but for overall high recovery efficiency, the matrix blocks must exchange oil and gas with the surrounding fracture system. The geological aspects of such matrix-block/fracture systems are difficult to quantify because their dimensions and fracture characteristics cannot be easily discerned from cores and logs. Descriptions of nearby outcrops of the reservoir formation can often be helpful in understanding the macrodimensions of the matrix-block/fracture system.

A number of technical papers have explored aspects of the geology/fluid-flow interactions of such matrix-block/fracture network carbonate reservoirs. Firoozabadi and coworkers[32–37] have developed theories, made calculations, and performed experiments to explore aspects of these types of reservoirs. Saidi[30] has discussed the physical phenomena affecting the performance of the Haft Kel field (Iran) and analyzed its performance; more discussion of the Haft Kel field is found in the Case Studies section of this chapter.

12.5 General Immiscible Gas/Oil Displacement Techniques

In this section, the general technical features of the various immiscible gas injection projects are discussed.

12.5.1 Types of Gas-Injection Operations. Immiscible gas injection is usually classified as either crestal or pattern, depending on the location of the gas injection wells. The same physical principles of oil displacement apply to either type of operation; however, the overall objectives, type of field selected, and analytical procedures for predicting reservoir performance vary considerably by gas injection method.

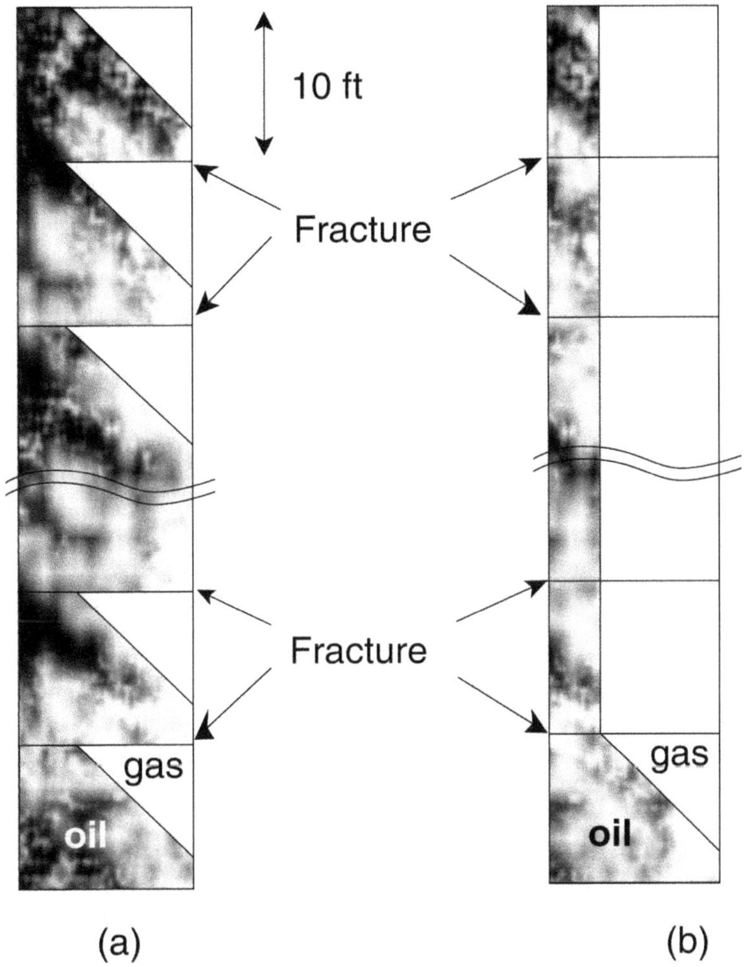

Fig. 12.8—Schematic of oil saturation profiles (dark shading) from stacks of matrix blocks: (a) without capillary continuity and (b) with capillary continuity bounded by vertical and horizontal fractures.[30]

Crestal Gas Injection. Crestal gas injection, sometimes called external or gas-cap injection, uses injection wells in higher structural positions, usually in the primary or secondary gas cap. This manner of injection is generally used in reservoirs with significant structural relief or thick oil columns with good vertical permeability. Injection wells are positioned to provide good areal distribution and to obtain maximum benefit of gravity drainage. The number of injection wells required for a specific reservoir depends on the injectivity of individual wells and the distribution needed to maximize the volume of the oil column contacted.

Crestal injection, when applicable, is superior to pattern injection because of the benefits of gravity drainage. In addition, crestal injection, if conducted at gravity-stable rates—e.g., less than the critical rate (see Eq. 12.8 later in this chapter)—will result in greater volumetric sweep efficiency than pattern injection operations. There are many examples of ongoing crestal injection projects throughout the world, including some very large projects in the Middle East.

Pattern Gas Injection. Pattern gas injection, sometimes called dispersed or internal gas injection, consists of a geometric arrangement of injection wells for the purpose of uniformly distributing the injected gas throughout the oil-productive portions of the reservoir. In practice, injection-well/production-well arrays often vary from the conventional regular pattern configurations—e.g., five-spot, seven-spot, nine-spot (see the chapter on waterflooding in this section for more description of these patterns)—to irregular injection-well spacing. The selection of an injection arrangement is a function of reservoir structure, sand continuity, permeability and porosity levels and variations, and the number and relative locations of existing wells.

This method of injection has been applied to reservoirs having low structural relief, relatively homogeneous reservoirs with low permeabilities, and reservoirs with low vertical permeability. Many early immiscible gas-injection projects were of this type. The greater injection-well density results in pattern gas injection, rapid pressure and production response, and shortened reservoir depletion times.

There are several limitations to pattern-type gas injection. Little or no improvement in recovery is derived from structural position or gravity drainage because both injection and production wells are located in all areas of the reservoir. Low areal sweep efficiency results from gas override in thin stringers and by viscous fingering of gas caused by high flow velocities and adverse mobility ratios. High injection-well density increases installation and operating costs. Typical results of applying pattern injection in low-dip reservoirs are rapid gas breakthrough, high producing GORs, significant gas compression costs to reinject the gas into the reservoir, and an improved recovery of < 10% of original oil in place (OOIP). Note that gas inefficiently displaces oil in gas-swept areas. Attempts to subsequently waterflood such areas result in rapid water breakthrough and little, if any, additional oil displacement.

Few pattern gas injection projects have been implemented in recent years because this method is not as attractive economically as alternative methods for increasing oil recovery.

12.5.2 Optimum Time To Initiate Gas Injection Operations. The optimum time to begin gas injection is site specific and depends on a balance of risks, gas market availability, environmental considerations, and other factors that affect project economics. When only oil recovery and improvements in reservoir producing characteristics are considered, reservoir conditions for gas injection operations are usually more favorable when the reservoir is at or slightly below the oil bubblepoint pressure, unless the bubblepoint pressure is low compared with the initial reservoir pressure. Near the oil bubblepoint pressure, nonrecovered oil represents the smallest volume of stock-tank oil, oil relative permeability is high, and oil viscosity is low.

12.5.3 Efficiencies of Oil Recovery by Immiscible Gas Displacement. It is customary in most displacement processes to relate recovery efficiency to displacement efficiency and volumetric sweep efficiency. The product of these factors provides an estimate of recoverable oil expressed as a percentage of OOIP. Analytical procedures are available for evaluating each efficiency factor. For the purposes of this chapter, the two components describing the overall recovery efficiency are defined as follows:

1. Displacement efficiency is the percentage of oil in place within a totally swept reservoir rock volume that is recovered as a result of viscous displacement and gravity drainage processes.
2. Volumetric sweep efficiency is the percentage of the total rock or PV that is swept by gas. This factor is sometimes divided into horizontal and vertical components, with the product of the two components representing the volumetric sweep.

Recovery efficiencies increase with continued gas injection, but the rate of recovery diminishes after gas breakthrough occurs as the GOR increases. The overall result is that the ultimate oil recovery efficiency is a function of economic considerations, such as the cost of gas compression and the volume and availability of lean residue gas or potentially more expensive alternatives like N_2 from a nitrogen rejection plant.

12.6 Vertical or Gravity Drainage Gas Displacement

In this section, the primary manner in which the immiscible gas/oil displacement process has been used is discussed in qualitative terms. This is the use of gas injection high on structure to displace oil downdip toward the production wells that are completed low in the oil column. In

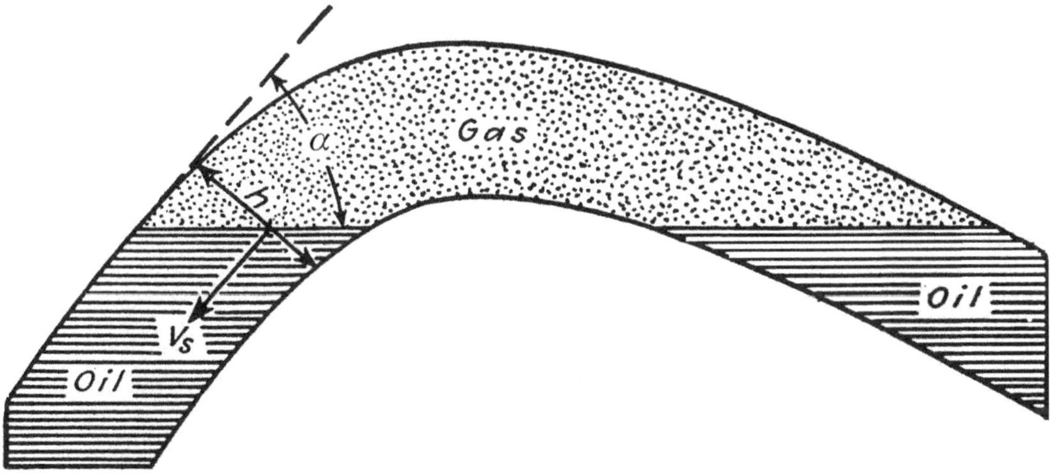

Fig. 12.9—Schematic cross-sectional view of anticlinal reservoir of thickness *h* and dip angle *α* with gas cap overlying oil column.[1]

many cases, an original gas cap was present, so the gas was injected into that gas-cap interval (see **Fig. 12.9** for cross-sectional view of anticlinal reservoir with gas cap over oil column with dip angle α and thickness h). In this situation, the force of gravity is at work, trying to stabilize the downward gas/oil displacement process by keeping the gas on top of the oil and counteracting the unstable gas/oil viscous displacement process. If the oil production rate is kept below the critical rate, then the gas/oil contact (GOC) will move downward at a uniform rate. In the next section, the simple engineering calculation techniques for estimating the rate for stable gravity drainage for a gas/oil system are discussed.

In normal oilfield operations, producing wells are drilled and perforated as far as possible from the gas cap, both vertically and laterally. Hence, many production wells may not have any of the gas cap directly over them initially. Until the gas cap extends over the whole of the oil column and its producing wells, because of the pressure differential between the gas cap and the area around the producing wells, the downdip oil production will result in the gas cap moving rapidly along the top of the structure to provide pressure support to the whole oil column. When this happens, if the reservoir is a dipping structure, the gas cap is described as "draping" over the oil column. If the oil reservoir is an anticlinal structure, then this process is often described as the gas cap having expanded to form an "umbrella" over the whole oil column. Once the gas-cap drape or umbrella has formed, the continued downward movement of the gas cap at each location can be stabilized. **Fig. 12.10** shows views of the Prudhoe Bay oil field (Alaskan North Slope) in which the original gas cap was offset to the northeast of most of the oil column and where there has been gas cap underrunning of the top of structure until it draped over much of the downdip oil column where the hundreds of producing wells were located.

There are likely to be local variations in the GOC caused by reservoir heterogeneities and near-wellbore pressure gradients. The most notable of these results in gas coning caused by high pressure gradients around the perforated interval of each wellbore. Here, the controlling factors are the oil and gas production rates, the distance from the top of the perforations to the overlying GOC, and the horizontal and vertical permeabilities.[38] In this situation, the presence of a small shale interval between the GOC and the top of the perforated interval can have a very beneficial effect on the maximum oil production rate before gas coning occurs (see **Fig. 12.11**). For a particular reservoir situation, gas-coning calculations are best made with a numerical reservoir simulation model.[38,39]

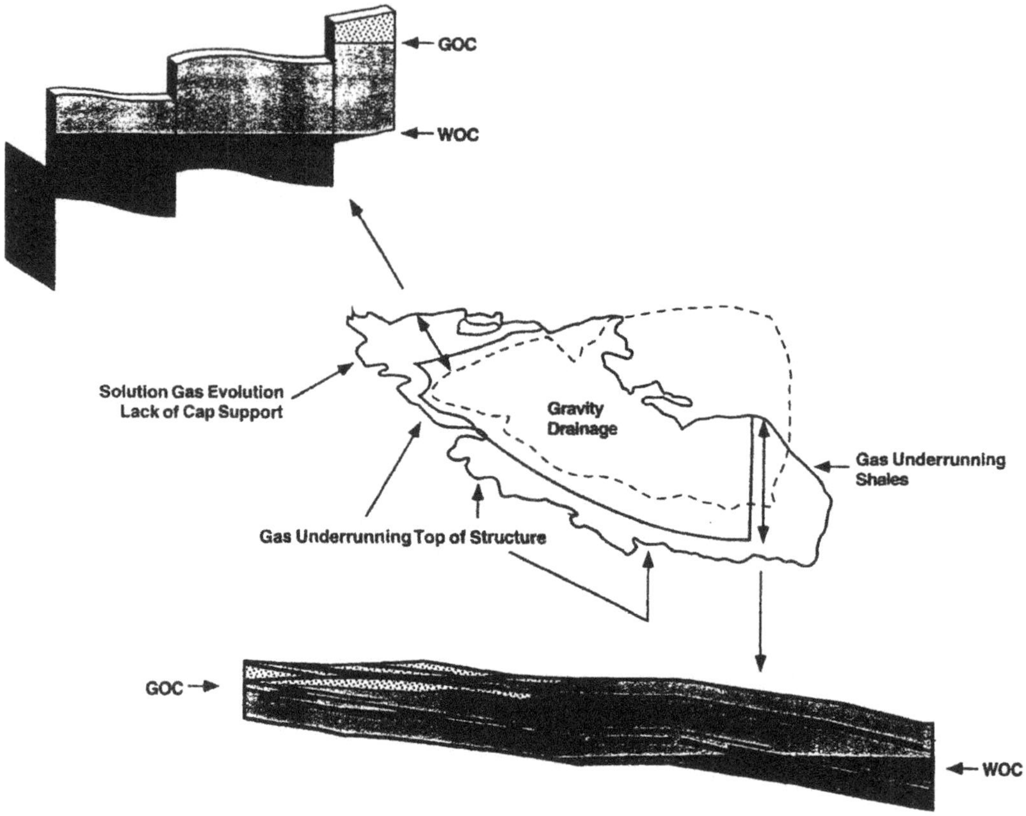

Fig. 12.10—Prudhoe Bay field: different natural depletion producing mechanisms in various areas of the Sadlerochit reservoir.[13]

The gas/oil gravity drainage process is complicated if the oil column is underlain by an aquifer because the aquifer will provide pressure support to the oil column in response to any decrease from original reservoir pressure caused by oil production. If the aquifer is very strong, it will invade the lower portions of the oil column and may provide almost barrel-for-barrel voidage replacement. In this case, the original gas cap may not expand much. If the aquifer is weak or if there is a tar mat at the oil/water contact (OWC) inhibiting water influx, then the gas cap will be the primary means of pressure support for the oil column and the reservoir will perform almost as if there were no aquifer present. A problem sometimes experienced with oil reservoirs with both overlying gas caps and underlying aquifers is that the near-wellbore coning behavior is more complicated. The reason is that gas-cap gas is coning downward toward the perforated interval and aquifer water is trying to cone upward toward the same perforated interval. If water cones first into the perforated interval, then the gas coning will be more severe because, with three-phase relative permeability effects, the near-wellbore pressure gradients are greater, which causes gas coning to occur at lower oil production rates.

12.7 Calculation Methods for Immiscible Gas Displacement

Techniques described in this chapter are classic methods for describing immiscible displacement assuming equilibrium between injected gas and displaced oil phases while accounting for differing physical characteristics of the fluids, the effects of reservoir heterogeneities, and injection/production well configurations. The reservoir is treated in terms of average properties for volume of rock, and production performance is described on the basis of an average well. Black-

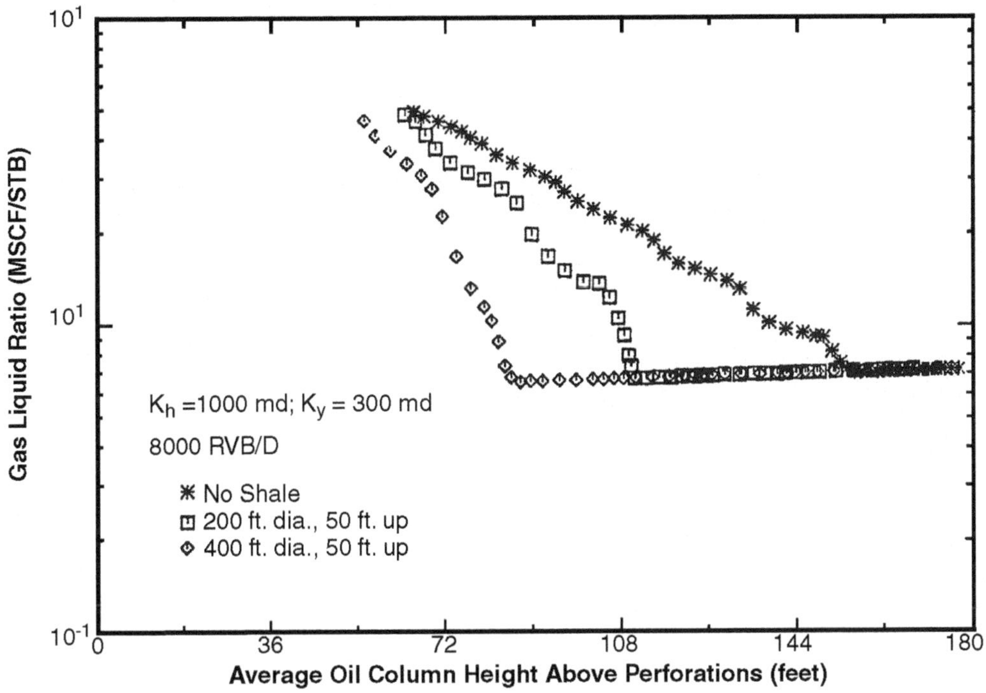

Fig. 12.11—Numerical simulation results of the effects of small shales on near-wellbore gas coning behavior.[38]

oil-type reservoir simulation models use essentially these same techniques but, by means of 1D, 2D, or 3D cell arrays, account for areal and vertical variations in rock and fluid properties, well-to-well gravity effects, and individual well characteristics. More complex compositional models account for nonequilibrium conditions between injected and displaced fluids and can be used to describe individual well streams in terms of the compositions of the produced fluids.

The increasing capability of desktop computers and the growing amount of affordable simulation software are making it possible to use numerical reservoir simulation more often. However, results obtained from simulation will be directly dependent on the quality of data to describe the reservoir rocks and fluids. It is also important to comprehend the physics of displacement to understand the simulation results and to identify incorrect results. The fundamentals of the displacement process presented in this section are intended to provide the background needed to produce good-quality predictions of oil recoveries.

12.7.1 Modifications of Displacement Equations. Applicability of the basic displacement equations to a given reservoir is governed by whether the underlying assumptions are reasonable. Several authors have reported modifications that eliminate the need to make certain assumptions. Modifications that take into consideration the swelling effects experienced from injection into an undersaturated reservoir and production of fluids from behind the gas front have been presented by Welge,[18] Kern,[19] Shreve and Welch,[40] and others. Jacoby and Berry,[41] Attra,[42] and others have presented equations and simple analytical procedures for calculating performance to account for some of the compositional interchange between the displacing gas and the reservoir oil.

These works are mentioned for completeness. If significant deviations from the basic assumptions of the Buckley-Leverett method are a concern, the more practical approach is to use numerical reservoir simulation to account for reservoir heterogeneities and gravity, capillary, and compositional effects. These simulators are discussed in the numerical simulation chapter in this section of the *Handbook.*

12.7.2 Methods for Evaluating Sweep Efficiency. Some techniques for estimating the volumetric, vertical, and areal sweep efficiency of an immiscible gas/oil displacement are discussed below.

History Matching. If there are sufficient data concerning the location of the gas front and oil recovery as a function of time, past reservoir performance can be used to calculate the volumetric sweep efficiency by dividing observed recovery at various times by the theoretical recovery determined from displacement efficiency calculations.

If there are adequate data to reliably describe spatial variations in reservoir rock and fluid characteristics, numerical reservoir simulation is the best way to predict sweep efficiency, particularly after the historical production and pressure data are matched. If data or sufficient economic justification to undertake a full numerical reservoir simulation study is lacking, the following methods are presented as useful for screening studies and in situations when more detailed studies are inappropriate.

Vertical Sweep Efficiency. Several authors have presented methods for determining vertical sweep efficiency based on statistical treatments of routine core analysis data. Some of the most frequently used methods are adaptations of the Stiles[43] method for evaluating the effect of permeability variations on waterflood performance. The same assumptions and calculation procedures may be used for immiscible gas/oil displacements. The relative permeability ratio used in such calculations is considered to be a constant equal to the relative permeability to gas at residual oil saturation ($k_{rg@Sor}$) divided by the relative permeability to oil at initial gas saturation ($k_{ro@Sgi}$).

Areal Sweep Efficiency. Several investigators have shown that areal sweep efficiency is primarily a function of injection/production well pattern arrangement, mobility ratio, and volume of displacing phase injected. Various studies have confirmed what would be expected intuitively, that areal sweep efficiency increases with the volume injected and with a lower mobility ratio. Data from model studies that show the influence of mobility ratio and displacement volume on areal sweep efficiency in a regular five-spot pattern are illustrated in **Fig. 12.12.**[44]

Thin models containing miscible fluids of varying viscosity were used to develop these area sweepout curves. These data are considered applicable to either water/oil or gas/oil displacement. These data are presented to aid in the understanding of the effect of some factors on the gas displacement mechanism and may prove useful in preliminary studies of a potential gas injection project to predict volumetric sweep. However, the quantitative applicability of laboratory data is inherently questionable because of uncertainties in model scaling, laboratory techniques, and associated simplifying assumptions regarding no vertical gas override effects or reservoir heterogeneities. The instability of the very unfavorable mobility ratio gas/oil displacement is most difficult to quantify in laboratory experiments. All these effects can cause a smaller sweepout efficiency than presented in Fig. 12.12. Nevertheless, laboratory model studies do offer a convenient means of making quantitative estimates when simulation is not practical or justified and injected gas remains dispersed in the reservoir.

When the laboratory data are used, the common practice is to calculate a mobility ratio using the viscosity and relative permeabilities of the oil ahead of the gas front and of the gas at the average saturation behind the displacing front.

12.7.3 Calculating Immiscible Gas Injection Performance. Numerical simulation represents the best way to predict the performance of immiscible gas injection if there are sufficient data to characterize the reservoir rocks and fluids adequately. Even simple 2D and 3D black-oil models provide insight into the more important aspects of oil recovery for reservoirs in which

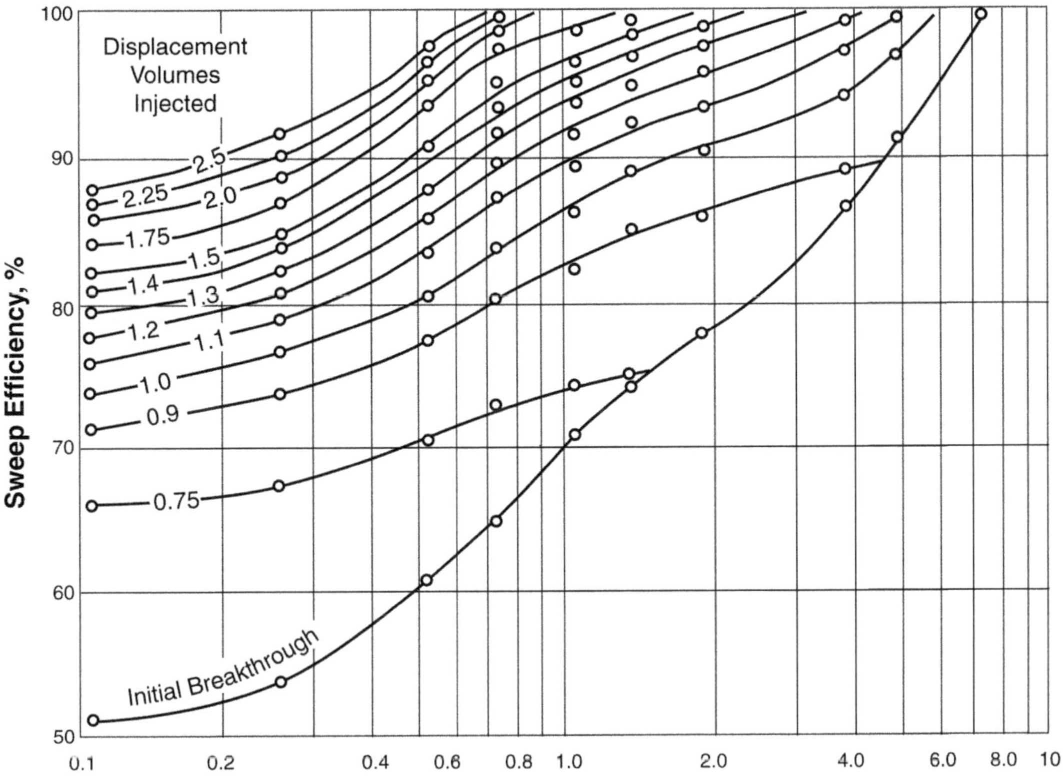

Fig. 12.12—Sweep efficiency as a function of mobility ratio.[44]

compositional effects are not a major concern. When adequate data are unavailable or when screening work is being done, simple models may suffice.

The immiscible displacement of oil by gas is described with fractional flow theory. Muskat[1] presented the basics of this theory more than 60 years ago. Since then, additional work has been done to develop various mathematical calculation methods based on fractional flow theory. A few of the more recent papers discussing these techniques are listed in the References section.[45–51]

As discussed above, pattern gas injection is seldom used now because waterflooding performance is much better in those types of reservoirs where pattern gas injection has historically been tried. Therefore, the remainder of this section discusses a simple model used for reservoirs in which stabilized gravity drainage controls the gas/oil displacement process and increases the ultimate oil recovery.[13]

Viscous, gravitational, and capillary forces and diffusion are involved in the displacement of oil by gas, complicating technical analysis of a particular reservoir if each of these forces and flow in all three dimensions are important. Fortunately, there are instances in which one force is dominant and only one dimension is involved in the rate-limiting step. In these circumstances, engineering solutions can be direct and simple. One such circumstance is that of thick reservoirs with high permeabilities.

In steeply dipping oil reservoirs containing sands with high vertical permeabilities, gravity drainage of the oil can be more effective than is calculated from the Buckley-Leverett assumptions alone, as illustrated in **Fig. 12.13**.[52] When sufficient vertical permeability exists, even at lower oil saturations, oil behind the gas front can continue to flow vertically downward

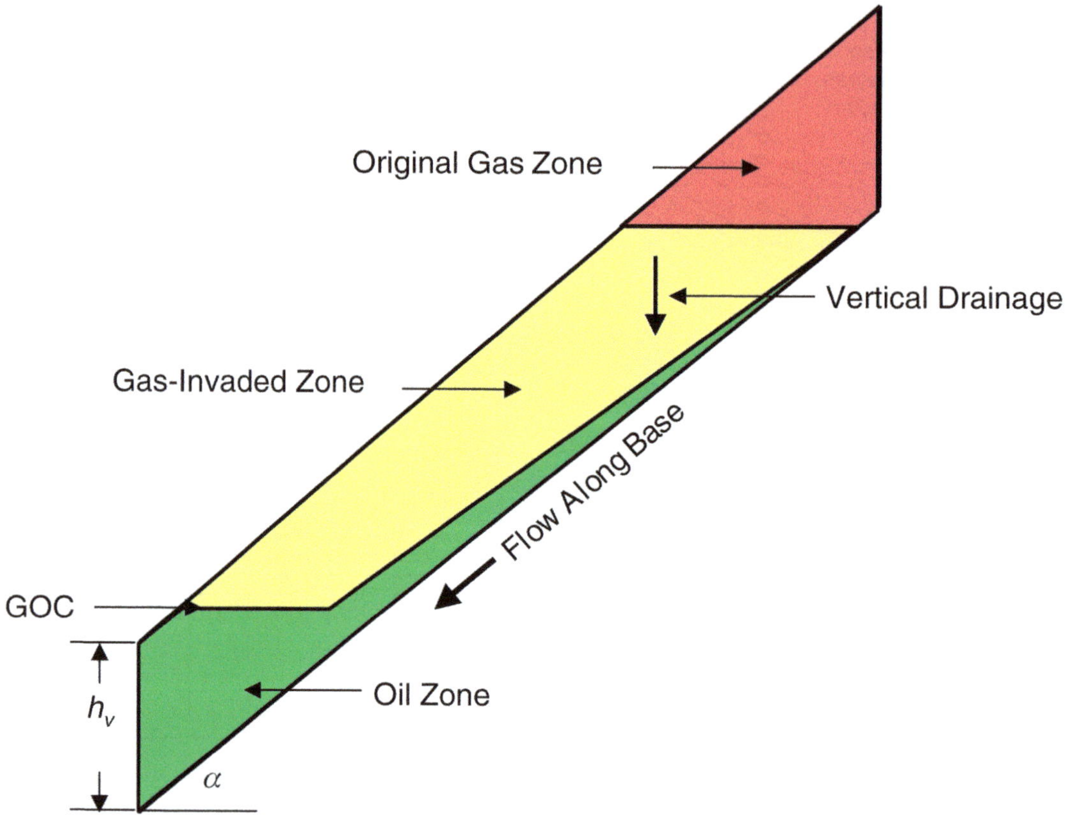

Fig. 12.13—Mechanisms of gravity drainage.

through the reservoir. Thus, the displacement process occurs in two steps. First, gas invades the originally oil-saturated sand as the GOC moves downdip because of oil production farther downdip. Second, oil drains vertically downward through the gas-invaded region and forms a thin layer with high oil saturation (with high k_{ro}) that drains along the base of the reservoir interval to the remaining downdip oil column.

Mathematical Model. A simple mathematical model can be used to describe the displacement of oil by gas drive and gravity drainage when the rate is less than one-half the critical rate. The critical rate is given by

$$\left(\frac{q_T}{A}\right)_{\text{critical}} = \frac{0.044 k \Delta\rho \sin\alpha}{\dfrac{\mu_o}{k_{ro}} - \dfrac{\mu_g}{k_{rg}}}, \quad (12.8)$$

where q_T = total volumetric flow rate through area A, ft^3/D, and k = permeability, darcies.

The first calculation determines the gas saturation just above the GOC by using Eq. 12.5, plotting F_g vs. S_g, and finding the tangent to the curve passing through the origin, as shown in Fig. 12.2. For ease of calculation, the GOC is assumed to move at a constant rate. The next calculation determines the quantity of oil that drains from the region invaded by gas in a given time increment. For ease of calculation, this region is divided into arbitrary lengths, and the amount of oil produced by vertical gravity drainage is calculated for the average time since passage of the gas front.

For vertical drainage of oil, the rate is given by Darcy's law, with the driving force proportional to the density difference between gas and oil. The assumption is made that resistance to flow of gas and capillary effects are negligible[53]:

$$u_{ov} = \frac{-0.044 k k_{ro} \Delta\rho}{\mu_o}, \quad (12.9)$$

where u_{ov} = vertical oil flow per unit area, res ft³/ft²-D.

From continuity considerations,

$$\left(\frac{dz}{dt}\right)_{S_o} = \frac{1}{\phi}\left(\frac{du_{ov}}{dS_o}\right), \quad (12.10)$$

where z = vertical distance, ft; t = time, days; ϕ = porosity, fraction; and S_o = oil saturation, fraction.

The rate of movement of a particular saturation, $(dz/dt)_{So}$, can be determined by plotting u_{ov} calculated from Eq. 12.9 vs. saturation, taking the slope to determine du_{ov}/dS_o and dividing by porosity, as indicated in Eq. 12.10. The amount of oil drained from each region since passage of the gas front can be calculated by graphical integration of the height vs. saturation plot.

As a first approximation, the time for oil to flow downdip in the thin layer along the base of the reservoir interval can be neglected, as can the volume of oil in this layer. If the displacement rate exceeds one-half the critical rate, oil tends to accumulate rather than flow away along the bottom of the reservoir. More accurate calculations also include consideration of the thickness of the gas/oil transition zone arising from capillary effects above this layer of oil, especially if the transition zone is < 10 ft thick.

Recoveries calculated by this technique are quite sensitive to the values of k_{ro} at low oil saturations.[10] Ways to extrapolate measured information are discussed next. First, conventional laboratory data can be extended to low oil saturations by plotting measured values of k_{ro} vs. $(S_o - S_{org}^*)/(1 - S_{wi} - S_{org}^*)$, in which S_{org}^* is the irreducible oil saturation in the presence of gas and connate water. S_{org}^* can be calculated by material balance for areas of a reservoir that have been invaded by gas if good data are available on GOC movement and oil recovery from the area. A second source is the oil saturation in an associated gas cap as determined in cores from that region. If the gas cap was originally filled with oil, drainage of oil over geologic time as gas migrates into the reservoir establishes an endpoint relict oil saturation. For instance, the Prudhoe Bay Sadlerochit reservoir was originally filled with oil. Gas then migrated into the reservoir several million years later, creating the gas cap.[54] Water-based mud cores from the gas cap interval showed an average routine core analysis oil saturation at discovery of 7% PV. The dip of the reservoir is 1 to 3°, but vertical permeabilities throughout the gas and oil columns are generally very high. Interestingly, oil saturations above small shale lenses in the gas cap averaged more than 7% PV, indicating that more time may be required to reach irreducible oil saturations when oil drainage is limited by the dip of this reservoir. A third source is drainage capillary pressure vs. saturation measurements. Experience has indicated that S_{org}^* should be < 10% PV and sometimes approaches zero. Although these endpoint saturations are seldom realized in the depletion time of a reservoir, it is important to have the correct value for predicting flow behavior and ultimate oil recovery. A benefit of even simple, multidimensional simulation models is that the inclusion of capillary effects controls the oil flow rate and conditions under which irreducible saturations are approached.

If a measured value of S_{org}^* is unavailable, a value is chosen to yield a straight line through the data, so for

TABLE 12.2—AVERAGE RESERVOIR PROPERTIES, HAWKINS FIELD

k, darcies	3.4
K_V, darcies	2.38
ϕ, fraction	0.279
S_{wi}, %	8
Dip angle, degrees	6
h_V, ft	49
L (along bedding planes), ft	3,500
p_r, psi	1,500
μ_o, cp	4.45
μ_g, cp	0.0185
ρ_o, lbm/ft^3	51.7
ρ_g, lbm/ft^3	5.35
u per unit area, ft^3/ft^2-D	0.0365

$$k_{rr} = \left(\frac{S_o - S_{org}^*}{1 - S_{wi} - S_{org}^*} \right)^n, \quad (12.11)$$

the slope of the line n should be ≈4 according to the theory of Corey *et al.*[55] but may be as large as 6 and as small as 2. Recovery data can be correlated by the dimensionless parameter $\frac{q_T \mu_o h_v}{k_v \Delta\rho L A}$, which is derived by dividing the time required for vertical drainage,

$$t_v = \frac{h_v}{\frac{k_v \Delta\rho}{\mu_o}}, \quad (12.12)$$

by the time required for flow along the bedding plane,

$$tu = \frac{LA}{q_T}, \quad (12.13)$$

where k_v = vertical permeability, darcies, and h_v = vertical thickness, ft.

Example Gas/Oil Gravity-Drainage Problem. The utility of this simple model can be illustrated by predicting recovery by gas drive and gravity drainage for an actual reservoir, in this case the Hawkins field in east Texas.[56]

Given: Average Hawkins Woodbine reservoir properties as presented in **Table 12.2 and Fig. 12.14.**

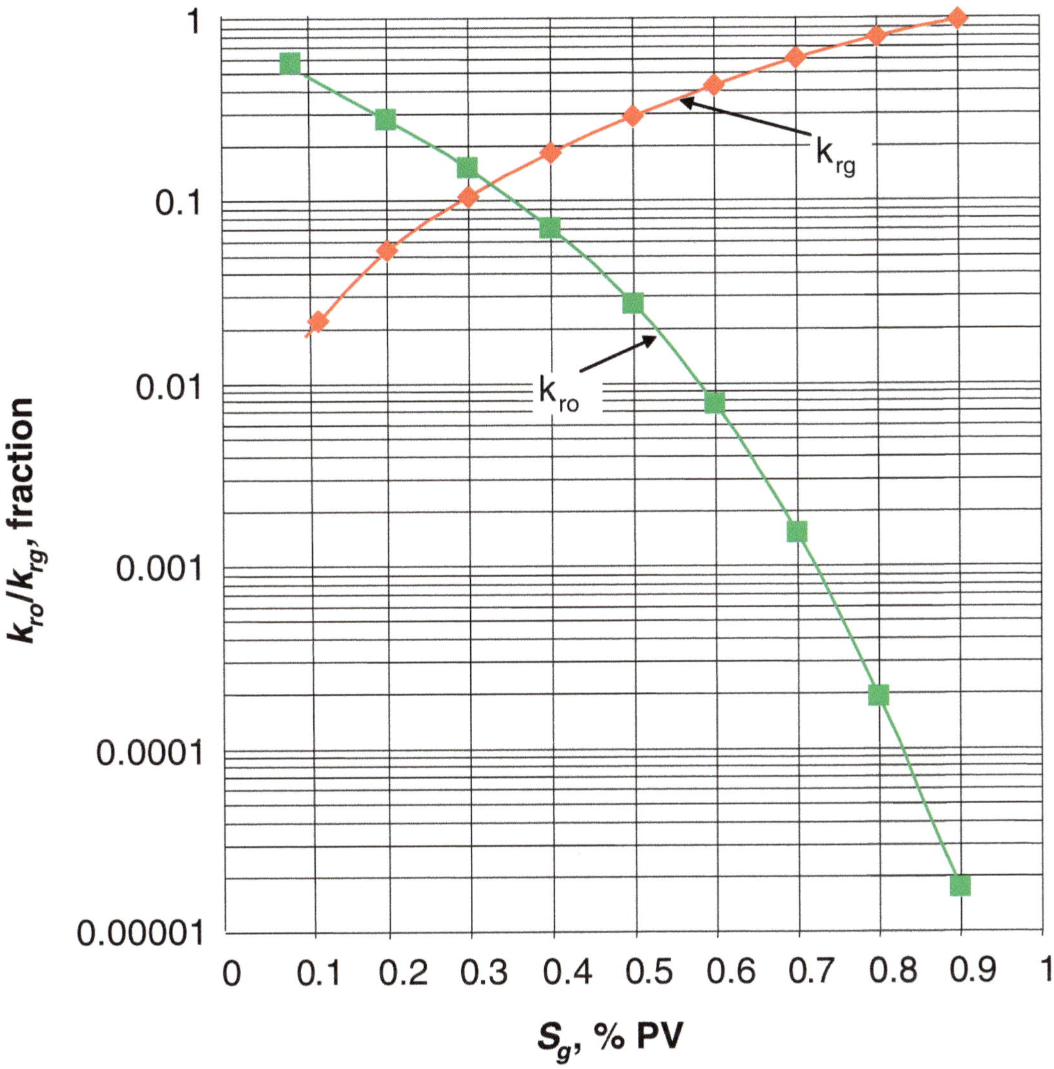

Fig. 12.14—Gas/oil relative permeability for Hawkins field.

Solution: With Eq. 12.8, the critical rate for the Hawkins field is calculated to be 0.173 ft^3/D-ft^2. The average actual rate, q_T/A, of 0.0365 ft^3/D-ft^2 (see Table 12.2) is 21% of the critical rate, and the simplified model should apply. With Eq. 12.5, the average gas saturation just above the GOC is found to be 46% PV by the Welge procedure, as shown in Fig. 12.2. The rate of frontal movement is

$$u_{\text{actual}} = \frac{q_T}{A\phi\,\Delta S} = \frac{(0.0365\ \text{ft}^3/D - \text{ft}^2)(365\ D/\text{yr})}{(0.279)(0.46\text{FracPV})} = 104\ \text{ft/yr}\ . \quad \text{(12.14)}$$

Time to gas breakthrough is 3,500/104 = 34 years. Recovery at breakthrough may now be estimated by dividing the reservoir into seven blocks, each 500 ft long and 49 ft thick. The average vertical movement of saturations in each block can be calculated from Eqs. 12.9 and 12.10. The relative permeability data for oil were extrapolated to low S_o values using the corre-

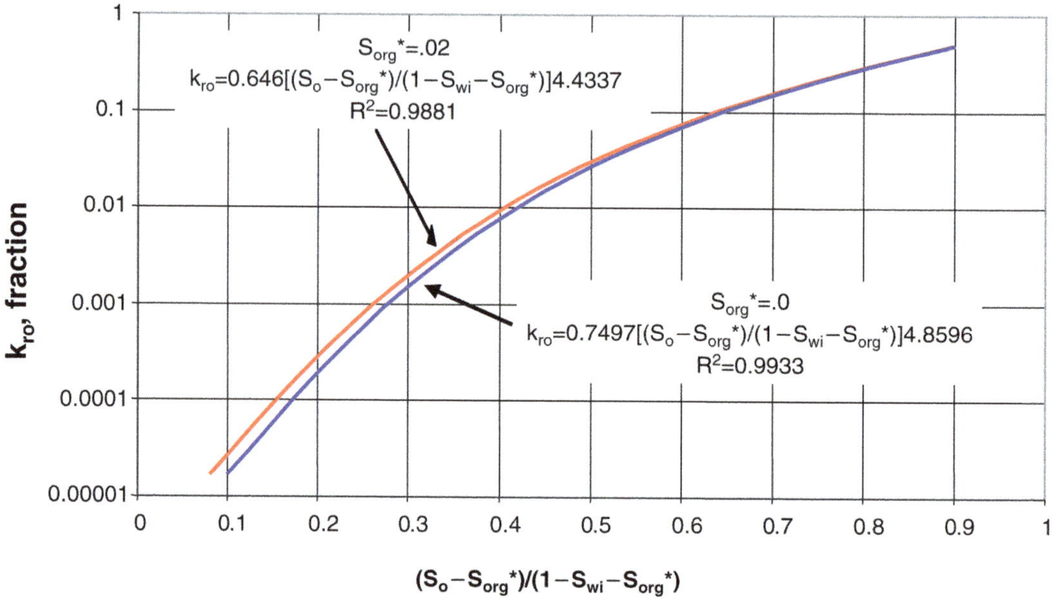

Fig. 12.15—Curve fit of laboratory k_{ro} vs. S_o data, Hawkins field.

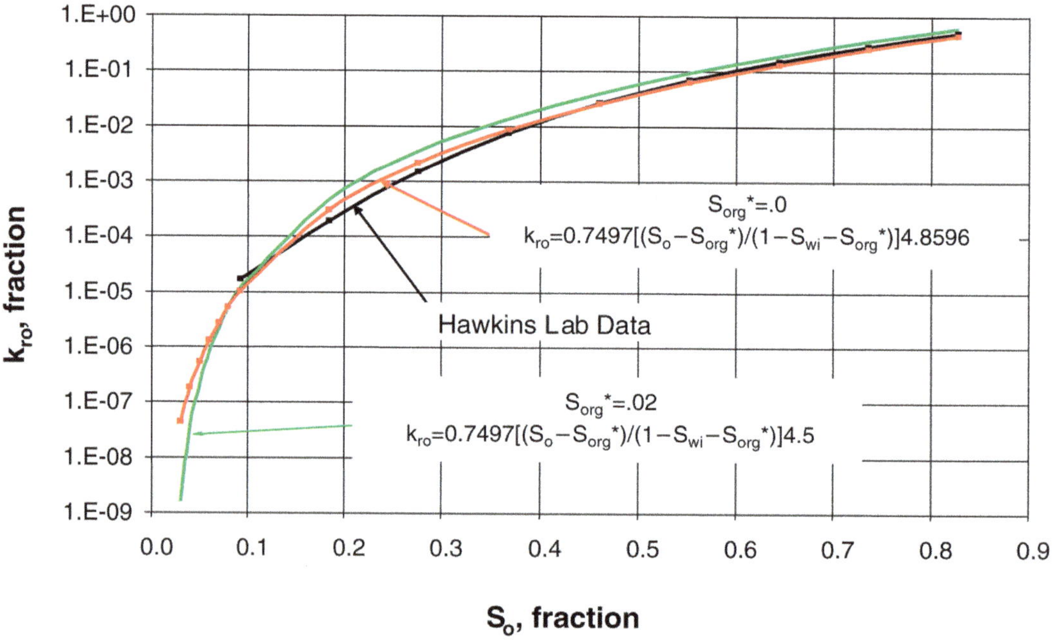

Fig. 12.16—Oil relative permeability projected to low S_o values, Hawkins field.

lation term $(S_o - S_{org}^*)/(1 - S_{wi} - S_{org}^*)$ discussed above. The resulting plots for two S_{org}^* assumptions are shown in **Fig. 12.15**.

In this case, $S_{org}^* = 0$ gave a slightly better fit. The resulting plot of k_{ro} vs. S_o is shown in **Fig. 12.16**. The laboratory data and a plot resulting from use of an S_{org}^* of 2% PV and an n of 4.5 are also shown to indicate the differences in k_{ro} at low S_o that result.

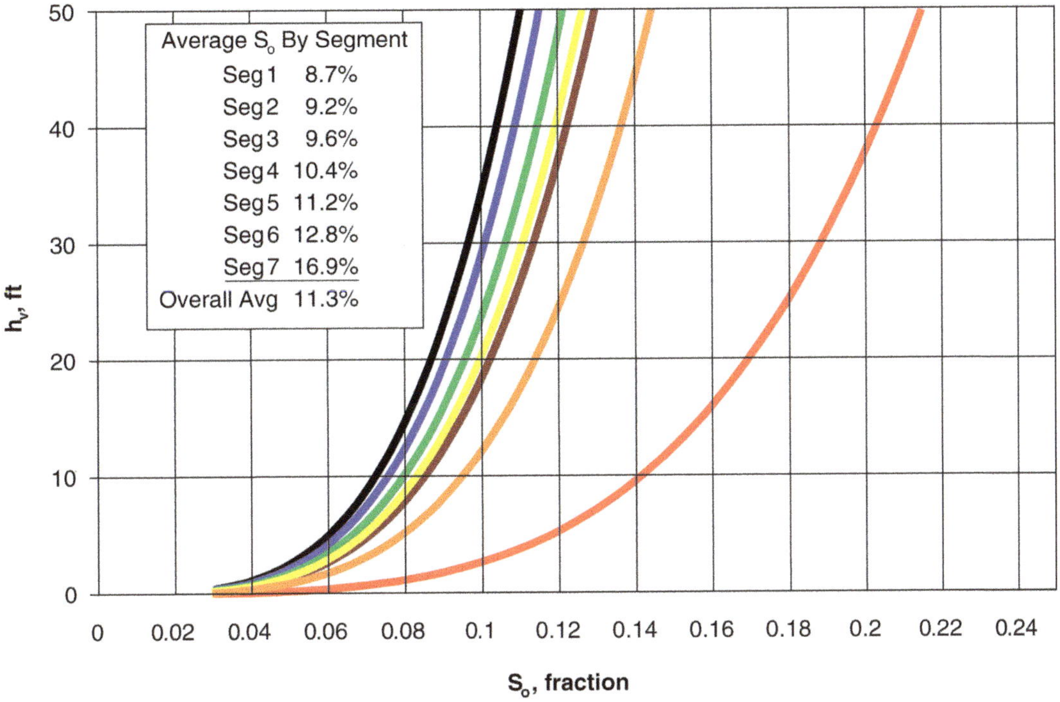

Fig. 12.17—Oil saturation profiles for different blocks, Hawkins field example problem.

The average time interval since passage of the gas front is calculated from the calculated rate of frontal movement of 104 ft/yr. The residual oil left in each block is determined by graphical integration of a plot of S_o vs. height, such as the ones shown in **Fig. 12.17**. Results of these calculations are shown in **Table 12.3**.

$$\text{Oil recovery to gas breakthrough} = \left[\frac{(1 - S_{wi} - S_{or})}{(1 - S_{wi})}\right] = \frac{1 - 0.080 - 0.113}{1 - 0.080} = 88\ \%\ \text{of OOIP.}$$

Comparisons With Field Data. A recovery of 87% of OOIP was observed in the Hawkins field for an area affected by an expanding gas cap.[56] The calculated recovery of 88% compares very favorably. A 2D two-phase reservoir simulation using similar relative permeabilities predicted 87% recovery at breakthrough. The recoveries observed in the field and predicted by models that permit flow in two dimensions are > 15% greater than those calculated by conventional 1D techniques that assume flow only along the bedding planes.

Model Summary. This section has shown how a simple gravity drainage model can be readily applied to predict recoveries by gas drive and gravity drainage when flow rates are less than one-half the critical rate and permeabilities in the vertical direction are high. Some applications of the model have been unsuccessful because of lower-than-expected vertical permeabilities. As a practical matter, the simple model should be used to predict reservoir behavior only when it can be shown to match history or when applied to a field analogous to one that the model fits.

TABLE 12.3—AVERAGE S_o IN EACH CALCULATION SEGMENT

Segment	End Location, ft	Breakthrough to Segment End, yr	Average Drain Time, yr	Days	Average S_o, %
1	3,500	33.7	31.3	11,406	8.7
2	3,000	28.8	26.4	9,651	9.2
3	2,500	24	21.6	7,897	9.6
4	2,000	19.2	16.8	6,142	10.4
5	1,500	14.4	12	4,387	11.2
6	1,000	9.6	7.2	2,632	12.8
7	500	4.8	2.4	877	16.9
				Average	11.3

12.8 Immiscible Gasflood Monitoring

There are a variety of methods for monitoring immiscible gas injection projects. Some apply both to the pattern type of gas injection projects and to the vertical gravity-drainage type of gas injection projects; others apply only to the gravity-drainage projects. In all cases, individual well production and pressure performance as a function of time must be recorded.

The most obvious monitoring method is to track the GORs of the individual production wells as a function of time. The GOR will be approximately flat at the oil's solution GOR until there is gas breakthrough. Then the GOR will climb. The timing of gas breakthrough and the rate of GOR climb will indicate how efficiently, or inefficiently, the gas/oil displacement is progressing. Field engineers should have made preliminary calculations, possibly using a numerical reservoir simulator, so that they have projections of what should be expected regarding gas breakthrough timing and GOR increases at the individual well locations (and as a function of the volume of gas injected at the individual injection wells).

The other methods concern primarily the monitoring of the vertical movement of the gas/oil interface (the current GOC) in gravity-drainage-type projects. Two techniques are generally used: cased-hole logging programs and monitor-well observations. Both help track the GOC movement as a function of time. By mapping the GOCs from the individual wells, engineers can determine whether the GOC is staying reasonably horizontal. By comparing the GOC movement as a function of time to the projected GOC behavior, engineers can determine how efficient the gas/oil displacement is and whether the project's expectations are being met.

For some reservoirs, other unique techniques can be used. For example, if a reservoir has a natural oil gravity variation as a function of depth, then the production wells' oil gravity can be tracked as a function of time.

Another technique is to take periodic gas samples and perform gas chromatograph analyses to determine the produced gas composition. To use this technique, baseline gas samples should be taken early and periodically from all wells. There are two circumstances in which gas chromatography is a useful tool for gas-injection project monitoring. The first is those projects in which flue gas (88% N_2, 12% CO_2) or pure N_2 is injected. In that type of project, it is important to track the BTU value of the gas from each well and how the nonhydrocarbon content of the produced or residue gas changes as a function of time. This is important with respect to the use of that gas for field fuel and the marketability of the residue gas stream.

Second, this technique can be important late in the life of a gas-injection project when wells are operating at very high GORs. This technique is useful for determining whether some of the very lean injection gas is breaking through into some of the wells without becoming saturated with light and intermediate hydrocarbon components from the residual oil phase in the gas-swept region above the current GOC. Such lean gas can actually strip intermediate hydrocarbon components from the produced oil in the field gas/oil separators. If there is no gas plant as part of the field facilities, then those hydrocarbons will be lost to the downstream owner of the gas plant that processes the field gas.

One other aspect of the monitoring activities is to track from which of the perforated intervals most of the gas flow is entering the wellbores. This can be accomplished with periodic spinner or temperature surveys. Depending on the nature of the reservoir interval, it may be possible to temporarily plug off some of the perforations to reduce wells' producing GOR.

The purpose of all these monitoring activities is to perform real-time analysis of reservoir performance and to consider any remedial actions. These actions include such alternatives as rebalancing the gas rates into the various injectors, rebalancing the flow rates of the various producers, and potentially drilling a few new injection or production wells into areas of the reservoir determined to require more drainage points or to improve the overall sweep efficiency.

12.9 Field Case Studies: Immiscible Gas Injection Examples

In this section, a number of field applications of immiscible gas injection are briefly reviewed. In each case, the following facts will be listed, to the extent that they are available.

- Field name.
- Structural closure.
- Rock type.
- Nature of pore system
- Average permeability.
- Source of injected gas.
- Field size.
- Presence of initial gas cap.
- Oil gravity.
- Initial reservoir pressure.
- Oil bubblepoint pressure.

Both successful and unsuccessful immiscible gas injection projects are discussed. A number of early successful and unsuccessful gas injection projects are summarized by Muskat in his 1949 classic book *Physical Principles of Oil Production.*[1] Most of this section describes successful projects because those have been presented in a number of SPE technical papers. Immiscible gas injection has been used in oil fields with a wide range of characteristics.

12.9.1 Early Successful and Unsuccessful Immiscible Gas Injection Projects. Muskat discusses four gas injection projects in the United States from the 1930s and 1940s.[1] Two of these projects were termed successes, and two were viewed as having poor response.

Cunningham Pool (Kansas). This 1,400-acre anticlinal 31 to 36°API oil field had a maximum closure of 75 ft and 53 producing wells. The reservoir is an oolitic limestone and had an initial gas cap. Field discovery was in 1932, and gas injection began in 1936 into three to five wells after the reservoir pressure had declined from 1,115 to 424 psi. The average reservoir properties were as follows: net thickness, 8 ft; porosity, 11% BV; and permeability, 105 md. Muskat termed this project a success because of its GOR history and concluded that "in spite of the thin pay section there has been effective gas segregation, so that the injected gas largely remained trapped in the reservoir and helped to sustain the oil saturation within the oil zone."[1]

Schuler Field (Arkansas). This 3,000-acre anticlinal 34°API oil field had a maximum closure of 135 ft and 146 wells drilled on 20-acre well spacing. This sandstone reservoir had a small initial gas cap. Field discovery was in 1937; gas injection began in 1941 into six wells at the crest of the structure and after the reservoir pressure had declined from 3,520 to about 1,550 psi. The average reservoir properties were as follows: net thickness, 0 to 70 ft; porosity,

17.6% BV; and permeability, 0 to 4,000 md (355-md average). Success was achieved because the reservoir pressure was stabilized, producing GORs were decreased, and produced gas was reinjected instead of flared. Because of proration limits on production, the allowed oil rate was being produced from a few optimally located downdip wells.

Grayburg Lime (West Texas). This 750-acre section of a west Texas anticlinal reservoir is a sandy dolomite and had an average net pay of 18 ft (in a gross thickness of 130 ft). Pay porosity ranged from 8 to 14% BV, and permeabilities ranged from 2 to 10 md. Oil gravities were in the 33 to 37°API range. In 1942, 1 of the 26 producing wells was converted to gas injection after the reservoir pressure had declined from 1,800 to 1,275 psi. Approximately a year after the start of gas injection, there was a sharp rise in the producing GOR. Because the oil production rate was considerably increased at about the same time as the GOR rise, Muskat reasons that, although the gas may have traveled rapidly through "a substantially continuous and intercommunicating fracture system," the GOR rise may have resulted from the inability of the reservoir matrix to supply oil to the fracture system at the higher oil rate.

Canal Field (California). This 1,100-acre dome oil field had a maximum closure of 150 ft with wells drilled on 20-acre well spacing. The reservoir is a sand of variable character with silt/shale streaks. There was no original gas cap; the original reservoir pressure was 3,550 psi; and the oil bubblepoint pressure was 2,800 psi. Porosity ranged from 15 to 32% BV with an average of 22% BV, and permeability ranged from 10 to 1,000 md with an average of 200 md. Field discovery was in 1937. Gas injection began in 1942 into a single crestal well, but within a year, two other injection wells located down the flanks of the structural axis were added. Within 6 months, the ethyl mercaptan tracer was spotted in one well, and within a year, two other wells showed tracer responses. It was concluded that "the appearance of the tracer at the producing wells definitely proves gas channeling through high-permeability streaks, rather than a uniform drive through the sand as a whole."[1]

In these four situations, the first two were deemed successful applications of immiscible gas injection; the last two were deemed failures. All these fields were relatively small compared with those discussed below. In small fields, there is less opportunity to optimize well placement and make changes in the course of the project life.

Hawkins Field (East Texas).[4,25,56,57] The Hawkins field contains > 1.3 billion bbl OOIP and 430 Bcf gas in the original gas cap. The reservoir consists of two high-quality sandstone intervals (27% BV porosity and 1- to 3-darcy permeability), the Lewisville and the Dexter; the Dexter, the better-quality sandstone, contains 70% of the oil. The structure caused by a deep-seated salt dome has 1,200 ft of closure and is extensively faulted (see structure map in **Fig. 12.18**). The reservoir is divided into two areas separated by a major fault. The eastern area contains 20% of the OOIP and 7% of the original gas cap gas and is underlain by the active Woodbine aquifer that covers much of east Texas. The western area contains the rest of the oil and gas. The western area has a tar mat that varies in thickness from 50 ft on the north to 100 ft on the south. This tar mat impeded aquifer influx until a decline in reservoir pressure resulted in water influx in the north that constituted a strong waterdrive that tilted the gas cap to the south, where aquifer influx did not occur. The average formation dip is 6 to 8°. Oil gravity averages 24.2°API gravity and varies somewhat vertically (21 to 26°API range). Oil viscosity averages about 3.7 cp (with values as high as 15 cp observed near the OWC on the east side), and its FVF is 1.22 RB/STB.

The 10,000-acre field, discovered in 1940, was developed on 20-acre well spacing. Unitization was completed on 1 January 1 1975 after 536 $\times$ 10^6 bbl of oil production. Before unitization, oil production was supported by gas-cap expansion and aquifer influx in the east. A small gas injection project was used in the west to stabilize the GOC and to prevent oil migration into the gas cap because of the aquifer influx from the north.

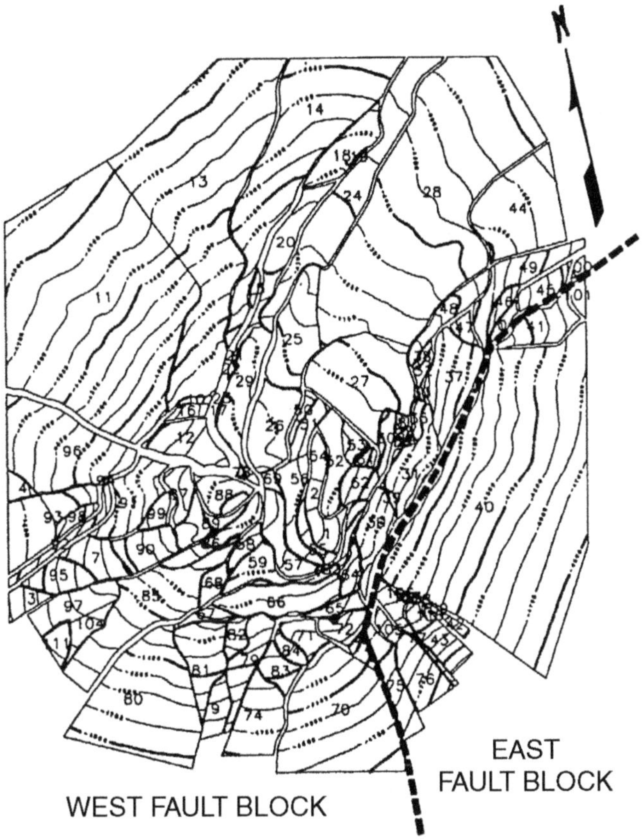

Fig. 12.18—Areal view of the Hawkins field. Top of reservoir structure showing major fault patterns.[25]

Extensive laboratory testing was conducted on reservoir core samples to quantify the ability of both water and gas to displace oil.[25] The results of these tests are shown in **Table 12.4**. These tests showed that gas was more efficient at displacing oil from the reservoir rock than water and that gas would recover at least 10% PV more oil. From the laboratory data, engineers calculated that, in the field, waterdrive would leave a residual oil saturation of 35% PV, whereas gas drive would leave an average residual oil saturation of 12% PV; the difference results from the lower density difference between the oil and water.

The field was unitized to facilitate the implementation of a gravity drainage project using crestal gas injection. Gas injection began in 1975. Two types of gases were injected. All produced gas, less fuel and shrinkage, was reinjected into the gas cap areas, and beginning in 1977, 120 MMcf/D of flue gas (88% N_2, 12% CO_2) generated at a nearby plant was also injected.[4] More recently, pure nitrogen from a cryogenic nitrogen rejection plant has been injected.

In 1987, a tertiary immiscible gas-drive process was started in the East Fault Block where the aquifer had invaded a large portion of the oil column. This tertiary process has been called the double displacement process (DDP).[25,57] In this process, the invading aquifer is being displaced to the original OWC so that the gas-drive gravity drainage process can remobilize much of the waterflood residual oil all the way down to this depth. Although the DDP is working, it is working more slowly than expected because of "higher viscosity oil (note the higher viscosity oil downdip discussed above), significant targeted oil volume found in lower-quality rock (in bypassed-oil zones), and lower-than-expected oil relative permeability."[57] With the success of the DDP in the east, a similar project was implemented in the west.

TABLE 12.4—SUMMARY OF LABORATORY TEST RESULTS ON HAWKINS FIELD CORE SAMPLES[25]

Plug	Porosity, % BV	Permeability, md k_a	Permeability, md k_{ocw}		Water Imbibition, % PV	Gas/Liquid Drainage, % PV
1	27.1	1,330	1,270		22.2	8.8
2	24.7	1,230	155		21.0	13.4
3	26.9	6,210	729		14.4	3.8
4	28.9	7,450	2,210		22.8	7.9
5	27.9	5,680	1,510		23.9	9.3
6	25.8	4,420	1,800		16.3	4.2
7	25.4	2,310	1,310		19.4	8.3
8	26.0	2,790	1,160		15.2	5.4
9	22.6	2,240	1,380		21.5	8.9
10	24.0	1,720	720		18.8	11.6
11	27.5	4,470	658		9.5	6.2
12	27.1	3,340	907		15.3	11.6
				Mean	18.4	8.3
			Gas/Liquid Drainage Only			
1	23.2	2,300	842			8.8
2	28.0	5,460	2,420			5.4
3	25.7	3,060	1,200			6.9
4	26.0	3,050	1,000			8.5
				Mean		7.4

Overall Hawkins' recovery efficiency from the gas-drive mechanism is about 87% in the gas-swept areas or > 20% better than estimated for the waterdrive process. Overall reservoir performance resulting from immiscible gas injection is considered excellent.

Prudhoe Bay Field (North Slope, Alaska).[13,27,38,54,58] The Prudhoe Bay field, the largest oil and gas field in North America, was discovered in 1968. The main Permo-Triassic reservoir is a thick deltaic high-quality sandstone deposit about 500 ft thick with porosities of 15 to 30% BV and permeabilities ranging from 50 to 3,000 md. The field contains > 20 × 10⁹ bbl of oil overlain by a 35-Tcf gas cap. The reservoir, a monocline structure dipping 1 to 2° to the south and southwest, is bounded on the north by major faults and on the east by a major lower Cretaceous truncation. The oil averages 27.6°API gravity and has an original solution GOR of about 735 scf/STB. Under much of the oil column area, there is a 20- to 60-ft-thick tar mat located above the OWC.

Because of its remote location, first production did not begin until 1977 after an oil pipeline across Alaska to the southern port of Valdez was built and extensive oilfield facilities were installed. The initial facilities had a gas-handling capacity of about 2 Bcf/D to separate and compress the produced gas for reinjection into the updip gas cap. For the regulated oil production rate of 1.2×10^6 BOPD, this was sufficient for more than the first decade of field operations. As of 2004, no gas pipeline has been built from the North Slope of Alaska.

Immiscible gas/oil displacement has been the production mechanism at work over much of the Prudhoe Bay oil column (Fig. 12.10), and waterflooding and miscible WAG processes have been used in the very downdip portions of the oil column. More recently, horizontal wells are being used nearer the GOC to exploit the thin oil columns that cannot be drained with existing wells. During the first decade of operations, the original gas cap had expanded to drape over much of the oil column area, but the producing GOR was kept low by perforating the production wells as far as possible, both vertically and laterally, from the gas cap. By the mid-1980s the engineers determined that maintaining the oil production rate depended on expanding the field's gas-handling capacity. Two major projects were undertaken. First, a central gas facility was constructed to strip the produced gas of many of the hydrocarbon components for sale as blendable NGLs and for use as miscible injectant. This project also increased the field gas handling capacity to 3.3 Bcf/D. The second project was a sequence of gashandling facility expansions that increased overall capacity to > 7.5 Bcf/D.

The reinjection of residue gas served two purposes. First, it maintained reservoir pressure, which helped increase oil recovery; the reservoir pressure declined only by ≈1,000 psi (from 4,300 to 3,300 psi) during the production of the first 10×10^9 bbl of oil. Second, the very lean residue gas has vaporized large amounts of hydrocarbon components from the relict oil saturation in the original gas cap[44] and from the remaining oil behind the gas/oil front. Simulations showed that the vaporization mechanism would contribute recovery of an additional 4 to 6 STB/MMcf of additional gas produced.[13] During the past 15 years, the average API gravity of the marketed Prudhoe Bay hydrocarbon liquids has increased by > 5°API.

Haldorsen *et al.*[58] have studied the efficiency of the Prudhoe Bay gravity drainage mechanism by means of laboratory experiments and field cased-hole log evaluations. They concluded, "The 'most likely' displacement efficiency, through the stochastic approach, was 68 percent after three years and 76 percent after 30 years of gravity drainage."

The ultimate Prudhoe Bay oil recovery has increased from initial estimates of 9.6 to 13×10^9 bbl oil, much of which is related to exploiting the immiscible gas/oil gravity drainage and oil stripping mechanisms. This has been accomplished by massive expansions of the gas handling facilities and extraction of the maximum volume of blendable NGLs from the produced gas stream. Application of the immiscible gas processes at Prudhoe Bay has been aided by the lack of alternative uses or markets for the produced gas.

Empire Abo Field (New Mexico).[26,39] Another example of the application of the immiscible gas/oil gravity drainage process is found at the Empire Abo field. This field covers ≈11,000 acres (12.5 miles long by 1.5 miles wide) and contains approximately 380 million STB of OOIP. This reservoir is a dolomitized reef structure (**Fig. 12.19**) with a dip angle of 10 to 20° from the crest toward the fore reef. The oil column is approximately 900 ft thick, but the average net pay is only 151 ft thick. The pore system of this reservoir is a network of vugs, fractures, and fissures because the primary pore system has been so altered by dolomitization; the average log-calculated porosity was 6.4% BV. Numerical simulations of field performance and routine core analysis data have indicated that the horizontal and vertical permeabilities are about equal. The Empire Abo field has a small initial gas cap (< 1% of the hydrocarbon PV), so the oil was gas saturated at the original reservoir pressure of 2,360 psia. The oil gravity was 43°API, and viscosity at reservoir conditions was approximately 0.4 cp.

The field was discovered in 1957 and unitized in 1973. At that time, high-GOR back-reef wells were shut in, only the downdip solution-GOR wells continued to produce, and gas-plant residue gas was reinjected into some of the former crestal production wells. The unit added 56 infill producers on 20-acre well spacing in the downdip area during the mid-1970s (see structure map in **Fig. 12.20**). In the late 1970s, 100 additional producers on 10-acre well spacing in the downdip area were drilled.

Because of the concentration of producing wells downdip, the gas cap draped downdip along the top of the reservoir. The Empire Abo field overall performance was excellent because of its very high vertical permeability; however, gas coning was a major issue, with the overall relatively low reservoir permeability of approximately 50 md. In 1979, Empire Abo

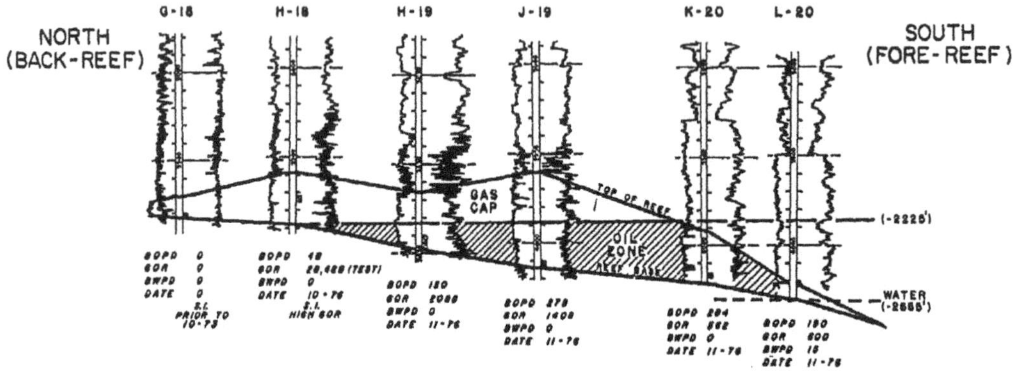

Fig. 12.19—Empire Abo field. Typical back-reef to fore-reef cross section, with openhole gamma ray/neutron logs.[26]

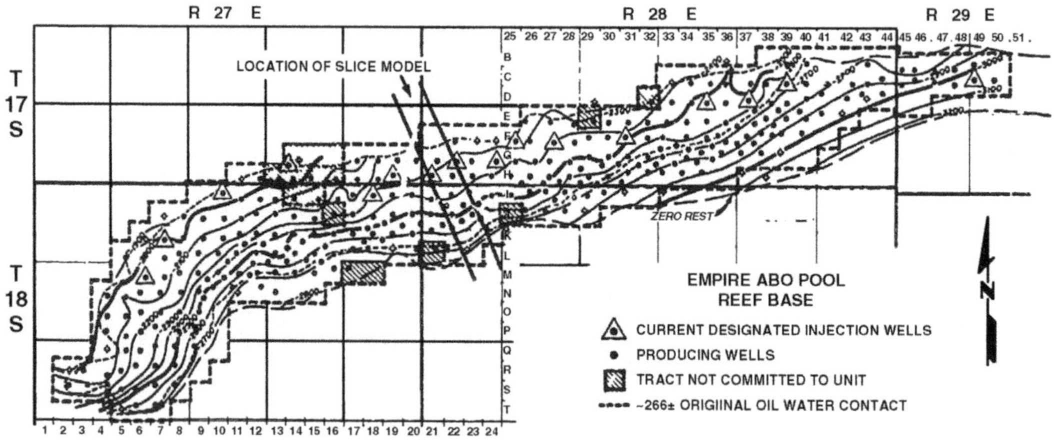

Fig. 12.20—Empire Abo field. Structure map on the base of the Abo reef.[26]

was the site of one of the first applications of horizontal wellbores to minimize gas coning.[59] The field is currently being blown down to recover as much gas as possible. On the basis of production data through 2002, on a stock-tank-oil basis, approximately 74% of the OOIP has been recovered from this field by application of the immiscible gas/oil gravity drainage process.

Heft Kel Field (Iran).[30] An example of the application of immiscible gas/oil displacement in a large Middle East matrix-block/fracture-system reservoir is the Heft Kel field. Its Asmari reservoir structure is a strongly folded anticline that is 20 miles long by 1.5 to 3 miles wide with an oil column thickness of approximately 2,000 ft. The most probable OOIP was slightly $> 7 \times 10^9$ STB with about 200 million STB in the fissures; numerical model history matching resulted in a value of 6.9×10^9 STB. The matrix block size determined from cores and flowmeter surveys varied from 8 to 14 ft. The numerical simulation model considered matrix permeabilities from 0.05 to 0.8 md. The overall horizontal and vertical permeabilities are approximately equal. There was an initial gas cap on the oil column. The oil gravity is approximately 37°API. The IFT at the bubblepoint pressure (1,412 psi and 116°F) is approximately 9 dynes/cm.

The field was discovered and put on production in 1928. It was produced on primary production from then until 1976 with a plateau rate of 200,000 BOPD for several early years. In 1976, gas injection began at a rate of 400 MMcf/D using gas from the nearby NIS gas dome. Recently, the field has been producing at approximately 35,000 BOPD.

Saidi[30] describes the many oil recovery mechanisms at work in this oil-wet reservoir as gravity drainage at constant IFT and reservoir pressure; oil swelling in the present gas-invaded zone because of the increase in reservoir pressure; oil swelling in the present oil zone through thermal convection/diffusion process; oil imbibition within the oil column; oil gravity drainage from the partially saturated blocks within the gas-invaded zone; and oil gravity drainage from the fully oil-saturated block in the oil zone and the blocks between that and the present GOC.

The flow behavior developed from the history match is that the oil-drainage performance follows that of stacks of discontinuous blocks, supporting practically no vertical capillary continuity between the matrix blocks (see Fig. 12.8).

By going to immiscible gas injection, oil recovery is increased by about 500×10^6 bbl by returning to the original reservoir pressure and could be increased by another 100×10^6 bbl if the reservoir pressure is increased an additional 100 psi because of the reduction in gas/oil IFT with increasing reservoir pressure.

Overall, the application of immiscible gas injection to the Haft Kel field has been considered a success. The estimated displacement efficiency by water was 17%, whereas that estimated for immiscible gas displacement was 32%.

Swanson River Field (Cook Inlet, Alaska).[24] A very different style of successful immiscible gas/oil displacement project is that applied to the Swanson River field's Hemlock reservoir. **Figs. 12.21 and 12.22** show an areal view of this reservoir and a type log through the Hemlock formation, respectively. This field is a north/south-trending anticlinal flexure about 6 miles long by 1 to 3 miles wide with as much as 600 ft of closure. The Hemlock formation consists of interbedded fine- to coarse-grained sandstone, conglomerate, siltstone, and coal, with numerous thin, impermeable, calcareous stringers of somewhat limited areal extent. Field experience has confirmed that these calcareous stringers are effective barriers to the vertical migration of fluids in the vicinity of producing wells. There are 10 Hemlock intervals, and the H1 through H5, H8, and H10 intervals have been engineered and managed separately (see Fig. 12.22).

The OOIP of this field is 435 million STB. This oil was very undersaturated at discovery in 1957, with an original reservoir pressure of 5,580 psi but with a bubblepoint pressure of only 1,350 psi. The oil is 37°API gravity, had an initial FVF of 1.21, and had a viscosity of 1.1 cp.

From these reservoir characteristics, it was clear that at the start of production the reservoir pressure would fall rapidly and that some form of pressure maintenance needed to be applied quickly. Fortunately for this very undersaturated oil field, a Tcf-sized dry gas field was discovered nearby for which there was no immediate or large outlets for its gas. A gas rental contract was signed between the two fields owners, and starting in 1966, 400 MMcf/D of gas was delivered for injection into crestal wells in the Swanson River Hemlock reservoir.

Laboratory studies indicated that a number of mechanisms were at work when methane gas displaced the low-bubblepoint-pressure Swanson River oil. First, if operating at a 5,000-psi pressure, the oil would swell from an FVF of 1.21 to an FVF of 1.80 on contact with the injected gas. Second, free gas would become saturated with intermediate hydrocarbon components from the fairly light (37°API) reservoir oil. Laboratory experiments show that breakthrough recovery efficiency exceeded 60%.

Although the Swanson River field has operated at high GORs (> 10 Mscf/STB) after the first several years of production, reservoir performance has been excellent, with 38% of the OOIP recovered through the first 10 years of gas injection. To show that the oil vaporization

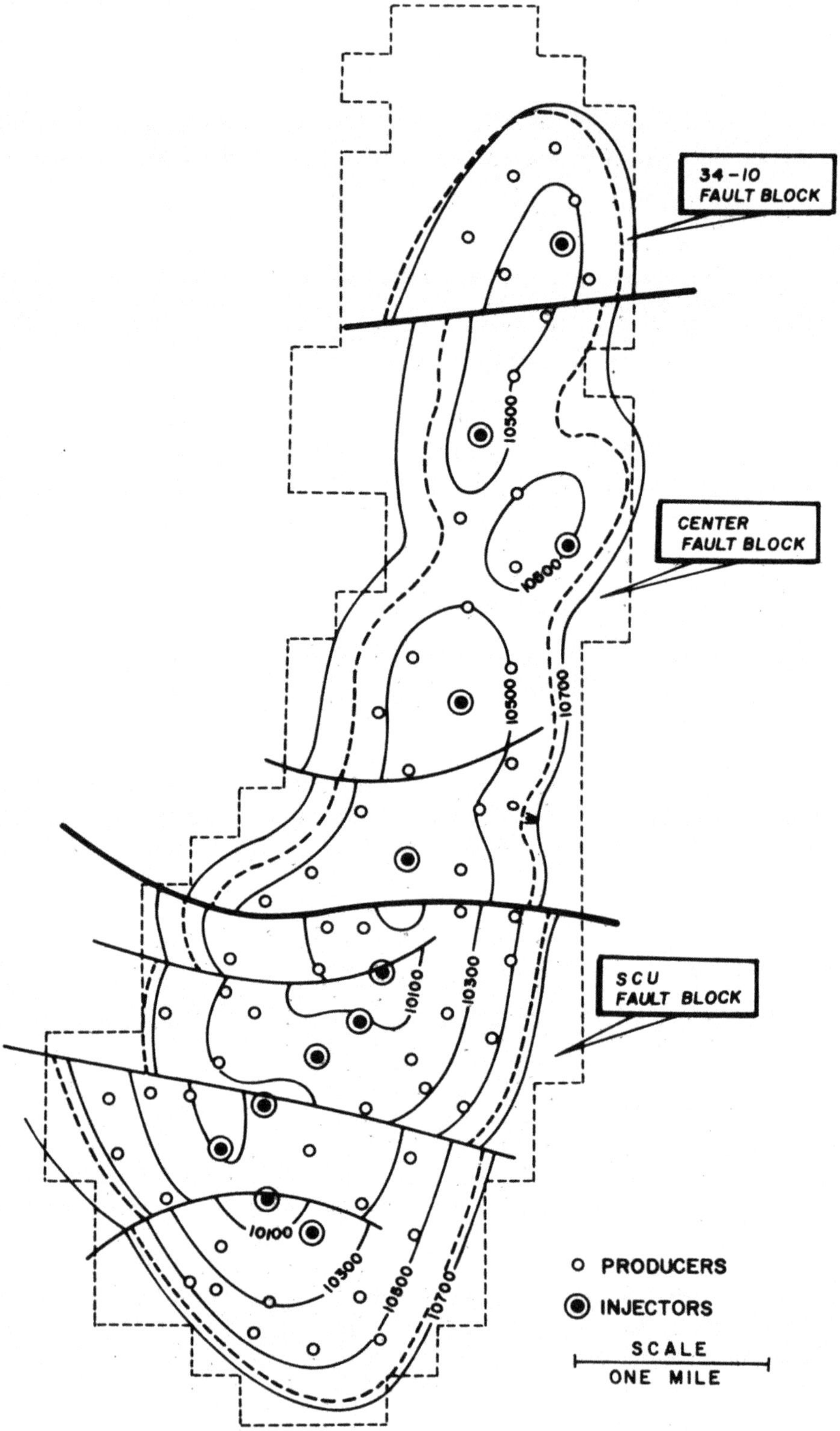

Fig. 12.21—Swanson River field. Contour map of top of Hemlock structure.[24]

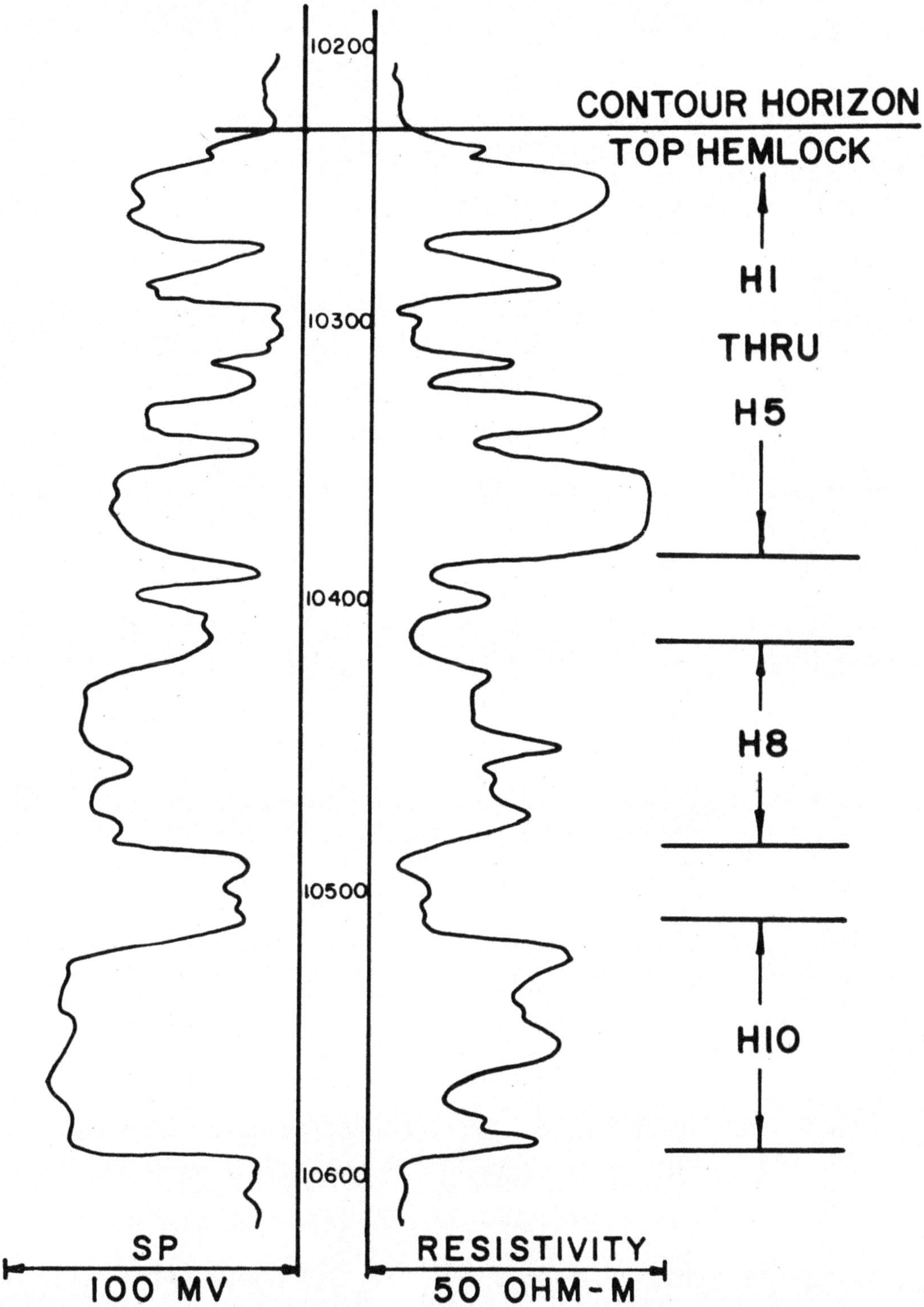

Fig. 12.22—Swanson River field. Typical electric log through Hemlock reservoir interval.[24]

mechanism is also at work in this reservoir, during these first 10 years of production after the start of gas injection, the produced oil gravity increased from 37 to 40°API.

The Swanson River field is currently being blown down to recover as much remaining gas as possible. From production statistics through 2002 and an updated OOIP of 390 × 10^6 bbl, ultimate recovery from application of immiscible gas/oil displacement to this field is nearly 58% of OOIP. This has been achieved primarily through oil swelling and oil stripping, not vertical gas/oil gravity drainage.

Kuparuk River Field, Alaskan North Slope.[14,15] Another type of immiscible gas/oil project is that applied to some areas of the Kuparuk River oil field. The Kuparuk River field contains approximately 5.9 × 10^9 bbl of 24°API oil in two producing zones (zones A and C). The oil was slightly undersaturated at discovery (3,300-psi original reservoir pressure, with oil 300 to 500 psi undersaturated). The shallower C sand is more permeable than the A sand, but the A sand contains approximately 62% of the OOIP. The reservoir covers > 200 sq miles.

The field was discovered in the late 1960s and went on production in 1981. The maximum rate has exceeded 300,000 BOPD, producing from 42 drillsites with wells on 160-acre well spacing. The challenges in this field were that there was no initial gas cap on the reservoir into which to reinject the produced gas; the reservoir dip is very slight, and the reservoir intervals were not very thick, so gas/oil gravity drainage could not efficiently occur; the gas could not be flared; and there was no off-site location to which the residue gas could be sent. For these reasons, the gas had to be reinjected into one of the reservoir intervals for storage and to maintain the oil production rates. The challenge for the field engineers was how to reinject the residue gas in the most efficient way, given the various constraints.

After a few years of gas injection into the A sand in one area of the field in which the offsetting producers soon experienced rapidly increasing GORs, the engineers decided to go to an immiscible WAG injection process.[14] WAG could be applied at minimal cost and with few operational complications because most of the field was already being waterflooded. Calculations indicated that there were three beneficial effects to this approach. First, there would be some swelling of the oil because of the free gas. Second, the residual oil saturation could potentially be reduced by the presence of trapped gas. Third, WAG injection was used to reduce the high mobility of the gas by means of three-phase relative permeability effects (simultaneously having mobile gas, oil, and water in the pore system); also, a tapered WAG scheme helped in this regard. The overall effect was that oil recovery could potentially be increased by 1 to 3% of OOIP without resulting in significant gas cycling problems. To date, immiscible WAG injection has worked as expected and has been a satisfactory solution to the Kuparuk River gas disposal problem.

12.10 Miscellaneous

12.10.1 Gas Sources for Immiscible Gas Injection Projects. The first consideration in any immiscible gas injection project is where to get the volume of gas necessary for the project. Historically, produced and processed residue gas from that particular oil field has been used. This is the most satisfactory solution if the economics of the additional oil recovery justify deferred gas sales. In some locations, this is not an issue because there is no current market outlet for the produced gas. Generally, reinjection of this local gas supply is sufficient to nearly maintain the current reservoir pressure.

The next best alternative in some locations is to develop deeper gas horizons as a gas supply. This is particularly true in the Middle East where massive volumes of gas are often found in the deeper formations, such as the Khuff, underlying some major oil reservoirs. The third alternative is to look to nearby fields for a source of gas; this alternative has been used at the Swanson River field in Alaska[22] and the Heft Kel and other Iranian fields.[30]

If there is no source for lean hydrocarbon gas, then possibly flue gas (88% N_2, 12% CO_2) or nitrogen might be used. Both of these options require that a plant be constructed at the field to generate these gases in the large volumes needed. This approach requires the economics of the project to justify the large capital expenditure for such facilities, additional operating costs, and future impact of the produced gas becoming increasingly contaminated with nonhydrocarbon components with time.

Sometimes, the solution to the gas supply problem is to use a combination of sources to provide the required volume of gas. At the Hawkins field, both residue hydrocarbon gas and flue gas initially were injected. More recently, nitrogen from a cryogenic nitrogen rejection plant has been injected.[4,25,57]

12.10.2 Use of Horizontal Wells in Immiscible Gas Injection Projects. Horizontal wells have not been discussed in this chapter because the technology to drill horizontally has developed rapidly since the mid-1980s and such wells have not had widespread use in historic immiscible gas injection projects. Horizontal wells are particularly suited for use in gravity drainage immiscible gas injection projects because they maximize the distance between the producing wells' perforations and the overlying gas cap and because such wells minimize the pressure gradients in the near-wellbore region (the cause of near-wellbore gas coning). At the Prudhoe Bay field, a large number of horizontal wells have been drilled for a variety of purposes, including the two mentioned above. For these reasons, in future immiscible gas gravity drainage projects, it is logical to consider using horizontal production wells. The chapter on fluid flow and well analysis in this section of the *Handbook* shows how to predict and to interpret the performance of horizontal wells.

12.10.3 Operating Procedures for Thin Oil Columns. One consideration in immiscible gas gravity drainage projects is the challenge of maximizing oil recovery from thin oil columns. The thin oil column may be what is found initially,[29] or in most cases, as the gas cap expands, it is all that is left to produce late in the life of the project. The field engineers have to monitor individual well performance and overall reservoir performance closely to optimize production under these circumstances. Obviously, if new wells are drilled, they should be either carefully targeted horizontal wells or wells with very limited perforated intervals.

A related consideration is when there is a thin oil column sandwiched between the expanding gas cap and the underlying aquifer.[29] In this situation, well perforations must be chosen to maximize recovery and to minimize the production of both gas and water. Although coning simulations with numerical reservoir simulators will provide insights into the best approach for a particular reservoir situation, actual field experience is necessary to optimize the operations.

12.11 Summary and Conclusions

In this chapter, the technical aspects of immiscible gas/oil displacement have been described, and several field case studies have been presented. The conclusions concerning immiscible gas/oil displacement are listed below:

1. Immiscible gas/oil viscous displacement is an inefficient oil displacement process because gas is a highly mobile fluid.

2. Gas-oil capillary pressure data indicate that in many situations the residual oil saturation to gas displacement is significantly lower than the residual oil saturation to water displacement.

3. The immiscible gas/oil process becomes efficient and desirable when gravity works to keep the very-low-density gas on top of the higher-density oil and/or there is significant mass transfer of components from the oil to the gas.

4. The most successful immiscible gas/oil injection projects are the vertical gravity drainage projects in which gas is injected into the crestal primary or secondary gas cap, with the oil wells producing from as far downdip as possible to maximize this distance from the gas cap

both vertically and laterally. To maximize the efficiency of this approach, the overall oil production rate has to be restricted to the critical displacement rate.

5. One gas/oil compositional mass-transfer effect is oil swelling. If an oil field contains a very undersaturated oil, then oil swelling by contact with the injected gas can be a very significant effect. However, if a reservoir has an original gas cap, the oil swelling effect is minimal because the oil is already fully saturated or nearly saturated with gas.

6. The other gas/oil compositional mass-transfer effect is stripping or vaporization of intermediate hydrocarbon components from the oil by the lean injected gas. The importance of this effect increases as the producing GOR rises. Toward the end of the life of an immiscible gas injection project, the stripping effect can contribute many of the liquid hydrocarbons produced in the surface facilities and associated gas plants. This effect occurs with all types of oils but is more significant for lighter, or higher API gravity, oils.

7. A few immiscible gas injection field projects have been undertaken that are not vertical gas/oil gravity drainage projects but in which compositional effects have led to project success. An excellent example of this approach is the Swanson River field in Alaska.

8. Gas coning into producing wellbores' perforated intervals occurs with thin oil columns or as the gas/oil interface moves downward. Horizontal wells are a method of further reducing the height of the remaining oil column by lowering pressure drawdown and thus minimizing the effects of gas coning.

9. Numerical reservoir simulators are the best tool to evaluate all the technical aspects of an immiscible gas injection project, either historical performance and/or projections of future performance. Simple mathematical techniques have been developed to analyze some types of immiscible gas/oil displacements.

Nomenclature

A = cross-sectional area, ft^2
F_g = fractional gas flow, fraction
h = thickness of oil zone normal to direction of dip, ft
h_v = vertical thickness, ft
k = permeability, darcies
k_h = permeability in horizongal direction, darcies
k_{ro} = relative permeability to oil, fraction
k_{rg} = relative permeability to gas, fraction
k_v = permeability in vertical direction, darcies
L = distance along the bedding plane, ft
M = mobility ratio, $k_{rg}\mu_o/k_{ro}\mu_g$
N_g = oil production, fraction
P_c = gas/oil capillary pressure, psi
q_T = total flow rate through area A, ft^3/D
S_{cw} = connate water saturation, fraction PV
S_g = gas saturation, fraction PV
S_o = oil saturation, fraction PV
$S_{org}{}^*$ = irreducible oil saturation in presence of gas, fraction PV
S_{wi} = initial water saturation, fraction PV
t = time, days
t_D = dimensionless time
t_v = time for vertical drainage, days
t_u = time for drainage along bedding plane, days
u = flow per unit area, ft^3/ft^2-D

vs = downdip free fall of oil, ft/D
V = volume of gas injected, res bbl
z = vertical distance, ft
α = angle of dip (positive downdip), degrees
μ_g = gas viscosity, cp
μ_o = oil viscosity, cp
ρ_g = gas density, lbm/ft^3
ρ_o = oil density, lbm/ft^3
ϕ = porosity, fraction

References

1. Muskat, M.: *Physical Principles of Oil Production,* McGraw-Hill Book Co. Inc., New York City (1949) 470–502.
2. Cotter, W.H.: "Twenty-Three Years of Gas Injection Into a Highly Undersaturated Crude Reservoir," *JPT* (April 1962) 361.
3. Shehabi, J.A.N.: "Effective Displacement of Oil by Gas Injection in a Preferentially Oil-Wet, Low-Dip Reservoir," *JPT* (December 1979) 1605.
4. Kuehm, H.G.: "Hawkins Inert Gas Plant: Design and Early Operation," paper SPE 6793 presented at the 1977 SPE Annual Technical Conference and Exhibition, Denver, 9–12 October.
5. Katz, D.L., Monroe, R.R., and Tanner, R.P.: "Surface Tension of Crude Oils Containing Dissolved Gases," *Pet. Tech.* (September 1943) 1.
6. Hough, E.W. and Warren, H.G.: "Correlation of Interfacial Tension of Hydrocarbons," *SPEJ* (December 1966) 345.
7. Katz, D.L. and Firoozabadi, A.: "Predicting Phase Behavior of Condensate/Crude-Oil Systems Using Methane Interaction Coefficients," *JPT* (November 1978) 1649–1655.
8. Firoozabadi, A. *et al.:* "Surface Tension of Reservoir Crude-Oil/Gas Systems Recognizing the Asphalt in the Heavy Fraction," *SPERE* (February 1988) 265.
9. Broseta, D. and Ragil, K.: "Parachors in Terms of Critical Temperature, Critical Pressure and Acentric Factor," paper SPE 30784 presented at the 1995 SPE Annual Technical Conference and Exhibition, Dallas, 22–25 October.
10. Hagoort, J.: "Oil Recovery by Gravity Drainage," *SPEJ* (June 1980) 139.
11. Johnston, E.F., Bossler, C.P., and Naumann, V.O.: "Calculation of Relative Permeability From Displacement Experiments," *Trans.*, AIME (1959) **216**, 61.
12. *Miscible Processes,* Reprint Series, SPE, Richardson, Texas (1957) **8**, 111–114.
13. Simon, A.D. and Petersen, E.J.: "Reservoir Management of the Prudhoe Bay Field," paper SPE 38847 presented at the 1997 SPE Annual Technical Conference and Exhibition, San Antonio, Texas, 5–8 October.
14. Ma, T.D. and Youngren, G.K.: "Performance of Immiscible Water-Alternating-Gas (IWAG) Injection at Kuparuk River Unit, North Slope, Alaska," paper SPE 28602 presented at the 1994 SPE Annual Technical Conference and Exhibition, New Orleans, 25–28 September.
15. Stoisits, R.F., Scherer, P.W., and Schmidt, S.E.: "Gas Optimization at the Kuparuk River Field," SPE 28467 presented at the 1994 SPE Annual Technical Conference and Exhibition, New Orleans, 25–28 September.
16. Aziz, K. and Settari, A.: *Petroleum Reservoir Simulation,* Applied Science Publishers Ltd., London (1979) 30–38.
17. Buckley, S.E. and Leverett, M.C.: "Mechanism of Fluid Displacement in Sands," *Trans.*, AIME (1942) **146**, 107.
18. Welge, H.J.: "A Simplified Method for Computing Oil Recovery by Gas or Water Drive," *Trans.,* AIME (1952) **195**, 91.
19. Kern, L.R.: "Displacement Mechanism in Multi-well Systems," *Trans.,* AIME (1952) **195**.
20. Craft, B.C. and Hawkins, M.F.: *Applied Petroleum Reservoir Engineering*, Prentice-Hall Inc., Englewood Cliffs, NJ (1959) 370.

21. Anders, E.L. Jr.: "Mile Six Pool: An Evaluation of Recovery Efficiency," *JPT* (November 1953) 279; *Trans.*, AIME, **198**.
22. *Miscible Processes,* Reprint Series, SPE, Richardson, Texas (1965) **8**, 197.
23. *Miscible Processes,* Reprint Series, SPE, Richardson, Texas (1965) **8**, 205.
24. Young, R.E., Fairfield, W.H., and Dykstra, H.: "Performance of a High-Pressure Gas Injection Project, Swanson River Field, Alaska," *JPT* (February 1977) 99.
25. Carlson, L.O.: "Performance of Hawkins Field Unit Under Gas Drive–Pressure Maintenance Operations and Development of an Enhanced Oil Recovery Project," paper SPE 17324 presented at the 1988 SPE/DOE Enhanced Oil Recovery Symposium, Tulsa, 17–20 April.
26. Christianson, S.H.: "Performance and Unitization of the Empire Abo Pool," paper SPE 6384 presented at the 1977 Permian Basin Oil and Gas Recovery Conference, Midland, Texas, 10–11 March.
27. Metz, W.P. and Elliott, R.A.: "Gas Handling Expansion Facilities at Prudhoe Bay, Alaska," paper SPE 22113 presented at the 1991 SPE International Arctic Technology Conference, Anchorage, 29–31 May.
28. Richardson, J.G., *et al.:* "The Effect of Small, Discontinuous Shales on Oil Recovery," *JPT* (November 1978) 1531.
29. Selamat, S., Goh, S.T., and Lee, K.S.: "Seligi Depletion Management," paper SPE 57251 presented at the 1999 SPE Asia Pacific Improved Oil Recovery Conference, Kuala Lumpur, 25–26 October.
30. Saidi, A.M.: "Twenty Years of Gas Injection History Into Well-Fractured Haft Kel Field (Iran)," paper SPE 35309 presented at the 1996 SPE International Petroleum Conference and Exhibition, Villahermosa, Mexico, 5–7 March.
31. O'Neill, N.: "Fahud Field Review: A Switch From Water to Gas Injection," *JPT* (May 1988) 609.
32. Horie, T., Firoozabadi, A., and Ishimoto, K.: "Laboratory Studies of Capillary Interaction in Fracture/Matrix Systems," *SPERE* (June 1990) 784.
33. Firoozabadi, A. and Hauge, J.: "Capillary Pressure in Fractured Porous Media," *JPT* (August 1990) 353.
34. Firoozabadi, A., Ishimoto, K., and Dindoruk, B.: "Theory of Reinfiltration in Fractured Porous Media: Part 2—Two-Dimensional Model," paper SPE 21798 presented at the 1991 SPE Rocky Mountain Regional Meeting and Low-Permeability Reservoirs Symposium, Denver, 15–17 April.
35. Dindoruk, B. and Firoozabadi, A.: "Crossflow in Fractured/Layered Media Incorporating Gravity, Viscous, and Phase Behavior Effects: Part I—Formulation and Features in Layered Media," paper SPE 35457 presented at the 1996 SPE/DOE Symposium on Improved Oil Recovery, Tulsa, 21–24 April.
36. Dindoruk, B. and Firoozabadi, A.: "Crossflow in Fractured/Layered Media Incorporating Gravity, Viscous, and Phase Behavior Effects: Part II—Features in Fractured Media," paper SPE 35458 presented at the 1996 SPE/DOE Symposium on Improved Oil Recovery, Tulsa, 21–24 April.
37. Ghorayeb, K. and Firoozabadi, A.: "Numerical Study of Natural Convection and Diffusion in Fractured Porous Media," *SPEJ* (March 2000) 12.
38. Addington, D.V.: "An Approach to Gas Coning Correlations for the Prudhoe Bay Field," *JPT* (November 1981) 2267.
39. Killough, J.E. and Foster, H.P. Jr.: "Reservoir Simulation of the Empire Abo Field: The Use of Pseudos in a Multilayered System," *SPEJ* (October 1979) 279.
40. Shreve, D.R. and Welch, L.W. Jr.: "Gas Drive and Gravity Drainage Analysis for Pressure Maintenance Operations," *JPT* (June 1956) 136; *Trans.,* AIME, **207**.
41. Jacoby, R.H. and Berry, V.J. Jr.: "A Method for Predicting Pressure Maintenance Performance for Reservoirs Producing Volatile Crude Oil," *JPT* (March 1958) 59; *Trans.,* AIME, **213**.
42. Attra, H.D.: "Nonequilibrium Gas Displacement Calculation," *SPEJ* (September 1961) 130; *Trans.,* AIME, **222**.
43. Stiles, W.E.: "Use of Permeability Distribution in Water Flood Calculations," *Trans.,* AIME (1949) **186**, 9.
44. Dyes, A.G., Caudle, B.H., and Erickson, R.A.: "Oil Production After Breakthrough—As Influenced by Mobility Ratio," *Trans.,* AIME (1954) **201**, 81.
45. Johns, R.T.: "Analytical Theory of Multicomponent Gas Drives with Two-Phase Mass Transfer," PhD dissertation, Stanford U., Palo Alto, California (1992).

46. Jessen, K., *et al.:* "Fast, Approximate Solutions for 1D Multicomponent Gas Injection Problems," paper SPE 56608 presented at the 1999 SPE Annual Technical Conference and Exhibition, Houston, 3–6 October.
47. Falls, A.H. and Schulte, W.M.: "Features of Three-Component, Three-Phase Displacement in Porous Media," *SPERE* (November 1992) 426.
48. Lake, L.W.: *Enhanced Oil Recovery,* first edition, Prentice-Hall Inc., Englewood Cliffs, NJ (1989).
49. Juanes, R. and Patzek, T.W.: "Relative Permeabilities in Co-Current Three-Phase Displacements with Gravity," paper SPE 83445 presented at the 2003 SPE Western Regional/AAPG Pacific Section Joint Meeting, Long Beach, California, 19–24 May.
50. Azevedo, A.V. *et al.:* "Nonuniqueness of Riemann Problems," *Zeitschrift für angewandte Mathematik und Physik* (1997) **47**, No. 6, 977.
51. Marchesin, D., Plohr, B., and Schecter, S.: "An Organizing Center for Wave Bifurcation in Multiphase Flow Models," *SIAM J. Appl. Math.* (October 1997) **57**, No. 5, 1189.
52. Richardson, J.G. and Blackwell, R.J.: "Use of Simple Mathematical Models for Predicting Reservoir Behavior," *JPT* (September 1971) 1145.
53. Cardwell, W.T. Jr. and Parsons, R.L.: "Gravity Drainage Theory," *Trans.,* AIME (1949) **179**, 199.
54. Erickson, J.W., and Sneider, S.M.: "Structural and Hydrocarbon Histories of the Ivishak (Sadlerochit) Reservoir, Prudhoe Bay Field," paper SPE 28574 presented at the 1994 SPE Annual Technical Conference and Exhibition, New Orleans, 25–28 September.
55. Corey, A.T. *et al.:* "Three-Phase Relative Permeability," *Trans.,* AIME (1956) **207**, 349.
56. King, R.L., Stiles, J.H. Jr., and Waggoner, J.M.: "A Reservoir Study of the Hawkins Woodbine Field," SPE 2972 presented at the 1970 SPE Annual Meeting, Houston, 4–7 October.
57. Langenberg, M.A., Henry, D.M., and Chlebana, M.R.: "Performance and Expansion Plans for the Double-Displacement Process in the Hawkins Field Unit," *SPERE* (November 1995) 301.
58. Haldorsen, H.H. *et al.:* "An Evaluation of the Prudhoe Bay Gravity Drainage Mechanism by Complementary Techniques," paper SPE 13651 presented at the 1985 SPE California Regional Meeting, Bakersfield, California, 27–29 March.
59. Stramp, R.L.: "The Use of Horizontal Drainholes in the Empire Abo Unit," paper SPE 9221 presented at the 1980 SPE Annual Technical Conference and Exhibition, Dallas, 21–24 September.

SI Metric Conversion Factors

acre	×	4.046 863	E + 03	$= m^2$
bbl	×	1.589 873	E − 01	$= m^3$
cp	×	1.0*	E − 03	= Pa·s
dyne	×	1.0*	E − 05	= N
ft	×	3.048*	E − 01	= m
ft^2	×	9.290 304*	E − 02	$= m^2$
ft^3	×	2.831 685	E − 02	$= m^3$
°F		(°F − 32)/1.8		= °C
in.	×	2.54*	E + 01	= cm
lbm	×	4.535 924	E − 01	= kg
psi	×	6.894 757	E + 00	= kPa

*Conversion factor is exact.

Chapter 13
Polymers, Gels, Foams, and Resins

Robert D. Sydansk, Sydansk Consulting Services

13.1 Introduction

This chapter provides an overview of selected chemical systems and technologies that promote improved conformance during light oil-recovery operations. These conformance-improvement systems and technologies include fluid systems for use during oil-recovery flooding operations in which the fluids promote sweep improvement and mobility control (e.g., polymer waterflooding) and oilfield conformance-improvement treatment systems (e.g., "small-volume" gel treatments). A conformance-improvement fluid system for promoting flood sweep improvement and mobility control involves injecting a volume of an oil-recovery fluid that constitutes a significant fraction of the reservoir pore volume. The volume of an oil-recovery flooding system that is applied for sweep improvement is usually greater than 5% of the reservoir and/or well-pattern pore volume. Conformance-improvement treatment systems normally are of a relatively small volume and usually are used to treat the near-wellbore region or a relatively small fracture volume within the reservoir.

This chapter provides an overview of the following conformance-improvement systems: polymer waterflooding and treatments, gel treatments, foam flooding and treatments, and resin treatments. This chapter does not discuss improving conformance during oilfield drilling and stimulations operations for which the application of conformance-improvement technologies also can be quite advantageous.

Conformance is a measure of the uniformity of the flood front of the injected drive fluid during an oil-recovery flooding operation and the uniformity vertically and areally of the flood front as it is being propagated through an oil reservoir. If there were perfect conformance in a perfect regular five-spot well pattern during an oil-recovery flooding operation, the flood front would reach all four of the offset producers at the same time, and the flood front would reach the entire vertical interval of all four of the producing wells at the same time. Of course, there never has been a reservoir that has exhibited perfect conformance during an oil-recovery flooding operation. The issues that must be considered are how imperfect is the conformance for a given flooding operation in an oil field, and what is the economic or other beneficial rate of return if a conformance-improvement flood or treatment is implemented.

Improved conformance during an oil-recovery operation will result in incremental and/or accelerated oil production and/or will result in reduced oil-production operating costs. Properly designed and executed conformance-improvement flooding or treatments will improve the effectiveness, efficiency, and profitability of an oil-recovery operation, regardless of whether the oil-recovery operation is primary production, secondary waterflooding, or tertiary-mode flooding.

This chapter contains seven major sections. Sec. 13.2 discusses conformance improvement and the benefits imparted by the technologies and treatment systems reviewed in this chapter and imparted during light oil-recovery flooding operations. Sec. 13.3 provides an overview of the conformance problems that flooding technologies and treatment systems are intended to reduce or eliminate. Sec. 13.4 briefly discusses disproportionate permeability reduction. Secs. 13.5 through 13.8 review, respectively, polymer waterflooding and treatments, gel treatments, foam flooding and treatments, and resin treatments.

13.2 Conformance Improvement

Improving conformance, in its original and most limited definition, is synonymous with improving the drive-fluid sweep efficiency during an oil-recovery flooding operation. Improving the conformance and/or sweep efficiency for any given oil-recovery drive fluid during a reservoir flooding operation involves improving one, or both, of two components of flood sweep efficiency: vertical and areal sweep efficiency.

The volumetric sweep efficiency of a given oil-recovery drive fluid during a flooding operation within a reservoir or well pattern is defined as

$$E_V = E_A E_I, \qquad (13.1)$$

where E_A is the areal sweep efficiency, and E_I is the vertical sweep efficiency.[1,2] For Eq. 13.1 to be strictly correct, the geological layers or strata of the reservoir must be uniform in terms of porosity, thickness, and oil saturation.[2]

Strictly speaking for a real reservoir, volumetric sweep efficiency is more precisely defined as

$$E_V = E_P E_I, \qquad (13.2)$$

where E_P is the pattern sweep efficiency,[2] which is the areal sweep efficiency for a reservoir with variations in thickness, porosity, and oil saturation. Areal sweep efficiency is defined as

$$E_A = \frac{A_f}{A_t}, \qquad (13.3)$$

where A_f is the area contacted by the oil-recovery displacement fluid, and A_t is the total reservoir area under consideration.

Vertical sweep efficiency is defined as

$$E_I = \frac{A_V}{A_{tV}}, \qquad (13.4)$$

where A_V is the reservoir vertical cross section contacted by the oil-recovery displacement fluid, and A_{tV} is the total reservoir vertical cross section.

For any given oil-recovery drive/displacement fluid, poor sweep efficiency often results primarily from spatial variation and/or heterogeneity in the permeability (fluid flow capacity) of the reservoir rock. Poor vertical conformance and poor vertical sweep efficiency in matrix rock (unfractured) reservoirs usually result primarily from geological strata of differing permeability overlying one another in a reservoir. Conformance treatments to improve poor vertical sweep profiles and/or to shut off competing water or gas production, emanating from a subset of geological strata, are referred to as profile modification treatments. The Dykstra-Parsons coefficient is a widely used measure of the vertical permeability heterogeneity of an oil-producing reservoir and is discussed in Ref. 1.

For any given oil reservoir, poor sweep efficiency that results from flooding with an oil-recovery drive fluid is aggravated as the viscosity of the drive fluid decreases. Within any reservoir with a given degree of permeability heterogeneity, as the viscosity of the drive/displacement fluid of an oil-recovery flooding operation decreases, the degree of viscous fingering and the associated poor sweep efficiency increases. The mathematical and engineering term that relates the viscosity of the oil-recovery drive fluid to conformance and sweep efficiency is "mobility ratio."

Mobility ratio is defined as

$$M = \frac{\lambda_D}{\lambda_d}, \quad \text{(13.5)}$$

where λ_D is the mobility of the oil-recovery displacement fluid phase, and λ_d is the mobility of displaced fluid phase. In this chapter, the mobility of the displaced fluid phase is the mobility of the reservoir oil phase. Eq. 13.5 holds for a piston-like oil-recovery flooding operation in which the flood front is sharp. Mobility is defined as

$$\lambda_i = \frac{k_i}{\mu_i}, \quad \text{(13.6)}$$

where k_i is the permeability to phase i, and μ_i is the viscosity of phase i.[2] Thus, as the viscosity of the oil-recovery displacement/drive fluid is increased in a reservoir with a given degree of permeability heterogeneity, the sweep efficiency and the degree of the oil-recovery flood conformance are improved. See Ref. 3 for the definition of mobility ratio in the case in which a waterflood does not exhibit piston-like displacement.

When the sweep efficiency and the degree of conformance are improved during an oil-recovery flooding operation, the rate at which the reservoir oil is recovered is increased, and the amount of oil-recovery drive fluid, which must be coproduced for a given oil recovery factor, is decreased. Reducing the amount of oil-recovery drive fluid (e.g., water) that must be coproduced for the attainment of a given oil-recovery factor reduces the operating and production costs associated with producing a given amount of oil. It also often reduces certain environmental liabilities, including the production of excessive and unnecessary amounts of saline reservoir brines that can contain toxic heavy-metal ions and the production of excessive and unnecessary volumes of possibly environmentally unfriendly secondary or tertiary flooding oil-recovery drive fluids.

For the most part, conformance-improvement flooding operations and treatments do not decrease residual oil saturation. However, there has been a contention made in the literature that polymer flooding can reduce residual oil saturation under certain circumstances. This contention is discussed briefly later in this chapter. Also, by virtue of the fact that surfactants are incorporated into foams of foam-flooding operations, foam flooding can, in principle, reduce residual oil saturation. However, oilfield foams that are applied for mobility control are believed to function primarily by improving flood sweep efficiency. For the most part, conformance-improvement treatments accelerate oil production and/or delay premature economic abandonment of wells, well patterns, and fields, and can do so while conducting normal primary, secondary, or tertiary oil-production operations.

As previously noted, conformance-improvement floods and treatments do not normally promote reductions in residual oil saturation. Therefore, conformance-improvement operations should be limited to well patterns or reservoirs with a substantial and economically viable amount of moveable oil that can be recovered as a result of conducting the conformance flood or treatment.

As originally and purely defined, conformance-improvement flooding operations and treatments involve improving the uniformity of the flood front of an injected drive fluid during an oil-recovery flooding operation. In addition to the original definition, the working definition of conformance-improvement treatments now includes treatments applied to production wells to shut off excessive, deleterious, and competing coproduction of water or gas coming from a source other than the producing oil-formation interval. Examples of such coproduction are the coning of water from an underlying aquifer and the coning of gas from an overlying gas cap. Thus, using this expanded working definition of conformance-improvement treatments, conformance treatments can include water or gas shutoff treatments that are applied to production wells during primary oil-recovery operations. In this chapter, we use the modern and expanded definition of conformance improvement. Ref. 4 presents a high-level overview of the use of polymers, gels, foams, and resins as water and gas shutoff treatments.

The vast majority of conformance-improvement treatments function by reducing the permeability and fluid-flow capacity of the offending and treated reservoir high-permeability flow paths, channels, and conduits. The only practical exception to this is acid stimulation treatments to improve the wellbore flow profiles in injection or production wells. Discussion of near-wellbore conformance-improvement treatments using acid stimulation is beyond the scope of this chapter. Although this chapter is focused primarily on conformance improvement for application to light oil reservoirs, many of the conformance-improvement technologies discussed can be applied during gas and heavy-oil recovery operations (i.e., gel and resin water-shutoff treatments).

In summary, conformance improvement facilitates improved sweep efficiency of an oil-recovery flooding operation and/or reduces the coproduction of water and gas, which impede the full production potential of any given production well. All the polymers, gels, foams, and resins discussed in this chapter are used for conformance improvement in oil or gas producing operations.

13.3 Conformance Problems

Conformance problems can be divided into six categories:

- Poor sweep efficiency and/or excessive coproduction of the oil-recovery drive fluid in a relatively homogeneous matrix-rock (unfractured) reservoir resulting from poor mobility control and/or oil-recovery drive-fluid fingering.
- Poor sweep efficiency and/or excessive coproduction of the oil-recovery drive fluid in a matrix-rock reservoir resulting from substantial permeability variation and heterogeneity.
- Poor sweep efficiency and/or excessive coproduction of the oil-recovery drive fluid occurring in a naturally fractured reservoir.
- Water or gas coning.
- Excessive and competing water or gas production emanating from a casing leak.
- Excessive and competing water or gas production emanating from flow behind pipe.

The remediation, or partial remediation, of the first conformance problem is exemplified by a mobility-control polymer flood conducted in a reservoir containing a viscous oil and/or a reservoir that is characterized as being relatively homogeneous.

13.3.1 Key Distinctions. The first of two key distinctions relating to conformance problems is to differentiate between vertical conformance problems and areal conformance problems.[5] Vertical conformance problems, which are probably the most pervasive and most easily remedied conformance problems in matrix-rock (unfractured) reservoirs, are commonly manifested by geological strata of differing permeability overlying one another. In matrix-rock (unfractured) reservoirs, areal conformance problems, also referred to as "directional" high-permeability trends, can exist. Such conformance problems, which as a whole are usually less often treated by the technologies discussed in this chapter, are normally best addressed (by technologies of this chapter) through the application of a mobility-control flood such as a polymer waterflood. Areal conformance problems in matrix rock oil reservoirs are often more effectively remedied through well-pattern alignment strategies, which are beyond the scope of this chapter.

Within the conformance-problem category involving vertical matrix-rock problems and involving reservoir geological strata of differing permeabilities, there exists another aspect of this critical conformance-problem distinction that needs to always be made when considering technologies for reducing vertical conformance problems. This distinction is whether the geological strata of differing permeability are in fluid and pressure communication with each other.[5–7] That is, is there vertical permeability communication between the zones or are there impermeable layers (e.g., a shale layer) separating the geological strata? If these geological strata are not in vertical fluid communication throughout the reservoir or the well pattern to be treated, then this conformance problem can be remedied or alleviated simply by reducing the injectivity into the high-permeability strata at the injection well or by reducing the productivity from the high-permeability strata at the production well. This problem can often be treated cost effectively in the wellbore or the near-wellbore environment with mechanical packer systems, tubing patches, sandpacks, squeeze cementing, near-wellbore polymer-gel treatments, or resins. When such a treatment involves the placement of a chemical fluid-flow shutoff material (e.g., a gel or resin) in the offending strata surrounding a radial-flow well of a matrix-rock reservoir, then it is imperative (as is discussed in more detail in Sec. 13.6) that the treatment be placed selectively only in the offending geological strata and that none of the treatment shutoff material be placed in the oil-producing strata.[6–8] This type of treatment for improving vertical conformance is referred to by some as a profile modification treatment.

If vertical pressure communication and fluid crossflow exists between the geological strata, then the oil-recovery sweep efficiency problem and/or the associated excessive drive-fluid production problem cannot be remedied effectively with a wellbore operation or by a near-wellbore blocking agent treatment.[5–7] As **Fig. 13.1** shows, when a conformance-treatment blocking agent is placed near wellbore in the high-permeability geological strata at either the production or injection well, the conformance-improvement gains are short lived in terms of improved sweep efficiency and/or reduced rate of the excessive oil-recovery drive fluid (e.g., water during waterflooding) production. If the blocking agent is placed selectively in the high-permeability strata near wellbore to the injection well, the subsequently injected oil-recovery drive fluid will be injected into, and flow through, the low-permeability strata for a relatively short distance until it flows beyond the radius of the blocking agent. At this point, the oil-recovery drive fluid will tend to rapidly crossflow into the high-permeability strata where the fluid flow resistance is less. Other than very early in the life of a flooding operation, the near-wellbore volume of the low-permeability strata is likely already swept of its mobile oil saturation. In this case, little, or often no, sweep improvement or incremental oil production is gained from the placement of the blocking agent in the near-wellbore volume of the high-permeability strata.

If the blocking agent is placed selectively in the high-permeability strata near wellbore to the production well when crossflow between the reservoir strata occurs, a relatively small, and often uneconomic, volume of incremental oil production and a short-lived reduction in the undesirable high rate of the oil-recovery drive fluid production are realized after the treatment. After placing the blocking agent near wellbore in the high-permeability strata, the oil-recovery drive fluid will flow from the high-permeability strata to the low-permeability strata at a point just beyond the outer radius of the emplaced blocking agent.

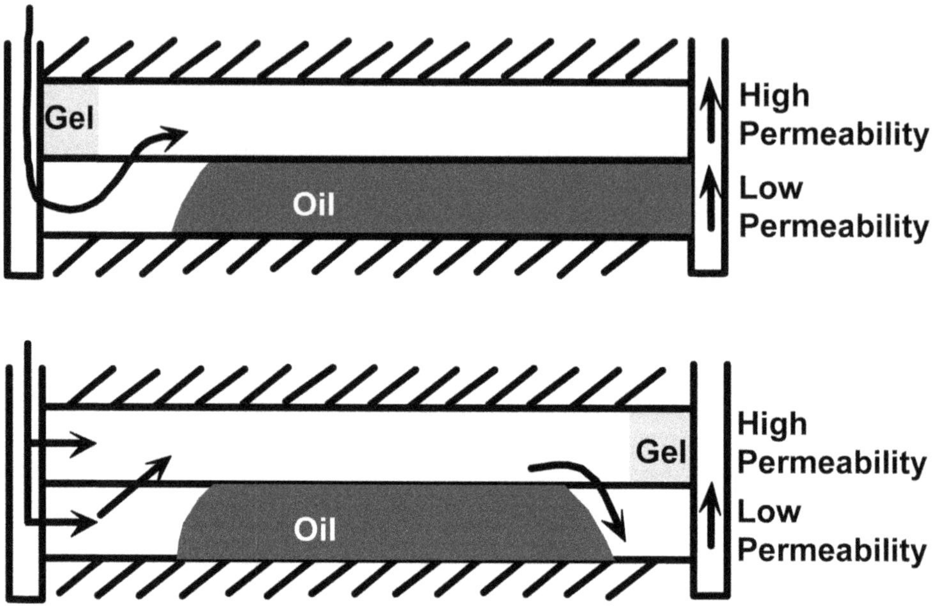

Fig. 13.1—Near-wellbore blocking agent placed in the high-permeability strata of a matrix rock reservoir exhibiting crossflow and suffering from a conformance problem.

Thus, when crossflow exists between the geological strata, when radial flow exists, and when the reservoir is undergoing an oil-recovery flooding operation, the selective placement of a blocking agent at, or near, the wellbore in the high-permeability strata of a matrix-rock reservoir renders little or no significant sweep improvement or reduction in the deleterious coproduction of the oil-recovery drive fluid (e.g., water during waterflooding).

If a matrix-rock reservoir with crossflow between geological strata and with radial-flow production is to be treated successfully with a blocking-agent conformance treatment, it must be treated such that the blocking agent is placed selectively deep in the reservoir in the high permeability strata.[6] The technical and economic feasibility of successfully applying water-shutoff treatments to this type of conformance problem has been questioned.[7] On the other hand, there are some reports in the literature, as exemplified by Ref. 9, that certain specialized polymer microgels have been applied through injection wells in the form of large volume conformance-improvement gel treatments that are intended to treat deeply into "matrix rock" reservoirs with crossflow between the reservoir geological strata.

A better strategy for rendering conformance improvement in layered reservoirs of matrix rock reservoirs where crossflow exists would be to use a mobility-control flood, such as a polymer flood. When flooding with a viscosity-enhancing mobility-control drive fluid, more of the injected drive fluid will be injected into, and flow through, the lower permeability and more poorly sweep geological reservoir strata. In this case, the strategy will result in accelerated oil production and reduced production of the oil-recovery drive fluid.

The second key conformance-problem distinction is whether the high-permeability flow path of the conformance problem is simply a high-permeability flow path through unfractured matrix rock or is a high-permeability anomaly, such as a fracture.[5] For this chapter, the cut off between a high-permeability flow path in matrix reservoir rock and a high permeability anomaly is the equivalent of about two Darcies in a sandstone reservoir.[5] High-permeability anomalies within a reservoir can include fractures (both natural and hydraulically induced), fracture networks, faults, joints, solution channels, interconnected vugular porosity, caverns, cobble layers, course sand strata, rubblized zones, and localized matrix reservoir rock with permeabilities greater than two Darcies. Reservoir fractures tend to be the most often encountered high-permeability anomaly. At depths greater than about 4,000 ft, fractures tend to be vertical in orientation and promote areal conformance problems.[10] At depths less than about 2,000 ft, fractures tend to be horizontal in orientation and can cause serious vertical conformance problems.

The distinction between conformance problems involving high-permeability flow paths through matrix reservoir rock and high-permeability anomalies is very important to the successful application of a number of technologies used to improve conformance. As discussed in Sec. 13.6, differentiating between these two conformance-problem regimes is critical to the success of the most widely applied polymer-gel treatment technologies because different versions of these polymer-gel technologies are normally required to treat these two different problems successfully. A polymer flood, which is applied to conformance problems involving solely matrix-rock permeability variation within a given well pattern or reservoir, is more likely to be successful than the same polymer flood that is applied to a similar well pattern or reservoir in which the conformance problem is dominated by high-permeability anomalies such as a carbonate well pattern or reservoir with numerous and extensive large solution channels. Classical mobility-control foam flooding is an inefficient option for use in a reservoir with high-permeability anomalies, such as an extensive and highly conductive fracture network.

Because the true nature of vugular-porosity conformance problems has often not been fully appreciated by many petroleum engineers, there have been a number of polymer-gel conformance treatment failures when treating vugular-porosity conformance problems. As **Fig. 13.2** depicts, the true and original definition of vugular porosity is relatively small voids (smaller than caverns) that exist randomly in matrix reservoir rock (especially carbonate reservoirs) where the vugular voids are not interconnected. If this is truly the vugular-porosity conformance problem that has been encountered in a given instance, then a matrix-rock conformance treatment is normally required. If, however, the conformance problem is dominated by large and extensive solution channels in the matrix reservoir rock (i.e., tubular flow pathways of often greater than ⅛-in. diameter), a high-permeability anomaly type of conformance treatment is required. The chances that a matrix-rock type of polymer-gel conformance treatment will be successful are remote when encountering reasonably large solutions channels. As is often the case when vugular-porosity conformance problems are encountered, the vugular porosity is actually vugs that are interconnected with solution channels.[5] If this is the actual nature of the vugular-porosity conformance problem, a high-permeability anomaly polymer-gel treatment is required. Failure to make the proper distinction between these two types of vugular-porosity problems can spell doom for a polymer-gel conformance-improvement treatment that is applied to such a vugular-porosity problem. When vugular-porosity conformance problems are encountered in those situations that the vugs are not interconnected, then a high-permeability anomaly polymer-gel treatment will not perform as expected and will not remedy this particular vugular-porosity conformance problem. Likewise, when vugular-porosity conformance problems are encountered in those situations that the vugs are interconnected, the application of a matrix rock polymer-gel conformance treatment will not be well suited for remedying such a vugular-porosity conformance problem.

13.3.2 How Conformance Problems Are Manifested. An alternate means of categorizing oilfield conformance problems is by the way conformance problems manifest themselves, such as by poor sweep efficiency during oil-recovery flooding operations, excessive and deleterious competing water coproduction, excessive and deleterious competing gas coproduction, coning and cusping, casing leaks, or water or gas flow behind pipe.

Two distinct types of water production exist. The first type, usually occurring later in the life of a waterflood, is water that is coproduced during oil/water fractional flow in reservoir matrix rock. When the production rate of this water is reduced, there will a proportional reduction in the oil-production rate. The second type of water production directly competes with oil

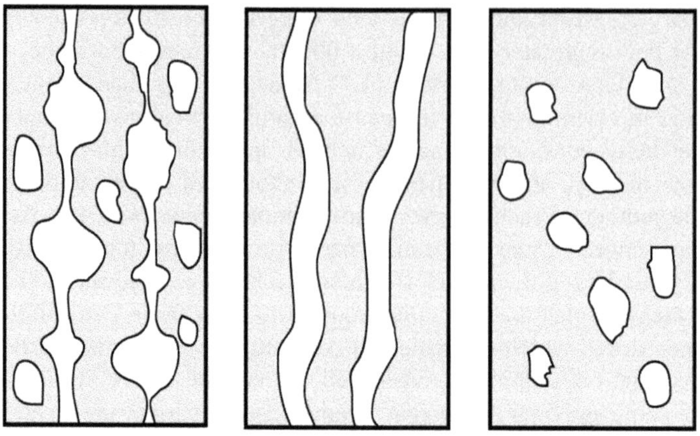

Fig. 13.2—Vugular porosity and solution channels.

production. This water often flows to the production wellbore via a flow path separate from that of the oil (e.g., water coning or a fracture emanating directly from a water injection well to the production well). For the second type of water-production problem, reducing water production can often lead to a greater pressure drawdown and/or an increase in the oil production rate. Thus, reducing the production of the second type of water production should be the objective of conformance-improvement floods and of water-shutoff treatments with gels, foams, and resins.[7]

A number of sources/causes of excessive and deleterious coproduction of water or gas exist:

• Early water or gas breakthrough during flooding operations resulting from poor sweep efficiency caused by vertical permeability variation in matrix-rock reservoirs (i.e., geological stratification).

• Early water or gas breakthrough during flooding operations resulting from poor sweep efficiency caused by variation in areal permeability in matrix-rock reservoirs.

• Early water or gas breakthrough caused by poor sweep efficiency that results from oil-recovery drive-fluid viscous fingering, where the viscous fingering is caused by an unfavorable mobility ratio between the oil-recovery displacement fluid and the reservoir oil.

• Fracture communication between the injector and producer (either extending fully or partially between wells).

• Fracture networks (with and with out directional trends).

• 2D coning via fractures.

• 3D coning via unfractured matrix reservoir rock.

• Cusping.

• Flow behind pipe.

• Casing leaks.

Coning and cusping can involve either water or gas. Cusping involves the production of aquifer water that flows to the production well through an inclined geological strata or zone, or gas-cap gas that flows to the production well through an inclined geological strata. In large part because of the relatively low viscosity and associated high mobility of gas, gas cusping tends to occur more easily than water cusping.

There are two distinctly different types and mechanisms of coning as it relates to conformance treatments such as water or gas shutoff coning treatments with gels. 2D coning occurs when water cones up, or gas cones down, to the production well's producing interval through vertical fractures or a fracture network. Conformance-treatment blocking agents, such as gels, can be used effectively and profitably to reduce such water or gas coning. 3D coning occurs when water cones up or gas cones down through matrix reservoir rock to the production well's producing interval. The use of conformance-treatment blocking agents, such as gels, has a very low probability of success when applied to a 3D coning problem.[7]

When large flow conduits with apertures substantially greater than approximately $^1/_{16}$ in. are the cause of flow behind pipe and the cause of the deleterious water or gas production, then the use of Portland cement is often favored for remedying such problems. The use of Portland cement for these purposes is beyond the scope of this chapter.

13.3.3 Implementing Conformance Treatments and Diagnosing Conformance Problems. In general, a conformance-improvement treatment (e.g., a gel treatment) to improve sweep efficiency and to generate incremental oil production is applied most effectively from the injection well side. When implementing a conformance-improvement treatment to reduce production operating costs by reducing the rate of competing water or gas production, these treatments usually are applied most effectively from the production well side. Treatments for both improving sweep and reducing excessive water and/or gas coproduction during gas or supercritical-liquid (e.g., CO_2) flooding operations in naturally fractured reservoirs normally are applied most effectively from the injection well side.[5]

The nature of the reservoir conformance problem to be addressed through the application of polymers, gels, foams, or resins needs to be diagnosed or deduced correctly or substantial negative consequences can occur.[5,7] A detailed discussion of how to properly and effectively diagnose and/or deduce reservoir conformance problems and flood sweep-efficiency problems is beyond the scope of this chapter. Refs. 5, 7, and 11 enumerate a number of techniques for diagnosing conformance problems and excessive water- and gas-production problems. Among the techniques discussed in these references are the use of interwell chemical and radioactive tracers, simple injectivity/productivity calculations to determine if fluid flow around a wellbore is radial or linear in nature, wellbore production and injection logs, various other logging tools, and pulse and pressure transient testing. An important element in successfully implementing a water or gas shutoff treatment is to determine at the onset, or at least hypothesize, the "plumbing" of the reservoir flow path of the excess water or gas production from its source to the production wellbore.[5] Production water/oil ratio (WOR) diagnostic plots[11] have also been used to help diagnose conformance problems. WOR diagnostic plots should be used in conjunction with another independent conformance-problem diagnostic tool, because many diagnostic plots can be interpreted in more than one way.

There historically has been a trend whereby petroleum engineers, when first considering the application of a conformance-improvement treatment in a new field, tend to underestimate the permeability and fluid-flow capacity of the high-permeability channels and flow paths within the reservoir to be treated.[5] This has contributed significantly to the low success rate of first-time conformance-improvement treatments being applied in a new field by an inexperienced petroleum engineer.

13.4 Disproportionate Permeability Reduction

Disproportionate permeability reduction (DPR) is a phenomenon whereby many water-soluble polymers and many polymer gels reduce the permeability to water flow to a greater extent than to oil or gas flow.[12–26] Most of the early work on, and application of, DPR involved fluid flow in reservoir matrix rock. More recently, water-shutoff chromium(III)-carboxylate/acrylamide-polymer (CC/AP) gels for use within fractures have been reported to impart DPR in gel-filled fractures.[27] However, because these relatively strong fracture-problem gels significantly reduce simultaneously the permeability to oil flow in fractures, these gels are better characterized as total-fluid-flow-shutoff gels and not DPR water-shutoff gels.

DPR is also referred to as relative permeability modification (RPM). However, some practitioners of this technology make the following subtle distinction. They tend to reserve the term

DPR for relatively strong polymer gels that impart a large degree of disproportionate permeability reduction and a relatively large reduction in water permeability. These practitioners reserve the term RPM for systems, such as solutions of water-soluble polymers or relatively weak gels, that impart more subtle disproportionate permeability reduction and more subtle reductions in water permeability. As used in this chapter when referring to water-shutoff treatments, the terms DPR and RPM are synonymous.

Alternatively, DPR conformance-improvement treatments, which involve relatively strong gels, can be successfully applied in hydraulically or naturally fractured reservoirs. In this case, the gel is placed and functions within the matrix rock that is adjacent to the fractures.[28]

DPR has the most value when used in water-shutoff/reduction treatments that are applied to production wells. DPR has little or no value for use during sweep-improvement treatments that are applied from the injection-well side.

A distinction needs to be made regarding classical and relatively strong (not DPR) polymer-gel water-shutoff (total-fluid-flow-shutoff) treatments that are placed within fractures that surround production wells. Such gel treatments tend to be placed, or place themselves, so that they will selectively reside in the water-producing fractures where the gels effectively block water flow. Such gel treatments (when properly designed and executed) tend to reduce water production without substantially reducing oil production. Here, the selective water shutoff results from selective placement of the gel in the water-producing fractures and not by the DPR mechanism.

The ability of acrylamide polymers to impart DPR to water flow in porous media was recognized as early as 1964 by Sandiford[26] and in 1973 by White *et al.*[12] The mechanism by which polymers and gels impart DPR and RPM effects has been the subject of a number of investigations. Refs. 29 through 39 are representative of these investigations. More recently, a set of plausible mechanisms have been proposed that explain how CC/AP polymer gels impart DPR.[40]

The application of DPR conformance-improvement technologies for water-shutoff (and/or water-reduction) purposes is not a panacea.[4,23,24,41–44] The successful application of bullheaded DPR long-term water-shutoff/reduction treatments, which involve radial flow in matrix reservoir (unfractured) rock and where the drawdown pressure on the producing interval is not increased after the gel treatment, is limited to when the following conditions are met:

• A conformance problem exists in a matrix rock reservoir involving differing geological strata.

• No fluid crossflow can occur within the reservoir between the water and the oil or gas producing geological strata.[44]

• The water strata is producing at an undesirably high water cut, and the oil or gas strata will produce for the economic life of the water-shutoff treatment at 100% oil or gas cut.

Possible exceptions to these limitations are as follows. First, if the DPR treatment induces an increase in the drawdown pressure on the producing interval, the DPR treatment may promote increased oil production. Second, if the DPR treatment material in the presence of oil flow breaks down, or is otherwise inactivated (with respect to its water-blocking ability), then selective water shutoff can occur over a wider range of excessive water-production problems.

For applications in radial-flow matrix-reservoir rock, commercially available DPR water-shutoff/reduction treatments usually attempt to reduce the permeability to oil or gas flow by a factor of two or less in the treated reservoir volume and reduce the permeability to water flow by a factor on the order of ten or more. Variability in performance of these systems has often led to erratic results when trying to accomplish this objective. To date, commercially available DPR water-shutoff/reduction treatments for application in radial-flow matrix-rock reservoirs have been based, almost exclusively, on the use of either solutions of water-soluble polymers or relatively "weak" gels.

For application to fractured wells, a DPR scheme for treating excess water production is discussed in the DPR subsection of Sec. 13.6.1. This scheme relies on placing a relatively strong DPR water-shutoff gel in the matrix rock that is adjacent to the fractures.[28]

DPR water-shutoff treatments are of no practical value [in terms of providing long-term (i.e., years) water shutoff] when applied to a single zone reservoir that is producing at a high water cut in the radial flow mode from a matrix rock reservoir. This is because a relative-permeability water block will form just beyond the outermost radial penetration of the DPR water-shutoff treatment.[44] What is less obvious is that for the same basic reason when producing from matrix rock reservoirs in the radial flow mode, DPR water-shutoff treatments are not effective at promoting long-term water shutoff/reduction anytime the oil-producing zones are producing at a finite water cut or crossflow exists between the oil- and water-producing zones.

Another issue with DPR water-shutoff treatments is that for such treatments, which are based on the use of solutions of water-soluble polymers or relatively "weak" gels, their water-shutoff performance is erratic and not highly reproducible in both the laboratory and field settings.[45] For DPR water-shutoff treatments, the restoration (or near restoration) of the oil permeability following the placement of the treatment in matrix reservoir rock can be quite slow.[45] An important point regarding DPR water-shutoff treatments is that DPR imparted in the treated volume of the matrix reservoir rock does not necessarily correspond in the field to a proportionate reduction in the water production rate.

An additional important distinction that needs to be made is to whether the DPR treatment is being applied for long-term (i.e., years) or short-term (i.e., hours to months) water-shutoff purposes. For the relative-permeability water-block reason discussed previously, many DPR water-shutoff treatments will render short-term or transient water-shutoff/reduction during treatments that are applied in the field. The water-block problem goes a long way in explaining why so many DPR water-shutoff treatments have failed to provide long-term water shutoff and have tended to render only short-term water shutoff. There are scenarios for limited reservoir conditions, where DPR treatments that render only transient water shutoff, can be engineered to be economically attractive and profitable.

The reason that bullheadable DPR water-shutoff treatments have created such a great interest in the oil industry is that they do not require the use of mechanical zone isolation when applied to layered matrix-rock reservoirs. Mechanical zone isolation often requires costly workover operations. Use of mechanical zone isolation during water-shutoff-treatment placement is not normally feasible when the well possesses a slotted-liner or gravel-pack completion or when the well involves a subsea tieback flow line.

Historically, a large number of ineffective DPR (RPM) water-shutoff treatments have been conducted. The high failure rate of DPR water-shutoff/reduction treatments has resulted from a combination of overexpectations by operators regarding DPR water-shutoff treatments, overselling of DPR water-shutoff treatments by oilfield service companies, and failure to recognize the constraints to the successful application of DPR water-shutoff treatments within matrix-rock (unfractured) reservoirs. However, DPR treatments remain one of the few options available to successfully treat excessive water-production problems in matrix rock reservoirs for which mechanical zone isolation is not possible or practical during treatment fluid placement. At the time of the writing of this chapter, the investigation, development, and exploitation of DPR conformance-improvement technologies were being actively pursued by petroleum industry R&D efforts.

13.5 Polymers

Early application of polymers for use during oilfield conformance-improvement operations was focused on improving volumetric sweep efficiency of waterfloods. More recently, polymers have been used extensively in DPR and RPM treatments for water shutoff and in conformance-improvement polymer-gel treatments. Most of this section focuses on the use of polymers in polymer waterflooding operations. Sec. 13.5.8 reviews the use of polymers in DPR and RPM treatments that are applied for "selective" water-shutoff/reduction.

13.5.1 Polymer Fundamentals. Polymers are large molecules and chemical entities referred to as macromolecules. Polymer molecules are the resultant chemical specie when a large number of relatively small and repeating molecular entities, called monomers, are joined together chemically. The chemical process of joining together the monomers and forming polymer molecules is referred to as the polymerization reaction process. Polymers, both natural and man made, have numerous beneficial uses and applications in modern society (everything from wood, to plastics, to man-made thickening agents added to milk shakes). Polymers can come in pure solid and liquid forms. Some polymers can be dissolved in liquids. This chapter is limited to polymers that can be dissolved (or dispersed) in an aqueous solution and that usually increase the viscosity of the aqueous solution.

Basic Oilfield Polymer Types. There are two fundamentally different types of water-soluble and viscosity-enhancing polymer chemistries that have been used during polymer waterflooding and conformance-improvement treatments. The first type is biopolymers, such as Xanthan gum polymer. The second type is man-made synthetic polymers, such as acrylamide-based polymers.

Biopolymers of the type used in conformance improvement are polysaccharides (poly sugars) in which the monomer chemical linkages of the polymer backbone are glycoside linkages, involving carbon-oxygen-carbon chemical bonds. In aqueous solution, the multistranded molecular complexes of xanthan polymer are fairly rigid molecular species, causing the polymer molecules to take on an extended molecular conformation.

Synthetic polymers of the type used in conformance improvement are usually highly flexible molecules in which the polymer backbone consists of a relatively chemically stable carbon molecular chain with single and flexible carbon-carbon bonds. Pendant water-soluble chemical groups (e.g., amide groups) on the molecule render the polymer molecule to be soluble in water.

Synthetic polymers have emerged to become the predominant and preferred polymer type for use in commercial oilfield conformance-improvement operations because of the inherent chemical and biological stability of synthetic polymers, along with injectivity and cost issues.

Historically, the two types of biopolymers primarily used in polymer waterflooding have been xanthan and scleroglucan polymers. The only synthetic polymers that have been used extensively during polymer waterflooding (and in polymer-gel conformance treatments) are those based on acrylamide-polymer chemistry.

Viscosity Enhancement and Permeability Reduction. Water-soluble polymers used in conformance-improvement operations operate by reducing fluid mobility by increasing viscosity of the oil-recovery drive fluid (primarily the flood water) and/or by reducing permeability, by which the polymers, directly or indirectly, act as a fluid-flow blocking agent. Reducing permeability is the conformance-improvement mechanism by which conformance treatments operate when polymers and polymer gels are used for imparting DPR. Polymers used in waterfloods often have a secondary component in their conformance-improvement mechanism that involves permeability reduction within the flooded volume of a matrix rock reservoir.

Chemistry of Polymers Used in Conformance Improvement. The chemistry of polymers used in conformance improvement is reviewed before discussing polymer waterflooding and DPR polymer treatments because several of the polymer chemistries that are reviewed are used in both of these oilfield conformance-improvement applications. In fact, some of the polymer chemistries discussed in this section are also used in conformance-improvement polymer gels (see Sec. 13.6).

Biopolymers. In addition to being environmentally friendly and readily available, advantages of biopolymers are their relative insensitivity to salinity and mechanical shear degradation. The two major concerns relating to the use of biopolymers are their susceptibility to biological and chemical degradation, and injectability issues resulting from cell debris that usu-

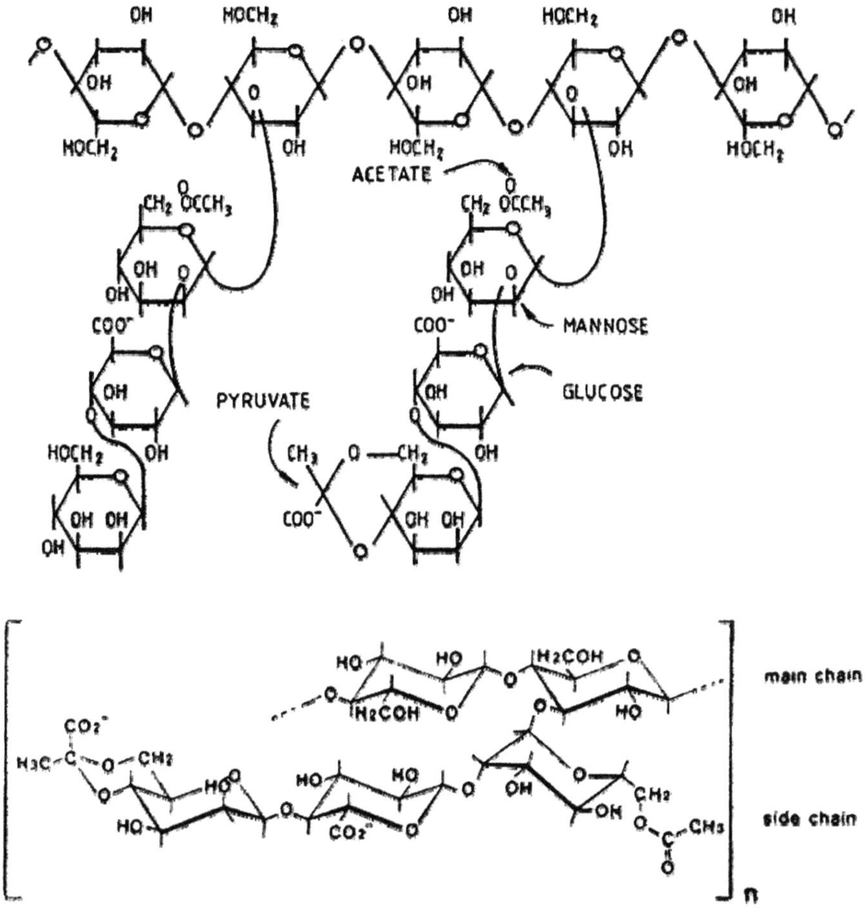

Fig. 13.3—Chemical structure of a xanthan polymer molecule.

ally remain in the biopolymer solutions that are derived from microorganism fermentation processes.

Xanthan[46] has been the most widely used biopolymer for polymer waterflooding. **Fig. 13.3** shows the chemical structure of the xanthan biopolymer molecule.[46] For a xanthan molecule with a molecular weight of 4 million daltons (atomic mass units), the xanthan molecule comprises on the order of 20,000 repeating sugar monomer units. Xanthan polymer is derived from a microorganism fermentation process that usually leaves a substantial amount of cell debris in the final polymer solution.

Flooding reservoir matrix rock with fully filtered xanthan polymer solutions tends to result in much less permeability reduction than the comparable flooding with appropriate acrylamide-polymer solutions. Xanthan polymers and the resultant solution viscosity are relatively insensitive to the salinity of the brine into which the Xanthan polymers are dissolved, and Xanthan polymers tend to be relatively insensitive to mechanical shear degradation. Xanthan polymers are quite susceptible to biological degradation.

Scleroglucan, with a triple-stranded molecular configuration, has been suggested to be a biopolymer possessing more favorable stability and performance properties for use during high-temperature polymer waterflooding (e.g., 195°F).[47,48]

Synthetic Polymers. Acrylamide polymers have emerged to be the most widely used synthetic polymer family for application in polymer flooding and in polymer and polymer-gel confor-

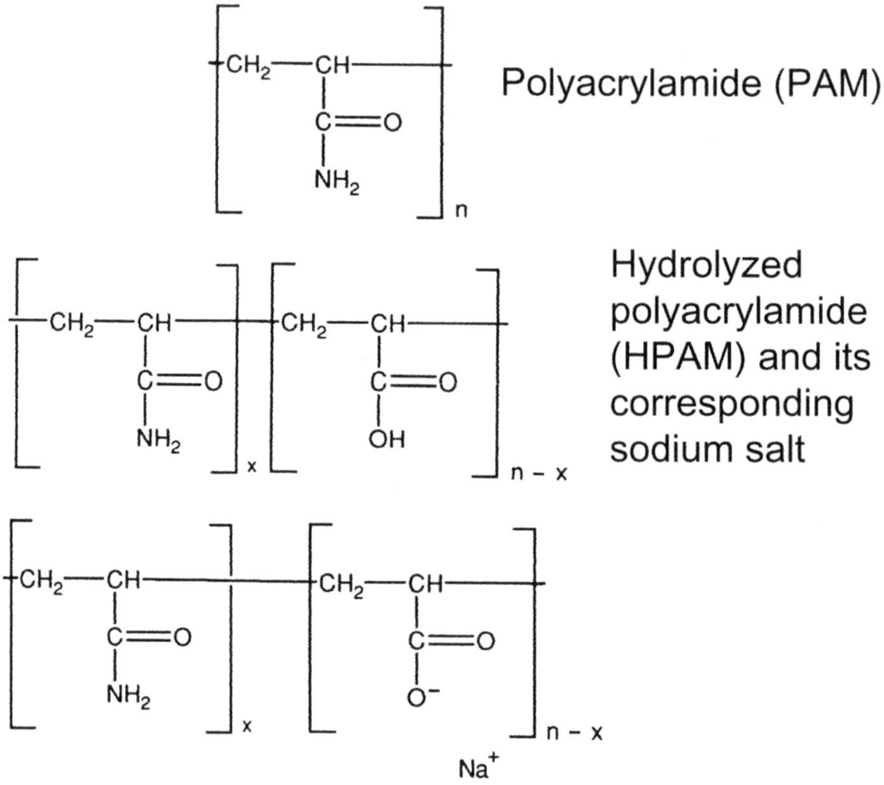

Fig. 13.4—Chemical structure of PAM and HPAM polymer molecules.

mance-improvement treatments. This has come about in large part because of cost and availability issues and because of the favorable chemical robustness and biological stability.

Polyacrylamide (PAM) is the simplest and most basic form of acrylamide polymers. **Fig. 13.4** shows the chemical structure of polyacrylamide and partially hydrolyzed polyacrylamide. For polyacrylamide with a molecular weight of 7 million daltons (a representative molecular weight of an acrylamide polymer to be used in polymer waterflooding), the value of n in Fig. 13.4 and the number of repeating monomer units is on the order of 100,000. When all other factors are equal and when dissolved in brine with a relatively low salinity, polyacrylamide is not as good a viscosity-enhancing agent and is not propagated as well through sand reservoirs when compared to partially hydrolyzed polyacrylamide (HPAM). Because pure polyacrylamide is slightly positively charged (cationic) in an acidic or "neutral" pH environment, polyacrylamide tends to adsorb onto reservoir rock surfaces, especially sands and sandstone pore surfaces. For these reasons, partially HPAM is most often favored for use in polymer flooding.

When polyacrylamide is manufactured commercially, it normally contains 1 to 2 mole percent hydrolyzed (carboxylate) content that is inadvertently imparted during the manufacturing process. This carboxylate "impurity" in many of the commercial polyacrylamide polymers is enough carboxylate content to render such polyacrylamide to be a good candidate for use in conformance-improvement polymer gels that involve chemical crosslinking reactions occurring through the polymer's carboxylate groups. A specialized polyacrylamide polymer is available commercially that contains essentially no carboxylate groups (less than 1 carboxylate groups in 1,000 acrylamide groups). Such polyacrylamide is referred to as ultra-low hydrolysis polyacrylamide. To manufacture ultra-low-hydrolysis polyacrylamide, normal acrylamide monomer feed stock is polymerized as usual, except the polymerization conditions (pH and temperature) are

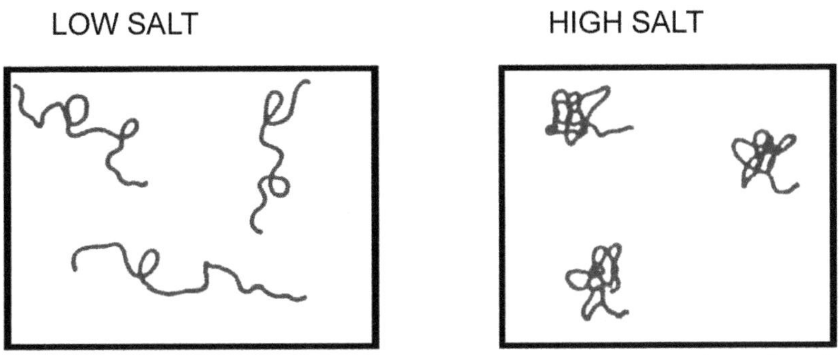

Fig. 13.5—Schematic of the effect of solution ionic strength on the molecular conformation of flexible coil polyelectrolyte molecules such as HPAM.

more tightly controlled. Polyacrylamide is normally not referred to as HPAM until the carboxylate content exceeds approximately 2 mole percent.

HPAM is the most widely employed water-soluble polymer for use in both polymer waterflooding and in oilfield conformance polymer-gel treatments. As mentioned previously and as compared with polyacrylamide, HPAM polymer tends to be a better viscosity-enhancing agent in low-salinity brines and tends to adsorb less onto the rock surfaces of reservoirs that are good polymer-waterflooding candidates. Thus, the use of the salt form of HPAM is favored over the use of straight polyacrylamide in most polymer-flooding applications. A number of practitioners of polymer flooding believe that because of the salt sensitivity of HPAM, this polymer performs best during polymer flooding conducted in reservoirs with low-salinity reservoir brines. However, there have been some instances in which HPAM, when dissolved in a fresh flooding brine, has performed well when flooded in a reservoir with a saline brine. Ref. 49 cites such an example.

A 30% hydrolysis level within polyacrylamide is near the optimum in terms of simultaneously promoting maximum viscosity enhancement of the polymer solution and minimizing polymer adsorption onto reservoir rock surfaces during most polymer waterfloods. For crosslinked polymer gels, the optimum hydrolysis level of 5 to 10 mole percent (in this case) simultaneously maximizes gel strength and minimizes unproductive intramolecular crosslinking.

As Fig. 13.4 shows, HPAM can come in two forms as it relates to the chemistry of the carboxylate groups. The carboxylate groups can be in the acid or salt form. For use in polymer waterflooding and in polymer gels, HPAM is almost always used in the sodium salt form. Unless specifically stated otherwise in this chapter, when referring to HPAM, we are referring to HPAM with its carboxylate groups in the sodium salt form.

In low salinity brines, the electrostatic charge repulsion between carboxylate groups of HPAM molecule tends to cause the flexible-backbone polymer to assume a distended tertiary conformational form that is a more effective for enhancing aqueous-solution viscosity than is the more balled-up form occurring for such polymers dissolved in a high salinity brine and the balled-up molecular conformational form of unhydrolyzed polyacrylamide. **Fig. 13.5** depicts the stretched out and more effective viscosity-enhancing molecular form of HPAM that exists in a low-salinity aqueous environment. Fig. 13.5 also shows the balled-up conformational form that HPAM assumes in a high salinity brine environment. High salinity causes the electrostatic fields around the carboxylate groups to shrink substantially and allows the HPAM molecule to assume a more balled-up form because of the elimination of a high degree of electrostatic repulsion between the negatively charged carboxylate groups on the polymer's backbone. The balled-up polymer form does not generate nearly as much viscosity as the distended form in otherwise comparable polymer-waterflood solutions.

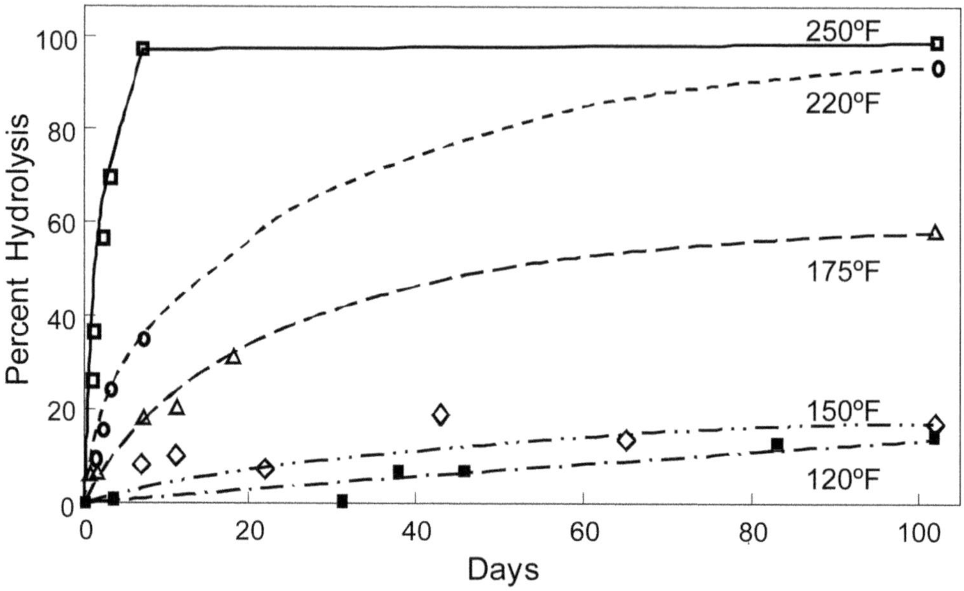

Fig. 13.6—Degree of hydrolysis vs. time at various temperatures for 1,000 ppm unhydrolyzed PAM in 5% salinity.

Hydrolyzed acrylamide groups, or equivalently termed carboxylate groups, can be introduced into polyacrylamide polymers by several means. First, polyacrylamide that is dissolved in aqueous solution can be reacted with caustic material, such as sodium hydroxide, to convert a portion of the polymer's pendant amide groups to carboxylate groups. This form of HPAM is referred to as partially hydrolyzed polyacrylamide. Second, during the polymerization process, acrylamide monomers can be copolymerized with acrylate monomers to form HPAM. This form of HPAM is referred to as being a copolymer of acrylamide and acrylate. All polyacrylamides and all commercially available HPAMs, when heated in aqueous solution, slowly undergo an autohydrolysis reaction in which a portion of the acrylamide polymer's pendant amide groups spontaneously hydrolyzes to carboxylate groups. The final degree of carboxylate content that is attainable within a polyacrylamide molecule increases with temperature, but does not reach 100 mole% carboxylate groups. That is, the acrylamide polymer cannot be converted in aqueous solution at high temperature (under reservoir conditions) to pure polyacrylate by means of the autohydrolysis reaction. The autohydrolysis reaction of acrylamide polymers is both acid and base catalyzed.

In high-temperature reservoirs after polyacrylamides or HPAMs autohydrolyze to sufficiently high levels, hardness ions, such as calcium or magnesium, in the reservoir brine cause the polymer to undergo a phase change, precipitate, and cause the polymer to lose most of its viscosity-enhancing function.[46,50] This outcome is the major limitation of acrylamide-polymer flooding in high temperature reservoirs. **Fig. 13.6** shows the degree of polymer hydrolysis vs. time at various selected temperatures for 1,000 ppm PAM polymer dissolved in a brine of 5% salinity.[46]

Copolymers containing 2-acrylamido-2-methyl-propanesulfonic acid (AMPS) monomers and acrylamide monomers have been suggested to form acrylamide polymers for use in polymer waterflooding of high-temperature (e.g., 200°F) and high-salinity reservoirs where the AMPS copolymer's performance and stability will be somewhat better than comparable HPAM.

Copolymers of vinylpyrrolidone and acrylamide, along with ter-polymers of vinylpyrrolidone, acrylamide, and acrylate, have been reported to be candidate polymers for use in

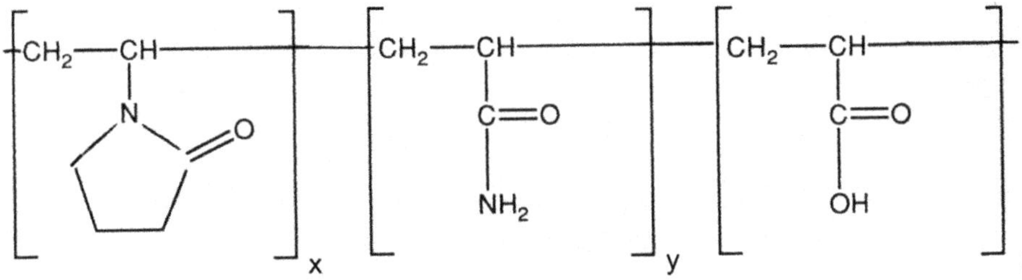

Fig. 13.7—Chemical structure of a ter-polymer poly (vinylpyrrolidone-acrylamide-acrylate) molecule.

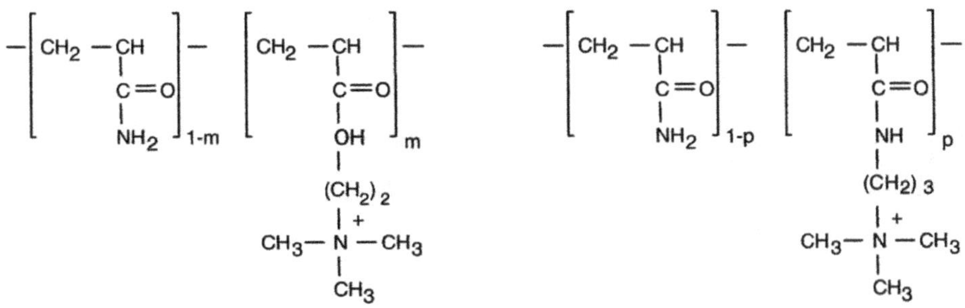

Fig. 13.8—Chemical structure of two cationic acrylamide polymers for possible use in the bridging-adsorption selective placement process for conformance gel treatments.[53]

polymer floods and conformance polymer-gel treatments that are to be applied to high-temperature reservoirs with harsh environments. Certain vinylpyrrolidone polymers were reported to not precipitate from seawater after aging for six years at 250°F.[51] Potential concerns regarding these co- and ter-polymers of vinylpyrrolidone are their relatively high cost as compared with more conventional acrylamide polymers and the relatively low molecular weight of the commercially available forms of these co and ter polymers.

Fig. 13.7 shows the general chemical structure of ter-polymers of vinylpyrrolidone, acrylamide, and acrylate. The primary beneficial function of the incorporated vinylpyrrolidone monomer into an acrylamide polymer is that it prevents the acrylamide monomer content of the polymer from autohydrolyzing at high temperatures to the excessively high levels of hydrolysis whereby the polymer would become susceptible to precipitating out of solution when the polymer encounters hardness divalent ions.

Cationic polyacrylamides are acrylamide polymers that have positively charged chemical groups attached to at least some of the polymer's pendant amide groups or acrylamide polymers that have been copolymerized with monomers containing positively charged pendant groups. These polymers have an exceptionally strong tendency to adsorb onto reservoir rock surfaces, especially sand and sandstone surfaces.

Cationic acrylamide polymers find specialized applications in conjunction with a variety of conformance-improvement treatments. These applications include use as polymer-anchoring agents to help promote conformance polymer-gel adsorption onto reservoir rock surfaces,[52] "bridging-adsorption"[53] and/or "flow-induced-adsorption"[54] polymers for injection before a conformance gel treatment to purportedly promote the selective placement of the gel treatment during the bullheaded treatment-placement mode, and polymer for use in certain polymer DPR conformance treatments.[42] Water-soluble cationic acrylamide polymers come in a wide variety of forms and chemistries. **Fig. 13.8** shows the chemical structure of two cationic acrylamide polymers that have been studied for use in the bridging-adsorption phenomenon.[53]

13.5.2 Benefits of Applying Conformance-Improvement Polymer Technologies. The application of oilfield polymer technologies, in the form of polymer waterflooding and polymer DPR treatments (and as polymer-gel treatments), can promote conformance improvement during oil-recovery-flooding and oil-production operations. They do so by the following means.

• Improve sweep efficiency—The application of polymer waterflooding and polymer DPR treatments promote more effective economic use of injected oil-recovery drive fluids, such as water during waterflooding. DPR treatments can also be used to reduce the amount of injected oil-recovery drive fluid that must be coproduced to yield a given oil-recovery factor.

• Accelerate production—Successful polymer waterflooding and polymer DPR treatments accelerate oil production during a waterflood or other oil-recovery flooding operations by reducing the amount of injected oil-recovery drive fluid that must be coproduced to attain a given level of oil recovery.

• Promotes incremental oil production—Polymer waterflooding and DPR treatments rarely reduce waterflood residual oil saturations. However, they do promote incremental oil recovery by increasing the amount of oil production before reaching the economic WOR limit of a production well, well pattern, or field during a waterflood or other oil-recovery flooding operation.

• Extend economic lives—Polymer waterflooding and DPR treatments can extend the economic lives of production wells, well patterns, and fields by increasing oil cuts as a function time and deferring the time when the economic WOR limit of a well, well pattern, or field is reached.

13.5.3 Polymer Waterflooding. When conducting a polymer waterflood, a high-molecular-weight and viscosity-enhancing polymer is added to the water of the waterflood to decrease the mobility of the flood water and, as a consequence, improve the sweep efficiency of the waterflood. The primary purpose of adding polymer to most polymer waterfloods is to increase the viscosity of the flood water; however, polymer addition to the flood water in many instances also imparts a secondary permeability-reduction component. Polymer waterflooding is normally applied when the waterflood mobility ratio is high or the heterogeneity of the reservoir is high. **Fig. 13.9** shows the polymer waterflooding process.[55]

How Polymer Flooding Improves Recovery. Polymer waterflooding promotes improved sweep efficiency by improving the mobility ratio. Improved sweep efficiency imparted during polymer flooding is primarily accomplished by increasing the viscosity of the waterflood drive fluid. Conventional wisdom states that polymer waterflooding does not reduce irreducible oil saturation (residual oil saturation to waterflooding);[46,55,56] however, at least one paper[57] has called this contention into question when flooding with selected acrylamide polymers.

Polymer Solutions. The principal beneficial property of polymer solutions for use in flooding oil reservoirs is the aqueous solution's enhanced viscosity. Aqueous polymer solutions that are used for conformance-improvement flooding normally exhibit non-Newtonian viscosity properties.

Viscosity of Polymer Solutions. The viscosity of a polymer solution is a measure of how "thick" a fluid is. For example, molasses is characterized as being "thicker" and more viscous than water. The viscosity of a fluid or solution may, in general terms, be defined as the solution's resistance to being sheared or as the resistance of a fluid mass to change its form.

Fluid viscosity, μ, is defined as

$$\mu = \frac{\tau}{\dot{\gamma}}, \quad (13.7)$$

where τ is shear stress and $\dot{\gamma}$ is shear rate.

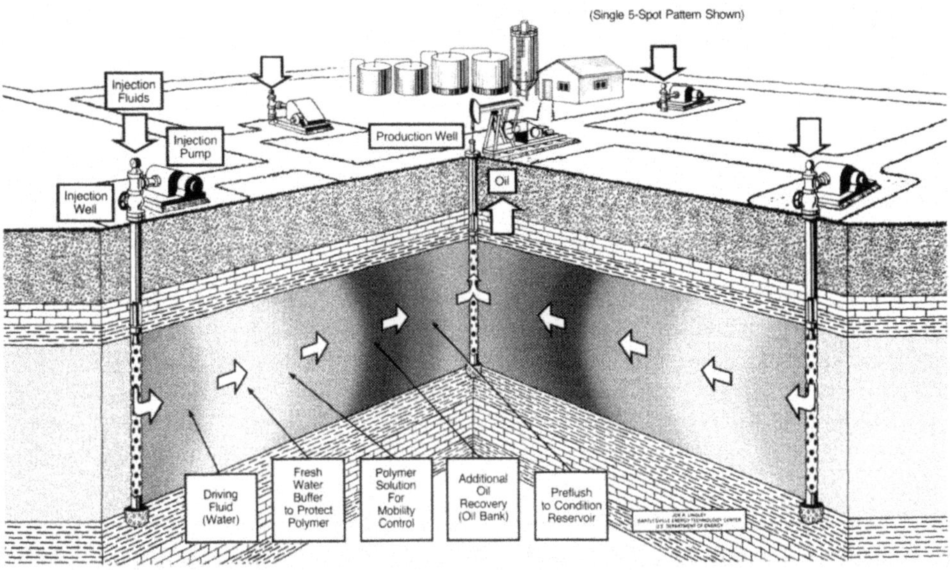

Fig. 13.9—Schematic of the polymer waterflooding process. The method shown requires a preflush to condition the reservoir, the injection of a polymer solution for mobility control to minimize channeling, and a driving fluid (water) to move the polymer solution and resulting oil bank to production wells. Mobility ratio is improved and flow through more permeable channels is reduced, resulting in increased volumetric sweep.[55]

Many common fluids, such as water and motor oils, exhibit Newtonian viscosity. For fluids with Newtonian viscosities, the fluid's viscosity is independent of the shear rate that the fluid is experiencing. That is, the value of the viscosity of a Newtonian fluid at a given temperature is a single value that is independent of shear rate.

The viscosity-enhancing power of a polymer is related to the size and extension of the polymer molecule in a particular aqueous solution.[46] For a number of reasons, the viscosity of a polymer solution that is measured in a viscometer and the effective viscosity of the polymer solution that is measured during flow through porous reservoir matrix rock often have different values.

To predict the viscosity-enhancing power of a polymer in a given solution, the polymer's intrinsic viscosity, $[\eta]$, can be measured by

$$[\eta] = \mathrm{Lim}_{c \to 0} \frac{\eta - \eta_s}{c\eta_s}, \quad \text{.......} \quad (13.8)$$

where c is polymer concentration, η is polymer solution viscosity, and η_s is solvent viscosity. Intrinsic viscosity is obtained by determining the value $\mathrm{Lim}_{c\to 0}(\eta\text{-}\eta_s)/c\eta_s)$ that is obtained from the plot of $(\eta-\eta_s)/c\eta_s)$ vs. polymer concentration and extrapolating the plotted data back to zero polymer concentration. See Chap. 3 of Ref. 46 for more details regarding intrinsic viscosity of polymer solutions. For a given polymer in an aqueous solution, the intrinsic viscosity for the polymer increases with polymer molecular weight (MW) according to the Mark-Houwink equation:

$$[\eta] = K'Mp^a, \quad \text{.......} \quad (13.9)$$

where K' and a are polymer-specific constants, and M_P is the polymer molecular weight. See Chap. 3 of Ref. 46 for more details regarding the Mark-Houwink equation.

The empirical Flory equation[58] can be used to estimate the mean end-to-end distance of a polymer in solution. The Flory equation is

$$d_p = 8\left(M_p[\eta]\right)^{1/3}, \quad (13.10)$$

where d_p is in Angstroms (10^{-10} m), and $[\eta]$ is in dl/g.

MW and Size. When all other factors are equal (such as polymer type and the brine solution into which the polymer is dissolved), as the MW of the polymer increases, the size of the polymer increases. As the size of the polymer increases, so does the polymer's viscosity enhancing ability when dissolved in a given brine. On the negative side, as the MW of a polymer increases, the propensity for the polymer to be retained during transport through matrix reservoir rock is increased, and the propensity for the polymer to exhibit injectivity problems is increased.

Because polymers used in polymer waterflooding are polydispersed in MW, polymer MW distribution is an important factor relating to how a given polymer will function during a polymer flood. Unfortunately, good MW distribution data are not readily and widely available for the polymers that are normally used in polymer flooding, because the determination of a polymer's MW distribution is relatively expensive and time consuming.

Fig. 13.10 shows the MW distribution for a typical HPAM polymer sample used in polymer flooding. The high MW tail of the MW distribution is quite significant. The small number of polymer molecules in the polymer's MW distribution have a disproportionately large effect on the viscosity-enhancing power of the polymer; are the polymer molecules that are first and most easily degraded by mechanical shear under intermediate- to high-shear flow conditions; are the polymer molecules that will be first and most easily retained during polymer transport through reservoir matrix rock; and are the polymer molecules that are most prone to causing injectivity damage.

Fig. 13.11 shows a series of MW distributions for a family of HPAM polymer samples. MAR-1 through MAR-9 denote sample numbers 1 through 9. When applied in a relatively high-permeability reservoir in which polymer retention during polymer transport and polymer injectivity are not major issues, polymers with higher molecular weights and narrower MW weight distributions perform relatively more effectively as viscosity-enhancing agents during a polymer waterflood.[46,59]

For a polymer that is dissolved in a given solvent, polymer MW is proportional to molecular size. For several illustrative polymers used in polymer waterflooding, their molecular size, as related to MW, is as follows. A 30% HPAM polymer of approximately 4×10^6 dalton (atomic mass units) MW dissolved in a good solvent is expected to be fibril in form and to have a diameter of 0.7 to 2.5 μm and a backbone chain length greater than 10 μm.[46,60] The hydrodynamic length of a xanthan molecule commonly used in polymer flooding has been reported to be approximately 1.5 μm.[46,61] The MW of such a xanthan molecule is approximately 4×10^6 daltons.

Rheology. The non-Newtonian viscosity of polymer solutions used in polymer waterflooding normally exhibit shear-thinning behavior when subjected to sufficiently high shear rates (but not at low shear rates). The viscosity of a Newtonian fluid does not vary with the shear rate to which the fluid is subjected. For a shear-thinning fluid, the apparent viscosity of the fluid decreases as the fluid experiences increasing shear rates.

Figs. 13.12 and 13.13 show the shear-thinning viscosity behavior of two polymers of the type used in polymer flooding. In Fig. 13.12, at low shear rates (< 0.1 s^{-1}), the viscosity behav-

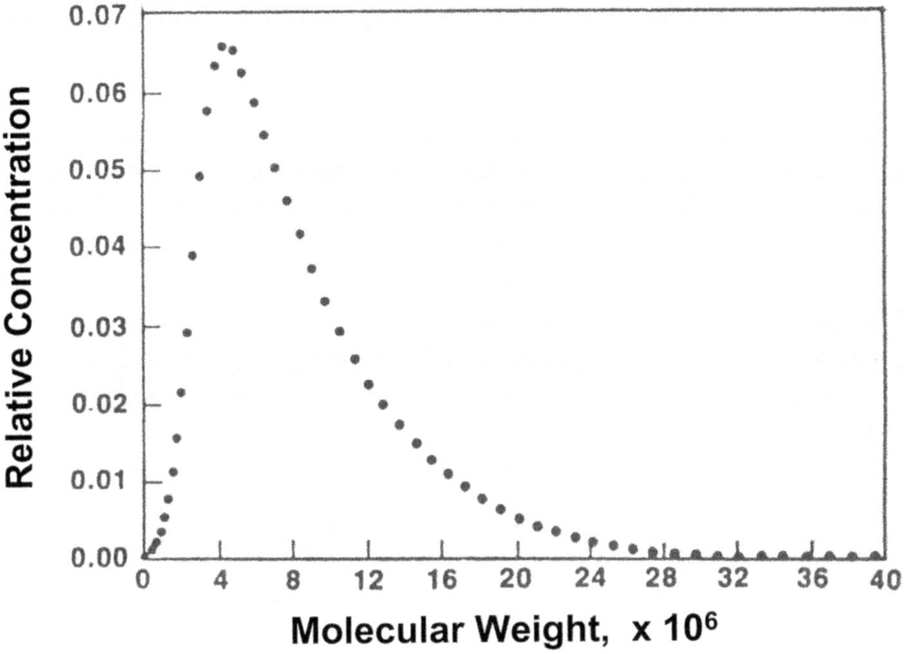

Fig. 13.10—Molecular weight distribution of a typical HPAM polymer used in polymer flooding.

ior of the polymer solutions at all polymer concentrations is invariant with shear rate and, thus, is Newtonian. At shear rates exceeding 1.0 s^{-1}, the viscosity of the polymer solutions decreases with increasing shear rate, thereby exhibiting shear-thinning viscosity behavior. The shear-thinning viscosity behavior of the polymer in Fig. 13.12 becomes relatively less pronounced as the concentration of the polymer in solution decreases. This trend is observed for all polymers that are used in polymer flooding. The shear-thinning viscosity reduction behavior results from the water-soluble high-MW polymers becoming uncoiled and unentangled when they are aligned and elongated in the fluid-flow shear field under sufficiently high shear-rate (~1 to 100 s^{-1}) conditions. When the polymers are aligned and unentangled by the shear field in this shear-rate range, the polymers become less effective viscosity enhancing agents. At the low shear rates of Fig. 13.12 (< 0.1 s^{-1}), the shear field is not strong enough to appreciably uncoil and untangle the polymer molecules. The viscosity is invariant over this shear-rate range, and the viscosity behavior is characterized as being Newtonian over this shear-rate range.

As expected, the apparent viscosity at any given shear rate increases as polymer concentration increases. For the studied AMPS polymer, Fig. 13.13 shows the dramatic and undesirable effect that increasing the salt concentration in the makeup water has on reducing the viscosity of the polymer solution at any given shear rate. Similar detrimental effects of increasing salt concentration are observed in polymer solutions of high-MW HPAM.[55] Note also the trend, which is quite generalized, that as the salt concentration of the polymer solution increases, the degree of shear thinning of the polymer solution decreases.

The shear-thinning viscosity behavior of these polymer solutions is favorable because the shear rate experienced by the polymer in the vast majority of the reservoir is usually quite low (approximately 1 to 5 s^{-1}) and is a shear rate at which the polymer exhibits near maximum viscosity.[55] In the near-wellbore region, share rates are often in the shear-thinning range for the polymer (e.g., 1 to 100 s^{-1}). This polymer shear thinning is fortuitous because the viscosity reduction improves the injectivity of the polymer solution.

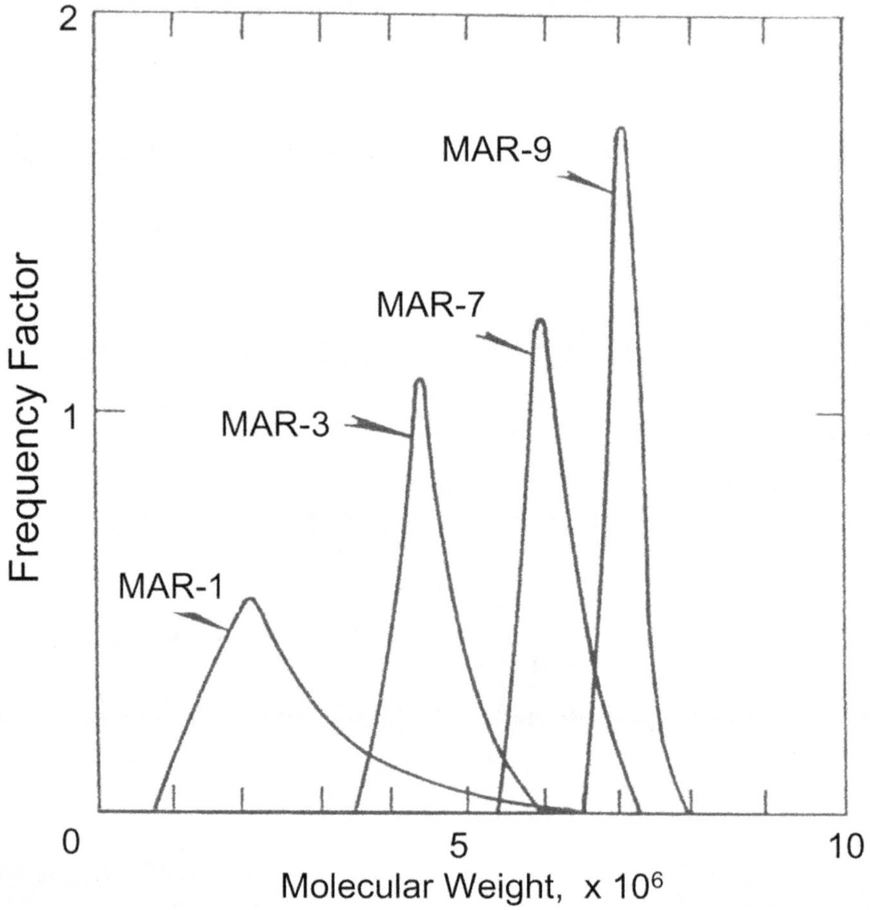

Fig. 13.11—Molecular weight distribution of a series of HPAM polymer samples.

The mathematical equation that describes the viscosity vs. shear-rate behavior (of the type of Figs. 13.12 and 13.13) for oilfield polymer solutions over the shear-rate range of approximately 1 to 100 s^{-1} is the power-law model that takes the form of

$$\eta = K(\dot{\gamma})^{n-1}, \qquad (13.11)$$

where K and n are, respectively, the power-law coefficient and exponent, and $\dot{\gamma}$ is shear rate. For polymer-flood fluids that are shear thinning, the value of the power-law coefficient, n, ranges between 0 and 1 and equals 1 for these fluids when they are Newtonian. The viscosity behavior of a polymer solution becomes more shear thinning as the value of the power-law exponent, n, decreases. The numerical value of the viscosity and the power-law constant, K, become equal when the value of the shear rate, $\dot{\gamma}$, equals 1.

As it relates to polymer solutions of polymer flooding, the power-law viscosity model is only applicable over a limited range of shear rates. For a description of other analytical mathematical expressions for describing polymer-solution viscosity vs. shear rate, especially expressions covering a wider range of shear rates, and for discussions on the viscoelastic properties and the extensional and elongational flow properties of high-MW polymer solutions, see Chap.

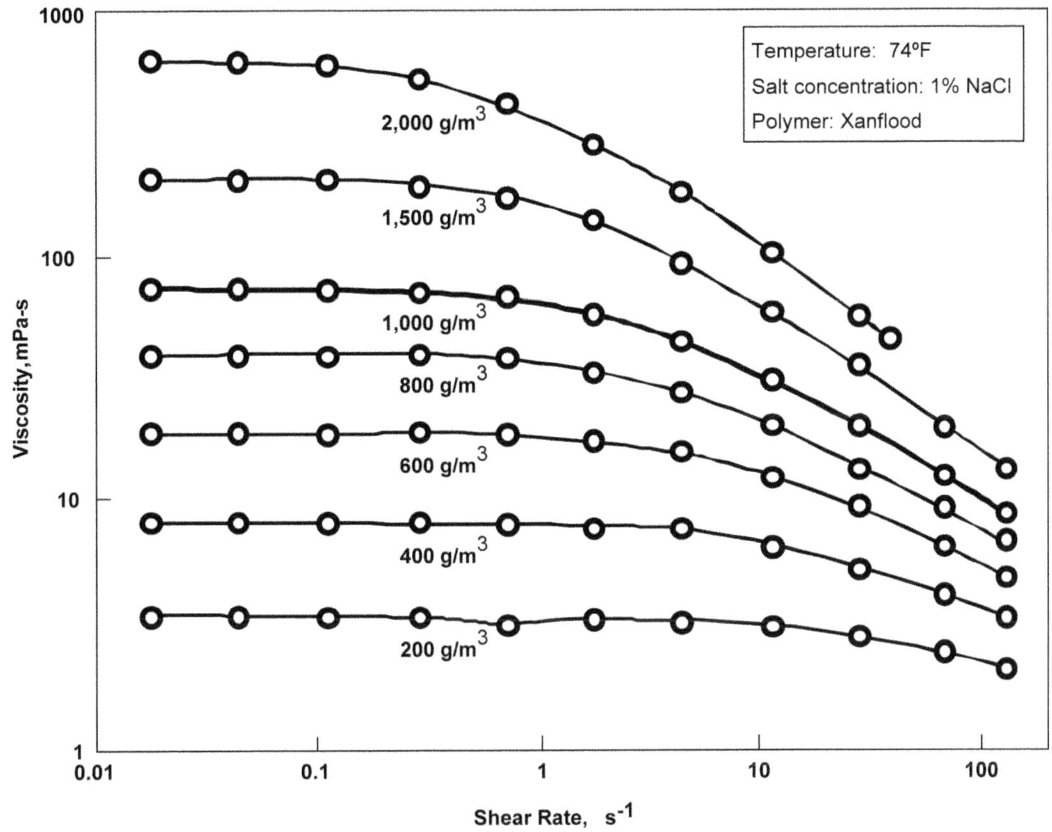

Fig. 13.12—Polymer solution viscosity vs. shear rate as a function of xanthan polymer concentration.

3 in Ref. 46. Extensional viscosity, which occurs under very high shear-rate conditions, can lead to pronounced increases in the apparent viscosity of polymer solutions and often leads to mechanical shear degradation of high-MW water-soluble polymers. The only location in a reservoir that a polymer solution is likely to experience extensional-viscosity conditions of any consequence is near wellbore to an injection or production well.

"Apparent viscosity" or "effective viscosity" refers to the viscosity of a polymer solution for which the viscosity is determined during flow of the polymer solution through porous media. This is discussed further later in this section.

Measuring Viscosity. When measuring the viscosity of polymer solutions to be used in polymer waterflooding, the use of standard laboratory steady-shear viscometers is often quite satisfactory. In addition to the use of conventional viscometers, the screen factor (SF) device has been used extensively to measure viscosity properties of polymer solutions that are used in polymer flooding. **Fig 13.14** shows the SF device. The SF "viscometer" consists of a small fluid reservoir in the glass unit that is in fluid communication above several wire-mesh screens, often three to five 100-mesh stainless-steel screens. A fluid sample of fixed volume is placed in the fluid reservoir, and the time is recorded for the fixed volume of fluid to flow through the screens under the influence of gravity. The SF value for a given polymer is the time it takes the fixed volume of polymer solution to flow through the screen viscometer divided by the time it takes the fixed volume of the solvent brine to flow through the screen viscometer. The SF value of a polymer solution is quite sensitive to the nature of the high-MW tail of the polymer's MW distribution. Some practitioners suggest that the SF value better correlates with

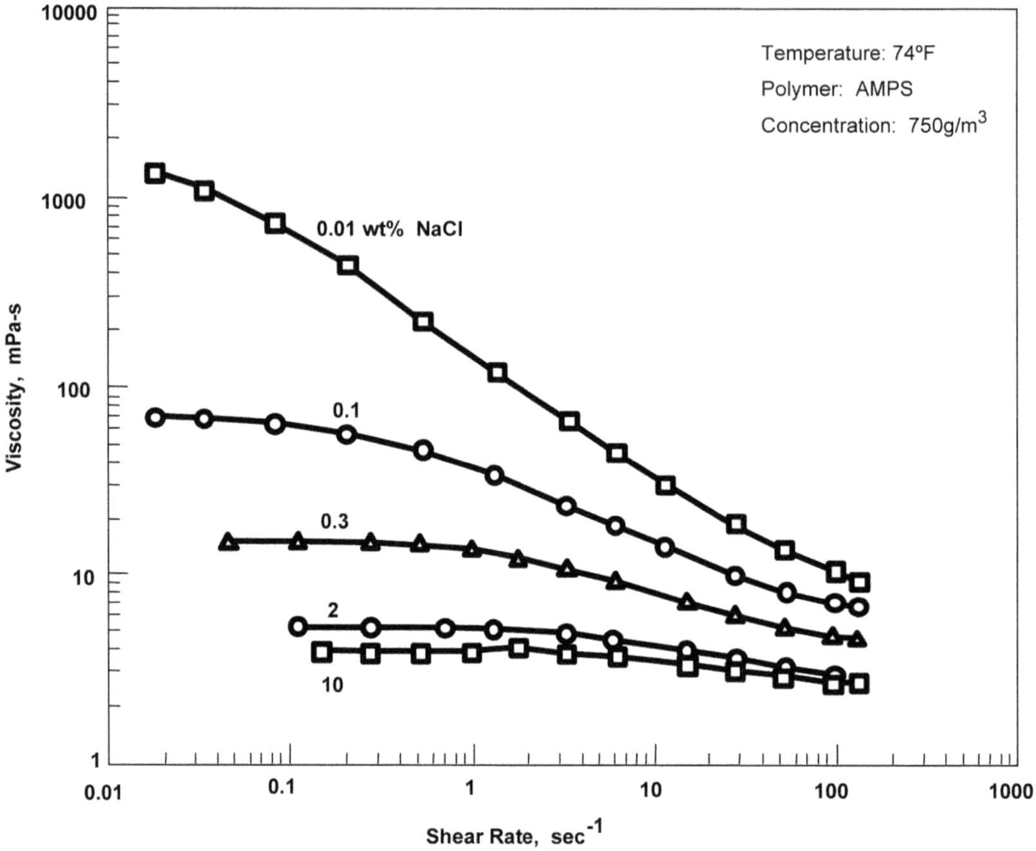

Fig. 13.13—Polymer solution viscosity vs. salinity as a function of AMPS polymer concentration.

mobility and permeability reduction exhibited by the polymer solution as it is propagated through matrix reservoir rock. However, this contention is not universally agreed on. The SF measurement is a simple, straightforward, and useful qualitative viscosity characterization of polymer solutions for use in polymer flooding.[46,51]

Effects of Salt, Hardness, and pH. The effects of salt and hardness on polymer-flood biopolymers are of relatively small consequence at lower temperatures (< 170°F), as compared with the effects on HPAMs polymers that are used in polymer flooding at the same reservoir temperature. Salt insensitivity is one of the attractive features of polymer-flood biopolymers such as xanthan. Likewise, pH within the range likely to be encountered in low-temperature (< 140°F) oil reservoirs is of relatively small consequence to the viscosity and mobility-control properties of xanthan polymer.

The effect of salt and hardness on the viscosity and mobility-control function of polymer-flood HPAM, and similar and related synthetic polymers, is quite significant and can be very deleterious. Cations of dissolved salts reduce the electrostatic repulsion of the negatively charged hydrolyzed carboxylate pendant groups on the polymer backbone of HPAM. Cations do this by screening and collapsing the local negatively charged double layer formed around the carboxylate species. The degree of collapse of the negatively charged electrostatic fields surrounding the polymer's carboxylate groups increases with increasing salt concentrations; and at constant salt concentration, with increasing charge of the cations of the salt. As the electrostatic fields surrounding the polymer's carboxylate groups collapse, the electrostatic repulsive

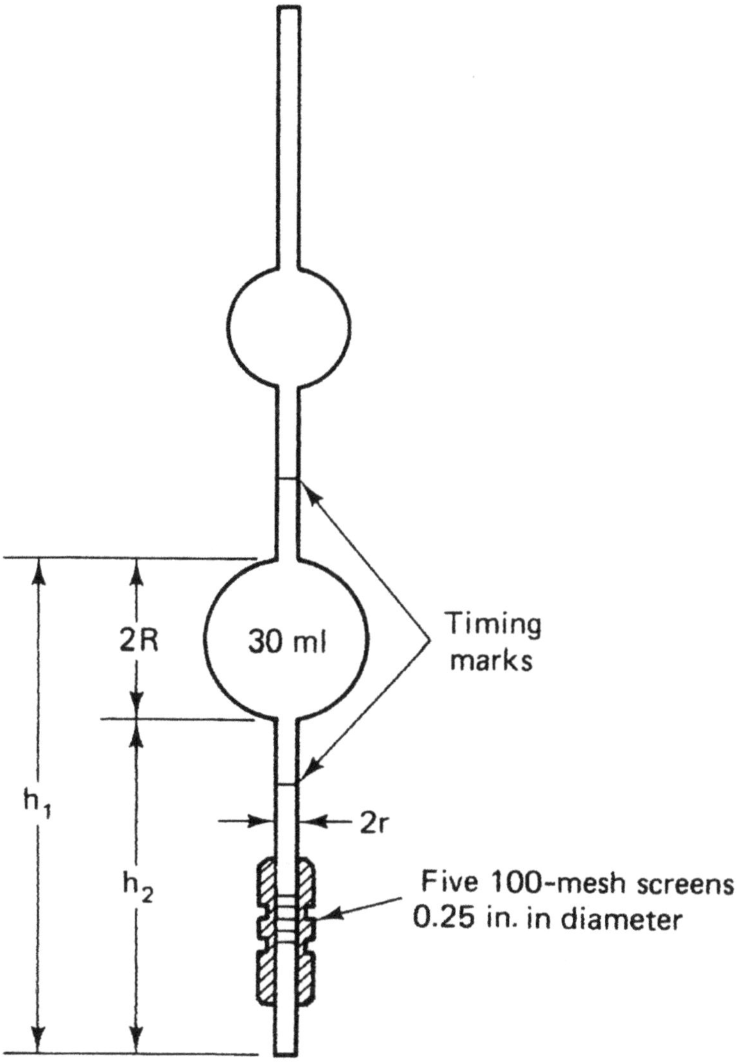

Fig. 13.14—Screen factor device.

forces that promote polymer backbone-chain distension decrease. As Fig. 13.13 shows, this leads to substantial reduction in polymer-solution viscosity. As a rule of thumb, the polymer-solution viscosity decreases by a factor of 10 for every factor of 10 increase in NaCl concentration.[55] The negative impact of divalent hardness ions, such as Ca^{++} and Mg^{++}, are much more deleterious at the same concentration than monovalent ions, such as Na^{+} and K^{+}.

As the concentration hardness cations, such as Ca^{++}, in the brine of a HPAM solution increases, the polymer becomes relatively more sensitive to mechanical shear degradation. The effect of pH on the viscosity of ionic HPAM can be significant. Decreasing the solution pH tends to convert the ionic salt form of the polymer's carboxylate groups to the relatively non-ionic carboxylic acid form of carboxylate groups. This diminishes the electrostatic repulsion of the ionic carboxylate groups along the polymer's backbone and leads to less distention of the polymer molecule and to less viscosity-enhancing power for the polymer in a low pH solution. For a studied hydrolyzed-polyacrylamide polymer solution, its viscosity decreased by a factor of approximately four when the pH of the polymer solution was decreased from 9.8 to 4.[46]

Flow in Porous Media. Polymer solutions used in waterfloods must be able to be transported successfully and effectively through the reservoir. Thus, the manner in which polymer solution flows through porous rock and the associated polymer interaction with the pore walls of matrix reservoir rock are important aspects regarding the attainment of the technical and economic success of a polymer flood.

Polymer Transport. Polymer retention during flow through reservoir matrix rock is discussed in an upcoming subsection. Polymer retention by adsorption and entrapment retards the rate of polymer propagation.

Inaccessible and Excluded Pore Volume. Accelerating the rate of polymer propagation, as compared with the rate of an inert chemical tracer dissolved in the injected polymer solution, is the inaccessible pore volume (IPV) phenomenon. Two explanations for, and contributions to, the IPV phenomenon have been reported. The first IPV explanation is that the large size of the polymer molecules prevents entry into smaller and dead-end pores. This promotes propagation of the polymer molecules faster than an inert chemical tracer because the polymer flows only through the larger-pore flow paths.

The second IPV explanation is the wall-exclusion effect. It is hypothesized that polymer molecules flow and concentrate in the center of the pore-level flow channels of matrix reservoir rock because the polymer molecule flow and the free tumbling of the polymer molecules are excluded from the near-surface volume of the pore walls. Such flow behavior accelerates the rate of polymer propagation through the porous media relative to the rate of propagation of an inert chemical tracer.[46,55,62]

Mobility Reduction. When polymer in solution flows through reservoir matrix rock, it imposes a mobility reduction that is the primary conformance-improvement benefit of polymer waterflooding. The mobility reduction can be imparted by one of two distinctly different mechanisms. First, the polymer can cause an increase in the viscosity of the brine being flooded through the porous media. This is normally the desired effect when flooding with polymer solutions for mobility control. The second mechanism reduces the permeability of the reservoir matrix rock. One measure of mobility reduction imparted by polymer-solution flow is the resistance factor, R_f, which is defined as

$$R_f = \frac{\lambda_w}{\lambda_p}, \qquad (13.12)$$

where λ_w is the mobility of the solvent of the polymer solution, and λ_p is the mobility of the polymer solution. When the polymer solution imparts no permeability reduction and for measurements made at ambient temperature,

$$R_f = \mu_{\text{eff}}, \qquad (13.13)$$

where μ_{eff} is the effective viscosity of the polymer solution as it flows through the reservoir matrix rock. Alternatively, for a single-phase polymer solution flowing through matrix reservoir rock at a given temperature and there is no imparted permeability reduction,

$$R_f = \frac{\mu_{\text{eff}}}{\mu_w}, \qquad (13.14)$$

where μ_w is the viscosity of the brine in which the polymer is dissolved.

Permeability Reduction. Polymer flow through reservoir matrix rock can cause permeability reduction. A measure of the polymer-induced permeability reduction is the residual resistance factor, R_{rf}:

$$R_{rf} = \frac{k_b}{k_a}, \quad \text{(13.15)}$$

where k_b is brine permeability measured before polymer flooding, and k_a is brine permeability measured after polymer flooding.

Permeability reduction induced by polymers tends to be greater in lower permeability reservoir rock. This is, in general, counterproductive. Polymers, especially HPAMs, that undergo even a small amount of mechanical shear degradation often lose much of their permeability reduction propensity because the relatively small number of exceptionally large molecules of a given polymer MW distribution (especially for many HPAMs) are the first polymer molecules to be shear degraded. These large molecules contribute disproportionately to permeability reduction.

Extensional Flow. As mentioned in the discussion on rheology, when flexible, coiled, high-MW polymers, such as HPAM, are forced to flow through matrix reservoir rock at exceptionally high rates and experience exceptionally high-flow shear fields, the polymer can enter extensional and elongational flow at which point the polymer solution's apparent viscosity can rise rapidly. In this flow regime, the polymer is also often mechanically shear degraded. Solutions of well-designed polymer floods are likely to experience extensional flow of noticeable consequence only in certain instances in the very near-wellbore region adjacent to the injection or production well.

Polymer Retention. Polymer retention will often profoundly affect the technical and economic success of a polymer-flooding project. The amount of oil that will be recovered per pound of polymer injected is inversely related to polymer retention.

Retention for a given polymer during a polymer flood increases as the permeability decreases, increases as the polymer molecular weight increases, increases as the clay content in the reservoir rock increases, usually decreases as oil wetness increases, tends to increase in sand and sandstone reservoirs with decreasing anionic charge and increasing cationic charge of the polymer's pendant groups, and has been reported to increase at times in the presence of crude oil.

Polymer retention should be determined carefully, or at least estimated carefully, before initiating a polymer waterflood. Polymer retention for a given polymer flood is normally best estimated by conducting flooding experiments in reservoir rock with reservoir fluids at reservoir temperature.

"Field-measured values of retention range from 7 to 150 μg of polymer per cm^3 of bulk volume, with a desirable retention level being less than approximately 20 μg/cm^3."[46] Laboratory measurements of polymer retention in reservoir rock are usually reported as mass of polymer adsorbed per unit mass of rock, Γ, and is usually reported as μg/g of polymer adsorbed onto reservoir rock. Frequently, it is preferred to have polymer retention reported in terms of mass of polymer adsorbed per unit volume of reservoir rock, Γ_v, or, more specifically, in terms of pounds of polymer adsorbed per acre-foot of reservoir, lbm/acre-ft. To convert from Γ to Γ_v,

$$\Gamma_v = 2.7194\Gamma(1-\phi)\rho_{RG}, \quad \text{(13.16)}$$

where ϕ is porosity, ρ_{RG} is the density of the reservoir rock grains (no pore space included), Γ_v is in units of lbm/acre-ft, and Γ is in units of µg/g.[46] Polymer retention, as measured during field projects, has been reported to range from 20 to 400 lbm polymer/acre-ft bulk volume, with desirable retention reported to be less than 50 lbm/acre-ft.[63]

Adsorption. Polymer adsorption results primarily from physical adsorption and not chemisorption.[46] Polymer adsorption is often the major cause of polymer retention.[46]

Mechanical Entrapment. Mechanical entrapment of polymer during propagation through reservoir porous media results from the larger polymer molecules becoming lodged in narrow flow channels (e.g., pore throats). Gogarty found that the HPAM polymers, under the conditions of his flooding experiment studies, had an effective size between 0.4 and 2 µm.[64] There are several significant consequences of mechanical entrapment:[46] permeability reduction, loss of the entrapped polymer's favorable viscosity enhancing functionality beyond the point of entrapment, loss of the largest of the polymer molecules first has a disproportionately large negative impact during the remainder of the polymer flood on viscosity and mobility-control properties, and loss of a disproportionately large portion of its viscosity and mobility control functionality relatively soon after the polymer solution is injected into a reservoir.

Hydrodynamic Retention. Hydrodynamic retention is the least understood and least well defined retention mechanism.[46] Polymer retention can increase as the flow rate of the polymer solution through reservoir matrix rock increases.[46] Hydrodynamic retention is thought to normally be a relatively small contributor to the total polymer retention during a polymer flood.[46] This retention mechanism is more significant in lower permeability reservoir rock. Hydrodynamic retention is thought to result from polymer molecules becoming temporarily trapped in stagnant flow regimes by hydrodynamic drag forces.[46]

Precipitation. Polymer precipitation from solution, especially in the presence of high reservoir brine salinity, is another source of polymer retention. Precipitation is especially problematic when flooding with HPAM in high-temperature reservoirs with formation waters containing hardness divalent cations.

Polymer Degradation. A decrease in the average molecular weight of the polymer can be caused by chemical, biological, mechanical, or thermal degradation.[46,55] Polymer stability, the inverse of degradation, should be evaluated and quantified under reservoir conditions in terms of a time span relevant to the lifetime of the polymer flood in question.

Chemical. Chemical free-radical species will degrade both biopolymers and synthetic polymers of polymer flooding. Free radicals cause chemical backbone scission of the polymer. Examples of free-radical sources that can be problematic for flooding and conformance-treatment polymers are free oxygen (O_2), hydrogen peroxide, sodium hypochlorite of bleach, and gel breakers such as ammonium peroxide. The combination of free oxygen and ferric ions is particularly problematic in causing oxygen free-radical degradation of polymer-flood polymers, especially of acrylamide polymers. Another source of polymer-degrading free radicals is free-radical or free-radical-precursor impurities within the polymer that are induced, in this case, during the manufacturing process. Polymer-degradation problems, caused by low levels of free radicals, are most problematic when conducting high-temperature (> 150°F) polymer-flooding experiments, especially during the high-temperature laboratory evaluation of polymers for high-temperature flooding. A countervailing phenomenon relating to free-radical chemical degradation is that oil reservoirs tend to quite rapidly neutralize and consume chemical free-radical species.

Two procedures are recommended for removing free oxygen from polymer-solution samples to be used during laboratory evaluation of polymer solutions for high-temperature applications. The first procedure uses a glass ampoule, which is glass-blown sealed after oxygen removal, and high-quality vacuum to reduce oxygen content to < 10 ppb. The second procedure consists of bubbling high-purity argon gas through the polymer solution. The need to deoxygenate polymer-solution samples in the laboratory during high-temperature testing is an aboveground laboratory artifact because polymer solutions that exist in most reservoirs are in an anaerobic and chemically reducing environment.

Hydrolysis reactions are important degradation reactions for both biopolymers and synthetic polymers; however, the hydrolysis reaction degrades the polymers in a much different manner for each of these two polymer types. Both acid- and base-catalyzed hydrolysis of the carbon-oxygen-carbon bonds of the backbone monomer chemical linkages of polysaccharide biopolymers cause polymer backbone scission and serious polymer MW degradation. The pH sensitivity of a biopolymer being used in a polymer flood needs to be considered. Serious polymer hydrolysis questions are raised if an acid-stimulation treatment contacts a previously placed conformance-improvement biopolymer gel treatment.

Hydrolysis (autohydrolysis) reactions of the amide pendant groups of the acrylamide polymers are of significant concern when such polymer is being flooded through a high-temperature reservoir that contains a significant concentration of hardness divalent ions in the formation water. If an acrylamide polymer, which is dissolved in a hardness-containing brine, autohydrolyzes to excessively high levels at high temperatures, the acrylamide polymer will undergo a phase change to an undissolved solid state that causes the polymer to precipitate. When this happens, the polymer loses its viscosity-enhancing properties. Technically, this type of acrylamide-polymer autohydrolysis is not polymer degradation, but simply leads to a phase change of the polymer from being dissolved in solution to being an undissolved solid specie.

Biological. Biological degradation is a serious potential problem for biopolymers, especially for use in shallow reservoirs and for the biopolymer as it resides in surface tanks and tubulars. For a properly designed acrylamide-polymer flood that uses solid polymer as the polymer source, potential biological degradation is essentially not an issue.

Mechanical. The form of mechanical degradation that is of most concern for polymers of polymer waterflooding is shear degradation. All dissolved polymers mechanically degrade if subjected to a sufficiently high-flow shear rate. During polymer flooding, deleteriously high-flow shear rates can exist in surface-injection equipment (valves, orifices, pumps, and tubing), at downhole constrictions (tubing orifices, perforations, or screens), and at the formation face of the injection well.

Xanthan biopolymer is usually not mechanically shear degraded under polymer-flood injection conditions. Under most radial-flow injection conditions, high-MW acrylamide polymers are quite susceptible to mechanical shear degradation. This is especially true if the flooding brine is high in hardness and salinity. When a water-soluble polymer encounters a sufficiently high-velocity flow field, both shear and elongational stress destroy the polymer solution's viscosity.[55] Maeker[65] and Seright[66] correlated permanent viscosity loss of a polymer solution to the product of the elongational stretch rate multiplied by the stretch length. The higher the MW of a given polymer, the more sensitive it is to mechanical shear degradation.

Thermal. All waterflooding polymers have an upper temperature limit above which they are no longer chemically stable, both with and without the addition of an appropriate thermal stabilizer package. This upper temperature limit varies with water chemistries of the polymer-dissolution and reservoir brines, polymer chemistry, manufacturer, and polymer lots from a given manufacturer. For the most part, the upper limit of thermal stability is fixed for a waterflood polymer obtained from a given manufacturer. It must be determined if the polymer to be used is thermally stable under reservoir conditions at the reservoir temperature of the polymer flood and that it will be sufficiently stable for the life of the polymer flood.

Stabilizers. Although once popular, the addition of chemical, biological, and thermal stabilizers to polymer-waterflooding solutions has lost a lot of its original attractiveness because of toxicity, environmental concern, effectiveness, and cost issues. A stabilizer that historically has been used widely as both a biological and thermal stabilizer for polymer-flood polymers is formaldehyde. Formaldehyde is now considered highly toxic and is highly regulated. Also, a number of the stabilizers used to protect against free-radical degradation can become, in themselves, polymer-degrading free radicals at high temperatures. In addition, early practitioners of chemical stabilizers did not fully appreciate chemical loss and chromatographic separations issues. If a chemical stabilizer is to be considered, it is prudent to proceed cautiously.

Polymer Injectivity. Polymer-solution injectivity is an important consideration for several reasons. First, the rate at which the polymer solution can be injected directly impacts the economics of a polymer-flood project. Second, routine injection-well cleanup jobs may be required if polymer or polymer-microgel damages injectivity. These cleanup jobs can detract from the polymer flood's economics and effectiveness. Injectivity decreases as polymer MW increases. Polymer-solution injectivity is more favorable when the polymer solution exhibits shear-thinning viscosity behavior.

13.5.4 When and Where Applicable. The following screening guidelines can be used to determine where polymer waterflooding is most applicable, in terms of reservoir properties.[46,67]

- Oil viscosity < 150 cp (preferably < 100 and > 10 cp) and API gravity > 15.
- Matrix-rock permeability > 10 md, with no maximum.
- Reservoir temperature: low temperatures are best (best at < 176°F; maximum of approximately 210°F).
- Water injectivity should be good with some spare capacity (hydraulic fracturing of injection wells may help).
- Reservoir clay content should be low.
- Low salinity of the injection and reservoir brines are preferable.

Polymer waterflooding has been conducted successfully in sandstone and carbonate matrix-rock reservoirs, fractured reservoirs, and in water-wet, mixed-wetting, and oil-wet reservoirs. For example, see Ref. 68.

13.5.5 Field Implementation. The following subsections focus on the field implementation of a polymer waterflood.

Flood Design. Working up a flood design is one of the first steps when implementing a polymer-waterflooding project.

Selecting a Polymer. When selecting a polymer for a polymer-waterflooding project, one should try to maximize, as best as possible, all the following polymer attributes. The polymer should[55]

- maximize the amount of viscosity enhancement and/or mobility reduction per unit cost.
- readily dissolve.
- propagate well and have low retention as transported through the reservoir.
- exhibit good shear stability.
- possess good chemical stability.
- have good biological stability.
- be thermally stable at reservoir temperature.
- possess acceptable injectivity properties.

Polymer Concentration. The optimum concentration of the polymer to be injected is a critical parameter in the design of an effective polymer-waterflooding project. The concentration of the injected polymer profoundly affects the cost, economics, and performance of a polymer-flooding project. The optimum concentration is a function of reservoir properties, the nature of the reservoir's conformance problems, and the business objective of the polymer flood. Business objectives of a polymer flood can include maximizing oil recovery, maximizing the rate of return on the cost of the polymer-flooding project, and minimizing the cost of the polymer flood.

Ref. 69 provides a literature review of 12 international polymer-flooding projects conducted between 1975 and 1992. The projects included both pilot and fieldwide projects. Ten of the polymer-flooding projects involved the use of acrylamide polymers, and two projects involved the use of xanthan polymers. All the flooding projects were conducted in reservoirs with temperatures less than 140°F. For the 12 international polymer-flooding projects, the median incremental oil recovery, as calculated from data in the paper, was 13% original oil in place (OOIP), and the range of incremental oil recoveries was reported to be 6 to 52% OOIP. The average pore volume (PV) of the polymer slug injected was calculated to be 51%, with the range of PVs being 21 to 100%.

These flooding parameter values are noteworthy when viewed in terms of comparable values reported in the paper, or calculated from data presented in the paper, for 128 polymer floods performed in the U.S. between 1980 and 1993. The median incremental oil production for the U.S. polymer floods was 4.9% OOIP. The average concentration of polymer injected in the U.S. projects was 460 ppm vs. 920 ppm for the international polymer floods. Ref. 69 states that it appears the amount of incremental oil production for the 12 international projects best correlates with the numerical value that is obtained from multiplying the PV of the polymer slug injected by the average concentration of the polymer injected during the polymer-flooding project.

On the basis of the 12 international polymer-flooding projects, 900-ppm polymer concentration in the polymer slug would be a good starting value in designing a polymer-waterflooding project. Working from this initial concentration, it should be determined, using appropriate engineering and evaluation tools, whether the optimum polymer concentration for the proposed polymer flood is actually higher or lower.

Sizing Volume Injected. At this writing, optimum sizing of the polymer-solution slug to be injected was one of the most controversial aspects of designing a polymer-waterflooding project. This design parameter profoundly affects the cost, economics, and performance of a polymer-flooding project. Underdesigning the size of the polymer slug injected has been thought to be a major cause for the disappointing performance of many polymer floods conducted in the U.S. Using the general arguments made in previous subsections, the optimum size of the polymer-solution slug during polymer flooding is a function of reservoir properties, nature of the reservoirs conformance problems, and the business objective of the polymer flood. Business objectives of a polymer flood can include maximizing oil recovery, maximizing the rate of return on the cost of the polymer-flooding project, or minimizing the cost of the polymer flood.

On the basis of 12 international polymer-flooding projects, a 50% PV slug of polymer solution would be a good starting value to use in designing a polymer-waterflooding project.[69] Working from this initial polymer-solution slug size, it should be determined, using appropriate engineering and evaluation tools, whether the optimum polymer-solution slug size for the proposed polymer flood is actually higher or lower. When attempting to design the optimum size of the polymer-solution slug, two important parameters that need to be accounted for are polymer retention and the rate and nature of the viscous fingering of the polymer-slug chase drive water into the polymer-solution slug.

Grading of Polymer Concentration. To overcome, or substantially reduce, the problem of viscous fingering of the polymer-slug chase drive water into the polymer slug, most polymer floods are designed with a tapered, decreasing-polymer-concentration chase slug beginning at or near the end of the design volume of the primary polymer-solution slug.

Timing. There is general agreement that, when all other factors are held constant, the earlier in the life of a waterflood that a polymer flood is initiate, the relatively more effective the polymer flood will be. Two factors offset the benefits of an early start. First, it is more difficult to definitively assess the oil-recovery potential and economic effectiveness of a polymer flood until definitive waterflood performance has been established. Second, reservoir description and associated reservoir conformance problems are often less well defined, especially in a new reservoir.

Suggested Steps for Designing a Polymer Flood. Lake[55] and Sorbie[46] suggest that the design and planning procedure for a polymer waterflood should entail the following elements.

- Screen the candidate reservoirs for both the technical and economic feasibility of performing a successful polymer waterflooding project.
- If appropriate and needed, improve the reservoir description.
- Select the polymer that should be used in the flooding project.
- When and where cost-effective, conduct laboratory studies under reservoir conditions to perform screening and compatibility tests on the polymer and polymer-solution core-flooding tests to determine polymer-solution flow properties and to estimate incremental oil recovery (on the scale of the core size used).
- Estimate the amount of polymer that will be required for the polymer flood.
- Design the polymer-injection facilities.
- When and where cost-effective, conduct a polymer-injectivity test and a field polymer-flood pilot test.
- If feasible and cost-effective, conduct reservoir simulation studies.
- Optimize the reservoir, operational, and economic performance of the polymer waterflooding project, such as optimizing well and pattern spacing, completion strategies, and injection rates.

One of the primary causes of failure for polymer-flood projects is that the reservoir description used was inaccurate.[46]

Polymer Injection. The polymer-injection facilities and the actual injection of the polymer solution are important aspects of a successful polymer-waterflooding operation.

Polymer Degradation. If the polymer is allowed to become mechanically shear degraded during surface mixing and pumping operations or during injection into the reservoir, the polymer will have lost a substantial amount, if not most, of its viscosity-enhancing and mobility-reducing power before leaving the injection-well near-wellbore region. The higher the rate that a polymer solution is injected across a unit area of injection surface, the greater the propensity for mechanical shear degradation of the polymer. It is most often the goal to inject the polymer solution as rapidly as possible without exceeding reservoir parting pressure. The strategy usually implemented is to use injection equipment and well completions that permit desired injection rates without substantial mechanical degradation.

To minimize mechanical shear degradation, the operator needs to use specialized pumping equipment, be sure that the polymer solution is not passed through any valves or orifices that cause high and damaging shear fields, and use special mixing and dilution equipment, such as the use of static mixers, to not shear degrade the polymer during mixing and dilution operations.

Well Completion. Injection wells of polymer floods are often openhole or gravel-packed completions. Hydraulically fracturing the injection well with short, wide fractures has been reported to have been used successfully to aid in injecting HPAM without excessive shear degradation.[68] The injection of polymer solutions through mandril completions, with associated flow of the polymer solution through mandril orifices, has proved to be detrimental in several instances.

Polymer Dissolution. Historically, the polymer, as supplied to (or near) the final wellsite, must be dissolved and/or diluted to some degree. The polymer dissolution or dilution process needs to be implemented so that the polymer is fully dissolved, at the proper concentration before injection into the reservoir, and dissolved or diluted in a manner that does not mechanically shear degrade the polymer. The dissolution or dilution process can range from a fairly simple process for broths of xanthan polymer to a quite difficult and technically challenging process for ultra-high-MW solid HPAM. A long dissolution time is an economic determent for a polymer flood, because it requires extra equipment and extra polymer-solution holding tanks.

Polymer Filtration. The polymer solution normally should be filtered before injection to ensure that it is readily injectable and does not unduly damage injectivity. This is a particularly critical and challenging issue when injecting biopolymer solutions, such as xanthan solutions, that are notorious for containing substantial amounts of cell debris from the polymer's fermentation process and cell debris that are difficult to remove fully with filters.

HPAM polymer, when obtained in the solid form and then dissolved in the field, should normally be filtered to remove microgels or undissolved "fisheyes." The amount of microgel and fisheye material in a given hydrolyzed acrylamide polymer varies from manufacturer to manufacturer and even with manufacturing lots from the same polymer manufacturer. These microgels and fisheyes of partially hydrated solid HPAM in solution often largely result from the final drying process in the manufacturing of the polymer, in which overheating, overdrying, and other factors cause some of the polymer molecules to crosslink together chemically during the drying process.

Parting Pressure. Downhole injection pressure should normally be kept below formation fracturing and parting pressures. If injection-well fracturing is required to obtain adequate polymer solution injectivity or to eliminate mechanical shear degradation, the injection wells are normally appropriately hydraulically fractured during a separate hydraulic fracturing operation.

Pilot Testing. It is recommended that a pilot test of a polymer-waterflooding design be implemented in one or several injection wells before a polymer-waterflooding project is implemented field wide and/or before implementing an expensive polymer-flood project. The primary objective of a pilot test is to assure that there will be adequate injectivity and that there will be no substantial injectivity issues. The polymer-solution injectivity trial can be as short as several days. Proper interpretation of a single-well polymer injectivity test can be difficult. If time and economics permit, a secondary objective of the pilot test is to demonstrate that the polymer flood will perform in the reservoir as expected in terms of mobilizing and recovering incremental oil. If it proves to be cost-effective, observation wells can be drilled near the injection well to observe, in a relatively short time frame, how the polymer flood can be expected to perform in the reservoir.

Issues Regarding Manufacturing of Polymers. Polymers used in polymer waterflooding can be manufactured and provided to the end user in one of four forms. First, polymers in solid particle form, the oldest form, are readily transported and stored. Polymers supplied in solid particles are a challenge to dissolve properly and fully. In addition, the polymer of solid particle polymers can be damaged in the drying process during polymer manufacturing, may contain undesirable chemicals that coat the polymer particles, and often contain undissolvable microgels that are not injectable into matrix reservoir rock and that damage the injectivity of the polymer-solution injection wells. The second form, concentrated (~10%) broths of aqueous polymer (especially biopolymers), is more easily dissolved in the field, but is more costly per pound of polymer to transport to the field. The third form is either aqueous emulsions that contain up to 35% or more active polymer or hydrocarbon-fluid suspensions/dispersions that contain ~50% active polymer. The challenge in using polymers supplied in this form is to routinely and consistently fully invert the emulsion/suspension in the field to permit the polymer to be dissolved fully in the flood water and to be fully effective during the polymer waterflood. Fourth, where economic and project scale permit, field-manufactured polymer, especially field-manufactured partially hydrolyzed polyacrylamide, can be an attractive option for providing polymer of high quality and exceptional performance characteristics.[59,68]

Quality Control. Quality control is an essential element for the successful implementation of a polymer-waterflooding project. A quality control program should include, but not be limited to, the following elements:

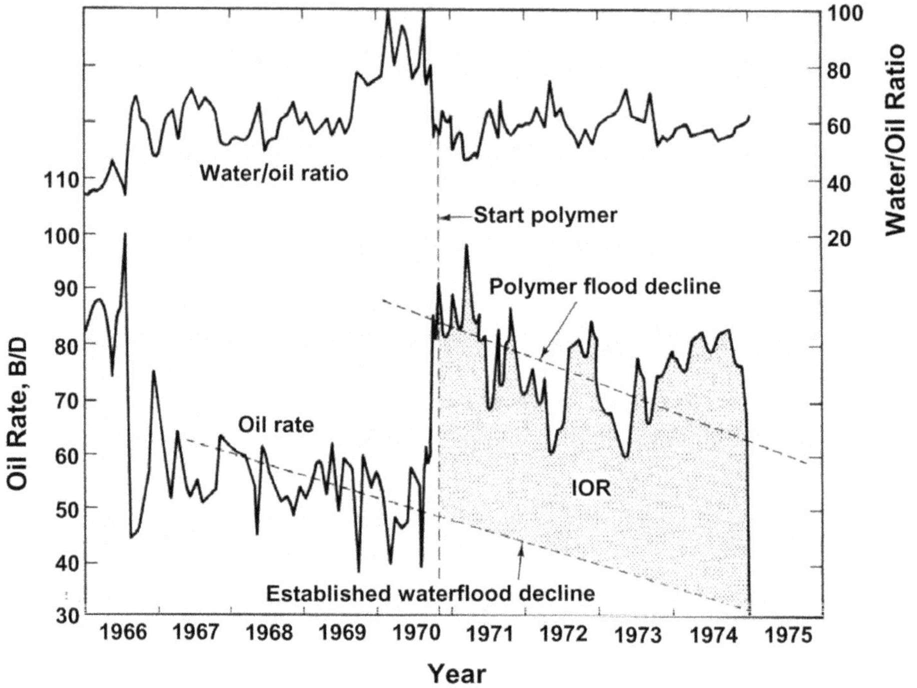

Fig. 13.15—Tertiary polymer flood response at the North Burbank unit, Osage County, Oklahoma.[55]

- Routine verification of the polymer concentration in the polymer as supplied and in the polymer solution injected.
- Routine determination of the viscosity and SFs of the injected polymer solution.
- Check of the filterability of the polymer solution to be injected.
- Check of the dissolution-rate properties of the polymer as supplied.
- Check of the polymer to be injected for complete dissolution.
- Periodical check of the thermal and chemical stability of the polymer by measuring the viscosity and SF of wellhead polymer samples that have been aged for appropriate periods of time at reservoir temperature under scrupulous anaerobic conditions in sealed glass ampoules. (See the recommended polymer-solution-sample deoxygenation procedure discussed previously in Sec. 13.5.3.)

13.5.6 Illustrative Field Results and Trends. This section provides a brief review of illustrative field applications of polymer waterflooding, along with a very brief discussion of trends in the field application of polymer waterflooding in the U.S.

Comprehensive Manning Survey. In 1983, Manning *et al.*[70] published a comprehensive and classic summary of the field results and performance of more than 250 polymer waterflooding projects and provided information relating to the early field applications of polymer waterflooding.

North Burbank Unit Flood. **Fig. 13.15** shows the incremental oil production response for the North Burbank polymer flood.[55]

Wyoming Polymer Floods. A polymer-waterflooding project that involved a large full-field flooding project at the North Oregon Basin field in Wyoming's mature Big Horn Basin oil-producing area was reported in 1986 to be producing 2,550 BOPD of incremental oil production. It was reported that this polymer-flooding project would recover ultimately more than 10 million bbl of incremental reserves from the mature North Oregon Basin field. The field project involved the flooding of both a fractured carbonate formation and a fractured sandstone formation with a polymer flood using partially hydrolyzed polyacrylamide. The polymer used in this flooding project was field manufactured at the North Oregon Basin field in a plant with a capacity of 23 million pounds of polymer per year.[68] This is an example of a successful polymer-flooding project using HPAM and conducted in naturally fractured sandstone and carbonate reservoirs where both the polymer dissolution brine and the reservoir brine were relatively saline and relatively hard (relatively high concentrations of divalent cations and anions).

Chinese Polymer Floods. The Chinese have reported on a number of polymer-flooding projects. During a pilot test of polymer flooding with HPAM in the 167°F Shuanghe reservoir of the Henan oil field in China, the incremental oil production was expected to approach 9.8% OOIP, and the polymer flood was expected to recover 0.7 bbl of oil per pound of polymer injected.[71]

It has been reported that polymer waterflooding, using HPAM, in the Daqing field in China has recovered cumulatively more than 300 million bbl of oil. This polymer waterflooding project was reported to be producing 70 million bbl/yr of oil in 2001. The cost of the oil from the polymer waterflooding project was stated to be U.S. $6.60/bbl. The field was reported to be producing at a rate 310% greater than that expected for waterflooding alone. Incremental oil production attributed to the Daqing polymer waterflood is projected to be in the range of 12% to 15% of the OOIP.

French Polymer Flood. In 1995, an update was reported on the French Courtenay polymer flood that was conducted in the secondary recovery mode in a shallow sand reservoir with a 40 cp viscosity oil and 86°F reservoir temperature. Waterflooding was not conducted in this reservoir because of the combination of thin, high-permeability sand channels and an unfavorable mobility ratio. Oil recovery for the polymer flood was 6.6% PV. The total cost of oil production from this polymer flood project was U.S. $12/bbl.[72]

Canadian Rapdan Polymer Flood. A 13-producer and 5-injector pilot of polymer waterflooding was conducted in 1986 in the 130°F and 110 md average permeability Upper Shaunavan formation of the Rapdan Unit in Saskatchewan, Canada. The polymer-flood pilot consisted of injecting 17% PV of 1,100 to 1,500 ppm polyacrylamide polymer solution. Polymer flooding was started after waterflooding. As a result of polymer flooding, oil production was reported to have increased from 410 BOPD at 18% oil cut to peak production of 1,100 BOPD at 36% oil cut.[73]

Review of Worldwide Polymer Floods. As Sec. 13.5.5 describes, a literature review was conducted on twelve international polymer floods, both pilots and fieldwide projects, that were conducted between 1975 and 1992.[69] All floods were conducted in reservoirs with a reservoir temperature of less than 140°F. In ten of the floods, partially hydrolyzed polyacrylamide was used as the polymer, and, in two of the floods, xanthan was used as the polymer. The international polymer floods recovered between 6 and 52% of the OOIP, compared with the medium recovery of 4.9% OOIP for 128 U.S. polymer floods. The general conclusion of Ref. 69 was that there tended to be a correlation between the polymer-flood incremental oil production and the total amount of polymer used in the polymer floods, as defined by PV of polymer injected multiplied by the average concentration of polymer injected.

Survey of U.S. Polymer-Flood Project Production. Ref. 74 reported the total U.S. enhanced oil recovery (EOR) oil production from polymer-waterflooding projects in 1998 was only 139 BOPD. This production figure was down from 21,000 BOPD in 1988 for U.S. EOR oil production resulting from polymer-waterflooding projects. In the same survey in 2000,[75] EOR production for polymer and chemical flooding were combined so that the EOR production figure for polymer flooding alone was not discernable. The reported combined U.S. oil production in 2000 resulting from chemical and polymer flooding was 1,600 BOPD.

In 2002,[76] EOR production from polymer flooding was reported to have fallen to zero, and it was reported that the number of polymer-flooding projects in the U.S. had fallen from 178 in 1986 to 4 in 2002. In the 2002 survey, 20 polymer-flooding projects that were being conducted outside the U.S. and Canada were listed. There is, at this writing, a trend in the U.S. toward less EOR polymer-flood oil production.

13.5.7 Additional Reading. Polymer waterflooding is thoroughly documented in Ref. 46 and nicely reviewed in Chap. 8 of Ref. 55. Also, see Chap. 5 of Ref. 2.

13.5.8 Polymers for Imparting DPR and RPM. In Sec. 13.4, a review was presented of the concepts, applicability, limitations, and desirability of the DPR phenomenon as it applies to conformance-improvement water-shutoff (and/or water-reduction) treatments.

As early as 1964, certain polymer-flood water-soluble polymers were known to impart DPR to water flow in reservoir rock that had been previously flooded with the polymer.[26] Although, in concept, water-soluble permeability-reducing polymers can be injected (using appropriate polymers and conditions) into matrix rock to reduce the absolute permeability to all fluids (including water, oil, and gas), the injection of water-soluble permeability-reducing polymers into matrix rock is most often performed to impart DPR.

Advantages and Issues. Bullhead injection of a simple aqueous solution containing a water-soluble polymer to treat conformance problems, such as excessive water production, in matrix rock (unfractured) reservoirs is a highly appealing concept. Most DPR polymers (also known as RPM polymers) are not usually highly exotic or costly. Thus, polymer-alone DPR treatments for reducing excessive water production are much simpler and less risky, in concept, than conducting the same task using a relatively strong total-shutoff gel, particularly a "strong" crosslinked polymer gel. A polymer solution alone poses less risk of totally sealing off the treated reservoir volume compared with injecting a water-shutoff gel. The chemical and operational aspects of injecting a polymer solution for water-shutoff/reduction purposes are substantially less complicated than injecting a comparable polymer gel.

However, polymer DPR water-shutoff and/or water-reduction treatments do have a number of significant limitations, in addition to those already discussed in Sec. 13.4. First, how fast will the treatment polymer be desorbed and flow back to the production well? Second, DPR polymer treatments for conformance improvement are normally successful only when applied to matrix rock oil reservoirs with a relatively low permeability (usually less than 1 Darcy). In addition, many existing DPR polymer conformance-improvement treatments are only applicable to sand and sandstone reservoirs. DPR polymer conformance-improvement treatments are not directly applicable within fractures and other high-permeability anomalies. Third, the amount of DPR, which is imparted by polymer systems available at this writing, is often quite small as compared with the amount of DPR that can be imparted by DPR polymer gels. Ref. 42 provides guidelines for well-candidate and chemical selection for use when considering the application of polymer DPR water-shutoff/reduction treatments. Fourth, the performance of polymer DPR treatments has been erratic in both the laboratory and field setting.

Mechanism for Imparting DPR. Although the mechanism by which polymers impart DPR to water flow in reservoir porous media is currently under active study, the basic mechanism is thought to involve polymer adsorption onto the pore body walls and/or retention at the pore throats.[31–36] In most cases, DPR polymers tend to decrease the relative permeability to water with little effect on the oil or gas relative permeability curve.[35,77,79]

Fig. 13.16, taken from Ref. 35, depicts the DPR and RPM effect on relative permeability curves imparted in a 4.8 Darcy sand pack that was flooded at 140°F with a 10 g/L solution of biopolymer Polysaccharide G in 10 g/L KCl brine. The figure shows, as a result of flooding the sand pack with the biopolymer solution, how the relative permeability curve to water was substantially reduced while the relative permeability curve to oil was relatively unaffected.

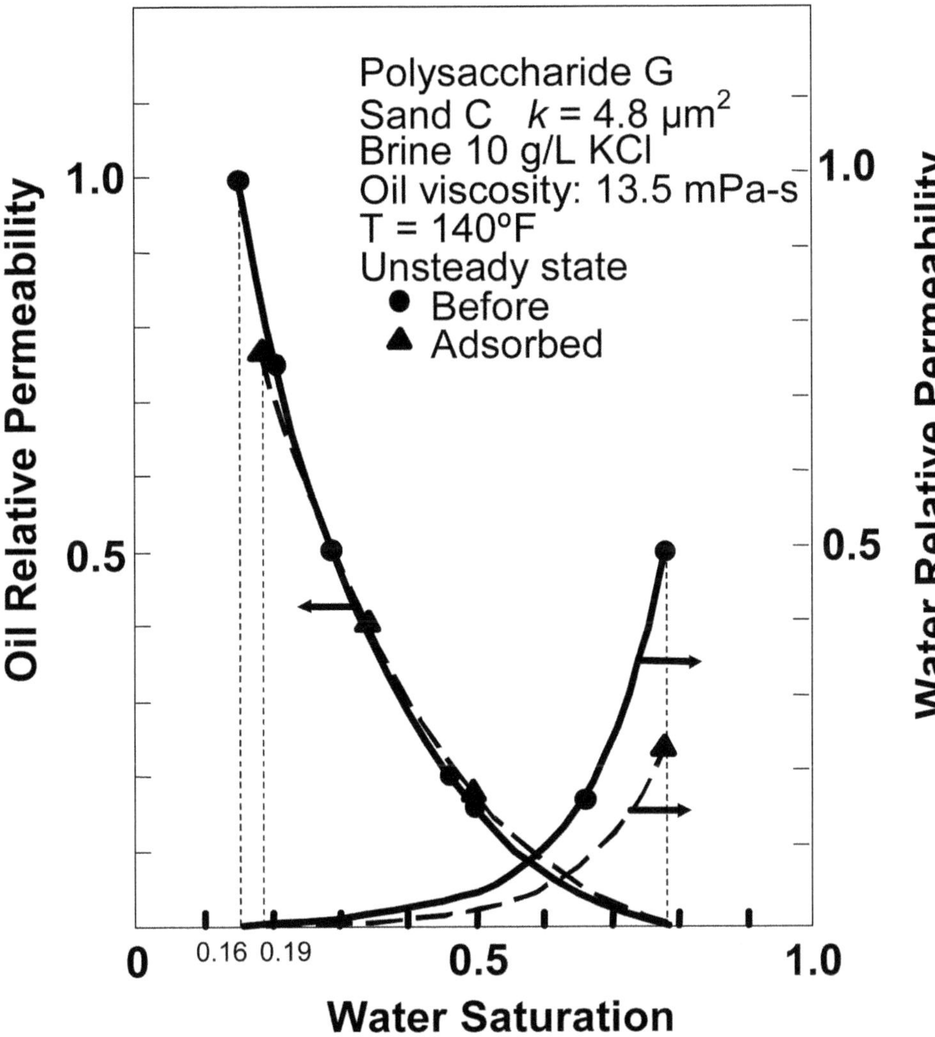

Fig. 13.16—Relative permeability modification effect induced in a sand pack by Polysaccharide G.[33]

Range of Applicability of DPR Polymer Treatments. While DPR polymer systems for imparting conformance improvement have been targeted primarily at sand and sandstone reservoirs, presumably being water wet, favorable DPR polymer effects have been observed when nonionic polyacrylamide was placed in various carbonate rocks with either water-wet and oil-wet conditions.[36] DPR polymer treatments have been applied to reservoirs with temperatures up to 225°F;[22] however, DPR polymer systems for use in conformance-improvement treatments are only applicable over a limited lower permeability range (approximately 5 md to hundreds of md in most cases). Because the upper permeability limit varies with the specific DPR polymer system and with specific reservoir rock lithologies, it is difficult to provide a universal upper permeability limit for the successful application of DPR polymer systems. The upper permeability limit for the successful application of any specific DPR polymer system is a treatment variable that should be scrutinized closely. The application of polymer DPR treatments to reservoir conformance problems involving flow channels with permeabilities exceeding the upper permeability limit of a DPR polymer is a major cause of failures for such polymer conformance treatments.

Illustrative Technologies and Field Applications. The earliest polymers (uncrosslinked) reported to have DPR properties used in water-shutoff/reduction treatments applied to production wells were polyacrylamides,[12,13,26] which is the same general type of water-soluble polymer that has been used extensively in polymer flooding for mobility-control and sweep-improvement purposes. In 1973, SPE literature reported on a proprietary and commercially available "brush" polymer for "selectively reducing water production."[14]

In 1988, Zaitoun and Kohler reported on how the adsorption of, respectively, polyacrylamide and a polysaccharide onto water-wet sand and sandstone promotes DPR to water flow. Also, it was reported how the adsorption of these polymers increase the irreducible water saturation.[35] The same research group reported on the development of two polyacrylamide-based DPR water-shutoff processes and the field application of one of the processes in an underground gas-storage facility.[77,79]

Gruenenfelder *et al.*[80] reported on the application of DPR polymer water-shutoff/reduction treatments to two gravel-packed wells of high-temperature (190 to 200°F) sand reservoirs of the U.S. Gulf Coast. The DPR polymer treatments involved the use of a nonionic triple-stranded polysaccharide biopolymer.

The application of DPR polymer water-shutoff/reduction treatments, involving an amphoteric polymer material, to five wells in a high-permeability and high-temperature (up to 225°F) sandstone reservoir in Indonesia has been reported.[22]

Refs. 34 and 42 discuss the use of cationic polyacrylamide as a candidate polymer for use in polymer DPR treatments to impart conformance improvement.

In 2001, Eoff *et al.* reviewed the structure and process optimization of a commercial "brush" polymer that has been used since 1973 in various forms and under various trade names as a RPM polymer in conjunction with conformance-improvement treatments.[17]

13.6 Gels

Gels are a fluid-based system to which, at least, some solid-like structural properties have been imparted. In other words, gels are a fluid-based system within which the base fluid has acquired at least some 3D solid-like structural properties. These structural properties are often elastic in nature. All of the conformance-improvement gels discussed in this section are aqueous-based materials. The term "gel" as used in this chapter (unless specifically noted otherwise) refers to classical, continuous, bulk, and "relatively strong" gel material and does not refer to discontinuous, dispersed, "relatively weak," microgel particles in an aqueous solution. Gels discussed in this chapter, when formed in a beaker for example, constitute a single and continuous gel mass throughout its entire volume within the beaker. The term "gelant" refers to a gel fluid before any appreciable crosslinking of the gel's chemical building blocks has occurred. The term gel refers to a gel fluid that has attained either partial or full chemical-crosslinking maturation. This chapter discusses polymer gels, as well as inorganic gels and monomer gels. Sec. 13.8 briefly discusses a conformance-improvement plugging material (i.e., a resin) that involves an organic-fluid-based gel.

An older definition of gel is "a jelly-like substance formed by the coagulation of a colloidal solution into a semisolid phase." In modern oilfield and technical literature, the term gel includes the elastic and semisolid material that results from chemically crosslinking together water-soluble polymers in an aqueous solution. Crosslinked-polymer gels can possess rigidity up to, and exceeding that of, Buna rubber. They contain polymer concentrations in the 150 to 100,000 ppm range (but more commonly 2,000 to 50,000 ppm and most commonly 3,000 to 12,000 ppm). Gels are often formulated with relatively inexpensive commodity polymers.

Gels have found broad application as oilfield fluid-flow blocking agents because gels are often an exceptionally cost-effective plugging and/or permeability-reducing agent for use in a number of different conformance-improvement treatments. Conformance-improvement gels are, for the most part, essentially nothing but water (often produced brine) with the remainder of the gel constituents incorporated as low concentrations of relatively inexpensive polymers and chemical crosslinking agents.

The use of conformance-improvement gels is an emerging technology that, at this writing, is still not well understood by petroleum engineers nor widely applied. Conformance-improvement gels are another conformance technology and engineering tool that should be added to the "petroleum engineering toolbox." However, gels are not a panacea for remedying reservoir and flooding conformance problems, but rather simply another tool that can be used to alleviate conformance problems when there is a good match between a conformance problem and the application of a particular gel technology.

13.6.1 Gel Conformance-Improvement Treatments. Gels operate, for the most part, either by diverting fluid flow from high-permeability, low-oil-saturation reservoir flow paths to low-permeability, high-oil-saturation flow paths, and thereby promoting improved flood sweep efficiency and incremental oil production, or by reducing oil-production operating costs by reducing excessive, deleterious, and competing coproduction of water or gas.

Oilfield gel conformance treatments can be applied in a number of forms including sweep improvement treatments, water shutoff treatments, gas shutoff treatments, zone abandonment treatments, squeeze and recompletion treatments, and water and gas coning treatments involving fractures and other linear-flow high-permeability reservoir anomalies. Gels are particularly effective for treating oil-production coning problems when the coning is occurring via linear flow in "vertical" fractures.[5,7]

When there is a good match between a given conformance problem and a particular gel technology, relatively large volumes of incremental oil production and/or substantial reductions in oil-production operating costs, by means of the shutting off of excessive, deleterious, and competing coproduction of water or gas, can be achieved profitably. Gel treatments are usually applied in the course of normal and ongoing primary, secondary, or tertiary oil-recovery operations, just as an operator would apply an acid or scale-inhibition treatment. Gel treatments are an emerging oilfield technology that can help extend the life of maturing oil reservoirs that are approaching their economic limit.

Background and Historical Information. Although conformance-improvement gel treatments have existed for a number of decades, their widespread use has only begun to emerge. Early oilfield gels tended to be stable and function well during testing and evaluation in the laboratory, but failed to be stable and to function downhole as intended because they lacked robust chemistries. Also, because of a lack of modern technology, many reservoir and flooding conformance problems were not understood, correctly depicted, or properly diagnosed. In addition, numerous individuals and organizations tended to make excessive claims about what early oilfield gel technologies could and would do. The success rate of these gel treatments was low and conducting such treatments was considered high risk. As a result, conformance-improvement gel technologies developed a somewhat bad reputation in the industry. Only recently has this reputation begun to improve. The information presented in this chapter can help petroleum engineers evaluate oilfield conformance gels and their field application on the basis of well-founded-scientific, sound-engineering, and field-performance merits.

How Gel Treatments Function. Gels used in conformance-improvement treatments reduce permeability in the reservoir treatment volume in which the gels are ultimately placed. Conformance gels do not function as viscosity-enhancing agents during oil-recovery flooding operations.

Although microgels (discussed in a later subsection of this section), which are colloidal-sized aggregates of suspended and noncontinuous gel particles, have, at times, been claimed to be viscosity-enhancing agents during their "propagation" through reservoir flow paths possessing high permeabilities, there is no substantiation in the petroleum engineering literature of any microgels substantially enhancing the viscosity of an aqueous oil-recovery drive fluid beyond the viscosity of the gel's polymer solution alone. In fact, microgel-gel-containing aqueous solutions tend to have lower viscosities than the polymer solutions from which they were derived. Permeability reduction is the dominant mechanism by which microgels impart conformance improvement. Oilfield microgels are also referred to, at times, as colloidal dispersion gels.

The more widely applied conformance-improvement gels are characterized as being bulk gels that function as permeability-reducing agents. Bulk gels have a continuous chemically crosslinked polymer network structure throughout the entire macroscopic scale of the gel. These gels are blocking and plugging agents to fluid flow in the reservoir volume in which they have been placed. For example, gels used in near-wellbore water-shutoff treatments of producing wells are selectively placed in a high-permeability strata (that are not in fluid or pressure communication with the other reservoir strata) to function purely as a plugging and blocking agents to fluid flow.

If a gel is placed some significant distance into the fractures of a naturally fractured reservoir surrounding an injection well, the gel not only functions as a blocking agent, but also as a diverting agent. These gels divert injected oil-recovery drive fluid from predominantly flowing through the high-permeability, low-oil-content fractures to predominantly flowing through the relatively low-permeability, high-oil-saturation matrix reservoir rock adjacent to the gel-filled and gel-treated fractures.

What Benefits Can Be Expected. For good gel-treatment well and well-pattern candidates, the following is a partial list of benefits that can be achieved.

- Generate incremental oil production through conformance improvement, and possibly generate large volumes of incremental oil production per unit cost of chemical expended.
- Substantially reduce oil-production operating expenses per unit cost of chemical expended.
- Reduce competing water production that can be unproductive, costly, and environmentally unfriendly.
- Reduce competing gas production that can be unproductive, excessive, and economically detrimental.
- Improve the performance of an ongoing oil-recovery operation.
- Reduce certain environmental liabilities by reducing the amount of excessive and unnecessary coproduction of environmentally unfriendly fluids, such as highly saline and hard reservoir brines.
- Extend the economic lives of marginal wells, well patterns, and oil fields.

Properties of an Ideal Gel System. An ideal conformance-improvement gel technology should be applicable as injection and production well treatments, as sweep improvement treatments, and as water and gas shutoff treatments. It should also be applicable to all reservoir mineralogies and lithologies and to a wide variety of reservoir and flooding-operation conformance problems. The ideal gel technology should be a single-fluid system and should possess a robust gel chemistry, which requires that it be insensitive to oilfield and reservoir environments and chemical interferences (especially H_2S and CO_2); be insensitive to all reservoir minerals and fluids; and be applicable over a broad pH range.

An ideal conformance-improvement gel technology should also involve a simple and straightforward gel-forming chemical system; be applicable over a broad range of reservoir temperatures; be stable over the long term; provide for a broad range of gel strengths, including rigid gels, and provide highly controllable and predictable gelation-delay onset times. For matrix-rock reservoir treatments, an ideal gel technology must include gelant solutions that are readily injectable into matrix reservoir rock. An ideal gel must be environmentally acceptable and friendly, be formulated with low concentrations of relatively inexpensive chemicals, and be formulated with readily available (preferably commodity) chemicals. Finally, it should reduce the permeability to water flow in matrix rock more than the permeability to oil and gas flow.

TABLE 13.1—CONFORMANCE-IMPROVEMENT GEL SYSTEMS AND CHEMISTRIES

Inorganic Gels	Silicate based		
	Al (III) based		
	Hydroxides of selected divalent and trivalent cations		
Organic polymer gels	Biopolymer gels		
	Synthetic (usually acrylamide-based) polymers	Metal crosslinking agents	Cr (VI) redox
			Cr (III) inorganic
			Cr (III) complex
			Al (III)
			Zr (IV)
		Organic crosslinking agents	Aldehydes
			Phenol-formaldehyde and derivatives
			Polyethyleneimine
	Monomer gels	Acrylamide monomer	
		Acrylate monomers	
		Phenolics	
	Lignosulfonate gels		
	Polymer self-induced gels		

Gel Systems and Chemistries. Oilfield conformance-improvement gels come in a wide range of forms and chemistries.[4] **Table 13.1** provides an overview of various conformance-improvement gels.

Chromium (III)-Carboxylate/Acrylamide-Polymer Gels. Widely applied as sweep-improvement treatments and as water and gas shutoff treatments, chromium (III)-carboxylate/acrylamide-polymer (CC/AP) gels[5,81–83] are aqueous acrylamide-polymer gels in which the chemical crosslinking agent is a chromium (III) carboxylate complex. CC/AP gels have an exceptionally robust gel chemistry and are highly insensitive to oilfield and reservoir interferences and environments. They are also applicable over an exceptionally broad pH range.[5] As a result, these gels, when properly formulated, are applicable to the acidic conditions associated with CO_2 flooding for which most earlier oilfield polymer gels did not function well. The chromium (III), as used in the crosslinking agent of this gel technology, is relatively nontoxic,[84] but it is highly regulated. This single-fluid gel technology provides a wide range of gel strengths and a wide range of controllable gelation-onset delay times. The gel technology is applicable over a broad range of reservoir temperatures and applicable to a broad range of conformance problems and reservoir mineralogies and lithologies. Chromic triacetate ($CrAc_3$) is the often-preferred crosslinking agent used with the CC/AP gel technology. A chemical gelation-rate-acceleration additive package, involving chromic trichloride, has been developed for use with the CC/AP gel technology. Two chemical means are available to retard the rate of gelation of CC/AP gels that are applied to high-temperature reservoirs. The first involves the use of low or ultra-low hydrolysis polyacrylamide and capitalizes on the slow formation of the required chemical crosslink sites on the polymer by means of autohydrolysis. The second chemical means is the addition of relatively strong carboxylate ligands, such as lactate or malonate, to the gelant solution.

Chromium (VI) Redox Gels. One of the early conformance-improvement gel technologies involved acrylamide polymers that were chemically crosslinked together using a chromium (VI) redox system.[85] This oilfield gel system has largely fallen from favor because of issues relating to the use of a crosslinking agent that contains toxic and carcinogenic chromium (VI) and because the crosslinking chemistry is rather complicated and subject to a number of oil-field interferences.

Aluminum Crosslinked Gels. Although other aluminum crosslinking agents have been developed and used in conformance-improvement gels, aluminum-citrate-crosslinked gels have historically been the most widely applied. The early widespread application of the aluminum-citrate gel technology was conducted in the sequential-injection mode, involving the repeated sequential injection of aqueous slugs containing, respectively, the polymer and the aluminum-citrated crosslinking agent. Sequential-injection aluminum-citrate gel technology is discussed in more detail later in this section. The application of conformance-gel treatments, involving the sequential injection of aqueous slugs containing the different chemicals that are required to form the gel in the reservoir, have largely fallen from favor.

More recently, aluminum-citrate acrylamide-polymer gels, which are formulated with low concentrations (200 to 1,200 ppm) of polymer and which are referred to as colloidal dispersion gels (CDGs), have been somewhat widely used as large-volume treatments applied through injection wells to "matrix rock" for improvement of waterflood sweep efficiency.[9,86] CDGs are a form of discontinuous microgel particles. The mechanism and means by which this particular gel technology generates incremental oil production is not fully understood.

Published laboratory studies have reported the following regarding aluminum-citrate CDGs. First, CDGs of acrylamide polymers crosslinked with aluminum citrate are not readily injectable into, and propagatable through, matrix rock of normal permeabilities (e.g., sandstone of < 1,000 md).[87,88] Ref. 88 discusses an experimental flooding study where an aluminum-citrate CDG was observed to not be readily propagatable, after two hours of aging, through matrix rock during a flooding experiment involving the injection of an aluminum-citrate colloidal dispersion gelant solution, containing 300 ppm high-MW HPAM, into a 700 md Berea sandstone core plug at 105°F and at a superficial velocity through the sandstone of 16 ft/d. Second, aluminum does not readily propagate through reservoir matrix rock.[87] Third, aluminum crosslinking of the polymer of CDGs normally occurs within several hours.[88] Fourth, aluminum-citrate CDGs do not preferentially enter high-permeability zones any more selectively than is dictated by Darcy's law. Fifth, aluminum-citrate CDGs do not viscosify water more than do the gel's polymer without the addition of the crosslinking chemical.[88] If aluminum-citrate CDGs are not readily injectable into and propagatable through matrix reservoir rock of normal permeabilities (sub-Darcy permeabilities), there are two possible explanations for the apparent success of a number of large-volume aluminum-citrate CDG treatments in terms of generating conformance improvement and generating incremental oil production when treating "matrix-rock conformance problems." First, the successfully treated "matrix rock" reservoirs may actually have been at least somewhat naturally or otherwise fractured. Second, the offending high-permeability strata within the successfully treated matrix-rock reservoirs may have possessed multi-Darcy permeabilities.

Gels Crosslinked With an Organic Crosslinker. There has been a long-standing desire within the oil industry to develop effective conformance-improvement polymer-gel technologies using benign organic chemical crosslinking agents that would impart carbon-carbon-bond chemical crosslinks between the gel polymer molecules. This would avoid the use of metal crosslinking agents and would result in exceptionally strong and stable polymer gels. However, with the possible exception of the polyethyleneimine-crosslinked gel technology briefly discussed later in this section, no such gel technology has been developed and reduced to commercial practice.

The majority of organically crosslinked polymer-gel technologies developed to date have been based on phenol-formaldehyde chemistries. These gels either use phenol and formaldehyde as the chemical crosslinking agent or use derivatives or precursors to phenol and formaldehyde. There have been attempts to identify and use less toxic and more environmentally friendly derivatives of phenol and formaldehyde as the crosslinking agents.[89]

Some of the most thermally stable polymer gels for use in high-temperature conformance-improvement treatments have been formulated with acrylamide polymers that are chemically crosslinked with organic crosslinking agents that are based on phenol-formaldehyde type chemistry.[90]

At this writing, an organically crosslinked gel technology that does not involve a phenol-formaldehyde type crosslinking chemistry was somewhat recently developed. The gel technology involves the use of a specially manufactured and derivatized acrylamide polymer and the use of polyethyleneimine as the crosslinking agent.[91] This gel technology is most readily applicable to reservoirs with temperatures exceeding approximately 180°F, and reports regarding the field application of this conformance polymer-gel technology have been favorable overall.

Biopolymer Gels. Gels based on the crosslinking of biopolymers with organic or inorganic crosslinking agents have been pursued. A popular conformance biopolymer-gel technology in the 1970s and early 1980s was based on gels of xanthan polymer crosslinked with an inorganic chromium (III) crosslinking agent.[92]

Monomer Gels. Conformance-improvement gels based on the in-situ polymerization of organic monomers to form polymers with and without the inclusion of crosslinking monomers have been developed and successfully applied. Early monomer-gel treatments were often based on the in-situ polymerization of the acrylamide monomer; however, this is seldom practiced currently because of toxicity and environmental concerns. Most modern monomer-gel technologies for oilfield application are based on the in-situ polymerization of less toxic acrylate monomers.[93]

An advantage of monomer gel technologies is the low water-like viscosity of the gelant solution. Disadvantages include cost and the sensitivity of the polymerization reaction to oilfield interference and environments. Care also needs to be taken to carefully distinguish between "linear" (uncrosslinked) and crosslinked oilfield monomer gels.

An older monomer gel technology involved gels formed from the reaction of formaldehyde with phenol.[94] A modern day concern with this gel technology is the toxicity and environmental issues associated with the use and handling in the field of formaldehyde, phenol, and/or their chemical derivatives.

Polymer Self-Induced Gels. A conformance-improvement biopolymer gel system has been reported that involves the injection of the polymer into the treated reservoir volume in the form of an alkaline high-pH solution. Once emplaced in the reservoir rock, the pH of the polymer solution is reduced by spontaneous or induced means. Following the reduction in the polymer solution's pH, the polymer solution spontaneously forms a gel.[95]

Inorganic Gels. A variety of inorganic gel technologies have been developed and applied over the years. These include gels based on, respectively, silicate or aluminum ion chemistries, along with gels of hydroxides of iron and magnesium.

Gels based on silicate chemistries were some of the earliest gel technologies applied for conformance improvement. A silicate gel is formed when a relatively high-pH aqueous solution, containing a sufficient concentration orthosilicate monomers or oligomers of orthosilicate, has its pH reduced or is exposed to hardness cations. Aqueous-based conformance-improvement silicate gels can be formed in a petroleum reservoir by either "internally catalyzed" or "externally catalyzed" means. Internally catalyzed silicate gels are formed by including in the aqueous gelant solution an acid-generating chemical that will spontaneously decrease the pH of the gelant solution when it is placed in the reservoir. Externally catalyzed silicate gels are usually formed by contacting the orthosilicate monomer or oligomer solution with a brine (e.g., reservoir brine) that contains a high level of harness cations (e.g., Ca^{++}). Internally catalyzed silicate gels are often favored for use in conformance-improvement treatments during oil-recovery operations. Externally catalyzed silicate gels are often used during drilling operations for lost circulation applications. Several potential concerns should be noted. Solution-aging, filterability, and quality-control issues can be a concern for silicate gelant solutions that are to be injected into matrix-rock reservoirs. There can be safety and environmental issues associated with the acid-generating chemical used in internally catalyzed silicate gel systems.

Silicate-based gels of conformance-improvement treatments have been applied successfully in Hungary.[96] A large-volume sodium-silicate gel treatment was reported to have been applied to an offshore Norwegian oil production well.[97] A large number of silicate-based conformance-improvement gel treatments have been applied worldwide.

The main advantage of inorganic gels is their environmentally friendly nature. Disadvantages historically have been that many of the inorganic gels were relatively weak gels, and a number of the inorganic gels do not provide good long-term fluid-shutoff performance. The latter is especially true for inorganic hydroxide "gels" that tend to convert over time to solutions containing ineffective oxide solids.

Gels Formulated With Synthetic Organic Polymers. Although a number of the early conformance-improvement gel technologies were based on inorganic gels and biopolymer gels, the recent trend has been toward the application of oilfield polymer gels based on the use and crosslinking of synthetic organic polymers, primarily acrylamide polymers.

Classification of Gel Treatment Types. Conformance-improvement gels can be classified in several manners. First, conformance-improvement gels can be classified as to whether they are intended to treat matrix-rock conformance problems, involving permeabilities less than roughly two Darcies, or treat reservoir high-permeability anomalies (usually fractures), involving permeabilities greater than roughly two Darcies. A subcategory of this classification is whether the gel treatments for treating matrix-rock conformance problems are to be placed in the near-wellbore environment (radial penetration of less than ~15 ft) or to be placed deeply in the far-wellbore environment (radial penetration of greater than ~15 ft). Although this classification scheme was originally developed for the CC/AP polymer-gel technology, the classification scheme is also generally applicable to all other conformance-improvement polymer-gel technologies.

Gels used to treat high-permeability-anomaly conformance problems are often treating fluid-flow problems involving linear flow, such as that occurring into fractures. Near-wellbore gel treatments placed in unfractured matrix rock are usually used to block radial fluid flow. Near-wellbore gel treatments that are placed in matrix rock are of relatively small volume, are often total-fluid-shutoff gel treatments, and are often relatively simple and straightforward to apply. These gel treatments usually have a relatively low risk factor if they can be placed properly and if the conformance problem is correctly diagnosed. In addition, these gel treatments can have high payout-to-cost ratios. Gels placed deep in matrix rock are large volume treatments that can be relatively costly and technically complex. In addition, these gel treatments can have a relatively high risk factor associated with them and tend to render a relatively low payout-to-cost ratio. The application of low-concentration aluminum-citrate CDG treatments that are applied via injection wells has been cited as a possible exception to this trend.

Second, the widely applied form of conformance-improvement gel treatments, namely crosslinked organic-polymer bulk gel treatments, can be classified as to whether when being injected from the wellbore into the reservoir, the gel fluid is a gelant solution or is a preformed or partially preformed gel fluid. In most instances, such as with the CC/AP gel technology, the polymer-gel fluid must be in its gelant form (i.e., before any initial microgels form) for the gel fluid to be readily injectable into matrix reservoir rock (such as sandstone having a permeability less than 1,000 md). However, for polymer gels to be selectively placed in high-permeability anomalies such as fractures, the gel should be designed so at least some initial gelation has occurred when the gel leaves the wellbore to assure that the gel will not substantially leak off from the high-permeability anomaly (e.g., a fracture) into the adjacent matrix reservoir rock. This is the key to properly formulated polymer gel being able to selectively treat fracture conformance problems, without substantially damaging the adjacent matrix reservoir rock.

Third, conformance-improvement organic-polymer gels are classified as to whether they are the more widely applied bulk gels possessing relatively high polymer concentration with a continuous crosslinked polymer-gel structure on the macro scale or they are microgels, alternately know as colloidal dispersion gels. Microgels have been purportedly used to treat deeply in matrix reservoir rock. Microgel solutions contain low polymer concentrations (usually < 1,300 ppm) and do not have a continuous crosslinked polymer gel structure on the macro scale. Examples of conformance-improvement microgels are low concentrations of acrylamide polymer crosslinked with aluminum citrate[9,86] and low concentrations of acrylamide polymer crosslinked with zirconium lactate.[98,99] As mentioned previously, the conformance-improvement mechanism by which microgels function when treating "matrix-rock" reservoirs is not fully understood.

The fourth classification relates to the injection mode of the gel chemicals. Early oilfield conformance gel technologies tended to be based on the sequential injection of fluids containing, respectively, two of the reactive chemicals of the gel's chemical make up. For example, the polymer would be injected in one aqueous fluid followed by injection of the crosslinking agent in a second fluid. This was done at that time because the gelation-delay onset time could not then be controlled and/or delayed sufficiently to permit the gel to be injected into the reservoir as desired and required. This strategy is flawed for two reasons. First, operational constraints almost always require a substantial inert fluid spacer be placed between the two reactive gel-forming solutions to prevent mixing of the reactive chemicals in the injection tubing. This practice essentially precludes formation of a gel in the near-wellbore region. Second, when the second reactive fluid begins to diffuse and/or finger into the first reactive fluid in the reservoir, gel will begin to form and tend to divert the second reactive fluid from further mixing. In general, this outcome leads to the highly inefficient use of injected chemicals. Most current oilfield gel technologies do not involve the sequential injection mode. Essentially all current technologies involve the injection of a single gelant solution that contains all of the gel chemical constituents. State-of-the-art single-fluid gel technologies have sufficient chemical gelation-onset-delay-time chemistries to allow proper placement.

A popular early polymer-gel technology for improving conformance within matrix rock reservoirs was the repeated sequential injection of aqueous slugs containing, respectively, HPAM and aluminum-citrate crosslinking agent.[100,101] The mechanism by which the sequential-injection aluminum-citrate gel treatments were originally purported to operate was the so-called "layering mechanism." This mechanism envisioned that repeated sequential injection of the gel-forming aqueous slugs would result in alternate adsorbed layers of crosslinking agent and polymer that would crosslink to form gel that would build out from the pore walls of matrix reservoir rock.[102] The layering mechanism was later recanted by its original proponents. Today, most polymer-gel treatments involving acrylamide polymers crosslinked with aluminum citrate are injected as a single-fluid.

Gel Parameters and Evaluation. This section focuses on important formula parameters and on temperature effects as they relate to gelation rate and gel strength of conformance-treatment polymer gels. Similar relationships often hold for other gel technologies. **Figs. 13.17 through 13.20** relate to gel formula parameters and the effect of temperature for a specific CC/AP gel formula. Other oilfield polymer-gel technologies tend to follow similar relationships. The gel formula of **Figs. 13.17 through 13.21** is a fracture-problem fluid-shutoff gel that has a rigid and soft Buna rubbery consistency. The gel was formulated in fresh water and contained

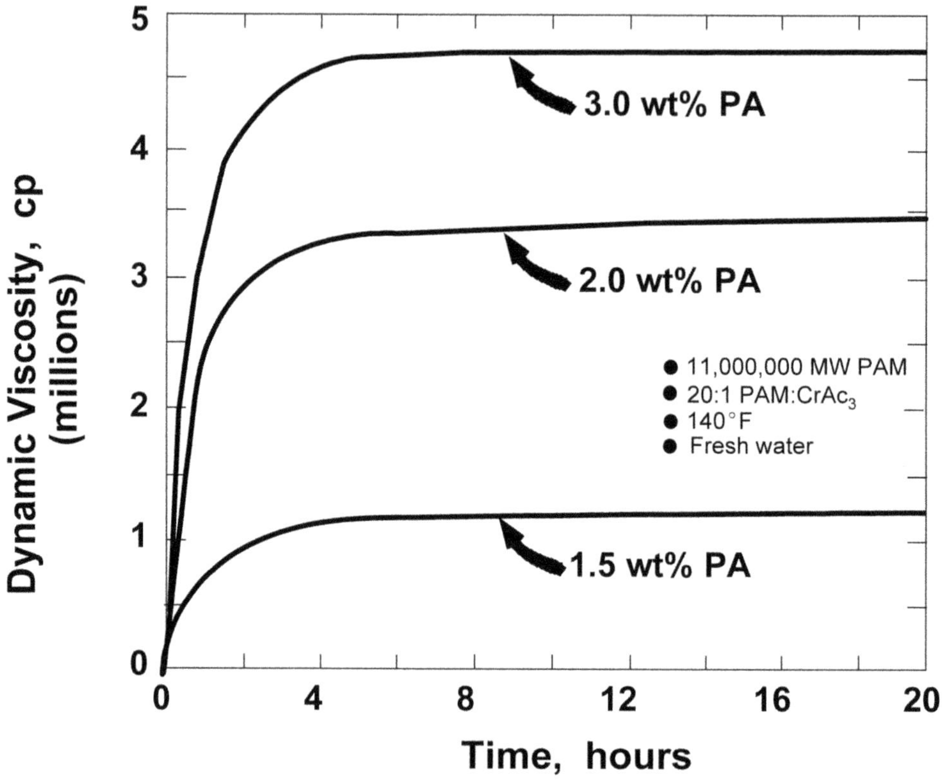

Fig. 13.17—Varying polymer concentration in a CC/AP gel.

2.0 wt% active polyacrylamide (PAM) polymer possessing 11 million MW and 2% hydrolysis. The polymer was chemically crosslinked together to form the gel using chromic triacetate ($CrAc_3$) as the crosslinking agent.

In **Figs. 13.17 through 13.22**, dynamic oscillatory viscometry was used to measure the elastic strength of the gel sample[103] as a function of aging time during the gel maturation process following the addition of the crosslinking agent to the gelant solution. The dynamic oscillatory viscosities were measured at an oscillation frequency of 0.1 radians per second.

These plots of dynamic oscillatory viscosity vs. gel aging time can be used to discern two important properties of any given gel sample. First, the equilibrium dynamic-oscillatory viscosity value that is eventually obtained for a given gel sample is a measure of the elastic strength of the gel. Second, the rate at which the dynamic oscillatory viscosity of the gel sample reaches the equilibrium viscosity value is a reflection of the gelation rate of the sample.

Polymer Concentration. As Fig. 13.17 shows, gel strength increases with polymer concentration when all other gel formula parameters are held constant. For the gels of this figure, the crosslinker loading was held constant at a 20:1 weight ratio of weight active polymer to weight active $CrAc_3$ crosslinking agent. The gel samples were aged at 140°F. The trend shown is a near universal relationship for all gel technologies. As the concentration is increased of the polymer, monomer, or fundamental chemical building block of the gel's solid-like structure (up to its solubility limit), gel strength increases.

Polymer Molecular Weight. When all other gel formula parameters are held constant, polymer-gel strength increases as the polymer molecular weight (MW) is increased. For Fig. 13.18, which depicts this trend, the crosslinking agent concentration within the gel samples was 1,000

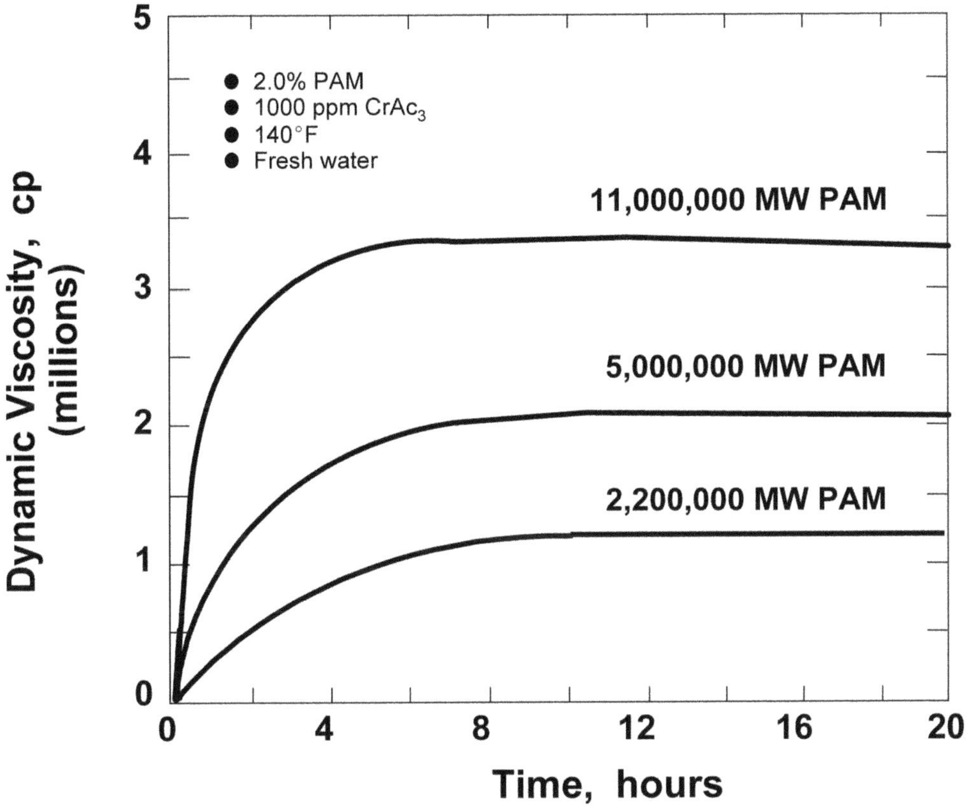

Fig. 13.18—Varying polymer MW in a CC/AP gel.

ppm $CrAc_3$. The trend shown explains why a gel formulated with a low-MW polymer needs a higher concentration of polymer to obtain the same strength of a comparable fracture-problem gel formulated with high-MW polymer. Comparable gels formulated with higher-MW polymers tend to be more elastic in nature.

Crosslinker Concentration. As Fig. 13.19 shows, when all other polymer-gel formula parameters are held constant, increasing the crosslinking agent concentration increases gel strength. Although not readily apparent in this figure, in general, increasing the crosslinking agent concentration also increases the gel's rate of gelation.

It should be emphasized that, for a given polymer gel, there usually is an optimum concentration for the crosslinking agent. For the samples of Fig. 13.19, that concentration is approximately 1,000 ppm $CrAc_3$ (or alternatively stated as a weight ratio of 20:1 active polymer to active crosslinking agent). The optimum crosslinking agent concentration is often specified in terms of the ratio of weight of active polymer to weight of active crosslinking agent (or, alternatively, in terms of the weight of the active polymer to the weight of metal ion in the crosslinking agent). Above a gel's optimum crosslinking-agent concentration, syneresis (expulsion of water from the gel) will occur. Below the optimum concentration, the gel is underoptimum in terms of gel strength. For most oilfield applications, the petroleum engineer wants to maximize effective gel strength to maximize economic performance per pound of polymer used in a given gel formula.

Syneresis results from shrinkage of the gel volume. Syneresis, a thermodynamic equilibrium phenomenon, results from excessive chemical attractive forces within the gel structure.

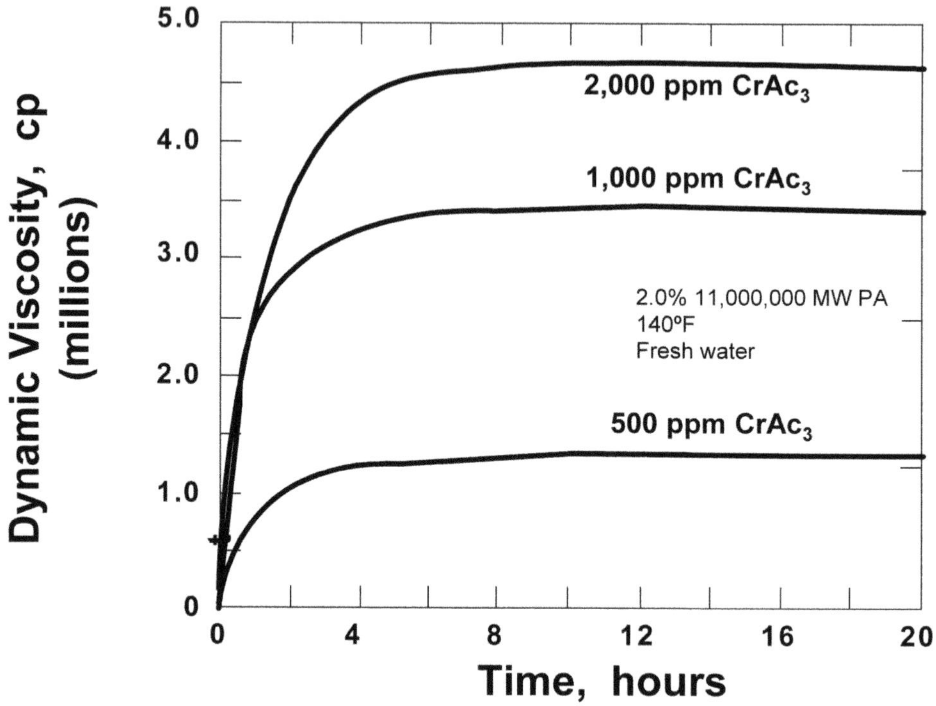

Fig. 13.19—Varying crosslinking agent concentration in a CC/AP gel.

Syneresis of a given polymer-gel formula usually results from one of two causes. First, syneresis for a given polymer-gel formula (fixed salinity and pH) can result from excessive crosslinking agent being incorporated into the gel formula. The second cause results from additional and excessive crosslinking chemical sites developing on the polymer over time, as exemplified by autohydrolysis of acrylamide polymers of CC/AP gels at high temperatures.

Syneresis is usually considered undesirable and unacceptable for gels to be used in near-wellbore treatments that are applied in matrix-rock reservoirs for total-fluid-shutoff purposes or in gels for sealing fluid flow in fractures. However, for gels, especially microgels, that are to be used in deeply placed gel treatments of matrix-rock reservoirs and to function where the microgel particles act as check valves in pore throats, gel syneresis may not be of significant concern.[104] Gel syneresis and gel degradation/degelation can be, and often are, distinctly different phenomena; however, both of these phenomena can result in expulsion of water from a gel.

The optimum concentration of the crosslinking agent for a given CC/AP polymer-gel formula decreases as the MW of the polymer increases, decreases proportionally with increasing polymer concentration (i.e., less pounds of crosslinker per pound of polymer), and increases with increasing temperature.

Temperature. Fig. 13.20 shows how the rate of gelation substantially increases with increasing temperature. For the CC/AP gel technology, the rate of gelation for a given gel formula approximately doubles with every 10°C increase in temperature.

Gelation Rate Acceleration. There are instances when applying conformance-improvement gel treatments that the optimum treatment design will call for the acceleration of the gelation rate of a given gel formula. This is especially true for gel treatments applied to low-temperature (e.g., subambient temperature) reservoirs. For example, for the CC/AP gel technology, a chemical gelation-rate-acceleration additive package involving the use as chromic trichloride as

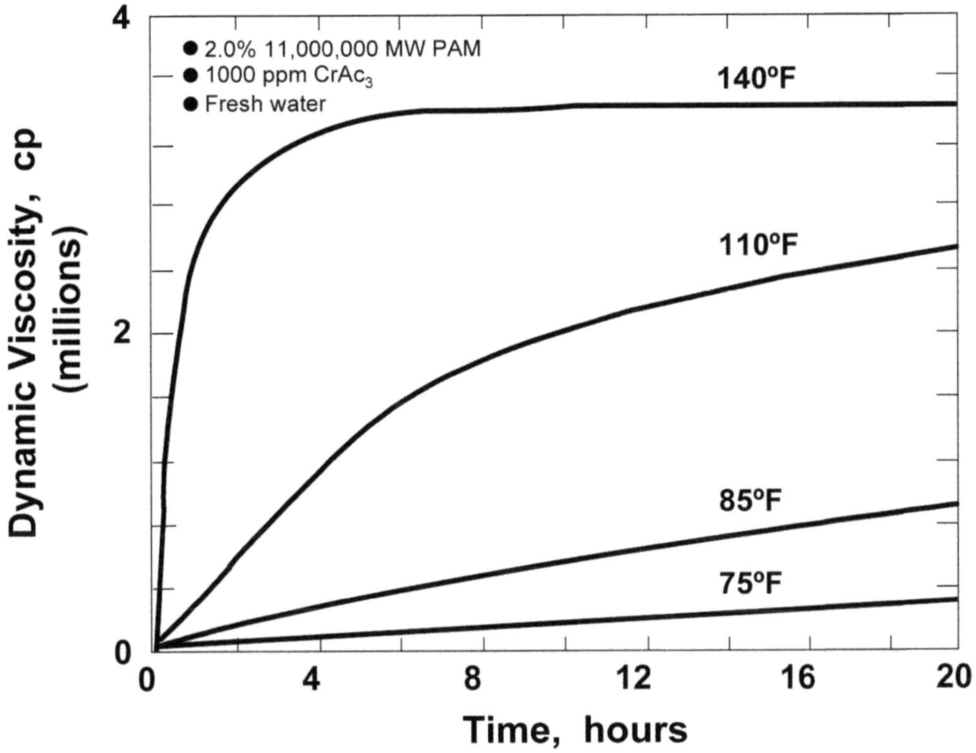

Fig. 13.20—Temperature effect on the gelation rate of a CC/AP gel.

the accelerating agent has been developed. Fig. 13.21 illustrates the use of this acceleration package with the CC/AP gel technology.

Gelation Rate Retardation. There are instances when applying oilfield gel treatments that the optimum treatment design will call for the retardation of the gelation onset time of a given gel formula. This is especially true for gel treatments that are to be applied to high-temperature reservoirs.

For example, for the CC/AP gel technology, a chemical gelation-rate-retardation additive package has been developed involving the use of strong carboxylate ligands as the gelation-rate retardation agent. Fig. 13.22 illustrates the use of this chemical gelation-rate-retardation additive package with the CC/AP gel technology. In the study of Fig. 13.22, the gel being investigated was a gel formula that was intended to impart total near-wellbore fluid shutoff in matrix-rock reservoirs at elevated temperatures. The gel was formulated in fresh water and contained 5.0 wt% active polyacrylamide (PAM) having a MW of ~250,000 daltons. Two versions of the PAM polymer were used in the study. The first PAM was 1.9 mole% hydrolyzed. The second PAM was ultra-low-hydrolysis PAM that had a hydrolysis level of < 0.1 mole%. The gelation-rate retardation agent used in the study of Fig. 13.22 was sodium lactate (NaLac). As the figure shows, the gelation rate was nearly instantaneous at 248°F when the gel was formulated with normal polyacrylamide (1.9% hydrolyzed) and no gelation-rate retardation agent was used. By using various combinations of gelation-rate chemical retarder addition and ultra-low-hydrolysis PAM in appropriately formulated gel recipes, the gelation onset delay time at 248°F could be extended in increments out to 23 hours. As the figure also shows, if the near-wellbore reservoir rock is cooled to 212°F (as can normally be easily done near wellbore through the use of ambient-temperature water injection), the gelation onset time could be extended to 68

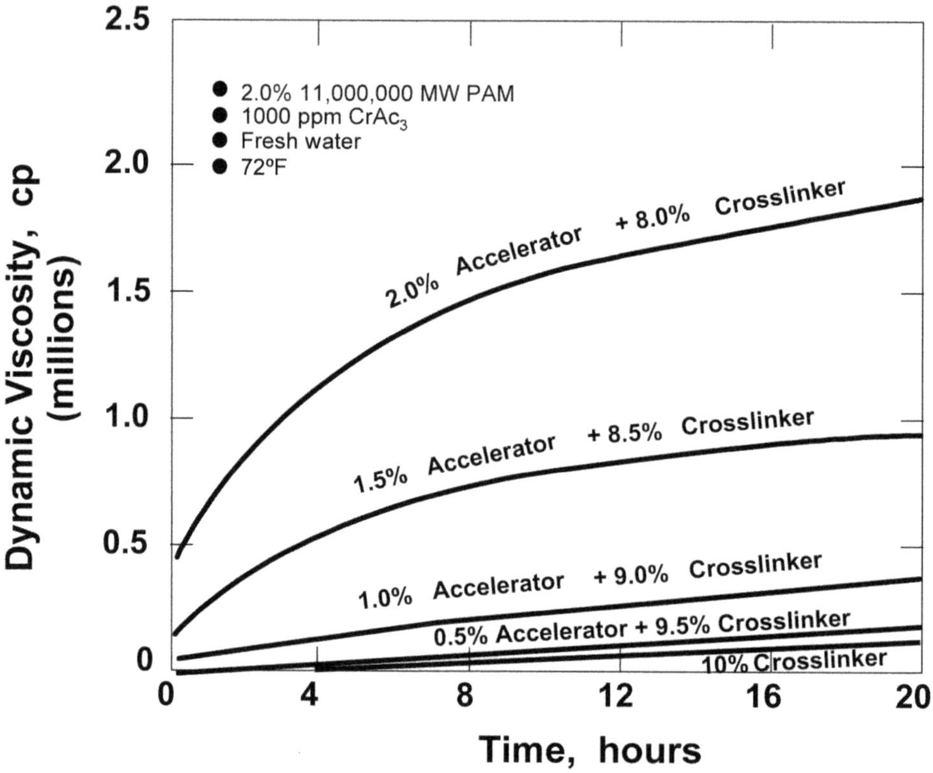

Fig. 13.21—Accelerating the gelation rate of a CC/AP gel.

hours. Appropriately formulated lactate-retarded CC/AP gels usually contain somewhat higher loadings of polymer and/or crosslinking agent. These increased gel chemical loadings are incorporated to counteract the slight weakening and destabilization of the gel that is imparted by the lactate addition.

Gel Strength. There are several different measurements of polymer-gel strength.[81] These gel strengths have at least some analogous correlations in most other gel technologies.

The first measurement is elastic strength. As discussed previously, dynamic oscillatory viscosity can be used to measure the elastic strength of a gel. In practical oilfield terms, the measure of the elastic strength of a gel relates to the resistance to physical deformation that a gel will exhibit while extruding through a constriction in its flow path, such as a constriction in a fracture flow path.

The second measurement is yield or failure strength. When this strength is exceeded, portions of the chemical gel structure are broken. An example of exceeding the gel yield or failure strength is the rupturing of the chemical bonds of the polymer's backbone structure of a shear-non-rehealable gel. This gel strength is measured by placing a mature gel sample in a large container and increasing container pressure until there is flow through a small orifice in the testing container. The yield or failure strength of a total-fluid-flow-shutoff and shear-non-rehealing gel formed and residing in a sandpack is measured by the differential pressure required to make the gel flow from the sandpack. The yield or failure strength of a polymer gel is often much larger than its elastic strength.

The third measurement is compressive strength. Polymer gels do not possess large compressive strengths compared to Portland cement. Thus, polymer gels should not be applied as

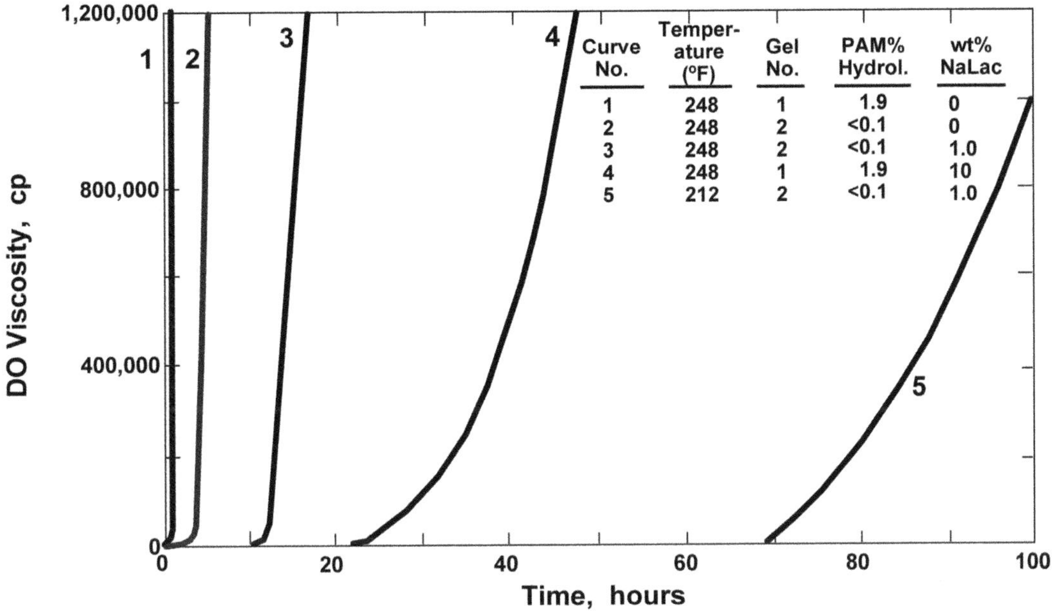

Fig. 13.22—Retarding the gelation rate of a CC/AP gel.

plugging material where substantial compressive strength is required. Adding solids to a polymer gel can greatly increase its compressive strength, especially at high solids loading.

The relative effectiveness of gels for plugging fluid flow through flow channels and flow paths, as measured by the differential pressure that causes the gel to breakdown and flow, increases as the size of the flow channel decreases. Gels that are placed as clear-fluid gelant solutions are particularly effective and strong plugging agents for use when plugging or reducing the fluid-flow capacity in microflow paths, such as is pores of matrix reservoir rock. Gels are especially effective at plugging small fluid-flow paths where solids-containing plugging agents, such as cement or even "microfine" cement, cannot be readily placed.

Gel Onset Time. There have been many and, at times, apparently conflicting means proposed to define and measure gelation onset times. There are two distinctly different and important times in the gel maturation process. The first is the initial onset of gel formation. The second is the time required to reach full gel strength. Because the time required to attain full gel strength is often reached asymptotically, pragmatically it is often better to measure the attainment of "near full gelation," where "near full gel strength" is quantitatively defined.

When determining the time of initial onset of gelation for a classical bulk polymer gel to be used to treat matrix reservoir rock, the investigator needs to determine when the first onset of microgel particle formation occurs. It is at this point that the gelant usually ceases to be readily injectable into sandstone of less than 1,000 md. Filtration testing, core flooding injectivity experiments, bottle testing, and dynamic oscillatory viscosity techniques have all been used successfully to determine the first onset of such gelation. For a bulk polymer gel, microgel formation usually occurs at a time just before the onset of first being able to visually detect any gel structure in the gelant solution. For classical bulk polymer gels, initial microgel formation constitutes the formation on a colloidal scale of the first discreet crosslinked polymer aggregates that are the precursor to the formation of the continuous macro-scale crosslinked polymer network of the bulk gel.

For crosslinked polymer gels, there are several different modes of the gelation onset delay and different ways these modes manifest themselves. First, there can be a substantial induction time before any gel formation, including microgel formation, can be detected. Second, and not mutually exclusive of the first mode, there can be "right angle gelation onsets." Here, once the gelation process starts, the final gel strength is reached very rapidly (almost instantaneously). Third, the gelation begins as soon as the crosslinking agent is added and proceeds over a protracted period of time. CC/AP gels widely applied to treat fracture conformance problems in reservoirs with temperatures < 150°F are of the third gelation-delay type mode.

Gel Bottle Testing. Although not a highly sophisticated and exacting quantitative technique, bottle testing (ampoule testing at high temperatures) has been used widely in both the laboratory R&D setting and the field quality-control setting to test and evaluate gels, especially polymer gels.

Bottle testing provides a highly cost-effective and straightforward technique to obtain a semiquantitative measure of gel strength and a semiquantitative measure of gelation rate. It is also a convenient means to evaluate the long-term stability of gels at a given test temperature. Bottle testing is often relatively easy to conduct in the field setting. Bottle testing in the laboratory provides a means to screen rapidly the performance of a large number of gel samples before selecting a few of the gels samples for more costly and rigorous testing, such as dynamic-oscillatory-viscosity and core-flooding experimental testing.

A polymer-gel bottle-test gel strength code has been developed and is now widely used (at times in various forms and with various modifications) for visually determining and evaluating, in a semiquantitative manner, the strength of polymer gels.[81] **Table 13.2** shows the gel strength code. The gel strength code is set up such that two observers, who view the same gel sample, could possibly assign to the sample a gel strength code that differs by one letter code. However, the gel strength code is designed such that it is "virtually not possible" for two reasonable observers to view the same gel sample and assign the gel sample a gel strength code that differs by two letter codes. The comparison of the strength of different gels using this bottle-testing gel-strength-code scheme should only be made when the same volume of gel sample is placed in a bottle or ampoule with the same size and geometric shape. When quoting a gel strength code, the size of the gel sample and the size of the bottle should be provided. Bottle testing at low to intermediate temperatures is often conducted with a 50 cm^3 gel sample placed in a 120 cm^3 (4 oz) wide-mouth bottle.

When conducting polymer-gel bottle testing, there are several pitfalls to avoid. When performing field quality control work involving polymer gels containing H_2S, a nitrogen blanket should be applied to prevent the H_2S from reacting with atmospheric oxygen and forming intermediate free-radical chemical species that attack and degrade the polymer gel. This deleterious reaction with free oxygen does not occur in anaerobic oil reservoirs (most oil reservoirs are anaerobic). Free oxygen in the injected water of the gel formula tends to be chemically reduced and deactivated quickly in the chemically reducing environment of most oil reservoirs. Consumption of free oxygen in the chemically reducing reservoir environment is generally much faster than the oxygen-induced gel degradation reaction.

When conducting high-temperature ampoule testing of this bottle-testing scheme, the gelant solutions need to be scrupulously deoxygenated to less than 10 ppb free O_2. There are two recommended procedures to deoxygenate the gelant solution of polymer-gel samples for high-temperature testing in the laboratory setting. The first procedure involves vacuum degassing to the point that the aqueous solution boils (ebulates) for a minute or so and then glassblowing the ampoule shut under vacuum. It is highly recommended that properly tempered heavy-walled (~4 mm thick) glass ampoules be used for this purpose and that the glass ampoules be placed in appropriate transparent safety containers when the gel samples are aged in an oven at the test temperature. The second effective procedure for effectively deoxygenating gelant samples (to < 10 ppb O_2) for high-temperature laboratory testing is to bubble high purity argon gas through the gelant solution. The need to deoxygenate gel samples during high-temperature test-

TABLE 13.2—BOTTLE-TESTING GEL STRENGTH CODES	
A	**No detectable gel formed.** The gelant solution appears to have the same viscosity (fluidity) as the original polymer solution, and no gel structure is visually detectable.
B	**Highly flowing gel.** The gel appears to be only slightly more viscous than the initial polymer solution.
C	**Flowing gel.** Most of the obviously detectable gel flows to the bottle cap on inversion of the bottle.
D	**Moderately flowing gel.** A small portion (approximately 5 to 15%) of the gel does not readily flow to the bottle cap on inversion—usually characterized as a "tonguing" gel (i.e., after hanging out of the bottle, the gel can be made to flow back into the bottle by slowly turning the bottle upright).
E	**Barely flowing gel.** The gel slowly flows to the bottle cap and/or a significant portion (>15%) of the gel does not flow to the bottle cap on inversion.
F	**Highly deformable nonflowing gel.** The gel does not flow to the bottle cap on inversion (gel flows to just short of reaching the bottle cap).
G	**Moderately deformable nonflowing gel.** The gel flows approximately halfway to the bottle cap on inversion.
H	**Slightly deformable nonflowing gel.** Only the gel surface slightly deforms on inversion.
I	**Rigid gel.** There is no gel-surface deformation on inversion.
J	**Ringing rigid gel.** A tuning-fork-like mechanical vibration can be felt after the bottle is mechanically tapped.

ing is a laboratory artifact because gels existing in a reservoir will normally be in an anaerobic environment.

Flow and Placement Considerations. The flow properties of a gelant or gel as it is being placed are important parameters.

Gelant and Gel Flow and Placement in Matrix Rock. To date, for all known gelant solutions used in conformance-improvement treatments (including polymer gelant solutions), these gelant solutions place themselves in all matrix-rock geological strata according to Darcy flow considerations and do so without any special selective placement in only the high-permeability strata and flow paths.

Mechanical Zone Isolation. Any placement of gel into, and the associated permeability reduction of, a low-permeability and/or high-oil-saturation strata in the near-wellbore region surrounding a radial-flow matrix-rock-reservoir well will almost always be counter productive to improving the conformance of that well.[7,8] Thus, when applying a gel treatment, especially a near-wellbore gel treatment, to treat a vertical conformance problem of a radial-flow well in a matrix rock reservoir, mechanical zone isolation must be used to assure that the gelant is injected only into the high-permeability and/or low-oil-saturation geological strata to be treated.[7,8]

Gel Treatments When Matrix Crossflow Exists. For a gel treatment to be successful when treating vertical conformance problems in matrix-rock reservoirs in which there is fluid-flow and pressure crossflow communication between geological strata of differing permeabilities, the gel treatment must be selectively placed deeply in the reservoir.[6,7] In addition, the gelant solution should approach unit mobility, or stated another way, the resistance factor equals 1.0 (i.e., the gelant solution should approach the viscosity of water).[6] The successful treatment of vertical conformance problems in matrix rock reservoirs of normal permeabilities (permeabilities < 1,000 md), where there is crossflow between the geological strata, requires the application of large-volume gel treatments for which the economic risks can be relatively high and the rates of returns can be relatively low. The large volume of gelant solution that needs to be injected begins to approach the volume of a traditional chemical flood, such as a polymer conformance-improvement waterflood. In addition, the requirement of being able to successfully propagate the gelant solution, or possibly microgels, deep into a matrix-rock reservoir (especially without damaging oil-productive zones) is a challenging task.

Gel Placement in Fractures. While no gel treatment fluid is available for the selective placement of gels (beyond Darcy flow considerations) into high-permeability flow channels of matrix reservoir rock of normal permeabilities, gels, especially polymer gels, can be routinely formulated for selective placement into fractures or into other high-permeability anomalies within a reservoir—such as fractures, solution channels, cobble packs, and rubblized zones. Such a polymer gel is normally designed so that the gelant solution is at least partially gelled when it leaves the wellbore, and initial microgels have formed. This initial gelation and the formation of microgel particles prevent the gel from substantially invading and damaging the matrix rock adjacent to high-permeability anomalies.[105] The next subsection of this chapter discusses a gel fluid-loss mechanism involving gel dehydration and expulsion of water from the gel into matrix reservoir rock adjacent to the gel-filled fracture. This gel dehydration does not involve the loss of gel from the fracture. For a properly designed polymer-gel treatment, gel leakoff from treated fractures is most often insignificant in terms of gel-treatment functionality. CC/AP fracture-problem polymer gels that are placed in such a manner are capable of selectively plugging the treated fracture volume effectively.[105]

Gel Extrusion Through Fractures. Because many of the most successful gel treatments have been applied as large-volume treatments to naturally fractured reservoirs and because the injection times of such gel treatments often exceed the injected gel's gelation onset time (often by a factor of 10 or more), much of these gels must be flowing and extruding through the fractures in a mature gel state.[106–108] Polymer gels used to treat fracture conformance problems have been shown, while extruding through fractures, to exhibit shear-thinning rheological behavior that correlates with gel superficial velocity and fracture width.[107] When extruding these gels through fractures at high velocities, the resultant pressure gradients within the fractures are insensitive to flow rate. This is a partial explanation for why these polymer gels have exhibited unexpectedly good injectivity into fractured formations. This explanation is not intuitive to many petroleum engineers.

Fracture-problem polymer gels of the type that are widely applied as sweep-improvement and water-shutoff treatments have a minimum pressure gradient that is required to mobilize the flow of the gel. This minimum pressure gradient for gel flow is proportional, over a broad range of fracture widths and differential pressures, to the inverse of the square of the fracture width.[107–109] The implications of this observation are extremely significant. One implication is that these polymer gels will tend to be selectively placed in the widest and most offending fractures when treating fracture conformance problems in naturally fractured reservoirs. A second implication is that fracture-problem gel water-shutoff treatments, which are applied to a naturally fractured reservoir, should be designed so that the drawdown pressure of normal production operations does not exceed minimum pressure gradient for gel flow. If the drawdown pressure exceeds the minimum pressure gradient for gel flow, any gel experiencing drawdown exceeding the minimum pressure gradient for gel flow will be mobilized and back produced. Of note, the pressure gradient in the intermediate- and far-wellbore region of most naturally fractured reservoirs during oil-recovery operations is quite small (often less than 5 psi/ft). For a widely applied fracture-problem CC/AP gel formula under the studied experimental conditions,

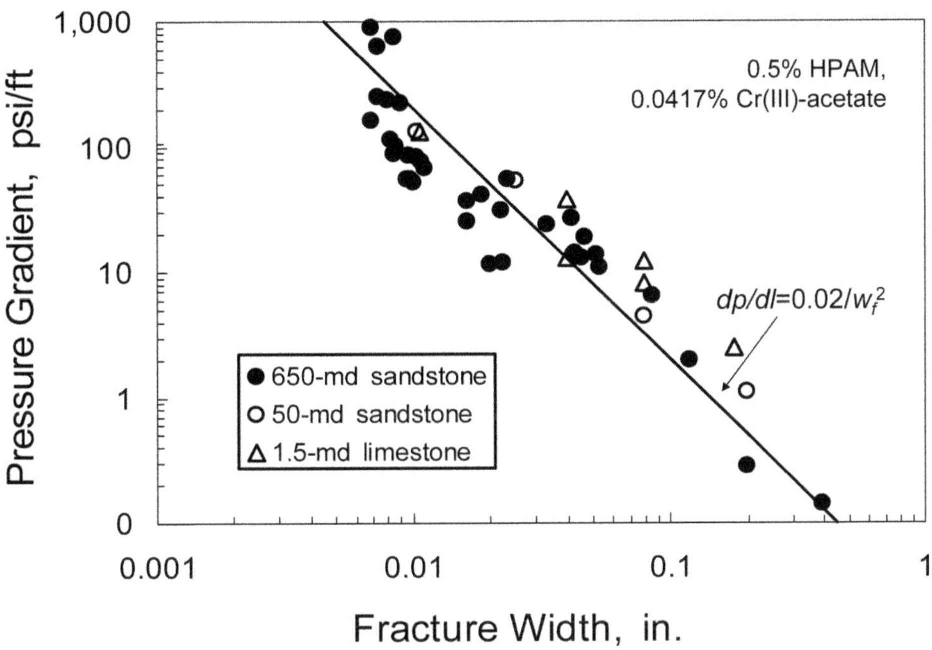

Fig. 13.23—Pressure gradients required to extrude a gel through open fractures.

the pressure gradient, d*p*/d*l*, required to extrude the gel from the studied fractures is described by the following mathematical equation:

$$dp/dl = 0.02/w_f^2, \qquad (13.17)$$

where w_f is the fracture width.[108,109] (See **Fig. 13.23.**) The data of Fig. 13.23 involved fracture widths ranging from 0.008 to 0.4 in. (0.2 to 10 mm) and pressure gradients from 0.1 to 1,000 psi/ft.

Aqueous polymer gels, being sponge like, can undergo dehydration while being propagated through fractures. Gel dehydration can occur any time a fracture-problem polymer gel experiences a differential pressure between the gel in the fracture and the adjacent permeable matrix reservoir rock. The rate of dehydration is not necessarily directly proportional to the differential pressure.[108,109] This gel dehydration is loss of water from the gel and not leakoff of the gel itself. Gel dehydration decreases the rate at which the gel propagates through a given fracture and strengthens the gel that resides within the fracture. As previously noted, polymer-gel strength increases as the concentration of polymer and crosslinking agent increases within the gel. For a fracture-problem CC/AP gel formula that has been widely applied in the field and under experimental conditions simulating such field applications, the gel dehydration rate, μ_l in ft/D (or alternatively ft³/ft²/D), has been described, as **Fig. 13.24** shows, by the empirical equation:

$$\mu_1 = 0.05t^{-0.55}, \qquad (13.18)$$

where t is time in days.[108] During the laboratory flooding-experiment study of Fig. 13.24, the facture width was 0.04 in. (1 mm), fracture lengths varied from 0.5 to 4 ft, fracture heights varied from 1.5 to 12 in., and injection fluxes in the fracture varied from 130 to 33,000 ft/D.

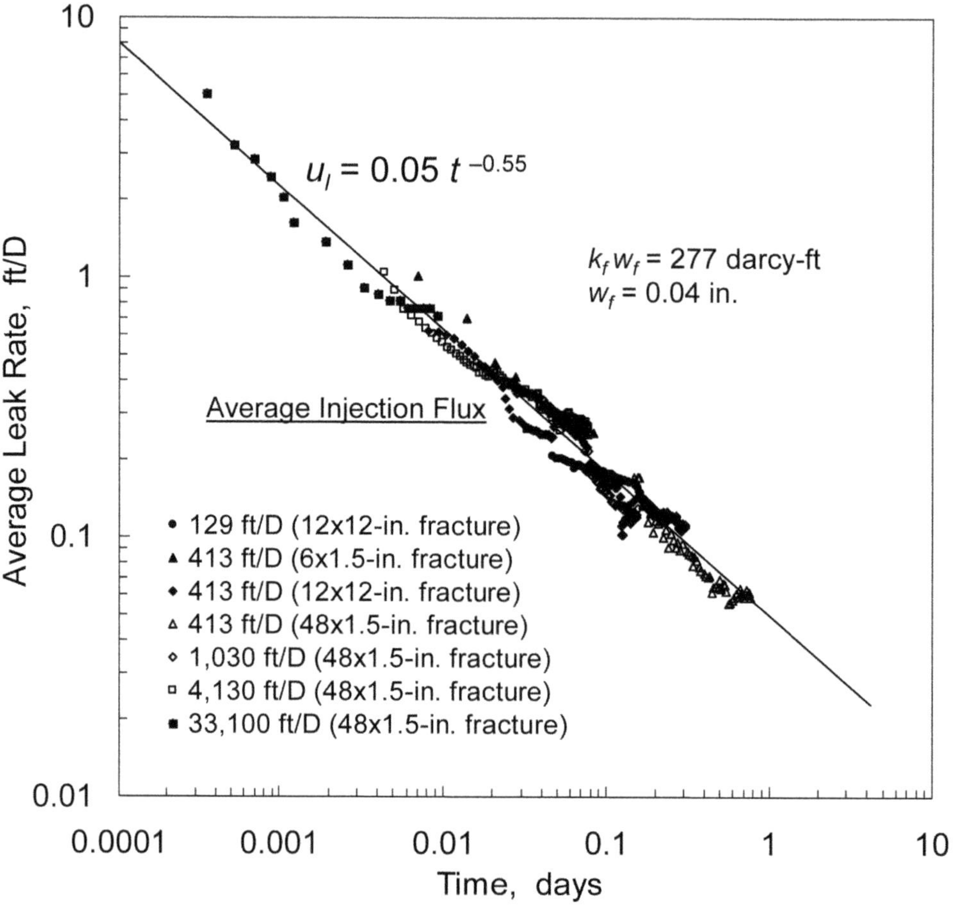

Fig. 13.24—Average gel-dehydration leakoff rate from seven fracture flooding experiments at different velocities.

The dehydration of fracture-problem polymer gels is the reason why if the objective is to inject the fracture-problem gel as deeply into a reservoir as possible, the gel should be injected as rapidly as feasible (without exceeding formation parting pressure). Conversely, to maximize the strength of the emplaced gel, the gel should be injected as slowly as feasible.

Gel Shear Rehealing. When a mature gel is exposed to a high shear-rate field and the gel structure is sheared, the gel may or may not be able to spontaneously reheal. Most polymer gels of the type used in hydraulic fracturing operations employ polymer crosslinking chemistries that impart shear-rehealing properties into the gel. That is, if the gel is subjected to a sufficient intensity shear flow field, the gel will temporarily shear degrade. The polymer chemical crosslinks will be temporarily broken, and the chemically crosslinked polymer molecules will temporarily separate in solution. However, on termination of the shear flow field, the gel and its chemical crosslinks of the polymer molecules will spontaneously reheal (for the most part). As a result, the gel will regain all (or nearly all) of its original gel strength. However, these gels are not normally good plugging agents for use in conformance-improvement treatments, especially for placement under high differential pressure conditions surrounding production wells. These gels, just like the "linear gels" resulting from high concentrations of uncrosslinked water-soluble polymers alone, tend to "slowly" flow under high differential pressure conditions. This is one possible shortcoming of the use of in-situ polymerization of monomers for conformance-improvement purposes when no crosslinking monomer is incorporated into the in-situ polymerization process.

Many conformance-treatment polymer-gel technologies, such as CC/AP gels, use polymer-crosslinking and polymer chemistries that do not lend themselves to shear rehealing. These gels have effectively no tendency to flow through constricting microflow paths, such as pore throats, when subjected to differential pressure. When these gels are subjected to even very high shear-rate fields, the gel crosslinking sites and/or multiplicity of crosslinked sites on any given polymer molecule do not permit the crosslinked polymer molecules to separate at the crosslinking sites. If the shear-rate flow or shear-stress conditions become exceptionally high, the gel's polymer backbone begins to experience scissions caused by mechanical shear degradation, which results in irreversible shear damage and mechanical degradation of the polymer gel.

Injection Pressure. As a general rule, the reservoir facture and/or parting pressure should not be exceeded during the injection of the gel treatment fluid. If reservoir fracture or parting pressure is unexpectedly and/or inadvertently exceeded when performing a gel treatment involving a relatively strong gel, normal practice calls for going to water injection at the same rate and pressure until the gel solution is displaced from the tubing. At worst in this situation, a minifracture will be created in the reservoir. On cessation of gel injection, the fracture will close, the gel will mature and should seal the fracture, and there will be little damage to the reservoir.

Hall Plots. Hall plots are often generated and analyzed during real-time placement of gel conformance-improvement treatments. At times, Hall plots of gel treatment placement have been creatively and unscientifically interpreted. The Hall plot was originally developed to analyze steady-state injectivity data for waterflood injection wells that are injecting into a single zone.[110–112] As normally used in conjunction with gel treatments and gel injection, $\Sigma p_{tf}\,\Delta t$ is plotted vs. cumulative gel volume injected W_i, where p_{tf} is the flowing wellhead pressure in psi, Δt is time in days, and W_i is cumulative injection volume in barrels. Under steady-state conditions, the slope of the Hall plot is

$$M_H = \frac{141.2\mu\left[\ln\left(r_e / r_w\right)\right] + s}{kh}, \quad (13.19)$$

where μ is viscosity in cp, r_e is the external reservoir radius in feet, r_w is the wellbore radius in feet, s is the skin factor, k is permeability in md, and h is formation height in feet. If a change in slope occurs in a Hall plot, all that the slope change can indicate, in the absence of some other and independent data, is that there has been a change in the well's injectivity. Without having other independent data, when a change occurs in the slope of a Hall plot, one cannot tell if the slope change was caused by a change in the well's skin factor, a change in the mobility (k/μ), or a change in the effective height of the well interval accepting the injection fluid.

Fig. 13.25 is a Hall plot illustrating an injectivity reduction after 5,000 bbl cumulative injection. In themselves, Hall plots, and any changes in their slopes, do not indicate selective placement of the gel in high-permeability channels during a gel treatment.[112]

Temperature Considerations and Limits. All gels, especially polymer gels, have a finite upper temperature limit above which the gels are not stable or functional. Significant progress has been made over the past two decades in increasing the upper temperature limit for the successful application of conformance-improvement gels. A continuation of this trend can be reasonably expected.

For CC/AP gel technology, the reservoir temperature limit for applying matrix-rock, near-wellbore total-shutoff treatments is reported to be 300°F.[5] Ref. 5 reported that the upper temperature limit for treating high-permeability-anomaly conformance problems (e.g., fractures)

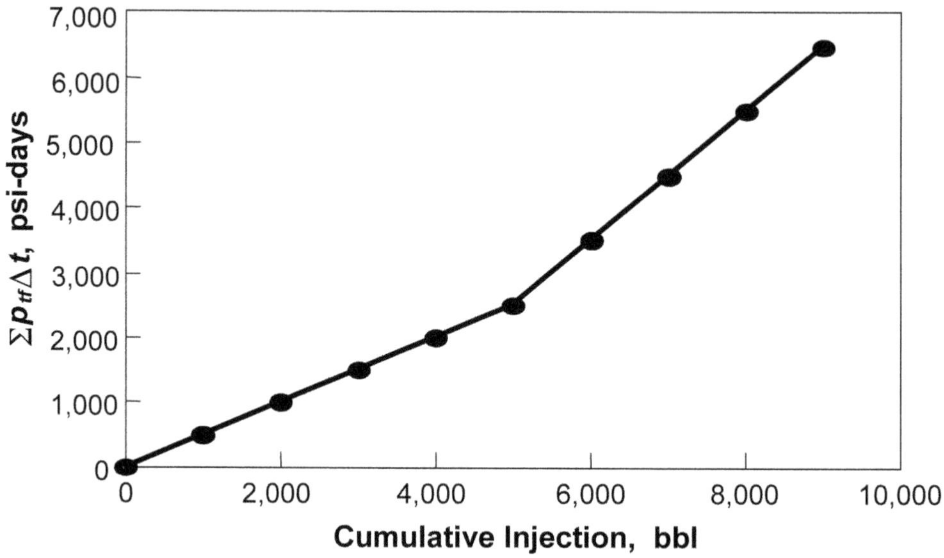

Fig. 13.25—Hall plot illustrating an injectivity decrease after 5,000 bbl cumulative injection.

with this gel technology can be up to approximately 270°F. Ref. 5 also discusses laboratory testing at 300°F that demonstrate how appropriately formulated CC/AP gels can essentially totally block flow to water in sandstone at high differential pressure conditions (1,000 psi per 3 in.) for an extended period of time (testing conducted for 23 days). **Fig. 13.26** is a photo of a CC/AP gel that remained stable, rigid, and clear after aging for 2.5 years at 300°F. This CC/AP gel is used for near wellbore, total-fluid-shutoff purposes in high-temperature matrix-rock reservoirs. For most polymer-gel technologies, such as the CC/AP gel technology, it is necessary to increase the polymer concentration within a given formula as temperature increases to maintain gel stability, performance, and strength similar to that of the gel formula at lower temperatures.

Conformance-improvement gels, consisting of an acrylamide polymer crosslinked with a set of organic crosslinking agents, have been reported to form strong and stable gels up to 350°F.[90] This is an interesting observation because, in both this and the CC/AP conformance gel technology, it is believed that it is not the crosslinking chemistry that is limiting high-temperature gel stability, but limitations in the thermal stability of the gel's organic polymer. Thus, one interpretation of these observations is that the acrylamide polymer used in the organic-crosslinked acrylamide-polymer gels was of a purer and more stable form.

The upper reservoir temperature limit, at which a given gel technology is stable and functional, is an interrelated function of polymer concentration and chemistry used in the gel, hardness divalent-ion concentrations within the gel's makeup water, polymer purity, and, if used, the chemical stabilizer package.[5] For organic polymer gels, stability is, in part, a function of the level of free-radical and free-radical-precursor chemical impurities in the polymer material itself. Free radicals cause organic-polymer backbone scission and associated polymer-gel degelation.

It is imperative to truly know what the upper reservoir temperature limit is for a conformance gel technology that is to be applied to a high-temperature reservoir and to not apply such a treatment at a temperature exceeding the temperature rating of the gel being used.

Treatable Conformance Problems. **Table 13.3** provides guidelines as to which conformance problems are attractive and unattractive to treat with polymer gels.

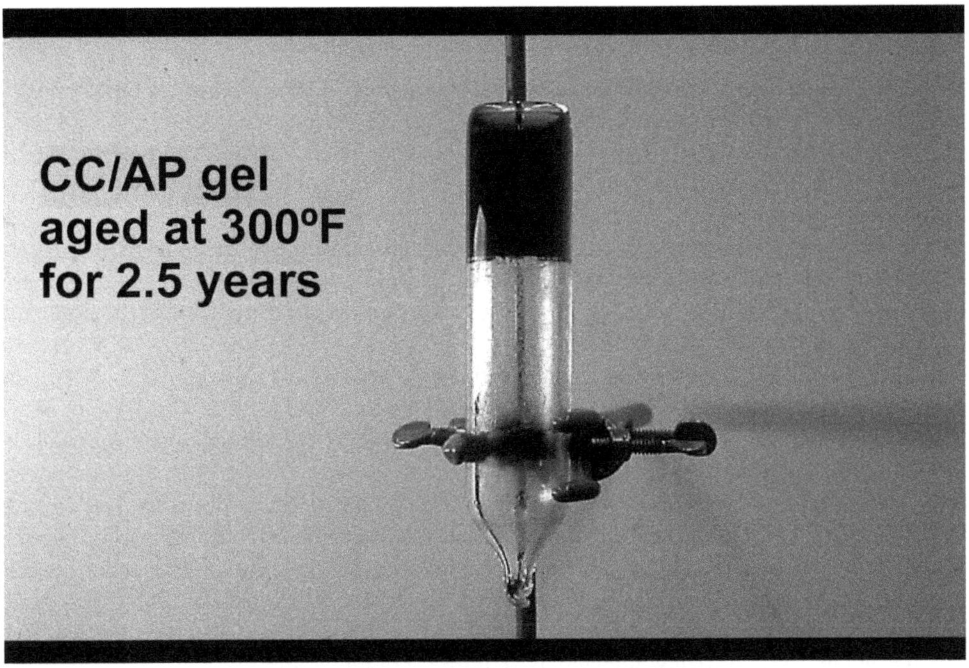

Fig. 13.26—CC/AP gel aged for 2.5 years at 300°F.

TABLE 13.3—CONFORMANCE PROBLEMS THAT ARE ATTRACTIVE TO TREAT WITH POLYMER GELS	
Matrix Conformance Problems	
Without crossflow	Yes
With crossflow	Challenging—must place very deeply
Fracture Conformance Problems	
Simple	Depends—case-by-case basis
Network—intermediate intensity and directional trends	Yes
Network—highly intense	Often not
Hydraulic	Yes
Coning	
Water and gas via fractures	Yes
Water and gas via matrix reservoir rock	No
Behind Pipe Channeling	Yes, for microflow channels
Casing Leaks	Yes, for microflow channels

Gel Treatment Design. The first step in designing a gel treatment is to correctly identify the nature of the conformance problem to be treated. This includes, during water- or gas-shut-off treatments, identifying the flow path of excessive water or gas production from its source to the production wellbore.

Gel Technology Selection. The following procedure for gel technology selection is highly generalized, and the procedure should be modified as dictated by the actual reservoir conformance problem to be treated. If a service company or a company specializing in conformance-treatment gels is to be involved, they should be consulted during each step of the selection process. A prerequisite is to eliminate all gel technologies, if any, that are prohibited by locally applicable safety or environmental regulations.

First, determine the type of problem that is to be treated. That is, whether it is a matrix-rock problem or a high-permeability-anomaly problem, such as fractures. If treating a matrix-rock problem, decide if you need a gel for treating near-wellbore or deeply in the reservoir. Determine how strong the gel needs to be. If the gel must be placed deep in the reservoir, economic constraints on the unit volume cost of gel pumped need to be carefully factored into the gel technology selection process. If the gel treatment will require a shut-in time following gel placement, decide on the shut-in strategy (i.e., what shut-in time range is acceptable) and chose a gel technology accordingly. Then, begin to select a gel technology that will function as required at the reservoir temperature being treated.

Next, if treating a high-permeability-anomaly (> 2 Darcy) conformance problem, select a gel-treatment fluid that can be injected into the reservoir in a mature or partially mature state. If treating matrix-rock conformance problems for which all the permeabilities are less than approximately 2 Darcy, select a gel technology that can be injected in the gelant state. After completing these steps in the gel-selection process, decide what the initial and near-full gelation times should be. Depending on the exact needs of the treatment design so far, predict the thermal history that the gelant will experience as it is being injected. If a precise thermal history for the gelant as it is being injected is required, computer thermal-simulation work may be needed.

Based on the outcome of the gel technology selection process so far, numerous conformance gel technologies may have already been ruled out. If so, the remaining selection process may be simpler. Next, the remaining gel technologies, which fit all the gel technology selection criteria, should be sorted through to select the gel technology that will perform the most effectively in treating the reservoir conformance problem and will meet the specialized needs of the operator who is applying the gel treatment. At this point, if economically justified, comparative laboratory studies may be conducted to help select with more certainty which of several gel technologies will perform most effectively in treating the reservoir conformance problem that is to be remedied.

Treatment Sizing. At this writing, technical sophistication relating to the method used to size the volumes of various conformance-improvement gel treatments needs improvement. The encouraging news is that there are active R&D programs that are specifically pursuing the development of a more rigorous scientific and engineering basis for use in the sizing of conformance-improvement gel treatments. Ref. 28 reports on a sound engineering basis for the sizing of polymer-gel treatments, injected as gelants, for treating excessive water production that results when a hydraulic fracture inadvertently extends down into an aquifer or into some other type of water-bearing geological strata. Ref. 28 also references a user-friendly computer program for use in sizing such gelant treatments and provides information on how the computer program can be downloaded.

Unfortunately, sizing of gel treatments remains highly empirical and is often based on the experience of operators and gel service companies. There are a number of empirical guidelines for sizing conformance gel treatments. When conducting near-wellbore gel treatments in matrix-rock reservoirs for total shutoff purposes, such as when using a CC/AP gel treatment for zone-abandonment or squeeze-and-recompletion treatment purposes, the rule of thumb is to inject a gel treatment volume that extends, on the average, 6 to 9 ft radially from the wellbore. This is a good guideline when treating reservoirs with elevated temperatures (> 150°F) and large drawdown pressures (> 300 psi). For lower temperatures and lower drawdown pressures, this rule of thumb can be relaxed somewhat. The depth of gel placement must be increased for gel tech-

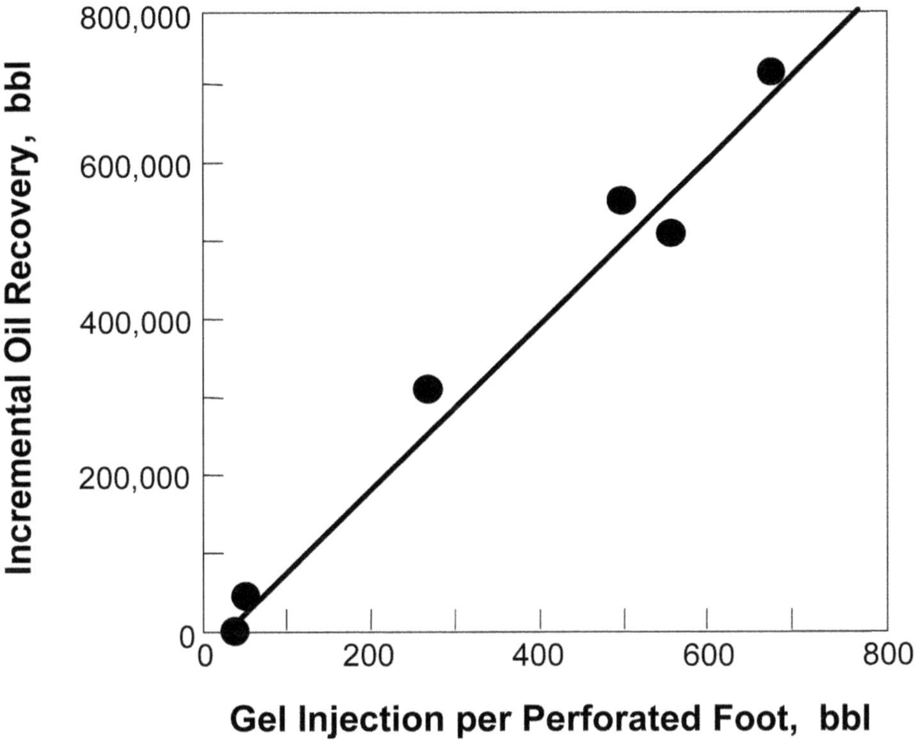

Fig. 13.27—Incremental oil production vs. volume of gel injected for CC/AP gel treatments applied to a series of fractured injection wells.

nologies that produce relatively weaker gel strengths. For any given gel treatment applied for total-fluid-shutoff application in the near-wellbore region in matrix reservoir rock, as the gel chemical loading is increased, the rule of thumb can again be relaxed (but do not treat, on the average, less than 3 ft radially). When using such higher chemical loadings in the gelant formula, the required depth of treatment penetration can be reduced, in part, because more chemical is available on the leading edge of gelant volume to be consumed and lost to the reservoir rock in the treated reservoir volume.

A series of CC/AP gel treatments were applied in an economically attractive manner as sweep-improvement treatments to waterflood injection wells of carbonate and sandstone reservoirs possessing fracture networks of intermediate intensity with directional trends. Gel treatment volumes ranged between 50 to 700 bbl of gel injected per perforated foot of reservoir pay zone.[106] **Fig. 13.27** shows how the incremental oil production for these treatments increased as the amount of gel injected per perforated foot increased. Gel treatment volumes for this series of treatments ranged up to 37,000 bbl of gel injected. The trend depicted in Fig. 13.27 cannot be extrapolated to an infinite volume of incremental oil production. This curve must eventually bend over and not exceed the recoverable amount of oil in any given reservoir. However, for the volumes of gel injected for this series of gel treatments, the volume of gel injected had not yet begun to approach the volume of gel required to cause the curve to bend over.

Gel treatment volumes of 25 to 50 bbl (100 bbl in specialized cases) of gel injection per perforated foot of reservoir producing interval have been reported to have been injected during successful CC/AP gel water-shutoff treatments of production wells of naturally fractured reservoirs.[5] One to two barrels of gel per foot of treated wellbore interval were injected during a series of total-shutoff CC/AP gel treatments that were successfully applied as chemical liner completions to the curved section of openhole short-radius horizontal wellbores.[113] These gel chemical-liner treatments were applied during the drilling of short-radius horizontal wells in a thin oil column. The gel chemical-liner treatments were applied to the curved section of the openhole borehole. Without the application of these chemical-liner gel treatments, excessive and uneconomic volumes of gas would have been produced into the short interval of openhole curved section of the borehole that extended up into the gas cap. For gel water-shutoff treatments to be applied to a vertical fracture that intersects a horizontal wellbore, it has been suggested that the gel needs to be placed only deep enough so that the gel will be able to prevent direct water entry into the borehole at the point of the fracture intersection.[114]

In general, an acceptable design strategy for sizing the volume of gel treatments applied to matrix rock (unfractured) reservoirs is to specify the radial distance that a gel treatment will extend, on the average, from the treated wellbore. The exact values of the required gel radial-penetration distance will be a strong function of, and vary widely with, the nature of the specific gel to be used, the nature of the reservoir-conformance problem, the nature of the reservoir properties, and the well's drawdown pressure. When conducting a gel treatment in a reservoir with a significant amount of fracturing, discussing radial depth of gel penetration into the reservoir from the wellbore is near meaningless, unless the fracture reservoir plumbing is quantitatively well characterized.

Injection Rate. When injecting fracture-problem gels, especially polymer gels that undergo gel dehydration during placement, the gel should be injected as rapidly as practical (without exceeding parting or fracture pressure) if the objective is to place the gel as deeply as possible into the fracture or fracture network. If the objective is to maximize the strength of the placed fracture-problem gel, then the gel should be injected as slowly as is feasible.[115]

In general, gel treatments should be injected as rapidly as feasible without exceeding reservoir parting or fracture pressure. For a gel with a given gelation onset time and gel that is to be injected into a matrix rock reservoir and for the situation of the gel-treatment pumping time possibly exceeding the gelation-onset time, maximizing the injection rate will maximize the amount of gel that can be injected within the gelation-onset-time constraint. Maximizing the injection rate will also reduce pumping time and costs.

A special word of caution needs to be made. If while pumping a gel treatment unexpectedly high or rapidly increasing injection pressure is encountered, normally a poor option to choose is to cut the pumping rate. When injecting a gel treatment, cutting the injection pump rate increases the residence time of the gel in the injection tubulars and results in a more structured gel being injected into the reservoir. Both of these outcomes are the opposite of what is needed if excessive injection pressure is being encountered. Also, many polymer-gel solutions are shear-thinning fluids. Thus, reducing the rate at which these gel solutions are being pumped results in a more viscous gel solution being pumped and injected. Again, this is the opposite of what is needed if excessive injection pressure is being encountered. The better options are to either stop gel injection and immediately clear the injection tubulars with water or reduce chemical loading in the injected gel.

Overdisplacement. The importance of the overdisplacement fluid itself and the volume of the overdisplacement fluid can range from negligible to profound. For example, the nature of the switch over to injecting water immediately following a large-volume weak gel treatment that is applied to an injection well as a sweep-improvement treatment is normally very straightforward and noncrucial. However, when applying a polymer-gel water-shutoff treatment to a naturally fractured reservoir surrounding a production well for which the excessive water production results from fracture channeling during a waterflood, the choice of the fluid and the volume injected of the overdisplacement fluid following gel injection is a crucial element of the treatment design and can have a major effect on treatment performance.

Balancing the following three (often opposing) requirements is critical to the success of this type of gel water-shutoff treatment. The first requirement is to displace the gel deep enough into the formation so that when the well is put back on normal production, the large near-wellbore drawdown differential pressure does not overlap some of the emplaced gel volume such that the large differential pressure will exceed the critical differential pressure for gel flow and, in turn, cause such emplaced gel to be back produced. The second requirement is that the gel is not excessively overdisplaced from the most offending and troublesome of fracture flow paths so as to permit undesired water production. The third requirement is that the gel is sufficiently overdisplaced so that the gel will not excessively block desired oil production through the fracture network to the production well. Many such successful polymer-gel water-shutoff treatments require post-treatment oil flow through fractures to the wellbore. At this writing, a lot of art is used in the selection of the overdisplacement fluid and its volume for this type of fracture-problem water-shutoff gel treatment.

The three basic varieties of overdisplacement fluids are water or brine (usually injection or produced water), a polymer solution (often the polymer solution of the gel without the addition of the crosslinking agent), and a hydrocarbon liquid (such as diesel or the reservoir crude oil). For water-shutoff gel treatments, the use of a liquid hydrocarbon has been advocated as a means to establish favorable relative permeability to oil flow in the near-wellbore environment. The pros and cons of the use of a hydrocarbon overdisplacement fluid during water-shutoff gel treatments have been vigorously debated among experts. The bottom line of these debates is that whether the use of a hydrocarbon fluid is favorable is reservoir specific and is relatively more advantageous when treating matrix-rock reservoirs. The use of a viscous polymer solution as the overdisplacement fluid helps to mitigate (or completely mitigates) the problem of a non-viscous brine overdisplacement fluid fingering into the gelant solution in the wellbore and near-wellbore environment, especially in a near-wellbore fracture environment. Additionally, there are numerous gel-treatment cases for which the use of a brine overdisplacement fluid is favored for functional, operational, safety, environmental, and/or economic reasons.

For total-fluid-flow-shutoff purposes in matrix rock, both the type of overdisplacement fluid used during a water- or gas-shutoff gel treatment and how much the gelant solution is over- or underdisplaced from the wellbore are critical. Gel will be left in the wellbore if underdisplaced. Overdisplacement can result in near-wellbore flow paths being opened to flow and essentially negating the value of the entire treatment. In general, the gel is underdisplaced, and some gel is left in the wellbore downhole near the production interval when treating matrix rock problems. Gel left in the wellbore will often spontaneously clean up during normal post-gel-treatment production, but, if not, can be cleaned/jetted out of the wellbore using coiled tubing.

Shut-in Time. After the placement of many gel treatments, the well needs to be shut in to allow the gelant solution to mature and set up. Post-treatment well shut in is mandatory following fluid-shutoff gel treatments applied to production wells in matrix rock reservoirs. Usually, the well needs to be shut in for the length of time it takes the gelant to reach near-full gel strength under the conditions encountered within the treated reservoir volume.

Returning a Well to Production. During the application of gel treatments to production wells for water- and gas-shutoff purposes, the manner in which the well is returned to production can have a major impact on the performance of the gel treatment. It is generally recommended to slowly return a gel-treated production well back to full production over a period of a day to several days. Obviously, the exact manner in which a production well should be slowly returned to full production is a strong function of the nature and the strength of the particular gel used and the nature of the reservoir being treated.

Injection Mode. There are two major modes of injecting conformance polymer-gel treatments. The single-fluid injection mode involves incorporating all the chemical components required to form the gel into a single gelant (pregel) solution. Essentially all modern conformance-improvement polymer-gel treatments are injected in this mode. As discussed in the Classification of Gel Treatment Types subsection of Sec. 13.6.1, many early polymer-gel treatments were pumped using the sequential-fluid injection mode.

Disproportionate Permeability Reduction. Sec. 13.4 reviewed the applicability, limitations, desirability, and fundamental concepts of DPR as it applies to conformance-improvement (including water and gas shutoff) treatments. It focused on the application of radial-flow DPR conformance-improvement treatments that are applied in matrix rock (unfractured) reservoirs. The development and successful application of conformance-improvement DPR gels is of substantial interest to the petroleum industry, as indicated in Refs. 4, 19 through 21, and 37 through 39. Gels that are used in matrix-rock radial-flow DPR gel treatments are typically characterized as being relatively weak gels.

A DPR gel treatment scheme exists[28] that often favors the use of relatively strong DPR polymer gels, for treating excessive water or gas production problems that occur in wells producing from a fractured oil-bearing formation. Such DPR fracture-problem gel treatments, which are conducted for water-shutoff purposes, are applied in the situation in which the oil-producing layer (producing 100% oil) is underlain or overlain by a water-producing zone and the vertical fractures provide the conduit for the undesired water production. The scheme was originally developed for, and is most directly applicable to, a production well that has been hydraulically fractured, where the hydraulic fracture has inadvertently extended down into, or up into, a water zone. Such gel DPR water-shutoff treatments require that the gel be placed in the matrix reservoir rock that is adjacent to water-producing fracture. A publicly available computer program can help design and size such a DPR water-shutoff gel treatment.[28] A strong CC/AP gel would be a good candidate for use in this type of DPR fluid-shutoff treatment. This DPR gel treatment scheme for imparting water-shutoff/reduction in fractured reservoirs is also applicable as production-well water-shutoff treatments that are applied to naturally fractured reservoirs in which the oil zone overlies or underlies a water zone or aquifer and when hydraulically fracturing through thin interbedded oil- and water-bearing geological strata.

13.6.2 Illustrative Field Results. The following examples illustrate field applications of various types of conformance-improvement gel treatments.

Fracture Problem Sweep-Improvement Treatments. **Fig. 13.28**[5,106] shows the type of production response that is possible when applying a polymer-gel treatment to a waterflood injection well to improve sweep efficiency. The sweep-improvement treatment involved the application of a CC/AP gel treatment. The figure shows the combined production-response of the four direct offsetting production wells to the gel-treated injection well. The gel treatment was applied for waterflood sweep-improvement purposes to the naturally fractured Embar carbonate formation surrounding Well O-7 of the highly mature SOB field in the Big Horn basin of Wyoming. The wide variations in WOR and oil production rate are quite common in many of the well patterns of this highly fractured reservoir. Ref. 106 provides more details regarding the 20,000 bbl gel treatment. The economics of applying this gel treatment were exceptionally good.

Fig. 13.29[5,116] shows seven years of incremental-oil-production response for the combined direct offset producing wells to CC/AP gel-treated injection well O-17 of the SOB field. This figure illustrates the type of treatment longevity that can be expected from CC/AP gel treatments. Well O-17 was treated in approximately the same manner and as part of the same gel treatment series cited for the Well O-7 gel treatment in Fig. 13.28.

Water Shutoff Treatments. **Fig. 13.30** shows the production response type that can occur when a gel treatment is applied to a production well to reduce excessive water production. In this case, a CC/AP gel treatment was applied to a production well for water-shutoff purposes. The gel treatment was applied to the 145°F fractured carbonate reservoir surrounding produc-

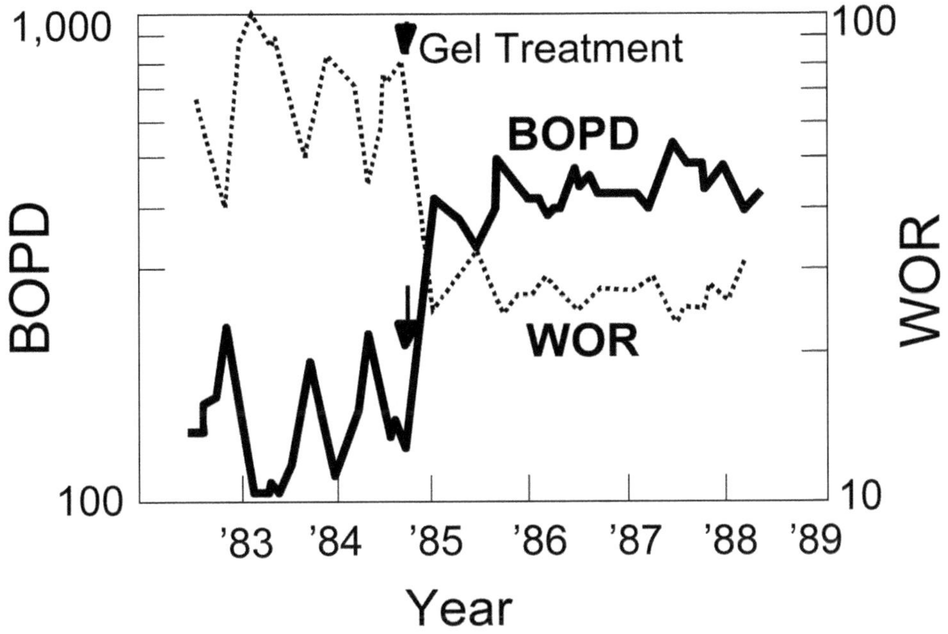

Fig. 13.28—Production response to the CC/AP gel treatment applied to injector O-7 in the SOB field.

tion well LSD N-17P in a field in Wyoming's Big Horn basin. The excessive water production, which was occurring during primary production, was thought to be edge water encroaching through fractures from a strong aquifer. Water production before the gel treatment was 5,000 BWPD. Following the treatment, the water production was reduced to 1,000 BWPD.[5] There was a short-lived (several months) modest oil-production increase following the treatment that paid out the cost of the gel treatment. The primary objective and benefit was the water shutoff.

More recently, a large number of CC/AP gel water-shutoff treatments have been successfully applied to often old and highly marginal production wells of the dolomitic Arbuckle formation in Kansas. The excessive and detrimental water production is believed to be aquifer water coning up to the production wells via fractures or some other type of high-permeability anomaly. Successful Kansas Arbuckle CC/AP gel water-shutoff treatments often reduce the water production rate by more than 90%.

Willhite and Pancake reported in 2004 that more than 250 CC/AP water-shutoff treatments had been applied to Kansas Arbuckle wells.[117] Their paper reported that incremental oil production was, in general, the business driver for conducting these gel treatments. The focus of their paper was on seven of these gel treatments that were studied in detail, where downhole pressure data was obtained before, during, and after the gel treatments. Highlights relating to these seven Arbuckle gel water-shutoff treatments are as follows:

- Water production was reduced in every well by 53 to 90%.
- Incremental oil production was obtained for 5 out of the 6 wells that were produced for 6 months after the gel treatment.
- Oil productivity indexes increased following the gel treatments.
- Incremental oil production increased with increasing volume of the gel treatment for the wells that were completed open hole.
- The duration of the treatment response is expected to be a function of the volume of the gelant injected.

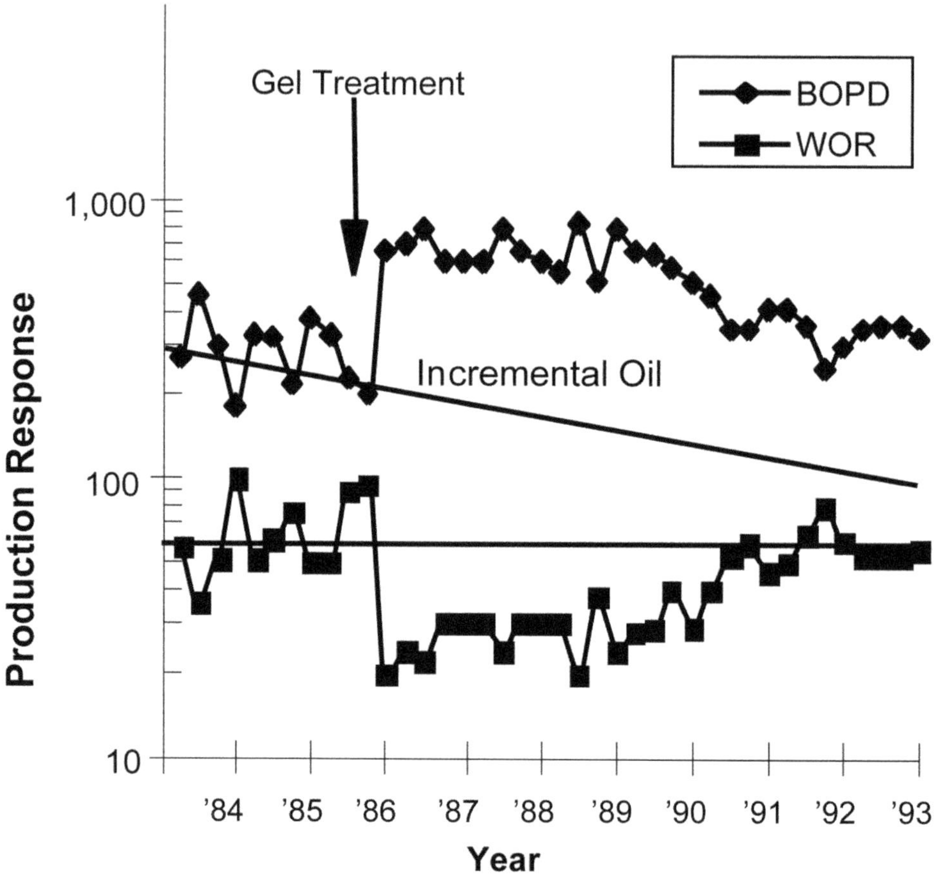

Fig. 13.29—Production response to the CC/AP gel treatment applied to injector O-17 in the SOB field (after Ref. 5 and Southwell[116]).

Rodney Reynolds reported on the performance of 300 CC/AP gel water-shutoff treatments that were applied in the Arbuckle formation of Kansas by more than 30 different operators.[118] The 250 CC/AP gel water-shutoff treatments of Ref. 117 are likely a subset of these 300 gel treatments. The following analyses are based on performance data that were obtained from 95 of the 300 treatments and then factored/proportioned up to the 300 CC/AP gel water-shutoff treatments. Reynolds reported the following performance attributes for the 300 Arbuckle water-shutoff treatments (based on the 95 gel treatments analyzed):

• Shut off 110,000,000 bbl of excessive and undesirable water production.

• Generated 1,600,000 bbl of incremental oil from these, in general, old and marginal wells.

• Cost of the gel pumped ranged from approximately U.S. $10 to $15 per bbl.

• Average treatment size was 2,600 bbl of gel fluid injected.

• For gel treatments with "average performance," payout times ranged from 3 to 6 months (based on incremental oil production only).

• Average incremental oil production from these "marginal" production wells was 5,500 bbl of oil.

• Operators reported reserve development costs for the incremental oil of U.S. $2 to $7 per bbl.

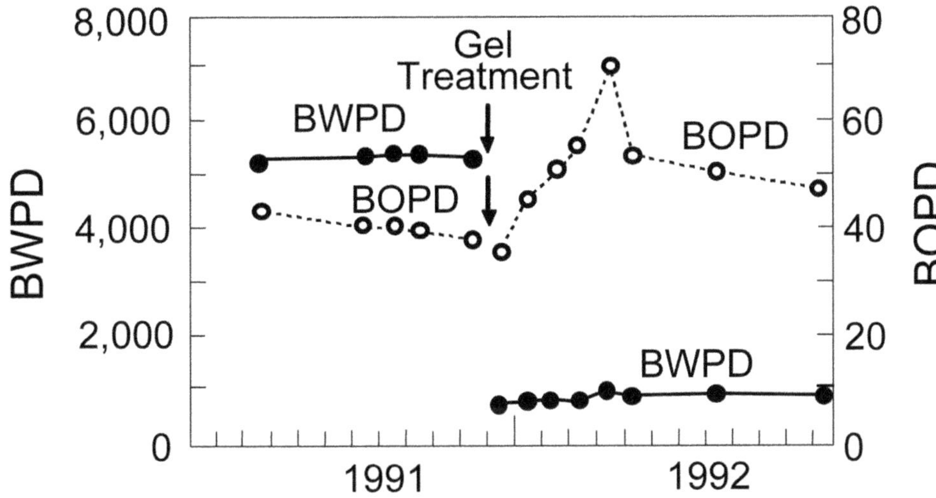

Fig. 13.30—Production response to the CC/AP gel water-shutoff treatment applied to production well LSD N-17P.

Gas Shutoff Treatments. Ref. 119 reported on 37 CC/AP gel gas-shutoff treatments that were applied to 31 production wells in the 190 to 220°F reservoir of Alaska's Prudhoe Bay field. It reported that these gas-shutoff gel squeeze treatments cost 75% of comparable Portland cement gas-shutoff squeeze treatments in this field and afforded a higher success rate. Ref. 119 also reported that these gel gas-shutoff treatments had been credited with a gross initial (one month of production) incremental-oil-production rate of 22,000 BOPD and that these gel treatments reduced gas production to 213 MMscf/D. Ref. 119 stated that "squeeze longevity has been greater than one year [to date] in some cases with drawdown pressures exceeding 1,500 psi."

Microgel Sweep-Improvement Treatments. Colloidal dispersion gels of acrylamide polymer crosslinked with aluminum citrate have been applied somewhat extensively in the U.S. Rocky Mountain region, especially in the Minnelusa sand formation. These large-volume gel treatments were applied to secondary and tertiary projects to promote in-depth permeability modification in highly heterogeneous "matrix" sandstone reservoirs. Ref. 9 reviews the recovery and economics of 29 such aluminum-citrate colloidal dispersion gel treatments. The polymer concentration of the aluminum-citrate colloidal dispersion gels was typically in the 200 to 1,200 ppm range. The mechanism by which these aluminum-citrate colloidal dispersion gel treatments function is not well understood, and technical issues relating to this technology are being debated.

Fractures Intersecting a Horizontal Well. Ref. 115 reviews the successful design and application of polymer-gel water-shutoff treatments that were applied using bullhead placement for treating excess water production from horizontal wellbores into which the excessive and competing water production was emanating from a "vertical" natural fracture that intersected the horizontal borehole and extended into an underlying aquifer. This is an emerging gel water-shutoff technology that is creating substantial interest because of the large number of expensive multilateral and/or subsea horizontal wells for which any water-shutoff treatments to be applied through these horizontal wells must be bullheaded due to their often openhole, multilateral, and/or subsea completions that do not readily permit the use of mechanical zone isolation.

Use During CO_2 Flooding. Borling[120] reported on successful conformance-improvement CC/AP gel treatments that were applied at the Wertz field CO_2 tertiary water alternating gas (WAG) flooding project in Wyoming's Wind River basin. He reviewed 10 injection-well gel treatments applied to a 165°F naturally fractured Tensleep sandstone reservoir. The following

TABLE 13.4—COMPARISON OF WYOMING INJECTION-WELL TREATMENTS IN CARBONATE VS. SANDSTONE FORMATIONS

Reservoir lithology	Carbonate	Sandstone
Number of treatments	11	6
Average incremental oil recovery per treatment, STBO	261,000	129,000
Average incremental oil recovery per unit of polymer injected, STBO/lbm	17.5	7.6

benefits were reported to have been derived from applying these gel treatments during the Wertz CO_2 flooding project in this naturally fractured reservoir. The gel treatments promoted incremental oil recoveries of up to 140,000 barrels per well pattern and increased oil production rates by 100 to 300 BOPD per well pattern. The gel treatments extended the economic lives of marginal well patterns by nearly two years. They reduced GORs and WORs, reduced gas and water cycling, reduced gas and water breakthrough times, and improved water and gas injection profiles. The CC/AP gel treatments reduced operating expenses, contributed substantially to the fieldwide decline-rate reduction in 1992 from 24 to 9%, were effective where conventional oilfield foams had failed, and had rapid payout times of often less than three months. The Wertz conformance-improvement gel treatments were responsible for recovering substantial reserves that would not have been otherwise recovered.

Hild and Wackowski[121] reported on 44 injection-well CC/AP gel treatments that were applied during 1994 through 1997 at the CO_2 miscible WAG flooding project of the Rangely Weber Sand unit in northwestern Colorado. These injection well treatments (average size ~10,000 bbl) had an 80% success rate and rendered an economic rate of return of 365%.

Gas Shutoff in an Openhole Gravel-Pack Completion. A "successful and selective" gas-shutoff treatment, using an organically crosslinked acrylamide-polymer gel, has been documented.[122] The gel treatment was applied to an offshore well in the Norwegian North Sea. The gel treatment was applied to a gravel-packed well penetrating a multilayer reservoir containing shaley sandstone strata. A temporary blocking gel of crosslinked hydroxypropyl guar was selectively placed to temporarily seal and protect the upper producing portion of the well, while the organically crosslinked acrylamide-polymer gel for shutting off the lower gas-producing strata was being injected.

Carbonate vs. Sandstone Reservoirs. **Table 13.4** shows a comparison of the production response and the economic performance of a series of large-volume (4,000 to 37,000 bbl) CC/AP gel treatments that were applied, between 1985 and 1988, as sweep-improvement treatments to naturally fractured injection wells in Wyoming's Big Horn basin. The averaged combined production responses of the four direct offsetting production wells to the treated injection wells are shown. For this set of treatments, 11 of the gel treatments were applied to the naturally fractured Embar carbonate formation and six of the gel treatments were applied to the naturally fractured Tensleep sandstone formation. The gel treatments performed well in both the carbonate and sandstone reservoirs. The difference in performance of the gel treatments in the two sets of reservoir mineralogies is believed to have resulted mostly from differences in fracture characteristics and not from differences in gel performance in the two types of mineralogical formations. The fracture network of the carbonate formation in this instance is believed to be more conducive to being successfully treated with gel sweep-improvement treatments.

Chemical Liner Completions. More than 100 successful CC/AP gel gas-shutoff treatments were applied as chemical liner completions and were performed during a drilling program involving the drilling of short-radius horizontal wellbores in the Yates field of Pecos County, Texas.[5,123,124] The gel chemical liner completions were applied to the openhole curved section of the horizontal wellbores. The lateral borehole of these wells penetrates a relatively thin oil column in a fractured dolomitic reservoir, where the oil column is overlain by a substantial gas cap. The openhole curved section of the short-radius borehole partially extended up into the gas cap. If the curved section of the boreholes were not sealed, the horizontal wells would produce excessive, and often uneconomic, high GORs. In 1995, the first 80 of these horizontal wellbores, which had been treated with the gel chemical liner treatments, were producing 12,500 BOPD and had recovered approximately six million cumulative barrels of crude oil.[5]

13.6.3 Miscellaneous Gel-Treatment Issues. ***Gels Complement Portland Cement.*** A number of petroleum engineers believe that oilfield gel treatments are a competing technology to Portland cement. Gels of oilfield conformance-improvement treatments are, in reality, a complementary technology to Portland cement. Where Portland cement functions well, oilfield gels do not, and vice versa.

Gels are effective plugging agents for use in blocking fluid flow in flow channels with small apertures, such as pore-level flow paths in matrix reservoir rock and in microannulae behind pipe, where such microannulae often exist between a primary cement job and the adjacent geological formation material.

However, the solids contained within Portland cement prevent cement slurries from penetrating any significant distance into matrix reservoir rock or other similar small-channel flow paths. In fact, squeeze cement water-shutoff treatments function primarily by squeezing off perforations or by breaking down the formation rock and placing the cement into the newly formed reservoir partings. Even so-called microfine cement does not propagate any significant distance into oil-reservoir matrix rock of normal permeabilities, such as sandstone formations with permeabilities less than 1,000 md.

If the goal is to treat to any significant depth into matrix reservoir rock or other small flow channels such a microannulae behind pipe, the placement and maturation of a solids-free gelant solution of an appropriate gel is the preferred technique for shutting off fluid flow in such instances. However, if the goal is to plug large flow channels, such as a perforation, a large void space behind pipe, or oilfield tubulars, then the use of cement is the preferred technology. Oilfield gels, in general, simply do not have the favorable compressive strength characteristics of Portland cement that are required to plug such large flow channels, especially if there is any significant differential pressure involved in the fluid flow to be blocked. However, if solids are incorporated into the gelant solution of an oilfield gel, especially a polymer gel, the compressive strength of the resultant gel can be dramatically increased. When the solids loading is significantly increased in the gel or a local volume of the gel, the compressive strength of the gel can approach that of Portland cement.

A problem faced by petroleum engineers is determining the size of the fluid-flow channels to be plugged and whether they should pump an appropriate gel treatment or a Portland cement squeeze job. When conducting a gel production-well water- or gas-shutoff treatment for which there may be a significant distribution in the size of the offending flow channels, there may be uncertainty in the size of the problematic flow channels or significant drawdown pressures involved. In these instances, a good strategy is to cap off a gel treatment with Portland cement.[125] The reverse order of first injecting cement and then gel is not recommended. The cement cap should not be pumped at rates that will exceed the downhole reservoir parting or fracture pressure.

Historical Trends. Part of the explanation for the historically low success rate of gel treatments applied for the first time in a new field by inexperienced petroleum engineers is that the permeability of offending reservoir flow channels tend to have permeabilities higher than initially anticipated.[5] Several reasons for this are the tendency to average permeabilities over relatively long logging intervals, ignoring what unrecovered core intervals might be suggesting during coring operations, and incomplete core, borehole, and reservoir volumetric sampling.

The success of gel-conformance treatments tends to be proportional to the involvement of the field operator in the well-candidate selection, the treatment design, execution, and quality control. This is especially true for first-time treatments in a new field by an operator who is not highly familiar with oilfield gel treatments.[5] An important aspect in explaining this observation is that the success of conformance-improvement and fluid-shutoff treatments is highly dependent on properly understanding and diagnosing the conformance and excess-fluid coproduction problems. These conformance problems are often best addressed and determined by the operator; however, the experience and capabilities of an oilfield-gel service company in determining and deducing the reservoir conformance problems for an operator should not be overlooked and underused.

Gel Treatment Elements That Must Be Successfully Executed. The successful application and execution of a sweep-improvement or fluid-shutoff gel treatment requires that all five of the following treatment elements be simultaneously implemented, otherwise, there is a high probability of failure.

- The conformance problem must be correctly identified.
- A proper and effective gel system must be selected.
- The gel treatment must be properly designed and sized.
- The gelant solution and/or gel must be properly applied and placed.
- The gel must function as intended downhole.

The success rate of any given gel sweep-improvement or fluid-shutoff treatment is often directly proportional to the operator's involvement in all the gel treatment elements. The implementation of successful gel treatments for sweep-improvement and fluid-shutoff purposes requires a high degree of teamwork between the field's operator and the service and technology providers.[5]

Prerequisites of Good Candidate Wells and Well Patterns. Because polymer-gel conformance treatments that are applied during oil-recovery production operations are, in fact, just treatment*s*, the well or well pattern to be treated must be suffering from a treatable conformance sweep problem or suffering from a treatable excessive fluid coproduction problem. Because gel treatments alone do not reduce microscopic-displacement residual oil saturation, any well pattern to be treated successfully must contain sufficient remaining recoverable oil to make the treatment economical. When performing a fluid-shutoff treatment, the production well must be producing an excessive amount of unproductive competing water or gas.

Attributes of Good Well Candidates. Good well candidates for the application of gel conformance-improvement treatments during oil-recovery operations[5] have the following attributes. Good injection-well candidates have some combination of early injectant breakthrough, an excessive injection capacity, a substantial movable oil saturation within the well pattern, and unexpectedly low oil recovery within the well pattern. Good production well candidates are characterized by some combination of high WOR or gas/oil ratio (GOR), excessive competing water or gas coproduction, a substantial movable oil saturation within the well pattern, unexpectedly low oil recovery, early water or gas breakthrough, and high producing levels within wells that are being pumped.

Quality Control Is Critical. There is a strong correlation between service companies and operators who conduct and/or insist on a strong quality control and quality assurance program during the application of conformance-improvement gel treatments and the success rate and the degree of benefits derived from the applied gel treatments.[5]

It is important for the operator to request, and to closely monitor and/or actually participate in, the quality control program for polymer-gel conformance treatments (or for any gel technology) if the operator expects to enjoy a high success rate for such chemical treatments.[5] This is especially true for the first application of a gel treatment in any given field. The quality control program should include, but not necessarily be limited to, assuring that the proper chemicals are being used in the actual gel formula of the treatment; before pumping the gel, formulating and testing the gel with the actual chemicals and water to be used; assuring complete dissolution of the gel chemicals before injection; when conducting matrix rock treatments, assuring that the gelant solution is injectable into the matrix reservoir rock without face plugging occurring; and taking gelant/gel solution samples regularly at or near the wellhead as the treatment is being pumped.

When performing nonroutine gel treatments using any gel, particularly for a new application or during a first-time treatment in a field, properly executed bottle testing, which is conducted in the field or at a nearby laboratory using actual field gel-formula chemicals and gel make-up water and which is conducted at reservoir temperature, is an especially powerful and effective quality control and quality assurance tool. Such testing provides a semiquantitative check on gelation rate, a semiquantitative check on final gel strength, and an indication of gel stability. In addition, such testing provides a degree of assurance that the proper chemicals are being used in the actual field gel formula, and that there are no chemical interferences involving the field make-up water that will interfere with the gel. Furthermore, such testing can provide a degree of assurance that the actual field recipe being used is the correct formula.

When conducting "bottle testing" of polymer gels at high temperatures (greater than ~180°F), the use of glass-blown sealed glass ampoules and appropriate associated safety procedures are recommended for both short-term and long-term testing. For such high-temperature gel ampoule testing, the gelant solution needs to be scrupulously deoxygenated (described in the Gel Bottle Testing subsection of Sec. 13.6.1) before placing the gel samples in the oven.

Limitations, Constraints, Pitfalls, and Risks. Polymer-gel treatments are not a panacea for rendering conformance improvement, but polymer gel treatments for sweep improvement and water and gas shutoff are a relatively new, emerging, and promising technology that should be added to the petroleum engineer's toolbox. An important constraint is that, unfortunately, gel sweep-improvement and fluid-shutoff treatments tend to be highly well, well pattern, and reservoir specific. An improperly designed or executed gel conformance-improvement treatment can reduce oil or gas production rates, reduce ultimate oil or gas recovery from the treated well or well pattern, cause injection or production operational problems, and, following poorly designed producing-well treatments, result in excessive back production of gel.

Common pitfalls of oilfield sweep-improvement and fluid-shutoff treatments include improper diagnosis of the conformance problem to be treated with gel; applying a gel treatment to a radial-flow matrix rock reservoir suffering from a vertical conformance problem without selectively placing the gel in only the high-permeability geological strata; inadequate and improper quality control; applying a gel treatment designed for matrix rock application to a high-permeability anomaly conformance problem, such as a fracture problem; and failure to use a chemically robust and adequately thermally stable gel technology or a sufficiently strong gel formula. Additional pitfalls include the use of too small of a gel-treatment volume, insufficient involvement by the technical staff of the field operator in the well-candidate selection and in the design and execution of the gel treatment, overexpectations of what DPR and RPM gel treatments can do, an incomplete understanding of how microgels function, incomplete dissolution of all the gel's makeup chemicals before gel-fluid injection, and failure to properly design production-well gel treatments, thereby encountering excessive back-production of the fluid-shutoff gel.

Guidelines for Most Effective Application. In general, gel sweep-improvement treatments for promoting incremental oil production are most advantageously applied to injection wells. Water coning through vertical fractures is a problem that often can be treated successfully with polymer gels. On the other hand, water coning through matrix reservoir rock is very difficult to

TABLE 13.5—EXCESS WATER PRODUCTION PROBLEMS AND TREATMENT CATEGORIES*[7]
Category A—Conventional treatments normally are an effective choice
1. Casing leaks without flow restrictions (apertures greater than approximately $^1/_{16}$ in.)
2. Flow behind pipe without flow restrictions (apertures greater than approximately $^1/_{16}$ in.)
3. Unfractured wells (injectors or producers) with effective barriers to crossflow
Category B—Treatments in which gelants normally are an effective choice
4. Casing leaks with flow restrictions (apertures less than approximately $^1/_{16}$ in.)
5. Flow behind pipe with flow restrictions (apertures less than approximately $^1/_{16}$ in.)
6. 2D coning through a hydraulic fracture from an aquifer
Category C—Treatments in which preformed or partially formed gels are injected
7. Natural-fracture system in communication with an aquifer
8. Faults or fractures crossing a deviated or horizontal well
9. Single fracture causing channeling between wells
10. Natural fracture system allowing channeling between wells
Category D—Difficult problems in which gel treatments should not be used
11. 3D matrix rock coning
12. Cusping
13. Channeling through strata (no fractures) with crossflow
*Categories are listed roughly in order of treatment difficulty.

treat successfully with gels. Generally, gel treatments that are applied for water and gas shutoff purposes are usually most advantageously applied to production wells, except as noted below. When applying gel-conformance treatments in conjunction with gas flooding (e.g., CO_2 flooding) in naturally fractured reservoirs, whether for sweep improvement or fluid shutoff, they are usually more advantageously applied to injection wells. When treating naturally fractured reservoirs with gels for sweep improvement or water shutoff, the best reservoir candidates are found when a fracture network exists of intermediate intensity (fracture spacing) with a directional trend for the most problematic fractures.

List of Water Shutoff Problems By Increasing Difficulty to Treat. Ref. 7 proposed a straightforward strategy for the use of polymer-gel treatments to solve excess water-production problems.[7] The strategy advocates that the easiest excess water-production problems to remedy should normally be attacked first. The paper advocates that conventional water-shutoff methods (e.g., cement and mechanical devices) should normally be applied first, where applicable. **Table 13.5**[7] provides a general ranking of water-production problems and treatment categories in order of increasing difficulty to treat successfully.

13.6.4 Emerging Trends and Issues. At this writing, there are a number of significant emerging trends and issues relating to gel treatments. An important emerging trend is the attempt to effectively exploit and capitalize on the DPR and RPM properties exhibited by many polymer gels. This emerging trend was discussed in a previous section and, in general terms, early in the chapter. Five other important emerging trends are discussed next.

Selective Placement. High on the oil industry wish list in the area of conformance-improvement treatments is to be able to bullhead treatments during gel injection, especially polymer-gel treatments when treating matrix-rock-strata (unfractured) radial-flow conformance problems, particularly when fluid crossflow between the strata does not occur. To do this, the gelant must be injected into and/or function selectively in only the high-permeability and/or water-bearing strata. Three of the more promising selective placement strategies under study are described here.

The use of bridging-adsorption and/or flow-induced-adsorption properties of certain water-soluble polymer macromolecules are being studied as a means to promote the selective placement of conformance-improvement gels into high-permeability reservoir geological strata and selectively into water-producing strata.[53,54,126] This scheme involves the pretreatment injection, under appropriate conditions, of a solution containing such polymers. The use of water-reactive diverting agents has been suggested as a means to selectively plug water-bearing strata.[127]

Dual injection of two fluids to impart selective gel placement has been suggested and applied in pilot tests; however, it is not yet a routine practice, except possibly by a few large service companies. The dual-injection scheme is as follows. One of the injected fluids is a nongel, nonreactive, and nondamaging fluid that is injected into the low-permeability geological strata. The second fluid is the gelant solution that is simultaneously injected into the strata to be treated. The two fluids must be pumped down the well via isolated flow conduits, such as being pumped down two separate tubing stings or being pumped down a tubing string and the tubing annulus. When mechanical zone isolation can be used effectively (e.g., a mechanical packer), this version of the dual injection scheme is readily applicable with existing technology. The real "plum" and emerging technology in this area is to be able to dual inject the gelant and the protecting fluid without the use of mechanical-zone isolation and without performing a well-workover operation. The fluid-injection rates must be precisely controlled and selected so that the two fluids are injected only into the targeted strata. In this case, the nongelant and protective fluid that is being injected into the high-oil-saturation geological strata is often a hydrocarbon fluid, such as diesel or crude oil. The application of dual injection of fluids, which is to be applied without the use of mechanical-zone isolation, is an advanced and sophisticated technique. Such dual injection is custom designed for the well to be treated and often requires substantial computer simulation during the design phase. In addition, sophisticated downhole-pressure monitoring and computer-aided fluid-injection control will likely be required.

Selective Stimulation of Low-Permeability Strata. Another strategy being developed for treating matrix-rock vertical conformance problems is to conduct a bullhead treatment followed by selective stimulation of the damaged high-oil-saturation strata. Methods of stimulation being considered and developed are ultradeep perforation techniques, hydraulic fracturing, and, in certain and limited instances, the use of a gel chemical breaker.

Foamed Gels. Foamed gels[4,83,128,129] provide the possibility of reducing the unit-volume cost of a given oilfield gel by replacing the bulk of the volume of the relatively expensive liquid phase of a gel with a relatively inexpensive gas phase. Foamed gels, in principal, would combine desirable features of foam-blocking agents and classical gels for use in conformance-improvement treatments, especially for use in the far-wellbore environment in which differential pressures are relatively low. The low density of foamed gels provides a driving force during placement in the reservoir for foamed gels to seek out different and, at times, more favorable flow paths than denser fluids, such as conventional aqueous gel fluids. This would be especially true when foamed gels are placed in highly conductive vertical fractures. Such selective placement could be particularly effective in reducing gas override, as occurs during CO_2 flooding in naturally fractured reservoirs of the U.S. Permian Basin and the Rocky Mountain region. Two countervailing issues relating to conformance foamed gels are that foamed gels are relatively more complex chemically and operationally compared with conventional gels, and their low density requires more pump horsepower to be expended during injection than conventional aqueous-based gelants of the same viscosity. Foamed gel has been applied as a conformance-improvement technology at the Rangely field CO_2 flooding project.[130,131]

Solids Addition. One of the drawbacks of many gels is their low compressive strengths that prevent effective use when encountering large differential pressures in large-aperture reservoir fluid-flow conduits, such as centimeter aperture fractures or solution channels. When solids are added to polymer gels, the compressive strength of the gel can be greatly increased and increased up to that of Portland cement when the gel is fully loaded with an appropriate solid.

There are numerous near-wellbore conformance problems that are best treated with gel. A small minority of these wells randomly and "unpredictably" require a plugging material with more compressive strength than the gel alone can provide. Previously when such unexpected well and conformance problems were encountered and detected during a gel treatment, a separate cement job had to be called out after, or during, the gel job. When an unexpected well and conformance problem is now encountered, appropriate solids can be added to the gel fluid as it is being pumped and only needs to be added near the end of gel fluid injection to be able to cost effectively plug the flow conduits that have unexpectedly wide apertures. This can be done cost effectively on the fly without the need to subsequently conduct an additional cement squeeze treatment. One of the keys to the effective addition of solids to conformance-improvement gels is to know how to precisely control the screen out of such solids addition during the placement of the gel treatment.

Gels Functional Only in the Presence of Water. Attempts have been made in the past to develop conformance-improvement gel technologies that are functional and active only in the presence of water, but inactive or inactivated in the presence of oil. There has been revived interest in developing such a gel technology.[127] One of the technical challenges that has to be overcome is that even when there is 100% oil flow in a matrix-rock, connate water also exists.

13.6.5 Gel Breakers. When an aqueous gel is contacted under appropriate conditions, chemical breakers can degrade the gel back to a low-viscosity and watery solution. Two possible reasons to use a breaker after a conformance-improvement gel treatment are to remove gel from the wellbore or perforations and to undo a gel treatment in the near-wellbore region if it was determined after its placement that the emplaced gel was not beneficial.

There are several reasons why a chemical breaker cannot be used to successfully and fully degrade a gel that has been placed deeply in either a matrix-rock or a fractured reservoir. First, successfully delivering the chemically reactive gel-breaker solution deeply in an oil reservoir is a daunting task. Second, and more fundamentally problematic, even if a chemical breaker solution could be 100% effective in the reservoir during its entire gel-breaking life, once injected into the reservoir, the gel-breaker solution would tend to wormhole through the emplaced gel. Thus, the chemical breaker would only be able to regain a small fraction of the pregel-treatment fluid-flow capacity within the gel-treated reservoir volume.

Many biopolymer gels and freshly placed inorganic gels can be chemically broken and reversed by contacting them with a strong acid solution. However, acids are usually ineffective at chemically breaking down metal-crosslinked synthetic-organic-polymer gels, such as metal-crosslinked acrylamide-polymer gels. Acrylamide-polymer gels can be chemically degraded back to a watery solution by contacting them with a free-radical chemical breaker, such as hydrogen peroxide, sodium hypochlorite of bleach, and ammonium peroxide. Free radicals chemically degrade polymer gels by a polymer-backbone scission mechanism.

Hydrogen peroxide is, in many instances, the most chemically powerful of the gel chemical breakers commercially available. However, its decomposition is catalyzed by tubular rust and many other oilfield substances, such that the injected hydrogen peroxide can be rendered essentially spent before it can be delivered to the downhole gel. The use of hydrogen peroxide may be favored when plastic-coated well tubulars have been used. Hydrogen peroxide is an extremely reactive chemical. It is advised to not inject concentrations exceeding 10% hydrogen peroxide; however, a concentration of less than 5% is ill advised, because ineffectively low concentration of hydrogen peroxide will often result downhole. Hydrogen peroxide decomposes to water and free oxygen during the gel-degradation process. The creation of oxygen in the wellbore and/or the reservoir after the hydrogen peroxide is injected raises safety issues that need to be addressed as part of the hydrogen peroxide selection process.

Bleach (containing sodium hypochlorite) is probably the most widely used material to chemically breakdown acrylamide-polymer gels. It is more chemically robust downhole than hydrogen peroxide. A note of caution: when hydrogen peroxide or bleach is used to break gels crosslinked with a chromium (III)-containing crosslinking agent, some of the chromium (III) will be converted, at least temporarily, to chromium (VI). Because any Cr (VI) that might be formed in the chemically reducing reservoir environment is rapidly converted back to relatively nontoxic Cr (III), field experience has shown that this is often only a theoretical concern. Most oil reservoirs are characterized as having a chemically reducing environment. It should be noted that certain types of metal-crosslinked polymer gels, under certain conditions, can be degelled when contacted with an aqueous solution containing a high concentration of either a caustic chemical (e.g., sodium hydroxide) or a strong ligand (e.g., oxalate).

If an effective delayed and single-fluid reversible-gel technology, especially a reversible-polymer-gel technology, were to be developed in which the gel chemical breaker or gel-breaking mechanism were chemically built directly into the gel structure itself, there would be numerous oilfield applications for such reversible gels. The delayed gel reversal/degelation time would need to be controllable.

If a water-soluble chemical breaker (breaker not built into the gel structure itself) were incorporated into an aqueous gel formula and the gel were placed downhole under a differential pressure (as usually is the case), then as the chemical breaker begins to break down the gel, the differential pressure would begin to squeeze water, including the dissolved breaker, out of the gel. Unfortunately, increasing breaker concentration is required as the concentration of the polymer increases in the gel. When a water-soluble breaker is being squeezed out of the partially broken gel, the opposite trend is occurring. Thus, the use of water-soluble chemical breakers incorporated into a single-fluid aqueous gel under differential pressure always results in an incomplete gel break and always leaves a significant gel residue. To date, the addition of a water-soluble chemical breaker into a single-fluid aqueous gel formula has not proven effective in fully degrading a gel when the gel is broken under differential pressure.

13.7 Foams

Currently, the three major applications of conformance-improvement oilfield foams are as a mobility-control agent during steamflooding, a mobility-control agent during CO_2 flooding, and gas-blocking/plugging agents placed around production wells, often applied in conjunction with a gas-flooding project.

Although the use of foams for oil-recovery applications has been actively considered and studied for more than forty years, widespread application of foams for improving oil recovery has not occurred to date. In the pioneering work of the late 1950s and through the early 1970s, foam was identified to be a promising candidate for improving mobility control and sweep efficiency of oil-recovery drive fluids, especially gas-drive fluids.[132–138] Early R&D personnel observed the following foam characteristics:

- Foams can be quite effective at reducing gas mobility.
- On the microscopic scale, the gas and liquid phases of foam flow separately through porous media with the liquid usually flowing as thin films or lamellae that are separated by gas bubbles.
- The pressure gradient during foam flow is proportional to the liquid flow rate, but quite independent of the gas flow rate.
- The macroscopic effective viscosity of foam during its flow in porous media is a function of the number and strength of the lamellae (alludes to the importance of foam texture and bubble size).

• Foams, at times, tend to promote larger mobility reductions in high-permeability porous media, as compared with lower-permeability porous media (an attractive property for improving conformance and reducing channeling).

• Foams might be good candidates for use as gas-flow blocking agents.

These early workers also noted that oil in porous media often tends to destabilize most aqueous foams and tends to harm oilfield foam performance. A number of the earliest oil-industry proponents of the use of foam hoped that foams would eventually lead to routine "air flooding" of reservoirs. This has not come to fruition.

The earliest study of foams for use during oil-recovery flooding operations attempted to capitalize on the ability of numerous aqueous-based foams to significantly reduce the mobility of gas flow during gas flooding and to be able to substantially improve oil-recovery flood sweep efficiency when flooding with a gas. In concept, foam flooding offers an alternative to polymer flooding. That is, foams can also provide mobility control during oil-recovery flooding operations. Early study and development focused on foam use for mobility-control purposes during oil-recovery flooding projects, especially during gas-flooding operations.

The focus of foam development and application has changed in more recent years. Two major factors have been largely responsible for promoting this change. First, it is unclear if foams (especially steam and natural gas foams) can be propagated distances of more than 100 ft in an oil reservoir because of the substantial minimum pressure gradient required for foam propagation and in view of the small pressure gradients that exist in most of the volume of matrix rock reservoirs. Second, economics now tend to favor small-volume chemical treatments (e.g., gel conformance-improvement treatments) over chemical-based improved oil-recovery flooding operations. Thus, the focus of oilfield foam development and application has shifted somewhat toward the use of foams as blocking/plugging agents that are part of relatively small volume treatments applied through production wells, especially for use as blocking agents to gas flow. The fluid-flow-blocking and permeability-reducing propensity of foams is one of the major factors hampering effective application of mobility-control foams (especially steam and natural gas foams) in the far-wellbore environment. The significant negative impact that crude oil often exerts on the desired performance of foams during mobility-control flooding also helped to shift the focus of oilfield foam use from mobility-control applications to fluid-flow-blocking treatments.

Conventional foams (i.e., not polymer-enhanced foams and foamed gels) are considered effective only when placed in matrix reservoir rock and are not applicable when placed in reservoir fracture channels with aperture widths on the order of greater than 0.5 mm. The application of foams for sweep-improvement and gas/water-blocking purposes is considered to be an advanced and nonroutine form of an oilfield conformance-improvement operation. It is recommended that the average petroleum engineer not undertake a foam conformance-improvement operation without in-house or commercially available technical support and/or without support from an organization that has expertise in conformance-improvement foam technologies. In addition, before implementing a foam conformance-improvement operation, it is usually necessary to perform a laboratory evaluation of the proposed foam formulation and the actual foam process to be used in the field.

13.7.1 Fundamentals and Science of Foams. ***General Nature of Foams.*** Bulk foam, as found in the head of a glass of beer or as found in association with cleaning solutions, is a metastable dispersion of a relatively large volume gas in a continuous liquid phase that constitutes a relatively small volume of the foam. An alternate definition of bulk foam is an "agglomeration of gas bubbles separated from each other by thin liquid films."[139] In most classical foams, the gas content is quite high (often 60 to 97% volume). In bulk form, such as in oilfield surface facilities and piping, foams are formed when gas contacts a liquid in the pres-

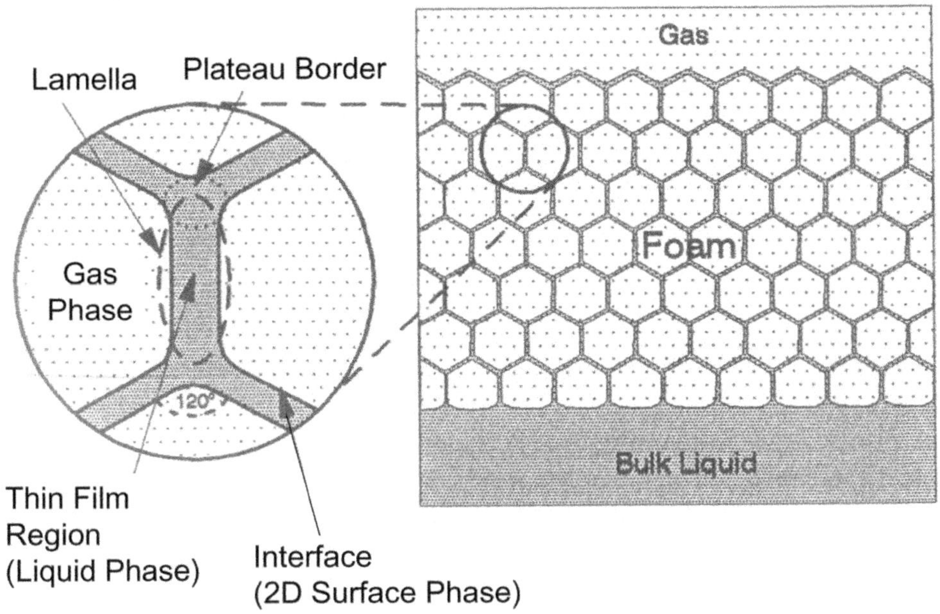

Fig. 13.31—Generalized 2D slice of a bulk foam system.

ence of mechanical agitation. As used herein, bulk foams are foams that exist in a container (e.g., a bottle or pipe) for which the volume of the container is much larger than the size of the individual foam gas bubbles.

Capillary processes control the formation and properties of foams in porous media. Foams for use in conformance improvement are dispersions of microgas bubbles usually with diameters/lengths ranging between 50 and 1000 μm. Foam in porous media exists as individual microgas bubbles in direct contact with the wetting fluid of the pore walls. These microgas bubbles are separated by liquid lamellae that bridge the pore walls and form a liquid partition on the pore scale between gas bubbles. Foam propagates in most matrix reservoir rock as a bubble train in which each gas bubble is separated from the next by a liquid lamellae film. In many instances, individual foam bubbles in reservoir matrix rock can be many pore bodies in length. Gauglitz *et al.* have defined foam structure in porous media as "a dispersion of gas in a continuous liquid phase with at least some gas flow paths made discontinuous by thin liquid films called lamellae."[140]

All foams discussed in this chapter and all foams that are used for conformance improvement have surfactants dissolved in the foam's liquid phase to stabilize the gas dispersion in the liquid. The gas phase of a foam can include both a classical gas and a supercritical gas, such as supercritical/dense CO_2. Except as specially noted, all foams discussed in this chapter that are used to impart oilfield conformance improvement are aqueous-based foams. This chapter is limited primarily to the discussion of surfactant-stabilized aqueous-based foams for use in improving conformance during oil-recovery operations.

Fig. 13.31 shows a 2D slice of a generalized bulk foam system.[141] The thin liquid films separating the foam gas bubbles are defined to be foam lamellae. The connection of the three lamellae of a gas bubble at a 120° angle is referred to as the Plateau border. In persistent bulk foams, spherical foam gas bubbles become transformed into foam cells, polyhedra separated by nearly flat thin liquid films. Such a foam is referred to as a dry foam. The polyhedra foam cells are almost, but not quite, regular dodecahedra. In three dimensions, four Plateau borders of a foam cell meet at a point at a tetrahedral angle of approximately 109°.[141]

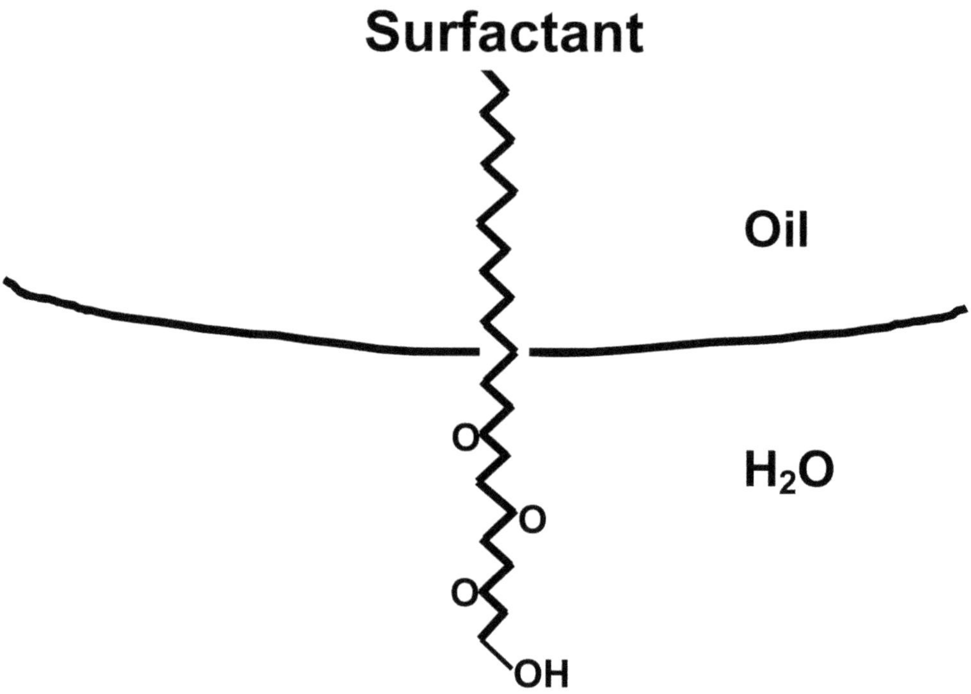

Fig. 13.32—Depiction of a polymer molecule residing at an oil/water interface.

Foams in porous media generally have bubbles that are as large as, or larger than, the pore bodies. Foam exists in reservoir-rock porous media as bubble trains where the Plateau border of the foam lamellae is formed at the pore wall and has, for static nonflowing foam in the pore body, an angle of about 90° between the liquid lamellae and the pore wall.

Foaming Agents. Surfactants are the necessary third ingredient required for the formation of the foams discussed in this chapter. An understanding of basic surfactant chemistry is essential when selecting a proper surfactant for a given oilfield foam application.

A surfactant molecule contains, within the same molecule, both a polar and nonpolar segment. The polar or hydrophilic segment of a surfactant molecule has a strong chemical affinity for water. The nonpolar or lipophilic segment has a strong chemical affinity for nonpolar hydrocarbon molecules. When water and oil or water and gas are in contact, surfactant molecules tend to partition to the oil/water or gas/water interface and reduce the interfacial tension of the interface. **Fig. 13.32** depicts a surfactant molecule residing at an oil/water interface. The partitioning of the surfactant molecule to the gas/water interface and the ensuing reduction of the interfacial tension is the primary mechanism through which surfactants stabilize dispersions of gas in water to form metastable foam.

Surfactants are classified into four types that are distinguished by the chemistry of the surfactant molecule's polar group.

• Anionics—The polar group of an anionic surfactant is a salt (or possibly an acid) where the polar anionic group is directly attached to the surfactant molecule and the counter and surface-inactive cation (often sodium) is strongly partitioned into the aqueous side of an oil/water or gas/water interface. Anionic surfactants are often used in oilfield foams because they are relatively good surfactants, generally resistant to retention, quite chemically stable, available on a commercial scale, and fairly inexpensive.

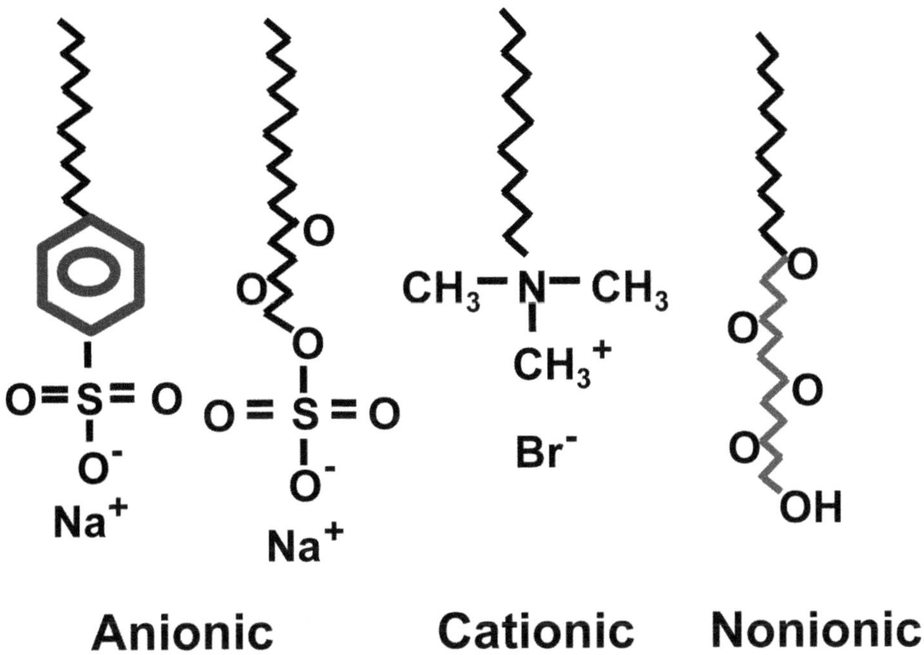

Fig. 13.33—Surfactant chemistry types.

• Cationics—The polar group of a cationic surfactant is a salt where the polar cationic group is directly attached to the surfactant molecule and the counter and surface-inactive anion is strongly partitioned into the aqueous side of an oil/water or gas/water interface. Cationic surfactants are infrequently used in oilfield foams because they tend to strongly adsorb on the surfaces of clays and sand and are relatively expensive.

• Nonionics—The polar group of a nonionic surfactant is not a salt, but rather a chemical specie, such as an alcohol, ether, or epoxy group, which promotes surfactant properties by imposing electronegativity contrast. Nonionic surfactants are less sensitive to high salinities and can be relatively inexpensive.

• Amphoterics—Amphoteric surfactants contain two or more characteristics of the previously listed chemical types of surfactants.

Fig. 13.33 illustrates the chemical structure of selected surfactants. Within any of the surfactant types, there can be substantial variations in their chemistries and performance. The chemistry, size, and degree of branching of the lipophilic segment of a surfactant molecule can have a major impact on foam-surfactant performance, just as the chemistry of the hydrophilic portion of a surfactant molecule can have. Even small and subtle differences in the lipophilic segment can alter surfactant properties dramatically. Most commercial surfactant products contain a distribution of surfactant types and sizes that adds further complexity of the surfactants used in conformance-improvement foams.

When using foam in conjunction with steam flooding or any elevated-reservoir-temperature application, it is important to choose a surfactant that will be thermally stable over the needed life of the foam in the reservoir. Historically, alpha-olefin surfactants and petroleum sulfonate surfactants have been most widely used in foams applied to high-temperature (> 170°F) reservoirs. Sulfate surfactants have been used at times in low-temperature (< 120°F) reservoirs.

Alpha-olefin sulfonates have emerged to be one of the most popular and widely employed surfactant chemistries for use in foams. This has resulted in large part because of their combined good foaming characteristics, relatively good salt tolerance, good thermal stability, availability, and relatively low cost. Mixtures of different surfactant chemistries have been suggested to provide advantages when formulating conformance foams.[142]

The use of fluorinated surfactants in foam formulas has shown some promise.[143] Fluorinated surfactants used with other surfactants have been reported to often improve the tolerance of the foam to oil.[144] Fluorinated surfactants have not been widely used in field applications of oilfield foams largely because of their relatively high cost.

Foam Properties. Several properties important to the characterization of bulk foam, as might exist in a bottle, are foam quality, foam texture, bubble size distribution, foam stability, and foam density. Foam quality is the volume percent gas within foam at a specified pressure and temperature. Foam qualities can exceed 97%. Bulk foams, having sufficiently high foam quality such that the foam cells are made of polyhedra liquid films, are referred to as dry foams.[141] Oilfield conformance-improvement foams typically have foam qualities in the range of 75 to 90%. When propagated through porous media, the mobility of many foams decreases as foam quality increases up to the upper limit of foam stability in terms of foam quality (an upper limit of often > 93% foam quality). When dealing with oilfield steam foams, steam quality refers to the mass fraction of water that is converted to steam.

Foam texture is a measure of the average gas bubble size. In general, as a foam texture becomes finer, the foam will have greater resistance to flow in matrix rock.

Bubble size distribution is a measure of the gas bubble size distribution in a foam. When holding all other variables constant, a bulk foam with a broad gas-bubble size distribution will be less stable because of gas diffusion from small to large gas bubbles. The imparted resistance to fluid flow in porous media by a foam will be higher when the bubble size is relatively homogeneous.[141]

Stability of an aqueous-based foam is dependent on the chemical and physical properties of the surfactant-stabilized water film separating the foam's gas bubbles. Foams are metastable entities; therefore, all foams will eventually break down. Foam breakdown is a result of the foam liquid films excessively thinning and rupturing with time and a result of gas diffusing from smaller bubbles into the larger bubbles, thus coarsening the foam's bubble size. External effects, such as contact with a foam breaker (e.g., oil or adverse salinities), contact with a hydrophobic surface, and local heating can break foam structure.

Factors affecting foam lamellae stability include gravity drainage, capillary suction, surface elasticity, viscosity (bulk and surface), electric double-layer repulsion, and steric repulsion.[141] The stability of foam residing in porous media evokes a whole series of additional considerations that are addressed in the next subsection of this chapter.

One of the attractive features of foams for use with gas-flooding operations is the relatively low effective density of foams. (As a countervailing note of reference, conformance-improvement foams formulated with supercritical CO_2 can attain densities exceeding the density of some crude oils.) The low-density feature has positive ramifications for foams used in both mobility-control flooding and for blocking fluid-flow. The low effective density causes the foam to be selectively placed higher in the reservoir interval where gas-flooding flow or gas production is most likely occurring.

For technical clarification, foam flow in porous media actually occurs as bubble trains of gas bubbles separated by liquid lamellae. Thus, strictly speaking, foam flow in porous media occurs as two phase flow—namely, gas bubble flow and liquid lamellae flow. In this more technically correct view, it is really the low density of the gas phase that promotes favored placement of the foam higher in the reservoir. During gas flooding, such as steam or CO_2 flooding, low-density foams used for mobility control are well suited to address and reduce the common problem of gas override that often precludes injectant oil-recovery gas from contacting the oil saturation lower in the reservoir vertical interval. Selective mobility-control by low-density foams in the upper portion of the reservoir will force more displacing fluid gas to contact oil-saturated sections lower in the reservoir.

The low density of foam used during a gas-blocking treatment will tend to drive the placement of the foam higher up in the reservoir interval where the offensive gas flow and production is most likely occurring. In this respect, foams for use in blocking-agent treatments are well suited to treat gas coning and gas cusping problems occurring at production wells. Also, gas override in a relative homogeneous reservoir with good vertical permeability causes excessive gas production in the upper interval of production wells. Low density gas-blocking foam helps favorable placement around such problem wells.

When considering the potential benefit of low density during foam placement of a conformance-improvement operation, the relative effects of gravity forces vs. viscous forces that are operating during the foam placement need to be carefully considered. That is, the horizontal differential pressure gradient vs. vertical differential pressure gradient that the foam will experience during its flow and/or placement in the reservoir needs to be evaluated.

Foams in Porous Media. Understanding how foams behave and perform in porous media is critical to the effective application of foams for conformance-improvement applications in matrix-rock reservoirs. How foam exists and functions in porous media is not always intuitively obvious on the basis of how foam behaves in bulk form (e.g., when existing in a bottle).

Properties. In addition to the properties of bulk foams, which for the most part, are applicable to foam that resides in porous media, there are two specialized properties of foams that reside in porous media.

In general, foams in matrix rock pores do not exist as a continuous interconnected liquid/film structure that contains gas bubbles, as is the case for a bulk foam. Foam in porous media exists as individual gas bubbles that are in direct contact with the wetting fluid of the pore walls. These in-situ microgas bubbles are separated by liquid lamellae that bridge the pore walls and form a liquid partition on the pore scale between the in-situ foam gas bubbles. Foam propagates in most matrix reservoir rock as a bubble train, where each gas bubble is separated from the next gas bubble by a liquid lamellae film. The length of a foam gas bubble in porous media is on the order of, or exceeds, one pore length.[139]

See the subsection on General Nature of Foams in Sec. 13.7.1 for additional discussion of the nature of foams in porous media. Foam stability and performance in porous media is strongly influenced by lamellae/pore-wall interactions. Foam texture in porous media is believed to most often be controlled by the porous media.

Mobility Reduction. Compared with the mobility of the gas phase from which a foam is formulated, the mobility of the resultant foam is dramatically reduced. Often, the mobility of the foam in foam-saturated reservoir matrix rock is less that the mobility of the aqueous phase alone that is used in the foam formula. For a given foam formula, there often is a general trend of decreasing foam mobility with increasing foam quality (gas content) up to the upper foam-quality stability limit.

Foam mobility reduction results from a combination of foam-induced permeability reduction and, on the macro scale, apparent foam-induced viscosity enhancement. Foam-induced mobility reduction is caused by at least two different mechanisms: the formation of, or an increase in, the trapped residual gas saturation; and increased resistance to flow of the gas phase resulting from the drag of propagating the foam-lamellae aqueous films through constricting pore bodies and, especially, through constricted pore throats.

As the texture of a foam becomes finer, the apparent viscosity of the foam increases, and the foam mobility decreases. This occurs because the number of foam lamellae films within a given volume of the porous rock has increased; thus, foam texture is an important variable in determining the amount of mobility reduction that will occur during foam flow.

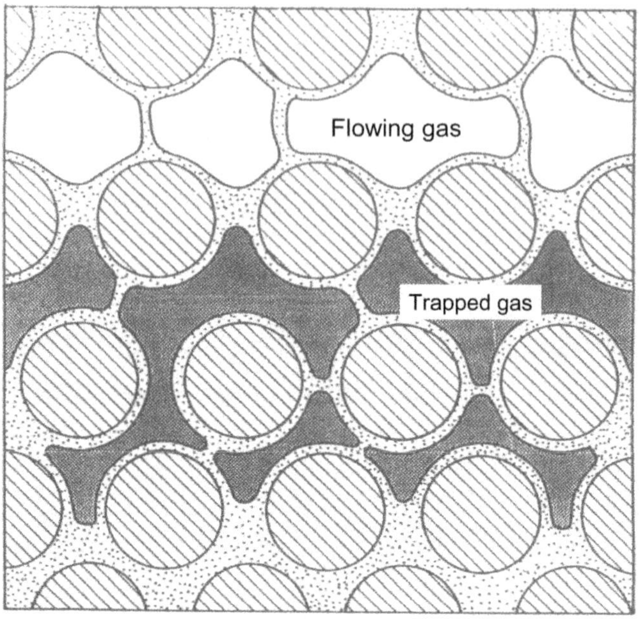

Fig. 13.34—Pore-level schematic of a flowing gas-bubble train and the trapped gas saturation during foam flooding in porous media. Flowing gas bubbles are unshaded; trapped gas is darkly shaded; liquid phase is dotted; and solid phase is hatched.

Rheology, Flow, and Transport. The apparent viscosity of foam (on the macro scale), as it is being propagated through porous media, frequently exhibits shear-thinning behavior. Such foams on the macro scale are a pseudoplastic fluid in porous media. Foams used in oilfield conformance treatments usually exhibit a yield stress. That is, the shear rate remains zero until a threshold stress and/or pressure gradient is reached, and, thereafter, foam flow begins.

Foam placement and saturation in water-wet reservoir matrix rock with normal permeabilities greatly reduces permeability to subsequent flow of gas. The reduction in gas permeability can be on the order of several hundred-fold. Following such foam placement in matrix reservoir rock, the relative permeability to water has often experienced little change.[145] The substantial selective reduction of the gas permeability has been attributed, in part, to a higher trapped gas saturation than occurred before foam flooding. Such foam trapped gas saturations have been reported to range between 10 and 70% PV, depending on surfactant type and the presence of oil during the foam flood.[145]

It is the drag and the resistance to flow of the lamellae through the pore structure that imparts much of the mobility reduction of foam flow. **Fig. 13.34** shows a schematic drawing of a flowing train of foam gas bubbles and the trapped gas saturation during foam flow through a foam-saturated porous media. Only a small percentage, typically 1 to 15%, of the foam gas saturation actually flows. The stationary portion of the foam blocks gas flow in intermediate and small-sized flow paths and lowers the effective permeability to gas flow. In the flowing portion of the foam, interactions of the foam gas bubbles and the interspaced lamellae determine the effective foam viscosity of the foam that flows in the largest pore flow paths. The resultant effective foam viscosity is most often larger than the effective viscosity of the water when water alone saturates the same flow channels.[145]

The rheology of flowing foam in porous media is controlled by the dynamics of foam generation and decay, in combination with the resulting foam texture and bubble size distribution of the equilibrium in-situ foam.[145] Foam flow in porous media is a complex process that involves a number of interacting microscopic foam events. Macroscopic manifestations of foam flow in porous media are the result of the combination of many pore-scale events that involve foam bubble evolution and foam bubble/lamellae pore-wall interactions during multiphase flow. Fully understanding the macroscopic manifestation of foam flow in porous media requires understanding the pore-level phenomena of foam flow.

Foam lamellae formation during foam flow in porous media results from a combination of three foam-generation mechanisms: snap off, division, and leave behind.[145] The snap-off mechanism is believed to be the dominant foam-generation mechanism during foam flow and transport in porous media.

During foam flow through porous media, foam destruction (decay and coalescence) is primarily brought on by capillary suction and gas diffusion. In certain instances, gravity can also contribute to foam decay when there is a significant density difference between the gas and liquid phases of the foam. The gas diffusion mechanism leads to coursing of the flowing foam and is normally of minor consequence for foam flow in porous media. Capillary suction coalescence is the dominant mechanism for lamellae breakage during foam flow in porous media and is strongly affected by the surfactant used in the foam.[145] If the lamellae of a foam can withstand the imposed capillary suction pressure, such a foam is termed a "strong foam."

Coalescence of flowing foam bubbles in porous media is more complicated than the coalescence of static bulk foam. A limiting capillary pressure, P_c*, exists above which foam coalescence is significant and below which coalescence in minimal. Limiting capillary pressure varies with gas flow rate, absolute permeability, and the surfactant used in the foam. The limiting capillary pressure of flowing foam in porous media is typically on the order of 0.44 psi (3 kPa).[145]

Many foams flowing in matrix sand-reservoir rock do so under steady flow at, or near, P_c*. Such foam flow occurs when the gas fractional flow rate is in the high range (i.e., high foam quality) and when the gas and liquid flow rates are fixed. In the P_c* foam-flow regime, the wetting-liquid (usually water) saturation is nearly constant and is independent of gas and liquid velocities over a wide range. This limiting wetting-phase saturation is thought to result from the constant P_c*.[145]

Foam flow in the limiting capillary pressure (P_c*) regime is quite interesting. When the liquid (water) velocity is held constant and the gas velocity is varied, the pressure drop is highly independent of the gas flow rate. Increasing the liquid velocity while holding the gas velocity constant usually results in a linearly increasing pressure drop. Increasing both the liquid and gas velocities, while holding the fractional flow constant, produces a linear response of pressure drop vs. total flow rate. Steady-state liquid and gas saturations are independent of gas fractional flow.[145]

Foam flow in the P_c* regime has a number of important ramifications. Under such foam-flow conditions, the aqueous phase saturation remains constant, as does the relative permeability of the water phase.[145]

However, not all foam flow occurs under the limiting-capillary-pressure regime. The P_c* regime for foam flow does not necessarily apply for low gas fractional flow (i.e., flow of low-quality foams). Osterloh and Jante[146] studied a wide range of flow rates and fractional flows (foam qualities) for foam flow in a 6,200 md sand pack at 302°F. During their flooding experiments involving a nitrogen foam and an alpha olefin-sulfonate-surfactant foam formula, they observed two foam-flow regimes when varying the foam quality. During flow of foam with foam qualities exceeding 94%, the pressure gradient was quite independent of gas velocity at a fixed liquid velocity and varied with liquid velocity to approximately the 0.33 power. The liquid saturation remained nearly constant. During flow of foam with lower foam qualities (< 94%), the opposite behavior was noted. The pressure gradient increased little with increased liquid velocities, but increased with gas velocity to the 0.31 power. It was suggested that the transition between these two foam flow regimes occurred at the point where the limiting capillary pressure was attained. Subsequently, Alvarez *et al.*[147] reported on what is described to be a unified model for steady-state foam flow at both high and low foam qualities and a model that helps reconcile apparently contradictory foam flow data for foam flow occurring in reservoir-rock porous media. The unified model is predicated on the contention that, in the high-foam-quality flow regime of Osterloh and Jante, foam flow behavior is dominated by capillary pressure and coalescence and that, in the low-foam-quality flow regime, foam flow behavior is dominated by bubble trapping and mobilization.

Steady-state foam flow refers to foam flow in a given length of porous media after foam has been propagated and formed throughout the entire length of the porous media in question, and the liquid saturation profile is nearly uniform throughout the entire length of the porous media. Transient foam flow in a given length of porous media refers to foam flow as foam is progressively being formed and propagated along the length of the porous media and, as the liquid saturation profile varies from low to high along the length of the porous media, in the direction of flow.

An alternative definition for "strong foam" to the one given previously is based on the continuity of foam within porous media. For a given volume of porous media that contains foam, a "continuous gas foam" exists when there is at least one continuous flow path along the length of the porous media that is uninterrupted by the existence a foam lamellae. A "discontinuous gas foam" exits when there is at least one foam lamellae along all the gas flow paths over the entire length of the porous media volume. Thus, when gas must flow through a discontinuous gas foam in a given porous media, the gas must mobilize and propagate (or possibly rupture) at least one foam lamellae. A strong foam is said to exist when a discontinuous gas foam occurs. A weak foam is said to exist when a continuous gas foam occurs.[139]

Chang and Grigg[148] have studied and reported on the effect of foam quality and flow rate on the imparted mobility reduction resulting from the steady state flow of dense CO_2 foam in porous media at reservoir-like temperature and pressure conditions. Over the range of foam qualities normally used in oil reservoirs and for the studied conditions and foam formula, CO_2 foam mobility was observed to increase with increasing flow rate and to decrease with increasing foam quality.

Magnetic resonance imaging has been reported to be a useful tool for high-resolution viewing of foam flow in selected porous media.[149]

Foams have been reported to have the very desirable feature, under certain conditions, of being able to reduce mobility to a greater extent in high-permeability porous media, as compared with lower-permeability porous media.[150–154]

Questions persist about the ability to propagate and place foams deep within matrix rock. One aspect of this concern is the often destabilizing effect of oil on foam transport. The next subsection discusses the effect of the presence of oil. Another aspect of this concern is the pressure gradient that is normally required for initiating and maintaining foam flow. Can foam flow be maintained in the far-wellbore regime where pressure gradients are inherently low?[155–157]

As a result, in part, of the low surface tension of CO_2, CO_2 foam is more easily propagated (than nitrogen, steam, and natural-gas foams) at the relatively small pressure gradients that exist in the far-wellbore region of most reservoirs.[139] Ref. 140 reports on laboratory results and literature references indicating that for dense/supercritical CO_2 foams, minimum pressure gradients in porous media of less than 1 psi/ft can exist for foam flow when flooding with strong CO_2 foams. However, under similar conditions, the minimum pressure gradients for the formation and flow of strong nitrogen foams are reported to be a factor of 20 psi/ft or greater. Viewed conversely, the relatively high minimum pressure gradient for foam flow in many instances can be advantageously used as the basis for foam treatments to block gas flow.

Another important aspect of the problem of deep foam placement is surfactant adsorption/retention. The upcoming subsection Surfactant Adsorption/Retention discusses adsorption and retention of surfactant during foam transport through matrix reservoir rock.

Effects of Oil and Wetting. Much has been published on the interaction of crude oil and foam within porous media—with much of this literature discussing negative interactions.[150,157–174] When oil contacts a foam, the oil often has a destabilizing effect.[141] The probability that oil will destabilize foams has been a major impediment to the widespread use of foams for oilfield conformance-improvement applications. The destabilizing effects of oil can range from minor to very deleterious. Ref. 159 reviews the mechanisms by which crude oil can destabilize foam in porous media. Foams are more stable in the presence of some crude oils as compared to others. In general, foams are more highly destabilized when contacted with lighter, lower viscosity crude oils.

The degree of sensitivity of foam to oil as foam flows through matrix reservoir rock depends on both the nature of the foam and the nature of the crude oil. Although many conformance-improvement foams are sensitive to oil contact, some foam formulas are quite resistant to destabilization by crude oil.

It has been suggested that even foams sensitive to crude oil can still be effective in matrix reservoir rock if the residual oil saturation is < 10%.[159] On the other hand, oil-sensitive foams will be significantly destabilized by the contact with the crude oil at higher oil saturations (e.g., 20% oil saturation). It has also been suggested that foam sensitivity to oil might be advantageously exploited. That is, by using foams to selectively reduce gas or water mobility in high-gas or water-saturation flow paths within an oil reservoir where the oil saturation is low, while the foam is simultaneously being destabilized and unable to reduce the oil mobility and productivity in the high-oil-saturation flow paths within the reservoir.

The general consensus of several investigators is that oil wetting is detrimental to foam stability and propagation in matrix reservoir rock; however, there is not universal agreement on this point.[139,159–162]

Surfactant Adsorption/Retention. The degree of surfactant adsorption/retention often can "make or break" the oil-recovery performance and the economics of a foam application. "Retention" is the combination of all other mechanisms, other than adsorption, that retards surfactant propagation during foam propagation through reservoir matrix rock.

Surfactant adsorption/retention under reservoir conditions should be one of the key factors considered and is one of the first parameters that should be examined and/or estimated when considering the application of foam injection for mobility-control purposes during a gas flooding operation.

The use of low-cost adsorption/retention sacrificial agents in the foam or such agents injected before the foam have been proposed as a means to alleviate the adsorption/retention problem.[175] Surfactant adsorption is often lower when foam is transported though an oil-wet reservoir.

Injection Mode. One of three distinctly different modes is used for injecting conformance-improvement foams: sequential injection, coinjection, or preformed foam created on the surface before injection. Sequential injection involves the alternate injection into the oil reservoir of the foam's gas and aqueous phases. Coinjection involves the coinjection into the reservoir of the foam's gas and liquid phases. Because of the substantial effective viscosities of foams and the associated poor injectivity of preformed foams, early applications of conformance-improvement foams tended to involve the sequential-injection or coinjection mode. Also, sequential-injection and coinjection are substantially simpler to implement in the field. Sequential injection also avoids tubular corrosion problems if the gas and the foaming-solution form a corrosive mixture, such as found in CO_2 foams.

The concept, which is supported by laboratory evidence, is that during the sequential or coinjection mode, foam will form in situ in the matrix reservoir rock. Supporting this contention is the expectation that low-viscosity and high-mobility gas will tend to finger into the aqueous foaming solution and generate the foam in situ.

However, there are two significant countering concerns. First, as the gas begins to finger into the aqueous solution and form foam in situ, the newly formed foam will substantially reduce subsequent gas fingering and divert subsequent gas flow away from the remaining aqueous foaming solution residing just ahead of the initially formed foam. This phenomenon results in ineffective and inefficient use of the injected foam chemicals and fluids in generating foam. Second, in intermediate and far wellbore locations, there may not be enough mechanical energy and/or differential pressure to generate foam in situ when using common foaming solutions. This is especially of concern for steam, nitrogen, and natural-gas foams.

Krause *et al.*[176] reported on relatively near-wellbore production-well foam treatments that were applied at the Prudhoe Bay field to reduce excessive GOR emanating from the production of reinjected natural gas. The first treatment involved the injection of the foaming solution into the reservoir, followed by a series of overflushes. It was thought that the subsequent production of gas through the emplaced foaming solution, in a similar manner to the sequential injection mode, would cause the generation of a gas-blocking foam in situ. The second foam gas-blocking treatment involved the sequential injection of the foaming solution and a slug of nitrogen. Neither of these first two foam gas-blocking treatments showed any post-treatment GOR decline. The third foam gas-blocking treatment was a nitrogen foam of 65% quality that was preformed at the surface before injection. This treatment significantly reduced GOR at the treated production well for several weeks. These results suggest that, for many applications of natural-gas and nitrogen conformance-improvement foams, foam injection using the preformed mode, as compared to the sequential injection or coinjection mode, will result in superior performance of the foam within the oil reservoir when conducting "near-wellbore" treatments. Unless compelling arguments for a specific application can be made to the contrary, foams for most applications of near and intermediate wellbore conformance-improvement treatments should be preformed at the surface before injection.

The sequential process, alternately known as the WAG process, of injecting sequentially and repeatedly alternating slugs of CO_2 and aqueous foaming solution is often favored when using CO_2 foam for mobility-control purposes during CO_2 flooding. This is because CO_2 dissolved in an aqueous surfactant solution forms carbonic acid that is corrosive to steel tubulars. Because of the low surface tension of CO_2, foam generation and propagation is much more feasible (than steam, nitrogen, or natural-gas foams) at realistic field pressure gradients that occur throughout the reservoir.[139]

Computer simulation studies have been reported to show that the optimal injection strategy for overcoming gas override during gas-flooding operations is the alternate/sequential injection of separate large slugs of gas and the foaming liquid at the maximum allowable fixed injection pressure.[177] This study was limited to foam injection into a homogeneous reservoir and did not account for any foam interactions with oil. The surfactant-alternating-gas-ameliorated (SAGA) injection mode for forming in situ mobility-control foam has been proposed for use when conducting large-volume WAG flooding projects in North Sea reservoirs.[178]

13.7.2 When, Where, and Why to Use Foams. The use of foams is most advantageously applied during gas flooding or for reducing gas coning and cusping in one of two manners. First, foams can be used to improve sweep efficiency and improve oil recovery during gas flooding (e.g., steam, CO_2, and hydrocarbon-miscible flooding). Such mobility-control foam is usually injected from the injection well side. Second, foams can be used as gas-blocking agents to reduce excessive, deleterious, and competing gas production. Such gas-blocking foam is most often placed from the production well side. Foams for use as both mobility-control and gas-blocking agents are attractive because they are relatively inexpensive on a unit-volume basis. The low unit-volume cost results from the combination of the bulk of the foam volume usually being a relatively low-cost gas, and the surfactant chemicals for the foaming solution are relatively inexpensive and used at relatively low concentrations.

13.7.3 Advantages and Disadvantages of Foams. There are a number of somewhat contrasting advantages and disadvantages for the use of foams for improving conformance during oil-recovery operations.

Advantages. The following is a list of the advantages of the use of foams.

- Foams are exceptionally effective in reducing gas mobility during gas flooding.
- Foams can be an effective gas-blocking agent.
- Foams are a conformance-improvement material that has the tendency, in numerous instances, to reduce permeability and mobility to a greater degree in higher permeability matrix reservoir rock.
- Foams are shear-thinning fluids resulting in relatively good injectivity and in more effective mobility control in the far-wellbore region where such mobility control is most needed.
- Foams possess low effective density that can often be exploited to help selectively place foams high in a reservoir thereby impeding problematic gas flow where it is most likely to occur.
- Foams are considered, in general, to be an environmentally friendly material for use in conformance-improvement operations.

Disadvantages. The following is a list of the disadvantages of the use of foams.

- Foams are a relatively complex technology, both chemically and operationally, to apply successfully.
- Oil tends to destabilize and deactivate many conformance-improvement foams.
- Many mobility-control foams (e.g., steam and natural gas foams) are difficult or impossible to propagate in the intermediate- to far-wellbore environment under the differential-pressure conditions encountered in most reservoirs.
- Surfactant adsorption/retention has a substantial negative impact on the performance and economics of mobility-control foams.
- Fluid-blocking (e.g., gas-blocking) foams used in production-well treatments have limited strength under high differential pressure conditions.
- Fluid-blocking (e.g., gas-blocking) foams are also limited by the inherent lack of long-term stability and the associated lack of long-term treatment effectiveness.
- The high viscosity and poor injectivity of preformed foams limit the application of this otherwise often favored foam-injection mode.
- The limited and sometimes poor ability to effectively form foam in situ in matrix reservoir rock during the coinjection or sequential injection of the foam's gas and liquid phases limit the effectiveness and the efficiency of the coinjection and sequential-injection modes for foam formation and placement in a reservoir.

13.7.4 Foams for Mobility Control. Foams, as a conformance-improvement technology for use during gas flooding (e.g., steam, CO_2, and miscible-gas flooding), have historically been most widely studied and applied when the foams are to be used in the form of a "viscosity-enhancing" mobility-control agent that is injected from the injection-well side. Because relatively large volumes of foam are required and because the foam must be propagated significant distances in the reservoir, applying foams for mobility control have proved technically and economically challenging.

Reducing Gas Channeling and Override. As initially discussed in the Foam Properties subsection of Sec. 13.7.1, the low effective density of most mobility-control foams, which are used during a gas flood such as a steam or CO_2 flood, provides a driving force for the foam to flow and be desirably placed in the upper reservoir vertical interval where the offending gas override is occurring and where the foam will be most effective at countering the negative impact of the gas override. Ref. 179 describes an "improved" surfactant-alternating-gas foam injection process to control gravity override during gas flooding projects.

Applications. *CO_2 Flooding.* CO_2 foams are considered to be an effective mobility-control agent candidate for use during CO_2 flooding to improve CO_2 sweep efficiency.[150,152,180–184] This includes the use of "foams" formulated with supercritical and dense CO_2. Surfactant selection and surfactant adsorption/retention losses are particularly critical parameters to the successful economic application of CO_2 foams during CO_2 flooding operations. The exploitation of relatively low-cost CO_2 foams formulated with surfactant concentrations below the critical micelle concentration has been suggested.[184] As discussed in the Injection Mode subsection of Sec. 13.7.1, the sequential or WAG injection of the CO_2 and the foaming solution is often preferred for the injection of mobility-control CO_2 foam.

Steam Flooding. Use of steam foams has been studied extensively and has been reduced to field practice as a technique to improve vertical and areal sweep efficiency and to reduce steam channeling and override during steam flooding that is being applied to shallow heavy-oil reservoirs. The steam foam process consists of adding surfactant, with and without the addition of a noncondensable gas, to the injected steam.[158,185] Based on the combined findings of theory, laboratory studies, and field performance, it has been determined that steam foams are normally more effective when a small amount of a noncondensable gas, such as nitrogen, is incorporated into a steam-foam formula. Steam foams have been used in conjunction with both continuous and cyclic steam injection.

As with foams for CO_2 flooding, the effectiveness and the economics of the steam-foam process are critically dependent on surfactant adsorption and retention. Unlike CO_2 foams, surfactant thermal stability is also a critical issue. Alpha-olefin sulfonates, along with petroleum sulfonates, are the surfactants that have been favored for use in conformance-improvement steam foams.[174] Borchardt and Strycker[186] have studied commercial olefin sulfonate surfactants to determine what the optimum chemistry should be in terms of favorable surfactant performance in foams for steam-flooding applications.

To mitigate the destabilizing effect of oil on steam foam, one proposed strategy is to inject a prefoam surfactant slug to mobilize the residual oil ahead of the steam foam.[185] Steam foams have been extensively applied in conjunction with the heavy-oil production operations in Kern County, California. The application of steam foam in Kern County, California has been considered a technical success, but its economic success is suspect.[185]

Miscible Gas Flooding. Although it would appear that foams would be well suited to impart mobility control and to improve sweep efficiency during miscible gas flooding (in a similar manner as during CO_2 and steam flooding), relatively few papers have appeared in the literature about this application of conformance-improvement foams, especially the actual field application of foams for use in conjunction with miscible-gas flooding. Mannhardt and Novosad[187] studied the adsorption of foaming surfactants to be used with hydrocarbon-miscible flooding in reservoirs with high salinities. Refs. 188 and 189 discuss the application of foams for use during hydrocarbon-miscible flooding in Canada.

Sizing Volume Injected. The volume of foam that should be injected during application, as a mobility-control agent in conjunction with gas flooding, is a subject that lacks good and sound engineering guidelines. Thus, the sizing of such foam applications must be a custom design based on previous experience with similar applications, and/or be based on empirical guidelines. It does not make sense to design the depth of foam placement to be greater than the distance the foam can propagate through the reservoir.

Polymer-Enhanced Foams. The addition of a water-soluble polymer to the foaming solution has been suggested as a means to increase stability of the foam, increase the effective viscosity and structure of foams, and improve the oil tolerance of foams.[4,157,190–193] The possible use of polymer-enhanced foams to treat fracture conformance problems has been suggested.[190] Polymer-enhanced foams formulated with high-MW acrylamide polymers have been noted to be rheologically shear-thinning fluids that substantially aid in the injectivity of preformed polymer-enhanced foams.

Potential disadvantages of the use of polymer-enhanced foams are reduced injectivity of preformed polymer-enhanced foams as compared with conventional foams, possible increased difficulty in propagating the polymer-enhanced foam through matrix reservoir rock, and somewhat increased operational and chemical complexity in applying polymer-enhanced foams as compared with conventional foams.

13.7.5 Foams as Blocking Agents. Because foam applications for mobility-control during gas flooding have proven technically challenging and marginally attractive, the recent focus has shifted somewhat to the application of relatively small volumes of foam that are placed as gas-blocking agents from the production well side. The application of foams as gas-blocking agents has been discussed and reviewed numerous times in the literature.[4,139,155,156,194,195]

Concept. Because foams are exceptionally effective at reducing gas permeability, they are good candidates for use in gas-blocking treatments that are placed relatively near to producing wellbores. The foam's low effective density results in the tendency for selective placement in the upper sections of the reservoir where gas, especially coning and cusping, is entering the wellbore. The obvious and major challenges that must be overcome to successfully apply foams as a gas-blocking agents are to assure that the emplaced blocking foam will have adequate strength and that the metastable foam will be stable long enough to result in attractive economics.

Laboratory Studies and Treatment Design. Treatment design and laboratory studies in support of applying foam gas-blocking treatments need to assure the following.

- The foam has adequate strength to function as intended under actual reservoir conditions, including adequate strength within the maximum permeability rock to be blocked/sealed and adequate strength to withstand the maximum differential pressure that will be encountered.
- The foam has adequate durability and stability to function as intended under actual reservoir conditions for at least the minimum intended economic life of the foam gas-blocking treatment, including adequate thermal stability, adequate stability to the reservoir oil and to the salinity of the reservoir brine, and adequate stability in the presence of the reservoir minerals and lithologies to be encountered.
- The foam can be emplaced and/or generated in the desired reservoir volume to be treated.

Foams for Reducing Gas Coning. The use of foams as blocking agents to reduce or eliminate gas coning in matrix-rock reservoirs has long been studied. One concept is to inject a foaming solution at, or near, the gas/oil contact (GOC). As gas cones down through the foaming solution, it is proposed that gas-blocking foam will form in situ and reduce or eliminate gas coning. More recently, it has been proposed that the foaming solution in this case should be a hydrocarbon solution. For a properly designed treatment, gravity forces will tend to promote the selective placement of the hydrocarbon foaming solution at the GOC, just as desired. Dalland and Hanssen reported on a laboratory study of this concept.[145]

Sizing Volume Injected. The volume of the emplaced foam required to successfully function as a gas-blocking treatment is smaller than the foam volume required for mobility-control purposes. In concept, the volume of the foam placed through a production well needs only be large enough to provide sufficient strength and durability to assure effective gas blockage. The volume and depth of foam placement for such a treatment will vary substantially with the foam formula used, the permeability and mineralogical nature of the reservoir volume being treated, and the drawdown pressure that the foam gas-blocking treatment will experience. At this writing, no exact treatment size guidelines for foam gas-blocking treatments can be provided. The sizing of such foam treatments needs to be determined on a well-by-well and treatment-by-treatment basis.

Polymer-Enhanced Foams and Foamed Gels. A trend, which has often occurred when applying foams as gas- or water-blocking treatments from the production well side, is that the initial application of a conventional foam resulted in some blocking of gas or water flow that lasted for only a relatively short time. Encouraged by the initial favorable gas- or water-blocking performance of conventional foam treatments, but pursuing and needing longer-term gas- or water-blocking performance, the application and exploitation of polymer-enhanced foams were next pursued.[4,191,192,196–201] Again encouraged by the improved gas- and water-flow blocking performance of the polymer-enhanced foam treatments, but still pursuing and needing even longer-term gas- and water-blocking performance, foamed gels were subsequently pursued.[4,197,200,201] The application of foamed gels as a blocking-agent material for use across a broad spectrum of conformance treatments has been suggested and studied.[4,83,129,201]

Ref. 131 reports on three large-volume (~40,000 res bbl) foamed-gel treatments that were applied for conformance-improvement purposes to injection wells of the Rangely field miscible CO_2 WAG project in northwestern Colorado. The three foamed-gel treatments were reported to have generated an incremental oil production rate of 155 BOPD. Each of the foamed-gel treatments induced stabilization in the pattern oil rate. The cost of the foamed-gel treatments was said to be 40 to 50% below that of comparable polymer-gel treatments that would have been conducted at the Rangely field.

13.7.6 Design Strategies for Field Application. Rossen[139] suggests the following design strategies for the field application of conformance-improvement foams. The initial steps are to characterize the field and its conformance problem; determine that there is sufficient recoverable oil to render the foam process economic; determine process goals (e.g., increase oil recovery or recovery rate, or reduce operating costs); and perform a preliminary economic evaluation of the foam project. Next, the surfactant to be used should be chosen by conducting wet-chemistry testing, conduct foam-property testing in porous media, and determining surfactant retention with reservoir core material, if possible. Finally, determine the injection strategy to be used.

13.7.7 Outlook. At this writing, the outlook for the use of foams extensively and routinely is not as encouraging as it was during the past two decades. Petroleum industry R&D was diminishing. The original interest in foams as mobility control agents has faded somewhat and interest in the use of conventional foams as fluid-flow blocking agents was also fading because foam fluid-flow blocking treatments are operationally and chemically relatively complex and polymer gels are considered by many petroleum engineers to be more effective, durable, and stronger.

13.7.8 Illustrative Field Results. The following are a series of illustrative field applications of foams for conformance-improvement purposes and/or a series of references to papers discussing such foam applications.

Foams Used During Steamflooding. In 1989, Hirasaki[158] reviewed early steam-foam-drive projects. In 1996, Patzek[203] reviewed the performance of seven steam-foam pilots conducted in California. Early and delayed production responses were discussed for these pilots. Ref. 204 reviews a steam-foam trial conducted at the Midway-Sunset field of California.

Foams Used During CO_2 Flooding. The design, results, and analysis of a two-year CO_2 foam field trial at the North Ward-Estes field in Texas have been documented.[205] The alternate injection of CO_2 and surfactant foaming solution was reported to have reduced injectivities by 40 to 85%. Gas production at an offset problematic production well decreased dramatically, while gas and oil production at the other offset producers increased, indicating favorable areal diversion. CO_2 foam application during this field trial was reported to have significantly improved CO_2 sweep efficiency and to have been economically successful.

Stephenson *et al.*[206] reported that a foam bank could only be propagated several meters from the wellbore during an extended foam test at the Joffre Viking miscible CO_2 flood in Canada, where the foaming aqueous solution and the CO_2 were coinjected.

Hoefner and Evans[207] reviewed four pattern-scale CO_2-foam field trials to determine the technical and economic potential for reducing channeling during CO_2 flooding. The field trials involved two different foaming surfactants, alternating and coinjection of the CO_2 and the aqueous foaming solution, and two field trials each in a San Andres carbonate reservoir of West Texas and one trial in a platform carbonate formation in southeast Utah. In all, 161,000 lbm of active foaming surfactant were injected, with one of the field trials lasting 18 months. The CO_2 foam treatments resulted in reduced CO_2 production and indications of increased oil production.

Ref. 208 reports on a 3,000-reservoir-bbl CO_2 foam treatment that was applied to reduce CO_2 channeling in the Wasson ODC Unit in Texas. The foam treatment was reported to have been a technical success, but an economic failure. Foam performance was noted to decrease with time and to be a treatment success issue.

Ref 182 provides a review of a four-year CO_2 foam mobility-control pilot test conducted in a dolomitic carbonate reservoir of the East Vacuum Grayburg/San-Andres Unit in New Mexico. During the CO_2 foam pilot, CO_2 mobility of the ongoing CO_2 WAG flooding operation was reduced, incremental oil production was noted at three of the eight offsetting producers, and gas cycling was significantly reduced.

Foam Gas-Blocking Treatments. Eight production wells in Nigeria were treated with foam gas-blocking treatments to reduce gas coning or cusping and to reduce excessive GOR.[196] The producing formation was thin and overlain by a gas gap. The reported set of applied nitrogen foam gas-blocking treatments included the use of polymer-enhanced foams, sequential and coinjection of the gas and foamer solution, and fluorosurfactants. The treatments were designed to place the foam barrier 5 to10 m radially from the wellbore. A volume of 600 to 800 bbl of foam solution was injected per treatment. The success rate of the foam gas-blocking treatments was reported to be 50+%. Results ranged from significant reductions in GOR that lasted for twelve months (reduced GOR from 7,000 to 2,000 scf/bbl and increased the oil production rate from 340 to 450 BOPD) to minor reductions in GOR that lasted only for a few weeks. The foam treatments were said to be easy to apply and relatively inexpensive. The preliminary conclusion based on this set of foam gas-blocking treatments is that for this type of foam treatment, the sequential injection mode is preferred over the coinjection mode. Four out of six foam gas-blocking treatments that used the polymer-enhanced foam formulas were successful, while neither of the foam gas-blocking treatments using the fluoro-surfactant foam formula were successful.

Ref. 209 presents a review of the application of two foam gas-blocking treatments that were applied to high GOR problems occurring at production wells in the North Sea. The first foam pilot test was applied to a sandstone formation of a production well in the Oseberg field in the Norwegian North Sea, where the high GOR production problem involved gas coning. The upper producing interval was selectively treated with a strong and "oil-resistant" foam. This Norwegian foam pilot was reported to have been successful (at a minimum, technically successful) and delayed the onset of gas coning by several months.

Foam ASP. The reported use in China of foamed alkaline-surfactant-polymer (ASP) flooding is noteworthy for several reasons.[210] First, the field trial of foam flooding in this manner was reported to be both a technical and economic success with the reported economics being quite attractive. Second, definitive positive reservoir and flooding responses were noted during this foam pilot test. Positive responses included substantial incremental oil production; substantial increases in the injection pressure; significantly lower GOR indicating reduced gas mobility, fingering, and channeling; and favorable changes in the produced water salinity which indicated improved sweep efficiency. Third (and of significant importance), substantial increases in injection pressure occurred during the pilot test.

The implications of the higher injection pressures are three fold. First, for an older waterflood operation in a mature field, this situation may require installing relatively expensive high-pressure injection equipment, injection lines, and wellheads. Second, even if the foam should happen to have an effective viscosity identical to water, it will require substantially more horsepower to inject the foam at the same rate as compared to waterflooding because of the lower density and lower wellbore hydrostatic pressure of the foam. Third, if the previous waterflood injection pressure was just below reservoir parting pressure, any increase in injection pressure would require lower injection rates that might pose a possible negative impact on the economics of the project.

On the positive side (although. at first, a foam-ASP flood may appear to be quite complex), if an oil operator is conducting either an ASP or a WAG flood and if the operator has excess natural gas being produced in the field, then flooding in the WAG mode, using the ASP solution as the water phase of the WAG flood, could prove to be relatively easy to implement and relatively attractive if the conformance of the flooding operation could be improved in a similar fashion as reported for the Chinese foam-ASP pilot test.

13.7.9 Suggested Additional Reading. The subject of foam use in the petroleum industry is covered in depth in Refs. 211 and 139. Ref. 212 provides a more recent review of experimental and modeling studies of conformance-improvement foam flow through porous media. Ref. 213 provides a review of published field experience relating to 30 applications of foam for conformance improvement in North Sea reservoirs. Ref. 202 presents an overview of the field application of foam for improved sweep efficiency and control of produced gas in the United States, the North Sea, and the Former Soviet Union.

13.8 Resins

The term "resin," as used in this chapter, refers to an organic, polymer-based, solid plastic material. Resins do not contain a significant amount of a solvent phase (as do gels), and resins are placed downhole in a liquid monomeric (or oligomeric) state and polymerized in situ to the mature solid state.

13.8.1 Oilfield Resins and Resin Treatments. Oilfield resins are exceptionally strong materials for use in blocking and plugging fluid flow in the wellbore and/or the very near-wellbore region.[4] The three classical oilfield resins discussed here have exceptionally good compressive strengths. Also, these three resins usually have good bonding strength to oil-free rock surfaces. Resins for fluid-shutoff purposes during squeeze treatments can normally only be placed in the wellbore, perforations, gravel packs, and other near-wellbore multi-Darcy flow channels.

Solid fillers, such as silica flour and calcium carbonate, are sometimes added to fluid-shutoff resin formulas to reduce cost, increase resin specific gravity, and/or provide higher temperature stability. If solid fillers are added to a fluid-shutoff resin formula, the solids should normally be added at the wellsite just before resin placement. The exception to this is solids addition to epoxy resins that is often conducted at the resin manufacturing facility. This is done because of the need for a high level of agitation while mixing the solids into the viscous immature epoxy-resin fluid.

In addition to resin treatments applied during a well's production phase, resin treatments have been successfully used as part of well completion strategies to improve conformance during the subsequent oil-recovery production phase of the well's later life. The use of resins for water control dates back at least to 1922.[214] The vast majority of the resin field applications, involving the use of the fluid-shutoff resins technologies that are described next, have used treatment volumes on the order of one to five barrels. At this writing, resin fluid-shutoff treatments were available from relatively few oilfield service companies.

Types and Chemistries. There are three resin chemistry types that have been studied and applied for use as fluid-shutoff treatments in wellbores and the very near-wellbore region. A fourth fluid-shutoff "resin" technology involves a blocking agent that is a cross between a resin and a gel.

Epoxies. Epoxy resins have been favored for use in conjunction with CO_2 flooding. Many of the early epoxy resin technologies were extremely sensitive to water and were "deactivated" while being placed. Wiper plugs are often placed in the tubing at the beginning and the end of the injected resin material. Some of the newer oilfield epoxy resin technologies are much less sensitive to water contact. In general, epoxy resins are the strongest (especially in terms of bonding strength) of the three oilfield resins discussed here. The kinetic rate of maturation of epoxy resins is somewhat touchy and subject to numerous possible interferences during resin placement as a consequence of the epoxy-resin maturation chemistry being based on free-radical chemistries.

Oilfield epoxy resins commercially available are applicable over a downhole temperature range (downhole temperature possibly attained by cooling) of 80 to 130°F and have an ultimate temperature stability of up to 400°F. Placement outside the wellbore is limited to fractures, behind-pipe channels, large vugs, and multi-Darcy matrix rock. Epoxy resins are applicable for shutting off fluid flow involving water, CO_2, and hydrocarbon gases. Epoxy resins are also applied for casing repair purposes. In general, epoxy resins have better bonding properties and strengths than phenolic or furan resins. The relatively fast curing times of epoxy resins result in more restrictive placement times than those of the other two fluid-shutoff resins.

Phenolics. Phenolic resins are well suited for use during steamflooding operations because of their good thermal stability. Phenolic resin treatments have been used to improve steam injection profiles by blocking near-wellbore steam thief zones. During resin treatment placement, immature phenolic resins are not highly sensitive to water and will not be deactivated by limited contact with water. The high viscosity of immature phenolic resins suppresses excessive mixing with water in the wellbore. The maturation chemical kinetics of phenolic resins is not highly subject to interferences during resin placement. Two types of catalysts/activators, caustic and acid, can be used to initiate the polymerization reaction. The maturation chemical kinetics for the more commonly used base-catalyzed phenolic resins is primarily a function of the amount of hydroxide ion incorporated into the resin formula. Phenolic resin maturation involves a chemical condensation polymerization reaction. Commercial base-catalyzed phenolic resins that were available at this writing were applicable over a downhole temperature range (downhole temperature possibly attained by cooling) of 90 to 170°F and applicable to an ultimate downhole temperature of up to 450°F. Acid-catalyzed phenolic resins are considered to be most ideally applied over a downhole temperature range of 100 to 130°F.

Primary uses of phenolic resins are for near-wellbore applications involve the plugging of perforations, behind pipe channels, gravel packs, and fractures. Phenolic resins are applicable for shutting-off fluid flow involving water, steam, CO_2, and hydrocarbon gases. In general, fluid-shutoff phenolic resins have slower maturation chemical-reaction (polymerization) rates than

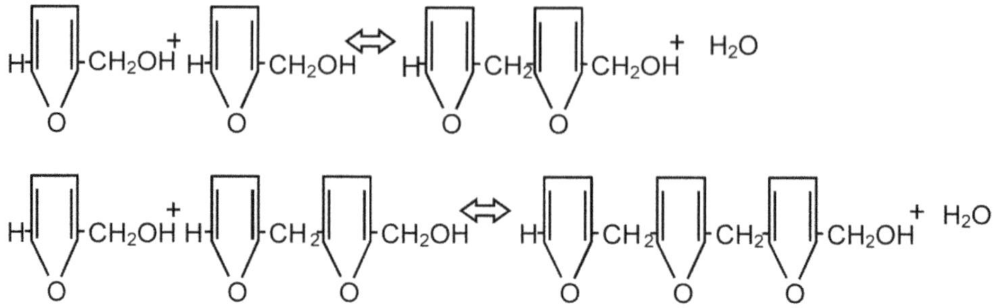

Repeated until obtained polymer

Fig. 13.35—Polymerization of furfuryl alcohol.

fluid-shutoff epoxy resins have. The bonding properties of phenolic resins are not as good as those of cement or epoxy resin.

Toxicity, safety, and environmental issues and regulations need to be carefully considered and reviewed before applying a phenolic resin treatment. The fundamental chemistry of phenolic resins is based on phenol/formaldehyde chemistry. Phenol and formaldehyde are both toxic. The use of phenolic resin treatments for oilfield fluid-shutoff purposes goes back to the 1960s.

Furans. Oilfield furan resins are, in general, weaker resins than epoxy and phenolic resins. Commonly available commercial furan resins are applicable over a downhole temperature range (downhole temperature possibly attained by cooling) of 60 to 350°F and are applicable to an ultimate downhole temperature of up to 700°F.[215,216] The penetration of furan resins into an oil reservoir is limited to fracture channels, vugs, and multi-Darcy matrix rock. Furan resins are applicable for shutting off fluid flow involving water, steam, CO_2, and hydrocarbon gases. Furan resins are also applicable for casing repair purposes. Primary uses of furan resins are for near-wellbore plugging of perforations, behind pipe channels, and fractures, along with their use for casing repairs. The use of furan resins is reported to be environmentally friendly.[216]

Oilfield furan resins are derived from the condensation polymerization reaction of furfuryl alcohol, as **Fig. 13.35** shows.[215,217] The polymerization reaction of furfuryl alcohol is acid catalyzed. Typically, furfuryl alcohol is obtained from the destructive distillation of corncobs. Furfuryl alcohol polymerizes to form a hard, black, chemically resistant, and thermally stable polymer resin.[217] Hardened furan resins have, in general, lower compressive strengths than epoxy and phenolic resins.

Modern oilfield furan resin treatments start with use and placement of furan oligomers or prepolymer chemical species (not the monomer specie). Use of oligomers of furfuryl alcohol reduces the reactivity of the polymerization reaction and greatly increases the controllability of the furan resin polymerization reaction.[215] Modern furan resin fluid-shutoff treatments are a single-fluid process in which an organic acid and an ester have been included in the resin formula.[215–217] The ester reacts with, and absorbs, the water formed during the condensation reaction of the furan resin polymerization process. The uptake of the water rendered by the condensation reaction drives the polymerization reaction further to completion and, thus, renders a more effective and stronger final furan resin. Some of the currently used furan resin fluid-shutoff formulas contain a small amount of a water-swelling polymer to prevent shrinkage of the furan resin after it is polymerized.

In general, furan resins provide better control over the resin maturation time and do so over a larger temperature range compared with the epoxy resin technology. The lower viscosity, in general, of immature furan resins, as compared with epoxy and phenolic resins, permits better penetration of furan resins, especially into reservoir matrix rock. Although not common, furan resins have been placed using coiled tubing.

Common Properties. The three resin technologies discussed previously have a number of features in common. They are all thermosetting and applicable to both sandstone and carbonate reservoirs. None of the three resin technologies are highly sensitive to the pH in the wellbore during resin placement or sensitive to H_2S. All three are not substantially degraded by conventional acid treatments performed after resin treatment placement. The three resins are more stable to acid than conventional Portland cement. Resins without solid fillers can flow into small flow channels where conventional Portland cement cannot.

All three resins types have good mechanical, bonding, and compressive strengths. Most fluid-shutoff resins have compressive strengths exceeding 1,000 psi, with some of these resins possessing compressive strengths up to, and exceeding, 20,000 psi.

The treatment volume of a high percentage of resin fluid-shutoff treatments placed downhole has only been in the range of one to five barrels (and most of these in the one to two barrel range). For all three of these resin technologies, the immature resin solution, as it is being placed downhole, is quite viscous (often qualitatively described as having approximately the consistency of molasses).

Crosslinked Styrene-Butadiene Block Copolymers. There is a conformance-improvement technology that involves a fluid-shutoff plugging agent that is a cross between a resin and a gel. The mature plugging agent contains approximately 10 to 20% styrene-butadiene block copolymer dissolved in an aromatic solvent, such as xylene, and the polymer has been chemically crosslinked using a peroxide free-radical crosslinking agent.[219]

This plugging-agent material is an example of a crosslinked polymer gel in which the gel solvent is an organic fluid. However, this plugging-agent technology also has the attributes of a resin that contains a substantial amount of added organic solvent. Because this plugging technology has many of the functional attributes of a resin (both in terms of application and final physical form) and because, like the three resins described previously, the crosslinked styrene-butadiene block copolymer gel technology is essentially an organic and nonaqueous blocking-agent technology, it is included in this resin technology section.

Crosslinked styrene-butadiene block copolymer gels/resins, as reported in the petroleum literature,[219] are not a fully rigid material. However, crosslinked styrene-butadiene block copolymer gels/resins are an effective "resin" material for use in correcting steam injection profiles during steamflooding.[219] This blocking-agent material is reportedly stable to 500°F. The thermally activated chemical crosslinking of the styrene-butadiene block copolymer is imparted by a peroxide agent that starts a free-radical crosslinking chemical chain reaction. Control of the gelation onset time at various temperatures is achieved by the proper selection of the chemistry of the peroxide crosslinking agent such that the chosen peroxide chemical decomposes to free radicals in an appropriate time frame at a specified temperature.

13.8.2 Problem Identification and Temperature Issues. Because of the limited and relatively small volumes of resin that are placed during a typical resin fluid-shutoff treatment, the correct identification of both the nature and location of the fluid-flow path to be shut off are critical to the successful application of resin conformance-improvement treatments.

There are three reasons that resin jobs are inherently small-volume treatments. First, the immature resin fluid as it is being placed downhole is quite viscous, so that, at best, the immature resin can only be placed very slowly into matrix reservoir rock. This situation is compounded by the fact that the resin fluid has a relatively short maturation time. Second, because of the highly reactive nature of the immature resin solutions, they cannot be propagated any significant distance into or through porous reservoir rock or any other type of restricted flow path without encountering serious chemical interferences to the resin polymerization reaction. Third, the high unit-volume cost usually precludes the application of large-volume resin treatments.

Temperature issues and the thermal history that a resin experiences during its placement to the final downhole location are critical to the successful application of resin fluid-shutoff treatments. Thermosetting resins have maturation/polymerization rates that are highly sensitive to temperature. If a resin were to inadvertently set up (as has occurred at times) in the dump bailer during placement or in an injection tubing string, this will be a costly outcome. If the resin sets up in an injection string, normally that injection string will have to be decommissioned.

Thus, it is critical that the temperature of the reservoir volume to be treated at the time of initial resin placement be known. It is also critical to know how the temperature of the resin will change with time during its placement downhole. Not correctly identifying the downhole temperature where the resin is to be placed and not knowing the thermal history of the resin while it is being placed are major causes of resin treatment failures.

13.8.3 Well Selection. The process of well-candidate selection for resin squeeze treatments is an important aspect of successfully applying resin fluid-shutoff treatments, and a resin treatment aspect that needs careful attention. In particular, the conformance problem of the well to be treated must be a problem for which only a few barrels of successfully placed resin will be sufficient to create the necessary flow barrier.

13.8.4 Placement. Originally, because of the small-volume of resin treatments, wireline dump bailers were widely used for placement. More recently, the majority of resin treatments have been placed through clean, uncorroded injection tubing. The advantages of using injection tubing are faster placement and the ability to attain large differential pressures during placement.

When treating reservoirs with elevated temperatures, the downhole target zone is often precooled. For phenolic and furan resin treatments, the injected cooling fluid is usually water. However, when performing a water-intolerant epoxy resin treatment, the bulk of the injected cooling fluid is usually water that is followed by a hydrocarbon fluid spacer (often xylene). If preresin-treatment cooling-fluid injection is conducted, it is critical to accurately know to what degree the downhole target zone has been cooled and how rapidly the downhole treatment volume will heat back up. The determination of such thermal information often requires the running of appropriate computer thermal-simulation programs and/or a high degree of experience with applying resins to the well type to be treated.

When placing fluid-shutoff resins using an injection string, wiper plugs are usually placed in the injection tubing at the beginning and the end of the resin injection. Historically, clean, uncorroded production or drill strings have been used. The use of such a relatively large internal-diameter injection string during resin injection has often been considered necessary because of the high viscosity of the resin material being injected. More recently, the use of coiled tubing has been successfully used in limited instances.[218]

It is mandatory that the resin be placed only in the wellbore interval intended for the resin treatment. As a result, the resin must normally be place with the use of mechanical zone isolation (e.g., use of mechanical packers). In the event any resin left in the wellbore needs to be removed, the resin material is usually drilled out using special drill bits. A major cause of failures for resin fluid-shutoff treatments is the improper placement downhole.[217]

13.8.5 Wellbore and Injection-String Condition. For several reasons, the condition of the wellbore and the injection string is a critical parameter to the success of resin fluid-shutoff treatments.

First, if the surfaces of the wellbore or the near-wellbore reservoir material to be contacted with the resin have a substantial oil-coating film, the resin will fail to bond to the reservoir material or wellbore hardware, and much of the resin treatment effectiveness is lost. For this reason, if oily solid surfaces are expected, a hydrocarbon-wash preflush is often performed. The hydrocarbon fluid used in such a preflush is usually xylene.

Second, corroded placement equipment or injection tubular strings can play havoc with the kinetics of the resin polymerization reaction, especially for epoxy resins.

Third, an oil-coated injection string can be detrimental to the polymerization reaction product of some resin formulas.

Fourth, when placing the resin through liners, screens, or course sands where these flow paths have previously become partially plugged with organic debris (e.g., asphaltenes) and inorganic debris (e.g., formation fines and scale), such previous plugging can prevent the resin from being fully placed into the desired treatment volume. These plugging materials should be removed before placing the resin treatment material. Jet washing is a common technique used to remove plugging material.

Fifth, in addition to water interfering with the maturation chemistry of certain oilfield resins, water in the injection string can cause physical problems during resin placement. Water can lead to "clumping" in which water becomes interspersed between blobs of resin material. The resulting course emulsion will often not flow readily and the ultimate strength of the placed resin will be decreased.

13.8.6 Advantages and Disadvantages. The advantages and disadvantages of the use of resins as small-volume fluid-shutoff treatments to remedy oilfield conformance problems are as follows.

Advantages. Resins possess good mechanical strength, good bonding strength, good thermal stability, and good chemical inertness (e.g., can acidize over resins).

Disadvantages. Resin fluid-shutoff treatments are constrained by limited resin penetration distances into the reservoir formation (because of a combination of high viscosity, being highly chemically reactive, and cost issues); are operationally and chemically relatively complicated; are somewhat tricky to successfully apply; are constrained by placement techniques and issues; are costly on a unit volume basis; have, in practice, usually been limited to shallow reservoir applications; and are considered to be a niche conformance-treatment technology.

13.8.7 Dos and Don'ts. The following items are a partial list of "do's" and "don'ts" as they apply to resin fluid-shutoff treatments.

Dos.

• Test the resin maturation and hardening rate onsite using the actual resin chemicals to be used and simulating the thermal history that the resin will experience during its placement.

• Before initiating a resin treatment, understand and be able to fully predict how the rate of resin maturation and hardening varies with the amount of catalyst and/or activator that is added to the resin formula.

• Know the age of the resin as it is received from the supplier. Older resins will often react faster than fresh ones because of an increased level of "self-polymerization" that occurs while the resin sits on the shelf during storage. (This is largely a concern for phenolic and furan resins.)

• Know, with a high degree of certainty, the thermal history that the resin will experience as it is placed and the rate at which the resin will mature and harden under this thermal history.

• Know, with a high degree of certainty, the exact location of the problem to be treated. (This is important because of the limited volume of resin that is to be placed.)

• Ensure that the injection tubing and placement equipment are free of rust and oil.

• Clean oil from the wellbore and reservoir location where the resin is to be placed and function.

• If economics permit, perform laboratory testing in reservoir or gravel-pack sandpacks if the resin is to be placed in such sands. Such testing, among other purposes, assures that the sands will not chemically interfere with the expected performance of the emplaced resin.

• Conduct, or have conducted, a good quality control and quality assurance program in conjunction with the application of resin squeeze treatments.

Don'ts.

• For most common epoxy resins, make sure water does not contact the epoxy resin while it is being placed and matured.

• For most common epoxy resins, don't inject the epoxy resin through injection tubing without the use of wiper plugs.

• Don't store the resin material above the recommended storage temperature before its use.

• Don't use aged resin material—always use fresh resin material.

13.8.8 Illustrative Field Results. ***Water-Encroachment Treatments.*** Ref. 215 reports on 26 production wells of the Kern River and San Ardo fields that were treated in 1990 and 1991 with furan resin jobs in which the treated production wells were suffering from water-encroachment problems. Most (24) of the wells were in the heavy-oil Kern River field of the Lower San Joaquin Valley in California. When the furan resin treatments were applied, the Kern River field was undergoing steamflooding. Of the production wells treated, 79% showed significant reductions in water production after the treatments. After drilling out the resin plug in the wellbore, the resin remaining in the treated perforations and in the treated formation was of a sufficient amount and strength to prevent water entry into the production wells.

Gravel Pack Resin Treatments. Phenolic-resin plugback treatments were applied to 32 wells with openhole gravel pack completions in the Midway-Sunset field in 1987.[220] These resin treatments were reported to have decreased water production by 5,900 BWPD and increased oil production by 256 BOPD. Total revenue and savings from the resin treatments were estimated to be U.S. $1 million, and the resin treatment project was reported to have paid out in 120 days. The phenolic resin treatments were applied successfully to downhole temperatures ranging from 100 to 200°F at this California cyclic-steam injection project.

Furan Resin Treatments in California. In 1980, Hess provided a brief review of 36 small-volume (1 to 5 bbl) furan resin treatments applied for a variety of purposes in California.[217] Of these 36 furan resin treatments, 27 were reported to have achieved "permanent" downhole plugging. The furan resin treatments were applied for a variety of purposes, including shutting off undesirable steam breakthrough in production wells, plugging thief zones in water and steam injection wells, shutting off bottomwater entry, and repairing casing and liner damage. The furan resin used in treating steam-stimulated and steamflood injection wells was reported to have held up through two years of service.

Epoxy Resin Treatment. A three-barrel epoxy resin treatment was applied to a Green Canyon production well in the Gulf of Mexico. The objective of this treatment was to shut off the lower perforated interval of the gravel-packed well. The temperature of the sand to be treated was 139°F. The wellbore had a deviation angle of 60° across the treated interval. The epoxy resin formula used was designed for a 4- to 6-hour pump time through coiled tubing and required a 24- to 48-hour shut-in time. Following the epoxy resin fluid-shutoff treatment, oil production increased from 240 BOPD before the resin job to 470 BOPD following the resin job. Following the resin treatment, the flowing tubing head pressure was reported to have increased dramatically.[218]

Steam Injector Profile Correction. More than 40 crosslinked styrene-butadiene block copolymer resin/gel treatments of approximately four-barrel volume have been successfully applied as injection-well profile correction treatments in conjunction with the steamflood conducted at the Kern River field in California.[219] It was reported that none of the resin/gel treatments showed any steam entry into the treated wellbore intervals after the steam-shutoff resin treatments. Payout times for these resin steam-shutoff treatments, based on steam cost savings alone, were reported to have averaged less than two months. These resin/gel treatments were applied with conventional oilfield surface and downhole equipment.

13.8.9 Historical Trend. Resin conformance fluid-shutoff treatments had been somewhat widely applied in previous years and had been relatively popular at that time. However, at the time of writing this chapter, relatively few resin fluid-shutoff treatments were being applied.

13.9 Summary

Conformance improvement encompasses improving drive-fluid sweep efficiency during oil-recovery flooding operations and the shutting off of excessive, deleterious, and competing coproduction of an extraneous fluid, such as water or gas. Conformance problems occur because of heterogeneity of reservoir permeability and/or viscosity and mobility-control attributes of the oil-recovery drive and reservoir fluids. A number of conformance problems and a number of key distinctions relating to conformance problems were discussed. Failure to correctly identify/diagnose the nature of an offending oilfield conformance problem and failure to adequately account for the key distinctions relating to the various conformance problems are major contributors to failures of conformance-improvement floods and treatments.

DPR is a phenomenon whereby a substantial number of, respectively, polymers and polymer gels of conformance-improvement treatments reduce the permeability to water flow to a greater extent than they reduce the permeability to oil and gas flow. The application of DPR technologies for water-shutoff purposes is not a panacea. The successful application of bullheaded DPR water-shutoff/reduction treatments (which are to be applied through production wells, involve radial flow in matrix rock, and for which the well's drawdown pressure is not increased post-treatment) is limited to the following conditions:

• A conformance problem exists in a matrix rock reservoir involving differing geological strata.

• No fluid crossflow exists within the reservoir between the water and the oil or gas producing geological strata.

• The water strata are producing at an uneconomically high water cut.

• The oil or gas strata is producing and will produce for the economic life of the water-shutoff treatment at 100% oil or gas cut.

Early use of water-soluble and viscosity-enhancing polymers during conformance-improvement operations was primarily for improving mobility control during waterflooding. Such polymer flooding was, and is, applied to improve the flood volumetric sweep efficiency of a waterflood by increasing the viscosity of the oil-recovery drive fluid. More recently, polymer application during conformance-improvement operations has additionally been used extensively in polymer-gel treatments and DPR polymer treatments for water-shutoff/reduction purposes. The science, technology, engineering, historical trends, and field application of polymers used in conformance improvement were reviewed.

Gels are a fluid-based system (usually relatively inexpensive) to which solid-like structural properties have been imparted. Gels, especially crosslinked-polymer gels, are effective, widely applied permeability-reducing and blocking/plugging agents for use in conformance-improvement treatments. Conformance-improvement gels are used to reduce or eliminate the fluid-flow capacity of high-permeability and/or water-producing flow paths within an oil or gas reservoir. Organic- and inorganic-based gels and the when, where, and how of successfully applying gel treatments were reviewed. The economics of polymer-gel treatments can be exceptionally attractive.

The use of conformance-improvement foams as mobility-control agents and as gas-flow blocking/plugging agents was reviewed. Foams are primarily used as mobility-control agents during gas-flooding operations or as gas-shutoff treatments. The fundamentals, history, science, technology, engineering, trends, and field application and performance of foams for conformance improvement were reviewed. There are some significant distinctions between foam properties and performance as they exist in bulk foam (e.g., foam in a bottle) and foam as it exists in reservoir-rock porous media. Foam flow in porous media tends to occur as gas-bubble trains, in which the individual gas bubbles are separated by thin-film liquid lamellae. An important aspect of foams in porous reservoir rock is the increased trapped gas saturation that the foams impart. Polymer-enhanced foams and foamed gels for conformance improvement were briefly discussed.

Resins for use in conformance-improvement treatments are an organic polymer-based plastic solid material. Such resin material normally does not contain a significant amount of a solvent phase. Resins are placed downhole in a liquid monomeric (or oligomeric) state and then polymerized to the mature solid resin state. Resins are an exceptionally strong material and an exceptionally chemically and thermally stable material for use in blocking and plugging fluid flow in the wellbore and/or the very near-wellbore region. The chemistry, technology, engineering, limitations, trends, cost concerns, special issues, and field applications of resins for use in conformance-improvement treatments were reviewed. Resin treatments have been used successfully in conjunction with steam flooding. Volumes of resin placed during a conformance treatment are typically on the order of one to five barrels.

Nomenclature

a = polymer-specific constant of the Mark-Houwink equation
A_f = area contacted by oil recovery displacement fluid, L^2
A_t = total reservoir area under consideration, L^2
A_V = reservoir vertical cross section contacted by oil recovery displacement fluid, L^2
A_{tV} = total reservoir vertical cross section, L^2
c = polymer concentration
d_p = mean end-to-end distance or size of a polymer molecule dissolved in solution, L
E_A = areal sweep efficiency
E_I = vertical sweep efficiency
E_P = pattern sweep efficiency
E_V = volumetric sweep efficiency
h = formation height, L, ft
k = permeability, L^2, md
K = power-law coefficient
K' = polymer-specific constant of the Mark-Houwink equation
k_a = permeability measured after polymer flooding, L^2
k_{av} = average permeability, L^2
k_b = permeability measured before polymer flooding, L^2
k_{hi} = high permeability, L^2
k_i = permeability of phase i
M = mobility ratio
M_P = polymer molecular weight
M_H = Hall plot slope
n = power-law exponent
P_c^* = limiting capillary pressure, m/Lt^2
p_{tf} = flowing wellhead pressure, m/Lt^2, psi
R_f = resistance factor
R_{rf} = residual resistance factor

r_e = external radius, L, ft
r_w = wellbore radius, L, ft
s = skin factor
t = time, t, days
w_f = fracture width/aperture, L
W_i = cumulative volume injected, bbl
ϕ = porosity
Γ = polymer retention in µg/g
Γ_v = polymer retention in lbm/AF
$\dot{\gamma}$ = shear rate
η = viscosity of the polymer solution
η_s = viscosity of the solvent
$[\eta]$ = intrinsic viscosity
ρ_{RG} = density of reservoir rock (no porosity included)
μ = viscosity
μ_{eff} = effective viscosity of a polymer solution
μ_i = viscosity of phase *i*
μ_l = average rate of gel dehydration
μ_w = viscosity of the brine into which polymer is dissolved
λ_D = mobility of the displacement phase
λ_d = mobility of the displaced phase
λ_i = mobility of phase *i*
λ_p = mobility of the polymer solution
λ_w = mobility of the brine solution
τ = shear stress

Abbreviations and Acronyms

AMPS = 2-acrylamido-2-methyl-propanesulfonic acid
ASP = alkaline surfactant polymer
BOPD = barrels oil per day
BWPD = barrels water per day
CC/AP = chromium (III)-carboxylate/acrylamide-polymer
CDG = colloidal dispersion gels
DPR = disproportionate permeability reduction
EOR = enhanced oil recovery
GOC = gas/oil contact
GOR = gas/oil ratio
HPAM = hydrolyzed polyacrylamide
IPV = inaccessible pore volume
MW = molecular weight
OOIP = original oil in place
PAM = polyacrylamide
PV = pore volume
RPM = relative permeability modification
SF = screen factor
WAG = water alternating gas
WOR = water/oil ratio

Acknowledgments

Special thanks to Craig Phelps, who provided much of the information on which the Resins section is based. The following individuals contributed significantly to the writing of this chapter by providing reviews of the chapter or selected sections of the chapter: Randy Seright, Paul Willhite, Malcolm Pitts, Bill Rossen, Jim Mack, Dwyann Dalrymple, Craig Phelps, Reid Grigg, Larry Eoff, and Ed Holstein.

References

1. Lake, L.W.: *Enhanced Oil Recovery*, Prentice Hall, Englewood Cliffs, New Jersey (1989) 189–200.
2. Green, D.W. and Willhite, G.P.: *Enhanced Oil Recovery*, Textbook Series, SPE, Richardson, Texas (1998) **6,** 73–75.
3. Willhite, G.P.: *Waterflooding*, Textbook Series, SPE, Richardson, Texas (1986) **3,** 87–89.
4. Kabir, A.H.: "Chemical Water & Gas Shutoff Technology—An Overview," paper SPE 72119 presented at the 2001 SPE Asia Pacific Improved Oil Recovery Conference, Kuala Lumpur, 8–9 October.
5. Sydansk, R.D. and Southwell, G.P.: "More Than 12 Years' Experience With a Successful Conformance-Control Polymer-Gel Technology," *SPEPF* (November 2000) 270.
6. Sorbie, K.S. and Seright, R.S.: "Gel Placement in Heterogeneous Systems with Crossflow," paper SPE 24192 presented the 1992 SPE/DOE Symposium on Enhanced Oil Recovery, Tulsa, 22–24 April.
7. Seright, R.S., Lane, R.H., and Sydansk, R.D.: "A Strategy for Attacking Excess Water Production," paper SPE 70067 presented at the 2001 SPE Permian Basin Oil and Gas Recovery Conference, Midland, Texas, 15–16 May.
8. Seright, R.S.: "Placement of Gels To Modify Injection Profiles," paper SPE 17332 presented at the 1988 SPE/DOE Enhanced Oil Recovery Symposium, Tulsa, 17–20 April.
9. Mack, J.C. and Smith, J.E.: "In-Depth Colloidal Dispersion Gels Improve Oil Recovery Efficiency," paper SPE 27780 presented at the 1994 SPE/DOE Symposium on Improved Oil Recovery, Tulsa, 17–20 April.
10. Martinez, S.J., Steanson, R.E., and Coulter, A.W.: "Formation Fracturing," *Petroleum Engineering Handbook*, H.B. Bradley (ed.), SPE, Richardson, Texas (1987) Chap. 55, 55–2.
11. Chan, K.S.: "Water Control Diagnostic Plots," paper SPE 30775 presented at the 1995 SPE Annual Technical Conference and Exhibition, Dallas, 22–25 October.
12. White, J.L., Goddard, J.E., and Phillips, H.M.: "Use of Polymers To Control Water Production in Oil Wells," *JPT* (February 1973) 143.
13. Sparlin, D.D.: "An Evaluation of Polyacrylamides for Reducing Water Production," *JPT* (August 1976) 906.
14. Weaver, J.D.: "A New Water-Oil Ratio Improvement Material," paper SPE 7574 presented at the 1978 SPE Annual Technical Conference and Exhibition, Houston, 1–3 October.
15. VanLandingham, J.V.: "Laboratory & Field Development of Dispersed Phase Polymer Systems for Water Control," paper SPE 8423 presented at the 1979 SPE Annual Technical Conference and Exhibition, Las Vegas, Nevada, 23–26 September.
16. Dunlap, D.D., Boles, J.L., and Novotny, R.J.: "Method for Improving Hydrocarbon/Water Ratios in Producing Wells," paper SPE 14822 presented at the 1986 SPE Symposium on Formation Damage Control, Lafayette, Louisiana, 26–27 February.
17. Eoff, L. *et al.*: "Structure and Process Optimization for the Use of a Polymeric Relative-Permeability Modifier in Conformance Control," paper SPE 64985 presented at the 2001 SPE International Symposium on Oilfield Chemistry, Houston, 13–16 February.
18. Schneider, F.N. and Owens, W.W.: "Steady-State Measurements of Relative Permeability for Polymer/Oil Systems," *SPEJ* (February 1982) 79.
19. Kohler, N. *et al.*: "Weak Gel Formulations for Selective Control of Water Production in High-Permeability and High-Temperature Production Wells," paper SPE 25225 presented at the 1993 SPE International Symposium on Oilfield Chemistry, New Orleans, 2–5 March.

20. Dovan, H.T. and Hutchins, R.D.: "New Polymer Technology for Water Control in Gas Wells," *SPEPF* (November 1994) 280.
21. Seright, R.S.: "Reduction of Gas and Water Permeabilities Using Gels," *SPEPF* (May 1995) 103.
22. Stanley, F.O. *et al.*: "Amphoteric Polymer Improves Hydrocarbon/Water Ratios in Producing Wells—An Indonesian Case Study," *SPEPF* (August 1997) 181.
23. Ligthelm, D.J.: "Water Shut Off in Gas Wells: Is There Scope for a Chemical Treatment?" paper SPE 68978 presented at the 2001 SPE European Formation Damage Conference, The Hague, 21–22 May.
24. Zaitoun, A. *et al.*: "Water Shutoff by Relative Permeability Modifiers: Lessons from Several Field Applications," paper SPE 56740 presented at the 1999 SPE Annual Technical Conference and Exhibition, Houston, 3–6 October.
25. Faber, M.J. *et al.*: "Water Shut-Off Field Experience With a Relative Permeability Modification System in the Marmul Field (Oman)," paper SPE 39633 presented at the 1998 SPE/DOE Improved Oil Recovery Symposium, Tulsa, 19–22 April.
26. Sandiford, B.B.: "Laboratory and Field Studies of Water Floods Using Polymer Solutions to Increase Oil Recovery," *JPT* (August 1964) 917.
27. Sydansk, R.D. *et al.*: "Characterization of Partially Formed Polymer Gels for Application to Fractured Production Wells for Water-Shutoff Purposes," *SPEPF* (August 2005) 240.
28. Seright, R.S., Liang, J., and Seldal, M.: "Sizing Gelant Treatments in Hydraulically Fractured Production Wells," *SPEPF* (November 1998) 223.
29. Seright, R.S. *et al.*: "Characterizing Disproportionate Permeability Reduction Using Synchrotron X-Ray Computed Microtomography," paper SPE 71508 presented at the 2001 SPE Annual Technical Conference and Exhibition, New Orleans, 30 September–3 October.
30. Nilsson, S., Stavland, A., and Jonsbraten, H.C.: "Mechanistic Study of Disproportionate Permeability Reduction," paper SPE 39635 presented at the 1998 SPE/DOE Improved Oil Recovery Symposium, Tulsa, 19–22 April.
31. Barreau, P. *et al.*: "Water Control in Producing Wells: Influence of an Adsorbed-Polymer Layer on Relative Permeabilities and Capillary Pressure," *SPERE* (November 1997) 234.
32. Mennella, A. *et al.*: "Pore-Scale Mechanism for Selective Permeability Reduction by Polymer Injection," paper SPE 39634 presented at the 1998 SPE/DOE Improved Recovery Symposium, Tulsa, 19–22 April.
33. Zitha, P.L.J., Vermolen, F.J., and Bruining, H.: "Modification of Two Phase Flow Properties by Adsorbed Polymers and Gels," paper SPE 54737 presented at the 1999 SPE European Formation Damage Conference, The Hague, 31 May–1 June.
34. Al-Sharji, H.H. *et al.*: "Disproportionate Permeability Reduction Due to Polymer Adsorption Entanglement," paper SPE 68972 presented at the 2001 SPE European Formation Damage Conference, The Hague, 21–22 May.
35. Zaitoun, A. and Kohler, N.: "Two-Phase Flow Through Porous Media: Effect of an Adsorbed Polymer Layer," paper SPE 18085 presented at the 1988 SPE Annual Technical Conference and Exhibition, Houston, 2–5 October.
36. Elmkies, P. *et al.*: "Further Investigations on Two-Phase Flow Property Modification by Polymers: Wettability Effects," paper SPE 64986 presented at the 2001 SPE International Symposium on Oilfield Chemistry, Houston, 13–16 February.
37. Thompson, K.E. and Fogler, H.S.: "Pore-Level Mechanisms for Altering Multiphase Permeability with Gels," *SPEJ* (September 1997) 350.
38. Liang, J.T., Sun, H., and Seright, R.S.: "Why Do Gels Reduce Water Permeability More Than Oil Permeability?" *SPERE* (November 1995) 282.
39. Grattoni, C.A., Jing, X.D., and Zimmerman, R.W.: "Disproportionate Permeability Reduction When a Silicate Gel Is Formed In-Situ to Control Water Production," paper SPE 69534 presented at the 2001 SPE Latin American and Caribbean Petroleum Engineering Conference, Buenos Aries, 25–28 March.
40. Seright, R.S., Prodanovic, M., and Lindquist, W.B.: "X-Ray Computed Microtomography Studies of Disproportionate Permeability Reduction," paper SPE 89393 presented at the 2004 SPE/DOE Symposium on Improved Oil Recovery, Tulsa, 17–21 April.
41. Stavland, A. *et al.*: "Disproportionate Permeability Reduction Is Not a Panacea," *SPEREE* (August 1998) 359.

42. Mennella, A. *et al.*: "Candidate and Chemical Selection Guidelines for Relative Permeability Modification (RPM) Treatments," *SPEPF* (August 2001) 181.
43. Liang, J.T., Lee, R.L., and Seright, R.S.: "Gel Placement in Production Wells," *SPEPF* (November 1993) 276; *Trans.,* AIME, **295.**
44. Botermans, C.W., Van Batenburg, D.W., and Bruining, J.: "Relative Permeability Modifiers: Myth or Reality?" paper SPE 68973 presented at the 2001 SPE European Formation Damage Conference, The Hague, 21–22 May.
45. Seright, R.S.: "'Clean' Up of Oil Zones after a Gel Treatment," paper SPE 92772 presented at the 2005 SPE International Symposium on Oilfield Chemistry, Houston, 2–4 February.
46. Sorbie, K.S.: *Polymer-Improved Oil Recovery*, Blackie, Glasgow and London (1991).
47. Davison, P. and Mentzer, E.: "Polymer Flooding in North Sea Reservoirs," *SPEJ* (June 1982) 353.
48. Rivenq, R.C., Donche, A., and Nolk, C.: "Improved Scleroglucan for Polymer Flooding Under Harsh Reservoir Conditions," *SPERE* (February 1992) 15.
49. Sohn, W.O., Maitin, B.K., and Votz,, H.: "Preconditioning Concepts in Polymer Flooding in High-Salinity Reservoirs: Laboratory Investigations and Case Histories," *SPERE* (November 1990) 503.
50. Moradi-Araghi, A. and Doe, P.H.: "Hydrolysis and Precipitation of Polyacrylamides in Hard Brines at Elevated Temperatures," *SPERE* (May 1987) 189.
51. Stahl, G.A., Moradi-Araghi, A., and Doe, P.H.: "High Temperature and Hardness Table 14. Copolymers of Vinylpyrrolidone and Acrylamide," *Water-Soluble Polymers for Petroleum Recovery*, G.A. Stahl and D.N. Schulz (eds.), Plenum Press, New York and London (1988) 121–130.
52. Mack, J.C.: "Improved Oil Recovery—Product to Process," paper SPE 7179 presented at the 1978 SPE Rocky Mountain Regional Meeting, Cody, Wyoming, 17–19 May.
53. Denys, K., Fichen, C., and Zaitoun, A.: "Bridging Adsorption of Cationic Polyacrylamides in Porous Media," paper SPE 64984 presented at the 2001 SPE International Symposium on Oilfield Chemistry, Houston, 13–16 February.
54. Chauveteau, G., Denys, K., and Zaitoun, A.: "New Insight on Polymer Adsorption Under High Flow Rates," paper SPE 75183 presented at the 2002 SPE/DOE Improved Oil Recovery Symposium, Tulsa, 13–17 April.
55. Lake, L.W.: *Enhanced Oil Recovery*, Prentice Hall, Englewood Cliffs, New Jersey (1989) 314–353.
56. *Improved Oil Recovery*, Interstate Oil Compaction Commission, Oklahoma City, Oklahoma (1983) 36.
57. Demin, W. *et al.*: "Viscous-Elastic Polymer Can Increase Microscale Displacement Efficiency in Cores," paper SPE 63227 presented at the 2000 SPE Annual Technical Conference and Exhibition, Dallas, 1–4 October.
58. Flory, P.J.: *Principles of Polymer Chemistry*, Cornell U. Press, Ithaca, New York (1953).
59. Argabright, P.A., Rhudy, J.S., and Phillips, B.L.: "Partially Hydrolyzed Polyacrylamides of Superior Flooding and Injection Properties," paper SPE 11208 presented at the 1982 SPE Annual Technical Conference and Exhibition, New Orleans, 26–29 September.
60. Herr, J.W. and Routson, W.G.: "Polymer Structure and Its Relationship to the Dilute Solution Properties of High Molecular Weight Polyacrylamide," paper SPE 5098 presented at the 1974 SPE Annual Meeting, Houston, 6–9 October.
61. Whitecombe, P.J. and Macosko, C.W: *J. Rheology* (1978) **22,** 493.
62. Duda, J.L., Klaus, E.E., and Fan, S.K.: "Influence of Polymer-Molecule/Wall Interactions on Mobility Control," *SPEJ* (October 1981) 613.
63. Lake, L.W.: "Chemical Flooding," *Petroleum Engineering Handbook*, H.B. Bradley (ed.), SPE, Richardson, Texas (1987) Chap. 47.
64. Gogarty, W.B.: "Mobility Control with Polymer Solutions," *SPEJ* (June 1967) 161.
65. Maerker, J.M.: "Mechanical Degradation of Partially Hydrolyzed Polyacrylamide Solutions in Unconsolidated Porous Media," *SPEJ* (August 1976) 172.
66. Seright, R.S.: "The Effects of Mechanical Degradation and Viscoelastic Behavior on Injectivity of Polyacrylamide Solutions," *SPEJ* (June 1983) 475.
67. Taber, J.J., Martin, F.D., and Seright, R.S.: "EOR Screening Criteria Revisited—Part 2: Applications and Impact of Oil Prices," *SPERE* (August 1997) 199.

68. DeHekker, T.G. *et al.*: "A Progress Report on Polymer-Augmented Waterflooding in Wyoming's North Oregon Basin and Byron Fields," paper SPE 14953 presented at the 1986 SPE/DOE Symposium on Enhanced Oil Recovery, Tulsa, 20–23 April.
69. De Bons, F.E. and Braun, R.W.: "Polymer Flooding: Still Viable IOR Technique," paper presented at the 1995 European IOR Symposium, Vienna, Austria, 15–17 May, 57–65.
70. Manning, R.K., Pope, G.A., and Lake, L.W.: "A Technical Survey of Polymer Flooding Projects," U.S. Department of Energy DOE/BETC/10327-19, Bartlesville, Oklahoma (September 1983).
71. Tielong, C. *et al.*: "A Pilot Test of Polymer Flooding in an Elevated-Temperature Reservoir," *SPEREE* (February 1998) 24.
72. Putz, A., Pedron, B., and Bazin, B.: "Commercial Polymer Injection in the Courtenay Field, 1993 Update," Geological Soc. Special Publication No. 84, New Developments in Improved Oil Recovery, H.J. De Haan (ed.), The Geographical Soc., London (1995) 239–249.
73. Pitts, M.J. *et al.*: "Polymer Flood of the Rapdan Pool," *SPERE* (August 1995) 183.
74. Moritis, G.: "EOR Production Up Slightly," *Oil & Gas J.* (20 April 1998) 49.
75. Moritis, G.: "EOR Weathers Low Oil Prices," *Oil & Gas J.* (20 March 2000) 39.
76. Moritis, G.: "Special Report—Enhanced Oil Recovery," *Oil & Gas J.* (15 April 2002) 43.
77. Zaitoun, A., Kohler, N., and Guerrini, Y.: "Improved Polyacrylamide Treatments for Water Control in Producing Wells," *JPT* (July 1991) 862.
78. Stavland, A. and Nilsson, S.: "Segregated Flow Is the Governing Mechanism of Disproportionate Permeability Reduction in Water and Gas Shutoff," paper SPE 71510 presented at the 2001 SPE Annual Technical Conference and Exhibition, New Orleans, 30 September–3 October.
79. Zaitoun, A. and Pichery, T.: "A Successful Polymer Treatment For Water Coning Abatement in Gas Storage Reservoir," paper SPE 71525 presented at the 2001 SPE Annual Technical Conference and Exhibition, New Orleans, 30 September–3 October.
80. Gruenenfelder, M.A. *et al.*: "Implementing New Permeability Selective Water Shutoff Polymer Technology in Offshore, Gravel-Packed Wells," paper SPE 27770 presented at the 1994 SPE/DOE Symposium on Improved Oil Recovery, Tulsa, 17–20 April.
81. Sydansk, R.D.: "A Newly Developed Chromium(III) Gel Technology," *SPERE* (August 1990) 346.
82. Sydansk, R.D.: "Acrylamide-Polymer/Chromium(III)-Carboxylate Gels for Near Wellbore Matrix Treatments," *SPE Advanced Technology Series* (April 1993) **1,** No. 1, 146–152.
83. Sydansk, R. D.: "Foam for Improving Sweep Efficiency in Subterranean Oil-Bearing Formations," U.S. Patent No. 5,105,884 (1992).
84. "Toxicological Profile for Chromium," U.S. Dept. of Health and Human Services, Public Health Service, Agency for Toxic Substances and Disease Registry, NTIS Report PB93-182434 (April 1993).
85. Purkaple, J.D. and Summers, L.E.: "Evaluation of Commercial Crosslinked Polyacrylamide Gel Systems for Injection Profile Modification," paper SPE 17331 presented at the 1988 SPE/DOE Enhanced Oil Recovery Symposium, Tulsa, 17–20 April.
86. Smith, J.E.: "Performance of 18 Polymers in Aluminum Citrate Colloidal Dispersion Gels," paper SPE 28989 presented at the 1995 SPE International Symposium on Oilfield Chemistry, San Antonio, Texas, 14–17 February.
87. Ranganathan, R. *et al.*: "An Experimental Study of the In Situ Gelation Behavior of a Polyacrylamide/Aluminum Citrate 'Colloidal Dispersion' Gel in a Porous Medium and Its Aggregate Growth During Gelation Reaction," paper SPE 37220 presented at the 1997 SPE International Symposium on Oilfield Chemistry, Houston, 18–21 February.
88. Seright, R.S.: "Improved Techniques for Fluid Diversion in Oil Recovery," second annual report, Contract No. DE-AC22-92BC14880, U.S. DOE, Washington, DC (March 1995).
89. Moradi-Araghi, A.: "Application of Low-Toxicity Crosslinking Systems in Production of Thermally Stable Gels," paper SPE 27826 presented at the 1994 SPE/DOE Symposium on Improved Oil Recovery, Tulsa, 17–20 April.
90. Dovan, H.T., Hutchins, R.D., and Sandiford, B.B.: "Delaying Gelation of Aqueous Polymers at Elevated Temperatures Using Novel Organic Crosslinkers," paper SPE 37246 presented at the 1997 SPE International Symposium on Oilfield Chemistry, Houston, 18–21 February.

91. Hardy, M. *et al.*: "The First Carbonate Field Application of a New Organically Crosslinked Water Shutoff Polymer System," paper SPE 50738 presented at the 1999 SPE International Symposium on Oilfield Chemistry, Houston, 16–19 February.
92. Avery, M.R., Burkholder, L.A., and Gruenenfelder, M.A.: "Use of Crosslinked Xanthan Gels in Actual Profile Modification Field Projects," paper SPE 14114 presented at the 1986 SPE International Meeting of Petroleum Engineering, Beijing, 17–20 March.
93. Dalrymple, D., Tarkington, J.T., and Hallock, J.: "A Gelation System for Conformance Technology," paper SPE 28503 presented at the 1994 SPE Annual Technical Conference and Exhibition, New Orleans, 25–28 September.
94. Jones, P.W. and Baker, R.O.: "Profile Control in Virginia Hills EOR Injectors," paper SPE 24193 presented at the 1992 SPE/DOE Symposium on Enhanced Oil Recovery, Tulsa, 22–24 April.
95. Vossoughi, S. and Buller, C.S.: "Polymer Modification by In-Situ Gelation With a Newly Discovered Biopolymer," *SPERE* (November 1991) 485.
96. Lakatos, I. *et al.*: "Application of Silicate-Based Well Treatment Techniques at the Hungarian Oil Fields," paper SPE 56739 presented at the 1999 SPE Annual Technical Conference and Exhibition, Houston, 3–6 October.
97. Rolfsvag, T.A. *et al.*: "Thin Gel Treatment of an Oil Producer at the Gullfaks Field: Results and Evaluation," paper SPE 35548 presented at the 1996 SPE European Production Operations Conference and Exhibition, Stavanger, 16–17 April.
98. "Controlling Gelation Time and Microgel Size for Water Shutoff," *JPT* (March 2001) 51.
99. Moffitt, P.D. *et al.*: "Development and Field Testing of a New Low Toxicity Polymer Crosslinking System," paper SPE 35173 presented at the 1996 SPE Permian Basin Oil and Gas Recovery Conference, Midland, Texas, 27–29 March.
100. Trantham, J.C. and Moffitt, P.D.: "North Burbank Unit 1,440-Acre Polymer Flood Project Design," paper SPE 10717 presented at the 1982 SPE/DOE Symposium on Enhanced Oil Recovery, Tulsa, 4–7 April.
101. Mack, J.C. and Warren, J.: "Performance and Operation of a Crosslinked Polymer Flood at Sage Spring Creek Unit A, Natrona County, Wyoming," *JPT* (July 1984) 1145.
102. Needham, R.B., Threlkeld, C.B., and Gall, J.W.: "Control of Water Mobility Using Polymers and Multivalent Cations," paper SPE 4747 presented at the 1974 SPE Improved Oil Recovery Symposium, Tulsa, 22–24 April.
103. Sydansk, R.D.: "A New Conformance-Improvement-Treatment Chromium(III) Gel Technology," paper SPE 17329 presented at the 1988 SPE/DOE Enhanced Oil Recovery Symposium, Tulsa, 17–20 April.
104. Bryant, S.L., Rabaioli, M.R., and Lockhart, T.P.: "Influence of Syneresis on Permeability Reduction by Polymer Gels," *SPEPF* (November, 1996) 209.
105. Seright, R.S.: "Gel Placement in Fracture Systems," *SPEPF* (November 1995) 241.
106. Sydansk, R.D. and Moore, P.E.: "Gel Conformance Treatments Increase Oil Production in Wyoming," *Oil & Gas J.* (20 January 1992) 40.
107. Seright, R.S.: "Use of Preformed Gels for Conformance Control in Fractured Systems," *SPEPF* (February 1997) 59.
108. Seright, R.S.: "Gel Propagation Through Fractures," *SPEPF* (November 2001) 225.
109. Seright, R.S.: "Mechanism for Gel Propagation Through Fractures," paper SPE 55628 presented at the 1999 SPE Rocky Mountain Regional Meeting, Gillette, Wyoming, 15–18 May.
110. Hall, H.N.: "How to Analyze Waterflood Injection Well Performance," *World Oil* (October 1963) 128.
111. Earlougher, R.C. Jr.: *Advances in Well Test Analysis*, SPE, Richardson, Texas (1997) 85–87.
112. Seright, R.S.: "First Annual Technical Progress Report—Improved Techniques for Fluid Diversion in Oil Recovery," DOE/BC/14880-5, U.S. DOE, Bartlesville, Oklahoma (December 1993) Chap. 6.
113. Southwell, G.P. and Posey, S.M.: "Application and Results of Acrylamide-Polymer/Chromium (III) Carboxylate Gels," paper SPE 27779 presented at the 1994 SPE/DOE Symposium on Improved Oil Recovery, Tulsa, 17–20 April.
114. O'Brien, W.J., Stratton, J.J. Jr., and Lane, R.H.: "Mechanistic Reservoir Modeling Improves Fissure Treatment Gel Design in Horizontal Injectors, Idd El Shargi North Dome Field, Qatar,"

paper SPE 56743 presented at the 1999 SPE Annual Technical Conference and Exhibition, Houston, 3–6 October.
115. Lane, R.H. and Seright, R.S.: "Gel Water Shutoff in Fractured and Faulted Horizontal Wells," paper SPE 65527 presented at the 2000 SPE/Petroleum Society of CIM International Conference on Horizontal Well Technology, Calgary, 6–8 November.
116. Southwell, G.P.: "Marathon Oil Company's Experience with Produced Water Control and Conformance Improvement Using Polymer Gels," *Proc., Forty-Sixth Annual Southwestern Petroleum Short Course,* Texas Tech U., Lubbock, Texas (April 1999) 353–365.
117. Willhite, G.P. and Pancake, R.E.: "Controlling Water Production Using Gelled Polymer Systems," paper SPE 89464 presented at the 2004 SPE/DOE Symposium on Improved Oil Recovery, Tulsa, April 17–21.
118. Reynolds, R.R.: "Gelled Polymer Treatments in Kansas Arbuckle Wells," Petroleum Technology Transfer Council—Technology Connections website http://www.nmcpttc.org/Case_Studies/GelPolymer/index.html (accessed February 2003).
119. Sanders, G.S., Chambers, M.J., and Lane, R.H.: "Successful Gas Shutoff With Polymer Gel Using Temperature Modeling and Selective Placement in the Prudhoe Bay Field," paper SPE 28502 presented at the 1994 SPE Annual Technical Conference and Exhibition, New Orleans, 25–28 September.
120. Borling, D.C.: "Injection Conformance Control Case Histories Using Gels at the Wertz Field CO_2 Tertiary Flood in Wyoming," paper SPE 27825 presented at the 1994 SPE/DOE Symposium on Improved Oil Recovery, Tulsa, 17–20 April.
121. Hild, G.P. and Wackowski, R.K.: "Results of the Injection Well Polymer Gel Treatment Program at the Rangely Weber Sand Unit, Rangely, Colorado," paper SPE 39612 presented at the 1998 SPE/DOE Improved Oil Recovery Symposium, Tulsa, 19–22 April.
122. Bach, T. *et al.*: "Polymer Sealant for Unwanted Gas in Openhole Gravel-Pack Completion," paper SPE 68975 presented at the 2001 SPE European Formation Damage Conference, The Hague, 21–22 May.
123. Southwell, G.P. and Posey, S.M.: "Applications and Results of Acrylamide-Polymer/Chromium (III)-Carboxylate Gels," paper SPE 27779 presented at the 1994 SPE/DOE Symposium in Improved Oil Recovery, Tulsa, 17–20 April.
124. Odorisio, V.G. and Curtis, S.C.: "Operational Advances From Field Application of Short-Radius Horizontal Drilling in the Yates Field Unit," paper SPE 24612 presented at the 1992 SPE Annual Technical Conference and Exhibition, Washington, DC, 4–7 October.
125. Lai, Q.J. *et al.*: "Gel-Cement Combination Squeezes For Gas Shutoff," paper SPE 54596 presented at the 1999 SPE Western Regional Meeting, Anchorage, 26–28 May.
126. Zitha, P.L.J. and Botermans, C.W.: "Bridging Adsorption of Flexible Polymers in Low-Permeability Porous Media," *SPEPF* (February 1998) 15–20.
127. Thompson, K.E. and Fogler, H.S.: "A Study of the Diversion Mechanisms by Reactive Water Diverting Agent," *SPEPF* (May 1995) 130.
128. Wassmuth, F.R. *et al.*: "Screening and Coreflood Testing of Gel Foams To Control Excessive Gas Production in Oil Wells," *SPEREE* (June 2001) 187.
129. Miller, M.J. and Fogler, H.S.: "A Mechanistic Investigation of Waterflood Diversion Using Foamed Gels," *SPEPF* (February 1995) 62.
130. Friedmann, F. *et al.*: "Development and Testing of a Foam-Gel Technology to Improve Conformance of the Rangely CO_2 Flood," *SPEREE* (February 1999) 4.
131. Hughes, T.L. *et al.*: "Large-Volume Foam-Gel Treatments to Improve Conformance of the Rangely CO_2 Flood," *SPEREE* (February 1999) 14.
132. Fried, A.N.: "The Foam-Drive Process for Increasing the Recovery of Oil," Report of Investigation 5866, USBM (1961).
133. Bernard, G.G. and Holm, L.W.: "Effect of Foam on Permeability of Porous Media to Gas," *SPEJ* (September 1964) 267.
134. Bernard, G.G. and Holm, L.W.: "Effect of Foam on Trapped Gas Saturation and on Permeability of Porous Media to Water," *SPEJ* (December 1965) 295.
135. Bernard, G.G. and Holm, L.W.: "Model Study of Foam as a Sealant for Leaks in Gas Storage Reservoirs," *SPEJ* (March 1970) 9.

136. Holm, L.W.: "The Mechanism of Gas and Liquid Flow Through Porous Media in the Presence of Foam," *SPEJ* (December 1968) 359.
137. Holm, L.W.: "Foam Injection Test in the Siggins Field, Illinois," *JPT* (December 1970) 1499.
138. Mast, R.F.: "Microscopic Behavior of Foam in Porous Media," paper SPE 3997 presented at the 1972 SPE Annual Meeting, San Antonio, Texas, 8–11 October.
139. Rossen, W.R.: "Foams in Enhanced Oil Recovery," *Foams—Theory, Measurement, and Applications*, R.K. Prud'homme and S.A. Khan (ed.), Marcel Dekker Inc., New York (1996) 413–464.
140. Gauglitz, P.A. *et al.*: "Foam Generation in Porous Media," paper SPE 75177 presented at the 2002 SPE/DOE Improved Oil Recovery Symposium, Tulsa, 13–17 April.
141. Schramm, L.L. and Wassmuth, F.: "Foams: Basic Principles," *Foams: Fundamentals and Applications in the Petroleum Industry*, L.L. Schramm (ed.), Advances in Chemistry Series 242, American Chemical Soc., Washington, DC (1994) 3–45.
142. Llave, F.M. and Olsen, D.K.: "Use of Mixed Surfactants To Generate Foams for Mobility Control in Chemical Flooding," *SPERE* (May 1994) 125.
143. Dalland, M. and Hanssen, J.E.: "GOR-Control Foams: Demonstration of Oil-Based Foam Process Efficiency in a Physical Flow Model," paper SPE 50755 presented at the 1999 SPE International Symposium on Oilfield Chemistry, Houston, 16–19 February.
144. Mannhardt, K., Novosad, J.J., and Schramm, L.L.: "Comparative Evaluation of Foam Stability to Oil," *SPEREE* (February 2000) 23.
145. Kovscek, A.R. and Radke, C.J.: "Fundamentals of Foam Transport in Porous Media," *Foams: Fundamentals and Applications in the Petroleum Industry*, L.L. Schramm (ed.), Advances in Chemistry Series 242, American Chemical Soc., Washington, DC (1994) 115–163.
146. Osterloh, W.T. and Jante, M.J. Jr.: "Effects of Gas and Liquid Velocity on Steady-State Foam Flow at High Temperature," paper SPE 24179 presented at the 1992 SPE/DOE Symposium on Enhanced Oil Recovery, Tulsa, 22–24 April.
147. Alvarez, J.M., Rivas, H.J., and Rossen, W.R.: "Unified Model for Steady-State Foam Behavior at High and Low Foam Qualities," *SPEJ* (September 2001) 325.
148. Chang, S.-H. and Grigg, R.B.: "Effects of Foam Quality and Flow Rate on CO_2-Foam Behavior at Reservoir Temperature and Pressure," *SPEREE* (June 1999) 248.
149. Wassmuth, F.R., Green, K.A., and Randall, L.: "Details of In-Situ Foam Propagation Exposed With Magnetic Resonance Imaging," *SPEREE* (April 2001) 135.
150. Casteel, J.F. and Djabbarah, N.F.: "Sweep Improvement in CO_2 Flooding by Use of Foaming Agents," *SPERE* (November 1988) 1186.
151. Llave, F.M. *et al.*: "Foams as Mobility Control Agents for Oil Recovery by Gas Displacement," paper SPE 20245 presented at the 1990 SPE/DOE Symposium on Enhanced Oil Recovery, Tulsa, 22–25 April.
152. Tsau, J.-S., Yaghoobi, H., and Grigg, R.B.: "Smart Foam to Improve Oil Recovery in Heterogeneous Porous Media," paper SPE 39677 presented at the 1998 SPE/DOE Improved Oil Recovery Symposium, Tulsa, 19–22 April.
153. Tsau, J.-S., and Heller, J.P.: "How Can Selective Mobility Reduction of CO_2-Foam Assist in Reservoir Floods?" paper SPE 35168 presented at the 1996 SPE Permian Basin Oil and Gas Recovery Conference, Midland, Texas, 27–29 March.
154. Lee, H.O., Heller, J.P., and Hoefer A.M.W.: "Change in Apparent Viscosity of CO_2 Foam with Rock Permeability," *SPERE* (November 1991) 421.
155. Albrecht, R.A. and Marsden, S.S.: "Foams as Blocking Agents in Porous Media," *SPEJ* (March 1970) 51.
156. Rossen, W.R.: "Theories of Foam Mobilization Pressure Gradient," paper SPE 17358 presented at the 1988 SPE/DOE Enhanced Oil Recovery Symposium, Tulsa, 17–20 April.
157. Minssieux, L.: "Oil Displacement by Foams in Relation to Their Physical Properties in Porous Media," *JPT* (January 1974) 100.
158. Hirasaki, G.J.: "The Steam-Foam Process," *JPT* (May 1989) 449.
159. Schramm, L.L.: "Foam Sensitivity to Crude Oil in Porous Media," *Foams: Fundamentals and Applications in the Petroleum Industry*, L.L. Schramm (ed.), Advances in Chemistry Series 242, American Chemical Soc., Washington, DC (1994) 165–197.

160. Kanda, M. and Schechter, R.S.: "On the Mechanism of Foam Formation in Porous Media," paper SPE 6200 presented at the 1976 SPE Annual Technical Conference and Exhibition, New Orleans, 3–6 October.
161. Suffridge, F.E., Raterman, K.T., and Russell, G.C.: "Foam Performance Under Reservoir Conditions," paper SPE 19691 presented at the 1989 SPE Annual Technical Conference and Exhibition, San Antonio, Texas, 8–11 October.
162. Sanchez, J.M. and Hazlett, R.D.: "Foam Flow Through an Oil-Wet Porous Medium: A Laboratory Study," *SPERE* (February 1992) 91.
163. Raza, S.H.: "Foam in Porous Media: Characteristics and Potential Applications," *SPEJ* (December 1970) 328.
164. Friedmann, F. and Jensen, J.A.: "Some Parameters Influencing the Formation and Propagation of Foams in Porous Media," paper SPE 15087 presented at the 1986 SPE California Regional Meeting, Oakland, California, 2–4 April.
165. Jensen, J.A. and Friedmann, F.: "Physical and Chemical Effects of an Oil Phase on the Propagation of Foam in Porous Media," paper SPE 16375 presented at the 1987 SPE California Regional Meeting, Ventura, California, 8–10 April.
166. Lau, H.C. and O'Brien, S.M.: "Effect of Spreading and Nonspreading Oils on Foam Propagation Through Porous Media," *SPERE* (August 1988) 893.
167. Raterman, K.T.: "An Investigation of Oil Destabilization of Nitrogen Foams in Porous Media," paper SPE 19692 presented at the 1989 SPE Annual Technical Conference and Exhibition, San Antonio, Texas, 8–11 October.
168. Schramm, L.L. and Novosad, J.J.: "Micro-Visualization of Foam Interactions with a Crude Oil," *Colloids and Surfaces 46* (1990) 21.
169. Kuhlman, M.I.: "Visualizing the Effect of Light Oil on CO_2 Foams," *JPT* (July 1990) 902.
170. Manlowe, D.J. and Radke, C.J.: "A Pore-Level Investigation of Foam/Oil Interactions in Porous Media," *SPERE* (November 1990) 495.
171. Hornbrook, J.W., Castanier, L.M., and Pettit, P.A.: "Observation of Foam/Oil Interactions in a New, High-Resolution Micromodel," paper SPE 22631 presented at the 1991 SPE Annual Technical Conference and Exhibition, Dallas, 6–9 October.
172. Kristiansen, T.S. and Holt, T.: "Properties of Flowing Foam in Porous Media Containing Oil," paper SPE 24182 presented at the 1992 SPE/DOE Symposium on Enhanced Oil Recovery, Tulsa, 22–24 April.
173. Law, D.H.-S., Yang, Z.-M., and Stone, T.W.: "Effect of the Presence of Oil on Foam Performance: A Field Simulation Study," *SPERE* (May 1992) 228.
174. Hamida, F.M. *et al.*: "Further Characterization of Surfactants as Steamflood Additives," *In Situ* (1992) **16,** No. 2, 137.
175. Tsau, J.-S., Syahputra, A.E., and Grigg, R.B.: "Economic Evaluation of Surfactant Adsorption in CO_2 Foam Application," paper SPE 59365 presented at the 2000 SPE/DOE Improved Oil Recovery Symposium, Tulsa, 3–5 April.
176. Krause, R.E. *et al.*: "Foam Treatment of Producing Wells to Increase Oil Production at Prudhoe Bay," paper SPE 24191 presented at the 1992 SPE/DOE Symposium on Enhanced Oil Recovery, Tulsa, 22–24 April.
177. Shan, D. and Rossen, W.R.: "Optimal Injection Strategies for Foam IOR," paper SPE 75180 presented at the 2002 SPE/DOE Improved Oil Recovery Symposium, Tulsa, 13–17 April.
178. Hanssen, J.E. *et al.*: "SAGA Injection: A New Combination IOR Process for Stratified Reservoirs," *New Developments in Improved Oil Recovery*, H.J. De Haan (ed.), Geological Soc. Special Publication (1995) **84,** 111.
179. Shi, J.-X. and Rossen, W.R.: "Improved Surfactant-Alternating-Gas Foam Process to Control Gravity Override," paper SPE 39653 presented at the 1998 SPE/DOE Improved Oil Recovery Symposium, Tulsa, 19–22 April.
180. Heller, J.P.: "CO_2 Foams in Enhanced Oil Recovery," *Foams: Fundamentals and Applications in the Petroleum Industry*, L.L. Schramm (ed.), Advances in Chemistry Series 242, American Chemical Soc., Washington, DC (1994) 201–234.
181. Di Jullo, S.S. and Emanuel, A.S.: "Laboratory Study of Foaming Surfactant of CO_2 Mobility Control," *SPERE* (May 1989) 136.

182. Martin, F.D., Stevens, J.E., and Harpole, K.J.: "CO_2-Foam Field Test at the East Vacuum Grayburg/San Andres Unit," *SPERE* (November 1995) 266.
183. Prieditis, J. and Paulett, G.S.: "CO_2-Foam Mobility Tests at Reservoir Conditions in San Andres Cores," paper SPE 24178 presented at the 1992 SPE/DOE Symposium on Enhanced Oil Recovery, Tulsa, 22–24 April.
184. Kuhlman, M.I. *et al.*: "CO_2 Foam With Surfactants Used Below Their Critical Micelle Concentrations," *SPERE* (November 1992) 445.
185. Isaacs, E.E., Ivory, J., and Green, M.K.: "Steam-Foams for Heavy Oil and Bitumen Recovery," *Foams: Fundamentals and Applications in the Petroleum Industry*, L.L. Schramm (ed.), Advances in Chemistry Series 242, American Chemical Soc., Washington, DC (1994) 235–258.
186. Borchardt, J.K. and Strycker, A.R.: "Olefin Sulfonate for High Temperature Steam Mobility Control: Structure—Property Correlations," paper SPE 37219 presented at the 1997 SPE International Symposium on Oilfield Chemistry, Houston, 18–21 February.
187. Mannhardt, K. and Novosad, J.J.: "Adsorption of Foam-Forming Surfactants for Hydrocarbon-Miscible Flooding at High Salinities," *Foams: Fundamentals and Applications in the Petroleum Industry*, L.L. Schramm (ed.), Advances in Chemistry Series 242, American Chemical Soc., Washington, DC (1994) 259–316.
188. Chad, J., Matsalla, P., and Novosad, J.J.: "Foam Forming Surfactants in Pembiana/Ostracod 'G' Poll," paper CIM 88-39-40 presented at the 1988 CIM Petroleum Soc. Technical Meeting, Calgary, 12–16 June.
189. Liu, P.C. and Besserer, G.J.: "Application of Foam Injection in Triassic Pool, Canada: Laboratory and Field Test Results," paper SPE 18080 presented at the 1988 SPE Annual Technical Conference and Exhibition, Houston, 2–5 October.
190. Sydansk, R.D.: "Polymer-Enhanced Foams Part 1: Laboratory Development and Evaluation," *SPE Advanced Technology Series* (April 1994) **2,** No. 2, 150–159.
191. Zhu, T. *et al.*: "Foams for Mobility Control and Improved Sweep Efficiency in Gas Flooding," paper SPE 39680 presented at the 1998 SPE Improved Oil Recovery Symposium, Tulsa, 19–22 April.
192. Ye, Z. *et al.*: "Laboratory Study on Profile Modification by Foamed Polymer Solution," paper CIM 97-128 presented at the 1997 Petroleum Soc. Annual Technical Meeting, Alberta, Canada, 8–11 June.
193. Romero, C., Alvarez, J.M., and Muller, A.J.: "Micromodel Studies of Polymer-Enhanced Foam Flow Through Porous Media," paper SPE 75179 presented at the 2002 SPE/DOE Improved Oil Recovery Symposium, Tulsa, 13–17 April.
194. Hanssen, J.E. and Dalland, M.: "Gas-Blocking Foams," *Foams: Fundamentals and Applications in the Petroleum Industry*, L.L. Schramm (ed.), Advances in Chemistry Series 242, American Chemical Soc., Washington, DC (1994) 319–353.
195. Hanssen, J.E. and Dalland, M.: "Foam Barriers for Thin Oil Rims: Gas Blockage at Reservoir Conditions," paper presented at the 1991 European IOR Symposium, Stavanger, 21–23 May.
196. Chukwueke, V.O., Bouts, M.N., and van Dijkum, C.E.: "Gas Shut-Off Foam Treatments," paper SPE 39650 presented at the 1998 SPE Improved Oil Recovery Symposium, Tulsa, 19–22 April.
197. Wassmuth, F.R. *et al.*: "Screening and Coreflood Testing of Gel Foams to Control Excessive Gas Production in Oil Wells," *SPEREE* (June 2001) 187.
198. Dalland, M. and Hanssen, J.E.: "Enhanced Foams for Efficient Gas Influx Control," paper SPE 37217 presented at the 1997 SPE International Symposium on Oilfield Chemistry, Houston, 18–24 February.
199. Aarra, M.G., Ormehaug, P.A., and Skauge, A.: "Foams for GOR Control – Improved Stability by Polymer Additives," paper 010 presented at the 1997 SPE European Symposium on Improved Oil Recovery, The Hague, 20–22 October.
200. Thach, S. *et al.*: "Matrix Gas Shut-Off in Hydraulically Fractured Wells Using Polymer-Foams," paper SPE 36616 presented at the 1996 SPE Annual Technical Conference and Exhibition, Denver, 6–9 October.
201. Zhdanov, S.A. *et al.*: "Application of Foam for Gas and Water Shut-Off: Review of Field Experience," paper SPE 36914 presented at the 1996 SPE European Petroleum Conference, Milan, Italy, 22–24 October.

202. Isaacs, E.E. *et al.*: "Conformance Improvement by Using High Temperature Foams and Gels," paper SPE 23754 presented at the 1992 SPE Latin America Petroleum Engineering Conference, Caracas, 8–11 March.
203. Patzek, T.W.: "Field Applications of Steam Foam for Mobility Improvement and Profile Control," *SPERE* (May 1996) 79.
204. Gauglitz, P.A. *et al.*: "Field Optimization of Steam/Foam for Profile Control: Midway-Sunset 26C," paper SPE 25781 presented at the 1993 SPE International Thermal Operations Symposium, Bakersfield, California, 8–10 February.
205. Chou, S.I. *et al.*: "CO_2 Foam Field Trial at North Ward-Estes," paper SPE 24643 presented at the 1992 SPE Annual Technical Conference and Exhibition, Washington, DC, 4–7 October.
206. Stephenson, D.J., Graham, A.G., and Luhning, R.W.: "Mobility Control Experience in the Joffre Viking Miscible CO_2 Flood," *SPERE* (August 1993) 183.
207. Hoefner, M.L. and Evans, E.M.: "CO_2 Foam: Results From Four Development Field Trials," *SPERE* (November 1995) 273.
208. Henry, R.L. *et al.*: "Field Test of Foam to Reduce CO_2 Cycling," paper SPE 35402 presented at the 1996 SPE/DOE Symposium on Improved Oil Recovery, Tulsa, 21–24 April.
209. Surguchev, L.M. and Hanssen, J.E.: "Foam Application in North Sea Reservoirs, I: Design and Technical Support of Field Trials," paper SPE 35371 presented at the 1996 SPE/DOE Symposium on Improved Oil Recovery, Tulsa, 21–24 April.
210. Demin, W. *et al.*: "First Ultra-Low Interfacial Tension Foam Flood Test Is Successful," paper SPE 71491 presented at the 2001 SPE Annual Technical Conference and Exhibition, New Orleans, 30 September–3 October.
211. *Foams: Fundamentals and Applications in the Petroleum Industry*, L.L. Schramm (ed.), Advances in Chemistry Series 242, American Chemical Society, Washington, DC (1994).
212. Nguyen, Q.P. *et al.*: "Experimental and Modeling Studies of Foam in Porous Media: A Review," paper SPE 58799 presented at the 2000 SPE International Symposium on Formation Damage Control, Lafayette, Louisiana, 23–24 February.
213. Hanssen, J.E., Holt, T., and Surguchev, L.M.: "Foam Processes: An Assessment of Their Potential in North Sea Reservoirs Based on a Critical Evaluation of Current Field Experience," paper SPE 27768 presented at the 1994 SPE/DOE Symposium on Improved Oil Recovery, Tulsa, 17–20 April.
214. Barnes, R.M.: "Notes on the Use of Resin for Excluding Water From Oil Wells," *Annual Report of the State Oil and Gas Supervisor of California,* California Division of Oil and Gas (1922–1923) **12,** 3.
215. Littlefield, B.A., Fader, P.D., and Surles, B.W.: "Case Histories of New Low-Cost Fluid Isolation Technology," paper SPE 24802 presented at the 1992 SPE Annual Technical Conference and Exhibition, Washington, DC, 4–7 October.
216. Fader, P.D. *et al.*: "New Low-Cost Resin System for Sand and Water Control," paper SPE 24051 presented at the 1992 Western Regional Meeting, Bakersfield, California, 30 March–1 April.
217. Hess, P.H.: "One-Step Furfuryl Alcohol Process for Formation Plugging," *JPT* (October 1980) 1834.
218. "Coiled-Tubing Resin Squeeze to Mitigate Water Production," *JPT* (June 1998) 41.
219. Morgenthaler, L.N. and Schultz, H.A.: "A Novel Process for Profile Control in Thermal Recovery Projects," paper SPE 28504 presented at the 1994 SPE Annual Technical Conference and Exhibition, New Orleans, 25–28 September.
220. Peavy, M.A.: "Successful Water Control in Openhole Gravel-Packed Completions Within a TEOR Environment," paper SPE 22778 presented at the 1991 SPE Annual Technical Conference and Exhibition, Dallas, 6–9 October.
221. Demin, W. *et al.*: "Experience Learned After Producing More Than 300 Million Barrels of Oil by Polymer Flooding in Daqing Oil Field," paper 77693 presented at the 2002 Annual Technical Conference and Exhibition, San Antonio, Texas, 29 September–2 October.

SI Metric Conversion Factors

acre-ft	×	1.233 489	E + 03	= m^3
Å	×	1.0*	E − 01	= nm
bbl	×	1.589 873	E − 01	= m^3
cp	×	1.0*	E − 03	= Pa·s
ft	×	3.048*	E − 01	= m
ft^2	×	9.290 304*	E − 02	= m^2
ft^3	×	2.831 685	E − 02	= m^3
°F		(°F − 32)/1.8		= °C
in.	×	2.54*	E + 00	= cm
$in.^3$	×	1.638 706	E + 01	= cm^3
lbm	×	4.535 924	E − 01	= kg
psi	×	6.894 757	E + 00	= kPa

*Conversion factor is exact.

Chapter 14
Miscible Processes

E.D. Holstein, Consultant and **Fred I. Stalkup,** PetroTel Inc.

14.1 Introduction

Miscible injection is a proven, economically viable process that significantly increases oil recovery from many different types of reservoirs. Fieldwide projects have been implemented in fields around the world, with most of these projects being onshore North American fields. Many of these projects are quite mature, making the recovery and production-rate benefits well established. As a result, the ability to predict recovery levels, rate improvements, costs, and resulting economics can now be considered proven and reliable. The purpose of this chapter is to introduce some fundamental concepts about miscible displacement, suggest some methods of predicting the benefits of miscible injection, and present a few field examples that demonstrate what has been accomplished with miscible injection.

14.2 Introductory Concepts

To put miscible flooding into perspective, it is instructive to compare the performance of a miscible flood to that of a waterflood. Although it is impossible to define a "typical" flood, the simplistic example shown in **Fig. 14.1** introduces the physics of the process and illustrates the level of incremental recovery and sweep often achieved with miscible flooding. This example is based on simulation results for the Means Lower San Andres reservoir in west Texas.[1]

The upper diagram in Fig. 14.1 shows a pattern element in the reservoir (with an injector on the left and a producer on the right) near the end of a waterflood. The initial average oil saturation, S_{oi}, of 70% at the start of the waterflood was reduced to an average of 44%, S_{orwf}, at the end. The 44% accounts for much higher saturations near the edges of the pattern and in lower-permeability strata not contacted by water. The waterflood recovers 37% of the original oil in place (OOIP) and sweeps approximately 80% of the reservoir.

The average S_{orwf} in those regions swept by water is 38%. The oil that remains in the water-swept part of the reservoir is trapped as a discontinuous phase within the pore space. The primary goal of a miscible flood is to recover part of this trapped residual oil. However, solvent can sometimes also displace oil from upper regions of the reservoir not swept or poorly swept by water because of gravity-driven water "slumping."

The lower diagram in Fig. 14.1 shows the same reservoir element near the completion of a miscible flood started near the end of the waterflood. In practice, a solvent flood is often initiated before completing the waterflood—or, in some cases, even before starting the waterflood.

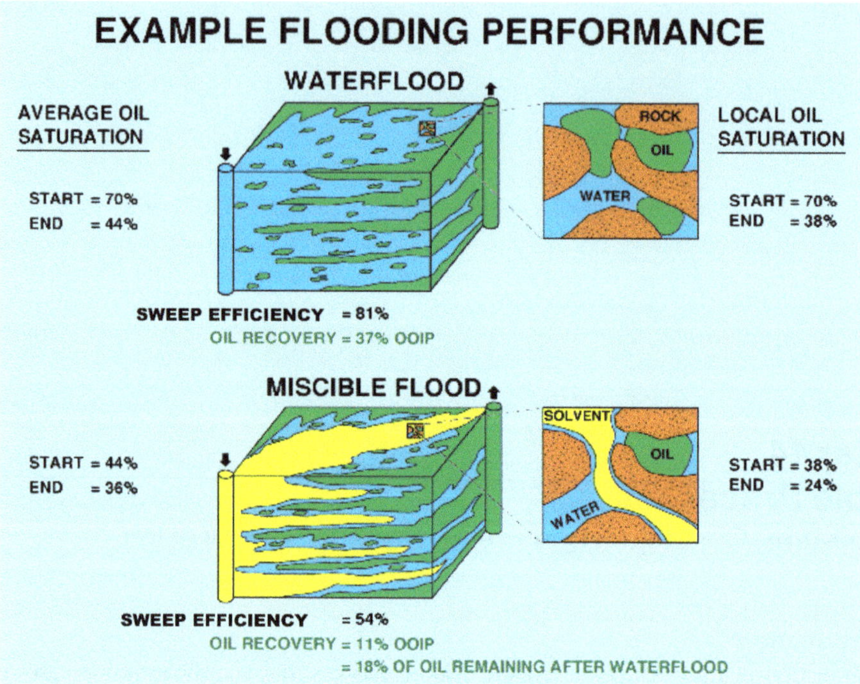

Fig. 14.1—Example flooding performance.[2]

The schematic shows that solvent sweeps only part of the reservoir previously swept by water, and only a portion of the residual oil in the solvent-swept regions is recovered. The average oil saturation after the miscible flood (S_{orm}) of 36% includes (1) higher oil saturations remaining in regions near the edge of the pattern element and in lower-permeability strata bypassed by both water and solvent and (2) residual oil in that part of the reservoir swept by water but not by solvent. The miscible flood recovered an incremental 11% OOIP over and above waterflooding results. Expressed another way, the incremental recovery was 18% of the oil remaining after waterflooding.

The solvent sweeps approximately 50% of the pattern, compared to approximately 80% for water. In regions swept by solvent, oil saturation is reduced from an average value of 38% to 24%. The average oil saturation of 24% in solvent-swept regions accounts for oil saturations lower than 24% near the injector, as well as higher saturations near the producer. At the pore level, solvent displaces some, but not all, of the waterflood residual oil.

Three solvent-injection strategies commonly used in commercial miscible-flooding applications are slug injection, water-alternating-gas (WAG) injection, and gravity-stable injection. The slug process usually involves continuous injection of approximately 0.2 to 0.4 hydrocarbon pore volumes (HCPV) of solvent that, in turn, is displaced by water or dry solvent.[2] The WAG process involves alternately injecting small volumes (0.01–0.04 HCPV) of water and solvent.[2] The total amount of solvent injected usually ranges from 0.2 to 0.6 HCPV.[3] As with the slug process, the final drive fluid is usually water or dry solvent. It is commonly accepted that alternate injection of small slugs of water decreases solvent mobility and leads to increased solvent sweep efficiency.[4] Experience with many projects has indicated that field-management processes improve with time, and additional volumes of injectant can be justified to further increase recovery.

For some pinnacle reefs and steeply dipping reservoirs with high vertical communication, it is advantageous to inject less-dense solvent at the top of the reservoir in a gravity-stable dis-

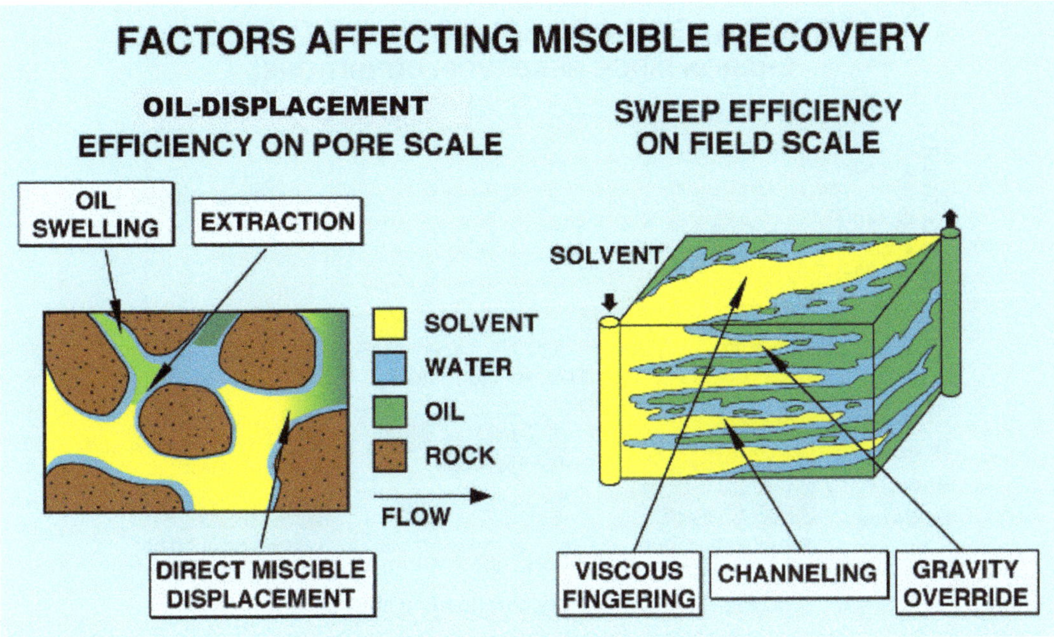

Fig. 14.2—Factors affecting miscible recovery.[2]

placement process. Solvent sweep efficiency and oil recovery are quite high, provided that there is sufficient vertical continuity.[5–9]

The two major factors that affect the performance of a miscible flood are oil-displacement efficiency at the pore level and sweep efficiency on the field scale. Oil displacement can be explained using the schematic on the left side of **Fig. 14.2**, which shows solvent flowing from left to right through a pore space. The displacement process involves several mechanisms.[2,10,11] One is direct miscible displacement of oil by solvent along higher-permeability pore paths. Additionally, part of the oil initially bypassed (on the pore level) by solvent can later be recovered through oil swelling that occurs as solvent dissolves in the oil, or by extraction of oil into solvent. Swelling and extraction take place as solvent continues to flow past the initially bypassed oil. These can be significant mechanisms in field processes and together may account for as much as 20 to 30% of the total incremental recovery. Oil-displacement efficiency is affected by solvent composition and pressure. Solvents can be designed that give very high displacement efficiencies at the pore level.[1,12–17]

The right side of Fig. 14.2 shows that, on a field scale, sweep efficiency is affected by viscous fingering and solvent channeling through high-permeability streaks. Gravity override can sometimes occur because solvent is usually less dense than the oil it is displacing.[18,19] When vertical communication is high, solvent tends to gravity segregate to the top of a reservoir unit and sweep only the upper part of that zone. Although gravity override can be a problem in reservoirs having good vertical communication (such as Judy Creek[20] and Prudhoe Bay[21]), it is not usually a serious problem for west Texas carbonates,[22] which tend to be more stratified and have poor vertical communication.

As discussed in specific field examples, sweep efficiency on the field scale is usually the single most important factor affecting performance of a miscible flood. Sweep efficiency can be increased to some extent by reducing well spacing, increasing injection rate, reconfiguring well patterns, increasing solvent-bank sizes, and modifying the ratio of injected water to injected solvent (WAG ratio).

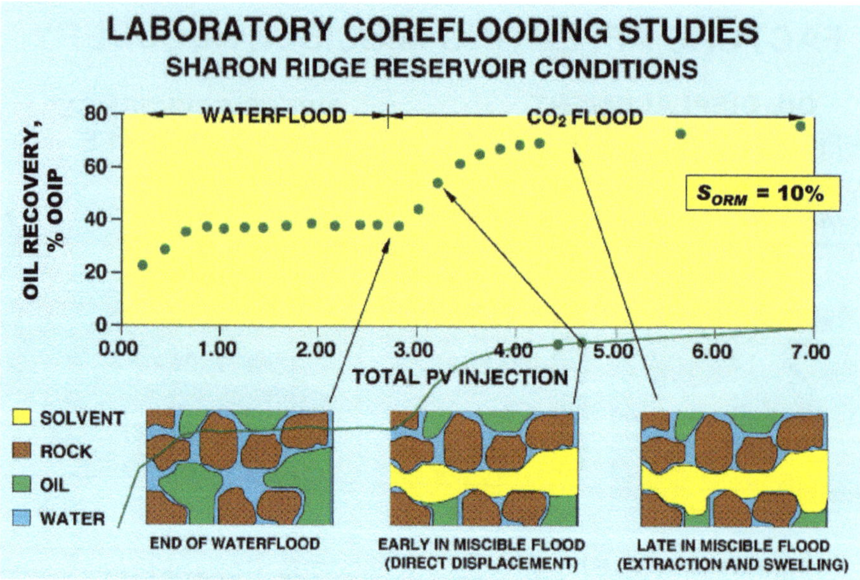

Fig. 14.3—Laboratory corefooding studies.[2]

Fig. 14.3 presents part of a considerable body of laboratory evidence that solvent effectively displaces oil from contacted regions of the reservoir. The graph of oil recovery as a function of total pore volumes of fluid injected shows the results of a laboratory coreflood conducted at conditions corresponding to the Sharon Ridge reservoir in west Texas. The waterflood recovered approximately 40% OOIP. A CO_2 flood that followed increased oil recovery to approximately 80% OOIP, demonstrating that CO_2 can displace a large portion of the residual oil remaining after a waterflood. S_{orm} was 10%; the WAG ratio for the miscible flood was 1.

The schematics at the bottom of Fig. 14.3 illustrate the pore-level recovery mechanisms discussed earlier (Fig. 14.2). At the end of the waterflood, residual oil is a discontinuous phase that occupies approximately 40% of the pore space. Early in the miscible flood [3.0 to 3.5 total pore volumes (PV) injected], some of this oil has been miscibly displaced by solvent from the higher-permeability flow path (on the pore scale). However, some oil also has been initially bypassed by solvent. Note that this bypassing at the pore level is much different from solvent bypassing, which can occur at the field scale because of larger-scale reservoir heterogeneities. As depicted in the schematic corresponding to late in the flood (to 7.0 total PV injected), part of this locally bypassed oil is subsequently recovered by extraction and swelling that takes place as solvent continues to flow past the bypassed oil. In this case, approximately 30% of the total amount of oil recovered by the CO_2 flood was recovered by extraction and swelling.

14.3 Designing a Miscible Flood

14.3.1 Determining Miscibility. True miscible displacement implies that injected and displaced phases mix in all proportions without forming interfaces or two phases. The single-phase condition also implies that the solvent eventually displaces all resident oil from the pore space that it invades. Although some fluids, such as propane, fulfill this definition, most solvents available for oilfield use form two distinct phases over a broad range of mixtures and pressures when combined with reservoir oils. However, when the same solvent displaces oil at reservoir temperature and above a suitably high pressure in a long, small-diameter (slim) sand-

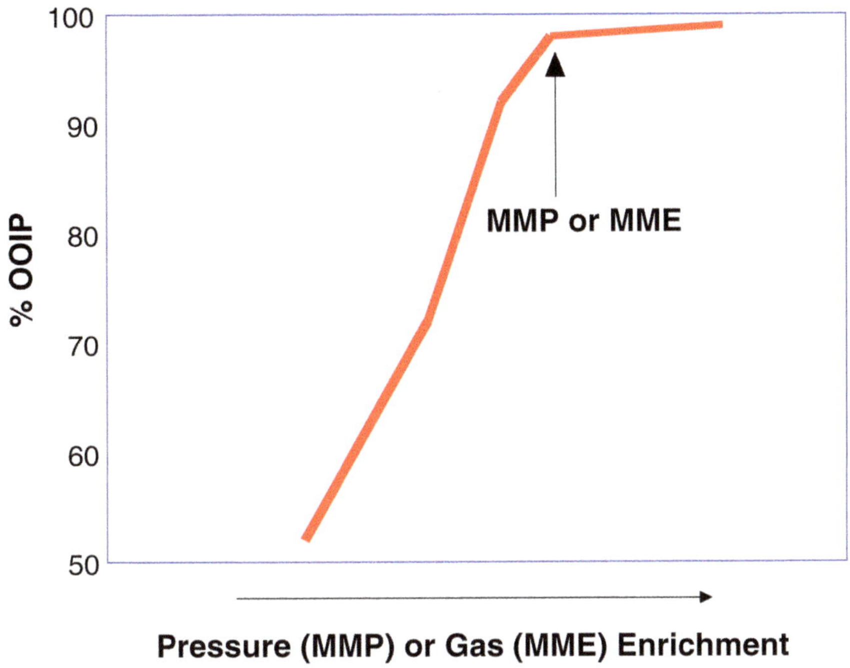

Fig. 14.4—Slimtube displacements used to determine MMP or MME.

packed tube, a miscible-like displacement occurs. Slimtube experiments are designed to make the displacement essentially 1D with 100% volumetric sweep by the solvent front.

Fig. 14.4 shows a series of hypothetical slimtube experiments. In these experiments, the solvent displaces oil from the fully oil-saturated slimtube at several pressures. Oil recovery is shown after 1.2 PV of injection for each pressure. Oil recovery increases with pressure up to approximately 95 to 98% and then increases very little thereafter. The pressure at which the break in the recovery curve occurs is said to be the minimum miscibility pressure (MMP). If the displacements had been conducted at constant pressure and with increasing enrichment by components such as ethane, propane, and butane, the break over would have been at the minimum miscibility enrichment (MME). Above the MMP or MME, the displacement is said to be "multiple-contact" or "dynamically" miscible. The increasing recovery with pressure or solvent enrichment results from in-situ mass transfer of components between solvent and resident oil. Each pressure increase produces an equilibrium mixture that becomes compositionally similar at the MMP or MME. Methane, methane enriched with C_2–C_4 hydrocarbons, CO_2, N_2, and flue solvent will all give compositionally enhanced displacements under the right conditions of pressure, temperature, and oil composition. The MMP or MME can be significantly different for each of these solvents.

A slimtube is not representative of the performance in reservoir rock because it does not account for the effects of factors such as gravity segregation and reservoir heterogeneity on volumetric sweep. **Fig. 14.5** shows how ethane enrichment of methane affected oil recovery in an experiment conducted by Chang *et al.*[23] with live reservoir fluid in a 34.6-in.-long, relatively homogeneous Bentheimer sandstone core. The experimental conditions resulted in a single, gravity-overriding tongue of solvent in the core.[23] Oil recovery continued to increase from enrichment levels well below to well above the MME determined from slimtube displacements without evidencing a pronounced break over. In other words, compositional enhancement continued to increase recovery well above the slimtube MME for this gravity-dominated displacement.

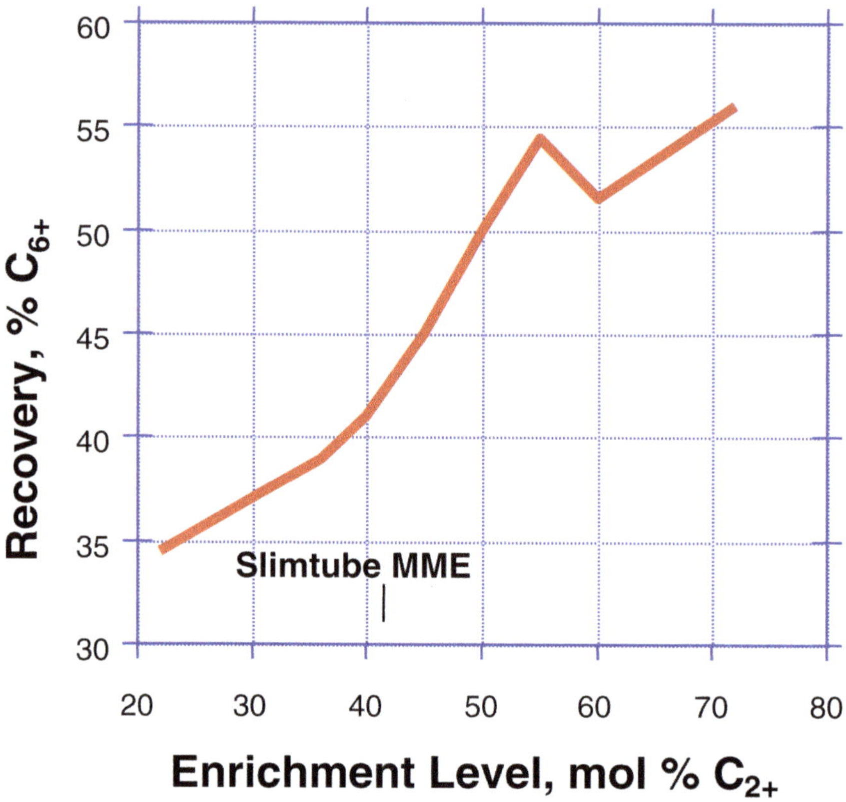

Fig. 14.5—Enriched gasflood with gravity tongue in outcrop core (after Ref. 23).

Jerauld[24] reported a similar finding by use of a compositional reservoir simulation of a one-fourth nine-spot pattern element (**Fig. 14.6**). Recovery continued to increase from well below to well above the simulator-predicted MME of 0.65.

14.3.2 Choosing a Candidate. A decision to implement a miscible flood in a particular field will usually consist of a sequential approach.

First is the screening stage. Data in the literature allow a reasonable estimate of MMP or MME, S_{orm}, amount of solvent required, and operating costs. This information is adequate to determine if a reservoir is a candidate.

MMP and MME Guidelines and Correlations. **Table 14.1** gives some rough guidelines for achieving MMP or MME with different injection solvents. For hydrocarbon solvent enriched with the C_2–C_4 hydrocarbons, MMP in the range 1,500 to 3,000 psi might be expected for mid-API-gravity oils, depending on oil and solvent compositions and reservoir temperature.

The vaporizing-solvent process (see later sections) is applicable to high-gravity oils. MMP typically is greater than 3,500 psi and is usually greater than 4,500 psi (and can be much greater).

CO_2 flooding is applicable with medium-gravity oils. At temperatures less than approximately 125°F, MMP can be as low as 1,200 psi. MMP increases with temperature.

Fig. 14.7[25] shows an approximate correlation for CO_2 flooding MMP. **Fig. 14.8**[14] shows one graph for a similar correlation that was developed for condensing-solvent-drive MMP. It also appears to be useful for condensing/vaporizing drives. Other graphs are given in Ref. 14.

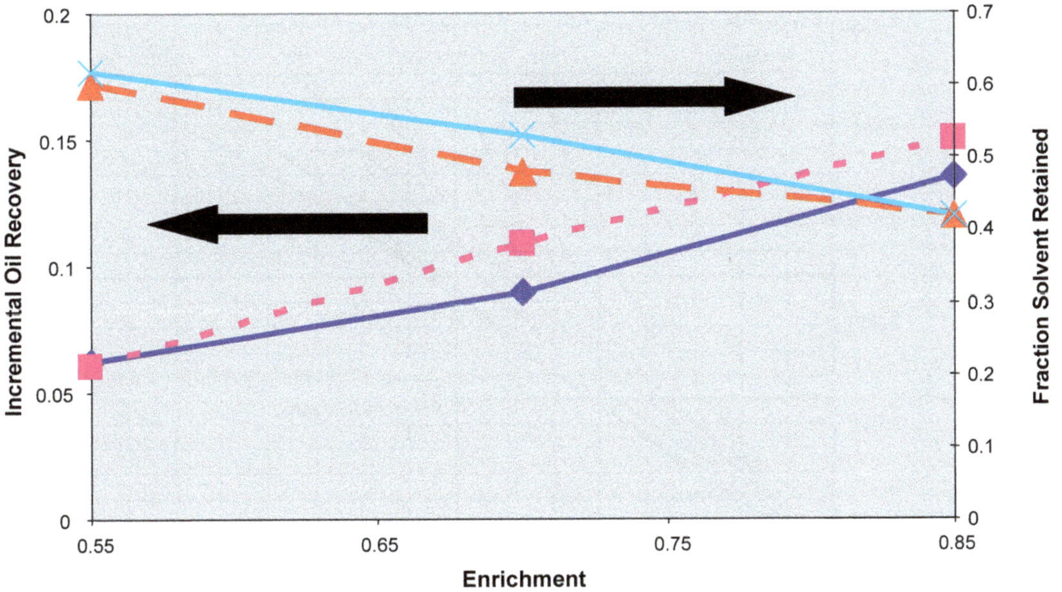

Fig. 14.6—Effect of enrichment on recovery in a reservoir study. After Jerauld (solid lines are the reference model, and dashed lines are the scaleup model).[24]

TABLE 14.1—ROUGH GUIDELINES FOR MMPs WITH DIFFERENT SOLVENTS

Process	Injection Gas	Miscibility Pressure (psi)	Typical Depth (ft)
Enriched gas	Natural gas enriched with C_2, C_3, C_4	1,500–3,000; >26°API	>2,000
Vaporizing gas	Natural gas; nitrogen; flue gas	>3,500; >40°API	>5,000
Carbon dioxide	CO_2	>1,200; >26°API	>2,000

In field projects in which the displacement was above either the MMP or the MME, residual oil saturation determined by coring behind the solvent front varied from approximately 3 to 10% PV.[26]

Second, a more thorough assessment may require acquiring laboratory data on S_{orwf}, S_{orm}, and MMP or MME. Analog data from nearby fields may be adequate for these values and may indeed be adequate for evaluating a project without additional work. Occasionally, some type of field pilot may be thought necessary to address such questions as displacement and sweep efficiencies (most small pilots do not produce results to reliably predict these factors), injectivity of miscible fluids, and existence of fractures or very-high-permeability layers that would prevent the miscible fluids from contacting a significant volume of the reservoir.

Third, some type of simulation will normally be done to incorporate reservoir and fluid characteristics unique to the field in question. The type of simulator used will depend on the

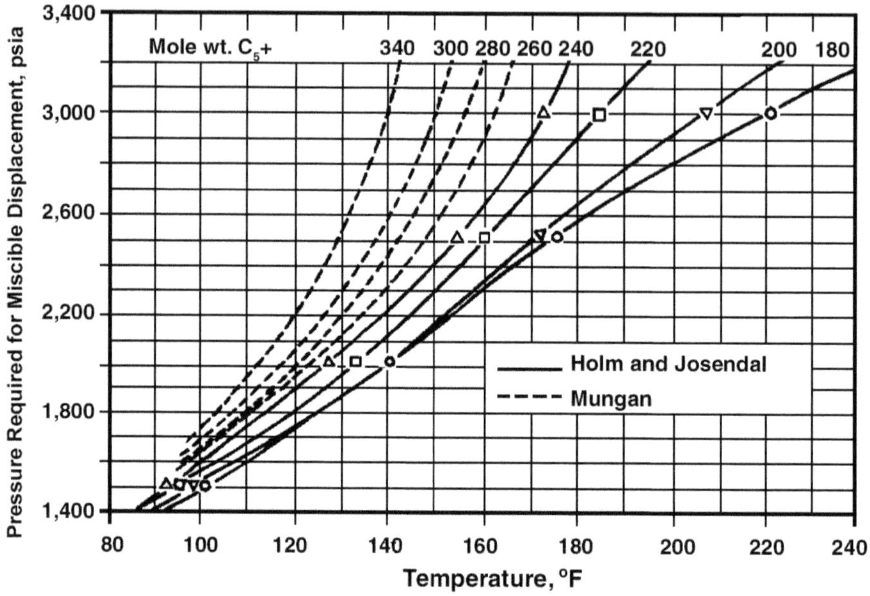

Fig. 14.7—Correlation for CO_2-flood MMP (after Refs. 25 and 26).

amount and quality of the characterization data available and the perceived risks of the project that justify the costs of the various types of simulation that can be done.

Several factors should be considered in assessing the economic viability of a miscible project:

• What miscible fluids are available, and what is the corresponding MMP or MME?

• Miscible fluids commonly considered are hydrocarbon solvents such as enriched methane, CO_2, N_2, and, less often, exhaust or flue gases. Assessments of hydrocarbon solvents should include near-term lost revenue because of delays in sales (if any) and the ultimate amount of solvent that will remain in the reservoir at abandonment. MMP and MME will also narrow the number of solvents that may be applicable to a specific field.

• Is the MMP sufficiently below overburden pressure, or are adequate enrichment fluids available for the field in question?

• In many instances, the MMP for a given solvent may exceed the overburden pressure of a formation. Where MMP is less than overburden, the question becomes whether a high-enough injection rate can be achieved to satisfy a reasonable project life.

• Are near-miscible recoveries high enough to support a project?

• As discussed earlier, significant additional recovery may be possible without reaching the slimtube MMP.

• What is the incremental recovery vs. the solvent slug size (**Fig. 14.9**)?

• Numerical simulations can provide sufficient insight to evaluate the economics of a project. For most projects, slug size can be refined further during the actual flood (usually increased) when actual performance can be used to modify initial projections.

• Which WAG ratio will be most effective?

• Simulations can give a good initial estimate and will be good enough for defining the costs of a WAG project. Several different WAG schemes have been used in practice. These include:

• An initial slug of miscible solvent followed by a low WAG ratio tapering to a high ratio.

• A constant WAG ratio that tapers to a high ratio near the end of the project.

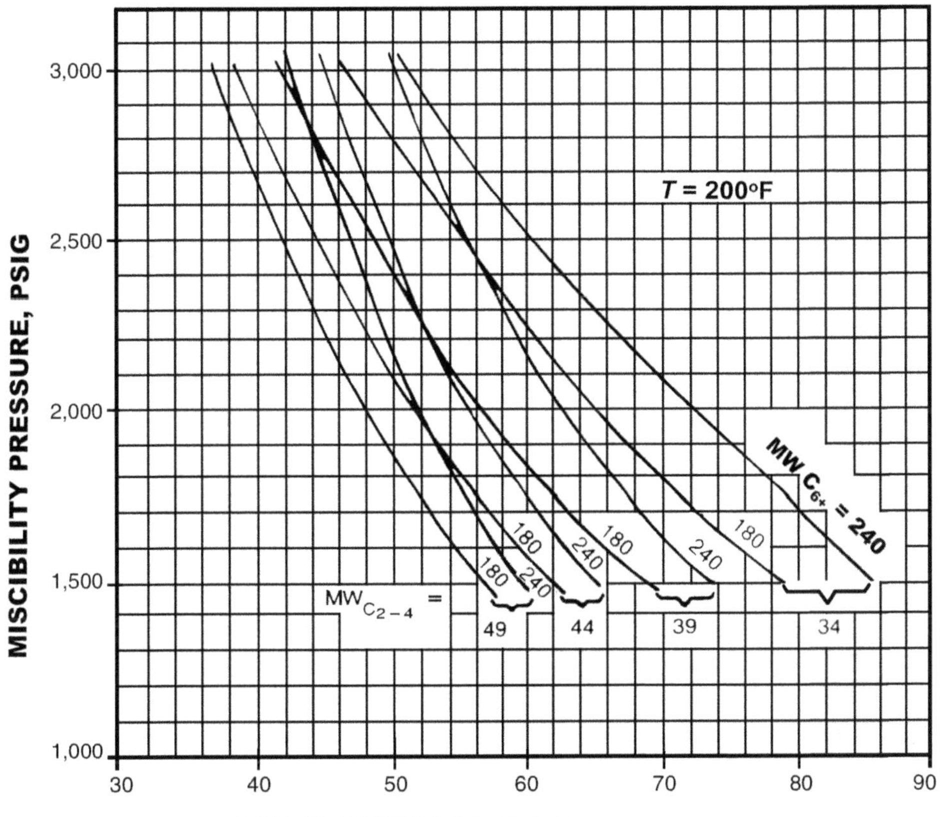

Fig. 14.8—Correlation for enriched-gas-drive MMP (after Ref. 14).

• Monthly adjustments in individual-pattern WAG ratios based on the observed performance of offset producers.

• In all cases, surveillance practices after the start of a project include periodic (monthly to quarterly) studies of the gas/oil-ratio (GOR) and water/oil-ratio (WOR) trends in producing wells and the indicated adjustments needed in WAG ratios in offsetting injectors to achieve desired performance.

• Will water- and solvent-injection rates change?

• Several projects have experienced reduced water injectivity (20% or more) after the WAG process was started.[27]

• Some projects have experienced solvent injectivity higher than water, while others have seen significant decreases.[27]

• Such decreases in injectivity may significantly affect project economics.

• Will separation of the solvent from produced fluids be necessary?

• This is a problem usually associated with CO_2 and N_2 projects. Breakthrough of solvents (which occurs early and grows with time) will contaminate produced hydrocarbon gases. Separation is required to remove contaminants before sale. The investment and operating cost of separation facilities and compression for reinjecting the recovered miscible materials should be included in the economic assessment of the project.

• In some instances, the amount of hydrocarbon gases produced may not justify its recovery, and reinjection of the total solvent stream may be a more practical solution.

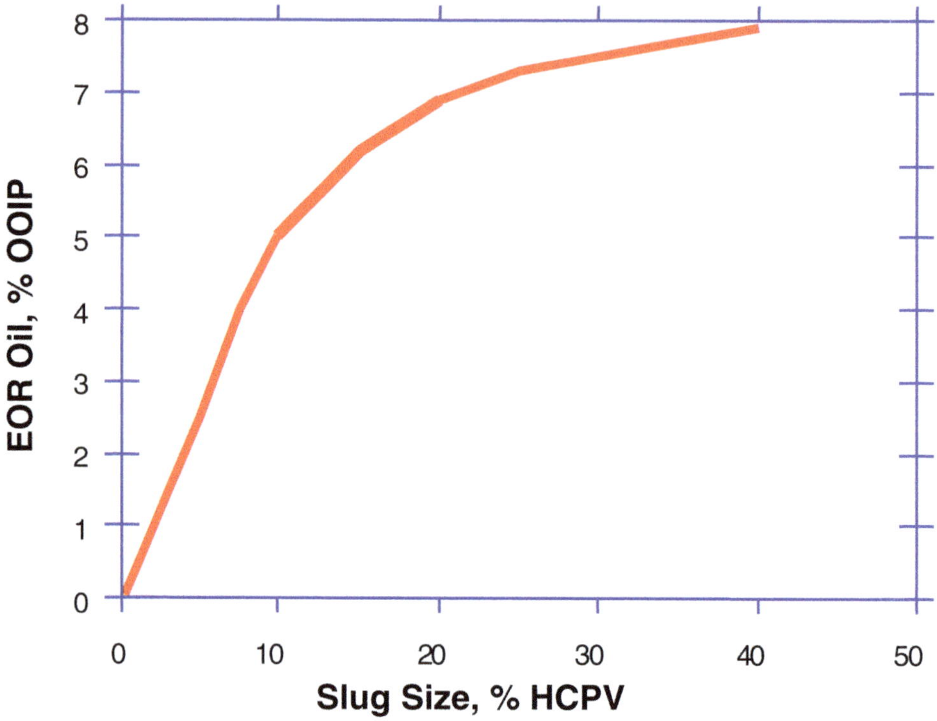

Fig. 14.9—Example of the relationship between incremental oil and slug size.

• How much solvent must be purchased, and how much should be recycled?

• Simulation results should provide sufficient insight on the rates of solvent breakthrough to answer this question. Predicted purchased volumes are generally in the 40 to 50% range of the total injection required, with the remainder being recycled solvent.

• Initial estimates of solvent needs are usually in the 40 to 60% range of HCPV. This figure tends to increase over the project life as reservoir management practices improve sweep and reduce costs, thus sustaining the economic viability of the project longer than originally forecast.

• What is the time from first expenditure to first incremental production (**Fig. 14.10**)?

• Capital outlays for equipment, drilling, and modifications to existing facilities will be made before the project start.

• In addition, experience shows that there is a delay from the time solvent injection starts to the first significant production response. This delay roughly corresponds to injection of 0.05 to 0.1 HCPV of solvent. In most cases, most of the purchased solvent is injected before significant recovery of incremental oil. Such a delay, of course, is inevitable because waterflood residual oil that is displaced by the solvent front has to travel from injector to producer. Often, there is no delay because incremental increases in oil production result from immediate improvements made to operations. There also can be a substantial delay before the peak incremental oil rate is attained, amounting very roughly to 0.1 to 0.2 HCPV of solvent injected. Project economics should include the impact of this timing.

• Another important feature of field tests that is consistent with mechanistic concepts and simulations is that solvent breakthrough usually occurs concurrently with the first production of the incremental-process oil or shortly thereafter. This signifies only a small, clean oil bank ahead of the advancing solvent front because of solvent fingering caused by adverse mobility ratios, gravity override, or permeability stratification. Much of the banked-up oil is located

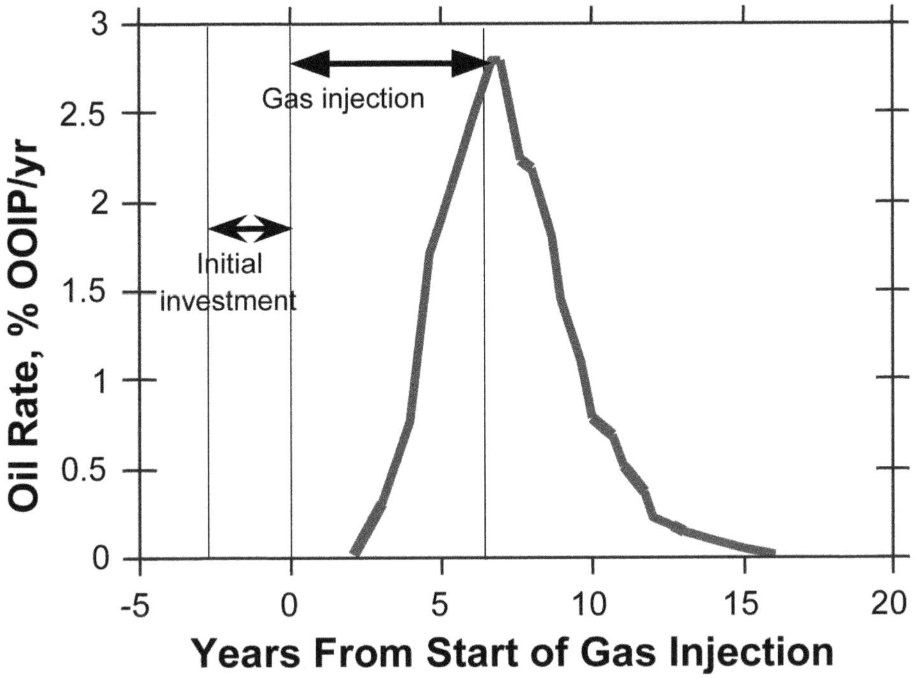

Fig. 14.10—Example of the relationship between initial investments, start of gas injection, and incremental oil production.

around the sides of these fingers and will be recovered with additional solvent injection (see Figs. 14.1 and 14.2).

• Are additional wells needed?

• Many projects have included the drilling of infill wells to provide more effective injectors, better volumetric sweep, and the productivity needed for good project economics. This applies to relatively low-cost environments in mature operating areas and has not been necessary or practical when wellbores are in good condition; in other, higher-cost operating areas; or where the reservoir characteristics did not require additional development.

14.4 Compositional Numerical Simulation

Prediction of a miscible flood is best done with a compositional reservoir simulator. The simulation must be able to predict the phase behavior as well as the sweep behavior in the reservoir to forecast such quantities as incremental oil recovery, miscible-solvent requirement, and solvent utilization efficiency and to optimize such variables as solvent composition, operating pressure, slug size, WAG ratio, injection-well placement, and injection rate. These topics are discussed in more detail in subsequent sections.

14.4.1 Phase Behavior. Methods for characterizing the reservoir and injected fluids vary from several approximation methods to rigorous component analysis.

Representation of Phase Behavior With Ternary and Pseudoternary Diagrams. Ternary diagrams and pseudoternary diagrams have been used for decades to visualize conceptually the phase behavior of injection-fluid/crude-oil systems. This is done by representing multicomponent fluids or mixtures by three pseudocomponents and then plotting fluid compositions in the interior of an equilateral triangle with apexes that represent 100% of each pseudocomponent and where the side opposite an apex represents 0% of that pseudocomponent. Usually, the three pseudocomponents represent a fraction of low-molecular-weight materials, a fraction of

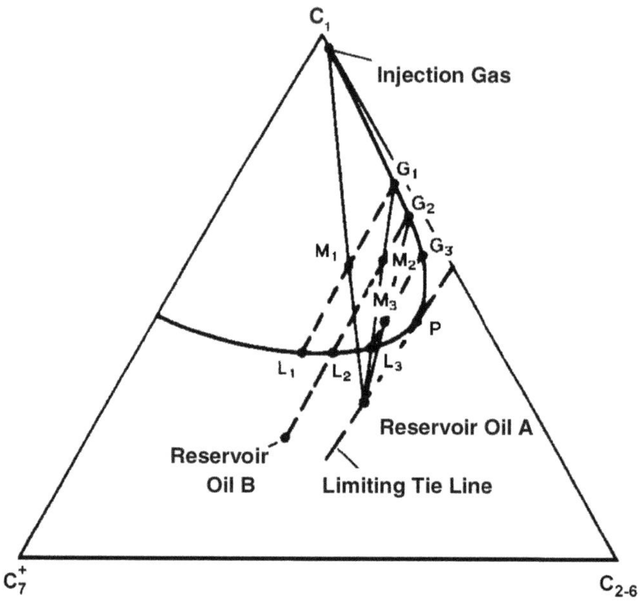

Fig. 14.11—The vaporizing-gas-drive process (after Ref. 4).

intermediate-molecular-weight materials, and a fraction of the higher-molecular-weight materials. For example, the low-molecular-weight fraction might include methane and nitrogen and perhaps CO_2 if CO_2 is the primary injection solvent. The intermediate-molecular-weight pseudocomponent might include the C_2–C_6 hydrocarbons and perhaps CO_2 if the CO_2 is a constituent of an otherwise hydrocarbon injection solvent. The higher-molecular-weight pseudocomponent in this scheme would be the leftover C_7+ fraction. There is no particularly "right" way to divide a fluid into three pseudocomponents. Different injection processes may be better represented by one type of grouping vs. another, and different groupings may give somewhat different insights into phase-behavior mechanisms.

Pseudoternary diagrams apply rigorously only to true ternary systems, and a strictly ternary analogy may give a somewhat misleading view of the mass-transfer mechanisms that result in compositional enhancement. Even so, a pseudoternary diagram is still a useful way to represent some complex phase-behavior concepts that are not so easily visualized otherwise.

A ternary diagram represents phase behavior at a constant temperature and pressure. **Fig. 14.11** is a traditional pseudoternary-diagram representation of phase behavior for the pseudocomponents C_1, $C_{2\text{–}6}$, and C_{7+}. It has the following characteristics.

There is a bubblepoint curve representing oil mixtures at their bubblepoint and a dewpoint curve representing solvent mixtures at their dewpoint. These curves come together at a critical composition. The dashed lines, called tie lines, connect liquid and solvent compositions that are in equilibrium. All mixtures inside the region bounded by the bubblepoint and dewpoint curves consist of two phases. The tie line that passes through a given mixture composition gives the equilibrium solvent and liquid compositions for this mixture where it intersects the dewpoint and bubblepoint curves. The tie line that just passes through the critical composition is called the critical tie line. All mixtures outside the region bounded by the dewpoint and bubblepoint curves are single phase.

The Vaporizing-Solvent-Drive Process. Fig. 14.11 shows how compositions change in situ when a lean injection solvent displaces an oil represented by point A, whose composition lies just to the right of the limiting tie line.

The injection solvent identified on the C_1–C_{2-6} side of the triangle has a high methane content. When this solvent mixes with the reservoir fluid, an overall composition such as M1 may result. M1 is in the two-phase region of the diagram and consists of dewpoint solvent G1 and bubblepoint liquid L1. As more solvent is injected, solvent G1, formed in the first contact, is pushed ahead, where it contacts fresh reservoir fluid. Upon mixing, solvent G1 and the reservoir fluid form another overall mixture, M2, which consists of dewpoint solvent, G2, and bubblepoint liquid, L2. Solvent G2 then flows ahead and contacts fresh reservoir fluid, forming an overall mixture M3, which consists of dewpoint solvent G3 and bubblepoint liquid L3. In this way, solvent at the displacing front is progressively enriched to the critical composition P, which is first-contact miscible with the reservoir fluid.

In the vaporizing-solvent-drive process, compositional enhancement occurs by the injection solvent vaporizing intermediate-molecular-weight hydrocarbons from the oil and enriching the composition at the solvent front. For pseudoternary phase behavior, as long as the reservoir-fluid composition lies to the right of the limiting tie line through the critical point, multiple-contact miscibility can be achieved. If oil composition should lie to the left of the critical tie line, solvent enrichment can occur only to the composition of dewpoint solvent lying on the tie line that can be extended to pass through the oil composition. For example, if reservoir oil B were being displaced by the injection solvent, enrichment of the solvent front could occur only up to dewpoint solvent G2. Although multiple-contact miscibility is not achieved, efficient immiscible vaporization may still occur (depending on the compositions actually achieved) and should be evaluated for effectiveness.

The mechanism for compositional enhancement described above can be effective for other solvents besides lean hydrocarbon solvent. N_2 and flue-gas solvent can give compositionally enhanced displacements by this mechanism, although with different MMPs. CO_2 also achieves compositionally enhanced displacement by a similar mechanism, although for lower temperatures the mechanism may be one of liquid/liquid extraction rather than vaporization.

The conceptual argument given above indicates that compositions in the solvent/oil transition zone lie along the dewpoint curve until the critical composition is reached so that no transition-zone compositions lie inside the two-phase region. Because of mixing caused by diffusion and reservoir flow mechanisms, this is not the case in practice. Transition-zone compositions cut into the two-phase region, causing some solvent-flood residual oil to be left behind the solvent front.

The Condensing-Solvent-Drive Process. **Fig. 14.12** shows another mechanism whereby compositional enhancement can occur in a solvent flood. The oil is a bubblepoint liquid lying on the bubblepoint curve. For this phase behavior, an injection solvent that has composition A on the C_1–C_{2-6} side of the triangle is completely miscible with the oil because the line connecting oil with composition A passes through only the single-phase region. However, according to the traditional ternary-diagram view of phase behavior, solvents that have compositions between A and B can develop multicontact miscibility in the following manner.

Solvent B lies just to the right of the critical tie line. When this solvent mixes with the original oil, an overall composition such as M1 may result, which is in the two-phase region. M1 consists of dewpoint solvent G1 and bubblepoint liquid L1, which are in equilibrium. Injection of fresh injection solvent B pushes the equilibrium solvent G1 ahead and contacts residual oil around the injector that now has composition L1. B mixes with L1 and forms a new overall composition in the two-phase region, M2, which splits into a new equilibrium solvent G2 and equilibrium liquid L2. Continued injection of solvent B pushes the new solvent G2 out of the way, and B mixes with liquid L2 to form a new overall mixture M3. Thus, solvent B successively contacts oil around the injector and enriches the oil along the bubblepoint curve by condensation of the intermediate-molecular-weight hydrocarbons that were used to enrich C_1–C_{2-6} mixtures to composition B. This enrichment of the oil proceeds until the enriched oil reach-

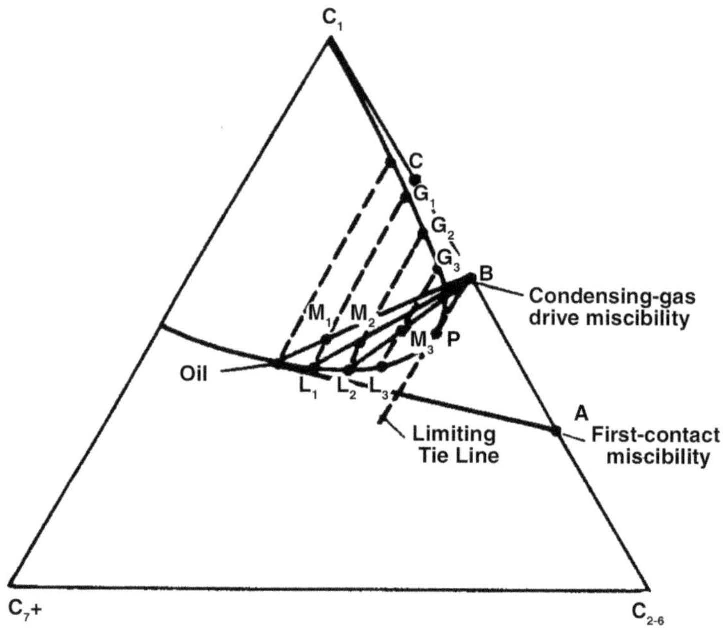

Fig. 14.12—The condensing-gas-drive process (after Ref. 4).

es the critical composition P at the critical point. Composition P is then first-contact miscible with solvent B. According to this concept, there is a transition zone of contiguously miscible liquid compositions from the original oil composition to the critical composition.

This method of solvent flooding historically was called the condensing-solvent-drive process because condensation of the intermediate-molecular-weight hydrocarbons into the oil was thought to be the mechanism responsible for the development of multicontact miscibility. In this process, miscibility is generated and propagated through the porous medium at the rear of the transition zone.

According to the pseudoternary diagram of Fig. 14.12, if the composition of injection solvent were to the left of the critical tie line, the displacement would be immiscible because the oil could never be enriched to the critical composition. For example, if solvent C were injected, the oil could be enriched only to composition L1 on the tie line that passes through C when the tie line is extended. Further contact of L1 with C only gives new overall mixtures that are on the tie line, so equilibrium solvent G1 on the tie line ends up displacing oil L1, and G1 and L1 are immiscible. The criterion for condensing-solvent-drive multiple-contact miscibility is that the injection-solvent composition must lie to the right of the critical tie line on the ternary diagram.

For most oils, however, the mechanism described previously for an enriched-solvent displacement is too simplified. Compositional enhancement occurs by a mixed mechanism that has both vaporizing and condensing features, as described below.

The Condensing/Vaporizing Process. **Fig. 14.13** shows a pseudoternary diagram for a condensing/vaporizing displacement in which the equilibrium vapor and liquid compositions were calculated for a simulated slimtube displacement with a compositional simulator. The two-phase region has an hourglass shape. There is what looks like a condensing lobe in which vapor and liquid are trying to come together at a critical composition similar to condensing-solvent drive. However, before this happens, vapor and liquid compositions begin to diverge at the trailing part of the displacement.

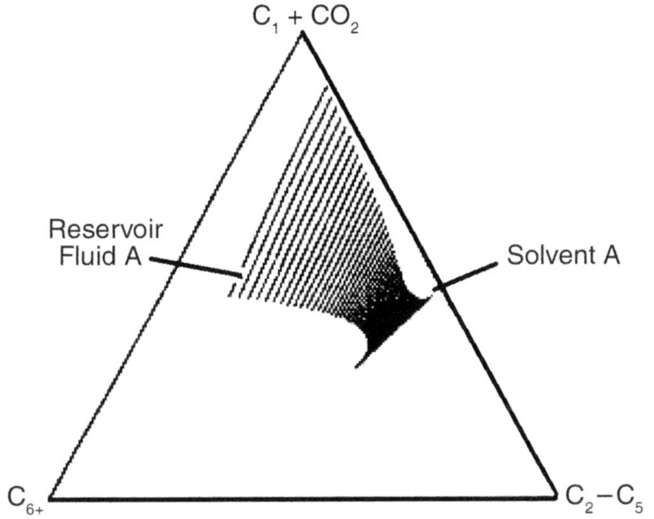

Fig. 14.13—The condensing/vaporizing process (after Ref. 27).

Zick[16] was first to explain this behavior. He deduced that in addition to condensation of intermediate-molecular-weight hydrocarbons from the injection solvent, vaporization of mid-range hydrocarbons from the oil also played an important role. Zick explained this mechanism in the following way:

The easiest way to understand the condensing/vaporizing mechanism is to consider an oil/ solvent system composed of essentially four groups of components. The first group consists of the lean components, such as methane, nitrogen, and carbon dioxide, which usually have equilibrium K-values greater than one. The second group consists of the light intermediate components, such as ethane, propane, and butane, which are the enriching components present in the injection solvent. The third group contains the middle intermediates, which are present in the oil but not significantly present in the injection solvent. These are components that can be vaporized from the oil. The lightest component in this group typically ranges from butane to decane, depending on the injection solvent composition. The heaviest component in this group cannot be defined precisely, but it might be around C_{30}. The fourth group consists of everything else, i.e., those heavy components in the oil which are very difficult to vaporize.[16]

When the enriched solvent comes into contact with the oil, the light intermediates condense from the solvent into the oil, making the oil lighter. The equilibrium solvent is more mobile than the oil, so it moves ahead and is replaced by fresh injection solvent, from which more light intermediates condense, making the oil even lighter. If this kept occurring until the oil was light enough to be miscible with the injection solvent, it would constitute the condensing-solvent-drive mechanism. However, this is unlikely to occur within a real reservoir oil. As the light intermediates are condensing from the injection solvent into the oil, the middle intermediates are being stripped from the oil into the solvent. Because the injection solvent contains none of these middle intermediates, they cannot be replenished in the oil. After a few contacts between the oil and the injection solvent, the oil becomes essentially saturated in the light intermediates, but it continues to lower the middle intermediates, which are stripped out and carried ahead by the mobile solvent phase. The light intermediates of the injection solvent cannot substitute for the middle intermediates the oil is losing. So after the first few contacts make the oil lighter by net condensation of intermediates, subsequent contacts make the oil heavier by net vaporization of intermediates. Once this begins to occur, the oil no longer has a chance of becoming miscible with the solvent. Ultimately, all the middle intermediates are removed, and the residual oil will be very heavy, containing only the heaviest, nonvolatile fraction and the components present in the injection solvent.

If the mechanism stopped there, a considerable amount of oil would remain unrecovered. However, there are further steps to the mechanism. Consider the oil in place slightly downstream from the injection point. The first solvent it will see will not be the injection solvent, but the equilibrium solvent. This relatively lean solvent essentially will be injection solvent that has lost most of its light intermediates and picked up a very small amount of middle intermediates. There will be very little mass transfer between this solvent and the fresh oil. The solvent that follows, however, will be richer. Eventually, the solvent that comes through will be solvent that has passed over oil that was saturated in the light intermediates. Therefore, this solvent will have approximately the same amount of light intermediates as the injection solvent. However, it will also contain a small amount of middle intermediates that it stripped from the oil over which it passed. Thus, it actually will be a little richer than the original injection solvent. The oil that sees this solvent will receive slightly more condensable intermediates than did the oil just upstream. Before the vaporization process takes over and again makes it heavier, this oil will become slightly lighter than the upstream oil had become.

This process continues farther downstream. The farther downstream, the richer the solvent that eventually comes through because it will have passed over an increasing amount of residual oil, allowing it to pick up increasing amounts of middle intermediates. This is beginning to sound like the vaporizing-solvent-drive mechanism, in which a lean injection solvent passes over an oil rich in intermediates, vaporizing the intermediates and becoming richer and richer until it becomes rich enough to be miscible with the original oil. There is a significant difference, however. The solvent in the condensing/vaporizing mechanism does not become rich enough to be miscible with the original oil. The original oil does not have to be rich in intermediates, nor does it even have to be undersaturated, both of which are necessary conditions for developing a vaporizing-solvent-drive mechanism. Instead, the solvent develops only enough richness by the vaporization part of the mechanism so that it nearly generates a condensing-solvent-drive mechanism with the original oil. The intermediates that were originally present in the solvent, plus those that were stripped from the oil, condense when the solvent encounters fresh oil downstream. This condensation proceeds in a manner very much like the condensing-solvent-drive mechanism. A sharp transition zone develops and propagates, and multicontact miscibility is almost achieved before the condensation process reverts to the vaporization process. The vaporization results in a trail of residual oil being left behind the moving transition zone, although the saturation level of the residual oil supplies subsequent solvent with the middle intermediates necessary to continue the propagation of the transition zone. The intermediates are vaporized from the residual oil, carried upstream into and beyond the transition zone, condensed there, and again become part of the residual oil after the transition zone has passed.

The condensing region is at the leading edge of the enriched-solvent displacement. The vaporizing region, with a small saturation of residual oil, is at the trailing end. In between is the sharp, two-phase transition zone, the two phases of which are almost—but not quite—miscible. The propagation of the sharp transition zone results in a very efficient, "apparently miscible" displacement, even though miscibility is not actually developed (except possibly, and only speculatively, as the displacement front travels to infinite distances, relative to dispersion length scales, downstream of the injection point). The sharpness of the transition zone deteriorates rapidly as either the pressure or the enrichment of the injection solvent falls below some critical value, resulting in the reduced displacement efficiencies typical of immiscible displacements.

Mixing by diffusion and flow mechanisms affects how close the mixed mechanism gets to multicontact miscibility. The greater the mixing, the farther vapor and liquid compositions remain apart in the neck of the hourglass, which results in larger solvent-flood residual oil.

In all the processes just described, equilibrium solvent and oil properties become more similar as compositions become more similar and approach the critical composition. This causes the interfacial tension between the solvent and oil to decrease as the solvent and oil compositions become more alike. This in turn causes the capillary number for oil displacing solvent to increase.

The capillary number is defined as

$$N_{ca} = \frac{k \left| \Delta p_g \right|}{\sigma}, \quad (14.1)$$

where k = permeability, Δp_g = pressure gradient through the displacing phase, and σ = interfacial tension.

Below a threshold value of capillary number, solvent-flood residual oil and solvent/oil relative permeability remain unchanged. Above the threshold, residual oil begins to decrease with increasing capillary number, and the solvent and oil relative permeability curves begin to straighten, ultimately becoming straight lines at very high values of the capillary number. These changes may have contradictory effects on a displacement. For a given equilibrium oil and solvent composition near the critical composition, solvent-flood residual oil saturation may be a bit lower than it would because of phase-behavior effects alone. However, solvent mobility may be higher because of the more favorable solvent relative permeability, which may result in somewhat poor sweep. The overall effect needs to be evaluated with a simulator that accounts for changing capillary number.

Prediction of Phase Behavior With Equations of State. In practice, vapor/liquid reservoir phase behavior is calculated by an equation of state (EOS). Refer to the chapter on Thermodynamics and Phase Behavior in the General Engineering section of this *Handbook* for more detail on EOSs. The two most common EOSs that have been used for oil-recovery solvent-injection processes are the Peng-Robinson EOS[28] and the Soave-Redlick-Kwong EOS.[29] Of the two, the Peng-Robinson EOS seems to be the one most often cited in the literature and is the one discussed in some detail in the remainder of this section. The Soave-Redlick-Kwong EOS is used in a similar manner to predict solvent/oil phase behavior.

Peng and Robinson originally proposed the two-parameter EOS shown next for a pure component:

$$P = \frac{RT}{v - b} - \frac{a(T)}{v(v + b) + b(v - b)}, \quad (14.2)$$

$$a(T) = 0.45724\left(\frac{R^2 T_C^2}{p_C}\right)\left[1 + \kappa\left(1 - \sqrt{\frac{T}{T_C}}\right)\right]^2, \quad (14.3)$$

$$\kappa = 0.37464 + 1.54226\omega - 0.26992\omega^2, \quad (14.4)$$

$$\text{and } b = 0.07780\frac{RT_C}{p_C}, \quad (14.5)$$

where ω = the component acentric factor, T_c = component critical temperature, and p_c = component critical pressure.

For heavier components, where $\omega > 0.49$, the following equation is recommended:

$$\kappa = 0.3796 + 1.485\omega - 0.1644\omega^2 = 0.01667\omega^3 \text{ . } \qquad (14.6)$$

The constants in Eqs. 14.3 and 14.5 are often designated Ω_a and Ω_b.

Eq. 14.2 represents continuous fluid behavior from the solvent to liquid state, and it can be rewritten as

$$Z^3 - (1 - B)Z^2 + (A - 3B^2 - 2B)Z - (AB - B^2 - B^3) = 0, \qquad (14.7)$$

$$\text{where } A = 0.45724 \frac{P}{p_C}\left(\frac{T_C}{T}\right)^2 \left[1 + \kappa\left(1 - \sqrt{T/T_C}\right)\right]^2 \qquad (14.8)$$

$$\text{and } B = 0.07780\left(\frac{P}{p_C}\right)\left(\frac{T_C}{T}\right) \text{ . } \qquad (14.9)$$

Jhaveri and Youngren[30] adapted a procedure used by Peneloux *et al.*[31] and modified the original Eq. 14.2 to include a third parameter to allow more-accurate volumetric predictions, which is recommended for solvent/oil simulations. The third parameter does not change the vapor/liquid equilibrium conditions determined by the unmodified, two-parameter equation. Instead, it modifies the phase volumes by making a translation along the volume axis. Eqs. 14.10 and 14.11 give the modified three-parameter equation:

$$(Z + C)^3 - (1 - B)(Z + C)^2 + (A - 3B^2 - 2B)(Z + C) - (AB - B^2 - B^3) = 0, \qquad (14.10)$$

$$C = sB, \qquad (14.11)$$

where s is the volumetric shift parameter.

For mixtures:

$$A = \sum_{i=1}^{n} \sum_{j=1}^{n} x_i x_j A_{ij}, \qquad (14.12)$$

$$A_{ij} = \left(1 - \partial_{ij}\right)\sqrt{A_i A_j}, \qquad (14.13)$$

$$B = \sum_{i=1}^{n} x_i B_i, \qquad (14.14)$$

$$\text{and } C = \sum_{i=1}^{n} x_i C_i \text{ . } \qquad (14.15)$$

In Eq. 14.13, ∂_{ij} is the binary interaction coefficient that characterizes the binary formed by components i and j. Eqs. 14.11 through 14.14 apply both to pure components and to lumped pseudocomponents that represent two or more pure components in complex mixtures.

The following expression derived from thermodynamic relationships and the EOS allows calculation of the fugacity, f_j, of component j in a mixture:

$$\ln\left(\frac{f_j}{x_j P}\right) = \frac{B_j}{B}(Z-1) - \ln(Z-B) - \frac{A}{2B\sqrt{2}}\left(\frac{2\sum_i x_i A_{ij}}{A} - \frac{B_j}{B}\right)\ln\left(\frac{Z+2.414B}{Z-0.414B}\right). \quad \text{.... (14.16)}$$

Thus, by satisfying the equilibrium condition $f_j^L = f_j^V$, vapor/liquid equilibrium ratios can be calculated, and flash calculations can be made to calculate the compositions of vapor and liquid in equilibrium, molar splits, and volumes.

Solution of the EOS does not calculate phase viscosities directly. This is done from some external calculation once the phase compositions and densities are known. A commonly used calculation for liquid-mixture viscosity is the Lohrenz-Bray-Clark method, which requires the critical volumes of each component or pseudocomponent in the mixture.[32] Refer to chapters in the General Engineering section for more information on estimating viscosities for gas and oil.

To use Eqs. 14.2 through 14.16 for calculating the phase behavior and properties of solvent compositional processes in oil recovery, the following steps must be taken to "characterize" the fluid system in question:

- Analyze the oil composition. This can be done by distillation or chromatographic methods. An extended analysis through at least C_{25+} is preferred. The advantage of distillation is that molecular weight, boiling point, and density can be measured on the distillation cuts.
- Represent the multicomponent reservoir fluid by an appropriate division into pure components and pseudocomponents. Pure components through C_5 plus three to five pseudocomponents usually will suffice. It may be possible to reduce the number of pure components and pseudocomponents further by combining similar components.
- Make an initial assignment of critical pressure and temperature, acentric factor, critical volume (or critical compressibility), volumetric shift parameter, and interaction parameters for each component and pseudocomponent.
- Tune the above properties for the pseudocomponents by comparing predicted phase behavior and properties with suitable experimental data.

Methods for dividing into pseudocomponents and estimating critical properties, shift parameters, and binary interaction coefficients are described in detail in Ref. 33.

Because of the approximations inherent in an EOS as well as the approximations required to represent a multicomponent reservoir fluid in a tractable form, it should be expected that phase-behavior properties and equilibrium compositions predicted with an EOS will depart from measured values over the range of composition and pressure conditions anticipated in a reservoir simulation. For this reason, additional adjustment of EOS parameters will be required for predictions to represent experimental measurements adequately. These adjustments usually are made by regression.

Reservoir oils usually are subjected to routine pressure/volume/temperature (PVT) experiments that give the volumetric and phase-behavior information necessary for predicting conventional recovery methods such as solution solvent drive or waterflooding. Experiments such as constant-composition expansion, differential liberation, constant-volume depletion, and separator tests provide black-oil properties. Other PVT experiments are more specific for solvent injection. These include swelling tests and multiple-contact experiments.

The swelling experiment is sometimes called a pressure-composition diagram determination. Injection solvent is added to reservoir oil in increments to give mixtures that contain increasing amounts of injection solvent. After each addition of solvent, the saturation pressure is measured at reservoir temperature. Overall composition of these mixtures ranges from that of black oil to compositions up to and beyond near-critical conditions (i.e., overall compositions that traverse a range from bubblepoint to dewpoint mixtures at reservoir temperature). Thus, the swelling experiment provides some PVT and phase-equilibrium information on mixture ranges that might reflect compositions as solvent displaces oil through the reservoir. It provides information on the saturation pressure of injection-solvent/oil mixtures, the swelling or increase in oil formation volume factor as solvent is added, the composition of the critical mixture, and the liquid saturation vs. pressure in the two-phase region of the diagram.

Multiple-contact tests seek to simulate the solvent/oil multiple contacting that occurs in a reservoir. A forward multicontact experiment tries to simulate multicontacting in a vaporizing-solvent drive. A reverse multicontact experiment tries to simulate the multicontacting that occurs in a purely condensing-solvent drive. The experiments give information concerning equilibrium-phase volumes and compositions.

In a reverse-contact experiment, the PVT cell is charged with the reservoir fluid at the desired pressure and temperature, and an increment of injection solvent is added sufficient to form a two-phase mixture (or a three-phase mixture in some tests). The phases are allowed to equilibrate, and phase volumes are measured. The solvent phase is then displaced from the cell, and oil and solvent compositions are measured. The procedure is repeated, with injection of a new increment of injection solvent introduced into the cell to contact equilibrium oil left after the first contact.

In the forward-contact experiment, the oil phase is displaced after the first contact, and the remaining equilibrium-solvent phase in the cell is contacted with a fresh increment of reservoir oil.

The objective of tuning is to ensure that the EOS predicts fluid properties and phase equilibrium compositions accurately over the range of pressure, temperature (if this varies), and composition that one expects to encounter in a simulation. If the simulation is for a solvent compositional process, then at a minimum the EOS should predict properties and phase equilibrium for the range of injection-solvent/oil mixtures and pressures encountered in the simulation study. It also should predict adequately for any black-oil conditions expected in the simulation (e.g., waterflooding or pressure depletion before solvent injection) and for the separator conditions expected.

Pedersen *et al.*[34] observed that an EOS tuned to match a specific set of data may not give reliable predictions for other data not included in the tuning process. However, when both sets of data are included in the tuning process, the prediction for either one may not be quite as good as for tuning against these data individually.

It seems prudent that at a minimum, there should be differential depletion data, separator tests, and swell data to tune an EOS against for making solvent-compositional simulations. Swelling tests are necessary when near-critical compositions are expected in the simulation, and it is necessary for the swell tests to explore this composition region. Swell tests with several different injected-solvent compositions might be warranted if optimization of the solvent composition is an objective of the simulation study.

The value added by multiple-contact tests is unclear. These are the most difficult and expensive of the experiments discussed earlier, yet they provide direct measurements of vapor/liquid equilibrium compositions and molar splits for a composition path that at least crudely mimics the development of compositions at the leading or trailing edges of the solvent/oil transition zone, which is, of course, what the simulator is trying to calculate. However, for the condensing/vaporizing process, multiple-contact experiments do not give compositions that are very near the critical point.

Although they are difficult and expensive to run, slimtube tests give a direct verification of the ability of the EOS to predict MMP or MME. If the EOS after regression of parameters does not predict slimtube MMP or MME, further adjustment of parameters is required.

For additional information on EOS, refer to the chapter on Phase Behavior in the General Engineering section of this *Handbook*.

14.5 Prediction of Compositionally Enhanced Solvent Flood Behavior

The compositional reservoir simulator is the preferred simulator for predicting compositionally enhanced solvent flood reservoir behavior. This simulator calculates the flow in up to three dimensions of solvent, oil, and water phases as well as n components in the solvent and oil phases. It also computes the phase equilibrium of the oil and solvent phases (i.e., the equilibrium compositions and relative volumes of the solvent and oil phases) in each gridblock of the simulator. In addition, it computes solvent- and oil-phase densities. The equilibrium compositions and densities are calculated with an EOS. From knowledge of the phase compositions and densities, solvent and oil viscosity and other properties such as interfacial tension are estimated from correlations. See the chapter on Reservoir Simulation in the Reservoir Engineering and Petrophysics section for more information on compositional reservoir simulation.

A compositional simulator is the most mechanistically accurate simulator for solvent compositional processes. When the EOS is tuned properly to appropriate experimental data, it computes realistic phase behavior. Thus, the appropriate phase behavior for flooding with enriched hydrocarbon solvent, lean hydrocarbon solvent, N_2, and CO_2 all can be taken into account. Compositional simulators predict the effect of changing pressure and injection-solvent composition on a displacement without the need to enter approximations into the simulator for these effects (except as the EOS itself is an approximation). The compositional simulator is capable of computing realistic behavior when pressure is well below the MMP of the injection solvent, is near but still below the MMP, or is well above the MMP. For this reason, it is ideally suited to study optimum operating conditions.

In addition to these advantages, a compositional simulation, to a large degree, removes the need for a user-defined miscible-flood residual oil saturation, as it naturally computes the amount of residual oil left after the interaction of phase behavior and dispersion and distributes this residual saturation realistically as a varying saturation instead of an input, constant saturation.

A compositional simulation can have other aspects of mechanistic reality besides phase behavior. The mechanisms of molecular diffusion and convective dispersion may be included in the equations solved by the simulator. Although grid-refinement sensitivity (described later), or numerical dispersion, may dwarf the effects of these mechanisms in many simulations, they may be important to include in the finely gridded reference simulations (also described later).

Another physical mechanism that can be included in compositional simulations is the effect of interfacial tension (IFT) on solvent/oil relative permeability and capillary pressure. Although one cannot readily foresee the impact of a particular mechanism in the complex compositional simulation of solvent flooding, inclusion of the IFT mechanism seems prudent.

When an appropriate relative permeability treatment is included, compositional simulation predicts realistic solvent trapping, especially the trapping of solvent by crossflowing oil. Oil crossflow into a solvent-swept zone immiscibly displaces the solvent in a compositional simulation and leaves the solvent as a residual saturation consistent with the phase behavior.

The primary disadvantages of a compositional simulator are the degree of grid refinement often required to compute oil recovery with satisfactory accuracy and the computing time required for fine-grid simulations. These factors generally preclude using a compositional simulator directly for full-field simulations unless some kind of scaling-up technique is used to transfer the information developed from fine-grid reference-model simulations on a limited reservoir scale to coarse-grid simulations on the full-field-model scale. The predicted benefit of compositionally enhanced solvent flooding can be substantially in error if the simulation is made directly with a full-field model with typical coarse grids. This is illustrated by **Fig. 14.14**, which shows the results of an enriched-solvent-drive reservoir study.[24] In this figure, simulations were made for two one-fourth nine-spot models that represented the same reservoir description. One model had a fine grid (30×30×31 cells in the *x*-, *y*-, and *z*- directions); the other had the same grid as that used in the full-field model (5×5×17). The incremental recov-

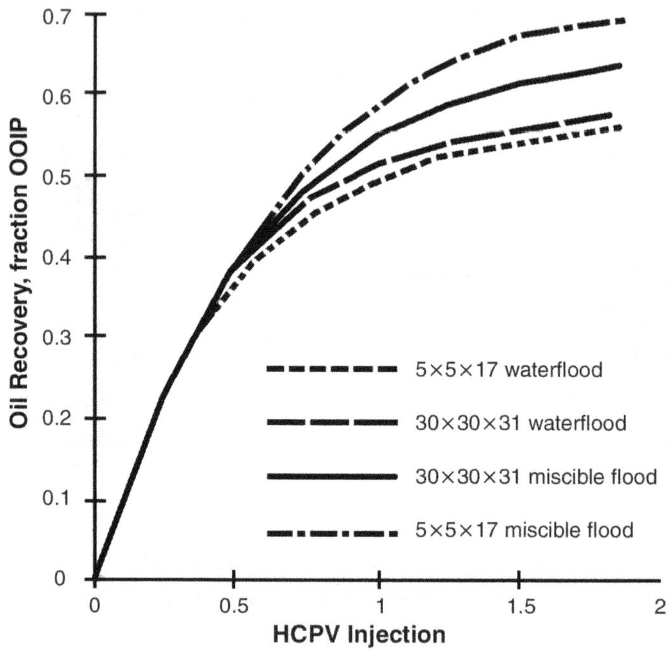

Fig. 14.14—Predictions with reference model and corresponding model with full-field grid size (after Ref. 24).

ery in this figure is the difference between solvent-flood and waterflood simulations in each model. The direct full-field simulation overpredicted incremental recovery by a factor of two.

There also are some additional data requirements for predicting solvent trapping and solvent relative permeability hysteresis that are not found in black-oil waterflood simulations. Solvent trapping, reference models, and scaleup are discussed next.

14.5.1 Solvent Trapping and Solvent Relative Permeability Hysteresis. In most compositionally enhanced solvent displacements, some of the solvent will be trapped permanently in the reservoir and will not be produced. This happens when water is used to drive a solvent slug and the oil displaced by the solvent. Solvent is trapped by advancing water much like oil is left as a residual in a waterflood. Solvent also can be trapped by oil that crossflows into a previously solvent-swept zone.

Laboratory data indicate that there is little dependence of the trapped solvent saturation on whether water or oil is the trapping phase.[35] These data also show that the magnitude of the trapped solvent saturation is insensitive to whether the measurement is made at reservoir or ambient conditions, on core plugs or composites of core plugs, or on native state or extracted cores.

The magnitude of the trapped-solvent saturation depends on the magnitude of the maximum solvent saturation present when the solvent is trapped. This is illustrated in **Fig. 14.15**, which shows data for Prudhoe Bay cores.[35] In these experiments, various initial solvent saturations were established before the cores were flooded with either water or oil. There is a substantial degree of scatter in the data, but the trapped-solvent saturation clearly depends on the initial solvent saturation. Moreover, within the data scatter, the trapping is nearly independent of the liquid phase doing the trapping and even independent of a small oil saturation if it should happen to be present as a third phase.

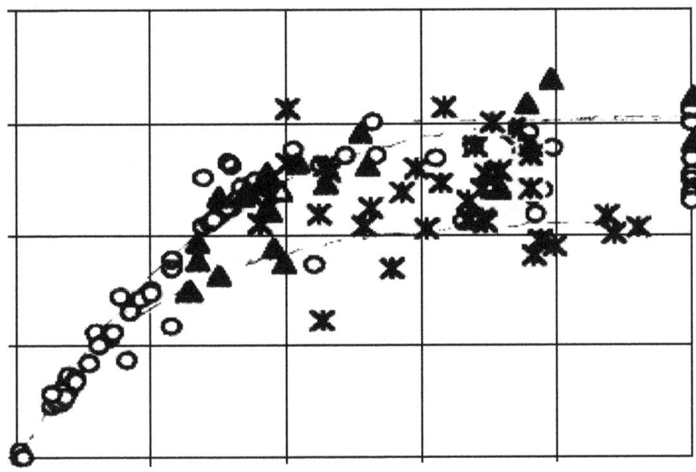

Fig. 14.15—Trapped-gas saturation in Prudhoe Bay cores (after Ref. 35).

According to Jerauld,[35] the relationship between trapped-solvent saturation and the maximum solvent saturation at the location at which trapping occurs is generally well represented by a "zero slope" adaptation of the Land curve:

$$S_{gt} = \frac{S_g^{max}}{1 + \left(1 / S_{gr}^{max} - 1\right)\left(S_{gr}^{max}\right)^{1/\left(1 - S_{gr}^{max}\right)}} . \qquad (14.17)$$

In this equation, S_{gr}^{max} is the solvent trapped when the rock is 100% solvent saturated, and S_g^{max} is the maximum solvent saturation at the location where trapping occurs. According to Eq. 14.17, if solvent subsequently is mobilized from this location only to be trapped again when the flowing saturation is lower, the retrapped solvent still attains the trapped saturation achieved at the previous maximum saturation. If on subsequent remobilization the solvent saturation should reach a higher value than the previous maximum, the trapped saturation will attain a new and higher value, according to Eq. 14.17.

The maximum trapped-solvent saturation, S_{gr}^{max}, when the rock initially is 100% solvent saturated depends strongly on both porosity and clay content, or microporosity. This is illustrated in **Fig. 14.16** for data from various sandstones, which show a generally increasing trend for S_{gr}^{max} with decreasing porosity.[35]

Because of the nature of solvent trapping, an important mechanism to be accounted for in WAG flooding is solvent relative permeability hysteresis. This is illustrated in **Fig. 14.17**. Consider solvent injection for the first WAG cycle. Solvent is the nonwetting phase, and solvent injection is a drainage process. The solvent relative permeability depends only on the solvent saturation. The solvent may be displacing only oil in the presence of connate water for a secondary flood, only water in the presence of waterflood residual oil for a tertiary flood, or both oil and water for a partially waterflooded reservoir. All these situations are approximated by the same solvent primary-drainage curve, and on the first WAG cycle the solvent relative permeabilities follow the primary-drainage curve AE. If at the end of the first solvent cycle a

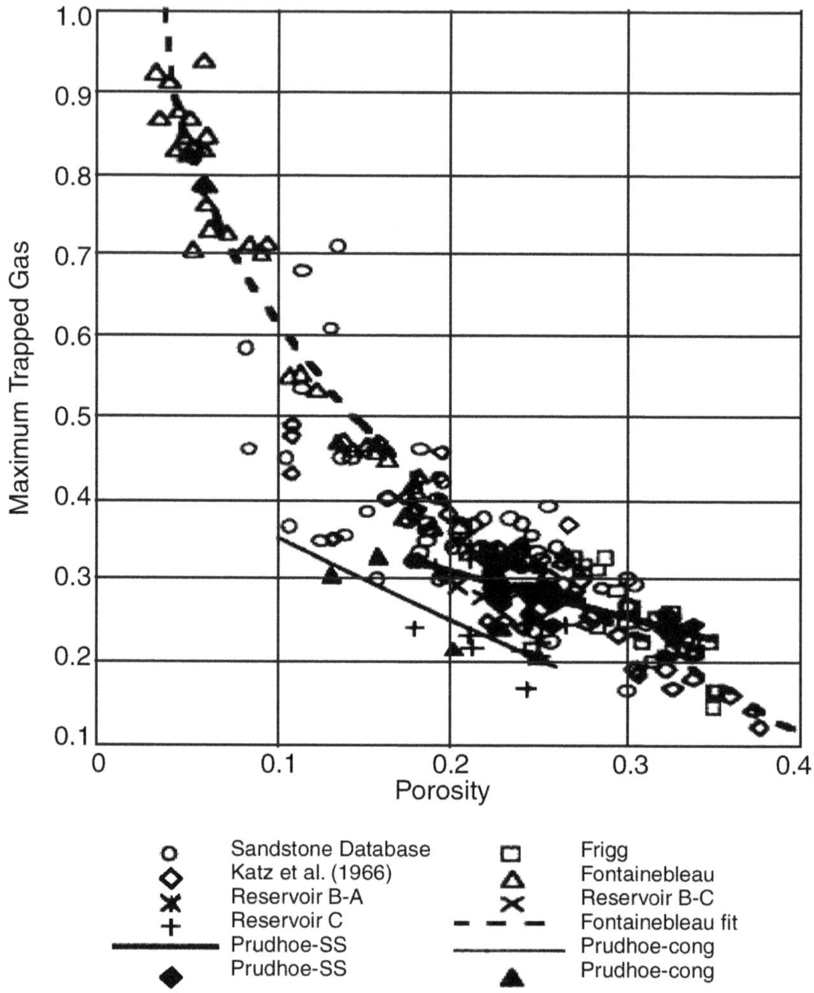

Fig. 14.16—Maximum trapped-gas saturation for different sandstones (after Ref. 35).

volume of solvent has been injected so that point B on the primary-drainage curve has been reached at a given location in the reservoir, the solvent saturation at this location will be S_{gB}.

Now consider water injection on the first WAG cycle. Assume that the solvent saturation at the location in question remains S_{gB}. (Actually, the saturation may increase somewhat as water at first displaces solvent past this location.) When water reaches this location, it will drive the solvent down to a trapped saturation, S_{gtD}, at point D according to the trapped-solvent- vs. maximum-solvent-saturation relationship for the rock. This is an imbibition process, and the solvent relative permeability follows the curve BD.

When solvent is injected on the second WAG cycle, the solvent relative permeability follows the curve DB because Land's evidence demonstrates that imbibition relative permeability often is nearly reversible.[36] It is important to take this hysteresis into account in a WAG simulation because the imbibition relative permeability is substantially less than the primary-drainage relative permeability and will cause the mobility ratio to be lower and the displacement more effective than would be the case with primary-drainage relative permeability only.

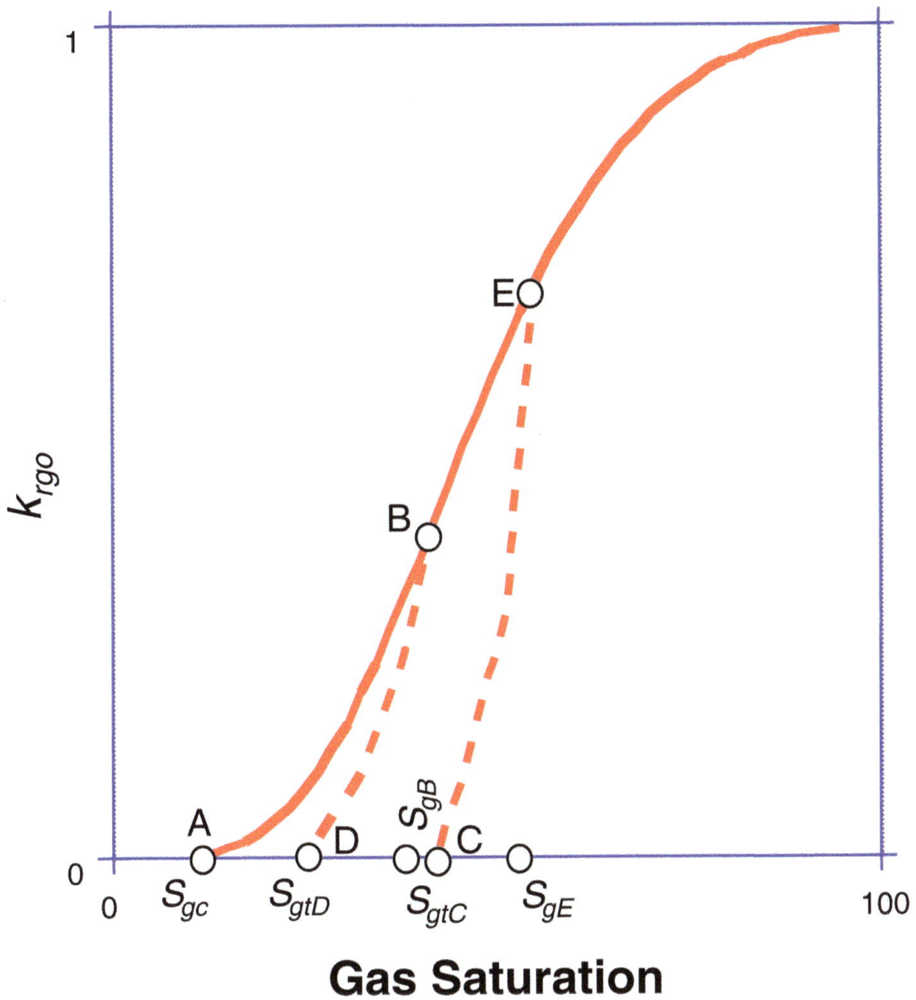

Fig. 14.17—Gas relative permeability hysteresis.

If the solvent saturation at the location in question never reaches S_{gB}, solvent relative permeability will stay on the curve BD during the subsequent second-WAG-cycle water slug. If such a large solvent slug is injected that S_{gB} is exceeded at this location, solvent relative permeability will once again follow the primary-drainage curve, perhaps to point E, and attain a new maximum solvent saturation at this location, S_{gE}. Then, on the subsequent water cycle, the solvent relative permeability will follow a new imbibition curve, EC, and solvent will be trapped at a new trapped-solvent saturation, S_{gtC}, according to the trapped-solvent- vs. maximum-solvent-saturation relationship.

14.5.2 Fine-Grid Reference Models. Grid-refinement sensitivity is an extremely troublesome problem in many compositionally enhanced solvent simulations. The problem manifests itself by the predicted behavior changing as the grid is refined (i.e., as the gridblocks become smaller and smaller). This behavior can be caused by truncation error or numerical dispersion that results from representing derivatives by finite differences; by the inability to accurately resolve the size of solvent tongues or fingers with large gridblocks; and by the inability to represent

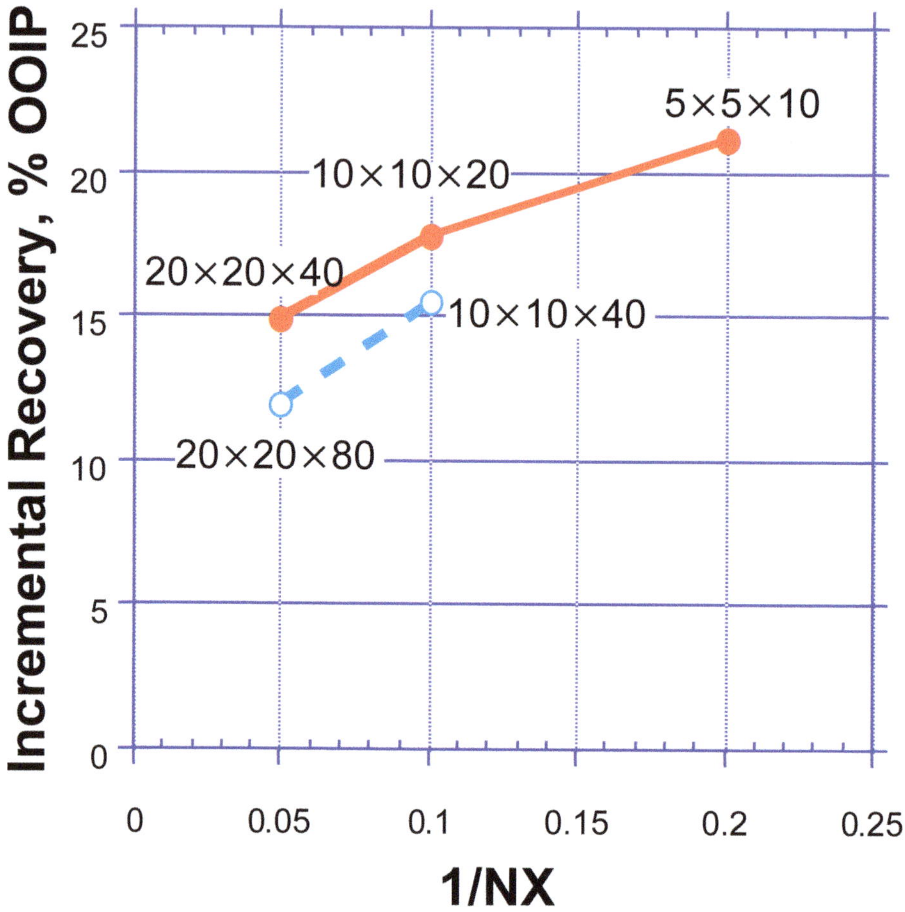

Fig. 14.18—Example of grid-refinement sensitivity.

with large gridblocks some features of reservoir description that have an important effect on solvent sweep, such as discontinuous shales, thin high-permeability strata, or thief zones.

Fig. 14.18 shows the incremental recovery computed for two different 3D models, one representing one-eighth of a nine-spot pattern, the other representing one-fourth of a nine-spot. Each model had a different geostatistical distribution of correlated permeability with scattered, discontinuous shales represented by zero vertical permeability between gridblocks. Permeability and porosity were scaled up by the renormalization method from the model with the smallest gridblocks to the other models.[37]

The base model for the one-eighth nine-spot has a grid of 20×20×40. Gridblocks were 93 ft on a side and 1 ft thick. The gridding of the one-fourth nine-spot model was 20×20×80, with gridblocks also 93 ft on a side and 1 ft thick.

Incremental recovery in this figure is plotted vs. 1/NX, where 1/NX is the dimensionless *x*-direction gridblock size. However, in this problem the dimensionless gridblock sizes in the other two directions also vary directly with the *x*-direction gridblock size. It is apparent that as the gridblock size is refined, the predicted incremental recovery decreases for what is supposed to be the same reservoir problem.

Fig. 14.18 illustrates the importance of minimizing grid-refinement error and explicitly including reservoir-description details that affect flow in an important way. Generally, minimizing the error from grid refinement and accounting for important reservoir-description details adequately requires small gridblocks. Layers that are 1 ft or no more than a few feet thick and have at least 20 to 40 lateral gridblocks between wells are desirable. Unfortunately, such fine gridding is not feasible for full-field simulations, for most 3D simulations of a single pattern, or perhaps even for some 3D repeating elements of a pattern. Because of this, field predictions need to be made in two steps—with reference models that can be gridded finely enough to accomplish the objectives summarized above, and with scaleup models that incorporate the information derived from reference models into field predictions that account for fieldwide reservoir description, multiple patterns, and operating realities and constraints.

Although it is desirable to make 3D reference-model simulations gridded so finely that the computed answer is adequately close to the converged answer, the discussion above shows that in general, it may not be feasible to do this. A reasonable alternative may be to make finely gridded 2D cross-section simulations to study the grid-refinement issue because for many problems, grid refinement has a larger effect on the computed outcome than the areal effects captured by a coarser-gridded 3D model. Variable-width 2D cross sections sometimes adequately represent the behavior of 3D pattern-segment models with the same fine gridding. In these cross sections, the width is smaller near the injector and producer and increases in the interwell region. This causes flow rate to be greatest near the wells and lowest midway between wells, as it would in a 3D displacement. Even when a fine-grid cross section does not realistically model a fine-grid 3D displacement, it still may predict incremental recovery better than a simulation in a more coarsely gridded 3D model. Moreover, 2D cross-section simulations are well suited for scaleup with the segment and streamline/streamtube models discussed in the next section.

A potential procedure for developing a 3D reference model is first to make a 3D simulation of a pattern element with the finest-grid refinement that is practical. Then, well-to-well cross sections are taken from this model, and the cross sections are refined further. Pseudoproperties are developed for the original cross sections (see Sec. 14.5.4) that predict the performance of the more finely gridded cross sections. Then, these pseudoproperties are used in the moderately gridded 3D model to approximate the effect of further grid refinement.[24]

14.5.3 Scaleup To Predict Full-Field Behavior. The objective of scaleup is to take the behavior predicted from detailed, fine-grid reference models that at best represent only a few wells and a tiny part of the reservoir and transfer it to a model that attempts to represent many wells and the integrated behavior of the entire compositionally enhanced solvent flood (or at least a significant portion of it). Three scaleup methods are (1) development of pseudorelative permeability relations and pseudoproperties for use in coarse-grid, full-field, numerical reservoir simulators; (2) development of segment models that estimate the behavior of pattern-repeating elements and add up the behavior of all the segments of a pattern, as well as the behavior of all the patterns; and (3) development of areal streamtube or streamline models that incorporate 2D vertical cross-section solutions into the streamtubes or streamlines and integrate behavior for all the streamtubes or streamlines.

14.5.4 Pseudorelative Permeability Relations and Pseudoproperties. Ref. 24 is a good example of the application of this method. Several reference models describe different areas of the field. Water/oil, solvent/oil, and solvent/water pseudorelative permeability relations are developed, along with pseudotrapped-solvent and solvent-flood residual oil values, so that the relevant behavior of the reference models is reproduced by corresponding models that have the same coarse grids as the full-field model. The coarse-grid models, of course, represent the same parts of the full-field model that the reference models represent. The pseudorelative permeability relations may be developed by trial and error, or they may be estimated by various methods from the fluid-flow and saturation values in the reference-model gridblocks.

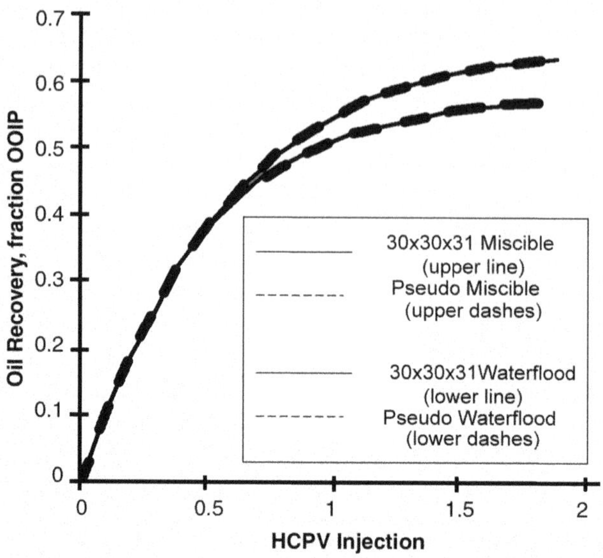

Fig. 14.19—Example of predicting reference-model behavior in a corresponding model with full-field-model grid size using pseudorelative permeability and pseudoproperties (after Ref. 24).

In practice, the pseudorelative permeability relations and pseudoproperties reproduce behavior of one particular fine-grid simulation (e.g., a particular slug size, WAG ratio, injection-solvent composition, start of injection relative to waterflooding, etc.). Of course, an objective of full-field-model simulations is to study different operating scenarios to optimize the flood; to have much utility, the pseudoproperties must predict behavior for conditions other than those of the particular fine-grid simulation that was used to derive them.

Fig. 14.19 shows how well behavior in one of the coarse-grid models of Ref. 24 reproduced the behavior of the corresponding fine-grid model for both the waterflood and the compositionally enhanced solvent flood. The fine-grid and corresponding coarse-grid models represented one-fourth of a nine-spot pattern. The grid of the reference model was 30×30×31, which contained layers that ranged from 4 to 8 ft thick. This 3D reference model already contained pseudorelationships that had been derived from cross sections with 1-ft layers by the method described previously. The grid of the coarse-grid model was 5×5×17. Both the timing and level of waterflood and WAG-flood oil recovery are predicted very well.

Although the pseudorelations were developed for a single slug size, WAG ratio, and solvent composition in the reference-model simulation, they reproduced the effect of these variables surprisingly well in subsequent coarse-grid simulations, as seen in **Figs. 14.6 and 14.20.** The reader is cautioned that this degree of predictability of the pseudorelations may result from the fact that the coarse-grid model still has a relatively large number of gridblocks, especially in the vertical direction, for a full-field model, and thus still retains a large degree of description detail and ability to resolve solvent tongues.

14.5.5 Scaleup With Segment Models. When there are too many patterns to allow a sufficient number of gridblocks between wells for adequate scaleup and predictions by the pseudofunction method, segment-model scaleup may be a suitable choice. This was the case for the enriched-solvent flood at Prudhoe Bay, where there are more than 200 enhanced oil recovery (EOR) and waterflood patterns.[38–40]

The segment model is best suited for regular, repeating patterns. Segment-model scaleup divides each pattern into injector/producer segments, as illustrated in **Fig. 14.21**, for a nine-spot

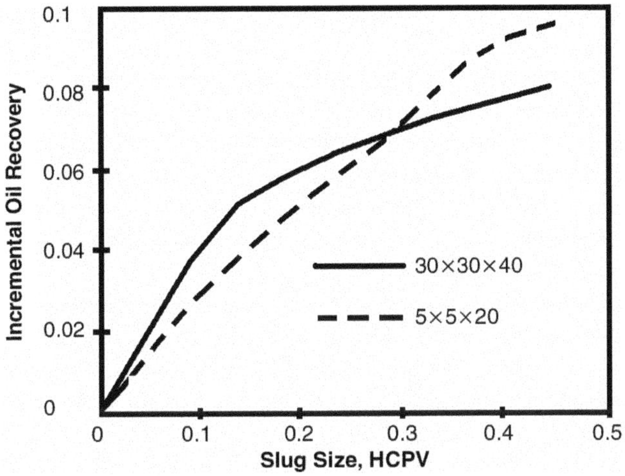

Fig. 14.20—Comparison of reference-model and full-field-model predictions (after Ref. 24).

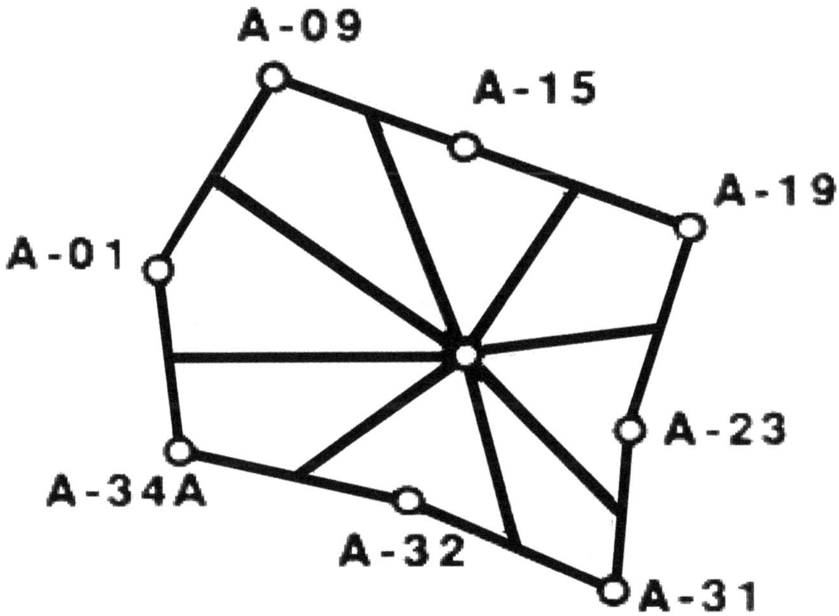

Fig. 14.21—Dividing a pattern into segments (after Ref. 42).

pattern. The dimensionless behavior of each segment is computed using information from suitable reference-model simulations (e.g., dimensionless recovery and solvent production vs. dimensionless injection such as HCPV or displaceable pore volumes). Time rating is accomplished with assumed rates, rates calculated by some other model, or actual injection and production data.

The preferred method for dividing the patterns into segments is with a streamline model that honors actual injection and production rates. Such a model calculates the streamline pattern from injectors to producers, and from this the no-flow boundaries can be determined. Typically, the streamlines for a waterflood are used. Such a model does not assume a balanced, closed boundary system and is able to account for the effects of unbalanced patterns, reservoir heterogeneities, and faulting.

Of course, in an actual flood, the streamlines change with time as they are influenced by injection and production rates and fluid-mobility changes, and dividing a pattern into segments on the basis of a snapshot of streamline distribution is an approximation. It also may be prudent to adjust segments further according to engineering judgment.

A simpler, and more approximate, method for dividing into segments is by geometric construction. Segments are constructed by bisecting the area surrounding the producing wells and connecting these segments to the appropriate injector.

Using the reference-model curves directly can become cumbersome quickly because of such factors as different solvent slug sizes injected into each segment (because of different degrees of throughput into each segment), different amounts of water preinjection, and differing WAG ratios. Reference simulations need to be run for all these variations, and schemes must be developed to interpolate between the various reference curves.

Wingard and Redmond[41] proposed a novel way of transforming the dimensionless performance computed by the reference model by making an analogy with the behavior of a series of stirred tanks. In their procedure, segment performance is calculated over a series of timesteps. The solvent injected during a given timestep mobilizes a given amount of incremental oil and creates a given amount of returned solvent that was not effective in mobilizing oil. The incremental oil mobilized during each timestep and the returned solvent are then produced from each segment according to the dimensionless performance of the reference model. The Wingard and Redmond method calculates the incremental oil recovered by solvent injection vs. time. The total recovery is obtained by superimposing this incremental recovery on a waterflood calculation.

Wingard and Redmond[41] give equations that (1) represent the type of information shown in Fig. 14.9 and derived from reference-model simulations for incremental recovery vs. slug size, (2) represent the increment of returned solvent for each increment of solvent slug injected, and (3) represent the production of each mobilized-oil increment according to the dimensionless injection vs. production performance derived from the reference-model simulations.

The advantage of the segment model is its simplicity and tractability for a large number of patterns. Its major drawback is the inflexibility in accounting for drastic changes in streamline patterns and, thus, changing segment volumes and creation of new segments as wells are converted from producer to injector (or vice versa), as patterns are reconfigured, as wells are recompleted or shut in, or as new wells are drilled. Wingard and Redman discuss these issues and propose approximations to make.

In addition, if solvent-flood performance depends on factors other than just slug size, such as throughput rate, WAG ratio, degree of prior waterflooding, and changing pressure level, approximations must be made to account for these factors.

14.5.6 Scaleup With Streamtube and Streamline Models. In many respects, scaleup with areal streamtube models is similar to scaleup with segment models. First, segments are assigned from injectors to producers—in this case, the streamtubes. The streamtubes are defined with a special model that calculates streamlines.[42–44] Typically, the pressure distribution is solved on an underlying grid for a given distribution of total mobility, and the streamlines are calculated for that pressure distribution. **Fig. 14.22** illustrates streamlines calculated for 2D areal flow, no-flow boundaries between wells, and the resulting streamtubes for several wells. Refer to the chapter on Reservoir Simulation in the Reservoir Engineering and Petrophysics section for more information on streamtube models.

Emanuel *et al.*[45] describe a procedure for superimposing fine-grid reference-model solvent-flood simulations on 2D areal streamtubes as follows:

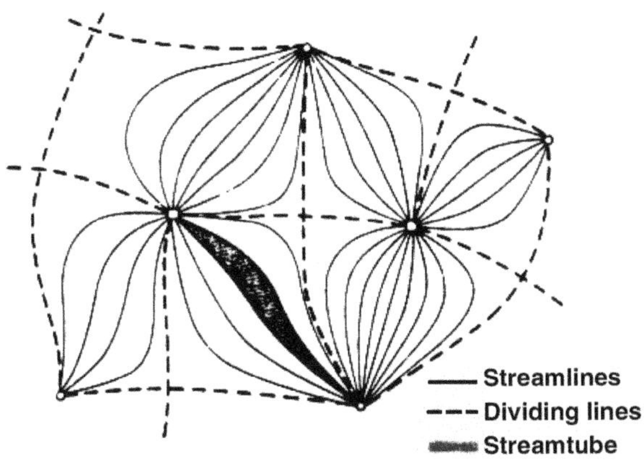

Fig. 14.22—Scaleup with streamtube or streamline models.

• Construct a detailed fine-grid geostatistical cross-section reference model that characterizes the permeability and porosity heterogeneity and correlation between wells in the reservoir area of interest. The wells chosen should typify the flow path of the displacing fluid. This selection is judgmental and will depend on reservoir characteristics. The reference model should be highly detailed in the vertical direction to represent measured log or core data as closely as possible. Layers should be 1 or 2 ft thick. The number of gridblocks between wells should be 20 to 100 depending on computational tractability. Although only one gridblock wide, the cross section should be of variable width to represent the shape of the streamtube. This geometry is intended to model the transition from radial flow near the wells to more linear flow midway through the pattern.

• Simulate the behavior of the process of interest in the reference model. The results of these simulations are reduced to correlations of phase fractional flow at the producer and the total mobility vs. distance from injector to producer as a function of pore volumes or HCPV injected.

• Map the fractional-flow solution onto each streamtube by (a) determining the total mobility in each tube for the cumulative HCPV injected into each tube, (b) allocating injected volume to each streamtube according to its total mobility (the fluid rate for each injection well can be specified or calculated from the imposed pressure drop and total resistance to flow of all the streamtubes), (c) calculating the incremental HCPV injected into each streamtube for the timestep selected, (d) determining the fractional phases of the produced fluid for the cumulative HCPV injected according to correlations, and (e) summing up the contributions from all the tubes connected to a producing well.

The primary advantage of streamtube models over segment models is that the effect of well-rate changes and changing mobility on the streamlines and resulting areal sweep can be accounted for, provided that the streamlines and streamtubes are updated with time. Even if they are not updated, such a model still gives an estimate of the effect of short streamlines and long streamlines on areal sweep, provided that areal sweep is dominated by viscous forces (i.e., a sufficiently high viscosity/gravity ratio) or by areal heterogeneity. In addition, throughput rates can be estimated from the total mobility of the streamtubes rather than being assumed as in segment models. Otherwise, limitations are similar to those encountered with segment models.

Giordano *et al.*[46] propose a novel method of mapping the behavior computed by a reference model directly on the streamlines. Their method makes an analogy between oil mobilization and solvent trapping with tracer adsorption and desorption. They convert the incremental solvent EOR performance computed for a 2D cross-section reference model to 1D tracer-model equations that represent adsorption and desorption of the tracer as it flows through the reservoir:

$$\frac{\partial\left(C_i + D_i\right)}{\partial t} + \frac{v_{xs}}{\phi_i}\left(\frac{\partial F_i}{\partial x_s}\right) = 0, \qquad (14.18)$$

where C_i and A_i are the flowing and adsorbed concentrations of component i, x_s is the distance along the streamline from the injection well, and ϕ_i is the accessible pore volume of component i, which is used as a parameter to scale breakthrough times.

Giordano *et al.*[46] present equations that, for a series of timesteps, (1) calculate the solvent that is effective in mobilizing incremental oil and leaves it trapped according to an adsorption curve; (2) calculate additional ineffective solvent that is left trapped according to another adsorption curve; (3) calculate production of the remaining solvent that is not adsorbed; and (4) calculate the oil mobilized by a given increment of solvent injected according to a desorption curve. Injected water is calculated as the difference between the total injection rate and the injected-solvent rate; produced water is the difference between the total production rate and the sum of the produced solvent, EOR oil, and waterflood oil.

The power of the Giordano *et al.*[46] method is its ability to account for the effect of well conversions and shut-ins, infill drilling, and changing well rates on streamline patterns, as well as its ability to avoid the complications of having to recalculate streamtubes. Its limitations, similar to those for streamtube models, are in the complexity of accounting for changes in performance caused by changes in WAG ratio, throughput rate, pressure level, injected-solvent composition, and other operating conditions.

14.6 Field Examples

References included in this chapter[19–21,38–40,47–75] present performance information on many different miscible projects. A few examples of these projects are included in this section to illustrate the types of performances that have been observed with the use of CO_2, hydrocarbon, and N_2 solvents in varying types of reservoirs.

14.6.1 CO_2 Projects. There have been more CO_2 projects than any other type of miscible flood. The three examples reviewed in this section are considered typical of such applications.

SACROC Four-Pattern Flood.[1,47] This project has been completed. It was thoroughly waterflooded before starting miscible injection. This sequence allows a straightforward evaluation of increased recovery because of miscible displacement.

Fig. 14.23 shows the oil-production rate for the end of the waterflood and the miscible flood. Actual field data are represented by the solid curve, and the forecast decline curve for a continuing waterflood is shown as the dotted curve. The difference between the actual field rate and the forecast waterflood decline represents increased recovery resulting from the miscible project (shaded area); the amount is given in MMSTB. Additional reservoir performance data, including primary plus secondary (P + S), miscible, and total recovery, are given in the upper-right-hand box as a percent OOIP. These data are given in terms of cumulative recovery to date as well as projected ultimate recovery.

After completing the waterflood in 1981, the CO_2 flood was initiated with the same wells and injection patterns. The four-pattern area encompasses 600 acres and 19 MMSTB (3.0 million m^3) OOIP. The well pattern is an inverted nine-spot with 40-acre well spacing. Shortly after starting CO_2 injection, there was an increase in oil-production rate. The EOR of 1.7 MMSTB (0.3 million m^3) is equivalent to 9% OOIP, which, when added to primary plus secondary recovery (57% OOIP), gives a total recovery of 66% OOIP. Net CO_2 use was 3.2 Mscf/STB of increased recovery (570 std m^3/m^3).

This project demonstrated that incremental oil can be recovered by a miscible flood after an efficient waterflood. In this case, water injectivity after CO_2 injection was higher than during the waterflood, thus enabling oil to be recovered more quickly.

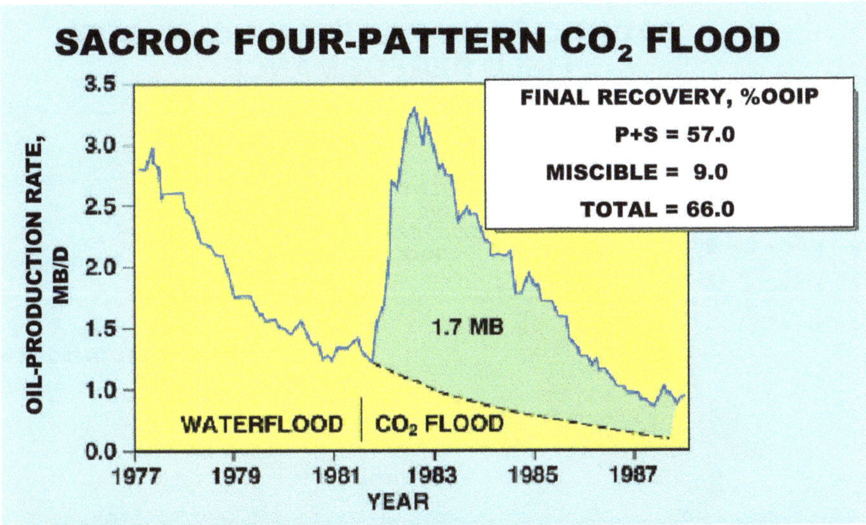

Fig. 14.23—Production history, SACROC four-pattern pilot. (From Healy, Holstein, and Batycky: "Status of Miscible Flooding Technology," *Proc.*, 14th World Petroleum Congress, Improved Recovery and Heavy Oil, 1994. © John Wiley & Sons Limited. Reproduced with permission.)

Fig. 14.24 illustrates the comparison of actual miscible flood performance to that predicted with a four-component compositional simulator. The Todd-Longstaff mixing model[67] was used to account for viscous fingering, and phase behavior was represented by a pseudoternary diagram. Two major empirically based physical parameters, S_{orm} and a viscous-fingering parameter, were used to model local displacement and sweep efficiencies. S_{orm} was based on laboratory displacement tests using representative samples of reservoir rock and fluids. The first step in simulation was to history match the waterflood. This enabled fine-tuning of the reservoir-description model. The compositional simulator was then used to calculate performance of the miscible flood without further adjustment to any match parameters.

As shown in Fig. 14.24, the simulation of cumulative oil recovery vs. cumulative injection for the miscible flood agrees reasonably well with actual field results. The produced water/oil ratio from the simulation is also in reasonable agreement with field results. Waterflood sweep efficiency was 74%, and the miscible flood sweep efficiency was 44%. These sweep efficiencies were determined from analysis of the simulation studies.

Means San Andres Unit.[1,7,12,21,48] This field, located in the eastern edge of the Central Basin Platform of the Permian Basin, produces primarily from the Permian-aged San Andres formation. It was discovered in 1934; waterflooding began in 1963. The field was developed initially on 40-acre spacing and subsequently drilled to 20-acre spacing after the start of waterflooding. The flooding pattern was first peripheral, then a three-to-one line drive, and finally an inverted nine-spot that proved most efficient for this reservoir. Reservoir characteristics are porosity of 9%, permeability of 20 md, S_{wi} of 35%, T_r of 95°F, net-to-gross ratio of 0.18, oil gravity of 29°API, and μ_{oi} of 6 cp.

An oil viscosity of 6 cp makes the waterflood mobility ratio relatively high. From pressure cores and laboratory corefloods, waterflood residual oil saturation was estimated to be 34% of pore volume. A CO_2 miscible project was evaluated with laboratory investigations, field pilots, and reservoir simulations. The pilot tests indicated that CO_2 could successfully mobilize the waterflood residual oil. Even though it is difficult to determine the governing mechanisms for

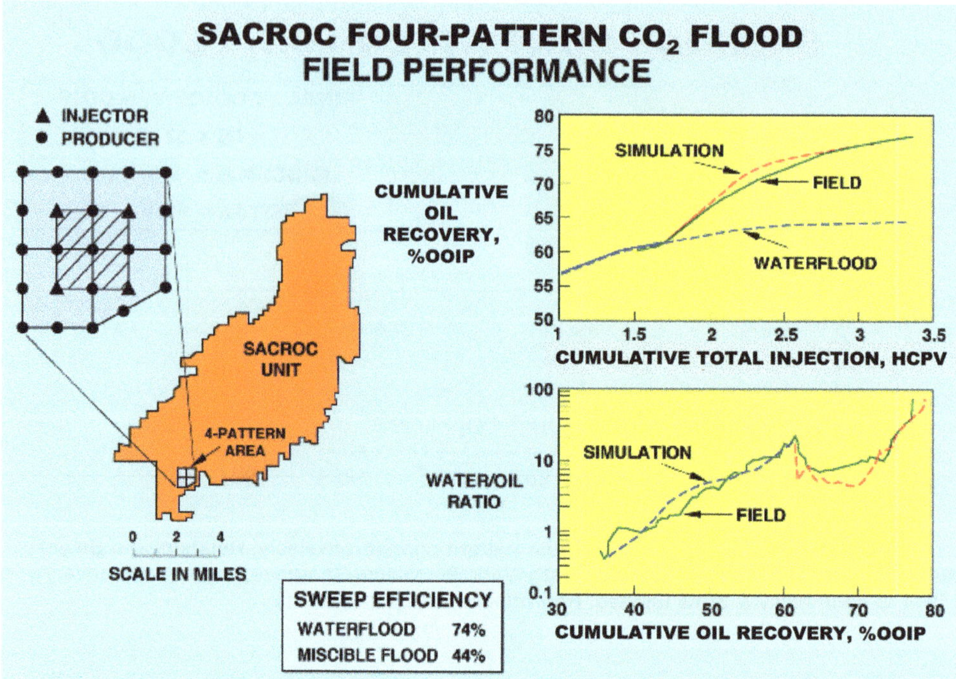

Fig. 14.24—SACROC pattern area, performance, and simulator match. (From Healy, Holstein, and Batycky: "Status of Miscible Flooding Technology," *Proc.,* 14th World Petroleum Congress, Improved Recovery and Heavy Oil, 1994. © John Wiley & Sons Limited. Reproduced with permission.)

improved oil recovery, it appears that after the initial direct displacement of oil by the solvent bank, lighter components of the remaining oil are recovered by extraction.

The original CO_2 project of 167 patterns on approximately 6,700 acres (which contained 82% of OOIP) was expanded to 7,830 acres as evaluation of performance indicated additional prospective areas. Factors affecting process design were the oil viscosity of 6 cp, high MMP, and low formation parting pressure that make operating pressure a critical factor. On the basis of the MMP estimation of 1,850 to 2,300 psi by slimtube experiments and the formation parting pressure of approximately 2,800 psi, a 2,000-psi operating pressure was selected.

Assessment of the economic viability of CO_2 miscible flooding was based on pattern-element simulations for representative project areas that were then used in a scaleup program to forecast total project incremental recovery. A 2:1 WAG ratio and primary CO_2 slug size of 0.40 HCPV were selected as optimum. Updated simulations after gaining operating experience indicated that a CO_2 slug size of 0.60 HCPV was better.

Results of the infill-drilling program and CO_2 flood combined for a total unit oil production increase from approximately 8,500 B/D in 1983 to approximately 16,000 B/D in 1987, as illustrated in **Fig. 14.25**. Much effort has been made to distinguish between the contributions of infill drilling, improved waterflooding, and miscible displacement. Originally incremental oil recovery resulting from infill drilling was projected to be 5.3% OOIP, while the incremental recovery resulting from CO_2 flooding was to be 7.1% OOIP. These recovery estimates have increased over time as a result of an effective reservoir-management program. Current estimates of recovery resulting from primary and waterflooding methods exceed 30% of OOIP, and incremental recovery resulting from the miscible CO_2 flood is more than 15% of the OOIP.

Utilization of new infill wells for injectors helped minimize downhole mechanical problems. A continuous injection-well profiling program is maintained for flood-management pur-

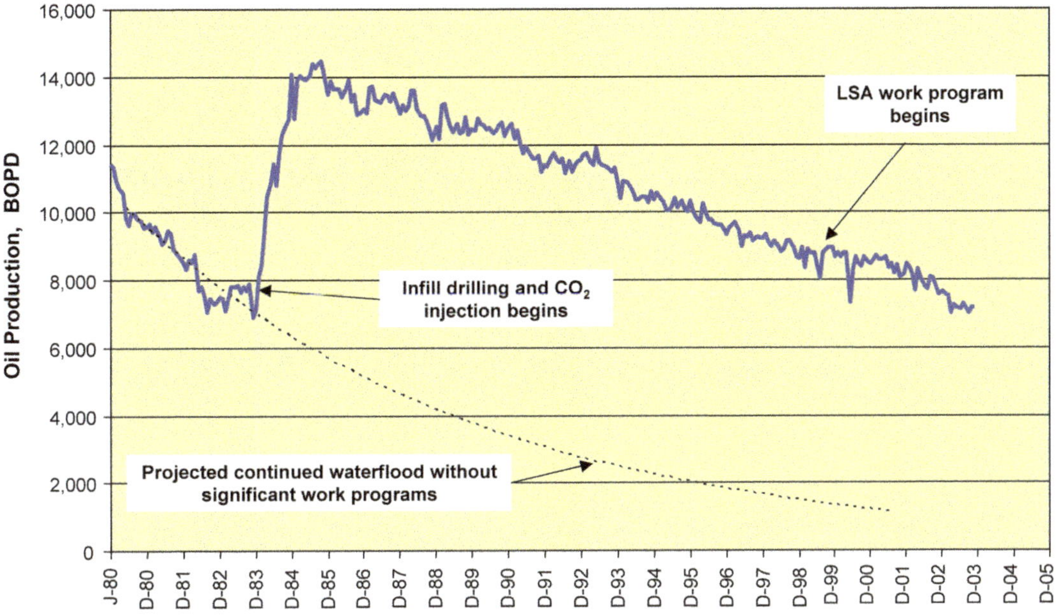

Fig. 14.25—Means San Andres miscible project performance.

poses. Increasing the WAG frequency minimized gas breakthrough between some WAG injectors and offsetting producers experiencing rapid gas breakthrough. While a detailed history-matching simulation of the test did not indicate solvent channeling through known, high-permeability leached layers to be a problem, all other indicators suggested otherwise. Dealing with leached pathways continues to be a challenge. History matching also indicated some loss of CO_2 into the basal water zone.

Several production enhancements have improved field and miscible project performance. First, a 360-acre, nine-pattern pilot was implemented in the North Dome to evaluate the Lower San Andres (LSA) potential. Results showed that additional reserves could be captured from this deeper horizon, although produced-water volumes exceeded initial projections and limited near-term LSA development because of facility constraints. Once water-handling issues were addressed, 59 additional wells were deepened to the LSA in 1992. Performance of these wells provided more insight into factors affecting reservoir performance and resulted in the deepening of an additional 81 injectors and producers and upgrading of facilities to handle more water and gas. As of this writing, the deepening of 97 production wells to the LSA has resulted in a 75% increase in unit oil production and a significant increase in reserves.

Several different types of profile modifications were attempted throughout the 1990s. Early foam and polymer treatments were discontinued because of limited, short-term benefits. Preliminary results from a recent conformance program indicate the possibility to mechanically isolate mature intervals and redirect CO_2 into oil-bearing intervals that would otherwise remain uncontacted.

The miscible-project performance is exceeding previous recovery projections. To better characterize the reservoir and improve business decisions for the asset, a detailed geologic study incorporating engineering and geologic data was used to provide the framework for 3D, three-phase reservoir simulation. Benefits of the study include increasing OOIP by 40%, identifying the potential in the residual oil zone found below the observed oil/water contact in the LSA,

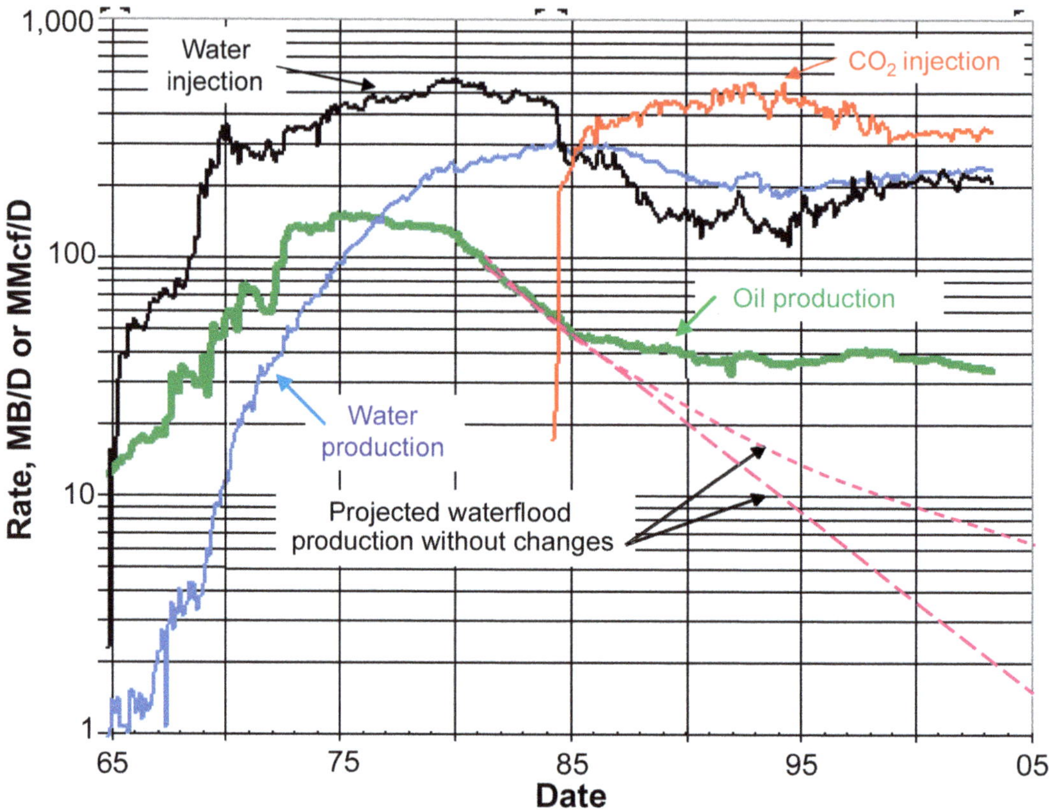

Fig. 14.26—Production performance of the Denver Unit miscible project.

and gaining a better understanding of reservoir continuity using flow units identified with sequence stratigraphy.

Future possibilities for the miscible project include expanding the CO_2-flood project on the basis of the geologic study, continuing the mechanical-isolation program to maximize sweep efficiency, and fine tuning other programs such as varying WAG ratios to further optimize flood performance and enhance profitability.

Denver Unit.[1,49–52] The Wasson Denver Unit CO_2 flood, started in 1983, is one of the larger industry CO_2 projects [28,000 acres, 2.1 BSTB (0.33 billion m^3) OOIP]. No new wells were drilled initially for this project; however, there was significant reconfiguration of the inverted nine-spot patterns (20-acre well spacing) being used in the waterflood preceding miscible injection. Unit performance is shown in **Fig. 14.26** for the period beginning with the waterflood through the first 19 years of miscible CO_2 injection. The reservoir was depressured from 3,200 to 2,200 psi to reduce the amount of trapped CO_2. Oil response occurred after approximately 6 to 8 months. Unit oil-production rates have been sustained since the start of CO_2 injection as a result of response to miscible injection and to the continuing efforts of reservoir-management practices that identify more patterns to miscible flood and ways to improve volumetric sweep with well workovers and conversions. The first CO_2 production occurred almost simultaneously with incremental oil production.

There are uncertainties in the continued waterflood curve because of the usual difficulties in estimating waterflood decline and additional uncertainties introduced as a result of pattern reconfiguration and other modifications that may have affected future waterflood performance as well as miscible recovery.

WAG Ratio. Different WAG ratios were implemented in different areas of the field to determine the most effective method. In the "Continuous Area," CO_2 was injected continuously for approximately 7 years, and then some patterns were converted to 1:1 WAG to reduce CO_2 producing rates. Oil rates were sustained after WAG started.

In the "WAG Area," HCPV injection rate was maintained at a level comparable to the Continuous Area. The WAG ratio was approximately 1:1. Incremental production response was poorer than in the Continuous Area, with a maximum of only about 17% of the waterflood oil rate at the start of CO_2 injection. In addition, there was about a 30% loss of water injectivity, and injection pressures exceeded fracturing pressure on water cycles. The area was converted to a line drive in 1988.

As a result of the experience described above, a "Hybrid Process" was applied in a final area of the field to capture the early response of continuous injection and the long-term gas management of WAG. In this process, CO_2 is injected continuously for 4 to 6 years, followed by 1:1 WAG, until a 60% (or larger) HCPV volume of CO_2 is injected. The final phase will be continuous water injection.

The project has performed well overall. There were a few problems in the western part of the field, where the WAG process was used. Water injection at the desired rates was difficult, and solvent was lost to the gas cap in a limited portion of the reservoir. Neither of these was a complete surprise because the operator recognized both as potential problems during the design phase of the project. The slug process used in the eastern part of the field has performed well, and an increase in the CO_2 slug size is being considered.

A fully compositional, fieldwide simulation model is being used to match field and individual-well performance. The simulator is then used to identify locations (which may require infill drilling or horizontal wells) for project expansion, which wells to shut in or return to production, where solvent losses are occurring, and needed changes in WAG ratios. Opportunities for infill drilling and pattern conversion were implemented and added several million barrels of recoverable oil.

The original estimated CO_2 slug size of 0.4 to 0.6 HCPV has now been increased to 0.72%. The current estimated ultimate EOR is 16.7% OOIP. Continued improvements in reservoir management may improve this outlook.

14.6.2 Enriched-Hydrocarbon Projects. While applied less frequently, these types of projects have been very successful when adequate supplies of methane and enriching fluids are available and profitable to inject.

Prudhoe Bay Field.[20,35,38–41,68–73] The permotriassic-aged Ivishak (also known as Sadlerochit) reservoir, the largest producing horizon in the field, is a series of clastic zones ranging from near-shore marine deposits in the lower sections to sandstones and conglomeratic braided-stream deposits in the remaining, more productive units that contain most of the OOIP. The productive area is 225 square miles. Production began in 1977.

The reservoir is a structural stratigraphic trap consisting of a faulted, south- and southwestward-dipping, 1 to 2° homocline. The main and western areas contain gas caps with different gas/oil contacts (GOCs). Both have an oil/water contact (OWC). The reservoir has an average thickness of more than 500 ft (1600 m). Average rock and fluid properties include a porosity of 22%, S_{wi} of 35%, a net-to-gross ratio of 0.87, an oil gravity of 28°API, and an oil viscosity of 0.8 cp. Permeability averages 500 md and ranges from <100 md in the western area and the deeper deltaic sands up to several darcies in some of the open-framework conglomerate deposits. While the reservoir is connected to a sizeable aquifer, rock properties degrade rapidly

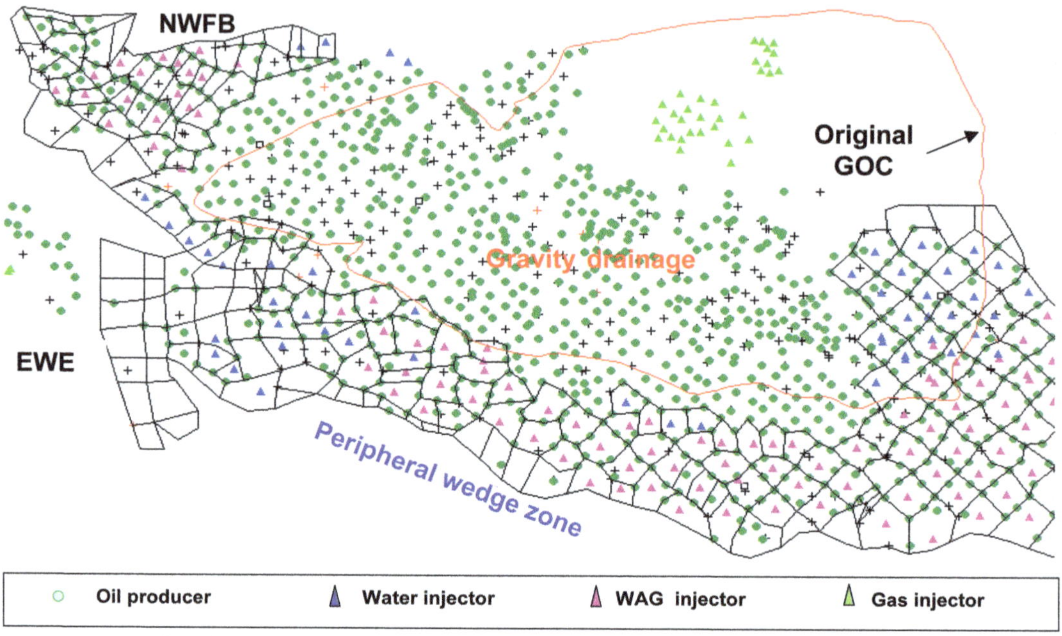

Fig. 14.27—Prudhoe Bay field project areas and recovery processes.

off structure, and the light oil column is separated from the aquifer in portions of the field by a heavy-oil tar mat.

The primary-depletion mechanisms were gravity drainage below the sizable gas cap, a very weak waterdrive, and a potential solution-gas drive. The initial spacing of 320 acres per well was quickly reduced to 160 acres per well and then further reduced to 80 acres per well to sustain production levels and contact more of the reservoir.

Waterflooding (planned as part of the original development plan) was started in conjunction with the 80-acre infill-drilling program. Source water was obtained from the Beaufort Sea. Inverted seven- and nine-spot injection patterns were used in areas of the oil column not overlain by the gas cap.

The gas-cap cycling project was begun in 1977. Gas-handling facilities subsequently were expanded three times and reached a capacity throughput of 8.0 Bcf/D by the late 1990s.

An enriched-hydrocarbon miscible (methane with propane and butane added) WAG injection project was initiated in 1983 with the Flowstation 3 Injection Project. This was expanded to additional areas of the field in 1987. **Fig. 14.27** shows the type of recovery mechanisms being managed in various parts of the field.

Miscible Flood Specifics. The method for optimizing the WAG ratio has shifted with time. Initially, all patterns received a solvent injection (miscible injectant, or MI) of 1% of total pore volume per year. This scheme was changed to a tiered allocation process whereby the more efficient patterns received more MI. When implementation of most miscible-injection patterns (149) was completed, the allocation scheme was changed again to direct MI to the most-efficient patterns (on the basis of a detailed analysis of field-performance data) using a binary method. All patterns are ranked in order of efficiency (production per volume of returned MI, or RMI), and solvent is allocated on a WAG ratio of 1 based on a pattern's throughput until available MI is assigned. The less-efficient patterns are shut in until another reallocation is made. A workover to correct well problems such as flow behind pipe or to open new zones or decline in another pattern's performance can result in a shut-in pattern being ranked higher and

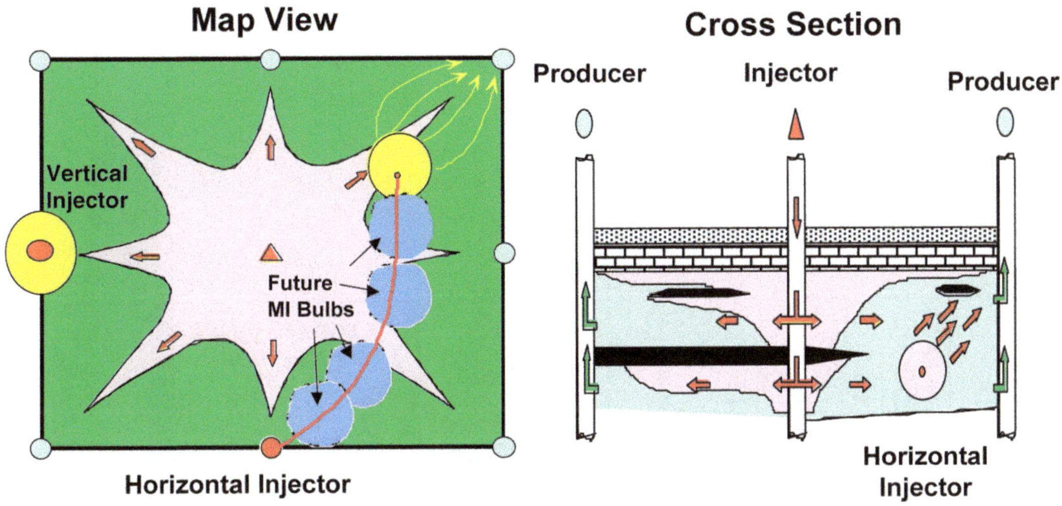

Fig. 14.28—MIST concept.[73]

placed on the active list. Gas samples are routinely collected from production wells to measure RMI as part of the MI allocation process.

Estimated ultimate recovery in the main part of the field is more than 60% OOIP and 80% original condensate in place. Of the total oil recovery, the miscible contribution is 10% OOIP in affected areas. Because the miscible project was started early in field history, a primary-waterflood base production curve was not established to provide an estimate of incremental recovery because of miscible injection. The miscible contribution is based on saturation changes measured by logs run in observation wells and simulations that match actual performance. Also, both tracer and log-inject-log tests have been conducted, and specialized core data have been obtained to measure and improve the effectiveness of the WAG miscible project.

Waterflood and WAG pattern recoveries can be improved with more focus on the management of individual injector/producer pairs within the floods. The objective is to ensure better vertical and areal distribution of the injectant. Patterns with the best performance have recovered in excess of 70% of the OOIP, while the poorer patterns have recovered less than 50% OOIP.

Miscible Injectant Stimulation Treatment (MIST) Concept.[72] As would be anticipated in a high-quality sand with good vertical permeability, the miscible-recovery process has been dominated by gravity forces. Rapid gravity segregation of MI away from the injection well prevents the MI from contacting a significant portion of the target waterflood residual oil, as illustrated in **Fig. 14.28**. As shown on the right, only a small portion of the reservoir is contacted in clean sand intervals typical of much of the miscible flood area. As shown on the left, shale lenses tend to mitigate gravity override, resulting in more of the reservoir being contacted. Both vertical and lateral MIST processes are being implemented to improve volumetric sweep. A vertical MIST process involves completing a production well at the bottom of a thick, continuous, watered-out interval. A large slug of solvent is injected, followed by a small slug of chase water. The solvent sweeps rock not contacted by previous solvent injection.

In the lateral MIST process, horizontal wells are placed in the bottom of the overridden portion of the oil column to distribute injectant laterally (**Fig. 14.29**). A series of injection cycles (referred to as MI bulbs) can be scheduled along the horizontal wellbore to sweep more of the oil column in the pattern. After an MI bulb is complete, the perforations are isolated with a packer or sand, and new perforations are added for the next bulb.

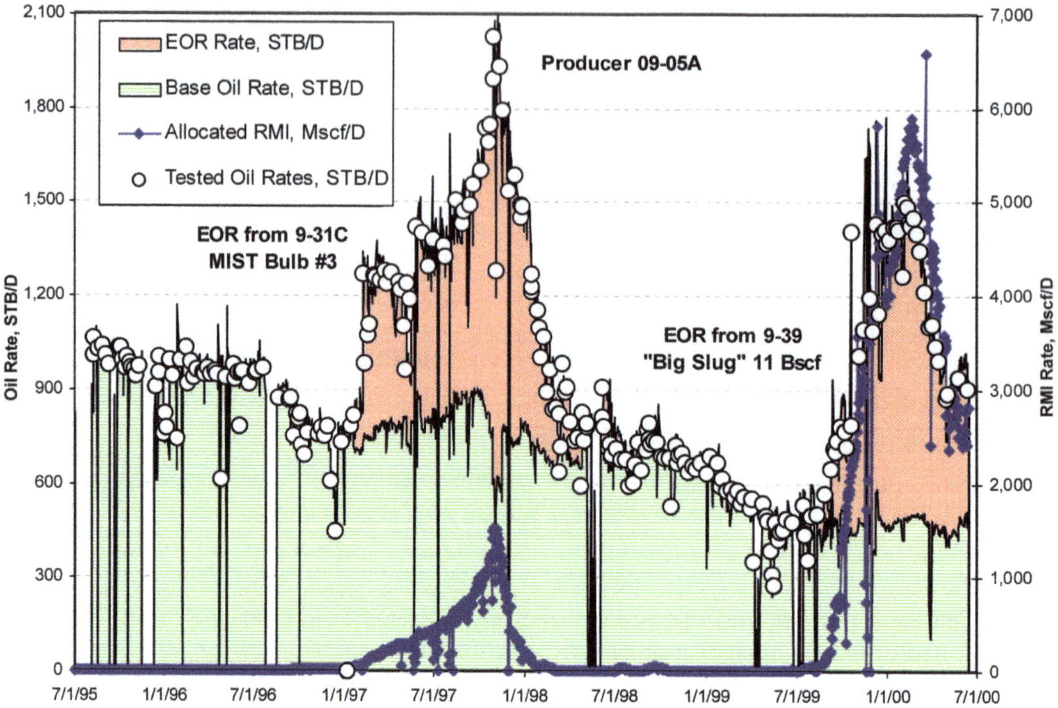

Fig. 14.29—Results of MIST injection.[73]

Fig. 14.29 shows the production history of one producer in a lateral MIST pattern. There are distinctly recognizable production increases because of miscible injection. The first response was injection in MIST injector 9-31C, where production doubled above the base rate. The second response was injection in conventional WAG injector 9-39.

The correlation between response and RMI is consistent with compositional-simulation runs that show that the rapid response is caused by vapor-phase transport. The more subdued response that occurs when liquid oil is displaced and banked by MI is more difficult to discern from other factors affecting producing rate.

Piercement Salt Dome Field. This field is composed of upper- and lower-faulted, unconsolidated sands that dip away from the salt dome at 65 to 85°. Porosity averages 26.5%, and permeability is 1.3 darcies. Oil gravity is 38°API.

This project is operated as a gravity-stable hydrocarbon miscible flood. The injection rate corresponds to a velocity of roughly one-half the critical velocity required for gravity-stable operations. A volume, corresponding to 17% PV, of enriched gas [natural-gas liquids (NGLs) plus solution gas] was injected, followed by injection of solution gas alone. When injection is completed, blowdown of the gas cap is expected to recover approximately 90% of the enriched gas and a significant portion of the injected solution gas, thus reducing the effective cost of the solvent.

Constant pressure is being maintained to improve recovery by eliminating shrinkage of oil over the course of displacement. Coreflood experiments gave recoveries similar to those predicted by the simulations. Miscible residual oil saturation in corefloods was 7% of PV. Slimtube experiments also were carried out to determine the MME required to achieve miscibility at a given pressure level. An MMP was then selected consistent with the volume of enriching NGL available for the project.

Primary production occurred through gas-cap expansion. Miscible-gas injection began after a short primary-production period. Estimated ultimate recovery is 50% of OOIP for primary recovery in both sands, 74% for total recovery after miscible flood in the lower sand, and 86% total recovery after miscible flood in the upper sand. These recovery levels were determined by tracking gas fronts with pulsed neutron capture (PNC) logging and performing material-balance calculations. These recovery levels are consistent with predictions based on simulations. To date, conformance has been excellent, with field recoveries quite similar to those seen in corefloods.

Routine pressure measurements, PNC logs, and pressure-transient testing are used to monitor reservoir performance and contact movements and to identify areas of good and poor communication. Pressure was initially allowed to decline to slightly above the MMP and was then maintained by scheduling injection volumes equal to production. Pressure maintenance became a challenge because of increasing gas/oil ratio (GOR) that resulted in water-cut increases and reservoir pressures below the MMP in some areas. Pressures were increased and maintained by curtailing production from high-GOR wells.

Pressure communication between injectors and producers has been good in the upper reservoir. In the lower reservoir, pressure communication between wells has been sporadic because of faults and shale barriers that act as baffles between injectors and producers.

14.6.3 Nitrogen Projects. ***Jay Field.***[74,75] Discovered in 1970, this field produces primarily from a Jurassic-aged Smackover carbonate that is heavily dolomitized and has complex lithology. The entire pay interval was cored in virtually all the wells to provide an accurate geologic description and aid in unitization efforts. The productive area is 14,415 acres, average depth is 15,400 ft subsea (SS), porosity is 14%, permeability is 35 md, S_{wi} is 12.7%, oil gravity is 51° API, oil viscosity is 0.18 cp, P_{ri} is 7,850 psi, T_r is 285°F, and net-to-gross ratio is 0.27. The dip is 3°.

The field was unitized in 1973, and waterflooding began 4 days later to arrest pressure decline. Miscible N_2 injection was started in 1981. The MMP for many solvents and the light oil in this reservoir is well below 7,000 psi. N_2 was selected over methane and CO_2 because of cost and supply considerations. Delay in methane sales was unattractive. Using CO_2 would have required a long pipeline from central Mississippi with accompanying costs and right-of-way complications.

The field was developed initially with 89 wells on 160-acre spacing. Selective infill drilling later in poorer sections of the reservoir (both areally and vertically) improved the sweep of injected water. This reduced the average spacing to 140 acres per well.

The waterflood was implemented with a 3-to-1 line-drive pattern using low-salinity water from a water-source well. As-produced water is also injected. Peak water-injection rates reached 250,000 B/D.

Nitrogen is produced by three air-separation units. Produced nitrogen is recovered with cryogenic units and reinjected. The injection rate peaked at 86 MMcf/D. WAG injection is used; the WAG ratio varies by pattern, as dictated by ongoing surveillance of producer oil rate and GOR performance. Current plans call for the injection of a 0.4 HCPV bank.

The oil-production rate reached 100,000 BOPD in 1973 and was sustained at or above that level through 1979. Currently, the field is producing 10,500 BOPD at a 95% water cut. The gas production of 80 Mcf/D is approximately 75% N_2. Reservoir pressure has been maintained at 7,500 psi. At this pressure, even high-water-cut wells continue to flow, negating the need for artificial lift.

Estimated ultimate recovery is 60% of OOIP, with recoveries approaching 70% in the upper section of the reservoir. Primary recovery resulting from fluid expansion and solution-gas drive was projected to be 17% of OOIP. Waterflooding increased this by 60%, and the misci-

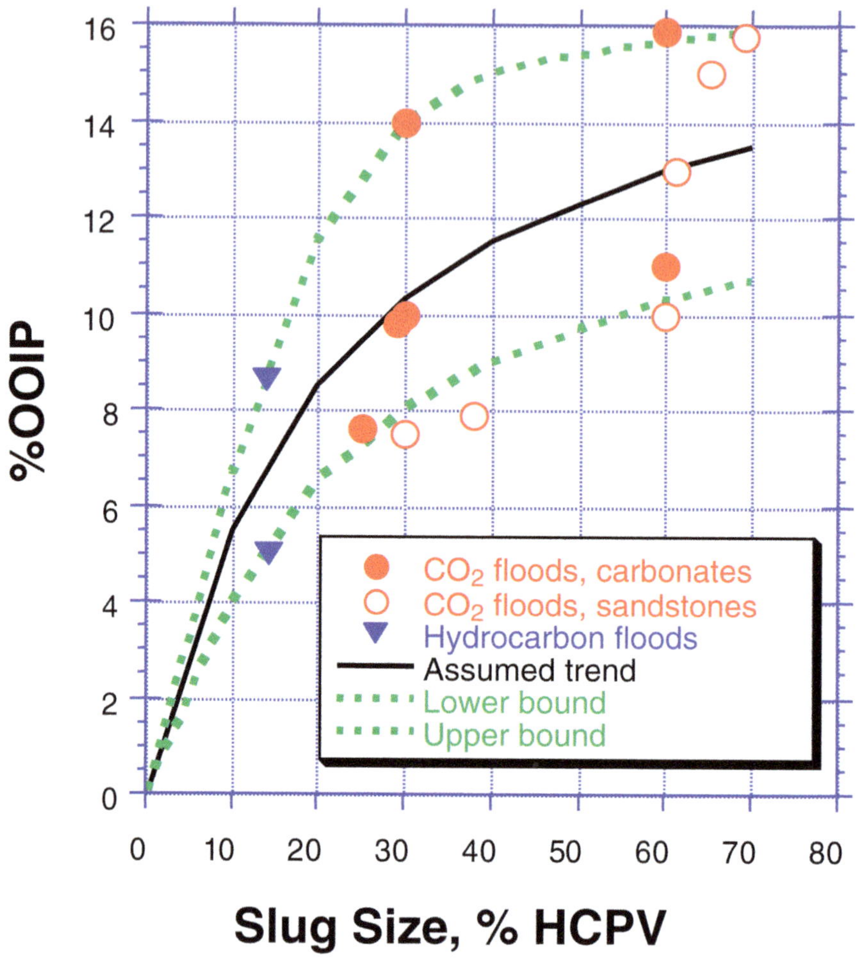

Fig. 14.30—Expected incremental ultimate recovery in selected field projects.

ble project will add another 7 to 10% of OOIP, with recovery in the upper part of the reservoir approaching 13%.

The reservoir surveillance program includes monitoring the WAG ratios, injection-to-withdrawal ratios, and profitability of each pattern. Tools used in the surveillance processes include a history-matched, fully compositional fieldwide simulation model based on an updated geologic model that includes sequence stratigraphy, geostatistical methods, and a styolite model. This finely gridded model includes an updated fluid characterization. These refinements enabled well-by-well matches of production and pressure data for all 137 wells in the field. The updated model has been used to optimize N_2 distribution and investigate several operational changes that are scheduled for implementation and will extend field life by more than 10 years.

14.6.4 Overall Industry Experience. Fig. 14.30 shows how incremental ultimate recovery increased with total solvent slug size for a few projects for which these data were available. The interpretations given in this figure are the author's and are not necessarily those of the project operators. Many of the projects represented in this figure are ongoing, and the ultimate incremental recovery is an estimate.

The contribution of incremental oil production from gas-injection projects (mostly miscible) has continued to grow since 1984, as reported biennially by the *Oil & Gas Journal*. **Fig. 14.31**

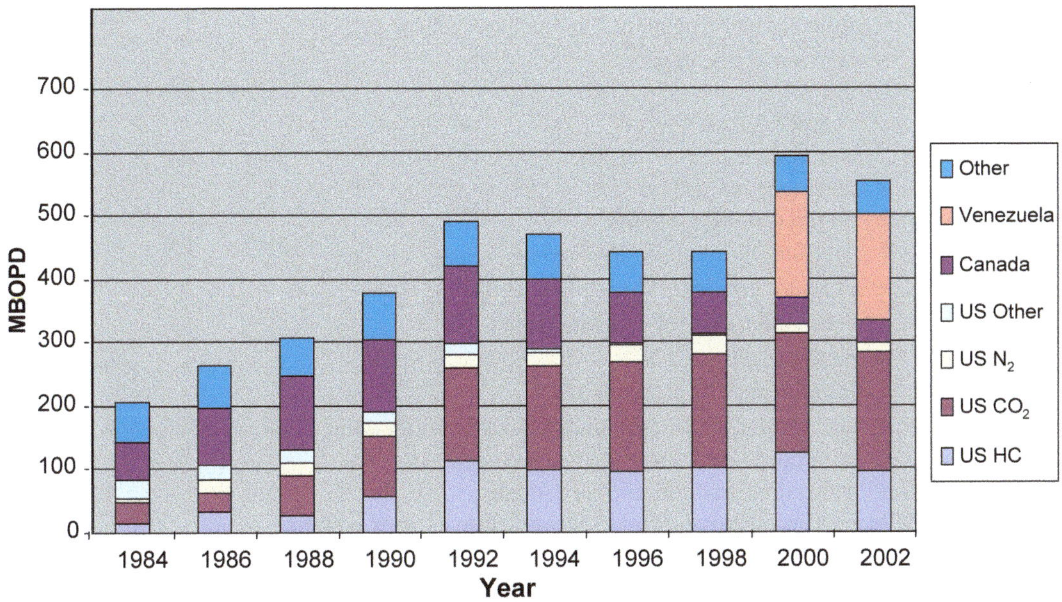

Fig. 14.31—World incremental oil production caused by gas injection (adapted from biennial reports, *Oil & Gas J.*).

shows the volumes reported in this reference. In 2002, total incremental production was nearly 550,000 BOPD. Approximately 60% of the total comes from hydrocarbon miscible projects, and most of the remainder comes from CO_2 miscible projects in the U.S. The advent of the significant contributions from the projects (hydrocarbon miscible) in Venezuela beginning in 2000 indicates an ongoing effort to identify and implement miscible projects around the world.

14.7 Summary

Miscible injection has been applied successfully in many reservoirs. The resulting experience has made it possible to reliably predict the economic viability of new projects in other reservoirs. This chapter contains some general guidelines that should suffice in screening studies of the applicability of a miscible process to a given reservoir or field. In addition, more-detailed information on phase behavior and compositional simulation has been discussed to provide guidance on how to make a more-detailed assessment of miscible flooding. Several field examples are presented to illustrate how CO_2, enriched hydrocarbons, and N_2 solvents have been used to increase oil recoveries significantly. Finally, proper assessment of the application of a miscible project should include the timing of capital outlays for project implementation, the timing of solvent injection and production response, changes in injectivity, and the costs and need to reinject produced solvent.

Nomenclature

a = constant
A = defined with Eq. 14.8
b = constant
B = defined with Eq. 14.9
C = defined with Eq. 14.11
D = adsorbed concentration
f = fugacity
f_j^L = fugacity of component j in the liquid phase, psi

f_j^V = fugacity of component *j* in the vapor phase, psi
F = flowing concentration
g = gas
i = component *i*
j = component *j*
k = permeability, md
n = number of components
N_{ca} = capillary number, dimensionless
Δp_g = pressure gradient through the displacing phase, psi
p_c = component critical pressure, psia
P = pressure, psia
P_{ri} = initial reservoir pressure, psi
R = universal gas constant, units consistent with other equation parameters
s = volumetric shift parameter, dimensionless
S = saturation, fraction of pore volume
S_g^{max} = maximum solvent saturation, fraction of pore volume
S_{gr}^{max} = solvent trapped when rock is 100% solvent saturated
S_{gt} = trapped gas saturation, fraction of pore volume
S_{oi} = initial oil saturation, fraction of pore volume
S_{orm} = miscible flood residual oil saturation, fraction of pore volume
S_{orwf} = final waterflood residual oil saturation, fraction of pore volume
S_{wi} = initial water saturation, fraction of pore volume
t = trapped
T = temperature, °R
T_c = critical temperature, °R
T_r = reduced temperature, T/T_c
v = volume, cubic ft
v_{xs} = fluid velocity along a streamline, ft/sec
x = distance along the streamline, ft
x,y,z = simulator grid directions
Z = fluid compressibility factor, dimensionless
ω = component acentric factor, dimensionless
ϕ = accessible pore volume
κ = define by Eq. 14.6
σ = interfacial tension, dynes/cm
∂_{ij} = binary interaction coefficient
μ_{oi} = oil viscosity, cp
Ω_a = constant in Eq. 14.3
Ω_b = constant in Eq. 14.5

Acknowledgments
The authors would like to express their appreciation to members of ExxonMobil Production Co. for their help in describing the performance of the Means San Andres Unit miscible flood performance and the members of Oxy Petroleum for their help in describing the Denver Unit miscible flood performance.

References

1. Healy, R.N., Holstein, E.D., and Batycky, J.P.: "Status of Miscible Flooding Technology," *Proc.*, 14th World Petroleum Congress, Improved Recovery and Heavy Oil, John Wiley and Sons, Vol. 7, No. 3, 407–16.
2. Hadlow, R.E.: "Update of Industry Experience with CO_2 Injection," paper SPE 24928 presented at the 1992 SPE Annual Technical Conference and Exhibition, Washington, DC, 4–7 October.
3. Brock, W.R. and Bryan, L.A.: "Summary Results of CO_2 EOR Field Tests, 1972–1987," paper SPE 18977 presented at the 1989 SPE Joint Rocky Mountain Regional/Low Permeability Reservoirs Symposium, Denver, 6–8 March.
4. Young, M.N. and Martin, W.E.: "The Wizard Lake D-3A Pool Miscible Floods," *J. Cdn. Pet. Tech.* (April–June 1980) 65.
5. Backmeyer, L.A. et *al.:* "The Tertiary Extension of the Wizard Lake D-3A Pool Miscible Flood," paper SPE 13271 presented at the 1984 SPE Annual Technical Conference and Exhibition, Houston, 16–19 September.
6. DaSie, W.J. and Guo, D.S.: "Assessment of a Vertical Hydrocarbon Miscible Flood in the Westpem Nisku D Reef," *SPERE* (May 1990) 147; *Trans.*, AIME, **289.**
7. Johnston, J.R.: "Weeks Island Gravity Stable CO_2 Pilot," paper SPE/DOE 17351 presented at the 1988 SPE/DOE Enhanced Oil Recovery Symposium, Tulsa, 17–20 April.
8. Tiffin, D.L. and Kremesec, V.J.: "A Mechanistic Study of Gravity-Assisted CO_2 Flooding," paper SPE 14895 presented at the 1986 SPE/DOE Symposium on Enhanced Oil Recovery, Tulsa, 20–23 April.
9. Shyeh-Yung, J.-G.J.: "Mechanisms of Miscible Oil Recovery: Effects of Pressure on Miscible and Near-Miscible Displacements of Oil by Carbon Dioxide," paper SPE 22651 presented at the 1991 SPE Annual Technical Conference and Exhibition, Dallas, 6–9 October.
10. Stern, D.: "Mechanisms of Miscible Oil Recovery: Effects of Pore-Level Fluid Distribution," paper SPE 22652 presented at the 1991 SPE Annual Technical Conference and Exhibition, Dallas, 6–9 October.
11. Koch, H.A. and Hutchinson, C.A. Jr.: "Miscible Displacements of Reservoir Oil Using Flue Gas," *Trans.*, AIME (1958) **213,** 7.
12. Clark, N.J. *et al.:* "Miscible Drive—Its Theory and Application," *JPT* (June 1958) 11.
13. Hutchinson, C.A. Jr. and Braun, P.H.: "Phase Relations of Miscible Displacement in Oil Recovery, *AIChE J.* (1961) **7,** No. 1, 64.
14. Benham, A.L., Dowden, W.E., and Kunzman, W.J.: "Miscible Fluid Displacement—Prediction of Miscibility," *Trans.*, AIME (1960) **219,** 229.
15. Wu, R.S. *et al.:* "Enriched Gas Displacement: Design of Solvent Compositions," *J. Cdn. Pet. Tech.* (May–June 1986) 55.
16. Zick, A.A.: "Combined Condensing/Vaporizing Mechanism in the Displacement of Oil by Enriched Gases," paper SPE 15493 presented at the 1986 SPE Annual Technical Conference and Exhibition, New Orleans, 5–8 October.
17. Stone, H.L.: "Vertical Conformance in an Alternating Water-Miscible Gas Flood," paper SPE 11130 presented at the 1982 SPE Annual Technical Conference and Exhibition, New Orleans, 26–29 September.
18. Jenkins, M.K.: "An Analytical Model for Water/Gas Miscible Displacements," paper SPE/DOE 12632 presented at the 1984 SPE/DOE Symposium on Enhanced Oil Recovery, Tulsa, 15–18 April.
19. Pritchard, D.W.L. *et al.:* "Reservoir Surveillance Impacts Management of the Judy Creek Hydrocarbon Miscible Flood," paper SPE/DOE 20228 presented at the 1990 SPE/DOE Enhanced Oil Recovery Symposium, Tulsa, 22–25 April.
20. Dawson, A.G., Jackson, D.D., and Buskirk, D.L.: "Impact of Solvent Injection Strategy and Reservoir Description on Hydrocarbon Miscible EOR for the Prudhoe Bay Unit, Alaska," paper SPE 19657 presented at the 1989 SPE Annual Technical Conference and Exhibition, San Antonio, Texas, 8–11 October.

21. Magruder, J.B., Stiles, L.H., and Yelverton, T.D.: "Review of the Means San Andres Unit CO_2 Tertiary Project," *JPT* (May 1990) 638; *Trans.,* AIME, **289.**
22. Moritis, G.: "EOR Increases 24% Worldwide; Claims 10% of U.S. Production," *Oil & Gas J.* (20 April 1992) 51.
23. Chang, H.L. *et al.:* "The Effects of Injectant-Enrichment Level on Oil Recovery in Horizontal, Gravity-Tongue-Dominated Enriched-Gas Drives," paper SPE 26084 presented at the 1993 SPE Western Regional Meeting, Anchorage, 26–28 May.
24. Jerauld, G.R.: "A Case Study in Scaleup for Multicontact Miscible Hydrocarbon Gas Injection," *SPEREE* (December 1998) 575.
25. Mungan, N.: "Carbon Dioxide Flooding—Fundamentals," *J. Cdn. Pet. Tech.* (January–March 1981) 87.
26. Stalkup, F.I.: *Miscible Displacement,* Monograph Series, SPE, Dallas (1983) **8.**
27. Rogers, J.E. and Grigg, R.B.: "A Literature Analysis of the WAG Injectivity Abnormalities in the CO_2 Process," *SPEREE* (October 2001) 375.
28. Peng, D.Y. and Robinson, D.B.: "A New Two-Constant Equation of State," *Ind. Eng. Chem. Fundam.* (1976) **15,** No. 1, 59.
29. Soave, G.: "Equilibrium Constants from a Modified Redlick-Kwong Equation of State," *Chem. Eng. Sci.* (1972) **27,** 1197.
30. Jhaveri, B.S. and Youngren, G.K.: "Three-Parameter Modification of the Peng-Robinson Equation of State To Improve Volumetric Predictions," *SPERE* (August 1988) 1033.
31. Peneloux, A., Rauzy, E., and Freze, R.: "A Consistent Correction for Redlich-Kwong-Soave Volumes," *Fluid Phase Equilibria* (1982) **8,** 7.
32. Lohrenz, J., Bray, B.G., and Clark, C.R.: "Calculating Viscosities of Reservoir Fluids From Their Composition," *JPT* (October 1964) 1171; *Trans.,* AIME, **231.**
33. Whitson, C.H. and Brule, M.R.: *Phase Behavior,* Monograph Series, SPE, Richardson, Texas (2000) **20.**
34. Pedersen, K.S., Thomassen, P., and Fredenslund, A.: "Thermodynamics of Petroleum Mixtures Containing Heavy Hydrocarbons. 1. Phase Envelope Calculations by Use of the Soave-Redlich-Kwong Equation of State," *Ind. Eng. Chem. Process Des. Dev.* (1984) **23,** No. 1, 163.
35. Jerauld, G.R.: "Gas-Oil Relative Permeability of Prudhoe Bay," paper SPE 35718 presented at the 1996 SPE Western Regional Meeting, Anchorage, 22–24 May.
36. Land, C.S.: "Comparison of Calculated with Experimental Imbibition Relative Permeability," *Trans.,* AIME (1971) **251.**
37. Christie, M.A. *et al.:* "A Renormalisation-Based Upscaling Technique for WAG Floods in Heterogeneous Reservoirs," paper presented at the 1995 SPE Symposium on Reservoir Simulation, San Antonio, Texas, 12–15 February.
38. Simon, A.D. and Petersen, E.J.: "Reservoir Management of the Prudhoe Bay Field," paper SPE 38847 presented at the 1997 SPE Annual Technical Conference and Exhibition, San Antonio, Texas, 5–8 October.
39. Szabo, J.D. and Meyers, K.O.: "Prudhoe Bay: Development History and Future Potential," paper SPE 26053 presented at the 1993 SPE Western Regional Meeting, Anchorage, 26–28 May.
40. McGuire, P.L. and Stalkup, F.I.: "Performance Analysis and Optimization of the Prudhoe Bay Miscible Gas Project," paper SPE 22398 presented at the 1992 SPE International Meeting on Petroleum Engineering, Beijing, 24–27 March.
41. Wingard, J.S. and Redman, R.S.: "A Full-Field Forecasting Tool for the Combined Water/ Miscible Gas Flood at Prudhoe Bay," paper SPE 28632 presented at the 1994 SPE Annual Technical Conference and Exhibition, New Orleans, 25–28 September.
42. Martin, J.C. and Wegner, R.E.: "Numerical Solution of Multiphase, Two-Dimensional Incompressible Flow Using Stream-tube Relationships," *SPEJ* (October 1979) 313; *Trans.,* AIME, **267.**
43. Thiele, M.R., Blunt, M.J., and Orr, F.M. Jr.: "A New Technique for Predicting Flow in Heterogeneous Systems Using Streamtubes," paper SPE/DOE 27834 presented at the 1994 SPE/DOE Symposium on Improved Oil Recovery, Tulsa, 17–20 April.
44. Batycky, R.P., Blunt, M.J., and Thiele, M.R.: "A 3D Field-Scale Streamline-Based Reservoir Simulator," *SPERE* (November 1997) 246.

45. Emanuel, A.S. *et al.*: "Reservoir Performance Prediction Methods Based on Fractal Geostatistics," *SPERE* (August 1991) 311.
46. Giordano, R.M., Redman, R.S., and Bratvedt, F.: "A New Approach to Forecasting Miscible WAG Performance at the Field Scale," paper SPE 36712 presented at the 1996 SPE Annual Technical Conference and Exhibition, Denver, 6–9 October.
47. Langston, M.V., Hoadley, S.F., and Young, D.N.: "Definitive CO_2 Flooding Response in the SACROC Unit," paper SPE/DOE 17321 presented at the 1988 SPE/DOE Enhanced Oil Recovery Symposium, Tulsa, 17–20 April.
48. Stiles, L.H. and Magruder, J.B.: "Reservoir Management in the Means San Andres Unit," *JPT* (April 1992) 469; *Trans.,* AIME, **293.**
49. Tanner, C.S. *et al.*: "Production Performance of the Wasson Denver Unit CO_2 Flood," paper SPE/DOE 24156 presented at the 1992 SPE/DOE Symposium on Enhanced Oil Recovery, Tulsa, 22–24 April.
50. Kittridge, M.G.: "Quantitative CO_2 Flood Monitoring, Denver Unit, Wasson (San Andres) Field," paper SPE 24644 presented at the 1992 SPE Annual Technical Conference and Exhibition, Washington, DC, 4–7 October.
51. Hsu, C.-F., Morell, J.I., and Falls, A.H.: "Field-Scale CO_2 Flood Simulations and Their Impact on the Performance of the Wasson Denver Unit," paper SPE 29116 presented at the 1995 SPE Symposium on Reservoir Simulation, San Antonio, Texas, 12–15 February.
52. Thai, B.N. *et al.*: "Denver Unit Infill Drilling and Pattern Reconfiguration Program," paper SPE 59548 presented at the 2000 SPE Permian Basin Oil and Gas Recovery Conference, Midland, Texas, 21–23 March.
53. Harpole, K.J. and Hallenbeck, L.D.: "East Vacuum Grayburg San Andres Unit CO_2 Flood Ten Year Performance Review: Evolution of a Reservoir Management Strategy and Results of WAG Optimization," paper SPE 36710 presented at the 1996 SPE Annual Technical Conference and Exhibition, Denver, 6–9 October.
54. Ring, J.N. and Smith, D.J.: "An Overview of the North Ward Estes CO_2 Flood," paper SPE 30729 presented at the 1995 SPE Annual Technical Conference and Exhibition, Dallas, 22–25 October.
55. Brokmeyer, R.J., Borling, D.C., and Pierson, W.T.: "Lost Soldier Tensleep CO_2 Tertiary Project, Performance Case History: Bairoil, Wyoming," paper SPE 35191 presented at the 1996 SPE Permian Basin Oil and Gas Recovery Conference, Midland, Texas, 27–29 March.
56. Flanders, W.A. and DePauw, R.M.: "Update Case History: Performance of the Twofreds Tertiary CO_2 Project," SPE 26614 presented at the 1993 SPE Annual Technical Conference and Exhibition, Houston, 3–6 October.
57. Reinbold, E.W. *et al.*: "Early Performance and Evaluation of the Kuparuk Hydrocarbon Miscible Flood," paper SPE 24930 presented at the 1992 SPE Annual Technical Conference and Exhibition, Washington, DC, 4–7 October.
58. Winter, B.T. and Edwards, K.A.: "Reservoir Management and Optimization of the Mitsue Gilwood Sand Unit #1 Horizontal Hydrocarbon Miscible Flood," paper SPE 30725 presented at the 1995 SPE Annual Technical Conference and Exhibition, Dallas, 22–25 October.
59. Pritchard, D.W.L. and Nieman, R.E.: "Improving Oil Recovery Through WAG Cycle Optimization in a Gravity-Override-Dominated Miscible Flood," paper SPE/DOE 24181 presented at the 1992 SPE/DOE Symposium on Enhanced Oil Recovery, Tulsa, 22–24 April.
60. Hoolahan, S.P. *et al.*: "Kuparuk Large Scale Enhanced Oil Recovery Project," paper SPE 35698 presented at the 1996 SPE Western Regional Meeting, Anchorage, 22–24 May.
61. Bellavance, J.F.R.: "Dollarhide Devonian CO_2 Flood: Project Performance Review 10 Years Later," paper SPE 35190 presented at the 1996 SPE Permian Basin Oil and Gas Recovery Conference, Midland, Texas, 27–29 March.
62. Pittaway, K.R. and Rosato, R.J.: "The Ford Geraldine Unit CO_2 Flood—Update 1990," *SPERE* (1991) 410.
63. Reitzel, G.A. and Callow, G.O.: "Pool Description and Performance Analysis Leads to Understanding Golden Spike's Miscible Flood," *J. Cdn. Pet. Tech.* (April–June 1977) **16,** No. 2, 39.

64. Lee, K.H. and El-Saleh, M.M.: "A Full-Field Numerical Modeling Study for the Ford Geraldine Unit CO_2 Flood," paper SPE/DOE 20227 presented at the 1990 SPE/DOE Symposium on Enhanced Oil Recovery, Tulsa, 22–25 April.
65. Phillips, L.A., McPherson, J.L., and Leibrecht, R.J.: "CO_2 Flood: Design and Initial Operations, Ford Geraldine (Delaware Sand) Unit," paper SPE 12197 presented at the 1983 SPE Annual Technical Conference and Exhibition, San Francisco, 3–5 October.
66. Pittaway, K.R. and Runyan, E.E.: "The Ford Geraldine Unit CO_2 Flood: Operating History," *SPEPE* (August 1990) 333.
67. Todd, M.R. and Longstaff, W.J.: "The Development, Testing, and Application of a Numerical Simulator for Predicting Miscible Flood Performance," *JPT* (July 1972) 874; *Trans.,* AIME, **253.**
68. Erickson, J.W. and Sneider, R.M.: "Structural and Hydrocarbon Histories of the Ivishak (Sadlerochit) Reservoir, Prudhoe Bay Field," paper SPE 28574 presented at the 1994 SPE Annual Technical Conference and Exhibition, New Orleans, 25–28 September.
69. Sneider, R.M. and Erickson, J.W.: "Rock Types, Depositional History, and Diagenetic Effects—Sadlerochit Reservoir, Prudhoe Bay Field," paper SPE 28575 presented at the 1994 SPE Annual Technical Conference and Exhibition, New Orleans, 25–28 September.
70. Weaver, J.W. and Uldrich, D.O.: "Optimizing Solvent Allocation in the Prudhoe Bay Miscible Gas Project," paper SPE 54615 presented at the 1999 SPE Western Regional Meeting, Anchorage, 26–28 May.
71. Cockin, A.P. *et al.*: "Analysis of a Single-Well Chemical Tracer Test To Measure the Residual Oil Saturation to a Hydrocarbon Miscible Gas Flood at Prudhoe Bay," *SPEREE* (December 2000) 544.
72. McGuire, P.L., Spence, A.P., and Redman, R.S.: "Performance Evaluation of a Mature Miscible Gasflood at Prudhoe Bay," *SPEREE* (August 2001) 318.
73. McGuire, P.L. and Holt, B.M.: "Unconventional Miscible EOR Experience at Prudhoe Bay: A Project Summary," *SPEREE* (February 2003) 17.
74. Langston, E.P. and Shirer, J.A.: "Performance of Jay/LEC Fields Unit Under Mature Waterflood and Early Tertiary Operations," *JPT* (February 1985) 261.
75. Lawrence, J.J. *et al.*: "Jay Nitrogen Tertiary Recovery Study: Managing a Mature Field," paper SPE 78527 presented at the 2002 Abu Dhabi International Petroleum Exhibition and Conference, Abu Dhabi, UAE, 13–16 October.

SI Metric Conversion Factors

acre ×	4.046 873	E + 03	= m^2
°API	141.5/(131.5 + °API)		= g/cm^3
bbl ×	1.589 873	E − 01	= m^3
cp ×	1.0*	E − 03	= Pa·s
dyne ×	1.0*	E − 02	= mN
ft ×	3.048*	E − 01	= m
ft^3 ×	2.831 685	E − 02	= m^3
°F	(°F − 32)/1.8		= °C
in. ×	2.54*	E + 00	= cm
psi ×	6.894 757	E + 00	= kPa

*Conversion factor is exact.

Chapter 15
Thermal Recovery by Steam Injection

Jeff Jones, SPE, Nations Petroleum

15.1 Steam

The most common method used to enhance oil production over primary rates is water injection, commonly referred to as secondary oil recovery. Common practice in the industry is to refer to all other methods as tertiary enhanced oil recovery. According to Prats,[1] thermal enhanced oil recovery (TEOR) is a family of tertiary processes defined as "any process in which heat is introduced intentionally into a subsurface accumulation of organic compounds for the purpose of recovering fuels through wells." By far, the most common vehicle used to inject heat is saturated steam. Hot water and heated gasses have been tried, but none are as effective as quality steam. According to a 2000 *Oil and Gas Journal* survey,[2] steam enhanced oil recovery projects account for 417,675 barrels of oil per day (BOPD), or 56% of the total for all tertiary enhanced recovery methods. That production rate has been essentially flat for more than 15 years. Hydrocarbon gas injection and CO_2 gas injection are the only other significant contributors and amount to only 17 and 24%, respectively. This chapter refers to the general process as steam enhanced oil recovery (SEOR).

15.1.1 Properties of Saturated Steam. Like other substances, water can exist in the form of a solid (ice), as a liquid (water), or as a gas (loosely called steam). SEOR processes are concerned with the liquid and gas phases, and the change from one phase to the other. The phase change region, in which water coexists as liquid and gas, is where our interest lies when considering steam for use in the oil field. The term "steam" is an imprecise designation because it refers to a water liquid/gas system that can exist from 32°F to any higher temperature; from 0.1 psia to any higher pressure; and from nearly all liquid to 100% gas. Steam quality refers to the phase change region of liquid to gas and is defined as

$$f_s = \frac{m_v}{m_v + m_l} \quad \text{(15.1)}$$

Heat capacity is expressed in units of Btu/(lbm-°F). A "Btu" is defined as the amount of heat required to raise 1 lbm of water from 60 to 61°F. All liquids and solids are compared to pure water, which has the highest heat capacity of any substance at 1 Btu/(lbm-°F). By calculating a

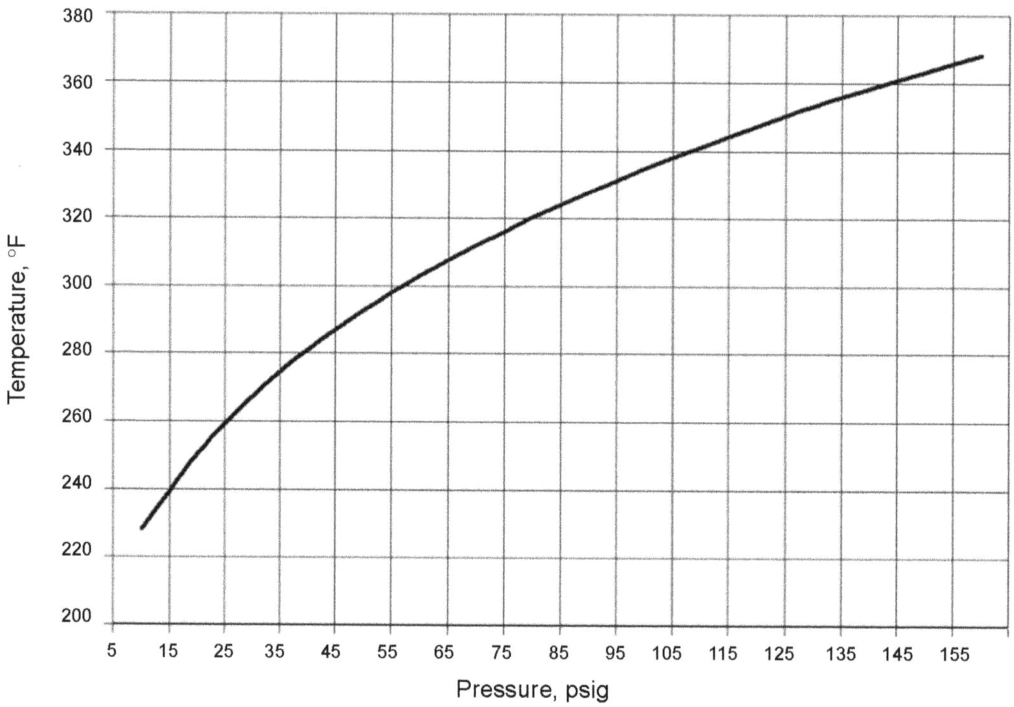

Fig. 15.1—Saturation steam temperature and pressure.

ratio of the heat capacity of water divided by that of another substance, a convenient fraction called "specific heat" is obtained. Notice that petroleum has a specific heat of 0.5, or half that of water, and sandstone is only 20% of water on a per pound basis. No other liquid or gas carries as much heat per pound as water. Also, the temperature range at which this high heat carrying performance is achieved, 34 to 700°F, is ideal for many processes, including SEOR.

Enthalpy is a useful property defined by an arbitrary combination of other properties and is not a true form of energy. The absolute value of enthalpy is of no practical value. Changes in enthalpy are extremely useful, however, and are the basis for SEOR energy calculations. The total enthalpy held by each pound of liquid water at any temperature is called sensible heat, h_f. The heat input, which produces a change of state from liquid to gas without a change of temperature, is called the "latent heat of evaporation" and is shown by h_{fv}. The total heat, h_v, in each pound of 100% quality or saturated steam is the sum of these two, $h_v = h_f + h_{fv}$.

In the phase change or "saturation" region, steam can only exist at one temperature for a given pressure regardless of quality or latent heat content, as shown in **Fig. 15.1**. Steam increases in volume as latent heat increases, as in **Fig. 15.2**. This is a useful property in displacing oil in an SEOR process. The volume occupied by 1 lbm of steam at any pressure is its specific volume in ft³/lbm and is represented by υ_s. Values for these thermal properties of water are published widely.[3,4] **Fig. 15.3** is a chart of the phase change region. Equations have been derived that approximate the values to acceptable accuracy for most SEOR calculations.

Simple versions that are accurate to within a few percent in the normal pressure ranges encountered in SEOR projects are

$$T_s = 116.79 p_s^{0.2229} \text{, °F ;} \quad (15.2)$$

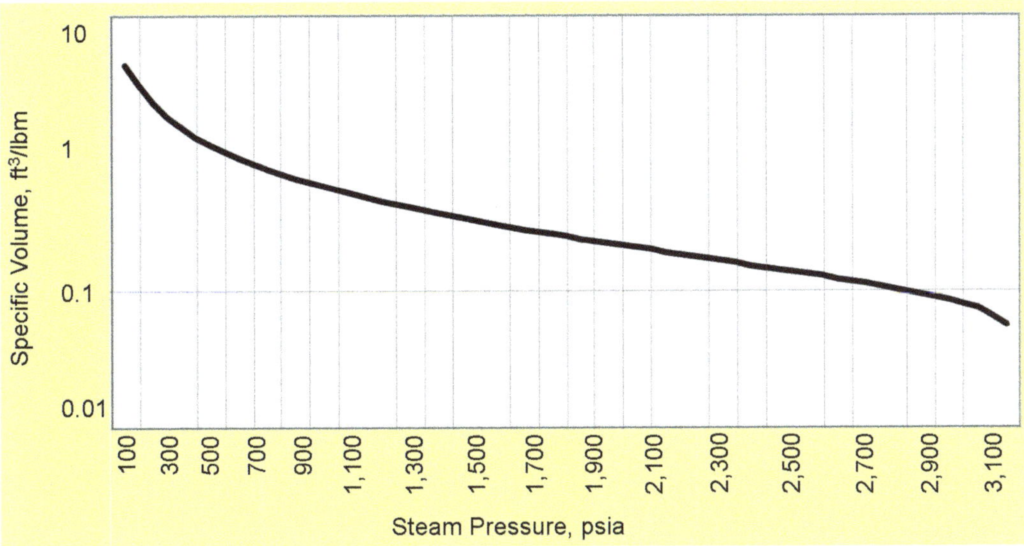

Fig. 15.2—Steam specific volume vs. saturation pressure.

$$v_s = 0.02 + f_s\left(\frac{443}{p_s} - 0.02\right), \text{ ft}^3/\text{lbm}\ ; \quad (15.3)$$

$$\rho_s = 5.06e^{0.000359}p_s - 5, \text{ lbm/ft}^3\ ; \quad (15.4)$$

$$h_f = 91p_s^{0.2574}, \text{ Btu/lbm}\ ; \quad (15.5)$$

$$h_{fv} = 1{,}318p_s^{-0.08774}, \text{ Btu/lbm}\ ; \quad (15.6)$$

$$h_v = 1{,}119p_s^{0.01267}, \text{ Btu/lbm}\ ; \quad (15.7)$$

and

$$h_{f_s} = \frac{f_s(1{,}318)}{p_s^{0.08774}} + 91p_s^{0.2574}, \text{ Btu/lbm}\ . \quad (15.8)$$

These are recommended for hand calculations or simple analytical equations. There are more precise versions in the literature[4], but for most purposes, Eqs. 15.2 through 15.8 are more accu-ate than any other available information that goes into SEOR calculations and are more than adequate for most calculations. The more accurate equations are cumbersome and lend themselves to computer applications. However, in a computer application, a lookup table is easy to create and is much more accurate than even the advanced analytical equations.

15.2 Reservoir Heating

The basis for reservoir heat-transfer calculations is traced back to the published solution[5] to a mechanical engineering problem. Marx and Langenheim[6] were the first to publish an adaptation of this solution to an SEOR process. They assumed that the equations for temperature response

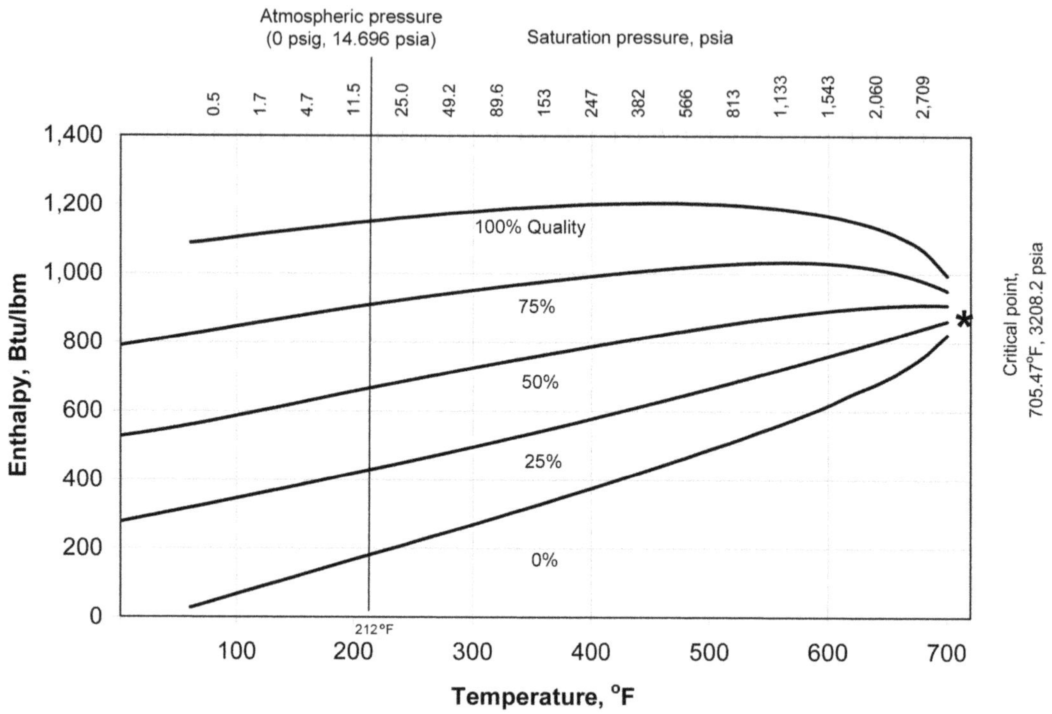

Fig. 15.3—Enthalpy of wet steam as a function of saturation pressure, temperature, and steam quality.

in a thin plate, backed in perfect contact to a semi-infinite solid after sudden exposure to constant-heat input, were analogous to steam injection into an oil-bearing reservoir.

Fig. 15.4 shows the temperature distribution anticipated in this model. The entire flood zone is at steam temperature upstream of the steam front, and the reservoir downstream of the steam front is at initial reservoir temperature. The true temperature profile is much less defined.

Their equation for the heated area, A_t, over time, t, is

$$A_t = \left[\frac{\dot{Q}_i M_R h_t \alpha}{4 \lambda_S^2 \Delta T} \right] G(t_D), \text{ ft}^2 \text{ .} \quad (15.9)$$

$G(t_D)$ from the mechanical engineering problem is a function of dimensionless time, t_D.

$$G(t_D) = e^{t_D} erfc\sqrt{t_D} + 2\sqrt{\frac{t_D}{\pi}} - 1, \quad (15.10)$$

where t_D is time of injection, t, multiplied by a few reservoir properties.

$$t_D = 4\left(\frac{M_S}{M_R}\right)^2 \frac{\alpha_S}{h^2} t, \quad (15.11)$$

and the complementary error function, $erfc(x)$, is common in heat conduction calculations. Several publications have tables of values for $G(t_D)$ vs. time, but van Lookeren[7] offers a simple equation with sufficient accuracy for most calculations, which is written as

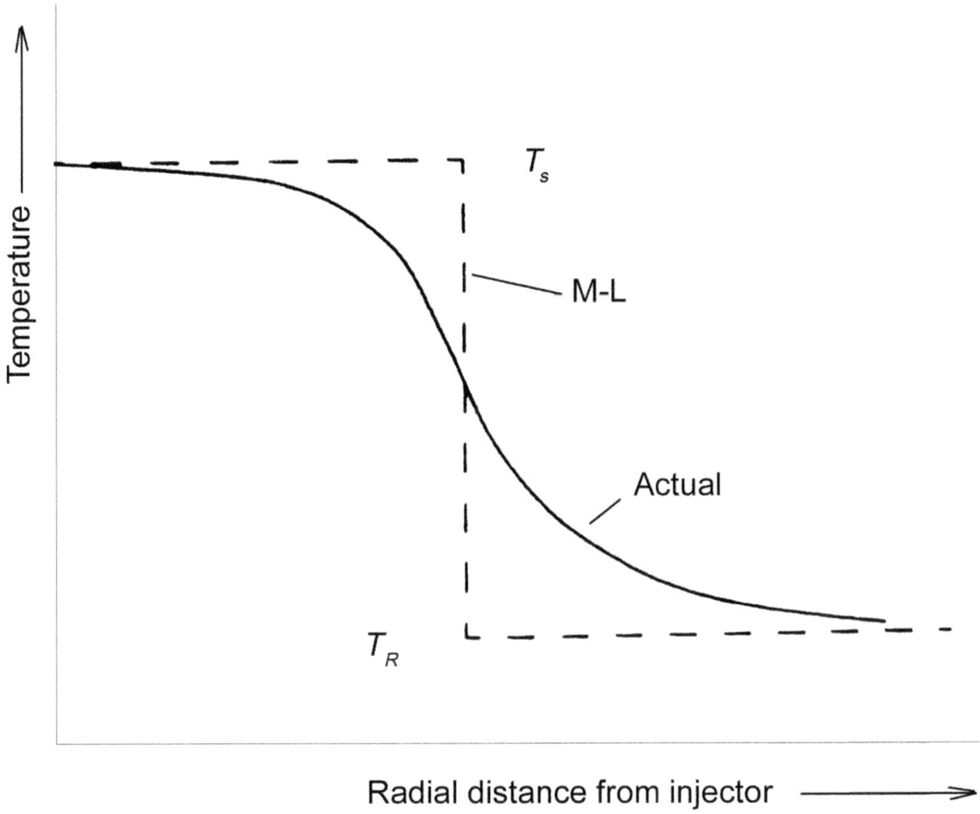

Fig. 15.4—Temperature profile resulting from convection and conduction caused by piston displacement of steam (after Marx and Langenheim[6]).

$$G(t_D) = \frac{1}{(1 + 0.85\sqrt{t_D})} \quad \text{.......} \quad (15.12)$$

Other useful relationships derived from the Marx and Langenheim equations are heat loss to the adjacent formations,

$$\dot{Q}_l = \dot{Q}_i \left[2\sqrt{\frac{t_D}{\pi}} - G(t_D) \right] ; \quad \text{.......} \quad (15.13)$$

rate of heated zone growth,

$$\frac{dA}{dt} = \frac{\dot{Q}_i}{43{,}560 \Delta T_i M_R h_t} \left[G(t_D) - 2\sqrt{\frac{t_D}{\pi}} + 1 \right] ; \quad \text{.......} \quad (15.14)$$

heat remaining in the reservoir,

$$Q = \frac{\dot{Q}_i M_R^2 h_t^2}{4\alpha_s M_S^2} G(t_D) ; \quad \text{.......} \quad (15.15)$$

cumulative heat loss to the adjacent formations,

$$Q_l = \dot{Q}_i t - Q \; ; \quad \text{(15.16)}$$

and reservoir efficiency or fraction of injected heat remaining in the reservoir,

$$E_h = \frac{Q}{\dot{Q}_i t} = \frac{G(t_D)}{t_D} \; . \quad \text{(15.17)}$$

Two constants appear in the equations that need definitions. M_R is the volumetric heat capacity of the composite formation including rock and fluids.

$$M_R = (1-\phi)M_\sigma + \phi S_o M_o + \phi S_w M_w + \phi S_g \left[f_s M_g + (1+f_s)\left(\frac{\rho_s}{\Delta T} + \rho_s C_w \right) \right] . \quad \text{(15.18)}$$

Note that there are two gas components: inert gas represented by the volumetric heat capacity, M_g, and steam represented by two terms, one owing to the latent heat of vaporization and one to the sensible heat.

Thermal diffusivity, α, is the ratio of the thermal conductivity to the volumetric heat capacity,

$$\alpha = \frac{\lambda}{\rho C} \; . \quad \text{(15.19)}$$

Farouq Ali[8] showed it is appropriate to use average values for M and α when the thermal properties of various layers of formation and over/underburden differ significantly.

Ramey[9] and Prats[10] showed that Eq. 15.17, reservoir heating efficiency, E_h, is independent of reservoir and steam zone geometry. Further the relationships apply to any form of heat transport (convection, conduction, or even radiation) in the plane of the reservoir and when the heat transport to adjacent formations is three-dimensional (3D). Assuming the reservoir properties are constant and vary little with changing temperature, E_h is solely a function of dimensionless time.

15.3 Steam Zone Growth

Reservoir heating by steam injection translates into the formation of a swept zone of some shape and an oil bank consisting of a migrating zone containing the displaced oil. Two general concepts have been proposed to describe this process.

15.3.1 Viscous Displacement. This is an extension of early water flooding concepts in which the displacement front is considered to be an advancing vertical front, the plane of which is normal to the bedding plane and extending vertically from the top to the bottom of a homogeneous flow section. Displacement of oil in this model is piston-like and is directly proportional to injection rate. The equations in the Reservoir Heating section, Sec. 15.2, have been used to calculate steamfloods as frontal advance floods.

Mandl and Volek,[11] followed by a slight modification by Myhill and Stegemeier,[12] contributed the concept of a critical time beyond which the vapor front stagnates and heat is carried only by hot water traveling through the condensation front. Dimensionless critical time, t_{cD}, is defined by

$$e^{t_{cD}} erfc\sqrt{t_{cD}} \equiv 1 - f_{hv}, \quad (15.20)$$

where f_{hv} is the fraction of heat injected as latent heat.

$$f_{hv} = \left(\frac{1 + C_w \Delta T}{f_s h_{fv}}\right)^{-1}, \quad (15.21)$$

and f_s and h_{fv} are at reservoir conditions.

$$E_h = \frac{1}{t_D}\left\{G(t_D) + (1 - f_{hv})\frac{U(t_D - t_{cD})}{\sqrt{\pi}}\left[2\sqrt{t_D} - 2(1 - f_{hv})\sqrt{t_{D-t_{cD}}} - \int_0^{t_{cD}} \frac{e^x erfc\sqrt{x}dx}{\sqrt{t_D - x}} - \sqrt{\pi}G(t_D)\right]\right\}. \quad (15.22)$$

The Marx and Langenheim equation (Eq. 15.10) is an upper bound to Eq. 15.22, as shown in **Fig. 15.5**.

15.3.2 Bypass. These models address the severe buoyancy forces that pertain when steam vapor, a gas, is injected into a liquid filled reservoir. In bypass models, the fronts are not vertical; the steam migrates quickly to the top of the reservoir. Van Lookeren[7] describes the severity of this override with a buoyancy factor.

$$A_{RD} = \left(\frac{\mu_s w_s}{\pi(\rho_o - \rho_s) g h_n^2 k k_s \rho_s}\right)^{0.5}. \quad (15.23)$$

Average steam zone thickness is

$$h_s \cong 0.5 h_n A_{RD}. \quad (15.24)$$

The shape of the steam zone is described by

$$h_s \cong h_n A_{RD}\left(\ln\frac{r_e}{r} - 0.5 + 0.5\frac{r^2}{r_e^2}\right)^{0.5}. \quad (15.25)$$

Note that steam zone thickness is not dependent on net zone thickness in Eqs. 15.24 and 15.25.

Fig. 15.6 shows the calculated injector liquid level as a function of time and five values of A_{RD}. A_{RD} is normally smaller than 0.6 in field projects; thus, the steam zone does not cover the vertical zone except for very thin reservoirs.

Neuman[13] addresses the bypass or steam override concept by basically assuming that injected steam quickly rises to the top of the zone and spreads out evenly. Thickness of the steam zone, h_s, is related to

$$\Delta h_s = \frac{4(M\sqrt{\alpha})_s C_w \Delta T}{h_v M_s}\sqrt{\frac{\Delta t}{\pi}}. \quad (15.26)$$

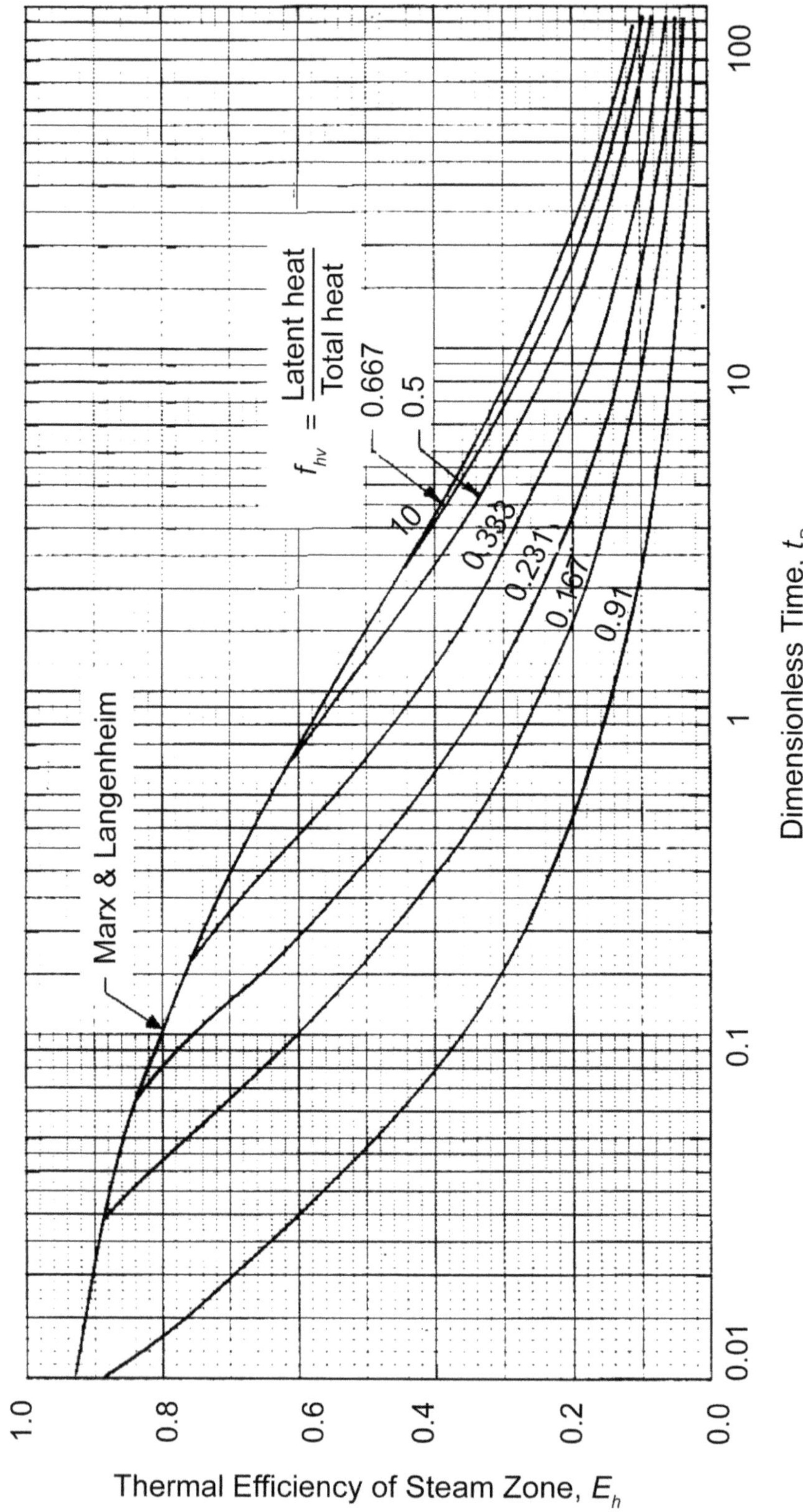

Fig. 15.5—Fraction of injected steam remaining in steam zone.[11]

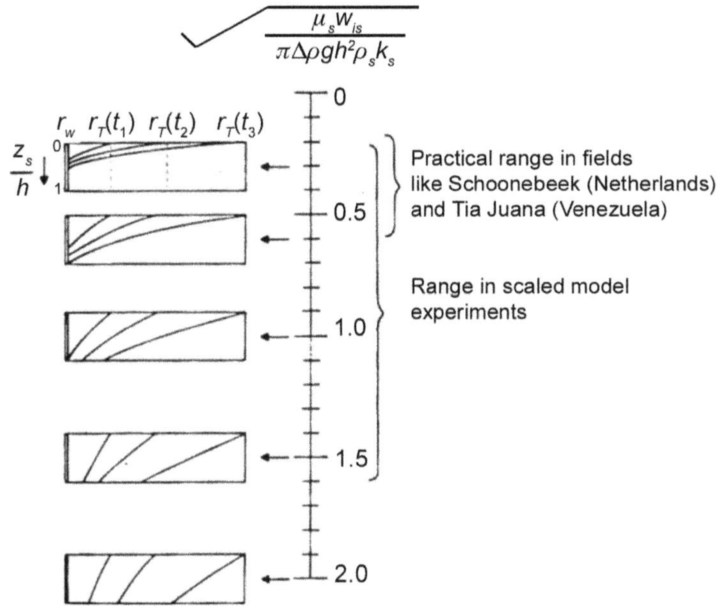

Fig. 15.6—Effect of the ratio of viscous to gravity forces on the location of steam displacement fronts.[7]

15.3.3 Combination. Field projects usually include features of both frontal advance and bypass processes. Owens and Ziegler[14] propose an analytical model that calculates the contribution of each process. They calculate total oil rate, q_o, as the sum of gravity drainage oil rate, q_{og}, after Clossman,[15] and viscous oil rate, q_{ov}, as

$$q_{ot} = q_{ov} + q_{og}, \quad (15.27)$$

$$q_{ov} = \frac{\frac{X}{Y}\left(\exp^{(-q_{ot} Y h_h)} - 1\right)(p_e - p_w)}{q_{ot}}, \quad (15.28)$$

and

$$q_{og} = \frac{\frac{X}{Y}\left(h_e(0.433\Delta\gamma)\right)\left\{1 + \left(\frac{1}{Y q_{ot} h_e}\right)\left(\exp^{(-Y q_o h_e)} - 1\right)\right\}}{q_{ot}}, \quad (15.29)$$

where

$$X = \frac{0.001127(2\pi)k\rho_s}{\mu_s \rho_o\left(\ln \frac{r_e}{r_w} - 0.5\right)}, \quad (15.30)$$

$$Y = \frac{m^*}{\alpha A \phi (S_{oi} S_{or})}, \quad (15.31)$$

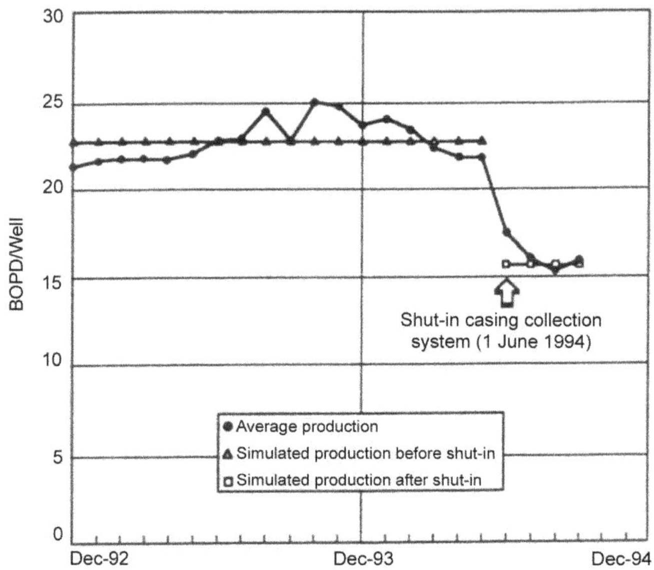

Fig. 15.7—Effect of shutting in casing vapor recovery system on a selected steamflood project.[14]

$$h_{e(n+1)} = h_{e(n)} \frac{q_o \Delta t}{7{,}758 m^* A\phi (S_{oi} \quad S_{or})}, \quad \text{(15.32)}$$

and

$$m^* = \frac{1}{2}\left\{1 + \left[1 - \frac{1}{2\ln\left(\frac{r_e}{r_w}\right) - 1}\right]^{0.5}\right\}. \quad \text{(15.33)}$$

This series of equations can be solved in a computer spreadsheet. The authors report that the viscous component of oil rate in the Kern River Ten-Pattern is significant and varied from 66% early in the project to 53% later. The method is also used to show the effect of operating with (supporting q_{ov}) and without (inhibiting q_{ov}) a casing vapor recovery system. **Fig. 15.7** shows the impact of shutting in the casing vapor recovery on a selected project. This is not to say that shutting in casing vapor recovery will have this profound negative impact on every steamflood. Notice that this method does not account for steam injection rate. A project that is significantly over-injected (high-pressure drawdown) would be expected to respond as in Fig. 15.7, while a project with more controlled steam rate (modest-pressure drawdown) may show little, if any, reduction in production rate.

Kimber *et al.*[16] found with a physical model that there is a viscous component of steamflooding that has potential to add significant oil recovery. **Fig. 15.8** shows that there is an optimal steam quality for a steamflood and that some amount of overinjection may be beneficial also. Ward and Sharpe[17] studied the subject at the Kern River field using a numerical simulator and found that shutting in the casing vents resulted in a permanent 18% loss in oil production rate as shown in **Fig. 15.9**. Oil production could be restored if either the vents were reopened or a depleted zone in the wellbore were perforated, allowing the excess steam to vent by crossflow.

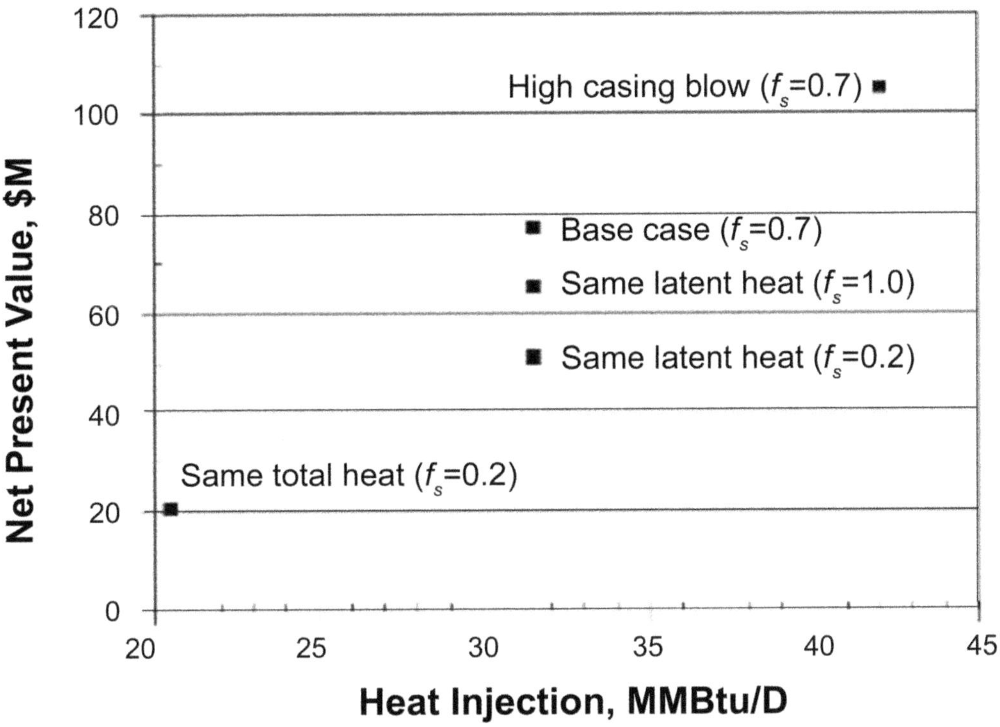

Fig. 15.8—The effect of latent heat injection rate on strategy economics.[16]

15.4 Steamflood Design

15.4.1 Candidate Selection. Screening criteria for identification of steamflood candidates have been published for many years. **Table 15.1** shows the screening guides from five different sources.[18–22]

It is obvious from Table 15.1 that there is a finite envelope of properties that define successful candidates. However, within that envelope there is a relatively wide spread of values for the indicators. The reason for this is that each reservoir is unique and success is a function of a combination of all of the screening criteria plus a myriad of other considerations. The authors of the papers (Refs. 18 through 22) typically offer linear regression equations to generate an indicator for a specific reservoir. The most recent version is by Donaldson,[23] which is written as

$$F_{os} = 0.0384 + 0.643\phi S_o + 6.48 \times 10^{-6}\left(\frac{kh_n}{\mu_{oi}}\right)D/(q_{i_s} A^{0.5}) \; . \qquad (15.34)$$

Doscher and Ghassemi[24] showed that there is an upper oil viscosity barrier that makes economic recovery of very viscous oils very difficult by conventional SEOR methods. Using scaled physical models, they found that for oils with a viscosity much greater than that in the Kern River or Midway Sunset fields in California, the steam/oil ratio for conventional steamflood or cyclic steam stimulation was uneconomic. The 14 and 12° API lines on **Fig. 15.10** show the general upper bound range for oil viscosity in a successful conventional steamflood. They proposed a correlation equation to estimate the limiting viscosity. The correlation is a function of steam and oil viscosities.

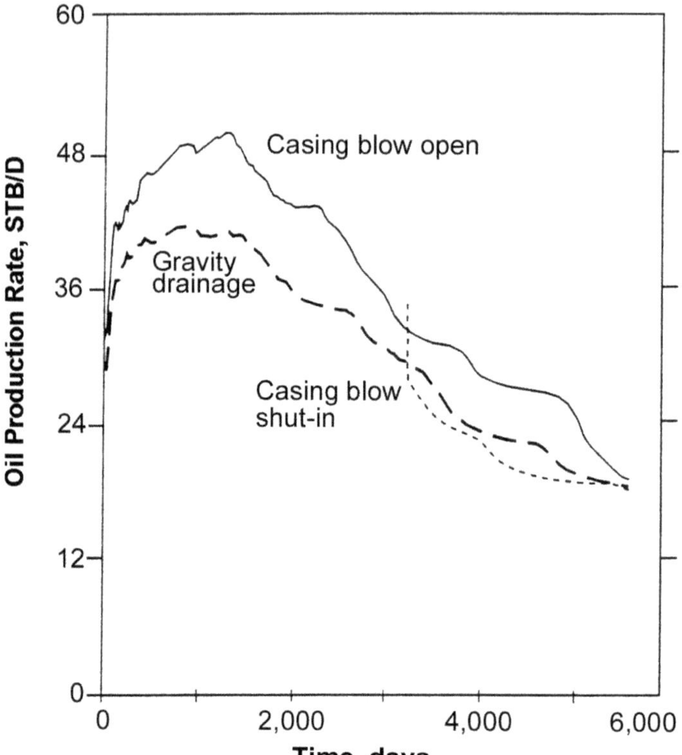

Fig. 15.9—Results of a numerical simulation showing 18% loss in production resulting from shutting in casing vapor recovery.[17]

TABLE 15.1—SEOR PROJECT SCREENING CRITERIA[23]

Reference	Ref. 18	Ref. 19	Ref. 20	Ref. 21	Ref. 22
Depth, ft	<3,000	<2,500	<3,000	>20	>400
Net thickness, h, ft	>20	>30	>30	>20	>10
Porosity, ϕ	>0.2	>0.3	>0.3		>0.2
Oil saturation, S_o		>0.5			>0.4
$\phi \times S_o$	>0.1		0.15–0.22	>0.0	>0.8
Oil gravity, API	10–34	8–40	12–15	<25	<36
Permeability, k, md	>250	>1,000	>1,000		
Oil viscosity, μ, cp	<15,000	200–1000	<1,000	>20	
Transmissibility, hk/μ, md-ft/cp	>5	>50			
Initial pressure, psia	<1,500				
Pattern size, acre		<10			

$$F_{os} = 5.12\left[f_s\left(\frac{\mu_s}{\mu_{oi}}\right)\right]^{0.57} \quad \text{. ......} \quad (15.35)$$

Other critical requirements exist such as reservoir continuity between injector and producer and barriers to contain steam from uncontrolled migration to nonsteam-bearing intervals. Qualitatively, one looks for relatively shallow, low pressure, low temperature, thick, high porosity, highly permeable reservoirs with medium to high saturation of high viscosity liquid oil.

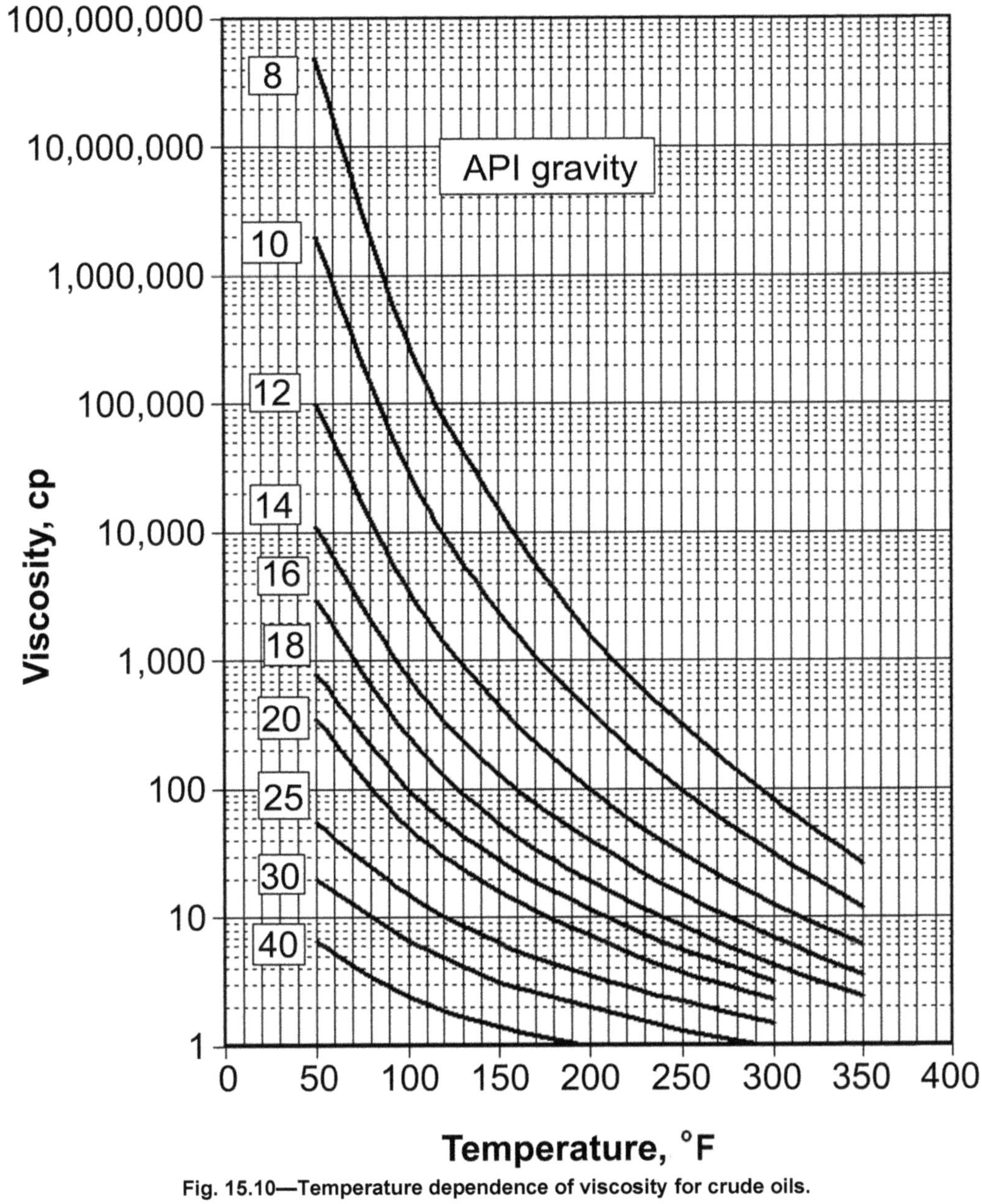

Fig. 15.10—Temperature dependence of viscosity for crude oils.

All of the following applies to oil sands with mobile oil. The SEOR process in reservoirs that are fractured during the steam injection process are too complicated to be calculated with analytical models.

15.4.2 Analogy. The simplest way to design an SEOR project is by analogy. If there is an analogous project in the same field or in a field with enough similarities to the one in question, simply using the design and results from the former may be adequate. **Fig. 15.11** from a paper by Greaser *et al.*[25] in the Kern River field, California is an example of the use of analogy. In that project there are thousands of steamflood injection zones. Texaco made good use of the

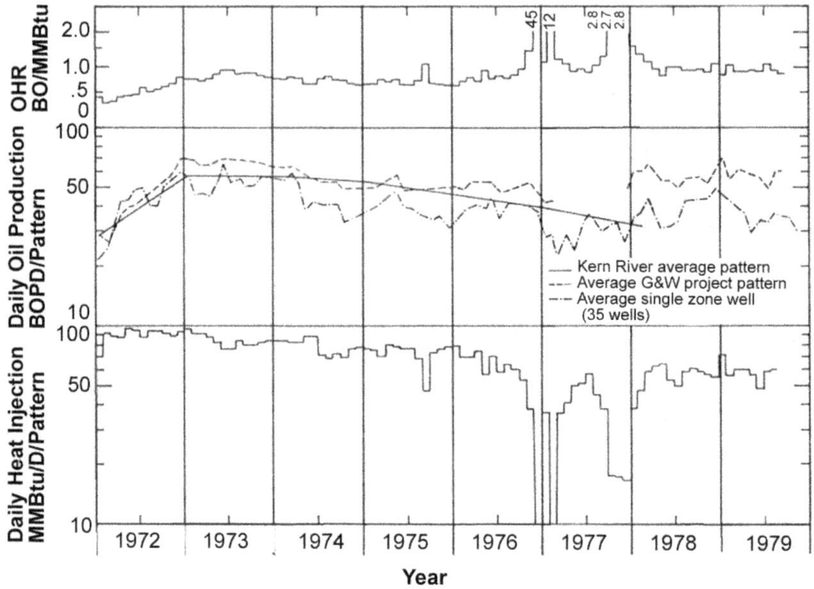

Fig. 15.11—Kern River "average pattern" used as an analogy to evaluate field performance.[25]

huge amount of steamflood performance data by calculating an "average pattern response" and using that as the standard to design and monitor future flood patterns.

15.4.3 Recovery Mechanisms. Waterflooding, successful in reservoirs with low viscosity/high mobility crude oil, is extremely inefficient in reservoirs with low mobility/high viscosity crude oil. Three dimensionless flow parameters help us understand the problem.

$$\frac{Lg\Delta\rho\cos\theta}{\Delta p},\ \frac{\sigma\cos\theta_c}{\sqrt{k\Delta p}},\ \text{and}\ \frac{k\Delta p}{\mu Lu}\ . \qquad (15.36)$$

These are, respectively, the ratio of gravity forces to applied pressure; the ratio of capillary forces to applied pressure; and the ratio of applied pressure to viscous forces. The ability to modify the variables in the above relationships is indicative of the potential success of extracting oil from the reservoir.

In the gravity ratio, the main controllable parameter is the distance between the injector and producer. The only other potential variable is density difference between injectant and the crude oil, which is large but does not change appreciably.

$\sigma\cos\theta_c$ in the capillary pressure-applied pressure ratio represents the interfacial tension of the crude oil in the rock pore and can be varied. The obvious goal is to reduce it to zero. Several approaches have been tried, including adding emulsified chemicals in the steam to form a single-phase water/oil emulsion, which has lower viscosity and lower interfacial tension than the crude alone. Many field trials have been tried over the years but have yet to prove economical.

The viscosity parameter in the third ratio has been the most important in designing a successful SEOR project. Other methods of reducing oil viscosity, such as miscible solvent injection, have been tried but have proven to be far inferior to heat injection. Fig. 15.10 shows the viscosity vs. temperature relationship for a few representative crude oils. Raising the temperature of the oil from the typical reservoir temperature of less than 100°F to approximately

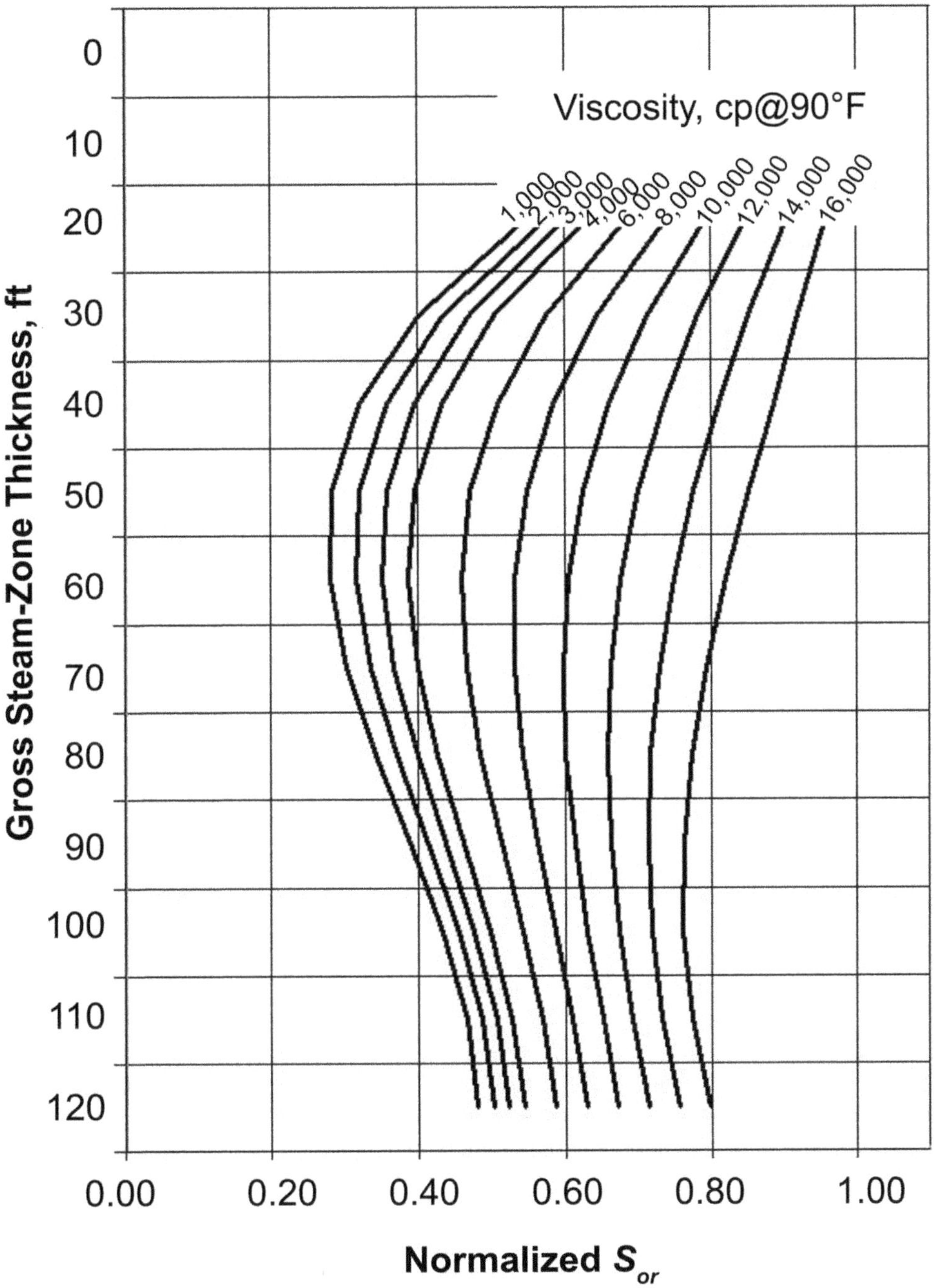

Fig. 15.12—Graph of a dimensionless function F_{Sor}, which relates residual oil saturation to gross section thickness and presteam oil viscosity.

300°F gives a viscosity reduction of orders of magnitude. Of all of the potential heat-carrying media, water-based steam is inexpensive and universally available and has the highest heat-carrying capacity of any compound; it exists at the ideal 300 to 500°F temperature range to achieve optimal oil viscosity reduction.

Texaco[26] published a correlation that estimates the residual oil saturation to steam based on the huge database generated over the decades of steamflood in the Kern River field. They found the residual saturation is a function of gross zone thickness and presteam oil viscosity. There appears to be an optimal zone thickness, above and below, where residual oil is less attractive. Residual oil saturation continuously increases for increasing viscosity, reaching a point of immobile oil at some high value. Their data was based on an average initial oil saturation of 55%. **Fig. 15.12** shows their correlation, normalized for any initial oil saturation. Short of actual core or laboratory data, that figure can be used to estimate residual oil saturation by finding the value of the factor for the desired initial oil saturation and zone thickness and then multiplying that factor by initial oil saturation. The equations presented next provide an alternative method for obtaining the same information.

$$S_{or} = F_{S_{or}} S_{oi}, \quad (15.37)$$

where

$$S_{or_b} = \frac{(49.9876 - 1.626h_t + 0.02155h_g^2 - 0.0000823h_t^3)}{50}, \quad (15.38)$$

$$S_{or_a} = \frac{(50.5357 - 0.20877h_t - 0.000260989h_t^2 + 0.00000991h_t^3)}{50}, \quad (15.39)$$

and

$$F_{S_{or}} = S_{or_a} + S_{oi}\left(S_{or_b} - S_{or_a}\right)\left(\frac{\mu_{oi}}{15{,}000}\right). \quad (15.40)$$

This is instructive as to the qualitative relationships between the parameters, but caution should be used when applying it quantitatively outside of the Kern River field. Also, the authors explain that they had no information for zones thicker than 120 ft or oil viscosity greater than 16,000 cp.

Other mechanisms include formation compaction, in-situ steam distillation generated solvent banks, and formation and fluid heat-induced swelling, but they are rarely necessary to supplement the previous ones in adequately describing the SEOR process. Of these secondary mechanisms, compaction has had the most significant impact on SEOR. The most notable regions to benefit are the Bolivar Coast, western Venezuela, and the Long Beach area near Los Angeles, California. The Bolivar Coast has been using cyclic steam stimulation (CSS) since the early 1960s. Rattia and Farouq Ali[27] published a study that concluded that formation compaction enhanced process efficiency in CSS but harmed process efficiency in steamflood.

15.5 Design Calculations: Viscous Displacement Models

15.5.1 The Marx-Langenheim[6] and Mandl-Volek[11] (M-V) Models. These steam zone growth models have often been used to calculate cumulative oil recovery and steam zone size over time. They are a simple way to get a quick estimate of project viability. Volume of steam zone is proportional to the fraction of heat remaining in the steam zone, E_h.

$$V_s = \frac{Q_i E_h}{43{,}560 M_R \Delta T} \quad \text{(15.41)}$$

It follows that oil displaced from the steam zone is

$$N_d = 7{,}758\phi \frac{h_n}{h_t}\left(S_{oi} - S_{or}\right)E_c V_s \quad \text{(15.42)}$$

E_c is an arbitrary "capture" factor that is inserted to "scale" invariably optimistic oil volume to realistic values. This factor is best determined by history matching the equation to field project and normally has a value from 0.7 to 1.0. This represents a serious limitation in calculating oil recovery because the calculation predicts the highest oil production rate at the beginning of the project.

This method is most useful in calculating steam zone size and extent. It is less useful for calculating oil rates and recovery because it does not account for the terms in the ratios in Eq. 15.36.

The calculation allows any value for steam injection rate and calculates oil rate. Information on practical steam rates must be found before practical results can be derived. Either method can be done by hand or in a computer spreadsheet.

15.5.2 The Jones[28] Model. The Jones model extended the M-V model by accounting for the third dimensionless factor in Eq. 15.2 and by honoring the oil in place. The model modifies the former model by calculating the delayed oil response to a growing steam chest using the third term in Eq. 15.36, which shows that well spacing and oil viscosity are important parameters. Also addressed is the often-present depletion in the form of in-situ gas and depletion gas cap, both of which must be filled with steam before oil can be displaced. M-V steam zone growth rate is converted to an oil production rate by multiplying by three dimensionless factors.

$$q_{o_h} = q_{oD} A_{cD} V_{oD} V_{pD}, \quad \text{(15.43)}$$

where

$$q_{oD} = \frac{N_{d(n)} - N_{d(n-1)}}{\Delta t}, \quad \text{(15.44)}$$

$$A_{cD} = \left\{ \frac{A_t}{A\left[0.11 \ln\left(\frac{\mu_{oi}}{100}\right)\right]^{0.5}} \right\}^2, \quad \text{(15.45)}$$

$$V_{oD} = \left(1 - \frac{N_p S_{oi}}{N(S_{oi} - S_{ors})}\right)^{0.5}, \quad \text{(15.46)}$$

$$V_{pD} = \left(\frac{5.62 V_s}{43{,}560 h_s \phi S_g}\right)^2, \quad \text{(15.47)}$$

$$N = \frac{43{,}560 A h_n \phi S_{oi}}{5.62}, \quad (15.48)$$

and

$$N_P = \sum_{t=0}^{t=n} q_{oh} \,. \quad (15.49)$$

The method can be calculated in a spreadsheet by first calculating a displaced oil rate from Eq. 15.44 for a time period, Δt; $\frac{1}{12}$ year is convenient. That volume is then multiplied by the dimensionless modifiers in Eqs. 15.45 through 15.47.

A_{cD} accounts for the viscosity of the oil and the size of the well spacing and has these restrictions: $0 \leq A_{cD} \leq 1.0$, and $A_{cD} = 1.0$ at $\mu_{oi} \leq 100$ cp.

V_{pD} accounts for reservoir fill-up that must occur because of in-situ gas saturation or depleted gas zones before steam zone growth can begin to displace oil. Restrictions are $0 \leq V_{pD} \leq 1.0$, and $V_{pD} = 1.0$ at $S_g = 0$.

The M-V method allows the steam zone to grow to an indefinite size for an indefinite time. Oil recovery can amount to more than original oil in place. V_{oD} is used to limit predicted oil recovery to some fraction of the original oil in place with this limit; $0 \leq V_{oD} \leq 1.0$.

Fig. 15.13 shows the results of the M-S viscous displacement model with and without the Jones corrections for a hypothetical steamflood. In this example Kern River California reservoir properties were used. A small pattern size was chosen to illustrate that the M-V displacement rate is independent of oil in place and pattern size. At 75 months, the M-V displacement was 132% of original oil in place, while the Jones model corrected it to 91%. This calculation was done to an unrealistic terminal point, but if the flood were terminated at 48 months when instantaneous oil/steam ratio had fallen to below 0.2 then recovery calculated by Jones would be 77%.

This method is very good for estimating oil rates, especially early in a steamflood and for incremental recovery. It has been used in many projects worldwide with success. As with the M-V method, this calculation allows any value for steam injection rate and calculates oil rate. Information on practical steam rates must be found before practical results can be derived. Because this method produces best results if calculated in small time steps (i.e., 10 days), it is best programmed on a computer.

15.6 Design Calculations: Gravity Drainage Models

15.6.1 The Neuman[13] Model. The Neuman method is very useful for calculating post-steam breakthrough performance and for heat management in calculating required steam injection rates. It does not address the presteam breakthrough period. Van der Knaap[29] shows that the Neuman method can be derived from the Mandl-Volek method, and they are completely compatible. Note that Neuman's paper[13] contains typographical errors in Eqs. D-2 and D-3. They are corrected next.

Steam-zone thickness is calculated as a function of the time lapse between the time of prediction, t, and the time when the location reached steam temperature, τ.

$$h_s = \left(\frac{4 k C_w \Delta T}{M_s h_{fv}} \right) \sqrt{\frac{t - \tau}{\pi \alpha_S}} \,. \quad (15.50)$$

An estimation of the steam-injection rate and the time required to achieve the required complete coverage of the project area with a steam zone is given by

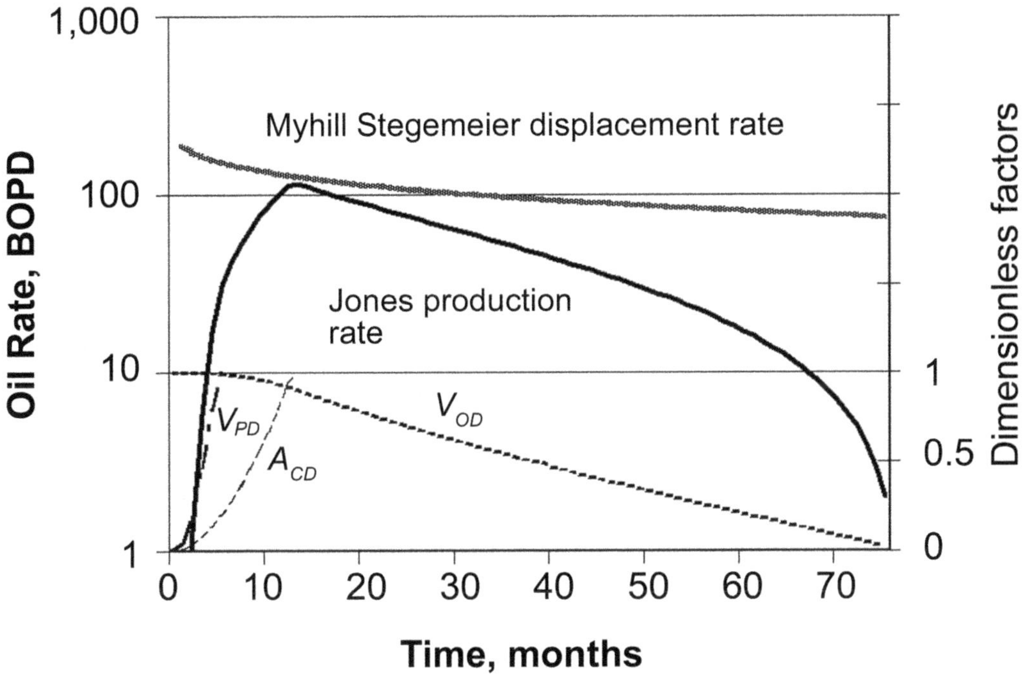

Fig. 15.13—Graphical presentation of Eq. 15.39.[28]

$$q_{i_S} = \frac{Ak\Delta T}{5.615\rho_w h_{fv} f_s (1 - f_p)\sqrt{\frac{\alpha_S t}{\pi}}} \text{ .} \quad (15.51)$$

Because everything in Eq. 15.51 is a constant except time (t), an estimate of required rate is easy to derive by simply changing the time. Caution should be used because this time/rate is divorced from reservoir reaction, primarily to viscous forces during these early stages of a flood; therefore, impractical results are easily obtained.

The steam-zone volume is

$$V_{sz} = f_s(1 - f_p)q_{i_s}\frac{\rho_w C_w}{M_s}t \text{ .} \quad (15.52)$$

Oil displaced from both the steam zone and the adjacent hot-water zone, heated by condensate convection and by conductive heat lost from the steam zone, is

$$V_o = (1 - f_b)\phi(S_o - S_{ors})f_s(1 - f_p)q_{i_S}\frac{\rho_w C_w}{M_s}t, \quad (15.53)$$

where oil rate from the hot-water zone is expressed as a fraction (f_b) of the oil from the steam zone.

$$f_b = \frac{f(T)\sqrt{\pi}(S_o - S_{orw})M_s h_{fv}}{2(S_o - S_{ors})M_w C_w \Delta T}, \quad (15.54)$$

and

$$f(T) = 0.8864 - 0.9288\left(\frac{T^* - T_R}{\Delta T}\right) + \frac{0.333}{\left(\frac{T^* - T_R}{\Delta T}\right)}. \quad (15.55)$$

Once steam covers the entire area, steam injection can be reduced by

$$\dot{Q}_i(n = t^*) = \dot{Q}_i(n = 0)\frac{2}{\pi}\sin^{-1}\left(\frac{t^*}{t}\right)^{0.5}, \quad (15.56)$$

where

$$\dot{Q}_i(n) = 350 q_{i_s}(n) f_s (1 - f_p). \quad (15.57)$$

This method can be done by hand but is easier to use if programmed on a computer spreadsheet.

15.6.2 The Vogel[30] Model. Similar to the Neuman method, the Vogel method is very useful for post-steam breakthrough heat-management calculations. It does not address early steam-flood performance. Vogel first postulates that a producing well in a California-type reservoir with thick, dipping zones and gas caps that are incapable of maintaining pressure will produce at a maximum rate regardless of excess steam injection over some optimal rate. Thus, the steam rate can be calculated to maintain a steam chest by replacing lost heat and the voidage left by produced fluids. To use this method, a good estimate of a hot well producing rate is necessary, but it is very useful for optimizing existing projects with established producing rates.

A modified interpretation of the M-V method is used to calculate heat necessary to overcome reservoir heat loss. Vogel uses the following equations to calculate the steam injection rate in barrels per day.

$$q_{i_s} = (q_{ls} + q_{p_s})h_{f_s} + q_{\text{excess}_s}, \quad (15.58)$$

where

$$q_{l_s} = \frac{2A\lambda_S \Delta T}{0.319 \times 10^6 \sqrt{\pi \alpha_S t}}, \quad (15.59)$$

and

$$q_{p_S} = \frac{(5.615)M_s \Delta T(q_{oh} + q_{oc})}{0.319 \times 10^6 \phi (S_{oi} - S_{or})}. \quad (15.60)$$

q_{excess_s} is a value added that accounts for surface and wellbore and miscellaneous heat losses.

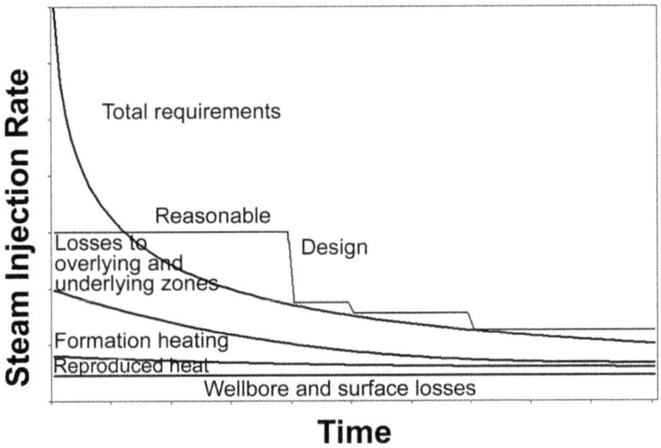

Fig. 15.14—Typical steam-injection rate schedule for gravity-dominated steam displacement.[30]

Fig. 15.14 shows the main benefit provided by the Neuman and Vogel methods. They are very useful in managing steam-injection requirements that continuously reduce as the project matures.

These equations apply only after steam has overlaid the reservoir and oil production has peaked because oil rate is known and steam rate is unknown in the equations. This method is easily done by hand but can also be programmed on a computer spreadsheet.

15.6.3 Discussion of Steam Rate Reduction. Kumar and Ziegler[31] investigated the issue of steam rate reduction schedule using a numerical reservoir simulator. The Neuman and Vogel methods previously described are based on analytical heat balances and do not directly predict oil production rates. **Fig. 15.15** shows the effect on oil production rate determined using the simulator as a result of three steam-injection schedules: constant rate, linear constant reduction, and Neuman reduction schedule. **Fig. 15.16** shows the steam injection for each of these schedules. The constant rate schedule results in the highest oil-production rate but at the cost of high late-steam rates. The linear reduction schedule yields nearly the same oil for a lot less steam. Neuman results in arrested steam-zone growth, but the severe early reduction in steam results in an equally severe loss in oil rate.

Table 15.2 shows that the linear rate-reduction compromise is economically superior to either of the other two schedules. Because the injection rate reduction changes net salable oil at various times in the project life, the authors used discount factors to properly value the oil production stream for each case. They termed this time-weighted oil volume "discounted net present barrels of oil."

15.7 Design Calculations: Other Models

15.7.1 The Gomaa[32] Model. Gomaa approached the problem by first doing a history match on a Kern River field steamflood using a numerical simulator then extracting an analytical solution from the simulator results.

Because the method is totally a function of a history match to the Kern River field, caution must be used in applying it to other projects. However, because it is a graphical method with few calculations, it offers ease in estimating project performance. After achieving a history match, Gomaa did several parametric simulations that allowed generation of several graphs that are, in turn, used to estimate project performance. Important parameters for the method are initial reservoir pressure (< 100 psia); steam injection pressure (< 200 psia); reservoir thickness

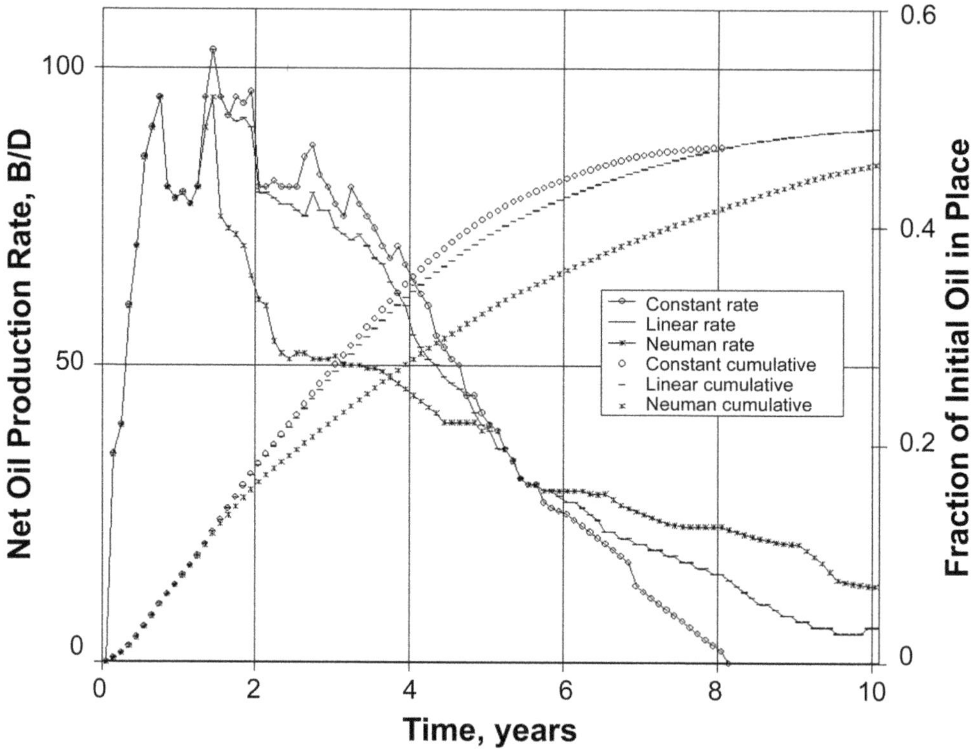

Fig. 15.15—Effect of steam injection schedule on oil production.[31]

(10 to 300 ft); mobile oil saturation (0.05 to 0.60); porosity (0.21 to 0.35); and heat injection rate (0.05 to 0.6 MM Btu/D/acre-ft). The steps for the procedure are outlined next.

- Provide downhole steam quality, pressure, and injection rate and determine enthalpy, h_{fs}.
- Calculate the heat injection rate,

$$\dot{Q}_{s_v} = 350 \times 10^{-6} \frac{q_{i_s} h_{f_s}}{A h_t} \ . \qquad (15.61)$$

- Determine vertical heat loss, Q_l, from **Fig. 15.17**.
- Determine the heat utilization factor from

$$f_h = 0.793 + 1.074 f_s - 1.737 f_s^2 + 0.628 f_s^3 \ . \qquad (15.62)$$

For a series of time steps (t, days), calculate the effective heat injection rate and the cumulative heat injected with

$$Q = \Delta t \dot{Q}_{s_v} f_h (1 - \dot{Q}_l), \qquad (15.63)$$

and

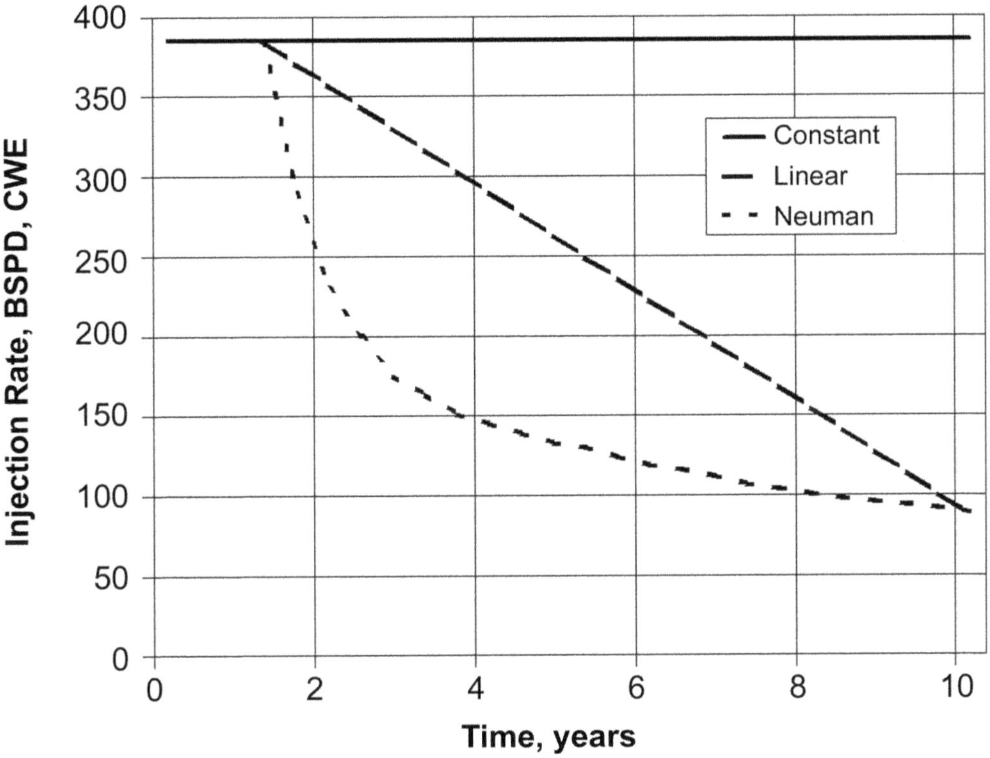

Fig. 15.16—Steam injection schedule.[31]

TABLE 15.2—KUMAR AND ZIEGLER[31] DISCOUNTED NET PRESENT BARRELS OF OIL (CONFINED MODEL)			
	Discount Rate		
Injection Schedule	0%	5%	10%
Constant rate	19,290	16,780	14,690
Linear reduction	19,770	16,960	14,690
Neuman reduction	18,360	15,410	13,120

$$\sum Q = \sum_n \dot{Q}_{s_{vn}} f_{h_n} (1 - \dot{Q}_{\ln}) \Delta t_n \; . \qquad (15.64)$$

• Estimate the mobile oil saturation,

$$S_{om} = S_{oi} - S_{ors} \; . \qquad (15.65)$$

• Use **Fig. 15.18** to determine the oil recovery factor, f_R.
• Calculate the initial mobile oil in place with

$$N_m = \frac{7{,}758 \phi S_{om} h_n A}{B_o} \; . \qquad (15.66)$$

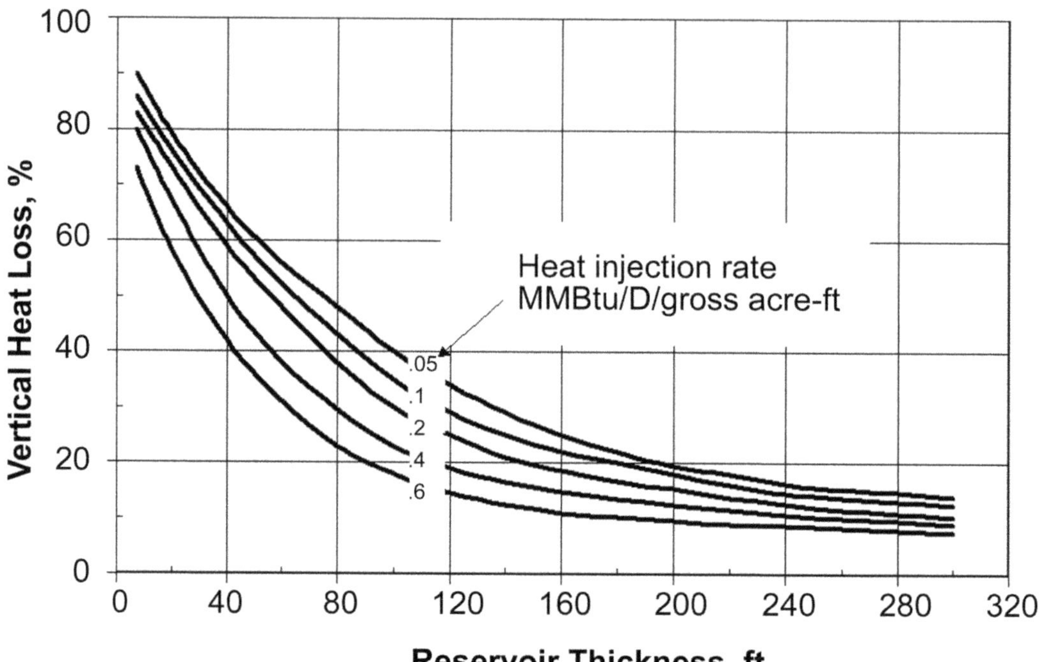

Fig. 15.17—Heat loss to overlying and underlying strata.[32]

- Calculate the oil recovery at time (t),

$$N_p = N_m f_R \text{ . } \qquad (15.67)$$

- Calculate the cumulative oil/steam ratio,

$$F_{os} = \frac{N_p}{\Delta t q_{i_s}} \text{ . } \qquad (15.68)$$

Repeat these steps for a period equal to the expected life of a steamflood in convenient time steps (i.e., 365 days) to estimate steamflood performance. This calculation method is easily done by hand but can be programmed in a computer spreadsheet.

15.8 Cyclic Steam Stimulation Design

Prats[1] defines stimulation as "any operation (not involving perforating or recompleting) carried out with the intent of increasing the post-treatment production rate without changing the driving forces in the reservoir." Periodic injection of steam into a producing well, alternating with a production cycle, has many features of this definition but also has many features that distinguish it as a true enhanced recovery mechanism. The primary benefit of the process is true stimulation—near wellbore reduction of flow resistance, viscosity reduction. However, there are EOR benefits of high-temperature gas dissolution, wetability changes, and relative permeability hysteresis (water flows into the reservoir easier than it flows out). Fortunately, calculating the temperature history of the wellbore, tracking the water/oil saturation history and the oil viscosity reduction is adequate to estimate the oil production response to the process.

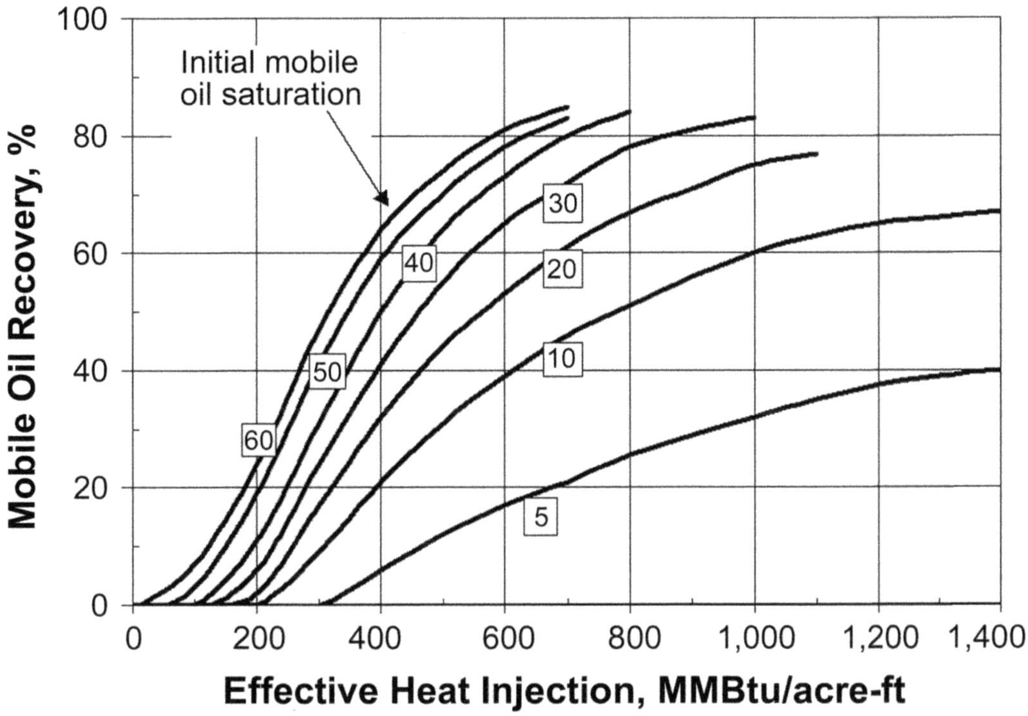

Fig. 15.18—Steamflood oil recovery as a function of effective heat injected and mobile oil saturation.[32]

15.9 Design Calculations

Steamflood design is simple compared to cyclic steam stimulation (CSS) design. Whereas steamflood reaches equilibrium and can be represented by a set of steady-state equations for much of its life, the CSS process is one of constantly changing conditions. First there is the injection phase, which is relatively so short that it is a total transition period. Then during the soaking period, steam vapor condenses and temperature begins to fall. The producing period is in a constant state of flux as testified by the constantly changing producing rates. Relative permeability curves, which can typically be ignored in steamflood calculations, become very important to CSS.

In spite of these problems, there are several desktop calculations that give a good representation of what can be expected from CSS. Probably the simplest representation of the process is by Owens and Suter,[33]

$$\frac{q_{oh}}{q_{oi}} = \frac{\mu_{oi}}{\mu_{oh}} . \qquad (15.69)$$

This simply indicates the productivity ratio resulting from steam temperature-induced oil-viscosity reduction. No attempt was made to calculate how the reservoir got the peak temperature, but once the well is steamed and placed on production, the authors propose that the operator can simply watch leadline temperature and accurately predict the production history of the production period prior to the next cycle.

15.9.1 The Boberg and Lantz[34] Method. The referenced paper describes the definitive work that serves as the basis of virtually all subsequent analytical analyses of CSS. They first calcu-

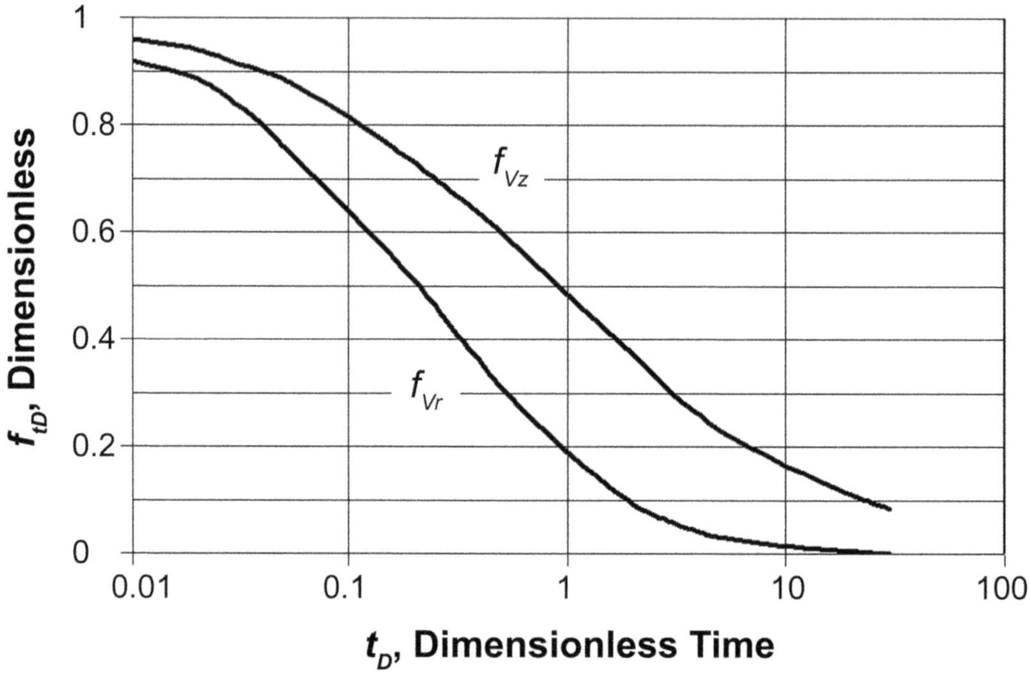

Fig. 15.19—Vertical and radial heat-loss factors for Eq. 15.65.[34]

late the reservoir temperature distribution resulting during the injection period. Eq. 15.9 is used to calculate the area of the processed zone that is heated to T_i. Then, the well is placed on production and temperature of the heated volume, which is assumed to remain constant and begins to fall by conduction to the surrounding cold reservoir rock and by hot fluid production. The average temperature in the hot zone is

$$\bar{T} = T_R + \Delta T\left[f_{Vr} f_{Vz}(1 - f_{pD}) - f_{pD}\right], \quad \text{(15.70)}$$

where f_{Vr} and f_{Vz} are unit solutions of component conduction in the radial and vertical directions, respectively. They can be estimated from **Fig. 15.19** or from

$$f_{Vr} = 0.92e^{\frac{(\ln t_D + 4.6)^2}{-13.5}}, \quad \text{(15.71)}$$

and

$$f_{Vz} = 0.96e^{\frac{(\ln t_D + 4.4)^2}{-27}}. \quad \text{(15.72)}$$

The term f_{pD} accounts for heat removed with produced fluids.

$$f_{pD} = \frac{1}{2Q}\int_0^t \dot{Q}_p dt, \quad \text{(15.73)}$$

TABLE 15.3—BOBERG AND LANTZ[33] RADIAL FLOW FACTORS

Flow Conditions	F_1	F_2
Steady state	$\dfrac{\ln\left(\dfrac{r_h}{r_w}\right)+S_h}{\ln\left(\dfrac{r_e}{r_w}\right)+S}$	$\dfrac{\ln\left(\dfrac{r_e}{r_w}\right)}{\ln\left(\dfrac{r_e}{r_w}\right)+S}$
Semisteady state	$\dfrac{\ln\left(\dfrac{r_h}{r_w}\right)-\dfrac{r_h^2}{2r_e^2}+S_h}{\ln\left(\dfrac{r_e}{r_w}\right)-\dfrac{1}{2}+S}$	$\dfrac{\ln\left(\dfrac{r_e}{r_h}\right)-\dfrac{1}{2}+\dfrac{r_h^2}{2r_e^2}}{\ln\left(\dfrac{r_e}{r_w}\right)-\dfrac{1}{2}+S}$

and

$$\dot{Q}_p=\left[5.615\left(q_oM_o+q_{wh}M_w+q_sM_w+\frac{q_s\rho_wh_{fv}}{\Delta T}\right)+(10^3q_{gh}M_g)\right]\Delta T\ . \quad (15.74)$$

The subscript, *h,* indicates that the properties should be for fluids from the hot zone at the sand face. The model does not predict steam, gas, or water producing rates, which must be estimated from some other source. Oil production rates are given by a method similar to Eq. 15.69, which is written as

$$\frac{q_{oh}}{q_{oi}}=\frac{J_h}{J}, \quad (15.75)$$

and

$$\frac{J_h}{J}=\frac{1}{(\mu_{oh}/\mu_{oi})F_1+F_2}\ . \quad (15.76)$$

F_1 and F_2 are radial flow factors for which Boberg and Lantz give expressions in **Table 15.3**. Note that the production rate is a function of only two variables—oil viscosity and the heated radius.

The method can be calculated by hand for a very few time steps, but it is much easier to use if programmed into a spreadsheet.

15.9.2 The Towson and Boberg[35] Model. The Boberg and Lantz method assumes that there is significant reservoir energy to produce oil under primary conditions. Because many CSS candidates have only gravity forces and initial viscosity is high, there is no significant primary production. Many California reservoirs have free liquid surfaces in the oil zones with a gas oil interface at atmospheric pressure. Towson and Boberg extended the former work to cover this situation. Eq. 15.70 is used to calculate the heated zone temperature from which oil viscosity is estimated. Then, gravity drainage oil rate may be calculated.

$$q_{o_h} = \frac{\pi(1.127)\rho_o g k k_{ro}(h_h^2 - h_w^2)}{\mu_{o_h}\left(\ln \frac{r_e}{r_w} - \frac{1}{2}\right)} \quad \text{(15.77)}$$

h_h must be computed for each time step during the production cycle by first calculating the average hot-zone fluid level.

$$\Delta h_h = \frac{\dot{Q}_o \Delta t(5.615)}{\pi\left(r_h^2 - r_w^2\right)\phi(S_{oi} - S_{or})} \quad \text{(15.78)}$$

Now the fluid level at the heated zone radius is

$$h_h = 4\Delta h_h\left(\ln \frac{r_h}{r_w} - \frac{1}{2}\right) - \sqrt{16\left(\Delta h_h\right)^2\left(\ln \frac{r_e}{r_w} - \frac{1}{2}\right)^2 - h_w^2 - 8\left(\Delta h_h\right)^2\left(\ln \frac{r_e}{r_w} - \frac{1}{2}\right)} \quad \text{(15.79)}$$

This procedure can be hand calculated but is much easier to use if a computer spreadsheet is used.

15.9.3 The Jones[36] Method. Jones took a similar approach to Towson and Boberg[35] in calculating oil rates as a function of gravity forces alone. He extended the model by also calculating heated-zone water rate. Information on relative permeability is necessary to accomplish this. Further, recognizing that Towson and Boberg and other similar models commonly over-predict oil production, he limited the vertical size of the zone that is invaded with steam using a version of Eq. 15.24. This phenomenon is easily demonstrated by running a downhole temperature survey following a steam cycle. Then, because cold oil sand is still exposed in the wellbore, another set of equations similar to Eq. 15.77 is used to calculate oil and water from the cold zone. Using this modification, fluid rates can be matched quite well without need of a scaling factor to reduce predicted oil rates to realistic levels.

A convenient parameter to track, when trying to history-match a field steam cycle with this model, is produced fluid temperature that represents a combination of cold/hot oil and water.

$$T_p = \frac{T_R(q_{oc}C_o + q_{wc}C_w) + T_h(q_{oh}C_o + q_{wh}C_w)}{(q_{oc} + q_{oh})C_o + (q_{wc} + q_{wc})C_w} \quad \text{(15.80)}$$

This method does not lend itself to hand calculation and should be programmed on a computer.

Because steam only enters a small fraction of the sandface in a thick interval as in California oil fields, there is opportunity to improve performance of a steam cycle by using packers or other methods to divert steam into more of the oil zone.

15.10 Process Optimization

There are always the operational questions of how much steam should be injected during a cycle; what rate should steam be injected; when should a well be resteamed; etc. Jones[36] reported the results of the use of the model previously described to history-match a massive 20-year, 1,500-well cyclic steam project in the Potter Sand in the Midway Sunset field, California. He then used the history-match information to do a long-life parametric study of the process. **Table 15.4** lists the conclusions for this particular application. This is, however, not a common practice. There are so many variables that the results from a single well or even a small group of wells cannot be used for a meaningful history match. Further, cyclic steam is easy to apply

TABLE 15.4—OPTIMIZED CYCLIC STEAM STIMULATION PARAMETERS FOR A SELECTED PROJECT USING THE JONES[36] MODEL

Parameter	Model Results
Cycle life	150–180 days
Cycle size	9,000 bbl
Injection rate	1,000 B/D
Well spacing	5/8 acre/well
Soak time	Short as possible
Steam quality	High as possible

in the field and is relatively inexpensive, so most operators simply start immediately with a field trial. Very little is published on optimizing CSS.

It is generally true of CSS that soak time should be as short as possible and that steam quality should be as high as possible. Further, efforts should be made to divert steam out of depleted zones and gas caps and into as much good oil-saturated sand as possible.

There are generally two reasons to apply CSS. First, there is the obvious stimulation of economic oil production immediately from the well. Second, because of the time delay in oil response from the initiation of steam injection into a continuous steam injector in a steamflood project, CSS concentrated in the steamflood zone is often used to accelerate project response.

15.10.1 Cumulative Average Daily Profit Method. Because process optimization is ultimately an economic decision, a resteaming decision can be based on the Rivero and Heintz[37] cumulative average daily profit (CADP) method. **Fig. 15.20** shows a graphical representation of how to use this method. When steam is injected into a producer, profits of the cycle are driven negative because of the cost of the steam, costs to prepare the well for steaming, and lost production as a result of the well being shut down. Once the well is put back on production, the oil rate will peak, and daily cash flow will be at a relative high. Concurrently, CADP for the cycle will begin to increase as the daily production begins to pay for the injection costs. As the well continues to produce, the oil rate gradually falls, as does daily profit. CADP hopefully soon becomes positive, then continues to increase until it reaches a value equal to the daily cash flow. It is at this point that the well should be recycled because cash flow for the next day's production will fall below the CADP.

Although instructive as a concept for picking resteaming time, actual field application of the method is practically impossible because of ever-present problems in gathering precise enough well production gauges and in collecting all of the necessary economic data in a timely manner. Also, because the method is divorced from the reservoir process, it may lead to short-term economic decisions that damage the reservoir.

15.10.2 Sequential CSS Method. In a large CSS project, one needs a way to decide which well to steam and in what sequence. McBean[38] and Jones and Cawthon[39] presented a sequential CSS method that ensures that all wells will be stimulated in a timely manner and takes advantage of the interwell stimulation often observed.

By steaming wells in a sequential manner from downdip to updip as shown in **Fig. 15.21**, they observed not only the oil response from the steamed wells but also some response from offset wells caused by a mini-steamflood. Kuo *et al.*[40] found in numerical simulations that small cycles in closely spaced wells are preferable in this process. Field experience in the sequential CSS project confirmed that finding with wells drilled on ⅝ acre (0.25 ha) spacing.

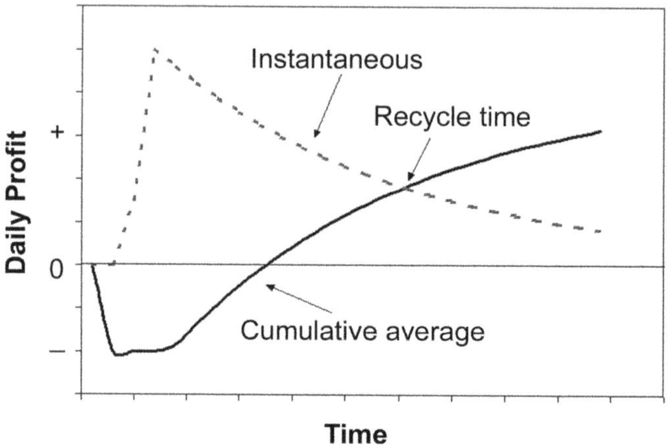

Fig. 15.20—Graphical representation of the cumulative average daily profit method of the recycle determination method.[37]

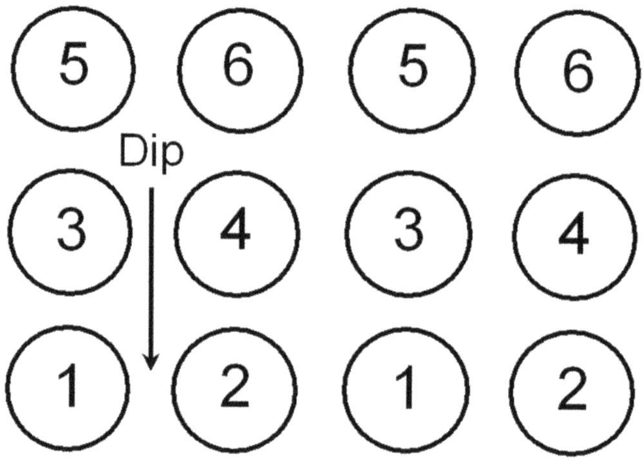

Fig. 15.21—Well steaming schedule for the sequential steam method of cyclic steam design.[39]

15.11 Steam Delivery Systems

15.11.1 Heat Loss in Surface Distribution Piping. For any SEOR process, no matter how efficient, the major cost is always that of generating the process steam. Whether the product of oilfield steam generators, industrial boilers, or electrical/steam cogeneration plants, steam must be delivered through a network of pipes and through pipes down a wellbore to the oil-bearing formation. It is imperative that the unavoidable heat losses in this distribution system be minimized with some type of insulating system.

The basic equation for heat loss is

$$\dot{Q}_{ls} = \frac{(T_s - T_A)}{R_h} \quad \text{(15.81)}$$

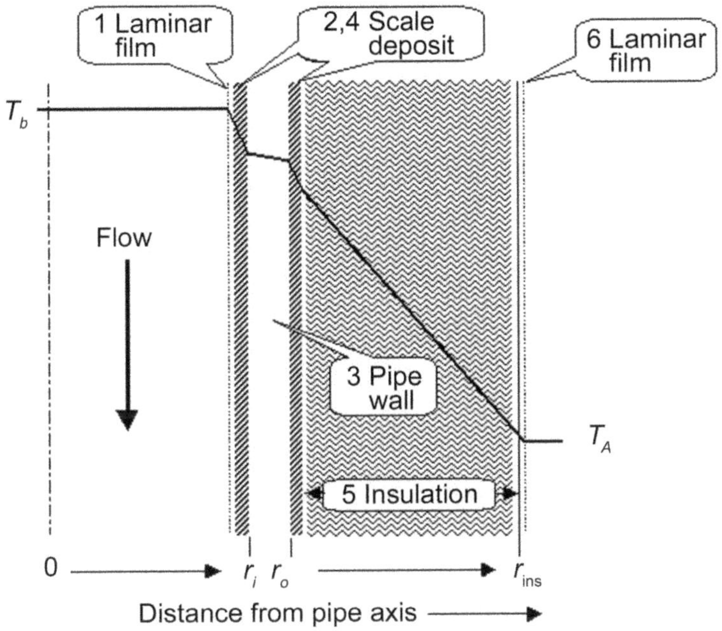

Fig. 15.22—Schematic diagram of resistance to heat transfer and of temperature profile in a suspended surface pipe.

The rate of heat loss in surface lines is usually calculated at steady-state conditions because transients disappear quickly in surface pipes. Thermal resistance to heat loss for that system is

$$R_h = \frac{1}{2\pi}\left[\frac{1}{h_f r_i} + \frac{1}{h_{p_i} r_i} + \frac{1}{\lambda_p}\ln\frac{r_o}{r_i} + \frac{1}{h_{p_o} r_o} + \frac{1}{\lambda_{\text{ins}}}\ln\frac{r_{\text{ins}}}{r_o} + \frac{1}{(h_{fc}+h_r)r_{\text{ins}}}\right]. \quad \text{..........} \ (15.82)$$

The terms in Eq. 15.82 are the coefficients of heat transfer for each of the layers of an insulated pipe as shown in **Fig. 15.22**. They are, from left to right: conduction in the laminar layer in contact with the pipe wall; conduction in the scale or other solid coating on the inside pipe wall; conduction in the pipe wall; conduction in the scale or other coating on the outside pipe wall; conduction in the insulation; and convection and radiation from the outer surface of the insulation. **Table 15.5** shows thermal conductivity for various materials. Refer to appropriate textbooks[41,42] for more in depth information.

In the previous calculation, h_r is the coefficient of radiant heat transfer for the outermost surface of the system; in this case, it is for the insulation. It is common practice to cover insulation with a thin sheath of bright aluminum, mainly for protection from weather and from mechanical abuse. A side benefit is that the bright surface has low radiant emissivity that, combined with low surface temperature, results in negligible radiant heat loss, so this term is often ignored. If the pipe is uninsulated, the term applies.

Note that for every system there is an optimum insulation thickness. Adding more insulation above this optimum will not result in more heat savings. This is because there are two competing effects; the rate of heat loss decreases with increasing insulation thickness, but heat loss increases as the exposed surface area increases.

Coefficients of Heat Transfer. The following are useful in most cases, but the reader should refer to appropriate[41,42] texts for more complicated systems.

Value of h_f. For condensing steam, the coefficient is large, and it is generally adequate to use

TABLE 15.5—THERMAL PROPERTIES FOR SELECTED MATERIALS

Material	λ, Btu/(hr-ft-°F)	ε, Dimensionless	Surface Temp.,°F
Black body emitter		1.00	all
Aluminum, polished	118	0.040–0.060	20–1,000
Carbon steel	28		
Rusty, mill finish		0.55–0.60	200–1,000
Stainless steel	9.4		
316, mill finish		0.28–0.66	100–1,000
Earth, wet	1.5		
Earth, dry	0.30		
Silica	0.83		
Concrete	1.05		
Glass wool insulation	0.022		

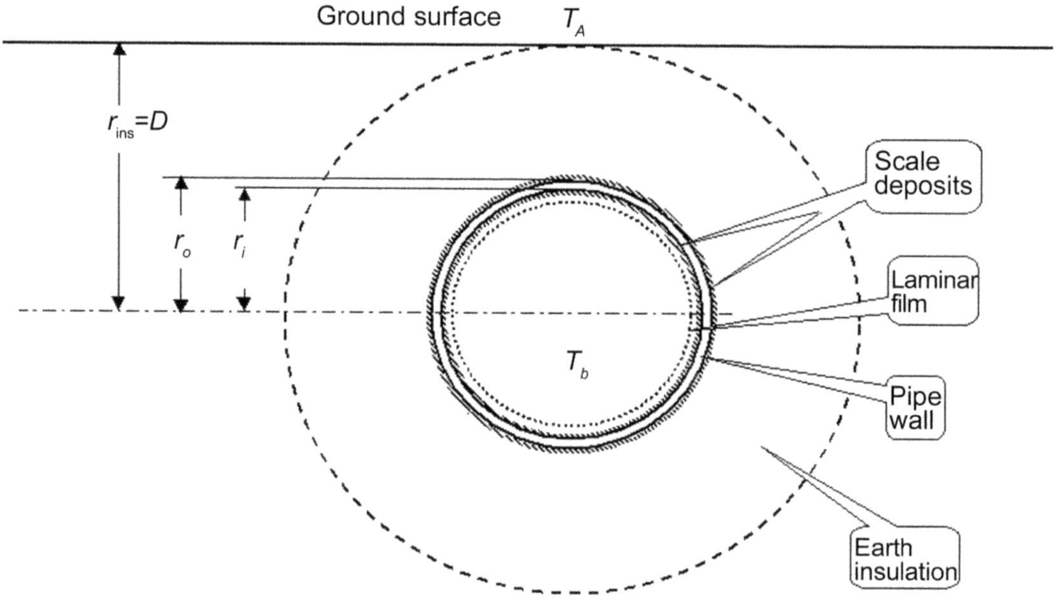

Fig. 15.23—Schematic diagram of idealized thermal resistances in buried surface lines.

$$h_f = 48{,}000, \quad (15.83)$$

when flow is turbulent (which is most of the time) and is determined by

$$N_{\mathrm{Re}} > 2{,}100, \quad (15.84)$$

where

$$N_{\mathrm{Re}} = \left[\frac{\pi}{2}(0.0616)\right]\frac{2i_w \rho_{w,\,sc}}{\pi \mu_s r_i}. \quad (15.85)$$

Because $\mu_s \sim 0.018$ cp for typical oilfield steam temperatures, turbulence will prevail at

$$i_w > 9.85 r_i \;. \qquad (15.86)$$

Value of h_{pi} and h_{po}. These are seldom known, and the terms are usually ignored for steam distribution lines. Actually, these deposits outside and, if not too thick and firmly attached, inside the pipe are desirable because they result in resistance to heat conduction. If they are present but no values are known, McAdams[41] recommends a value of 48,000 Btu/(ft²-D-°F).

Value of h_{fc}. McAdams[41] offers the next equation to calculate the coefficient of forced convection at the outer surface of a pipe system in air.

$$h_{fc} = \frac{0.12\lambda_a}{r_e} N_{\mathrm{Re}}^{0.6}, \qquad (15.87)$$

for $1{,}000 < N_{\mathrm{Re}} < 50{,}000$, where

$$N_{\mathrm{Re}} = 4{,}364 \frac{r_e v_w \rho_{a,\,sc}}{\mu_a} \;. \qquad (15.88)$$

Value of h_r. In the following relationship for the coefficient of radiant heat transfer, temperatures must be expressed in °Rankine, which is °F + 460.

$$h_r = 1.712 \times 10^{-9} \varepsilon (920 + T_o + T_a)\left[(T_o + 460)^2 + (T_a = 460)^2\right] \;. \qquad (15.89)$$

Emissivity, ε, of various materials is listed in Table 15.5.

Buried Lines. A special case of insulated lines is pipes buried in the earth. See **Fig. 15.23.** Eq. 15.82 applies, except for two modifications,[43] which are

$$\ln\left(\frac{r_{\mathrm{ins}}}{r_o}\right) = \cos h^{-1}\left(\frac{D}{r_o}\right), \qquad (15.90)$$

and

$$h_{fc} = \infty \;. \qquad (15.91)$$

Heat loss rate is very high for short-term injection for buried pipes, even in dry soil, so this is not recommended in cyclic steam projects. If the soil contains moisture, the losses are even greater.

15.11.2 Heat Loss in Wells. Heat loss in wells never reaches a steady-state condition. It begins at a very high rate when the well casing is suddenly heated by initial steam injection, then continually decreases in rate as the surrounding earth is heated. For long term continuous steam injection over a period of years, wellbore heat loss becomes relatively small. Conversely, for intermittent cyclic steam injection, the heat-loss rate will always be relatively high because the surrounding earth is never appreciably heated. Eq. 15.82 still applies but is complicated by the ambient (earth) temperature increasing with depth because of geothermal gradient and by the "insulation," earth again, having high conductivity and practically infinite thickness. The latter property results in the thermal resistance being time dependent. **Fig. 15.24** is a schematic depiction of typical elements that contribute to the resistance to heat flow which is described by

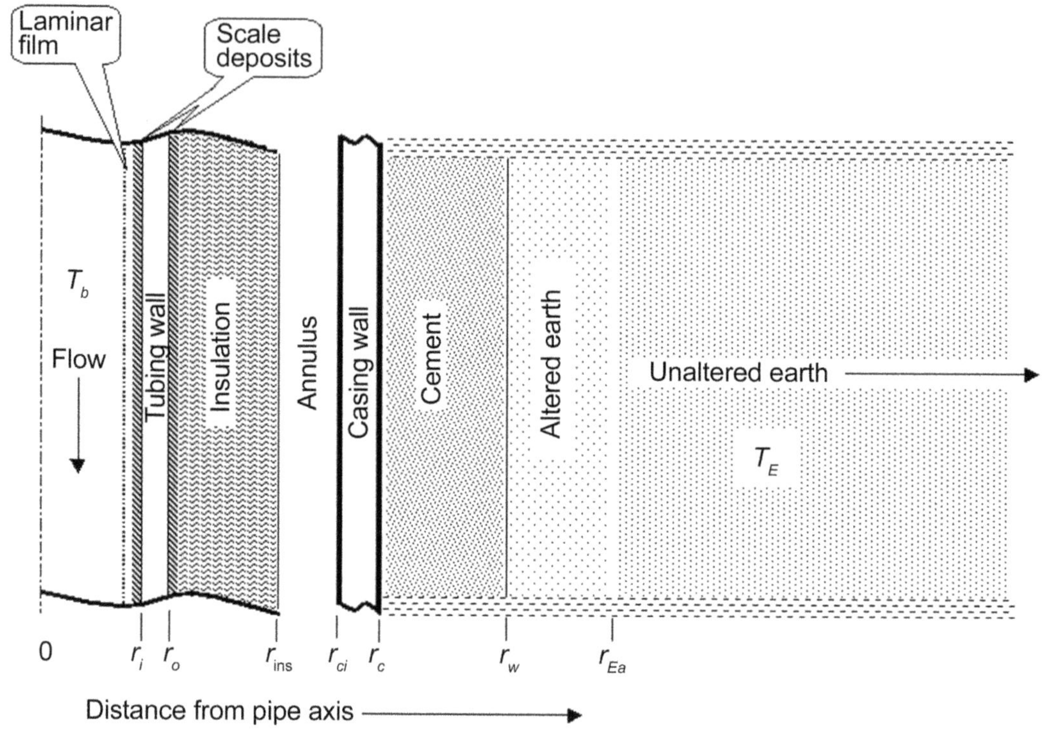

Fig. 15.24—Schematic diagram of resistance to heat transfer in wellbores.

$$R_h = \frac{1}{2\pi}\left[\frac{1}{h_{ft}r_i} + \frac{1}{h_{p_i}r_i} + \frac{1}{\lambda_p}\ln\frac{r_o}{r_i} + \frac{1}{h_{po}r_o} + \frac{1}{\lambda_{ins}}\ln\frac{r_{ins}}{r_o} + \frac{1}{h_{rc,\,an}r_{ins}} + \frac{1}{\lambda_p}\ln\frac{r_{co}}{r_{ci}} + \frac{1}{\lambda_{cem}}\ln\frac{r_w}{r_{co}} + \frac{1}{\lambda_{Ea}}\ln\frac{r_{Ea}}{r_w} + \frac{f(t_D)}{\lambda_E}\right], \quad (15.92)$$

where the first five terms are similar to those in Eq. 15.82. The last five terms are the resistances in radiation and convection in the casing annulus, in the casing, in the cement, in the altered earth zone (dried earth because of high temperatures), and in the time dependent loss to the earth. If other resistance zones can be identified, such as coatings in the casing or scale deposits, etc., terms should be added for them. Every system should be analyzed according to the elements included, such as wells with no insulation on the tubing, wells with no tubing at all, or simply injection down the casing. All of the additional terms can be determined with equations previously presented except for the coefficient of heat transfer in the annulus, $h_{rc,an}$ and f_{tD}.

Heat loss is a serious problem in cyclic steam stimulation because the wellbore and surface lines are never heated to steady-state conditions. **Fig. 15.25** shows the results of Eq. 15.92 for several steam-injection rates in a typical Kern River field producer. It demonstrates that because steam is at a relatively constant temperature heat loss rate, it is a constant and injection should be done at the highest practical rate. **Fig. 15.26** shows the benefit of insulating the casing from contact with steam for short duration injection as in a steam cycle. Conversely, **Fig. 15.27**[44] shows that for long-term injection, as in a steamflood injector, and for shallow wells, as encountered in the San Joaquin Valley oil fields, there is no benefit from insulating the

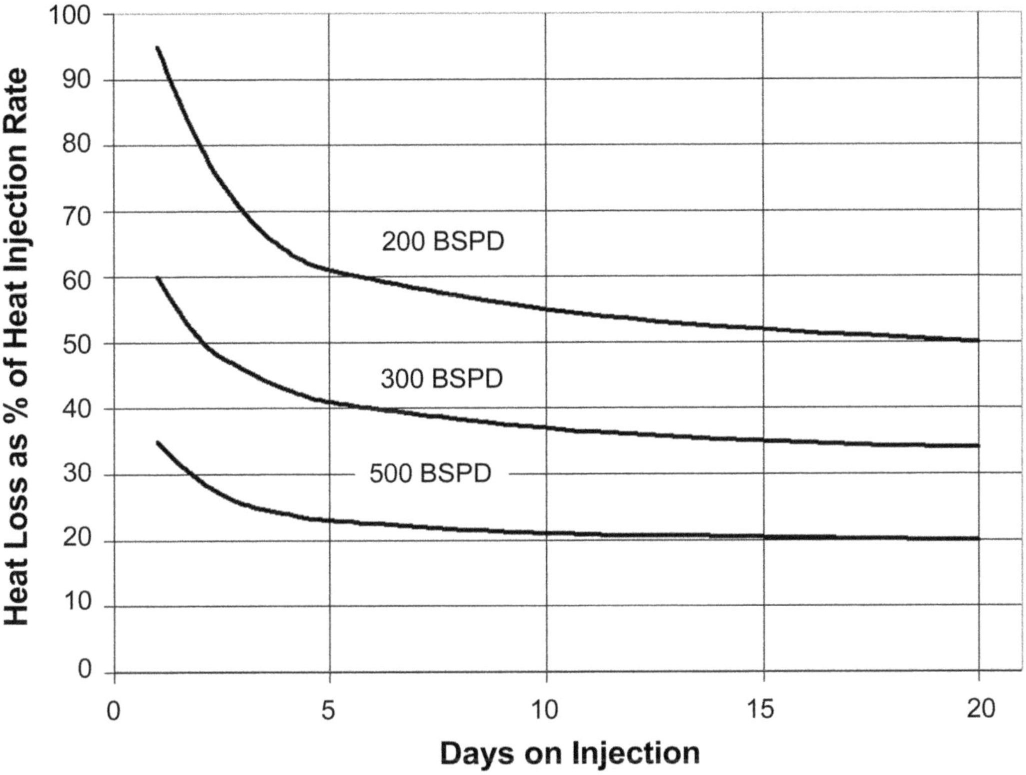

Fig. 15.25—Heat loss for several steam-injection rates in a typical California cyclic steam well. Rate should be as high as possible to minimize heat loss.

casing. Thus, it is possible to drill inexpensive slimhole injectors, completed simply with a tubing string, and not appreciably increase heat loss over the life of a project.

Value of $h_{rc,an}$. In an air-filled annulus operating under free convection, the coefficient of heat transfer for radiation and convection is given by Willhite.[43]

$$h_{rc,\,an} = 4.11 \times 10^{-8}\left[\frac{1}{\varepsilon_{\text{ins}}} + \frac{r_{\text{ins}}}{r_{ci}}\left(\frac{1}{\varepsilon_{ci}} - 1\right)\right] \cdot F(T_{\text{ins}}, T_{ci}) + \frac{1}{r_e}\frac{\lambda_{a,\,a}}{\ln\left(\frac{r_{ci}}{r_{\text{ins}}}\right)}, \quad (15.93)$$

where

$$\lambda_{a,\,a} = 0.049\lambda_a N\text{Gr}^{0.333} N\text{Pr}^{0.047} \;. \quad (15.94)$$

The Grashof number is

$$N_{\text{Gr}} = \left[7.12 \times 10^7 \frac{g_c}{g}\right] \cdot \frac{(r_{ci} - r_{\text{ins}})^3 g\rho_{an}^2 \beta_{an}(T_{\text{ins}} - T_{ci})}{g_c \mu_{an}^2} \;; \quad (15.95)$$

the Prandtl number is

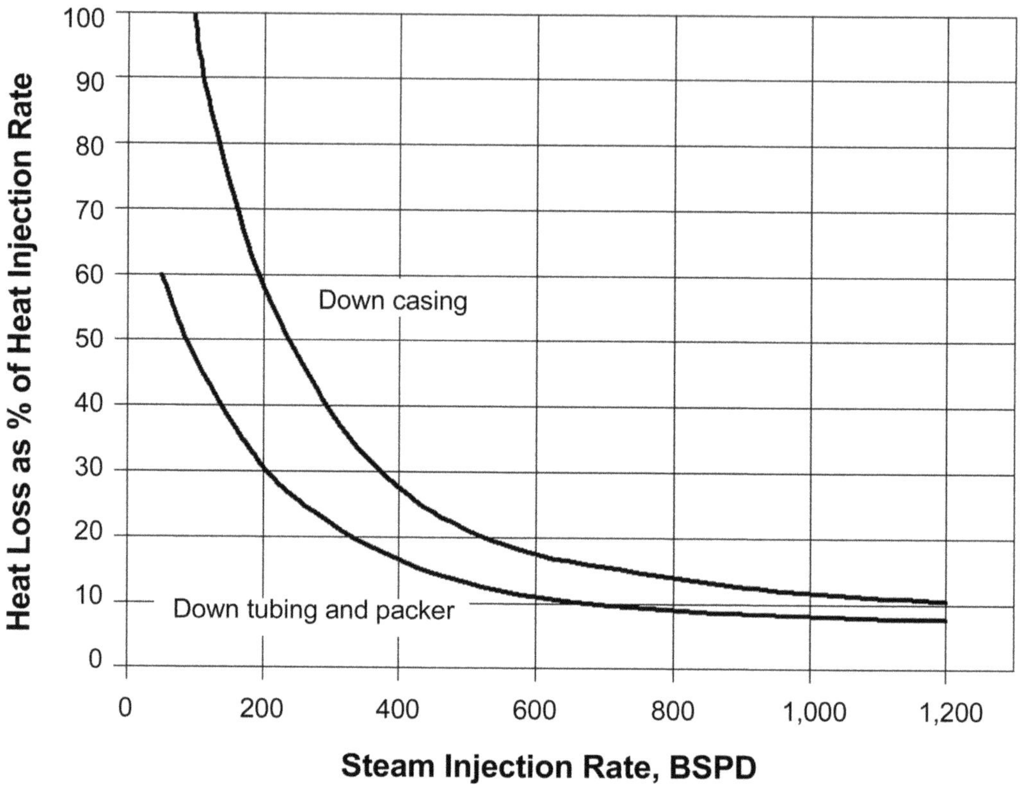

Fig. 15.26—Heat loss for cyclic steam in a typical California cyclic steam well after 10 days with and without tubing and a packer.

$$N_{Pr} = 58.1 \frac{C_{an} \mu_{an}}{\lambda_{an}} ; \quad \text{(15.96)}$$

and the temperature function is

$$F(T_{ins}, T_{ci}) = \left[(T_{ins} + 460)^2 + (T_{ci} + 460)^2 \right] \cdot (920 + T_{ins} + T_{ci}) . \quad \text{(15.97)}$$

The temperatures for use in Eq. 15.97 are proportional to the fractional thermal resistance between the outer tubing surface, in this case the insulation, T_{ins}, and the casing inner wall, T_{ci}. They can be estimated by

$$T_{ins} = T_b \frac{T_b T_A}{2\pi R_h} \frac{1}{h_{ft} r_i} + \frac{1}{h_{p_i} r_i} + \frac{1}{\lambda_p} \ln \frac{r_o}{r_i} + \frac{1}{h p_o r_o} + \frac{1}{\lambda_{ins}} \ln \frac{r_{ins}}{r_o} , \quad \text{(15.98)}$$

and

$$T_{ci} = T_A - \frac{T_b - T_A}{2\pi R_h} \left[\frac{1}{\lambda_p} \ln \frac{r_{co}}{r_{ci}} + \frac{1}{\lambda_{cem}} \ln \frac{r_w}{r_{co}} + \frac{1}{\lambda_{Ea}} \ln \frac{r_{Ea}}{r_w} + \frac{f(t_D)}{\lambda_E} \right] . \quad \text{(15.99)}$$

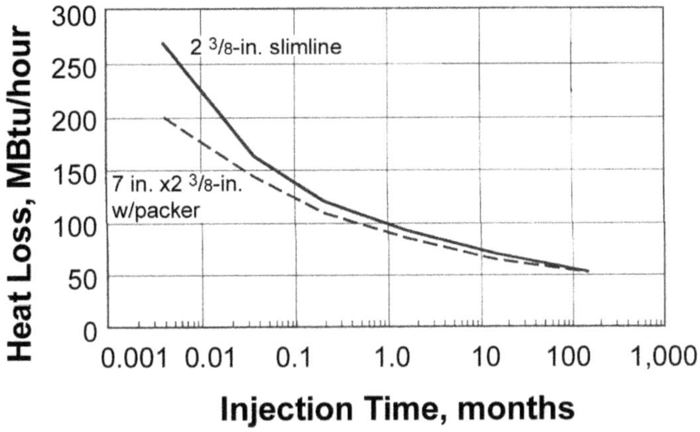

Fig. 15.27—Heat loss for a steamflood injector with a conventional casing, tubing, and packer completion compared to a slimhole completion.[44]

Value of $f(t_D)$. This is a function of dimensionless time.

$$t_D = \frac{\alpha_E t}{r_{Ea}^2} \quad \text{.......... (15.100)}$$

The radius in the denominator is the radius of the outermost element in contact with the reservoir, which is the heat altered zone in this case. Willhite[43] gives a table of values for $f(t_D)$, for t_D <100. A reasonable estimate can be derived from

$$f(t_D) \cong 3.53 e^{\frac{\left(\ln(t_D) - 7.66\right)^2}{-44.85}} \quad \text{.......... (15.101)}$$

Ramey[44] gives a calculation of $f(t_D)$, for $t_D \geq 100$.

$$f(t_D) \cong 0.5 \ln(t_D) + 0.403 \quad \text{.......... (15.102)}$$

Because R_h, T_{ins}, T_{ci}, and $f(t_D)$ are interrelated by nonlinear expressions, they must be solved by an iterative trial-and-error procedure.[43]

15.12 Heat Management

While always an implicit goal in SEOR processes, overall process heat management became a topic in the literature in the mid-1980s. The growth of the discipline has closely followed the development of the personal computer and computer applications. **Fig. 15.28**[46] is a graphical representation of the major components of a heat balance that must be performed to properly manage a SEOR process. Ziegler *et al.*[47] published a very good summary of a method of implementing the principle. In essence, the operator must establish an iterative process that continues for the duration of the project and that continuously collects and analyzes pertinent data. Based on that analysis, the operator then makes appropriate midcourse adjustments to optimize the project. The process is actually a complete project optimization method but has adopted the name "heat management" because steam generating costs tower over any other cost and are

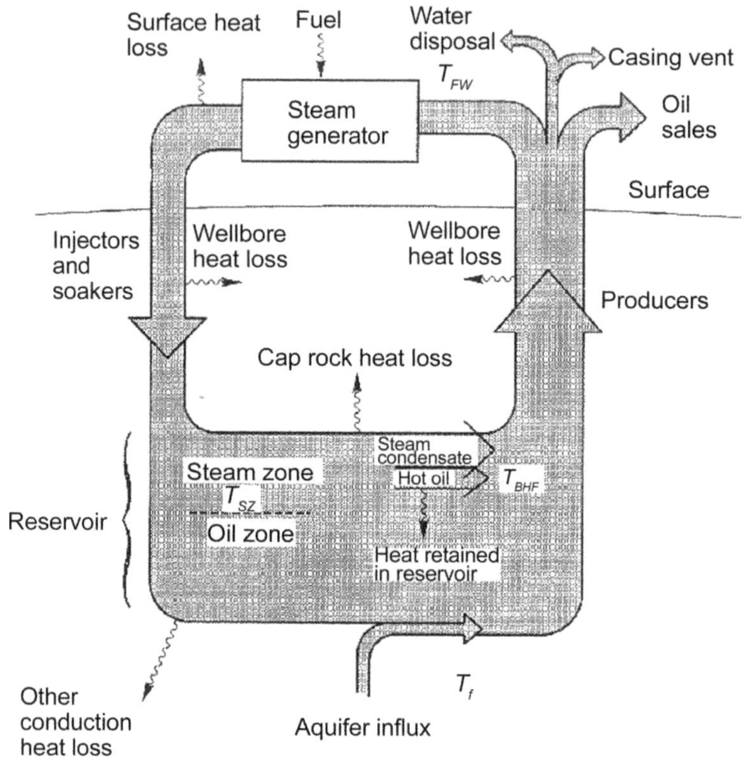

Fig. 15.28—Schematic diagram of the components of heat management for SEOR projects.[46]

even several times larger than the initial substantial capital investment in a steamflood. There are three basic parts of the method, as shown in **Fig. 15.29**,[47] which are data gathering, data monitoring, and adjustments to the process.

15.12.1 Data Gathering. As its name implies, data gathering consists of compiling regular and typically large amounts of information on the operation. Data from producing wells, injection wells, observation wells, and surface facilities are compiled and stored in computer databases. **Table 15.6** was derived from Ziegler[47] and indicates the information currently considered necessary for California steamfloods. In addition to the types of data collected, the operator must also specify a regular schedule for collection.

All of the data are from existing and necessary components of the system except for the observation wells. Although it is tempting to economize by eliminating them, that is false economy. Information they provide on the change in temperature and gas/oil saturation over time is critical in maintaining process efficiency. The data must be stored in computer databases that are accessible to desktop PCs.

15.12.2 Data Monitoring. As with any oil production operation, there are daily and weekly tasks that must be done. However, in heat management, the operator must schedule formal project reviews on a much longer term such as a semiannual or annual basis. **Table 15.7**[46] is an example review schedule. During these reviews, the project performance data is compared to the original design (using methods described in previous sections in this chapter), and adjustments are made accordingly.

Key calculations are heat and mass balances to quantify the terms shown in Fig. 15.28.

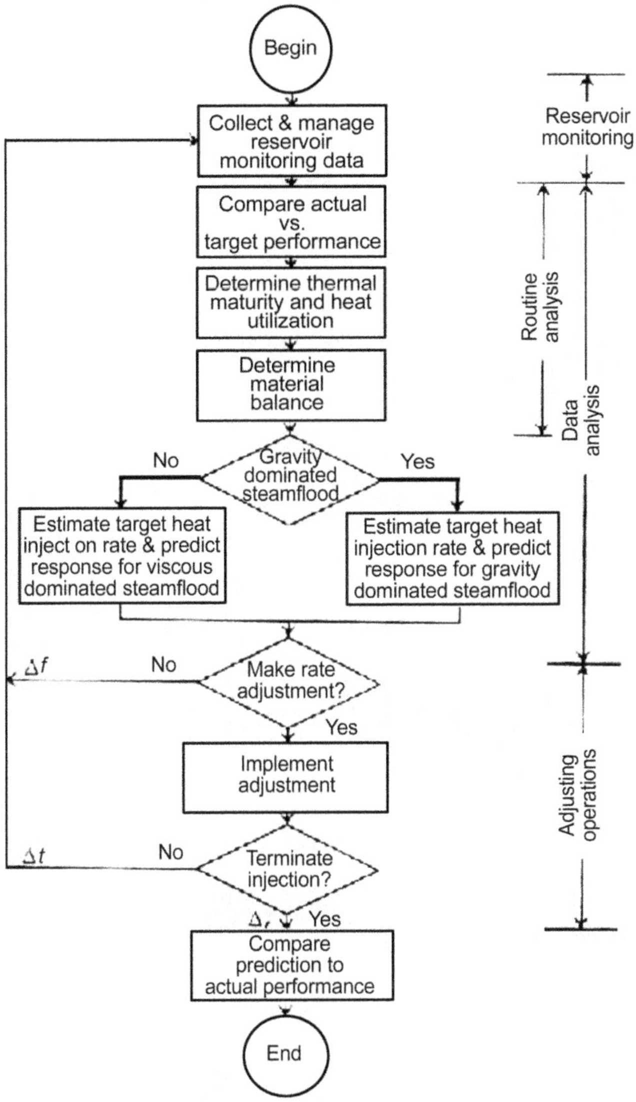

Fig. 15.29—Heat management flowchart.[47]

$$m_i + m_{(o/w)\text{influx}} = m_{(o/w)\text{prod}} + m_{\text{casing\_blow}} + m_{\text{influx}} + m_{Z(\text{accumulation})}, \quad (15.103)$$

and

$$Q_i + Q_{(o/w)\text{influx}} = Q_{(o/w)\text{prod}} + Q_{\text{casing\_blow}} + Q_{\text{loss}} + Q_{\text{influx}} + Q_{z(\text{accumulation})}. \quad (15.104)$$

Table 15.8[47] is an example of a list of metrics that should be established to aid in making heat management decisions. Note the use of the term "produced oil/fuel" ratio, F_{of}, in the metrics table. This is in place of the more common "produced oil/steam injected" ratio, F_{os}. The reason for using the former is that steamfloods often are operated at steam qualities that vary significantly from the normal 70 to 80%. In these cases, F_{os} would be a misleading indicator,

TABLE 15.6—EXAMPLE STEAMFLOOD MONITORING DATA PLAN FOR HEAT MANAGEMENT[47]

Source	Data Type	Collection Frequency	Presentation Format for Routine Analysis	Derived Information for Data Analysis
Injection wells	Rate	Bi-weekly	Rate, bubble, contour plots	Performance update, energy, material balance
	Quality	Annual, selected wells	Heat-injection map	Steam quality distribution, energy balance
	RA and spinner profiles	1–3 years	Injection profile	Remedial potential, compliance
	Shut-in pressure	6 months–1 year	Depth vs. pressure, areal map	Steam-zone pressure trends, maximum heating
	PNC log	Nonroutine	Saturation, shale maps	Screening for VE potential
	Pressure/temperature profile	Nonroutine	Heat injection log	Zonal heat allocation
	Pressure transient	Nonroutine	Log/log analysis	P, k, skin, steam-zone volume, boundary
Observation wells	Temperature profile	1–2 months	Temperature profile, average temperature map	Average temperature, heat rate, steam zone
	PNC log	6 months–1 year	Gas saturation trace, x-sections	S_g-sweep efficiency, steam-zone growth
	C/O log	Annual	Oil saturation trace	S_o, S_w recovery factor, allocation
Producing wells	Liquid rates	1–2 months	Rate, bubble, contour map	Performance update, material balance
	Effluent rate	Annual, selected wells	Bubble map	Steam breakthrough, energy balance
	Casing pressure	Monthly	Bubble, contour map	Steam breakthrough, pressure drawdown
	Flowline temperature	Monthly	Bubble, contour map	Steamflood response, steam-zone temperature
	Lead-line temperature	Annual, selected wells	Bubble, contour map	Reservoir heating, steam breakthrough
	F-Scott	Quarterly, if applicable	Zone rate, bubble, contour map	Zonal oil-production allocation
	Isolation test	Nonroutine	Performance x-section	Zonal oil/water/gas allocation
	Tracers	Nonroutine	Areal sweep map	Reservoir heterogeneity, areal sweep efficiency
Casing collection settings	Liquid rates	Daily	CCS effluent vs. steam injection rate	Energy balance
	Setting pressure	Daily	Pressure vs. oil rate plot	Effect of casing pressure on oil production
	Gas composition	Nonroutine	Lap report	Steam, reservoir interaction
Steam generator	Water rate	Daily		Mass delivery
	Steam quality	Daily		Heat delivery
	Fuel rate	Daily		Fuel consumption

TABLE 15.7—EXAMPLE DATA ANALYSIS SCHEDULE FOR STEAMFLOOD HEAT MANAGEMENT[47]

Steps	Data Requirements	Evaluation Methods	Evaluation Outcomes	Possible Adjustments
1. Compare actual performance to prediction	Oil/water/steam rates, OFR, steam production, flowline temperature, current OPEX, oil price, TOW surveys	Compare actual and predicted reservoir performance and profitability	Identify gaps between actual and predicted	
2. Determine heat balance, thermal maturity and heat utilization	Heat injected: rate, quality, temperature, pressure and wellbore profiles	Comparison of actual vs. target injection	Estimate steamfront geometry	Changes to delivery system
	Heat stored in reservoir: temperature profile and PNC logs	Gravity, viscous-force ratio	Estimate process efficiency	Adjust injection intervals in injection wells
	Heat produced: oil/water rates, casing effluent, flowline temperature	Thermal maturity	Estimate process efficiency	Change cyclic steam frequency in pattern producers
		Heat utilization		
3. Determine material balance	Production data: oil, water, and steam	Liquid balance vs. time	Voidage and influx estimates	Increase well productivity
	Injection data: steam and water	Material balance		Infill drilling
				Control Influx
4. Determine target heat injection rate	Reservoir properties	A. Neuman and Vogel analogy simulation	Current steam injection requirement	A. Steam rates
A. Gravity dominated	Fluid properties	B. Mandl-Volek and Jones analogy, simulation	Steam-injection schedule	B. Steam rates and qualities
B. Viscous dominated	OPEX forecast		Production forecast	Return to step one in six months
	Oil price forecast		Profitability forecast	

and an in/out energy factor would be a better benchmark to use.[48] Either of these is useful as a project metric, but the user must be aware of the limitations. **Fig. 15.30** shows the F_{of} for three different steam injection schedules.[31] Using only F_{of} as the criteria, The Neuman steam-injection schedule (Fig. 15.15) seems to be far superior to the other two. However, Fig. 15.14 shows that the linear steam rate reduction probably gives the best overall economic performance. The Neuman reduction rate is designed to optimize F_{of}, but severe early steam reduction results in lower oil rates and cumulative production.

15.12.3 Adjustments to the Process. Once the project has been reviewed and compared to the various benchmarks, it is either on or ahead of schedule, thus requiring no changes, or is not performing as expected and is in need of a midcourse correction. Table 15.7 lists common changes. There are only a few operational areas that can be changed.

TABLE 15.8—EXAMPLE HEAT MANAGEMENT METRICS[47]

Economic Metrics	Tracking Frequency	Internal Data Source	External Data Source
Oil production rate	Monthly	Company database	Commercial production data service
Instantaneous oil/fuel ratio	Monthly	Company database	Commercial production data service (with steam quality estimate)
Cumulative oil recovery	Monthly	Company database	Commercial production data service (with OOIP estimate)
Fuel cost: $/bbl produced oil	Monthly	Company database	Commercial oil industry reporting service
Total operating cost: $/bbl produced oil	Monthly	Company database	Commercial oil industry reporting service

Process Metrics	Tracking Frequency	Recommended Goal
Status of company database	Quarterly	100%
Number of heat analyses per year per project	Quarterly	Measured against target set by team
Percentage of time on heat management	Quarterly	An initial increase at startup of project declining as project matures

Fig. 15.31[30] is an example of a Midway Sunset, California, steamflood that benefited from heat management. The operator determined that the wells were producing at their maximum rate but that steam injection was much higher than required. Based on this analysis, the steam rate was redirected and then reduced with improved oil production, thus greatly improving project economics and ultimate recovery.

Fig. 15.32[47] is another example of heat management techniques being used to improve an in-progress steamflood. This Kern River, California, project was found to be overinjected, and some producers were not completed in the proper zones. Steam rate was redirected in new injectors, and some producing wells were recompleted; later, steam rate was reduced, resulting in lower steam cost, increased oil production, and reduced water production. At the time of publication, the operator was faced with an apparent opportunity for another heat management study and adjustment to the process.

15.13 Horizontal Well Applications in Steam Injection—S. M. Farouq Ali, Consultant and Jeff Jones

Horizontal wells are being employed in innovative ways in steam injection operations to permit commercial exploitation of reservoirs that are considered unfavorable for steam, such as very viscous oils and bitumen and heavy oil formations with bottomwater, etc. Numerous papers have explored steam injection using horizontal- vertical-well combinations by use of scaled physical models or numerical simulators. For example, Chang, Farouq Ali, and George[49] used scaled models to study five-spot steamfloods, finding that for their experimental conditions, a horizontal steam injector and a horizontal producer yielded the highest recovery. **Fig. 15.33** shows a comparison of oil recoveries for various combinations of horizontal and vertical wells and for four different cases: homogeneous formation, 10% bottomwater (% of oil zone thickness), 50% bottomwater, and homogeneous formation with 10% pore volume solvent injection before steam. Huang and Hight[50] carried out numerical simulations of a variety of hypothetical

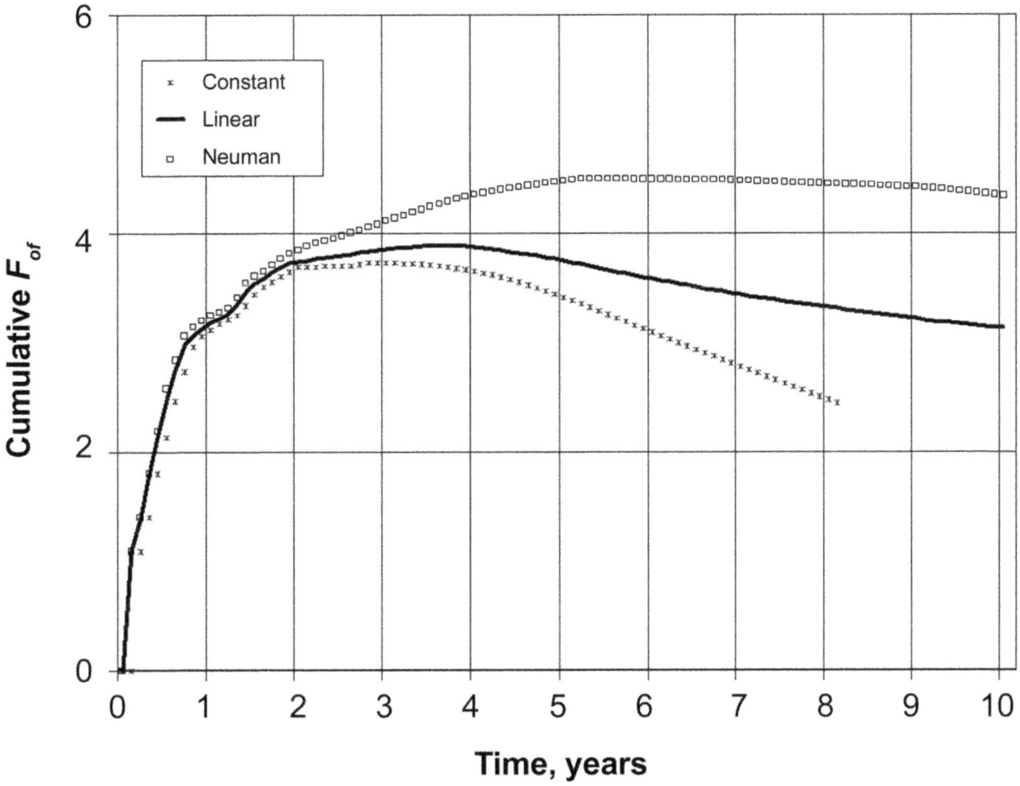

Fig. 15.30—Produced oil/fuel burned ratio for three steam-injection scenarios.[31]

situations involving horizontal and vertical wells. A few field-tested applications of horizontal wells are briefly described next.

15.13.1 Vertical Injectors, Multilateral Producer: Kern Hot Plate Test. Dietrich[51] described the Kern Horizontal-Well Pilot, in a high viscosity (about 5,000 cp) oil reservoir, shown in **Fig. 15.34**. Key data are given in **Table 15.9**. Eight vertical injectors were used to inject steam at an average rate of 1,900 B/D (CWE). The producer consisted of 8 branches of ultra-short radius horizontal wells. The project operated for about 14 months. The production response from the wells averaged 120 B/D. The project was not economic (steam-oil ratio of 15) because of unfavorable placement of the horizontal producer and other factors.

15.13.2 Vertical Injectors and Horizontal Producer With Bottomwater: Tangleflags Project. Jespersen and Fontaine[52] described the Tangleflags project that utilizes a system of vertical injectors and horizontal producers. There are 11 producers. Table 15.9 gives the key data for the project. **Fig. 15.35** shows the first well, together with the estimated sizes of the steam zones around the vertical injectors at three different times. The original horizontal well (400 m [1,312 ft] in length, but only 107 m [351 ft] were open to flow) was drilled for primary production, but the water cut became prohibitive within months, at which time it was decided to inject steam into the offset vertical wells. As a result, the water cut declined rapidly and stayed at a low value. The well produced almost 2.5 million bbl of oil. It should be noted that a small, discontinuous primary gas cap is present over the field. Temperature surveys showed that the injected steam migrated into the gas cap and during the early part of the project, exerted a

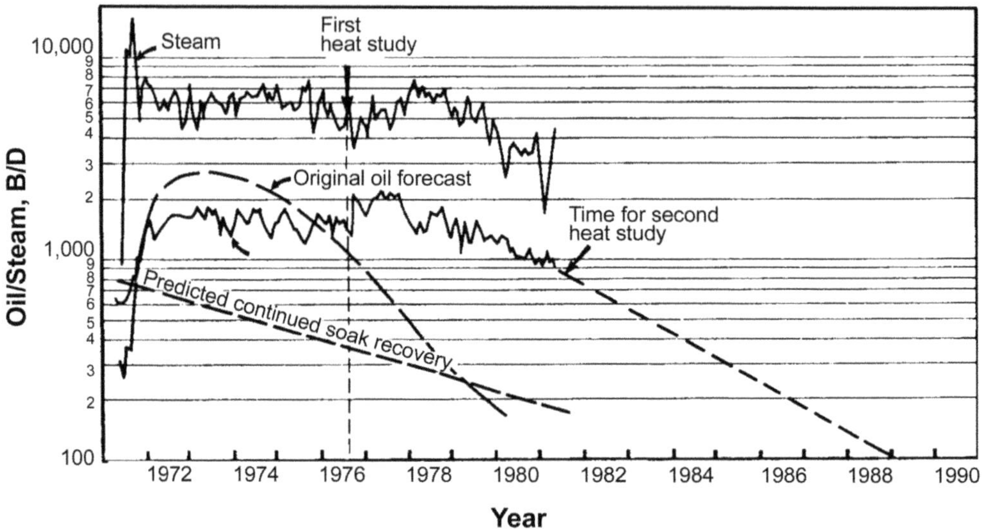

Fig. 15.31—Midway Sunset field example of heat management results on an in-progress steamflood.[30]

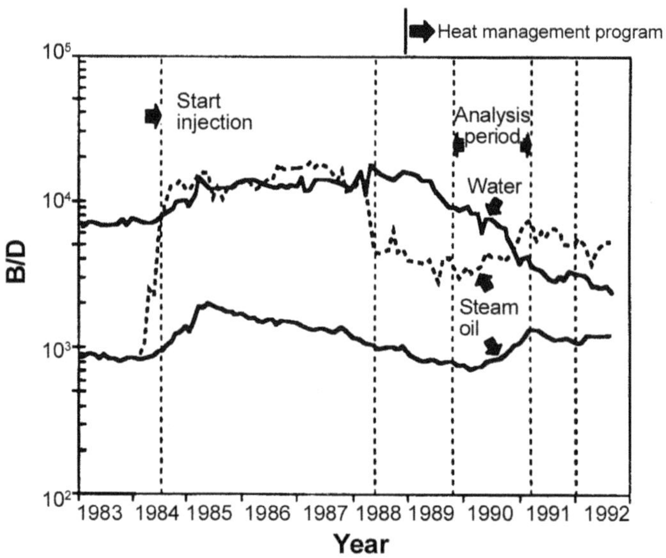

Fig. 15.32—Kern River field example of heat management results on an in-progress steamflood.[47]

downward drive, restricting the advance of the oil/water contact. At a later stage, steam injection rate was decreased and fluid flow was gravity controlled.

There are several other steam injection operations like Tangleflags. If the vertical permeability is high, and gas cap absent or small, an operation utilizing horizontal producers and vertical injectors may be viable.

15.13.3 Steam-Assisted Gravity Drainage (SAGD). SAGD is an outstanding example of a steam injection process devised for a specific type of heavy oil reservoir utilizing horizontal wells. Several variations of the basic process have been developed, and are being tested. The

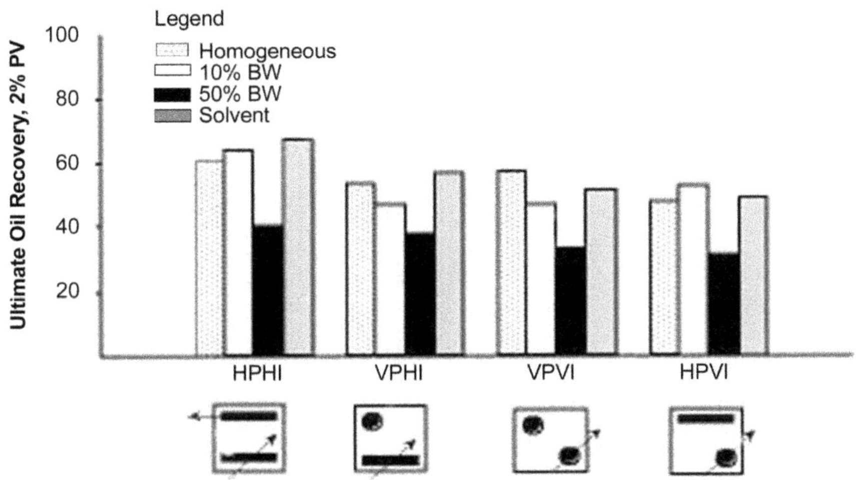

Fig. 15.33—Comparison of the various injection-production strategies on ultimate oil recovery, for the different bottomwater cases investigated (after Chang, George, and Farouq Ali[49]).

TABLE 15.9—RESERVOIR PROPERTIES IN HORIZONTAL WELL PROJECTS

Project	Kern	Tangleflags	Midway Sunset Monarch	Cymric/McKittrick Tulare	Duri
Depth, ft	500	1,475	1,000–1,400	900	450–700
Average thickness, ft	77	89	20–300	40	>40
Reservoir dip, degrees			60–70	20–65	10–15
Oil gravity, °API	18–20		14	10–14	21
Oil viscosity, cp	5,000	13,000	6,000	3,000–10,000	400
Reservoir temperature, °F	110	66	Mature cyclic ~150	Mature flood ~150	Mature flood ~150
Porosity, %	26	33	30	32–34	32
Horizontal permeability, md	300–700	~4000	200–1,500	400–7,000	500–2,000
Vertical permeability, md	60–470				
Oil saturation, %	50	50	80	55–65	55
Initial pressure, psig		591		75–150	
Solution gas/oil ratio, scf/STB		62			

original process, as developed by Butler, McNab, and Lo[53] in 1979, utilizes two parallel horizontal wells in a vertical plane: the injector being the upper well and the producer the lower well (**Fig. 15.36**, taken from Butler[54]). If the oil/bitumen mobility is initially very low, steam is circulated in both wells for conduction heating of the oil around the wells. Steam is then injected into the upper well, while producing the lower well. As a result, steam rises forming a steam chamber with oil flowing at the sides of the chamber and condensate flowing inside the chamber, as shown in Fig. 15.36. This is an idealized situation. Other flow regime may occur depending on the oil and formation properties. For the simplest case, the oil production rate q_{oh}, in B/D per ft well length, is given by (multiplier "2" indicates flow from two sides of the steam chamber)

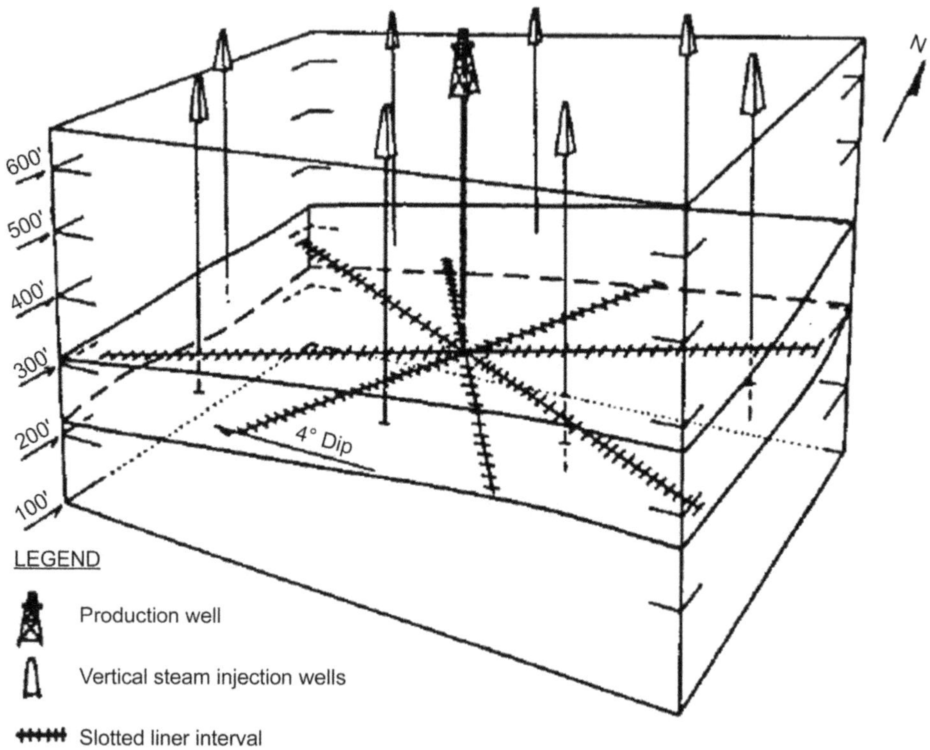

Fig. 15.34—Schematic of Kern River Hot Plate pilot injection and production wells.[51]

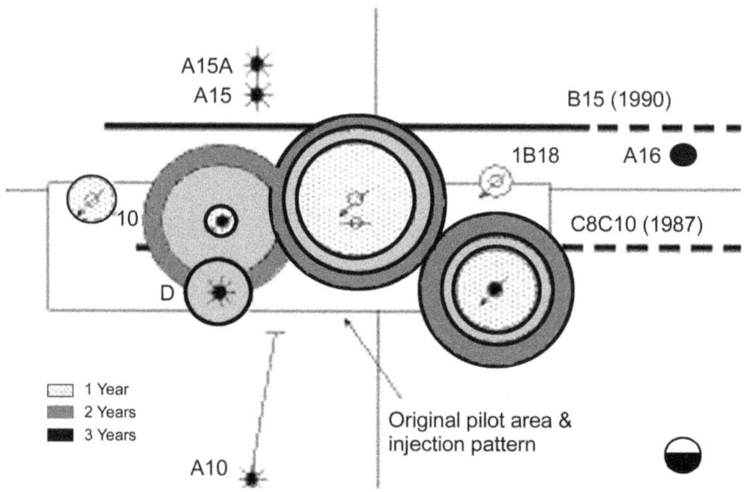

Fig. 15.35—Tangleflags project bubble map showing horizontal producer and vertical injectors and the growth of the steam zone over time (after Jesperson and Fontaine[52]).

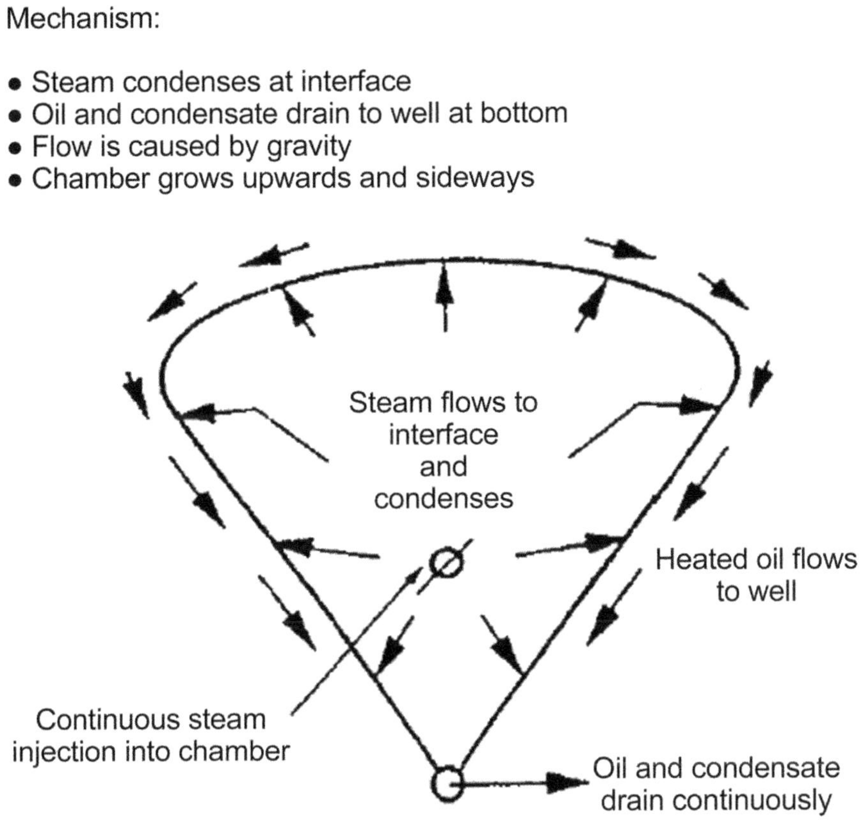

Fig. 15.36—Conceptual diagram of the SAGD process (after Butler[54]).

$$q_{oh} = 2(0.295)\sqrt{\frac{2kk_{ro}\alpha\phi(S_o - S_{ors})h_h}{m^* v_{os}}}, \qquad (15.105)$$

where the kinematic viscosity of oil (in centistokes) at the steam temperature, T_s, is given by v_s, and that at any other temperature, T, is given by

$$\left(\frac{v_{os}}{v}\right) = \left(\frac{T - T_R}{T_s - T_R}\right)^{m^*}, \qquad (15.106)$$

where m is derived from the viscosity-temperature relationship of the oil.

Eq. 15.106 predicts rates of 0.1 to 0.7 B/D per ft for a horizontal well for an oil viscosity of 100,000 cp. For example, a 2,000-ft long well may be expected to produce about 800 B/D at a steam temperature of 400°F. The theory has been verified by laboratory experiments. Field results to-date have been encouraging. One commercial project (EnCana's Foster Creek Project), consisting of 22 well pairs, has been in operation since October 2001. The current steam/oil ratios are averaging 2.5 bbl oil/bbl of steam. Earlier field tests of SAGD in Athabasca tar sands were successful at a depth of about 600 ft, which is too deep for surface mining and not deep enough for high-pressure steam injection.

SAGD is a complex process because gravity flow strongly relies on a high vertical permeability. The initial oil mobility determines the vertical spacing of the two wells. In a million cp

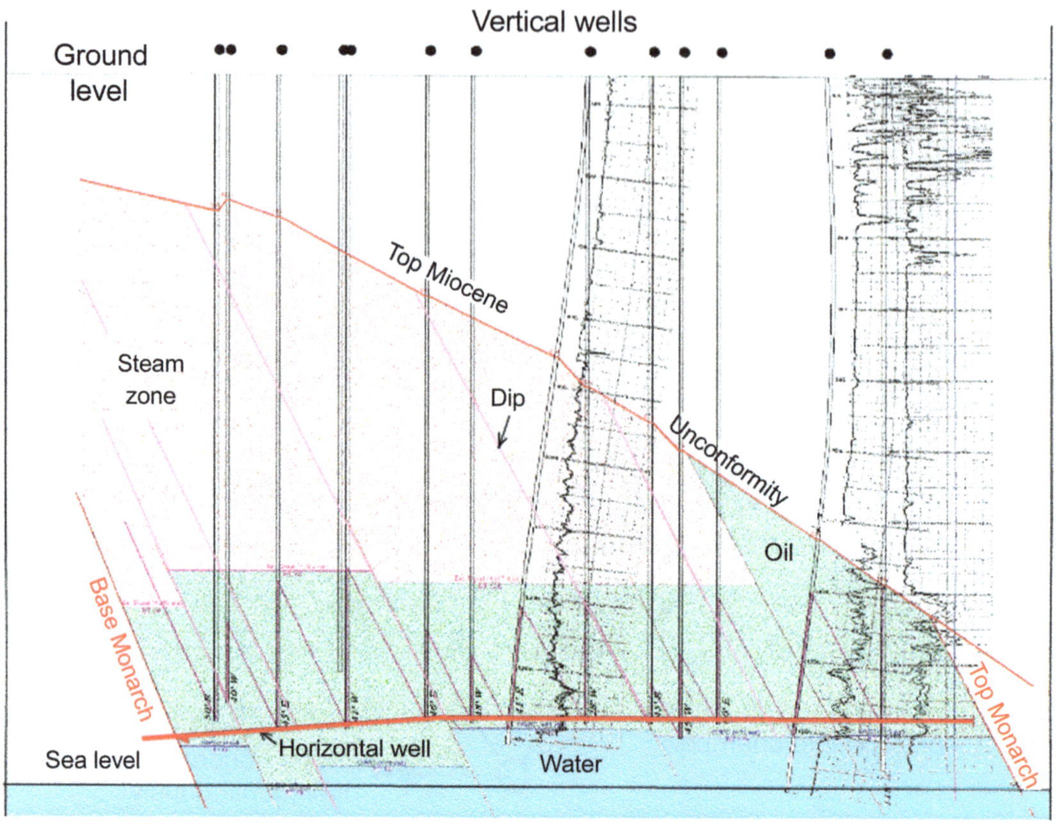

Fig. 15.37—Cross section of Midway Sunset Monarch sand cyclic steam project showing oil in bottom zone wedges bypassed by vertical wells but recovered by horizontal well.

tar sand, the spacing would be 5 to 6 m [16 to 20 ft]. It is also important that the steam chamber be sealed. There is no steam migration to offset vertical wells. In California, SAGD failed to achieve commercial success because of relatively high initial mobility of oil, as well as other reasons.

One variation of SAGD is known as single-well SAGD. Here, insulated tubing is used to inject steam into a single horizontal well, with production from the annulus. This process was successful in a few cases but generally failed. Another variation (Vapex) utilizes a suitable solvent (such as ethane, propane, etc.) instead of steam and is being field tested.

15.13.4 Cyclic Steam Stimulation and Other Uses of Horizontal Wells. Horizontal wells are being used for cyclic steam stimulation, much in the same manner as vertical wells, in Cold Lake, Alberta, and in Eastern Venezuela. In Cold Lake, horizontal well lengths are of the order of 3,000 ft, and steam slugs are very large (100,000 to 200,000 bbl). The resulting steam/oil ratios are 2 to 4, and performance is superior to SAGD continuous steamflood.[55] In Venezuela, wells are around 1,000 ft in length, and the steam slugs are of the order of 50,000 bbl. Response is good, with average stimulation ratios of 2 to 4 over a production period of 6 months.

In California, horizontal wells have been used sparingly in thermal projects, mainly because of sand control problems but also because of the fact that vertical wells are very cost effective, and it is difficult for horizontal wells to be competitive. Horizontal wells have successfully been incorporated into existing thermal projects to supplement conventional vertical-well performance for both Midway Sunset field cyclic steam[56,57] and Cymric/McKittrick field

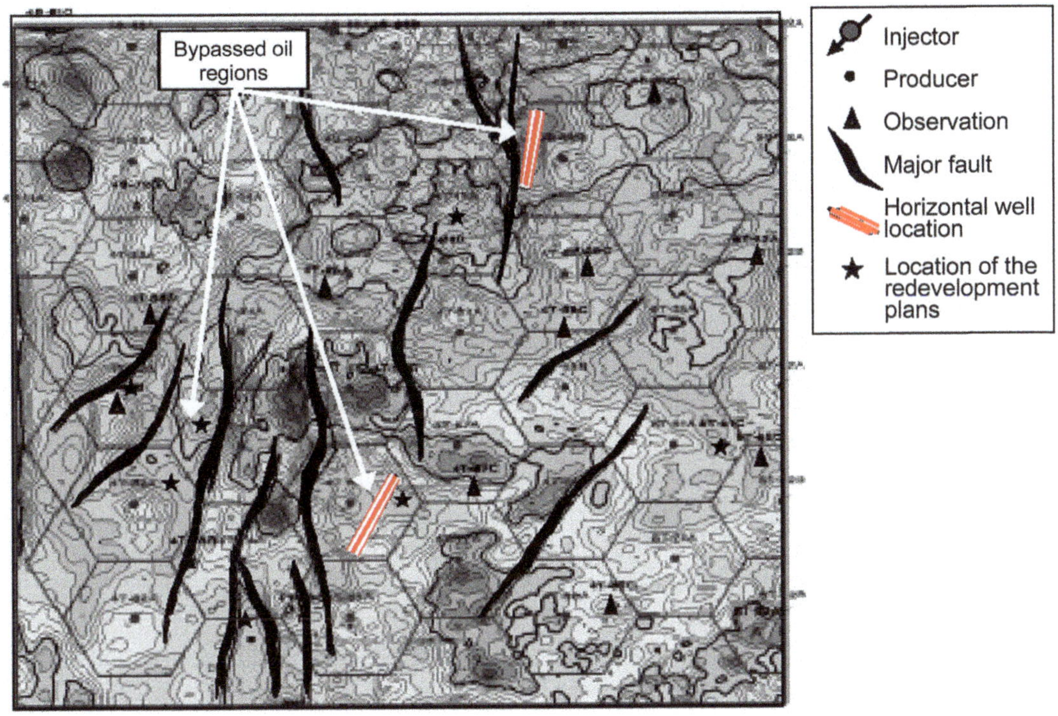

Fig. 15.38—Bypassed oil map of a zone in the Duri field and possible horizontal well locations.[59]

steamfloods.[58] In the Duri field, Indonesia, horizontal wells are also being used to supplement conventional vertical-well steamflood.[59] Table 15.9 shows reservoir data for these three areas. All of these latter applications are accessing bypassed oil in mature projects for which vertical wells would be impractical. **Fig. 15.37** is a schematic drawing showing the wedges of oil sand in the Midway Sunset cyclic steam project at the very bottom of the zones that cannot be accessed by the existing vertical wells but will be drained by the horizontal wells. **Fig. 15.38**[59] shows the result of a seismic study of the mature Duri steamflood. Highlighted on the map are pockets of oil that were bypassed by the flood because of proximity to a fault. The operators are strongly considering a suite of horizontal wells to access these reserves.

Nomenclature

A = area, sq ft [m^2]
A_{cD} = dimensionless factor defined by Eq. 15.45
A_{RD} = dimensionless buoyancy defined in Eq. 15.23
A_t = time-dependent heated area, sq ft [m^2]
B_o = oil formation volume factor, RB/STB [res m^3/stock-tank m^3]
C = isobaric specific heat
C_{an} = isobaric specific heat of annular fluid, Btu/(lbm-°F) [kJ/kg·K]
C_o = isobaric specific heat of oil, Btu/(lbm-°F) [kJ/kg·K]
C_w = isobaric specific heat of water, Btu/(lbm-°F) [kJ/kg·K]
D = depth below surface, ft [m]
$erfc(x)$ = complementary error function
E = efficiency
E_c = fraction of oil displaced that is produced
E_h = heat efficiency—fraction of injected heat remaining in reservoir

f = volumetric fraction of noncondensable gas in vapor phase
f_b = hot-water zone oil-rate factor defined by Eq. 15.54
f_h = heat utilization factor defined by Eq. 15.55
f_{hv} = fraction of heat injected as latent heat
f_p = fraction of heat injected that is produced
f_{pD} = heat loss factor caused by hot fluid production
f_R = oil recovery factor defined by Eq. 15.14
f_s = steam quality
$f_{(tD)}$ = function of dimensionless time defined by Eq. 15.102
$f(T)$ = temperature function defined in Eq. 15.56
f_{Vr} = conductive heat loss factor caused by radial conduction
f_{Vz} = conductive heat loss factor caused by vertical conduction
F_1, F_2 = constants defined in Table 15.2
F_{fo} = ratio of fuel burned to produced oil, B/B [m^3/m^3]
F_{of} = produce oil/fuel burned ratio, B/B [m^3/m^3]
F_{os} = produce oil/injected steam ratio, B/B [m^3/m^3]
F_{Sor} = residual oil factor used in Eq. 15.37
$F(T_{ins}, T_{ei})$ = temperature function defined by Eq. 15.97
g = gravity acceleration constant, 32.174 ft/sec^2 [9.8067 m/s^2]
g_c = conversion factor in Newton's second law of motion, 32.174 lbm-ft/lbf-s^2 [1.0 kg·m/N·s^2]
$G(t_D)$ = function defined by Eq. 15.10
h = enthalpy per unit mass, Btu/lbm [kJ/kg]
h_e = fluid level in reservoir at external boundary, ft [m]
h_f = enthalpy of liquid portion of saturated steam, Btu/lbm [kJ/kg]
h_{fc} = forced convection coefficient of heat transfer, Btu/(sq ft-D-°F) [kJ/m^2·d·K]
h_{fs} = enthalpy of < 100% quality saturated steam, Btu/lbm [kJ/kg]
h_{ft} = film coefficient of heat transfer, Btu/(sq ft-D-°F) [kJ/m^2·d·K]
h_{fv} = enthalpy of vapor portion of saturated steam, Btu/lbm [kJ/kg]
h_h = fluid level in stimulated reservoir, ft [m]
h_n = net reservoir thickness, ft [m]
h_{pi} = film coefficient of heat transfer at pipe inner radius, Btu/(sq ft-D-°F) [kJ/m^2·d·K]
h_{po} = film coefficient of heat transfer at pipe outer radius, Btu/(sq ft-D-°F) [kJ/m^2·d·K]
h_r = coefficient of radiant heat transfer for the outermost surface, Btu/(sq ft-D-°F) [kJ/m^2·d·K]
$h_{rc,an}$ = radiant/convection heat transfer coefficient in well annulus, Btu/(sq ft-D-°F) [kJ/m^2·d·K]
h_s = steam zone thickness, ft [m]
h_t = gross reservoir thickness, ft [m]
h_v = enthalpy of 100% quality (saturated) saturated steam, Btu/lbm [kJ/kg]
h_w = fluid level in cold wellbore, ft [m]
i_w = cold water equivalent steam injection rate, B/D [m^3/d]
J = productivity of a cold well, B/psi-D [m^3/kPa·d]
J_h = productivity of a stimulated well, B/psi-D [m^3/kPa·d]

k = reservoir permeability, md [μm^3]
k_{ro} = relative permeability to oil
k_{rs} = relative permeability to steam
L = distance between wells, ft [m]
m = mass, lbm [kg]
m^* = exponent in Eqs. 15.33, 15.105, and 15.106
$m_{\text{casing\_blow}}$ = mass (gas) extracted from system, lbm [kg]
m_i = mass injection, lbm [kg]
m_{influx} = mass exiting system, lbm [kg]
m_l = mass of liquid, lbm [kg]
$m_{(o/w)\text{influx}}$ = mass flowing into system, lbm [kg]
$m_{(o/w)\text{prod}}$ = mass (fluid) extracted from system, lbm [kg]
m_v = mass of vapor, lbm [kg]
$m_{Z(\text{accum})}$ = mass accumulating in system, lbm [kg]
M_g = volumetric heat capacity of gas, Btu/(ft^3-°F) [kJ/m^3·K]
M_o = volumetric heat capacity of oil, Btu/(ft^3-°F) [kJ/m^3·K]
M_R = volumetric heat capacity of the reservoir, Btu/(ft^3-°F) [kJ/m^3·K]
M_s = volumetric heat capacity of steam zone, Btu/(ft^3-°F) [kJ/m^3·K]
M_S = volumetric heat capacity of surrounding formation, Btu/(ft^3-°F) [kJ/m^3·K]
M_w = volumetric heat capacity of water, Btu/(ft^3-°F) [kJ/m^3·K]
M_σ = volumetric heat capacity of reservoir rocks, Btu/(ft^3-°F) [kJ/m^3·K]
n = index of time increment
N = initial oil in place, B [m^3]
N_d = oil displacement rate, B/D [m^3/d]
N_{Gr} = Grashof number
N_m = initial mobile oil in place, B [m^3]
N_p = cumulative oil produced, B [m^3]
N_{Pr} = Prandtl number
N_{Re} = Reynolds number
p = atmospheric pressure, psia [kPa]
p_e = external boundary pressure, psia [kPa]
p_s = steam pressure, psia [kPa]
p_w = wellbore pressure, psia [kPa]
q_{excess_s} = steam injection rate to make up for surface heat losses, B/D [m^3/d]
q_{gh} = hot gas production rate, Mcf/D [std m^3/d]
q_{is} = reproduced steam rate, B/D [m^3/d]
q_{iso} = initial steam injection rate, B/D [m^3/d]
q_{ls} = steam injection rate to make up for reservoir heat losses, B/D [m^3/d]
q_{oc} = cold oil production rate, B/D [m^3/d]
q_{oD} = displaced oil rate defined in Eq. 15.49, B/D [m^3/d]
q_{og} = oil production rate owing to gravity displacement, B/D [m^3/d]
q_{oh} = hot oil production rate, B/D [m^3/d]
q_{oi} = initial oil production rate, B/D [m^3/d]
q_{ot} = total oil production rate, B/D [m^3/d]
q_{ov} = oil production rate because of viscous displacement, B/D [m^3/d]
q_{ps} = steam rate to replace reservoir volume of produced oil, B/D [m^3/d]

q_{wc} = cold water production rate, B/D [m³/d]
q_{wh} = hot water production rate, B/D [m³/d]
Q = amount of injected heat remaining in reservoir, Btu [kJ]
$Q_{\text{casing\_blow}}$ = heat removed with produced gas, Btu [kJ]
Q_i = total heat injected, Btu [kJ]
Q_{influx} = heat leaving system, Btu [kJ]
Q_l = heat lost in reservoir, Btu [kJ]
Q_{ls} = surface piping heat loss/unit length, Btu/ft [kJ/m]
Q_{ot} = cumulative oil recovery at time (t), B/D [m³/d]
$Q_{(o/w)\text{influx}}$ = heat flowing into system, Btu [kJ]
$Q_{(o/w)\text{prod}}$ = heat removed with produced liquids, Btu [kL]
$Q_{z(\text{accum})}$ = heat accumulating in system, Btu [kJ]
$\dot{Q}_i$ = heat injection rate, Btu/D [kJ/d]
$\dot{Q}l$ = heat loss rate, Btu/D [kJ/d]
$\dot{Q}p$ = heat removed with produced fluids, Btu/D [kJ/d]
$\dot{Q}sv$ = volumetric heat injection rate, MMBtu/D/acre-ft [kJ/m³]
r = radius of reservoir, ft [m]
r_{ci} = casing internal radius, ft [m]
r_{co} = outer casing radius, ft [m]
r_e = external radius of heated zone, ft [m]
r_{Ea} = altered radius in earth around wellbore, ft [m]
r_h = radius of heated or steam zone, ft [m]
r_i = inside pipe radius, ft [m]
r_{ins} = insulation external radius, ft [m]
r_o = outside pipe radius, ft [m]
r_w = radius of well, ft [m]
R_h = overall specific thermal resistance, °F-ft-D/Btu [K·m·d/kJ]
S = skin factor before stimulation
S_g = gas saturation fraction
S_h = skin factor after stimulation
S_o = oil saturation
S_{oi} = initial oil saturation fraction
S_{om} = mobile oil saturation fraction
S_{or} = residual oil saturation fraction
S_{ora}, S_{orb} = terms used in Eq. 15.40
S_{ors} = residual oil saturation to steam fraction
S_{orw} = residual oil saturation to water fraction
S_w = water saturation fraction
t = time, D [d]
t^* = time at which steam injection rate reduction is to begin, D [d]
t_{cD} = critical dimensionless time
t_D = dimensionless time
T = average temperature in heated reservoir, °F
T^* = temperature above which oil saturation is reduced to S_{orw}, °F
T_a = air temperature, °F
T_A = ambient temperature, °F

T_b = bulk fluid temperature, °F
T_{BHF} = bottomhole fluid temperature, °F
T_{ci} = temperature of casing wall, °F
T_E = temperature of earth, °F
T_{FW} = steam generator feedwater temperature, °F
T_h = temperature in stimulated zone, °F
T_i = influx water temperature, °F
T_o = temperature of outer surface, °F
T_p = produced fluid temperature, °F
T_R = unaffected reservoir temperature, °F
T_s = steam temperature, °F
T_{SZ} = steam zone temperature, °F
u = volumetric flux, ft³/sq ft-D [m/d]
U = unit function equals 1 for $t_D - t_{cD} > 0$, 0 for $t_D - t_{cD} < 0$ in Eq. 15.22
V_{oD} = dimensionless factor defined by Eq. 15.42
V_{pD} = dimensionless factor defined by Eq. 15.43
V_s = steam zone volume, acre ft [m³]
V_{sz} = steam zone volume per Neuman in Eq. 15.52, B [m³]
w_{st} = mass flow rate of dry steam, lbm/D [kg/d]
x = distance along the x ordinate
X = factor defined in Eq. 15.30
Y = factor defined in Eq. 15.31
α = thermal diffusivity of reservoir, ft²/D [m²/d]
α_E = thermal diffusivity of earth, ft²/D [m²/d]
α_s = thermal diffusivity of surrounding formation, ft²/D [m²/d]
β_1 = thermal volumetric expansion coefficient, 1/°F [1/K]
β_{an} = volumetric thermal expansion coefficient of gas in annulus, 1/°F [1/K]
γ = specific gravity
Δ = increment or decrement
Δh_h = change in stimulated zone fluid level, ft [m]
Δt = time steps, D [d]
ΔT = steam temperature/reservoir temperature, T_s/T_R, °F
$\Delta\gamma$ = oil/steam specific gravity difference
$\Delta\rho$ = density difference between water and oil, lbm/ft³ [kg/m³]
ε = emissivity
ε_{ci} = radiant emissivity of casing wall
ε_{ins} = radiant emissivity of insulation outer surface
θ_c = wetting contact angle, deg (°) [rad]
θ = formation dip angle, deg (°) [rad]
λ = thermal conductivity, Btu/(ft-D-°F) [kJ/m·d·K]
λ_a = thermal conductivity of air, Btu/(ft-D-°F) [kJ/m·d·K]
$\lambda_{a,a}$ = thermal conductivity of air in well annulus, Btu/(ft-D-°F) [kJ/m·d·K]
λ_E = thermal conductivity of unaltered earth, cp [Pa·s]
λ_{Ea} = thermal conductivity of altered earth, cp [Pa·s]
λ_{ins} = thermal conductivity of insulation, cp [Pa·s]
λ_p = thermal conductivity of pipe, cp [Pa·s]

λ_S = thermal conductivity of surrounding formation, cp [Pa·s]
μ = viscosity, cp [Pa·s]
μ_a = viscosity of air, cp [Pa·s]
μ_{an} = viscosity of well annulus gas, cp [Pa·s]
μ_{oh} = hot oil viscosity, cp [Pa·s]
μ_{oi} = initial oil viscosity, cp [Pa·s]
μ_s = steam viscosity, cp [Pa·s]
π = constant pi, 3.141
ρ = density, lbm/ft^3 [kg/m^3]
$\rho_{a,sc}$ = density of air, lbm/ft^3 [kg/m^3]
ρ_{an} = density of well annulus gas, lbm/ft^3 [kg/m^3]
ρ_o = density of oil, lbm/ft^3 [kg/m^3]
ρ_s = density of dry steam, lbm/ft^3 [kg/m^3]
ρ_w = density of water, lbm/ft^3 [kg/m^3]
$\rho_{w,sc}$ = density of water at standard conditions, 62.4 lbm/ft^3 [662.69 kg/m^3]
σ = interfacial tension, oil/water, dyne/cm [mN/m]
τ = time since location in reservoir reached steam temperature, D [d]
v_s = steam specific volume, ft^3/lbm [m^3/kg]
v_w = wind velocity, miles/hr [km/h]
ϕ = porosity

Subscripts

a = air
cem = cement
d = displaced
ins = insulation
o/w = oil/water
prod = production

Superscripts

$*$ = special value as in t^* and T^*
x = integration variable in Eq. 15.22

References

1. Prats, M.: *Thermal Recovery,* Monograph Series, SPE, Richardson, Texas (1982) **7,** 1.
2. Moritis, G.: "EOR Weathers Low Oil Prices," *Oil & Gas J.* (March 2000) 39.
3. Keenan, J.H. *et al.*: *Steam Tables—Thermodynamic Properties of Water Including Vapor, Liquid, and Solid Phases,* John Wiley & Sons Inc., New York City (1969) 2.
4. Chien, S.: "Empirical Correlations of Saturated Steam Properties," paper SPE 20319 presented at the 1991 SPE Asia-Pacific Conference, Perth, Australia, 4–7 November.
5. Carslaw, H.S. and Jaeger, J.C.: *Conduction of Heat in Solids,* Oxford U. Press, Amen House, London (1950) 373.
6. Marx, J.W. and Langenheim, R.H.: "Reservoir Heating by Hot Fluid Injection," *Trans.,* AIME (1959) **216,** 312.
7. van Lookeren, J.: "Calculation Methods for Linear and Radial Steam Flow in Oil Reservoirs," *SPEJ* (June 1983) 427.
8. Farouq Ali, S.M.: *Oil Recovery by Steam Injection,* Producers Publishing Co., Bradford, Pennsylvania (1970) 51.

9. Ramey, H.J. Jr.: "Discussion of Reservoir Heating by Hot Fluid Injection," *Trans.,* AIME (1959) **216,** 364.
10. Prats, M.: "The Heat Efficiency of Thermal Recovery Processes," *JPT* (March 1969) 323; *Trans.,* AIME, **246**.
11. Mandl, G. and Volek, C.W.: "Heat and Mass Transport in Steam Drive Processes," *SPEJ* (March 1969) 59; Reprint Series, SPE, Richardson, Texas (1972) **10,** 27.
12. Myhill, N.A. and Stegemeier, G.L.: "Steamdrive Correlation and Prediction," *JPT* (February 1978) 173.
13. Neuman, C.H.: "A Gravity Override Model of Steamdrive," *JPT* (January 1985) 163.
14. Owens, B.K. and Ziegler, V.M.: "An Oil Production Model for a Well Producing by Both Gravity Drainage and Viscous Flow From a Mature Steamflood," paper SPE 29656 presented at the 1995 SPE Western Regional Meeting, Bakersfield, California, 8–10 March.
15. Closmann, P.J.: "A Simplified Gravity Drainage Oil Production Model for Mature Steamdrives," paper SPE 25790 presented at the 1993 SPE International Thermal Operations Symposium, Bakersfield, California, 8–10 February.
16. Kimber, K.D. *et al.:* "The Role of Latent Heat in Heat Management of Mature Steamfloods," paper SPE 29659 presented at the 1995 SPE Western Regional Meeting, Bakersfield, California, 8–10 March.
17. Ward, R.C. and Sharpe, H.N.: "Mechanistic Modeling of Casing Blow Shut-In Effects and Strategies for the Recovery of Lost Production," paper SPE 37561 presented at the 1997 SPE International Thermal Operations Symposium, Bakersfield, California, 10–12 February.
18. Hayes, H.J. *et al.: Enhanced Oil Recovery,* National Petroleum Council, Industry and Advisory to the U.S. Department of the Interior, Washington DC (1984).
19. Iyoho, A.W.: "Selecting Enhanced Oil Recovery Processes," *World Oil* (November 1978) 61.
20. Farouq Ali, S.M.: "Current Status of Steam Injection as a Heavy Oil Recovery Method," *J. of Canadian Petroleum Technology* (January–March 1974) 1.
21. Venkatesh, E.S., Venkatesh, V.E., and Menzie, D.E.: "PDS Data Base Aids Selection of EOR Method," *Oil & Gas J.* (June 1998) 63.
22. Chu, C.: "State-of-the-Art Review of Steamflood Field Projects," paper SPE 11733 presented at the 1983 SPE California Regional Meeting, Ventura, California, 23–25 March.
23. Donaldson, A.B. and Donaldson, J.E.: "Dimensional Analysis of the Criteria for Steam Injection," paper SPE 38303 presented at the 1997 SPE Western Regional Meeting, Long Beach, California, 25–27 June.
24. Doscher, T.M. and Ghassemi, F.: "Limitations on the Oil/Steam Ratio for Truly Viscous Crudes," *JPT* (July 1984) 1123.
25. Greaser, G.R. and Shore, R.A.: "Steamflood Performance in the Kern River Field," paper SPE 8834 presented at the 1980 SPE/DOE Symposium on Enhanced Oil Recovery, Tulsa, 20–23 April.
26. Restine, J.L.: "On the Effect of Viscosity and Sand Thickness on Kern River Field, Single-Zone, Steamflood Performance," paper SPE 21526 presented at the 1991 SPE International Thermal Operations Symposium, Bakersfield, California, 7-8 February.
27. Rattia, A. and Farouq Ali, S.M.: "Effects of Formation Compaction on Steam Injection Response," paper SPE 10323 presented at the 1981 SPE Annual Technical Conference and Exhibition, San Antonio, Texas, 5–7 October.
28. Jones, J.: "Steamdrive Model for Hand-Held Programmable Calculators," *JPT* (September 1981) 1583; *Trans.,* AIME, **271.**
29. van der Knaap, W.: "Physical Aspects of Some Steam Injection Theories, an Application," paper SPE 26293 available from SPE, Richardson, Texas (1993).
30. Vogel, J.V.: "Simplified Heat Calculations for Steamfloods," *JPT* (July 1984) 1127.
31. Kumar, M. and Ziegler, V.M.: "Injection Schedule and Production Strategies for Optimizing Steamflood Performance," *SPERE* (May 1993) 101.
32. Gomaa, E.E.: "Correlations for Predicting Oil Recovery by Steamflood," *JPT* (February 1980) 325.
33. Owens, W.D. and Suter, V.E.: "Steam Stimulation—Newest Form of Secondary Petroleum Recovery," *Oil & Gas J.* (April 1965) **90,** 82.
34. Boberg, T.C. and Lantz, R.B.: "Calculation of the Production Rate of a Thermally Stimulated Well," *JPT* (December 1966) 1613; *Trans.,* AIME, **237.**

35. Towson, D.E. and Boberg, T.C.: "Gravity Drainage in Thermally Stimulated Wells," *J. of Canadian Petroleum Technology* (October–December 1967) 130.
36. Jones, J.: "Why Cyclic Steam Predictive Models Get No Respect," *SPERE* (February 1992) 67.
37. Rivero, R.T. and Heintz, R.C.: "Resteaming Time Determination—Case History of a Steam-Soak Well in Midway Sunset," *JPT* (June 1975) 655.
38. McBean, W.N.: "Attic Oil Recovery by Steam Displacement," paper SPE 4170 presented at the 1972 SPE California Regional Meeting, Bakersfield, California, 8–10 November.
39. Jones, J. and Cawthon, G.J.: "Sequential Steam: An Engineered Cyclic Steaming Method," *JPT* (July 1990) 848–853, 901.
40. Kuo, C.H., Shain, S.A., and Phocas, D.M.: "A Gravity Drainage Model for the Steam-Soak Process," *SPEJ* (June 1970) 119; *Trans.,* AIME, **249.**
41. McAdams, W.H., Williams, G.C., and Smith, K.A.: "Standard Handbook for Mechanical Engineers," *Transmission of Heat by Conduction and Convection,* T. Baumeister and L. Marks (eds.), seventh edition, McGraw-Hill Book Co. Inc., New York City (1967) Chap. 4, 92.
42. Rohsenow, W.M. and Hartnett, J.P.: *Handbook of Heat Transfer*, McGraw-Hill Book Co. Inc., New York City (1973) 3.
43. Willhite, G.P.: "Overall Heat Transfer Coefficients on Steam and Hot Water Injection Wells," *JPT* (May 1967) 607.
44. Dennis, E.L.: "Project Design for Slimhole Steam Injectors in Thermal Recovery Projects as Compared to Conventional Steam Injectors," paper SPE 29629 presented at the 1995 SPE Western Regional Meeting, Bakersfield, California, 8–10 March.
45. Ramey, H.J. Jr.: "Wellbore Heat Transmission," *JPT* (April 1962) 427; *Trans.,* AIME, **225.**
46. van Dorp, J.J. and Roach, R.H.: "Steam Management in Composite Mature Steamfloods—Midway Sunset Field," paper SPE 29658 presented at the 1995 SPE Western Regional Meeting, Bakersfield, California, 8–10 March.
47. Ziegler, V.M. *et al.:* "Recommended Practices for Heat Management of Steamflood Projects," paper SPE 25808 presented at the 1993 SPE International Thermal Operations Symposium, Bakersfield, California, 8–10 February.
48. Jones, J.A.: "Discussion of Simulating a Steamflood at the Georgsdorf Field, Federal Republic of Germany," *JPT* (June 1985) 1069.
49. Chang, H.L., George, A.E., and Farouq Ali, S.M.: "Performance of Horizontal-Vertical Combinations for Steamflooding Bottomwater Formations," *J. Cdn. Pet. Tech.* (May 1992) 41.
50. Huang, W.S. and Hight, M.A.: "Evaluation of Steamflood Processes With Horizontal Wells," *SPERE* (February 1989) 69.
51. Dietrich, J.K.: "The Kern River Horizontal-Well Steam Pilot," *SPERE* (August 1988) 935.
52. Jesperson, P.J. and Fontaine, T.J.C.: "The Tangleflags North Pilot: A Horizontal Well Steamflood," *J. Cdn. Pet. Tech.* (May 1993) 52.
53. Butler, R.M., McNab, G.S., and Lo, H.Y.: "Theoretical Studies on the Gravity Drainage of Heavy Oil During In-Situ Steam Heating," *Cdn. J. of Chem. Eng.* (August 1981) **59,** 455 (paper presented in 1979).
54. Butler, R.M.: "A New Approach to the Modeling of Steam-Assisted Gravity Drainage," *J. Cdn. Pet. Tech.* (May/June 1985) 42.
55. Scott, G.R.: "Comparison of CSS and SAGD Performance in the Clearwater Formation at Cold Lake," paper SPE 79020 presented at the 2002 SPE International Thermal Operations and Heavy Oil Symposium and International Horizontal Well Technology Conference, Calgary, 4–7 November.
56. Chona, R.A. *et al.:* "Evaluation of a Horizontal Infill Well in a Mature Cyclic-Steam Project," paper SPE 37087 presented at the 1996 SPE International Conference on Horizontal Well Technology, Calgary, 18–20 November.
57. McKay, C., Jones, J., and Pomerene, J.: "Successful Horizontal Producers in Midway Sunset Thermal Operations," paper SPE 83479 presented at the 2003 SPE Western Regional/AAPG Pacific Section Meeting, Long Beach, California, 19–24 May.
58. Cline, V.J. and Basham, M.: "Improving Project Performance in a Heavy Oil Horizontal Well Project in the San Joaquin Valley, California," paper SPE 78981 presented at the 2002 SPE International Thermal Operations and Heavy Oil Symposium and International Horizontal Well Technology Conference, Calgary, 4–7 November.

59. Sigit, R. *et al.*: "Seismically Guided Bypassed Oil Identification in A Mature Steamflood Area," paper SPE 57261 presented at the 1999 SPE Asia Pacific Improved Oil Recovery Conference, Kuala Lumpur, 25–26 October.

General References

Prats, M.: *Thermal Recovery,* Monograph Series, SPE, Richardson, Texas (1982) **7.**

Green, D.W. and Willhite, G.P.: *Enhanced Oil Recovery,* Textbook Series, SPE, Richardson, Texas (1998) **6,** 301.

Hong, K.C.: *Steamflood Reservoir Management Thermal Enhanced Oil Recovery,* PennWell Books, Tulsa (1994).

Thermal Recovery Processes, Reprint Series, SPE, Richardson, Texas (1965) **7.**

Thermal Recovery Processes, Reprint Series, SPE, Richardson, Texas (1985) **7.**

Thermal Recovery Techniques, Reprint Series, SPE, Richardson, Texas (1972) **10.**

Lake, L.W.: *Enhanced Oil Recovery,* Prentice Hall, Indianapolis, Indiana (1989).

SI Metric Conversion Factors

acre	×	4.046 856	E − 01	= ha
acre-ft	×	1.233 489	E + 03	= m^3
°API		141.5/(131.5 + °API)		= g/cm^3
B	×	1.590	E − 01	= m^3
B/D	×	1.590	E − 01	= m^3/d
Btu	×	1.055 056	E + 00	= kJ
Btu/(ft-D-°F)	×	6.231	E + 00	= kJ/m·d·K
Btu/hr	×	2.930 711	E − 01	= W
Btu/lbm	×	2.326	E + 01	= kJ/kg
Btu/(lbm-°F)	×	4.187	E + 00	= kJ/kg·K
cp	×	1.0*	E − 03	= Pa s
D	×	1.0	E + 00	= d
ft	×	3.048*	E − 01	= m
ft^2	×	9.290 304*	E − 02	= m^2
ft^3	×	2.831 685	E − 02	= m^3
°F		(°F − 32)/1.8		= °C
lbm	×	4.536	E − 01	= kg
lbm/ft^3	×	1.062	E + 01	= kg/m^3
MM Btu/D/acre-ft	×	8.556	E + 02	= kJ/m^3
psi	×	6.894 757	E + 00	= kPa
psia	×	6.895	E + 00	= kPa

*Conversion factor is exact.

Chapter 16
In-Situ Combustion

William E. Brigham and **Louis Castanier,** Stanford U.

16.1 Introduction

In-situ combustion (ISC) is the oldest thermal-recovery technique. It has been used for more than nine decades with many economically successful projects. Nevertheless, it is regarded as a high-risk process by many primarily because of the many failures of early field tests. Most of those failures came from the application of a good process (ISC) to the wrong reservoirs or the poorest prospects. An objective of this chapter is to clarify the potential of ISC as an economically viable oil-recovery technique for a variety of reservoirs. This chapter is a summary containing a description of ISC, a discussion of laboratory screening techniques, an illustration of how to apply laboratory results to field design, a review of performance-prediction methods, a discussion of operational practices and problems, and an analysis of field results. For a more complete review, the work of Sarathi,[1] Prats,[2] and Burger *et al.*[3] should be consulted.

16.2 Process Description

ISC is basically injection of an oxidizing gas (air or oxygen-enriched air) to generate heat by burning a portion of resident oil. Most of the oil is driven toward the producers by a combination of gasdrive (from the combustion gases), steam, and waterdrive. This process is also called fire flooding to describe the movement of a burning front inside the reservoir. Based on the respective directions of front propagation and air flow, the process can be forward (when the combustion front advances in the same direction as the air flow) or reverse (when the front moves against the air flow).

16.2.1 Reverse Combustion. This process has been studied extensively in laboratories and tried in the field. The idea is that it could be a useful way to produce very heavy oils with high viscosity. In brief, it has not been successful economically for two major reasons.

First, combustion started at the producer results in hot produced fluids that often contain unreacted oxygen. These conditions require special, high-cost tubulars to protect against high temperatures and corrosion. More oxygen is required to propagate the front compared to forward combustion, thus increasing the major cost of operating an ISC project.

Second, unreacted, coke-like heavy ends will remain in the burned portion of the reservoir. At some time in the process, the coke will start to burn, and the process will revert to forward combustion with considerable heat generation but little oil production. This has occurred even in carefully controlled laboratory experiments.

In summary, reverse combustion has been found difficult to apply and economically unattractive.

16.2.2 Forward Combustion. Because only forward combustion is practiced in the field, we will only consider this case. Forward combustion can be further characterized as "dry" when only air or enriched air is injected or "wet" when air and water are coinjected.

Dry Combustion. The first step in dry forward ISC is to ignite the oil. In some cases, autoignition occurs when air injection begins if the reservoir temperature is fairly high and the oil is reasonably reactive. This often occurs in California reservoirs. Ignition has been induced with downhole gas burners, electrical heaters, and/or injection of pyrophoric agents or steam injection; it will be discussed in more detail later.

After ignition, the combustion front is propagated by a continuous flow of air. Rather than an underground fire, the front is propagated as a glow similar to the hot zone of a burning cigarette or hot coals in a barbecue. As the front progresses into the reservoir, several zones exist between injector and producer as a result of heat, mass transport, and chemical reactions. **Fig. 16.1**[4] is an idealized representation of the various zones and the resulting temperature and fluid-saturation distributions. In the field, there are transitions between zones; however, the concepts illustrated provide insight on the combustion process.

Zone Definitions. Starting from the injector, seven zones have been defined.

• A—The burned zone is the volume already burned. This zone is filled with air and may contain small amounts of residual unburned organic solids. Because it has been subjected to high temperatures, mineral alterations are possible. Because of the continuous airflow from the injector, the burned-zone temperature increases from injected-air temperature at the injector to combustion-front temperature at the combustion front.

• B—The combustion front is the highest temperature zone. It is very thin, often no more than several inches thick. It is in this region that oxygen combines with the fuel and high-temperature oxidation occurs. The products of the burning reactions are water and carbon oxides. The fuel is often misnamed coke. In fact, it is not pure carbon but a hydrocarbon with H/C atomic ratios ranging from approximately 0.6 to 2.0. This fuel is formed in the thermal-cracking zone just ahead of the front and is the product of cracking and pyrolisis, which is deposited on the rock matrix. The amount of fuel burned is an important parameter because it determines how much air must be injected to burn a certain volume of reservoir.

• C/D—The cracking/vaporization zone is downstream of the front. The crude is modified in this zone by the high temperature of the combustion process. The light ends vaporize and are transported downstream, where they condense and mix with the original crude. The heavy ends pyrolize, resulting in CO_2, CO, hydrocarbon gases, and solid organic fuel deposited on the rock.

• E—The steam plateau. This is the zone in which some of the hydrocarbon vapors condense. Most of those condense further downstream as the steam condenses. The steam plateau temperature depends on the partial pressure of the water in the gas phase. Depending on the temperature, the original oil may undergo a mild thermal cracking, often named visbreaking, that usually reduces oil viscosity.

• F—A water bank exists at the leading edge of the steam plateau, where the temperature is less than steam saturation temperature. This water bank decreases in temperature and saturation downstream, with a resulting increase in oil saturation.

• G—The oil bank. This zone contains most of the displaced oil, including most of the light ends that result from thermal cracking.

• H—Beyond these affected areas is the undisturbed original reservoir. Gas saturation will increase slightly in this area because of the high mobility of combustion gases.

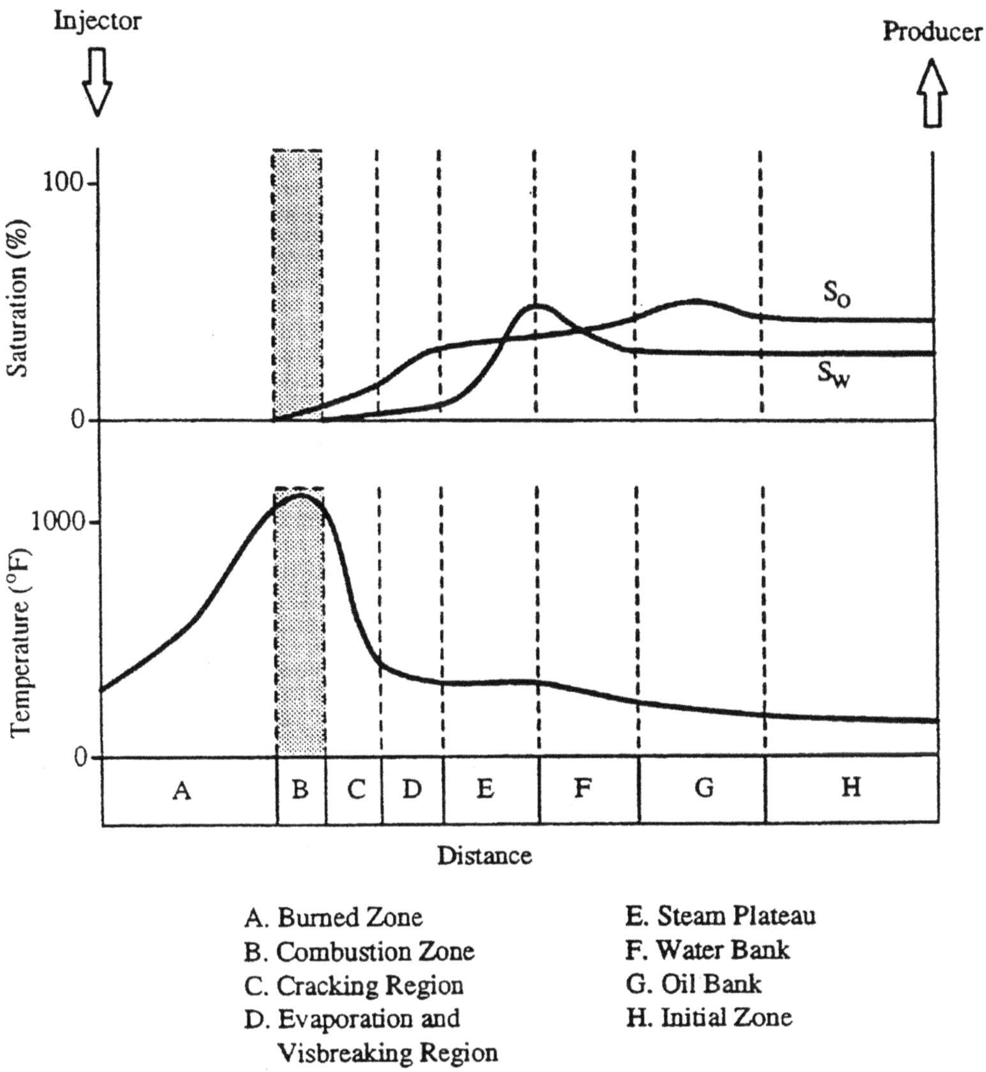

Fig. 16.1—Schematic diagram of temperature and saturation profiles and zones in ISC.[4]

Wet Combustion. A large amount of heat is stored in the burned zone during dry forward ISC, (Fig. 16.1) because the low heat capacity of air cannot transfer that heat efficiently. Water injected with the air can capture and advance more heat stored in the burned zone.

During the wet combustion process, injected water absorbs the heat from the burned zone, vaporizes, moves through the burning front, and expands the steam plateau. As a result, the steam plateau is extended. This results in faster heat movement and oil displacement.

Depending on the water/air ratio, wet combustion is classified as (1) incomplete when the water is converted into superheated steam and recovers only part of the heat from the burned zone, (2) normal when all the heat from the burned zone is recovered, and quenched or (3) super wet when the front temperature declines as a result of the injected water.

When operated properly, water-assisted combustion reduces the amount of fuel needed, resulting in increased oil recovery and decreased air requirements to heat a given volume of reservoir. Up to a 25% improvement in process efficiency can be achieved.[5] Determination of the optimum water/air ratio is difficult because of reservoir heterogeneities and gravity override

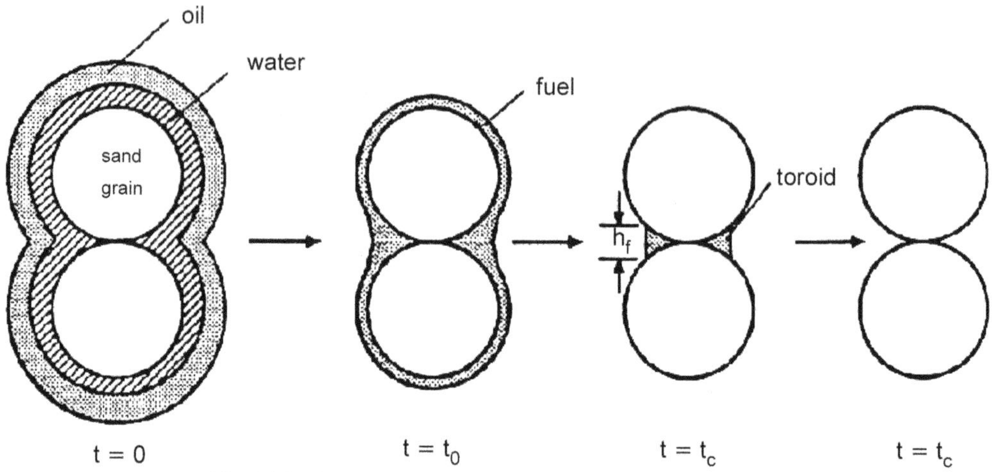

Fig. 16.2—Schematic diagram of varying fuel geometry.[6]

that can affect fluid movement and saturation distributions. Injecting too much water can result in an inefficient fire front, thus losing the benefits of the process.

Some authors recommend, as a best practice, injecting water at high rates to achieve "partially quenched combustion." This method has limited application. A high-temperature burn is preferred but is difficult to achieve with oils that are not highly reactive. Injecting large amounts of water can lower combustion temperatures, resulting in a greater fraction of oil burned and higher costs for oxygen. At the same time, these types of burns only partially oxidize the oil. This partial oxidation results in a much more viscous liquid, which in turn lowers the flow rate. So, in brief, if water injection is used, great care should be taken to assure that liquid water never reaches the high-temperature combustion front. A discussion of heat- and material-balance calculations that includes chemical reactions and the effect of injected air and water is presented later in some detail.

16.3 Laboratory Studies

ISC processes are largely a function of oil composition and rock mineralogy. The extent and nature of the chemical reactions between crude oil and injected air, as well as the heat generated, depend on the oil-matrix system. Laboratory studies, using crude and matrix from a prospective ISC project, should be performed before designing any field operation.

16.3.1 The Reactions. The chemical reactions associated with ISC are complex and numerous. They occur over a broad temperature range. Most researchers group them into three classes in ascending temperature ranges:

- Low-temperature oxidation (LTO)—heterogeneous gas/liquid reactions producing partially oxygenated compounds and few carbon oxides.
- Medium-temperature reactions—cracking and pyrolisis of hydrocarbons to form fuel.
- High-temperature oxidation (HTO)—heterogeneous H/C bond breaking reactions in which the fuel reacts with oxygen to form water and carbon oxides.

A more recent and more accurate kinetics model has been developed.[6] Only two reactions are used, but in addition, the geometry of the reacting residual fuel in the pore spaces is taken into account, as indicated in **Fig. 16.2**. The crude-oil oxidation consists of two stages: LTO forming an oxygenated hydrocarbon fuel, and high-temperature combustion of this fuel. A detailed description of the different reaction regimes is outside the scope of this *Handbook*; some practical comments on the role of LTO, however, are appropriate at this stage.

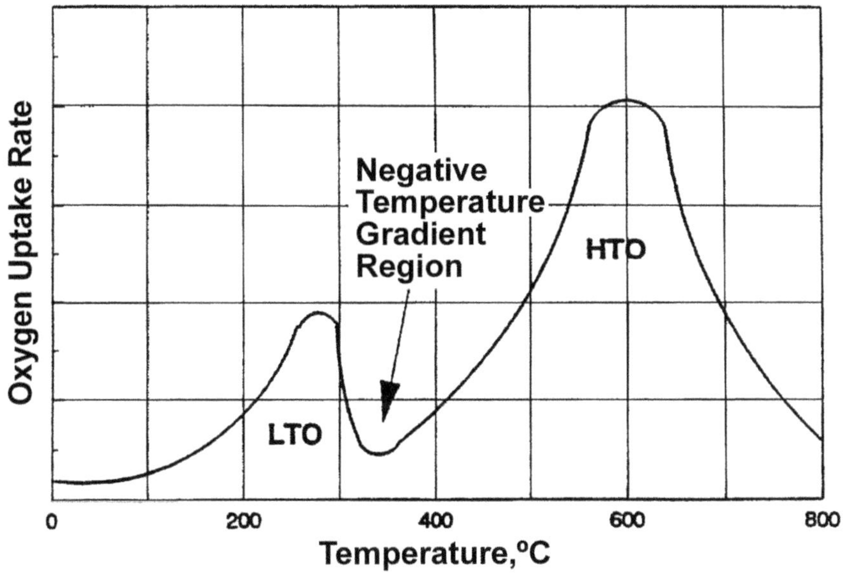

Fig. 16.3—Schematic of dry combustion temperature profile showing the general effect of temperature on oxygen uptake rate for heavy oils and the negative temperature gradient region.[8]

LTO can be described as oxygen addition to the crude oil. LTO yields water and oxygenated hydrocarbons such as ketones, alcohols, and peroxides. A good description of LTO can be found in Burger.[7] LTO generally increases original oil viscosity, boiling range, and density. LTO increases the amount of fuel and is promoted by low air flux in the oxidation zone. Poor crude oxidation characteristics can also play a role. In heavy-oil reservoirs (API gravity $< 20°$), LTO tends to be more pronounced when oxygen (rather than air) is injected in the reservoir.[8]

A U. of Calgary research team has shown that, for heavy oils, LTO reactions must be minimized. **Fig. 16.3** shows the oxygen uptake as the temperature of a typical heavy oil is raised linearly with time. Notice the negative temperature-gradient region, that is, the region in which oxygen-rate uptake decreases with temperature increase. If the temperature of the ISC process stays at or below the negative temperature-gradient region, the oil displacement efficiency will be very low. This is because LTO increases the oil viscosity and fuel content. The injected-air flux in a heavy-oil project should be maintained at a value well above the value needed to maintain the reactions in the high-temperature oxidation regime. LTO generally has almost no effect on light oils in terms of mobility or recovery despite the fact that light oils are more susceptible to LTO than heavy oils.

Fuel deposition determines the feasibility and economic success of a combustion project. It occurs at intermediate temperatures after the LTO reactions. Numerous studies have been conducted in an attempt to understand fuel formation and deposition at intermediate temperatures. The oil type and chemical structure determine the rate and extent of the different reactions. Catalytic effects from the matrix and/or injected solutions of metals may affect the type and amount of fuel formed. Again, all laboratory experiments must include not only the crude to be tested but also representative core material from the reservoir of interest.

16.3.2 Kinetics. Kinetics of combustion reactions can be defined by how fast the chemical reactions occur and how much of the oil is affected. It is important to study kinetics for several reasons:

- Characterization of oil reactivity.
- Determination of ignition conditions.

• Insight on the nature of the fuel and its combustion characteristics.
• Use of kinetic parameters as input for possible numerical simulation of the process.

Because crude oils contain hundreds of different compounds, it is impossible to accurately represent all the reactions occurring during ISC. Even if it were possible to detail all the reactions, the use of such information in numerical models would be impossible because of cost and computer limitations. Consequently, we will concentrate on useful simple models describing ISC reaction kinetics that have been published in the literature. Most studies use the Arrhenius expressions defined as follows. The model assumes a functional dependency on fuel concentration and oxygen partial pressure. It is given by

$$R_c = \frac{dC_f}{dt} = K p_{O_2}^a C_f^b, \qquad (16.1)$$

where R_C = reaction rate of the crude, kg/m³s; C_f = concentration of fuel, kg/m³; and p_{O_2} = oxygen partial pressure, Pa.

The exponent constants, a and b, are the orders of the reactions with respect to oxygen partial pressure (a) and fuel concentration (b). Data show that a ranges between 0.5 and 1.0, while b is close to 1.0. The reaction rate, K, is the Arrhenius constant, expressed as a function of temperature as follows:

$$K = A \exp\left(-E / RT\right), \qquad (16.2)$$

where A = Arrhenius rate constant; E = activation energy, kJ/mol; T = absolute temperature, °K; and R = 0.00831 universal gas constant, kJ/mol °K.

When using literature values, be careful because the parameters in Eqs. 16.1 and 16.2 vary depending on the system of units used.

A variety of experimental techniques can be used to determine the kinetics of ISC reactions. Among those are differential thermal analysis, thermogravimetric analysis, accelerating rate calorimetry, and effluent analysis. The reference list contains several descriptions of various methods and results.

The effluent analysis method, also called the ramped temperature method in Canada, is quantitative and consists of heating a sample of oil and rock while flowing oxygen (for oxidation) or nitrogen (for pyrolisis) through the sample. The kinetic parameters can be calculated from effluent gas evolution with temperature and chemical analysis of post-test cores. Details of the analysis techniques can be found in Refs. 3, 6, and 9.

16.3.3 Combustion-Tube Studies. Although the kinetic studies can provide useful insight on ISC reactions, combustion-tube experiments are mandatory to determine the parameters needed to design and implement field projects. These data are used to make predictions of field test performance. As Sarathi[1] points out, "Combustion-tube studies are the necessary first step in the design of an ISC project."

Combustion tubes aim at representing a small volume of the reservoir. They are usually packed with native reservoir cores or representative samples of matrix material and oil, placed in vertical position to minimize gravity effects and heated to reservoir temperature. Ignition is usually started at the top by electrical heaters, and the combustion front is propagated downward. This allows propagation of a combustion front and the associated chemical reactions at conditions close to those in a reservoir.

Temperature profiles, pressures, gas and liquid injection and production rates, and composition histories at the inlet and outlet are recorded. ISC tube runs are unscaled, and direct correlation of combustion-tube results to the field is not possible. However, as long as the runs are performed with reservoir rock and fluids at reservoir conditions, the reactions of fuel deposition and combustion will be similar in both tube and reservoir.

Tube runs will not provide information on ISC sweep efficiency. They adequately model the chemistry of the process, but not the flow behavior in the reservoir, and only partially model the heat-transfer processes. Flow behavior in the reservoir is affected by gravity override, well spacing, and geometry and reservoir heterogeneities, and tube runs cannot reproduce these phenomena. Heat transfer from the tube to the surroundings can be much higher than reservoir heat losses.

Two different schools of thought exist on this heat-transfer problem. Many experimenters use strip heaters around the tube to lessen the temperature gradient between the tube and the surroundings. This practice can, however, lead to overestimation of water/oil ratios in wet combustion if the strip heaters provide too much energy to the system, as they often do. Information on front cooling by injected water may also be masked by the heaters. As a result, the extent of the steam plateau may not be correct. Most of these types of experiments are bulky and time consuming and require extensive instrumentation.

The other solution is to increase the air flux and minimize heat losses by insulation alone. This may slightly overestimate air requirements and fuel content but is much simpler and easier to operate. As a result, it is widely used. Descriptions of various setups for combustion-tube studies have been provided.[5,6,10–12]

The information that can be acquired from tube runs includes:

- Fuel burned.
- Air required to burn a unit volume of reservoir.
- Atomic H/C ratio of burned fuel.
- Excess air and oxygen use.
- Air/fuel ratio.
- Oil recovery from the swept zone.
- Optimization of water/air ratio in wet combustion.
- Composition of produced fluids.
- Front temperature and stability.

This last piece of information is quite important in heavy oils to determine if the process is operating properly in the desired high-temperature regime. If high temperature cannot be achieved in ideal laboratory conditions, it is likely that field results would be worse.

Data Analysis. The following is a simple analysis of data from tube runs. It assumes that the combustion occurs at high temperature where the fuel exclusively combines with oxygen to produce water and carbon oxides. The stoichiometric equation[13] is then

$$CH_n + \left[\left(\frac{2m+1}{2m+2}\right) + \frac{n}{4}\right]O_2 = \left(\frac{m}{m+1}\right)CO_2 + \left(\frac{1}{m+1}\right)CO + \left(\frac{n}{2}\right)H_2O, \quad \text{(16.3)}$$

where n = hydrogen/carbon atomic ratio of fuel and m = CO_2/CO concentration ratio produced. The other symbols indicate the various components in the chemical-balance equation.

This equation is only an approximation of the process. It neglects LTO reactions, oxygen/mineral reactions, and water/organic fuel reactions. Alternate analysis when some of these reactions are important is detailed in Sarathi[1] based on work from Moore and Mehta.[10] Assuming Eq. 16.3 to be valid, the apparent H/C ratio, n, can be estimated from the concentration of exhaust gases and the injected oxygen concentration.[13]

$$n = \frac{4\left[O_2 - CO_2 - CO/2 - (O_2)_{prod}\right]}{CO_2 + CO}, \quad \text{(16.4)}$$

where $(O_2)_{prod}$ = oxygen concentration produced.

It is prudent to normalize the concentrations by making a balance on the nitrogen, which in these conditions can be considered inert. The basic chemical equation is then

$$CH_n + aO_2 + RaN_2 = bCO_2 + dCO + fO_2 + RaN_2, \quad (16.5)$$

where R is the molar ratio of nitrogen to oxygen in the feed gas, and a, b, d, and f are stoichiometric coefficients similar to those in Eq. 16.3.

The range of the ratio, n, for high-temperature reactions should be from 0.5 to 2. Calculation of an unusually high value of n indicates that LTO is important. In the very early stages of field projects, large n values are often observed because of the solubility of the combustion gases (particularly CO_2) in the oil.

Once n and m are known, the amount of air required to burn one unit weight of fuel is found from Eq. 16.3. The heat generated by burning a unit weight of fuel can be calculated by simple addition of the heat generated by each reaction, as described in the stoichiometric equation (Eq. 16.3). The calculation of heat produced must take into account the production of carbon monoxide. The following formula[13] can estimate heating values of fuels as a function of n and m.

$$H_c = \frac{m(174{,}000)}{(m+1)(n+12)} + \frac{52{,}500}{(m+1)(n+12)} + n\frac{61{,}500}{(n+12)}, \quad (16.6)$$

where H_c = heating value, Btu/lbm fuel. To convert to J/kg, multiply by 2,326.

The air required to burn a given volume of reservoir is, of course, a very important design parameter and one of the keys to the economics of the combustion process. This is calculated directly from the experimental data by dividing the amount of oxygen consumed by the volume swept during the tube run. The mass of fuel burned in a unit volume of reservoir can be calculated from the oxygen consumed by a unit volume and by applying Eq. 16.3. All other relevant parameters can be estimated.[13,14] It is prudent to perform multiple laboratory tube runs before field implementation.

16.4 Combining Material- and Heat-Balance Calculations

Many useful and reasonably accurate calculations can be made on ISC to predict the behavior of a proposed project. These ideas will be explained in the following diagrams and example calculations. They start with a very simple heat balance and are then extended to more closely represent what happens in the reservoir.

16.4.1 First Assumptions. Start by assuming that no combustion data are available to get an initial idea of the feasibility of a project. This preliminary work gives the engineer a sound basis to decide whether further work has economic promise.

Assume a sandstone formation with a porosity of 22%, a temperature of 100°F, a 24°API oil at a saturation of 65%, and an injection pressure of 300 psia. Also assume the CO_2/CO atomic ratio m to be

$$m = \frac{CO_2}{CO} = 20. \quad (16.7)$$

This is a reasonable ratio to assume, based on both laboratory and field experience.

TABLE 16.1—EFFECT OF TEMPERATURE ON H/C RATIO, 21.8° API CRUDE	
H/C Ratio, *n*	Combustion Temperature (°F)
0.40	1,000
0.80	800
1.40	700
2.40	600

Because there are no tube run data, generalized correlation curves[13] (**Figs. 16.4 and 16.5**) will be used to calculate expected results. From Fig. 16.4, the fuel availability, *W*, for 24°API crude is

$$W = 0.95 \text{ lbm } C/100 \text{ lbm rock.} \quad (16.8)$$

The apparent *H/C* atomic ratio (*n*) of the fuel is also needed. This is a function of the combustion-front temperature,[13] as shown in Fig. 16.5. Selected data from the graph are listed in **Table 16.1**. These data for 21.8°API crude are close enough to 24°API crude for initial estimates.

16.4.2 Calculate Initial Heat Balances and Temperatures. Start by assuming that all heat generated is used to heat the rock formation through which the combustion front has moved. This assumption is not accurate, but it simplifies the understanding of the mathematics and concepts involved in heat-balance calculations. A sketch of the temperature profile generated is shown in **Fig. 16.6**. Corrections to this heat balance calculation will be discussed later.

Assuming 1.0 ft³ of rock formation burned and a front temperature of 1,000°F, from Eq. 16.8 and Table 16.1,

$$\frac{\text{lbm fuel}}{100 \text{ lbm rock}} = \frac{\text{lbm carbon}}{100 \text{ lbm rock}} + \frac{\text{lbm hydrogen}}{100 \text{ lbm rock}}$$
$$= 0.95 + \frac{0.95(0.40)}{12} = 0.9817. \quad (16.9)$$

Quartz weighs approximately 164 lbm/ft³. The amount of fuel for a cubic foot of formation equals

$$\frac{(0.9817)(164)(1-0.22)}{100} = 1.256 \text{ lbm fuel/ft}^3\text{rock.} \quad (16.10)$$

Using the heat of combustion, Eq. 16.6, with the appropriate parameters, becomes

$$H_c = \frac{(20)(174{,}000)}{(21)(12.4)} + \frac{52{,}500}{(21)(12.4)} + \frac{(0.4)(61{,}500)}{12.4}$$
$$= 15{,}550 \text{ Btu/lbm fuel.} \quad (16.11)$$

Thus, the total heat generated is

$$\frac{\text{Heat generated}}{\text{ft}^3\text{rock}} = (15{,}550)(1.256) = 19{,}530\frac{\text{Btu}}{\text{ft}^3\text{rock}}. \quad (16.12)$$

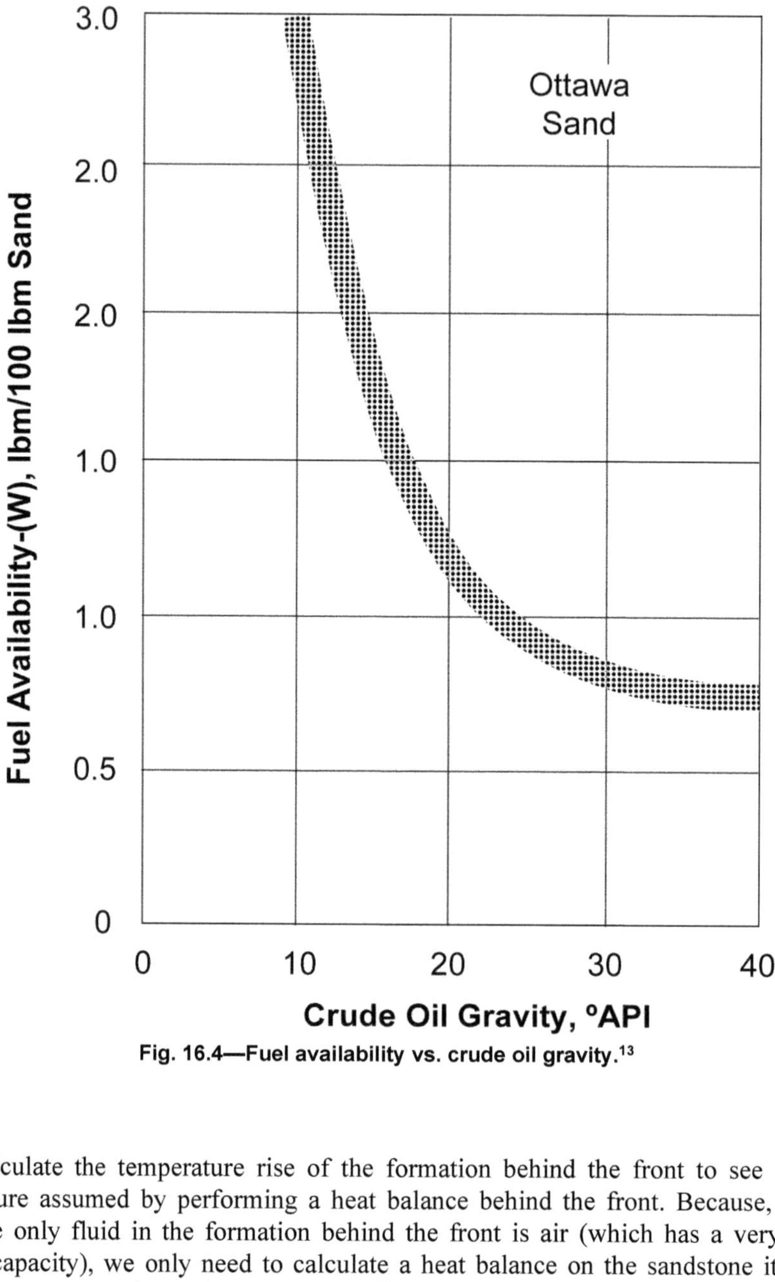

Fig. 16.4—Fuel availability vs. crude oil gravity.[13]

Next, calculate the temperature rise of the formation behind the front to see if it matches the temperature assumed by performing a heat balance behind the front. Because, for practical purposes, the only fluid in the formation behind the front is air (which has a very small volumetric heat capacity), we only need to calculate a heat balance on the sandstone itself. A good equation for average sandstone heat capacity is[13]

$$c_s = \frac{T_1 + 2{,}000}{10{,}000} + \frac{T_2 - T_1}{20{,}000} \quad \text{(16.13a)}$$

$$= 0.21 + \frac{1{,}000 - 100}{20{,}000} = 0.255 \text{ Btu/lbm-°F}, \quad \text{(16.13b)}$$

where T_1 = initial reservoir temperature, °F and T_2 = final reservoir temperature, °F.

From a heat-balance calculation, the reservoir sand temperature is as follows:

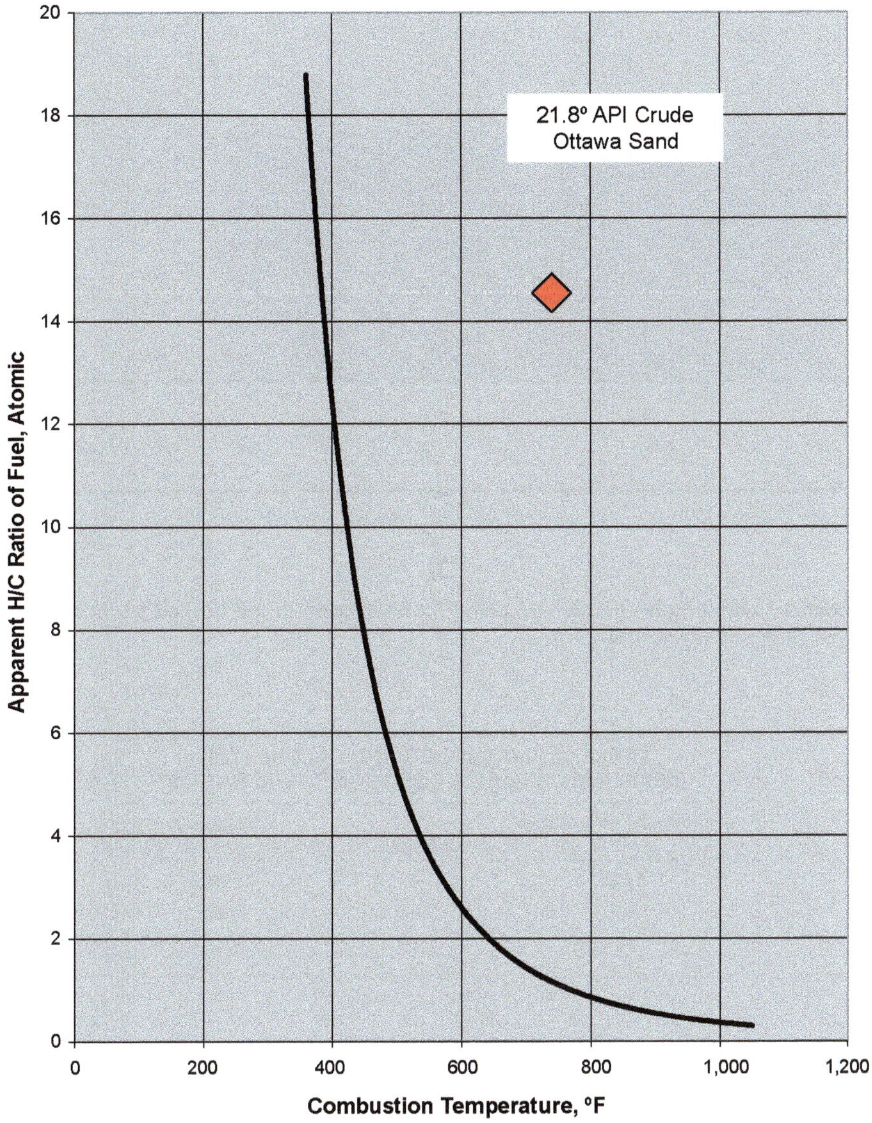

Fig. 16.5—H/C ratio vs. combustion temperature.[13]

$$T_2 - 100 = \frac{19{,}530}{(0.255)(164)(1-0.22)} = 598.7$$

$$\text{or } T_2 = 699°\text{F}. \qquad (16.14)$$

The result from Eq. 16.14 does not agree with the assumed temperature of 1,000°F. Calculations with other assumed temperatures result in the calculated temperature values shown in **Table 16.2**.

The tabular data are graphed as circles in **Fig. 16.7**. The two temperatures match at 801°F. This is the calculated combustion-front temperature if all the heat generated is used to heat the formation behind the combustion front.

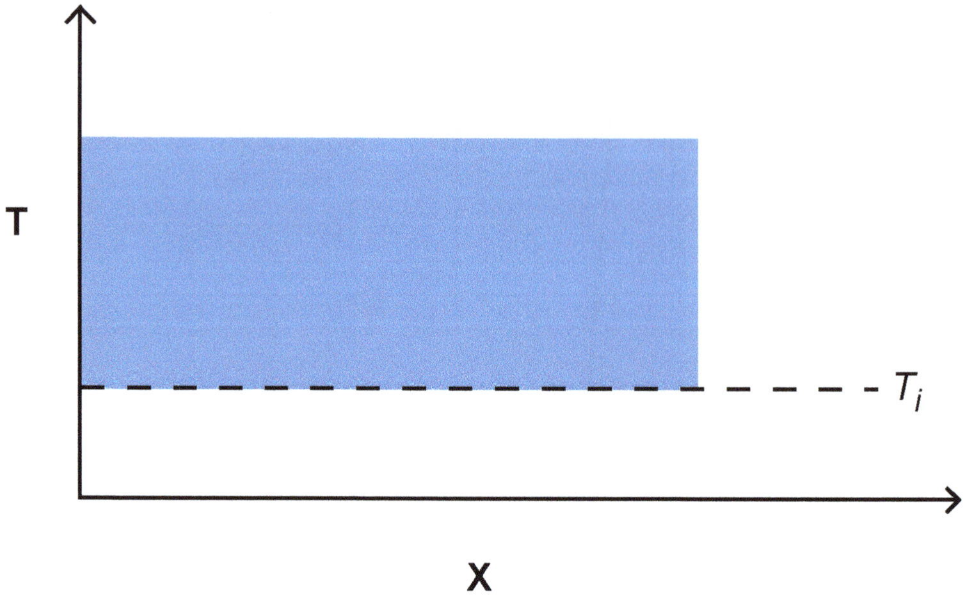

Fig. 16.6—Idealized temperature profile assuming all heat stays in the burned zone; *x* is the distance coordinate away from an injector at *x* = 0.

TABLE 16.2—ASSUMED AND CALCULATED TEMPERATURES AS A FUNCTION OF H/C RATIOS

Assumed Temperature (°F)	H/C Ratio	Calculated Temperature (°F)
1,000	0.40	699
800	0.80	802
700	1.40	939

16.4.3 Correction for Water of Combustion. These results do not include all the processes occurring in the reservoir. First, the water formed by combustion will condense beyond the combustion front, absorb some heat of combustion, and reduce the heat of the formation behind the combustion front. This effect can be calculated as follows.

From Eq. 16.9, 0.0317 lbm of H_2 are burned per 100 lbm of rock at 1,000°F. Assuming that a pound of steam will release 1,000 Btu when cooling from combustion temperature and condensing (this number is only approximately correct but is adequate for estimation purposes), the amount of heat carried forward by the steam is calculated with concepts similar to Eqs. 16.10 and 16.12:

$$\text{Heat carried ahead by steam} = 0.0317\left(\frac{18}{2}\right)\frac{(1{,}000)(164)(1-0.22)}{100}$$
$$= 365\ \text{Btu/ft}^3\ \text{rock burned.} \qquad (16.15)$$

In this equation, $^{18}/_{2}$ is the ratio of molecular weight of water and hydrogen. The heat given up by the steam is 1,000 Btu/lbm, and the other numbers are similar to those in Eq. 16.10. Thus, the calculated temperature is lower than it was in Eq. 16.14, as shown below:

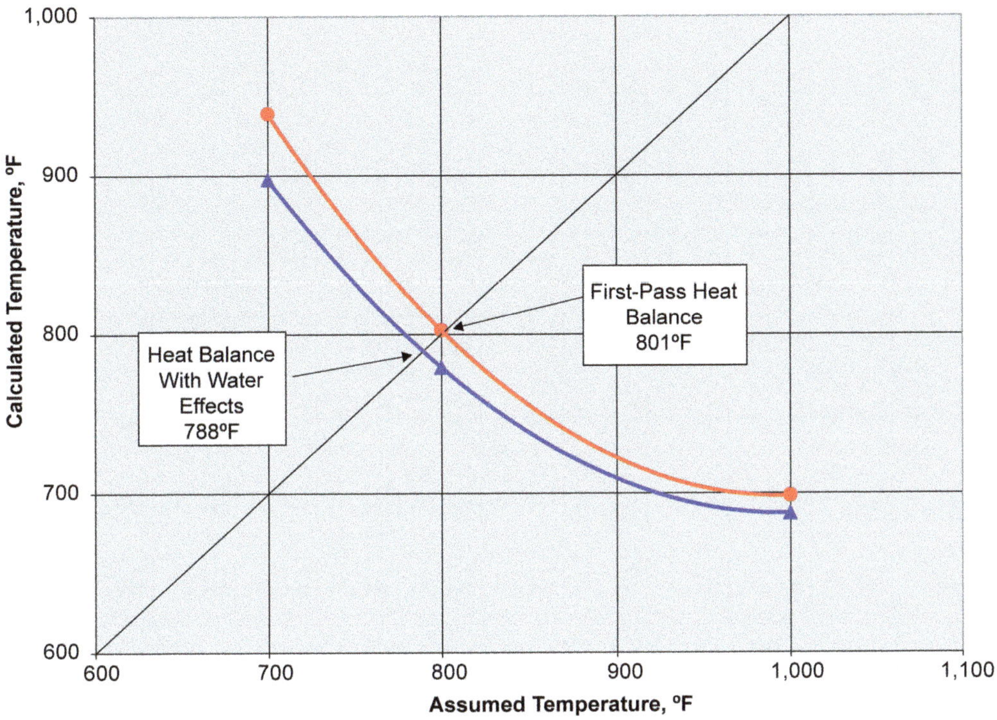

Fig. 16.7—Assumed and calculated combustion-zone temperatures.

$$T_2 - 100 = \frac{19{,}530 - 365}{(0.255)(164)(1-0.22)} = 587.5$$

$$\text{or } T_2 = 688\text{°F}. \qquad (16.16)$$

Other temperatures were calculated in a similar way, and the results, graphed as triangles in Fig. 16.7, show a corrected combustion temperature of 788°F. At this temperature, the H/C ratio is 0.85, as indicated in **Fig. 16.8.**

16.4.4 Calculating the Volume and Temperature of the Steam Plateau. No calculations on the steam plateau were necessary in the above calculations. The steam-plateau temperature and volume directly affect the volume of oil moved as a result of the combustion process. To calculate these terms, use the H/C ratio of 0.85 and calculate the partial pressure of the water as follows.

The fuel composition is $CH_{0.85}$. From Eq. 16.3, the moles of oxygen used per mole of fuel are

$$O_2 = \frac{2m+1}{2m+2} + \frac{n}{4} = \frac{2(20)+1}{2(20)+2} + \frac{0.85}{4}$$

$$= 1.189 \text{ mol } O_2 \; . \qquad (16.17)$$

Combustion products are calculated in a similar way:

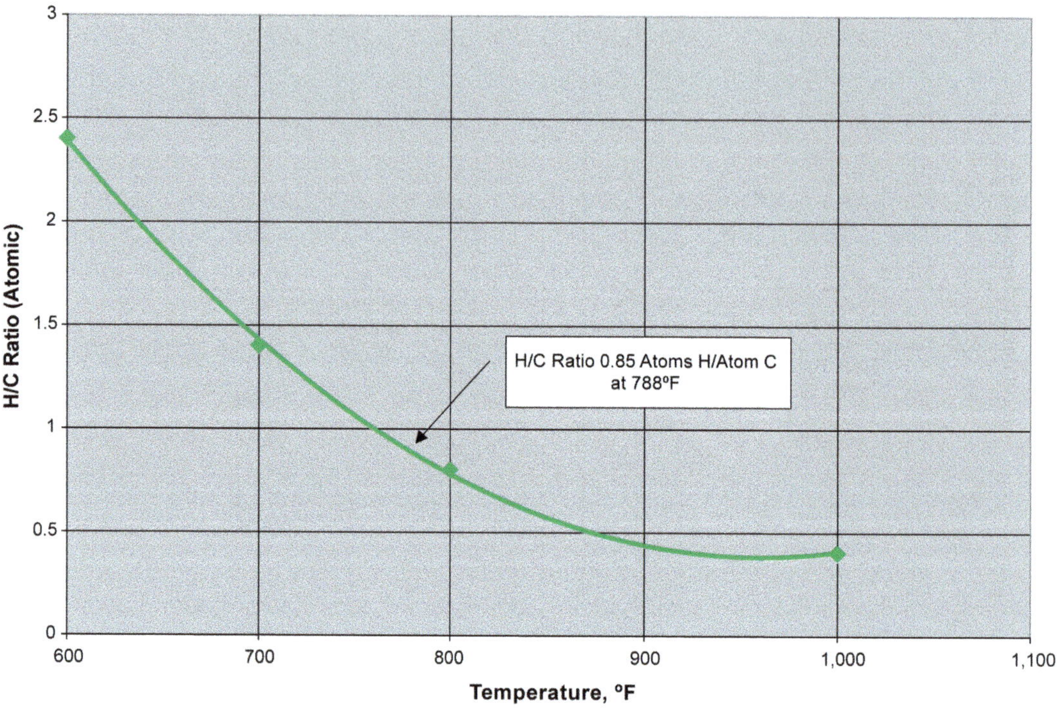

Fig. 16.8—H/C atomic ratio vs. combustion temperature.[13]

$$CO_2 = \frac{m}{m+1} = \frac{20}{21} = 0.952 \text{ mol } CO_2 . \quad (16.18)$$

$$CO = \frac{1}{m+1} = \frac{1}{21} = 0.048 \text{ mol CO}. \quad (16.19)$$

$$H_2O = \frac{n}{2} = \frac{0.85}{2} = 0.425 \text{ mol } H_2O. \quad (16.20)$$

$$N_2 = \left(\frac{79}{21}\right)(O_2) = \frac{79}{21}(1.189) = 4.473 \text{ mol } N_2 . \quad (16.21)$$

The operating pressure is 300 psia. The partial pressure of H_2O in the combustion gas is

$$p_{\text{water}} = \frac{0.425(300)}{0.952 + 0.048 + 0.425 + 4.473}$$
$$= 21.6 \text{ psia.}$$

From steam tables, the saturation temperature for 21.6 psia is 232°F. This is the temperature of the steam plateau.

The volume of the steam plateau is a function of the amount of H_2O formed. Knowing that there are 0.95 lbm C/100 lbm rock burned, and knowing from Fig. 16.8 that the H/C ratio is 0.85, an equation similar to Eq. 16.10 yields the amount of water formed per cubic foot of rock burned.

$$\text{Water formed} = 0.95\left(\frac{0.85}{12}\right)\left(\frac{18}{2}\right)\frac{(164)(1-0.22)}{100}$$

$$= 0.775\frac{\text{lbm H}_2\text{O formed}}{\text{ft}^3\text{rock burned}} \quad \text{......} \quad (16.22)$$

Thus, the total heat carried forward by the water formed is

$$\text{Heat carried by steam} = 0.775\,(1{,}000)$$

$$= 775\ \text{Btu/ft}^3\ \text{rock burned.} \quad \text{......} \quad (16.23)$$

Using Eq. 16.13a, the heat capacity of the formation is

$$c_s = 0.21 + \frac{232-100}{20{,}000} = 0.2166\ \text{Btu/lbm-°F} \quad \text{......} \quad (16.13c)$$

The amount of heat required to raise the temperature of a cubic foot of sand from 100 to 305°F, from a heat balance, is

$$\Delta H = \frac{0.2166\ \text{Btu}}{\text{lbm-°F}}\left[\frac{164(1-0.22)\ \text{lbm}}{\text{ft}^3}\right](232-100)\text{°F}$$

$$= 3{,}701\text{Btu/ft}^3 \quad \text{......} \quad (16.24)$$

Thus, the volume of rock heated by condensing steam is Eq. 16.23 divided by Eq. 16.24:

$$\text{Volume of steam heated rock} = \frac{775}{3{,}701} = \frac{0.209\ \text{ft}^3\ \text{steam zone}}{\text{ft}^3\ \text{rock burned}} \quad \text{......} \quad (16.25)$$

A sketch of the resulting temperature profile is shown in **Fig. 16.9**.

16.4.5 Calculating the Effects of Injected Air and Water. Further corrections are needed to increase the accuracy of the temperature profile in Fig. 16.9. Injected air will partially cool the burned zone, raise the temperature as it approaches the combustion front, and carry heat forward. This will have little effect on the combustion kinetics or the amount of heat generated by combustion, so, in essence, this amount of energy is merely carried forward to extend the size of the steam plateau.

A sketch of this idea is shown in **Fig. 16.10**. In this sketch, the area marked "1" is the temperature profile behind the burning front; Area 2 is the steam plateau, which is now larger than calculated before because of the heat carried forward by the combustion gases.

This temperature profile can be approximated as indicated in **Fig. 16.11**, where the profiles of the burned zone and steam plateau are treated as square waves that have been adjusted so that the total heat in Areas 1 and 2 is the same as in Fig. 16.10.

There are several reasons for using this square-wave concept. One is that it makes it easier to calculate heat losses to be expected from either a laboratory or field combustion operation using superposition calculations similar to those discussed by Ramey as seen in Prats.[2] The reference also indicate that the heat losses calculated with Fig. 16.11 are quite adequate.

When wet combustion is used, the temperature behind the front tends to change abruptly, as shown in Fig. 16.11. As a result, heat and material balances of the sort discussed next can be used to calculate the movement of the resulting cooling front, burn front, and steam plateau.

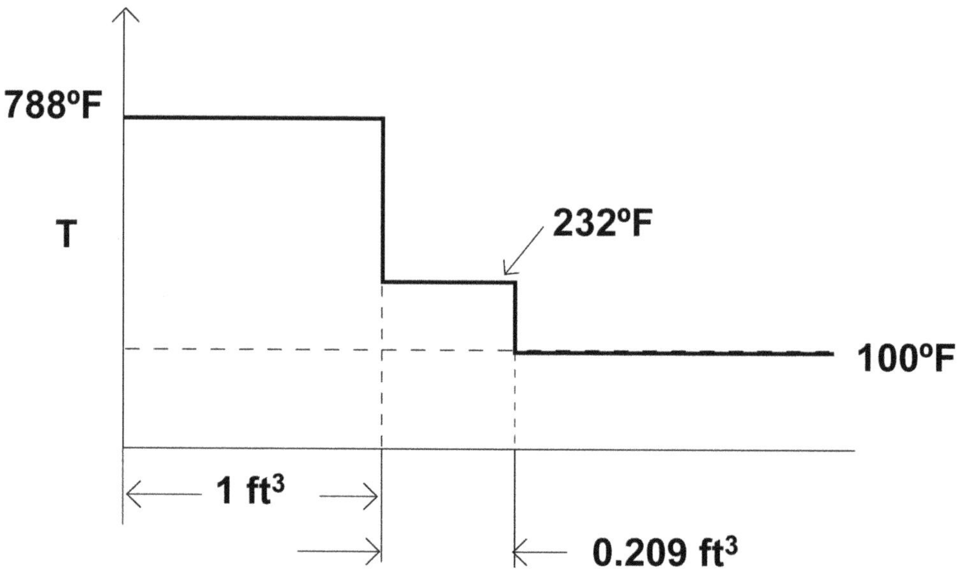

Fig. 16.9—Idealized temperature profile with steam plateau ahead of the combustion front.

The amount of air injected per cubic foot of rock burned and the heat capacity of air are needed to calculate this heat-transfer process for dry combustion. The volume of combustion gas should also theoretically be calculated, but normally this isn't necessary because its volume is nearly identical to the air volume. Further, its heat capacity is nearly the same—remember that most of the combustion gas is nitrogen (assuming that air is injected).

The moles of air injected are calculated by adding the O_2 from Eq. 16.17 to the N_2 from Eq. 16.21:

$$\text{Air injected} = 1.189 + 4.473 = \frac{5.662 \text{ mol air}}{\text{mol fuel}} . \qquad (16.26)$$

The heat capacity for air is 7.00 Btu/lbm mol-°F. With previously determined factors of 0.95 lbm of carbon burned for 100 lbm rock, 164(1 – 0.22) pounds of rock per cubic foot of rock, 12 lbm of carbon per mole, a combustion-zone temperature of 788°F (Fig. 16.7), and the results of Eq. 16.26, we obtain the amount of heat carried forward by the injected air as follows.

$$\frac{\text{Heat carried by air}}{\text{ft}^3 \text{ rock burned}} = \left(\frac{0.95}{100}\right)\left[\frac{(164)(1-0.22)}{12}\right](5.662)(7.00)(788-100)$$

$$= \frac{2{,}761 \text{ Btu}}{\text{ft}^3 \text{ rock burned}} . \qquad (16.27)$$

This heat extracted behind the burned zone is deposited into the steam plateau. The resulting size of the steam plateau can be calculated in a way similar to Eq. 16.25 by adding the heat carried by the combustion gas to that carried by the water as follows:

$$\text{Vol of steam heated rock} = \frac{775 + 2{,}761}{3{,}701} = 0.901 \text{ ft}^3 . \qquad (16.28)$$

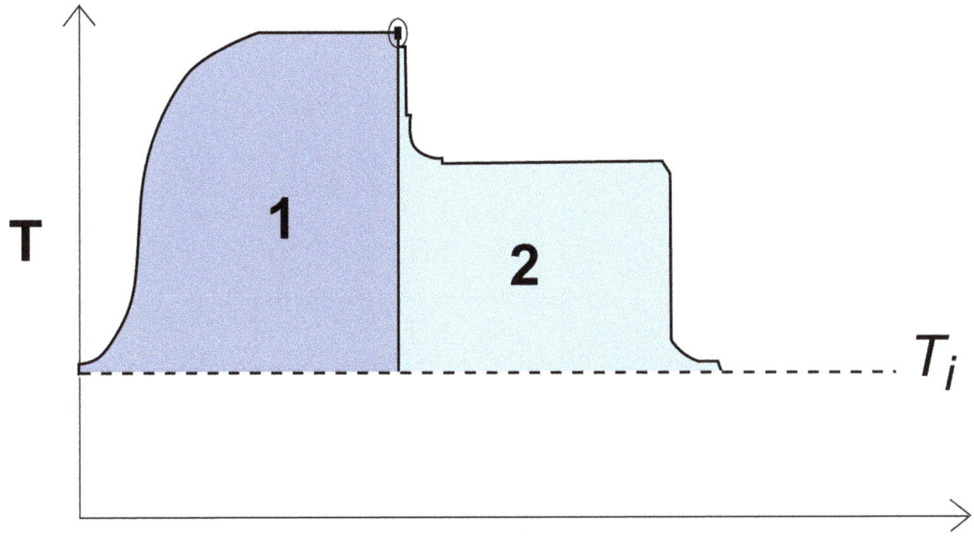

Fig. 16.10—Schematic of combustion temperature profile including cooling by injected air.

This calculation shows that the condensing steam front is far enough ahead of the combustion front to displace oil efficiently, and it is unnecessary to have the combustion front cover the entire reservoir to obtain good recovery. Recovery can be estimated by knowing the amount of fuel burned, by estimating the residual oil saturation in the steam plateau, and by estimating the volumetric sweep efficiency of the process.

16.4.6 Heat Losses. An estimate of heat losses using the superposition concepts seen in Prats,[2] based on Ramey's work, will make the calculations just presented more accurate. These estimates are particularly important if a laboratory heat balance indicates significant heat losses. The temperature profiles just calculated assuming no heat losses can be used to make a first estimate of heat losses and a recalculated steam-plateau size. This is a reasonable way to handle the heat balance. Because all heat transferred was assumed to be in the steam plateau, any reduction in that transferred heat because of losses will reduce the amount of heat in the plateau.

Because the size of the steam zone and the size of the calculated heat losses are interdependent, iterative calculations are necessary until the assumed and calculated heat balances match. This will usually require only two to three iterations to achieve this match.

Data used in the previous calculations were based on generalized predictions of combustion behavior (i.e., the amount of fuel per cubic feet of formation) and the H/C ratio of the fuel. If combustion tube runs are made in the laboratory, those parameters will be known and should be used in the calculations. In addition, accurate temperature and saturation profiles vs. time will allow reasonably accurate heat-balance calculations to determine the heat losses from the experiment. As an alternative, reasonable assumptions about the heat losses can be used to check the heat-balance calculations and indicate if there is significant experimental error.

Computer assisted tomography (CAT) scanner measurements produce the most accurate saturation histories. Accurate measurements of temperature profile and accurate oil, water, and gas production data also make it possible to estimate reasonable saturation histories. These are the major sources of error in the overall heat-balance calculations, but they are fairly small compared to the amount of heat stored in the hot matrix.

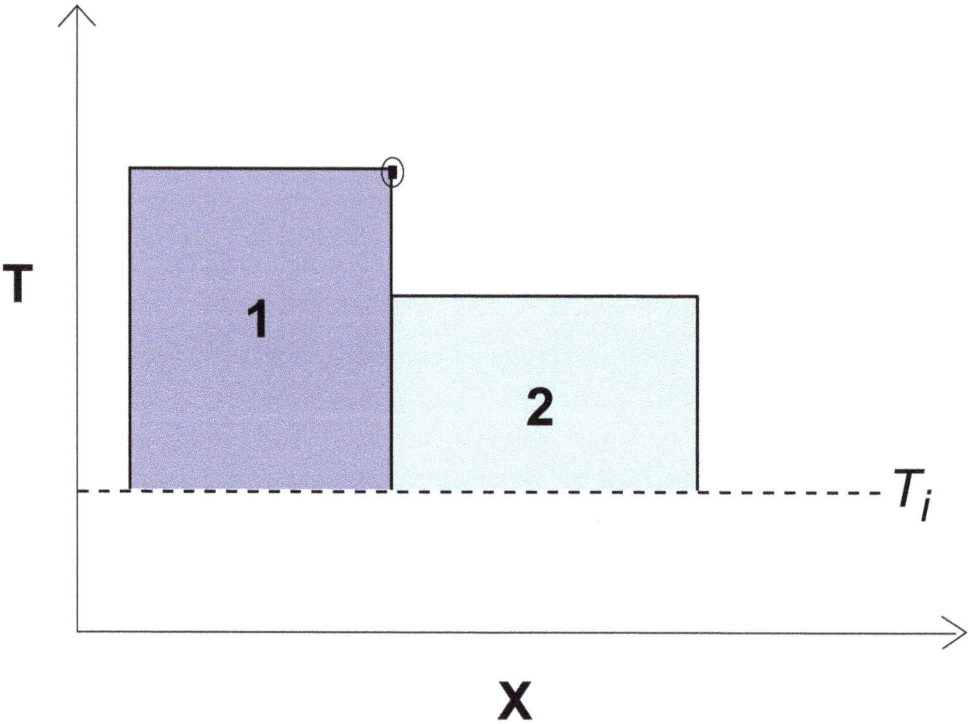

Fig. 16.11—Idealized temperature profile including injected-air cooling.

16.5 Design Considerations

Conditions favoring the use of ISC rather than steam include the following: (1) high reservoir pressure at which steam is inefficient, (2) potential for severe wellbore heat losses (i.e., depth, offshore, permafrost), (3) reservoir clay swelling in contact with fresh water, (4) limited water supply, and (5) environmental regulations prohibiting steam generation.

Like any other injection process, the design of ISC projects must consider injection-pressure limitations and reservoir flow resistance. These are especially important in heavy-oil reservoirs in which combustion must occur in the high-temperature regime to be successful. The minimum air flux needed to maintain high temperatures at the front is estimated to be 0.125 ft/D (0.04 m/d).[14] Because the burn-zone growth is directly proportional to the injected air, the maximum air-injection rate determines the minimum lifetime of the project. Ways to increase the air-injection rate are often needed, especially in heavy-oil reservoirs. They may include reduced well spacing, cyclic steaming of injectors and producers, and an increase in injection pressure. These factors will determine the compressor pressure and volume output.

There has been some controversy over whether ISC projects should be developed using patterns or line drives. Many early projects were started as pilots with a single injector. Usually, this practice ultimately resulted in a five-spot pattern. These pilots behaved contrary to plan, with the combustion front moving in only one direction because of permeability variations, gravity effects, well spacing differences, or a combination of these factors.

Attempts to correct the unbalanced flow included stimulating unresponsive wells and limiting withdrawal rates of wells that producing excessive volumes of combustion gas. Generally, these efforts did not have the desired effect.

In retrospect, this reservoir behavior makes sense. Once a combustion front is even slightly asymmetric, the higher temperature and, thus, higher mobility will cause greater flow in that direction. Thus, the flow will become more asymmetric, finally resulting in flow principally in only one direction.

Because it is often difficult to decide *a priori* which direction the major flow will take, operating plans should remain flexible until field performance indicates which injection scheme best uses the flow directions.

For the above reasons, many of the more successful ISC projects have been line-drive operations that start near the top of the reservoir and move downdip. In such an operation, the direction of the fire front is known. The operating engineers can then plan the completion and operating history in a rational way that will mirror the front movement and breakthrough history. This operating practice can be seen in most of the successful ISC field projects to be discussed later.

16.6 Performance Prediction

Predicting the production response to ISC has been the topic of various studies. Complete numerical simulation of ISC is difficult because of the complex reactions and the thin burning front that requires small gridblocks for representation. Simulators range from tank models to complex 3D simulators. In addition to simulation, empirical models, hybrid models, and correlation methods have been developed. A discussion of some of these methods follows.

The easiest method is essentially a tank balance,[14] adapted by Prats.[2] The oil and water produced are given by

$$N_p = \phi V_b (S_{oi} - S_f) + 0.4 S_{oi} (V_p - V_b) \quad (16.29)$$

and

$$W_p = V_b \phi (S_{wi} + S_{wf}), \quad (16.30)$$

where S_{oi} = initial oil saturation, fraction; S_f = oil saturation burned, fraction; V_b = volume burned, m^3; N_p = oil produced, m^3; W_p = water produced, m^3; ϕ = porosity, fraction; V_p = volume of the pattern, m^3; and S_{wf} = water saturation resulting from the combustion process, fraction.

If the volumes are in acre-ft and the production terms are in bbl, a multiplication factor of 7,758 must be used. The estimate of 40% of the oil produced coming from outside the burned volume is an empirical value based on experience. This is the 0.4 term in Eq. 16.29.

Fig. 16.12, presented by Gates and Ramey,[15] combines laboratory results and field observations from the Belridge ISC projects. It shows the effect of initial gas saturation on the oil-recovery history. Oil production rates and instantaneous air/oil ratios can be estimated from the slopes of the curves. At late times, the above two techniques give similar results.

Satman and Brigham[16] used data from dry combustion field tests to obtain two empirical correlations. Those are presented in **Figs. 16.13 and 16.14**. The terms in the ordinates are cumulative incremental oil produced (CIOP), original oil in place (OOIP), fuel burned (FB), and oil in place (OIP) at the start of the project.

In addition to original oil saturation S_o, thickness h, oil viscosity μ_o, and porosity ϕ, the abscissa includes cumulative air injected (CAI), and OIP at the start of the project, and fraction oxygen use. The second correlation, Fig. 16.14, is the most accurate except for oils with less than 10 cp original viscosity, where the first correlation must be used. These correlations were generated from pilot floods; thus, they would not be expected to be accurate for pattern flooding. However, the narrative in the previous section points out that pattern flooding is generally not the best way to operate an ISC project, and these correlations are expected to be reasonably accurate for line-drive projects.

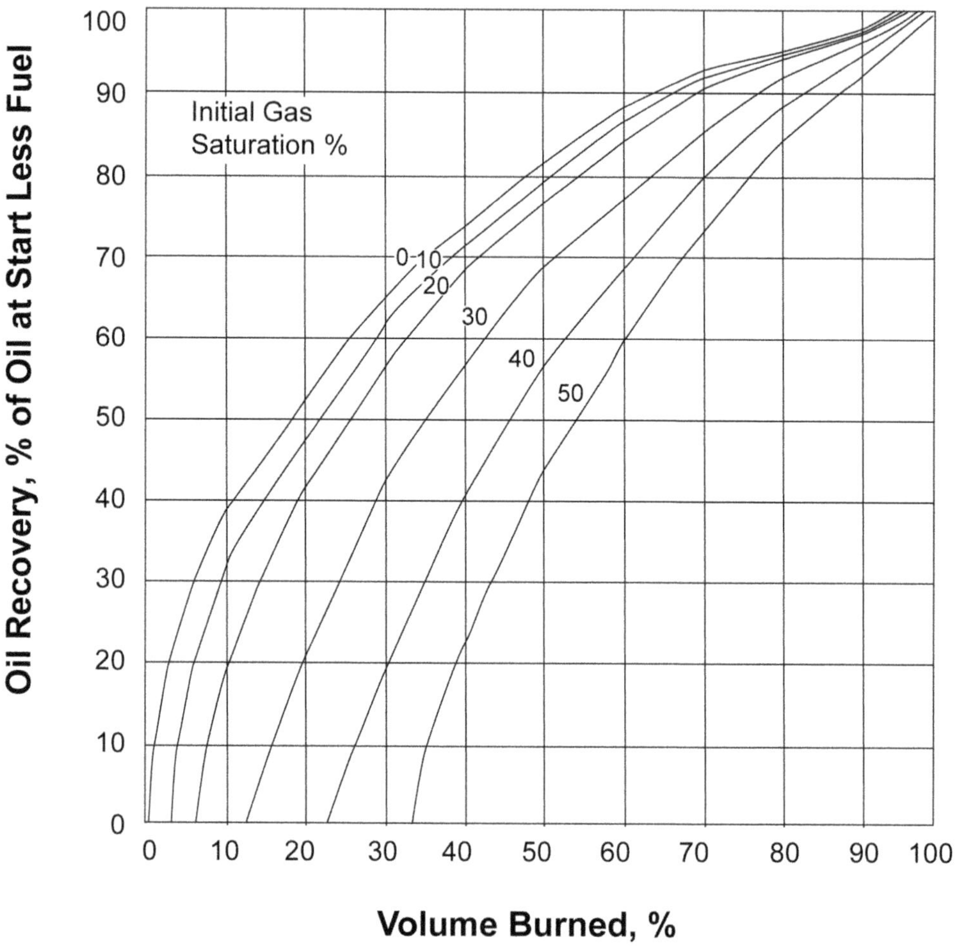

Fig. 16.12—Gates and Ramey method.[15]

16.7 Operating Practices

In addition to the standard field equipment for oil production, ISC requires particular attention to air compression, ignition, well design, completion, and production practices.

16.7.1 Compressors. Air-compression systems are critical to the success of any ISC field project. Past failures often can be traced to poor compressor design, faulty maintenance, or operating mistakes. A detailed discussion of compressors and sizing considerations appears in the Facilities and Construction Engineering volume of this *Handbook*. Other discussions are available in Sarathi.[1]

The factors to be considered when selecting compressors include peak air requirements, injection pressure, capital cost, power requirements, operation and maintenance costs, and other relevant technical and economic parameters specific to the field considered. Compressor terminology varies among manufacturers. It is best to obtain a complete description including compressor, driver, interstage cooling system, and all ancillary equipment, including control and safety systems from each vendor being consulted.

Air compression causes high temperatures because of the large heat capacity (c_p/c_v ratio). Compressor design must consider these high temperatures to ensure continuous, sustained oper-

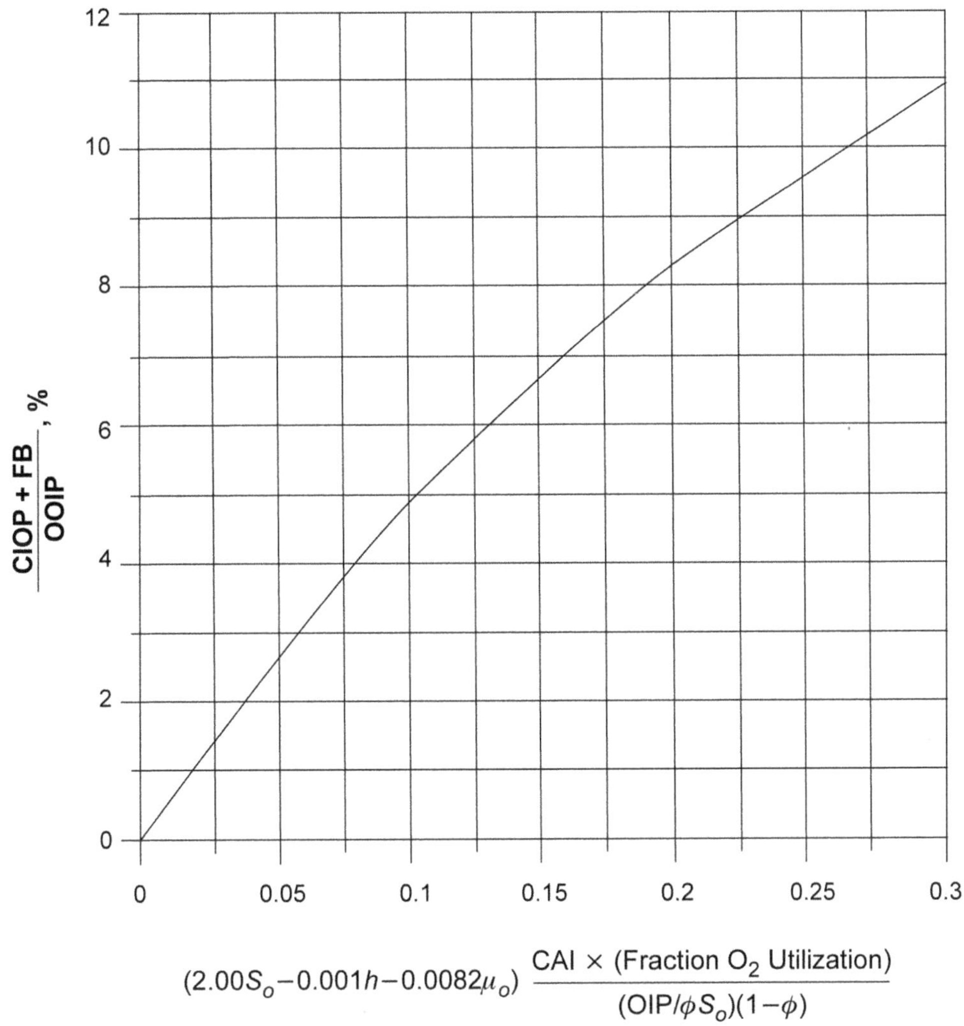

Fig. 16.13—First Satman correlation.[16]

ations free from the corrosive effects of air and the explosion hazards of some lubricating fluids. Mineral oils are not recommended. Synthetic lubricants withstand the higher temperatures and offer lower volatility and flammability than conventional lubricants.

16.7.2 Ignition. Ignition and maintenance of high combustion temperatures, especially in heavy-oil projects, are the most critical factors of an ISC project. Shallcross[17] presented a complete review of ignition methods. The following is a summary of this study.

Ignition can occur spontaneously if the oil is reactive, the reservoir temperature is high enough, and the reservoir is reasonably thick. Various models have been proposed to determine the time for spontaneous ignition.[18,19]

When spontaneous ignition does not occur or is not desired (i.e., in heavy oil reservoirs, where it is important to maintain high combustion temperatures), the most appropriate ignition method depends on the reservoir and the equipment available on site.

Downhole gas-fired burners allow good control of the temperature of injected gases and may be operated at a greater depth than other methods. The disadvantages include the need to

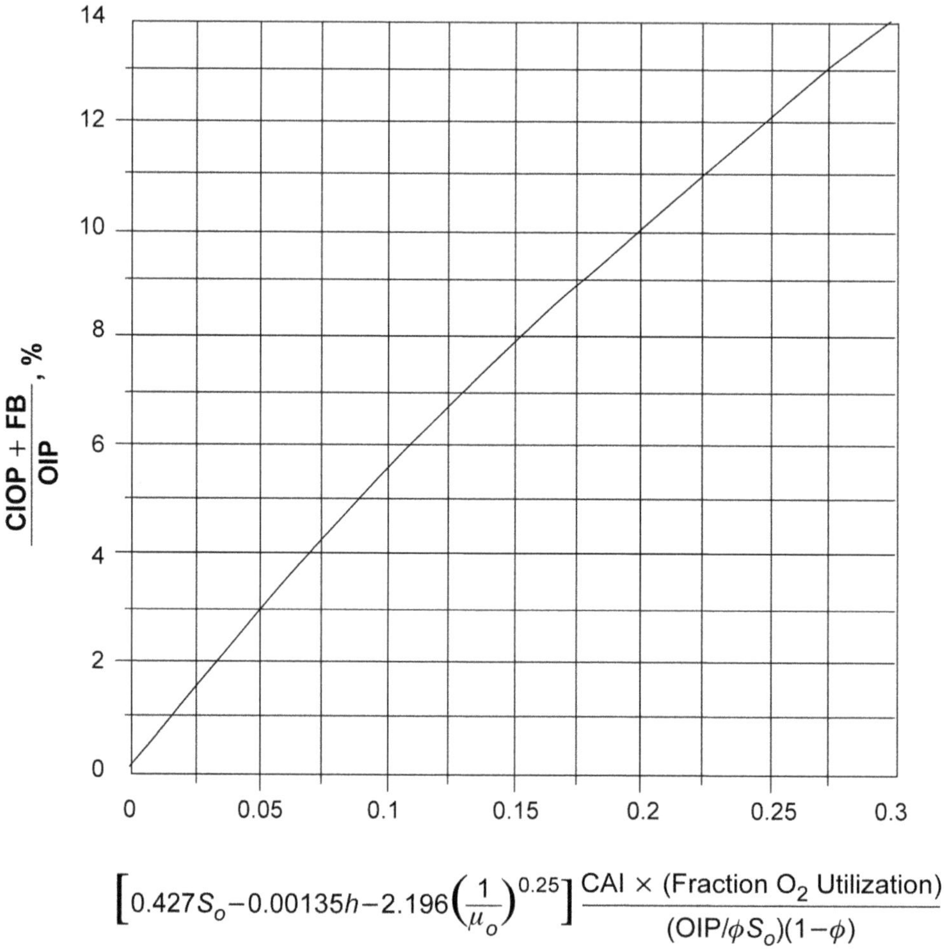

Fig. 16.14—Second Satman correlation.[16]

run multiple tubing strings in the injection wells. Some particulates such as soot may be carried into the formation if the gas does not burn cleanly.

Catalytic heaters run at lower temperatures but are sometimes prohibitively expensive. Electrical heaters can be lowered with a single cable, can provide excellent temperature control, and can be reused repeatedly. There is, however, a depth limitation because of electrical power losses in the cable.

Chemically enhanced ignition does not have a depth limitation but may require handling and storage of dangerous materials. Fuel packs are not recommended because of poor temperature control and nonuniform ignition across the entire reservoir thickness. Well damage from elevated temperatures and plugging by particulate matter may occur.

Steam may be used to locally increase reservoir temperature and facilitate auto ignition. It suffers from depth limitation because of wellbore heat losses, but when the conditions are right, it can be a very simple and effective method for ignition.

The References section includes details of design and implementation of the above methods.

16.7.3 Well Design and Completions. ISC wells must be designed to account for several factors amplified by the combustion, namely high temperature, corrosive environment, and sand and clay control. Safe operations should be the primary concern.

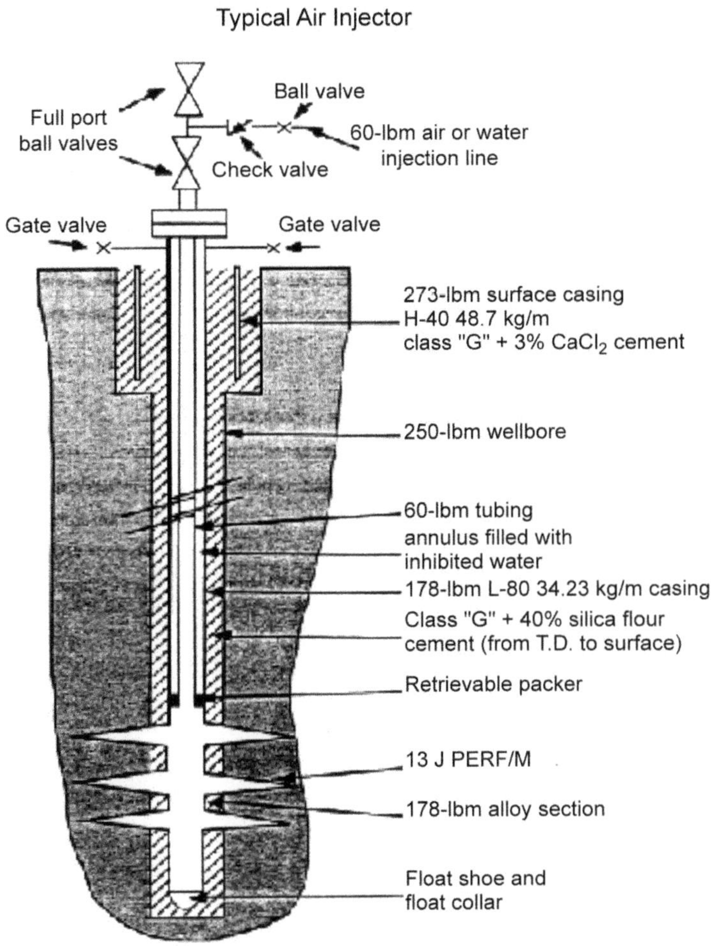

Fig. 16.15—Typical injection-well design.[1]

Typical well designs for injection and production are shown in **Figs. 16.15 and 16.16**. Completion type and design depends on the reservoir being considered. Laboratory testing for sand control and completions can help to determine the best completion technique for a given field. Care must be taken to cement the wells properly. There are cement formulations that are stable at high temperatures.[20] Openhole completions may be used in conjunction with slotted liners, screens, gravel packs, or various other sand and clay control methods. To maximize productivity, producing wells should be completed toward the bottom of the zone of interest to take advantage of gravity drainage and avoid hot gases as long as possible. Rat holes have been used successfully in certain heavy-oil combustion projects to increase the effect of gravity drainage.[21]

16.7.4 Injection and Production Practices. Safe air injection requires that the surface injection equipment and the injection well are free of hydrocarbons. All lubricants used in compression and downhole operations should be synthetic or nonhydrocarbon types. All equipment, tools, lines, tubing, work strings, and injection strings must be clean and hydrocarbon free. Personnel at all levels should be aware of the importance of preventing hydrocarbons in the injection wells. As a safety measure to protect injection wells if the compressor is shut down, a system to prevent backflow of oil from the formation must be present at every injection well.

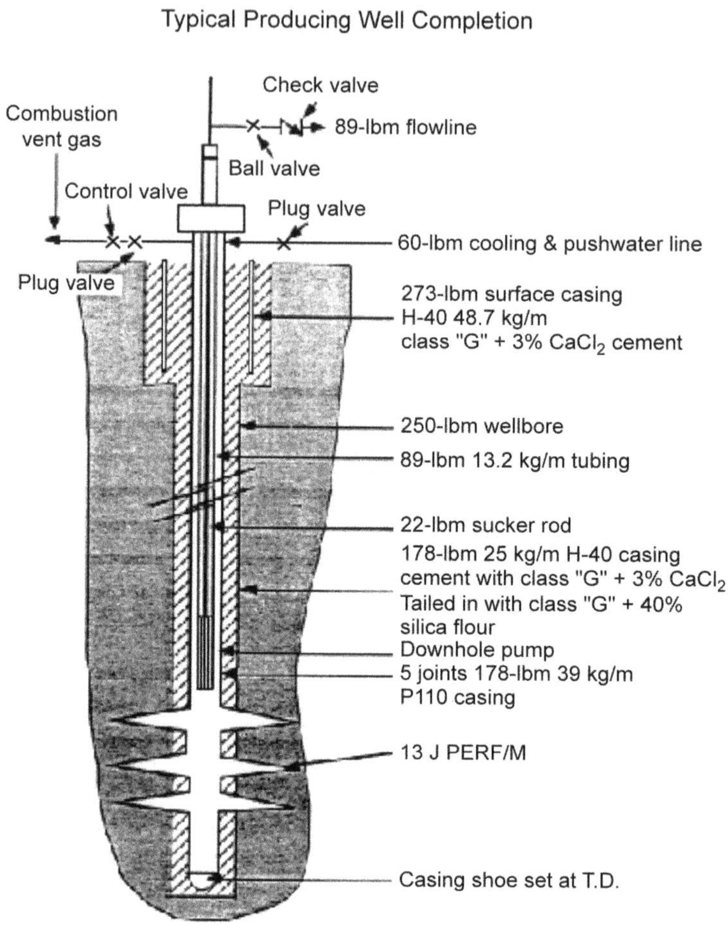

Fig. 16.16—Typical production-well design.[1]

Downhole temperatures in producing wells increase as displaced oil, hot water, and steam fronts reach the well. Producers are preserved by downhole cooling and proper material selection. **Fig. 16.17** provides an estimate of the water requirements to maintain bottomhole temperature no higher than 250°F as a function of oil and water production rate and formation flowing temperature. Significant additional oil recovery can be obtained from hot wells with downhole cooling, especially if the well is completed in the lower section of the producing zone to maximize gravity segregation in the reservoir. In many cases, after the combustion front has moved through the well, it is possible to convert the former producer to a new air injector, thus realizing significant cost reductions over the life of the project.

Monitoring is crucial for proper combustion operations. In addition to testing individual producers for oil and water rates, injected fluids must be measured. Also, produced gases must be measured and analyzed to determine the efficiency of the combustion operation. Downhole temperature measurements are essential to calculate the size and location of the burned zone. Flowline temperatures can indicate thermal stimulation or downhole problems.

Combustion projects generate waste water, flue gases, and pollutants from compression and oil-handling equipment. Local pollution disposal regulations must be consulted before designing any ISC operation.

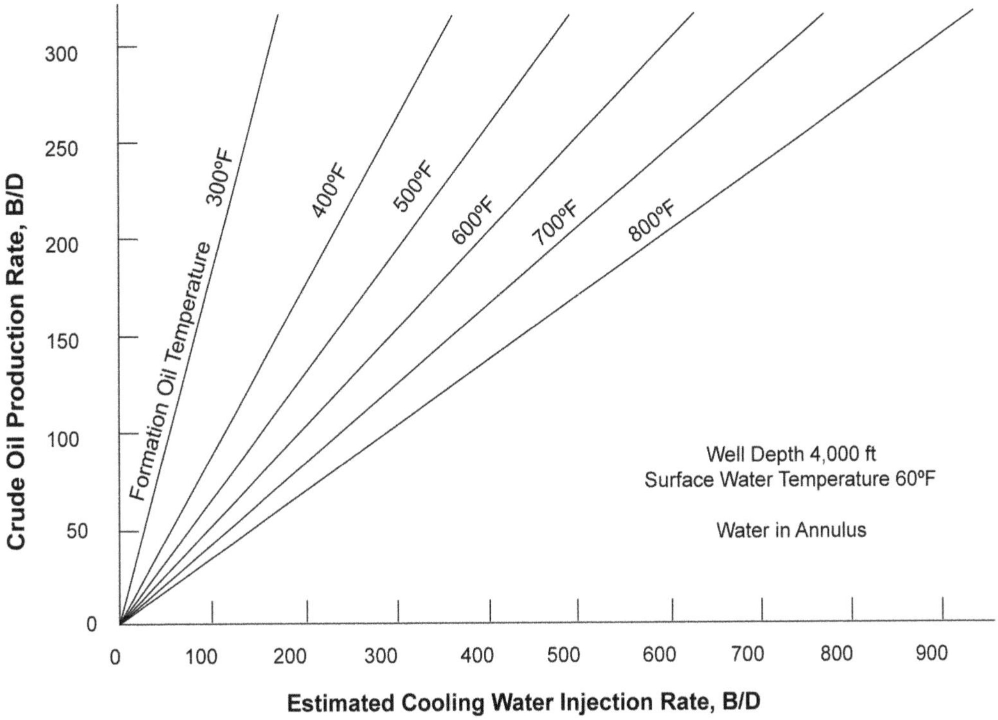

Fig. 16.17—Water needed to cool hot wells.[1]

In general, environmental problems are similar to those posed by steam injection. The produced water may contain H_2S and/or CO_2, which may require special handling and anticorrosion equipment. Flue gases may contain hydrocarbons, H_2S, CO_2, CO, and other trace amounts of sulfur gases. **Table 16.3**[1] summarizes the various pollution-control systems suitable for combustion projects and their recommended applications. Sarathi[1] also provides detailed descriptions of the various types of systems and their uses. Other problems that can be encountered are sand production, corrosion, emulsions, well failures, and compressor failures.

16.8 Field Experience

ISC has been used in the field since 1920. In the U.S., more than 230 projects have been implemented. Many of those were technically and economically successful. Unfavorable reservoir and fluid characteristics, poor design, and engineering or operational problems caused failures. Most of the failed projects were small pilot projects implemented in unfavorable reservoirs. Worldwide, combustion accounts for approximately 10% of the oil produced by thermal methods; 29 projects were active as of 1998.[1] Most of the projects outside of the U.S. are large heavy-oil projects, while the current trend in the U.S. is to use ISC in deep, lighter-oil reservoirs in which waterflooding and steamflooding are not effective. Brief comments on these projects follow.

16.8.1 Heavy Oils. If we define heavy oil as under 20°API, 19 projects using ISC were active in 1998. Some general comments are applicable:

- Most of these projects last a long time; projects initiated in the 1960s are still active. Economics of successful projects are favorable even as compared to steamflooding and waterflooding.[22]
- All the successful projects operate in the high-temperature mode.

TABLE 16.3—APPLICATION OF POLLUTION-CONTROL SYSTEMS TO A FIREFLOOD PROJECT[1]

Equipment	Gas Treated	Pollutant Removed	Method	Suggested Application
Flare stack	Flue gas	None	Vent to atmosphere	Flue gas meets air-quality regulations
Flare stack with flame burner	Flue gas	H/C$^{(1)}$,CO$^{(1)}$, S-gases$^{(1,2)}$	Burn	Flue gas with enough H/C to support combustion (>200 Btu/scf)
Thermal incinerator	Flue gas	H/C, CO, S-gases$^{(1,2)}$	Burn	Flue gas not suitable for a flare with burner or a catalytic incinerator (heat value > 85 Btu/scf but < 200 Btu/scf)
Catalytic incinerator (excess air)	Flue gas	H/C, CO, S-gases$^{(2)}$	Burn	Flue gas with heat value <85 Btu/scf
H_2S scrubber	Flue gas	$H_2S^{(4)}$	Chemical reaction	Flue gas containing H_2S but acceptable amounts of other S gases
SO_2 scrubber	Incinerator exhaust	$SO_2^{(4)}$	Chemical reaction	When H_2S removed from flue gas is inadequate or impractical

1. Removal efficiency may be poor.
2. S-gases are converted to SO_2.
3. May increase amount of CO for flue gas with heat values > Btu/Scf.
4. Typical removal efficiency is 90 to 95%.

• Gravity override and channeling do occur. Gravity drainage of the hot oil is an important mechanism and should be maximized. Frequently, improved production of oil continues after the air injection has been terminated.

• Line-drive projects, starting at the top of the reservoir and moving downward, exhibit superior performance compared to repeated pattern projects.

• Most of the projects failed when air injection was attempted in different layers of the reservoir at the same time. Air injectivity is a critical parameter, and injectivity contrast between layers is usually too difficult to overcome.

Because a detailed description of all or even a few of the projects is outside the scope of this narrative, the reader can find additional information on several current projects in the following references:

• Projects in Romania: Supplacu de Barcau is the world's largest combustion project; it started in 1964 and is operated in a line-drive mode from the top downward. Videle and Balaria are other ISC projects.[23,24]

• In India, Balol, which started as a pilot in 1990, was expanded to the whole field. Designed as a wet combustion project, the water injection rate had to be cut in half because of too much cooling. This project also was changed from patterns to updip line drive because of premature breakthrough in the producers.[25,26]

• Projects in Russia, Kazakhstan, and Azerbaijan are not very well described in Western literature.[27]

• The Albanian project of Kasnice is described in Marko *et al.*[28]

• Batrum in Canada is a successful project using horizontal wells as producers. The Eyehill field is another project with horizontal wells.[26,29] Wabaska also uses the same concept of horizontal wells as producers.[30] Horizontal wells in ISC operations have been successfully used in Canada since 1993.

Cyclic applications such as pressure-up and blowdown, have been described.[30,31] This operational technique allows production from very low mobility oil fields or tar sands in which fracturing or cyclic steaming are needed before air injection. The Wabaska project is a cyclical combustion project with horizontal wells. This type of pressure-up, blowdown technique has also been successfully implemented at Wolf Lake.[31] Air was injected until the front arrived at the producers. When the front reached a given producer, this well was shut down, and cooling water was circulated. When all the producers were shut down, injection was stopped, and the producers reopened to blow down the reservoir. This process was repeated for several cycles. Operating combustion in this fashion allows production from fields in which injectivity is low because of a high crude viscosity at reservoir conditions.

U.S. projects at Bellevue and Midway-Sunset have been described.[26] More details on the Midway-Sunset project can be found in Hoffmann.[32] Ramey *et al.*[21] describe the Belridge project as an economic success.

ISC in heavy-oil reservoirs has been successful in both the dry and wet modes. Dry combustion early in the life of the project is the preferred method to form the desired high-temperature regime. When the process is well established, moderate amounts of water can be added to improve efficiency. Quenched, or super-wet, combustion seems to have limited success except when used at the end of a field operation to scavenge the heat remaining in the rock.

Another operating variation includes the use of enriched air or pure oxygen. Oxygen-enriched combustion presents technical and economic advantages for reservoirs with high pressure or very low injectivity. It has been demonstrated successfully in the field.[33,34] The additional literature covering the special handling methods and additional precautions needed for enriched air injection is listed in the References section. Commercial application of the oxygen technology has been limited because of oil-price variations.

16.8.2 Light Oils. ISC is used in light-oil reservoirs for two different purposes:

- To reduce the viscosity of unconventional light oils such as Demjen in Hungary[35] or Niemangu in China.[1] In these cases, thermal effects are important. In the case of Demjen, a catalyst had to be injected to promote combustion. Iron was used to increase the amount of fuel burned because the light oil by itself was not depositing enough fuel to sustain combustion. The oil is parafinic and almost solid at reservoir temperature despite an API gravity of 32°.
- To produce from light-oil reservoirs in which waterflooding or other enhanced oil recovery methods are not attractive. Combustion is used to generate flue gases for reservoir pressure maintenance and production by gravity drainage. Thermal effects are only minor for this process. An interesting case is the West Hackberry double displacement process,[36] in which the gas cap is expanded for gravity drainage to recover residual oil after waterflood.

Combustion is successful in tight carbonate reservoirs located in the Dakotas, such as Medicine Pole Hill, Buffalo, west and south Buffalo,[37] and Horse Creek. In those cases, combustion allows exploitation of thin reservoirs with large well spacing.

16.9 Screening Guidelines

ISC is a complex process; it combines the effects of steamdrive, hydrocarbon miscible and immiscible flooding, immiscible gas drive, and hot and cold waterflooding. Because of its complexity, there is a misconception that combustion has a low probability of success. The truth is that combustion is an economically attractive, proven recovery process, capable of economically recovering a large fraction of the oil in place.

ISC can be applied to many different reservoirs. Some suggested screening guidelines are:

• Nature of the formation: The rock type is not important provided that the matrix/oil system is reactive enough to sustain combustion. As in any drive process, high-permeability streaks are detrimental. Swelling clays may be a problem in the steam-plateau area.

• Depth: The reservoir must be deep enough to ensure containment of the injected air in the reservoir. There is no depth limit, except that this may affect the injection pressure.

• Pressure: This will affect the economics of the process but does not affect the technical aspects of combustion.

• Temperature: Temperature will affect the propensity for autoignition but is otherwise not critical.

• Reservoir thickness: Thickness must be greater than 4 m (15 ft) (Sarathi[1]) to avoid excessive heat losses to surrounding formations. Thick formations may present sweep-efficiency problems because of gravity override.

• Permeability: This has to be sufficient to allow injection of air at the designed air flux. The air injectivity is especially important for heavy-oil reservoirs. Conditions are favorable when kh/μ is greater than 5md m/cp[3].

• Porosity and oil saturation: These must be large enough to allow economic oil recovery. The product, ϕS_o, should be more than approximately 0.08 for combustion to be economically successful.

• Oil gravity: This parameter is not critical. In-situ viscosity has to be small enough to allow air injection and resulting oil production at the design rate.

• Oil nature: In heavy-oil projects, the oil should be readily oxidizable at reservoir and rock matrix conditions. This relationship must be determined by laboratory experiments. The same laboratory experiments can also determine the amount of air needed to burn a given reservoir volume. This is key to the economics of the process.

16.10 Conclusions

ISC is applicable to a wide array of reservoirs. In fact, it is the only thermal method that can presently be applied to deep reservoirs, though deep downhole steam generation is being tested. It can be used at any stage of reservoir depletion; it can be used in special situations such as offshore or in Arctic regions. Because of the lack of heat losses at the surface and in the injection wells, it is the most thermally efficient thermal-recovery method. The injectant (air) is readily available. Combustion allows wider well spacing than steam; economic results are comparable to those of steam injection.

Several aspects of operating ISC projects must be considered. First is the large compression ratio and associated costs required to inject air into the formation. Second is the planning and design requirements for a combustion project; these are more difficult than for steam injection. Third is extensive laboratory work to assess fuel availability, air requirements, and burning characteristics of the crude that are required before designing ISC projects. Fourth is the high degree of technical sophistication and the careful monitoring needed to ensure proper operation of a project. Fifth is the limitation of numerical simulation and other techniques that makes predictions of recovery more difficult than most other enhanced oil recovery methods.

Considerable improvements in the application of ISC have been made since the early projects. New developments, such as application to light-oil reservoirs, and the use of horizontal wells are reviving interest in ISC. This process deserves consideration for many reservoirs, including those in hostile environments or those not amenable to other recovery methods.

Nomenclature

a, b, d, f = stoichiometric coefficients similar to those in Eq. 16.3

A = Arrhenius rate constant

c_p/c_v = heat capacity ratio

C_f = concentration of fuel, kg/m^3

E = activation energy, kJ/mol
H_c = heating value, Btu/lbm fuel
K = reaction rate
m = CO_2/CO atomic ratio
m = CO_2/CO concentration ratio produced
n = hydrogen/carbon atomic ratio of fuel
N_p = oil produced, m^3
$(O_2)_{prod}$ = oxygen concentration produced
p_{O_2} = oxygen partial pressure, Pa
R = 0.00831 universal gas constant, kJ/mol °K
R_C = reaction rate of crude, kg/m^3s
S_f = oil saturation burned, fraction
S_o = oil saturation, fraction
S_{oi} = initial oil saturation, fraction
S_{wf} = water saturation resulting from the combustion process, fraction
T = absolute temperature, °K
T_1 = initial reservoir temperature, °F
T_2 = final reservoir temperature, °F
V_b = volume burned, m^3
V_p = volume of the pattern, m^3
W = fuel availability
W_p = water produced, m^3
ϕ = porosity, fraction
μ = oil viscosity

References

1. Sarathi, P.: "In-Situ Combustion Handbook Principles And Practices," Report DOE/PC/91008-0374, OSTI ID 3175 (January 1999).
2. Prats, M. *Thermal Recovery,* Monograph Series, SPE, Richardson, Texas (1982) **7.**
3. Burger, J., Sourieau, P., and Combarnous, M.: *Thermal Methods Of Oil Recovery,* Editions Technip, Paris (1985; re-edited 1989).
4. Wu, C.H. and Fulton, F.P.: "Experimental Simulation of the Zones Preceding the Combustion Front of an In-Situ Combustion Process," *SPEJ* (March 1971) 38; *Trans.,* AIME, **251.**
5. Bousaid, S.: "Multiple-Quenched Fireflood Process Boosts Efficiency," *JPT* (November 1989) 1202.
6. Mamora, D. *et al.:* "Kinetics of in-situ combustion," Report No. DOE/BC/14600-51 (DE93000152) SUPRI TR 91, U.S. Dept. of Energy, Washington, DC (July 1993).
7. Burger, J.: "Chemical Aspects of In-Situ Combustion–Heat of Combustion and Kinetics," *SPEJ* (October 1972) 410; *Trans.,* AIME, **253.**
8. Moore, R.G.: "New Strategies for In-Situ Combustion," *J Cdn. Pet. Tech.* (December 1993) **32,** No. 10.
9. Fassihi, M.R., Brigham, W.E., and Ramey, H.J. Jr.: "Reaction Kinetics of In-Situ Combustion: Part 2—Modeling," *SPEJ* (August 1984) 408.
10. Moore, R.G. *et al.:* "A comparison of the laboratory in-situ combustion behaviour of Canadian oils," *Proc.,* 6th UNITAR Conference, Houston, February 1995.
11. Bardon, C. and Gadelle, C.: "Essais de laboratoire pour l'etude de la combustion in-situ," paper presented at the 1977 French Soviet Symposium on Enhanced Oil Recovery, Moscow, May.

12. Leaute, R.P. and Collyer, C.J.: "Laboratory studies of in situ combustion with cold lake crude," paper 5 presented at the 1984 Annual Conference on Upgrading Technology and Petroleum Recovery, Calgary.
13. Dew, J.N. and Martin, W.I.: "How to calculate air requirements for in-situ combustion," *Petroleum Engineer* (December 1964 and January 1965).
14. Nelson, T.W. and Mc Neil, J.S.: "How to engineer an in-situ combustion project," *Producer Monthly* (May 1961); *Oil & Gas J.* (5 June 1961).
15. Gates, C.J and Ramey H.J. Jr.: " A Method for Engineering In-Situ Combustion Oil Recovery Projects," *JPT* (February 1980) 285.
16. Brigham, W. and Satman, A.: "Recovery Correlations for In-Situ Combustion Field Projects and Application to Field Pilots," *JPT* (December 1980) 2132.
17. Shallcross, D.C.: "Devices and Methods for In-Situ Combustion Ignition," Report No. DOE/BC/14126-12 (DE 89000766), U.S. Dept. of Energy, Washington, DC (October 1989).
18. Burger, J.G.: "Spontaneous Ignition in Oil Reservoirs," *SPEJ* (April 1976) 73.
19. Tadema, H.J. and Weidjeima, J.: "Spontaneous ignition of oils," *Oil & Gas J.* **68,** No. 50.
20. Smith, D.K.: *Cementing,* Monograph Series, SPE, Richardson, Texas (1976) **4.**
21. Ramey, H.J. Jr. *et al.:* "Case History of South Belridge, California, In Situ Combustion Oil Recovery," paper SPE 24200 presented at the 1992 SPE/DOE Enhanced Oil Recovery Symposium, Tulsa, 22–24 April.
22. Ramage, W.E., Castanier, L.M., and Ramey, H.J. Jr.: "Economic Evaluation of Thermal Recovery Projects," paper SPE 16852 presented at the 1987 SPE Annual Technical Conference and Exhibition, Dallas, 27–30 September.
23. Machedon, V.: "Suplacu de Barcau field: 28 years of in situ combustion," *Proc.,* 7th European Symposium on Enhanced Oil Recovery, Moscow (1993).
24. Machedon, V.: "Romania: 30 years of experience in in-situ combustion," *Proc.,* NIPER/DOE Symposium on In-situ Combustion, Tulsa (21–22 April 1994); U.S. DOE Report NIPER/BDM-0086 (January 1995).
25. Roychaudurhi, S. *et al.:* "Experience with in-situ combustion pilot in presence of edge water," *Proc.,* 6th UNITAR Conference on Heavy Crudes and Tar Sands, Houston (12–17 February 1995).
26. Moritis, G. "EOR Oil Production Up Slightly," *Oil & Gas J.* (20 April 1998) 49.
27. Mamedov, Y.G. and Bocserman, A.A.: "Development of heavy oils and bitumen in the former Soviet Union and eastern and central Europe," *Proc.,* 6th UNITAR Conference on Heavy Crudes and Tar Sands, Houston (12–17 February 1995).
28. Marko, D.: "Oil production history in Albanian fields and their perspective," *Proc.,* 6th UNITAR Conference on Heavy Crudes and Tar Sands, Houston (12–17 February 1995).
29. Ames, B.G., Gramms, R.E., and Pebdani, F.N.: "Improved sweep efficiency through application of horizontal well technology in a mature combustion EOR," *Proc.,* NIPER/DOE Symposium on In-Situ Combustion, Tulsa (21–22 April 1994); U.S. DOE Report NIPER/BDM-0086 (January 1995).
30. Marjerrison, D.M. and Fassihi, M.R.: "Performance of Morgan Pressure Cycling In-Situ Combustion Project," paper SPE 27793 presented at the 1994 SPE/DOE Improved Oil Recovery Symposium, Tulsa, 17–20 April.
31. Hallam, R.J. and Donnelly, J.K.: "Pressure-Up Blow Down Combustion: A Channeled Reservoir Recovery Process," *SPE Advanced Technology Series* (April 1993) 153.
32. Hoffmann, S.J.: "Successful Application of In Situ Combustion to a Dipping Heavy Oil Reservoir Midway Sunset Field," paper 39639 presented at the 1998 SPE/DOE Symposium, Tulsa.
33. Hvizdos, L.J., Howard, J.V., and Roberts, G.W.: "Enhanced Oil Recovery Through Oxygen-Enriched In-Situ Combustion: Test Results From the Forest Hill Field in Texas," *JPT* (June 1983) 1061.
34. Moore, R.G. *et al.:* "New Insights Into Enriched-Air In-Situ Combustion," *JPT* (July 1990) 916; *Trans.,* AIME, **289.**
35. Racz, D. *et al.:* "Development of a thermocatalytic in situ combustion process in Hungary," *Proc.,* 1985 European Meeting on Improved Oil Recovery, Rome (April).
36. Gilham, T.H. *et al.:* "Keys To Increasing Production Via Air Injection in Gulf Coast Light Oil Reservoirs," paper SPE 38848 presented at the 1997 SPE Annual Technical Conference and Exhibition, San Antonio, Texas, 5–8 October.

37. Miller, R.J.: "Koch's experience with deep in-situ combustion in Williston basin," *Proc.*, NIPER/DOE Symposium on In-Situ Combustion, Tulsa (21–22 April 1994); U.S. DOE Report NIPER/BDM-0086 (January 1995).

General References

Alexander, J.D., Martin W.L., and Dew, J.N.: "Factors Affecting Fuel Availability and Composition During In-Situ Combustion," *JPT* (October 1962) 1154; *Trans.*, AIME, **225.**

Bahtia, A.K.: "Reservoir management of heavy oil reservoirs of North Gujarat," *Proc.*, 7th UNITAR Intl. Conference on Heavy Crude and Tar Sands, Beijing (1998).

Basham, M.: "Planning an in situ combustion project: Cymric field, California," *Proc.*, PTTC Meeting, Bakersfield, California (1999).

Burger, J. and Sahuquet, B.: "Laboratory Research on Wet Combustion," *JPT* (October 1973) 1137.

Carcoana, A. *et al.:* "Balayage et recuperation d'un gisement d'huile lourde par combustion in situ," *Proc.*, Colloque Int. sur l'exploitation des hydrocarbures, Paris (1975).

Cato, R.W. and Franka, W.A.: "Getty Oil Reports Fireflood Pilot is Successful Project," *Oil & Gas J.* (12 February 1968).

Chu, C.: "State-of-the-Art Review of Fireflood Field Projects," *JPT* (January 1982) 19.

Coats, K.H.: "In-Situ Combustion Model," *SPEJ* (December 1980) 533.

Counihan, T.M.A.: "A Successful In-Situ Combustion Pilot in the Midway Sunset Field, California," Paper SPE 6525 presented at the 1977 SPE California Regional Meeting, Bakersfield, California, 13–15 April.

Craig, F.F. Jr. and Parrish, D.R.A.: "A Multipilot Evaluation of the COFCAW Process," *JPT* (June 1974) 659.

Dietz, D.N.: "Wet Underground Combustion, State of the Art," *JPT* (May 1970) 605; *Trans.*, AIME, **249.**

Dietz, D.N. and Wijema, J.: "Reverse combustion seldom feasible," *Prod. Monthly* (May 1968).

Ejiogu, J.G. *et al.:* "Wet combustion—A tertiary recovery process for the Pembina Cardium Reservoir," *J. Cdn. Pet. Tech.* (July–September 1979).

Farouq Ali, S.M.: "A Current Appraisal of In-Situ Combustion Field Tests," *JPT* (April 1972) 477.

Fassihi, M.R.: "Analysis of fuel oxidation on in-situ combustion oil recovery," PhD thesis, Stanford U. (1981); U.S. DOE Report DOE/ET/12056-26, U.S. Dept. of Energy, Washington, DC (August 1982).

Fattahi, B.: "Pyramids Hills Project, Mt. Poso Field, California," Aera Energy, 1999 PTTC Meeting, Bakersfield, California.

Gadelle, C. *et al.:* "Heavy Oil Recovery by In-Situ Combustion—Two Field Cases in Rumania," *JPT* (November 1981) 2057.

Galas, C.M.F. and Ejlogu, G.C.: "Enhancement of In-Situ Combustion by Steam Stimulation of Production Wells," *SPEREE* (November 1993) 270.

Garon, A.M. and Wygal, R.J. Jr.: "A Laboratory Investigation of Fire-Water Flooding," *SPEJ* (December 1974) 537.

Gates, C.J. and Ramey, H.J. Jr.: "Field Results of South Belridge Thermal Recovery Experiment," *Trans.*, AIME (1958) **213,** 236.

Gates, C.F., Jung, K.D., and Surface, R.A.: "In-Situ Combustion in the Tulare Formation, South Belridge Field, Kern County, California," *JPT* (May 1978) 799; *Trans.*, AIME, **265.**

Gates, C.J. and Sklar, I.: "Combustion as a Primary Recovery Process – Midway Sunset Field," *JPT* (August 1971) 981; *Trans.*, AIME, **251.**

Joseph, C. and Pusch, W.H.: "A Field Comparison of Wet and Dry Combustion," *JPT* (September 1980) 1523.

Koch, R.L.: "Practical use of combustion drive at West Newport field," *Pet. Eng.* (January 1965).

Martin, W.L., Alexander, J.D., and Dew, J.N.: "Process Variables of In Situ Combustion," *Trans.,* AIME (1958) **213,** 28.

Meldau, R.F., Shipley, R.G., and Coats, K.H.: "Cyclical Gas/Steam Stimulation of Heavy-Oil Wells," *JPT* (October 1981) 1990.

Moss, J.T., White, P.D., and McNeil, J.S. Jr.: "In Situ Combustion Process—Results of a Five-Well Field Experiment in Southern Oklahoma," *Trans.,* AIME (1959) **216,** 55.

Olsen, D. and Sarathi, P.: "Field application of in- situ combustion," Report No. NIPER/BDM 0086, U.S. Dept. of Energy, Washington, DC (1994).

Showalter, W.E. and Maclean, A.M.: "Fireflood at Brea Olinda Field, Orange County, California," paper SPE 4763 presented at the 1974 SPE Improved Oil Recovery Symposium, Tulsa, 22–24 April.

Showalter, W.E.: "Combustion-Drive Tests," *SPEJ* (March 1963) 53; *Trans.,* AIME, **228.**

Terwilliger, P.L. *et al.:* "Fireflood of the P2-3 Sand Reservoir in the Miga Field of Eastern Venezuela," *JPT* (January 1975) 9.

Widmeyer, R.H. *et al.:* "The Charco Redondo Thermal Recovery Pilot," *JPT* (December 1977) 1522.

Williams, R.L., Jones, J.A., and Counihan, T.M.: "Expansion of a Successful In-Situ Combustion Pilot in Midway Sunset Field," paper SPE 16873 presented at the 1987 SPE Annual Technical Conference and Exhibition, Dallas, 27–30 September.

SI Metric Conversion Factors

°API		141.5/(131.5 + °API)		= g/cm³
bar	×	1.0*	E + 05	= Pa
bbl	×	1.589 873	E − 01	= m³
Btu	×	1.055 056	E + 00	= kJ
cp	×	1.0*	E − 03	= Pa·s
ft	×	3.048*	E − 01	= m
ft³	×	2.831 685	E − 02	= m³
°F		(°F − 32)/1.8		= °C
°F		(°F + 459.67)/1.8		= K
kW-hr	×	3.6*	E + 00	= J
lbm	×	4.535 924	E − 01	= kg
psi	×	6.894 757	E + 00	= kPa

*Conversion factor is exact.

Chapter 17
Reservoir Simulation

Rod P. Batycky, Marco R. Thiele, StreamSim Technologies Inc.; **K.H. Coats,** Coats Engineering Inc.; **Alan Grindheim, Dave Ponting,** Roxar Software Solutions; **John E. Killough,** Landmark Graphics; **Tony Settari,** U. of Calgary and Taurus Reservoir Solutions Ltd.; **L. Kent Thomas,** ConocoPhillips; **John Wallis,** Wallis Consulting Inc.; **J.W. Watts,** Consultant; and **Curtis H. Whitson,** Norwegian U. of Science and Technology and Pera

17.1 Introduction—K.H. Coats

The Merriam-Webster Dictionary defines *simulate* as assuming the appearance of without the reality. Simulation of petroleum reservoir performance refers to the construction and operation of a model whose behavior assumes the appearance of actual reservoir behavior. The model itself is either physical (for example, a laboratory sandpack) or mathematical. A mathematical model is a set of equations that, subject to certain assumptions, describes the physical processes active in the reservoir. Although the model itself obviously lacks the reality of the reservoir, the behavior of a valid model simulates—assumes the appearance of—the actual reservoir.

The purpose of simulation is estimation of field performance (e.g., oil recovery) under one or more producing schemes. Whereas the field can be produced only once, at considerable expense, a model can be produced or run many times at low expense over a short period of time. Observation of model results that represent different producing conditions aids selection of an optimal set of producing conditions for the reservoir.

The tools of reservoir simulation range from the intuition and judgment of the engineer to complex mathematical models requiring use of digital computers. The question is not whether to simulate, but rather which tool or method to use. This chapter concerns the numerical mathematical model requiring a digital computer. The Reservoir Simulation chapter in the 1987 edition of the *Petroleum Engineering Handbook*[1] included a general description of reservoir simulation models, a discussion related to how and why they are used, choice of different types of models for different-reservoir problems, and reliability of simulation results in the face of model assumptions and uncertainty in reservoir-fluid and rock-description parameters. That material is largely omitted here. Instead, this chapter attempts to summarize current practices and trends related to development and application of reservoir simulation models.

TABLE 17.1—SPE COMPARATIVE SOLUTION PROJECT PROBLEMS

SPE1	Three-phase black oil
	10×10×3 300-block grid 3,650-day depletion with gas injection
SPE2	Three-phase black oil
	10×1×15 150-block *r-z* grid 900-day single-well coning depletion
SPE3	Nine-component retrograde gas
	9×9×4 324-block grid
	5,480-day cycling and blowdown
SPE4	Cyclic steam injection and steam displacement of heavy oils
SPE5	Six-component volatile oil
	7×7×3 147-block grid
	20-year WAG injection
SPE6	Three-phase black oil
	Single-block and cross-sectional dual porosity with drainage and gas and water injection cases
SPE7	Three-phase black oil
	9×9×6 486-block grid with horizontal wells
	Eight 1,500-day injection-production cases
SPE8	Two-phase gas-oil black oil
	10×10×4 400-block grid
	Comparison of 2,500-day 400-block grid results with 20-block unstructured and locally refined grid results
SPE9	Three-phase black oil
	24×25×15 9,000-block 25-well grid with geostatistical description
	900-day depletion
SPE10	Model 1: Two-phase gas-oil case with a 2,000-block 100×1×20 grid and gas injection to 2000 days
	Model 2: Two-phase water-oil case with a 1.12-million block 60×220×85 grid and water injection to 2,000 days
	Both models have geostatistical descriptions

Models have been referred to by *type,* such as *black-oil, compositional, thermal, generalized,* or *IMPES, Implicit, Sequential, Adaptive Implicit,* or *single-porosity, dual-porosity,* and more. These *types* provide a confusing basis for discussing models; some refer to the application (e.g., thermal), others to the model formulation (e.g., implicit), and yet others to an attribute of the reservoir formation (e.g., dual-porosity). The historical trend, though irregular, has been and is toward the *generalized* model, which incorporates all the previously mentioned types and more. The generalized model, which represents most models in use and under development today, will be discussed in this chapter. Current model capabilities, recent developments, and trends will then be discussed in relation to this generalized model.

The 10 SPE Comparative Solution Project problems, SPE1 through SPE10,[2–11] are used for some examples below. **Table 17.1** gives a brief description of those problems.

17.1.1 The Generalized Model. Any reservoir simulator consists of $n + m$ equations for each of N active gridblocks comprising the reservoir. These equations represent conservation of mass of each of n components in each gridblock over a timestep Δt from t^n to t^{n+1}. The first n (primary) equations simply express conservation of mass for each of n components such as oil, gas, methane, CO_2, and water, denoted by subscript $I = 1,2,...,n$. In the thermal case, one of the "components" is energy and its equation expresses conservation of energy. An additional m (secondary or constraint) equations express constraints such as equal fugacities of each component in all phases where it is present, and the volume balance $S_w + S_o + S_g + S_{\text{solid}} = 1.0$, where S_{solid} represents any immobile phase such as precipitated solid salt or coke.

There must be $n + m$ variables (unknowns) corresponding to these $n + m$ equations. For example, consider the isothermal, three-phase, compositional case with all components present in all three phases. There are $m = 2n + 1$ constraint equations consisting of the volume balance and the $2n$ equations expressing equal fugacities of each component in all three phases, for a total of $n + m = 3n + 1$ equations. There are $3n + 1$ unknowns: p, S_w, S_o, S_g, and the $3(n - 1)$ independent mol fractions x_{ij}, where $i = 1,2,...,n - 1$; $j = 1,2,3$ denotes the three phases oil, gas, and water. For other cases, such as thermal, dual-porosity, and so on, the m constraint equations, the $n + m$ variables, and equal numbers of equations and unknowns can be defined for each gridblock.

Because the m constraint equations for a block involve unknowns only in the given block, they can be used to eliminate the m secondary variables from the block's n primary or conservation equations. Thus, in each block, only n primary equations in n unknowns need be considered in discussions of model formulation and the linear solver. The n unknowns are denoted by P_{i1}, P_{i2}, ..., P_{in}, where P_{in} is chosen as pressure p_i with no loss of generality. These primary variables may be chosen as any n independent variables from the many available variables: phase and overall mol fractions, mol numbers, saturations, p, and so on. Different authors choose different variables.[12–15] Any sensible choice of variables and ordering of the primary equations gives for each gridblock a set of n equations in n unknowns which is susceptible to normal Gaussian elimination without pivoting. The (Newton-Raphson) convergence rate for the model's timestep calculation is independent of the variable choice; the model speed (CPU time) is essentially independent of variable choice.

The Ith primary or conservation equation for block i is

$$M_{iI}^{n+1} - M_{iI}^{n} = \Delta t\left(\sum_{j=1}^{j=N} q_{ijI} - q_{iI}\right) I = 1,2,...n, \quad (17.1)$$

where M_{iI} is mass of component I in gridblock i, q_{ijI} is the interblock flow rate of component I from neighbor block j to block i, and q_{iI} is a well term. With transposition, this equation is represented by $f_{iI} = 0$, the Ith equation of gridblock i. All n equations $f_{iI} = 0$ for the block can be expressed as the vector equation $\mathbf{F}_i = 0$ where f_{iI} is the Ith element of the vector $\mathbf{F}_i$. Finally, the vector equation

$$\mathbf{F}\left(\mathbf{P}_1, \mathbf{P}_2, ..., \mathbf{P}_N\right) = 0 \quad (17.2)$$

represents the entire model, where the ith element of the vector $\mathbf{F}$ is $\mathbf{F}_i$. $\mathbf{F}$ is a function of the N vector unknowns $\mathbf{P}_i$, where the Ith scalar element of $\mathbf{P}_i$ is P_{iI}. Application of the Newton-Raphson method gives

$$\mathbf{F}^l + \delta\mathbf{F} = \mathbf{F}^l + A\delta\mathbf{P} = 0, \quad (17.3)$$

where $\delta \mathbf{P}$ is $\mathbf{P}^{l+1}-\mathbf{P}^{l}$ and the $N \times N$ matrix A represents the Jacobian $\partial\mathbf{F}/\partial\mathbf{P}$. The element A_{ij} of A is itself an $n \times n$ matrix $\partial\mathbf{F}_i/\partial\mathbf{P}_j$ with scalar elements $a_{rs} = \partial f_{ir}/\partial P_{js}$, r and s each $= 1,2,...,n$. Eq. 17.3 is solved by the model's linear solver. The matrix A is very sparse because A_{ij} is 0 unless block j is a neighbor of block i.

The calculations for a timestep consist of a number of Newton (*nonlinear* or *outer*) iterations terminated by satisfaction of specified convergence criteria. Each Newton iteration requires:

(a) Linearization of the constraint equations and conservation Eq. 17.1.

(b) Linear algebra to generate the A matrix coefficients.

(c) Iterative solution of Eq. 17.3 (*inner* or *linear* iterations).

(d) Use of the new iterate $\mathbf{P}^{l+1}$ to obtain from Eq. 17.1 the moles of each component in the gridblock.

(e) A flash to give phase compositions, densities, and saturations which allow generation of the A matrix coefficients for the next Newton iteration.

17.1.2 Model Formulations. A major portion of the model's total CPU time is often spent in the linear solver solution of Eq. 17.3. This CPU time in turn reflects the many multiply operations required. The model formulation has a large effect on the nature and expense of those multiplies.

Implicit vs. Explicit. The interblock flow term in Eq. 17.1,

$$q_{ijI} = T_{ij} \sum_{J=1}^{J=3} \lambda_{j} \rho_{J} x_{IJ} \left(\Delta p_{J} - \gamma_{J} \Delta Z\right), \quad \text{(17.4)}$$

uses phase mobilities, densities, and mol fractions evaluated at the upstream blocks. A gridblock is *implicit* in, say, the variable S_g if the new time level value S_g^{n+1} is used to evaluate interblock flow terms dependent upon it. The block is *explicit* in S_g if the old time level value S_g^{n} is used.

The Implicit Formulation. The implicit formulation[16] expresses interblock flow terms using implicit (new time level) values of all variables in all gridblocks. As a consequence, all nonzero A_{ij} elements of the A matrix of Eq. 17.3 are full $n \times n$ matrices. The resulting multiplies in the linear solver are then either matrix-matrix or matrix-vector multiplies, requiring work (number of scalar multiplies) of order n^3 or n^2, respectively.

The IMPES Formulation. Early papers[17–19] presented the basis of the IMPES (implicit pressure, explicit saturations) formulation for the black-oil case: take all variables in the interblock flow terms explicit, except for pressure, and eliminate all nonpressure variables from the linearized expressions for M_{il}^{n+1} in Eq. 17.1. The obvious extension to any type model with any number of components was presented later,[20] and numerous IMPES-type compositional models have been published.[13–15,21]

The model Eq. 17.3 can be written as:

$$A_{ii}\delta\mathbf{P}_i + \sum_{j \neq i} A_{ij}\delta\mathbf{P}_j = -\mathbf{F}_i^l \; i = 1,2,..., N \;. \quad \text{(17.5)}$$

If all variables but pressure are explicit in the interblock flow terms, then all entries but those in the last column of the $n \times n$ A_{ij} $(j \neq i)$ matrix are zero (recall, the nth variable in each gridblock, P_{in}, is pressure p_i). This allows elimination of all nonpressure variables and reduction of the vector Eq. 17.5 to the scalar equation in pressure only[22]:

$$a_{ii}\delta p_i + \sum_{j \neq i} a_{ij}\delta p_j = -f_i^l \; i = 1,2,\ldots, N \qquad (17.6)$$

or

$$A\delta\mathbf{P} = -\mathbf{F}^l, \qquad (17.7)$$

where A is now a scalar $N \times N$ matrix and the **P** and **F** vectors have N scalar elements p_i and f_i, respectively. The multiplications required in solution of the IMPES pressure Eq. 17.7 are scalar multiplications, requiring a small fraction of the work of the matrix-matrix and matrix-vector multiplications of the implicit formulation. Thus, the model CPU time per gridblock per Newton iteration for moderate or large n is much less for the IMPES formulation than for the implicit formulation.

The Sequential Formulation. The stability of the IMPES formulation for the two-phase water/oil case was improved by following the IMPES pressure equation solution with solution of a water saturation equation using implicit saturations (mobilities).[23] This concept was extended to the three-phase case and called the *sequential* formulation.[24] For each Newton iteration, this method requires solution of the IMPES pressure Eq. 17.7, followed by solution for two saturations from a similar equation where the A_{ij} elements of A are 2×2 matrices.

A sequential compositional model was described[15] and mentioned the desirability of a sequential implicit treatment of mol fractions in addition to saturations.

The Adaptive Implicit Formulation. The Adaptive Implicit Method (AIM)[25] uses different levels of implicitness in different blocks. In each gridblock, each of the n variables may be chosen explicit or implicit, independent of the choices in other gridblocks. The choices may change from one timestep to the next. This results in the same equation $A\delta\mathbf{P} = -\mathbf{F}^l$ as the Implicit formulation except that the elements A_{ij} of the A matrix are rectangular matrices of variable size. The numbers of rows and columns in A_{ij} equal the numbers of implicit variables in blocks i and j, respectively; all A_{ii} are square matrices. The CPU expense per Newton iteration of an AIM model lies between those of IMPES and Implicit models, tending toward the former as more blocks are taken implicit in pressure only.

Choice of Formulation. For a given problem, the previous four formulations generally give widely different CPU times. Generalizations regarding the best formulation have many exceptions. Arguably, the trend is or should be toward sole use of the AIM formulation. This is discussed in the Stable Step and Switching Criteria sections to follow. Current simulation studies use all of these formulations. The Implicit formulation is generally faster than IMPES for single-well coning studies, and for thermal and naturally fractured reservoir problems. For other problems, IMPES is generally faster than Implicit for moderate or large n (say, $n > 4$). Most participants used IMPES for SPE Comparative Solution Project problems SPE1, SPE3, SPE5, and SPE10. All participants used the Implicit formulation for SPE2, SPE4, SPE6, and SPE9. No participants in SPE1 through SPE10 used a Sequential model, and, with few exceptions, none used AIM.

A frequently stated generalization is that numerical dispersion error is significantly larger for Implicit than for IMPES formulations. Truncation error analysis[26] shows this error to be proportional to $\Delta x + u\Delta t$ for Implicit and $\Delta x - u\Delta t$ for IMPES. Real problem nonlinearities and heterogeneity render the analysis approximate and the generalization of limited merit. For example, **Figs. 17.1 through 17.3** show virtually identical Implicit and IMPES results for the black-oil 9,000-block SPE9 and 2,000-block gas/oil SPE10 problems. For SPE9 (SPE10), the average timestep was 67 (9.7) times larger for Implicit than for IMPES. The percentage of total CPU

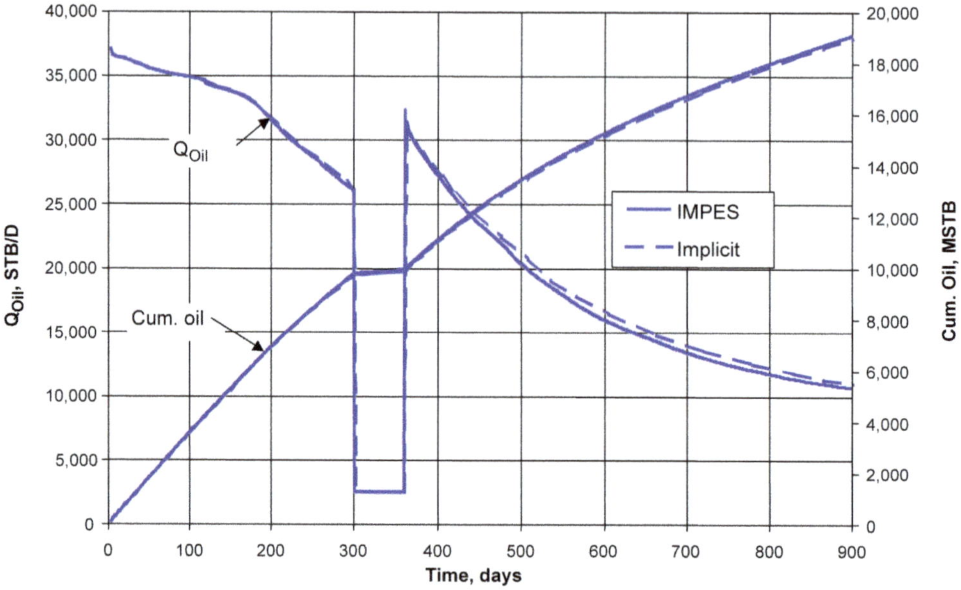

Fig. 17.1—IMPES vs. implicit oil rate and cumulative oil production, SPE9.

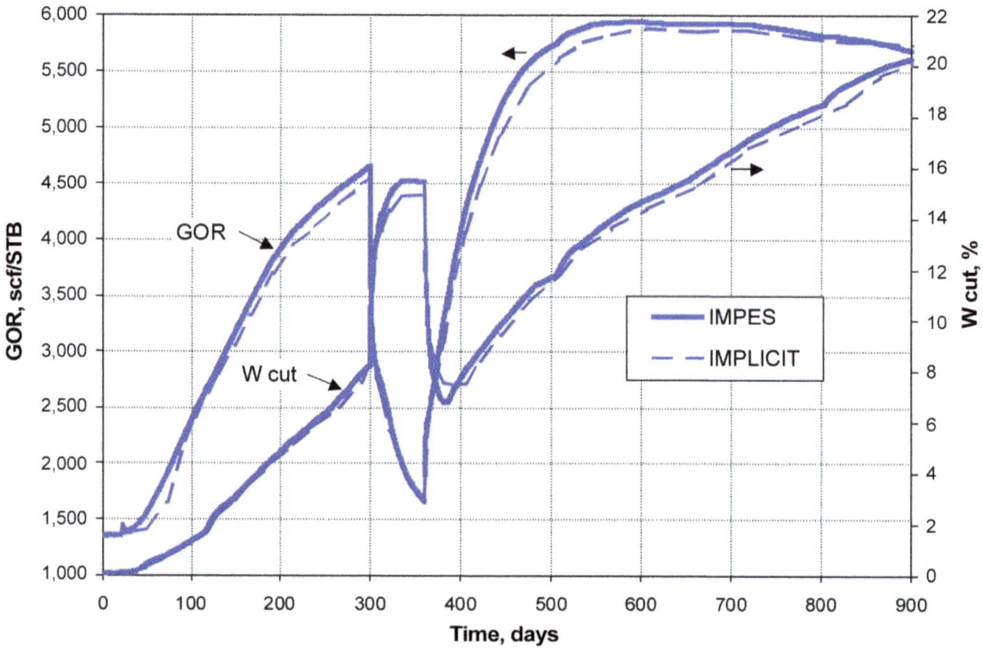

Fig. 17.2—IMPES vs. implicit GOR and water cut, SPE9.

time spent in the linear solver for IMPES (Implicit) was 23.7 (57.3) for SPE9 and 35.4 (73.4) for SPE10.

Implementation is an important factor in the relative efficiencies of different formulations. For a given problem, different models using the same formulation can give widely different

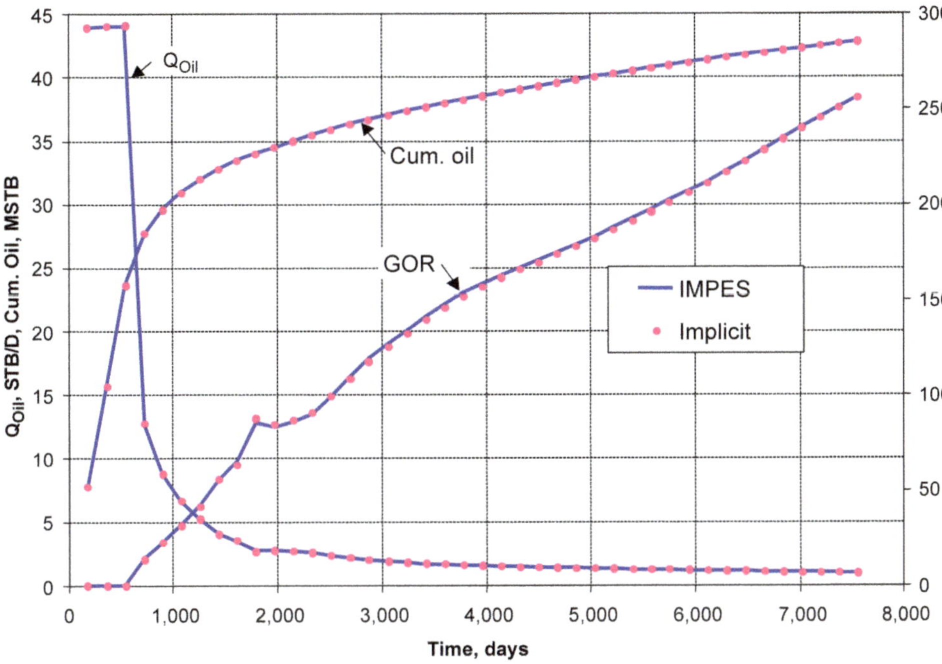

Fig. 17.3—IMPES vs. implicit oil rate, cumulative oil production, and GOR, SPE10 Model 1.

CPU times. For example, the IMPES CPU times reported by different participants in the six-component compositional SPE5 problem[6] differed by a factor of over 50.

17.1.3 Advances in Model Formulations. The IMPES formulation was improved by concepts of relaxed volume,[13–15] better choice of variables,[13] and "adaptive" flash calculations.[13]

Relaxed Volume. The relaxed volume concept relates to the timestep calculation Steps (d) and (e) given previously. Step (d) gives the mass of each component in the gridblock, M_I^{l+1}, which in turn gives overall composition $\{z_I\}^{l+1}$. The Step (e) flash then gives phase amounts and densities which in turn give new iterate S_w, S_o, and S_g values. These saturations do not sum to 1.0 because of the nonlinear nature of the conservation Eq. 17.1. If the saturations are altered (e.g., divide each by their sum) to exactly satisfy the volume balance $\Sigma_J S_J = 1$, then an incremental (timestep) mass-balance error occurs. If the saturations are not altered, then mass is conserved but there is a volume-balance error $\Sigma_J S_J - 1$. The authors[13–15] chose to preserve mass and *carry forward* the volume balance error from iterate to iterate and step to step. The volume balance going into iterate $l + 1$ is $\Sigma_J \delta S_J = 1 - \Sigma_J S_J^l$. This in effect conserves both mass and volume because there is no permanent or accumulating volume error—only that of the given timestep. Equally important, there is no need to iterate out the volume error to a "tight" tolerance, and Newton iterations and model CPU are reduced. In contrast, the previous or historical IMPES procedure reset saturations to preserve volume and iterated out the mass-balance error. Because the latter error was not carried forward, more Newton iteration (and CPU time) was required to keep the permanent, accumulating mass balance error tolerably low. This use of relaxed volume with carryover also reduces Newton iterations and CPU time in the Implicit formulation.[21]

This discussion implies some fundamental advantage of preserving mass and iterating out volume error as opposed to preserving volume and iterating out mass error. In the writer's opinion, that is not true provided the error is carried forward in both cases. The Newton iteration requirement and CPU time should be similar if "equivalent" mass and volume error tolerances are used as convergence criteria.

Variable Choice. The linear algebra required to reduce the gridblock's n conservation equations to the IMPES pressure equation is influenced by the choice of variables. The influence is absent for black oil, moderate for "moderate" n and up to a factor of three for large n (say, > 15).[22] The choices of p and mol fractions $\{z_i\}$[13] or mol numbers[14,15] are better than the choice of p, saturations, and phase mol fractions[12] for large n. The effect of this variable choice on total CPU time is often small because the affected work is often a small part of total CPU time. This IMPES reduction is absent in the Implicit formulation and the last of the above variable choices is arguably preferable.[22]

Adaptive Flash Calculations.[13] The work of EOS flash calculations, including the generation of fugacities and their derivatives, can significantly affect model efficiency when the linear solver does not dominate total CPU time. There may be little need to perform (most of) that work in a gridblock when p and composition are changing slowly. Use of internal, intelligent criteria dictating when that work is needed can significantly reduce the total-run flash calculation CPU time.[13] This is similar in principle to the AIM selection of explicit variables for gridblocks which are quiescent in respect to throughput ratio.

17.1.4 Stable Timestep and Switching Criteria. This topic relates to the observation that lower run turnaround time can increase benefits from a reservoir study allotted a budgeted time period. As a corollary, time spent in repeated runs fighting model instabilities or time-stepping is counterproductive. While many factors affect this run time, it always equals the product (CPU time/step) × (number of timesteps). The first factor is "large" and the second "small" for the Implicit formulation, and conversely for the IMPES formulation. IMPES is a conditionally stable formulation requiring that $\Delta t < \Delta t^*$ to prevent oscillations and error growth, where Δt^* is maximum stable timestep. The conditional stability stems from the explicit treatment of nonpressure variables in the interblock flow terms. Mathematicians performed stability analyses for constant-coefficient difference equations bearing some resemblance to IMPES. Authors in our industry extended and applied their results to derive expressions for Δt^*, in particular,[27]

$$\Delta t^* = \frac{V_p}{f_g'\left(|q_x| + |q_y| + |q_z|\right) + 2P_{cgo}'\psi\left(T_x + T_y + T_z\right)} \quad \text{(17.8)}$$

for the black-oil 3D case of gas/oil flow. This shows that stable step Δt^* is dependent upon flow rates, phase mobility, and capillary pressure derivatives, which of course vary with time and from one gridblock to another. Thus, at a given timestep, there are block-dependent stable step values Δt^*_i, where $1 < i < N$, and the IMPES stable step is Min(i) Δt^*_i. An IMPES model using this internally determined stable step will run stably but may suffer from the weakest-link principle. As an extreme example, consider a 500,000-gridblock problem where, over a 100-day period, the Δt^*_i value is 0.01 day for one block and > 30 days for all other blocks. The IMPES model will require 10,000 timesteps over the 100-day period.

In the AIM formulation, the stable step Δt^*_i depends upon the number and identities of variables chosen explicit in block i; theoretically, $\Delta t^*_i = \infty$ if all block i variables are chosen implicit. In the previous example, all nonpressure variables could be chosen implicit in the block where $\Delta t^*_i = 0.01$ and explicit in all other blocks. The AIM model would then require CPU time/step essentially no greater than the IMPES model but would require only three timesteps for the 100-day period.

Numerous papers[28–33] address the problem of determining expressions for the Δt^*_i for use internally as switching criteria to select block variables as explicit or implicit in the AIM model. The stability analyses involved are complex and may be impractically complex when allowing the implicit vs. explicit variable choice to include all permutations (in number and identity) of the n variables. The most reliable and efficient AIM models in the future will stem from continuing research leading to the following: (a) Δt^*_i estimates which are "accurate," and (b) implicit vs. explicit variable choices, block by block, which are near-optimal[34] and minimize total CPU time, (CPU time/step) × (number of steps).

17.1.5 The Linear Solver. Preconditioned Orthomin[35] is the most widely used method for iterative solution of Eqs. 17.3 or 17.7. Nested Factorization (NF)[36] and incomplete LU factorization [ILU(n)][37] are the two most widely used preconditioners. The term "LU factorization" refers to the factoring of the matrix A into the product of a lower triangular matrix L and an upper triangular matrix U. That is an expensive operation but is straightforward, involving only Gaussian elimination. The term "ILU(n)" denotes incomplete LU factorization, where only limited fill-in is allowed and n is the "order of fill."[37] NF performs exceptionally well when transmissibilities associated with a particular direction (in a structured grid) dominate those in other directions uniformly throughout the grid. In general, ILU(n) or red-black ILU(n)[38] [RBILU (n)] is less sensitive than NF to ordering of the blocks and spatial variation of the direction of dominant transmissibilities. In addition, RBILU(n) or ILU(n) have the parameter n (order of allowed infill) which can be increased as needed to solve problems of any difficulty.

A literature search and discussions with numerous developers and users have failed to establish consensus on whether NF or ILU preconditioning is better. Some are strong advocates of one method and others are just as adamantly supportive of the other. But many find, like this writer, that the better method is problem-dependent and it is difficult to find a reliable *a priori* indicator for making an up-front choice. In the writer's experience, (a) when NF works well, it is faster than ILU methods, (b) RBILU(0) with no residual constraint is frequently the best of the ILU variants and a good default choice, and (c) in some cases, global residual constraint with the ILU or RBILU method is beneficial.

17.1.6 Cartesian Grids and Reservoir Definition. For many years, simulation used orthogonal Cartesian grids. In the past 15 years, numerous papers have described local grid refinement and various non-Cartesian grids, as discussed in the Gridding section. These papers show that non-Cartesian grids can reduce grid-orientation effects and provide definition and accuracy near wells, faults, highly heterogeneous areas, and so on more efficiently than Cartesian grids. The premise that Cartesian grids cannot provide required accuracy efficiently in these respects has come to be accepted as a fact. In addition, advances in geophysics have led to geostatistical description of permeability and porosity on a fine scale once unimaginable. Increasingly, our papers include examples using thousands of gridblocks for two- or few-well "patterns," in part to reflect these geostatistical descriptions. The purpose of this section is to show, using a few examples, that Cartesian grids can provide adequate accuracy and reservoir and near-well definition efficiently in some cases, even without local grid refinement. No generalizations from the examples used are intended. For the most part, the examples are taken from the literature.

SPE7 is an x-direction horizontal well problem with a 9 × 9 × 6 Cartesian grid representing a 2,700 × 2,700 × 160-ft reservoir section. The specified block Δy values decrease from 620 to 60 ft at the well, presumably to increase near-well definition and accuracy of results. The Δx are uniformly 300 ft. **Fig. 17.4** compares Case 1A results for the SPE7 grid with results using uniform areal spacing $\Delta x = \Delta y = 300$ ft. The near-well y-direction refinement of the specified grid has no effect and is not necessary in this problem.

SPE8 is a gas/oil problem with one gas injector and two producers on the corners of a 5,000 × 5,000 × 325-ft square reservoir. A 10 × 10 × 4 Cartesian grid with uniform $\Delta x = \Delta y =$ 500 ft is specified. Five participants compared their results for that grid with results from their

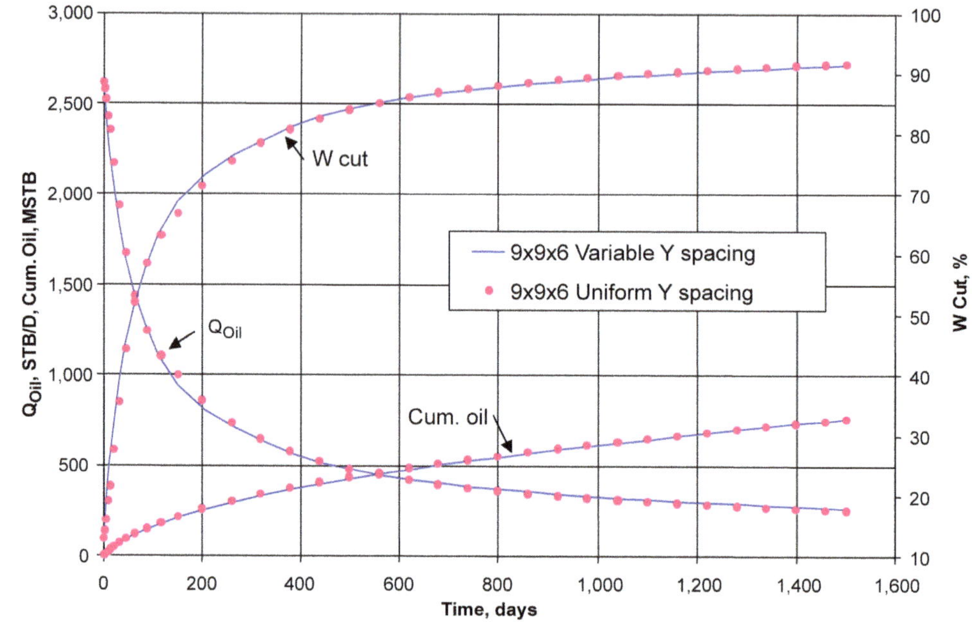

Fig. 17.4—Effect of near-well grid refinement, SPE7 Case 1A.

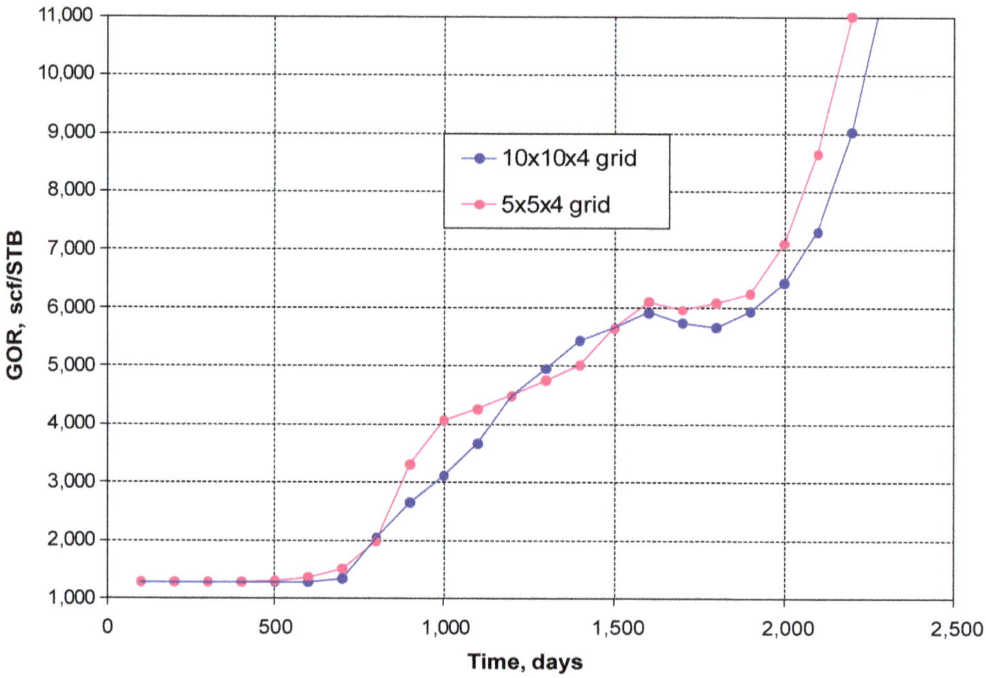

Fig. 17.5—Effects of coarser grids on producing GOR.

(areally) locally refined or unstructured grids. They showed good agreement for grids having approximately four times fewer blocks than the 10 × 10 × 4 grid. **Fig. 17.5** shows equally good agreement for a 5 × 5 × 4 ($\Delta x = \Delta y =$ 1,000 ft) Cartesian grid with no local refinement.

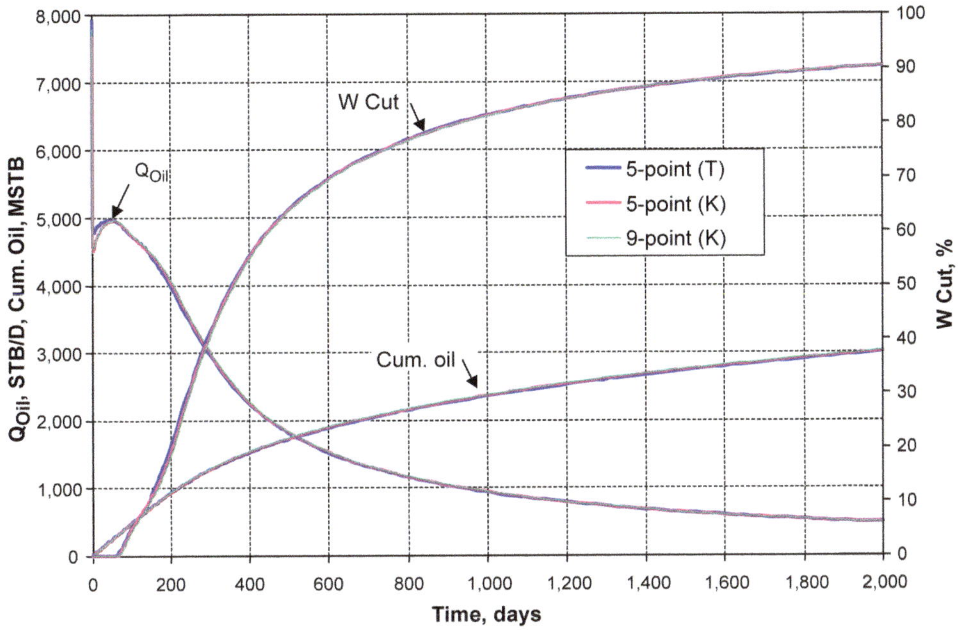

Fig. 17.6—9-point vs. 5-point, SPE10 Model 2.

SPE10 (Model 1) is a 2D cross-sectional gas/oil problem with a geostatistical permeability distribution given on a 100 × 1 × 20 Cartesian fine grid. Coarse-grid submittals included results using upscaling and local grid refinement. A homogeneous 5 × 1 × 5 Cartesian grid with no alteration of relative permeability matched the 100 × 1 × 20 results nearly exactly.[11]

SPE10 (Model 2) is a 3D water/oil problem with a 1.122 million-cell geostatistical grid. Some coarse grid submittals included sophisticated upscaling and gridding techniques with no pseudoization of relative permeability and grids from 4,810 to 70,224 blocks. Others used simple flow-based upscaling to 75- to 2,000-block Cartesian grids with moderate k_r changes. In general, the latter submittals showed the best agreement with the fine-grid solution.[11]

Numerous papers show that non-Cartesian grids can significantly reduce the grid-orientation effects of Cartesian grids. However, most of the examples used to study those effects are highly adverse mobility ratio displacements in homogeneous, horizontal reservoirs. In reservoirs with more normal fluid mobilities, areal fluid movement is more strongly affected by heterogeneity and/or gravity forces associated with reservoir structure (variable dip), and grid-orientation effects tend toward a second-order effect. As an example, the SPE10 (Model 2) water/oil problem reservoir is highly heterogeneous. **Fig. 17.6** compares five-point and nine-point field results for an upscaled 28 × 55 × 85 Cartesian grid. The close agreement indicates an absence of grid-orientation effects even though the unfavorable oil/water viscosity ratio is 10 and there is no dip.

Example 17.1. **Table 17.2** gives data for Example 17.1, a ¼ five-spot, vertical-well problem. **Fig. 17.7** shows two block-centered grids (a) and (b) used for this type of problem. The four-fold smaller well blocks of grid (b) provide finer well definition and presumably increase the accuracy of results. **Fig. 17.8** shows the identical results for 10 × 10 grid (a) and 11 × 11 grid (b). Results are nearly identical for the 5 × 5 grid (a) and 6 × 6 grid (b), and **Fig 17.9** shows insignificant difference between 3 × 3 grid (a) and 4 × 4 grid (b) results. The grid (b) doubles the grid (a) IMPES run CPU time but contributes no greater accuracy. Well-index ef-

TABLE 17.2—EXAMPLE 1 DATA

$^1/_4$ 40-acre five-spot with no dip and 180-ft pay thickness

Incompressible oil and water properties:

ρ_o = 43 lbm/ft^3 μ_o = 3 cp R_s = 0

ρ_w = 64 lbm/ft^3 μ_w = 0.3 cp

Grid: $N_x = N_y$ $\Delta x = \Delta y$ N_z = 15 Δz = 15 ft

ϕ=0.2 k_x = 1000 md k_z = 100 md

$S_{wc} = S_{orw}$ = 0.2 k_{rwro} = 1.0 k_{rocw} = 1.0

$k_{rw} = k_{rwro}[(S_w - S_{wc})/(1.0 - S_w - S_{orw}^2)]$, k_{rw}<1.0

$k_{ro} = k_{rocw}[(1.0 - S_w - S_{orw})/(1.0 - S_{wc} - S_{orw}^2)]$

Water injection well at $x = y = 0$ is completed in all 15 layers and injects 2,550.7 STB/D water

Production well at $x = y$ = 660 ft is completed in top 5 layers (upper $^1/_3$ of pay) and flows on pressure constraint against a bottomhole wellbore pressure of 4,000 psia

Discussion of runs for this problem refers to various grids by their $N_x \times N_y$ dimensions because N_z = 15 in all cases

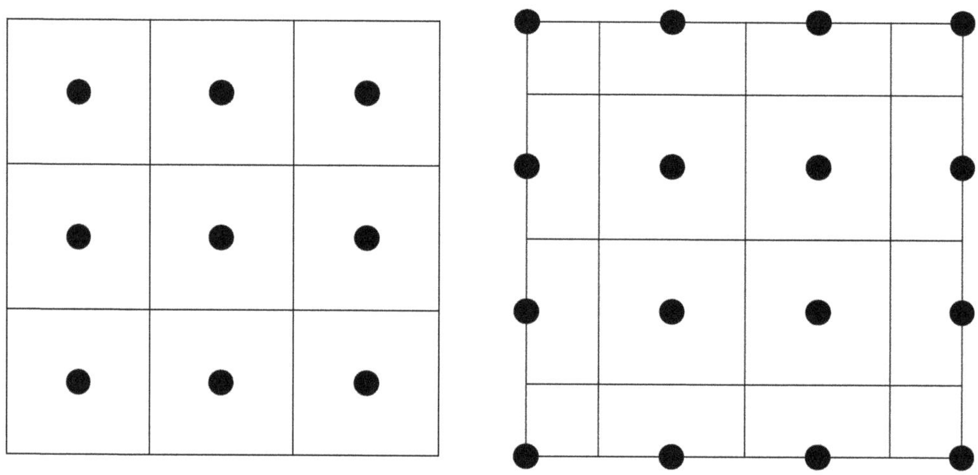

(a) (b)

Fig. 17.7—Block-centered 3 × 3 and 4 × 4 Cartesian grids used in Example 17.1.

fects are not important here. When they are, a single one-layer single-phase run can be made to determine the index correction factor for grid (a) wells located in the corners of their gridblocks.

Fig. 17.10 shows a small effect of grid refinement on Example 17.1 results for grids from 20 × 20 to 3 × 3. The results indicate little need to enhance near-well definition by unstructured grids or by grid refinement (global or local) for grids finer than 3 × 3 for this problem.

Example 17.2. Flexible non-Cartesian grids are shown to significantly reduce the required number of gridblocks.[39] An example provided[40] was water/oil coning in a horizontal well in a 600 × 300 × 230 m homogeneous reservoir. Results were: (a) a 25,823-block 31 × 17

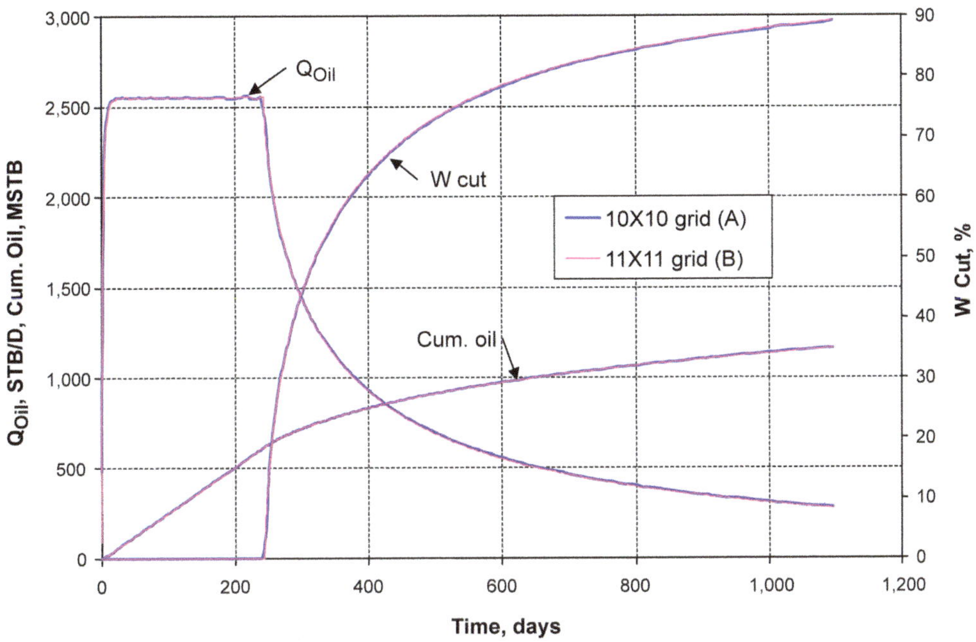

Fig. 17.8—Effect of (a) 10 × 10 vs. (b) 11 × 11 grid on Example 17.1 results.

× 49 Cartesian grid was required to obtain a converged solution, and (b) a 3D 2,066-block CVFE unstructured grid gave correct results. **Table 17.3** gives data for Example 17.2, a similar problem. **Fig 17.11** compares Example 17.2 results for 60 × 31 × 48 and 10 × 7 × 9 Cartesian grids. The 630-block coarse Cartesian grid results here agree as well with the Cartesian 60 × 31 × 48 fine-grid results as the reported 2,066-block CVFE results agree with the 31 × 17 × 49 Cartesian fine-grid results.

Non-Cartesian grids are argued to define irregular reservoir boundaries more efficiently than Cartesian grids. This is not necessarily true. For over 30 years, many models have used active-block coding. While the Cartesian grid extends past boundaries to numerous inactive blocks, those inactive blocks are dropped by the model and require no computer storage or CPU. These numerous inactive blocks pose a problem only for models, if any, that do not use active-block coding.

17.2 Linear Solver—John Wallis and J.W. Watts

The linear equation solver is an important component in a reservoir simulator. It is used in the Newton step to solve the discretized nonlinear partial differential equations. These equations describe mass balances on the individual components treated in the model. For nonisothermal problems, an energy balance is added to the system. The matrix problem involves solving $Ax = b$, where A is typically a large sparse matrix, b is the right-side vector, and x is the vector of unknowns. In the IMPES formulation, there is a single unknown per cell pressure. In the fully implicit formulation, there is a fixed number n of unknowns per cell where $n \geq 2$. In the adaptive implicit formulation, there is a variable number of unknowns per cell. In most formulations, pressure is an unknown for each cell. The matrix A typically has associated well constraint equations and well variables and may be partitioned in block 2 × 2 form as

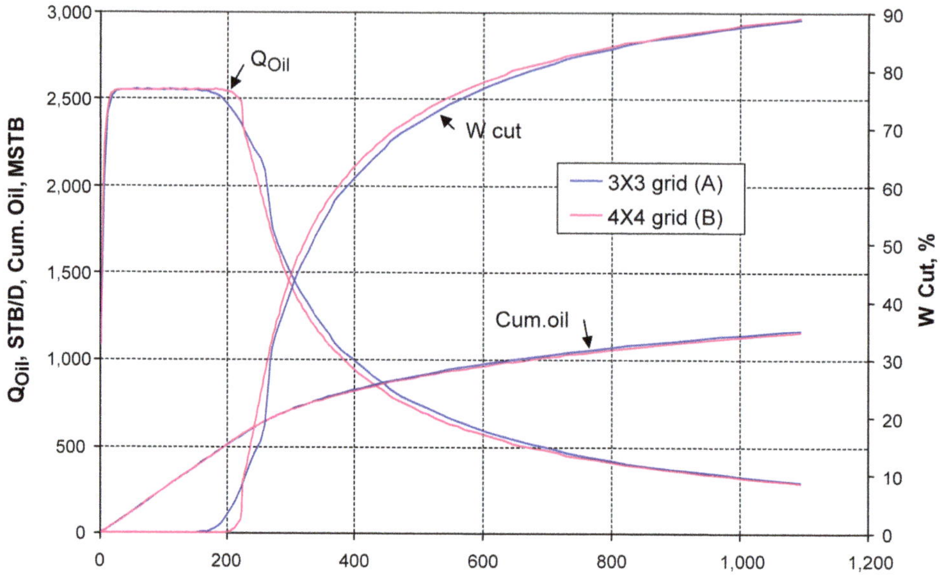

Fig. 17.9—Effect of (a) 3 × 3 vs. (b) 4 × 4 grid on Example 17.1 results.

$$\begin{bmatrix} A_{ww} & A_{wr} \\ A_{rw} & A_{rr} \end{bmatrix} \begin{bmatrix} x_w \\ x_R \end{bmatrix} = \begin{bmatrix} b_w \\ b_R \end{bmatrix}, \quad (17.9)$$

where x_w is the well variable-solution vector and x_R is the reservoir variable-solution vector. The matrix A_{ww} is often diagonal. In this case, the well variables may be directly eliminated, and the iterative solution is on the implicitly defined matrix system

$$(A_{RR} - A_{Rw}A_{ww}^{-1}A_{wR})x_R = \tilde{b}_R, \text{ where } \tilde{b}_R = b_R - A_{Rw}A_{ww}^{-1}b_w \,. \quad (17.10)$$

The well variables are then obtained by back substitution as

$$x_w = A_{ww}^{-1}(b_w - A_{wR}x_R) \,. \quad (17.11)$$

If A is large, solution of the matrix equations is impractical using direct methods such as Gaussian elimination because of computer storage or CPU time requirements. Iterative solution based on projection onto Krylov subspaces is typically used. These Krylov subspaces are spaces spanned by vectors of the form $p(A)v$, where p is a polynomial. Basically, these techniques approximate $A^{-1}b$ by $p(A)b$. The commonly used methods for constructing $p(A)b$ are Orthomin[36] and GMRES.[41] Both methods minimize the residual norm over all vectors in $span\{b, Ab, A^2b, \ldots, A^{m-1}b\}$ at iteration m. They should yield identical results. From a practical standpoint, it does not matter which is used.

A technique known as preconditioning can improve both the efficiency (speed in a typical problem) and robustness (ability to solve a wide range of problems at least reasonably well) of Orthomin or GMRES. Preconditioning involves transforming the original matrix system into one with the same solution that is easier to solve. As a rule, the robustness of the iterative

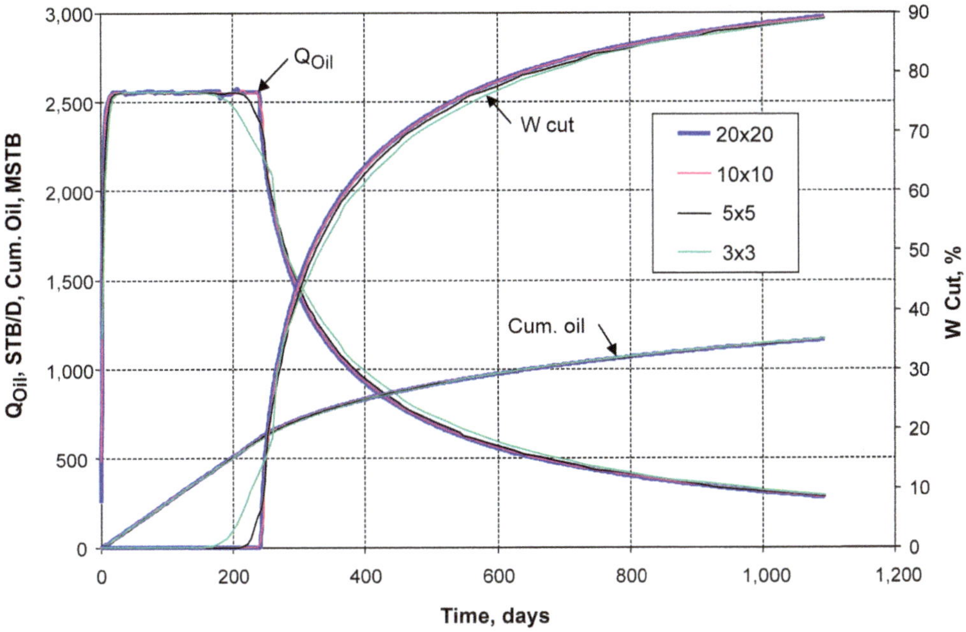

Fig. 17.10—Effects of Cartesian grid coarsening, Example 17.1 results.

TABLE 17.3—EXAMPLE 2 DATA
Water/oil coning problem with a 300-m *x*-direction horizontal well
Reservoir dimensions: 600×300×230 m
Pay zone thickness: 35 m
Aquifer thickness: 195 m
Producer: x = 150 m to 450 m, y = 150 m, z = 10 m, rate = 1,315 STB/D liquid
Water injector: x = 300 m, y = 150 m, z = 140 m, rate = 1,315 STB/D water
$k_x = k_y = 360$ md $k_z = 60$ md $\phi = 0.21$
Incompressible water and oil, $B_o = B_w = 1.0$ RB/STB, $R_s = 0$
$\rho_o = 53.04$ lbm/ft^3 $\mu_o = 1.2$ cp
$\rho_w = 64.27$ lbm/ft^3 $\mu_w = 0.52$ cp
$k_{rwro} = 0.22$ $k_{rocw} = 1.0$ $S_{wc} = 0.296$ $S_{orw} = 0.31$
$k_{rw} = k_{rwro} S_{wn}^2$ (k_{rw}<1) $k_{ro} = k_{rocw} S_{on}^2$
$P_{cwo} = 2.4*[(1.0-S_w)/(1.0-S_{wc}^7)]$ psi
Water/oil contact at z = 35 m (P_{cwo} = 0)

scheme is far more dependent on the preconditioning than on the specific Krylov subspace accelerator used. The preconditioner M is a matrix that approximates A, and has the property that the linear systems of the form $Mx = b$ are easily and inexpensively solved. For most linear solvers the following preconditioned system is solved:

$$AM^{-1}y = b, \quad \text{where } x = M^{-1}y \,.$$

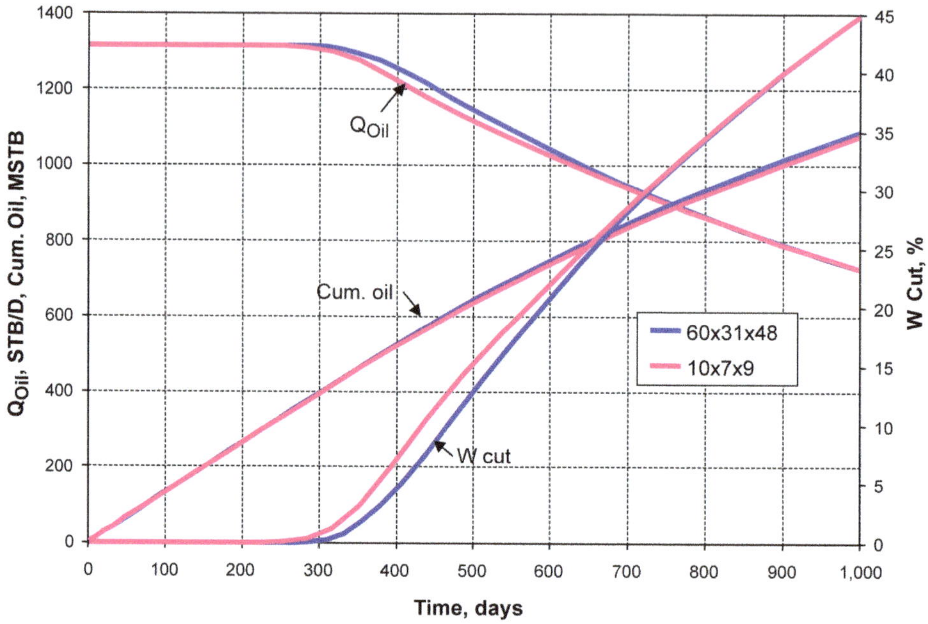

Fig. 17.11—Fine vs. coarse Cartesian grids, Example 17.2.

The preconditioned ORTHOMIN(k) algorithm, which retains the last k A-orthogonal direction vectors, is given by

1. Compute $r_0 = b - Ax_0$. Set $p_0 = M^{-1}r_0$.
2. For $j = 0,1,2, \ldots,$ until convergence Do :
3. $\alpha_j = \dfrac{(r_j, Ap_j)}{(Ap_j, Ap_j)}$
4. $x_{j+1} = x_j + \alpha_j p_j$
5. $r_{j+1} = r_j - \alpha_j Ap_j$
6. $\beta_{ij} = -\dfrac{(AM^{-1}r_{j+1}, Ap_i)}{(Ap_i, Ap_i)}$, for $i = j - k + 1, \ldots, j$
7. $p_{j+1} = M^{-1}r_{j+1} + \sum_{i=j-k+1}^{j} \beta_{ij} p_i$
8. End Do

Because of the nature of the reservoir simulation equations, only certain preconditioners are effective in solving them. Reservoirs are typically shaped like pancakes, being much broader than they are thick. This geometry leads to strong vertical connectivity. Some preconditioners exploit this property. The most commonly used such preconditioner is NF.[37] The convergence of NF is sensitive to the cell ordering. Best results are usually obtained by ordering the cells first along the direction of highest transmissibility and then successively along directions of decreasing transmissibility. This nearly always means that NF should be ordered first in the vertical direction.

The other commonly used preconditioners are incomplete lower triangular/upper triangular factorizations of the matrix, or ILU(n), where n is the level of infill that is retained during the elimination process. Performance of these can be improved by using a red-black checkerboard ordering (also called D4 ordering) of the nodes.[42] Red-black ordering on a five-point (in 2D) or seven-point (in 3D) grid leads to direct elimination of the unknowns at the red cells, leaving a system containing only the unknowns at the black cells. The result is a halving of the number of unknowns. An ILU preconditioner using red-black ordering with zero infill on the reduced system is referred to as RBILU(0) and is the most frequently-used form of ILU.

In IMPES models with either no faults or vertical faults only, RBILU(0) or ILU(1) combined with *z*-line additive corrections,[43,44] typically converges very rapidly. NF can also be used effectively in cases involving vertical faults and pinchouts because the matrix retains the structure required by NF. Nonvertical faults interrupt the matrix structure that makes red-black orderings attractive. In models containing them, ILU is the method of choice. Nine-point discretizations also cause problems for red-black orderings, but cause no difficulty for NF.

NF and RBILU(0) are commonly used in implicit models. Another very effective approach exploits the fact that pressure is the "stiff" variable. The Combinative[45] or CPR[44] method is a two-step preconditioning that extracts a pressure equation from the implicit matrix. It iteratively solves for a pressure correction at each iteration, uses the pressure correction to form a new residual, applies an inexpensive implicit preconditioning such as diagonal scaling or line Gauss-Seidel to the new residual, and then uses the sum of the two steps as the approximate solution. In compositional models, this two-step method can be much faster than one-step methods.

Many models include well-constraint equations that add well pressures to the set of unknowns. A simple but effective way of dealing with these equations is to order them first in the global matrix and then eliminate the well pressures from the set of unknowns.[46] In this approximate elimination, any infill terms are either column-summed or row-summed into the main diagonal of the reservoir matrix, which is then factored using NF. The reservoir matrix problem is then solved iteratively and the well variables are obtained by back substitution.

Parallel iterative solution typically uses a domain decomposition approach in which the grid is partitioned into domains that contain approximately the same number of cells. The partitioning should be done such that coupling is strongest between cells within a domain. As a result, domains normally are groups of columns of cells. One way to accelerate the iteration is to color the domains in red-black fashion and apply an NF-type procedure in which the outer level of nesting is the coupling between domains and either NF or an ILU variation is used to factor the individual domains.[47,48]

The solution of matrices arising from unstructured grids typically involves some variant of ILU with cell orderings such as Reverse Cuthill-McKee (RCMK)[49] or Minimum Degree Fill (MDF).[50,51]

17.3 Gridding in Reservoir Simulation—Dave Ponting

17.3.1 Introduction. The aim of gridding in reservoir simulation is to turn the geological model of the field into a discrete system on which the fluid-flow equations can be solved.

The basic structure of an oil reservoir is a set of geological horizons representing bedding planes. The reservoir may contain faults, at which the strata are displaced. It is usually possible to identify many more layers in the geological model than it is practical to include in reservoir flow simulation, so some upscaling of rock properties will normally be carried out. Even after this process, the geology to be represented is rarely homogeneous at the scale of the simulation grid.

Two related issues are involved in choosing a grid for reservoir simulation: the accuracy with which the geological description of the reservoir is matched, and the discretization of the flow equations. In a classical finite-difference scheme, the point values of pressures and saturations are used as solution variables, and the differential operators that appear in the fluid-flow equations may be expanded as difference expressions of these point values to some order. An alternative approach is to use an *integral finite-difference*[52] or *finite-volume*[53] method in which the fluid-flow equations are integrated over a set of cell volumes. This yields a set of equations in which the mass conservation conditions for the fluid in the simulation cell volumes are related to the flows through the interfaces between those cell volumes. Rock properties such as porosity are assumed constant over the cell or controlled volume. This yields a discretization scheme which is conservative (each outflow from one cell is an inflow to another) and for which the fluid in place may be obtained straightforwardly. The mass conservation equations for a timestep from T to $T + \Delta T$ then become:

$$V_{pa}^{T+\Delta T} \cdot m_{ca}^{T+\Delta T} - V_{pa}^{T} \cdot m_{ca}^{T} = \Delta T \cdot (Q_{ca} + \textstyle\sum_b \sum_p F_{cpab}), \quad (17.12)$$

where V_{pa} is the pore volume of cell a, m_{ca} is the density of conserved component c in cell a, Q_{ca} is the injection or production rate of component c because of wells, and F_{cpab} is the flow rate of component c in phase p from cell a to its neighbor b. In general, the flows F_{cpab} may involve the solution values of a number of cells, the number of cells involved defining the *stencil* of the numerical scheme. The linear pressure dependence of flows given by Darcy's law leads to an expression of the type:

$$F_{cpab} = \textstyle\sum_x T_{ax} M_{cpax} \Delta\Phi_{pax} . \quad (17.13)$$

M_{cpax} is the mobility of component c in phase p for the contribution to the flow between a and x, given by $x_{cp}.K_{rp}/\mu_p$, where x_{cp} is the concentration of component c in phase p, K_{rp} is the relative permeability of phase p, and μ_p is the viscosity of phase p. This is often set to an *upstream* value of the mobility, depending upon the sign of the potential difference.

$\Delta\Phi_{pax}$ is the potential difference of phase p between cell a and cell x, which includes pressure, gravity and capillary pressure contributions:

$$\Delta\Phi_{pax} = P_a - P_x - g\rho_p \cdot (d_a - d_x) + P_{cpa} - P_{cpx} . \quad (17.14)$$

The constant coefficients of mobility and potential difference products, T_{ax}, are commonly termed the *transmissibilities.*

When the flows between two cells a and b can be expressed as a function of the solution values in just those two cells, so that the summation over cells includes just $x = b$, the flow expression takes a *two-point* form. The flow expression then takes a simple form:

$$F_{cpab} = T_{ap} M_{cpab} \Delta\Phi_{pab} . \quad (17.15)$$

When solution values from other cells are required, the flow takes a *multipoint* form.[54]

Other options for discretization are available, such as Galerkin finite elements[55–57] and mixed finite-element.[58] It is sometimes possible to cast a finite-element Galerkin discretization into the upstreamed transmissibility-based form.[57]

17.3.2 Regular Cartesian Grids. A simple 3D grid is the regular *Cartesian* grid (**Fig. 17.12**). Cells in such a grid may be simply identified using their (i,j,k) index values.

Each of the grid elements will be assigned a single permeability or porosity value. In this case, it is possible to obtain the transmissibility value as a harmonic average:

$$T_{ap} = 1/(1/T_a + 1/T_b),\ T_a = K_a \cdot A_a/(d_a/2),\ T_b = K_b \cdot A_b/(d_b/2), \quad (17.16)$$

where cell b is the neighbor to cell a in some direction and K is the cell permeability in that direction. A is the area of the cell orthogonal to the direction of flow, and d the dimension of

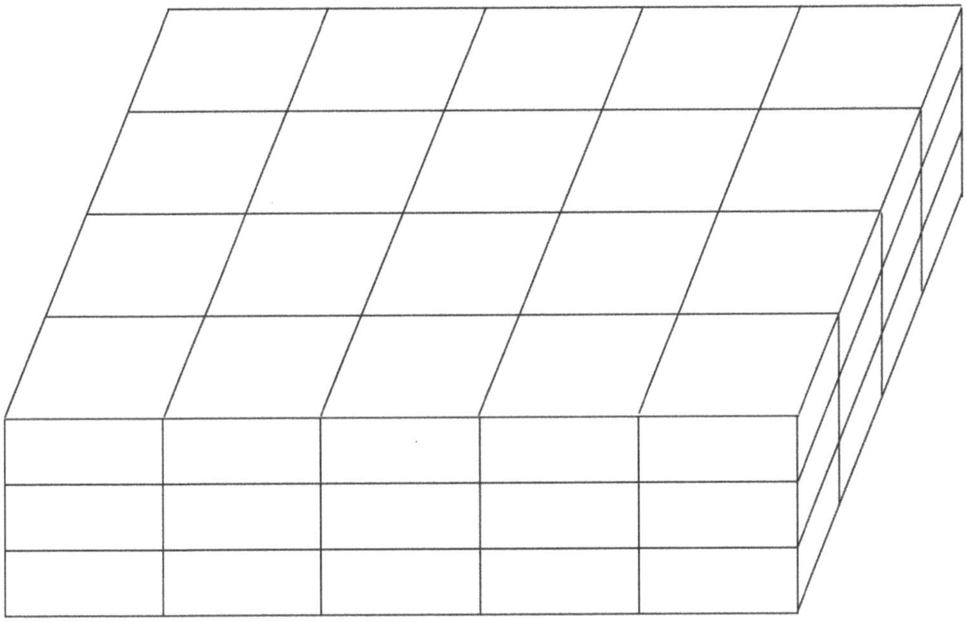

Fig. 17.12—Regular Cartesian grids.

the cell in that direction. Such a two-point transmissibility assumes a permeability tensor with primary axes aligned along the grid axes.

Although regular grids are normally defined in normal Cartesian coordinates, it is also possible to use an (r, ϕ, z) radial system.[52] The resulting grid is *cylindrical* and is important for the special case of near-well studies dominated by radial inflow. For a 3D system, regular grids yield seven-point schemes, in which the flow equations for a cell involve solution values for just the cell and its six neighbors. Not all the elements in the grid need represent active solution variables in the simulation. Some cells may be inactive, representing volumes of the reservoir with zero porosity. Such inactive cells are usually compressed out of reservoir simulation solution arrays prior to the memory and time-intensive flow solution stage, and enable reservoirs with irregular boundaries to be represented within extended simulation grids.

The horizons that delimit rock strata are generally not horizontal, but are dipped, curved, or faulted. Unless extremely fine, a true regular grid that is orthogonal in all three axes will be unable to assign rock properties accurately to cell volumes. Such a layer-cake structure can be used, but will generally misalign property values (**Fig. 17.13**) in which the orthogonal grid provides a rather poor match to the dipping strata represented by the shaded layers. However, it is possible that improving computer power will bring such *rasterized* grids to a level of refinement at which a sufficiently good representation may be obtained.

Dip-Normal Geometry. A simple variation of a regular grid, in which the regular grid is rotated to bring the layers of cells into alignment with the bedding planes. Such a description would only suit a reservoir with a single, constant angle of dip. As geological descriptions have improved, fewer and fewer model reservoirs are found to fit this simple pattern, and something more flexible is required.

Block-Center Geometry. A simple model in which transmissibility between blocks is calculated on the basis of linear interpolation between the center values of the cells. This is a simple way of representing variable dip, but is difficult to represent graphically in a consistent way. Pore volumes are calculated on the basis of a series of flat regular cells with variable depths

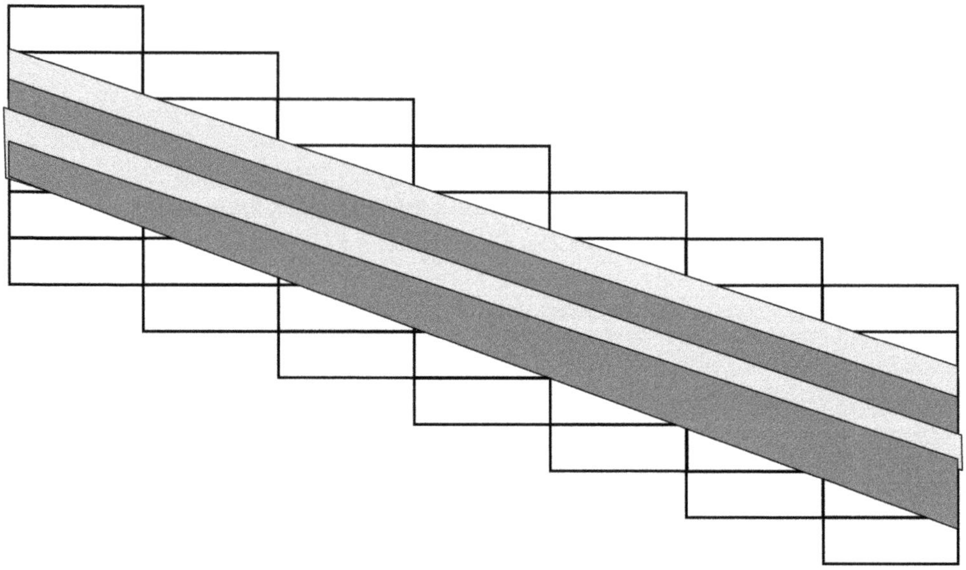

Fig. 17.13—Orthogonal grid used to represent dip.

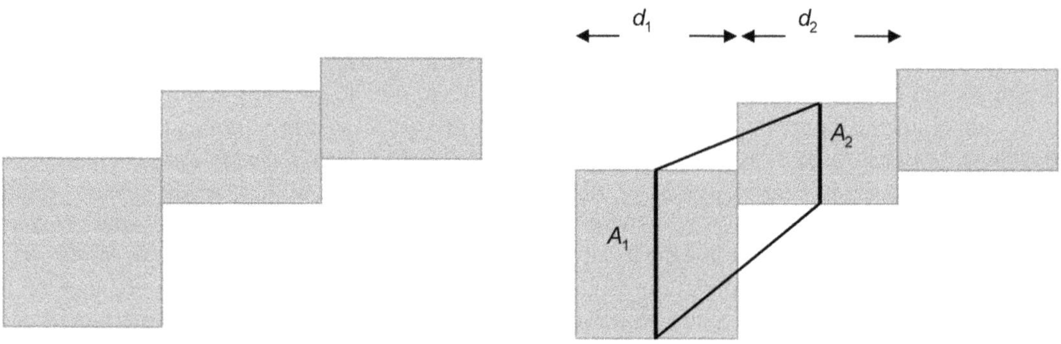

Fig. 17.14—(a) Obtaining pore volumes; (b) obtaining transmissibilities.

(**Fig. 17.14a**), but transmissibilities are calculated on the basis of interpolated values (Fig. 17.14b). The areal grid is rectangular.

Thus, for the pair of cells illustrated,

$$T = K \cdot A / d, \text{ where } K / d = 2 / (d_1 / K_1 + d_2 / K_2), \quad (17.17)$$

where A is the average area over which flow occurs and c is a dip correction given by $\cos^2\theta$, where θ is the angle of dip of a line joining the cell centers to the horizontal. Such a block-center option is suitable for unfaulted reservoirs and is commonly supplied as a simulator option.

17.3.3 Hexahedral Grids. Further improvements in geological modeling threw an emphasis on describing faults, and made it important to distinguish depth displacements due to dip and faulting. This is difficult in block centre geometry in which the cell is positioned by its centre depth and Δx, Δy, Δz dimensions. To define faulting more precisely it is useful to define the position of grid cell by its corner point locations. A hexahedral shape with eight corners and

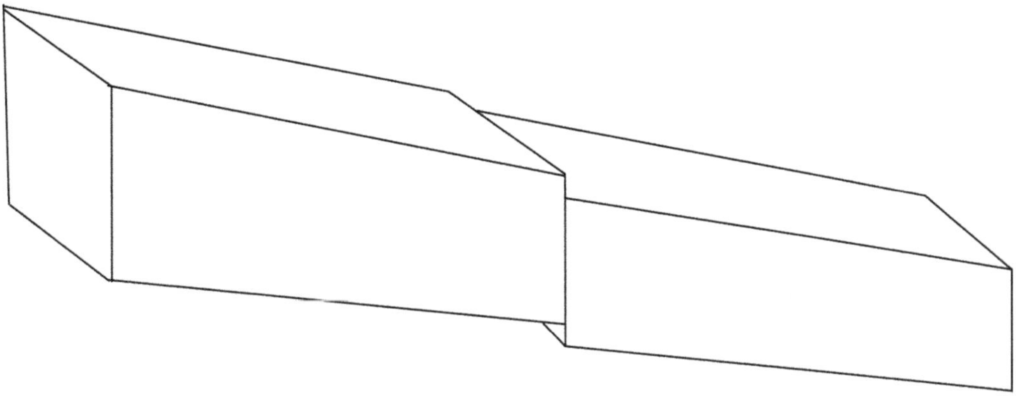

Fig. 17.15—Hexahedral grid system.

bilinear planes as surfaces then describes the cell geometry. Faults, both vertical and inclined, may be described precisely (**Fig. 17.15**). Such grids are often called *corner-point grids.*

In both the dipped and general hexahedral grids, the orthogonality of a completely rectangular grid no longer exists, and the result is that the two-point property of the flows between the cells is lost—the flow between cell a and cell b is not just a function of the solution values of cells and *a* and *b*.[53,59–62] Typically, the result is a 27-point scheme in three dimensions. However, if the grid distortion is mild, it may be possible to ignore some additional couplings and use a low-order transmissibility scheme. This is normally done for extra couplings introduced by dip angles, which are often small.

Although this corner-point description handles the fault issue, the basic coordinate system remains a regular grid (i.e., the grid is structured). Fitting such a basically regular system to the irregular shapes of a reservoir remains a difficulty that may be solved in two basic ways—either by distorting the grid and fitting the cells into the geometry, or by truncating the grid to the reservoir position.

17.3.4 Multiple-Domain Hexahedral Grids. In some cases, a single structured grid system cannot match the overall structure of a reservoir, so a block grid or domain-based grid is used.[63] This consists of a number of subgrids, each with a local regular (i,j,k) structure, but linked together to model the entire reservoir. The block hexahedral system gives rise to multiple (i,j,k) indexing systems—(i,j,k,l), where the l index specifies the grid system. These comprise a series of regular grids. Such regular gridding systems have advantages for upscaling and downscaling—for example, a natural coarsening of a regular grid may be simply defined by grouping sets of coordinates in each direction.

17.3.5 Grid Refinement. A common requirement in reservoir simulation is an increased level of detail around an item of interest such as a well. This is frequently obtained in structured grids by *local grid refinement*, replacing a set of cells in the original grid by a finer grid (**Fig. 17.16**). The inserted grid may be Cartesian (center) or radial (upper left). Local refinement may be regarded as a form of multiple domain structured grid, in that it consists of a number of linked structured grids. Flows at the edges of local refinements generally take a multipoint form.[54]

17.3.6 Unstructured Grids. The problems involved in using a regular structured grid to represent reservoir geometry can be avoided by using an *unstructured* grid. This is constructed around a set of solution points that need have no particular indexing scheme. These points may

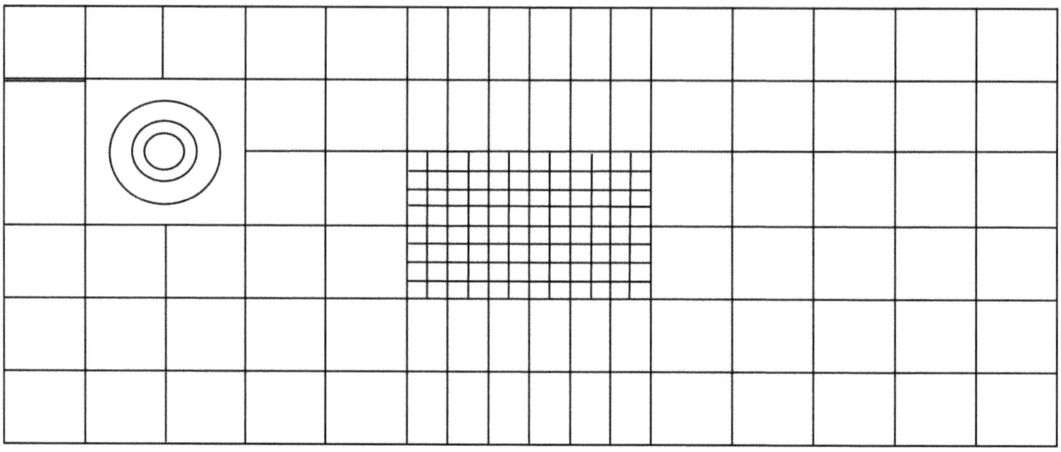

Fig. 17.16—Local grid refinement.

be triangulated into a mesh of triangles or tetrahedrons. A control volume is constructed around the nodes of the resulting mesh to define the simulator cell volumes. The perpendicular bisector (PEBI) method introduced into reservoir simulation by Heinemann[64,65] used a technique also known as a *Voronoi* grid.[58,66] Starting from any set of solution points, the PEBI cell volumes are defined by the perpendicular bisection planes between these points. The resulting control volume is defined by the perpendicular planes—it is the set of points closer to the node than any other. This is shown in **Fig 17.17**, in which the bisectors to the heavy lines joining the solution points enclose the control volume, represented by the shaded area. The grid is locally orthogonal, and the desirable property of two point flows is obtained. The actual cell volumes may have a variety of shapes, depending on the exact placement of the solution points, but are typically hexagonal in two dimensions. Grid refinement occurs naturally in areas where solution points are closely spaced.

The two-point property is not naturally preserved in anisotropic reservoirs, although it can be regained by transforming to a K-orthogonal grid in which the geometry is transformed so that $\mathbf{K}_n$ is parallel to the vector joining the solution nodes, where **K** is the permeability tensor and **n** is the normal to the cell volume surface.[67] For nonisotropic cases in which the grid is not K-orthogonal, the flows will be functions of the solution values in more than two cells, as in the general hexahedral case.

An unstructured grid may be defined in two dimensions, and then applied to each layer of a reservoir model, so that a typical cell is a hexagonal prism. This is sometimes termed a 2½D unstructured grid.[68] Alternatively, a full 3D unstructured grid may be defined. The 3D approach is most effective when applied to model a local structure such as a branched well.

Unstructured grids yield an elegant and flexible grid description. However, the ability to identify cells by a simple set of indices is lost, and items such as wells need to be positioned in true space terms. The systems of linear equations generated by unstructured grids are also commonly regarded as more difficult to solve than those produced by structured grids. However, it may be more true to say that optimal solution schemes are simpler to find for structured grids, where the row and plane order provides a natural *ab initio* solution variable-ordering scheme.

17.3.7 Truncated Regular Grids.[69] The truncated approach fits in well with the rectangular grids used in geological modeling. A simple rectangular grid is always used in the areal direction, but faults may subdivide the rock volume in a given column. The areal grid is not

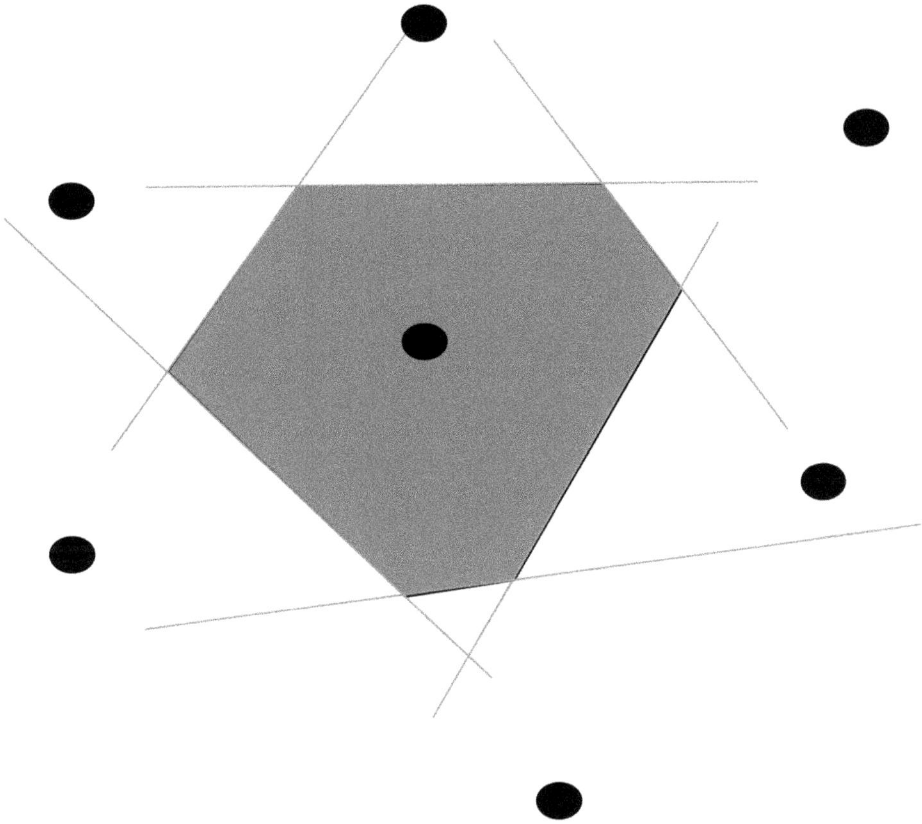

Fig. 17.17—Unstructured grids may result in hexagonal cells.

modified to match the faults. Thus the two marked volumes in **Fig. 17.18** represent different cells, but may have the same *i, j* indices, so this creates a multiple-domain grid. A disadvantage is the more complex shape of cells at the edge of the grid. Transmissibilities for such cells may need to be calculated numerically. Apart from the truncated cells, all the grid cells are hexahedra that are rectangular in plan.

17.3.8 Other Gridding Systems. ***Triangular or Tetrahedral Grids.*** The underlying solution points of a PEBI mesh can be linked together into a Delaunay triangulation. In 2D, this creates triangles, and in 3D it creates tetrahedra. One option would clearly be to use triangular or tetrahedral cells directly and associate cell volumes with these. This technique is rather rarely used in reservoir simulation. Partly this may be historical, but the Delaunay triangulation is rather less stable under grid changes than a Voronoi grid, and triangulation can more often lead to "sliver" cells with a high surface area but a small volume.

Curvilinear Grid Systems. In some special cases, a transformed coordinate system may be used, based around an expected flow pattern. Such grids are not well adapted to represent geological data, and have been used less frequently as more detailed reservoir descriptions have become available.

17.3.9 Future Directions. Two themes emerge from current trends in reservoir simulation gridding. The increasing sophistication of data preparation and solver technology indicates a move towards unstructured grids as a general method of solving the flow equations for a given reser-

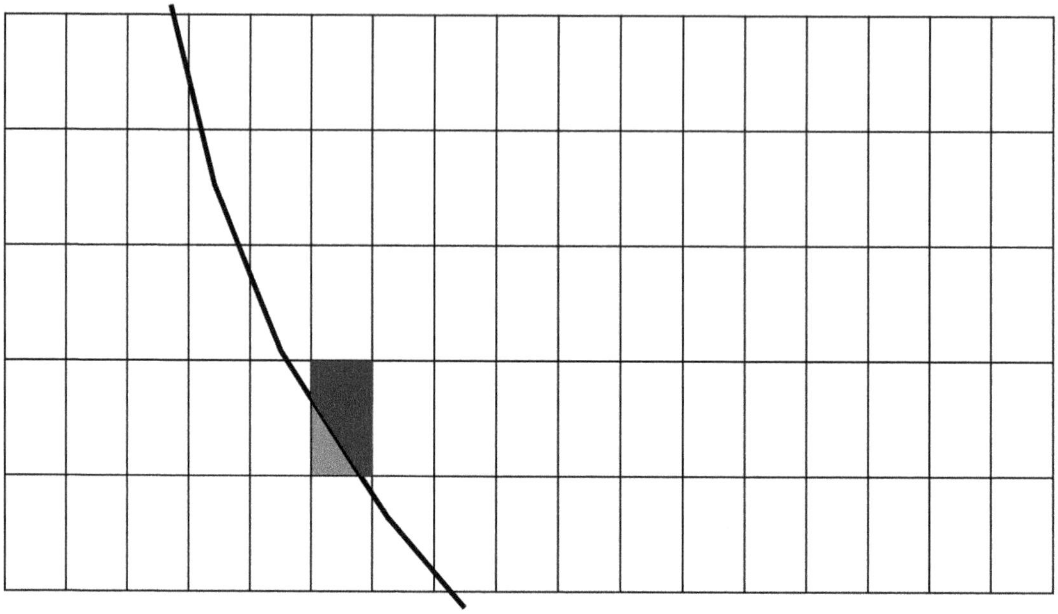

Fig. 17.18—A fault creating two cell volumes in a truncated grid.

voir simulation problem. On the other hand, reservoir simulation is increasingly seen as part of a decision-making process rather than an isolated activity, so the ability to map easily onto the generally regular data structures used in seismic and geological modeling becomes an important issue. In this role, structured grids may have advantages of simplicity and scalability.

An ideal is to separate the construction of the flow-simulation grid from the description of the reservoir geometry. This ties in with a further ideal, inherent in many discretization schemes, that the scale of the simulation grid should be below the scale of the problem structure.

For more complex shape-dominated problems, the unstructured approach looks general and flexible, providing that the data-handling and cell-identification methods can be moved to true *x,y,z* space preprocessing software.

17.4 Upscaling of Grid Properties—Alan Grindheim

17.4.1 Definition. Upscaling, or homogenization, is substituting a heterogeneous property region consisting of fine grid cells with an equivalent* homogeneous region made up of a single coarse-grid cell with an effective property value. It is performed for each of the cells in the coarse grid and for each of the grid properties needed in the flow-simulation model. Therefore, the upscaling process is essentially an averaging procedure in which the static and dynamic characteristics of a fine-scale model are to be approximated by that of a coarse-scale model. A conceptual illustration of the upscaling process is shown in **Fig. 17.19**.

17.4.2 Can Upscaling Be Avoided? Typically, 3D geological models contain detailed descriptions of the reservoir that can be hard to capture properly with a significantly coarser model. Therefore, it would be preferable if upscaling could be avoided. Currently, an average-sized flow simulation model consists of approximately 100,000 active grid cells. This is to ensure that the CPU consumption of a simulation run will be reasonable (i.e., within practical limits).

* Either volume or flux vice, depending on the type of property that is to be upscaled.

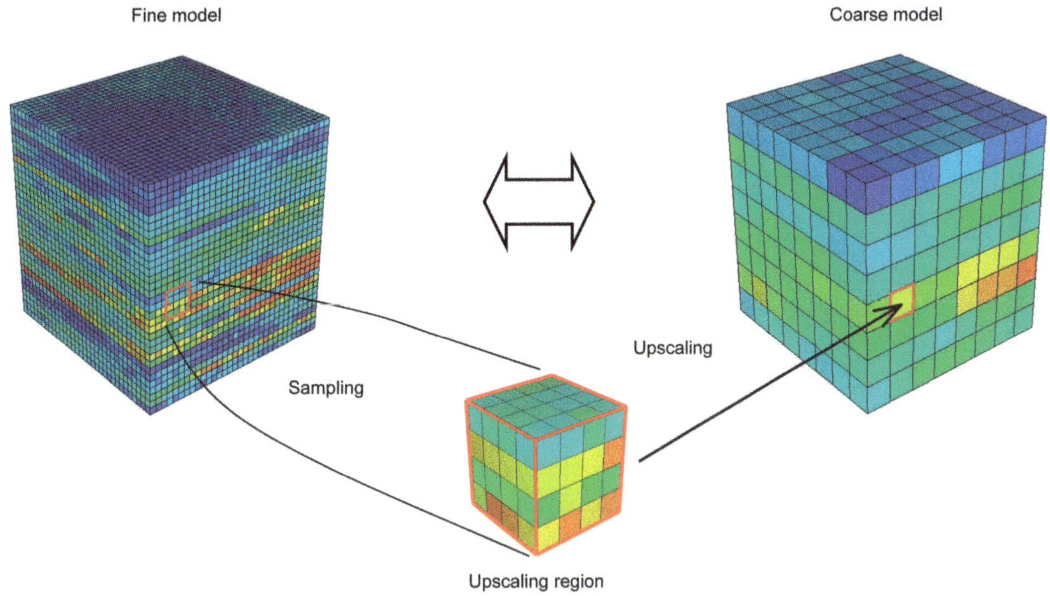

Fig. 17.19—Concept of upscaling.

Because a typical 3D geological model may consist of approximately 10 million active grid cells, it is obviously infeasible to run fluid-flow simulations directly on the geological model. Hence, upscaling is a required part of current reservoir modeling workflows.

Seen through the eyes of the geologist, the upscaling task may be a painful experience because all the geological details that were put into the model seem to be lost in the process. For a reservoir engineer, on the other hand, effective properties might be all that matter.

For volumetric (additive) properties such as porosity and saturation, the effective flow-cell value is simply given by the bulk and pore volume weighted arithmetic average, respectively, of the geo cells inside it. For the permeability, which is intrinsic (nonadditive) by nature, no such simple averaging method exists. The complexity one needs to take into account when upscaling permeability is considerable; therefore, all current techniques provide only an approximation of the true effective cell permeability. This approximation may range from very good to very poor, depending on the complexity of the fine-scale permeability distribution as well as the upscaling method used.

17.4.3 Upscaling Techniques for Absolute Permeability. Homogenization of absolute permeability does not have an exact analytical solution, except for in a few idealized cases. The challenge of computing an accurate effective permeability has resulted in a large number of upscaling techniques. These techniques range from simple statistical averages to advanced numerical methods.

Tensor methods are the most accurate techniques available for computing the effective cell permeability. These are based on solving a second-order elliptic differential equation describing single-phase, incompressible, steady-state fluid flow in a region without sources and sinks (i.e., wells). Some flow-based methods may provide a *full* permeability tensor. However, because most multiphase flow simulators can only handle a diagonal permeability tensor because of the use of a seven-point stencil in 3D, diagonal tensor methods are most frequently used whether directly or indirectly (through a diagonalization of a full tensor). For a diagonal tensor, only the effective permeability in the principal directions of flow (x, y, and z) will be nonzero.

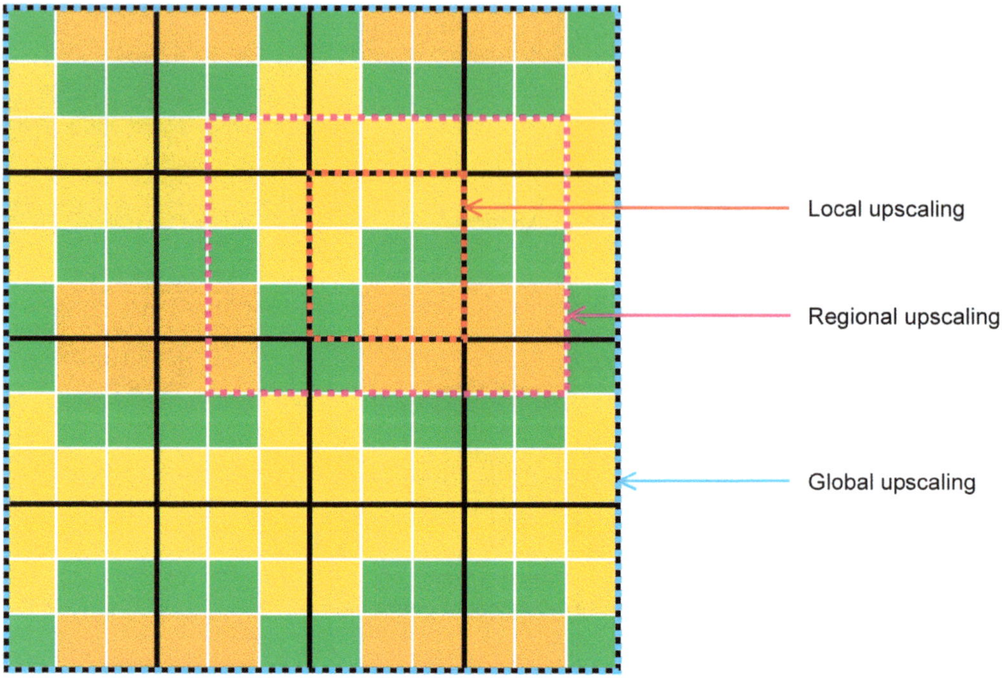

Fig. 17.20—Upscaling schemes and the size of the computational region (geo grid in white, flow grid in black).

The flow equation is usually discretized with a finite-difference scheme, although finite-element methods are also applied occasionally. To compute all the directional components of the permeability tensor, the discretization and solution of the flow equation must be performed for each of the principal flow directions (i.e., three separate single-phase simulations need to be performed). Each simulation involves the iterative solution of a linear equation system (typically, the linear solver is a conjugate gradient method, preconditioned by incomplete Cholesky or LU factorization). The unknowns in this equation system are the geo-cell pressures inside the flow cell, whereas known quantities are the geo-cell dimensions and permeabilities, as well as the pressure conditions along the faces of the flow cell. When the numerical solution of the fine-scale pressure distribution has converged, an effective permeability is computed by equating expressions for the flux through the heterogeneous geo cells with the flux through the equivalent homogeneous flow cell using some form of Darcy's law.

The pressure field is usually solved locally—that is, for one flow cell at a time. However, as discussed in the next subsection, the size of the *computational* region may not necessarily be limited to that of the upscaling region (i.e., the flow cell).

17.4.4 Upscaling Schemes for Absolute Permeability. Based on the size of the computational region, the single-phase upscaling process may either be described as local, regional, or global. With local upscaling techniques, the computational region is identical to the upscaling region (i.e., only geo cells inside the flow cell are considered in the upscaling computations). For regional upscaling, the computational region is expanded beyond that of the flow cell to include a buffer region of neighboring geo cells. In the case of global upscaling, the computational region is that of the entire geo model. **Fig. 17.20** provides a schematic drawing of how the computational region varies with the different upscaling schemes. These are further discussed in the subsections that follow.

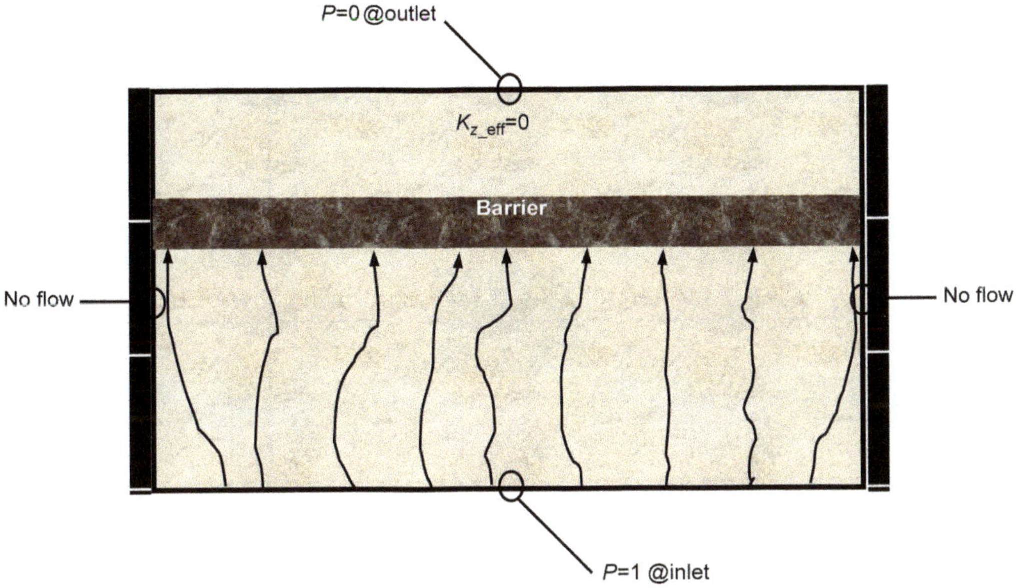

Fig. 17.21—The sealed sides boundary conditions (in *z*-direction).

It should be noted that the different upscaling schemes are only relevant when considering flow-based (tensor) methods. It is also important to realize that even though the computational region may vary according to the scheme used, the upscaling region remains unchanged and is of course defined by the flow cell, as in the case of the simple, analytical upscaling techniques.

17.4.5 Local Upscaling. Because it used to be too time-consuming to compute the fine-scale pressure field for the complete geo grid in a single operation, the flow-based methods have traditionally been restricted to solving the pressure field locally—that is, for a single flow cell at a time. Hence, the effective cell permeability is computed separately and independently of the other flow cells, which may or may not be correct depending on how representative the imposed pressure conditions along the faces of the flow cell are.

Different types of artificial boundary conditions for the flow cell have been suggested over the years, all with the objective of providing as good an approximation of the real boundary conditions as possible. An important design criterion for the artificial boundary conditions is the conservation of flux in and out of the flow cell.

The first type of boundary conditions proposed for the local solution of the pressure equation was published by Warren and Price in 1961.[70] Their approach is to impose a constant pressure gradient in a selected direction of flow by specifying a pressure of 1 on the inflow face and a pressure of 0 on the outflow face. By allowing no flow to pass through the sides of the cell, all fluxes are forced to go in the principal direction of flow. Therefore, this type of boundary conditions is often referred to as the no-flow or sealed-sides boundary conditions. The sealed-sides boundary conditions are graphically illustrated in **Fig. 17.21** for flow in the vertical direction (here in the case of a flow cell containing a barrier).

The choice of boundary conditions emulates the way core permeability is measured in the lab. This is hardly a coincidence. As in the coreflood experiment, the local numerical flow simulation is in effect 1D because the cell faces parallel to the main flow direction are sealed. This implies that the estimated effective permeability will be scalar. Hence, the maximum number of directional permeability components that can be obtained with this type of boundary conditions is three, one for each of the principal directions of flow. In practice, the diagonal

Fig. 17.22—The fine-scale permeabilities are symmetric on the scale of the flow cell (geo grid in white, flow grid in black).

permeability tensor is derived by setting up the boundary conditions for *x, y,* and *z* directions, respectively, in three independent single-phase simulations.

As documented in Refs. 71 and 72, a tensor technique based on the sealed-sides boundary conditions tends to bias the estimated effective permeability toward a low value. The physical implication of this is most clearly seen in the case of a bimodal permeability system of sand and shale. This is because the sealed-sides method consistently underestimates the reservoir flow characteristics by thickening shale barriers and narrowing sand channels. The latter effect also has a tendency of disconnecting stacked sand channels.

Take, for example, the flow illustrated by Fig. 17.21. Because the barrier extends across the entire length of the local upscaling region, the resulting effective permeability (in the *z*-direction) will be zero. For vertical flow, the result, therefore, is a thickening of the shale in the flow model equal to the thickness of the flow cell. Depending on which factors that affect fluid flow in the region of the cell, this may or may not be a representative value for that particular flow cell.

Strictly speaking, the sealed-sides boundary conditions are only valid if no wells are present and the flow cells are symmetric in each direction of the grid as illustrated in 2D by **Fig. 17.22**. Hence, the sealed-sides boundary conditions assume that the flow cell is surrounded by mirror images of itself.

By the end of the 1980s, 3D geological models had started to appear more regularly on the modeling scene. This resulted in a new demand for advanced upscaling. In this renewed effort, two alternative boundary conditions for solving the local pressure solution in a flow-based method were suggested more or less at the same time. One was based on linear boundary conditions, the other on periodic boundary conditions.

The use of linear boundary conditions in flow-based upscaling was suggested by Guerillot *et al.*[73] in 1989 and Samier[74] in 1990 to enable the computation of a full-permeability tensor. Instead of setting the flow through the sides of the cell to zero, the pressure along the sides is

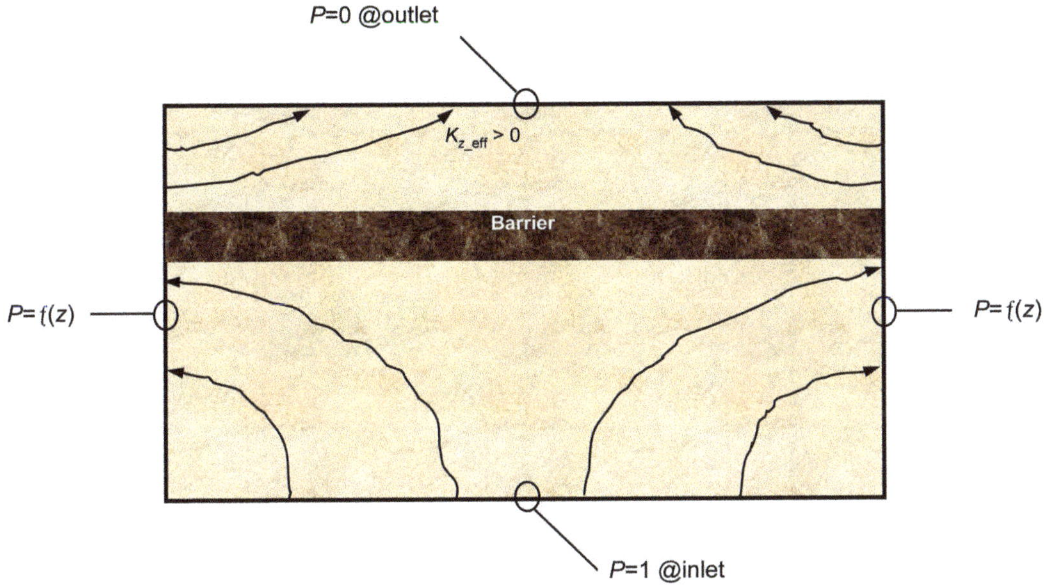

Fig. 17.23—The open sides boundary conditions (in z-direction).

allowed to vary in a linear fashion that matches the constant pressure on the two cell faces perpendicular to the flow. Hence, the imposed pressure gradient is still constant, but the flow is allowed to enter and leave the cell at any point along the sides parallel to the main flow direction. Therefore, this type of boundary conditions is also referred to as the open-sides boundary conditions. The situation is graphically illustrated in **Fig. 17.23** for flow in the vertical direction (here in the case of a flow cell containing a barrier).

As with the sealed-sides boundary conditions, three independent single-phase simulations, with the main flow direction in *x, y,* and *z,* respectively, are needed to yield all of the components of the permeability tensor. With open-sides boundary conditions, however, also the off-diagonal components will generally be nonzero. Hence, unlike the sealed-sides boundary conditions where the effective permeability is limited to that of a diagonal tensor, the open-sides boundary conditions, as previously mentioned, give a full permeability tensor. The resulting full tensor may be either symmetric or nonsymmetric depending on the properties of the method under consideration.

As documented in Refs. 75 and 76, a tensor technique based on the open-sides boundary conditions tends to bias the estimated effective permeability toward a high value. The physical implication of this is most clearly seen in the case of a bimodal permeability system of sand and shale. This is because the open-sides method consistently overestimates the reservoir flow characteristics by narrowing shale barriers and thickening sand channels. The latter effect also has a tendency of connecting isolated sand channels.

Take, for example, the situation illustrated by Fig. 17.23. Even though the barrier extends across the entire length of the local upscaling region, the resulting effective permeability (in the *z*-direction) will be significantly larger than zero. For vertical flow, the result is therefore a narrowing of the shale in the flow model equal to the horizontal dimensions of the flow cell. Depending on which factors affect fluid flow in the region of the cell, this may or may not be a representative value for that particular flow cell.

The use of linear boundary conditions has its origin in the effective medium theory,[77] which states that any region of permeability behaves as if embedded within the average medi-

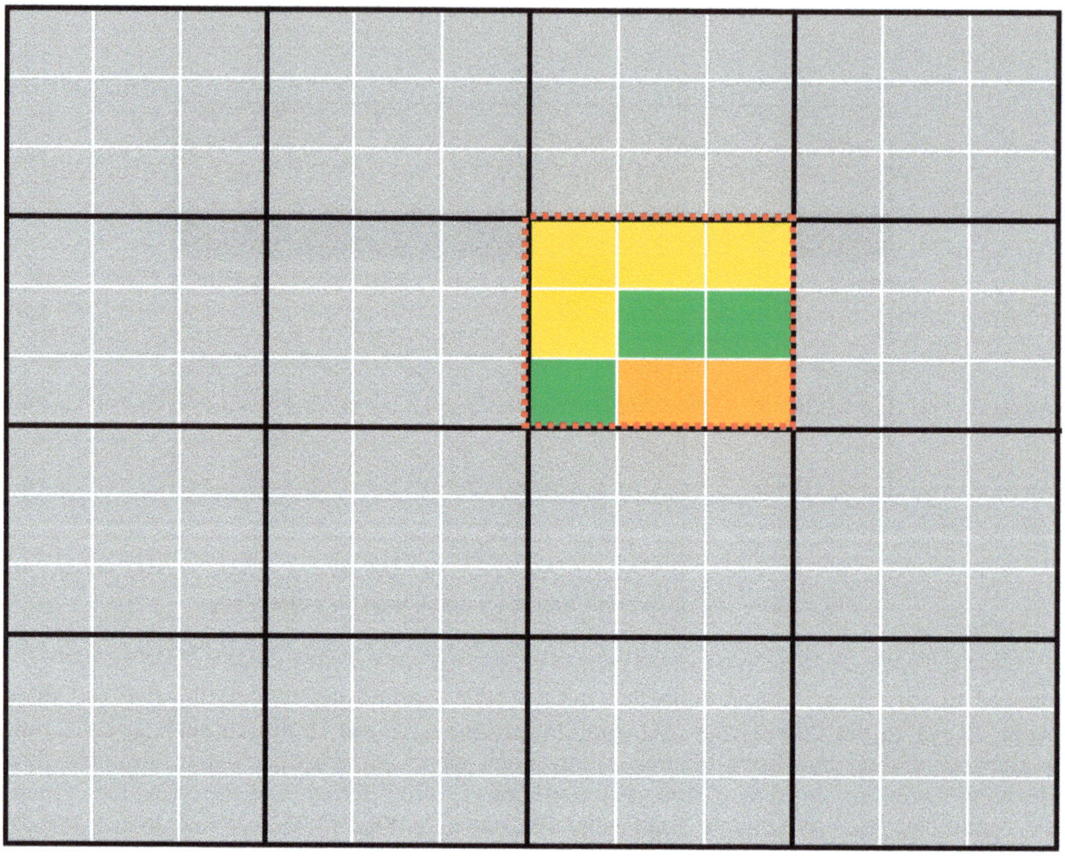

Fig. 17.24—The fine-scale properties are uniform and nonzero outside the upscaling region (the geo cells outside the upscaling region are drawn in gray to reflect that the average value need not be known).

um. Strictly speaking, these boundary conditions are therefore only valid if the neighboring flow cells are of a uniform, nonzero permeability. This is illustrated in 2D by **Fig. 17.24**.

The use of periodic boundary conditions originates from the volume averaging theory, and its use in flow-based upscaling was first introduced by Durlofsky and Chung in 1990 and by Durlofsky in 1991.[75,76] Durlofsky used periodic boundary conditions, together with Darcy's law and the classic requirement of flux conservation, to derive a full permeability tensor. A somewhat different approach, which also uses a periodic pressure field around the flow cell, was proposed by Øistein Bøe *et al.* in 1994.[78,79] This uses a weak form of Darcy's law to prove that periodic boundary conditions result in a full permeability tensor that is both symmetric and positive definite. The Norsk Hydro tensor method is based on the conservation of dissipation (mechanical energy per unit weight of fluid), although it turns out that fluxes are conserved as well.

A simplistic illustration of the periodic boundary conditions is given in **Fig. 17.25** for flow in the vertical direction (here in the case of a flow cell containing a barrier).

Although the periodic boundary conditions generally result in an effective permeability that is higher than that computed with the sealed-sides boundary conditions, the effective vertical permeability for the upscaling region illustrated in Fig. 17.25 will also be zero.

Strictly speaking, the periodic boundary conditions are only valid if no wells are present and the fine-scale medium is periodic on the scale of the flow cells (i.e., the fine-scale proper-

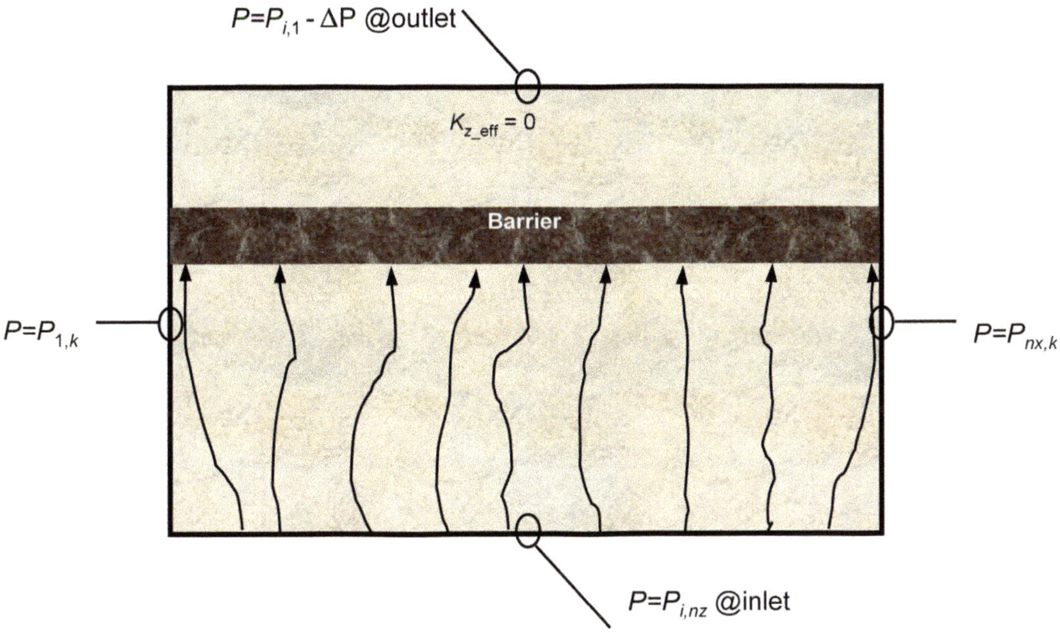

Fig. 17.25—The periodic boundary conditions in *z*-direction ($P_{1,k} = P_{nx,k}$, $P_{i,1} = P_{i,nz}$).

ty distribution inside each flow cell must be identical). This is illustrated in 2D by **Fig. 17.26**. Please note that if a medium is symmetric on the scale of ΔL, then it will be periodic on the scale of $2\Delta L$.

The relative performance of the tensor methods that is caused by the various boundary conditions has proven to be of considerable interest. As it happens, the sealed-sides method provides a lower bound and the open-sides method an upper bound of the effective permeability. The periodic-based method turns out to give an effective permeability estimate that generally lies in between the two previous methods.

With regard to the outer bounds of effective permeability, it is well known that the harmonic and arithmetic means provide the absolute lower and upper limit of the effective permeability, respectively. It is less known that the uncertainty range in the effective permeability may be narrowed using the composite averages. In fact, it may be mathematically proven that the harmonic-arithmetic average provides a closer lower limit than the pure harmonic mean, whereas the arithmetic-harmonic average provides a closer upper limit than the pure arithmetic mean (truly valid only for regular grids). In this context, it is important to realize that the two flow-based methods (sealed and open sides) provide an even narrower uncertainty band for the effective permeability, but at the expense of increased CPU time.

The relative performance of the most important local upscaling techniques is shown in **Fig. 17.27**. Fig. 17.27 also indicates the "inner" uncertainty range of the true effective permeability for the sake of comparison.

17.4.6 Regional Upscaling. Regional upscaling is applied to reduce the influence of the artificial boundary conditions on the effective permeability estimate by moving the boundary of the computational region away from the flow cell. This implies that the influence of neighboring geo cells is taken into account in addition to the geo cells inside the flow cell. In other words, regional upscaling represents an expansion of the local computational region outside the volume of the flow cell. The size of the so-called buffer or skin around each flow cell is usually

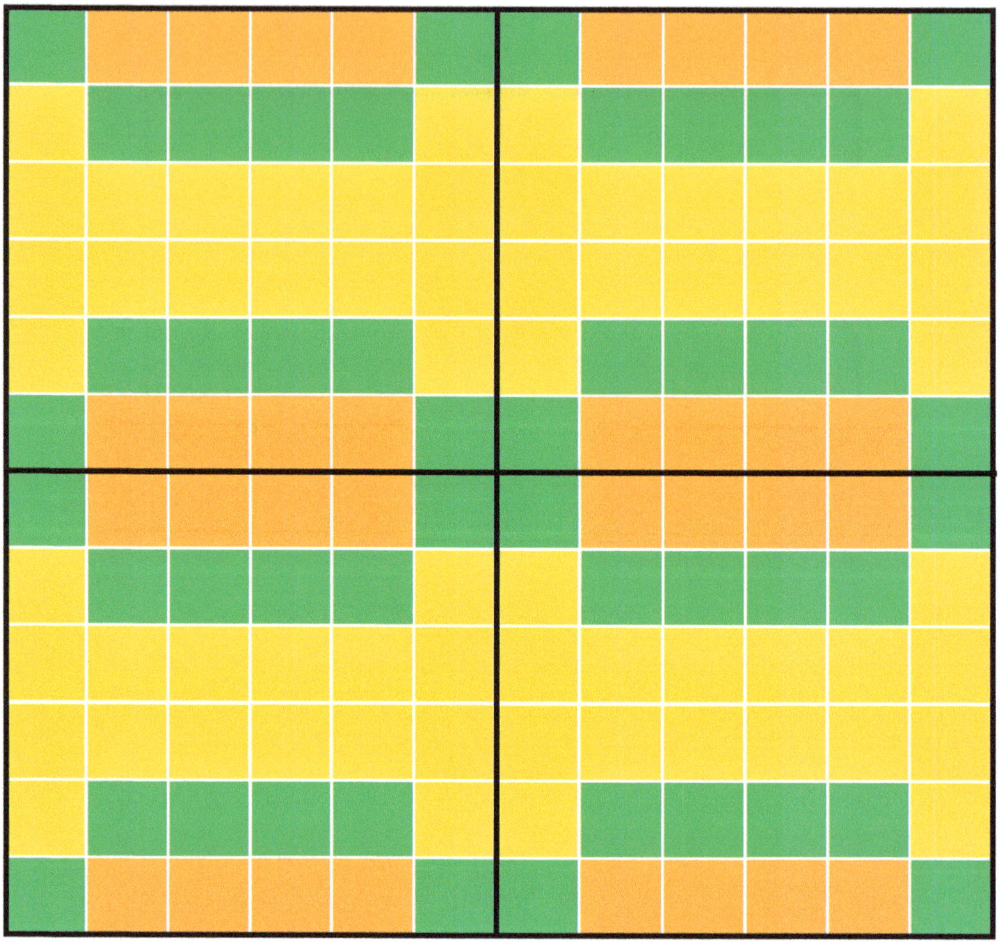

Fig. 17.26—The fine-scale properties are periodic on the scale of the flow cell (geo grid in white, flow grid in black).

given in number of neighboring geo cells to either side of the flow cell and must be specified by the modeler for each of the three coordinate directions.

The permeability estimate of a regional upscaling method will improve as the size of the buffer region increases, and it will ultimately be equal to the "true" effective permeability when the buffer size has reached the boundaries of the geo model for all three directions. The gain in accuracy is largest in the beginning (i.e., for small buffer values). This is illustrated by **Fig. 17.28**, showing the behavior of the lower- and upper-bound tensor techniques in the case of increasing buffer size. Please note that in Fig. 17.28, the outer bounds are shown to be symmetric around the "true" effective permeability. Generally, this is not the case.

17.4.7 Global Upscaling. Strictly, the fine-scale pressure field must be determined for the entire geo grid simultaneously to compute "exact" effective permeabilities for the flow cells. In the past, however, this has been too CPU-intensive to be performed in practice. With the introduction of new and promising solution algorithms such as the Output Least Squares (OSL) method, global upscaling schemes can now be realized. In the paper by Holden and Nielsen,[80] the OSL method is used to minimize the difference in pressure, as well as velocity, between the geo and flow grids in an iterative process. Because the CPU consumption of the applied

$$|\leftarrow \quad K_{\text{true}} \quad \rightarrow|$$

$$K_{\text{har}} \leq K_{h\text{-}a} \leq K_{nbc} \leq K_{pbc} \leq K_{lbc} \leq K_{\text{a-h}} \leq K_{\text{ari}}$$

Fig. 17.27—The relative performance of local upscaling techniques.

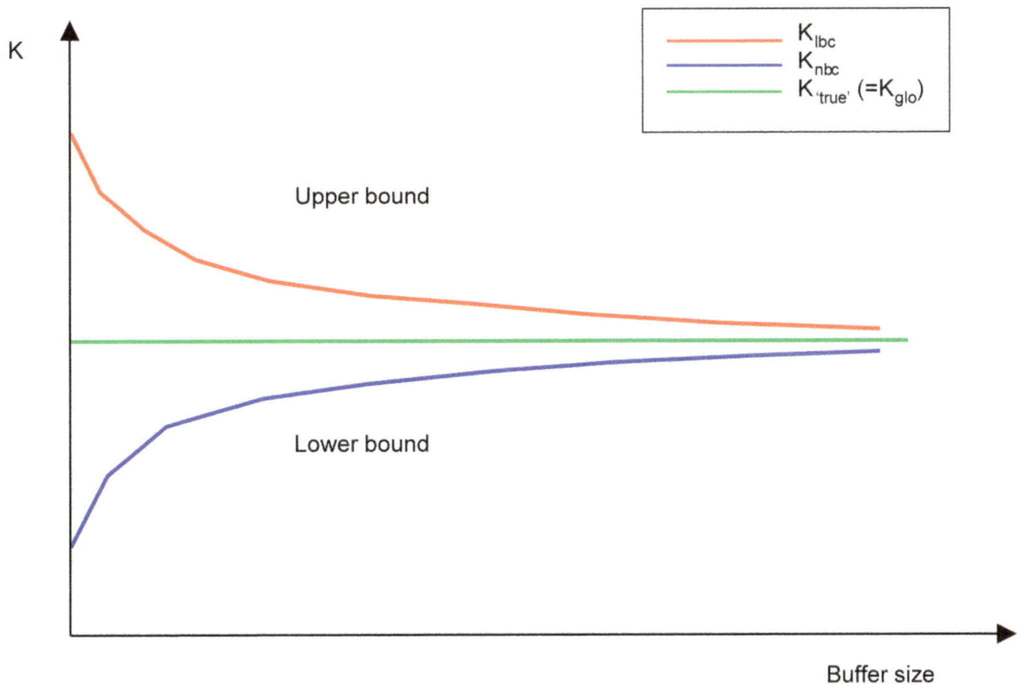

Fig. 17.28—The outer bounds of effective permeability and their behavior with increasing buffer size.

equation solver is proportional to the number of geo cells, a global solution will use approximately the same amount of computational time as the sum of all the local computations. Therefore, the new global upscaling scheme is just as fast as any local method.

An obvious advantage with the global upscaling approach is that one avoids using artificial boundary conditions around the upscaling region (i.e., instead of guessing what the boundary conditions for the flow cells might be, the pressure conditions surrounding the cells are explicitly known). Another important benefit is that a poor separation of scales in the upscaling will no longer occur because the size of the computational region is the same as the geo model.

Although still in its research stage, global upscaling has much potential for improving today's permeability estimators, especially for models containing a complex facies architecture with large permeability contrasts between facies. In fact, according to Holden and Nielsen,[80] preliminary results show an improvement factor of 10 in some cases.

Still, as discussed by Holden and Nielsen,[80] the global upscaling approach is not enough to ensure maximum accuracy in the modeling of the effective permeability. Because the value of the effective permeability is influenced by changes in the pressure field, the flow-cell permeabilities should strictly be recomputed by the global method for every timestep taken by the multiphase flow simulator. In practice, though, it might be good enough to update the effective permeability field whenever a significant change occurs as a result of altering the well configuration or production/injection rates and so on. Hence, the ultimate upscaling scheme for the absolute permeability might be the one that is coupled with the multiphase flow simulator and automatically updates the absolute effective permeability field for each timestep. With the current computer power and the lack of proper integration between the geological model and the simulation model, this is hardly achievable yet.

17.4.8 Best Practice Guidelines. As may be understood from the previous sections, there exists no single upscaling method for absolute permeability that is superior to all other methods in all situations, at least not until it has been fully established that the global upscaling scheme represents the ultimate method of choice. Selecting the proper upscaling method from the many available choices can be quite a challenge. The choice of sophistication in the upscaling method generally depends on one or several of the following factors:

- The complexity of the fine-scale permeability distribution (i.e., the geo model).
- The degree of upscaling that needs to be performed (i.e., the coarsening factor).
- The number of permeability realizations that need to be upscaled.
- The time available to the project for performing upscaling.
- The intended use of the flow model.

Because an exact validation of the upscaling process cannot be performed unless a multiphase flow simulation is carried out on the geo model itself, two alternative upscaling approaches for identifying the proper homogenization method are presented here.

The Absolute Upscaling Approach. This approach assumes that there exists a way to properly validate the *absolute* performance of an upscaling method without resorting to an extremely time-consuming (if at all feasible) finite-volume simulation of the geo model. As documented in Ref. 81, streamline simulation offers a very efficient way of validating the performance of upscaling methods. The validation process is carried out by first running a streamline simulation on the geo model to compute the reference solution. Then a streamline simulation is run on the flow model for each of the upscaling methods that are to be evaluated. The simulated performance of the various upscaling methods is then compared to that of the geo model. The validation of upscaling methods is best done under single-phase flow conditions to avoid introducing other model parameters (e.g., relative permeabilities and associated rock types) that also need to be upscaled in one way or the other. A higher confidence may also be obtained for the validation process if the actual well pattern is used in the streamline simulations.

Using the previously described validation scheme, the modeler may choose to evaluate any upscaling method until one with a satisfactory performance is found. Still, a more systematic way of identifying the optimum upscaling method is desirable. With the absolute upscaling approach that is presented here, the modeler is offered a multistep procedure that is to be terminated as soon as a satisfactory upscaling method has been identified. The recommended procedure involves the following steps:

1. Compute the upper and lower bounds of the effective permeability using the arithmetic-harmonic (or pure arithmetic) and harmonic-arithmetic (or pure harmonic) average techniques, respectively. Being of the analytical type, these methods are very fast and will provide a first quantification of the upscaling uncertainty. Validate the performance of the two composite methods against that of the geo model using a single-phase streamline simulator with the actual well pattern.

2. If the performance of any of the two methods in Step 1 is within an acceptable range of the geo model, then terminate the procedure and choose the appropriate method. If, on the other hand, the performance of both methods is unsatisfactory because of the complexity of the geo model, then use the upper (open-sides) and lower (sealed-sides) bound diagonal tensor methods to narrow the uncertainty in the flow-model performance. Validate the performance of the two tensor methods against that of the reference solution.

3. If the performance of any of the methods in Step 2 is within an acceptable range of the reference solution, then terminate the procedure and choose the appropriate method. If, on the other hand, the performance of both methods is unsatisfactory, then the following alternatives may be worth considering:

a. If time allows, refine or coarsen the flow grid (whatever is best) to achieve a better separation of the length scales. Then repeat the upscaling of the outer bounds (in Step 2) and redo the validation to check if the performance of either method has improved.

b. Apply a tensor method with periodic or semi-open-boundary conditions [the semi-open boundary conditions alternative is available in some applications using a multiplier between 0 (sealed) and 1 (open) to the side faces of the computational region]. As previously mentioned, this should result in an intermediate estimate of the effective permeability tensor and therefore provide a flow model performance that lies somewhere in between the two methods in Step 2.

c. Select the best of the two tensor methods in Step 2 and convert the local method to a regional method using a buffer region of modest size. Validate its performance. If necessary, repeat this step using an increasingly larger buffer region until a satisfactory performance of the flow model is obtained.

d. If for some reason none of the previous alternatives are an option, then one needs to apply the method that best satisfies the wanted flow behavior of a given cell or cells in a given region. In other words, a combination of the outer bound techniques within one and the same model may be a fourth alternative.

As previously mentioned, the method using open-sides boundary conditions is a good estimator of sand continuity and quality, whereas the sealed-sides boundary conditions method is better at detecting the presence and effect of barriers.

Consider a long horizontal oil producer in the Troll West Gas Province that is protected against coning from the overlying gas cap by a calcite barrier just above the well. If the vertical grid resolution in the Troll full-field model was such that one could apply the open-sides boundary conditions technique on the cells containing the well, and the sealed-sides boundary conditions technique on the cells containing the barrier, then this would be the optimum local upscaling approach.

However, if both a segment of the well trajectory and a segment of the calcite are present inside the same flow cell, then the open-sides boundary conditions technique will give a good estimate of the well's PI but result in a much too early gas breakthrough, whereas the sealed-sides boundary conditions method will better capture the effect of the calcite but give a too low estimate of the well's PI. If this is the case, one needs to consider applying one of the alternatives A or C (alternative B will, in this particular example, give the same result as the sealed-sides method).

The multistep procedure of the absolute upscaling approach is graphically illustrated in **Fig. 17.29**.

The Relative Upscaling Approach. This approach acknowledges the fact that an exact validation of the upscaling results cannot be achieved in practice. Therefore, instead of trying to validate the absolute performance of an upscaling method, the approach diagnoses the relative performance of outer bound methods using the actual multiphase finite-volume simulator. This implies that a full black-oil simulation is run on the flow model for each of the upscaling methods that are to be evaluated. The deviation in the simulated performance between outer bound

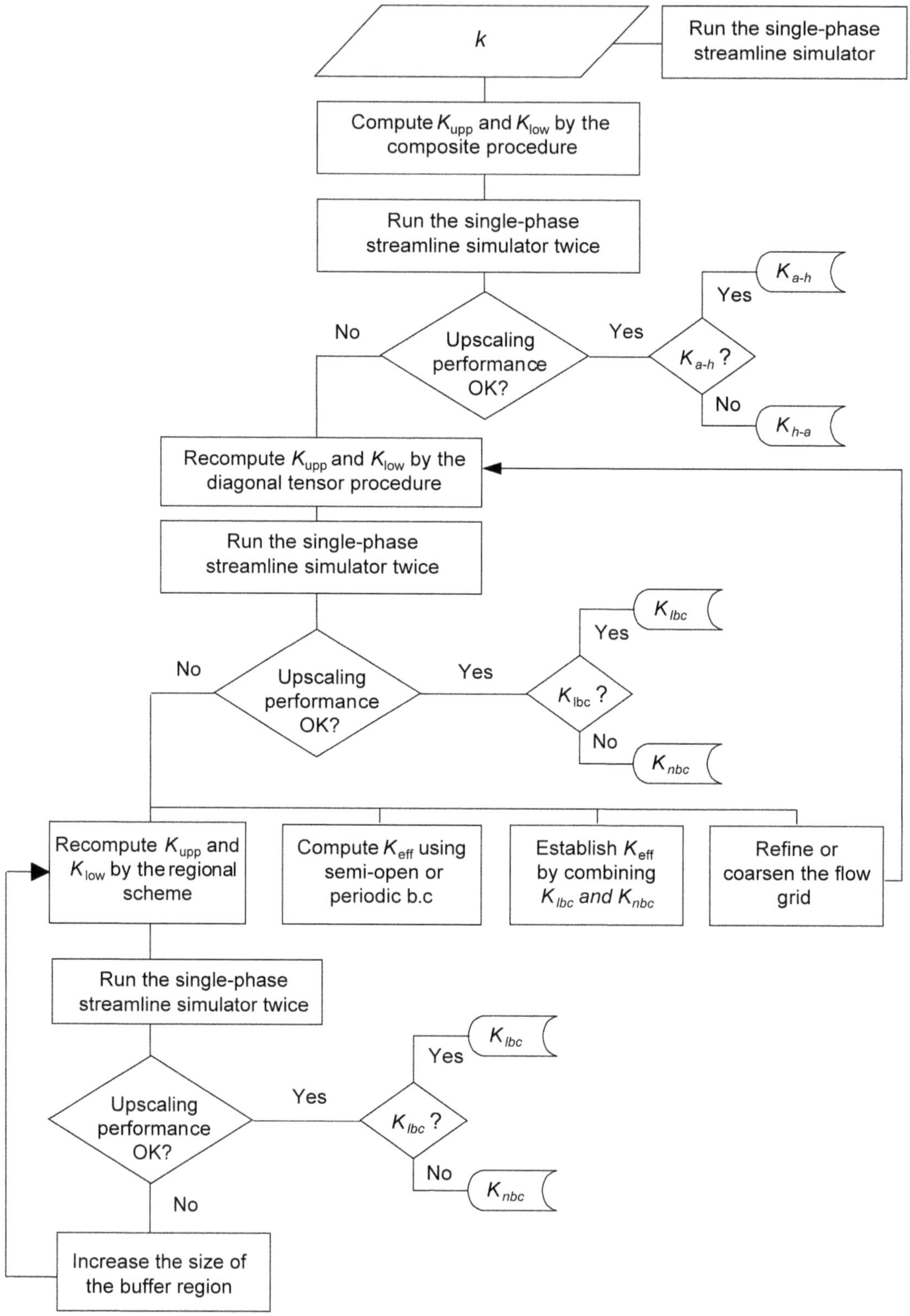

Fig. 17.29—Flow chart for the absolute upscaling approach.

methods will then reflect the part of the model uncertainty that originates from the upscaling process itself. To ensure a high degree of relevance in the diagnostics, it is important that the test simulations contain a representative description of the actual flow model.

If the project is pressed for time, the simulations may be skipped altogether in favor of a faster, although less robust, way of performing the diagnostics. Instead of analyzing simulation profiles, a normalized difference parameter may be computed on a cell by cell basis using the formula

$$\partial = (K_{\text{upp}} - K_{\text{low}}) / \left[(K_{\text{upp}} + K_{\text{low}}) / 2 \right] . \qquad (17.18)$$

The relative upscaling approach that is presented here utilizes a multistep procedure that applies outer bound methods of increasing accuracy until the best possible upscaling method can be identified. The recommended procedure involves the following steps:

1. Compute the upper and lower bounds of the effective permeability using the arithmetic-harmonic (or pure arithmetic) and harmonic-arithmetic (or pure harmonic) average techniques, respectively. Being of the analytical type, these methods are very fast and will provide a first quantification of the upscaling uncertainty. Run the finite-volume simulator for each of the two composite methods (or compute a grid-based difference parameter) and perform the diagnostics.

2. If the performance gap between the two methods in Step 1 is acceptable (small), then terminate the procedure and choose either of the two. If the deviation in performance is unsatisfactory because of the complexity of the geo model, then use the upper (open-sides) and lower (sealed-sides) bound diagonal tensor methods to narrow the uncertainty in the flow model performance. Run the finite-volume simulator for each of the two tensor methods (or compute a grid-based difference parameter) and perform the diagnostics.

3. If this reduces the upscaling uncertainty to within acceptable limits, then either of the two diagonal tensor methods may be used to provide the final permeability field for the flow model. If the deviation in the performance is unsatisfactory (large), then the following alternatives may be worth considering:

1. If time allows, refine or coarsen the flow grid (whatever is the best) to achieve a better separation of the length scales. Then repeat the upscaling of the outer bounds (in Step 2) and rerun the simulations (or recompute the difference parameter) to check if the performance gap (upscaling uncertainty) has narrowed.

2. Apply a tensor method with semi-open or periodic boundary conditions. As previously mentioned, this should result in an intermediate estimate of the effective permeability tensor and hence provide a flow model performance that lies somewhere in between the two methods in Step 2.

3. Convert the two local tensor methods in Step 2 into regional methods using a buffer region of modest size. Rerun the finite-volume simulator (or recompute the difference parameter) and check the performance gap. If necessary, repeat this step using an increasingly larger buffer region until the upscaling uncertainty reaches acceptable limits, at least as far as practically possible.

4. If for some reason none of the previous alternatives are an option, then one needs to apply the method that best satisfies the wanted flow behavior of a given cell or cells in a given region. In other words, a combination of the outer bound techniques within one and the same model may be a fourth alternative.

The multistep procedure of the relative upscaling approach is graphically illustrated in **Fig. 17.30**.

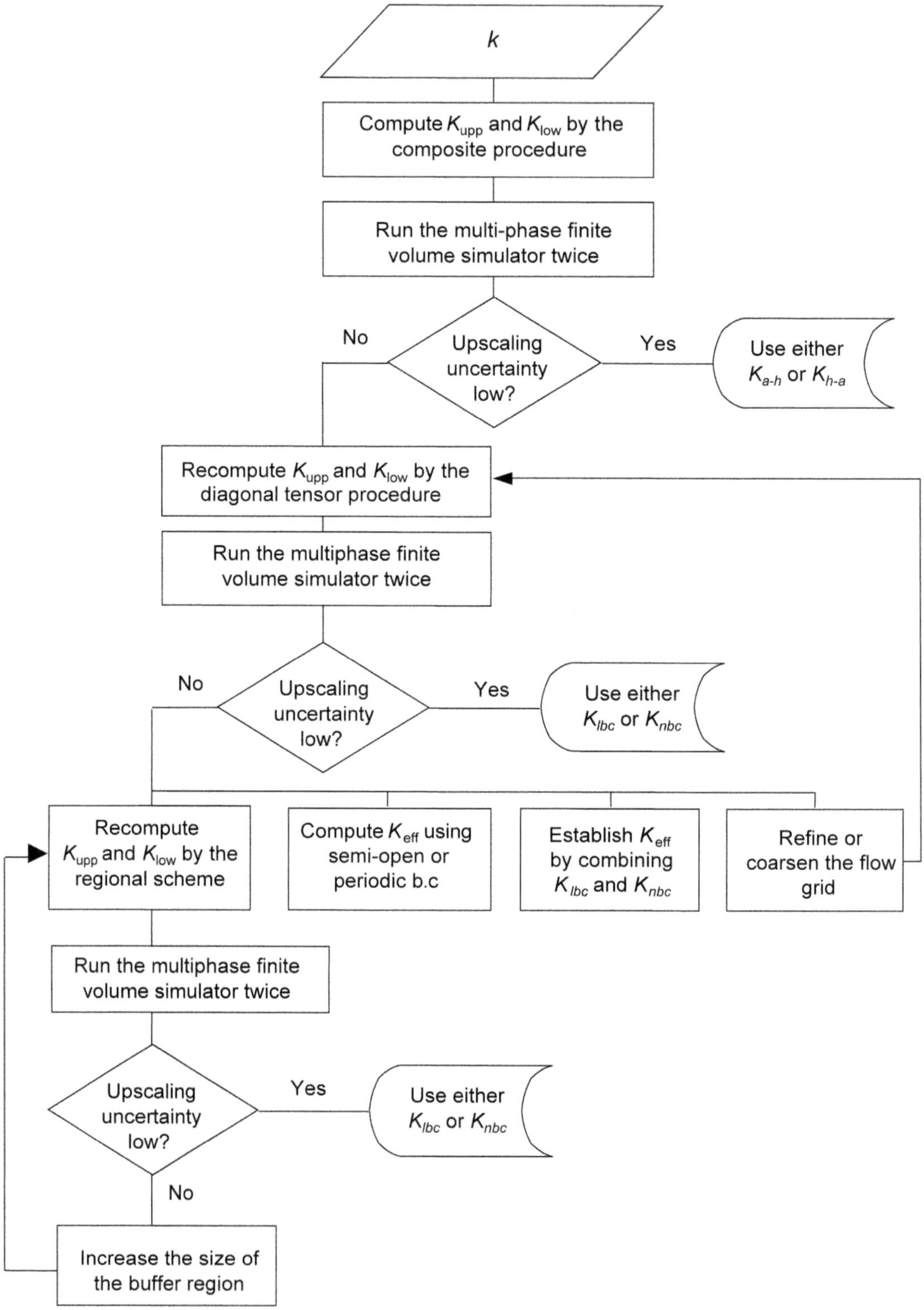

Fig. 17.30—Flow chart for the relative upscaling approach.

17.5 Streamline Simulation—Rod P. Batycky and Marco R. Thiele

17.5.1 Introduction. Streamline-based flow simulation differentiates itself from cell-based simulation techniques such as finite-differences and finite-elements in that phase saturations and components are transported along a flow-based grid defined by streamlines (or streamtubes) rather than moved from cell-to-cell. This difference allows streamlines to be extremely efficient in solving large, heterogeneous models if key assumptions in the formulation are met by the physical system being simulated (see below). Specifically, large relates to the number of active grid cells.

Streamlines represent a snapshot of the instantaneous flow field and thereby produce data such as drainage/irrigation regions associated with producing/injecting wells and flow rate allocation between injector/producer pairs that are not easily determined by other simulation techniques. The computational speed and novel solution data available have made streamlines an important, complementary approach to traditional simulation approaches to perform sensitivity runs, quantify the impact of upscaling algorithms used to move models from the geomodeling scale to the simulation scale, visualize the flow field, perform more reliable full-field simulations where sector models would normally be used, enable the ranking of predicted field behavior of given multiple production scenarios and input parameters, evaluate the efficiency of injectors and producers, reduce turnaround time in history matching, and perform other established reservoir engineering tasks. A comprehensive overview on streamline-based flow simulation has recently been presented by Thiele.[82]

17.5.2 Applicability of Streamline Simulation. The power of streamline simulation lies in its simplicity. The main objective is to capture how injected reservoir volumes (usually water and/or gas) displace resident reservoir volumes given well locations, well rates, reservoir geometry, and geological description. One of the key underlying assumptions in streamline simulation is that the system be close to incompressibility. This decouples saturations from the underlying pressure field and allows each streamline to be treated as being independent from the streamlines next to it.

Many fields under waterflooding or other pressure maintenance schemes are excellent candidates for streamline modeling and have been successfully modeled in this way.[83,84] Forecast simulations under the assumption of voidage replacement are another good example where streamlines can be very effective. Even miscible gas injection schemes have been successfully modeled.[85,86] At high pressures, the displacement of resident oil by gas is primarily an issue of simulating local sweep efficiency and channeling, something streamlines are designed to model without incurring numerical difficulties associated with other formulations.[87] An early application to streamdrive projects was presented by Emanuel.[88] Crane and Blunt[89] used streamlines to model solute transport. More recently, streamlines have been shown to be very effective in modeling fractured reservoirs using a dual porosity formulation.[90]

17.5.3 Historical. Streamlines have been in the petroleum literature as early as Muskat and Wyckoff's 1934 paper.[91] In 1937, Muskat presented the governing analytical solutions for the stream function and the potential function for 2D domains using the assumption of incompressible flow.[92] Since then, streamlines and streamtubes have received repeated attention as a way to numerically predict the movement of fluids, even after the advent of finite-difference methods in the early 1960s. Important early contributions were made by Fay and Pratts,[93] Higgins and Leighton,[94,95] Bommer and Schechter,[96] Martin and Wegner,[97] Lake *et al.*,[98] and Emanuel *et al.*[99]

In the early 1990s, streamlines were revived because advances in geological modeling techniques were producing models that were too large for finite differences to simulate in an acceptable time frame. For streamlines to be applicable to real field cases, important advances were made that extended streamlines to 3D using a time-of-flight variable[87,100–103] allowed for streamlines to be periodically updated[87,104,105] and included gravity.[106]

17.5.4 Mathematics of the Streamline Method. The streamline method and the underlying mathematics for incompressible multiphase flow are briefly outlined here. For a detailed discussion as well as additional references describing streamline methods, see Batycky *et al.*,[87] Batycky,[103] and Blunt *et al.*[107]

Governing IMPES Equations. The streamline method is an IMPES-type formulation with the pressure field solved for implicitly and the oil/gas/water saturations solved for explicitly along streamlines. The governing equation for pressure, *P*, for multiphase incompressible flow without capillary or diffusion effects is given by

$$\nabla \cdot \sum_{j=1}^{n_p} \frac{\vec{\vec{k}}\, k_{rj}}{\mu_j}\left(\nabla \cdot P + \rho_j \vec{g} D \right) = 0, \quad (17.19)$$

where D is the depth below datum, g is gravitational acceleration constant, k is the permeability tensor, k_{rj} is the relative permeability, μ_j is viscosity, and ρ_j is the phase density of phase j. The total velocity, $\vec{u}_t$, is derived from the 3D solution to the pressure equation and application of Darcy's law. The explicit material-balance equation for each incompressible phase j is then given by

$$\phi \frac{\partial S_j}{\partial t} + \vec{u}_t \cdot \nabla f_j + \nabla \cdot \vec{G}_j = 0. \quad (17.20)$$

Each phase fractional flow, f_j, is given by

$$f_j = \frac{k_{rj}}{\mu_j} \frac{1}{\sum_{j=1}^{n_p} k_{rj} / \mu_j}, \quad (17.21)$$

and the phase velocity resulting from gravity effects because of phase density differences is given by

$$\vec{G}_j = \vec{\vec{k}} \cdot g f_j \nabla D \sum_{i=1}^{n_p} k_{ri} (\rho_i - \rho_j) / \mu_i . \quad (17.22)$$

The difference between finite-difference simulation and streamline simulation is the way the explicit material balance equations (Eqs. 17.20 through 17.22) are solved. In finite difference, the material balance equations are solved between gridblocks, whereas in streamline simulation the material balance equations are solved along streamlines. How this is done is explained next.

Solution to the Transport Equation. In a standard finite-difference method, Eq. 17.20 is discretized and solved on the underlying grid on which the pressure field is computed. The solution to Eq. 17.20 is governed by the grid CFL condition, which can lead to prohibitively small timestep sizes, particularly for models with high permeability contrasts and/or high local flow velocities. With streamlines, this grid CFL limit is avoided completely by solving Eq. 17.20 along each streamline using a time-of-flight (TOF) coordinate transform.[106]

Streamlines are traced from sources to sinks based on the underlying total velocity field. As each streamline is traced, compute the TOF along the streamline, which is defined as

$$\tau = \int_0^s \frac{\phi(\delta)}{\left| \vec{u}_t(\delta) \right|} d\delta \quad (17.23)$$

and leads to the definition

$$\left| \vec{u}_t \right| \frac{\partial}{\partial \delta} \equiv \vec{u}_t \cdot \nabla = \phi \frac{\partial}{\partial \tau} \ . \quad (17.24)$$

Using Eq. 17.19, rewrite Eq. 17.20 as

$$\frac{\partial S_j}{\partial t} + \frac{\partial f_j}{\partial \tau} + \frac{1}{\phi} \nabla \cdot \vec{G}_j = 0. \quad (17.25)$$

Because the gravity term is not aligned along a streamline direction, Eq. 17.25 is split into two parts (operator splitting), giving two 1D equations. The convective portion of the material-balance equation along streamlines is given by

$$\frac{\partial S_j^c}{\partial t} + \frac{\partial f_j}{\partial \tau} = 0, \quad (17.26)$$

while the portion resulting from phase-density differences solved along gravity lines is given by

$$\frac{\partial S_j^g}{\partial t} + \frac{1}{\phi} \nabla \cdot \vec{G}_j = 0. \quad (17.27)$$

Both Eqs. 17.26 and 17.27 represent 1D equations that are solved using standard finite-difference numerical techniques. There are still CFL limits that restrict timestep sizes in these equations, but these are local to each streamline or gravity line, rather then at the 3D grid level.

Timestepping. In field-scale displacements, the streamline paths change with time because of the changing fluid distributions and the changing well conditions. As a result, the total velocity field is periodically updated, and new streamlines are recomputed to reflect the nonlinear nature of the displacement.

To move the 3D saturation distribution forward in time between successive streamline distributions from time T_i to $T_{i+1} = T_i + dT_i$, the algorithm pictured in **Fig. 17.31** is used.

The basic algorithm for streamline-based flow simulation is as follows: (1) Given initial conditions (i.e., pressures and saturations for each active cell in the system) and well conditions, the pressure is solved implicitly for each cell, as is done in conventional finite-difference methods (Eq. 17.19). (2) With the pressures known, the total velocity for each cell interface can be determined using Darcy's Law. The total velocity is then used to trace streamlines using Pollock's algorithm.[100] (3) 1D mass conservation equations are then solved along each streamline, independently of each other (Eq. 17.26). The initial conditions for the streamlines are obtained by a mapping from the underlying 3D grid onto each streamline. The mass-transport problem is marched forward in time along each streamline for a pre-specified global timestep dT_i, and then the solution is mapped back onto the 3D grid. Gravity is included by

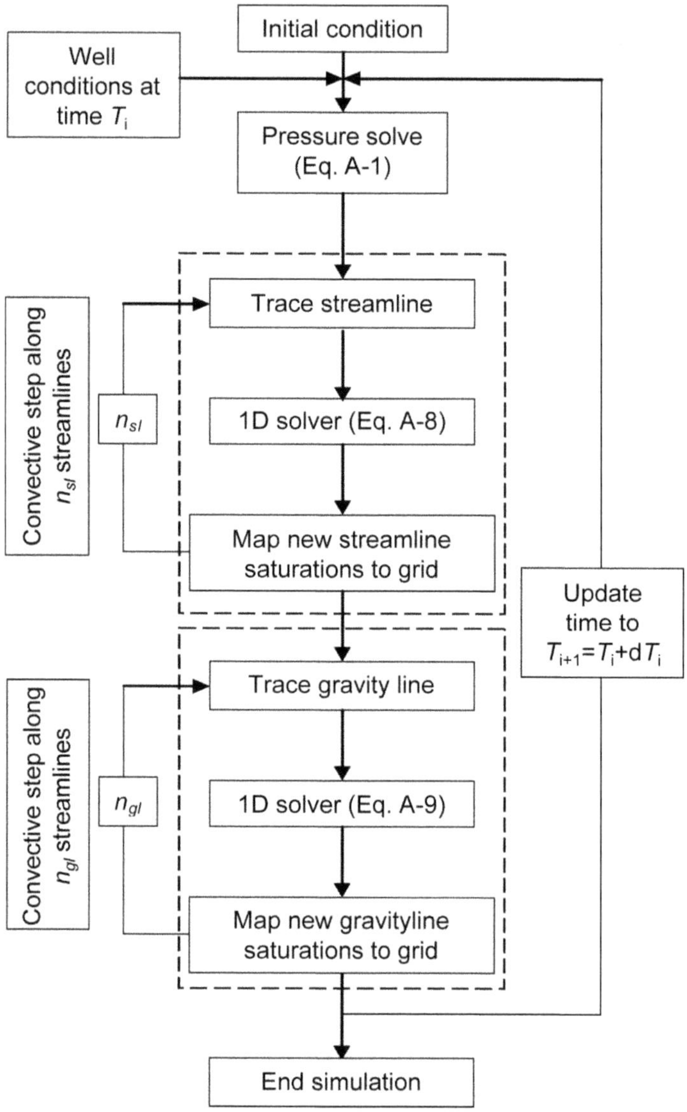

Fig. 17.31—Flow chart showing the basic algorithm for streamline-based flow simulation.

considering a vertical segregation step along gravity lines after movement along all streamlines (Eq. 17.27). While simple in its approach, important details must be considered. In particular:

1. The algorithm is similar to an IMPES approach, in that the pressure is solved implicitly for a new time level n+1 assuming saturations at level n. The saturations at time n are given by mapping back solutions from each streamline onto the 3D grid at the previous timestep. Because of the implicit nature of the pressure solution, there is no limitation on the timestep to reach n+1. However, for compressible systems numerical convergence problems might limit the actual size of the timestep. This is no different than in FD simulation.

2. The tracing of the streamlines using Pollock's algorithm assumes Cartesian cells. Nonorthogonal corner-point cells require an isoparametric transformation for tracing streamlines.[108]

3. For incompressible systems, streamline will start at injection wells and end at production wells. For compressible systems, streamline can start/end anywhere in the system, because any

gridblock in the system might act as a source (volume expansion) or a sink (volume contraction). Multiphase gravity effects can give rise to circulation cells for both incompressible and compressible systems.

4. Initial launching of streamlines from wells can be proportional to the total flux at the wells, though this will in general leave many cells in the system without a streamline passing through them. For missed cells, tracing begins at the center of the missed cell and then traced backward until a source is encountered. If a cell does not have a streamline pass through it, then it is not possible to assign an updated saturation back to that cell.

5. In practice, it is not possible to have all streamlines carry the same flux and ensure at least one streamline per cell. Thus, streamlines do not carry the same flux. Furthermore, for incompressible problems the flux along each streamline is a constant, while for compressible systems it is not.

6. The tracing of streamlines using the TOF variable produces a highly irregular 1D grid along each streamline. To numerically solve the 1D problem efficiently, the 1D grid must be regularized, solved using an implicit approach, or regridded in some way to allow for a more efficient solution.

7. The tracing of the streamlines relies on an accurate solution of the velocity field. Excessive distortions of the grid (nonorthogonal) or a pressure solution that has not been solved to a small enough tolerance can cause problems in tracing streamline paths.

17.5.5 The Computational Efficiency of Streamlines. One advantage of streamline simulation over more traditional approaches is its inherent efficiency, both in terms of memory and computational speed. Specifically, streamline-based simulation can exhibit a near-linear scaling in run times as a function of active cells in the model. Memory efficiency is a result of two key aspects of the formula: streamline-based simulation is an IMPES-type method and therefore involves only the implicit solution of pressure, and tracing of streamlines and solution of the relevant transport problem along each streamline is done sequentially. Only one streamline needs to be kept in memory at any given time.

Computational speed, on the other hand, is achieved because the transport problem is decoupled from the 3D grid and instead solved along each streamline. Because transport along streamlines is 1D, they can be solved efficiently. Because the number of streamlines increases linearly with the number of active cells, and streamlines only need to be updated infrequently, the computational time exhibits a near-linear scaling with increasing number of gridblocks

The number of global timesteps is related to how often the flow field (streamlines) requires updating. Specifically, changing flow paths are a function of heterogeneity, mobility changes, gravity, and changing well conditions. For many practical problems, it is the changing well rates that introduced the greatest impact on a changing flow field and is therefore the limiting factor in deciding on global timestep sizes. Grouping well events into semiyearly or yearly intervals and assuming that the streamlines remain unchanged over each period is reasonable. This is why field simulations with 30- to 40-year histories are successfully and routinely simulated with 1-year timesteps.[84]

A good example to demonstrate the efficiency of SL simulation is Model 2 of the 10th SPE comparative solution project.[12] The total run time, T, of any streamline simulation is approximately proportional to

$$T \propto \sum_{1}^{n_{ts}} \left(t^{\text{solver}} + \sum_{1}^{n_{sl}} t_j^{sl} \right), \text{ where}$$

n_{ts} number of timesteps (number of streamline updates)
t^{solver} time required to solve for the global pressure field ($Ax = b$) at each timestep
n_{sl} number of streamlines at each timestep
t_j^{sl} time to solve transport equation for each streamline .... (17.28)

A near-linear scaling arises because:

1. The number of timesteps (streamline updates) is independent of the model size, heterogeneity, and any other geometrical description of the 3D model. It is mainly a function of the number of well events and the actual displacement physics. For the SPE10 problem in **Fig. 17.32**, all cases were run with the exact same number of streamline updates—24.
2. An efficient pressure solver is expected to have a near-linear behavior as well.[109]
3. The number of streamlines tends to increase linearly with the number of gridblocks, all else being equal. Fig. 17.32 illustrates this behavior.
4. The time to solve the transport problem along each streamline can be made efficient by regularizing the underlying TOF grid and choosing the number of nodes to use along each streamline regardless of the size of the underlying 3D grid.

The linear behavior with model size is the main reason why streamline simulation is so useful in modeling large systems. In FDs, finer models not only cause smaller timesteps because of smaller gridblocks but usually face problems because of increased heterogeneity as finer models tend to have wider permeability and porosity distributions. The usual workaround is to use an implicit or adaptive-implicit formulation, but for large problems these solutions can become prohibitively expensive, both in terms of CPU time and memory.

17.5.6 Novel Data Produced by Streamlines. Streamlines produce new data not available with conventional simulators. Because streamlines start at a source and end in a sink, it is possible to determine which injectors are (or which part of an aquifer is) supporting a particular producer, and exactly by how much. A high water cut in a producing well can therefore be traced back to specific injection wells or boundaries with water influx. Conversely, it is possible to determine just how much volume from a particular injection well is contributing to the producers it is supporting—particularly valuable information when trying to balance patterns (**Fig. 17.33**) or optimize water injection over a field.

Streamlines can also identify the reservoir volume associated with any well in the system, because a block traversed by a streamline attached to a particular well will belong to that well's drainage volume. It is therefore possible to divide the reservoir into dynamically defined drainage zones attached to wells (Fig. 17.33). Properties normally associated with reservoir volumes can now be expressed on a per-well basis, such as oil in place, water in place, and average pressure, just to mention a few.

The most successful uses of new data produced by streamlines are in the area of waterflood management and reservoir surveillance,[110,111] and in the area of history matching.[112–115]

17.5.7 Applications of Streamlines. Streamlines are a powerful complementary tool to more traditional simulation techniques, and they are expected to play an important part in optimizing field production and management in the future. Specifically, streamlines can be used to:

1. Validate upscaling techniques by allowing to generate reference solutions of fine-scale models.[81]
2. Efficiently perform parametric studies.

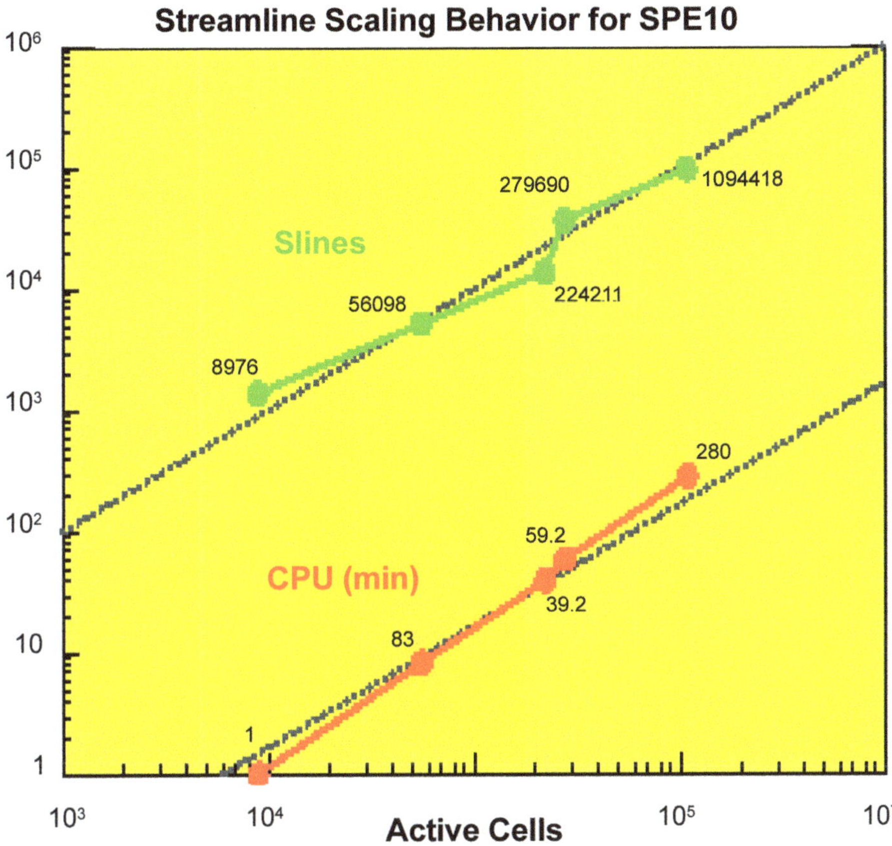

Fig. 17.32—Example of linear scaling of run time and number of streamlines as a function of active cells for SPE Comparative Solution Project 10 using 3DSL, a commercial streamline simulator, on a PIII 866MHz PC.

3. Visualize flow.
4. Balance patterns.
5. Determine efficiency of injectors and producers using data provided by streamlines.
6. Aid in history matching.
7. Enable ranking of production scenarios/geological models.
8. Optimize and manage field injection/production.
9. Conduct reservoir surveillance.

It is important to underline that the theory on which streamline simulation rests is firmly rooted in the incompressible formulation of exact voidage replacement. Thus, streamline simulation is particularly powerful for modeling systems that are not a strong function of absolute pressure, but are instead governed by a pressure gradient. In addition, the strong assumptions of independence between streamlines favors modeling displacements that are not a strong function of diffusive phenomena, such as capillary pressure, transverse diffusion, or compressibility. For example, streamline simulation offers little or no advantage over conventional simulation for modeling primary production. This is because the main feature of modeling primary production is to accurately capture the pressure decline over time, not the movement of a saturation front.

Fig. 17.33—Streamlines automatically allow the determination of the allocation of flow between wells by summing the flux of all streamlines associated with a particular well, well pair, or group of wells. Using this information and the visual display of streamlines allows patterns to be balanced correctly and efficiently. From left to right: rates are progressively changed to yield a balanced pattern.

17.5.8 The Future of Streamline Simulation. The next few years are expected to bring a further maturing and extended application of streamline-based flow-simulation technology. It is reasonable to expect that most companies using conventional simulation technology today will in one form or another use SL simulation in their future work. What remains uncertain is whether new user groups, such as geologists and geophysicists, will adopt the technology to bring a dynamic flow component to their analysis. Developments in the following areas are currently under way: the use of streamlines in conditioning static reservoir models to production data, extension of streamline simulation to compositional models, tracing of streamlines in structurally complex reservoirs, modeling of dual-porosity/dual-permeability models, and parallelization of streamline numerics for the solution of large models.

17.6 Simulation of Geomechanics—Tony Settari

Historically, much of the simulation has accounted for rock mechanics by simple use of a time-invariant rock compressibility c_R, spatially constant or variable. In reality, rock mechanics is intimately coupled with fluid flow in two aspects. First, the porosity changes are a direct result of the deformation of the skeleton, which is a complex function of both pressure and stress (effective stress state). Second, it is well known that the permeability of the media is also a function of effective stresses in the reservoir. Therefore, rigorous reservoir simulation should include simultaneous solution of multiphase flow and stresses as well as the appropriate dependencies between these processes. While these couplings physically exist to some extent in all reservoirs, they can be often ignored or approximated when the reservoir behaves elastically. However, the changes in porosity and permeability are more pronounced when rock failure occurs, such as in compacting reservoirs or in high-pressure injection operations, and these processes require use of more complex, coupled geomechanical modeling.

17.6.1 Types of Coupling. There are essentially two main types of coupling between reservoir flow and stress: pore-volume coupling and flow-properties coupling. In the past, the first led to the development of various simplified "compaction" modeling techniques, while the latter is reflected in the "pressure-dependent permeability" options available in many simulators. In a coupled geomechanical model, both can be treated more rigorously.

Pore-Volume Coupling. The porosity in the reservoir model is traditionally treated as a function of pressure via rock compressibility:

$$\phi = f(p)\left(= \phi^0\left[1 + c_R(p - p^0)\right]\right), \quad (17.29)$$

and changes in block pore volume V_p are computed as $V_p = V_b^0\phi$, where V_b^0 is the block (bulk) volume. In reality, pore volume changes are a result of complex interaction of fluid (pore) pressure, stresses acting on the element of the rock and temperature. The deformation of the rock solid (also called the skeleton) caused by combination of stress and pressure changes results in changes in the bulk volume of an element V_b, which is computed at any conditions as

$$V_b = V_b^0(1 - \varepsilon_v), \quad (17.30)$$

where ε_v is the volumetric strain at these conditions, and V_b^0 is the bulk volume at reference conditions at which the volumetric strain = 0. Then the true porosity is given by

$$\phi = V_p / V_b, \quad (17.31)$$

where V_p is the pore volume of the element. In Eq. 17.31, both pore volume and bulk volume are variable; therefore, true porosity and pore volumes are both functions of pressure, temperature and stress:

$$\phi = V_p / V_b = f(p, T, \sigma),\ V_p = g(p, T, \sigma), \quad (17.32)$$

which shows the coupling between fluid flow and geomechanics (stress modeling).

In stress modeling, the changes in volumetric strain and porosity are calculated from complex constitutive relations of the material, which define both the stress-strain and volumetric behavior.[116,117] To compute pressure changes correctly in the reservoir simulator, it is necessary to force the changes in pore volume to be the same as computed by the stress model, which is the essence of the "volume coupling." This can be achieved in two ways. The rigorous solution is for the reservoir model to recalculate the block sizes based on the stress solution, and use the true porosity. However, reservoir simulators do not allow for modifying the bulk volume. In this case, one can define a pseudoporosity $\phi^* = V_p / V_b^0$, which will give the correct pore volume.[118] In either case, the usual treatment of porosity by rock compressibility must be replaced by the coupling with stress-strain solution. The porosity changes become more complex when failure of the skeleton is reached either in compression (rock compaction) or in shear.

Flow-Properties Coupling. The primary mechanism is the dependence of permeability on stress, usually of the form

$$k = f(\sigma_{\text{eff}}) = f(\sigma_{\text{avg}} - p),\ \sigma_{\text{avg}} = (\sigma_x + \sigma_y + \sigma_z)/3. \quad (17.33)$$

This relationship is well established for many types of reservoirs[119] and generally becomes stronger as permeability decreases. It is dominant (and more complex) in fractured reservoirs where stress-dependent fracture aperture and reopening/creating fractures under injection can cause large, anisotropic changes. The tensor character of permeability may be important in these applications. In soft formations and unconsolidated sands, deformations can lead to porosity dilation, which will also lead to permeability enhancement. On reloading, there is a hysteretic effect. Two to three orders of magnitude enhancements in permeability can occur because of injection at fracture pressure (e.g., in microfractured rock[120] or coal seams).

The permeability changes are a function of some measure of effective stress, but are often simplified (and laboratory data reported) as a function of pressure. The creation of new fractures can cause a transition from initially single-porosity media to dual-porosity, and in such a case will also have an effect on relative permeabilities.

17.6.2 Modeling of Reservoir Compaction and/or Dilation. Modeling reservoir deformation is of considerable importance in soft and/or thick reservoirs where the results of compaction may provide an important production mechanism, cause well failures, and/or cause ground subsidence or heave with environmental consequences. Review of the compaction mechanics and its consequences for field development is found in Ref. 121. Initial approach to modeling compaction was based on modifications of reservoir models.[122–125] The common feature of such reservoir compaction models is that the compaction is treated as a 1D problem (uniaxial strain) by assuming that (a) only vertical deformations take place, and (b) each vertical column of blocks deforms independently. In such models, the porosity changes are calculated by modifying the conventional compressibility c_R as a function of pressure only, in the form of "compaction tables." The tables are based on results of uniaxial strain laboratory experiments, and the stress problem is not solved. The compaction of the reservoir is then obtained analytically assuming uniaxial deformation. The relation between reservoir compaction and surface subsidence can be then obtained by an independent solution of a stress problem using the computed compaction as a boundary condition. Dilation (increase of porosity) is an important geomechanical mechanism occurring during steam injection into unconsolidated sands. This process has been also modeled by the "compaction-dilation" tables.[126] In chalk reservoirs, increased water saturation in waterflooding causes weakening of the rock and therefore S_w is an additional variable in Eq. 17.32.[127]

The major drawback of the use of compaction tables is that the dependence on stress indicated in Eq. 17.32 must be either ignored, or the change of stress must be estimated in terms of change in pressure. A more accurate modeling approach is to couple in some fashion the reservoir simulator with stress-strain (geomechanics) solution. Such models typically combine the solution of the multiphase flow in the reservoir and elastoplastic solution of the deformations in a much larger domain including the reservoir, sideburden, underburden, and overburden. The majority of coupled models use the iterative coupling[128]; the different variations and their shortcomings are discussed next.

Coupled models have much larger computing requirements compared with a reservoir model of the same reservoir, primarily for two reasons: first, the larger number of unknowns per gridblock, and second, the stress-solution grid must be usually much larger laterally than the reservoir grid to eliminate the effect of boundary conditions as well as extend up to surface to provide subsidence solution. Combination of these factors leads to computing times typically of one order of magnitude larger compared with conventional simulation, and even more if elastoplastic solution is required for the stresses.

17.6.3 Modeling of Stress-Dependent Flow Properties. The primary flow-dependent property is permeability, and the problems to model its dependency in Eq. 17.33 are similar to modeling the pore-volume coupling (Eq. 17.32). However, the problem is somewhat easier because stress-dependent permeability (or transmissibility) does not affect mass-balance formulation.

Again, the traditional approach is to use tables of k vs. pressure in an uncoupled model. However, the problem remains one of replacing the dependency on effective stress by one on pressure only. Even in a single-phase, single-porosity gas-flow case, different assumptions about the stress change during depletion can lead to large errors in well decline.[129] Different strategies for converting the stress-dependent data to pressure tables are based on local constrainment assumptions.[130]

In coupled models, the permeability dependency can be usually computed explicitly on a timestep basis, and "loose" coupling can be used. In fact, a "coupled" model that deals only with flow properties coupling and ignores the pore-volume coupling can be run successfully even if the stress solution is done in larger intervals of time compared to the reservoir. Such models have been used extensively to study permeability changes in waterfloods, particularly in fractured or jointed media.[131,132] Here, the advantage of coupled modeling is in its capability to predict the permeability changes from the geomechanics of reopening of fractures or failure (dilation) of joints.[133] The development of anisotropy is dictated by the orientation of fractures or faults, and requires a "full tensor" treatment of transmissibilities in the flow model. In the stress strain model, different methods of pseudoizing the fracture/joint networks into a continuum are used, which include predicting permeability as a function of effective stress and/or strain. While the need for the tensor transmissibilities in such models has been recognized,[134] in injection processes dual-porosity media can be created. Therefore, reservoir description may be changing in time because of geomechanics; this aspect has been ignored in coupled models to date.

The same principles can be also applied to model hydraulically induced fractures being represented by dynamically changing transmissibility multipliers in the potential fracture plane.[135] The effective stress dependency (as opposed to pressure in an uncoupled model) allows capturing the changes of the fracture propagation pressure with time, which can be large, in particular in steam injection. Another application is the prediction of production/injection-induced slippage on faults, which can induce communication between reservoir fault blocks and/or seismicity.

17.6.4 Types of Coupled Models. Coupled models can be either fully coupled (i.e., all unknowns solved simultaneously) or modular (reservoir simulator and stress code). In the latter case, different coupling strategies can be used, with consequences for running speed and accuracy. The majority of coupled models use a conventional finite-difference reservoir simulator coupled with a finite-element (FEM) stress simulator. However, attempts have been made to develop fully coupled FEM codes,[136] and a fully coupled geomechanics was implemented in a commercial model using a finite-difference stress solution.[137] Considering the proliferation and sophistication of the geomechanics codes available outside the petroleum engineering, the modular approach offers the best solution.[119,127] Generally, the reservoir simulator is the "host" or "master." Commercial stress simulators are, in principle, easy to couple to it (in particular if only permeability coupling is considered).

Because of the extreme complexity and large computing requirements of coupled models, different simplifications have been developed. The main types of models (in the order of increasing rigorousness, but also computing time) are as follows:

One-Way Coupling. Pressure and temperature changes are passed from the reservoir code to the geomechanics module, but no information is passed back on timestep basis. The geomechanics does not improve the flow solution, but the model can be useful for predictions of wellbore stability for infill drilling, fracturing pressures, and so on. Manual adjustment of "compaction tables" is possible manually through restarts. Such manually coupled solutions[138] of the stress problem (at intervals of time) or one-way coupling[139] were often used in early coupled modeling. The method can be satisfactory when the reservoir fluid system is highly compressible (i.e., in gas reservoirs), but can lead to errors when the porosity is strongly coupled to flow.

Loose Coupling. Reservoir and geomechanics modules are run sequentially on a timestep basis, passing converged solutions of flow and stress variables to each other. Pore volumes and permeabilities in the flow model are computed as a function of p, T, and σ_{avg} with stress variables lagged a timestep. However, the relationships are "distilled" into tables similar to the "compaction tables," but now a function of effective stress. The advantages are functional similarity to the uncoupled reservoirs with "compaction tables," no need for iteration during timestep, and the possibility of updating the stress solution less frequently than the reservoir solution. However, complex constitutive models of the solid (e.g., plasticity) may be difficult to represent.

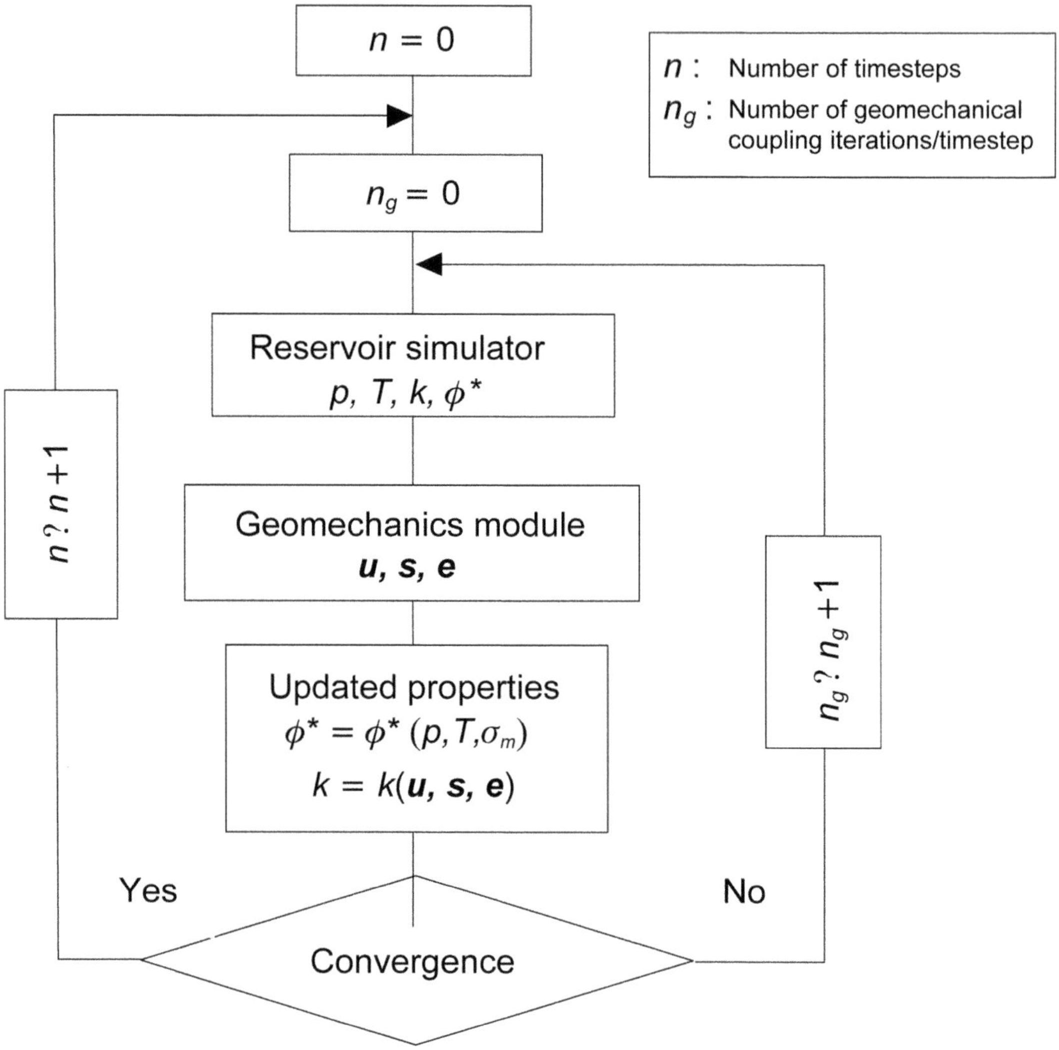

Fig. 17.34—Schematic of the iterative coupling algorithm.

Iterative Coupling. This method is shown schematically in **Fig. 17.34**. Iteration is carried out between the reservoir and stress solution at every timestep until the pore volumes and permeabilities calculated from the stress model and those used by the reservoir model agree. In each iteration, the previous guess of the V_p^{n+1} for the end of the timestep is used to converge the flow solution, and the changes of p and T over timestep are then used to solve for the new deformations and stresses, which in turn provide updated estimate of V_p^{n+1}. The changes of permeability are also iterated on. Therefore, each "geomechanical" iteration costs the equivalent of a timestep solution of the previous methods. The original formulation of the coupling iteration[128] is always convergent, and its efficiency has been recently improved.[140]

The iterative coupling, when converged, is equivalent to a fully coupled code while it retains flexibility. In many problems, it is not necessary to fully converge the timestep, and if used with only one iteration per timestep, the method is similar to the loose coupling, except that the porosity is determined directly by the constitutive model of the solid rather than by a simplified relationship.

Full Coupling. This requires simultaneous formulation of the flow and stress variables and therefore results in larger matrices. The advantage is that consistent approach to discretization can be used, and the model is integrated from the point of view of code development. However, it is very costly to redevelop all the features of the physics and numerics now readily available in stress codes. Moreover, in published fully coupled models, the approach for solving the resulting matrix problem has been to partition the matrix in the same fashion as in the iterative coupling[141] and to apply the geomechanical iteration at the matrix-solution level. Therefore, the fully coupled formulation, which results in larger, strongly nonlinear matrix equations, does not reduce the difficulty of the problem, and it may need to use geomechanical iteration in the solution process as the best strategy. These aspects need further study.

17.6.5 Future Trends and Needs. As a result of much larger computing requirements, coupled models lag behind conventional models in the size of the problems that can be currently handled. Therefore, they are a prime candidate for the use of massively parallel hardware and will require large future development effort in parallelization. Because of the increased understanding of the complexities of the geomechanics, the current trend is also toward more strongly coupled models with fewer simplifications. This further increases the computing requirements.

Given that not all problems require use of geomechanics, and the cost of the study may increase dramatically, it is important to be able to identify when coupled simulation is needed, and what approximations can be made without compromising the answers. There are no simple rules, but there is a growing need to conduct a "screening" process at an early stage of a reservoir study to determine if geomechanics is an issue. This process requires an integration of reservoir, production, and completion engineering data as well as field experience.

Finally, coupled geomechanical modeling is the future tool for truly integrated reservoir management. Conventional reservoir simulation studies ignore numerous constraints placed on the development scenarios from the point of view of drilling, completion, and operations. These constraints can be incorporated into coupled models, and additional modules can be integrated (e.g., long-term wellbore stability and sand production predictions, subsidence management, 4D seismic interpretation, and so on[142]).

17.7 Pressure/Volume/Temperature (PVT) Treatment—Curtis H. Whitson

The PVT treatment of fluids in reservoir simulation describes the phase behavior of gas, oil, and water at reservoir and surface conditions. Phase behavior of a mixture with known composition consists of defining the number of phases, phase amounts, phase compositions, phase properties (molecular weight, density, and viscosity), and the interfacial tension between phases. In addition to defining the phase behavior of mixtures at a specific reservoir pressure, knowing the derivatives of all phase properties with respect to pressure and composition is important in reservoir simulation.

The calculation of phase behavior in a reservoir model can be made in one of two ways—using a "black-oil" approach[22,143–146] based on simple interpolation of PVT properties as a function of pressure, or using a "compositional" approach based on a thermodynamically-consistent model such as a cubic EOS.[14,15,22] With either approach, the PVT quantities required by a reservoir simulator are essentially the same. Modern reservoir simulators[14,22] are usually written with a general compositional formulation, whereas black-oil PVT properties are converted internally to a two-component "compositional" model; the two components are surface gas and surface oil.

A reservoir simulator keeps track of overall composition in each computational grid cell as a function of time. A grid cell's overall composition changes because of cell-to-cell fluxes and because of production or injection in wells. The phase fluxes and component movement within the reservoir are greatly affected by phase behavior (e.g., the mobility of each phase and which components are carried in each phase). The surface products from wells are dependent on the phase behavior of produced wellstreams, but at conditions far removed from reservoir pressure and temperature.

In most reservoir simulators, the water phase and water component are treated in a simplified manner—namely, that the water component does not partition into the hydrocarbon phases, and the hydrocarbon components do not partition into the water phase; the term "hydrocarbon" also includes N_2, CO_2, and H_2S. Because of relatively high CO_2 solubility in water, and the potential importance of CO_2 "accounting" in CO_2 floods, some compositional models allow CO_2 partitioning in the water phase.[147]

Conceptually and computationally, it is feasible to allow complete partitioning of all components in all phases in reservoir simulation. The partitioning related to the water phase and the water component could be treated with simple pressure-dependent tables, or even with an EOS model. The water-related *K*-values have simple composition dependence that would make EOS-based fugacity updates almost trivial. The main problem would be treating the impact of changing salinity on water-related *K*-values. From a practical point of view, however, modeling water-related component partitioning will have a marginal impact on reservoir performance. Hereafter, only phase behavior of nonaqueous phases will be discussed.

17.7.1 Number of Phases and Phase Type. The starting point for PVT calculations in a model grid cell is to determine if the overall composition is single-phase or two-phase at the current pressure estimate. If a cell is single-phase, the phase "type" (gas or oil) may also be needed to select the proper relative permeability and capillary pressure curves. For a black-oil PVT model (i.e., any model with simple pressure-dependent *K*-values), determining the number of phases and phase identification is trivial.

For EOS models, two methods can be used for establishing how many phases exist: the Michelsen[148] phase stability test or a saturation pressure calculation.[148] The stability test is relatively "slow" because good *K*-value estimates are not used to initiate the calculation, whereas a saturation pressure can be quite fast because good *K*-value estimates are often available. A stability test is more reliable in the sense of giving a conclusive result, whereas a near-critical saturation pressure test may not be found[149] or converge correctly to a nontrivial solution. Another advantage of using saturation pressure is that it gives a consistent method to define phase type for single-phase cells, something that is not provided by the stability test. The choice of which method to use depends on the tradeoff between speed and reliability. Both methods, if detecting two phases, give excellent starting *K*-values to initiate the two-phase flash calculation.

17.7.2 Two-Phase Flash. Having established that two phases exist in a cell, one must perform a flash calculation. The flash calculation uses the overall moles n and molar composition z_i to determine the molar amounts of each phase (n_g and n_o), and the phase compositions (y_i and x_i). For simple pressure-dependent *K*-value models, the Rachford-Rice[150,151] (RR) procedure is used to find molar amounts and compositions,

$$\sum_{i=1}^{N}(y_i - x_i) = \sum_{i=1}^{N}\frac{z_i(K_i - 1)}{1 + \alpha_g(K_i - 1)} = 0, \qquad (17.34)$$

solving for $\alpha_g = n_g / (n_g + n_o)$, with $x_i = z_i / [1 + \alpha_g(K_i - 1)]$ and $y_i = K_i x_i$.[152]

For an EOS model, using the RR solution can be used, but an additional constraint must be satisfied; the fugacity of each component must be the same in each phase,[153] $f_{gi} = f_{oi}$. Using an initial set of K_i estimates, the RR equation is solved. Resulting compositions y and x are then used to calculate fugacities f_{gi} and f_{oi}. If the constraint $f_{gi} = f_{oi}$ is not satisfied for all components, a new set of K-values is estimated using some form of a successive substitution (SS) update $K_{inew} = K_{iold}(f_{oi} / f_{gi})$.

For reservoir simulation, a Newton solution of the flash equation should be much faster than SS methods. Including the equal-fugacity constraint within the set of nonlinear equations used to solve for model pressures "automatically" provides the first Newton update of the flash equation.[153] Usually only one additional Newton iteration of the flash equation is needed to converge the fugacity constraints to required tolerance.

In summary, K-values alone determine the phase amounts and phase compositions (from the Rachford-Rice equation). The EOS model guarantees rigorous thermodynamic consistency through the equal-fugacity constraint, ensuring that K-values properly account for pressure and composition dependence of the phase split and component partitioning.

Having completed the flash calculation in a cell, the following information is known: phase moles, phase masses, phase densities, phase volumes (saturations), and phase compositions. The viscosity and gas/oil IFT can be calculated using compositional correlations,[152] or interpolated from pressure-dependent tables for a black-oil model. The impact of PVT properties on reservoir performance will consider individual PVT properties and their impact on individual terms in the flow equations.

17.7.3 Density. Phase molar amounts, n_g and n_o, are converted to phase volumes using phase molecular weights and densities: $V_g = n_g M_g / \rho_g$ and $V_o = n_o M_o / \rho_o$, where $S_g = V_g / V_{\text{block}}$ and $S_o = V_o / V_{\text{block}}$. Saturations determine relative permeabilities, which can have a dramatic impact on phase fluxes. Typically, $k_r \sim S^n$, where $n \sim 2$ to 4. For $n = 3$, a 5% error in density results in a 15% error in k_r.

For reservoirs with sufficiently high vertical permeability, gravity will often play an important role in recovery. Gravity segregation is caused by vertical fluxes driven by the potential terms $\nabla \rho_g = d(p_g + \rho_g gz + P_{cgo}) / dz$, $\nabla \rho_o = d(p_o + \rho_o gz) / dz$, and $\nabla \rho_w = d(p_w + \rho_w gz + P_{cwo}) / dz$. Errors in densities have a direct impact on gravity segregation. It is actually the density difference between flowing phases that determines the magnitude of gravity segregation (i.e., $\Delta\rho_{wo} = \rho_w - \rho_o$, $\Delta\rho_{go} = \rho_o - \rho_g$, and $\Delta\rho_{wg} = \rho_w - \rho_g$). Likewise, the condition for equilibrium initialization $\nabla \rho_g = \nabla \rho_o = \nabla \rho_w = 0$ results in saturation-depth distributions, which must honor input $\Delta\rho_{go} g = P_{cgo}(S_g)$ and $\Delta\rho_{wo} g = P_{cwo}(S_w)$ relationships.

A +5% error in oil density (e.g., 630 instead of 600 kg/m^3) and a –5% in gas density (e.g., 380 instead of 400 kg/m^3) results in gravity segregation potential ($\Delta\rho_{go} = \rho_o - \rho_g$) in error by 25% (250 instead of 200 kg/m^3). Similarly, in the initialization of a water/oil system, the inadvertent use of $\rho_w = 62.5$ lbm/ft^3 instead of the correct value of 70.0 lbm/ft^3 for a high-salinity brine could easily cause a 25% error in initial oil in place for a low-permeability (large capillary transition zone), low-API oil.

In black-oil models, it is particularly easy to model densities incorrectly because of the inadvertent use of "arbitrary" surface densities. The reservoir phase densities are not input directly in a black-oil model. Instead, they are calculated using the formation volume factors (FVF) B_o, B_g, and B_w, solution gas/oil and oil/gas ratios R_s and r_s, and surface densities ρ_{gs}, ρ_{os}, and ρ_{ws}, based on the relations $\rho_g = (\rho_{gs} + \rho_{os} r_s) / B_g$, $\rho_o = (\rho_{os} + R_s \rho_{gs}) / B_o$, and $\rho_w = \rho_{ws} / B_w$. Because reservoir densities are not usually tabulated as output by reservoir simulators, the user may not know how "wrong" the densities might be.

A well-tuned EOS using volume shift factors[154] should always predict reservoir and surface densities within 2 to 4%, and often within 1 to 2%. The author's experience has found it generally unnecessary to use different volume shift factors for reservoir and surface calculations unless all densities are needed with accuracies of 1 to 2%. The use of either Peng-Robinson[155] or Soave-Redlich-Kwong[156] EOS with volume-shift factors provides densities as good or better than the more-complicated ZJRK[157] EOS model; in this author's opinion, the ZJRK no longer has a role in reservoir simulation, as supported by decreasing industrial use (almost to nonexistent) during the past 5 years.

17.7.4 Component Partitioning. The partitioning of components in gas and oil phases, as dictated by *K*-values, is important for properly describing vaporization and condensation. Vaporization is the process of "stock-tank oil" (STO) components (C_{6+}) moving from the oil phase to the gas phase, while condensation is the process of intermediate (C_3–C_{10+}) components moving from the gas phase to the oil phase. Retrograde condensation is particularly important in the depletion of gas condensate reservoirs, where in-situ retrograde condensation results in decreasing surface liquid yields.[22,152,158]

For most gas-injection projects, vaporization will play an important role. Proper modeling of vaporization usually requires a compositional model that uses a detailed description of the heavier (C_{7+}) components, typically three to five fractions.[22,159,160] The lightest fractions (C_7 to C_{12}) will typically vaporize 80 to 100%, while intermediate heavies (C_{13} to C_{20}) may vaporize in varying amounts from 30 to 90%, and the heaviest components C_{20+} may vaporize from 0 to 50%. The degree of vaporization depends primarily on the local pressure and composition of displacing gas. Cook *et al.*[144] suggested a modification of the black-oil model that allows saturated properties to change as a function of how much injection gas has contacted a given cell, thereby allowing reasonable description of a vaporization-dominated process. Tang and Zick[161] propose a limited three-component compositional model that allows accurate description of developed miscibility for use in models in which grid-related dispersion may result in underpredicted conditions of miscibility using an EOS-based simulator.

In some gas-injection processes, the injection gas may be enriched (1) during injection by adding NGLs (C_3 to C_5 components) and/or (2) within the reservoir by multiple contacts with reservoir oil, which gradually enriches the injection gas with intermediate components C_3 through C_{10}. For either type of enrichment, the vaporization process can become extremely efficient and result in a near-miscible or miscible displacement with near-100% recovery of the oil. When such a displacement results, it is often associated with the development of a complex process whereby the near-miscible front consists of upstream vaporization and downstream condensation—the "condensing/vaporizing" gas drive mechanism first described by Zick.[162]

For an EOS model to properly describe complex phase behavior related to vaporization, condensation, and near-miscible mechanisms, special PVT experimental data should be measured and used to tune the EOS. For immiscible processes with significant vaporization, a single-cell multicontact PVT test is useful, quantifying the degree of vaporization in terms of oil volumetric stripping and gas compositional changes. For near-miscible or miscible processes, a swelling-type test which has the following features is strongly recommended: (1) five to seven mixtures of reservoir oil and injection gas are used, with two to three mixtures being bubble-points and two to three mixtures being dewpoints; (2) a constant composition expansion is conducted on each mixture, where saturation pressure and the oil relative volumes are measured and reported; and (3) a single equilibrium "flash" somewhat near the critical point is used for obtaining a set of equilibrium gas and oil compositions (*K*-values).

Proper EOS tuning of the complex phase behavior measured in such an experiment is difficult, and requires the ability to match near-critical volumetric and compositional changes. Such predictions are almost never available *a priori* with an EOS model that has been tuned only to simple depletion data. Small but important modifications of the heavy-end properties and binary interaction parameters are usually required to obtain a satisfactory match of near-critical PVT data provided by this special "swelling" test. A tuned EOS model that is able to match all PVT data, including near-critical phase behavior, has a good chance of properly predicting true multicontact near-miscible/miscible behavior.[152,162]

17.7.5 Viscosity. Gas viscosities are typically estimated by correlation[152] within 5 to 10%, and they are almost never measured experimentally. Such accuracy is adequate for most applications, and gas viscosities seldom vary beyond the range of 0.02 to 0.03 cp. For gas-injection processes at high pressure or near-miscible displacements, gas viscosities can range from 0.03 to 0.1 cp. Gas-viscosity correlations are not usually accurate for this higher range of viscosity, and errors up to 20 to 30% may be expected; compositional viscosity correlations also have the same level of accuracy.

Oil viscosities are notoriously inaccurate based on correlations,[152] at best being within 10 to 30%, but often in error by 50% or more. Oil viscosities should always be measured and used to tune a viscosity model. A minimum requirement would be measurement of stock-tank oil viscosities, and normally live-oil viscosities are available from a differential liberation test.

For gas condensates, oil viscosities are almost never measured. This may be a serious problem if condensate blockage has a significant impact on well deliverability.[163] Condensate viscosities are difficult to measure because retrograde condensate volumes can be very small. The use of a separator condensate sample is recommended. Bring it to reservoir conditions (T and P) for density and viscosity measurements, then tune the viscosity model to these data. This approach is feasible for any gas condensate; it is not expensive, and it is better than relying on untuned compositional viscosity correlations.

Our experience has been that the compositional viscosity correlation by Lorenz, Bray, and Clark[164] (LBC) is not predictive, and it is highly dependent on accurate density predictions. Tuning the heavy-component critical volumes and, sometimes, careful modification of the higher-order constants in the LBC equation, provide the required accuracy for gas and oil viscosities in most reservoir processes. Unfortunately, modification of the LBC correlation to match a limited number of viscosities (in a narrow range of pressure and composition typical of depletion experiments) can lead to unphysical viscosity predictions at conditions developed during a gas injection process.

The Pedersen *et al.*[165] compositional viscosity correlation, though more complicated and CPU-intensive than the LBC correlation, has quite good predictive capabilities. The correlation is based on a corresponding states formulation in which methane viscosity is the reference compound.

17.7.6 Gas/Oil Interfacial Tension (IFT). Gas/oil IFTs may be used in a reservoir simulator[13] to modify input capillary pressures P_{cgo}, relative permeabilities k_{rog} and k_{rg}, and residual oil saturation S_{org}. As IFT decreases, P_{cgo} decreases proportionately. Relative permeabilities and residual oil saturation change only marginally until IFT reaches a fairly low value (e.g., 1 mN/m), whereas at lower IFTs, the relative permeability-related properties change approximately as log(IFT).

Practically, IFT impact on capillary pressure is limited to fractured reservoirs[166] in which displacement efficiencies may be strongly linked to the balance of gravity and capillary forces on a relatively small distance scale (e.g., block heights of 1 to 10 m).

IFT impact on relative permeabilities and residual saturations may have an impact on some gas-injection processes, though near-miscible and miscible processes have minimal IFT dependence because they are usually dominated by the strong compositional effects (vaporization and condensation) that result in near-100% recoveries. In fact, displacements that are miscible should, by definition, be independent of the relative permeabilities and residual oil saturation.

IFT impact on near-well relative permeabilities in gas condensate reservoirs can, together with high velocities, result in large capillary numbers, which have the tendency to "straighten" relative permeabilities and improve flow performance. It has been shown[167,168] in a number of recent publications that this effect can have an important impact on gas-condensate well deliverabilities. Neglecting IFT and velocity dependencies of relative permeability can lead to overly conservative prediction of well deliverability (i.e., overprediction of condensate blockage).

17.7.7 Black-Oil PVT Models. Black-oil PVT properties are generated in one of two ways. For low- to medium-GOR oils (< 150 Sm^3/Sm^3), a traditional differential liberation experiment (DLE) is used, with corrections for separator flash to calculate oil FVF B_o and solution GOR R_s, as well as the gas FVF B_g.[152] This approach assumes the reservoir gas contains unsubstantial amounts of condensate in solution, with solution oil/gas ratio $r_s \sim 0$.

The more common and general approach to generating black-oil PVT properties uses an EOS model to simulate a depletion-type PVT experiment (differential liberation, constant volume depletion, or constant composition expansion), with the equilibrium gas and equilibrium oil at each stage in the depletion being individually processed to surface conditions to provide the four black-oil properties B_o, R_s, B_g, and r_s.[22,156] For highly volatile oils, the EOS method gives substantially different and improved black-oil properties compared with the traditional laboratory DLE/separator-corrected approach.

The conversion of black-oil PVT data (R_s, B_o, r_s, and B_g) to a compositional model[144] uses K-values of surface gas and oil pseudo "components" $K_{gs} = (R_s + C_{os}) \ / \ (1 + r_s C_{os}) \ / \ R_s$ and $K_{os} = r_s(R_s + C_{os}) \ / \ (1 + r_s C_{os})$, with $C_{os} = (RT_{sc} \ / \ p_{sc})(\rho_{os} \ / \ M_{os})$. The reservoir-phase densities are calculated from $\rho_g = (\rho_{gs} + \rho_{os} r_s) \ / \ B_g$ and $\rho_o = (\rho_{os} + R_s \, \rho_{gs}) \ / \ B_o$, while phase molecular weights are given by $M_g = (M_{gs} + r_s M_{os} C_{os})/(1 + r_s C_{os})$ and $M_o = (R_s M_{gs} + M_{os} C_{os}) \ / \ (R_s + C_{os})$. Viscosities and gas/oil IFTs are interpolated directly from input tables.

Coats *et al.*,[22] Coats *et al.*,[143] and Fevang *et al.*[145] have shown that black-oil models can be used for practically any type of reservoir produced by depletion or waterflooding, including reservoirs with large compositional gradients. Some issues require special treatment for complex fluids systems, including fluid initialization and the method for generating black-oil tables. In a nutshell, the recommended procedures[145] are to generate the black-oil tables with an EOS model using the fluids with the highest saturation pressure (e.g., at the gas/oil contact) and to initialize with solution GOR (R_s and r_s) vs. depth—not saturation pressure vs. depth.

A common problem with black-oil models is the calculation of "negative compressibility,"[23] meaning that a small pressure drop results in a reduction in total (gas + oil) volume. Another problem is physical extrapolation of saturated PVT properties to saturation pressures higher than given in the input table (e.g., caused by gas injection, gravity segregation in undersaturated reservoirs, or near-well behavior during rate reductions).

When can the black-oil PVT treatment not be used? Basically, for any gas-injection process with significant component partitioning that changes during the displacement.[145,159] This would include processes with high-pressure vaporization using lean gas, condensation from enriched injection gas, and developed-miscibility processes such as the condensing/vaporizing mechanism. Surprisingly, a black-oil treatment is sometimes adequate even for complex gas injection problems, though it is not usually known *a priori.* To check the validity of a black-oil model in a gas injection project, the reservoir process should first be simulated with a compositional model, and preferably a relevant 3D model that captures all important flow characteristics.

17.7.8 Equation-of-State Models. The most common EOS used in reservoir simulation are the PR and the SRK models. Both models have two constants, a and b. Each constant must be calculated for each component based on component critical properties (T_c and p_c) and acentric factor (ω).

The PR EOS has two versions—the original 1976 version[155] and the modified 1979 version[169]; the latter uses a third-degree polynomial expression for the correction term to constant a. For some systems, the difference in equilibrium calculations for the two PR EOS models is significant.

The Peneloux[154] volume shift factors should always be used with two-constant EOS models to ensure accurate oil and gas densities. The volume shift factors have no impact on calculated *K*-values or compositions, only densities. As mentioned earlier, the ZJRK[157] EOS is outdated and was used before the volume-shift method was introduced in 1980, with complex correction functions to constants *a* and *b* to improve liquid density predictions.

Binary interaction parameters (BIPs) k_{ij} are important for adjusting predictions of equilibrium properties (*K*-values and compositions).[170] These parameters represent a correction of the form $(1 - k_{ij})$ to the $a_i a_j$ term in the quadratic mixing rule for EOS constant *a*. BIPs can be positive or negative; they are symmetric ($k_{ij} = k_{ji}$); they are usually ~0 for most hydrocarbon-hydrocarbon pairs, except C_1 to C_{7+} pairs which may reach values as high as 0.25; and they are generally close to 0.1 for nonhydrocarbon (N_2, CO_2, H_2S)-hydrocarbon pairs.

17.7.9 Three-Phase PVT Behavior. Three-phase (L1-L2-V) behavior is an occasional but serious problem for EOS-based compositional models. The third phase (L2) is usually a light liquid and typically appears at low temperatures (< 140°F) in gas-injection processes using CO_2 or NGL-enriched gas.[171] Physically, three phases may actually exist, and the EOS model is correctly predicting the behavior. Sometimes a three-phase system is predicted without one physically existing; this may result for lower temperatures when the heaviest component properties are improperly modified to fit two-phase gas/oil PVT data.

For reservoir simulators, the three-phase problem is caused by the EOS formulation "allowing" only two phases. If three phases actually exist, the two-phase flash may find any of the three possible two-phase combinations: L1-V, L2-V, or L1-L2. These false two-phase solutions may indicate a single-phase condition, or they may actually appear as a valid solution (meeting the equal fugacity constraints). Unfortunately, the reservoir model, in a given cell during a given timestep, may flip-flop between two of the possible solutions, resulting in unstable behavior because the pressure solution is not continuous from one two-phase solution to the other. Models may have to simply give up because of repeated timestep reductions, which result from the inadequacy of the EOS two-phase model handling a three-phase condition.[172]

17.7.10 Surface Phase Behavior. In compositional simulation, the surface calculations are usually made using multistage separation with an EOS model, with fixed *K*-values for each separator, or using so-called "gas plant" factors, which define the fraction of a wellstream component that is processed into the stock-tank oil.

For black-oil models, the surface separation is "built in" to the PVT tables. Consequently, if the surface process changes during a model run, all black-oil PVT tables must be reentered at the appropriate time. This also requires that vertical flow performance (VFP) tables be reentered because surface rate and volume ratio nodes change with the process change.

It is difficult to use traditional black-oil models for fields with various well groups that have significantly different processing facilities.[143]

17.7.11 Thermal Model PVT Requirements. Additional PVT requirements for thermal processes such as steamflooding include quantifying the temperature dependence of *K*-values, densities, and viscosities. Water-phase behavior of liquid and steam must also be defined in terms of pressure and temperature. Water-hydrocarbon phase behavior is still assumed to be simple, without water/hydrocarbon component partitioning.

An EOS model can be tuned to distillation data for improving the predictive capabilities of *K*-value dependence; otherwise, a simple correlation of the form $K_i = a_i \exp(-b_i T) / p$ may be used for distillable components.[173] Using distillation data is an indirect approach to defining *K*-value behavior, and it is used in lieu of high-temperature gas/oil/water phase-behavior experiments, which are not usually available. Oil viscosities in thermal processes may be difficult to correlate with a compositional correlation, so empirical correlations may be used instead.

17.7.12 Fluid Initialization. As with rock and other petrophysical properties such as permeability and porosity, a reservoir simulator model must also initialize the spatial distribution of initial fluids. For an EOS-based model, the initial molar compositions are defined, z_i(x,y,z). For a black-oil model, the initial solution gas-oil ratio R_s and solution oil-gas ratio r_s ratio are defined, R_s(x,y,z) and r_s(x,y,z); sometimes saturation pressures are used instead, p_b(x,y,z) and p_d (x,y,z), but this is not recommended.[26] Specifying a saturated gas-oil contact (GOC) is also a means to initialize fluids vertically in a reservoir simulator, where solution GOR R_s (and bubblepoint) are assumed constant below the GOC, while solution OGR r_s decreases upwards from the GOC to honor the model-imposed assumption that reservoir pressure equals dewpoint pressure, $p(z) = p_d(z)$.

Another fluid initialization data might include temperature T(x,y,z). Some black-oil models allow spatial variation of stock-tank oil density, γ_{API}(x,y,z), where black-oil properties are correlated in multidimensional tables as a function of pressure and γ_{API}. Across barriers such as faults and sealing shales, discrete PVT model data may be defined, such as EOS parameters or black-oil tables; such "discontinuous" fluid descriptions may cause physical and model incompatibilities if fluids mix in the reservoir or at the surface.

A typical problem with initialization is that the specified fluid distribution, initial pressure specifications, and fluid contacts lead to fluid movement when the model is turned on (without production or injection). Initial fluid movement may be unimportant without significantly changing the user-specified fluid system; serious inconsistencies may lead to time-zero flow that has an impact on model performance.

17.8 High-Performance Computing and Reservoir Simulation—John E. Killough

The motivation for high-performance computing in reservoir simulation has always existed. From the earliest simulation models, computing resources have been severely taxed simply because the level of complexity desired by the engineer almost always exceeded the speed and memory of the hardware. The high-speed vector processors such as the Cray of the late 1970s and early 1980s led to orders-of-magnitude improvement in speed of computation and led to production models of several hundred thousand cells. The relief brought by these models, unfortunately, was short-lived. The desire for increased physics of compositional modeling and the introduction of geostatistically/structurally based geological models led to increases in computational complexity even beyond the large-scale models of the vector processors. Tens of millions of cells with complete reservoir parameters now became available for use by the engineer. Although upscaling provided a tool to dramatically reduce model sizes, the inherent assumptions of the upscaling techniques left a strong desire by the engineer to incorporate all of the available data in studies. The only available solution to this problem became the subdivision of the model into small segments and the use of parallel computers for reservoir simulation. The recent introduction of fast, low-cost commodity hardware based on the INTEL architecture has led to a revolution in higher-performance computing based on clusters.

17.8.1 The Explosion of Data and the Chasm of Scale. From the earliest reservoir simulation models of the 1960s with only a few tens of finite-difference cells, reservoir simulation had progressed by several orders of magnitude in the early 1990s to hundreds of thousands of cells. Unfortunately, these fine-scale models still were far from the scale of the data. This chasm of scale was further exacerbated by the introduction of geocellular models with geostatistically derived attributes. With these geostatistically based models, it was possible to generate geological descriptions for the reservoir simulation with literally tens of millions of finite-difference cells. Although upscaling offered some relief to this problem, the resultant assumptions required often left the engineer wondering how far from reality the resultant solutions had been driven. If the additional degree of freedom of uncertainty of the reservoir data is added to this problem, then the number and size of simulations becomes unlimited. This implies that the need for further improvements in high-performance computing will always exist for reservoir simulation. Parallel computing offers one approach to overcome the problem of the explosion of information through the use of fine-scale models with limited or no upscaling.

17.8.2 Characteristics of High-Performance Computing. The basis of high-performance computing is the use of specialized hardware to achieve computational speeds that are much faster than conventional computers. This idea of "speedup" is encapsulated in what is known as Amdahl's Law:

$$\text{Speedup} = 1/(\text{scalar} + \text{special}/n),$$

where "scalar" is the fraction of the program which is scalar and "special" is the fraction of computations performed on specialized hardware (one minus scalar). Simply stated, Amdahl's law indicates that if there is a piece of specialized hardware that is n times faster than the scalar processor(s) in a computer or group of computers, it must be used a large fraction of the time to have a significant impact on overall performance. The earlier supercomputers primarily used specialized hardware known as vector processors for the speedup. More recently, parallel computers based on connecting tens to hundreds of processors have formed the basis of high-performance computers. The importance of the serial or single-processor computer cannot be overemphasized, however. The introduction of the reduced instruction set computer (RISC) with "superscalar" performance in 1990 spelled the end of the dominance of the Cray supercomputer in the high-performance reservoir-simulation market because of a rapid decrease in the price/performance of the new hardware. Similarly, the recent introduction of low-cost, commodity hardware based on the INTEL chipset has led to significant improvements in price/performance. The application of this hardware to reservoir simulation has been a natural evolution in the application of the lowest-cost, highest-performance computers to reservoir simulation.

17.8.3 The Parallel Reservoir Simulator. The earlier high-performance computing hardware such as the Cray Computer required that modifications be made to the reservoir simulator programs to efficiently use the hardware.[174,175] The modifications known as vectorization led to different approaches in the data organization. One of the best examples of this is given in Refs. 176 and 177, in which a compositional model's data structure is reorganized so that like phases in the finite-difference gridblocks of the reservoir are grouped. The notion behind parallel reservoir simulation is similar; however, in this case, the reservoir grid is subdivided into 3D blocks or domains. The computational work for each of the domains is then performed in parallel by a separate CPU of one of the computers in the parallel hardware. Profiles of computing workload for a typical reservoir simulation model often show tens, if not hundreds, of subroutines that are involved in a substantial portion of the calculations. Because of this, major reprogramming of reservoir simulation models is required to achieve high parallel efficiencies. As Ref. 178 points out, there are numerous obstacles to parallelization for reservoir simulation: recursive nature of the linear equation solver, load imbalance caused by reservoir heterogeneity and/or physics, and diverse data structure of well and facility management routines.

Several papers[179–189] discuss techniques that have been used to bring about efficient parallel reservoir simulations. Each of these addresses the solutions to parallelization and, in particular, the obstacles mentioned previously in various ways. One of these simulators uses local grid refinement for parallelization.[179] The main concept of parallelization with local grid refinement (LGR) is that LGR is not necessarily used to add additional cells to the model but simply to facilitate the subdivision of the grid. With LGR, the same simulation program can be used to perform simulations either serially on a single processor or in parallel on multiple processors simply through data manipulation. In addition, a great deal of flexibility for domain decomposition exists, which can lead to enhanced load balance for parallel simulations.

Parallelization using LGR involves assigning each grid to a processor. The same processor can be assigned to many grids or in the limit; each grid can be assigned to a separate processor. Variation of processor assignment and the grid refinement can dramatically affect parallel performance. Different preconditioners for the parallel linear equation solvers can also have a dramatic effect on parallel efficiency. Finally, the flexibility of LGR and processor assignment can be used to achieve improved parallel efficiency. For parallel domain decomposition, the base grid may be totally refined into subgrids so that all gridblocks in the base grid become inactive. Each grid is assigned to a processor so that only a portion of the model resides on a given processor. In this manner, the simulator memory requirements become scaleable. That is, as the number of processors is increased, the grid assigned to a given processor becomes smaller. Alternatively, only a portion of the base or root grid may be refined; the remaining unrefined cells in the root grid are then assigned to a processor. Because of this use of local grid refinement, only an extremely small number of global arrays are required on each processor —for example, the pore-volume array.

The existence of horizontal wells in most modern simulations requires that no restriction be placed on the placement of wells relative to grid and/or processor boundaries in a parallel simulator. To alleviate any difficulty with this, one approach is for well information to be passed by a broadcast among all processors.[179] The overhead for broadcasting completion level information for all wells is included in the following results and appears to have little overall effect on parallel efficiency. Tubing hydraulics calculations for a predictive well-management routine have also been parallelized. These calculations are computationally intensive enough to support gather/scatter-type operations. For example, in a case with 1,500 producing wells, more than 90% of the computational workload for well management was in the tubing hydraulics calculations. Parallelization of only these calculations has shown a substantial improvement in parallel efficiency for several test cases for the overall well-management routines.

17.8.4 The Parallel Linear-Equation Solver. A typical parallel linear solver involves composite grid decomposition and is based on the work of Wallis and Nolen.[190] At each level of the composite grid, a red-black or block Jacobi ordering of the grid is used. Red-black refers to an ordering of the grids or domains. For example, if the grids form an equal areal subdivision of an areally square model, the red and black grids would appear as a checkerboard when viewed from the top of the grid. For a 1D or strip decomposition of the model, the red-black coloring would appear as alternately colored stripes when viewed perpendicular to the decomposition direction. An approximate block LU factorization approach is used for preconditioning and involves, for example, an L-factor for the red-black ordering as follows:

$$L = \begin{pmatrix} A_1 & & & & & & \\ & A_2 & & & & & \\ & & A_3 & & & & \\ & & & A_4 & & & \\ Q_1 & R_2 & & & B_1 & & \\ & Q_2 & R_3 & & & B_2 & \\ & & Q_3 & R_4 & & & B_3 \end{pmatrix} \quad \text{(17.35)}$$

The A_i blocks are ordered first and represent the red grids or domains, while the B_i blocks are the black grids and are ordered second. The diagonals of the B_i submatrices are modified from the original coefficients to maintain rowsum constraints (see Ref. 190). This approximate factorization leads to independent solutions on each processor for the first the red A_i and then followed by the black B_i grid matrices. For the block Jacobi approach, only a single color exists, and all grids at a given level are solved independently. As described in Ref. 179, parallelization of this preconditioning can be achieved in a manner almost identical to the technique used for the coefficient routines. Ref. 47 provides detail of a similar parallel solver developed independently, but almost simultaneously, by Burrows, Ponting, and Wood. As described previously, each grid is assigned to a processor. These may be the same or different processors. If all active grids lie on the same level, the composite grid solver performs all red or black preconditioning solutions independently [i.e., either block Jacobi (all red) or red-black orderings are used]. If grids are assigned to different levels, each grid of the same level and same color (Jacobi being all red) is solved simultaneously, with communication being performed after the entire solution of all grids on a level is completed. Factorization is performed so that rowsum constraints are also imposed. The flexibility of this solver allows the easy implementation of block red-black algorithms, or, as a special case, a red-black linear decomposition ("strip") decomposition could be used.

17.8.5 Parallel Computers and Message-Passing Interface. With the rapid changes in technology for parallel computing, it is difficult to provide the state-of-the-art because this appears to change daily; however, current classification practice divides parallel computers into two distinct categories: shared and distributed memory. Shared-memory systems allow all of the memory associated with the parallel computer to be accessed by all of the processors. For UNIX systems, the memory sizes currently can range up to several tens of billion of bytes with hundreds of processors. For systems based on PC products, the shared-memory systems usually are limited to a few billion bytes of storage and fewer than 10 processors. (This may change with the recently introduced 64-bit processor Itanium.) The distributed memory systems allow access to local memory only by a small number of processors that reside on a "node" of the parallel system. Each of the "nodes" in the distributed memory parallel system often communicates with other nodes by a high-speed switch, although conventional ethernet communications are possible for some applications. Either of these parallel computers is capable of delivering several hundred million floating point operations per second (MFLOPS) for each node, or several billions of floating point operations per second (GFLOPS) for applications that can use the parallel hardware. Parallel PC hardware can be further subdivided into operating systems: Windows- and LINUX-based systems. Current trends indicated that it is likely that the Windows-based systems will dominate the engineer's desktop because of the wide availability of engineering/business applications available for this operating system. LINUX-based systems will likely be limited to clusters of servers to perform computer-intensive work such as parallel simulations.

To perform calculations in parallel, data must be passed in the form of messages among processors of the parallel machine. Message-passing interface (MPI) has been primarily used for parallelization on all classifications of parallel machines because of the wide availability and portability of MPI. For shared-memory computers (Sun and SGI for example), other forms of message passing exist such as the recently popular OpenMP. There are generally three forms of message passing for a distributed memory system: synchronous, asynchronous, and global operations. Synchronous message passing requires that all messages be completed before computations can proceed. Asynchronous, on the other hand, allows some overlap of processing and communications. Global message-passing operations distribute data from single processors to one processor, or vice-versa. Often some form of arithmetic function, such as summation, is performed as part of the global concatenation operation. These functions are particularly useful for evaluation of reduction functions such as dot products, or for distribution of data from a single node to all nodes in the parallel partition. Global communications are normally optimized for a particular architecture and are much more efficient than a simple set of send-receive operations for achieving the same results.

17.8.6 Application of Parallel Computing. The rationalization of parallel reservoir simulation is based on two concepts: the ability to run larger models, or simply the ability to improve turnaround time on existing models. A typical larger-size model would be considered to have more than 1 million gridblocks. Results for a 1-million-cell, black-oil, implicit model show scaling is almost perfect from 1 to 18 processors with only a small degradation up to 36 processors (speedup = 32) on an SGI shared-memory parallel computer. Although this is a somewhat idealized example, it does point out the excellent scaling possible because of low message-passing overhead for a parallel simulator using domain decomposition. Results for a real-world compositional example with 3.4 million cells have been obtained on an IBM SP parallel computer with 108 processors. For this case, with 12 to 27 processors, the problem scales extremely well with a speedup of a factor of 2 for slightly more than doubling the number of processors. The scaling falls off slightly from 27 to 54 processors with only a factor of 1.7, but from 54 to 108 processors, this scaling is maintained at a factor of 1.7. Overall, the speedup from 12 to 108 processors was a very good factor of 5.6. If perfect scaling on 12 processors is assumed, this indicates a factor of 68 overall for 107 processors for a parallel efficiency of approximately 64%. The total elapsed time for the 108 processor case was approximately 12 hours, indicating that overnight turnaround for this 3.4-million-cell model could be achieved. Another example application of parallel simulation with more than 1 million gridblocks was recently presented in a black-oil simulation study of complex water encroachment in a large carbonate reservoir in Saudi Arabia.[189]

17.8.7 The Future of High-Performance Reservoir Simulation. It is clear that the uptake of parallel reservoir simulation has been limited by the hurdle of cost and the robustness of the technology. The recent advances in commodity hardware based on the INTEL architecture have provided a significant boost, however. Currently, the use of parallel INTEL-based clusters using inexpensive high-speed switches has lowered the entry point for significant parallel simulation by an order of magnitude. Just as the UNIX workstation caused a revolution in reservoir simulation, this new lost-cost hardware will likely bring parallel computing to the engineer's desktop in the near future. The remaining hurdles for parallel simulation then are limited to enhancing parallel simulation technology. In particular, emphasis must be placed on load balancing and efficient linear equation solvers. Load balancing refers to the fact that all processors must perform the same amount of work to achieve high parallel efficiency. If only a few processors perform most of the work for a parallel simulation involving large numbers of processors, then poor parallel efficiency results—a case known as load imbalance. The allocation of optimization techniques to solve the load balancing problem offers some promise.[191] An area of particular importance for load balancing is the coupled surface network/reservoir simulator. For this case, the network often dominates the simulation by almost an order of magnitude; therefore, to achieve reasonable parallel efficiency, the imbalance caused by the serial surface network calculations must be overcome. One approach to this is mentioned previously, the parallelization of the tubing hydraulics calculations; however, considerable additional parallelization effort will be required, especially if more than a few processors are to be used efficiently. Load imbalance is also brought about by complex geologies and complex locally refined grids. It is likely that the introduction of unstructured grids and associated load balancing using techniques such as "Hilbert space-filling curves"[192] may well lead to the solution of this problem. Finally, the linear equation solver as the heart of the parallel simulator must be enhanced to provide efficient and robust solutions. Solver degradation often results from the use of large numbers of domains (processors) or a poor choice of the decomposition geometry.[179] Although the previous examples show good parallel performance, future models with large, implicit systems must be routinely solved on large numbers of processors. Multigrid-like solvers may show the best promise for this improvement.[183,193]

17.9 Reservoir Simulation Applications—L. Kent Thomas

Reservoir simulation is a widely used tool for making decisions on the development of new fields, the location of infill wells, and the implementation of enhanced recovery projects. It is the focal point of an integrated effort of geosciences, petrophysics, reservoir, production and facilities engineering, computer science, and economics. Geoscientists using seismic, well-log, outcrop analog data and mathematical models are able to develop geological models containing millions of cells. These models characterize complex geological features including faults, pinchouts, shales, and channels. Simulation of the reservoir at the fine geologic scale, however, is usually not undertaken except in limited cases. Generally, the fine-scaled geological model is partially integrated or "upscaled" to a coarse-grid model, which is computationally more tractable. The grid of the upscaled model is designed to capture the principal geological features of the geologic model to simulate the fluid flow. The grid may also be designed to capture the effects of complex wells. In the upscaling process, the laboratory relative-permeability and capillary-pressure functions may be upscaled to "pseudofunctions." These pseudofunctions attempt to capture fluid-flow behavior that is lost because of the integration of fine-scale geologic features in the upscaling process. Phase-behavior treatment can range from simple black-oil PVT to compositional and thermal processes.

The reservoir simulation model may either be used directly to forecast the performance of a new reservoir or adjusted so that it reasonably models the historical behavior of an existing reservoir and wells. This adjustment process is called history matching. Programs called "pre-processors" and "post-processors" enable the engineer to prepare data, manipulate the model, and view results. Once a history-matched model is obtained, then forecasts are made under a variety of operating conditions. These results are combined together with economic models to enable the engineer to make decisions concerning the operation of the reservoir.

17.9.1 Development of the Geological Model. A sound understanding of the structural elements of the reservoir and the depositional environment under which the sediments were deposited is critical to the development of an accurate geologic model. Today, the geologic model is frequently constructed as a numerical representation of the reservoir and adjacent aquifer and is referred to as a static, or geocellular, model. This model provides the vehicle to capture and combine the seismic structural interpretation and well petrophysical data in a numerically consistent way with known depositional characteristics.[194,195] Petrophysical properties such as porosity, permeability, and water saturation can be distributed throughout the interwell 3D volume using various techniques, many of which rely on geostatistics.[196] Efforts are also underway to condition these numerical, static models with production[197] and well-test[198] data to further reduce geologic uncertainty. The construction of a geocellular model represents a significant collaborative effort between geoscientists, petrophysicists, and reservoir engineers.

Geocellular models today may consist of over 25 to 50 million cells on large and/or geologically complex reservoirs. The ability to build static geologic models of this magnitude has outstripped the reservoir engineer's ability to simulate an equal number of cells in a full physics reservoir simulator (and will continue to do so). Classical development geologic efforts have focused on defining and describing the reservoir geology using 2D maps, which depict the most likely interpretation of the depositional environment and the variability of the reservoir parameters between wells. These interpretations have historically been referred to as "deterministic" reservoir descriptions. With the advent of geocellular models and the application of such technologies as geostatistics, it is now possible for geoscientists to generate multiple reservoir descriptions for the reservoir engineer to simulate. In some cases, one of these descriptions may be selected to represent the "deterministic" model. Regardless if one or several static models are handed over for reservoir simulation, it is generally necessary to reduce the cell count to run the problem with existing reservoir simulators. Significant effort is being spent improving techniques to reduce the number of reservoir cells in the areal and vertical dimension while maintaining the essential geologic character that impacts the recovery process under consideration. This approach is referred to as upscaling, and it will be discussed in greater detail in the following section. To date, the largest reservoir simulators consist of reservoir descriptions of 2 million grid cells and are run using massively parallel processing power.

17.9.2 Upscaling Geological Model to Reservoir Flow Model. Geological models, which contain the complex structural features of large oil and gas reservoirs, commonly have tens of millions of cells. These models, which contain pinchouts, faults, and other significant information including lithology and facies distributions, are upscaled in both the vertical and areal directions to tens or hundreds of thousands of cells for reservoir simulation.[199]

Several upscaling methods have been developed over the last several years including analytical techniques, local flow-based methods, and global flow-based methods. Analytical methods use arithmetic, harmonic, power law, and geometric averaging to calculate effective properties for each reservoir model gridblock. Local flow-based methods calculate effective gridblock properties by performing single-phase flow simulations in each direction across the upscaled block.[200] The diagonal permeability tensor is calculated by sealing the boundaries perpendicular to the applied pressure gradient. The full-permeability tensor can be calculated in a similar manner by leaving the boundaries normal to the imposed pressure gradient open and applying periodic boundary conditions. Global flow-based methods use pressure gradients across the entire field subject to a specific set of wells to calculate the permeability tensor. Local and global flow-based techniques can be used to compute upscaled transmissibilities directly.

17.9.3 Inclusion of Faults in Reservoir Flow Model. Faults and pinchouts of geological layers are incorporated in geological models to capture the complex geometry of many reservoirs. This information is then upscaled into the reservoir model, and it results in both neighbor and non-neighbor connections across the faults and non-neighbor connections across the pinchouts. In Cartesian coordinates, the trace of a fault may need to be represented by a "stair-stepped" line, while a somewhat better representation of faults can be made with corner-point grids. PEBI grids, which will be discussed subsequently, are best suited to accurately model fault geometry.

Models for calculating the transmissibility across the fault and parallel to the fault have been developed based on fault type, displacement, geochemical deposition, and whether open joints occur along the fault.[201] In general, transmissibilities across a fault can be at least an order of magnitude lower than those parallel to the fault. Inclusion of this information in a reservoir model is frequently a key parameter in reservoir description.

A recent paper describes the analysis that was performed to calculate fluid flow through conductive faults in the Khafji oil field in the Arabian Gulf.[202] Two sandstone reservoirs separated by a thick continuous shale are both connected to the same large aquifer and had the same initial WOC. The top reservoir has edgewater drive, while the deeper reservoir is bottomwater drive. Early water breakthrough in the upper sand was determined to be a function of supplemental water influx from the aquifer of the lower sand through conductive faults.

17.9.4 Development of Pseudofunctions for Multiphase Flow. Pseudorelative permeability curves are developed for upscaled reservoir models to match multiphase fluid flow at the fine-grid level. Several methods for performing these calculations have been presented in the literature.[203] In the "10th SPE Comparative Solution Project: A Comparison of Upscaling Techniques," the fine-scale geological model was chosen to be sufficiently detailed such that it would be difficult to run the fine grid and use classical pseudoization methods.[12] Several participants, however, used some level of fine-grid simulation to develop pseudorelative permeability curves, with two of the participants adjusting the exponents of the Corey equations to effect a reasonable match. This approach can be done manually or with an automated history-matching algorithm.

17.9.5 Gridding Techniques. The majority of reservoir simulation studies conducted today use Cartesian or corner-point structured grids with some application of local grid refinement to evaluate infill well locations or to more accurately calculate water and/or gas coning in a well. In a structured grid, cell locations are specified using a 3D, *i, j, k,* indexing system. This allows for ready access either numerically or visually, using pre- and post-processing software, to multilayer well information or data and calculated results at any physical location in the reservoir model.

A more flexible approach for modeling reservoirs with complex geometries that still relies on structured gridding was presented by Jenny *et al.*[63] Here, a hexahedral multiblock grid is used which is unstructured globally, but maintains an *i, j, k* structure on each subgrid.

PEBI grids[58] are now being used on a limited basis to simulate reservoirs with complex geological features that have been developed with nonconventional wells to maximize recovery.[204] These grids are unstructured and are described internally in a simulator with a 1D index, *i,* that ranges from one to the number of nonzero pore volume cells in a model. Evaluation of simulator input and results relies heavily on pre- and post-processing software that allows the user to visually look at the model and make changes during the history-matching phase of a study.

17.9.6 Simulation of Nonconventional and Intelligent Wells. Nonconventional wells are routinely used to maximize production rates and ultimate recovery in oil and gas reservoirs. Wells in this category include deviated, horizontal, horizontal and vertical multilaterals, and multilateral fishbone designs. This latter well type is especially effective in low-permeability or heavy-oil reservoirs.

Simulation of nonconventional wells can be approached in several ways. First, the productivity of each perforation in a conventional model can be approximated by applying the appropriate skin and Peaceman's equation.[205] Second, simulation grids can be constructed that closely follow the well path and allow a more accurate calculation of well rates.[206] Another approach, which is quite appealing, is based on a semi-analytical method.[207] It results in a good approximation for PIs in nonconventional wells and incorporates the near-wellbore skin because of heterogeneity in this region as well as mechanical skin. This method, which is very efficient, can also include wellbore hydraulic effects.

Nonconventional wells coupled with intelligent completions can be used to improve sweep efficiency and optimize recovery.[208] One example of this technology is the use of surface-controlled, downhole-adjustable chokes, which can be used to apply different pressure drawdowns to separate zones along the well. This allows a more uniform inflow in the well and control of early water or gas breakthrough. Real-time measurements of wellbore pressures and temperatures are being made for use in conjunction with PLT tests for inflow performance analysis.

17.9.7 Integrated Reservoir and Surface Facilities Models. Integration of reservoir and surface facilities simulation can result in improved production forecasts and allows optimization of the surface facilities structure and operating conditions. An integrated reservoir, well flow string, and surface network model of the Prudhoe Bay oil field was built and successfully applied to a facility optimization study.[209] Production costs as a result of this effort were reduced by defining the optimum number of separator stages and their connections, and defining the optimum separator operating conditions and by using excess capacity in the Prudhoe Bay facilities to process production from satellite fields. Procedures for the simultaneous solution of the reservoir and surface pipeline network flow equations are described in Refs. 210 and 211.

In the Ekofisk field in the Norwegian sector of the North Sea, integrated reservoir and facilities simulations have been made to optimize throughput in existing surface facilities and to forecast production from planned expansion of current facilities.[212] This optimization project has resulted in sustained high production of approximately 300,000 STB/D over the last several years. Another important aspect in the management of this field is the inclusion of compaction logic in the model based on both stress and water saturation changes during depletion and waterflooding.[213] Treatment of geomechanical effects in stress-sensitive reservoirs has received increased attention throughout the industry in recent years.

17.9.8 Simulation of Multiple Reservoirs. Simulation of multiple fields producing into a common production facility is routinely practiced today to capture the interplay between well deliverability, water and gas injection, operating constraints, and contract rates. In the J-Block area fields, in the U.K. sector of the North Sea, an integrated reservoir study was conducted that included a gas condensate reservoir with gas cycling that was simultaneously modeled with volatile oil reservoirs.[214] The fields were developed with a single platform and one subsea manifold completion and a combination of vertical and horizontal wells. Four separate PVT regions were used to describe the fluid behavior. The integrated model used in this study results in an efficient reservoir management tool for making development and operating decisions.

Another example of reservoir management of multiple fields with shared facilities is the Gannet cluster, located in the U.K. sector of the North Sea, which connects four fields.[215] Wells from one of the fields are directly linked to the production platform, and the other three fields are subsea tiebacks to the platform. Three of the four fields are oil fields and the fourth is a gas field. An integrated model was built to simulate the interaction of the subsurface and surface processes. The well-management objective on this project was to maximize hydrocarbon recovery while simultaneously meeting a long-term gas contract.

17.9.9 Use of Larger Models. The maximum practical model size has increased from tens of thousands to hundreds of thousands of cells at essentially a linear rate vs. time during the last decade. This trend has developed as a result of the dramatic increase in computer hardware speed accompanied with larger memory and cache. Both high-speed UNIX workstations and high-end PCs are used for reservoir simulation, with a close race developing between the two platforms in regard to run times. Additional advances in computing speed for megamodels have been achieved using parallel hardware along with the necessary developments in model software, discussed in a previous section. An example application of this technology was recently presented in a simulation study of complex water encroachment in a large carbonate reservoir in Saudi Arabia.[189]

17.9.10 History Matching and Production Forecasting. Once a reservoir simulation model has been constructed, the validity of the model is examined by using it to simulate the performance of a field under past operating conditions. This is usually done by specifying historical controlling rates, such as oil rate in an oil reservoir vs. time, and then making a comparison of the nonspecified performance such as GOR, WOR, and reservoir pressure with measured data. If there are significant differences between the calculated performance and the known performance of the well/reservoir system, then adjustments to the reservoir simulation model are made to reduce this difference. This process is called history matching. These adjustments should be made in a geologically consistent manner.[198] Modification of those parameters that have the highest degree of uncertainty will give the maximum reduction in the error. The history-matching process should be approached in a consistent manner to minimize the effort.[216] In addition to well rates and bottomhole pressures, and reservoir pressures measured at the time the well is drilled, production logs, long-term pressure gauges, and time-lapse seismic data enable the engineer to better constrain the model during the history-matching process. Time-lapse (4D) seismic[217,218] is becoming an integral part of the field performance monitoring and history matching. Streamline models together with reservoir simulators[219,197] can be used to improve the history-matching process, especially in waterflood operations. Tools to assist in the history-matching process consist of the use of parallel computers, sensitivity analysis, and gradient techniques.[220,221]

Once a history match is obtained, then forecasts of future well/reservoir performance under various operating scenarios are made. Models of multiphase flow in the wellbore and production lines are used to constrain the production rate. These models may include subsea completions[222] with very long gathering lines or complex surface facilities with reinjection of produced fluids.[223] Because of the uncertainty in the geological and reservoir simulation models for new fields, often multiple forecasts[223] with different reservoir parameters are made to determine the uncertainty in the forecasts. Multiple history-matched models based on multiple geological models,[224] and experimental design[225] may also be used to characterize the uncertainty in production forecasts.

Nomenclature

A = a matrix
B = formation volume factor, reservoir volume/surface volume
b = right-side vector
C_{os} = equivalent surface gas volume for a unit volume of surface oil
d = dimension in direction of flow
D = depth below datum, ft
D = future time, time units
f = fractional flow
f_i = fugacity of component i
F = flow rate, volume/time unit
F, F_i, f_{il} = see Eq. 17.2
f_g = gas-phase fractional flow, $\lambda_g / (\lambda_g + \lambda_o)$
f_g' = $\partial f_g / \partial S_g$
g = acceleration constant (length/length/time) or gas phase, depending on use
G = phase velocity
k = permeability, md
k_{ij} = binary interaction parameters
k_r = relative permeability, fraction
K = permeability, md
K_i = equilibrium constant for component i
K_r = relative permeability, fraction
k_{rocw} = relative permeability of oil at S_{wc}, $S_g = 0$, fraction
k_{rwro} = relative permeability of water at residual oil, fraction
L = length, feet
L1 = liquid type 1
L2 = liquid type 2

m = density, mass/unit volume
M = mass, mobility, molecular weight, or preconditioner matrix depending on use
n = number of components or number of moles
n_{st} = number of streamlines
n_{ts} = number of timesteps
N = number of active gridblocks
o = oil phase
p = polynomial
p_J = phase J pressure, psi
P = pressure, psi
P_c = capillary pressure, psi
P_{cgo}' = dP_{cgo} / dS_g
P_i = vector of primary unknowns for block i
P_{ii} = ith scalar element of P_i
q_{ii} = well production rate of component i from block i, volume/time
q_{Iji} = interblock flow rate of component I from block j to block i, volume/time
q_x = total volumetric flow rate in the x direction, volume/time
Q = production or injection volume
r_s = solution gas-condensate ratio, volume/volume
R_s = solution gas, scf/STB oil or nm^3/nm^3 oil
S = saturation
S_{on} = $(1.0 - S_w - S_{orw})/(1.0 - S_{wc} - S_{orw})$, fraction
S_{orw} = residual oil saturation after water displacement
S_{wc} = connate water saturation, fraction
S_{wn} = $(S_w - S_{wc})/(1.0 - S_{wc} - S_{orw})$, fraction
t^{solver} = time to solve for a global pressure field
t^{st} = time to solve transport equation for each streamline
T = time, time units or transmissibility, md·ft/cp
T_{ij} = transmissibility connecting blocks i and j, md·ft/cp
u = pore velocity, v/ϕ
u_t = total velocity
v = Darcy or superficial velocity
V = vapor phase or volume
V_p = pore volume, bbl
w = water phase
x = an unknown vector or concentration of a component in the liquid phase
x,y,z = Cartesian coordinates
x_i = concentration of component i in oil phase
$x_{i,J}$ = mol fraction of component i in phase J
y = an unknown vector
y_i = concentration of component i in the gas phase
z = gas deviation factor
z_i = overall mol fraction of component i in a gridblock
Z = depth to gridblock center, measured positive downward, ft or m
α = mole fraction vapor phase
β = see Linear Solver section, Step 6
γ = phase density expressed as psi/ft or bar/m

δ = $\delta X = X^{l+1} - X^l$, where l is the Newton iterate index and X is any quantity
Δp = pressure difference, psi
ΔQ = potential difference
Δt = timestep
Δt^* = maximum stable IMPES timestep
Δx, Δy, Δz = gridblock dimensions
ΔZ = $Z_j - Z_i$
θ = dip angle of line joining cell centers, degrees
λ = phase mobility, k_r / μ
μ = viscosity, cp
ρ = density, mass/volume or molar density, mols/volume unless noted otherwise
Σ = sum
τ = time-of-flight along a streamline
ϕ = porosity, fraction
ψ = $\lambda_o \lambda_g / (\lambda_o + \lambda_g)$
ω' = acentric factor
∇ = potential term
∂ = normalized difference parameter

Subscripts

a = cell designation
$a\text{-}h$ = arithmetic-harmonic average
ari = arithmetic average
b = adjacent cell designation
c = component designation
eff = effective
g = gas phase
h = horizontal
$h\text{-}a$ = harmonic-arithmetic average
har = harmonic average
i = component number, index counter, or initial condition
j = index counter
J = phase number
k = index counter
lbc = linear (open sides) boundary conditions
low = lower bound
nbc = no-flow (sealed sides) boundary conditions
o = oil phase
p = phase designation
pb = periodic boundary conditions
r = residual
R = reservoir
s = surface
sc = standard conditions
true = true
upp = upper bound
v = vertical
w = water phase or well
x = cell designation
x,y,z = Cartesian x,y,z directions

z_eff = effective in z direction

Superscripts

0 = reference (in compressibility equation)
1 = Newton iterate index
n = timestep number

References

1. *The Merriam-Webster Dictionary,* New Revised Edition, Merriam-Webster, New York City (2004).
2. Coats, K.H.: "Reservoir Simulation," *Petroleum Engineering Handbook,* H.B. Bradley (ed.), SPE, Richardson, Texas (1987) Chap. 48.
3. Odeh, A.S.: "Comparison of Solutions to a Three-Dimensional Black-Oil Reservoir Simulation Problem," *JPT* (January 1981) 13.
4. Weinstein, H.G., Chappelear, J.E., and Nolen, J.S.: "Second Comparative Solution Project: A Three-Phase Coning Study," *JPT* (March 1986) 345.
5. Kenyon, D.E. and Behie, G.A.: "Third SPE Comparative Solution Project: Gas Cycling of Retrograde Condensate Reservoirs," *JPT* (August 1987) 981.
6. Aziz, K., Ramesh, A.B., and Woo, P.T.: "Fourth SPE Comparative Solution Project: Comparison of Steam Injection Simulators," *JPT* (December 1987) 1576.
7. Killough, J.E. and Kossack, C.A: "Fifth Comparative Solution Project: Evaluation of Miscible Flood Simulators," paper SPE 16000 presented at the 1987 SPE Symposium on Reservoir Simulation, San Antonio, Texas, 1–4 February.
8. Firoozabadi, A. and Thomas, L.K.: "Sixth SPE Comparative Solution Project: Dual-Porosity Simulators," *JPT* (June 1990) 710.
9. Nghiem, L., Collins, D.A., and Sharma, R.: "Seventh SPE Comparative Solution Project: Modeling of Horizontal Wells in Reservoir Simulation," paper SPE 21221 presented at the 1991 SPE Symposium on Reservoir Simulation, Anaheim, California, 17–20 February.
10. Quandalle, P.: "Eighth SPE Comparative Solution Project: Gridding Techniques in Reservoir Simulation," paper SPE 25263 presented at the 1993 SPE Symposium on Reservoir Simulation, New Orleans, 28 February–3 March.
11. Killough, J.E.: "Ninth SPE Comparative Solution Project: A Reexamination of Black-Oil Simulation," paper SPE 29110 presented at the 1995 SPE Symposium on Reservoir Simulation, San Antonio, Texas, 12–15 February.
12. Christie, M.A. and Blunt, M.J. "Tenth SPE Comparative Solution Project: A Comparison of Upscaling Techniques," *SPEREE* (August 2001) 308.
13. Coats, K.H.: "An Equation of State Compositional Model," *SPEJ* (October 1980) 363.
14. Young, L.C. and Stephenson, R.E.: "A Generalized Compositional Approach for Reservoir Simulation," *SPEJ* (October 1983) 727.
15. Acs, G., Doleschall, S., and Farkas, E.: "General Purpose Compositional Model," *SPEJ* (August 1985) 543.
16. Watts, J.W.: "A Compositional Formulation of the Pressure and Saturation Equations," *SPERE* (May 1986) 243.
17. Blair, P.M. and Weinaug, C.F.: "Solution of Two-Phase Flow Problems Using Implicit Difference Equations," *SPEJ* (December 1969) 417; *Trans.*, AIME, **246.**
18. Stone, H.L. and Garder, A.O. Jr.: "Analysis of Gas-Cap or Dissolved-Gas Drive Reservoirs," *SPEJ* (1961) 92; *Trans.,* AIME, **222.**
19. Sheldon, J.W., Harris, C.D., and Bavly, D.: "A Method for General Reservoir Behavior Simulation on a Digital Computer," paper SPE 1521-G presented at the 1960 SPE Annual Meeting, Denver, 2–5 October.
20. Fagin, R.G. and Stewart, C.H.: "A New Approach to the Two-Dimensional Multiphase Reservoir Simulator," *SPEJ* (June 1966) 175; *Trans.,* AIME, **237.**

21. Coats, K.H.: "Reservoir Simulation: A General Model Formulation and Associated Physical/Numerical Sources of Instability," *Boundary and Interior Layers—Computational and Asymptotic Methods,* J.J.H. Miller (ed.), Boole Press, Dublin, Ireland (1980).
22. Coats, K.H., Thomas, L.K., and Pierson, R.G.: "Compositional and Black Oil Reservoir Simulation," *SPEREE* (August 1998) 372.
23. Coats, K.H.: "A Note on IMPES and Some IMPES-Based Simulation Models," *SPEJ* (September 2000) 245.
24. MacDonald, R.C. and Coats, K.H.: "Methods for Numerical Simulation of Water and Gas Coning," *SPEJ* (December 1970) 425; *Trans.,* AIME, **249.**
25. Spillette, A.G., Hillestad, J.G., and Stone, H.L.: "A High-Stability Sequential Solution Approach to Reservoir Simulation," paper SPE 4542 presented at the 1973 SPE Annual Meeting, Las Vegas, Nevada, 30 September–3 October.
26. Thomas, G.W. and Thurnau, D.H.: "Reservoir Simulation Using an Adaptive Implicit Method," *SPEJ* (October 1983) 759.
27. Lantz, R.B.: "Quantitative Evaluation of Numerical Diffusion (Truncation Error)," *SPEJ* (September 1971) 315; *Trans.*, AIME, **251.**
28. Todd, M.R., O'Dell, P.M., and Hirasaki, G.J.: "Methods for Increased Accuracy in Numerical Reservoir Simulators," *SPEJ* (December 1972) 515; *Trans.*, AIME, **253.**
29. Fung, L.S.-K., Collins, D.A., and Nghiem, L.X.: "An Adaptive-Implicit Switching Criterion Based on Numerical Stability Analysis," *SPERE* (February 1989) 45.
30. Russell, T.F.: "Stability Analysis and Switching Criteria for Adaptive Implicit Methods Based on the CFL Condition," paper SPE 18416 presented at the 1989 SPE Symposium on Reservoir Simulation, Houston, 6–8 February.
31. Grabenstetter, J. *et al.*: "Stability-Based Switching Criterion for Adaptive-Implicit Compositional Reservoir Simulation," paper SPE 21225 presented at the 1991 SPE Symposium on Reservoir Simulation, Anaheim, California, 17–20 February.
32. Watts, J.W. and Rame, M.: "An Algebraic Approach to the Adaptive Implicit Method," paper SPE 51900 presented at the 1999 SPE Reservoir Simulation Symposium, Houston, 14–17 February.
33. Coats, K.H.: "IMPES Stability: The CFL Limit," paper SPE 66345 presented at the 2001 SPE Reservoir Simulation Symposium, Houston, 11–14 February.
34. Coats, K.H.: "IMPES Stability: The Stable Step," paper SPE 69225 presented at the 2001 SPE Reservoir Simulation Symposium, Houston, 11–14 February.
35. Young, L.C. and Russell, T.F.: "Implementation of an Adaptive Implicit Method," paper SPE 25245 presented at the 1993 SPE Symposium on Reservoir Simulation, New Orleans, 28 February–3 March.
36. Vinsome, P.K.W.: "Orthomin, an Iterative Method for Solving Sparse Sets of Simultaneous Linear Equations," paper SPE 5729 presented at the 1976 SPE Symposium of Numerical Reservoir Simulation of Reservoir Performance, Los Angeles, 19–20 February.
37. Appleyard, J.R. and Cheshire, I.M.: "Nested Factorization," paper SPE 12264 presented at the 1983 SPE Reservoir Simulation Symposium, San Francisco, 16–18 November.
38. Watts, J.W. III: "A Conjugate Gradient-Truncated Direct Method for the Iterative Solution of the Reservoir Simulation Pressure Equation," *SPEJ* (June 1981) 345.
39. Forsyth, P.A. and Sammon, P.H.: "Local Mesh Refinement and Modeling of Faults and Pinchouts," *SPEFE* (June 1986) 275.
40. Verma, S. and Aziz, K.: "A Control Volume Scheme for Flexible Grids in Reservoir Simulation," paper SPE 37999 presented at the 1997 SPE Reservoir Simulation Symposium, Dallas, 8–11 June.
41. Saad, Y. and Schultz, M.H.: "GMRES: A Generalized Minimal Residual Algorithm for Solving Nonsymmetric Linear Systems," *SIAM Journal on Scientific and Statistical Computing* (1986) 856.
42. Tan, T.B.S. and Letkeman, J.P.: "Application of D4 Ordering and Minimization in an Effective Partial Matrix Inverse Iterative Method," paper SPE 10493 presented at the 1982 SPE Symposium on Reservoir Simulation, New Orleans, 31 January–3 February.
43. Watts, J.W.: "An Iterative Matrix Solution Method Suitable for Anisotropic Problems," *SPEJ* (March 1971) 47; *Trans.*, AIME, **251.**

44. Wallis, J.R.: "Incomplete Gaussian Elimination as a Preconditioning for Generalized Conjugate Gradient Acceleration," paper SPE 12265 presented at the 1983 SPE Reservoir Simulation Symposium, San Francisco, 15–18 November.
45. Behie, A. and Vinsome, P.K.W.: "Block Iterative Methods for Fully Implicit Reservoir Simulation," *SPEJ* (October 1982) 658.
46. Holmes, J.A.: "Enhancements to the Strongly Coupled, Fully Implicit Well Model: Wellbore Crossflow Modeling and Collective Well Control," paper SPE 12259 presented at the 1983 SPE Reservoir Simulation Symposium, San Francisco, 15–18 November.
47. Burrows, R., Ponting, D., and Wood, L.: "Parallel Reservoir Simulation with Nested Factorization," presented at the 1996 European Conference on the Mathematics of Oil Recovery, Leoben, Austria, 3–6 September.
48. Killough, J.E. *et al.*: "Parallelization of a General-Purpose Reservoir Simulator," presented at the 1996 European Conference on the Mathematics of Oil Recovery, Leoben, Austria, 3–6 September.
49. Cuthill, E. and McKee, J.: "Reducing the Bandwidth of Sparse Symmetric Matrices," *Proc., 24th National Conference of the Assoc. for Computing Machinery*, Brandon Press, Princeton, New Jersey (1969) 157–172.
50. D'Azevedo, E.F., Forsyth, P.A., and Tang, W.-P.: "An Automatic Ordering Method for Incomplete Factorization Iterative Solvers," paper SPE 21226 presented at the 1991 Symposium on Reservoir Simulation, Anaheim, California, 17–20 February.
51. D'Azevedo, E.F., Forsyth, P.A., and Tang, W.-P.: "Ordering Methods for Preconditioned Conjugate Gradient Methods Applied to Unstructured Grid Problems," *SIAM Journal of Matrix Analysis and Applications* (July 1992) 944.
52. Aziz, K. and Settari, A.: *Petroleum Reservoir Simulation* (1979) ISBN 0–9730614-0-5, reprinted 2002, Blitzprint Ltd., Calgary.
53. Rozon, B.J.: "A Generalized Finite Volume Discretization Method for Reservoir Simulation," paper SPE 18414 presented at the 1989 SPE Symposium on Reservoir Simulation, Houston, 6–8 February.
54. Aavatsmark, I. *et al.*: "MPFA for Faults and Local Refinements With Application to Field Simulations," *Proc.*, 2000 European Conference on the Mathematics of Oil Recovery, Baveno, Italy, 5–8 September.
55. Forsyth, P.A.: "A Control-Volume Finite-Element Method for Local Mesh Refinement in Thermal Reservoir Simulation," *SPERE* (November 1990) 561.
56. Herge, T.M., Dalen, V., and Henriquez, A.: "Generalized Transmissibilities for Distorted Grids in Reservoir Simulation," paper SPE 15622 presented at the 1986 SPE Annual Technical Conference and Exhibition, New Orleans, 5–8 October.
57. Young, L.C.: "Rigorous Treatment of Distorted Grids in 3D," paper SPE 51899 presented at the 1999 SPE Reservoir Simulation Symposium, Houston, 14–17 February.
58. Aziz, K.: "Reservoir Simulation Grids: Opportunities and Problems," paper SPE 25233 presented at the 1993 SPE Symposium on Reservoir Simulation, New Orleans, 28 February–3 March.
59. Aavatsmark, I., Barkve, T., and Mannseth, T.: "Control Volume Discretization Methods for 3D Quadrilateral Grids in Inhomogeneous, Anisotropic Reservoirs," *SPEJ* (June 1998) 146.
60. Edwards, M.G. and Rogers, C.F.: "A Flux Continuous Scheme for the Full Tensor Pressure Equation," *Proc.,* 1994 European Conference on the Mathematics of Oil Recovery, Roros, Norway, 7–10 June.
61. Lee, S.H., Tchelepi, H., and DeChant, L.J.: "Implementation of a Flux-Continuous Finite Difference Method for Stratigraphic, Hexahedron Grids," paper SPE 51901 presented at the 1999 SPE Reservoir Simulation Symposium, Houston, 14–17 February 1999.
62. Ponting, D.K.: "Corner Point Geometry in Reservoir Simulation," *Proc.,* First European Conference on the Mathematics of Oil Recovery, 1989.
63. Jenny, P. *et al.*: "Modeling Flow in Geometrically Complex Reservoirs Using Hexahedral Multiblock Grids," *SPEJ* (June 2002) 149.
64. Heinemann, Z.E. and Brand, C.W.: "Gridding Techniques in Reservoir Simulation," *Proc.,* First Intl. Forum on Reservoir Simulation, Alpbach, Austria (1988) 339.
65. Heinemann, Z.E. *et al.*: "Modeling Reservoir Geometry With Irregular Grids," *SPERE* (May 1991) 225.

66. Palagi, C.L. and Aziz, K.: "Use of Voroni Grid in Reservoir Simulation," paper SPE 22889 presented at the 1991 SPE Annual Technical Conference and Exhibition, Dallas, 6–9 October.
67. Gunasekera, D.L., Cox, J., and Lindsey, P.: The Generation and Application of K-Orthogonal Grid Systems," paper SPE 37998 presented at the 1997 SPE Symposium on Reservoir Simulation, Dallas, 9–11 June.
68. Verma, S. and Aziz, K.: "Two- and Three-Dimensional Flexible Grids for Reservoir Simulation," *Proc.,* 1996 European Conference on the Mathematics of Oil Recovery, Leoben, Austria, 3–6 September.
69. Lasseter, T.J.: "A New Approach for the Efficient Construction of 3D Geological Models for Reservoir Applications," *Proc.,* Eighth European Conference on the Mathematics of Oil Recovery, Freiberg, Germany (2002).
70. Warren, J.E. and Price, H.S.: "Flow in Heterogeneous Porous Media," *SPEJ* (September 1961) 153; *Trans.,* AIME, **222.**
71. King, M.J. *et al.:* "Application of Novel Upscaling Approaches to the Magnus and Andrew Reservoirs," paper SPE 50643 presented at the 1998 SPE European Petroleum Conference, The Hague, 20–22 October.
72. King, M.J. and Mansfield, M.: "Flow Simulation of Geological Models," *SPEREE* (August 1999) 351.
73. Guerillot, D. *et al.:* "An Integrated Model for Computer Aided Reservoir Description: From Outcrop Study to Fluid Flow Simulations," paper presented at the 1989 IOR Symposium, Budapest, Hungary.
74. Samier, P.: "A Finite-Element Method for Calculating Transmissibilities," *Proc.,* 1990 European Conference on the Mathematics of Oil Recovery, Arles, France, 11–14 September.
75. Durlofsky, L.J. and Chung, E.Y.: "Effective Permeability of Heterogeneous Reservoir Regions," *Proc.,* 1990 European Conference on the Mathematics of Oil Recovery, Arles, France, 11–14 September.
76. Durlofsky, L.J.: "Numerical Calculation of Equivalent Grid Block Permeability Tensors for Heterogeneous Porous Media," *Water Resources Research* (1991) **27,** 699.
77. Kirkpatrick, S.: "Classical Transport in Disordered Media: Scaling and Effective-Medium Theories," *Phys. Rev. Lett.* (December 1971).
78. Tjølsen, C., Bøe, Ø., and Damsleth, E.: "Homogenisation of Absolute Permeability—Does it Affect the Final Results?" *Proc.,* Second Lerkendal Petroleum Engineering Workshop (1992).
79. Bøe, Ø.: "Analysis of an Upscaling Method Based on Conservation of Dissipation," *Transport in Porous Media* (1994) **77,** 17.
80. Holden, L. and Nielsen B.F.: *Global Upscaling of Permeability in Heterogeneous Reservoirs: The Output Least Squares (OLS) Method,* Norwegian Computing Centre, November 1998.
81. Samier, P. *et al.:* "Applications of Streamline Simulations to Reservoir Studies," *SPEREE* (August 2002) 324.
82. Thiele, M.R.: "Streamline Simulation," keynote address at the 2001 Intl. Forum on Reservoir Simulation, Schloss Fuschl, Austria, 3–7 September.
83. Lolomari, T. *et al.:* "The Use of Streamline Simulation in Reservoir Management: Methodology and Case Studies," paper SPE 63157 presented at the 2000 SPE Annual Technical Conference and Exhibition, Dallas, 1–4 October.
84. Baker, R.O. *et al.:* "Full-Field Modeling Using Streamline-Based Simulation: 4 Case Studies" paper SPE 66405 presented at the 2001 SPE Reservoir Simulation Symposium, Houston, 11–14 February.
85. Thiele, M.R., Batycky, R.P., and Blunt, M.J.: "A Streamline-Based 3D Field-Scale Compositional Reservoir Simulator," paper SPE 38889 presented at the 1997 SPE Annual Technical Conference and Exhibition, San Antonio, Texas, 5–8 October.
86. Thiele, M.R., Batycky, R.P., and Thomas, L.K.: "Miscible WAG Simulations Using Streamlines," paper presented at the 2002 European Conference on the Mathematics of Oil Recovery, Freiberg, Germany, 3–6 September.
87. Batycky, R.P., Blunt, M.J., and Thiele, M.R.: "A 3D Field Scale Streamline-Based Reservoir Simulator," *SPERE* (November 1997) 246.

88. Emanuel, A.S.: "Development of an Analytical Streamtube Model for Estimating Stream-Drive Performance," paper SPE 21756 presented at the 1991 SPE Western Regional Meeting, Long Beach, California, 20–22 March.
89. Crane, M.J. and Blunt, M.J.: "Streamline-Based Simulation of Solute Transport," *Water Resources Research* (October 1999) **35,** No. 10, 3061.
90. Di Donato, G., Huang, W., and Blunt, M.J.: "Streamline-Based Dual-Porosity Simulation of Fractured Reservoirs," paper SPE 84036 presented at the 2003 SPE Annual Technical Conference and Exhibition, Denver, 5–8 October.
91. Muskat, M. and Wyckoff, R.: "Theoretical Analysis of Waterflooding Networks," *Trans.,* AIME (1934) **107.**
92. Muskat, M.: *Flow of Homogeneous Fluids,* Intl. Human Resources Development Corp., Boston, Massachusetts (1937 and 1982).
93. Fay, C.H. and Prats, M.: "The Application of Numerical Methods to Cycling and Flooding Problems," *Proc.,* Third World Petroleum Congress, The Hague, 28 May–6 June (1951).
94. Higgins, R.V. and Leighton, A.J.: "A Computer Method to Calculate Two-Phase Flow in Any Irregular Bounded Porous Medium," *JPT* (June 1962) 679.
95. Higgins, R.V. and Leighton, A.J.: "Computer Predictions of Water Drive of Oil and Gas Mixtures Through Irregularly Bounded Porous Media—Three Phase Flow," *JPT* (September 1962) 1048.
96. Bommer, M.P. and Schechter, R.S.: "Mathematical Modeling of In-Situ Uranium Leaching," *SPEJ* (December 1979) 393.
97. Martin, J.C. and Wegner, R.E.: "Numerical Solution of Multiphase, Two-Dimensional Incompressible Flow Using Streamtube Relationships," *SPEJ* (October 1979) 313.
98. Lake, L.W., Johnston, J.R., and Stegemeier, G.L.: "Simulation and Performance Prediction of Large-Scale Surfactant/Polymer Project," *SPEJ* (December 1981) 731.
99. Emanuel, A.S. *et al.:* "Reservoir Performance Prediction Methods Based on Fractal Geostatistics," *SPERE* (August 1989) 311.
100. Pollock, D.W.: "Semianalytical Computation of Path Lines for Finite-Difference Models," *Ground Water* (November–December 1988) **26,** No. 6, 743.
101. King, M.J. *et al.:* "Rapid Evaluation of the Impact of Heterogeneity on Miscible Gas Injection," paper SPE 26079 presented at the 1993 SPE Western Regional Meeting, Anchorage, 26–28 May.
102. Datta-Gupta, A. and King, M.J.: "A Semi-Analytic Approach to Tracer Flow Modeling in Heterogeneous Permeable Media," *Advances in Water Resources* (1995) **18,** No. 1, 9.
103. Batycky, R.P.: "A Three-Dimensional Two-Phase Field Scale Streamline Simulator," PhD dissertation, Stanford U., Stanford, California (1997).
104. Renard, G.: "A 2D Reservoir Streamtube EOR Model with Periodic Automatic Regeneration of Streamlines," *In Situ* (1990) **14,** No. 2, 175.
105. Thiele, M.R. *et al.:* "Simulating Flow in Heterogeneous Systems Using Streamtubes and Streamlines," *SPERE* (February 1996) 5.
106. Bratvedt, F., Gimse, T., and Tegnander, C.: "Streamline Computations for Porous Media Flow Including Gravity," *Transport in Porous Media* (October 1996) **25,** No. 1, 63.
107. Blunt, M.J., Liu, K., and Thiele, M.R.: "A Generalized Streamline Method to Predict Reservoir Flow," *Petroleum Geoscience* (1996) **2,** 256.
108. Prévost, M., Edwards, M.G., and Blunt, M.J.: "Streamline Tracing on Curvilinear Structured and Unstructured Grids," *SPEJ* (June 2002) 139.
109. Thiele, M. and Batycky, R.P.: "Water Injection Optimization Using a Streamline-Based Workflow," SPE paper 84080 presented at the 2003 SPE Annual Technical Conference and Exhibition, Denver, 5–8 October.
110. Batycky, R.P. *et al.:* "Revisiting Reservoir Flood Surveillance Methods Using Streamlines," SPE paper 95402 presented at the 2005 SPE Annual Technical Conference and Exhibition, Dallas, 9–12 October.
111. Wang, Y. and Kovscek, A.R.: "Streamline Approach for History Matching Production Data," *SPEJ* (December 2000) 353.
112. Agarwal, B. and Blunt, M.J.: "Streamline-Based Method With Full-Physics Forward Simulation for History-Matching Performance Data of a North Sea Field," *SPEJ* (June 2003) 171.
113. Caers, J. *et al.:* "A Geostatistical Approach to Streamline-Based History Matching," *SPEJ* (September 2002) 250.

114. Fenwick, D.H. *et al.:* "Reconciling Prior Geologic Information With Production Data Using Streamlines—Application to a Giant Middle-Eastern Oil Field," SPE paper 95940 presented at the 2005 Annual Technical Conference and Exhibition, Dallas, 9–12 October.
115. Stüben, K.: "Algebraic Multigrid (AMG): An Introduction with Applications," *Multigrid,* U. Trottenberg, C.W. Oosterlee and A. Schüller (eds.), Academic Press, New York City (2000) appendix.
116. Darve, F.: *Geomaterials: Constitutive Equations and Modeling,* Elsevier Applied Science, New York City (1990).
117. Coussy, O.: *Mechanics of Porous Continua,* John Wiley and Sons, Chichester, England (1995).
118. Settari, A. and Walters, D.A.: "Advances in Coupled Geomechanical and Reservoir Modeling With Applications to Reservoir Compaction," *SPEJ* (September 2001) 334.
119. Lorenz, J.C.: "Stress Sensitive Reservoirs," *JPT* (January 1999) 61.
120. Settari, A. *et al.:* "Brine Disposal Into a Tight Stress Sensitive Formation at Fracturing Conditions: Design and Field Experience," *SPEREE* (April 1999) 186.
121. Settari, A.: "Reservoir Compaction," *JPT* (August 2002) 62.
122. Finol, A. and Farouq Ali, S.M.: "Numerical Simulation of Oil Production With Simultaneous Ground Subsidence," *SPEJ* (October 1975) 411.
123. Merle, H.A. *et al.:* "The Bachaquero Study—A Composite Analysis of the Behavior of a Compaction Drive/Solution Gas Drive Reservoir," *JPT* (September 1976) 1107.
124. Rattia, A.J. and Farouq Ali, S.M.: "Effect of Formation Compaction on Steam Injection Response," paper SPE 10323 presented at the 1981 SPE Annual Technical Conference and Exhibition, San Antonio, Texas, 5–7 October.
125. Chase, C.A. and Dietrich, J.K.: "Compaction Within the South Belridge Diatomite," *SPERE* (November 1989) 422.
126. Beattie, C.I., Boberg, T.C., and McNab, G.S.: "Reservoir Simulation of Cyclic Steam Stimulation in the Cold Lake Oil Sands," *SPERE* (1991) 200.
127. Chin, L.Y. and Thomas, L.K.: "Fully Coupled Analysis of Improved Oil Recovery by Reservoir Compaction," paper SPE 56753 presented at the 1999 SPE Annual Technical Conference and Exhibition, Houston, 3–6 October.
128. Settari, A. and Mourits, F.M.: "A Coupled Reservoir and Geomechanical Simulation System," *SPEJ* (September 1998) 219.
129. Settari, A. and Price, H. S.: "Simulation of Hydraulic Fracturing in Low-Permeability Reservoirs," *SPEJ* (April 1984) 141.
130. Settari, A.: "How To Approximate Effects of Geomechanics in Conventional Reservoir Simulation" paper SPE 97155 presented at the 2005 SPE Annual Technical Conference and Exhibition, Dallas, 9–12 October.
131. Heffer, K.J. *et al.:* "The Influence of Natural Fractures, Faults and Earth Stresses on Reservoir Performance—Geomechanical Analysis by Numerical Modelling," *Proc.,* Third Intl. Conference on North Sea Oil and Gas Reservoirs, Trondheim, Norway (1992) 201–211.
132. Jin, M., Somerville, J., and Smart, B.G.D.: "Coupled Reservoir Simulation Applied to the Management of Production Induced Stress-Sensitivity," paper SPE 64790 presented at the 2000 SPE International Oil and Gas Conference and Exhibition in China, Beijing, 7–10 November.
133. "Coupled Thermo-Hydro-Mechanical Processes of Fractured Media," *Developments in Geotechnical Engineering,* O. Stephansson, L. Jing, and S.-F. Tsang (eds.), Elsevier, New York City (1996) 575.
134. Koutsabeloulis, N.C. and Hope, S.A.: "'Coupled' Stress/Fluid/Thermal Multiphase Reservoir Simulation Studies Incorporating Rock Mechanics," paper SPE 47393 presented at the 1998 SPE/ISRM Eurock 98, Trondheim, Norway, 8–10 July.
135. Settari, A.: "Coupled Fracture and Reservoir Modeling," *Proc.,* Advanced Hydraulic Fracturing Workshop, 2000 North American Rock Mechanics Conference, Seattle, Washington (2000).
136. Lewis, R.W. and Schrefler, B.A.: *The Finite Element Method in the Deformation and Consolidation of Porous Media,* John Wiley and Sons, New York City (1987) 344.
137. Stone, T. *et al.:* "Fully Coupled Geomechanics in a Commercial Reservoir Simulator," paper SPE 65107 presented at the 2000 SPE European Petroleum Conference, Paris, 24–25 October.
138. Sulak, R.M., Thomas, L.K., and Boade, R.R.: "3D Reservoir Simulation of Ekofisk Compaction Drive," *JPT* (October 1991) 1272.

139. Fredrich, J.T. *et al.*: One-Way Coupled Reservoir-Geomechanical Modeling of the Lost Hills Oil Field, California," *Proc.*, 38th U.S. Rock Mechanics Symposium, Washington, DC (2001).
140. Tran, D., Settari, A., and Nghiem, L.: "New Iterative Coupling Between a Reservoir Simulator and a Geomechanics Module," paper SPE 78192 presented at the 2002 SPE/ISRM Rock Mechanics Conference, Irving, Texas, 20–23 October.
141. Dean, R.H : "A Poroelastic Multicomponent Reservoir Simulator," paper SPE 62806 available from SPE, Richardson, Texas (2000).
142. Settari, A., Walters, D.A., and Behie, G.A.: "Use of Coupled Reservoir and Geomechanical Modeling for Integrated Reservoir Analysis and Management," *J. Cdn. Pet. Tech.* (2001) **40,** No. 12, 55.
143. Coats, B.K., Nolen, J.S., and Sherman, A.H.: "Application and Accuracy of Two Component Characterizations of Reservoir Oils for Numerical Modeling Purposes," paper SPE 17614 presented at the 1988 SPE International Meeting on Petroleum Engineering, Tianjin, China, 1–4 November.
144. Cook, R.E., Jacoby, R.H., and Ramesh, A.B.: "A Beta-Type Reservoir Simulator for Approximating Compositional Effects During Gas Injection," *SPEJ* (October 1974) 471; *Trans.,* AIME, **257.**
145. Fevang, Ø. and Whitson, C.H.: "Modeling Gas-Condensate Well Deliverability," *SPERE* (November 1996) 221.
146. Spivak, A. and Dixon, T.N.: "Simulation of Gas-Condensate Reservoirs," paper SPE 4271 presented at the 1973 SPE Annual Meeting, Houston, 10–12 January.
147. Chang, Y.-B., Coats, B.K., and Nolen, J.S.: "A Compositional Model for CO_2 Floods Including CO_2 Solubility in Water," paper SPE 35164 presented at the 1996 SPE Permian Basin Oil and Gas Recovery Conference, Midland, Texas, 27–29 March.
148. Michelsen, M.L.: "Saturation Point Calculations," *Fluid Phase Equilibria* (1985) **23,** 181.
149. Michelsen, M.L.: "The Isothermal Flash Problem. Part I. Stability," *Fluid Phase Equilibria* (1982) **9,** 1.
150. Muskat, M. and McDowell, J.M.: "An Electrical Computer for Solving Phase Equilibrium Problems," *Trans.,* AIME (1949) **186,** 291.
151. Rachford, H.H. and Rice, J.D.: "Procedure for Use of Electrical Digital Computers in Calculating Flash Vaporization Hydrocarbon Equilibrium," *JPT* (October 1952) 19; *Trans.,* AIME, **195.**
152. Whitson, C.H. and Brule, M.R.: *Phase Behavior*, Monograph Series, SPE, Richardson, Texas (2000) **20.**
153. Michelsen, M.L.: "The Isothermal Flash Problem. Part II. Phase-Split Calculation," *Fluid Phase Equilibria* (1982) **9,** 21.
154. Peneloux, A., Rauzy, E., and Freze, R.: "A Consistent Correction for Redlich-Kwong-Soave Volumes," *Fluid Phase Equilibria* (1982) **8,** 7.
155. Peng, D.Y. and Robinson, D.B.: "A New-Constant Equation of State," *Ind. Eng. Chem. Fund.* (1976) **15,** No. 1, 59.
156. Soave, G.: "Equilibrium Constants from a Modified Redlich-Kwong Equation of State," *Chem. Eng. Sci.* (1972) **27,** No. 6, 1197.
157. Zudkevitch, D. and Joffe, J.: "Correlation and Prediction of Vapor-Liquid Equilibrium with the Redlich-Kwong Equation of State," *AIChE J.* (1970) **16,** 112.
158. Whitson, C.H. and Torp, S.B.: "Evaluating Constant Volume Depletion Data," *JPT* (March 1983) 610; *Trans.,* AIME, **275.**
159. Cook, A.B. *et al.*: "The Role of Vaporization in High Percentage Oil Recovery by Pressure Maintenance," *JPT* (February 1967) 245.
160. Whitson, C.H.: "Characterizing Hydrocarbon Plus Fractions," *SPEJ* (August 1983) 683; *Trans.,* AIME, **275.**
161. Tang, D.E. and Zick, A.A.: "A New Limited-Compositional Reservoir Simulator," paper SPE 25255 presented at the 1993 SPE Symposium on Reservoir Simulation, New Orleans, 28 February–3 March.
162. Zick, A.A.: "A Combined Condensing/Vaporizing Mechanism in the Displacement of Oil by Enriched Gases," paper SPE 15493 presented at the 1986 SPE Annual Technical Conference and Exhibition, New Orleans, 5–8 October.

163. Fevang, Ø., Singh, K., and Whitson, C.H.: "Guidelines for Choosing Compositional and Black-Oil Models for Volatile Oil and Gas-Condensate Reservoirs," paper SPE 63087 presented at the 2000 SPE Annual Technical Conference and Exhibition, Dallas, 1–4 October.
164. Lohrenz, J., Bray, B.G., and Clark, C.R.: "Calculating Viscosities of Reservoir Fluids From Their Compositions," *JPT* (October 1964) 1171; *Trans.*, AIME, **231.**
165. Pedersen, K.S. and Fredenslund, A.: "An Improved Corresponding States Model for the Prediction of Oil and Gas Viscosities and Thermal Conductivities," *Chem. Eng. Sci.* (1987) **42,** 182.
166. Saidi, A.: "Twenty Years of Gas Injection History Into Well-Fractured Haft Kel Field (Iran)," paper SPE 35309 presented at the 1996 SPE International Petroleum Conference and Exhibition, Villahermosa, Mexico, 5–7 March.
167. Whitson, C.H., Fevang, Ø., and Saevareid, A.: "Gas Condensate Relative Permeability for Well Calculations," paper SPE 56476 presented at the 1999 SPE Annual Technical Conference and Exhibition, Houston, 3–6 October.
168. Mott, R.E., Cable, A.S., and Spearing, M.C.: "Measurements of Relative Permeabilities for Calculating Gas-Condensate Well Deliverability," *SPEREE* (December 2000) 473.
169. Robinson, D.B. and Peng, D.Y.: "The Characterization of the Heptanes and Heavier Fractions," *Research Report 28,* Gas Producers Assn., Tulsa (1978).
170. Katz, D.L. and Firoozabadi, A.: "Predicting Phase Behavior of Condensate/Crude-Oil Systems Using Methane Interaction Coefficients," *JPT* (November 1978) 1649; *Trans.*, AIME, **265.**
171. Shelton, J.L. and Yarborough, L.: "Multiple-Phase Behavior in Porous Media During CO_2 or Rich-Gas Flooding," *JPT* (September 1977) 1171.
172. Ngheim, L.X. and Li, Y.K.: "Effect of Phase Behavior on CO_2 Displacement Efficiency at Low Temperatures: Model Studies With an Equation of State," *SPERE* (July 1986) 414.
173. Coats, K.H.: "Simulation of Steamflooding With Distillation and Solution Gas," *SPEJ* (October 1976) 235.
174. Killough, J.E.: "The Use of Vector Processors in Reservoir Simulation," paper SPE 7673 presented at the 1979 SPE Symposium on Reservoir Simulation, Denver, 1–2 February.
175. Nolen, J.S., Kuba, D.W., and Kascic, M.J. Jr.: "Application of Vector Processors to the Solution of Finite Difference Equations," paper SPE 7675 presented at the 1979 SPE Symposium on Reservoir Simulation, Denver, 1–2 February.
176. Killough, J.E. and Levesque, J.M.: "Reservoir Simulation and the In-house Vector Processor: Experience for the First Year," paper SPE 10521 presented at the 1982 SPE Symposium on Reservoir Simulation, New Orleans, 31 January–3 February.
177. Young, L.C.: "Equation of State Compositional Modeling on Vector Processors," paper SPE 16023 presented at the 1987 SPE Symposium on Reservoir Simulation, San Antonio, Texas, 1–4 February.
178. Killough, J.E.: "Is Parallel Computing Ready for Reservoir Simulation?" paper SPE 26634 presented at the 1993 SPE Annual Technical Conference and Exhibition, Houston, 3–6 October.
179. Killough, J.E. *et al.:* "A General-Purpose Parallel Reservoir Simulator," presented at the 1996 European Conference on the Mathematics of Oil Recovery, Leoben, Austria, 3–6 September.
180. van Daalen, D.T. *et al.:* "The Parallelization of BOSIM, Shell's Black/Volatile Oil Reservoir Simulator," *Proc. of the First IMA/SPE European Conference on the Mathematics of Oil Recovery*, Oxford U. (1990).
181. Killough, J.E. and Bhogeswara, R.: "Simulation of Compositional Reservoir Phenomena on a Distributed Memory Parallel Computer," *JPT* (November 1991) 1368.
182. Wheeler, J.A. and Smith, R.A.: "Reservoir Simulation on a Hypercube," *SPERE* (1990) 544.
183. Rutledge, J.M. *et al.:* "The Use of a Massively Parallel SIMD Computer for Reservoir Simulation," paper SPE 21213 presented at the 1991 SPE Symposium on Reservoir Simulation, Anaheim, California, 17–20 February.
184. Gautam S. *et al.:* "Parallel Computing Alters Approaches, Raises Integration Challenges in Reservoir Modeling," *Oil & Gas J.* (20 May 1996) 48.
185. Tchelepi, H.A.: "Practical Use of Scale Up and Parallel Reservoir Simulation Technologies in Field Studies," paper SPE 38886 presented at the 1997 SPE Annual Technical Conference and Exhibition, San Antonio, Texas, 5–8 October.
186. Chien, M.C.H. *et al.:* "A Scalable Parallel Multi-Purpose Reservoir Simulator," paper SPE 37976 presented at the 1997 SPE Reservoir Simulation Symposium, Dallas, 8–11 June.

187. Shiralkar, G.S. *et al.:* "Falcon: A Production-Quality Distributed-Memory Reservoir Simulator," paper SPE 37975 presented at the 1997 SPE Reservoir Simulation Symposium, Dallas, 8–11 June.
188. Wang, P. *et al.:* "A New Generation EOS Compositional Reservoir Simulator: Part I—Formulation and Discretization," paper SPE 37979 presented at the 1997 SPE Reservoir Simulation Symposium, Dallas, 8–11 June.
189. Pavlas, E.J. Jr.: "MPP Simulation of Complex Water Encroachment in a Large Carbonate Reservoir in Saudi Arabia," paper SPE 71628 presented at the 2001 SPE Annual Technical Conference and Exhibition, New Orleans, 30 September–3 October.
190. Wallis, J.R. and Nolen, J.S.: "Efficient Linear Solution of Locally Refined Grids Using Algebraic Multilevel Approximate Factorizations," paper SPE 25239 presented at the 1993 SPE Symposium on Reservoir Simulation, New Orleans, 28 February–3 March.
191. Song, T.: "A Load-Balancing Technique for Reservoir Simulation Based on Dantzig's Transportation Model," MS thesis, U. of Houston, Houston (1996).
192. *Reference Manual for Zoltan,* Los Alamos Natl. Laboratory, Los Alamos, New Mexico (1998).
193. Bogeswara, R. and Killough, J.E.: "Parallel Linear Solvers for Reservoir Simulation: A Generic Approach for Existing and Emerging Computer Architectures," paper SPE 25240 presented at the 1993 SPE Symposium on Reservoir Simulation, New Orleans, 28 February–3 March.
194. Guerberoff, D. *et al.:* "Lateral Delineation of Sandstone Bodies Guided by Seismic and Petrophysical Data Using Geocellular Model: Canadon Seco Formation, San Jorge Basin, Argentina," paper SPE 69487 presented at the 2001 SPE Latin American and Caribbean Petroleum Engineering Conference, Buenos Aires, 25–28 March.
195. Agarwal, B. *et al.:* "Reservoir Characterization of Ekofisk Field: A Giant, Fractured Chalk Reservoir in the Norwegian North Sea," *SPEREE* (December 2000) 534.
196. Caers, J., Avseth, P., and Mukerji, T.: "Geostatistical Integration of Rock Physics, Seismic Amplitudes, and Geological Models in North Sea Turbidite Systems," paper SPE 71321 presented at the 2001 SPE Annual Technical Conference and Exhibition, New Orleans, 30 September–3 October.
197. Al Qassab, H.M. *et al.:* "Conditioning Integrated Geological Models to Dynamic Flow Data of Giant Saudi Arabian Reservoir," paper SPE 71319 presented at the 2001 SPE Annual Technical Conference and Exhibition, New Orleans, 30 September–3 October.
198. Raghavan, R. *et al.:* "Integration of Geology, Geophysics, and Numerical Simulation in the Interpretation of a Well Test in a Fluvial Reservoir," paper SPE 62983 presented at the 2000 SPE Annual Technical Conference and Exhibition, Dallas, 1–4 October.
199. Durlofsky, L.J.: "Upscaling and gridding of fine-scale geological models for flow simulation," *Proc.,* 2005 Intl. Forum on Reservoir Simulation, Stresa, Italy, 20–25 June.
200. Peaceman, D.W.: "Effective Transmissibilities of a Gridblock by Upscaling—Comparison of Direct Methods With Renormalization," *SPEJ* (September 1997) 338.
201. Flodin, E.A. *et al.:* "Representation of Fault Zone Permeability in Reservoir Flow Models," paper SPE 71617 presented at the 2001 SPE Annual Technical Conference and Exhibition, New Orleans, 30 September–3 October.
202. Nishikiori, N. and Hayashida, Y.: "Investigation of Fluid Conductive Faults and Modeling of Complex Water Influx in the Khadji Oil Field, Arabian Gulf," *SPEREE* (October 2000) 401.
203. Barker, J.W. and Thikcan, S.: "A Critical Review of the Use of Pseudorelative Permeabilities for Upscaling," *SPERE* (May 1997) 109.
204. Beckner, B.L. *et al.:* "EMpower: New Reservoir Simulation System," paper SPE 68116 presented at the 2001 SPE Middle Ease Oil Show, Bahrain, 17–20 March.
205. Peaceman, D.W.: "Interpretation of Well-Block Pressures in Numerical Reservoir Simulation with Nonsquare Grid Blocks and Anisotropic Permeability," *SPEJ* (June 1983) 531.
206. Mlacnik, M.J. and Heinemann, Z.E.: "Using Well Windows in Full Field Reservoir Simulation," *SPEREE* (December 2003) 275.
207. Wolfsteiner, C., Durlofsky, L.J., and Aziz, K.: "Approximate Model for Productivity of Nonconventional Wells in Heterogeneous Reservoirs," *SPEJ* (June 2000) 218.
208. Valvante, P.H., Durlosky, L.J., and Aziz, K.: "Semi-Analytical Modeling of the Performance of Intelligent Well Completions," paper SPE 66368 presented at the 2001 SPE Reservoir Simulation Symposium, Houston, 11–14 February.

209. Litvak, M.L. *et al.:* "Integration of Prudhoe Bay Surface Pipeline Network and Full Field Reservoir Models," paper SPE 38885 presented at the 1997 SPE Annual Technical Conference and Exhibition, San Antonio, Texas, 5–8 October.
210. Fang, W.Y. and Lo, K.K.: "A Generalized Well-Management Scheme for Reservoir Simulation," *SPERE* (May 1996) 116.
211. Litvak, M.L. and Wang, C.H.: "Simplified Phase-Equilibrium Calculations in Integrated Reservoir and Surface-Pipeline-Network Models," *SPEJ* (June 2000) 236.
212. Hermansen, H. *et al.:* "Twenty-Five Years of Ekofisk Reservoir Management," paper SPE 38927 presented at the 1997 SPE Annual Technical Conference and Exhibition, San Antonio, Texas, 5–8 October.
213. Syltc, J.E. *et al.:* "Water Induced Compaction in the Ekofisk Field," paper SPE 56426 presented at the 1999 SPE Annual Technical Conference and Exhibition, Houston, 3–6 October.
214. Ashton, K. *et al.:* "Judy/Joanne Field Development," paper SPE 49128 presented at the 1998 SPE Annual Technical Conference and Exhibition, New Orleans, 27–30 September.
215. Pieters, J. and Por, G.J.A.: "Total System Modeling—A Tool for Effective Reservoir Management of Multiple Fields With Shared Facilities," paper SPE 30442 presented at the 1995 SPE Offshore Europe Conference, Aberdeen, 5–8 September.
216. Williams, M.A. and Keating, J.F.: "The Stratigraphic Method: A Structured Approach to History Matching Complex Simulation Models," paper SPE 38014 presented at the 1997 SPE Reservoir Simulation Symposium, Dallas, 8–11 June.
217. van Ditzhuijzen, R., Oldenziel, T., and van Kruijsdijk, C.P.J.W.: "Geological Parameterization of a Reservoir Model for History Matching Incorporating Time-Lapse Seismic Based on a Case Study of the Statfjord Field," paper SPE 71318 presented at the 2001 SPE Annual Technical Conference and Exhibition, New Orleans, 30 September–3 October.
218. Bogan, C., Johnson, D., Litvak, M., and Stauber, D.: "Building Reservoir Models Based on 4D Seismic and Well Data in Gulf of Mexico Oil Fields," paper SPE 84370 presented at the 2003 SPE Annual Technical Conference and Exhibition, Denver, 5–8 October.
219. Milliken, W.J., Emanuel, A.S., and Chakravarty, A.: "Applications of 3D Streamline Simulation to Assist History Matching," paper SPE 63155 presented at the 2000 SPE Annual Technical Conference and Exhibition, Dallas, 1–4 October.
220. Schulze-Riegert, R.W. *et al.:* "Optimization Methods for History Matching of Complex Reservoir," paper SPE 66393 presented at the 2001 SPE Reservoir Simulation Symposium, Houston, 11–14 February.
221. Grussaute, T. and Gouel, P.: "Computer Aided History Matching of a Real Field Case," paper SPE 50642 presented at the 1998 SPE European Petroleum Conference, The Hague, 20–22 October.
222. Zapata, V.J. *et al.:* "Advances in Tightly Coupled Reservoir/Wellbore/Surface-Network Simulation," *SPEREE* (April 2001) 114.
223. Gorell, S. and Bassett, R.: "Trends in Reservoir Simulation: Big Models, Scalable Models? Will You Please Make Up Your Mind?" paper SPE 71596 presented at the 2001 SPE Annual Technical Conference and Exhibition, New Orleans, 30 September–3 October.
224. Zabalza-Mezghani, I., Mezghani, M., and Blanc, G.: "Constraining Reservoir Facies Models to Dynamic Data-Impact of Spatial Distribution Uncertainty on Production Forecasts," paper SPE 71335 presented at the 2001 SPE Annual Technical Conference and Exhibition, New Orleans, 30 September–3 October.
225. Manceau, E. *et al.:* "Combination of Experimental Design and Joint Modeling Methods for Quantifying the Risk Associated with Deterministic and Stochastic Uncertainties—An Integrated Test Study," paper SPE 71620 presented at the 2001 SPE Annual Technical Conference and Exhibition, New Orleans, 30 September–3 October.

SI Metric Conversion Factors

bbl	×	1.589 873	E − 01	= m^3
cp	×	1.0*	E − 03	= Pa·s
dyne	×	1.0*	E − 02	= mN
ft	×	3.048*	E − 01	= m

ft^3	×	2.831 685	E − 02	= m^3
°F		(°F − 32)/1.8		= °C
lbm	×	4.535 924	E − 01	= kg

*Conversion factor is exact.

Chapter 18
Estimation of Primary Reserves of Crude Oil, Natural Gas, and Condensate

Ron Harrell, SPE, Ryder Scott Co. and **Chap Cronquist,** SPE, Consultant

18.1 Introduction

Most exploration and production companies' assets consist of the resources and reserves owned by that entity. Resources typically are classified as either contingent (discovered but presently uneconomic) or undiscovered, but their quantification is outside the scope of this work. This chapter discusses primary reserves, which are discovered quantities of hydrocarbons that can be produced at a profit and are classified by level of uncertainty. The discussion of estimated reserves in this chapter excludes limitations imposed by the terms of ownership.

Since publication of the first two petroleum handbooks,[1,2] the international petroleum industry has increased its understanding of the need to develop more reliable estimates of petroleum reserves and to quantify the uncertainty associated with the classifications of reserves. Furthermore, the regulatory authorities of many countries, particularly of the U.S.A., have found it necessary to accommodate the increasing internationalization of the industry and to manage their increasing involvement and influence in the industry.

With this global expansion of the petroleum industry has come ongoing technological development that provides better tools and techniques for analyzing reservoirs and reservoir fluids as well as greater understanding of how reservoir geology affects reservoir performance. In developing mathematical models that replicate the geologic environment, the reservoir engineer must incorporate all available hard technical data and work closely with and use fully the experience of multidisciplinary teams of geophysicists, geologists, petrophysicists, and other specialists. The engineer must develop a working knowledge of the skills contributed by each member of the reserves assessment team and apply the professional judgment of each team member to the estimation and classification of reserves.

The estimation of reserves is more than just a periodic, statutory calculating and reporting of company assets (although those are important functions); it is an essential element of investment planning and resource management for every prudent operator. Estimating reserves begins with identifying a drillable prospect, and it continues while the prospect is developed and placed on production, and thereafter as warranted by well and/or reservoir performance, new geologic data, competitor (offset) operations, unitization, contract renegotiation, improved technology, and/or changing economic conditions.[3]

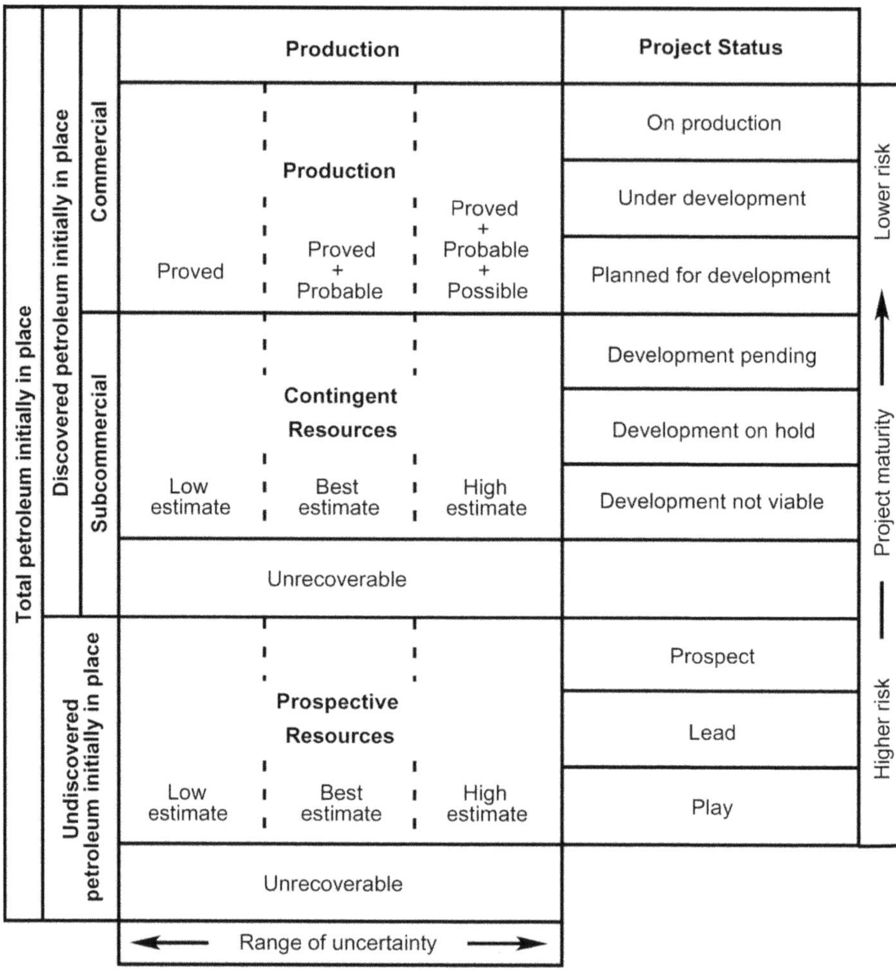

Fig. 18.1—Resource classification system, showing possible project status categories (after Ross[4]).

Fig. 18.1[4] illustrates the entire spectrum of in-place and recoverable hydrocarbons, the total resource base that is contained within the subsurface of the Earth. The vertical scale represents the total resource base, including cumulative production, reserves, and the unrecoverable portions of the discovered and economic quantities, as well as the recoverable and unrecoverable portions of the two categories of resources.

The horizontal scale reflects the increasing degrees of uncertainty of reserves quantities, showing (left to right) reserves classifications of proved through possible and resource-estimates categories of low to high. A project status column on the far right contains terms typically used to describe the stage of exploration and development associated with the various degrees of uncertainty. The classifications of proved plus probable and proved plus probable plus possible reserves are consistent with those used with probabilistic methods of reserves estimation and classification.

Fig. 18.2 adapts the so-called McKelvey[5] box to show the relationship between USGS reserves classifications and those of the 1997 Society of Petroleum Engineers/World Petroleum Congress (SPE/WPC),[6] which are quoted in a subsequent section. USGS (McKelvey[5]) classifications of identified (discovered) reserves (measured, indicated, and inferred) are approximately equivalent to the 1997 SPE/WPC classifications of proved, probable, and possible, respectively.

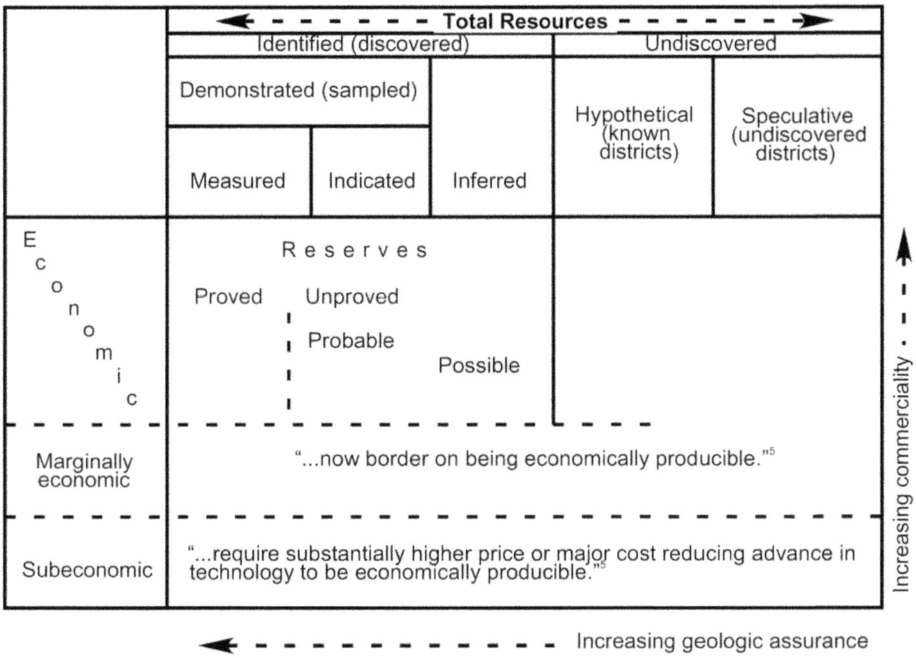

Fig. 18.2—The McKelvey box, illustrating the relationship between the U.S. Geological Survey (USGS) classification of resources and the 1997 SPE/WPC classifications of reserves. (After McKelvey.[5])

The integration of this concept with the 2001 SPE/WPC/AAPG (American Assn. of Petroleum Geologists) resource definitions[4] led to the development of Fig 18.1.

Reserves are the quantities that remain to be commercially produced as of a given date, under stated economic and operating conditions. Ultimate reserves are the sum of cumulative production and quantities (reserves) yet to be commercially produced. Terms such as "remaining reserves" and "recoverable reserves" are redundant. Terms such as "exploratory reserves," "geologic reserves," "potential reserves," and "technical reserves" often are used in conversation, but are misleading. The use of such terms is discouraged.

For this chapter discussion, reserves are quantities that are available for sale following lease or platform separation. Oil and condensate quantities do not reflect any reduction for field losses or field use. Natural-gas reserves typically are reported net after field condensate removal, but no consideration is given herein to gas volume reduction resulting from the extraction of natural-gas liquids or for the removal of contaminants. (See McMichael and Spencer[7] for a discussion of these issues.)

18.2 Reserves Definitions

18.2.1 1997 SPE/WPC Petroleum Reserves Definitions.*

Reserves are those quantities of petroleum which are anticipated to be commercially recovered from known accumulations from a given date forward. All reserves estimates involve some degree of uncertainty. The uncertainty depends chiefly on the amount of reliable geologic and engineering data available at the time of the estimate and the interpretation of these data. The relative degree of uncertainty may be conveyed by placing reserves into one of two principal classifications, either proved or unproved. Unproved reserves are

* Definitions here are quoted from Ref. 6. Cronquist[3] has provided comments regarding interpretation of some of the terms used in these definitions. **New definitions** were approved in 2007.

less certain to be recovered than proved reserves and may be further subclassified as probable and possible reserves to denote progressively increasing uncertainty regarding their recoverability.

The intent of the SPE and WPC in approving additional classifications beyond proved reserves is to facilitate consistency among professionals using such terms. In presenting these definitions, neither organization is recommending public disclosure of reserves classified as unproved. Public disclosure of the quantities classified as unproved reserves is left to the discretion of the countries or companies involved.

Estimation of reserves is done under conditions of uncertainty. The method of estimation is called *deterministic* [our emphasis] if a single best estimate of reserves is made based on known geological, engineering, and economic data. The method of estimation is called *probabilistic* [our emphasis] when the known geological, engineering, and economic data are used to generate a range of estimates and their associated probabilities. Identifying reserves as proved, probable, and possible has been the most frequent classification method and gives an indication of the probability of recovery. Because of potential differences in uncertainty, caution should be exercised when aggregating reserves of different classifications.

Reserves estimates will generally be revised as additional geologic or engineering data become available or as economic conditions change. Reserves do not include quantities of petroleum being held in inventory, and they may be reduced for usage or processing losses if required for financial reporting.

Reserves may be attributed to either natural energy or improved recovery methods. Improved recovery methods include all methods for supplementing natural energy or altering natural forces in the reservoir to increase ultimate recovery. Examples of such methods are pressure maintenance, cycling, waterflooding, thermal methods, chemical flooding, and the use of miscible and immiscible displacement fluids. Other improved recovery methods may be developed in the future as petroleum technology continues to evolve.

Proved Reserves. Proved reserves are those quantities of petroleum which, by analysis of geological and engineering data, can be estimated with reasonable certainty to be commercially recoverable, from a given date forward, from known reservoirs and under current economic conditions, operating methods, and government regulations. Proved reserves can be categorized as developed or undeveloped.

If deterministic methods are used, the term reasonable certainty is intended to express a high degree of confidence that the quantities will be recovered. If probabilistic methods are used, there should be at least a 90% probability that the quantities actually recovered will equal or exceed the estimate.

Establishment of current economic conditions should include relevant historical petroleum prices and associated costs and may involve an averaging period that is consistent with the purpose of the reserves estimate, appropriate contract obligations, corporate procedures, and government regulations involved in reporting these reserves.

In general, reserves are considered proved if the commercial producibility of the reservoir is supported by actual production or formation tests. In this context, the term "proved" refers to the actual quantities of petroleum reserves and not just the productivity of the well or reservoir. In certain cases, proved reserves may be assigned on the basis of well logs and/or core analysis that indicate the subject reservoir is hydrocarbon-bearing and is analogous to reservoirs in the same area that are producing or have demonstrated the ability to produce on formation tests.

The area of the reservoir considered as proved includes (1) the area delineated by drilling and defined by fluid contacts, if any, and (2) the undrilled portions of the reservoir that can reasonably be judged as commercially productive on the basis of available geological and engineering data. In the absence of data on fluid contacts, the lowest known occurrence of hydrocarbons controls the proved limit unless otherwise indicated by definitive geological, engineering, or performance data.

Reserves may be classified as proved if facilities to process and transport those reserves to market are operational at the time of the estimate or there is a reasonable expectation that such facilities will be installed. Reserves in undeveloped locations may be classified as proved undeveloped provided (1) the locations are direct offsets to wells that have indicated commercial production in the objective formation, (2) it is reasonably certain such locations are within the known proved productive limits of the objective formation, (3) the locations conform to existing well spacing regulations where applicable, and (4) it is reasonably certain the locations will be developed. Reserves from other locations are classified as proved undeveloped only where interpretations of geological and engineering data from wells indicate with reasonable certainty that the objective formation is laterally continuous and contains commercially recoverable petroleum at locations beyond direct offsets.

Reserves that are to be produced through the application of established improved recovery methods are included in the proved classification when (1) successful testing by a pilot project or favorable response of an installed program in the same or an analogous reservoir with similar rock and fluid properties provides support for the analysis on which the project was based, and (2) it is reasonably certain that the project will proceed. Reserves to be recovered by improved recovery methods that have yet to be established through commercially successful applications are included in the proved classification only (1) after a favorable production response from the subject reservoir from either (a) a representative pilot or (b) an installed program where the response provides support for the analysis on which the project is based and (2) it is reasonably certain the project will proceed.

Unproved Reserves. Unproved reserves are based on geologic and/or engineering data similar to those used in estimates of proved reserves; but technical, contractual, economic, or regulatory uncertainties preclude such reserves being classified as proved. Unproved reserves may be further classified as probable reserves and possible reserves.

Unproved reserves may be estimated assuming future economic conditions different from those prevailing at the time of the estimate. The effect of possible future improvements in economic conditions and technological developments can be expressed by allocating appropriate quantities of reserves to the probable and possible classifications.

Probable Reserves. Probable reserves are those unproved reserves which analysis of geological and engineering data suggests are more likely than not to be recoverable. In this context, when probabilistic methods are used, there should be at least a 50% probability that the quantities actually recovered will equal or exceed the sum of estimated proved plus probable reserves.

In general, probable reserves may include (1) reserves anticipated to be proved by normal step-out drilling where subsurface control is inadequate to classify these reserves as proved, (2) reserves in formations that appear to be productive based on well log characteristics but lack core data or definitive tests and which are not analogous to producing or proved reserves in the area, (3) incremental reserves attributable to infill drilling that could have been classified as proved if closer statutory spacing had been approved at the time of the estimate, (4) reserves attributable to improved recovery methods that have been established by repeated commercially successful applications when (a) a project or pilot is planned but not in operation and (b) rock, fluid, and reservoir characteristics appear favorable for commercial application, (5) reserves in an area of the formation that appears to be separated from the proved area by faulting and the geologic interpretation indicates the subject area is structurally higher than the proved area, (6) reserves attributable to a future workover, treatment, retreatment, change of equipment, or other mechanical procedures, where such procedure has not been proved successful in wells which exhibit similar behavior in analogous reservoirs, and (7) incremental reserves in proved reservoirs where an alternative interpretation of performance or volumetric data indicates more reserves than can be classified as proved.

Possible Reserves. Possible reserves are those unproved reserves which analysis of geological and engineering data suggests are less likely to be recoverable than probable reserves. In this context, when probabilistic methods are used, there should be at least a 10% probability that the quantities actually recovered will equal or exceed the sum of estimated proved plus probable plus possible reserves.

In general, possible reserves may include (1) reserves which, based on geological interpretations, could possibly exist beyond areas classified as probable, (2) reserves in formations that appear to be petroleum-bearing based on log and core analysis but may not be productive at commercial rates, (3) incremental reserves attributed to infill drilling that are subject to technical uncertainty, (4) reserves attributed to improved recovery methods when (a) a project or pilot is planned but not in operation and (b) rock, fluid, and reservoir characteristics are such that a reasonable doubt exists that the project will be commercial, and (5) reserves in an area of the formation that appears to be separated from the proved area by faulting and geological interpretation indicates the subject area is structurally lower than the proved area.

Reserves Status Categories. Reserves status categories define the development and producing status of wells and reservoirs.

Developed Reserves. Developed reserves are expected to be recovered from existing wells including reserves behind pipe. Improved recovery reserves are considered developed only after the necessary equipment has been installed, or when the costs to do so are relatively minor. Developed reserves may be subcategorized as producing or nonproducing.

Producing. Reserves subcategorized as producing are expected to be recovered from completion intervals that are open and producing at the time of the estimate. Improved recovery reserves are considered producing only after the improved recovery project is in operation.

Nonproducing. Reserves subcategorized as nonproducing include shut-in and behind-pipe reserves. Shut-in reserves are expected to be recovered from (1) completion intervals that are open at the time of the estimate but that have not started producing, (2) wells which were shut in for market conditions or pipeline connections or (3) wells not capable of production for mechanical reasons. Behind-pipe reserves are expected to be recovered from zones in existing wells that will require additional completion work or future recompletion prior to the start of production.

Undeveloped Reserves. Undeveloped reserves are expected to be recovered (1) from new wells on undrilled acreage, (2) from deepening existing wells to a different reservoir, or (3) where a relatively large expenditure is required to (a) recomplete an existing well or (b) install production or transportation facilities for primary or improved recovery projects.

The 1997 SPE/WPC definitions[6] quoted above evolved over many years through the efforts of several organizations. The Soc. of Petroleum Evaluation Engineers (SPEE) contributed significantly to the 1981 and 1987 SPE Petroleum Reserves definitions.[8,9] The 1981 definitions refer only to proved reserves; the 1987 definitions introduce the concept of unproved reserves and the subclassifications of probable and possible reserves.

The 1997 SPE/WPC definitions recognize both the deterministic and probabilistic methods and establish relative standards for each. The SPE and the WPC continue to work together to improve these definitions in light of several unresolved ambiguities. Such ambiguities include reconciliation of the terms "reasonable certainty" and "at least a 90% probability" used to describe proved reserves. For deterministic estimates, quantities for each of the three reserves categories of proved, probable, and possible are estimated, and the evaluator is cautioned not to add these quantities together because of the differing degrees of uncertainty associated with each. Conversely, the probabilistic approach requires the addition of "proved plus probable" or "proved plus probable plus possible" categories to comply with the stated numerical levels of uncertainty. Further, definitions place certain specified limitations on the "lowest known occurrence of hydrocarbons," average prices, and conformance to regulatory well spacing (where applicable), thus reducing the variables that are subject to classic probabilistic reserves assessment.

18.2.2 1978 U.S. Securities and Exchange Commission (SEC) Oil and Gas Reserves Definitions.[1] The U.S. SEC Regulation S X Rule 4 10 paragraph (a)[10] defines proved reserves as follows:*

Proved Oil and Gas Reserves. Proved oil and gas reserves are the estimated quantities of crude oil, natural gas, and natural gas liquids which geological and engineering data demonstrate with reasonable certainty to be recoverable in future years from known reservoirs under existing economic and operating conditions; i.e., prices and costs as of the date the estimate is made. Prices include consideration of changes in existing prices provided only by contractual arrangements, but not on escalations based upon future conditions.

1. Reservoirs are considered proved if economic producibility is supported by either actual production or conclusive formation test. The area of a reservoir considered proved includes:

(A) that portion delineated by drilling and defined by gas/oil and/or oil/water contacts, if any; and

(B) the immediately adjoining portions not yet drilled, but which can be reasonably judged as economically productive on the basis of available geological and engineering data. In the absence of information on fluid contacts, the lowest known structural occurrence of hydrocarbons controls the lower proved limit of the reservoir.

2. Reserves which can be produced economically through application of improved recovery techniques (such as fluid injection) are included in the "proved" classification when successful testing by a pilot project, or the operation of an installed program in the reservoir, provides support for the engineering analysis on which the project or program was based.

3. Estimates of proved reserves do not include the following:

(A) oil that may become available from known reservoirs but is classified separately as "indicated additional reserves";

(B) crude oil, natural gas, and natural gas liquids, the recovery of which is subject to reasonable doubt because of uncertainty as to geology, reservoir characteristics, or economic factors;

(C) crude oil, natural gas, and natural gas liquids, that may occur in undrilled prospects; and

(D) crude oil, natural gas, and natural gas liquids, that may be recovered from oil shales, coal,† gilsonite and other such sources.

Proved Developed Oil and Gas Reserves. Proved developed oil and gas reserves are reserves that can be expected to be recovered through existing wells with existing equipment and operating methods. Additional oil and gas expected to be obtained through the application of fluid injection or other improved recovery techniques for supplementing the natural forces and mechanisms of primary recovery should be included as "proved developed reserves" only after testing by a pilot project or after the operation of an installed program has confirmed through production response that increased recovery will be achieved.

Proved Undeveloped Oil and Gas Reserves. Proved undeveloped oil and gas reserves are reserves that are expected to be recovered from new wells on undrilled acreage, or from existing wells where a relatively

*Subsection formatting in this section follows that of the SEC.

†SAB12, following, modifies this ruling.

major expenditure is required for recompletion. Reserves on undrilled acreage shall be limited to those drilling units offsetting productive units that are reasonably certain of production when drilled. Proved reserves for other undrilled units can be claimed only where it can be demonstrated with certainty that there is continuity of production from the existing productive formation. Under no circumstances should estimates for proved undeveloped reserves be attributable to any acreage for which an application of fluid injection or other improved recovery technique is contemplated, unless such techniques have been proved effective by actual tests in the area and in the same reservoir.

18.2.3 U.S. SEC Staff Accounting Bulletins (SABs). Certain SABs published after Regulation S X[10] concern the application of financial accounting and disclosure rules for oil and gas producing activities. In particular, the following interpretations extracted from a recent SEC compilation of SABs[11,12,13] set forth the Commission staff's view on specific questions pertaining to proved oil and gas reserves:

Economic producibility of estimated proved reserves can be supported to the satisfaction of the Office of Engineering if geological and engineering data demonstrate with reasonable certainty that those reserves can be recovered in future years under existing economic and operating conditions. The relative importance of the many pieces of geological and engineering data which should be evaluated when classifying reserves cannot be identified in advance. In certain instances, proved reserves may be assigned to reservoirs on the basis of a combination of electrical and other type logs and core analyses which indicate the reservoirs are analogous to similar reservoirs in the same field which are producing or have demonstrated the ability to produce on a formation test.[11]

In determining whether "proved undeveloped reserves" encompass acreage on which fluid injection (or other improved recovery technique) is contemplated, is it appropriate to distinguish between (i) fluid injection used for pressure maintenance during the early life of a field and (ii) fluid injection used to effect secondary recovery when a field is in the late stages of depletion? The Office of Engineering believes that the distinction identified in the above question may be appropriate in a few limited circumstances, such as in the case of certain fields in the North Sea. The staff will review estimates of proved reserves attributable to fluid injection in the light of the strength of the evidence presented by the registrant in support of a contention that enhanced recovery will be achieved.[11]

Companies should report reserves of natural gas liquids which are net to their leasehold interest; i.e., that portion recovered in a processing plant and allocated to the leasehold interest. It may be appropriate in the case of natural gas liquids not clearly attributable to leasehold interests ownership to follow instruction (b) of Item 2(b)(3) of Regulation S K and report such reserves separately and describe the nature of the ownership.[11]

The staff believes that since coalbed methane gas can be recovered from coal in its natural and original location, it should be included in proved reserves, provided that it complies in all other respects with the definition of proved oil and gas reserves as specified in Rule 4 10(a)(2) including the requirement that methane production be economical at current prices, costs, (net of the tax credit) and existing operating conditions.[12]

SAB statements are not rules or interpretations of the Commission, nor are they published as bearing the Commission's official approval. They represent interpretations and practices followed by the U.S. SEC's Div. of Corporation Finance and Office of the Chief Accountant in administering the disclosure requirements of the U.S. securities laws.

18.2.4 Unproved Oil and Gas Reserves. U.S. SEC Regulation S K[13] prohibits the disclosure of estimated quantities of probable and possible reserves of oil and gas and any estimated value thereof in any documents publicly filed with the Commission.

18.2.5 Website Release. In a 21 February 2001 website release,[14] the U.S. SEC addressed several topics relative to the reporting of proved reserves. No changes were made to the 1978 definitions, but the U.S. SEC engineering staff modified its application and interpretation of those definitions in light of the widespread technological advances in over the previous 20 years. The most significant of these application changes are discussed below.

18.2.6 Significance of the 1997 SPE/WPC and the 1978 U.S. SEC Reserves Definitions. The 1997 SPE/WPC definitions[6] have been proposed as the technical standards for the international petroleum industry. Regulatory authorities worldwide have been encouraged to use these definitions as much they reasonably can for their specific purposes.

When the U.S. SEC definitions were approved (1978), virtually all regulated companies were in the U.S.A. and most reported reserves were located in North America, neither of which is now the case. Also, virtually all natural gas then was sold through long-term contracts with a defined pricing structure. Oil prices were less volatile then, compared with price swings seen over the past 10 to 15 years. Privatization of numerous national or state-owned oil and gas companies, with the sale of their securities within the U.S.A., has greatly enlarged the effective reach and importance of the U.S. SEC. The U.S. SEC's petroleum-engineering staff has been increased in recent years to enable the agency to monitor more effectively the reserves-reporting activities of an increasing number of oil and gas companies subject to the jurisdiction of the agency.

18.2.7 1997 SPE/WPC and 1978 U.S. SEC Definitions Comparison. ***Reasonable Certainty.*** The SPE/WPC and the U.S. SEC definitions both use "reasonable certainty" to describe the controlling condition for proved reserves, but neither offers quantification of the term. Most engineers now accept that reasonable certainty indicates relatively high confidence. The 1997 SPE/WPC definitions require that for proved reserves estimated using probabilistic methods there be at least 90% probability (a "P90" estimate) that the quantity recovered will equal or exceed the estimated quantity. At least 50% probability (a "P50" estimate) is required for proved plus probable reserves estimated using probabilistic methods.

Proved reserves estimated using deterministic methods seldom, if ever, will meet a P90 requirement. Most volumetric estimates use *average* porosity, *average* water saturation, and recovery efficiencies (REs) that are consistent with expected reservoir drive mechanisms and operating conditions.

If an oil/water contact (OWC) or gas/water contact (GWC) is known, the resultant volumetric reserves estimate logically may be described as the "most likely" estimate, which might approach a P50 (probabilistic) estimate. For another example, when using a conservative RE because of an unknown drive mechanism and a lowest known limit of oil (LKO) or gas (LKG), the equivalent confidence level may be between P50 and P90.

For performance estimates using trend analysis, the engineer typically will extrapolate a "best fit" from the historical information, which essentially reflects a P50 estimate.[3] Extrapolations could be prepared for P90, P50, and P10 estimates, but this seldom is done.

In summary, reasonable certainty for deterministic estimates will represent confidence levels approaching P50 in most instances, but will seldom approach the P90 level as defined in the 1997 SPE/WPC definitions.

Known Accumulations. The 1997 SPE/WPC definitions use the term *known accumulations*, which the U.S. SEC definitions refer to as *known reservoirs*. In the industry, a known accumulation is an underground collection of moveable petroleum—one or more reservoirs confirmed through the drilling and gathering of reservoir data from one or more wells. A known accumulation must be considered commercial before reserves of any classification may be assigned.

Economic Conditions. The U.S. SEC definitions[10] specify that proved reserves are to be reported consistent with "existing economic and operating conditions; i.e., prices and costs as of the date the estimate is made." This is intended to be the contract price for the sale of oil and gas as of the report date, which typically is 31 December. For products sold in a spot market, the pricing available on the report date should be used. Operating costs typically are costs averaged over the preceding 12 months, unless a change material to the estimate has occurred during the preceding 6 months. In the estimation of proved reserves, the 1997 SPE/WPC definitions allow the use of a historical oil and gas price averaged over a time period that is "consistent with the purpose of the estimate."

According to the 1997 SPE/WPC definitions, probable and possible reserves also must be considered commercial before they can be classified, but they may be based on more favorable economic conditions than those existing at the effective date of the estimate.

Development Status. The U.S. SEC continues to enforce the 1978 definitions regarding undeveloped well locations, requiring that the proved classification be given only to those locations beyond one offset from a productive well where "it can be demonstrated with *certainty* that there is continuity of production" [emphasis added] from the existing productive formation. A subsequent clarification released by the U.S. SEC in 2001[14] stated that "there is no mitigating modifier for the word certainty."

These provisions are troublesome in at least three regards. First, the term "offset" refers to regulatory-controlled well spacing in North America and to few areas, if any, elsewhere. Second, and more importantly, the term "certainty" is used here by the U.S. SEC to describe undeveloped reserves, whereas the controlling term in the first sentence of the U.S. SEC definitions is "reasonable certainty." Third, the expression "continuity of production" is undefined, but is used in a context of a higher standard than reservoir continuity.

The 1997 SPE/WPC definitions leave the classification of reserves from undrilled locations to the discretion of the evaluator using good engineering practices.

Requirement for Flow Testing. Both the U.S. SEC definitions and the 1997 SPE/WPC definitions require evidence of producibility through actual production or conclusive formation tests. Both definitions also permit exceptions in certain cases. The U.S. SEC may grant an exception for a definitive flow test for a particular reservoir if core *and* log data indicate commercial productivity and the reservoir is analogous to one or more flow-tested or producing reservoir(s) in the same field. The 1997 SPE/WPC definitions have a similar but less restrictive flow-test exemption that requires favorable core *and/or* log data for the subject reservoir compared to an analogous reservoir(s) in the same area that has demonstrated commercial productivity.

There is increasing interest by evaluators using both the U.S. SEC and the 1997 SPE/WPC reserves definitions in certain instances, particularly for the deepwater areas of the U.S. Gulf of Mexico, where physical well testing is impractical because of costs and environmental concerns. Many of these discoveries are not close to other fields, but may be characterized by thick sections of highly permeable sandstones saturated with high-API-gravity, low-viscosity crudes and exhibiting reservoir pressures in excess of 8,000 psia. In such scenarios, flow calculations can confirm production rates that are considerably higher than the minimum commercial rates, and are adequate for producers to make commercial decisions about facility sizing and project sanctioning. The evaluator determines on a case-specific basis whether such indirect data are adequate to confirm commercial flow rates (without a physical flow test) in consideration of the applicable definitions.

Areal Extent of a Reservoir. The U.S. SEC definitions mandate the use of a recognized OWC, gas/oil contact (GOC), and GWC to define structural limits of proved reserves. In the absence of such contacts, limits are imposed by the LKO or LKG, which typically are defined as the subsea depth of the base of the (permeable) reservoir, as recorded on well logs. Only in recent years has the U.S. SEC begun to consider indirect measurements or calculations of controlling contacts. The U.S. SEC's acceptance of indirect indicators such as pressure/depth calculations, seismic information, pressure-gradient calculations, and capillary-pressure computations has been only in exceptional cases in which the quality of the data was excellent and the presentation of the interpretation was compelling. In their acceptance of a limiting contact below the base of the lowest logged interval, the U.S. SEC engineering staff may further consider the materiality of the quantity of reserves added through the use of pressure-gradient data. They provide no definition of materiality. Also, agreement between two or more indirect measurements would be required as the basis for establishing proved reserves. Such agreement also applies to 1997 SPE/WPC definitions.

Enhanced-Recovery Reserves. The U.S. SEC traditionally has required that the proved designation be limited to reservoirs where enhanced-recovery (improved-recovery) methods have been demonstrated through a successful pilot project or an installed program "in the reservoir." This requirement was reinterpreted in 2001[14] to allow successful, operating analog reservoirs in the same geologic area to be used as support for assigning proved enhanced-recovery reserves. The subject reservoir should have reservoir and fluid parameters that are equal to or more favorable than those of the analog reservoir(s). The 1997 SPE/WPC definitions also permit the use of successful analogs in the area as the basis for establishing proved enhanced-recovery reserves.

Other Differences. The 1997 SPE/WPC and 1978 U.S. SEC definitions contain many other wording differences regarding proved reserves, but those cited above are of major importance. For questions of interpretation or application, however, the U.S. SEC typically will respond to inquiries by interested parties.

18.2.8 Other Reserves Definitions. Other petroleum-producing countries and regulatory authorities have promulgated petroleum reserves definitions that are recognized and based on years of development and sound engineering judgment, but none—however important in their sphere of influence—with such far-reaching consequences as the 1978 U.S. SEC and the 1997 SPE/WPC definitions. (See Cronquist[3] for a summary of many of these other definitions.)

18.3 Calculation Procedures

Although reserves estimates for known accumulations historically have used deterministic calculation procedures, the 1997 SPE/WPC definitions allow either deterministic or probabilistic procedures. Each of these is discussed briefly in the next two sections. Thereafter—except for another section on probabilistic procedures near the end—the chapter will focus on deterministic procedures because they still are more widely used. Both procedures need the same basic data and equations.

18.3.1 Deterministic Procedures. Deterministic calculations of oil and/or gas initially in place (O/GIP) and reserves are based on best estimates of the true values of pertinent parameters, although it is recognized that there may be considerable uncertainty in such values. Reserves calculated using such procedures are classified subjectively on the basis of professional judgments of the uncertainty in each reserve estimate and/or of pertinent regulatory and/or corporate guidelines.

18.3.2 Probabilistic Procedures. Probabilistic procedures recognize that uncertainties in input data and equations to calculate reserves may be significant. Accordingly, each input parameter uses a reasonable range of values, from which a set of reserves calculations is made. Reserves calculated using this procedure are classified on the basis of specified percentile rankings of reserves estimates within the calculated set and/or of pertinent regulatory and/or corporate guidelines.

18.4 Estimation Methods

Methods to estimate reserves may be categorized as either static or dynamic. Static methods typically are used before production is initiated in a subject reservoir, and include analogy methods and volumetric methods. Computer simulation that is used before production initiation is considered a static method. Dynamic methods might be used after sustained production has been initiated, and include production trend analysis, material-balance calculations, and computer simulation. Because dynamic methods typically consider well and/or reservoir performance, they generally are considered more reliable than static methods. A specific reserves estimate might involve one or more such methods. What method (or methods) are used depends on several factors, including production history of the area, if any; stage of development on the date of the estimate; geologic complexity; quality and quantity of data; maturity of production for the subject reservoir; and the purpose of the estimate. Each estimate should be corroborated using an alternate, preferably independent, method.

18.4.1 Analogy Methods. Analogy methods typically are used to estimate ultimate recovery—or unit recovery factors—of oil and/or gas for *undrilled locations* and to supplement volumetric methods of estimating reserves in the early stages of development and production. The analogy method assumes that the analogous reservoir or well is comparable to the subject reservoir or well in those aspects that control ultimate recovery of oil and/or gas. The method's weakness is that this assumption's validity cannot be determined until the subject reservoir or well has been produced long enough to estimate reserves using dynamic methods.

In some scenarios, analogy may be the *only* feasible method until there are sufficient pressure and/or production data for a reliable analysis of performance. Such scenarios include areas of widely spaced development, where subsurface information might be too sparse to facilitate reliable volumetric mapping, and reservoirs where log, core, and/or test data are insufficient for reliable characterization.

There are two broad categories of analogy methods: analytical and statistical. Regardless of the method used, however, analogous and subject reservoirs should be similar in their structural configuration; lithology and depositional environment of the reservoir rock; nature and degree of principal heterogeneity; average net thickness and ratio of net pay to gross pay; petrophysics of the rock/fluid system; initial pressure and temperature; reservoir-fluid properties and drive mechanism; spatial relationship between free gas (if any), oil, and aquifer at initial conditions ("stacked" vs. "en echelon"); well spacing; and economic scenario. Because all these reservoir aspects seldom, if ever, are similar, judicious compensating adjustments from analogous to subject reservoir usually are needed. Such adjustments require considerable local knowledge and reservoir engineering experience.

Analytical (analogy) methods include using recovery *factors* (e.g., STBO/acre-ft of reservoir) or recovery *efficiencies* (percent recovery) from analogous reservoirs to estimate oil and/or gas reserves for wells or reservoirs being studied. Basic rock and/or fluid parameters—porosity, water saturation, and formation volume factor (FVF)—may be used with, or as modifiers to, these recovery factors or recovery efficiencies. For example, the recovery factor observed in an analogous reservoir (F_{RA}) might be adjusted by:

$$F_{RS} = F_{RA}(\phi S_{hi})_S / (\phi S_{hi})_A, \qquad (18.1)$$

where the terms $(\phi S_{hi})_S$ and $(\phi S_{hi})_A$ refer to subject and analogous reservoirs, respectively.

Depending on circumstances, more complex relations might be appropriate, such as one or more of the factorial groups in Eqs. 18.14 and 18.15, which are discussed in the Recovery Efficiency section below.

TABLE 18.1—AGGREGATE RECOVERY EFFICIENCY FOR GROUPS OF OIL RESERVOIRS IN VARIOUS REGIONS OF THE U.S.A. (AFTER CRONQUIST[3])

State or Region	Dominant Lithology	Main Drive Mechanism	Recovery Efficiency[a]
Northern Louisiana	Carbonate[b]	Pressure depletion	0.15
Southern Louisiana	Sandstone[c]	Waterdrive	0.51[d]
Oklahoma	Sandstone[e]	Waterflood	0.34
Pennsylvania	Sandstone[f]	Waterflood	0.20
Permian Basin	Carbonate[g]	Waterflood	0.39[h]
Texas Gulf Coast	Sandstone[i]	Waterflood	0.47[j]
Texas Panhandle	Carbonates[k]	Pressure depletion	0.23
West Virginia	Sandstone[l]	Pressure depletion	0.21

[a]Calculated from API data.[15] Almost all fields included produced "light" oil. [b]Mostly Lower Cretaceous/Jurassic, porosity = 5 to 20%, permeability = 1 to 500 md. [c]Chiefly Miocene, porosity = 15 to 35%, permeability = 50 to 2,500 md. [d]Estimated ultimate recovery efficiency from a few high-quality giant fields = 56 to 66%. [e]Mostly Pennsylvanian, porosity = 5 to 20%, permeability = 5 to 100 md. [f]Mostly poor-quality Devonian graywackes. [g]Chiefly Permian shelf carbonates, porosity = 5 to 20%, permeability = 1 to 100 md. [h]Estimated ultimate recovery efficiency from some of the better floods = 50 to 60%. [i]Chiefly lower Cenozoic, porosity = 15 to 35%, permeability = 50 to 2,500 md. [j]Estimated ultimate recovery efficiency from several high-quality giant fields = 52 to 61%, with a few approaching 70 to 74%. [k]Mostly Upper Paleozoic, porosity = 5 to 15%, permeability = 1 to 100 md. [l]Middle Paleozoic, porosity = 5 to 15%, permeability = 1 to 100 md.

In many areas, estimates of rock properties—porosity, initial water saturation, and net pay—made from wireline logs are subject to considerable uncertainty. Without core data or definitive formation tests, analogy may be the only method available initially to estimate reserves. Analogies can be drawn from mature reservoirs in comparable geologic and engineering settings. Several examples are provided for mature areas in the U.S.A.

Using American Petroleum Institute (API) data,[15] *aggregate* REs were estimated[3] for groups of oil reservoirs in various regions in the U.S.A. (**Table 18.1**). (Comparable data on natural gas reservoirs in the U.S.A. have not been published.) From Table 18.1, one might note (1) for northern and southern Louisiana, for example, the respective influence of lithology and drive mechanism on RE; (2) for Oklahoma and Pennsylvania, for example, the significant difference in waterflood (ultimate) RE, which probably results from the Pennsylvania waterfloods' significantly poorer rock quality; (3) for southern Louisiana and the Texas Gulf Coast, for example, the reasonable agreement in waterdrive RE. The slightly higher-quality sands in the Louisiana Gulf Coast probably cause small differences.

From a review of these (admittedly limited) data, it appears that variations in aggregate oil RE between regions are attributable principally to differences in reservoir rock/fluid systems and reservoir drive mechanisms. SPEE has published additional, useful summary discussions of regional variations in rock quality and reservoir drive mechanism for various areas in the U.S.A.[16]

In 1984, the API published estimates of oil initially in place (OIP) and primary RE for 533 reservoirs in the U.S.A. and estimates of primary and waterflood RE for 230 reservoirs in the U.S.A.[17] (**Table 18.2**). Estimates such as these can be developed for other areas using detailed data published annually by most U.S.A. state oil and gas regulatory commissions, the Alberta Energy Resource Conservation Board (ERCB) [which since 1995 has been part of the Alberta Energy and Utilities Board (EUB)], the Saskatchewan Dept. of Energy and Mines, various regional geological societies, and other such agencies. This type of regional data might be useful

TABLE 18.2—OIL INITIALLY IN PLACE AND PRIMARY-RECOVERY FACTORS FOR 533 U.S.A. RESERVOIRS (AFTER REF. 17)

			OIP, STB/NAF		Recovery, STB/NAF		
Lithology	State or Region	Number of Reservoirs	Range	Average	Range	Average	Average RE
			Solution Gas Drive				
Sandstone	California	51	328 to 2,146	1,148	55 to 644	273	0.24
	Louisiana	17	360 to 1,540	1,167	110 to 613	315	0.27
	Oklahoma	41	386 to 1,030	604	40 to 227	117	0.19
	NW Texas	24	Not reported	578	Not reported	86	0.15
	SE Texas	38	Not reported	827	Not reported	205	0.25
	West Virginia	13	546 to 882	678	100 to 250	143	0.21
	Wyoming	23	386 to 1,135	692	82 to 361	176	0.25
Carbonate	All	100	59 to 1,067	526	5 to 190	93	0.18
			Natural Waterdrive				
Sandstone	California	9	508 to 1,770	1,020	169 to 568	367	0.36
	Louisiana	42	535 to 1,960	1,203	219 to 1,470	722	0.60
	Texas	101	547 to 1,801	1,167	125 to 1,094	633	0.54
	Wyoming	27	Not reported	860	Not reported	313	0.36
Carbonate	All	47	85 to 705	265	29 to 370	117	0.44

in making *preliminary* estimates of reserves, pending development of specific data for the well/reservoir in question.

Information presented in Tables 18.1 and 18.2 is included only as reference information and should be considered only in the absence of definitive data available from specific wells or reservoirs.

Statistical (analogy) methods include using per-well recoveries of oil and/or gas from analogous wells in the same producing trend or in analogous geologic settings to estimate recoveries from wells being studied. Two types of statistical (analogy) methods are discussed here: isoultimate recovery maps and analysis of observed frequency distributions of ultimate recovery. Additional comments are provided in the Recovery Efficiency section below.

When analogy methods are used to estimate reserves for individual wells, the analogous and subject wells should be similar regarding well completion, including stimulation type (if any); production method; initial absolute open-flow potential for gas wells; initial potential and/or productivity index for oil wells; well spacing; and economic limit.

In some producing areas, ultimate recovery of oil or gas from individual wells may be controlled by geologic trends such as depositional environment, intensity of fracturing, or degree of diagenesis. In such cases, an isoultimate recovery map can be made by posting and contouring estimated ultimate recovery from individual wells in the area of interest. Such a map can be used to estimate reserves for undrilled tracts, *but one should use this technique cautiously!* Nongeologic factors might control ultimate recovery of oil or gas.

Different completion and stimulation procedures may yield different ultimate recoveries from individual wells. For example, for wells in several areas of the U.S.A., a correlation has been observed between size of fracture treatment and ultimate recovery.[18] Thus, before relying on isoultimate recovery maps, one should determine whether there is a statistically valid relation between ultimate recovery and the completion or stimulation method or another parameter. Wells in the area of interest may be capable of draining more than a "spacing unit." In this case, wells that are adjacent to undrilled tracts and those placed on production early in reservoir life may exhibit higher ultimate recoveries than wells in central locations and those placed on production late in reservoir life. One should investigate these possibilities before using an isoultimate recovery map to estimate reserves for undrilled tracts.

An alternative to isoultimate recovery maps is the use of "bubble maps," in which the size of the circle drawn around a wellbore is proportional to the parameter or measurement being compared.

Factors such as well spacing, completion technique, operating cost, and operating procedure can affect ultimate recovery of oil and/or gas significantly. If there are significant differences in these factors from one property to the next, be careful in making comparisons or statistical analyses of wells between such properties.

If other factors are more or less the same, the RE of oil and/or gas generally will be greater in areas with low operating costs and closely spaced wells than in areas with high operating costs and widely spaced wells. Most statistical data on RE in the U.S.A. have come from operations with onshore properties, which historically have had relatively low operating costs and closely spaced wells. These are not typical conditions in other areas of the world, especially offshore areas; thus, one should exercise caution when using REs determined from U.S.A. reservoirs to estimate REs from otherwise analogous reservoirs elsewhere.

18.4.2 Volumetric Methods. Volumetric methods to estimate reserves generally are used early in the life of a reservoir, before there are sufficient production and/or pressure data to use the performance method, and for behind-pipe zones, which might not be placed on production until the current completion zone is abandoned. Although volumetric methods are the most widely used methods for estimating reserves, results from their use might be subject to considerable uncertainty, depending on the geologic setting and the amount and quality of geologic and engineering data. Thus, it is recommended that an evaluator *compare reserves estimated by volumetric methods against well and reservoir performance at the earliest practical stage of production and make adjustments as warranted.*

Volumetric methods for estimating reserves involve three steps:

1. Use volumetric mapping or another procedure to determine net volume of the reservoir.

2. Determine rock/fluid parameters to calculate unit volumes of oil, gas, and/or condensate initially in place (O/G/CIP).

3. Estimate REs for oil, gas, and/or condensate. Step 1 may involve the use of gross-rock-volume isopach maps and a net-to-gross net-pay ratio to obtain the net volume of the reservoir.

For oil reservoirs, initial reserves of oil and solution gas can be calculated using Eq. 18.2. and Eq. 18.4, respectively. (Remember, however, that all calculations of reserves must be considered *estimates*, and are accurate to no more than two significant figures.)

$$N_{Ri} = N_i E_{Ro} = \left[7{,}758\phi_o(1 - S_{wo})A_o h_{no} / B_{oi}\right] E_{Ro}, \quad \text{.............................. (18.2)}$$

where 7,758 = bbl/acre-ft (but if units in Eqs. 18.2 and 18.3 are ha and m, this constant becomes 1.0).

For analyses where $A_o h_{no}$ is determined from planimetry of isopach maps:

$$N_{Ri} = N_i E_{Ro} = \left[7{,}758\phi_o(1 - S_{wo})V_{to}R_{ngo} / B_{oi}\right]E_{Ro} \quad \text{(18.3)}$$

Eq. 18.3 assumes a *gross*-pay isopach and would be modified for a *net*-pay isopach by setting $R_{ngo} = 1$.

Initial reserves of solution gas can be calculated by:

$$G_{RSi} = G_{Si}E_{Rg} = N_i R_{si} E_{Rg} \,. \quad \text{(18.4)}$$

For gas reservoirs—either gas caps or nonassociated gas—initial gas and condensate reserves can be calculated by:

$$G_{RFi} = G_{Fi}E_{Rg} = \left[43{,}560\phi_g(1 - S_{wg})A_g h_{ng} / B_{gi}\right]E_{Rg} \quad \text{(18.5)}$$

and

$$C_{Ri} = C_i E_{Rc} = G_{Fi}R_{ci}E_{Rc}, \quad \text{(18.6)}$$

where 43,560 = ft^3/acre-ft (but if units in Eq. 18.5 are ha and m, this constant becomes 1.0). Gas initially in place (GIP) may, alternatively, be calculated with an equation analogous to Eq. 18.3.

Reservoir Maps. Most volumetric methods begin with determining the bulk reservoir volume that contains hydrocarbons. This usually involves preparing structure maps of the top and base of the reservoir and an isopach map of the reservoir. The reliability of results derived from the construction of reservoir maps is directly related to the geologic complexity of the reservoir, the number of wells drilled, the quality and extent of seismic data, information gathered through the drilling and completion of wells, and the evaluator's ability to accurately integrate all data into such maps.

Structure maps should be clearly marked to indicate reservoir limits related to faulting, fluid contacts, and/or facies changes, as well as to indicate all wellbores that penetrate the subject formation. The location of relevant seismic lines and shot points should be shown on all reservoir maps. Subsea depths (ft or m) typically are used to represent formation depths and fluid contacts. Structure maps prepared by explorationists often relate more to structural "markers" or log features that are recognizable over large areas. These lithological interval "tops" may be used to better define structural influences that may control hydrocarbon trapping forces and depositional trends; however, the reserves geologist should prepare reservoir structure maps that describe the top (and sometimes the base) of the reservoir interval in the mapped formation. Both logs and core data can be used in defining the mapped interval. Contouring may be hand drawn or accomplished using geologic mapping software.

The construction of isopach maps should incorporate the same level of detail as do structure maps, as described above. Net pay should be clearly marked for each well penetration. Failure to calculate gross reservoir volumes properly can lead to serious errors in the preparation of reservoir development plans, unwise expenditure of capital budget amounts, and seriously distorted estimates of reserves. Some of these concerns are discussed further in the Pitfalls section of this chapter.

Net Pay. Estimating net pay is one of the most important steps in volumetric mapping, but unfortunately, it also is one of its most subjective steps. Among the factors that influence it are the amount and quality of log, core, and test data; the nature of the rock/fluid system; and the anticipated drive/recovery mechanism.[19–26] In the following discussion, the term "net pay" refers to "true vertical net pay"—logged net pay thickness that is corrected for borehole inclination and formation dip. [The correction procedure assumes that volumetric mapping is based on vertical projection of dipping formations onto a horizontal (mapping) plane. For steeply dipping formations, kh so derived must be corrected for calculations of fluid flow parallel to the formation bedding surface.]

Net pay in a given reservoir might be determined for different purposes,[25] for each of which the procedure and results might be different. For example:

• In an initial well-log evaluation to determine whether to run production casing and attempt a completion, net pay typically includes only intervals that are judged likely to contribute to well inflow at commercial rates. The log analyst might estimate net pay conservatively to minimize the monetary risk of a subcommercial completion.

• When evaluating a reservoir to determine total hydrocarbons in place (e.g., as an independent check on material-balance calculations), net pay usually includes all hydrocarbon-bearing intervals that are likely to contribute to the "energy balance." Net pay for this purpose generally will be more than that estimated during initial completion.

• When considering a waterflood, net pay should include only intervals considered "floodable," a criterion that implies interwell continuity. Also, the methods used to define floodable intervals are subjective and may exclude intervals that will contribute to recovery by imbibition. Net pay for waterflooding might be less than for initial well-log evaluation or for determining total hydrocarbons in place.

• If a reservoir is to be unitized and net pay is part of the unitization formula, the determination of net pay may be subject to arbitrary rules to ensure "uniformity," which might require determination of net pay that is unrelated to the considerations above.

To facilitate visualization of pertinent data, prepare composite logs for all wells in the reservoir that is under study. Regional and individual variations are expected, but generally the two steps are:

1. Determine "gross interval" by establishing the top and bottom of the zone of interest. In sand/shale sequences, the inflection points of the spontaneous, or self, potential or the gamma ray curves usually are considered zone boundaries. For carbonates, establishing the zone boundaries might involve using one or more of the porosity curves to establish a minimum porosity (porosity cutoff) or using a combination of logs to determine lithologic top and bottom. (Choice of minimum porosity typically is quite subjective and is related to the minimum permeability that is considered productive in the area, which depends on petrophysics, reservoir fluid, and drive mechanism.)

2. Exclude nonpay intervals within the gross interval on the basis of maximum shaliness, minimum porosity, maximum water saturation, specified degree of reversal in either the spontaneous-potential or the gamma ray curve, or a combination of these.

The Petroleum Soc. of the Canadian Inst. of Mining[26] provides a list of "generally accepted (minimum) cutoffs" (**Table 18.3**). Local cutoff data for other regions are available in sources such as the Schlumberger* Well Evaluation Conference (WEC) publications for various regions (e.g., Ref. 27). Engineers should use published cutoff data only as general guidelines, however, and only when information for a specific reservoir or project is unavailable. There may be circumstances for which no minimum permeability, porosity, and/or saturation cutoffs are appropriate.

Historically, the objective of initially estimating net pay has been to determine which intervals in each well zone could be expected to contribute to fluid flow into the wellbore under the anticipated method of well and/or reservoir operation. (This is not a proposed definition of net pay; indeed, there is more than one recognized definition of net pay.). The criteria for determining net pay under primary drive should include fluid mobility (k_e/μ) and pressure

* This reference is not intended to be limiting and does not imply endorsement.

TABLE 18.3—GENERALLY ACCEPTED MINIMUM CUTOFFS[26]

Permeability[a]	
Medium- to high-gravity oil	1.0 md
Wet gas	0.5 md[b]
Porosity[c]	
Carbonates	2 to 4%
Sandstones	7 to 10%
Heavy oil (sandstones)	26 to 28%
Water saturation (maximum)	50 to 60%

[a]Unstressed, horizontal permeability from measurements on cores.
[b]As low as 0.1 md for dry gas under special circumstances.
[c]Unstressed, from cores.

differential.[24] Permeability data, however, seldom are available on all wells in a reservoir. Typically, permeability is estimated using correlations between porosity and permeability measured in cores from the subject reservoir, empirical correlations based on global data, low-frequency acoustic log interpretation,** and/or nuclear magnetic resonance (NMR) log response.

In some cases (e.g., shaly, laminated, sand reservoirs and many carbonate reservoirs), it may be difficult, at best, to determine net pay with an acceptable degree of confidence. [For examples of log response in shaly, laminated, sand reservoirs (known as low-resistivity, low-contrast), see publications of the Houston Geological Soc.,[28] the New Orleans Geological Soc.,[29] and the Rocky Mountain Assn. of Geologists.[30]]

A net-pay isopach made with considerable interpretation uncertainty might have to be revised frequently when interpretive procedures are revised or when new data become available. In this situation, it might be desirable to map gross pay and apply an average net-to-gross ratio to account for nonproductive rock. The net-to-gross ratio can be revised as warranted, thereby avoiding the need to remap the entire reservoir; however, this procedure is not appropriate if there are significant spatial variations in the net-to-gross ratio.

Depending on circumstances, it might be appropriate to classify net pay as proved, probable, or possible to recognize the degree of uncertainty in such estimates and to provide a basis for classifying reserves associated with each estimate. Regardless of the method used to determine net pay, the porosity and water-saturation values used in the volumetric equations (Eqs. 18.2, 18.3, and 18.5) should be consistent with the cutoff values used to determine net pay.

Volumetric-Mapping Example. **Figs. 18.3 through 18.6** illustrate several mapping and reserves-classification principles. Fig. 18.3 is a structure map of an oil reservoir along the upthrown side of a west/northwest dipping fault. A well log and core taken in Well 1 indicate that the sandstone section is oil-filled throughout the logged interval, thereby defining the highest known oil (HKO) at 10,500 ft subsea (ss) and LKO at 10,550 ft ss. A flow test indicated commercial rates of oil production. Pressure/volume/temperature (PVT) data indicate that the oil is gas-saturated. It is assumed that three wells (Well 1 and Locations A and B) are needed to effectively drain the oil reservoir, although B is contingent on the geologic interpretation after drilling A.

Although the production test in Well 1 does not indicate so, there is a possibility of a GOC just updip from the top of its logged interval, and of an OWC just below the base of the logged interval. Accordingly, *on the basis of available information at the conclusion of the logging and testing of Well 2,* only the reservoir volume between 10,500 and 10,550 ft ss can

** Personal communication with E.C. Thomas, Bayou Petrophysics, 23 April 2002.

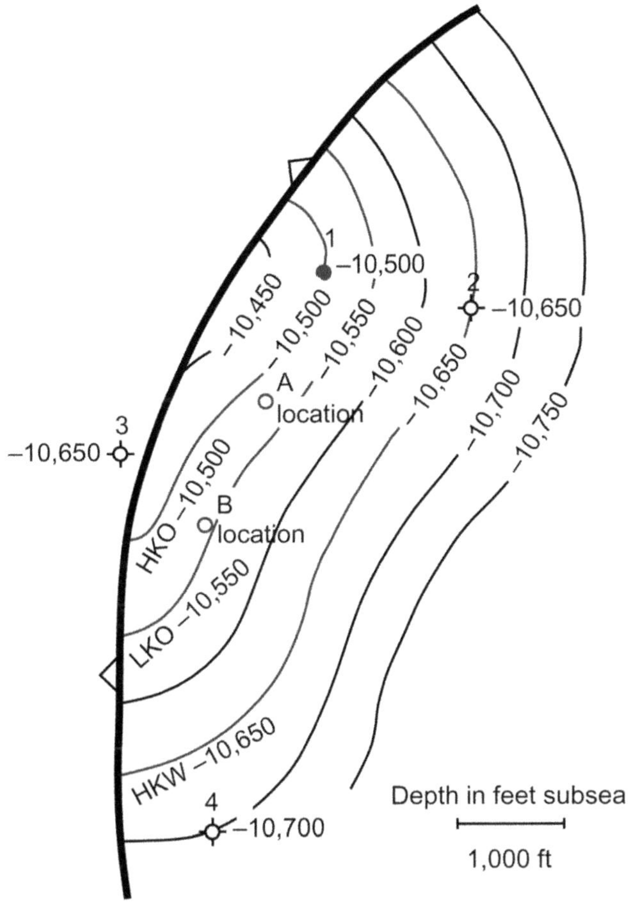

Fig. 18.3—Structure map on top of porosity.

be considered to contain proved reserves—by either the 1978 U.S. SEC or the 1997 SPE/WPC reserves definitions. Conversely, if the crude oil is undersaturated, the volume of the reservoir above the HKO may be considered to be oil-filled. (This proved volume might need to be reduced further because of lateral limitations imposed by optimum well spacing and geological uncertainty. See the later discussion in the Reserves Classifications section in this chapter.)

Fig. 18.4 is a "net sand" isopach of a channel sand, and it is based on an interpretation of the geologic and geophysical data available at the mapping date. This interpretation is supported by regional studies and the 2D seismic lines indicated on the net sand isopach. Note that using mechanically derived contours when contouring the net sand thickness in each of the wellbores would yield an entirely different interpretation, which illustrates the importance of using all available information, including subsurface data, geophysical interpretations, and experiential/analog data from the area.

Fig. 18.5 integrates Figs. 18.3 and 18.4 as a net oil isopach map of the reservoir volume between HKO and LKO, using the spatially correct "Wharton's method"[31] of contouring the "wedge edges." "Wedging" of the net pay contours northwest of Well 1 and southeast of Well 3 reflects truncation of the net (oil) sand contours on Fig. 18.4 by the fault. Wedging in the crescent-shaped area on the northwest side reflects truncation of HKO by the fault. Wedging

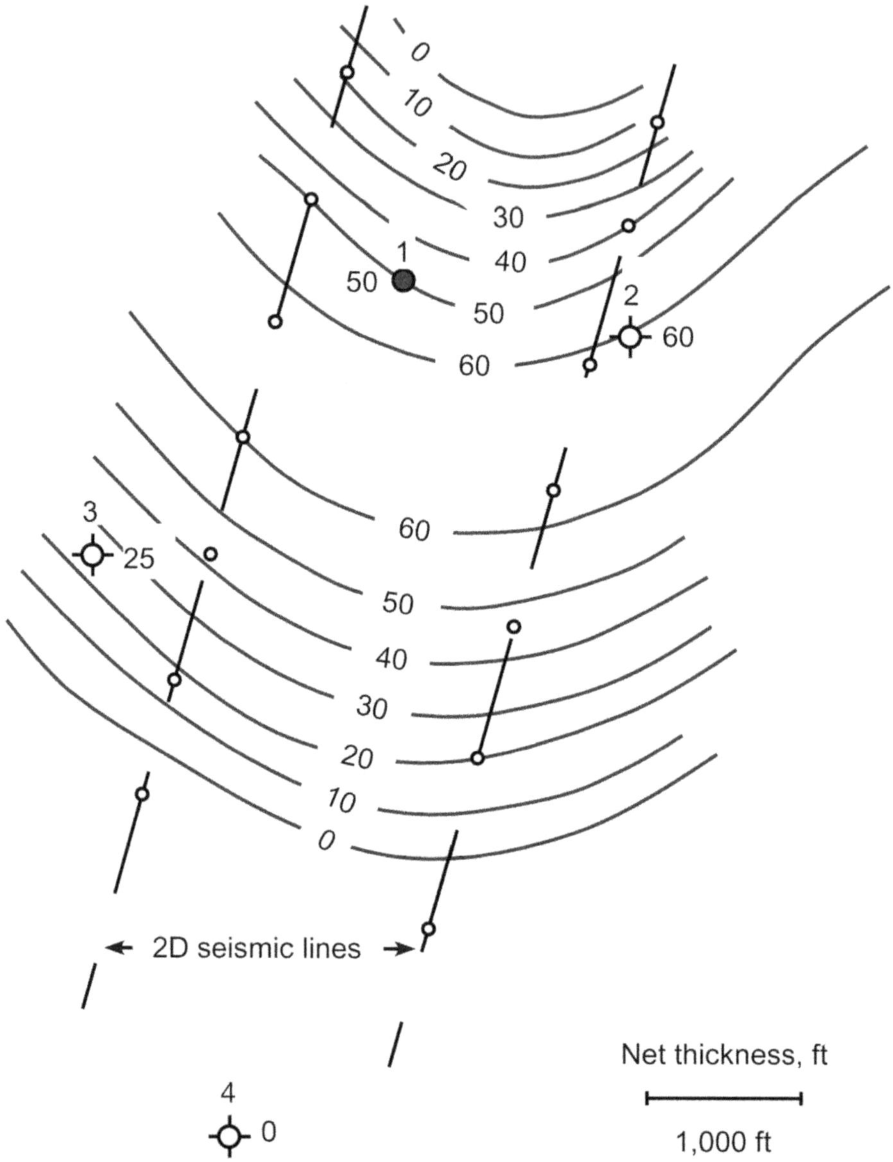

Fig. 18.4—Isopach map of total net sand.

on the southeast side reflects loss of structure between HKO and LKO, whereas thinning of the net oil contours to the northeast and southwest reflects thinning of the net (oil) sand.

Fig. 18.6 is an isopach map of the entire reservoir volume that is interpreted to exist above highest known water (HKW) at 10,650 ft ss, as seen in Well 2. By assuming the reservoir volume above HKO at 10,500 ft ss to be oil-filled, rather than (possibly) partially gas-filled, oil reserves in this volume might be considered to be proved plus probable plus possible (3P).

A map similar to Fig. 18.6 could be prepared for a proved-plus-probable (2P) interpretation, using an arbitrary downdip limit of 10,600 ft ss, which is the midpoint between LKO and HKW. If 3D seismic or other data are available to define or otherwise indicate an OWC, an alternate 2P interpretation based on this limit might be appropriate, as discussed on the section on reservoir limits.

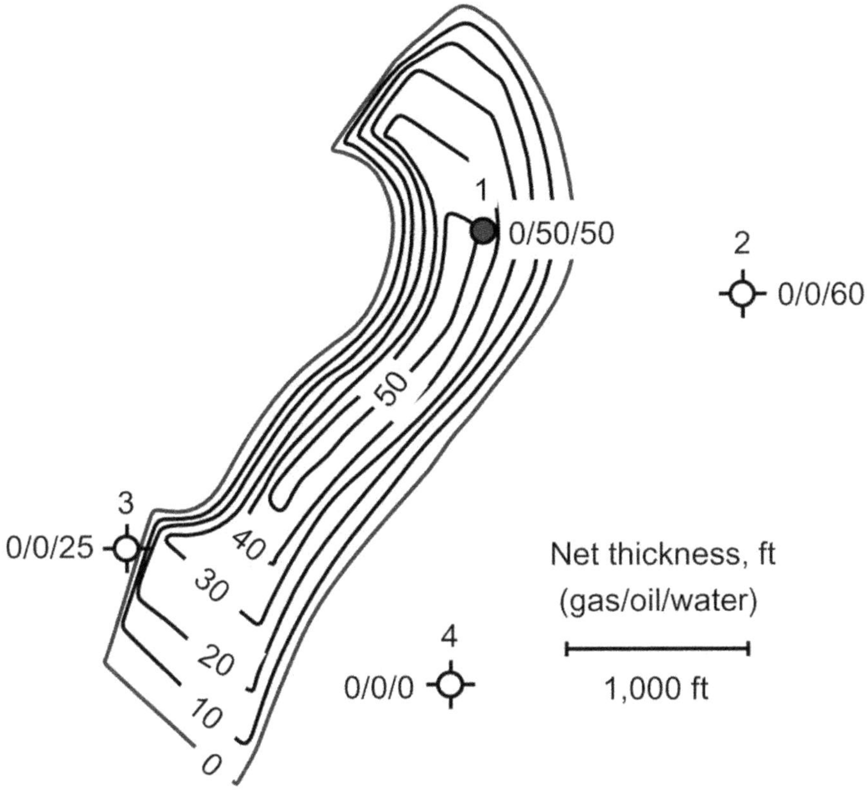

Fig. 18.5—Isopach map of total net (oil) sand.

Reservoir volumes for the proved case (P1) reserves usually are calculated directly by planimetry of the proved volume defined by an isopach map. Incremental probable (P2) reservoir volumes typically are the result of subtracting P1 volumes from the isopach of 2P volumes. Similarly, incremental possible (P3) reservoir volumes are the result of subtracting the 2P isopach volume from the 3P isopach volume.

It may be appropriate to clarify the terms P1, P2, and P3 at this point and to distinguish these from the seemingly similar terms 1P, 2P, and 3P as most often used in reservoir engineering vernacular. As stated previously, P1, P2, and P3 reflect reserves quantities for the classifications of proved, probable, and possible, respectively. The term 1P is synonymous with P1 as it applies to proved reserves only. The term 2P is a cumulative expression for the sum of proved and probable reserves; the term 3P embraces the sum of proved, probable, and possible reserves.

Seismic information, especially when "calibrated" through comparison to well log information, is especially important when well-control information is sparse and when faulting has not been defined through fault cuts recognized in wellbores. One should be alert to all circumstances in which reservoir extent may be limited by faulting, permeability changes, depositional discontinuities, or any other condition that might interrupt fluid flow. Individual reservoir compartments might be difficult to define without performance information, especially pressure information.

Typically, faults with a throw of less than the thickness of the productive formation are assumed to be nonsealing. Caution! One should not assume this without supporting evidence, which should be evaluated over time by a skilled, experienced engineer.

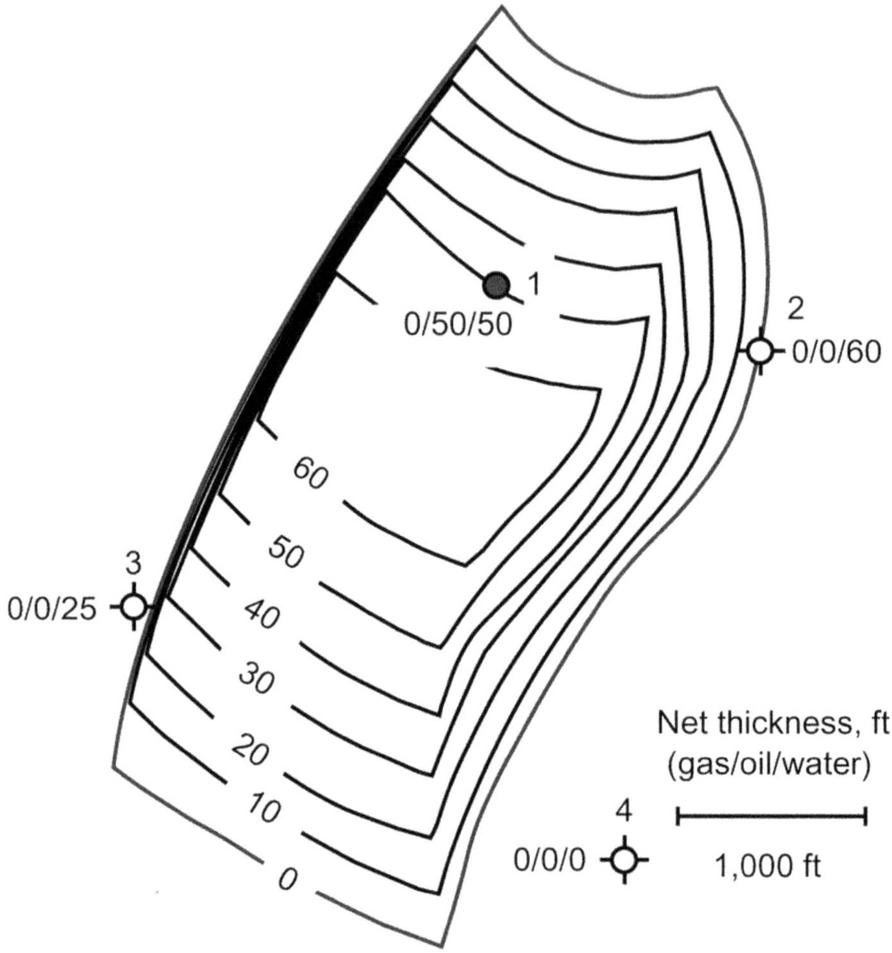

Fig. 18.6—Net oil isopach above HKW.

Isopach maps are designed to determine one or more measures of reservoir volume, and for a single-phase oil reservoir, they might include:

1. Bulk reservoir volume above an observed OWC or LKO. A net-to-gross pay ratio is applied to the bulk reservoir volume to yield net oil reservoir volume.
2. Net reservoir volume determined from the contouring of net oil pay thickness on the basis of log analysis and/or core analysis.
3. Isopore volume maps that reflect the product of net pay and wellbore weighted average porosity, $h_n\phi$.
4. Oil pore-volume (PV) maps that reflect the product of net pay, porosity, and oil saturation, $h_n\phi(1-S_w)$.

For an oil reservoir with a gas cap, similar maps might be prepared, which include:

1. Bulk volume of the reservoir above the GOC, and that between the GOC and the OWC (or LKO), with net-to-gross pay ratios applied separately to the oil and gas portions of the reservoir.
2. Net reservoir volumes from the contouring of net oil and net gas pays on the basis of log and/or core analyses.
3. Isopore volume maps that reflect the product of net oil and net gas pays and wellbore weighted average porosity $h_n\phi$ for each of the oil- and gas-filled portions of the reservoir.

4. PV maps of the oil- and gas-saturated portions of the reservoir that reflect the products of net pay, porosity, and oil saturation and of net pay, porosity, and gas saturation, respectively.

In each set of isopach map design objectives, each subsequent objective requires more data and effort to complete. The amount and quality of the available data and the purpose of the reserves estimate are important in determining what type of maps to construct.

For hand-drawn isopach maps, reservoir volumes are calculated using manually operated planimeters. If computer software is used in map preparation, reservoir volumes are generated within that software. All reservoir maps, whether manually or computer-prepared, should be constructed without bias, should be consistent with the geology of the area, and should use sound mathematical and spatial fundamentals.

Computer-Aided Mapping. Many commercial software programs are available to help prepare reservoir maps and calculate ranges and distributions of reservoir parameters. Such programs range from relatively simple, 2D contouring utilities to fully integrated applications that help construct and visualize the complex distribution of various reservoir parameters and attributes in 3D geostatistical models. Few reservoir engineers are trained to use the more complex software to describe the structural and stratigraphic complexities of a reservoir. Similarly, not all geologists and geophysicists are aware of the reservoir engineer's concerns about the uncertainty of reservoir parameters in absolute terms and in the distribution of these parameters throughout the reservoir model.

For this and other reasons, the reservoir engineer must work closely with the geologists and/or geophysicists who have contributed data and interpretations to the models and have made critical decisions about the construction details. The quality, completeness, and unbiased use of the data are critical to the evaluation team's ability to make decisions about in-place hydrocarbons, compartmentalization, expected recovery mechanisms, and REs that are consistent with the development and production plan approved for the project. These considerations enable reliable quantification and classification of reserves.

The phrase "computer-*aided* mapping" firmly conveys that all such programs are useful tools, but cannot substitute for the judgment and experience of their users. All the basic reservoir-mapping principles and reserves definitions must be integrated in all reservoir maps, whether hand-drawn or computer-created.

Areal Assignments. Frequently, available geologic information is inadequate to prepare volumetric maps, especially with discovery wells. In such cases, an estimated productive area may need to be "assigned" to the completion interval and other productive intervals where reserves are attributed. Typically, such areal assignments are used only until adequate geologic or performance data become available.

Initial areal assignments should reflect the expected drainage area for the well and should use all available analogous, geologic, and test information. The product of the expected drainage area or spacing pattern and two-thirds to three-fourths of the apparent wellbore thickness often is used to account for possible reservoir thinning away from the wellbore.

Reservoir Limits. In general, there are two types of reservoir limits, external and internal. External limits define the zero-isopach contour for either oil or free gas. Internal limits may affect the flow of fluids in the reservoir. The following discussion is related to external limits, which usually are major faults and/or fluid/fluid contacts. Internal limits are discussed in a later section.

In the absence of observation through wellbores, several methods might be used to calculate an OWC/GWC, including pressure/depth, 3D seismic mapping, pressure gradients, and capillary pressure. These are listed in order of their commonly recognized degree of reliability, but their true reliability is a direct function of the extent and quality of the data from which the limits are calculated.

Well A, wet	
Depth ss, ft	BHP, psia
6,075	2,662
6,091	2,669
6,108	2,677
6,220	2,725
6,232	2,731

Well B, discovery	
Depth ss, ft	BHP, psia
5,771	2,602
5,778	2,604
5,785	2,608
5,800	2,610
5,806	2,612
5,813	2,614

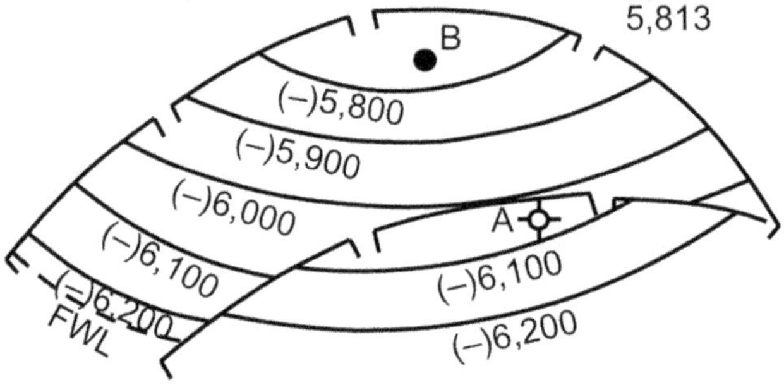

Fig. 18.7—Structure on top of productive sand (after Dake[32]).

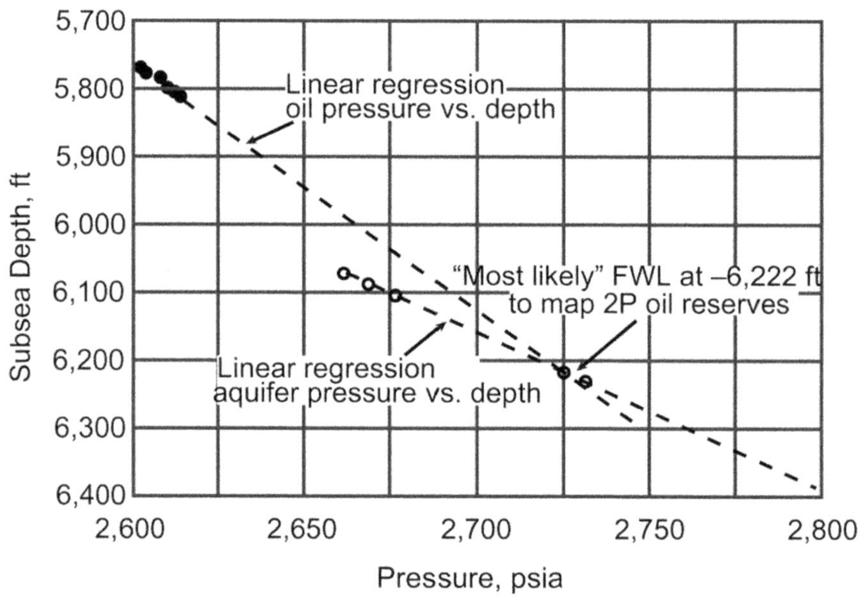

Fig. 18.8—BHP vs. depth, illustrating the pressure/depth method for estimating the FWL (after Cronquist[3]).

Pressure/Depth Method. If there are sufficient accurate subsurface pressure measurements at a series of depths within a given reservoir, such measurements can be used to calculate definitively the depth of a GOC, GWC, and/or an OWC. The principle is illustrated in **Figs. 18.7**[32] **and 18.8.**[3]

Fig. 18.7 is a structure map on the top of a productive sand, and it shows the locations of an oil completion (Well B) and a wet well (Well A). Well B encountered commercial oil throughout the logged interval. Well A apparently was in the aquifer. Static bottomhole pres-

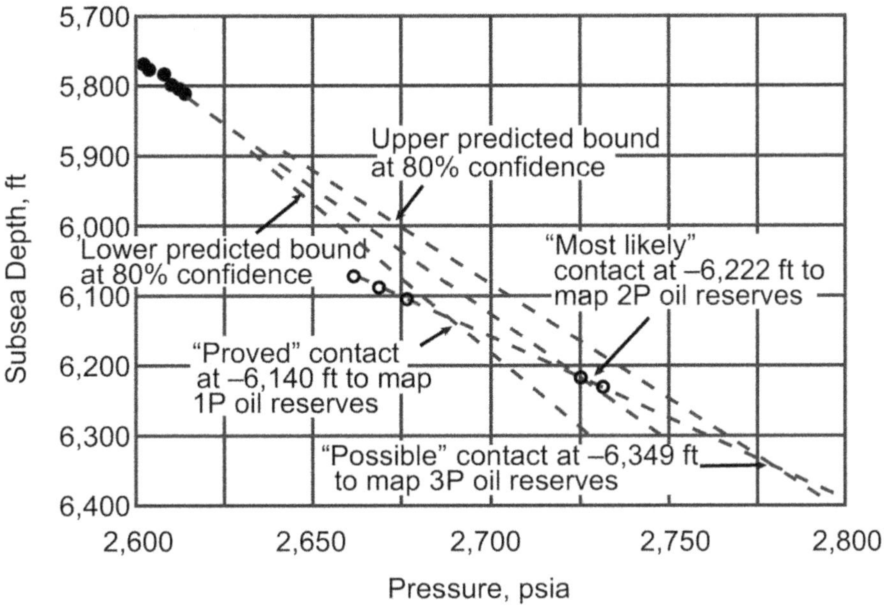

Fig. 18.9—BHP vs. depth, including the upper and lower bounds of predictions at 80% confidence and illustrating the pressure/depth method for estimating proved, most likely, and possible FWLs (after Cronquist[3]).

sures (BHP), measured in the fluid columns of each well using a wireline formation tester (WLFT), are tabulated in Fig. 18.7.

From linear regression through each data set, a "most likely" free-water level (FWL) is estimated at 6,222 ft ss, which may be used to map proved plus probable oil reserves.[3] This interpretation is based on several considerations:

• The backflow periods during the WLFTs in the oil column were sufficient to reconnect the oil in the filtrate-invaded zone,[33] thereby ensuring valid BHP data.

• The oil gradient calculated from the regression is consistent with the reservoir oil density measured from PVT analysis.

• From available subsurface data, the fault separating the two wells at the eastern end of the structure appears to die out to the southwest, thereby supporting pressure communication between the two wells through the aquifer.

• Regional data support the assumption of constant oil density between LKO and the estimated FWL.

If appropriate, this analysis can be expanded through a probabilistic approach.[3] From the regression results, the upper and lower bounds at 80% confidence can be used to estimate P10 and P90 projections of oil BHP vs. depth, respectively (**Fig. 18.9**). (At 80% confidence, oil gradients were calculated to range from 0.24 to 0.31 psi/ft, which is considered reasonable in light of other uncertainties in the data. Because the regression coefficient for the aquifer pressures is 0.9999, the projection of aquifer BHP vs. depth was considered error-free.) This analysis leads to a proved OWC at 6,140 ft ss, a most likely OWC at 6,222 ft ss, and a possible OWC at 6,349 ft ss. (Given the low density of the reservoir oil and the high quality of the reservoir rock in this example, the FWL is considered a reasonable approximation to the commercial OWC.) Note that the geologic interpretation (Fig. 18.7) would support an OWC at approximately 6,300 ft ss. Were there a significant disparity between the geologic and engineering interpretations, this analysis would have to be reviewed.

In more general applications of the pressure/depth method, be careful when estimating FWLs and commercial OWCs in cases where the density of the reservoir oil approaches that of the interstitial water in the aquifer, and/or the reservoir rock is of such low quality that there might be a significant vertical distance between the FWL and the 100% water level.[3]

3D Seismic Mapping. The widespread use of 3D seismic methods has increased confidence in the guiding of development drilling and defining of reservoir size and extent, especially in certain geographic locations. One should be cautious in assigning proved reserves to the entire area indicated to be productive, unless there is corroborating support from other measurements and unless such areas can be drained with the anticipated development plan. To classify gas as proved reserves on the basis of the analysis of seismic bright spots or flat spots in a faulted accumulation, Robertson[34] (quoted here) recommends that all the following conditions be met:

1. The flat spot and/or bright spot is clearly visible in the 3D seismic data.
2. The spatial mapping of the flat spot and/or downdip edge of the bright spot fits a structural contour, which usually will be the spill point of the reservoir.
3. A well penetrates the GWC in one fault block of the reservoir, so logs, pressure data, and test data provide a direct and unambiguous tie between the GWC in the well and the seismic flat spot and/or downdip edge of the bright spot; i.e., the borehole proves that there is producible gas, not residual gas, down to the seismic indicators of the GWC.
4. A well in another fault block penetrates the reservoir updip from the GWC.
5. This second well proves gas down to a lowest known depth, and pressure data show that this gas is in communication with the gas in the first fault block.
6. The seismic flat spot and/or downdip edge of the bright spot in the second fault block lies below the lowest known gas in the second well and is spatially continuous with and at the same depth as these seismic indicators in the first fault block.

Robertson further says that "[i]f all these conditions are met, the gas in the second fault block between the lowest known occurrence in the well and the seismic flat spot and/or downdip edge of the bright spot can reasonably be judged to be proved."[34] Authors' caution! This interpretation might not be accepted within all regulatory jurisdictions.

In exploration, good practice mandates the need for at least one well penetration into a seismic bright spot (amplitude anomaly) to demonstrate commercial producibility before the prospective resources in a pool defined by an amplitude anomaly can be classified as reserves.

There are recognized circumstances in which geophysical interpretations that are made using surveys designed for exploration can be considered reliable in determining reservoir thickness, certain petrophysical parameters, and fluid contacts, but these circumstances are few and currently apply minimally, if at all, to establishing proved reserves quantities. Using 3D surveys for reservoir evaluation increases opportunities for developing confidence in geophysical techniques to help quantify reservoir parameters and limits.

4D seismic analysis—a time sequence of 3D surveys—is an emerging technology that is beginning to be applied more widely to monitor the flow of hydrocarbons in a reservoir.[35–38] 4D seismic interpretation analysis can benefit the understanding of reservoir compartmentalization and RE, but its greatest application might be in optimized reservoir management.

Pressure-Gradient Method. The pressure-gradient method for calculating a hydrocarbon/water contact is similar to that described in the *Pressure/Depth* section in this chapter, but it relies on the computation of fluid gradients from fluid analyses, rather than gradients calculated by repeat pressure measurements at different depths in a common reservoir. Extrapolation over a large depth interval will magnify slight errors in computing fluid gradients. Accordingly, one should not use the pressure gradient method to produce definitive results, but rather to provide approximations that will require confirmation by other sources.

The density of reservoir oil *at the bubblepoint* (lbm/bbl) can be calculated by[39]:

TABLE 18.4—PRESSURE-GRADIENT METHOD

Case	Updip Fluid	Downdip Fluid	GOC or FWL 1P	2P	3P
GOC	Gas	Oil	C	B	A
OWC (FWL)	Oil	Water	A	B	C
GWC (FWL)	Gas	Water	A	B	C

A, B, and C correspond to data points in Fig. 18.10.

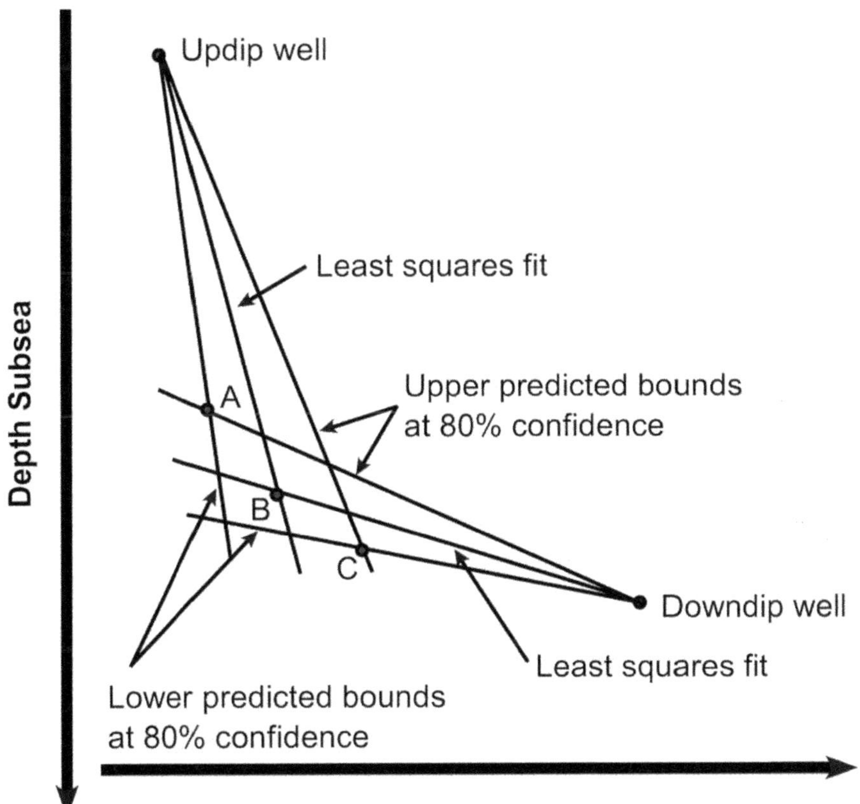

Fig. 18.10—Pressure-gradient method (after Cronquist[3]).

$$\rho_{or} = \left(350\gamma_{os} + 0.076 R_{sb}\gamma_g\right) / B_{ob} \text{ . } \qquad (18.7)$$

Oil density calculated from Eq. 18.7 can be converted to oil gradient (in psi/ft) by multiplying it by 0.00124. For calculations of density and gradient for undersaturated oil, see McCain.[39] Water gradients can be calculated from salinity or other compositional data.

The method discussed here and that in the Pressure Gradient section can be generalized to estimate the depths of 1P, 2P, and 3P GOCs and/or FWLs. The procedure is summarized in **Table 18.4** and illustrated by **Fig. 18.10**.[3]

Capillary-Pressure Method. The capillary-pressure method to calculate the depth of an OWC or GWC is based on the principle that *in a common reservoir* the family of capillary-pressure curves for various rock types in the reservoir has a common FWL. The capillary-pressure method involves several equations[40–42]:

$$F_g \approx \left[\ln \left(5.21 k_a^{0.1254} / \phi\right)\right]^2 / 2.303, \quad (18.8)$$

$$P_{dm} \approx 937.8 / \left(\phi k_a^{0.3406}\right), \quad (18.9)$$

$$\log\left(P_{cm}\right) \approx (-) F_g / \ln\left(1 - S_w\right) + \log\left(P_{dm}\right), \quad (18.10)$$

$$P_{cR} = P_{cL}\left(\sigma_R \cos\theta_R\right) / \left(\sigma_L \cos\theta_L\right), \quad (18.11)$$

and

$$Z_{fw} = P_{cR} / \left(\rho_w - \rho_o\right). \quad (18.12)$$

In Eqs. 18.8 through 18.12, the operator ≈ denotes correlations, whereas = is reserved for analytical equations.

The capillary-pressure method has several limitations (discussed later) and involves four steps:

1. For each well log level for which porosity and permeability are known, calculate F_g using Eq. 18.8 and P_{dm} using Eq. 18.9. [The permeability value k_a in Eqs. 18.8 and 18.9 is absolute (air) permeability (md) and must be estimated independently from log-determined permeability, and typically is computed by logging-company interpretation programs, which assume that log-calculated water saturation is irreducible water saturation.]
2. Use the F_g and P_{dm} values and the log-determined value of S_w at the same level to calculate P_{cm} for that level, using Eq. 18.10.
3. Use Eq. 18.11 to convert from mercury/air capillary pressure—the basis for the Thomeer[43] curves—to the fluid system of interest.
4. Use Eq. 18.12 to calculate the vertical distance from the level investigated to the FWL, using the fluid densities in the reservoir being evaluated. To reduce statistical scatter, make such calculations for multiple points within the interval being evaluated.

Fig. 18.11 illustrates the basic principle of the capillary-pressure method, showing a good-quality sand layer underlain by a poor-quality sand layer in a hypothetical well that logged LKO to the base of the poor-quality sand. The presence of variable-quality sand layers enhances the method's reliability because more than one capillary-pressure curve must be fitted in the calculations.

There are several limitations to the capillary-pressure method:

1. The well-log interval of interest must be within the part of the hydrocarbon/water transition zone (HWTZ) that exhibits a significant water-saturation vs. depth gradient.
2. There must be a robust correlation between permeability and porosity.
3. The Archie equation, frequently used in logging programs to calculate S_w, may not be the "correct" model for the log interval of interest because the value for S_w that is computed from log analysis is a function of the saturation and cementation exponents, each of which is a function of lithology, which might vary over the logged interval.
4. Reservoir fluid densities should not vary significantly between the observation well(s) and the FWL.
5. Although the FWL may reasonably approximate the initial hydrocarbon/water contact in reservoirs with good-quality rocks containing free gas or light oil, in reservoirs with poor-quali-

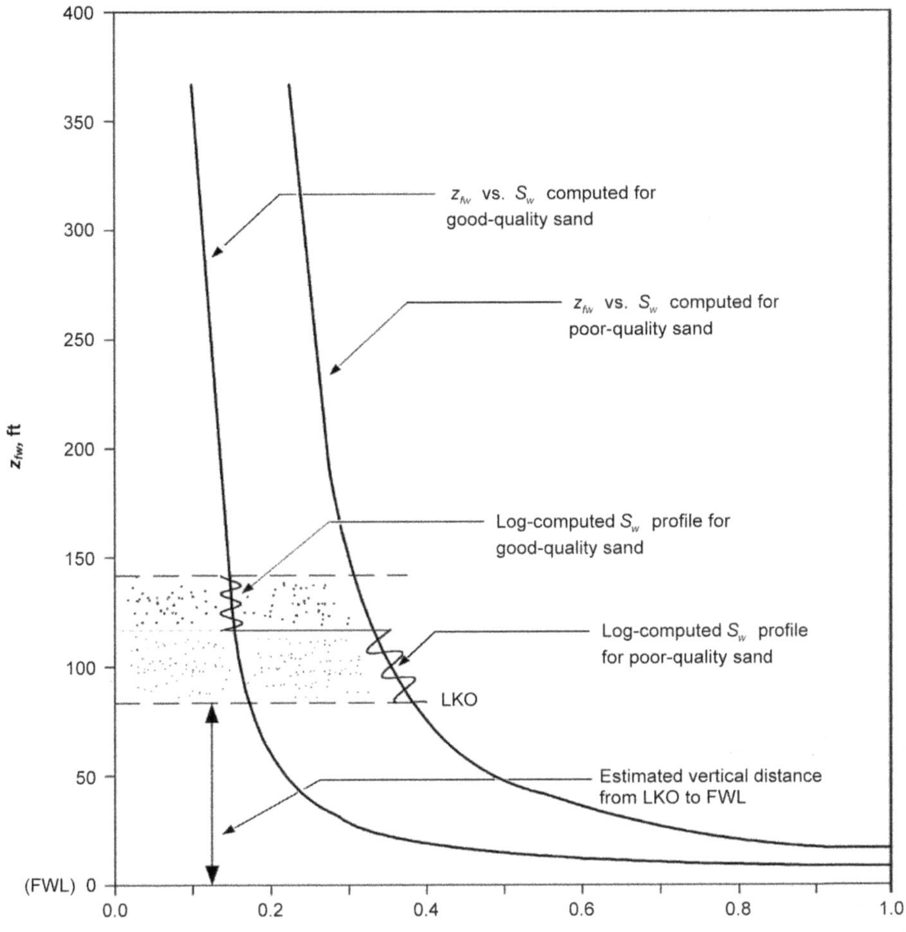

Fig. 18.11—Capillary pressure chart that illustrates the shift in capillary pressure across the boundary between good- and poor-quality sand (after Cronquist[3]).

ty rocks and/or those containing heavy oil, the commercial hydrocarbon/water contact may be significantly shallower than the FWL.

Internal Limits. In addition to external limits discussed in a previous section, there might be internal limits that could significantly affect RE. Two of these internal limits are internal faulting and depositional/diagenetic discontinuities.

Internal faults can be partial or complete barriers to fluid flow along all or part of their length. Under pressure gradients imposed by production, such barriers might respond differently from how they would under static conditions. Usually internal faults exhibit smaller vertical displacements than do bounding faults. In many cases, such faulting might be undetectable using existing well-control, or even seismic, data. Furthermore, internal faults might only become apparent after a period of sustained production, as happened in the Cormorant field in the North Sea[44] and in other fields in that area.

Depositional discontinuities internal to a reservoir might adversely affect RE. In the U.S.A. Gulf Coast, for example, in composite reservoirs that are producing by strong waterdrive, high-permeability (channel) sands typically flood out in preference to low-permeability (fringe) sands.[45] Such occurrences are common in comparable geologic and reservoir settings.

Several methods are used to detect internal limits, including repeat formation tester (RFT) depth/pressure traverses, differences in hydrocarbon isotopes, $^{87}Sr/^{86}Sr$ isotope ratios,[46] and differences in saturation pressure.

Volumetric Parameters. In addition to those determined from geologic mapping, volumetric parameters include petrophysical data (ϕ, S_w) and formation volume factors, all of which should be averaged over the reservoir; initial pressure and formation temperature, which are arguments for formation volume factors and other parameters; and REs, which are discussed in a subsequent section.

Petrophysical data used in Eqs. 18.2, 18.3, and 18.5 can be determined by several methods, including well logging, transient pressure testing, and routine and special core analysis, each of which is discussed in detail in separate chapters of the *Handbook.* Accordingly, only brief comments are provided here.

Porosity. Although reservoir porosity is estimable from well logs, it is good practice to calibrate log-determined porosity against core data, corrected for compressibility effects, if appropriate. As denoted in Eqs. 18.2, 18.3, and 18.5, porosity in the oil zone might differ from that in the gas zone. Depending on circumstances, procedures to estimate net pay (e.g., cutoff porosity and water saturation) might differ between the two zones. Accordingly, average porosity in each of the two zones should reflect such differences.

Water Saturation. In Eqs. 18.2, 18.3, and 18.5, the saturation terms S_{wo} and S_{wg} should reflect average initial water saturation—which might be greater than the irreducible saturation—in the oil zone and gas zone, respectively. Depending on the petrophysics and geometry of the reservoir, some of the accumulation might be in an oil/water or gas/water transition zone. This might reasonably be expected in reservoirs with low-relief, heavy oil, poor-quality rock, or a combination of these. In any event, the saturation terms in Eqs. 18.2, 18.3, and 18.5 should be consistent with the procedure(s) used to determine net pay and map the accumulation.

Compressibility Effects. For accumulations in geopressured reservoirs or friable, unconsolidated sands, it might be appropriate to correct properties measured at ambient conditions to those expected at reservoir conditions. For example, irreducible water saturation at reservoir conditions typically will be greater than that measured under ambient conditions in the same rock sample, and porosity and permeability will be less. Low-permeability rock, such as is seen in tight gas reservoirs, can have permeabilities of as much as two orders of magnitude less at reservoir conditions, compared to those measured at ambient conditions. (Geopressured reservoirs are discussed further in the Special Problems section of this chapter.)

Formation Volume Factor. In general, FVFs [expressed as reservoir volume/standard volume (RV/SV)] are a function of composition of the reservoir fluid, reservoir pressure and temperature, and surface separator(s) temperature(s) and pressure(s). As discussed in subsequent sections, such factors may be measured in a PVT cell; estimated from empirical correlations; or, for nonretrograde reservoir gases, calculated from the composition of the reservoir gas. The FVFs in Eqs. 18.2, 18.3, and 18.5 should reflect initial reservoir conditions and separator conditions that are anticipated over reservoir life.

Oil FVFs at a specified reservoir temperature and over a specified range of reservoir pressures and surface separator conditions can be measured in a PVT cell in a laboratory. Fluid samples used for these measurements are taken from the wellhead during carefully controlled flow and test conditions, or from bottomhole samples after careful conditioning of the sampled well. Such procedures are covered in detail by McCain[39] and elsewhere in the *Handbook.*

If PVT data are unavailable, oil FVFs at initial conditions can be estimated from empirical correlations. Since publication of Standing's correlation[47] in 1947—still considered industry standard—there have been at least 10 such correlations published. Each yields slightly different results, and it might not be apparent which correlation is best for a specific application.[3]

Gas reservoirs are either retrograde or nonretrograde. For retrograde gas reservoirs that are to be produced by pressure depletion below the dewpoint pressure, consider obtaining separator samples and running a PVT analysis at reservoir temperature and over an appropriate pressure range. For nonretrograde gas reservoirs, empirical correlations usually provide sufficient accuracy.

Gas Deviation Factor. The gas deviation factor—also called the supercompressibility factor or the *z*-factor—reflects the nonideal PVT behavior of real gases. Many engineers use this factor for the analysis of nonassociated gas reservoirs, as discussed later. The *z*-factor is a function of fluid composition, pressure, and temperature. Mathematically, it is the volume of a specific molar quantity of real gas, divided by the volume of the same molar quantity of an ideal gas, both measured at the same pressure and temperature:

$$z = V_R / V_I \quad (18.13)$$

To ensure accuracy, the *z*-factor should be measured in a PVT laboratory, using a gas sample from the reservoir being analyzed; however, the *z*-factor estimated using empirical correlations is acceptably accurate for most engineering calculations.

For nonretrograde gases, the reservoir fluid composition does not change, regardless of reservoir pressure changes, and the initial *z*-factor can be used for calculations over the reservoir life. For retrograde gases, however, the composition of the reservoir fluid—and the *z*-factor—will change if reservoir pressure goes below the dewpoint pressure, under which scenario the two-phase *z*-factor must be used. Correlations are available to estimate the two-phase *z*-factor, but using PVT data might be preferable.

Initial Pressure. The initial reservoir pressure is needed to determine the initial FVFs, and it is the basis for material balance calculations of O/GIP. An accurate estimate of initial reservoir pressure is particularly important for geopressured reservoirs and undersaturated oil reservoirs, especially those for which material balance calculations might be necessary.

Reservoir Temperature. Reservoir temperature also is needed to determine the initial FVFs. Accurate determination of reservoir temperature is especially important for reservoir fluids anticipated to be close to critical conditions.

Recovery Efficiency. Estimation of RE is one of the most difficult aspects of volumetric estimation of reserves and—unless there is considerable local experience and/or good analogs—is fraught with uncertainty. In approximate order of importance, RE for developed oil/gas reservoirs depends on drive mechanism, formation quality, degree and type of reservoir heterogeneity, well spacing, fluid type, operating and completion methods, reservoir geometry and well locations, and economics. The interaction between these factors—some of which are discussed below—is complex enough to preclude quantitative analysis during the static phase of reserves estimation, and to allow only qualitative guidelines to be provided here.

Depending on geologic setting, drive mechanism, and reservoir geometry, RE might be strongly influenced by well spacing. Such scenarios include thin oil columns producing by bottom waterdrive; heavy oil, in general; low-permeability gas reservoirs; and reservoirs with discontinuities.

Oil Reservoirs. The viscosity of the reservoir oil strongly influences RE for oil reservoirs, and roughly correlates to the stock-tank oil gravity. Reservoir oils are broadly classified as either heavy (stock-tank gravity less than approximately 25°API) or light (stock-tank gravity greater than approximately 25°API).

For light oils that are producing by solution gas drive with insignificant gravity segregation, REs from reasonably homogeneous reservoirs generally range from 5 to 35% OIP, depending on formation quality, initial solution gas/oil ratio (GOR), and stock-tank gravity.

For heavy oils that are producing by solution gas drive, REs from reasonably homogeneous reservoirs generally range from < 5 to 20% OIP, depending on production mechanism, forma-

TABLE 18.5—RECOVERY EFFICIENCY OF OIL FROM SOLUTION GAS DRIVE RESERVOIRS FOR PRODUCTION FROM BUBBLEPOINT PRESSURE TO ABANDONMENT PRESSURE EQUAL TO 10% OF BUBBLEPOINT PRESSURE[a,b]

		Sandstones			Carbonates		
Solution GOR	Stock-Tank Gravity, °API	Maximum[c]	Average[d]	Minimum[e]	Maximum[f]	Average[g]	Minimum[h]
60	15	0.128	0.086	0.026	0.280	0.040	0.006
	30	0.213	0.152	0.087	0.328	0.099	0.029
	50	0.342	0.248	0.169	0.390	0.186	0.080
200	15	0.133	0.088	0.033	0.275	0.045	0.009
	30	0.222	0.152	0.084	0.323	0.098	0.026
	50	0.374	0.264	0.176	0.398	0.193	0.074
600	15	0.180	0.113	0.060	0.266	0.069	0.019
	30	0.243	0.151	0.084	0.300	0.096	(0.025)[i]
	50	0.356	0.230	0.138	0.361	0.151	(0.043)
1,000	15	n/a[j]	n/a	n/a	n/a	n/a	n/a
	30	0.344	0.212	0.126	0.326	0.132	(0.040)
	50	0.337	0.202	0.116	0.318	0.120	(0.031)
2,000	15	n/a	n/a	n/a	n/a	n/a	n/a
	30	n/a	n/a	n/a	n/a	n/a	n/a
	50	0.407	0.248	0.156	0.328	(0.145)	(0.050)

[a]After Arps and Roberts.[48] [b]Use 0.75 as a preliminary estimate of RE of solution gas for reservoir pressure depleted to approximately 10% of bubblepoint pressure. [c]Maximum RE usually associated with unconsolidated or friable, well-sorted sands with porosity greater than approximately 25%. [d]Average RE usually associated with moderately cemented sandstones with porosity between approximately 15 and 25%. [e]Minimum RE usually associated with highly cemented sandstones with porosity less than approximately 15%. [f]Maximum RE calculated using average k_g/k_o data from 26 samples of Permian dolomite from west Texas. Because average primary RE from these reservoirs typically is less than that tabulated here, the results here should be used with caution. [g]Average RE usually associated with intergranular, slightly vugular limestones and dolomites. [h]Minimum RE usually associated with highly vugular or fractured carbonates, or with fractured cherts. [i]Values in parentheses are estimates made from calculations discontinued before reaching abandonment pressure. [j]An unlikely combination of solution GOR and stock-tank gravity; therefore, calculations were not made.

tion quality, initial solution GOR, and stock-tank gravity. Additional comments about heavy oil are provided in the Special Problems section of this chapter.

Table 18.5 is based on analytical calculations for an assumed range of oil properties and gas/oil relative permeability characteristics.[48] The calculations assume pressure depletion from the bubblepoint pressure to 10% of that pressure. For reservoirs developed on wide well spacing, especially those in poor-quality formations, it is conjectural whether average reservoir pressure can be reduced to such levels at commercial production rates. The calculations for Table 18.5 take no account of gravity segregation, which might increase RE; of possible heterogeneities, which might cause incomplete drainage; or of initially undersaturated reservoirs, where expansion of the rock/fluid system will contribute to overall RE when reservoir pressure is reduced from initial to bubblepoint pressure. Nevertheless, REs in Table 18.5 reasonably agree with many REs observed in the field.

Eqs. 18.14 and 18.15, published by the API in 1967, result from regression analyses of *observed* RE of oil vs. various rock/fluid properties.[49] Eq. 18.14 was developed from 80 U.S.A. reservoirs that had produced by solution gas drive with no initial gas cap. Eq. 18.15 was developed from 70 U.S.A. reservoirs that had produced by waterdrive.

$$\left(E_{Ro}\right)_{sg} \approx 0.41815\left[\phi\left(1-S_w\right)/B_{ob}\right]^{0.1611}\left(k/\mu_{ob}\right)^{0.0979} S_w^{0.3722}\left(p_b/p_a\right)^{0.1741} \quad \text{(18.14)}$$

$$\left(E_{Ro}\right)_{wd} \approx 0.54898\left[\phi\left(1-S_w\right)/B_{oi}\right]^{0.0422}\left(k\mu_{wi}/\mu_{oi}\right)^{0.077}S_w^{-0.1903}\left(p_i/p_a\right)^{-0.2159} \quad \text{....... (18.15)}$$

As with any correlation, these equations reflect the best fit through large data sets; there might be substantial errors when they are applied to a specific case. Accordingly, the user should seek independent corroboration.

Despite disclaimers[50] regarding the reliability of these correlations, their applicability has been corroborated in enough scenarios to warrant continued, but cautious, application. Estimates based on these correlations can be adjusted when more information becomes available. Early life performance data might be helpful in determining the drive mechanism (which correlation to use) or might lead to a better analysis of reservoir fluid properties (bubblepoints and/or formation volume factors).

The nature of water influx (bottomwater vs. edge water) apparently was not considered in the development of Eq. 18.15. In bottom waterdrive reservoirs, suboptimal well spacing might cause significantly lower RE than indicated by that correlation. (For additional comments regarding bottom waterdrive, see Cronquist.[3])

Generalized application of these correlations to estimate REs outside onshore North America is limited by the absence of terms for well spacing and heterogeneity and by the economic scenario under which the reservoirs were operated. Data for these correlations were derived from onshore, U.S.A. reservoirs that generally were developed on relatively close spacing; thus, the reservoirs probably were produced under semisteady-state conditions, under which well spacing might not be a significant parameter. Also, with close well spacing, the effects of reservoir heterogeneity tend to be minimal. Operating costs and economic limits for onshore U.S.A. reservoirs tend to be significantly less than are typical elsewhere, such that REs tends to be greater.

Gravity Segregation. In nature, there always is some gravity segregation of free gas and oil during production of oil by solution gas drive. The relative importance of gravity segregation can be shown by a gravity number N_G:

$$N_G = k_o\left(\rho_o - \rho_g\right)\sin\alpha/\mu_o \quad \text{.. (18.16)}$$

The N_G and RE data in **Table 18.6**[3] were compiled from published studies of actual reservoirs, and they may be useful for guidance in making a *preliminary* estimate of the relative importance of gravity segregation. These data do not show a correlation between N_G and oil RE; however, they do indicate that, in an oil reservoir with a N_G greater than approximately 15, significant gravity segregation is a reasonable expectation.[3]

Gas was injected into some of these reservoirs. The term m_i indicates the volume injected as a multiple of initial gas-cap volume. In the presence of gravity segregation, gas-cap injection reasonably is expected to have the same effect as a large initial gas cap.

Solution Gas. Industry literature rarely discusses RE of solution gas from oil reservoirs. This is because of the historically low economic value of solution gas and the expense of commercializing it (installation of multistage compressors). For oil reservoirs produced by strong waterdrive, such that abandonment pressure is not significantly less than initial bubblepoint pressure, RE of solution gas should be comparable to that for the oil in the same reservoir. For oil reservoirs produced by solution gas drive, such that abandonment pressure approaches atmospheric, as a *first approximation,* RE of solution gas reasonably can be estimated at approximately 75% solution gas initially in place (SGIP).

Volatile Oils. In composition and PVT behavior, volatile oils are transitional between black oil and retrograde gas, and their FVFs from differential liberation are greater than 2.0 RB/STB.

TABLE 18.6—COMPARISON OF N_G AND RE FOR SEVERAL OIL FIELDS PRODUCING UNDER GRAVITY DRIVE (AFTER CRONQUIST[3])

Field (Pool), Location	Gravity Number	RE % OIP
Elk basin (Tensleep)[a] Wyoming, U.S.A.	15[b]	61[c]
Midway Sunset (Lakeview), California, U.S.A.	41	62[d]
Mile Six, Peru	47	62[e]
Oklahoma City, Oklahoma, U.S.A.	30[b]	37
Shuler (Jones), Arkansas, U.S.A.	340[b]	60[f]

[a]Reportedly, reservoir is oil-wet. [b]Estimated from published data. [c]$m = 0$; $G_{Fi} = 110$ Bscf. [d]$m = 0$; $m_i = 0$. [e]$m = 0$; $m_i > 8.7$. [f]$m = 0.027$; $m_i = 14.8$. 1962 estimated ultimate recovery (70%) includes effects of water injection.

*Eq. 18.16 is not dimensionless and hence is not meant as an absolute indicator of the importance of gravity drive. Values calculated from Eq. 18.16 and those in this table should be used in a relative sense only.

Their initial solution GOR ranges from approximately 1,750 to 3,200 scf/STB, and their stock-tank gravity ranges from approximately 40 to 60°API.

A significant volume of stock-tank liquids may be produced as condensate from the liberated solution gas. For a specific reservoir, the producing GOR and FVF strongly depend on separator conditions. Use the *flash* FVF that reflects anticipated separator conditions—not the differential liberation factor—to calculate OIP and reserves.

REs of volatile oils by primary drive mechanisms are comparable to those observed for light oils. Unless there are compelling data to the contrary, Eqs. 18.14 and/or 18.15 may be used for a preliminary estimate of RE. Depending on the anticipated drive mechanism, volatile-oil/primary-drive-mechanism reservoirs frequently are operated under high-pressure gas injection. Cronquist[51] discusses several such projects.

Depending on circumstances, some or all of the wellhead production from accumulations of volatile oils might be processed through an off-site plant. One should take care to allocate products properly between lease and plant because ownership of each may differ.

Volatile oils sometimes are classified—erroneously—as retrograde gases. The reservoir temperature in such accumulations may be only slightly less than critical temperature, and a small error in measuring formation temperature can cause laboratory PVT analysis to be conducted at an incorrect temperature. Also, because producing GOR depends on separator conditions, recombination ratios might be incorrect, causing the laboratory fluid to be nonrepresentative.

Gas Reservoirs. RE of gas from nonassociated gas reservoirs is influenced by the ratio of abandonment to initial reservoir pressure and the degree to which the accumulation has been invaded by water from the aquifer, among other factors. In general, RE can be calculated by

$$\left(E_{Rg}\right)_{pwd} = \left(1 - R_{pz}\right) + \left(R_{pz} E_V E_D\right), \quad (18.17)$$

where

$$E_D = \left(1 - S_{wi} - S_{gr}\right) / \left(1 - S_{wi}\right), \quad (18.18)$$

and where S_{gr} can be approximated by

$$S_{gr} \approx 0.62 - 1.3\phi \; . \quad (18.19)$$

If there is no water influx, $E_V = 0$ and Eq. 18.17 becomes

$$\left(E_{Rg}\right)_{pd} = \left(1 - R_{pz}\right) = 1 - p_a z_i / p_i z_a \,. \qquad (18.20)$$

For moderate- to high-permeability reservoirs that are producing by pressure depletion with no water influx, an RE of 85% GIP frequently is assumed, pending observation of well and/or reservoir performance. In many such reservoirs, multiple-stage compression may be warranted, and REs exceeding 95% can be achieved. Caution! Without actual installation of such equipment and/or operating data to demonstrate the commerciality of such an installation, such REs should not be assumed in booking proved reserves. For poor- to low-permeability reservoirs, however, RE might be substantially less than 85%.

If there is a strong waterdrive (where $R_{pz} \approx 1.0$), Eq. 18.17 reduces to

$$\left(E_{Rg}\right)_{swd} \approx E_V E_D, \qquad (18.21)$$

and gas RE can be approximated by the product of the volumetric sweep efficiency, E_V, and the displacement efficiency, E_D.

RE of gas from partial waterdrive gas reservoirs is rate-sensitive.[52,53] Some engineers advocate producing such reservoirs at high rates to minimize abandonment pressure, thereby maximizing RE of GIP.[32] Depending on the nature of permeability distribution in a reservoir, well spacing and location, and the strength of the aquifer, however, high production rates might cause irregular water encroachment, which could cause lower ultimate recovery than if the reservoir were produced at a lower rate.[3,54]

Volumetric sweep efficiency is influenced by reservoir geometry, spatial distribution of permeability, horizontal/vertical permeability ratio, well spacing, and well positioning, among other factors. For example, in thin gas accumulations underlain by water, the volumetric sweep efficiency will be significantly less than in thick gas accumulations underlain by water. Accordingly, closer well spacing—if commercially feasible—might be required for efficient drainage of such thin gas accumulations. Regardless of well spacing, a high horizontal/vertical permeability ratio generally will be conducive to effective sweep efficiency in bottom waterdrive gas reservoirs. Also, sweep efficiency by bottom waterdrive generally will be higher in reservoirs where permeability increases toward the top of the reservoir (as in offshore barrier bars) than in ones where permeability decreases toward the top (as in alluvial channels). Well positioning in the reservoir also is important. A long, narrow reservoir that is developed by a single well at one end might not drain effectively, requiring additional wells to achieve a reasonable RE.

Gas-Cap Gas. Estimating RE of gas from gas caps can be more complex than might be inferred from Eqs. 18.17 through 18.21. In addition to the factors discussed in the Gas Reservoirs section of this chapter, RE of gas from gas caps also depends on the thickness and spatial relation of the gas cap with respect to the oil column and on the completion and production practices for wells that are completed in the oil column. In general, estimation of RE of gas from gas caps using volumetric methods might be subject to considerable error, and such reserves should be checked against those estimated from well performance as soon as clearly defined, reliable production trends have been established.

In reservoirs with thin gas caps that are completely underlain by the oil column, making gas-well completions may be infeasible and, regardless of the drive mechanism, most of the gas-cap gas might be produced by downward coning into oil-well completions. In a strong-waterdrive reservoir, this scenario could lead to the presence of substantial volumes of unswept gas at crestal locations. In a depletion drive reservoir, if the oil-well completions can be operated to relatively low economic limits, RE of gas should be commensurate with that calculated from Eq. 18.20.

In reservoirs where the gas cap is en echelon to the oil column and where gas-well completions are feasible, RE of gas should be commensurate with that expected from a nonassociated free-gas accumulation.

Condensate. Condensate is defined here as a petroleum liquid consisting mostly of pentanes and heavier hydrocarbons that is in the gas (vapor) phase under initial reservoir conditions and that condenses to the liquid phase when the gas is produced through surface separation equipment operating under ambient conditions on a lease. (The definition of "condensate" has been controversial among OPEC countries. Less restrictive definitions that make it difficult to distinguish between condensate and light oil have been proposed.)

RE of condensate from gas reservoirs depends on composition of the reservoir gas, reservoir abandonment pressure relative to the dewpoint pressure, RE of gas from the same reservoir, and separation equipment used in the lease, among other factors. For nonretrograde reservoirs, RE of condensate should be comparable to that of the gas in the same reservoir, but either way, separator conditions might influence RE. For example, high-pressure gas wells—those with flowing tubing pressure (FTP) greater than approximately 1,000 psia—typically are produced through low-temperature separators (LTS). If FTP gradually decreases below approximately 1,000 psia because of pressure depletion, LTS equipment becomes progressively less efficient, which may lead to condensate carry-over.

For retrograde reservoirs, if PVT data are not available, the correlations in Eqs. 18.22 through 18.26[55] may be used to calculate condensate/gas ratio (CGR) as a function of reservoir pressure, dewpoint pressure, and initial CGR. Caution! These correlations are based on limited data.

$$R_c \approx R_{cd}\left[1 + nD\left(1 - p/p_d\right)\right]^{-1/n}, \quad (18.22)$$

where R_{cd} = the CGR at the dewpoint pressure, STB/MMscf, and n and D are defined as follows:

$$\text{If } R_{cd} \le 15, \text{ then } n \approx 24.5 - 1.37R_{cd}\,. \quad (18.23)$$

$$\text{If } R_{cd} > 15, \text{ then } n \approx 3.91\left[1 + 2.71\left(R_{cd} - 15\right)\right]^{-0.129}. \quad (18.24)$$

$$\text{If } R_{cd} \le 150, \text{ then } D \approx 4.05 + 0.0099\left(R_{cd}\right)^{1.48}. \quad (18.25)$$

$$\text{If } R_{cd} > 150, \text{ then } D \approx 24.3 \ln\left(R_{cd}\right) - 101. \quad (18.26)$$

If retrograde gas reservoirs are produced to reservoir pressures that are less than the dewpoint pressure, retrograde condensation might cause irreversible loss of condensate in the reservoir pore space. The loss will be proportionately greater for richer gases.

To illustrate this point, *producing* CGR vs. reservoir pressure (as calculated using Eqs. 18.22 through 18.26) is plotted on **Fig. 18.12** as dimensionless CGR (R_c/R_{cd}) vs. dimensionless pressure (p/p_d). For example, if a reservoir with an initial CGR of 100 STB/MMscf is produced to a reservoir pressure that is 80% of dewpoint pressure, the producing CGR is estimated from Fig. 18.12 to be 40 STB/MMscf.

Eqs. 18.22 through 18.26 also are used here to calculate *ultimate* CGR for reservoirs produced to 10% of dewpoint pressure:

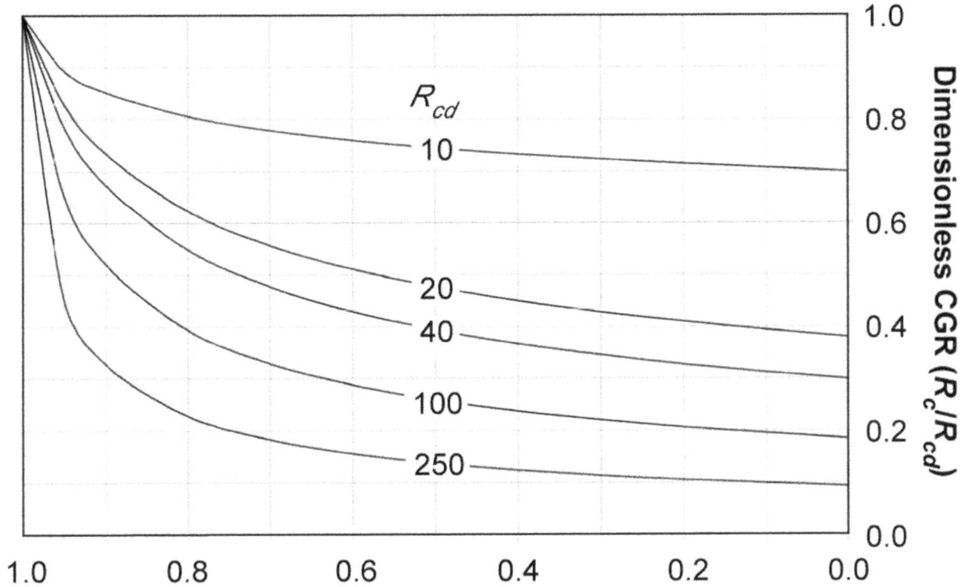

Fig. 18.12—Dimensionless CGR vs. dimensionless pressure (after Cronquist[3]).

$$R_{cuD} \approx 2.01 R_{cd}^{-0.412}, \quad \text{(18.27)}$$

where R_{cuD} = dimensionless ultimate condensate production ($C_{pu}/G_{pu} R_{ci}$). (In developing Eq. 18.27, it was necessary to assume z = 1, which may introduce intolerable errors.) For example, for a gas reservoir with an initial CGR of 100 STB/MMscf produced from dewpoint pressure to 10% of dewpoint pressure, from Eq. 18.27 the cumulative (ultimate) CGR is calculated to be 30 STB/MMscf.

These simple guidelines are adequate for retrograde gas reservoirs in good- formations; however, in poor-quality formations, estimating reserves from such reservoirs is a major problem. Producing such reservoirs by pressure depletion at reservoir pressures below the dewpoint pressure causes retrograde liquids to condense in the pore space. In the interwell areas of the reservoir, such condensation typically builds up to no more than a few percent of initial pore space, but around wellbores, it can build up to saturation levels that can substantially reduce the effective permeability to gas. Depending on circumstances, well productivity might be severely reduced, possibly causing premature abandonment.

A major difficulty continues to be reliable prediction of the depletion performance of such reservoirs, especially those with "rich gases" (high initial CGR) in low-permeability formations. Some of the problems are being overcome through advances in simulation technology, but as yet there are no simple guidelines in the open literature.

Classification of Volumetric Reserves. Classification of volumetric reserves may involve both updip/downdip limits and areal assignments.

Updip/Downdip Limits. The 1997 SPE/WPC[6] and the 1978 U.S. SEC[10] reserves definitions both limit proved reserves to a volume no deeper than GWC or OWC. The 1997 SPE/WPC definitions state: "In the absence of data on fluid contacts, the lowest known occurrence of hydrocarbons controls the proved limit unless otherwise indicated by definitive geological, engineering or performance data." Their use of "geological" embraces all facets of geology, including geophysics. Similarly, the U.S. SEC definitions state: "In the absence of information on fluid contacts, the lowest known structural occurrence of hydrocarbons controls the lower proved limit of the reservoir."

Until areal and vertical reservoir limits are defined by the drilling of enough wells, the location or depth of an OWC or a GWC may be subject to much uncertainty. The extent of this uncertainty often can be narrowed to the vertical distance between the HKW in an off–structure dry hole and the LKO or LKG that is demonstrated by an upstructure well. Without other information to define the hydrocarbon/water contact, LKO/LKG depths typically are used to limit the downdip volume of proved oil and gas reserves.

A deterministic approach might arbitrarily assign incremental probable reserves to the portion of the reservoir between the LKO/LKG and the midpoint between that depth and the HKW, and it might assign incremental possible reserves to the interval between this midpoint and the HKW.

A probabilistic approach in one case might ascribe various confidence values to different levels between the LKO/LKG and the HKW, or in another case might consider that the LKO/LKG and the HKW define the endpoints of a probabilistic distribution and use Monte Carlo analysis to calculate appropriate probabilities for 2P and 3P reserves. In the first case, for instance, the reservoir volume above the LKO/LKG might be assigned a 90% (P90) value, and the maximum possible reservoir size down to the HKW might be given a 5 to 10% (P5, P10) value. An intermediate depth between the LKO/LKG and the HKW can be established to define the 50% (P50) confidence level.

The six simplified reservoir cross sections included in **Fig. 18.13** establish some basic guidelines in reserves classification, but do not include all potential variations. All examples in Fig. 18.13 assume adequate geologic data to support the structural interpretation. See Cronquist[3] for additional guidelines.

Part A of Fig. 18.13 illustrates a one-well reservoir with the well interval fully saturated with oil or gas, but without direct evidence of a downdip GOC, GWC, OWC, or HKW. No other data are available. For gas or undersaturated crude, proved (P1) reserves are limited to the "lowest known occurrence" or the depth established by the base of the formation as seen in the well log. Some have assumed that one sand thickness below the LKG or LKO contains proved (P1) or probable (P2) reserves, even though little geologic data or logic typically supports this assumption. Possible (P3) reserves sometimes are assigned to the reservoir volume represented by one additional sand thickness below the assumed P2 limit, again often with little basis.

Part B of Fig. 18.13 is similar to part A, but includes data from a downdip wet well located below the LKO/LKG contact. This wet well establishes the HKW, which is considered equivalent to the deepest (and largest) possible accumulation of gas or oil in the reservoir. As in part A of Fig. 18.13, the volume of gas or oil above the LKG/LKO may be considered P1 reserves; the volume below the LKG/LGO, but above the HKW, often is split into P2 and P3 reserves.

Part C of Fig. 18.13 is similar to part A, but the depth and location of a reservoir structural spill point (SP) can be mapped. The SP depth establishes a maximum size of a hydrocarbon accumulation in much the same way as does the dry hole shown in part B of Fig. 18.13. The interval between the LKG/LKO and the SP may be subdivided into P2 and P3 intervals, assuming no other information to the contrary.

Part D of Fig. 18.13 is identical to part C in structural attributes, but illustrates a further complication for a saturated oil reservoir that might have a gas cap with a GOC that potentially is as deep as the HKO shown at the top of the logged interval. The engineer judges what portion of the reservoir volume above the HKO should be allocated to oil and/or gas. In any event, only the reservoir portion limited by the LKO and HKO can be classified as P1 reserves

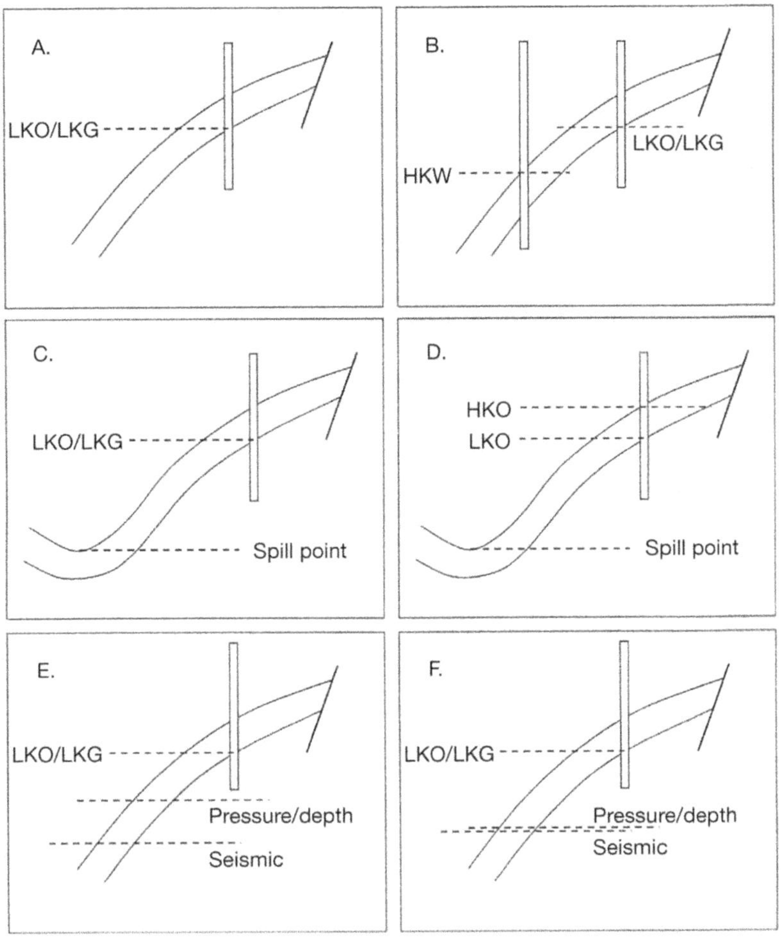

Fig. 18.13—Estimation of reservoir limits for classification of reserves.

in most circumstances. This limitation might have a significant impact on the value of the reservoir because oil usually has a higher value than natural gas.

Part E of Fig. 18.13 is a simplified example to guide where seismic interpretations lead to a downdip limit below that calculated through the use of pressure/depth data. Although using pressure and seismic data lead to different conclusions, they can be useful in defining P2 and P3 reserves allocations. In some cases, the quality of the pressure/depth data may warrant assigning P1 reserves to the calculated hydrocarbon/water contact, especially if the 3D seismic interpretation supports a larger reserves quantity.

Part F of Fig. 18.13 is an idealized case in which a reservoir limit below the LKG/LKO is corroborated by both pressure and seismic interpretations, perhaps enough to support the designation of P1 reserves to the lower depth. Such a conclusion would be a function of the amount and quality of the pressure and seismic data.

These six examples are useful, but they are only guidelines. The reserves engineer should consider fully the quantity and quality of all data used in reserves quantifications and classification and look beyond the specifics of any individual reserves determination for guidance obtainable from area analogs.

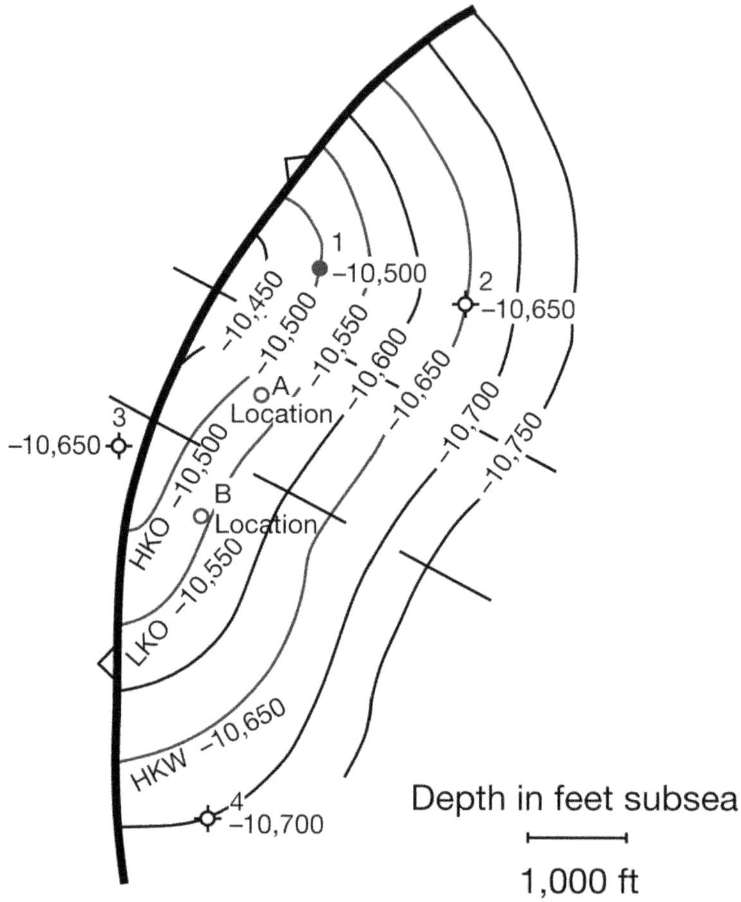

Fig. 18.14—Structure map on top of porosity.

Areal Assignments. **Fig. 18.14** is identical to Fig. 18.3, except for the two dashed lines that suggest assignment of different classifications reserves from an areal perspective. Consistent with the Fig. 18.3 discussion, Locations A and B in Fig. 18.14 are required for effective drainage and depletion of the reservoir. Well 1 is classified as proved producing. Location A can be classified as proved (undeveloped) on the basis of *reasonable certainty* of the geologic interpretation and the adequacy of well control.

Location B cannot be classified as proved, given the uncertainty introduced by reservoir sand being absent in Well 4 and the structural uncertainty regarding the position of an LKO or an OWC. The southwest boundary of the interpreted channel sand is not defined and has been inferred through seismic interpretation (see Fig 18.4). Most would classify the reservoir volume southwest of the southern line and above 10,600 ft ss as probable undeveloped. Note that all or part of the reservoir volume above the HKO may be assumed to be oil-filled in the P2 or P3 cases.

18.4.3 Performance Methods. Performance methods are used after a field, reservoir, or well has been on sustained production long enough to develop a trend of pressure and/or production data that can be analyzed, usually mathematically, to estimate O/GIP and/or future production. The analysis may involve material-balance calculations, computer simulation, or "fitting" historical trends of production rates [water/oil ratio (WOR), gas/oil ratio, water/gas ratio (WGR), condensate/gas ratio, and/or pressure], or some combination of these.

Material Balance. Classic material balance (CMB) is a computational procedure by which the fluid properties and pressure/performance history of a *reservoir* are averaged, thereby treating the reservoir as though it were a tank. Computer simulation (discussed in a subsequent section) uses material-balance principles, but segments the reservoir into numerous cells or gridblocks, rather than treating it as a tank. Also, simulation programs can be used to forecast production.

CMB equations are formulated to calculate volumes of O/GIP. These methods can be used when there are sufficient historical reservoir pressure and production data to perform reliable calculations. Computer simulation might seem to have rendered the CMB obsolete, but not so. For many scenarios, the CMB remains an essential adjunct to computer simulation.[56,57] Indeed, the CMB sometimes is the *only practical approach* to reservoir analysis, such as in the following cases:

• In highly fractured reservoirs, which defy characterization needed for reliable computer simulation.

• In fractured/vugular carbonates where lost circulation problems preclude complete penetration and, accordingly, adequate characterization of the reservoir.

• In reservoirs (typically offshore-gas) that are developed from clusters of wells from widely spaced platforms, a scenario in which subsurface data may be inadequate to characterize the reservoir.

For reliable material-balance calculations, the reservoir must have reached semisteady-state conditions (pressure transients must have affected the entire hydrocarbon fluid system and contiguous aquifer, if any). Depending on reservoir fluid and formation properties and on reservoir drive mechanism, this could require cumulative production of as much as 5 to 10% of the O/GIP.

Reliable application of the material balance method requires accurate historical production data for *all* fluids (oil, gas, and water), accurate historical (static) bottomhole pressure data, and PVT data representative of *initial* reservoir conditions.

Limitations for CMB include reservoirs with strong waterdrives or large gas caps that maintain reservoir pressure close to initial pressure, a scenario in which reservoir-pressure errors cause erratic calculations of the expansion terms and, consequently, unstable solutions (discussed in the next section); thin, areally extensive, and/or low-permeability reservoirs that have wide areal variations in reservoir pressure and saturation conditions that cannot be averaged reliably; and volatile oils, in which the volumes of stock-tank liquids separated from the solution gas are unaccounted for. This last limitation might be overcome using procedures discussed by Walsh.[58]

For gas reservoirs that are producing by pressure depletion, a graphical form of the material-balance equation (discussed in a subsequent section) can be used to estimate GIP *and* reserves. For oil reservoirs, the material-balance equation can be used to estimate O/GIP and likely drive mechanism. To estimate *reserves* from oil reservoirs, however, the material-balance equation must be used in a form that includes appropriate relative-permeability data and operating constraints in a predictive mode adapted for computer simulation. Otherwise, REs can be estimated by analogy.

Equations. In "expanded" form (including water influx), the material-balance equation may be stated as oil-and-gas production (plus water) equals expansion of oil and free gas initially in place plus water influx. Assuming an initial gas cap and, at this time, ignoring compressibility of PV and interstitial water, this relation may be written as:

PRODUCTION = EXPANSION + INFLUX

GAS CAP + OIL COLUMN + AQUIFER = GAS CAP + OIL COLUMN + AQUIFER

TOTAL GAS − SOLUTION GAS

$$N_p R_p B_g - N_p R_{si} B_g + N_p B_t + W_p B_w = \frac{N_i m B_{ti}\left(B_g - B_{gi}\right)}{B_{gi}} + N_i\left(B_t - B_{ti}\right) + W_e \quad (18.28)$$

and gathering terms,

$$N_p\left[B_t + B_g(R_p - R_{si})\right] + W_p B_w = N_i\left[(B_t - B_{ti}) + mB_{ti}(B_g - B_{gi})/B_{gi}\right] + W_e \ . \quad (18.29)$$

In Eqs. 18.28 and 18.29, units on both sides of the equal sign are RB. Also, units for B_g in these and subsequent equations in this section are RB/scf (whereas in Eq. 18.5, B_g is in Rcf/scf).

Depending on the magnitude of pore and water compressibility compared to overall system compressibility, it might be desirable to include compressibility terms for oil material balance calculations, both above and below the bubblepoint pressure. If compressibility terms were included for both the initial gas cap and the oil column, Eq. 18.29 would be written:

$$N_p\left[B_t + B_g\left(R_p - R_{si}\right)\right] + W_p B_w = N_i\left[\left(B_t - B_{ti}\right) + (1+m)\left(c_p + c_w S_w\right)B_{ti}\left(p_i - p\right)/S_o + mB_{ti}\left(B_g - B_{gi}\right)/B_{gi}\right] + W_e, \quad (18.30)$$

where the pore compressibility c_p = the *integrated* value between p_i and p, and where p = the pressure at which Eq. 18.30 is evaluated.

There are three unknowns in Eqs. 18.29 and 18.30: OIP, N_i; size of the initial gas cap, m; and water influx, W_e. It is rarely possible to solve for all three with reasonable reliability,[59] but depending on circumstances, it might be possible to eliminate m or W_e, or both. For example, if PVT data indicate an undersaturated oil reservoir, m may be eliminated, and if geologic and/or well-performance data indicate no water influx, W_e may be eliminated. In general, however, the engineer might be faced with determining all three unknowns. In practice, after appropriate modifications, Eq. 18.29 or Eq. 18.30 is solved repeatedly and successively, using the cumulative production and pressure data corresponding to each time until stable solutions are achieved.

One approach to solving for all three unknowns is to group the terms in Eq. 18.30 to reflect the production and expansion[60]:

$$F_{pR} = N_p\left[B_t + B_g\left(R_p - R_{si}\right)\right] + W_p B_w \ . \quad (18.31)$$

$$E_o = (B_o - B_{oi}) + (R_{si} - R_s)B_g - (B_t - B_{ti}) \; . \qquad (18.32)$$

$$E_c = (1 + m)(c_p + c_w S_w)B_{ti}(p_i - p) / S_o \; . \qquad (18.33)$$

$$E_g = B_g - B_{gi} \; . \qquad (18.34)$$

(E_c, which accounts for expansion of PV and interstitial water in the gas cap, in addition to that in the oil column, is taken from Dake.[56] Havlena and Odeh[60] neglected to include a compressibility term for reservoirs with an initial gas cap.) Substituting these new terms in Eq. 18.30 yields:

$$F_{pR} = N_i\left[E_o + (1 + m)E_c + mB_{ti}E_g / B_{gi}\right] + W_e \; . \qquad (18.35)$$

Depending on pressure/production history and subsurface data, it might not be apparent in the early stages of the analysis whether a reservoir being studied had an initial gas cap or whether there is water influx, or both. Eq. 18.35 can be used in various forms to estimate the amount of O/GIP and the most likely reservoir-drive mechanism.

For example, if there is no initial gas cap, $m = 0$, and if there is no water influx, W_pB_w is dropped from Eq. 18.31. After dropping m and W_e, Eq. 18.35 reduces to

$$F_{pR} = N_i(E_o + E_c) \; . \qquad (18.36)$$

Without a gas cap and water influx, a plot of F_{pR} vs. ($E_o + E_c$) should be a straight line passing through the origin, as shown in **Fig. 18.15**. (For consolidated formations in normally pressured reservoirs, the compressibility term E_c may be ignored, and the plot would be of F_{pR} vs. E_o.) The points represent successive solutions of the fluid production and expansion terms, up to the date of evaluation. The origin is a *must* point; thus, one has a fixed point to guide the straight-line plot. The first few points might be erratic and may be ignored in constructing the straight line. As implied by Eq. 18.36, the slope of the line is $F_{pR}/(E_o + E_c)$, dimensionally equal to RB/(RB/STB), which equals N_i (OIP, in STB).

There are three other possibilities: no gas cap and water influx, gas cap and no water influx, and gas cap and water influx. Each requires appropriate modifications to Eqs. 18.31 through 18.36 and the plotting of different groups. Havlena and Odeh,[60,61] Dake,[56] Tehrani,[59] Sills,[62] and Cronquist[3] offer detailed discussion—and criticism—of these procedures.

p/z vs. Cumulative Wellhead Gas. For nonassociated gas reservoirs where there is no (or insignificant) aquifer influx, and where PV compressibility is negligible, the volume of the initial gas-bearing pore space typically is *assumed* to remain constant over life. For these *volumetric* gas reservoirs, the material-balance equation can be written as:

$$p / z = p_i / z_i - G_p p_i / G_{Fi} z_i \; . \qquad (18.37)$$

For such reservoirs, most engineers use a plot of p/z vs. G_p to monitor performance and estimate GIP and reserves, as shown by **Fig. 18.16**. In theory, such plots should be for a *reservoir*, which may involve averaging static BHPs for more than one well. Caution! G_p is cumulative *wellhead* gas, not just separator or pipeline gas, and should include the gaseous equivalent of stock-tank condensate (GEC).

Depending on the geologic setting and initial well spacing, however, it might not readily be apparent which wells are in a common reservoir, in which case it might be appropriate to plot

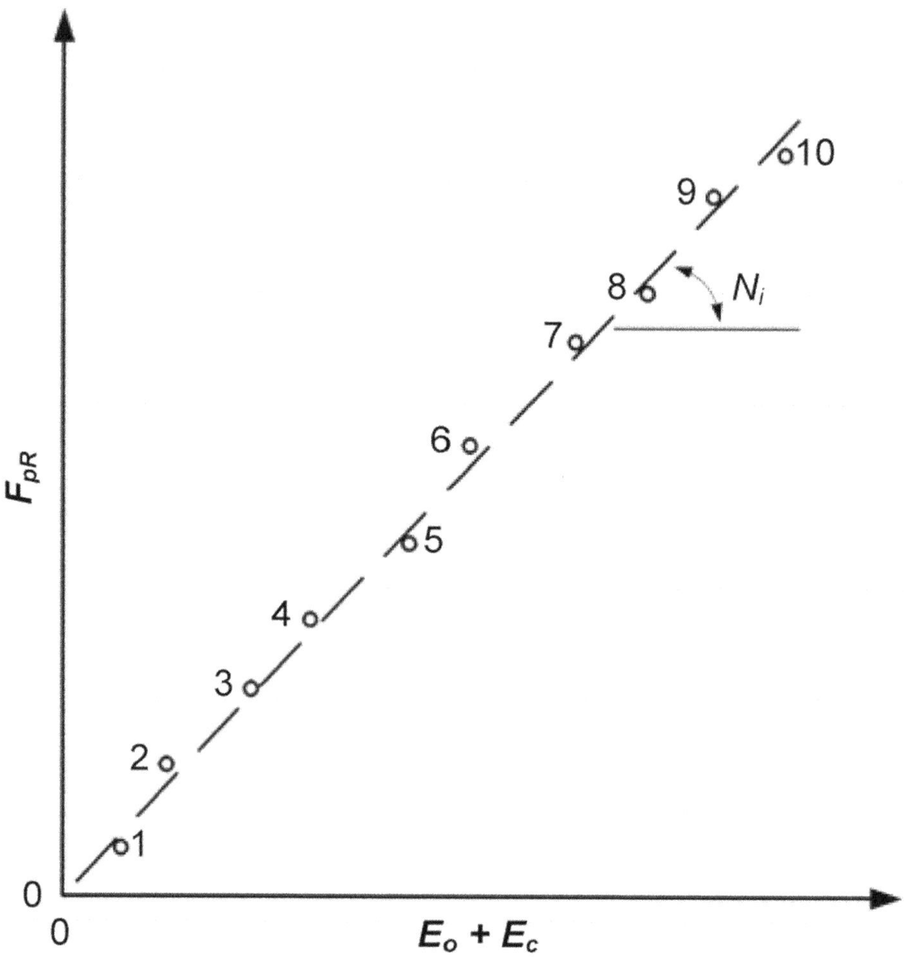

Fig. 18.15—Fluid production vs. compressibility (no gas cap and no water influx). (After Havlena and Odeh.[60])

initial static BHPs for the suspect wells vs. time on a common scale. It is good practice to maintain plots of p/z vs. G_p for each well until *performance* indicates which wells are in a common reservoir; however, even then it might be desirable to maintain a plot for each well because such plots might help to determine the drainage area of each well and to identify areas for infill drilling. In competitive reservoirs, well plots should be maintained over reservoir life.

Plots of p/z vs. G_p are among the most widely used reservoir engineering tools, as well as one of the most widely misused tools, as discussed in the Pitfalls section of this chapter.

Computer Simulation. Computer simulation is based, in part, on material balances, and can involve analysis of an entire reservoir, segments of a reservoir, or a single well. When such analysis includes a good-quality history match, future production forecasts might be considered proved reserves, depending on the constraints imposed on the forecast; however, if computer simulation is used to forecast production, considering such forecasts to be proved reserves is conjectural without a history match. More detailed comments regarding computer simulation are provided in a subsequent section.

Production/Decline Trend Analysis. Whereas material-balance methods focus on reservoirs, production/decline trend analysis (P/DTA) focuses on wells or aggregates of wells. In that

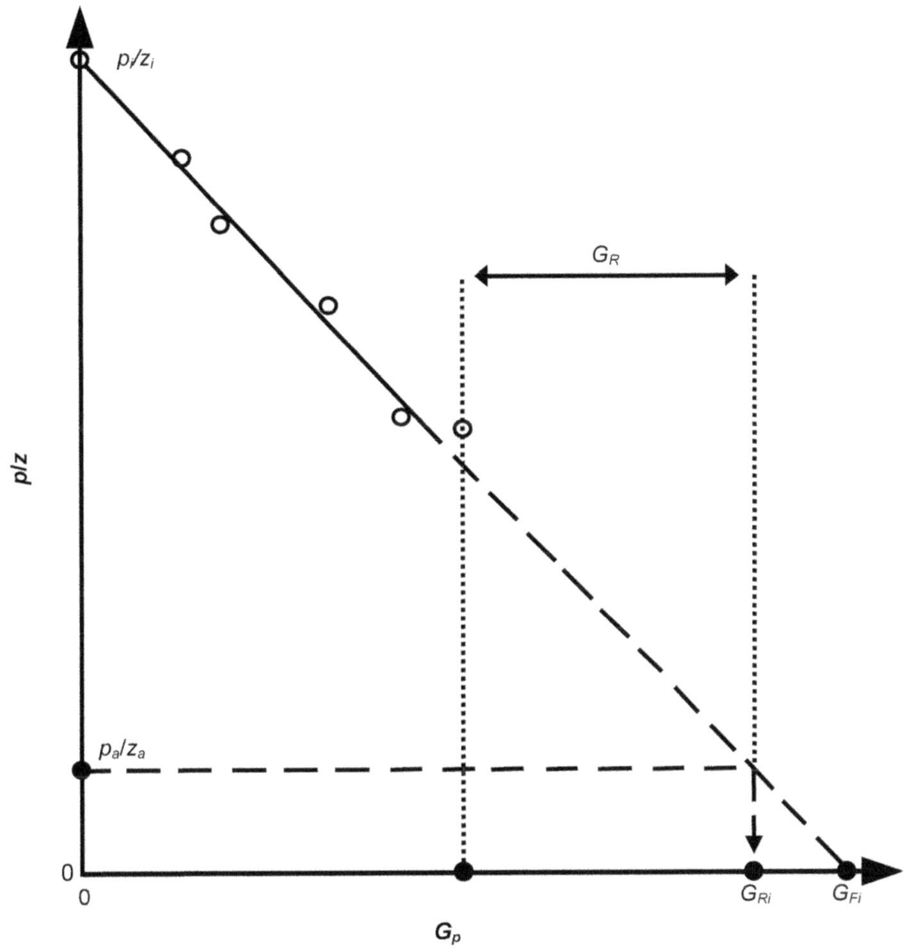

Fig. 18.16—Reservoir pressure vs. cumulative gas production for a volumetric reservoir.

sense, they are two fundamentally different, independent methods for estimating reserves. Depending on circumstances, their results might not agree, a situation that warrants further investigation. Pending resolution, the smaller of the reserves estimates might be classified as proved, and the larger one as proved plus probable or proved plus probable plus possible, depending on the degree of uncertainty.

Economic Limit. "Economic limit" generally refers to one or the other of the following:

• Producing characteristic(s) that typically are used locally to establish the practical limit of production—maximum WOR or maximum GOR—which are identified here as economic limit *conditions.*

• The minimum production rate at which income from such production is less than the cost of continued operation. This is known as the economic limit *production rate* or, simply, the economic limit.

The economic limit may refer to a single well; an aggregate of wells, a lease or unit, or other economic aggregation or financial grouping; or a production facility for an aggregate of wells with processing equipment that must be operated as a unit (e.g., offshore production platforms or fluid injection projects).

The economic limit generally is defined as the production *rate* at which the net revenue to the operator's working interest that is attributable to production from a well, aggregate, or facility equals the "out-of-pocket" cost to operate the well, aggregate, or facility. Net revenue is gross revenue, less production and ad valorem taxes, royalties, transportation, and treating expense, if any. Out-of-pocket costs—sometimes called direct operating costs—are costs that would be saved if the well or facility were shut in (e.g., costs for power and materials). Accordingly, such costs include labor only if shutting in the well or facility would save that cost. State and federal income tax and corporate overhead usually are excluded from out-of-pocket costs.

The economic limit for an oil well can be calculated by

$$q_{oel} = \frac{C_D}{\left[\left(1 - F_{ro}\right)p_o + R\left(1 - F_{rg}\right)p_g\right] - \left(T_P + T_A + C_{TP}\right)} \quad (18.38)$$

Application of production and ad valorem taxes varies from one jurisdiction to another, and Eq. 18.38 should be viewed only as an example.

Wellbore/Mechanical Problem vs. Reservoir Performance. A performance/decline trend in a well might, in some instances, be caused by a wellbore and/or mechanical problem, rather than by the performance of the well and/or reservoir. Extrapolation of such trends can lead to estimates that do not reflect production that might actually be recovered if the wellbore and/or mechanical problem were remedied.

Single Wells vs. Aggregates. Historical production data on single wells might not always be available, and reserves estimates might have to be made using production data aggregated to leases or to reservoir units that contain more than one well. Even if single-well data are available, such production might be volumes allocated from a common battery on the basis of periodic well tests.

In aggregates, various wells might be in different stages of decline and might have different GORs and/or different WORs. Remedial operations, infill drilling, and modifications to production equipment might have influenced the production history and may be expected to influence future trends. An evaluator should become familiar with past, current, and anticipated future operations before attempting to estimate reserves from production trends from multiwell aggregates. Routine lease operations that affect well performance might be reported only informally, if at all. Frequently, pumpers or lease foremen maintain records of well tests, pressures, and equipment changes in "daily gauge reports" that generally are not available elsewhere. Depending on circumstances, the engineer responsible for reserves estimates should consider making an on-site visit with field personnel to determine the nature of field operations and the possible influence of those operations on production rate.

The Pitfalls section of this chapter discusses some factors affecting production forecast reliability.

Production Curtailment. Production of oil and/or gas might be curtailed for many reasons, including capacity limitations on pipelines or plants, market restrictions, inability to handle all produced water or gas, limitations on treating facilities, contract arrangements, regulatory limits, and governmental interference. Here are several examples:

- If wellhead gas is being processed through a gasoline plant, total gas production might be limited by plant capacity. Wellhead gas may periodically bypass a processing plant if the relationship between oil and gas prices makes the sale of unprocessed gas more financially attractive.
- Production from low-pressure gas wells might have to be compressed before being sold, and such production might be constrained by available compressor capacity.
- Production from high-WOR oil wells might have to be limited because of limited capacity to separate, treat, and dispose of produced water.

Thus, before analyzing historical production trends, especially on multiwell aggregates, the engineer should determine whether there has been curtailment of production during the period being analyzed and whether there are facilities constraints that might cause curtailment in the future.

Regarding possible future curtailments, the reserves engineer should ensure that extrapolations of historical trends to estimate reserves do not generate future fluid production forecasts that are inconsistent with existing or anticipated processing capacity. For example, tacit in a future oil and gas production forecast from a waterdrive reservoir is the idea that, after breakthrough, water production will gradually increase until wells reach the economic limit; however, such forecasts might not be realistic because of physical limitations to handling the produced water.

Performance Indicators. One or more of the performance indicators of a well might exhibit a trend before the production rate of the principal product begins to decline. Depending on reservoir type and drive mechanism, these performance indicators include fractional flow of oil (f_o), WOR, WGR, GOR, CGR, BHP, FTP, and shut-in tubing pressure (SITP).

Several performance methods have been used to estimate oil reserves from individual wells in waterdrive reservoirs. Two of the most widely used methods involve plotting either f_o vs. N_p or WOR vs. N_p. These and other methods are discussed briefly here.

In waterdrive reservoirs, after breakthrough of water in individual wells, semilog plots of f_o vs. N_p for each well might exhibit linear trends, as illustrated by **Fig. 18.17**. Usually, the trend can be extrapolated to f_o at economic limit conditions to estimate reserves. The f_o at the economic limit will be governed by the total liquid capacity (oil plus water) of the well and the completion equipment used to produce the well.

For such wells produced at an approximately constant oil rate after breakthrough of water, the logarithm of the rate of total liquid production (oil plus water) typically increases linearly with time, as illustrated by **Fig. 18.18**. This behavior has been observed for high-capacity oil wells in waterdrive reservoirs being produced by gas lift. Oil reserves can be estimated by extrapolating the total liquids rate (oil plus water) to the maximum capacity of the well. Plots such as Figs 18.17 and 18.18 should be prepared for each well that is producing significant water. In using plots such as Fig. 18.18 to estimate reserves, note that the production rate of oil will decrease after the production rate of total liquids increases to the maximum capacity of the well. As discussed in the Rate/Time Trends section, following, the nature of the decrease will depend on the trend of f_o vs. N_p.

In many cases, a linear trend of f_o vs. N_p might turn downward as f_o becomes small. The downturn reportedly occurs at smaller f_o values for heavy (viscous) oils than it does for light oils, as illustrated by **Fig. 18.19**.[63]

In areas with high-permeability sands and a favorable water/oil-mobility ratio (e.g., offshore Nigeria), a Cartesian plot of f_o vs. N_p, rather than a semilog plot, might provide a more reliable basis for extrapolation.

Occasionally, the completion interval may extend across a shale break, as shown schematically by part A of **Fig. 18.20**. In many cases in bottom waterdrive reservoirs with such completions, discontinuities have been observed in the trend of f_o vs. N_p, as shown by part B of Fig. 18.20. After water breakthrough (point "a" on both figures), the trend of f_o vs. N_p decreases while the perforations below the shale break water out (point "b"). The trend stabilizes until the encroaching water rises above the shale break (point "c"). After water breakthrough into the perforations above the shale break, the trend again declines. The rate of decline of f_o after the stabilization period might be greater or less than the rate of decline before the stabilization period, as discussed below. During period b–c, f_o should be *approximately:*

$$f_o = \frac{k_o h_{n1}}{\mu_o \left(k_o h_{n1} / \mu_o + k_w h_{n2} / \mu_w \right)} \quad \text{.......... (18.39)}$$

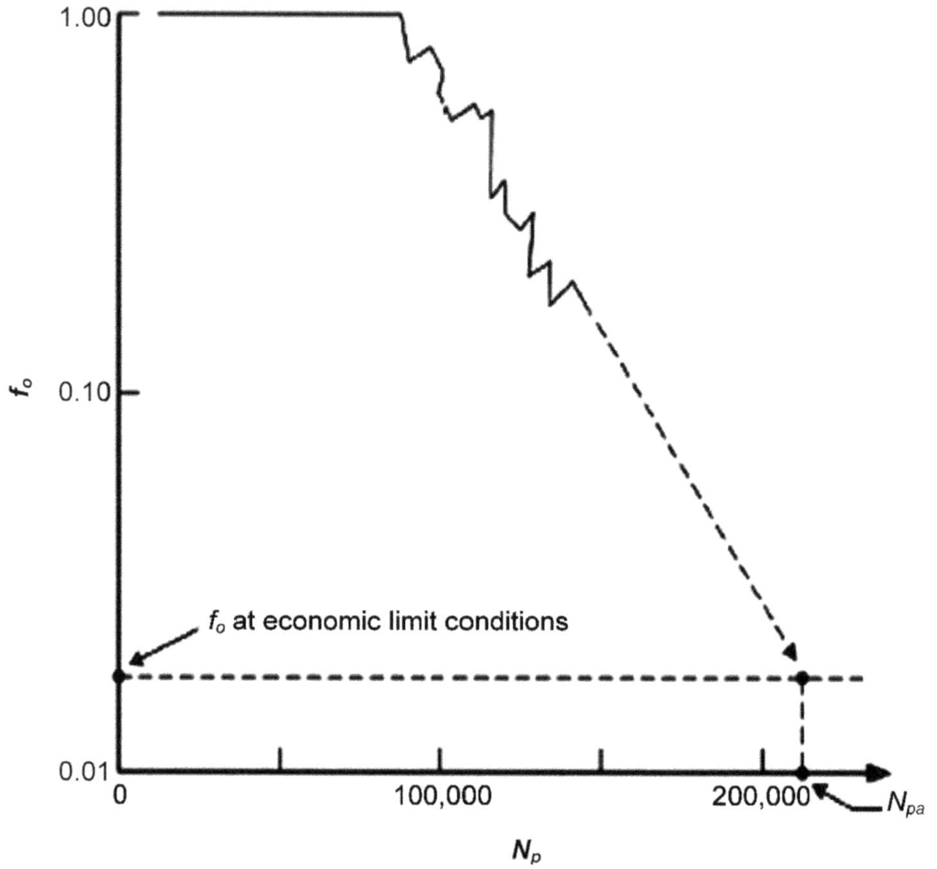

Fig. 18.17—Fraction of oil in total fluids vs. cumulative oil production for a *typical well* that is producing by waterdrive (after Cronquist[3]).

If such behavior is observed (trend a–b–c) and if well-completion geometry can be verified by Eq. 18.39, then, as a *first approximation:*

$$N_{pu} = N_{pc}\left(1 + h_{n1} / h_{n2}\right) . \qquad (18.40)$$

If $h_{n1} > h_{n2}$, the rate of decline in f_o after the stabilization period usually will be less than the rate of decline before the stabilization period. Conversely, if $h_{n1} < h_{n2}$, the rate of decline after stabilization will be more than before.

In a scenario like that shown by Fig. 18.20, extrapolation of trend a–b could lead to significant underestimation of reserves. Against such a possibility, one should check reserves from such an extrapolation against a reasonable volumetric estimate. In general, reserves estimated from any such extrapolation should be checked against those estimated from volumetric mapping and/or should be compared with reserves from analogous wells.

Trends of WOR vs. N_p also can be used to estimate reserves from waterdrive wells. In many scenarios, semilog plots of such data tend to become linear at WORs greater than approximately 1. In these cases, semilog plots of (WOR + 1) vs. N_p tend to be linear at WORs of less than 1, which might help to define trends at low WOR values.[64] Be careful in using such plots, however, without first examining the nature of the fractional flow trend, as discussed in the Rate/Time Trends section, following.

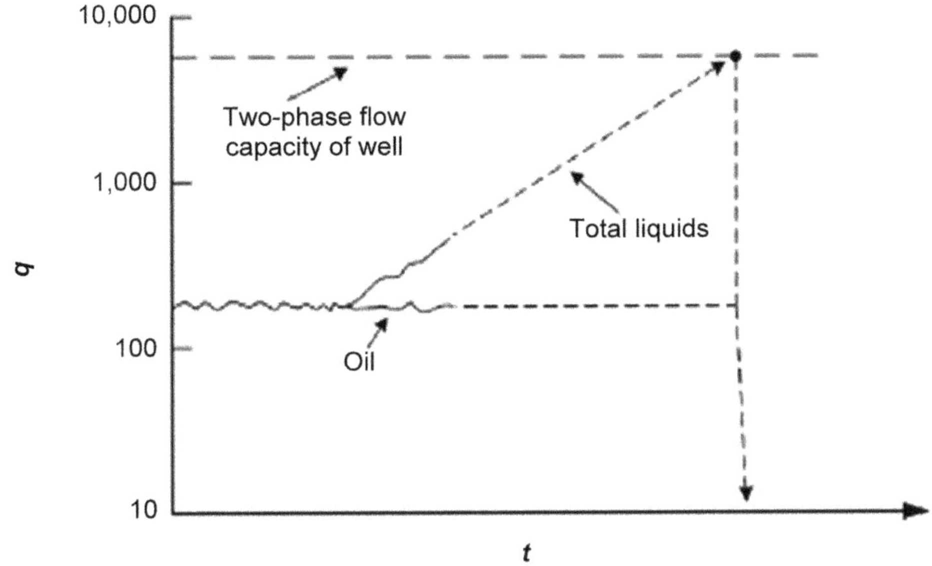

Fig. 18.18—Total liquids production vs. time for a typical waterdrive oil well (after Cronquist[3]).

Determining the correct performance plot to estimate reserves in waterdrive reservoirs is highly empirical. Many of the published procedures are based on the observation that a significant part of a semilog plot of (k_{rw}/k_{ro}) vs. S_w is linear. The floodout performance of wells and reservoirs, however, is governed not only by relative permeability, but also by formation heterogeneity and gravity/viscous forces, among other factors; thus, consider different types of plots, determining the best plot type by experience in analogous scenarios. Remember, however, that the best plot at low water cuts might not be the best plot at high water cuts.

Formation heterogeneity can have a significant and, typically, unpredictable influence on the WOR trend. For example, in naturally fractured reservoirs, the WOR trends in specific wells have varied sharply as a function of cumulative production and/or production rate.

Rate/Time Trends. Historically, engineers have used what have come to be known as "Arps equations" to analyze behavior and predict future production of wells. Three types of equations have been used: hyperbolic, harmonic, and exponential.[65] The rate/time equation for hyperbolic declines is:

$$q_2 = q_1\left(1 + ba_1t\right)^{-1/b} \quad \text{(18.41)}$$

Note that t, which should be written as Δt, is the *incremental* time $(t_2 - t_1)$ required for the production rate to decline from q_1 to q_2.

After integrating with respect to time, Eq. 18.41 becomes the rate-cumulative equation for hyperbolic declines:

$$\Delta Q_p = q_1^b\left[q_1^{(1-b)} - q_2^{(1-b)}\right]/\left[a_1(1-b)\right], \quad \text{(18.42)}$$

where ΔQ_p = denotes incremental cumulative production, ΔG_p or ΔN_p, as the production rate declines from q_1 to q_2. Eq. 18.42 may be solved to determine the *incremental* time for the production rate to decline from q_2 to the economic limit rate, q_{el}:

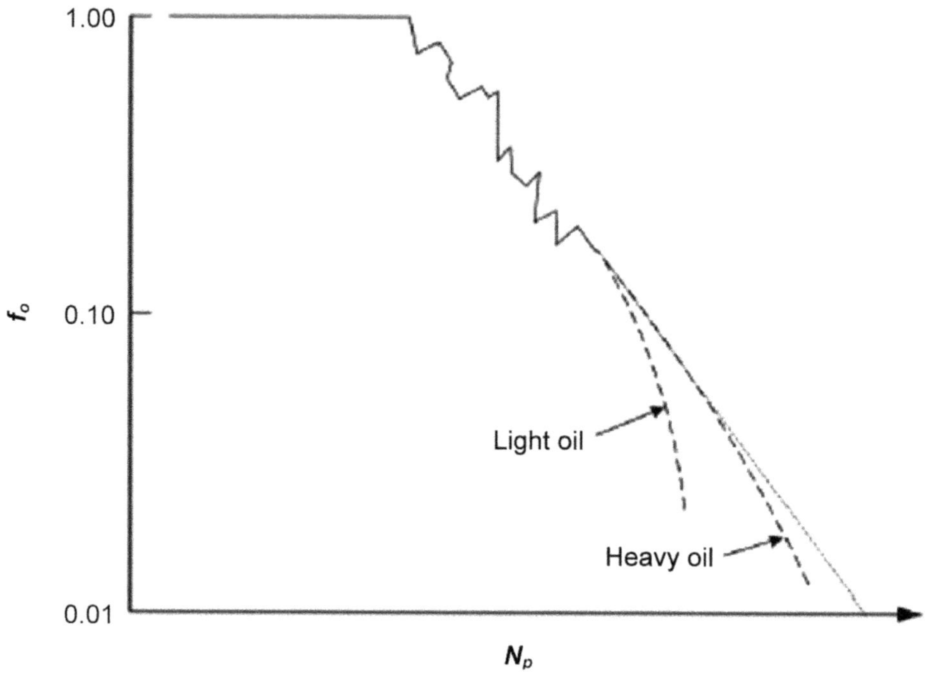

Fig. 18.19—Observed gas RE vs *kh* (after Brons[63]).

$$t_{el} = \left[(q_2 / q_{el})^b - 1 \right] / a_1 b \quad \text{.......... (18.43)}$$

The harmonic and exponential equations are special cases of a hyperbolic equation, where b is equal to 1 and 0, respectively. The three types of equations are compared graphically in **Fig. 18.21**,[65] which shows Cartesian and semilog plots of rate vs. time and rate vs. cumulative production for b values of 0, 0.5, and 1.0. This range of b values reflects semisteady-state, cylindrical flow of a single fluid with small, constant compressibility.[66] This is a reasonable approximation for many well/reservoir scenarios, but it is not universally applicable, and b values greater than 1 have been observed in many areas.[67]

Historically, the Arps equations have been considered as empirical fits to observed trends of production rate vs. time (or vs. cumulative production). As discussed by numerous authors, such equations should, in theory, be used only to fit the constant terminal pressure period of production—the semisteady-state period during which flowing BHP is maintained at a constant value; however, in practice, this operating condition is seldom maintained over long periods of time, which contributes to uncertainties in reserves estimation based on decline trend analysis.

As implied by Fig. 18.19, if the production rate of total liquids is maintained at a constant rate (a common practice), then a linear, *semilog* trend of f_o vs. N_p reflects a harmonic decline. In this scenario, a linear trend of log (WOR + 1) vs. cumulative oil may be observed, which also may be extrapolated to estimate reserves. Linear, *Cartesian* trends of f_o vs. N_p, however, have sometimes been observed, as mentioned previously, and such trends reflect exponential declines. Apparently linear trends of log (WOR + 1) vs. cumulative oil might be observed at low WORs, but they typically depart from linearity at high WORs and should not be extrapolated to estimate reserves.

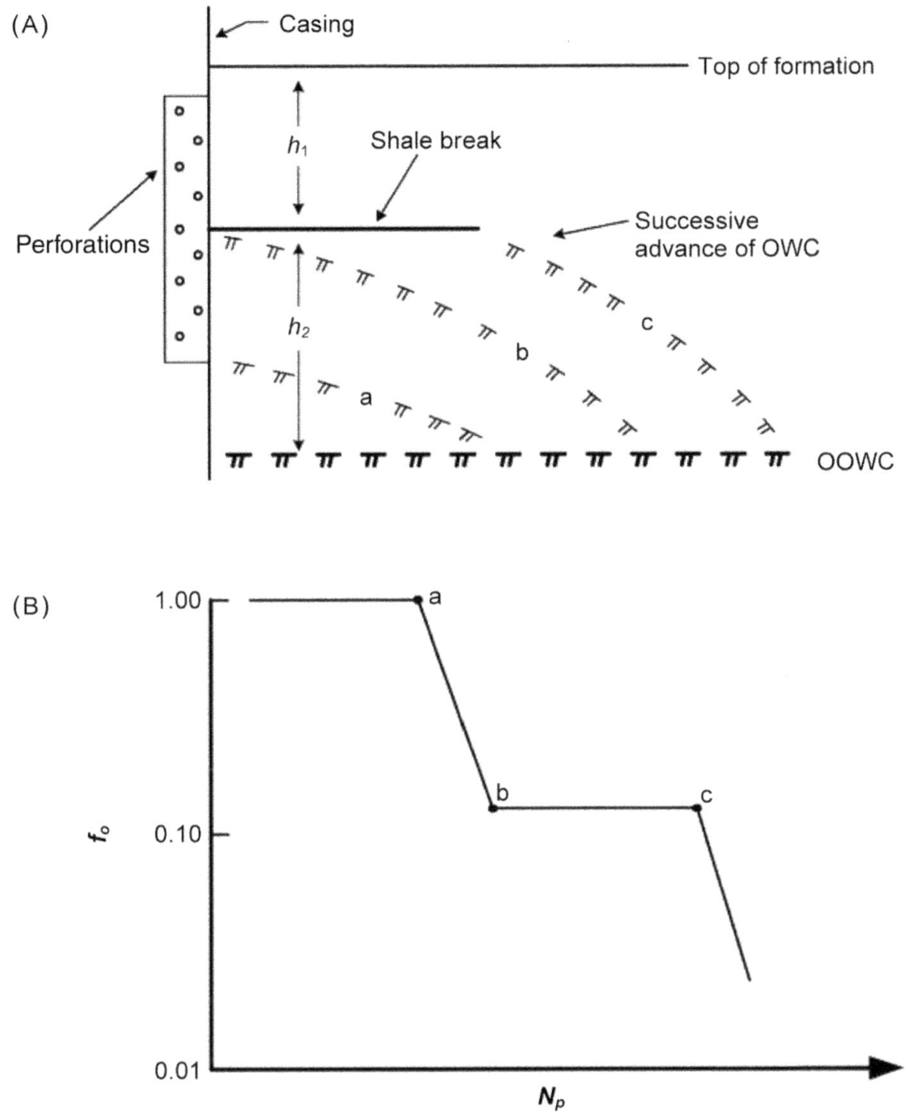

Fig. 18.20—Typical water encroachment and fractional flow behavior for a well in a bottom waterdrive oil reservoir with a shale break in the completion interval (after Cronquist[3]). OOWC = original oil/water contact.

Analysis and forward projection of rate/time trends (or rate/cumulative production or other trends) typically are handled with in-house or commercial software, as discussed in the next section.

Analysis, Forecasting, and Evaluation Software. In addition to in-house and other proprietary software, numerous commercial software products are available for analyzing historic trends and projecting future oil and gas production, or reserves. Typically, such products include a graphics module and an evaluation module. The graphics module is used to analyze historic trends and project future oil and gas production. These projections may be inserted into the evaluation module, together with prices, costs, taxes, ownership, and other pertinent data, to

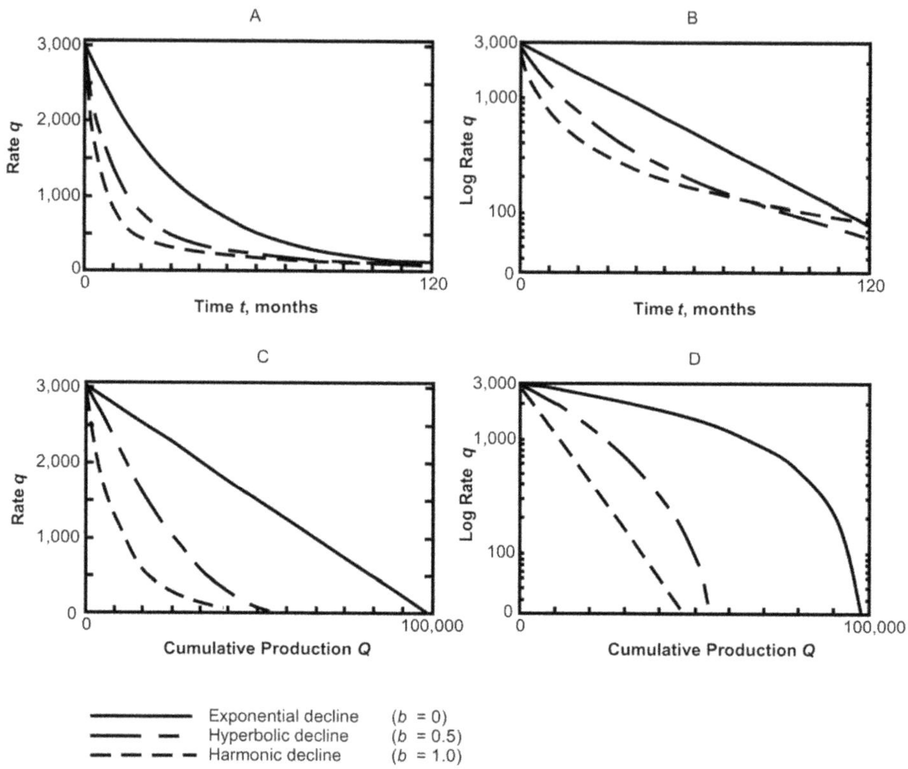

Fig. 18.21—Three types of production-decline curves (after Arps[65]).

calculate the economic limit and the reserves of oil, gas, and condensate, and to generate a cash flow projection.

The graphics module can be used either to curve fit a historic trend of production and generate a production forecast of reserves on the basis of projection of the fitted trend, or to generate a production forecast of reserves that has been calculated by analogy and/or volumetric methods. Three curve-fitting/-projection parameters are most important: the initial decline rate D_i, the hyperbolic exponent b, and the terminal effective decline rate a.

Procedures for curve fitting depend on the production maturity of the well(s) being analyzed. For example, the production trend for an immature well might be so erratic as to require considerable local knowledge to ensure a curve fit that is consistent with analogous wells/reservoirs. If such a well is in the transient stage, apparent b factors from curve fitting might not represent of the semisteady-state stage.[66,68,69] Projection of such curve fits might yield substantial overestimates of reserves. It might be appropriate to ignore the curve fit and impose a decline trend consistent with analogous wells in the area. The reasonableness of this procedure can be tested with volumetric and/or simulation analysis. In any event, the criterion of reasonable certainty should be observed in estimating proved reserves.

For wells in an advanced stage of production and in areas where there are good analogy wells, the appropriate b factors and terminal effective decline rates may be readily apparent.

For scenarios in which there is no production history, reserves can be estimated by analogy and/or volumetric methods, as discussed previously. The graphics module then can be used to compute a production forecast, given initial and final production rates, using any of the curve types (exponential, hyperbolic, or harmonic) that typically are programmed into the software. A constant rate period can be projected, followed by a prescribed decline period. Analogs are useful in selecting the decline curve parameters.

An evaluator should not automatically accept the best fit calculated by the software; rather, fully review and analyze the data it displays. Production histories can be affected by operational downtime, equipment modifications, market conditions, weather, and many other conditions that might distort the productive capacity of the property being reviewed. One should preview the pressure history, where relevant, to detect erroneous data. There might be a need to acquire missing data or to contact the operator for additional information.

The projection or forecast never should be made until the *engineering analysis* has been completed. Good engineering practice might require overriding the software-generated hyperbolic projection with an exponential projection.

Most graphics modules allow the user to exclude certain data points and to include all or a portion of the well history. One should use this feature cautiously and ensure that only erroneous or invalid data points are excluded.

Decline curves typically are used to forecast future production rates for properties that have some established performance history. In addition to monthly rates of oil and gas production, reservoir and surface pressure data and various fluid ratios can be analyzed and used in preparing reliable production forecasts. In most cases, the primary production stream volumes are forecast, and secondary products are estimated using constant or variable indices—primarily GOR or CGR. Gas and/or liquid handling constraints often are not considered in such projections. Iterative methods[70] are available to assist in forecasting oil production from a mature waterflood where there are surface or subsurface limitations on water handling, and in gas lift oil operations where gas-handling equipment imposes limitations.

Reconciliation of Estimates. For each well or reservoir, the reserves engineer should attempt to reconcile the estimates by comparing ultimate recovery estimated using P/DTA to ultimate recovery estimated using volumetric methods. (This reconciliation also should include results from computer simulation, as appropriate, but this is not specifically included in this discussion.) Consider the following possibilities:

1. If ultimate recovery estimated using volumetric methods is significantly greater than that from P/DTA, then

(a) The performance data might be representative only of the transient period (when the pressure sink around wells is still expanding) and not of semisteady-state conditions (when the drainage area has stabilized), which usually better represents well performance.

(b) The volumetric parameters might not be representative of the well or reservoir.

(c) Wells might need to be stimulated or equipped with higher-capacity equipment.

(d) Infill wells might be needed.

(e) Or, more than one of the foregoing factors may be relevant.

2. If ultimate recovery estimated using volumetric methods is significantly less than that from P/DTA, then

(a) The volumetric parameters might not be representative of the well or reservoir.

(b) The mapped area or volume might be too small and additional development might be warranted.

(c) Or, both factors may be relevant.

3. Pending determination of the reason for the difference in ultimate recovery estimated using the two methods, assign proved reserves only to those quantities demonstrated to be recoverable from performance methods.

As discussed earlier, reservoir heterogeneities often have an adverse effect on the recovery of oil and/or gas. A significant difference between ultimate recovery estimated using volumetric methods and that from P/DTA might indicate inefficient drainage caused by reservoir heterogeneities.

Computer Simulation. Computer simulation is a sophisticated method for analyzing well/reservoir performance and/or projecting future production (reserves). The widespread commercial availability of simulation software and the greatly expanded capabilities of personal computers have increased use of this technology.

Engineers use models of varying complexity to guide reservoir development and production programs and to compare and evaluate well spacing, rate sensitivity, and improved recovery potential, as well as other scenarios. It is important to realize, however, that although computer simulation is widely used, these methods typically oversimplify representation of the spatial distribution of reservoir properties.[71–73] Computer simulation results typically are sensitive to changes in reservoir characterization, and the degree of sensitivity may render moot an estimate of ultimate recovery that is based on this method.

Accordingly, check such estimates against simple models, as discussed by Richardson and Blackwell,[74] and against actual performance in analogous reservoirs. Saleri and Toronyi[71] observe that "determination of ultimate recovery through simulation is an unattainable goal, particularly for heterogeneous reservoirs that are coarsely gridded." Jacks[72] notes that "an estimate of ultimate recovery from any single simulation is subject to considerable uncertainty, especially if it is developed early in the life of a reservoir." Many forecasts of future performance have been significantly in error, despite the users' best efforts at reservoir characterization and history matching.[75] In some cases, the reservoir characterization and history matching were sound, but the field development was not consistent with that modeled. Continuous monitoring of performance, then, is essential in reserve estimation by computer simulation—indeed, by any method!

Computer Simulation vs. Reserves Definitions. Despite its limitations, computer simulation may provide the only acceptable answer to analysis of complex reservoirs, but such use must be consistent with the purpose(s) for which a specific reserves estimate is made. For example, the presumed "most likely" case—consistent with proved plus probable reserves—is the case typically modeled by computer simulation.[73] Such a case might not be appropriate for reporting reserves under U.S. SEC guidelines.

To ensure that simulation results will comply with SEC reserves definitions, either configure the model using the observed hydrocarbon/water contact or imposing the lowest known occurrences of oil or gas, or, if the constructed model does not comply with proved reserves definitions, make necessary adjustments to the model results to estimate the recovery that would approximate recovery reached using a model that meets proved reserves requirements. Many model runs might be needed to accomplish this. Manually adjusting the output data causes some of the rigorous nature of the simulation process to not be realized fully.[73]

One should not construct a reservoir model solely to comply with proved reserves definitions unless warranted by special, and perhaps limited, circumstances. Constructing a (3P) model may be best initially, with the ability to modify it to assist in estimating proved or proved plus probable reserves.

Reservoir models that do not comply with the limitations inherent in proved reserves definitions can help in understanding the hydrocarbon REs likely with each reserves category.[73] Sensitivities of various reservoir characteristics (e.g., relative permeability, drive mechanisms, and PV compressibilities) can be investigated to determine a range of recoveries that may comply with the "reasonable certainty" aspect of proved reserves.

In history matching, a fit typically is described as "poor," "good," or "excellent." Such subjective terms mean little in establishing a confidence level in predictions of future production (reserves). However, if a history match can be obtained that replicates individual well producing rates and pressures with reasonable tolerance, without illogical adjustments, the model might be appropriate for the estimating proved reserves.

Palke and Rietz[73] observe (in part) that

• Although reservoir simulation is a sophisticated technique, it does not always produce reliable or applicable results suitable for reserves estimation.

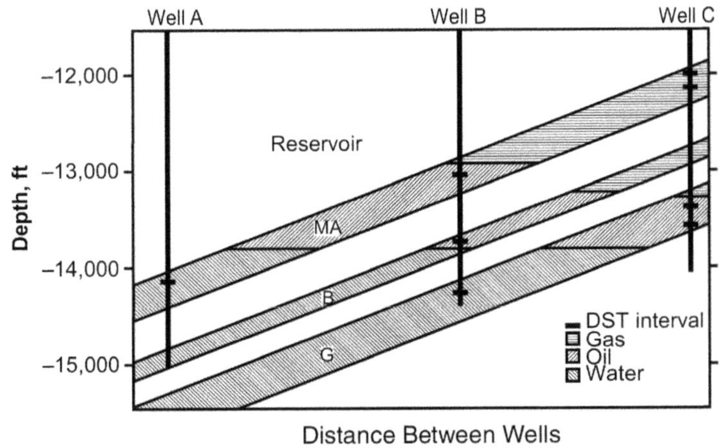

Fig. 18.22—Schematic cross section with fluid contacts and DSTs (after Rasor[76]).

• Reservoir simulation should be used to improve the understanding of a reservoir, but should not be used to circumvent the intent of U.S. SEC reserves definitions.

• Although reservoir simulation increasingly is being used as a tool for reserves estimation, the results do not necessarily constitute U.S. SEC reserves estimates.

• Models that do not comply with U.S. SEC reserves definitions can be modified to comply, but this process may be difficult.

• Results from models that are not consistent with U.S. SEC reserves definitions can be used through the alteration of the simulation output itself, which requires a great deal of simulation output and might provide less rigorous solutions.

• For immature reservoirs, simulation is useful primarily for estimating the oil and/or gas RE, and for testing the limits in terms of uncertain parameters (permeability, aquifer support, O/GIP).

• Models of mature reservoirs should exhibit reasonable history matches before they are accepted for reserves purposes.

• The uniqueness and quality of the history match affect the confidence to be placed in a model's ability to predict future performance, and thus dictate the model's appropriate usage in the process of estimating reserves.

Example Study. Rasor[76] presents a field study comparing the results of a simulation study of a three-reservoir field with those of a classic reservoir-engineering approach. He uses a 3D compositional model and assumes gas processing to maximize liquid recovery. Laboratory PVT and special core analysis data are matched. The classic procedure is limited to a traditional volumetric study incorporating estimated REs based on the fluid and formation properties and the experience of the estimator. **Fig. 18.22** is a schematic cross section of the three reservoirs, MA, B, and G, shallow to deep, respectively.

OWCs for reservoirs MA and G were calculated using the pressure/depth relationships illustrated in **Figs. 18.23 and 18.24**. The GOC in B is based on PVT analysis, **Fig. 18.25**. **Table 18.7** compares the results from the classic and simulation methods. The REs of oil and gas from reservoirs B and G are in reasonably close agreement. The oil REs for MA range from 24 to 42%. Without other information or suitable analogies, an engineer might assign a proved (P1) recovery of 24% OIP and a proved plus probable (2P) recovery of 42% OIP.

The condensate REs are significantly different for all three reservoirs. In this situation, without reasonable explanation for the differences or compelling data to the contrary, the lower REs would be appropriate for booking proved reserves.

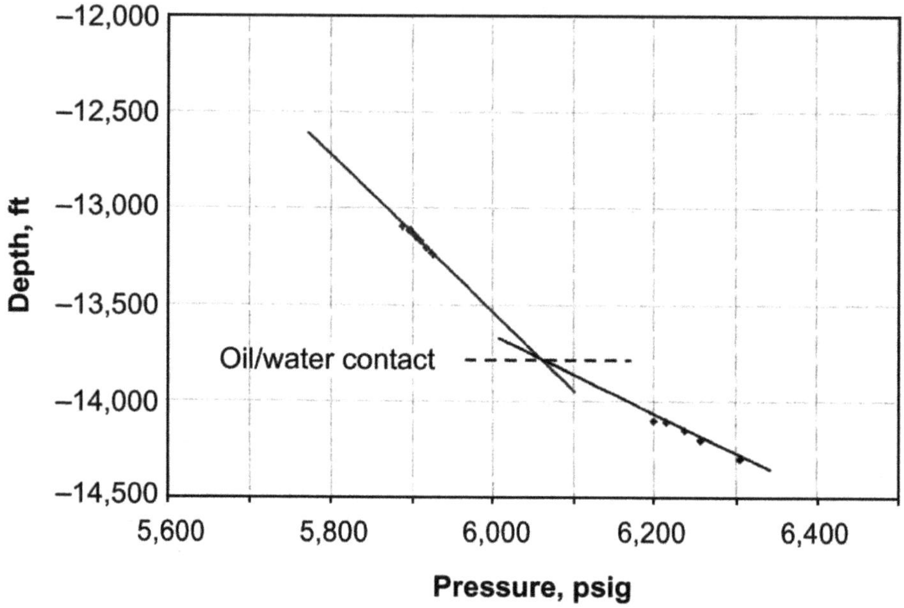

Fig. 18.23—Reservoir MA fluid gradients (after Rasor[76]).

This study illustrates the range in results produced by the two methods—classic engineering and reservoir simulation—by a competent evaluator using a reasonable amount of data. It is a real-world example of the uncertainties of relying on computer simulation, without history matching, for the estimation of proved reserves.

18.5 Special Problems*

Some operating and/or reservoir scenarios are so complex as to defy conventional analysis and to require a high degree of skill and ingenuity. Because of the substantial degree of uncertainty, such scenarios may be better suited for probabilistic procedures than for conventional scenarios.

18.5.1 Coalbed Methane. Chap. 6 of the Emerging Technologies section of the *Handbook* discusses the development of coal deposits in the U.S.A. and around the world for recovery of natural gas. Such natural gas is predominantly methane, but it also may contain small amounts of ethane, carbon dioxide, and nitrogen.

Coalbed methane (CBM) is adsorbed onto the coal surfaces exposed through the matrix microporosity and the naturally occurring fracture or cleat system. This cleat system typically is water-filled, often with fresh or slightly saline water, but may also contain some free gas.

Calculation of gas in place for a unit volume of the coal layers being developed does not follow the "porous media" approach of determining effective porosity, saturations, pressures, temperatures, and gas quality. Instead, the gas in place is measured physically through the recovery of coal samples, the number and distribution of which are important to the estimation of total gas in place pertinent to the property being evaluated.

Cored samples are transferred carefully from the core barrel to canisters, which are sealed immediately and transported to an analysis laboratory. In analysis, two measurements are taken. First, free gas in the canister is measured, and then the coal sample is crushed and the

* Section sequence is not intended to convey authors' views of relative importance.

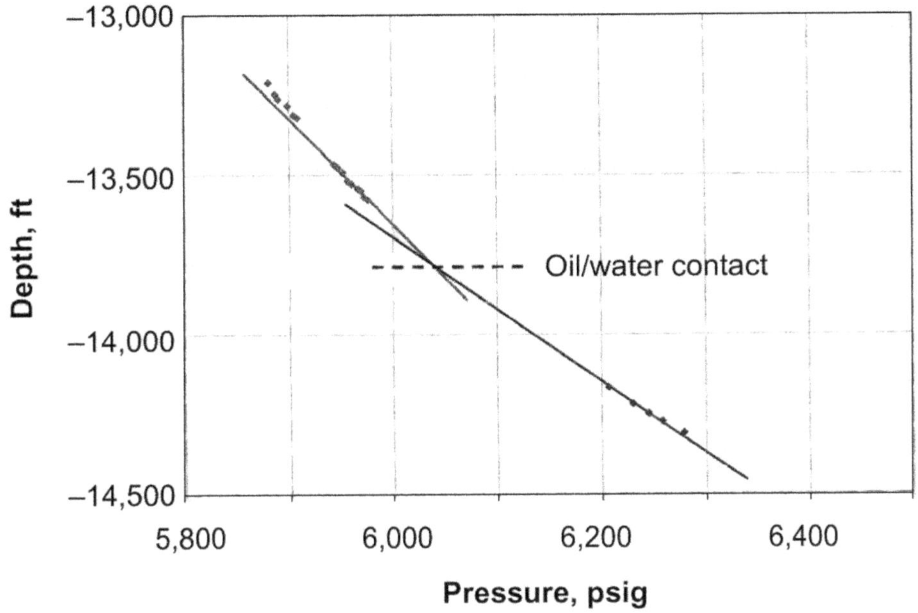

Fig. 18.24—Reservoir G fluid gradients (after Rasor[76]).

liberated gas measured. These two measurements are combined with an estimate of gas lost during the core recovery operation. The lost gas volume is estimated as a function of the coal type and depth of burial and other factors. Total gas in place is calculated as the product of the unit gas in place—considering areal variations—and the mapped volume of the coal seams being developed.

A volumetric estimate of GIP (scf) in coalbed reservoirs can be calculated by[77]

$$G_i = Ah\left[\frac{43{,}560\phi_f\left(1 - S_{wfi}\right)}{B_{gi}} + 1{,}359C_{gi}\rho_c\left(1 - f_a - f_m\right)\right]. \qquad (18.44)$$

The first term in the square brackets of Eq. 18.44 is used to calculate the gas volume contained in the interconnected fracture or cleat system (if any) and is identical to the term used for porous media reservoirs. The second term in the square brackets is used to calculate the adsorbed gas in the coal matrix. The adsorbed gas quantity results from laboratory measurements of the adsorbed gas in a unit of dry, ash-free coal and other coal-quality factors.

In most cases, cleat porosity is water-filled, so that the free gas therein essentially is zero. Most engineers ignore the gas volume in solution in the water.

Most of the data on which the reservoir engineer must rely is gathered through core analysis, fluid analyses, and well tests. **Table 18.8** presents certain pertinent data items and primary sources for each item.[77]

Seismic data historically has not been used in CBM project analysis because most of these projects have been in areas that had an abundance of subsurface data obtained as a result of underground mining (e.g., the Black Warrior basin) or oil and gas wells drilled to deeper objectives (e.g., the San Juan basin). As the industry advances into areas with a dearth of existing subsurface information, seismic information is expected to become more important in determining reservoir extent and structure.

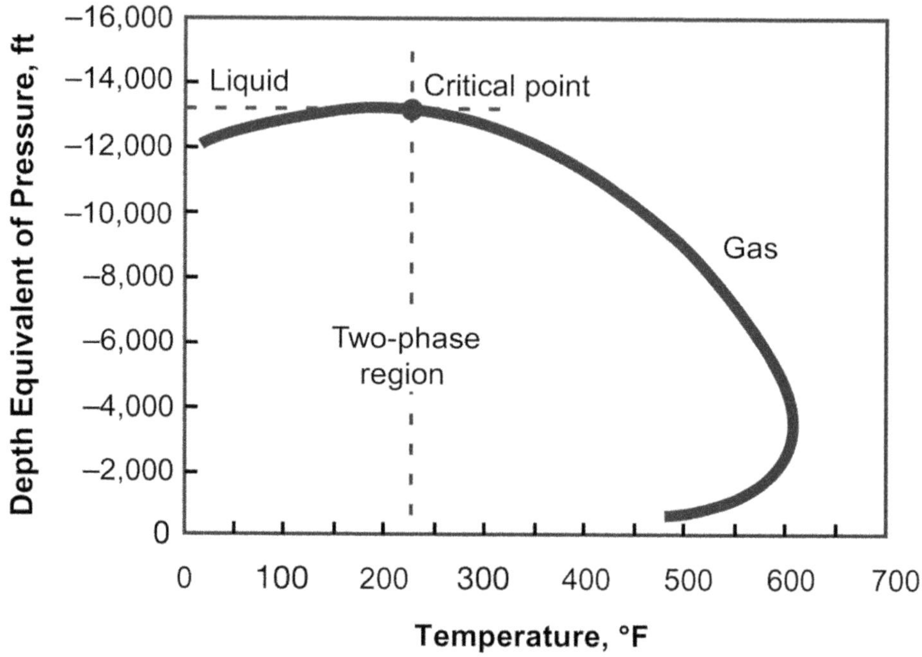

Fig. 18.25—Reservoir B PVT analysis (after Rasor[76]).

TABLE 18.7—COMPARISON OF RECOVERY EFFICIENCIES—CLASSIC AND SIMULATION METHODS (AFTER RASOR [76])

Classic Reservoir-Engineering Case				
Reservoir	Oil	Solution Gas	Gas-Cap Gas	Condensate
MA	0.24	0.73	0.77	0.13
B	0.36	0.64	0.76	0.25
G	0.36	0.64	0.76	0.25
Simulation-Prediction Case				
Reservoir	Oil	Solution Gas	Gas-Cap Gas	Condensate
MA	0.42	0.62	0.73	0.88
B	0.38	0.68	0.76	0.42
G	0.38	0.68	0.76	0.42

REs, typically up to 75% GIP, are related to well density, the degree of naturally occurring fractures, the effectiveness of wellbore hydraulic fracturing programs, and the ability to "dewater" the reservoir to reduce the reservoir pressure to a level where desorption can be effective. Laboratory measurements can be used to develop composited desorption isotherms, which are useful in estimating the rate of gas liberation while reservoir pressure is reduced.

Proved reserves can be assigned to an area where wells have been drilled and have demonstrated that commercial gas rates can be maintained. Well spacing in the U.S.A. ranges from approximately 40 to 160 acres per well. For coalbed projects in areas remote from comparable analog operations, the time to confirm commerciality may be as long as several years. Some projects dewater quickly, allowing commercial gas rates to be attained early; other projects

TABLE 18.8—SOURCES OF DATA FOR ANALYZING AND ESTIMATING RESERVES FROM COALBED METHANE RESERVOIRS (AFTER ZUBER[77])

Data Item	Primary Source(s)
Permeability	Well test
Initial pressure	Well test
Initial water saturation	Well test
Gas desorption pressure	Well test
Pore compressibility	Well test
Desorption isotherm	Core test
Sorbed gas content	Core test
Desorption time	Core test
Relative permeability	Production data and core test
Porosity	Core test
Net pay thickness	Well logs and core test
Temperature	Well log
Gas PVT properties	Gas analysis
Water PVT properties	Water analysis
Completion effectiveness	Well test
Well drainage area (spacing)	Geologic description

might prove to be noncommercial because of dewatering failure. A cluster of wells might need to create a pressure sink large enough to overcome the influx of water from a large aquifer. High permeability, together with a large aquifer, might create enough water influx to cause project failure.

Fig. 18.26 illustrates a typical individual well decline curve exhibiting a 2-year period of dewatering that is characterized by increasing gas production rates. An exponential trend has been drawn through the approximate 1-year decline period.

Confidence in the forecast would increase if there were nearby analog wells with more production history supporting the exponential projection. Lacking such support, however, the projection should be confirmed through volumetric means before booking the forecast volumes as proved reserves. Many (perhaps most) coalbed wells producing from coal that has low to moderate permeability will exhibit a wide range of hyperbolic declines, underscoring the need for suitable analogs.

Type curves (production vs. time) from successful analog operations are the most useful tools for predicting the production profile and reserves for completed wells. As in traditional reserves estimation, volumetric reserves estimates should be checked against performance-driven reserves estimates.

Assigning of proved undeveloped reserves to coalbed projects usually should be restricted to the "one-offset" limitation imposed by the 1978 U.S. SEC definitions, unless the engineer can demonstrate "certainty of production" beyond the one-offset location. The 1997 SPE/WPC definitions may, in some circumstances, permit a larger area to be classified as proved, but one should be cautious until both the presence of coal of commercial thickness and adequate permeability are determined with reasonable certainty.

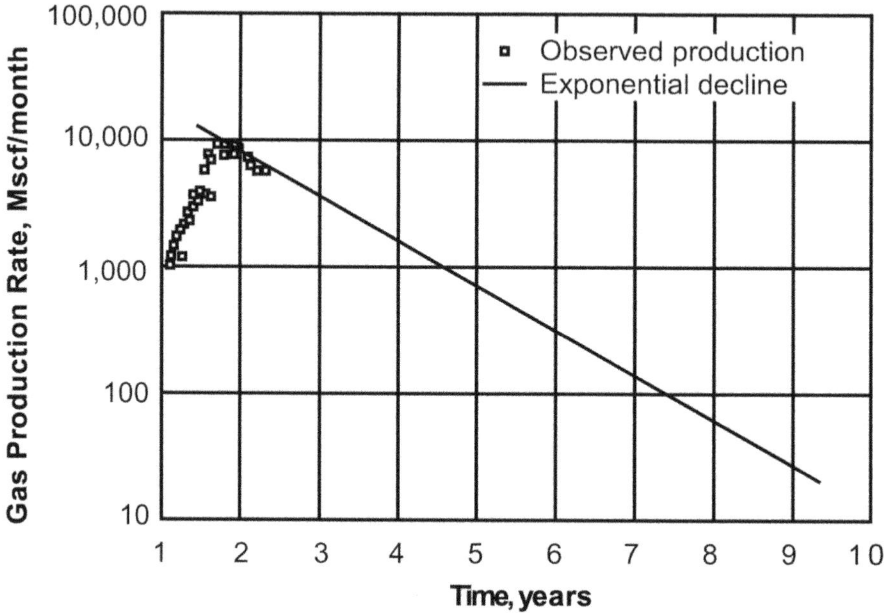

Fig. 18.26—Long-term forecast using exponential decline (after Zuber[77]).

Probable and possible reserves typically are assigned to acreage at increasing distances from the commercially developed portion of the project.

18.5.2 Fractured Reservoirs. Fractured reservoirs have been observed in most producing areas of the world, in igneous/metamorphic rocks, sandstones, carbonates, shales, and cherts. The two broad categories of fractured reservoirs are those with a porous matrix and those with a nonporous matrix.[78] In the porous matrix type (the more common), most of the hydrocarbons are stored in the matrix porosity, and the fractures serve as the principal flow conduits. Such reservoirs typically are identified as "dual-porosity" systems. Examples include many of the Iranian fields, Ekofisk (North Sea), Palm Valley (Australia), and Spraberry (Texas, U.S.A.). Some cherts exhibit dual porosity and have significant storage capacity in the matrix but that contributes little to reserves. Fractured reservoirs with a nonporous matrix occur in fractured igneous and/or metamorphic rocks, fractured shales, and fractured cherts. Such reservoirs frequently are associated with basement rocks.[79] Examples include the Bach Ho field (offshore Vietnam), the Augila field (Libya), the Edison field (California, U.S.A.), the Big Sandy gas field (Kentucky, U.S.A.), and the Santa Maria basin fields (California).

When they occur in carbonates, fractures tend to facilitate extensive leaching and diagenesis, which may lead to the development of vugular, sometimes karstic, porosity. Examples include the Albion Scipio trend (Michigan, U.S.A.) and the Rospo Mare field (offshore Italy).

Fractured reservoirs pose formidable difficulties for estimating reserves. These difficulties are attributable to the heterogeneity of the reservoir formation, which causes substantial uncertainties in estimates of O/GIP and RE. Because of uncertainties in determining the flow characteristics of dual-porosity systems, estimates of reserves using volumetric methods are subject to substantial uncertainty. When feasible, compare such estimates with observed recovery in analogous reservoirs.

In general, the following scenarios cause problems:

• Boreholes frequently are severely washed out, making log interpretation difficult or impossible.

TABLE 18.9—RECOVERY FACTORS IN NATURALLY FRACTURED RESERVOIRS (AFTER AGUILERA[80])

	Recovery Efficiency, %					
Drive Mechanism	Oil Reservoirs Storage Ratio			Gas Reservoirs Storage Ratio		
	A[a]	B[b]	C[c]	A[a]	B[b]	C[c]
Solution gas (*depletion*)	10 to 20	20 to 30	30 to 35	70 to 80	80 to 90	>90
Solution gas plus gas-cap expansion plus water influx	35 to 45	45[d] to 55	55 to 65			
Strong waterdrive				15 to 25	25 to 35	35 to 45

[a]Storage mostly in matrix, with small amount in fractures. [b]Storage about 50/50 in matrix and fractures. [c]All storage in fractures.
[d]In some areas, 25% might be a more appropriate lower limit (personal communication with Ed Holstein, 2 March 2002).

- Core recovery frequently is fragmental, at best.
- Even in good-quality boreholes, detection of fractures and measurement of fracture porosity using logging devices is highly empirical, although significant improvements have been made using formation imaging tools.
- In accumulations with a severe loss of circulation, operators typically stop drilling at the top of the reservoir section, a practice that, while necessary for safe operations, precludes characterization of the objective section.
- Well performance frequently is strongly influenced by proximity to major fractures, which can extend surprising distances. Because of this, be extremely cautious in assigning reserves to undrilled tracts that offset tracts at a mature stage of production.*
- Although transient pressure analysis provides useful data, applying modern interpretation techniques mandates using highly accurate quartz pressure transducers.
- The accuracy of type curve matching depends on the accuracy of the mathematical model used for the type curves. An invalid model cannot yield a valid interpretation. Even if the model is valid, analysis of results might not provide unique answers.
- In pressure-depletion reservoirs, the rate/time performance of wells typically is hyperbolic. The behavior of an average well might be used to estimate reserves, but one should expect wide variation in performance between wells.
- In reservoirs producing by pressure depletion, the early performance of wells typically is characterized by relatively rapid decline in the production rate, which usually is caused by transient pressure behavior. Reserves cannot be estimated with any degree of confidence using decline trend analysis until wells have passed through the transient pressure period and settled into semisteady-state conditions.

Aguilera[80] provides guidelines for estimating RE in fractured reservoirs, classifying fractured reservoirs by pore type and "storage ratio"—the relative amount of storage in matrix vs. fractures. **Table 18.9** is adapted from Aguilera's Tables 2 and 3.

Aguilera does not state so, but "oil" presumably means light oil—stock-tank gravities greater than approximately 25°API. Also, Aguilera does not mention the influence of well spacing on RE, which could be a major factor, depending on the nature of the fracture system.

18.5.3 Natural Gas From Fractured Shale. As reported in mid-2000, natural gas produced from shale in the U.S.A. has grown to be[81] approximately 1.6% (0.3 Tcf annually) of total gas production. The first commercial production of natural gas from shale was developed to supply

* Personal communication with Ed Holstein, retired, Exxon, 2 March 2002.

TABLE 18.10—U.S.A. SHALE GAS RESOURCES (AFTER REF. 82)

Basin	State(s)	Major Shale-Bearing Formation or Group	Basin Area	Total Organic Carbon (TOC)	Thermal Maturity	Shale Gas-in-Place		Estimated Recoverable Shale Gas	
			square mileage	%	% R_o	Tcf	Resource	Tcf	Resource
Appalachian	Ohio Kentucky New York Pennsylvania West Virginia Virginia	Ohio shale	160,000	0 to 4.5	0.4 to 1.3	225.0 to 248.0[a]	1980 NPC estimate[84] 1992 NPC estimate[85]	14.5 to 27.5	1980 NPC estimate[84] 1992 NPC estimate[85]
Michigan	Michigan Indiana Ohio	Antrim shale	122,000	1.0 to 20.0	0.4 to 0.6	35.0 to 76.0	1980 NPC estimate[84] 1992 NPC estimate[85]	11.0 to 18.9	1992 NPC estimate[85] 1995 USGS estimate[86]
Illinois	Illinois Indiana Kentucky	New Albany shale	53,000	1.0 to 25.0	0.4 to 1.0	86.0 to 160.0	1980 NPC estimate[84] 1992 NPC estimate[85]	1.9 to 19.2	1992 NPC estimate[85] 1995 USGS estimate[86]
Fort Worth	Texas	Barnett shale	4,200[b]	4.5	1.0 to 1.3	NA		3.4	Schmoker[87] Kuuskraa[88]
San Juan	Colorado New Mexico	Lewis shale	1,100[b]	0.45 to 2.5	1.6 to 1.88	96.8	1999 Williams estimate[89]	NA	

[a]Black shales only. [b]Play area only. NPC = National Petroleum Council.

gas to the town of Fredonia, New York, U.S.A., in the late 1820s, predating Col. Drake's first oil well by almost 40 years.

By 1979, some 60 Bcf/year was being produced from wells in the Appalachian (Ohio) basin. Production from the Antrim shale (Michigan basin) began in the mid-1980s and by 1994 had surpassed production of the Appalachian basin. Three other U.S.A. basins—San Juan (Colorado, New Mexico), Fort Worth (Texas), and Illinois (Illinois, Indiana, Kentucky) currently are producing from shale. Total U.S.A. gas-bearing shale resources[82–88] are shown in **Table 18.10**.

In a 1992 report,[89] the Gas Research Inst. (GRI) describes the Antrim shale as being organically rich and says that the majority of the in-place gas was sorbed into the organic constituent. Productivity is achieved by reducing reservoir pressure, and is maximized by hydraulic fracturing to access and connect the wellbore to the natural microfractures and other permeability pathways.

As in coalbed methane reservoirs, the naturally occurring fracture system usually is water-filled, requiring artificial lift equipment to dewater the wells to reduce the bottomhole pressure to a level consistent with maximum gas desorption and production.

Shale gas volumes initially in place (scf) can be calculated by:

$$G_i = 1{,}359 A h_s \rho C_{gi}, \quad (18.45)$$

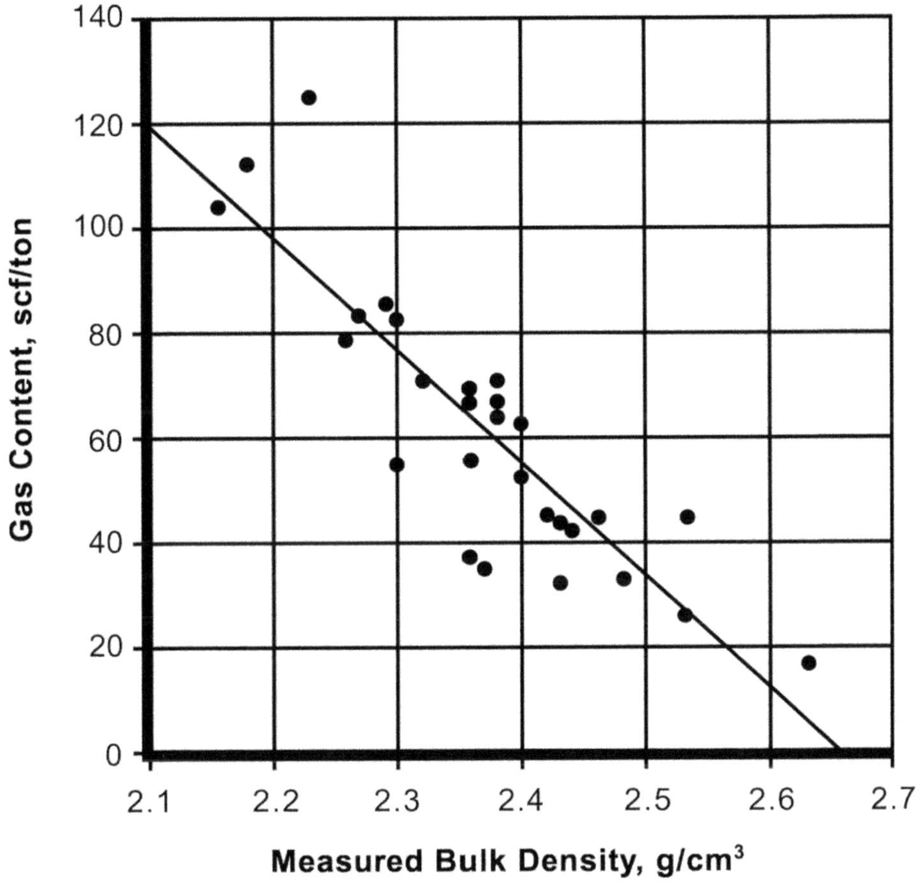

Fig. 18.27—Calculation of gas content from bulk density measurements from log, Antrim shale (after Ref. 82).

where 1,359 is a conversion factor to convert volume (acre-ft), shale density (g/cm^3), and gas content (scf/ton) to scf gas in place.

Shale density and gas content can be measured directly through core analysis or indirectly through well logs, using correlations established between gas content and shale bulk density. Core samples are taken and preserved to minimize the release of original gas in place. The free gas in the core sample canisters plus the gas that is released during core crushing is measured in the laboratory. This volume of gas may be adjusted to account for the volume estimated to have been lost during core retrieval. The analysis procedures are similar to those used for CBM.

Gas-content/shale-density correlations are an outgrowth of studies[90] in which the laboratory-measured shale densities and total organic content (TOC) of the samples were compared and related in a linear correlation. Similarly, gas content was found to have a linear relationship with TOC.

This research thus leads to the ability to measure bulk density from well logs and use this information to directly estimate gas content and log-derived gas in place.

Fig. 18.27[82] is an example of the relationship between gas content and shale density for the Antrim shale in a defined area.

Gas REs vary widely and are related to many factors, including completion efficiency, reservoir pressure, water removal efficiency, and well spacing. The recoveries[82] shown in **Table**

TABLE 18.11—KEY PROPERTIES FOR PRODUCTIVE GAS SHALES (AFTER REF. 82)

	Basin and Producing Shale				
	Fort Worth Barnett	Appalachian Ohio	Michigan Antrim	Illinois New Albany	San Juan Lewis
Depth, ft	6,500 to 8,500	2,000 to 5,000	600 to 2,200	500 to 2,000	3,000 to 6,000
Gross thickness, ft	200 to 300	300 to 1,000	160	180	500 to 1,900
Net thickness, ft	50 to 100	30 to 100	70 to 120	50 to 100	200 to 300
Bottomhole temperature, °F	200	100	75	80 to 105	130 to 170
TOC, %	4.50	0.00 to 4.70	1.00 to 20.00	1.00 to 25.00	0.45 to 2.50
Thermal maturity, % R_o	1.00 to 1.30	0.40 to 1.30	0.40 to 0.60	0.4 to 1.00	1.60 to 1.88
Total porosity, %	4.0 to 5.0	4.70	9.00	10.00 to 14.00	3.00 to 5.50
Gas-filled porosity, %	2.5	2.0	4.0	5.0	1.0 to 3.5
Water-filled porosity, %	1.9	2.5 to 3.0	4.0	4.0 to 8.0	1.0 to 2.0
kh, md-ft	0.01 to 2.00	0.15 to 50.00	1.00 to 5,000.00	NA	6.00 to 400.00
Gas content, scf/ton	300 to 350	60 to 100	40 to 100	40 to 80	15 to 45
Adsorbed gas, %	20	50	70	40 to 60	60 to 85
Reservoir pressure, psia	3,000 to 4,000	500 to 2,000	400	300 to 600	1,000 to 1,500
Pressure gradient, psi/ft	0.43 to 0.44	0.15 to 0.40	0.35	0.43	0.20 to 0.25
Well costs, U.S. $1,000	450 to 600	200 to 300	180 to 250	125 to 150	250 to 300
Completion costs, U.S. $1,000	100 to 150	25 to 50	25 to 50	25	100 to 300
Water production, BWPD	0	0	5 to 500	5 to 500	0
Gas production, Mcf/D per well	100 to 1,000	30 to 500	40 to 500	10 to 50	100 to 200
Well spacing, acre	80 to 160	40 to 160	40 to 160	80	80 to 320
Recovery factors, %	8 to 15	10 to 20	20 to 60	10 to 20	5 to 15
Gas-in-place, Bcf per section	30 to 40	5 to 10	6 to 15	7 to 10	8 to 50

18.11 range from 5 to 60% GIP, but probably average approximately 40% GIP in the Michigan basin (Antrim). A typical well production profile is shown in **Fig. 18.28**.

The initial dewatering period of approximately 1 year is characterized by diminishing water production and increasing gas production. Following perhaps a year of relatively constant production, a decline rate of approximately 6% per year is typical for a Michigan-basin Antrim well. Most wells exhibit exponential decline during their economic life.

The booking of proved reserves must be delayed until the production rate reaches a commercial level and/or there is ample evidence from nearby analog wells. Undeveloped locations may be classified as proved if these locations are directly adjacent to commercial wells (1978 U.S. SEC definitions). Additional locations may be classified as proved under the 1997 SPE/WPC definitions[6] if there is compelling evidence from nearby analogs and if the continuity of favorable reservoir conditions is reasonably certain.

18.5.4 "Tight" Gas Reservoirs. As defined by the U.S. Federal Energy Regulatory Commission (U.S. FERC), low-permeability ("tight") gas reservoirs have an average in-situ permeability of 0.1 md or less. Others have placed the upper limit at 1 md. Such accumulations in the U.S.A. contain substantial resources. Estimates of ultimate recovery from these resources vary widely and depend chiefly on assumptions of wellhead gas price.

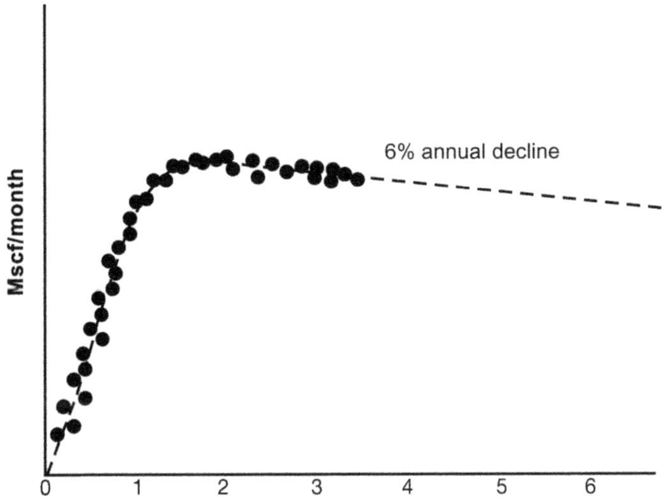

Fig. 18.28—Typical production graph for an Antrim shale well.

Methods for estimating gas reserves in moderate- to high-permeability reservoirs are unreliable in very-low-permeability reservoirs. The unreliability can be attributed to the geologic setting in which these reservoirs occur and the completion methods required to make them commercial. In general, their geologic setting is characterized by a high degree of permeability heterogeneity; lateral discontinuities in apparently blanket sands; stratigraphic, rather than structural, traps; and complex mineralogy, frequently with high-grain-density minerals randomly dispersed throughout the section as well as water-sensitive clays.

These attributes make it very difficult to determine porosity and interstitial water saturation by conventional log and core analysis.[91–95] Petrophysical properties measured at ambient conditions (e.g., k and S_w) differ substantially from those at reservoir conditions, and corrections for formation compressibility are subject to considerable uncertainty. For example, permeabilities can be as much as two orders of magnitude greater at ambient conditions than at reservoir conditions. These problems and poor lateral continuity lead to substantial uncertainties in volumetric estimates of GIP. In many cases, it is impossible to distinguish between commercial and noncommercial intervals from log analysis alone. Drillstem tests rarely provide useful information because formations often are damaged during drilling.

Massive hydraulic fracturing usually is required to obtain commercial flow rates. Despite more than 35 years of experience with fracturing technology, however, the industry remains unable to design a treatment and predict the results with high confidence when relying solely on analytical methods.[96,97] Typically, operators rely on analytical models coupled with analogy.

Along the U.S. Gulf Coast, tight-gas accumulations frequently occur in geopressured sections.[98,99] In this environment, understanding the influence of reservoir stress on rock properties is important for differentiating between productive and nonproductive formations.

Over the life of a well completed in low-permeability gas reservoirs, the gas production rate typically exhibits a hyperbolic decline,[100,101] with apparent b values generally >1. In addition to decline curve analysis, empirical log-log rate/time models might provide useful short-term information for such wells—before the onset of significant pressure depletion. The following equations, developed in the 1980s for more than 2,500 then-new wells in the U.S. Rocky Mountains, have been used to estimate flow rates (Eqs. 18.46 and 18.47) and near-term reserves (Eqs. 18.48 and 18.49) for damaged fracture flow[102]:

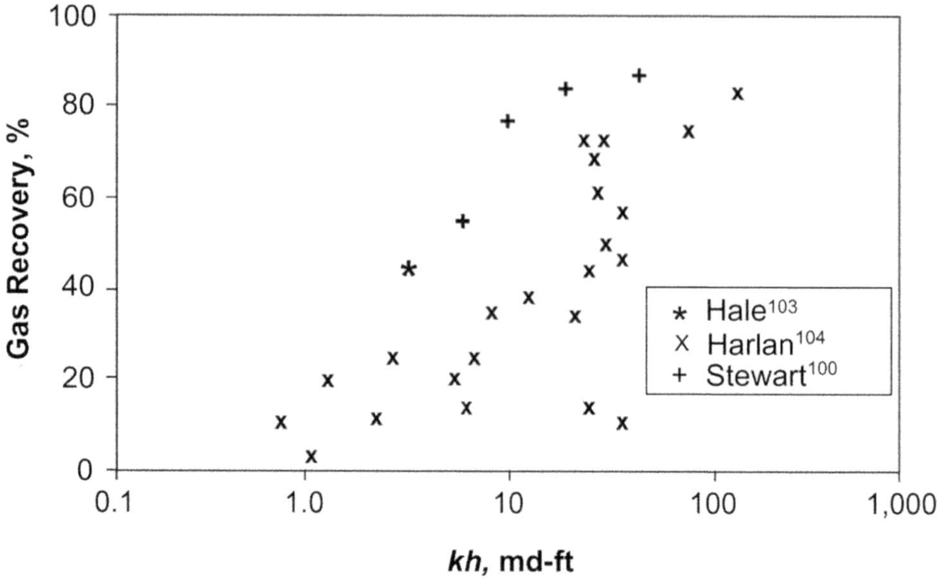

Fig. 18.29—Observed gas RE as related to *kh*. (After Cronquist.[3])

$$q_g = K_A t^n, \quad (18.46)$$

$$q_g = K_A / \left(K_B + t^{0.5}\right), \quad (18.47)$$

$$G_p = K_A^{n+1} / (n+1), \quad (18.48)$$

and

$$G_p = 2K_A K_B \ln\left[K_B / \left(K_B + t^{0.5}\right)\right] + 2K_A t^{0.5}, \quad (18.49)$$

where K_A and K_B are coefficients calculated during fitting these equations.

For an undamaged, fractured well, initial values of n should equal approximately –0.5. Because of damage, however, initial n values as small as –0.15 had been observed in the wells studied; the average was –0.34. With the onset of depletion, n decreases to –1.0 or more.[103,104]

Eqs. 18.46 and 18.48 can be used to account for the effects of a damaged fracture by using the field-observed value of n for each such well. Depending on circumstances, however, Eqs. 18.47 and 18.49 might provide a better fit to the observed data.

RE of gas from some of these reservoirs vs. $k_g h$ is shown in **Fig. 18.29**. Well spacing ranges from approximately 160 to 320 acres. For *kh* of less than approximately 50 md-ft, there is a decrease in RE, albeit erratically so.

Because of the high degree of permeability heterogeneity, drainage areas of individual wells vary widely. In the Green River basin (U.S.A.), for example, effective drainage areas reportedly have ranged from approximately 100 to 640 acres.[105] Depending on economics, such situations can offer opportunities for significant increase in reserves by infill drilling.

18.5.5 Heavy Oil. Discovered resources of heavy and extraheavy crude oil are estimated to be approximately 4,600 billion bbl, two-thirds of which are in Canada and Venezuela.[106] Bitumen

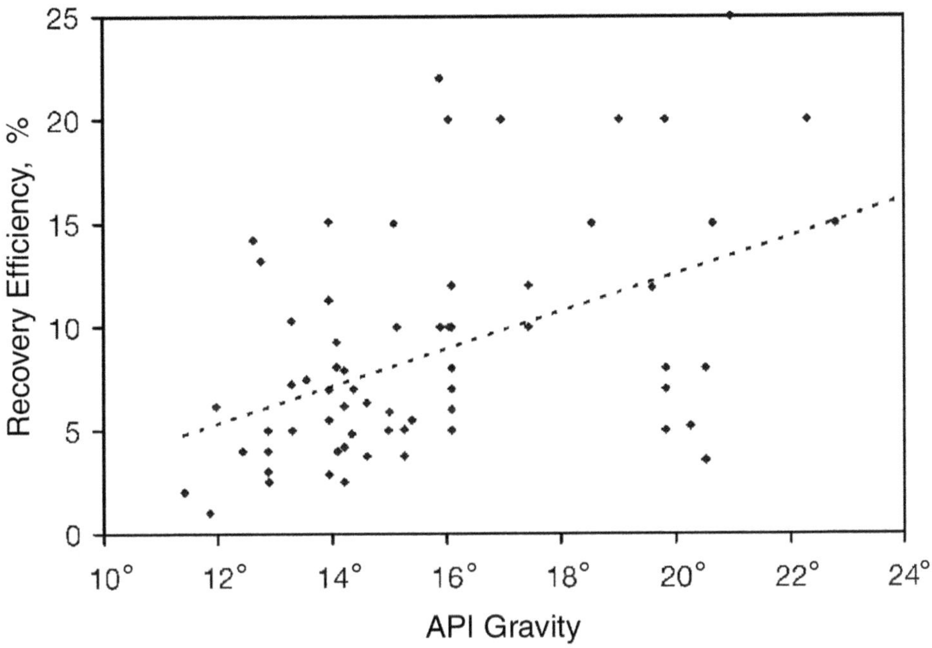

Fig. 18.30—Observed oil RE as related to oil gravity for 69 pools in Alberta and Saskatchewan (after Cronquist[3]).

and tar sands are excluded from this estimate. Published data on RE from this resource by primary drive mechanisms are sparse. Meyer and Mitchell[107] estimated worldwide ultimate recovery from heavy and extraheavy crude oils to be 476 billion bbl, which is 10% of the Briggs *et al.*[106] estimate of the discovered resource initially in place. Estimated primary RE ranges from 8 to 12% OIP for the Orinoco area of Venezuela, where stock-tank gravities range from 8 to 13°API.[108] Estimated primary RE ranges from 3 to 8% OIP[109] for the Lloydminster area of western Canada, where stock-tank gravities range from 13 to 17°API.

Primary RE vs. API gravity for heavy crude oil "pools" in Alberta and Saskatchewan* is plotted on **Fig. 18.30**. Data are from 69 "pools" with OIP > 10^6 m^3 (6.3 million STB). The trend line is not weighted by resource size and is shown only for reference; the regression coefficient, 0.21, is too small to infer a statistically significant correlation between RE and oil gravity. Of interest, however, is that the trend line is consistent with REs shown on Table 18.5 for "average" sandstone for 15°API and slightly-higher-gravity oils. Also interesting (lack of correlation notwithstanding) is the large number of heavy-oil reservoirs with significantly greater REs than would be predicted using "conventional" solution gas drive calculations. Reportedly, such REs are attributed to two mechanisms: the simultaneous production of oil and sand known as "cold production,"[110] and "foamy oil."[111]

Although there are reports of REs that range from 5 to 20% OIP,[110,112] no general correlations are available that relate specific rock/fluid properties and REs for heavy oil; thus, for volumetric methods, reserves engineers typically rely on analogy. The performance of wells in heavy oil reservoirs is erratic, however, and is influenced by varying production practices, varying volumes of sand production, and frequent downtime, among other factors, so that analogy estimates are subject to considerable uncertainty.

* Personal communication with Rajneesh Kumar, Ryder Scott Co., 2 April 2002.

Reserves estimates based on performance also are subject to considerable uncertainty. Production rates for single wells usually are erratic, thereby precluding meaningful trend analysis. Many engineers generate normalized production curves from groups of wells producing from zones comparable to those being analyzed.

In summary, the producing mechanisms for heavy oil are poorly understood; an optimum production strategy has yet to be developed; *a priori* prediction of the efficiency of the production mechanisms for heavy oil currently is impossible. Although progress is being made on computer modeling,[113] it may be several years before sufficient data are compiled for reliable estimates of RE and/or reserves from heavy oil.

18.5.6 Thin Oil Columns. Thin oil columns overlain by free gas and underlain by water pose difficult problems in well spacing and completion method, production policy, and reserves estimation. In this context, "thin" is a relative term. Whether an oil column is considered thin depends on costs to drill and produce the accumulation. For example, in the Bream field (Australia Bass Strait, 230 ft water depth), 44 ft was considered thin,[114] whereas in the Troll field (offshore Norway, 980 ft water depth), 79 ft was considered thin.[115] Onshore U.S.A., 20 ft is considered thin. Irrgang[116] takes a pragmatic approach, defining thin oil columns as those that "will cone both water and gas when produced at commercial rates."

The overall RE of oil from such accumulations can be influenced by well spacing and completion method and by gas-cap management policy. Economics, though, tend to be controlled by individual well recoveries and production capacities, rather than by average RE from the reservoir; thus, development planning focuses on the economics of individual wells—the cost to drill, complete, and operate vs. the oil rate/time profile. Ultimate oil recovery from individual wells tends to be controlled by a number of factors, including the gross thickness of the oil column and the horizontal/vertical permeability ratio in the well's drainage area. This ratio might vary significantly over the areal extent of the reservoir, depending on the depositional environment of the reservoir rock. Experience with such reservoirs indicates that this parameter typically is underestimated, causing underestimates of oil recovery. Localized shale breaks might contribute to suppressed coning of gas and/or water if wells can be completed to take advantage of these heterogeneities.

From limited data from conventional well completions in several such fields in Australia, Irrgang[116] developed the relation

$$N_{paw} = \text{fn}\left[\phi\left(1 - S_w - S_{or}\right)k_H h_t^{2.5} R_{ng}^{1.5} / \mu_o B_o\right], \quad (18.50)$$

where the bracketed term is a correlating parameter.

Irrgang does not provide details on estimation of k_H; the median permeability probably is appropriate. Irrgang has observed, "a higher power may be appropriate for permeability—possibly even 2."* The vertical/horizontal permeability ratio, k_V / k_H, influences volumetric sweep efficiency in bottom waterdrive reservoirs. The absence of this term in Irrgang's correlation is puzzling, but might be because of measurement difficulties.

Depending on the water/oil mobility ratio and the horizontal/vertical permeability ratio, oil wells completed in this type of accumulation might exhibit coning of the overlying gas and/or coning of the underlying water early in life. These phenomena might cause rapidly increasing GOR or WOR and relatively short economic life; thus, how efficiently this type of accumulation can be exploited depends on the degree to which premature coning of gas and/or water can be avoided by appropriate completion methods and production practices. In one of the earliest published analyses of this problem, Van Lookeren[117] advocated perforating below the

---

* Personal communication with H.H. Irrgang, Command Petroleum Holdings (December 1994).

TABLE 18.12—POTENTIAL FIELD ANALOGS FOR ESTIMATING OIL RECOVERY FROM THIN OIL COLUMNS (AFTER IRRGANG[116])

Parameter	Field and Lithology			
	Bream[a] sandstone	South Pepper[a] sandstone	Chookoo[b] sandstone	Taylor[b] sandstone
Porosity, fraction	0.21	0.21	0.17	0.08 to 0.16
Net/gross ratio	0.65	0.96	0.90	0.40 to 1.00
Permeability, md	2,900	80 to 4,800	500	100 to 2,000
k_V/k_H	0.01	0.01 to 0.20	Not reported	Not reported
Oil viscosity, cp	0.24	0.48	0.21	0.26
Oil column, m	13.5	9.7	3.7	0 to 11.3
Gas column, m	105[c]	18	19[c]	33
Well spacing, m[d]	300 to 800	300 to 800	600 to 1,200	700 to 1,500
Oil recovery, 10^6 bbl/well	2.00	1.50	0.26	0.09

[a]Offshore. [b]Onshore. [c]Produced gas injected into gas cap. [d]Authors' estimate.

initial OWC to minimize gas coning; however, the simple isotropic model used in his analysis essentially negates the practical application of this approach. In the last few years, horizontal drainholes have been used to develop these accumulations.[118] Because this technology is still evolving, consider apparent successes in analogous reservoirs with caution.

Determination of optimum well spacing and estimation of oil reserves in such reservoirs is subject to substantial uncertainty, at least until a reasonably well-defined performance trend has been established for each well. Before performance trends are established, however, reserves typically are estimated using a combination of volumetric mapping and analogy or analytical methods. In this context, computer simulation can be extremely useful in establishing sensitivity of RE to various assumed scenarios, thereby helping to determine optimum well spacing and commerciality. Potential analogs are provided in **Table 18.12**.[116]

A critical review of the more than 50 years of literature[119–125] makes apparent that the industry has yet to develop a general treatment of coning that includes the influences of gas cap, aquifer influx, and other relevant parameters. For example, some authors investigate the problem of coning in the presence of an inactive aquifer, which is analogous to the classic coning problem first discussed by Muskat and Wyckoff,[119] whereas others investigate it in the presence of an active aquifer. Clearly, the critical rate to avoid water coning would be less in the presence of an active aquifer than in the presence of an inactive aquifer, other factors being the same.

In addition to aquifer strength, another critical parameter to apply the correlations in Refs. 119 through 125 is the horizontal/vertical permeability ratio over each well's drainage area. Laboratory measurements of vertical and horizontal permeability of small core samples are inadequate for estimating this parameter. In theory, vertical interference testing or vertical pulse testing can determine this parameter, as discussed by Earlougher,[126] but the test procedure involves two sets of perforations separated by a packer, an expense operators might be reluctant to incur. Another possible approach for wells exhibiting coning is computer simulation to establish the horizontal/vertical permeability ratio that yields an acceptable match to observed behavior. Whether results from a few such wells would apply to all wells in the reservoir depends on the depositional environment of the reservoir formation and the degree of lateral heterogeneity; however, it is unlikely. It might be more practical to test wells at gradually increasing rates to determine a maximum rate at which each well can be produced without coning.

In the presence of a strong aquifer and a gas cap, the combination of water encroachment and gas-cap coning might cause displacement of part of the oil column into the gas cap. Depending on the size of the initial gas cap and the degree of gas-cap voidage, significant volumes of oil might be lost. In some cases, this loss might be minimized or avoided by injecting the produced free gas into the gas cap to maintain constant gas-cap volume.

18.5.7 Geopressured Oil/Gas. The term "geopressure," introduced in the late 1950s by Charles Stuart of Shell Oil Co., refers to reservoir fluid pressure that significantly exceeds hydrostatic pressure (which is 0.4 to 0.5 psi/ft of depth), possibly approaching overburden pressure (approximately 1.0 psi/ft). Geopressured accumulations have been observed in many areas of the world.

Geologic Setting. In regressive tertiary basins (the geologic setting for most geopressured accumulations), such pressures in sand/shale sequences generally are attributed to undercompaction of thick sequences of marine shales. Reservoirs in this depositional sequence tend to be geologically complex and exhibit producing mechanisms that are not well understood. Both of these factors cause considerable uncertainty in reserves estimates at all stages of development/production and reservoir maturity. Geologic complexity contributes to uncertainty in estimates of O/GIP that are based on volumetric mapping. Poorly understood producing mechanisms contribute to uncertainty in estimates of reserves that are based on pressure/production performance. Each aspect is discussed below.

Geopressured reservoirs frequently are associated with substantial faulting and complex stratigraphy, which can make correlation, structural interpretation, and volumetric mapping subject to considerable uncertainty.

The resistivity of interstitial water in geopressured sections may approach that of fresh water, which may suppress the SP log. Under these conditions, it might be difficult to estimate net pay unless a gamma ray log also has been run. In addition, the relatively fresh waters frequently encountered in geopressured sections complicate interpretation of resistivity logs, especially in shaly sands. Cases have been reported in which reserves were booked on the basis of high resistivity observed in porous sands that later investigation proved bore fresh water.

Drive Mechanism(s). As discussed in the Material Balance section of this chapter, if gas production is attributed to gas expansion only, a plot of p/z vs. G_p should be a straight line. Because geologists considered them to be closed accumulations, during the early years of exploitation it was assumed that geopressured gas reservoirs would produce by pressure depletion and exhibit linear plots of p/z vs. G_p. Although this was observed to be true in many cases, it is not universally true. The p/z vs. G_p plots for many geopressured reservoirs initially appear to be linear, but curve downward as reservoir pressure approaches hydrostatic pressure. Extrapolation of the initial part of such a plot might yield an estimate of GIP that is approximately twice that estimated using volumetric methods. The anomalously low initial slope of the p/z vs. G_p plot has been attributed to several phenomena, including PV compression, expansion of interstitial water, and partial waterdrive. The downward curvature of the p/z vs. G_p plot has been attributed to other factors, including depletion of a limited protoshale water aquifer[127] and rock collapse.[128] [The American Geological Inst. (AGI) defines shale as an "indurated (hardened)...sedimentary rock formed by the consolidation...of clay."[129] Because geopressures in tertiary basins generally are attributed to undercompaction, the term protoshale is adopted here to make that distinction.]

Producing mechanisms in a geopressured gas reservoir might include gas expansion; compressibility of the reservoir PV; expansion of the interstitial water; water influx because of water expansion from a contiguous aquifer; water influx because of dewatering of interbedded protoshale; and/or evolution of natural gas dissolved in interstitial and aquifer water. Any or all of these mechanisms may be active at various stages in the life of a geopressured gas reservoir. Pressure/production data typically are insufficiently diagnostic to distinguish one mechanism from another, so that there may be considerable uncertainty in analysis of historical data and estimation of reserves.

There is disagreement regarding the relative importance of these mechanisms, especially compressibility of reservoir PV[130] and water influx from interbedded protoshale.[131–133] Because it is difficult to analyze geopressure mechanisms separately for a specific reservoir, many engineers use Eq. 18.51 to make an aggregate adjustment to the p/z vs. G_p plot[134]:

$$p/z\left[\Delta p\left(c_w S_w + c_p\right)/\left(1-S_w\right)\right] = p_i/z_i - G_p p_i/G_{Fi} z_i \text{ . } \quad (18.51)$$

Eq. 18.51 differs from Eq. 18.37 by inclusion of a p/z adjustment factor, which is the left-side square-bracketed term. Eq. 18.51 sometimes is simplified by adjusting the apparent gas in place (AGIP)—that estimated by extrapolation of the initial part of the p/z vs. G_p plot—by multiplying the AGIP by the gas-compressibility/effective-compressibility ratio.

Both methods assume that PV compressibility remains constant over the life of the reservoir being evaluated, which is contrary to the findings of numerous investigators. In addition, neither accounts for possible water encroachment.

Regardless of the method used to adjust the p/z vs. G_p plot, always check a reserves estimate so derived against analogies and/or a volumetric estimate for the same well.

Analytical Methods. Analytical methods outlined in the literature typically require more information than usually is available. As an alternative, a method was proposed[3] that parallels that of Havlena and Odeh.[60] Under this method, Eq. 18.28 can be written for a gas reservoir as

$$G_p B_g + W_p B_w = G_{Fi}\left(B_g - B_{gi}\right) + G_{Fi} B_{gi} \Delta p\left(c_p + c_w S_w\right)/\left(1-S_w\right) + W_e \text{ . } \quad (18.52)$$

Define

$$F_{pR} = G_p B_g + W_p B_w, \quad (18.53)$$

and

$$E_g = B_g - B_{gi} \text{ . } \quad (18.54)$$

Substituting Eqs. 18.53 and 18.54 into Eq. 18.52 leads to

$$\begin{aligned} F_{pR} &= G_{Fi} E_g + G_{Fi} B_{gi} \Delta p\left(c_p + c_w S_w\right)/\left(1-S_w\right) + W_e \\ &= G_{Fi}\left[E_g + B_{gi} \Delta p\left(c_p + c_w S_w\right)/\left(1-S_w\right)\right] + W_e \text{ . } \end{aligned} \quad (18.55)$$

Divide by the gas-expansion and rock/fluid-compression term in brackets:

$$\begin{aligned} &\frac{F_{pR}}{E_g + B_{gi} \Delta p\left(c_p + c_w S_w\right)/\left(1-S_w\right)} \\ &= G_{Fi} + C \frac{W_e}{E_g + B_{gi} \Delta p\left(c_p + c_w S_w\right)/\left(1-S_w\right)} \text{ . } \end{aligned} \quad (18.56)$$

If the water-influx term and the rock/fluid expansion/compression terms are estimated correctly, a plot of the left-side term vs. the fractional part of the second right-side term of Eq. 18.56 will be a straight line. The y intercept should be equal to G_{Fi}. The slope of the line should equal C, the water-influx constant. Note that c_p is a function of pressure and is c_p (p) integrated over a change in net overburden pressure that corresponds to the value $p_i - p$.

The water-influx term probably will be the most difficult term to evaluate because water influx in a given reservoir could be attributable to expansion from a contiguous aquifer and/or to dewatering of interbedded protoshale. Favorable conditions for protoshale water influx include considerable interbedding of protoshale with the gas-bearing sand, a small contiguous aquifer, and a high initial fluid-pressure gradient. Opposite conditions would favor aquifer influx. Depending on the size and shape of the contiguous aquifer, W_e might be calculable using a limited linear aquifer model or a limited cylindrical aquifer model. If protoshale dewatering is suspected, a limited linear aquifer model might be more appropriate.

Geopressured gas reservoirs might exhibit retrograde behavior, a phenomenon discussed in the Condensate section of this chapter. Oil reservoirs are encountered less frequently than gas reservoirs in the geopressured section and rarely are discussed in the literature. Comments similar to those for geopressured gas reservoirs are appropriate regarding drive mechanism in geopressured oil reservoirs. Depending on circumstances, an approach analogous to that presented in Eqs. 18.52 through 18.56 might be appropriate for geopressured oil reservoirs.

PV Compressibility. On the basis of numerous studies of the influence of reservoir pressure on PV compressibility,[135–144] it seems apparent that PV compressibility of porous rocks depends on the stress conditions in the reservoir, decreases as stress increases, decreases as rocks become more consolidated, and might increase as temperature increases.

There appears to be no correlation between compressibility and rock properties that is generally valid across a broad spectrum of lithologies and pressures. Hall's[145] correlation between compressibility and porosity—still widely cited—covers only a narrow range of stress conditions and apparently reflects only data from well-consolidated rocks.

Reportedly, some geopressured sands have compressibilities approaching those usually associated with consolidated rock[146]; however, these data apparently were measured on rock samples taken from geopressured aquifers, rather than from hydrocarbon reservoirs. In the high temperatures usually associated with geopressured environments, sandstones undergo rapid diagenesis that can cause a geologically young rock to become tightly cemented. This is more likely to occur in aquifers (where the interstitial water is mobile) than in hydrocarbon reservoirs (where the interstitial water is immobile). Expect these tightly cemented sandstones to be less compressible than relatively uncemented sands; accordingly, measure compressibility on samples taken from the hydrocarbon-bearing zone, not from the aquifer. Take great care when using compressibility data from rocks that appear similar to the zone of interest or that have comparable porosity and permeability.

In the absence of laboratory data, the following correlation can be used to estimate PV compressibility[147]:

$$c_p \approx A\left[K_1 p_{ob} - K_2 p_i + K_3 (p_i - p) - B\right]^C + D, \quad (18.57)$$

where A, B, C, D, K_1, K_2, and K_3 depend on rock properties, as shown in **Table 18.13**.

During pressure reduction of reservoir fluids, the resultant stresses on reservoir rocks differ from those on core samples during hydrostatic testing in the laboratory. In the subsurface, when production reduces reservoir fluid pressure, the weight of the overburden compacts the reservoir rock, which uniaxially reduces the bulk volume of the rock and, consequently, reduces PV. This process can be replicated in the laboratory, but such tests require special

TABLE 18.13—CONSTANTS TO ESTIMATE PV COMPRESSIBILITY USING EQ. 18.57

Constant	Sandstones			Carbonates
	Unconsolidated	Friable	Consolidated	
A	–2.805 E–5	1.054 E–5	2.399 E–5	Not reported
B	300	500	300	Not reported
C	0.1395	–2.2250	0.0623	Not reported
D	1.183 E–4	–1.103 E–5	4.308 E–5	Not reported
K_1	0.95	0.90	0.85	0.85
K_2	0.95	0.90	0.80	0.85
K_3	0.75	0.60	0.45	0.55

equipment that is not used by most commercial laboratories. Most laboratory compressibility data are measured using hydrostatic stress, which can be related to reservoir stress by

$$p_n = K_1 p_{ob} - K_2 p_i + K_3 (p_i - p) . \qquad (18.58)$$

18.6 Probabilistic Procedures

18.6.1 Background. Experienced reservoir engineers know that uncertainty exists in geologic and engineering data and, consequently, in the results of calculations made with these data; however, the degree of uncertainty in most reservoir engineering calculations usually is not quantified.

Reserves estimates historically have been deterministic ("single-valued"), with the degree of uncertainty indicated by qualitative terms such as proved, probable, and possible. Additional information about the degree of uncertainty has been conveyed by describing producing category and development status (e.g., producing, behind-pipe, or not developed), as discussed in the Reserves Definitions section of this chapter.

Even now, deterministic estimates of reserves usually are considered appropriate in geologic settings and operating areas where there is substantial experience, and in fully developed, mature fields—situations having relatively little uncertainty; however, for new geologic settings (then, coalbed methane) and in new operating areas (then, the North Sea), the industry developed probabilistic procedures to estimate and classify reserves.[148,149] These procedures have been used to quantify potential range of reserves attributed to risky ventures and their degree of uncertainty.

Probabilistic procedures have long been used to assess exploratory ventures, but have not been widely used to assess production ventures. There are, however, several considerations and potential application areas for probabilistic procedures discussed below.

18.6.2 Probabilistic Classification of Reserves. Probabilistic classifications of reserves for a specific entity usually are based on the cumulative density function of the quantities calculated for the entity. (Cronquist[3] discusses basic principles of probabilistic methods.) As previously discussed, probabilistic classifications typically specify for proved reserves that

• There should be at least 90% probability that the quantities actually recovered will equal or exceed the estimate.

• There should be at least 50% probability that the quantities actually recovered will equal or exceed the sum of estimated proved plus probable reserves.

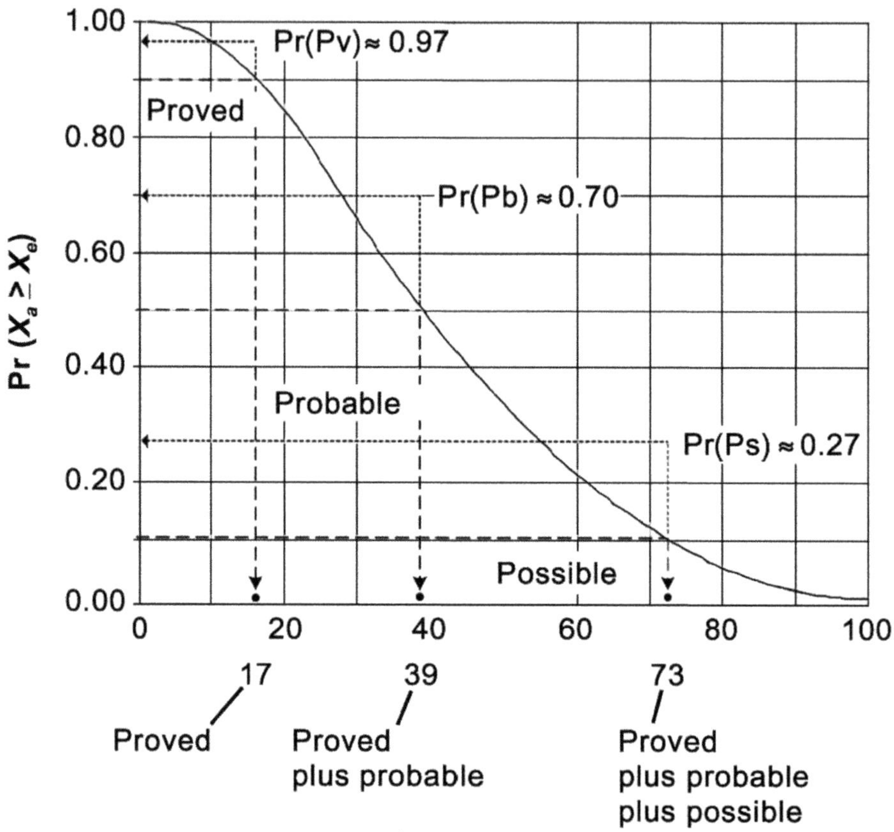

Fig. 18.31—Typical expectation curve for proved, proved plus probable, and proved plus probable plus possible reserves. X_a = the actual quantity of reserves and X_e = the estimated quantity of reserves.

• There should be at least 10% probability that the quantities actually recovered will equal or exceed the sum of estimated proved plus probable plus possible reserves.

Caution! Probabilistic notation and concepts discussed here should not be confused with those used by many exploration geologists [e.g., Rose,[150] in which probabilities refer to, among other considerations, *prospect reserves* distributions or *field size* distributions. In that context $P_{90\%}$ (Rose's notation) refers to *ultimate* reserves in a *yet to be discovered* accumulation. *After discovery*, reserves in such an accumulation might be estimated and classified using the procedures discussed here.]

For the expectation curve (EC) in **Fig. 18.31**, proved, proved plus probable, and proved plus probable plus possible quantities would be 17, 39, and 73, respectively.

The foregoing (probabilistic) definitions reflect a cumulative interpretation of the EC in Fig 18.31. An incremental interpretation is shown by the short dotted lines, which approximate the area under the EC having three rectangles with increments on the *x*-axis that are defined by the foregoing classifications; thus, for the EC in Fig. 18.31, it might be said that the probability of recovering proved reserves [Pr(Pv)] ≈ 0.97, the probability of recovering *incremental* probable reserves [Pr(Pb)] ≈ 0.70, and the probability of recovering *incremental* possible reserves [Pr (Ps)] ≈ 0.27.

Despite the 1997 SPE/WPC definitions' emphasis on a cumulative approach to probabilistic estimation and classification, an incremental approach might be more appropriate in situations that involve undrilled and/or unproved reserves where incremental expenditures and/or regulatory approval are required to bring such reserves on production. Specific guidelines for these situations are enumerated in separate paragraphs in the 1997 SPE/WPC reserves definitions for probable and possible reserves, respectively, and are preceded by the phrases "In general, probable reserves may include..." and "In general, possible reserves may include...."

18.6.3 Applications. Despite the (still) widespread use of deterministic methods for reserves estimation, there are several scenarios in which a probabilistic approach might be more appropriate.

Analogy/Statistical Methods. Historically, analogy/statistical methods have involved arithmetic averages of pertinent parameters, which have been considered "best estimates," a procedure that assumes that such parameters are (approximately) normally distributed; however, there are scenarios in which arithmetic averages might lead to significant bias in reserves estimates, especially those intended to conform to a P90 standard. The statistical frequency distribution of ultimate recovery—initial reserves—for wells in a common reservoir or geologic trend typically can be *approximated* with a log-normal (frequency) distribution. Also, the frequency distribution of geologic and/or engineering parameters (e.g., net pay) typically can be *approximated* with log-normal distributions. Log-normal distributions generally exhibit positive (right) skew. Depending on the degree of skew, the arithmetic average of such a distribution might be significantly greater than the intended "best estimate," which usually is considered the median of the distribution; thus, estimates based on arithmetic averages of pertinent parameters may (inadvertently) be biased on the high side. The more skewed the distribution, the greater the bias. With skewed distributions of pertinent parameters, consider using the median value for each such distribution, rather than the arithmetic average.[3]

These observations can be used to estimate reserves for new wells by analogy with data from older wells in the same trend. Before using this technique, however, determine whether per-well reserves are truly random, are not influenced by prior drainage, and are unrelated to a geologic or operating parameter. As drilling and production continue, update the analysis when reserves for existing and new wells are revised.

Random distributions of per-well reserves have been observed in many geologic/reservoir settings in the U.S.A. and elsewhere. In these areas, *at the time of the analysis,* there was no statistically valid correlation between ultimate production for individual wells and date of initial production, net pay, "frac" treatment, initial potential, or any other parameter for these same wells. In some of these areas, however, as infill drilling continued, correlations were observed between date of initial production and ultimate recovery; thus, in any mature area, one should be alert to similar possibilities.

Volumetric Methods. Depending on the geologic complexity, the stage of development, and the quality/quantity of subsurface and geophysical data, volumetric mapping might involve considerable uncertainty, including subsea depth of fluid/fluid contacts, location of bounding faults, position of sand pinchouts, and time/depth conversion(s) for seismic data, among other factors. Circumstances might make it appropriate to quantify one or more of these uncertainties using probabilistic procedures. There may be numerous perturbations and/or combinations, a subject that merits an entire chapter, but is only covered briefly here. (See Cronquist[3] for a more in-depth discussion of this topic.)

As discussed in the *Reservoir Limits* section, the subsea depth(s) of fluid/fluid contacts in a specific reservoir—GOC, GWC, and/or OWC—might not have been penetrated by wells, and there might be considerable disparity between analyses based on pressure, seismic, and/or capillary pressure data. Given such a scenario, it might be appropriate to define the range of uncertainty in subsea depth using a triangular probability distribution, with the mode of the distribution determined by the analysis procedure considered the most robust.

Performance Methods. The analysis of performance data might involve, for example, regression analysis (a least squares fit) of a production trend vs. time or vs. cumulative production. Historically, production projections based on such analyses have been classified as proved reserves. Regression analysis, however, yields a "fit" through the "means" of the analyzed trend. Probabilistically, such a fit is a P50, and reserves so calculated might, accordingly, be classified as proved plus probable. In a scenario where there is a long history of settled production and, consequently, a regression coefficient approaching unity, such a distinction is insignificant; however, where there is a relatively short, erratic production history and, consequently, a regression coefficient significantly less than unity, the distinction might be significant. Reserves engineers might consider a probabilistic analysis of the data.[3,151,152]

18.7 Pitfalls*[153]

A *pitfall* is a hidden or not easily recognized danger or difficulty that catches one unawares. In engineering, pitfalls are fallen into through mistakes from carelessness or ignorance. The potential petroleum engineering pitfalls below are described from the perspectives of an engineer preparing a reserves estimate and an engineer reviewing the work product of another.

Bias is not usually considered a pitfall, but rather is a conscious effort to "shape" a result for a specific purpose. An engineer must guard against all forms of external or internal bias and be careful to avoid the pitfalls that can affect the reliability of any reserves estimate. Some of the material presented here has been covered, in part, earlier in this chapter but is expanded here in light of the seriousness of these potential pitfalls.

18.7.1 Analogy/Statistical Methods. Assuming that reservoirs on trend and in the same geologic formation are analogous is a common pitfall. Depending on the geologic setting, depositional environment and, consequently, reservoir quality, might vary significantly over relatively short distances. For example, in a clastic deltaic environment both channel and fringe sands occur. Reservoirs in channel sands exhibit better quality than those in fringe sands and the two cannot be considered analogous; for the same drive mechanism, recovery factors in the former typically are significantly larger than those in the latter.

Statistical correlations—for recovery factor, for example—usually reflect the geologic, engineering, operational, and economic settings in which the correlated data were observed. Application of such correlations between disparate settings is a common pitfall and might lead to significant errors in reserves estimates.

Statistical correlations for PVT properties reflect the geologic setting from which the reservoir fluids originated.[3] Failure to recognize this is a common pitfall. For example, such correlations developed for fluids from California reservoirs might be inappropriate for estimating PVT properties of fluids from North-Sea reservoirs. Use of inappropriate correlations might cause significant errors in estimates of reserves of both gas and oil. Cronquist has provided additional guidelines regarding application of PVT correlations.[3]

18.7.2 Volumetric Methods. Pitfalls in volumetric methods include failing to integrate subsurface (well) data with seismic data, incorrectly constructing net-pay isopachs, and applying global correlations without adjustments for local conditions. Some of these pitfalls are discussed in the following paragraphs.

Structural Mapping. **Fig. 18.32** is a log section marked to indicate a structural top at 6,973 ft ss, top of first porosity at 7,018 ft ss, and an OWC at 7,133 ft ss. The three porosity intervals (at 7,018, 7,064, and 7,098 ft ss) might be part of a single pressure-connected reservoir, or they might be three separate reservoirs. Well log data alone usually are insufficient to resolve such uncertainties. If the single-reservoir scenario can be confirmed, a reservoir structure map incorporating the top of the first porosity (7,018 ft ss) and the OWC at 7,133 ft ss might be

* This section was inspired by a presentation by Harry Gaston Jr. at the 1992 Annual Convention of the SPEE on 13–16 June 1992 at Jackson Hole, Wyoming.

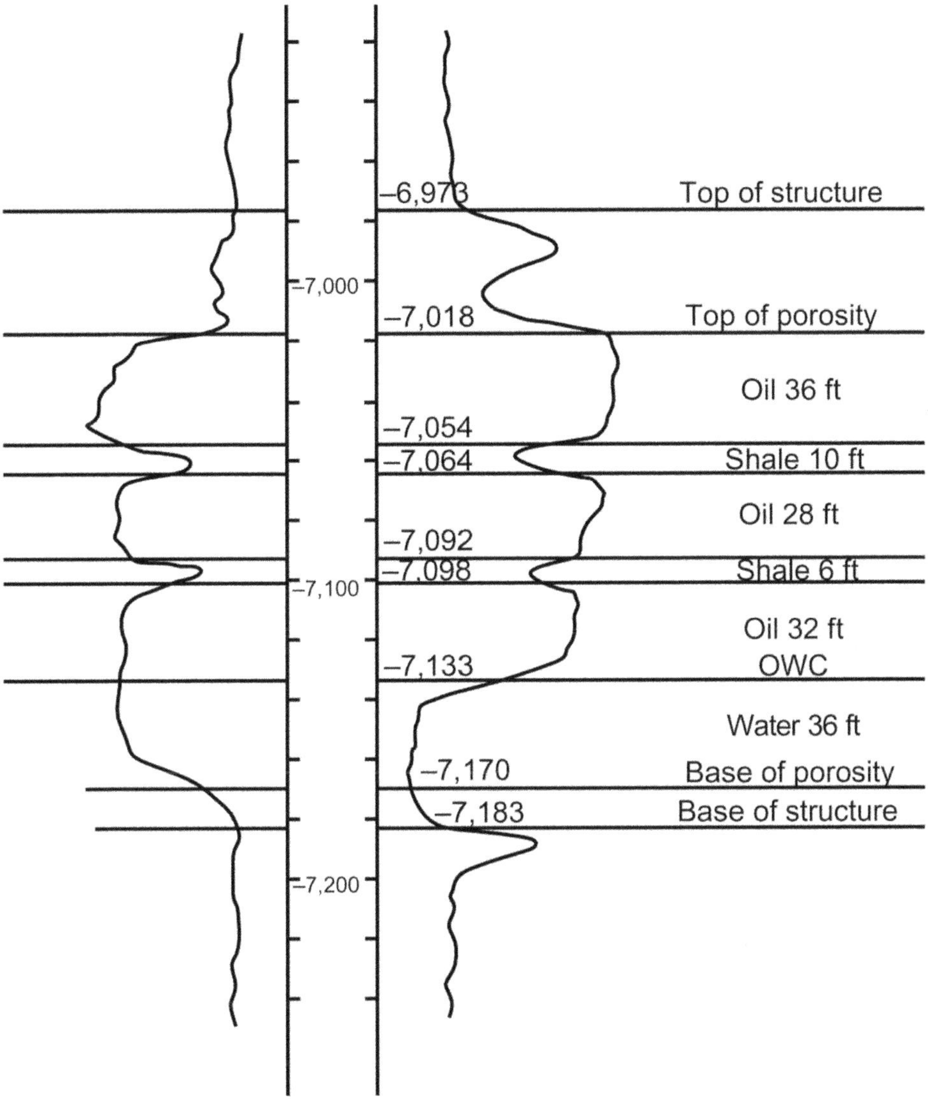

Fig. 18.32—Well log section (after Mistrot[154]).

appropriate. Net-pay isopach maps of each pay interval could be constructed using a common OWC.

An alternate geologic interpretation might indicate the likelihood of the continuation of the two 10-ft shale breaks over the area of the accumulation. Should this situation exist, it would be appropriate to recognize three separate reservoirs. The upper two would have P1 reserves limited to the LKO depths of 7,054 and 7,092 ft ss, respectively.

Additional wells might be required to confirm or deny the effectiveness of the intervening shale members. Assuming a single reservoir in similar circumstances might cause P1 reserves to be overstated significantly, should subsequent information support the presence of more than one reservoir. A structure map using a structural top of 6,973 ft ss, without adjusting for the distance between this top and the top of porosity, would lead to overstatement of the productive area and reserves through exaggeration of the reservoir area. **Fig. 18.33** illustrates the potential overstatement of reservoir extent, using lowest known hydrocarbons (LKH) limits for

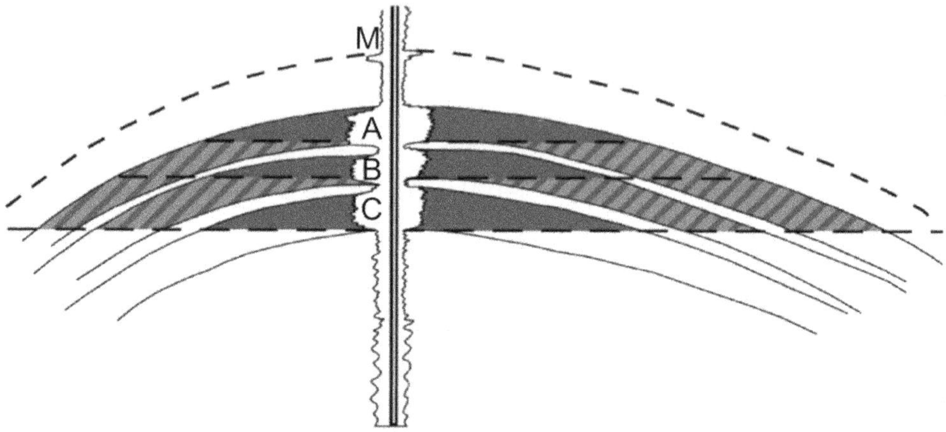

Fig. 18.33—Single-well cross section illustrating the possibility of three separate reservoirs.

three sand intervals, A, B, and C, that might be three individual reservoirs and the marker, M. Shaded areas above the individual LKH levels may be considered proved reserves. The crosshatched areas may be considered P2 or P3 reserves. Continued monitoring of performance history and/or development drilling would confirm the presence of a single accumulation or of two or three separate reservoirs.

Isopach Mapping. Isopach maps often are prepared improperly, which almost always leads to overstatement of O/GIP. **Fig. 18.34** illustrates a common pitfall of assuming a net-pay buildup updip from well control. Given that the maximum net sand observed in the immediate area is only 45 ft, there is no justification to assume such a buildup. If the dry hole to the north had a total net sand thickness greater than 45 ft, then a buildup in net pay to the observed maximum net sand might be justified. Also, there is uncertainty regarding the position of the fault bounding the northwest side of the reservoir, and the geologist's interpretation might have been optimistic.

Fig. 18.35 is similar to Fig. 18.34, but the area updip from the highest productive wells is interpreted to be no more than 45 ft thick. Also, the fault has been shifted to the southeast as a result of an alternate fault interpretation, which results in significantly less reservoir volume than the original interpretation. Which is correct? Is the difference from a pitfall or bias? One should be vigilant in discerning the differences in reserves estimates resulting from either bias or a pitfall, and should make appropriate revisions as warranted.

18.7.3 Performance Methods. Many pitfalls await the unwary who use performance methods to estimate reserves. Several of the more common pitfalls are discussed here: analysis of production decline trends to estimate reserves for wells still in the transient stage; extrapolation of performance indicators to unrealistic economic limit conditions; misuse of composite decline curves; physical life limitations; failure to recognize the effect of interference between wells; failure to reconcile results of volumetric and performance analyses; and miscalculation/misuse of p/z vs. G_p plots.

Analysis of Production Decline Trends During Transient Flow. An expanding drainage radius and (initially) steep, hyperbolic declines in production rate characterize transient flow from wells. Reserves estimates during this stage of production usually will be too low if they are based solely on decline-trend analysis, without appropriate adjustments. The problem is especially severe in shallow, low-permeability gas reservoirs, where wells may exhibit transient flow over a substantial part of their productive life.

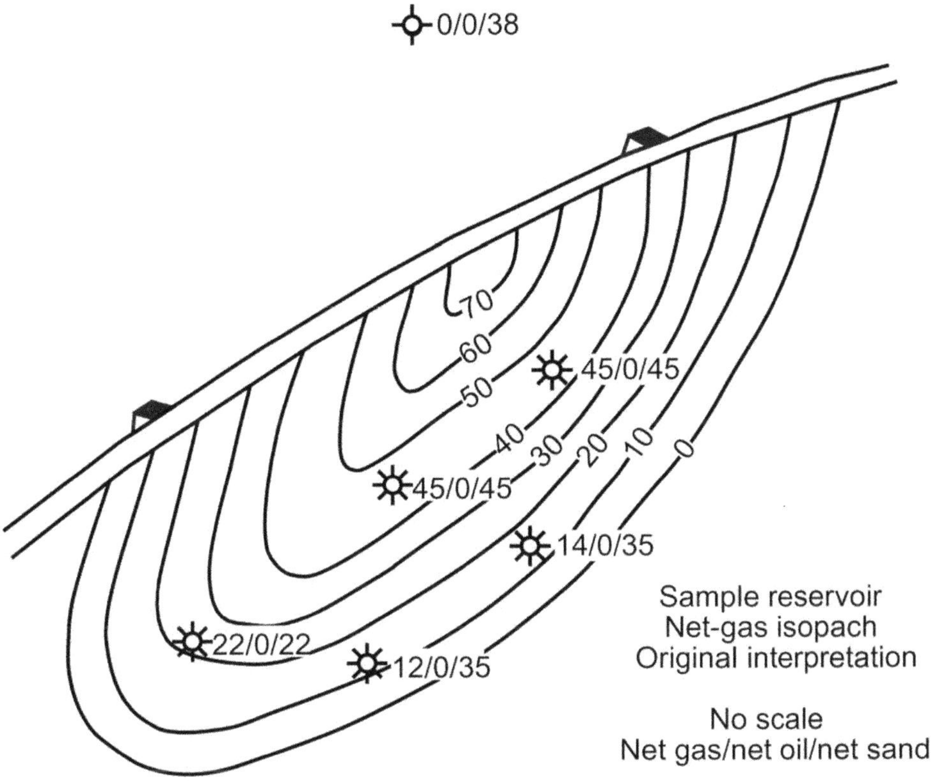

Fig. 18.34—Net gas isopach (after Gaston[153]).

Unrealistic Economic-Limit Conditions. Semilog plots of f_o vs. N_p commonly are used to estimate reserves for oil wells in waterdrive reservoirs. Such plots frequently are used before there is a clearly defined decline in oil production rate, which can be a pitfall. Plots of f_o vs. N_p can be extrapolated to a local average "f_o cutoff," but this cutoff might be too low for wells with low productivity. For example, in many areas, the average f_o cutoff is 0.01; however, if a well is capable of producing only 250 B/D (oil plus water) and the economic limit is 5 BOPD, then the f_o cutoff for this well should be 0.02, not 0.01.

Misuse of Composite Decline Curves. Projecting decline trends of composited production from multiple wells to estimate proved reserves often leads to an overstatement of reserves. Failure to thoroughly analyze the effects of operator efforts to sustain production rates, as well as of marketing limitation or transportation restriction influences, is a common pitfall. Production declines of individual wells and the composite production stream might have been mitigated by workovers, recompletions, new wells, stimulation, and/or compression equipment upgrade or other equipment and facility upgrades, the benefits of which might not be available during the forecast period to sustain the demonstrated production trend. The engineer should always try to evaluate the performance of individual wells or completions, even if the production history has been computed using an allocation of the composite production quantities.

Physical Life Limitations. The extrapolation of monthly production rates of oil and/or gas from a well can extend for 40, 60, or maybe 100 years, depending on the economic assumptions that are used. When estimating such reserves, consider the cost of drilling a replacement well and the replacement of infrastructure facilities. Some engineers arbitrarily limit future projection lives to 30 to 50 years, depending on the specific circumstances of each production area.

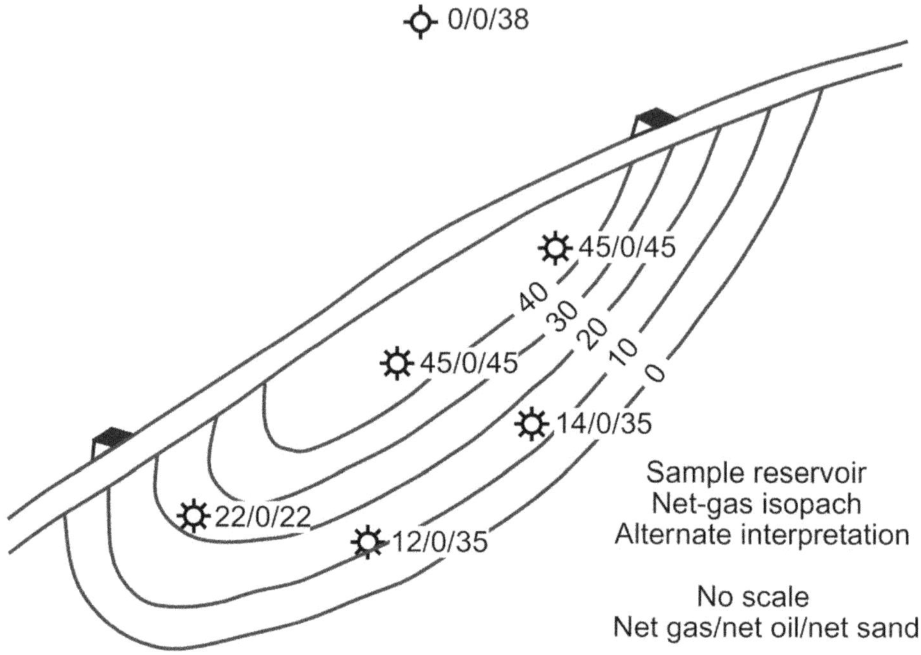

Fig. 18.35—Net gas isopach (after Gaston[153]).

Failure To Recognize the Effect of Interference Between Wells. Analyzing each well in a common reservoir often is appropriate for estimating reserves, particularly for a fully developed reservoir being produced under the control of a single operator; however, when individual wells are not analyzed, be aware of the potential for interwell interference to affect individual well-drainage patterns. Examples are the drilling of infill wells, which might increase reserves or serve only to accelerate rate, and increased production, from competitor-operated wells.

Failure To Reconcile Results of Volumetric and Performance Analyses. During the early life of a reservoir, performance-data analysis might suggest more or less reserves than are indicated by volumetric methods. Depending on the degree of difference and the economic consequences thereof, it might be appropriate to review the performance and volumetric data, as well as the analysis methods, to resolve these discrepancies. Such a review might reveal one or more causes (e.g., inappropriate net-pay cutoffs, invalid drainage-volume estimates, poor assumptions regarding drive mechanism, and/or unrealistic decline-trend projections).

Misuse and/or Improper Calculation of p/z vs. G_p Plots. Plots of p/z vs. G_p are powerful tools for analyzing the performance of gas reservoirs; however, there are several reservoir scenarios in which these plots are misused or misinterpreted, including geopressured reservoirs, partial waterdrive, conjectural economics of multistage compression, and plot miscalculation.

Geopressured Reservoirs. In geopressured reservoirs, plots of p/z vs. G_p frequently exhibit an anomalously low initial slope, followed by a steeper slope. The initial low slope might be caused by water influx and/or PV compressibility. The steeper slope, which might not appear until later in reservoir life, might be because of depletion of a limited aquifer. As discussed in the Special Problems section of this chapter, depending on the stage of production maturity, it might not be possible to determine the dominant drive mechanism in such reservoirs. In such case, either adjust plots of p/z vs. G_p or use an alternate material-balance method.

Partial Waterdrive. In many cases, a plot of p/z vs. G_p is insufficient to determine whether there is water influx into a gas reservoir under study.[3,139] Failure to recognize such influx may lead to substantial overestimates of gas reserves. The uncertainty in such cases might be resolved by using a procedure, discussed in the Geopressured Oil/Gas section of this chapter, that is based on the Havlena and Odeh method.[60] Ignoring the rock-compressibility term and rearranging Eq. 18.52 yields:

$$G_p B_g / \left(B_g - B_{gi}\right) = G_{Fi} + \left(W_e - W_p B_w\right) / \left(B_g - B_{gi}\right) . \quad (18.59)$$

If there is no water influx, the second right-side term equals zero and the left-side term vs. G_p should plot as a horizontal line, with the y-axis intercept equal to G_{Fi}. If there is water influx, however, the plot will be curved, with the degree and type of curvature dependent on the degree of water influx.

Conjectural Economics of Multistage Compression. In some operating and economic scenarios, RE of gas from volumetric reservoirs may exceed 95% GIP; however, such REs typically require multistage compression, the economics of which might be conjectural in the early stages of reservoir history. Be wary of assuming that installing such compression facilities can be justified economically, unless there is considerable experience in an analogous reservoir in a comparable operating and economic setting. On offshore platforms, for example, there may not be enough space to install such facilities, and major expansion may be infeasible, either structurally or economically.

Improper Calculations of p/z Values. Some of the more commonly seen problems associated with the determination of static reservoir pressures and related z-factor are:

- Nonrepresentative pressures obtained when some wells completed in the reservoir remain on production.
- Relying on shut-in wellhead pressures to calculate bottomhole pressures in the presence of water and hydrocarbon liquids.
- Shut-in time that is inadequate to achieve static pressure.
- Using inappropriate gas composition to compute z-factors.
- Not adjusting bottomhole pressures to a common subsea depth.

Each of these problems is discussed briefly in the following paragraphs.

Nonrepresentative pressures may occur when operators do not shut in certain wells because of high liquid ratios or other concerns. They allow these wells to continue to produce while other wells in the same reservoir are shut in for a pressure survey. Other operators might obtain static reservoir pressures in new wells being completed in a common reservoir at the same time when other reservoir wells continue to produce. Be cautious in relying on such pressure information in any circumstance wherein the reservoir is not allowed to approach an equilibrium pressure.

Shut-in wellhead pressures may be used to reliably calculate reservoir pressures only if the composition and relative volumes of the wellbore fluids are known. The range of error in the calculated values increases if there is a static liquid level (either condensate or water) in the wellbore at an unknown depth.

Inadequate shut-in time to achieve static reservoir pressure occurs commonly. Knowledge of actual shut-in time is important for assessing data reliability. Third-party engineers might not have full access to the operator's data. In many cases, such data may have been lost or misplaced because of transfers of interest. For example, a reserves engineer might have only publicly reported pressure data for use in reservoir analysis, and might not be aware of the length of the shut-in period before the pressure measurement or of the consequences of this lack of knowledge.

Using an inappropriate gas composition to calculate z-factors usually is the result of assuming a constant gas composition throughout the life of a gas reservoir. While this might be appropriate for a dry-gas reservoir, it could cause serious errors for a retrograde gas reservoir if static reservoir pressure is reduced to less than the dewpoint pressure during production. Ignoring the resultant change in composition of the reservoir gas in retrograde accumulations can cause errors in calculation of the z-factor, in-place hydrocarbons, and reserves.

Not adjusting bottomhole pressures to a common depth in a given reservoir often is attributed to time constraints or carelessness. Although this might cause minor errors in low-pressure, low-relief, dry-gas reservoirs, it might be critical for high-relief oil reservoirs.

Nomenclature

a = terminal effective decline rate
a_1 = decline rate, initial, 1/t
A = area of reservoir or accumulation, acre
A = constant (in Eq. 18.57 and Table 18.13 only)
A_g = area of gas cap or gas reservoir, acre or hectare
A_o = area of the oil zone, acre or hectare
b = hyperbolic decline exponent (same as n, used by earlier authors)
B = constant
B_g = formation volume factor, gas, Rcf/scf
B_{gi} = initial formation volume factor, gas, Rcf/scf or RB/scf
B_o = formation volume factor, oil, RB/STB
B_{ob} = formation volume factor at the bubblepoint, oil, RB/STB
B_{oi} = initial formation volume factor, oil, RB/STB
B_t = formation volume factor, total, RB/STB
B_{ti} = initial total formation volume factor, RB/STB
B_w = formation volume factor, water, RB/STB
c_p = compressibility, pore volume, vol/vol/psi
c_w = compressibility, water, vol/vol/psi
C = water-influx constant
C = constant (in Eq. 18.57 and Table 18.13 only)
C_D = direct operating cost, U.S. dollar/well
C_{gi} = initial sorbed gas concentration, scf/ton, dry, ash-free coal or shale
C_i = condensate (distillate) initially in place, STB
C_{pu} = ultimate condensate production, STB
C_{Ri} = initial condensate reserves, STB
C_{tp} = transportation costs, U.S. dollar/bbl
D = curve-fit coefficient
D = curve-fit coefficient (in Eq. 18.57 and Table 18.13 only)
D_i = initial decline rate
E_c = "expansion" of initial reservoir pore volume caused by compression effects on pore volume and interstitial water, vol/vol/psi
E_D = microscopic displacement efficiency, fraction
E_g = expansion of the initial gas cap, if one is present, RB/scf
E_o = expansion of a unit volume of oil and dissolved (solution) gas initially in place, RB/STB
E_{Rc} = recovery efficiency of condensate, fraction
E_{Rg} = recovery efficiency of gas, general, fraction
$(E_{Rg})_{pd}$ = recovery efficiency of gas attributable to pressure depletion
$(E_{Rg})_{pwd}$ = recovery efficiency of (free) gas attributable to partial waterdrive, fraction

$\left(E_{Rg}\right)_{swd}$ = recovery efficiency of (free) gas attributable to strong water drive, fraction
E_{Ro} = recovery efficiency of oil, general, fraction
$\left(E_{Ro}\right)_{sg}$ = recovery efficiency of oil attributable to solution gas drive, fraction
$\left(E_{Ro}\right)_{wd}$ = recovery efficiency of oil attributable to waterdrive, fraction
E_V = fraction of the initially gas bearing volume swept by the aquifer (volumetric sweep efficiency), fraction
f_a = average weight fraction of ash, fraction
f_m = average weight fraction of moisture, fraction
f_o = fractional flow of oil
F_g = Thomeer[43] parameter for capillary pressure curves
F_{pR} = volume of cumulative oil, gas, and water production, RB
F_{rg} = royalty on gas, fraction
F_{ro} = royalty on oil, fraction
F_{RA} = recovery factor, analogous reservoir
F_{RS} = recovery factor, subject reservoir
G_{Fi} = free gas initially in place, scf or m^3
G_i = gas-in-place at initial reservoir conditions, scf
G_p = cumulative gas production, scf
G_{pu} = ultimate cumulative gas production, scf
G_R = (remaining) gas reserves, scf
G_{RFi} = initial free gas reserves, scf
G_{Ri} = initial gas reserves, scf
G_{RSi} = initial solution gas reserves, scf
G_{Si} = solution gas initially in place, scf
h = thickness, ft
h_n = net pay thickness
h_{n1} = net thickness of upper (oil) zone, ft
h_{n2} = net thickness of the lower (oil) zone, ft
h_{ng} = average net thickness of gas cap or gas reservoir, ft or m
h_{no} = average net oil pay, ft or m
h_s = shale thickness, ft
h_t = gross oil column thickness
k = permeability, md
k_a = air permeability, md
k_e = effective permeability
k_g = effective gas permeability, md
k_H = horizontal permeability
k_o = effective oil permeability, md
k_{ro} = relative oil permeability, dimensionless
k_{rw} = relative oil permeability, dimensionless
k_V = vertical permeability, md
k_w = effective water permeability, md
K_A = coefficient
K_B = coefficient

K_1 = constant
K_2 = constant
K_3 = constant
m = ratio of initial gas cap volume to initial oil column volume, dimensionless
m_i = volume of gas injected, multiple of initial gas cap volume, dimensionless
n = variable
N_G = gravity number, dimensionless
N_i = oil initially in place, STB or m^3
N_p = cumulative oil production, STB
N_{pa} = cumulative oil production at abandonment, STB
N_{paw} = cumulative oil production, well, at abandonment, STB
N_{pc} = cumulative oil production at point "c," STB
N_{pu} = ultimate oil production, STB
N_{Ri} = initial oil reserves, STB
p = pressure, static reservoir, general, psia
p_a = abandonment pressure, psia
p_b = bubblepoint pressure, psia
p_d = dewpoint pressure, psia
p_g = wellhead price of gas, U.S. dollar/STB
p_i = initial reservoir pressure, psia
p_n = laboratory net (hydrostatic) pressure (confining pressure minus pore pressure), psia
p_o = wellhead price of oil, U.S. dollar/STB
p_{ob} = overburden pressure, psia
P_{cL} = capillary pressure at laboratory conditions, psia
P_{cm} = mercury (air) capillary pressure, psia
P_{cR} = capillary pressure at reservoir conditions, psia
P_{dm} = mercury (air) displacement pressure, psia
q = production rate, general, STB/month or scf/month
q_1 = production rate at the beginning of a period
q_2 = production rate at the end of a period
q_{el} = production rate at the economic limit, general, scf/D or STB/D
q_g = gas production rate, scf/month
q_{oel} = oil production rate at economic limit, BOPD
Q = cumulative production
R = gas/oil ratio, general, scf/STB
R_c = condensate/gas ratio, STB/MMscf
R_{cd} = condensate/gas ratio at dewpoint pressure, STB/MMscf
R_{ci} = initial condensate/gas ratio, STB/MMscf
R_{cuD} = cumulative condensate/gas ratio at abandonment divided by initial condensate/gas ratio, dimensionless
R_{ng} = net-to-gross pay ratio, dimensionless
R_{ngo} = average net-to-gross pay ratio in initially oil-bearing zone
R_o = vitrinite reflectance, %
R_p = cumulative (producing) gas/oil ratio, scf/STB
R_{pz} = ratio of abandonment p/z to initial p/z, or $p_a z_i / p_i z_a$

R_s = gas/oil ratio, ft^3/STB
R_{sb} = gas/oil ratio at bubblepoint pressure, scf/STB
R_{si} = initial solution gas/oil ratio, scf/STB
S_{gr} = residual gas saturation, fraction
S_{hi} = initial hydrocarbon saturation, fraction
S_o = oil saturation, fraction
S_{or} = residual oil saturation, fraction
S_w = water saturation, fraction
S_{wfi} = interconnected fracture water saturation, fraction
S_{wg} = water saturation in the free-gas zone, fraction
S_{wi} = initial water saturation, fraction
S_{wo} = water saturation in the oil zone, fraction
t = time, months or years
t_{el} = (incremental) time for production to decline from current rate to economic limit rate, days, months, or years
t_1 = time 1
t_2 = time 2
T_A = ad valorem tax, U.S. dollar/STB
T_P = production tax, U.S. dollar/STB
V_I = molar volume of ideal gas
V_R = molar volume of real gas
V_{to} = gross volume of initially oil-bearing rock, acre-ft
W_e = cumulative water influx, RB
W_p = cumulative water production, STB
X_a = the actual quantity of reserves
X_e = the estimated quantity of reserves
z = gas compressibility factor, general, dimensionless
z_a = gas compressibility factor at the economic limit, dimensionless
Z_{fw} = vertical distance above free-water level, ft
z_i = gas compressibility factor at initial conditions, dimensionless
α = dip angle, degree
γ_g = specific gravity of solution gas (air = 1.0), dimensionless
γ_{os} = specific gravity of stock-tank oil, dimensionless
ΔG_p = incremental cumulative gas production, scf
ΔN_p = incremental cumulative oil production, STB
Δp = pressure, incremental, psi
ΔQ_p = incremental cumulative production, ΔG_p or ΔN_p, as the production rate declines from q_1 to q_2
Δt = incremental time between t_1 and t_2
θ_L = contact angle at laboratory conditions, degree
θ_R = contact angle at reservoir conditions, degree
μ = viscosity, general, cp
μ_o = oil viscosity, general, cp
μ_{ob} = oil viscosity at the bubblepoint, cp
μ_{oi} = initial oil viscosity, cp
μ_w = viscosity, water, general, cp

μ_{wi} = initial water viscosity, cp
ρ = density, general, g/cm^3
ρ_c = density, coal, g/cm^3
ρ_g = density, gas, g/cm^3
ρ_o = density, oil, psi/ft or g/cm^3
ρ_{or} = density of reservoir oil, lbm/bbl
ρ_w = density, water, psi/ft
σ_L = interfacial tension at laboratory conditions, dyne/cm
σ_R = interfacial tension at reservoir conditions, dyne/cm
ϕ = porosity, general, fraction
ϕ_f = interconnected fracture (effective) porosity, fraction
ϕ_g = porosity in gas zone, fraction
ϕ_o = porosity in oil zone, fraction

References

1. *Petroleum Production Handbook,* T.C. Frick (ed.), McGraw-Hill Book Co. Inc., New York City (1962).
2. *Petroleum Engineering Handbook,* H.B. Bradley (ed.), SPE, Richardson, Texas (1987).
3. Cronquist, C.: *Estimation and Classification of Reserves of Crude Oil, Natural Gas, and Condensate,* SPE, Richardson, Texas (2001).
4. Ross, J.G.: "Petroleum Resources Classification and Definitions," *Guidelines for the Evaluation of Petroleum Reserves and Resources*, SPE, Richardson, Texas (2001) 25.
5. McKelvey, V.E.: "Mineral Resource Estimates and Public Policy," *American Scientist* (January–February 1972) **60,** 32.
6. *SPE/WPC Petroleum Reserves Definitions,* SPE, Richardson, Texas (1997).
7. McMichael, C. and Spencer, A.: "Operational Issues," *Guidelines for the Evaluation of Petroleum Reserves and Resources*, SPE, Richardson, Texas (2001) 25.
8. "Society Adopts Proved Reserves Definitions," *JPT* (November 1981) 2213.
9. "Reserves Definitions Approved," *JPT* (May 1987) 576.
10. "Regulation S-X, Rule 4-10," *Financial Accounting and Reporting for Oil and Gas Producing Activities Pursuant to the Federal Securities Laws and the Energy Policy and Conservation Act of 1975,* U.S. SEC, Washington, DC (1975).
11. *SEC Staff Accounting Bulletin: Codification of Staff Accounting Bulletins,* Topic 12 Oil and Gas Producing Activities, A.1. Estimates of Quantities of Proved Reserves, Questions 1–3 U.S. SEC, Washington, DC (19 May 2003).
12. *SEC Staff Accounting Bulletin: Codification of Staff Accounting Bulletins,* Topic 12 Oil and Gas Producing Activities, G. Inclusion of Methane Gas in Proved Reserves, U.S. SEC, Washington, DC (19 May 2003).
13. *SEC Staff Accounting Bulletin: Codification of Staff Accounting Bulletins,* Topic 12 Oil and Gas Producing Activities, A.3.b. Unproved Properties, U.S. SEC, Washington, DC (19 May 2003).
14. Website Release, www.sec.gov/divisions/corpfin/acctdisc.htm, U.S. SEC (21 February 2001).
15. *Reserves of Crude Oil, Natural Gas Liquids and Natural Gas in the United States and Canada as of December 31, 1979,* API, Washington, DC (1980) 34.
16. *Guidelines for Application of Petroleum Reserve Definitions*, second edition, Monograph Series, Soc. of Petroleum Evaluation Engineers, Houston (1998) 1.
17. *A Statistical Study of Recovery Efficiency*, API Bull. D14, second edition, API, Dallas (1984).
18. Fast, C.R., Holman, G.B., and Covlin, R.J.: "The Application of Massive Hydraulic Fracturing to the Tight Muddy 'J' Formation, Wattenberg Field, Colorado," *JPT* (January 1977) 10.
19. Havlena, D.: "Interpretation, Averaging, and Use of the Basic Geological-Engineering Data," *J. Cdn. Pet. Tech.* (October–December 1966) 236.

20. George, C.J. and Stiles, L.H.: "Improved Techniques for Evaluating Carbonate Waterfloods in West Texas," *JPT* (November 1978) 1547.
21. Sneider, R.M. *et al.*: "Predicting Reservoir Rock Geometry and Continuity in Pennsylvanian Reservoirs, Elk City Field, Oklahoma," *JPT* (July 1977) 851.
22. Wilhite, G.P.: *Waterflooding*, Textbook Series, SPE, Richardson, Texas (1986) 3.
23. Missman, R.A. and Jameson, J.: "An Evolving Description of a Fractured Carbonate Reservoir: The Lisburne Field, Prudhoe Bay, Alaska," *Proc.*, 1990 Archie Conference, Houston, 22–25 October, 204.
24. Cobb, W.M. and Marek, F.J.: "Net Pay Determination for Primary and Waterflood Depletion Mechanisms," paper SPE 48952 presented at the 1998 SPE Annual Technical Conference and Exhibition, New Orleans, 27–30 September.
25. Snyder, R.H.: "A Review of the Concepts and Methodology of Determining 'Net Pay,'" paper SPE 3609 presented at the 1971 SPE Annual Meeting, New Orleans, 3–6 October.
26. Determination of Oil and Gas Reserves, Petroleum Soc.-CIM, Calgary (1994).
27. *Venezuela Well Evaluation Conference,* Schlumberger Sorenco, C.A., Caracas (1997).
28. *Productive Low Resistivity Well Logs of the Offshore Gulf of Mexico,* Houston Geological Soc., Houston (1993).
29. *Productive Low Resistivity Well Logs of the Offshore Gulf of Mexico,* D. Moore (ed.), New Orleans Geological Soc., New Orleans (1993).
30. *Hydrocarbon Production from Low Contrast, Low Resistivity Reservoirs, Rocky Mountain and Mid-Continent Regions, Log Examples of Subtle Plays,* E.D. Dolly and J.C. Mullarky (eds.), Rocky Mountain Assn. of Geologists, Denver (1996).
31. Wharton, J.B.: "Isopachous Maps of Sand Reservoirs," *AAPG Bull.* (July 1948) **32**, 7.
32. Dake, L.P.: *The Practice of Reservoir Engineering*, Elsevier Science Publisher, Amsterdam (1994) **59**, 472
33. Elshahawi, H., Fathy, K., and Hiekali, S.: "Capillary Pressure and Rock Wettability Effects on Wireline Formation Tester Measurements," paper SPE 56712 presented at the 1999 SPE Annual Technical Conference and Exhibition, Houston, 3–6 October.
34. Robertson, J.D.: "Seismic Applications," *Guidelines for the Evaluation of Petroleum Reserves and Resources*, SPE/WPC/AAPG, SPE, Richardson, Texas (2001) 93.
35. King, G.: "4-D Seismic Improves Reservoir Management Decisions," *World Oil* (March 1996) 79.
36. Kristiansen, P. and Christie, P.: "Monitoring Foinaven Reservoir: Advances in 4-D Seismic," *World Oil* (November 1999) 71.
37. Hoover, A.R., Burkhart, T., and Flemings, P.B.: "Reservoir and Production Analysis of the K40 Sand, South Timbalier 295, Offshore Louisiana, With Comparison to Time-Lapse (4-D) Seismic Results," *AAPG Bull.* (October 1999) 1624.
38. Ashbaugh, J.P. and Flemings, P.B.: "Dynamic Reservoir Characterization of the ST295 Field (Offshore Louisiana): Reservoir Simulation, Acoustic Modeling, and Time-Lapse Seismic Refines Geologic Model and Illuminates Dynamic Behavior," SEG Workshop, Houston (1999) 5 November.
39. McCain, W.D. Jr.: *The Properties of Petroleum Fluids,* PennWell Publishing Co., Tulsa (1990).
40. Alger, R.P., Luffel, D.L., and Truman, R.B.: "New Unified Method of Integrating Core Capillary —Pressure Data With Well Logs," *SPEFE* (June 1989) 145.
41. Smith, D.: "Predicting a Downdip Water Level Using Capillary Pressure Relations," paper presented at the 1990 SPWLA Annual Logging Symposium, Lafayette, Louisiana, 24–27 June.
42. Hawkins, J.M., Luffel, D.L., and Harris, T.G.: "Capillary Pressure Model Predicts Distance to Gas/Water, Oil/Water Contact," *Oil & Gas J.* (18 January 1993) 39.
43. Thomeer, J.H.M.: "Introduction of a Pore Geometrical Factor Defined by the Capillary Pressure Curve," *JPT* (March 1960) 354; *Trans.,* AIME, **219**.
44. Stiles, J.H. Jr. and Mc Kee, J.W.: "Cormorant: Development of a Complex Field," *SPEFE* (December 1991) 427; *Trans.,* AIME, **291**.
45. Hartman, J.A. and Paynter, D.D.: "Drainage Anomalies in Gulf Coast Tertiary Sandstones," *JPT* (October 1979) 1313.

46. Mearns, E. and McBride, J.J.: "Hydrocarbon Filling History and Reservoir Continuity of Oil Fields Evaluated Using $^{87}Sr/^{86}Sr$ Isotope Ratio Variations in Formation Water, With Examples From the North Sea," *Petroleum Geoscience* (February 1999) 17.
47. Standing, M.B.: "A Pressure-Volume-Temperature Correlation for Mixtures of California Oils and Gases, *Drill. & Prod. Prac.*, API, Washington, DC (1947) 275.
48. Arps, J.J. and Roberts, T.G.: "The Effect of the Relative Permeability Ratio, the Oil-Gravity and the Solution Gas-Oil Ratio on the Primary Recovery From a Depletion Type Reservoir," *Trans.*, AIME (1955) **204,** 120.
49. Arps, J.J. *et al.: Bull. D-14, A Statistical Study of Recovery Efficiency,* API, Washington, DC (1967).
50. Doscher, T.M.: "Statistical Analysis Shows Crude Oil Recovery," *Oil & Gas J.* (29 October 1984) 61.
51. Cronquist, C.: "Volatile Oils: Fluid Characteristics, Reservoir Behavior and Production Facilities," *World Oil* (April 1979) 159.
52. Agarwal, R.G., Al-Hussainy, R., and Ramey, H.J.: "The Importance of Water Influx in Gas Reservoirs," *JPT* (November 1965) 1336; *Trans.*, AIME, **234.**
53. Cronquist, C.: "Determining Recovery From Partial Water Drive Reservoirs," *World Oil* (September 1980) 107.
54. Cronquist, C.: "Turtle Bayou 1936–83: Case History of a Major Gas Field in South Louisiana," *JPT* (November 1984) 1941; *Trans.*, AIME, **277.**
55. Garb, F.A.: "Property Evaluation with Hand-Held Computers: Part 9—Liquid Yield for Depletion Drive Gas Reservoirs and Gas Recovery Factors for Depletion Directed Water Displacement Reservoirs," *Pet. Eng. Intl.* (October 1978) 72.
56. Dake, L.P.: *Fundamentals of Reservoir Engineering,* Elsevier Science Publisher, New York City (1978).
57. Pletcher, J.L.: "Improvements to Reservoir Material Balance Methods," *SPEREE* (February 2002) 49.
58. Walsh, M.P.: "A Generalized Approach to Reservoir Material Balance Equations," *J. Cdn. Pet. Tech.* (January 1995) 55.
59. Tehrani, D.H.: "An Analysis of a Volumetric Balance Equation for Calculation of Oil in Place and Water Influx," *JPT* (September 1985) 1664.
60. Havlena, D. and Odeh, A.S.: "The Material Balance as an Equation of a Straight Line," *JPT* (August 1963) 896; *Trans.*, AIME, 228.
61. Havlena, D. and Odeh, A.S.: "The Material Balance as an Equation of a Straight Line—Part II, Field Cases," *JPT* (July 1964) 815; *Trans.*, AIME, **231.**
62. Sills, S.R.: "Improved Material-Balance Regression Analysis for Waterdrive Oil and Gas Reservoirs," *SPERE* (May 1996) 127.
63. Brons, F.: "On the Use and Misuse of Production Decline Curves," *API 801-39E,* API, Dallas (1963).
64. Purvis, R.A.: "Analysis of Production-Performance Graphs," *J. Cdn. Pet. Tech.* (July–August 1985) 44.
65. Arps, J.J.: "Analysis of Decline Curves," *Trans.*, AIME (1945) **160,** 228.
66. Fetkovich, M.J.: "Decline Curve Analysis Using Type Curves," *JPT* (June 1980) 1065; *Trans.*, AIME, **269**.
67. Long, D.R. and Davis, M.J.: "A New Approach to the Hyperbolic Curve," *JPT* (July 1988) 909.
68. Fetkovich, M.J. *et al.*: "Decline-Curve Analysis Using Type Curves—Case Histories," *SPEFE* (December 1987) 637; *Trans.*, AIME, **283**.
69. Fetkovich, M.J., Fetkovich, E.J., and Fetkovich, M.D.: "Useful Concepts for Decline Curve Forecasting, Reserve Estimation and Analysis," paper SPE 28628 presented at the 1994 SPE Annual Technical Conference and Exhibition, New Orleans, 25–28 September.
70. Wilson, S.J., Miller, M.R., Frazer, L.C. and Digert, S.A.: "Multiple Trend Forecasting Accounts for Field Constraints," paper SPE 39939 presented at the 1998 SPE Rocky Mountain Regional/Low Permeability Reservoir Symposium and Exhibition, Denver, 5–8 April.
71. Saleri, N.G. and Toronyi, R.M.: "Engineering Control in Reservoir Simulation: Part I," paper SPE 18305 presented at the 1988 SPE Annual Technical Conference and Exhibition, Houston, 2–5 October.

72. Jacks, H.H.: "Forecasting Future Performance," *Reservoir Simulation,* C.C. Mattox and R.L. Dalton (eds.), Monograph Series, SPE, Richardson, Texas (1990) **13**, 99–110.
73. Palke, M.R. and Rietz, D.C.: "The Adaptation of Reservoir Simulation Models for Use in Reserves Forecasting," paper SPE 71430 presented at the 2001 SPE Annual Technical Conference and Exhibition, New Orleans, 30 September–3 October.
74. Richardson, J.G. and Blackwell, R.J.: "Use of Simple Mathematical Models for Predicting Reservoir Behavior," *JPT* (September 1971) 1145; *Trans.*, AIME, **251**.
75. Trewin, N. and Bramwell, M.G.: "The Auk Field, Block 30/16, UK North Sea," *United Kingdom Oil and Gas Fields, 25 Years Commemorative Vol.,* I.L. Abbotts (ed.), Geological Soc. of London, London (1991) 227.
76. Rasor, R.: "Presentation Case—Reservoir Simulation Prediction," SPEE Forum, Soc. of Petroleum Evaluation Engineers, Houston (2000) 31 October–1 November.
77. Zuber, M.D.: "A Guide to Coalbed Methane Reservoir Engineering," Gas Research Inst. (Gas Technology Inst.), Chicago, GRI-93/0293 (1996).
78. Reiss, L.H.: *The Reservoir Engineering Aspects of Fractured Formations,* Editions Technip, Paris (1980).
79. P'an, C.H.: "Petroleum in Basement Rocks," *AAPG Bull.* (October 1982) 1597.
80. Aguilera, R.: "Recovery Factors and Reserves in Naturally Fractured Reservoirs," *J. Cdn. Pet. Tech.* (July 1999) 15.
81. "Gas Production from Fractured Shales: An Overview and Update," *Gas Tips* (June 2000).
82. "United States Fractured Shales Gas Resource Map," Gas Research Inst. (Gas Technology Inst.) (2000) http://griweb.gastechnology.org/pub/content/Jul/200007/11130/resources.html.
83. National Petroleum Council estimates, Volume III, Devonian Shales (1980).
84. National Petroleum Council Estimate, Volume II, Source and Supply (1992).
85. "1995 National Assessment of United States Oil and Gas Resources," USGS Survey Circular 1118, U.S. Government Printing Office, Washington, DC (1995).
86. Schmoker, J.W.: "Methodology for Assessing Continuous-type (Unconventional) Hydrocarbon Accumulations," *1995 National Assessment of United States Oil and Gas Resources—Results, Methodology, and Supporting Data,* D.L. Gautier, G.L. Dolton, K.I. Takahashi, and K.L. Varnes (eds.), USGS Digital Data Series DDS-30, release 2 (1996).
87. Kuuskraa, V.A. *et al.*: "Barnett Shale Rising Star in Fort Worth Basin," *Oil & Gas J.*, **96,** No. 21 (25 May 1998) 67–76.
88. Williams, P.: "San Juan Basin," *American Oil and Gas Investor* **19,** 6 (1999) 26–34.
89. Decker, D., Coates, J.-M.P., and Wicks, D.: "Stratigraphy, Gas Occurrence, Formation Evaluation and Fracture Characterization of the Antrim Shale, Michigan Basin," GRI, Chicago (December 1992).
90. Decker, A.H., Wicks, D.E., and Coates, J.-M.: "Gas Content Measurements and Log Based Correlations in the Antrim Shale," GRI-93/0293, GRI, Chicago (July 1993) 33–37.
91. Brown, C.A., Erbe, C.B., and Crafton, J.W.: "A Comprehensive Reservoir Model of the Low-Permeability Lewis Sands in the Hay Reservoir Area, Sweetwater County, Wyoming," paper SPE 10193 presented at the 1981 SPE Annual Technical Conference and Exhibition, San Antonio, 5–7 October.
92. Kukal, G.C. *et al.*: "Critical Problems Hindering Accurate Log Interpretation of Tight Gas Sand Reservoirs," paper SPE 11620 presented at the 1983 SPE/DOE Symposium on Low Permeability, Denver, 13–16 March.
93. Spencer, C.W.: "Geologic Aspects of Tight Gas Reservoirs in the Rocky Mountain Region," *JPT* (July 1985) 1308; *Trans.,* AIME, **279**.
94. Witherbee, L.J., Godfrey, R.D., and Dimelow, T.E.: "Predicting Turbidite-Contourite Reservoir Intervals in Tight Gas Sands: A Case Study From the Mancos B Sandstone," paper SPE 11609 presented at the 1983 SPE Symposium on Low Permeability, Denver, 13–16 March.
95. Lewis, D.J. and Perrin, J.D.: "Wilcox Formation Evaluation: Improved Procedures for Tight Gas-Sand Evaluation," *JPT* (June 1992) 724.
96. Kazemi, H.: "Low-Permeability Gas Sands," *JPT* (October 1982) 2229.
97. Veatch, R.W. Jr.: "Overview of Current Hydraulic Fracturing Design and Treatment Technology —Part 1," *JPT* (April 1983) 677.

98. Robinson, B.M., Holditch, S.A., and Lee, W.J.: "A Case Study of the Wilcox (Lobo) Trend in Webb and Zapata Counties, TX," *JPT* (December 1986) 1355; *Trans.*, AIME, **281**.
99. Lewis, D.J. and Perrin, J.D.: "Wilcox Formation Evaluation: Improved Procedures for Tight Gas-Sand Evaluation," *JPT* (June 1992) 724.
100. Stewart, P.R.: "Low-Permeability Gas Well Performance at Constant Pressure," *JPT* (September 1970) 1149.
101. Brown, C.A., Erbe, C.B., and Crafton, J.W.: "A Comprehensive Reservoir Model of the Low-Permeability Lewis Sands in the Hay Reservoir Area, Sweetwater County, Wyoming," paper SPE 10193 presented at the 1981 SPE Annual Technical Conference and Exhibition, San Antonio, Texas, 5–7 October.
102. Hale, B.W.: "Analysis of Tight Gas Well Production Histories in the Rocky Mountains," *SPERE* (July 1986) 310; *Trans.,* AIME, **281**.
103. Hale, B.W.: "A Type Curve Approach to Reserves for the Big Piney Gas Field," paper SPE 9840 presented at the 1981 SPE/DOE Symposium, Denver, 27–29 May.
104. Harlan, L.E.: "A Method for Determining Recovery Factors in Low-Permeability Gas Reservoirs," paper SPE 1555 presented at the 1966 SPE Annual Meeting, Dallas, 25 October.
105. Cipolla, C.L. and Kyte, D.G.: "Infill Drilling in the Moxa Arch: A Case History of the Frontier Formation," paper SPE 24909 presented at the 1992 SPE Annual Technical Conference and Exhibition, Washington, DC, 4–7 October.
106. Briggs, P.J. *et al.:* "Development of Heavy Oil Reservoirs," *JPT* (February 1988) 206.
107. Meyer, R.F. and Mitchell, R.W.: "A Perspective on Heavy and Extra Heavy Oil, Natural Bitumen, and Shale Oil," paper presented at the 1987 Twelfth World Petroleum Congresses, Houston.
108. Martinez, A.R.: "The Orinoco Oil Belt, Venezuela," *J. of Petroleum Geology* (1987) **10,** 125.
109. Adams, D.M.: "Experiences With Waterflooding Lloydminster Heavy-Oil Reservoirs," *JPT* (August 1982) 1643; *Trans.,* AIME, **273**.
110. Dusseault, M.: "Cold Production and Enhanced Oil Recovery," *J. Cdn. Pet. Tech.* (November 1993) 16.
111. Maini, B.B., Sarma, H.K., and George, A.E.: "Significance of Foamy-Oil Behaviour in Primary Production of Heavy Oils." *J. Cdn. Pet. Tech.* (1993) **32,** 6.
112. Maini, B.: "Foamy-Oil Flow," *JPT* (October 2001) 54.
113. Kumar, R., and Pooladi-Darvish, M.: "Solution-gas Drive in Heavy Oil: Field Prediction and Sensitivity Studies with Low Gas Relative Permeability," *J. Cdn. Pet. Tech.* (March 2002) 26.
114. Sulaiman, S. and Bretherton, T.A.: "Thin Oil Development in the Bream Field," paper SPE 19493 presented at the 1989 SPE Asia-Pacific Conference, Sydney, 13–15 September.
115. Seines, K., Lien, S.C., and Haug, B.T.: "Troll Horizontal Well Tests Demonstrate Large Production Potential From Thin Oil Zones," *SPERE* (May 1994) 133; *Trans.,* AIME, **297**.
116. Irrgang, H.H.: "Evaluation and Management of Thin Oil Column Reservoirs in Australia," *APEA J.* (1994) **1,** 64.
117. van Lookeren, J.: "Oil Production From Reservoirs With an Oil Layer Between Gas and Bottom Water in the Same Sand," *JPT* (March 1965) 354; *Trans,* AIME (1965) **234**, 354.
118. Sognesand, S.: "Reservoir Management of the Oseberg Field During Eight Years' Production," paper SPE 38555 presented at the 1997 SPE Offshore Europe Conference, Aberdeen, 9–10 September.
119. Muskat, M. and Wyckoff, R.D.: "An Approximate Theory of Water-Coning in Oil Production," *Trans.*, AIME (1935) **114**, 144.
120. Chaney, P.E. *et al.:* "How To Perforate Your Well To Prevent Water and Gas Coning," *Oil & Gas J.* (7 May 1956) 108.
121. Chierici, G.L., Ciucci, G.M., and Pizzi. G.: "A Systematic Study of Gas and Water Coning by Potentiometric Models," *JPT* (August 1964) 923; *Trans.,* AIME, **231**.
122. Bournazel, C. and Jeanson, B.: "Fast Water-Coning Evaluation Method," paper SPE 3628 presented at the 1971 SPE Annual Meeting, New Orleans, 3–6 October.
123. Kuo, M.C.T. and DesBrisay, C.L.: "A Simplified Method for Water Coning Predictions," paper SPE 12067 presented at the 1983 SPE Annual Technical Conference and Exhibition, San Francisco, 5–8 October.
124. Kuo, M.C.T.: "Correlations Rapidly Analyze Water Coning," *Oil & Gas J.* (October 1989) 77.

125. Hoyland, L.A., Papatzacos, P., and Skjaeveland, S.M.: "Critical Rate for Water Coning: Correlation and Analytical Solution," *SPERE* (November 1989) 495.
126. Earlougher, R.C. Jr.: *Advances in Well Test Analysis,* Monograph Series, SPE, Richardson, Texas (1977) **5.**
127. Duggan, J.O.: "The Anderson 'L'—An Abnormally Pressured Gas Reservoir in South Texas," *JPT* (February 1972) 132.
128. Harville, D.W. and Hawkins, M.F.: "Rock Compressibility and Failure as Reservoir Mechanisms in Geopressured Gas Reservoirs," *JPT* (December 1969) 1528.
129. *Glossary of Geology,* Gary, M., McAfee, R. Jr., and Wolf, C. (eds.), American Geological Inst., Washington, DC (1972).
130. Bernard, W.J.: "Reserves Estimation and Performance Prediction for Geopressured Gas Reservoirs," *J. of Petroleum Science and Engineering* (1987) 15.
131. Wallace, W.E.: "Water Production from Abnormally Pressured Gas Reservoirs in South Louisiana," *JPT* (August 1969) 969.
132. Bourgoyne, A.T.: "Shale Water as a Pressure Support Mechanism in Gas Reservoirs Having Abnormal Formation Pressure," *J. of Petroleum Science and Engineering* (1990) 305.
133. Chierici, G.L. *et al.:* "Water Drive From Interbedded Shale Compaction in Superpressured Gas Reservoirs—A Model Study," *JPT* (June 1978) 93.
134. Ramagost, B.P. and Farshad, F.F.: "*P/Z* Abnormally Pressured Gas Reservoirs," paper SPE 10125 presented at the 1981 SPE Annual Technical Conference and Exhibition, San Antonio, Texas, 5–7 October.
135. Geertsma, J.: "The Effect of Fluid Pressure Decline on Volumetric Changes of Porous Rocks," *Trans.,* AIME (1957) **210**, 331.
136. Fatt, I.: "Pore Volume Compressibilities of Sandstone Reservoir Rocks," *Trans*., AIME (1958) **213,** 362.
137. van der Knaap, W.: "Nonlinear Behavior of Elastic Porous Media," *Trans.,* AIME (1959) **216,** 179.
138. Dobrynin, V.M.: "Effect of Overburden Pressure on Some Properties of Sandstone," *SPEJ* (December 1962) 360; *Trans*., AIME, **225**.
139. Chierici, G.L., Pizzi, G., and Ciucci, G.M.: "Water Drive Gas Reservoirs: Uncertainty in Reserves Evaluation From Past History," *JPT* (February 1967) 237; *Trans.,* AIME, **240.**
140. von Gonten, W.D. and Choudhary, B.K.: "The Effect of Pressure and Temperature on Pore Volume Compressibility," paper SPE 2526 presented at the 1969 SPE Annual Meeting, Denver, 28 September–1 October.
141. Teeuw, D.: "Prediction of Formation Compaction From Laboratory Compressibility Data," *SPEJ* (1971) 263.
142. Newman, G.H.: "Pore-Volume Compressibility of Consolidated, Friable, and Unconsolidated Reservoir Rocks Under Hydrostatic Loading," *JPT* (February 1973) 129.
143. Jones, S.C.: "Two-Point Determinations of Permeability and PV vs. Net Confining Stress," *SPEFE* (March 1988) 235.
144. Yale, D.P. *et al.*: "Application of Variable Formation Compressibility for Improved Reservoir Analysis," paper SPE 26647 presented at the 1993 SPE Annual Technical Conference and Exhibition, Houston, 3–6 October.
145. Hall, H.N.: "Compressibility of Reservoir Rocks," *Trans.,* AIME (1953) **204,** 309.
146. Bernard, W.J.: "Reserves Estimation and Performance Prediction for Geopressured Gas Reservoirs," *J. of Petroleum Science and Engineering* (1987) 15.
147. Yale, D.P. *et al.:* "Application of Variable Formation Compressibility for Improved Reservoir Analysis," paper SPE 26647 presented at the 1993 SPE Annual Technical Conference and Exhibition, Houston, 3–6 October.
148. Dhir, R., Dern, R.R., and Mavor, M.J.: "Economic and Reserves Evaluation of Coalbed Methane Reservoirs," paper SPE 22024 presented at the 1991 SPE Hydrocarbon and Evaluation Symposium, Dallas, 11–12 April.
149. Keith, D.R., Wilson, D.C., and Gorsuch, D.P.: "Reserves Definitions—An Attempt at Consistency," paper SPE 15865 presented at the 1986 SPE European Petroleum Conference, London, 20–22 October.

150. Rose, P.R.: *Risk Analysis and Management of Petroleum Exploration Ventures,* AAPG, Tulsa (2001).
151. Huffman, C.H. and Thompson, R.S.: "Probability Ranges for Reserves Estimates From Decline Curve Analysis," paper SPE 28333 presented at the 1994 SPE Annual Technical Conference and Exhibition, New Orleans, 25–28 September.
152. Jochen, V.A. and Spivey, J.P.: "Probabilistic Reserves Estimation Using Decline Curve Analysis With the Bootstrap Method," paper SPE 36633 presented at the 1996 SPE Annual Technical Conference and Exhibition, Denver, 6–9 October.
153. Gaston, H.: "Pitfalls in Oil and Gas Evaluations," presented at 1992 SPEE Annual Convention, Jackson Hole, Wyoming, 15 June 1992.
154. Mistrot, Gus A.: *Reserves Estimating—Common Errors and How To Avoid Them*, SPE Short Course, SPE, Richardson, Texas (1998).

SI Metric Conversion Factors

acre	×	4.046 856	E − 01	= ha
acre-ft	×	1.233 489	E + 03	= m^3
°API		141.5/(131.5 + °API)		= g/cm^3
bbl	×	1.589 873	E − 01	= m^3
°F		(°F − 32)/1.8		= °C
ft	×	3.048*	E − 01	= m
ft^3	×	2.831 685	E − 02	= m^3
lbm	×	4.535 924	E − 01	= kg
mile	×	1.609 344*	E + 00	= km
psi	×	6.894 757	E + 00	= kPa
psia	×	6.894 757	E + 00	= kPa
square mile	×	2.589 988	E + 06	= m^2
ton	×	9.071 847	E − 01	= Mg

*Conversion factor is exact.

Chapter 19
Valuation of Oil and Gas Reserves

D.R. Long, Long Consultants Inc.

19.1 Introduction

This chapter describes the use of a reserves estimate to prepare an economic evaluation and perhaps then place a value on the reserves. This chapter often refers to a document titled *Perspectives on the Fair Market Value of Oil and Gas Interests*[1] published by the Society of Petroleum Evaluation Engineers (SPEE) in the spring of 2002. In this chapter, that document is referred to as the SPEE FMV document. To value reserves, the nature of the ownership must be considered. Reserves ownership is usually derived from contractual agreements that specify the obligations of the parties to those agreements for the payment of costs and the sharing of revenues. These agreements often include specific commitment obligations such as the drilling of wells. A common arrangement for such contracts is the oil and gas lease. Another common contractual structure is the production-sharing arrangement. Appendix A, which describes common types of oil and gas property interests, is from the SPEE FMV document.[1]

Sec. 19.8 discusses fair market value (FMV). While an FMV estimate is often not the main reason an economic evaluation is prepared, it is useful to keep in mind that it may be used for such a purpose. Like the estimation of reserves, the preparation of an economic evaluation has subjective elements. If the evaluation, inclusive of the reserves estimates and production schedules, is prepared in a manner anticipating an FMV estimate as the goal, the objective evaluator will be inclined to use appropriate judgment. One must be cautious, however, and be mindful of the applicable reserves definition. For example, a reserves definition that might be used for property acquisition might not meet the reporting requirements of the U.S. Securities and Exchange Commission (SEC).

The use of the word "valuation" in the title of this chapter is a carryover from the first handbook.[2] J.J. Arps wrote the valuation chapter, and reference to his name in this chapter is generally the same as a reference to the chapter he authored in that handbook. Arps' use of the term valuation seemed to have a relatively narrow scope. Valuation referred to the processes for the determination of FMV and embraced what he called an "analytical appraisal" method, which involved present worth discounting to address the time value of money. There are numerous references and comparative statements in this chapter about these methods because, despite a few variations that have occurred in the interim, enough is the same that presentation of the historical perspective is useful. Arps also included a section on "different concepts of valuation." Those methods have been discarded.

The analytical valuation methods embraced by Arps[2] are still in use, but today's business environment is quite different. Arps' chapter was for use only in the U.S. domestic market, but the marketplace has expanded vastly. Interest and discount rates are considerably higher, and there have been significant changes in the U.S. federal tax code. U.S. federal tax rates are lower now than in 1962, but other aspects of the law also have changed.

In 1962, the primary buyers for oil and gas properties were major oil and gas companies. Sellers were successful wildcatters or small oil and gas companies called independents. Today, the independents are the buyers and the majors are the sellers. There are many more independents today, and the major oil companies are undergoing consolidation. The marketplace for oil and gas properties involves a much greater number of transactions and is more competitive.

The properties tend to be different from those familiar to Arps in 1962. Today, gas reserves are much more likely to be the major source of projected future income. Ownership is more fractionated. In addition, discussion is much more likely to be about the rate of decline than about how many more years the rate will remain constant. Production rates are much more likely to be governed by the capacity of wells to produce than by artificial restraint imposed by regulation. Constant price projections are unlikely because there are futures markets for both oil and gas. The value offered for properties is more likely to include a significant component for reserves involving higher degrees of uncertainty.

One of the most significant changes in the oil and gas reserves evaluations is the use of the computer. In 1962, most cash flow projections were tabulated by hand on large ledger sheets. After plotting historical production curves by hand, forecasting was done by field to reduce the computational load. Modern petroleum economic evaluation software has eased laborious procedures tremendously. Forecasts are often done on a well-by-well basis, and sensitivity studies are much easier. The most valuable attributes of petroleum economic software are the scheduling tools and the summarization capabilities.

Despite the changes since 1962, some things remain the same. The offering price of the buyer is still very sensitive to the cost of money. The buyer's intent is to derive some measure of profit. No matter what form that profit takes in the buyer's figuring of the offering price, it must be of a sufficient magnitude to cover risks inherent in the industry. Another factor that remains the same is that the U.S. still has a federal income tax.

To facilitate meaningful discussion, some basic definitions need to be understood.

• Oil and Gas Reserves Study—A study of geologic and other information for estimating oil and gas reserves within the scope of a specified set of definitions. The definitions will usually include reserves classed as proved and will often categorize unproved reserves. Unproved reserves classes include probable and possible reserves. If reserves are estimated to support an economic evaluation, the reserves estimator will generate or describe associated schedules of future production. For a description of the proved, probable, and possible reserves classification system, see the chapter on estimating oil and gas reserves in this section of the *Handbook*.

• Oil and Gas Reserves Economic Evaluation—A report, often inclusive of an oil and gas reserves study and usually prepared by or under the direction of a petroleum engineer, that presents schedules of future net cash flow and discounted net cash flow based on the reserves and production schedules from an oil and gas reserves study. The reserves classes of the reserves study can be expected to flow through to the summary levels of the economic evaluation because such classifications are a tool for the assessment of uncertainty.

• Valuation—The process of determining value, whether it be for estimating FMV or for another purpose and whether it be derived with discounted cash flow projection methods or with other methods.

In this chapter, the term "reserves report" refers, sometimes collectively and at other times individually, to both a reserves study and an economic evaluation of reserves. The meaning should be clear from the context of the discussion. Each of the processes has specific obligations for objectivity and disclosure regarding methods. SPEE, an organization sometimes referred to as a sister organization to SPE but with no official relation, recently developed a set of Recommended Evaluation Practices (REPs).[3] These REPs and other information about SPEE, which is a strong promoter of ethical standards, are available at www.spee.org. SPEE also publishes *Guidelines for the Application of Petroleum Reserves Definitions*.[4] Sec. 19.8 includes excerpts from the SPEE FMV document,[1] which similarly presumes the preparation of an oil and gas reserves economic evaluation as described here.

In the following sections, the term valuation is not limited to use with procedures purporting to represent, or necessarily closely tied to, FMV. For example, the U.S. SEC stipulates the use of a 10% discount rate for certain financial reporting purposes. This must be considered a valuation process, but the result is primarily for comparative purposes and is not represented as FMV. A calculation to determine the collateral value of a property or properties (the loan value) would be considered a valuation but would logically be less than FMV.

The term "appraisal" is not used beyond this introduction. Appraisal was a term used by Arps.[2] It is most commonly used today with real estate. It has specific connotations that may not be consistent with methods used in the valuation of oil and gas reserves. Real estate appraisals are commonly made with comparative sales principles. The principle valuation techniques described in this chapter are based on anticipations of future revenue and conceptually bear little resemblance to the common real estate appraisal processes. The term appraisal is avoided in connection with the valuation of oil and gas reserves. There is a place in oil and gas valuation work for the comparative sale method.

A common application of valuation methods is to determine value for taxation. Jurisdictions will usually specify procedures somewhat differently than those contained in this chapter. Those differences often involve the way the cash flow projections are prepared. For example, it is sometimes specified that the valuation procedure ignore anticipated taxes to be levied by the taxing authority requiring the valuation. The taxing authority might specify a discount rate or method for arriving at the discount rate to be used in determining value. The evaluator should be knowledgeable about local practices.

The valuation procedures discussed in this chapter are classed properly as methods for the valuation of underlying assets, as opposed to methods for the valuation of a company. The value of a company is some combination of a perception of how well the organization's management will manage its assets for growth (or otherwise) and its debts and other liabilities. Individual property assets are, however, often more valuable when combined than when viewed separately.

In recognition of the international makeup of today's SPE, it is appropriate to write from an exclusively international perspective. Oil and gas companies around the world with roots in capitalistic societies operate on the basis of an incentive for profit from operation and a share of either the volume of oil and gas sold or some benefit from that sale. In the U.S., fractional interests are usually of a nature described in Appendix A. In other parts of the world, it is more common for the operator of the property to receive revenue on the basis of a more complex sharing arrangement made with a country's national government or other governmental faction. This chapter does not delve into the complexities of these sharing arrangements. The valuation circumstances described are generally specific to properties located in the U.S. An understanding of the principles should allow portability for universal application.

19.2 The Need for Economic Evaluations

Valuations of oil and gas properties are needed for many of the same reasons appraisals are needed for homes, cars, jewelry, or any other assets. Lenders require some type of valuation when assets are used as collateral for a loan. Taxes are often assessed on the basis of property value. Property values have a bearing on rates for insurance policies and settlements after loss, damage, or foreclosure. Economic evaluations are seldom made simply for curiosity. In most cases, they are needed for some business reason. Here is a listing of reasons economic evaluations are needed in the oil and gas industry:

• Project Evaluation—Proposals to drill wells, set platforms, or otherwise provide for the facilities required to produce or enhance the production of oil and gas require economic justification before implementation.

• Prioritize Opportunities—When presented with a set of investment opportunities, all of which pass certain minimum criteria and assume some restraint on available capital, discussion and theorization are never exhausted during the selection process. Economic analysis is part of the process.

• Acquisition and Divestiture—Buyers and sellers of properties need valuations to prepare for acquisition and divestiture transactions. The buyer's need for an evaluation is obvious. Sellers with a compulsion to seek the highest compensation usually prepare an economic evaluation to gauge the adequacy of an offer.

• Financing—Loans by banks or other institutions are directly collateralized by a company's property portfolio or a specified group of properties.

• Taxation—Governmental entities from the Internal Revenue Service (IRS) to local taxing authorities such as school districts want money, and the taxes are often determined on the basis of the value of the reserves.

• Securities Laws—In the U.S., the SEC requires the reporting of reserves information, which mainly comes from oil and gas reserves and economic evaluations.

• Legal Settlements—Some of the most dramatic arguments over property values occur in settlement proceedings involving foreclosure. Foreclosure is often associated with a condemnation case wherein lands are rendered unusable for oil and gas development because of confiscation for a public use that covers a wide area, such as a new water reservoir or recreational lake. Alternatively, a company going through bankruptcy proceedings might convince the court that creditors would be better served under a plan in which the company retained the interest. At one time, there were many court cases involving take-or-pay provisions of contracts between producers and pipeline companies.

• Allocation of Values—Value allocation can be part of legal settlements when issues are fundamentally divisive, but allocation is often an element when parties are trying to work together to promote common interests. Unitization is the process in which groups of small properties are united into a larger unit with interest holders of the smaller properties trading their interests for a smaller fraction of the new unit. Sometimes the process is the reverse and involves the distribution of interests. Either way, economic valuation is usually part of the process.

19.3 Evaluation Methods

19.3.1 Deterministic Cash Flow. The primary method of evaluation discussed in this chapter is the deterministic cash flow method. Projected schedules of quantities of reserves from a deterministic reserves study projected in selected future time frames (usually calendar years) are displayed in the results, but computations are often made monthly. Schedules of future prices and costs are projected for the same time frames. Volumes are multiplied by prices, and costs are deducted to estimate future net revenue (FNR). Present worth calculations to factor in the time value of money are applied to the projected cash flow stream, and the results are reported. A table presenting a suite of present worth values over a range of discount rates is usually included. Individual projections are usually made by well or ownership entity. Each projection is classed in a reserves category (i.e., proved, probable, or possible) and more specific subclasses are common. The results are summarized at the reserves category level.

This valuation procedure is based on the methods used to estimate reserves in the chapter on the estimation of oil and gas reserves in this section of the *Handbook*. Within the limits of the reserves category definitions, the parameters used in the estimation of reserves are generally average values. In the case of proved reserves, the results of the calculation are perceived to be best estimates. In the case of probable and possible reserves categories, the best estimate perception is inherently qualified with the perspective of "if they exist at all." Probable and possible reserves quantities have not been adjusted for risk.

19.3.2 Decision Trees. Decision tree analysis is a very useful tool to value estimated reserves deterministically. The result is a calculation of expected value or a range of discrete values. Sec. 19.6 illustrates this principle simply. The chapter on risk analysis and decision making in the Emerging and Peripheral Technologies section of this *Handbook* is a must-read chapter in connection with the application of the principles discussed here. Even if the methods are not physically used, the perspectives offered by both decision tree and Monte Carlo simulation must be considered. The chapter on general economics in the General Engineering section of this *Handbook* also presents a perspective on the use of decision trees for analysis.

19.3.3 Sensitivity Analysis. In this procedure, variables in a deterministic analysis are modified and results are recalculated to check the effect on the economic calculation. The results might be presented on a chart, or they might be expressed in a manner such as a one-dollar reduction in the projected gas prices reduces the FNR by 20%. The use of this type of procedure with decision trees has been a traditional evaluation tool with deterministic reserves valuation procedures.

19.3.4 Monte Carlo Simulation. Monte Carlo simulation is a method in which a frequency distribution representing what is perceived to be the full range of possible values for parameters is substituted for the single average values, instead of the use of single-value averages or estimates for parameters, as in a deterministic calculation. The frequency distribution for a parameter, such as porosity, might be shaped in a number of ways. Bell shapes, indicating a normal distribution, or triangular shapes are common. The value at the peak of the distribution's shape is the most likely value (the mode), and the higher values to the right and lower values to the left are defined as less likely. See the chapter on risk analysis and decision making in the Emerging and Peripheral Technologies section and the chapter on general economics in the General Engineering section of this *Handbook*.

In the case of oil and gas reserves economic evaluation, all the parameters used in the estimation of reserves quantities (such as porosity and water saturation) and the additional variables introduced in the economics of the evaluation process (such as prices, costs, and sometimes timing) are often represented as probability distributions. If a parameter, such as oil price, is locked into a contract, there might be no reason to assign a distribution to that parameter unless the model is intended to consider the default of a contracting party. The simulator is then allowed to run through what are often thousands of iterations. At each iteration, the simulator selects a value for each parameter from the frequency descriptions and computes and stores results. Because of the way the distributions are defined and the way the simulator takes its samples, the values for the individual parameters from the distributions that are closer to the mode value are selected more often. Occasionally, but not often, the simulator will select values at the extremes of the distributions. The method of sampling the described distributions is a math problem of its own.

A Monte Carlo simulator is not a numerical reservoir simulator. It is not a model of the depletion process of a reservoir. In numerical reservoir simulators, the reservoir is characterized in discrete segments and those segments are assigned values of porosity, permeability, fluid saturation, and pressure. The rocks and fluid properties are characterized, and the model is run most often with the intent of mathematically replicating the physical behavior of pressure and fluid flow within the formation during reservoir depletion or enhanced-recovery process. Input to a Monte Carlo simulation will require distributions for expected production rates and recovery factors. Those distributions might be determined with the aid of a reservoir simulator, but it is not expected (today) that the software be linked or integrated.

All the Monte Carlo generated results of reserves and cash flow projection would be incomprehensible; therefore, summary results, selected by the modeler, are reported. Output distributions often include only reserves quantities in total, FNR, and present worth values at a few discount rates. The results represent calculated distributions of expectations for reserves, FNR, and present worth (discounted FNR) at one or more discount rates. Depending on how the model is constructed, it may or may not be possible to distinguish between reserves and revenues derived from areas of the reservoir categorized (deterministically) as proved, probable, or possible.

Any (deterministic) model built in a spreadsheet can be converted to a Monte Carlo model by replacing some of the input values with probability distributions. Conversely, all the errors and pitfalls that might cause a deterministic model to lead to a false indication of value will similarly affect a Monte Carlo simulation. The construction of a Monte Carlo simulation model requires all the work and care that goes into a deterministic model and more.

19.3.5 Real Options. As applied to the evaluation of oil and gas reserves and associated opportunities, real options analysis is a new concept. In this *Handbook*, the use of option theory for the valuation of real options is mentioned in the chapter on risk analysis and decision making in the Emerging and Peripheral Technologies section and in the chapter on petroleum economics in the General Engineering section. Several SPE technical papers have addressed the subject in recent years.[5–11] Financial analysts use option-pricing models (OPMs) to value puts and calls on stocks. These stock options to sell or purchase at a specified price, at or before some specified time, always have some value of zero or greater. They are never negative because the option may be allowed to expire with no additional consequence to the holder. After the option is obtained, there is no further expectation of the possibility of financial loss. The value of the financial option stems from the opportunity to exercise the option immediately, plus the speculative potential that it can be held and exercised in the future for a greater financial gain. OPMs for valuing financial options consider the current stock price, the strike or exercise price associated with the particular option, the risk-free interest rate, the time to maturity of the option, and the variance in the stock price.

Real options represent investment opportunities outside the stock market that have some resemblance to stock options. The option to buy production from a lease in a future specified month at a specified price is essentially identical to a stock option. Most real options are not so ideal in nature and are often extremely complicated. Real options usually are not structured like the previous example, in which someone has contracted to purchase production. Real options are perhaps better thought of in oil and gas industry terms as anticipated opportunities and/or alternatives. Real options are more like the decisions we make in our everyday lives.

In most circumstances, real options will not have a specific exercise date, and many real options involve liability such that their value has the potential to be negative at a future date. The time frame associated with real options may be very long. Real option valuation (ROV) involves the application of the methods derived for the valuation of stock options to the valuation of real options. Real options that are perceived to be of value often revolve around management's perceived future opportunity to make an intelligent decision on the basis of the circumstances existing when the decision is made (i.e., an optional course in a future oil and gas property management decision).

In an attempt to acquire an oil and gas property, whether producing or not, many attach an additional value to being the operator. Sometimes that additional value has to do with cash flow projections and the opportunity to receive income from partners as overhead compensation [Council of Petroleum Accountants Socs. (COPAS) charges]. Perhaps more often, the additional value attached to being operator has to do with a greater degree of control. The operator is able to exercise options, and that flexibility has value. The operator has more real options than the nonoperator does.

In attempting an acquisition in a competitive market, incremental value frequently is added in recognition that the acquirer will have specific opportunities, in the future, to deviate from and optimize the exploitation plan on the basis of anticipated future events and new information or technology that will become available. It is unlikely that either the deterministic reserves evaluator or the Monte Carlo modeler has made an exhaustive accounting for all possible opportunities. The details behind all possible future opportunities cannot be fully anticipated. ROV is a tool that might be useful for quantifying such potential incremental value.

When considering an acquisition, future potential opportunities sometimes will not be of any specific nature. A common incentive behind oil and gas properties acquisition is the experience that with ownership there is additional value often attributed to serendipity. Perhaps real option analysis could be of assistance in quantifying such value. The use of option theory to quantify oil and gas industry investment opportunities is not a widely accepted method at this time.

19.3.6 Rules of Thumb, Yardsticks, and Comparative Sales. A number of "unit value processes" have been referred to as evaluation methods, but they are more properly described as screening tools. As a comparative measure between transactions, the price paid per barrel of reserves is often computed. Because most transactions involve reserves of both oil and gas in inconsistent proportions, a value per equivalent barrel is often computed, which necessitates the application of a relationship of equivalency between a thousand cubic feet of gas and a barrel of oil. The value of a thousand cubic feet of gas can easily vary over a broad range (more than 20%, often much more) only on the basis of quality issues, exclusive of transportation costs related to location or cost of production or processing to make marketable. Oil has its own issues, of which quality is only one, but the equivalency issue is only the beginning of the problem with trying to use unit value methods as a means of valuation. Such methods make no allowance for the timing or cost of recovery, there is no provision for transportation or other cost to market the production, and there is no recognition of what capital might be required to accomplish the recovery.

There is nothing wrong with making these comparisons. They are of interest, and sometimes the questions raised expose errors or facts that have not been, but should be, disclosed. If there is a comparative method that helps you become comfortable with the result of a valuation, use it. If it exposes a weakness in the valuation, fix the valuation.

Sometimes there are attempts to value oil and gas interests on a multiple of current income with the current income rate expressed in dollars per month or dollars per year. The value approximation might be expressed as 3 years or 36 months at current. The number of potential flaws in this type of valuation are so numerous and varied that an attempt to initiate a list seems futile. There are circumstances that require a determination of value in which the time and effort required is not warranted for even a minimum effort to generate a cash flow projection. Given a general knowledge of the character of the reserves and knowledge of valuation results of other properties of somewhat similar character, the multiple of current income approach may be the best of bad alternatives. In such circumstances, the income rate multiple might be selected on the basis of analogy with the valuations of other properties thought to be similar in character but for which a more rigorous valuation procedure has been appropriate.

Another valuation method is the use of the comparative sale. This is often the only practical tool and is particularly useful. Unlike the real estate market, it is usually impossible to locate a suitable basis for the comparison. The comparative sale method relies on the use of analogy to infer value. It is easy for the technique to be flawed, unless the valuation is done with complete understanding and with great care. For instance, it is common for the purchaser of an interest to attach an element of value that is not known to the observer. This attachment of added value occurs with the greatest frequency when small interests are traded but is not uncommon with transactions of any size. SPEE's FMV document labels these added values "strategic" values. An industrial user, for example, might purchase a fractional interest in a property near one of its facilities to provide an emergency fuel source in the event of a cutoff of its regular supply; therefore, it might pay a premium price for that interest. Given that the industrial user's needs are satisfied, someone valuing other interests in the same or nearby property could be mistaken in using that transaction as an analog. The SPEE FMV document suggests that the best use of comparative sale methods for oil and gas interests is in the valuation of nonproducing properties, but it points out that even then the process is more complicated than in the real estate business. It is not intended to imply that the use of the comparative sale method in the real estate market is easy, it is just much more difficult (perhaps less certain is a better perspective) when the value of minerals comes into play. Garb[12] discusses a variety of valuation methods, and his article is recommended reading.

19.4 The Nature of Reserves Reports and Discounted Cash Flow Schedules

While seldom listed as the preferred method by agencies requiring valuations (such as the U.S. IRS), the most practical valuation method for oil and gas properties is usually based on the expectations of future income. Reserves estimates and projections of future rates of production are used with expectations for oil and gas prices, items of burden, and costs to compute estimates, usually tabulated by calendar year, of future cash flow. **Table 19.1** presents the presentation form of the results of that calculation. Sometimes the form for the presentation is in greater detail, which results in the information being stacked on the page with the lines representing the calendar year period repeated two or three times. For valuation purposes, the projected cash flow stream is then analyzed in a variety of manners. Table 19.1 represents a projection for a single producing property in which the only production stream is oil. In most cases, a valuation is based on a summary level cash flow in which projected economics are derived from the expectation for the sale of produced oil, gas, and other substances such as sulfur. Table 19.1 illustrates certain elements to set the stage for later discussion.

Table 19.1 contains the results of a reserves and cash flow projection. Col. 1 indicates the time intervals associated with the individual lines. The intervals sometimes are indicated by their ending month (with the beginning of each interval being the end of the previous one), but the times are indicated by calendar year in Table 19.1. In Table 19.1, the beginning of the first interval is as of 1 September 2001. Col. 2 presents the full well stream (commonly referred to as the gross or $^8\!/_8$) production quantity schedule from the reserves report. (It is not production; it is a forecast of production. Terminology and proper expression in the economic evaluation business is a common problem.[13]) Col. 3 gives the production quantity schedules netted to the interest being evaluated by multiplying the gross quantities by the fractional revenue interest. The interest fractions will be displayed, unless the page is a summary. The revenue (income) interest is commonly less than the working (expense) interest because of royalty burdens or other sharing arrangements. Col. 4 presents the projected oil price schedules, which is projected to escalate in this example. Col. 5 is the product of the net oil quantity times the oil price. Col. 6 represents items of deduction for local taxes, and Col. 7 shows deductions for operation costs. Col. 8 presents capital cost (none is scheduled in this example), which is another deduction item. The values in Col. 9 represent the time interval's summary of revenue less items of deduction and the result of the net cash flow projection, which is computed by subtracting Cols. 6, 7, and 8 from Col. 5. Various items of identification and information are also presented. This identification information should include identification of the well or forecast entity and the reserves category. The present worth suite is also displayed. In Table 19.1, it is in the bottom right corner and is presented with discount rates from zero to 100%. Cols. 11 through

TABLE 19.1—PRODUCTION FORECAST AND ECONOMIC CALCULATION DEMONSTRATING PRESENT WORTH DISCOUNTING AS OF 1 SEPTEMBER 2001

1	2	3	4	5	6	7	8	9	10	11	12	13	14
Year	Oil Gross (thousand bbl)	Oil Net (thousand bbl)	Price ($/bbl)	Net Operating Revenue (U.S. $ thousand)	Local Taxes (U.S. $ thousand)	Lease Operating Expenses (U.S. $ thousand)	Capital (U.S. $ thousand)	Net Revenue (U.S. $ thousand)	Cumulative Net Revenue (U.S. $ thousand)	Time to Midpoint of Period (years)	12% Discount Factor	12% Discounted Present Worth (U.S. $ thousand)	Cumulative 12% Discounted Present Worth (U.S. $ thousand)
2001	165.000	14.025	24.00	336.600	24.235	132.000	0	180.365	180.365	0.167	0.9813	176.990	176.990
2002	222.024	18.872	24.72	466.517	33.589	190.000	0	242.928	423.292	0.833	0.9099	221.035	398.025
2003	196.000	16.660	25.46	424.190	30.542	184.000	0	209.649	632.941	1.833	0.8124	170.317	568.343
2004	173.000	14.705	26.23	385.645	27.766	180.000	0	177.879	810.820	2.833	0.7254	129.025	697.368
2005	153.000	13.005	27.01	351.294	25.293	176.000	0	150.001	960.820	3.833	0.6476	97.146	794.513
2006	136.000	11.560	27.82	321.629	23.157	172.000	0	126.472	1087.292	4.833	0.5782	73.132	867.645
2007	121.000	10.285	28.66	294.740	21.221	170.000	0	103.519	1190.811	5.833	0.5163	53.446	921.091
2008	107.000	9.095	29.52	268.457	19.329	168.000	0	81.128	1271.939	6.833	0.4610	37.398	958.489
2009	86.000	7.310	30.40	222.242	16.001	167.000	0	39.241	1311.179	7.833	0.4116	16.151	974.640
2010	77.000	6.545	31.31	204.954	14.757	166.000	0	24.197	1335.376	8.833	0.3675	8.892	983.532
2011	69.000	5.865	32.25	189.170	13.620	166.000	0	9.549	1344.926	9.833	0.3281	3.133	986.665

Working Interest Fraction	0.1000
Revenue Interest Fraction	0.0850

Present Worth Profile

15	16
Discount Rate (%)	Present Worth (U.S. $ thousand)
0	1344.926
8	1082.096
10	1032.035
12	986.665
15	926.165
20	841.722
30	716.545
40	629.006
60	515.711
100	398.536

Example oil well No. 1
Reserves category: Proved Developed Producing

13 are included to facilitate the following discussion of present worth discounting. They usually are not displayed in standard presentations.

Table 19.1 presents values and items that represent the projected cash flow for an existing well producing proved developed reserves. If the cost to drill and equip the well ($800M) were included in the capital costs as an initial outlay before production started, the discount factor would be 1.0 and the cumulative net revenue and discounted values would be reduced by that amount. Further, the indicated rate of return (ROR) for such an investment can be determined from the present worth table to be between 20 and 30%.

Discounting is a widely used method for accounting for the time value of money. It is not a method for evaluating risk. Like the loan shark, cash outlays bearing substantial risk of recoupment require the potential of a high ROR. The high ROR should be the apparent result of an analysis of risk, not the method. The method of risk analysis should be based on the element of risk: production rate, reserves, prices, and cost. Elevating the discount rate reduces the value of income in proportion to its distance into the future. If the reserves are correct but the beginning production rate of a new well is off by 25%, sensitivity analysis should be used to assess the risk or lower the forecasted rate so that the projection represents an expected value. Arbitrary discount rate adjustments should not be made. Despite sharing this view, the SPEE FMV document presents some material involving one elevated discount rate risk analysis method. If you are contesting a valuation instead of establishing value, it might be worth review.

The discount rate may be thought of as an inverted interest rate. As the discount rate increases, the value of future income is reduced. At a given discount rate, the farther in the future a dollar of income is expected to be received, the less it is computed to be worth. The present worth profile of Table 19.1 shows that as the discount rate increases, the value of the projected cash flow stream decreases. At a discount rate of zero, the present worth is the same as the total of the projected cash flow.

Not all discounting methods are the same. Continuous discounting at a rate of 12% and annual midpoint discounting at a rate of 12% do not yield the same results. The standard discounting method historically used in the oil and gas industry is midpoint discounting with either annual or calendar year time frames, as Table 19.1 illustrates. Today, with software that is more sophisticated and with some change in logic, other discounting methods frequently are used. A logical means of discounting is a method that recognizes when the cash from the sale of oil or gas is received by the owner. From that view, middle-of-the-production-month discounting would be too optimistic and so would end-of-month discounting. A retarded discounting method that recognizes that income is often received a month or two after associated quantities are produced is a possibility but is not a method known to be supported by existing software.

A common method of discounting computes an annual discount rate factor with the following equation:

$$D_f = \frac{1}{(1+r)^t}, \qquad (19.1)$$

where D_f = discount factor for the specific point in time, r = annual rate of discount expressed as a fraction, and t = time in years from the as of date to the point of reference for discounting. The value assigned to t determines if the discounting is midframe or end of frame and as suggested, could be set beyond the end of the frame. Table 19.1 illustrates calendar year midpoint discounting.

Example 19.1. Calculate the 12% discount factor for cash received at midyear in the fifth time interval of the projection in Col. 12 of Table 19.1.

Solution. The effective date is 1 September of 2001; therefore, there are 4 months following the "as of" date remaining in 2001 at the effective date. Then there are 3 full years of 12 months each (2002, 2003, and 2004). You must go 6 months into 2005 to get to the middle.

$t = 4 + 12 + 12 + 12 + 6 = 46$ months = 3.833 years.

$$D_f = \frac{1}{(1 + 0.12)^{3.833}} = 0.647636. \quad (19.1)$$

The value of D_f is expressed in Table 19.1 as 0.6476. Therefore, the discount FNR for the fifth year of the Table 19.1 projection is \$150,001 × 0.647636 = \$97,146. The sum of the discount factor net revenue for all years is referred to as the present worth of the income stream, but this amount should not be confused with the FMV. It is common to compute present worth values at multiple discount rates and to display the computed results as a set of numbers. The term present worth profile probably stems from the shape of the graph when the present worth values are plotted with the discount rates typically represented on the x-axis.

Other equations sometimes are used for discounting. Monthly discounting at the annual rate divided by 12 and envisioning monthly compounding is given by

$$D_f = \frac{1}{\left(1 + \frac{r}{12}\right)^{12t}} . \quad (19.2)$$

If the example is recalculated with Eq. 19.2, the discount factor is

$$D_f = \frac{1}{\left(1 + \frac{0.12}{12}\right)^{12 \times 3.833}} = 0.632727.$$

It is easy to see how the formula could be modified to discount daily or even hourly. Continuous discounting is accomplished with

$$D_f = \frac{1}{e^{rt}}, \quad (19.3)$$

where e is the base of the natural logarithm. Calculating the discount factor with Eq. 19.3,

$$D_f = \frac{1}{e^{0.12 \times 3.833}} = 0.631284.$$

The size of the time intervals used for scheduling is independent of the compounding rate envisioned by the discount factor formula. The only logic behind the use of discount rate formulas other than the annual equation is that the factors are more conservative. There is some obvious logic behind end-of-period discounting or even a retarded discounting, but the logic behind compounded discounting with fractional portions of a year is not clear.

Whatever use is made of an economic evaluation of reserves, an assessment of uncertainty is usually required. Acceptable reserves reports always include definitions given partially to indicate the estimator's opinion of certainty. A guide to the level of certainty generally is expressed by the reserves category in which the reserves quantity falls. Even though the confidence in reserves quantity is theoretically the same between proved reserves classes of proved producing and proved undeveloped, there is unquestionably more uncertainty in the timing for recovery for classes of reserves that are not physically on production.

Within a reserves class such as proved, there are well-recognized subclasses like proved developed producing, proved developed nonproducing, and proved undeveloped. Many evaluators maintain additional reserves classes added by individual evaluators for distinction. Reserves categories like "proved developed nonproducing shut-in" and "proved developed nonproducing behind pipe" are common in that they give the user of the report additional insight as to the nature of the reserves. SPEE includes a suggested extended classification system in Ref. 4. The list is potentially endless.

Regardless of the depth of the classification system, if the reserves estimate is made deterministically, which is still the most common method, the classification system cannot describe the uncertainty fully. An inspection of any of the common (deterministic) reserves definitions will reveal that the reserves class system is not based entirely on level of certainty. Even if certainty is the only criterion, the variations within a class may be significant. Probabilistic reserves estimating procedures, which yield values that quantify uncertainty, can be an important valuation tool. This discussion focuses on results from deterministic procedures. The nature of a reserves report is such that a true assessment of value (whether the report is prepared deterministically or probabilistically) can be made only with a thorough understanding of how the report was prepared and with an understanding of the nature of the properties. This assessment also must include judgment regarding the completeness, quality, and suitability of the availability of data, not only in the preparation of the reserves estimates but in the economic data as well. See Sec. 19.6. Assuming that a suitable production schedule has been provided with the reserves report, the key elements in generating the cash flow projections are discussed in the order in which they are encountered in Table 19.1.

19.4.1 Report Effective Date. The effective date is the time at which the elements that generate the cash flow projection start to be used. This is normally time zero for discounting purposes. The cumulative production (normally shown, although it is not included in the Table 19.1 example) will include the cumulative actual production plus estimated production for a stub period if production up to the effective date was not complete. For example, if the effective date of the report is at the end of the year, but production was available only through October when the production projection was made, the cumulative production will include an estimate of production for the last two months of the year. In preparation of the economic projection, the estimated future production will be cropped at the front end and used only from the effective date forward. A distinction is sometimes made between "effective date" and "as of date." Effective date is the time zero for discounting and scheduling, and as of date is the date through which information was available. For example, given a report prepared for estate tax purposes with the date of death and a time zero of 1 January and with that report being published on 1 July of the same year with all the information available at the time of report preparation, the effective date might be considered 1 January but the as of date is 1 July. There is no standard terminology.

19.4.2 Interest Position. This is intertwined with the subject of Appendix A. Generally, the interest position is defined in terms of an entity's fractional responsibility for costs, with a separate fraction defining its share of the proceeds from sales. There are sometimes financial, accounting, security, and/or business issues relative to whether an entity owns its fractional share of production or is merely the beneficiary of the revenue from sales. Such issues are ignored here, but Ref. 4 discusses the subject. The displayed interest positions are generally the ones applicable on the effective date of the report, and subsequent changes that might occur may or may not be displayed. Reversionary interest positions, if they exist, are not likely to be clearly identified in the petroleum economics schedules that resemble Table 19.1.

19.4.3 Prices. The prices used in generating the cash flows are often dictated by the report's intended user. The SEC sets forth its own requirements. Lending institutions generally have pricing guidelines as a part of their collateral valuation procedure. If the report is being prepared for the determination of FMV, the evaluation engineer will likely be the judge as to the view of the market place. For such purposes, it is not the evaluation engineer's place to be an economist and predict future prices of a global economy. The evaluator is attempting to reflect the view of the market. It is often convenient (and defensible) to rely on the indications by the New York Mercantile Exchange (NYMEX) futures market. One of the nice features of the NYMEX futures market is that it is available every business day. Other useful market indicators become available periodically, such as the State of Texas' annual budget. It is interesting to compare the numbers. Because of price volatility, the time over which any such comparison is valid might be very short or indeterminable.

NYMEX prices reflect benchmark prices for West Texas Intermediate Sweet crude and Henry Hub gas. While care must be used, it is usually possible to use historical prices over a brief period to arrive at suitable differentials to these NYMEX indexes and to then base future price projections appropriately. Ref. 7 is a valuable source for an industry perspective. Like the State of Texas budget, it is sometimes difficult to pin down the time frame to which the perspective applies. It sometimes seems that the NYMEX futures market is out of sync with the mainstream of the industry and with what a property purchaser would use as a guide for a decision on an acquisition. Different segments of the industry are not always synchronized.

The evaluator must take care that the prices used in the economic projection reflect certain other sales features. If transportation charges apply, they usually should be reflected in a reduced price, unless care is taken to be sure those transportation fees are reflected in the cost column. Also, while oil production and sales volume difference usually balance out over time (assuming a lack of theft or errors), gas sales quantities are commonly less than produced quantities because of lease use and shrinkage. The reserves engineer commonly forecasts wellhead production by necessity. Net produced and reserves quantities are reported conventionally as the fraction of the wellhead volume indicated by the revenue interest fraction. To get the arithmetic correct, it is often necessary to account for lease use and shrinkage losses (commonly as high as 6% or greater), gas liquids recovered, and the net back of liquids recovered in an off-site gas processing plant in the gas price.

To ensure that input variables and/or the economic model will project future cash flow suitably, it is a good idea to determine if historical produced quantities and prices can be used to duplicate historical revenue. In a presentation before the SPEE annual meeting in Park City,[14] Rick Riseden indicated the importance of being able to "follow the cash." His subject was the valuation of royalty interests where not all the operator's records might be available but the principle is the same. If you try to follow the cash and cannot, how can you predict future revenues with any accuracy?

19.4.4 Taxes. U.S. federal income taxes are not included in this item and the calculation of such taxes is not presented in this chapter, but Sec. 19.7 discusses the characteristic impact on value. This tax element is what is often referred to as local property taxes. In reality, the taxes normally included are all taxes other than federal and state income taxes. The nature of the tax structure varies from state to state. SPEE maintains a document[15] that presents a summary of the structure of these taxes by state. While the summary is quite helpful, caution is necessary.

Laws and rates change frequently, and some provisions of the laws might not be fully covered in the SPEE material. It is best to consult evaluators with local experience.

19.4.5 Operating Costs. It is common practice to use historical average costs for a preceding time period as the initial cost for projection purposes. Usually, the historical costs are calculated at a cost per well per month and projected forward in the same manner with future costs at the summary level declining as a direct function of the number of wells projected to be producing. Future costs typically are escalated at a rate commensurate with some perceived inflation rate thought to be consistent with the projection of oil and gas prices.

"Nonrecurring" costs are often removed from the historical averages before determining the parameters for projection of future costs. The costs projected in the economic evaluation are generally field-level expenses and exclude general and administrative (overhead) costs. To the extent a working interest owner is billed overhead charges by the operating company (COPAS charges are in that category), those charges are included in the operating costs of the nonoperating working interest owner.

There is much room for abuse, deception, and pitfalls concerning value on the operating cost side of the valuation process.[16] In the case of an offshore platform, it is obvious that there is a substantial base level expense that is not proportional to the number of producing wells. On a property producing a lot of water, the operating cost is more likely to be related to the amount of water being produced than the number of wells or amount of oil. Whether or not onshore operating costs can be expected to vary directly with the number of wells is a matter of circumstances and judgment. It is usually a mistake to eliminate costs classed as nonrecurring. It might be appropriate to spread such costs over a number of wells or make a judgment as to expected frequency of occurrence.

Some costs of operation are classed as "make marketable" charges. The most common of these is the cost to compress gas to the pressure required to enter the sales line, to reduce the water content, or to extract impurities such as carbon dioxide, nitrogen, and/or hydrogen sulfide to meet pipeline specifications. These charges are not handled in proportion to expense interest but are usually charged to the property owner in proportion to the revenue interests; thus, it is not universally true that royalty owners do not bear any share of the costs of operations.

19.4.6 Capital Costs. These costs are listed separately from operation costs because of the difference in their nature and because, if any after income tax economic calculation projection is to be made, there is a difference in the structure of their deductibility in net income computations. The capital costs most commonly included in petroleum economic projections are for the drilling of new wells and the workover of existing wells. Other common costs are for the addition of compression; installation, expansion, or addition of production collection and separation systems; the addition of artificial lift; and the installation of offshore platforms. Planned improved-recovery projects require additional capital outlays for facilities.

There occasionally are repetitive needs for relatively small, but potentially significant, amounts of capital that go unnoticed by the evaluator during reviews of historical expenditures. Such capital items are often associated with large units with consolidated facilities and are not captured in the "operating cost" category of historical expenditures or budgets. Engineers are accustomed to including capital for new wells, recompletions, and workovers, but they are sometimes unaware that the operating costs they are provided include only expensed portions of repetitive historical costs.

Reserves reports sometimes include income from sources other than from the sale of produced oil and gas. When the income is from the sale of byproducts such as sulfur, the treatment is relatively straightforward because it is a third sales revenue stream. If the income results from the processing of gas for the recovery of liquids, the economic model is often adjusted to reflect the associated quantities, but at other times, it is considered sufficient to make an adjustment of the gas price to reflect the elevated income level. A different approach might be required if reporting under U.S. SEC guidelines.

If the income results from providing a service such as disposing of produced saltwater for a fee paid by neighboring operators, the issue is less clear because the evaluation includes an income stream associated with a business other than the production of oil and gas. Evaluations have been observed that include, as a part of the petroleum economics, the operator's income (likely treated as a reduction or credit against other expenses) from COPAS payments received from others. The key to knowing the proper means of evaluation is to know the use of the report and any controlling regulations. Conspicuous disclosure of unusual circumstances is recommended.

Sometimes there are material exit costs such as well and offshore platform abandonments and site remediation. Future environmental liability also could be a significant cost. Economic evaluation of reserves should include such considerations; however, a report may present a fine perspective on specific underlying values but not be a complete expression of value. The preparers of reserves studies and evaluations usually have limited expertise in such matters. It is a good idea to read carefully the qualifications that are always contained in a thoughtfully prepared report or statement of opinion.

This section conveys the message that these projections of production and associated expense and income are constructed to be a model of the future and should include consideration of all parameters that bear on those expectations. A well-constructed model will include a specific value assignment to all the variables that might be included as a distribution should a Monte Carlo simulation be constructed later. As discussed in Sec. 19.4.3, it is a very good idea to be sure you can "follow the cash."

19.5 The Nature of Risk and Uncertainty

Forrest A. Garb is the author of the following text taken from the SPEE FMV document.[1]

> The petroleum industry has written extensively on the risk associated with exploration and development. Very little has been published, however, addressing the uncertainties associated with acquisition and operation of producing properties. The uncertainties associated with estimating the reserves and value of oil and gas producing properties are divided into three classifications. The technical uncertainty that the hydrocarbon volumes estimated do exist in the ground and that the recoverable amounts can be produced within the time frame projected, the economic uncertainty that product prices, operating costs, equipment costs, inflation, and market conditions will be in reasonable agreement with the assumptions used in the economic analysis, and the political uncertainty that world economics, international political stability, taxation, and regulations will not be significantly different than perceived at the time of the evaluation.
>
> 1. **Technical Uncertainty.** Technological uncertainty relates to whether or not the hydrocarbon volumes estimated actually exist in the ground and whether the reserves and recovery rates will be as projected. Technical uncertainty is strongly influenced by the length of time that the property has produced and the quantity and quality of the information available about that property. For example, porosity derived from well logs might truly measure the pore space of a rock, yet the logs may not identify how much of that pore space is interconnected. A Styrofoam™ cup is porous; however, the pores are not interconnected, otherwise the cup would leak. Because reservoirs depend on the pores to transmit oil or gas to the wellbore, pore space makes no contribution to reserves if, like the cup, it is not interconnected. In technical terms, the rock must have permeability. As the reservoir depletes during the production process, it is often possible to analyze its performance and, hopefully, confirm that its effective size is close to that indicated earlier through the use of logs, geological interpretations, and other data. Too often however, there are surprises. When prices for oil and gas are stable and/or expectations for future oil and gas prices fall within a narrow range, technical uncertainties (reservoir fluid content, productive area, porosity, net pay thickness, water saturation, producing mechanism, recovery factor, and producing rate) dominate the concerns. Somewhat stable industry costs and prices cause the estimation of reserves and producing rates to be the most significant inputs in the evaluation of a property.

2. **Economic Uncertainty.** Instability in oil and gas prices and questions about costs and taxation have, at times, caused economic uncertainty to take precedence over technical uncertainty. Prices at which oil and gas are expected to be sold are not the only variables that can affect the economic uncertainty. Capital, operating, and drilling costs react to increases and reductions in industry activity. Drilling costs during slow periods have been known to decline to as low as 50% of the costs prevalent during boom times. Operating costs do not necessarily follow changes in oil and gas prices as is frequently assumed in projections. Cost reductions, which may reflect the availability of equipment and services when the industry is operating below capacity (possibly due to low oil and gas prices), could disappear as quickly as they appeared and must be considered in the assessment of uncertainty. Inflation and interest rates on borrowed capital also add to uncertainty.

3. **Political Uncertainty.** Political uncertainty includes uncertainty regarding local and national taxes, environmental regulations, and global concerns. Oil prices have been significantly influenced by artificial supply restraints for almost all of the industry's life. There is the risk of nationalization, operational restrictions, and social unrest in foreign host countries. Exporting countries could restrict the exportation of crude produced within their boundaries or require the producer to sell that oil domestically at prices significantly below the world price. Revenues generated from local sales might not be easily repatriated by the producing company. In other circumstances, the terms of agreements and contracts might be subject to capricious interpretations or obligations.

A balance recognizing the element of romance and the possibility for better results than expected versus the uncertainty of predicted outcomes or the risk of loss is the target of an FMV analysis. The procedures for assessing risk and uncertainty include deterministic and probabilistic methodologies. Deterministic methods can be applied to yield one or a specific set of multiple results. Probabilistic methods target a description of all possible outcomes. The question, "What is the value of a property?" implies that there is a single answer. For most business transactions a single answer is required. The most frequently used procedure to assess the value of a property is deterministic.

If only a "most likely" case is developed, the fact that it is highly unlikely that this case will actually occur is ignored. The question, "What is the range of outcome possibilities for this property?" recognizes that there is no single answer and perhaps one should look to a probability theory as an aid in business decisions. Despite the limitations imposed, the thrust of this monograph is to address deterministic procedures. The consideration of alternate outcomes in the assessment of FMV is encouraged, whether it is through the use of multiple deterministic analyses or fully probabilistic procedures. Additional discussion of probabilistic procedures is included in the SPEE Monograph 1, second edition, dated October 1998.[4]

19.6 Methods to Adjust for Risk and Uncertainty

19.6.1 Application of Adjustment Factors for Risk and Uncertainty. For the economics presented in a reserves report to be useful in estimating value (loan value, exchange value, FMV, etc.), some method of adjustment may be needed so that the economics reflect an expected case. Garb[12] indicated "adjustment factors reflecting the uncertainties attributed to the different categories of reserves can be used to index or to equate their FNR stream of the equivalent of proved producing reserves." **Table 19.2** presents qualitative observed monetary value risk factors (MVRF). It replicates the table from Ref. 4 that summarizes part of the results of the SPEE annual parameters surveys and includes a range of factors grouped by deterministic reserves category. Table 19.2 presents value ranges generally greater than those presented by Garb,[12] but the intended application is similar and the ranges in Table 19.2 are more current. The table uses the term uncertainty, but risk would have been a better term.

The footnotes state the factors "assume risk factors reported were those applied against monetary values," but it is not precisely clear what the values apply to because of the way the surveys were conducted. They undoubtedly apply to deterministic reserves categories. Consider

TABLE 19.2—QUALITATIVE OBSERVED* MVRF RANGES (after *Guidelines for Application of Petroleum Reserves Definitions*)[4]

Deterministic Reserves Category	Adjustment Factors To Reflect Uncertainty** MVRF
Proved producing	0.85 to 1.00
Proved shut-in	0.65 to 1.00
Proved undeveloped	0.50 to 0.90
Proved behind-pipe	0.25 to 0.75
Probable behind-pipe	0.05 to 0.50
Probable undeveloped	0.00 to 0.40
Possible behind-pipe	0.00 to 0.25
Possible undeveloped	0.00 to 0.20

*The adjustment factors shown were derived from surveys of SPEE members and other presumed knowledgeable and informed persons in the industry over a period of years beginning in 1982. Comparison of the surveys shows little variation in the results over the period. These factors, which are included for reference and comparison, represent an approximated one standard deviation range, and assume risk factors reported were those applied against monetary values. The 1997 SPE/World Petroleum Congress definitions take a step toward the quantification of confidence levels associated with reserves estimates. The users of these data should be aware that, despite the noted consistency in the above adjustments factors over the years, reserves estimates that have been prepared with the "probabilistic option" provided in the new definitions may have a tendency to produce results at variance with the historical trend.

**These may also be considered chance of success (COS) factors expressed in decimal form.

TABLE 19.3—PROVED UNDEVELOPED CASH FLOW EXAMPLE A

Year	Revenue ($ thousand)	Taxes ($ thousand)	Expenses ($ thousand)	Capital ($ thousand)	Cash Flow ($ thousand)	Discounted Cash Flow ($ thousand)
1	5,000	320	1,350	8,000	(4,670)	(4,934)
2	4,300	430	850	0	3,020	2,356
3	3,500	350	700	0	2,450	1,620
4	3,000	311	620	0	2,069	1,159
5	2,500	250	500	0	1,750	831
6	2,000	200	400	0	1,400	563
7	1,500	150	350	0	1,000	341
Totals	21,800	2,011	4,770	8,000	7,019	1,936

the proved undeveloped reserves category and the simple cash flow schedule presented in **Table 19.3**.

The discount rate for the present worth calculation in the right column is not shown. Assume that the discount rate used to calculate that column is the discount rate that would represent FMV, if this were a cash flow summary for proved developed producing reserves. Table 19.2 shows a range of factors from 0.50 to 0.90 for the proved undeveloped reserves category. For this illustration, we used a factor of 0.70. Applying the 0.70 factor to the $1,936,000 present worth value results in an adjusted value of $1,355,000. Is this the correct application of the method? The answer depends significantly on precisely when the decision to spend the capital must be made.

TABLE 19.4—PROVED UNDEVELOPED CASH FLOW EXAMPLE B

Year	70% Revenue ($ thousand)	Taxes ($ thousand)	Expenses ($ thousand)	Capital ($ thousand)	Cash Flow ($ thousand)	Cash Flow ($ thousand)
1	3,500	224	1,350	8,000	(6,074)	(6,227)
2	3,010	301	850	0	1,859	1,450
3	2,450	245	700	0	1,505	995
4	2,100	218	620	0	1,262	707
5	1,750	175	500	0	1,075	510
6	1,400	140	400	0	860	346
7	1,050	105	350	0	595	203
Totals	15,260	1,408	4,770	8,000	1,082	(2,015)

TABLE 19.5—PROVED UNDEVELOPED CASH FLOW EXAMPLE C

Year	Revenue ($ thousand)	Taxes ($ thousand)	Expenses ($ thousand)	Capital ($ thousand)	Cash Flow ($ thousand)	Discounted Cash Flow ($ thousand)
1	0	0	0	8,000	(8,000)	(8,000)
2	5,000	320	1,350	0	3,330	2,598
3	4,300	430	850	0	3,020	1,997
4	3,500	350	700	0	2,450	1,373
5	3,000	311	620	0	2,069	982
6	2,500	250	500	0	1,750	704
7	2,000	200	400	0	1,400	477
8	1,500	150	350	0	1,000	289
Totals	21,800	2,011	4,770	8,000	7,019	420

What if the factor of 0.70 is applicable to reserves risk and all the capital must be committed up front? If the revenues were reduced to 70%, which could be expected to be the implication of a 0.70 factor applied to reserves, the values in Table 19.3 would appear as the values in **Table 19.4** and would have no value at the discount rate used. This type of analysis is essentially a decision tree calculation of expected results, given the circumstances described.

Another possibility is that the need for adjustment may have nothing to do with uncertainty regarding the reserves quantity. **Table 19.5** presents a cash flow example and present worth calculation that is the same as in Table 19.3, except that the receipt of revenues has been delayed by one year.

Again, the result is a value considerably less than the $1,355,000 suggested in Table 19.3 after application of the 0.70 factor to the discounted value. There are other possibilities that could cause an evaluator to make value adjustments independent of concerns about the confidence associated with reserves quantities. Another crucial consideration relative to Table 19.5 would be the possibility of overexpenditure of capital cost. There are almost an infinite number of possibilities, which is why probabilistic analysis and Monte Carlo simulation can be useful decision-making tools.

There are many aspects of uncertainty other than those associated with reserves quantities. Risk and uncertainties of all aspects should be addressed. It is unlikely that subjectivity can be eliminated from the evaluation. In making deterministic adjustments for risk, it is often helpful to think about how the distributions for variables might be described in a Monte Carlo simulation and to attempt to estimate what the most likely results would be. This will not help much in determining the breadth of the value range that might be revealed by Monte Carlo simulation, but it should help in coming up with values approximating the mode. If the deterministic method is the sole means of evaluation, it is hoped that the distribution of expected values will be narrow and the mode will be very meaningful.

19.7 The Effect of Federal Income Taxes on U.S. Acquisition Economics

This section shows how federal income taxes can affect the economic evaluation process and gives the reader some understanding of how after federal income tax (AFIT) and before federal income tax (BFIT) relate in the U.S. This section does not delve into the U.S. federal tax code and the mechanics of making after tax cash flow computations, which can be done with petroleum economic software packages. Any evaluator intending to make such an analysis should consult a tax professional.

The marketplace in which oil and gas interests are traded involves transactions in countries with income taxes at the federal and, sometimes, local levels. Cash flow calculations that have been described are BFIT. When a property is purchased, it is assumed that the buyer made the acquisition with the intent of making a profit. If one could determine precisely the nature of the bottom line economic return expectations of the buyer, it would help greatly in making estimates of FMV. If such expectations were known and universal, an FMV estimation formula could be devised that would be reasonably portable around the world. It is generally perceived by observers of the U.S. oil and gas property marketplace that the impact of income taxes is sufficiently small and the variations in other variables influencing perceived value are sufficiently large that it is not normally possible, from the data available on a specific transaction, to discern (back into) the AFIT (bottom line) economic parameters of the purchaser.

If BFIT rates of return that characterized market value are known and one is attempting to transport or compare the observations to another market, can the BFIT results be backed into an AFIT bottom line ROR? For example, when making computations relative to a company's weighted average cost of capital (WACC), the results are usually first expressed in terms of an AFIT ROR. The following equations are suggested for calculating WACC:

$$C_{at} = C_{bt} \times (1 - T_R) \qquad (19.4)$$

or

$$C_{bt} = \frac{C_{at}}{1 - T_R} \qquad (19.5)$$

The relationships expressed in Eqs. 19.4 and 19.5 do not work in the evaluation of project economics because the purchase price of the property is, over time, a deductible item as a reduction of income on which tax is paid (i.e., the statutory tax rate is not the effective tax rate). **Fig. 19.1** includes three computed present worth profiles that contain many assumptions but were constructed with the intent of illustrating theoretical differences in BFIT and AFIT (U.S.) rates of return. The federal income tax rate used to generate the AFIT profiles is 35%. In preparing the economics for one AFIT case, it was assumed that 70% of the purchase price was loaned to the buyer at an interest rate of 8% and that 90% of the revenue was dedicated to loan repayment. The other AFIT case was run with similar assumptions but without the loan. The economic life of the example is 19 years. The purchase price was deducted in the AFIT calculation with the units of depletion method.

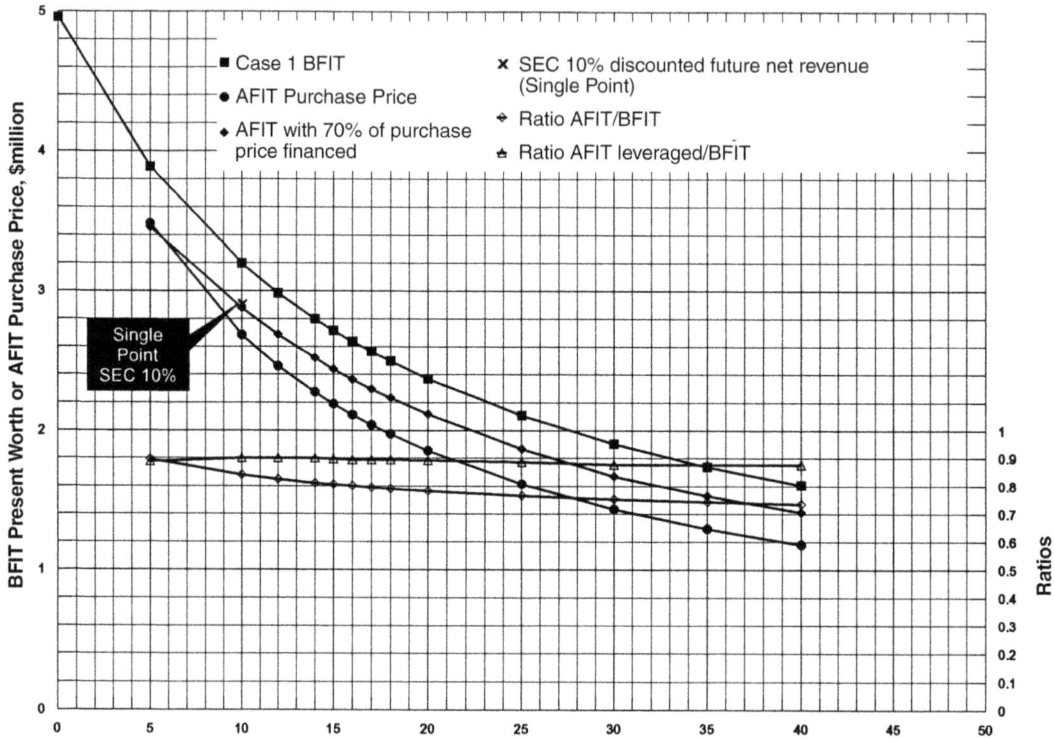

Fig. 19.1—Comparison of BFIT present worth and AFIT purchase price as functions of discount rates and ROR.

TABLE 19.6—BFIT AND AFIT RATE OF RETURN COMPARISION		
Discount Rate	BFIT Present Worth or AFIT Purchase Price (%)	Apparent Tax Burden Effect (%)
BFIT	15.4	
AFIT (leveraged)	12.0	3.4
AFIT (no leverage)	9.9	5.5

A specific purchase price can be represented by a horizontal line on Fig. 19.1 (ignore the posted lines on the graph that approach being horizontal for now and think about a horizontal line at a *y*-axis price of approximately $2,700,000). From Fig. 19.1, it appears that such a purchase could be represented by three different discount rates as shown in **Table 19.6**.

The burden imposed by income taxes in terms of the discount rate is shown to vary more than 2 percentage points simply based on one financing option. In the U.S., it is generally perceived that income taxes cause the spread between the BFIT discount rate and the AFIT purchase price to be approximately 4 to 6%. There is a price to pay for leveraging up the purchase price through financing in the manner described. The payment of interest reduces the total future net revenue and absolute profit in terms of total dollars.

Another way to compare the BFIT and AFIT calculations is by looking at ratios rather than spreads in discount rates. Fig. 19.1 plots the ratios to show how the benefits of leveraging with a loan improve, from a discount rate point of view, as the ROR increases. In the example, a loan interest rate of 8% and a tax rate of 35% are used. The leveraging effect stems from the deductibility of the interest from otherwise taxable revenues and the fact that the after tax ROR is greater than the interest rate times one minus the tax rate. [8 × (1−0.35) = 5.2%.] From Fig. 19.1, it is clear that below a 5% discount rate, the ratio of AFIT (no leverage) purchase price to the BFIT present worth is greater than the corresponding ratio with leverage, which indicates an undesirable condition for financing. If the financing costs more that the desired after tax ROR times one minus the statutory tax rate, there is no leveraging benefit. The ratios plotted on Fig. 19.1 show that as the ROR increases beyond 5%, so do the benefits of leveraging with a loan.

In the example presented by Fig. 19.1, AFIT leveraged case purchase prices bear a relatively constant ratio to the BFIT present worth values at approximately 0.9 over a wide range of values. This flatness isa usual occurrence and if it were to hold true over a wide range of conditions, it would be easy to approximate the AFIT leveraged purchase price by "haircutting" the BFIT present worth by 10%. This is only an observation at this time.

Fig. 19.1 includes the 10% discounted present worth parameters that would result from flat future pricing, as set forth by the U.S. SEC filings. This "single point" on Fig. 19.1 is posted on the vertical 10% line approximately half way between the lines connecting the Case 1 BFIT present worth values and the lines connecting the AFIT indicated purchase prices. The SEC data point plots essentially on top of the AFIT ROR at 10% (loan-leveraged case) at a purchase price of $2,900,000.

19.8 Fair Market Value

The FMV definition offered by SPEE[1] is "the price a willing buyer will pay and a willing seller will accept for a property, when the property is exposed to the market for a reasonable period and neither party has any compulsion to buy or to sell, and both have reasonable knowledge of relevant facts." The market for oil and gas properties is not static. Because it can be expected to change in the future, it is helpful to reflect on how valuations or perceptions of value have changed from the past. Perhaps what is most surprising is what little change there has been.

Table 19.7 describes the valuation methods embraced by Arps.[2] Arps' statements have been slightly modified to conform to today's terminology. Additional perspective offered by Arps is added in Italics. Each of these methods presumes that ROR is the applicable criterion (or hurdle) for valuation but that is not always the case, as is discussed later. The reason for the reverse order is that it is helpful to present the method that Arps considered "the most sophisticated approach" first so that the latter methods can be related to it. Today it is more common to refer to "present worth" as discounted future cash flow or discounted future net revenue, but there is no difference.

Any adjustments for risk suggested or implied in the parameters of Table 19.7 are management's judgment of general industry risk (perhaps inclusive of oil and gas price uncertainty) or experience and are not normally project specific unless the project is unusual in terms of a company's normal business. The risk or uncertainty associated with specific projects must be assessed on the basis of the specific merits of the individual parameters of those projects. Adjustments for project risk are discussed in Sec. 19.6.

There is debate about how overhead or general and administrative (G&A) expenses enter the cash flow projections. Some suggest that all costs and capital expenditures should be inflated to account for G&A; others argue that G&A should be a scheduled expense; and still others conclude it should not be considered at all. Some perceive that the elevation of the AFIT ROR above the cost of capital includes a component to cover G&A in addition to components for

TABLE 19.7—HISTORICAL AND MODERN VALUATION METHODS

Arps' Method Number	Arps' Description of Method	Modern Description of the Most Similar Method
3	A value equal to the present worth of the net cash flow after federal income taxes computed at an intermediate (6 to 9%) rate of interest. Arps referred to this as "the most sophisticated approach to the fair-market-value problem."	A price that yields an AFIT ROR equal to the cost of capital plus an elevation of that interest rate to cover management's perceptions of the need for profit and to account for industry risk. The AFIT ROR target is commonly in the very low double-digit range when expressed as a percent.
2	A value equal to the present worth of the net cash flow before federal taxes computed at a speculative (13 to 21%) rate of interest. This was a "not recommended method" because discounting tends to overvalue short life properties and undervalue long life properties.	The price equal to the discounted future net revenue using a discount rate determined to be applicable to BFIT cash flow projections based on observations of the market, or alternately, a discount rate sufficiently elevated above the AFIT ROR target (from Method 3) to account for income taxes. This elevation of the discount rate above the AFIT ROR target is typically on the order of 4 to 6%.
1	A value equal to a fraction (typically 0.67 to 0.75) of the present worth of the net cash flow before federal taxes computed at a safe (4 to 6%) rate of interest that approximates the cost of money. This method was "simple, easy to understand, and therefore widely used" according to Arps.	A price calculated using the value of discounted future net revenue at a discount factor equivalent to an AFIT ROR target, then adjusting (haircutting) that value using a factor to account for income taxes. The factor is typically in the range of 0.75 to 0.90.

profit and industry risk. In practice, cash flow schedules in the majority of reserves reports do not include overhead expenses except for those billed to nonoperating working interest owners by the operator of the property. Such charges are frequently referred to as COPAS charges because of the manner in which the rates are established.

Method 3 is used today essentially unchanged except that the discount rates are higher because of differences in the cost of money and industry risk. Method 3 is used often internally by companies to arrive at a price they might offer to a potential seller, but it is seldom used by outside parties to estimate value. There likely will be several more columns of detail in the economics than shown by Table 19.1. There are many details, and people want to see that detail if it has been judged to be worth that much work. The basic calculations are handled easily by most petroleum economic software packages, but one must take care to model the company's internal cash flow mechanics. The need for detail is partly because the method is used to investigate the leveraging benefits of financing. The benefits of the use of loan leverage to increase the offering price while maintaining the ROR were illustrated previously. Such leveraging has the disadvantage of reducing the undiscounted projected cash flow because of the payment of interest. If Method 3 were to be used for a purpose other than internal use, it probably would be for estimating FMV while speculating as to complexion of the likely acquirer. Similarly, a third party could be investigating how a recent change in tax rates or structure might affect those in the acquisition market.

There is always interest in the components of the AFIT ROR target discount rate. Generally, the components are expected to include the cost to borrow money and/or acquire capital, a margin to cover overhead and profit, and management's safety factor (referred to as industry risk component).

Method 2 uses an elevated discount rate to adjust for federal income tax effects. Arps classed this method as "not recommended."[2] The primary reason for Arps' criticism is that when the discount rate is elevated, the value of cash projected to flow at times more distant into the future is reduced more than warranted in comparison with Method 3. David and Hickman[17] wrote about a market for long-life reserves in 1990; however, others suggest that there is greater uncertainty related to those more distant projections of cash flow and that such reduction is warranted. When Arps wrote the 1962 chapter from a U.S. perspective, production levels were often regulated and projections of production and future cash flow were more often expected to be constant at existing levels for a significant time.

Method 2 is the most popular method today. Its modern popularity seems to stem from at least two sources. In 1982, SPEE began an annual publication that is currently titled *Survey of Economic Parameters Used in Property Evaluation.*[18] The questionnaire for that survey has inherently embraced Method 2 from inception. The publication has gained wide use and acclaim and is a source to those needing quantification of the discount factor required to apply Method 2. Another reason for Method 2's popularity is that it requires both a discount rate and a factor. Method 2 results in a simpler expression with a single parameter. Additionally, taxing authorities at state and county levels often require an expression of value in terms of the Method 2 discount rate, which might have had some bearing on the original construction of the SPEE questionnaire.

Method 1 uses discounted cash flow to estimate value, but a factor is applied to the discounted future net revenue to account for income taxes. In the Method 1 described by Arps, a "safe" discount rate was selected and value was determined by a factor that had a midrange of 0.71.[2] Arps' application of Method 1 used a discount rate lower than the rate used in Method 3. The modern application of Method 1 (although it is not used very much) uses the AFIT ROR target rate from Method 3 and a higher factor, the midrange of which is 0.825 from Table 19.1.

Because the modern application is to use a higher discount rate in comparison to the cost of money at the times being considered, a higher factor representing a smaller reduction in the discounted values is perhaps to be expected. Another factor that might bear on the difference in the factor is the U.S. federal income tax rate and structure. When Arps wrote the 1962 chapter, the income tax rate for corporations was 50% but there was a financial arrangement available known as an ABC transaction. Today the tax rate is lower and the ABC transaction was eliminated by Congress in the late 1960s. It is not clear how historical modification of tax rates and laws have affected market values.

FMV estimation has thus far focused on ROR considerations and the present worth of projected cash flow streams. That perspective is always important but is not necessarily the controlling factor. Other economic parameters that measure the nature of a contemplated investment also affect market value. The critical ones are payout, which is the length of time required for recovery of the investment (purchase price) from the projected cash flow stream, and return on investment (ROI), which is the projected future cash flow or FNR divided by the investment (in this case, the purchase price). Payout for oil and gas property acquisition is normally expected to be 5 years or less. Some suggest that this limitation is not a true evaluation parameter but the result of the application of other criteria

ROI is probably, or should be, a controlling parameter in many valuations. Arps,[2] in commenting on one of his own studies of 34 property transactions made during the postwar years in the Mid-Continent, Gulf Coast, and California, indicated "in none of these transactions did the total consideration exceed two-thirds of the undiscounted future net cash income before federal taxes." To be in line with this two-thirds observation by Arps, the ROI must be greater than 1.5. Many[12] set higher limits, as high as 2 or 3, while others find 1.7 to be an acceptable minimum. Sometimes a correlation is perceived between ROI and acceptable minimum ROR. ROI is a good indicator of the susceptibility of an investment to risk. If reserves (or any other parameter factored into the cash flow calculation) are missed by 25% on a project evaluated with an ROI of 2, the economic result will probably be tolerable. If the originally projected ROI was 1.5, the results will be embarrassingly poor. Low ROI acquisition opportunities are usually of quite short life and often seem attractive because of short payout times. Investors willing to take such risks need to have cash because the value of the asset diminishes so rapidly that if any of the purchase is financed (with the properties being acquired put up for collateral), the loan value will be only a small fraction of the purchase price. When ROI is the parameter controlling the purchase price, the ROR will often seem to be abnormally high.

Estimates of FMV are probably best made with the insight derived by observing the characteristics of marketplace transactions. Rather than direct use of comparative sale methods, as might be done in real estate, it is better to use ROR, ROI, and payout as tools in the comparative process. Even then, adjustments must be made that account for differences in circumstances such as tax credits, tax structure, production-sharing contract terms, and risks. The parameter used as a basis for comparison most often is the ROR. Discussions over a property's value, if the risk components can be dealt with, often are reduced to what ROR represents value. The SPEE FMV document[1] suggests that the applicable BFIT ROR in the U.S. correlates with the 30-year Treasury bond rate according to the following formula:

$$i_R = 1.72 \times (R_{tb} + 2.87), \quad (19.6)$$

as a percent where R_{tb} = 30-year U.S. Treasury bond rate expressed as a percent.

This relationship was derived from a 20-year review of SPEE's annual *Survey of Economic Parameters Used in Property Evaluation.*[18] The average BFIT ROR indicated by survey respondents in the summer of 2001 was 15.4%, and the mean was 15.0%. These survey results included participants outside SPEE. The membership was surveyed through an example problem to solicit opinions with a described set of circumstances to convey high levels of confidence. This survey resulted in a mode ROR of 19.2% in a narrow distribution. The results seem to be inconsistent, but the difference probably lies in perceptions of risk.

Nomenclature

C_{at} = weighted average cost of capital after federal income tax
C_{bt} = weighted average cost of capital before federal income tax
D_f = discount factor for a specific point in time
e = base of the natural logarithm
i_R = ROR before federal income tax
r = annual rate of discount expressed as a fraction or percent
R_{tb} = 30-year U.S. Treasury bond rate, expressed as a percent
t = time in years from the as of date to the point of reference for discounting
T_R = tax rate

References

1. *Perspectives on the Fair Market Value of Oil and Gas Interests,* D.R. Long (ed.), SPEE, Houston (2002).
2. Arps, J.J.: "Valuation of Oil and Gas Reserves," *Petroleum Production Handbook,* T.C. Frick and R.W. Taylor (eds.), SPE, Richardson, Texas (1962) **2,** Chap. 38.
3. *SPEE Recommended Evaluation Practice Program,* SPEE, www.spee.org/rep-main.htm.

4. *Guidelines for the Application of Petroleum Reserves Definitions,* second edition, Monograph Series, SPEE, Houston (1998) **1.**
5. Lohrenz, J. and Dickens, R.N.: "Option Theory for Evaluation of Oil and Gas Assets: The Upsides and Downsides," paper SPE 25837 presented at the 1993 SPE Hydrocarbon Economics and Evaluation Symposium, Dallas, 29–30 March.
6. Walkup, G.W. Jr. and Chatwin, R.E.: "Case Studies of a Real Option Approach to Asset Valuation in the Petroleum Industry," paper SPE 52952 presented at the 1999 SPE Hydrocarbon Economics and Evaluation Symposium, Dallas, 20–23 March.
7. Hooper, H.T. III and Rutherford, S.R.: "Real Options and Probabilistic Economics: Bridging the Gap," paper SPE 71408 presented at the 2001 SPE Annual Technical Conference and Exhibition, New Orleans, 30 September–3 October.
8. Luehrman, T.A.: "Extending the Influence of Real Options: Problems and Opportunities," paper SPE 71407 presented at the 2001 SPE Annual Technical Conference and Exhibition, New Orleans, 30 September–3 October.
9. Markland, J.T.: "Options Theory: A New Way Forward for Exploration and Engineering Economics?" paper SPE 24232 presented at the 1992 SPE Oil & Gas Economics, Finance, and Management Conference, London, 28–29 April.
10. Zettl, M.: "Extending the Option Pricing Theory for the Valuation of E&P Projects," paper SPE 62968 presented at the 2000 SPE Annual Technical Conference and Exhibition, Dallas, 1–4 October.
11. Faiz, S.: "Real-Options Application: From Successes in Asset Valuation to Challenges for an Enterprisewide Approach," *JPT* (January 2001) 42.
12. Garb, F.A.: "Which Fair-Market-Value Method Should You Use?" *JPT* (January 1990) 8.
13. Wright, J.D. and Thompson, R.S.: "A Comparative Analysis of 12 Economic Software Programs," paper SPE 68588 presented at the 2001 SPE Hydrocarbon Economics and Evaluation Symposium, Dallas, 2–3 April.
14. Riseden, R.: "Valuation of Royalty Interests," presented at the 2002 SPEE Annual Meeting, Park City, Utah, June.
15. "SPEE Oil & Gas Production Taxes, Section 29 and COPAS," SPEE, www.spee.org/productiontaxes.htm.
16. Doering, M.A.: "Acquisition Pitfalls: Operating Cost Forecasts," paper SPE 25841 presented at the 1993 SPE Hydrocarbon Economics and Evaluation Symposium, Dallas, 29–30 March.
17. David, A. and Hickman, T.S.: "What Is the Market Value of Long-Life Reserves?" paper SPE 20508 presented at the 1990 SPE Annual Technical Conference and Exhibition, New Orleans, 23–26 September.
18. "Survey of Economic Parameters Used in Property Evaluation," presented at the 2001 SPEE Annual Meeting, San Diego, California, June.

Appendix A—Description of Asset

This appendix is a reproduction of Appendix A from Ref. 1. It provides a general description of typical contractual arrangements for fractional ownership and/or beneficial rights.

A fundamental step in any property evaluation is a description of the asset. In the determination of FMV, the evaluator must be certain that the mineral interest is correctly described. The lease or production sharing agreement and any other significant issues relevant to ownership need to be reviewed and understood. If the asset is a piece of real estate, the physical dimensions and a complete description of the property are prepared along with a description of any structures that may be present. If the asset is the inventory of a retail business, then the stock must be counted and categorized. In these instances, the assets can be accurately described and quantified with little doubt as to their existence. In the oil and gas industry, however, the situation is somewhat different.

When evaluating oil and gas assets in the form of reserves, the process of describing the asset requires an estimate of the volume and quality of oil, gas, and by-product reserves that may be recovered in the future from subsurface reservoirs. These products are not available at the surface to be examined. One aspect of oil and gas property ownership is that ownership is usually of a fractional nature. Even if the property ownership is in the U.S., taxing authorities will somehow share in the revenue stream.

Types of Oil and Gas Property Ownership. The most common types of oil and gas property interests or ownership are

A. Working Interests

B. Royalties (landowner's, overriding, and sliding scale)

C. Net Profits Interests

D. Production Payments

E. Production Sharing Arrangements

The following discussion defines these types of contractual arrangements and, in addition, includes common variations of the basic oil and gas interest types.

In the United States and to a great degree in Canada, the minerals are deeded privately and, at least at the origin of the country's claim to lands, were transferred and/or deeded in tandem with ownership of the surface of the land. In almost all other parts of the world, the mineral ownership is the property of the government of the country and is separate from the private surface ownership, if any. All the indicated ownership types could be broadly classified as production sharing arrangements because all are strictly contractual in nature.

Today in the United States and Canada it is not uncommon to find that the surface and mineral interests have been severed through contractual agreement. Such severance occasionally (frequently) is the root of disputes when the mineral owners require access to the minerals through surface locations.

A. Working Interests. A working interest owner has an obligation to pay a share of costs. That share of costs is sometimes referred to as an expense interest. In some groups within the industry, the term working interest is synonymous with the term expense interest. Working interests receive a "revenue interest," also referred to as a net revenue interest (NRI) or division order interest. An NRI is a fractional interest in the gross revenue from the sales of oil, gas, and sometimes other products from a property. The NRI is normally less than expense interest due to royalty interest burdens.

Types of working interests are

1. Working Interest—This is the lessee's or operating interest (conventionally equal to the expense interest) under an oil and gas lease, which is also called a leasehold interest. A working interest is created by a lease agreement or a deed. Its primary characteristic is the right to enter the property to explore, drill, and conduct production operations. The working interest can be owned by one entity or partitioned between several entities as in a joint venture or unit operation. If there is more than one entity, one party is usually designated as the "operator." The other working interests, if any, are "nonoperating" working interests. The working interest fractions add up to 100%.

2. Carried Interest—A carried interest is a fraction of the total working interest in an oil and gas property that is not required to pay development and, in some cases, operating costs. The carried interest portion reflects the lack of participation in all or a portion of capital costs. The capital requirement normally attributable to the carried interest's working interest position is, through agreement, paid by others, normally by one or more of the other working interest owners putting up a disproportionate share. The classic reason for one of the working interest owners to carry the other is a means of buying one's way into participation. This type of buy-in is sometimes referred to as a "promote." In some provisions for carried interests, the owner of the carried interest does not receive any income until the other working interest owners that put up the disproportionate share receive the return of their capital investment or an otherwise agreed on sum of money, which might be considerably in excess of the disproportionate

amount put up to carry the carried interest. A carried interest may apply to one or more wells, the lease, or to the prospect.

3. Reversionary Interest—In some situations, a portion of the working interest (the expense part, the NRI part, or both) reverts to another party when a specified condition occurs. The carried interest described above is a form of a reversionary interest. The condition that triggers the reversion often is occasioned by a "payout" of an investment. There may be a change in the working interest, or an overriding royalty could become a working interest. Multiple reversionary interests with different triggers often apply to a single property.

4. Terminable Interest—A terminable interest is a working interest that terminates on the payment of specified charges from the production revenue. This is another description of a class of reversionary interest.

B. Royalty Interests.

1. Landowner's Royalty Interest—This is the landowner's share of production retained in the lease agreement. If the minerals have been severed from the surface, it is the interest retained by the mineral interest owner. This income is free from capital investment and from most operating expenses except for a pro rata share of production and severance taxes. Sometimes the lease requires the royalty to bear its share of power, substitute fuel, marketing costs, and other costs to make the produced oil and gas marketable substances. Other such costs include compression, dehydration, and the removal of impurities.

2. Overriding Royalty Interest (ORRI)—An overriding royalty interest is in addition to the landowner's royalty interest, and its terms and conditions are similar to the "basic lease" unless otherwise specified. An ORRI is carved out of the working interest. It usually represents a payment to a middleman for services rendered. The ORRI holder's benefit of ownership generally expires with the expiration of the lease to which it is associated.

3. Drillsite Royalty—A drillsite royalty is paid for the use of a parcel of land for drilling and production operations for directionally drilled wells when the completion interval is not under the drillsite tract.

4. Compensatory Royalty—This is also called an "offset" royalty. This is a payment from the lessee to the lessor in lieu of drilling an offset well. This usually occurs when the lease would require the drilling of an "offset" well, but the lessee does not consider the drilling of the required well to be economic.

5. Shut-in Royalty—A royalty is paid (usually in the form of a periodic fixed amount stipulated by the lease agreement) when a gas well, capable of economic production, is shut-in because of a lack of a market for the gas (e.g., pipeline is not yet installed).

6. Term Royalty—Any royalty on production from a lease that has a fixed duration in time (i.e., 20 years) is a term royalty. Most royalties are associated with leases that are evergreen (i.e., the lease continues as long as production is in paying quantities), and those royalties are not term royalties.

7. Minimum Royalty, Sliding Scale Royalty, Step-Scale Royalty—These royalties reflect minimum or varying royalty rates, which can depend on such variables as (1) total oil (or gas) rate, (2) average daily rate per producing well, and (3) the oil (or gas) price.

C. Net Profits Interests. A net profits interest is a share of the revenue from a well or property that bears a share of expenses and a share of the income, often very similar to a non-operated working interest. The holder of the net profits receives income when the operation to his position is profitable but has no (or a modified) cost burden when the income from the sale of production is insufficient to cover costs from operating a well or property. Net profits interests are carved out of the working interest. The net profits can be on an operating cost basis only or may also reflect the recovery of capital costs. The structure of net profits interests are often such that costs from periods of non-profitable or negative cash flow operations are paid from later periods of operations when the operations are profitable or have positive cash flow. One of the working interest owners, frequently the one from which the net profits interest was carved, administers the funds of the net profits interest account and, on the occasion of negative cash flow, at least temporarily absorbs the loss. Net profits interests may have tax (U.S. federal income tax) consequences that are identical to a working interest with similar associated expense and revenue interest fractions.

D. Production Payment. A production payment describes a financial arrangement wherein the owner of the production payment is entitled to a specified portion of the oil or gas production for a limited period of time or, as in most cases, until a specified condition is satisfied [normally related to when an amount of money (plus interest) has been received]. This income is free of production costs. A production payment is often used in financing an acquisition or a portion of the sales price to the seller with a bank loan. It is sometimes a method of "nonrecourse" financing, wherein the production payment holder can look only to the property to satisfy the loan.

There are two basic types of production payments, the "reserved" and the "carved out" payment. It is "reserved" when the owner of an interest assigns all or part of his interest and obtains an oil payment for a specified sum of money or a specified volume of production (barrels or MCF). A "carved out" payment is one in which a party assigns only a payment of a specified sum of money (usually with interest) out of a certain share of the grantor's production revenue. However, none of the grantor's interest in the property has been assigned, only a certain sum of money. A production payment may be "deferred" and not begin until after the operator has realized a certain sum of money, or until a specified period of time has elapsed.

E. Production Sharing Arrangements. Production sharing agreements can have terms similar to production payments. However, production payments are normally structured to satisfy a previous monetary commitment, while a production sharing arrangement is normally structured to compensate and provide incentives for an operator to perform under a contract to produce a government's minerals.

SI Metric Conversion Factors

bbl	× 1.589 873	E − 01	= m^3
ft^3	× 2.831 685	E − 02	= m^3

Chapter 20
Reservoir Management Programs

E.D. Holstein, Consultant and **E.G. Woods,** Consultant

20.1 Introduction

Reservoir management has been in place in most producing organizations for several years. Several authors[1–8] have described how reservoir management is structured; however, the type, quality, and consistency of programs vary. This chapter defines reservoir management and suggests how to maintain an effective, ongoing program that can be sustained and continually updated to represent the changing needs of an organization or resource.

Reservoir management consists of processes that require the interaction of technical, operating, and management groups for success. The complexity of the problem and size of the asset dictate the type and number of personnel assigned to the task. Commitments can vary from part-time assignments for technical and operating staff members to the full-time use of multifunctional and, in some instances, multiorganizational teams. Personnel changes, altered priorities, insufficient surveillance data, and lack of documentation, however, can reduce the effectiveness of reservoir management programs.

Methods for assessing the effectiveness of reservoir management programs, including identifying strengths and areas for improvement, are needed to approach the topic from a quality perspective (i.e., benchmark to an ideal, best-practice standard). Making these assessments on a systematic, regular basis can be effective in developing a common terminology that improves communication and in ensuring a comprehensive review and a more complete listing of improvement opportunities. Reservoir management assessments are also effective in providing a comparison with ideal or best practices that result in a more innovative environment and in establishing a method of documentation and measurement to determine how well reservoir management is being sustained despite changes in personnel and priorities. This chapter includes a method for assessing the quality of a reservoir management program.

20.2 Reservoir Management Processes

Fig. 20.1 illustrates reservoir management processes. The processes are divided into those stewarded by the reservoir management team (RMT) and those guided by the supervisors and managers associated with reservoir management who comprise the reservoir management leadership team (RMLT). The arrows in the RMT box show work flow and how data and opportunities are captured.

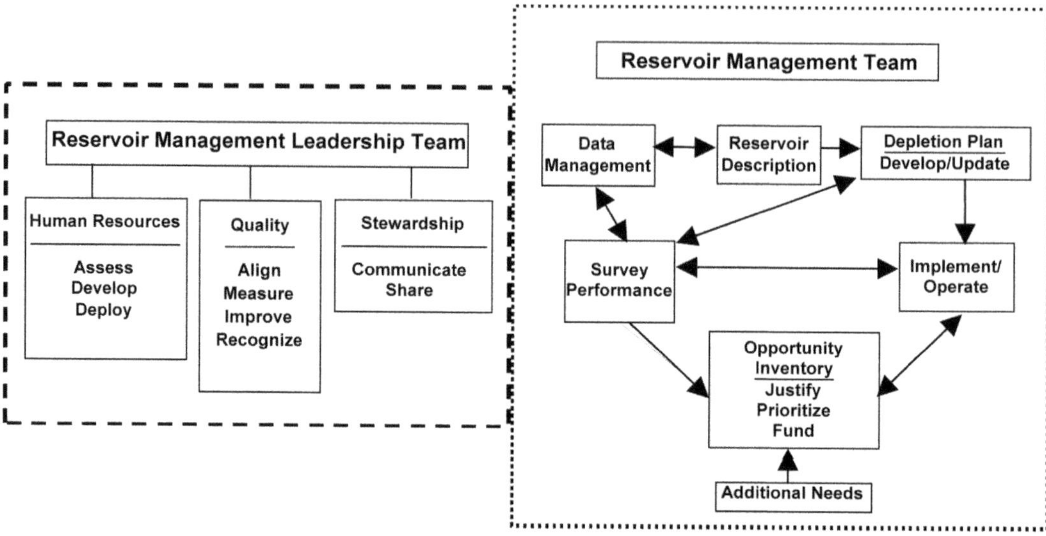

Fig. 20.1—Reservoir management processes.

20.2.1 Reservoir Management Team. The RMT members are the technical and operating personnel who perform reservoir management tasks. **Fig. 20.2** shows the skills represented by the team members. The RMT is a multifunctional team, and organizational structure should not be inferred from this illustration. A team with all the skills shown could be a permanent part of an organization; however, it is more likely that the team would meet on a regular basis with individual members being assembled as needed from groups within an organization. In-house skills may need to be supplemented with outside personnel. The objective of the team is to bring together the skills needed to describe the reservoir, prepare depletion plans (including economic justification of projects), drill the wells, design and maintain wellbores, design and maintain production equipment, and conduct the day-to-day operations of producing the field according to the depletion plan. The team also meets to provide the information needed to upgrade and improve the depletion plan. A major focus of the team is to obtain reliable data on a timely basis to analyze production performance.

Data Management. This process represents the organizing of raw and interpreted data into a readily accessible form. It is not intended to imply what type or quantity of data is needed. Those issues are addressed in other processes.

Data Captured. This information includes raw data such as seismic records, well logs, conventional and special core analyses, fluid analyses, static pressures, pressure-transient tests, flowing pressures, periodic well production tests, and monthly produced volumes of oil, gas, and water. Interpreted data could include seismic time maps, seismic conversion of time-to-depth maps, seismic attribute maps, log analyses, formation tops, structure and isopach maps, cross sections, geologic models, and simulation models.

How much information and how to capture this information varies with the size of the database, size of the resource, and the remaining life of the resource. Hand-kept records and hard copies of information may be adequate for small resources. However, digital databases should be considered for all resources for the systematic acquisition of data, the growing usability of software for data interpretation, and the value of having data available to individuals in a distributed network.

Quality Assurance. Processes for the timely capture and quality maintenance of data also should be established. Personnel may be required for this specific purpose. While this assign-

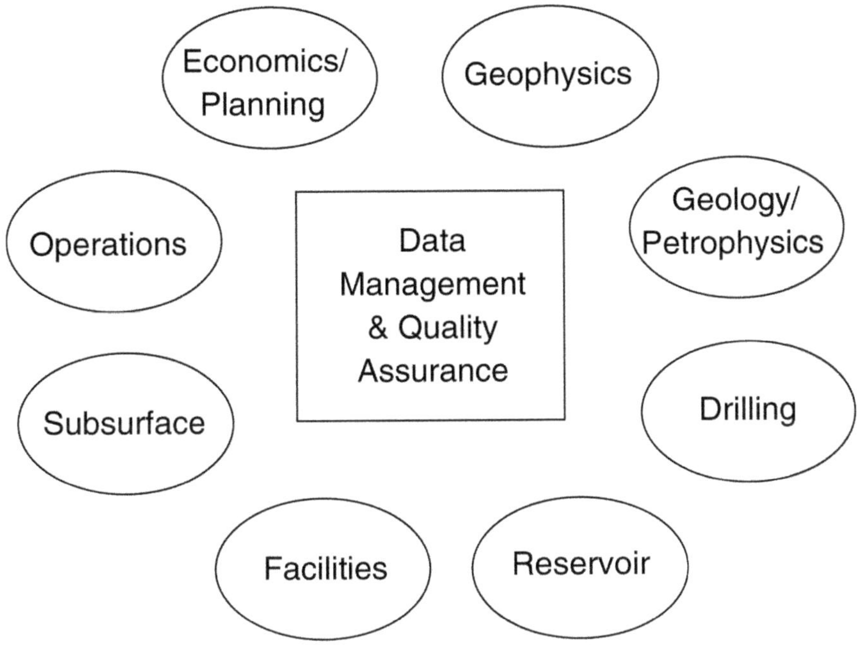

Fig. 20.2—Reservoir management team skills.

ment may be a drain on limited manpower, the benefits of readily available, high-quality data will save time spent in reorganizing, checking, and reinterpreting data each time a study is conducted. The time savings more than returns the cost of quality data capture. Studies of work output indicate that as much as 50% of the time spent on a project can be consumed by finding and organizing data that is not maintained in a readily accessible, high-quality format.

Reservoir Description. This process is the development of an up-to-date, detailed description of the reservoir that incorporates available data and technology into a fieldwide interpretation consistent with observed historical reservoir performance. Variations and risks in the description should be included. Again, the effort that goes into this description depends on the size of the remaining resource.

Geophysical, geological, and engineering interpretations are expected to produce information on the distribution of hydrocarbons in place and reserves. These interpretations include field and regional structure maps, including fluid-contact locations and the size of aquifers; isopach and porosity maps; the number of flow units or individual producing zones; the depositional environment including information on diagenetic changes and vertical and areal barriers to flow (or lack thereof); and variations in fluid saturations and permeabilities. The expected variability in these values should be included in these assessments. Descriptions from hand-drawn maps and correlations may suffice for small resources; however, in most cases, a geologic model is developed to capture these interpretations, with more complex models being needed for larger resources. The power of PCs and their software makes it more attractive to develop geologic models for all resources. (See the chapters on reservoir petroleum geology, geophysics fundamentals and outputs, and petrophysics in this section of the *Handbook* for information on geophysical, geological, and petrophysical interpretations and how they are used to develop a reservoir description. See the chapters on reservoir geophysics and geostatistics in the Emerging and Peripheral Technologies section of this *Handbook* for information on emerging technologies that can improve reservoir description.)

Original In-Place Volumes. Initially, the geologic model is used to estimate the amount and distribution of original in-place hydrocarbon volumes. These estimates include the range of uncertainties in rock properties, fluid saturations, and geologic interpretations and the resulting range in estimated in-place volumes.

Fluid-Flow Characteristics. The mapping of depositional environments, flow barriers, and flow test and core data aid in understanding the productivity and recovery trends in the reservoirs. This understanding is important in optimizing well placement and spacing and in recovery process selection.

Updating. Periodic collaboration between geoscientists and engineers is needed to include new seismic data and interpretations, well data, and performance characteristics into the geologic model. This work produces a better description of reservoir contents, reduces uncertainty, and establishes a basis for improved future development and reservoir operations.

Depletion Plan Development and Updating. Depletion plans define how to use primary drive mechanisms to deplete hydrocarbon resources and how, when, or if these mechanisms should be supplemented for additional recovery. The plan includes projected ultimate recoveries; producing rates of oil, gas, and water; and changes in reservoir pressure. As information is gained from field performance, the depletion plan is updated periodically to include any changes needed to better reflect how to optimize the depletion strategy. Specific parts of the plan include drilling schedules, well placement, individual well and total field offtake rates, total and well injection volumes, and wellbore utilization plans.

Determine the Primary Drive Mechanism. Determining the primary producing mechanism is the first step in selecting a depletion strategy. The chapters on oil reservoir primary drive mechanisms and gas reservoirs in this section of the *Handbook* describe primary drive mechanisms for oil and gas reservoirs, recovery potential, and methods to determine drive mechanism. This information can be used to project oil-, gas-, and water-producing rates, reservoir pressure trends, and ultimate recovery.

Well Placement and Well Density. The drilling and completion schedule and the number and placement of wells are a function of several factors:

• The total field offtake rate and individual well rates that can be sustained without ultimate loss of recovery.

• The number of wells and rates that are practical considering constraints such as the number of well slots on a drill pad or an offshore platform; the installed facility capacity; individual well capacities with the tubulars, completion techniques, and artificial lift used; and any regulatory limits on spacing and/or producing rates.

• The location of the wells for efficient drainage, that is, spaced evenly to contact all portions of the reservoir or targeted to specific areas because of reservoir geometry, quality variations, or invading water or gas. The amount of the reservoir that can be contacted is affected by limits on lateral drilling reach from existing platform or drill pads.

Infill drilling also should be considered at some point to sustain rates or contact portions of the reservoir inadequately drained with existing completions. The chapter on water injection in this section of the *Handbook* describes the benefits of infill drilling.

Horizontal/Multilateral Wells. Horizontal and multilateral wells increasingly are being used to contact more of the reservoir, to achieve higher rates with fewer wells, and to minimize coning of gas and/or water. Continuing technologic improvements are making horizontal wells an increasingly more cost-effective way to develop certain portions of a reservoir. The chapter on fluid flow through permeable media in this section of the *Handbook* covers performance characteristics of horizontal wells. It is now possible to simulate this performance with relatively simple models to determine the benefits of planning such wells for depletion of a reservoir. Horizontal wells are particularly effective in producing zones that have good vertical permeability, well-managed contact movements, and large drainage areas. These wells also have been

TABLE 20.1—SCREENING CRITERIA FOR SELECTING IMPROVED RECOVERY PROCESSES

Process	Rock Type	Average k md	Depth ft	Oil Viscosity cp	T_R °F
Waterflood	Either	>5*	NR**	<100	NR**
Immiscible					
Hydrocarbon	Either	>1000	NR**	<20	NR**
CO_2, N_2	Either	>1000	NR**	<20	NR**
Miscible					
High-pressure hydrocarbon	Either	All	>5,000	<5	NR**
Enriched hydrocarbon	Either	All	>3,000	<5	NR**
CO_2	Either	All	>3,000	<10	NR**
N_2	Either	All	>6,000	<5	NR**
Thermal					
Steam	Either	†	200 to 5,000	>20	NR**
Combustion	Sandstone	†	>1,000	NR**	NR**
Chemical					
Polymer	Either	>100	NR**	<40	<200
Alkaline	Sandstone	>100	NR**	<40‡	<200

*Can be <1 md for carbonates; **NR = no restriction; †kh/μ >100; ‡acid number>0.2.

effective in fractured formations in which the horizontal well intersects more fractures than a vertical wellbore and, therefore, drains a greater volume of the reservoir. Horizontal wells are not as well suited to zones with low vertical permeability, rapidly moving contacts, and small remaining volumes to be drained. Where vertical permeability is low, high-angle wells may be more appropriate to contact the vertical interval and still provide a large drainage area.

The Need for Improved Recovery Projects. Improved recovery mechanisms refer to the injection of fluid (i.e., water, water with additives, hydrocarbon gas, nonhydrocarbon gas, steam, or air for in-situ combustion) to augment or replace the primary drive mechanism. **Table 20.1** presents screening criteria for selecting improved recovery processes.

The chapters on water injection; immiscible gas injection; foam, polymer, and resin injection; miscible processes; steam; and in-situ combustion in this section of the *Handbook* describe the characteristics and potential for several injection schemes in oil reservoirs. The chapter on gas reservoirs in this section of the *Handbook* explains the benefits of injection into gas caps and gas reservoirs. Historically, a field development progressed from primary production to fluid injection, such as water or immiscible gas, and then, in some instances, a second type of fluid injection, such as a miscible injection project, was needed. It is now important to determine the need for injection projects as early as possible to minimize depletion times, provide space for necessary equipment, avoid retrofitting facilities, and avoid other costly intermediate steps.

Wellbore Utilization Plan. This plan identifies the intermediate and final drainage points in each zone and outlines how each well will be used to deplete each producing zone and reservoir in a field. The plan includes guidelines on when to rework completions to sustain production by avoiding unwanted, excessive gas or water volumes. Where multiple producing zones exist, the plan should describe the zones to be completed in each well and provide guidelines on when to recomplete and the sequence of those recompletions to provide efficient recovery in each zone and minimize overall depletion time of the total resource.

Data Acquisition. The depletion plan should include the type of data to be acquired during the development stages of the reservoir and during the early, middle, and final production phases. Such data plans should include the type and number of open- and cased-hole logs; number, location, and frequency of static and transit pressure tests; number and location of fluid samples; number and location of cores and analyses to be made; type, location, and frequency of production logs; frequency of individual well tests; and capture of monthly produced volumes of oil, gas, and water and monthly injected volumes. The chapters on petrophysics, production logs, single well chemical tracer testing, interwell tracer tests, and measurement of reservoir pressure in this section of the *Handbook* contain more information on these data.

Reservoir Models. Most depletion plans are based on some type of reservoir model.

Model Types. Reservoir models are basic tools for addressing reservoir management questions and issues. In selecting a model, it is normally desirable to select the simplest model that will give reliable results (i.e., selecting a model that will adequately discriminate among alternatives and lead to an optimal decision, although absolute results may not be precise). Several types of models of varying complexity are available that may be adequate for different uses. These models include analog, decline curve, analytical (material-balance, Darcy-law, Buckley-Leverett, pressure-transient), small numerical (well, cross-sectional, pattern-element, 3D-segment), and large-scale, full-field models. See the chapter on reservoir simulation in this section of the *Handbook* for details on building reservoir simulation models. Other chapters in this section also contain guidelines for models specific to the subject of the chapter.

Reservoir Issues. The first step in model selection is to identify the questions to be answered and their relative importance. The following issues must be addressed during this step.

- Exploration prospect forecast of oil, gas, and water production.
- Annual forecasts of oil, gas, and water production.
- Monthly tanker scheduling and storage requirements.
- Pressure maintenance requirements.
- Evaluation of alternative recovery processes: gas-cap expansion; natural waterdrive; and water, gas, or other fluid injection.
- Operational guidelines for pressure levels, injection volumes and distribution, and individual well and field total production targets.
- Well performance predictions: coning, artificial lift requirements.
- Stimulation evaluation.
- Gas- and water-handling requirements.
- The need for and timing of reservoir depressuring.

Model Description. A second consideration in model selection is deciding which primary forces will dominate reservoir performance. It must be determined whether viscous, gravity, or capillary forces, as reflected in coning, gas overrun, water underrun, or pressure drop, will dominate reservoir and well performance.

Model Data. Most models require at least some data describing fluid properties and reservoir description and may require multiphase flow (relative permeability and capillary pressure) and well performance [coning correlations, gas/oil ratio (GOR), water/oil ratio (WOR)] functions. Based on experience, certain simplifying assumptions may be acceptable. For example, if the reservoir description is dominated by a fining or coarsening upward depositional sequence, this may be more important than capturing the areal variation in reservoir description.

Case Design. Careful thought should be given to identifying cases to be run with the model to avoid running all combinations of the variables being studied, a number that can run in the several thousands for even modest-sized resources. In some cases, this may involve starting with a simple model to test the importance of some variables. For example, before building a full-field model, it can be helpful to build well, cross-sectional, 3D-segment, or pattern-element models.

Small Models. Coarse-grid 2D or 3D simulation models are useful for making a distributed volumetric balance in cases in which reservoir fluid migration and/or significant pressure gradients are issues. 3D models are useful for doing a multizone volumetric balance in which there is fluid migration between zones through commingled wellbores, along fault planes, or across fault planes by sand-on-sand contact. Also, these models are useful for regional aquifer modeling in which aquifers are irregular in shape, heterogeneous, or subject to pressure interference between fields.

Large Models. Large simulation models with more than 100,000 gridblocks are now constructed for many medium and large fields. The models usually are based on a detailed geologic model that may contain one million or more gridblocks. Methodologies are now in place to convert these geologic models into a more manageable reservoir model while retaining a good representation of the variation in reservoir characteristics. See the chapter on reservoir simulation in this section of the *Handbook* for more information on building and running reservoir models.

Implement and Operate. These are the specific activities required to implement and sustain the depletion plan.

Implementation Plan. This plan is developed with input from several engineering disciplines. The drilling engineers design drilling schedules, wellbore trajectories, and casing and cementing programs to locate wells at the required location for the lowest effective cost. The subsurface engineers design tubulars, completion techniques, stimulation processes, and artificial lift equipment required for target well rates. They also design workover procedures as needed. The facilities engineers design equipment needed initially to handle produced and injected fluids and provide plans for upgrades and additions when required. For offshore locations, the platform design engineers provide adequate sizing to accommodate wells and facilities within budget and physical restraints. The reservoir engineers assist in optimizing the drilling schedule to conform to available facility capacity and injection requirements, in providing well rate forecasts to aid well design, and in providing multiple production scenarios to aid facility design.

Need for Iteration. There is interdependence between depletion and implementation plans. Iteration normally is required to find the optimum combination of reservoir depletion objectives, equipment and completion constraints, and economic guidelines that result in optimum field development. For example, favorable mobility-ratio waterdrive reservoirs tend to have low sensitivity to rate of recovery; therefore, there is an economic tradeoff between the facility capacity and the time that the reservoir operates at a facility-constrained capacity (plateau rate).

Operating Plan. This plan guides operations personnel in implementing and accomplishing the depletion plan and can be strengthened by making operations part of the reservoir management team. Also, by establishing lines of direct communication between technical and operations staffs, early recognition of problem areas with individual well or facility performance will result in expedited solutions.

Survey of Performance. A periodic and systematic review of field performance, once the field has been placed on production, should be practiced. The types of activity vary with the type and size of resource and stage of depletion. At a minimum, frequent (daily to weekly) review of trends in individual well producing or injection rates, gas/oil ratios or gas/liquid ratios, water cut, wellhead pressure, and artificial lift performance should be required. Assuming the planned data are being captured, controlled for quality, and stored in a database, such reviews can be augmented by software packages that process data quickly, analyze performance trends, and generate exception reports.

Reservoir Performance Reviews. Surveillance includes periodic comparison (every 1 to 5 years) of performance (rates, pressures, recovery levels, injection volumes, contact movements, etc.) with the projections contained in the depletion plan. Significant deviations should trigger an update to the depletion plan to reflect the new information, to identify additional data needs, and to outline additional work programs needed to improve recovery and sustain rates. Reservoir reviews primarily should be an in-house assessment; however, the inclusion of experts not directly involved in day-to-day surveillance should be considered to provide another view and an independent source of ideas.

Rank, Justify, and Fund Opportunities. Opportunities that can increase the economic value of the reservoir will be an outgrowth of the initial development plan, the development and updating of the depletion plan, and surveillance activities. Assessing the economic benefits of each opportunity and obtaining the necessary management approvals is the next step.

Opportunities may involve investments in new wells, processing equipment, and improved recovery projects or changes to reservoir depletion strategies. Such assessments include recognition and evaluation of risks and benefits for programs to acquire data, improve recovery, increase production rates, or reduce operating costs. Outputs are a range of outcomes for each opportunity reflecting the associated risks for that opportunity and any alternative solutions. The natural outcome of such assessment is an up-to-date ranked list of opportunities. Outputs are recommendations to fund projects and/or implement changes to the depletion plan. The final phase is to obtain management approval for project funding consistent with budget and personnel resource constraints.

Additional Needs for Advancing Ideas. In the analysis of some opportunities, additional data, studies, or technology may be required to reduce the risks to an acceptable level. Additional data acquisition might include additional seismic, delineation drilling, or fluid analyses. Studies could include reinterpretation of seismic data or an in-depth reservoir study including complex geologic and reservoir simulation models. Acquiring technology could be by purchase or through liaison with a research organization in which the needs of the project are considered in planning research activities.

20.2.2 Reservoir Management Leadership Team. This group includes supervisors and managers responsible for allocating resources and creating an environment conducive to effective reservoir management.

Human Resources. The first objective of the RMLT is to assess and provide the personnel needed for successful reservoir management. This task involves the following procedures.

• Assess. Determine skills required to develop and manage a reservoir, and then judge the current staff's proficiency and need for improving such skills.

• Develop. Providing the training and work experiences needed to improve skills. Training includes in-house schools and seminars plus programs offered by third parties. Mentoring should be considered. This also includes the development of local training programs to meet specific needs not being met cost-effectively by more general programs.

• Deploy. Allocating supervisory, operations, and technical staff members consistent with established priorities and skill requirements to maximize the value of a resource. The process should consider the long-term benefits of skill development.

Achieving a Quality Program. The RMLT should have methods in place for determining the quality of the reservoir management program and how it can be improved to meet expectations. The quality model should be flexible to fit the needs of an organization. The needs most likely will vary from resource to resource depending on size, maturity, and perceived chance for developing additional investment opportunities.

• Align. There should be a process in place to ensure that all persons involved in reservoir management understand its importance, their role, and the quality objectives.

• Measure. Some process for systematically measuring the quality of a reservoir management program needs to be developed. Such a process would identify components of the program and the relative importance of each. A method of measuring how well each component is being accomplished needs to include the strengths in the current program that should be retained and areas that need improvement.

• Improve. Plans for improvements that lead to higher quality programs would include procedural issues, organizational structure changes, and the identification of skill needs. Such plans would also include goals for implementing the improvements.

• Recognize. A vital part of the quality program is the continuing recognition of both individual and team accomplishments. Whatever form of recognition is invoked, it offers an opportunity to reward good work and to reinforce the fact that management is dedicated to a good reservoir management program.

Stewardship. This process involves the periodic review and discussion of reservoir management activities and accomplishments with management and between groups.

• Communicate. These activities would involve various RMT and RMLT members and appropriate management. The primary purpose of these reviews is to describe reservoir management organization, future work plans, and historical performance of resources and to obtain management feedback on their expectations and support for continued reservoir management activities.

• Share. A corollary activity is the sharing of experiences between teams to identify best practices and company-wide improvement needs.

20.3 Measuring Reservoir Management Performance

Two questionnaires have been designed to survey and measure the quality of reservoir management performance. One is for RMT activities, and the second is for RMLT activities. **Tables 20.2 and 20.3** show suggested questions for the surveys.[9] The questions are a guideline of what could be included in the surveys.

The surveys are self-assessment tools for the RMT and RMLT groups. The strengths of the surveys are the systematic consideration of all reservoir management activities, improved communication and understanding of the multifunctional issues by the team, identification of strengths to be sustained and areas for improvement, and a method for measuring improvement with subsequent surveys.

20.3.1 Conducting a Survey. ***Participants.*** As dictated by the specific situation, operating, supervisory, and technical persons would be included in the assessment. In some instances, it may be desirable to have outside experts participate in the assessment to obtain different views on specific areas in which weaknesses are suspected.

Frequency. The size and importance of a resource affects the timing and frequency of surveys. Considering the effort and personnel required for an effective survey, annual or less frequent timing may be appropriate.

Questionnaire Format and Scoring. **Fig. 20.3** shows the suggested format for capturing assessments as they are made. Only the section of RMT questions on database processes is included in this example. This form would be developed for each major process of the RMT and RMLT.

A performance rating for each question is to be judged on a four-point scale ranging from "seldom equals" to "essentially always equals" as compared with a best-practice standard that is determined by the group doing the assessment. Best-practice standards tend to be somewhat subjective, but experience has shown there is good agreement among knowledgeable persons participating in surveys. Best practice will depend on the field or fields involved. A prime asset that represents a significant resource with a long remaining life would be held to different standards than smaller resources or mature fields in an advanced stage of development.

The four-point scale should be adequate for differentiating the quality for various activities. Scales with five to ten increments were considered to be too fine and beyond the capability of this process, while fewer increments would not allow enough difference between poor and excellent.

Each participant rates each question, the scores are tabulated, and the team works to reach consensus on each question. While the scoring system could be viewed as a numerical result (1 to 4) to determine an overall outcome for each field or resource reviewed, the descriptions are intended to indicate qualitative results that are more appropriate for determining strengths that

TABLE 20.2—RMT QUESTIONNAIRE

Determining the Condition of a Database.

Do the databases (hardcopy/computer):

1. Contain all existing geologic and geophysical data?
2. Contain all existing engineering and operational data?
3. Are these databases integrated?

Are the data:

4. Error free?
5. Easy to use? (Logically organized, clearly labeled with appropriate units?)

Are procedures/responsibilities effective for:

6. Inputting historical data?
7. Inputting newly acquired data?
8. Correcting errors in the data?

Describing the Reservoir.

Data Adequacy.

1. Is the core coverage adequate to describe the reservoir?
2. Is the modern log density adequate to describe the reservoir?
3. Is the geophysical coverage adequate to describe the reservoir?
4. Has the potential in 3D, 2D, or time-lapse seismic to improve the reservoir description been adequately evaluated?

Interpretation Quality.

1. Are maps and interpretations field wide, up to date, and internally consistent?
2. Are log analyses up to date and performed with the latest techniques?
3. Are geological/geophysical and reservoir engineering interpretations integrated and consistent with observed field behavior?
4. Do the reservoir descriptions use current technical concepts?
5. Are technology experts (persons with a high degree of skill and knowledge in reservoir description and management) from within and outside the company regularly consulted?

Maintaining the Description.

1. Do systems exist to keep interpretations current?
2. Is reservoir description documentation complete?

Determining the In-Place Resource.

1. Is there a suite of core measurements to provide confidence in log-determined rock properties and fluid saturations?
2. Are there sufficient fluid samples to provide confidence in pressure-volume-temperature relationships?
3. Does an efficient system exist for converting reservoir description to volumes?
4. Are all potentially valuable resources included in the assessment (oil, gas, gas plant liquids, sulfur, helium, CO_2).
5. Are the technical uncertainties quantified and their impact understood?
6. Is documentation of in-place volumes complete?

Depletion Plan Development and Updating.

Reservoir Performance.

1. Is there a reservoir performance model that incorporates up-to-date reservoir description and establishes performance expectations including identification of the primary drive mechanism, the need for improved recovery projects, producing rates of all fluids and ultimate recovery?
2. Is the model appropriate for the size and complexity of the resource?
3. Are model predictions consistent with performance data?
4. Are alternative development and depletion schemes considered?
5. Is the reservoir performance model sufficiently documented to provide explanation and continuity when team members change?

Wellbore Management.

1. Have the planned number, type, and placement of wells been documented adequately?
2. Do well plans include the evaluation of horizontal wells and/or high angle wells?
3. Has a wellbore utilization plan been developed and documented?

Updating.

1. Is there a plan to collect the necessary data on which to base an updated analysis and projection of reservoir performance?
2. Is there a plan for systemic, periodic updates to the depletion plan?
3. Does a reservoir performance analysis compare actual to expectations from the theoretical modeling work?
4. Are recommended adjustments to operating and surveillance plans made systematically?
5. Are the revised plans reviewed periodically with technical experts?

TABLE 20.2—RMT QUESTIONNAIRE (continued)

Implement and Operate.
1. Are specific work plans, including schedules and data needs, developed for each project initiated?
2. Are the work plans integrated across technical and operating functions?
3. Are responsibilities and leadership clearly defined?
4. Have experts had the opportunity to offer comments and suggestions?
5. Is there sufficient access to experts?
6. Does a written plan exist for providing operating guidelines to field personnel?
7. Do operating plans recognize uncertainties and provide sufficient flexibility?
8. Do field personnel have sufficient information to know key issues?
9. Do depletion planners and field personnel communicate on implementation of the plan?
10. Are data collected as planned?
11. Are sufficient data collected to ensure that the operations plan is being followed?
12. Are ongoing data collection costs and value understood and endorsed by appropriate management?
13. Is there a process in place to compare progress against the depletion plan and make any appropriate changes to depletion or implementation plans?

Survey of Performance.
1. Is there a written surveillance plan?
2. Does the written surveillance plan provide for obtaining needed data?
3. Are the key reservoir performance indicators identified in the plan?
4. Is input from field personnel included in the plan?
5. Do surveillance plans recognize uncertainties and provide sufficient flexibility?
6. Are periodic reviews of reservoir performance planned?
7. Are reviews coordinated with technical and operations personnel to obtain the necessary input?
8. Is management's endorsement of reviews obtained?

Justify, Prioritize, and Fund Opportunities.
Identifying Opportunities.
1. Does the process for identifying opportunities incorporate a field wide vs. wellbore analysis?
2. Does the process for identifying opportunities incorporate input from necessary technical and field functions?
3. Are ideas documented and maintained in an inventory system?
4. Is the inventory updated at appropriate intervals consistent with resource size and opportunity?
5. Are all possible opportunities captured in case improvements in technologies, economic, political, or risk factors occur?

Evaluating Opportunities.
1. Does an economic model exist that incorporates key issues such as costs, price forecasts, regulatory and legal boundaries, facility limitations, and/or royalty owner/host government issues?
2. Does the model forecast the economic drivers on the basis of specific field data?
3. Is the economic model sufficiently documented to provide explanation and continuity when team members change?
4. Are priorities evaluated on a timely basis with consistent criteria including long-term strategies?
5. Are the processes for obtaining expenditure authorization understood?
6. Are these processes documented?

Needs For Advancing Ideas.
1. Are plans formulated to acquire new information or analyses necessary to improve assessments of key opportunities and advance the idea?
2. Are these plans documented?
3. Are there plans to acquire new technology?
4. Are the costs of improving the assessments consistent with benefits?

are to be retained, barriers preventing execution of the process, and areas that require improvement. Also, some questions would have more importance than others and should be weighed accordingly.

Evaluation and Use of the Score. The initial assessment of a resource establishes a baseline and highlights areas of good practices and activities needing improvement. Subsequent reviews will show if targeted improvements are being accomplished while maintaining strengths. Over time, reviews also indicate where changes in emphasis are appropriate considering the changing maturity and needs of a resource. Another use of the results is the exchange of results

TABLE 20.3—RMLT QUESTIONNAIRE

Human Resources.

Assess.

1. Are the skills needed to manage the company's reservoirs systematically assessed?
2. Is the staff's current proficiency in those needed skills systematically assessed?
3. Are the assessment processes and assessments documented?
4. Is information collected to measure and improve the effectiveness of determining skills and training needs?

Develop.

1. Are training programs implemented on the basis of an assessment of the staff's proficiencies?
2. Is this training supplemented with coaching and developmental work assignments in reservoir management?
3. Are these development and training plans documented?
4. Is the effectiveness of the skill development processes assessed periodically?

Deploy.

1. Do criteria and processes exist for ensuring that technical staffs are employed consistent with business priorities and skills requirements?
2. Are the criteria periodically re-examined?
3. Are the criteria and processes documented?
4. Is information collected to measure and improve the effectiveness of this process?
5. Are the needs for outside expert skills identified and supplied?

Quality.

Align.

1. Are desired reservoir management goals and quality characteristics communicated to the organization?
2. Is multifunctional cooperation encouraged in forming reservoir management plans and goals?
3. Are the communication processes documented?
4. Is information collected to measure and improve the effectiveness of this process?

Measure.

1. Do processes exist for documenting reservoir management activities?
2. Are the critical components of reservoir management that need measuring periodically defined and documented?
3. Are the measures defined?
4. Are methods for collecting the data needed to measure the critical components established?
5. Are there efficient processes for analyzing the data?

Improve.

1. Do documented processes exist for identifying the needs (e.g., resources, skills, equipment, software) of people managing the reservoir?
2. Do documented processes exist that establish responsibilities for supplying the needs of people involved in managing the reservoir?
3. Are there established quality requirements?
4. Are near (1 to 2 years) and longer (> 5 years) term goals set for improving the reservoir management processes?
5. Are the strategies for achieving these goals established?
6. When setting goals, are business needs, current capabilities, and benchmarking data considered? (Benchmarking is a method of determining how your process compares with other company groups and/or competitors.)

Recognize.

1. Is there a process for recognizing individual and team accomplishments?
2. Is information collected to measure and improve the effectiveness of this process?

Stewardship.

1. Have all persons (customers) who will use the products of reservoir management been identified?
2. Have customer needs, quality characteristics, and use of product been identified?
3. Does a process exist for measuring and improving customer satisfaction?

Communicate.

1. Have the company's needs, including anticipated quality and use, been identified?
2. Does the RMLT take an active role in reviewing activities with management and gathering feedback?
3. Is the reporting process documented?
4. Is information collected to measure and improve the effectiveness of this process?

Share.

1. Are there processes for sharing approaches and lessons learned among various teams or groups that practice reservoir management?
2. Is this sharing process documented?
3. Does a process exist for identifying and communicating improvements in existing technology and new technology needed to achieve business goals?
4. Is information collected to measure and improve the effectiveness of this process?

	Seldom	Somewhat	Mostly	Essentially Always
Do the databases (hardcopy/computer)				
1. Contain all existing geologic and geophysical data?				
2. Contain all existing engineering and operational data?				
3. Are these databases integrated?				
Is the data				
4. Error free?				
5. Easy to use?				
Are procedures/responsibilities effective for				
6. Inputting historical data?				
7. Inputting newly acquired data?				
8. Correcting errors in the data?				
Strengths:				
Barriers:				
Areas for improvement:				

Fig. 20.3—Suggested format for capturing assessments.

between different RMT and RMLT groups with a focus on how strengths were accomplished and what has been implemented to improve the process.

20.4 Examples of Reservoir Management Benefits

In the following discussion of field histories, the reader should be aware that the small well spacing in some fields would not be optimal considering current technology and economic conditions. Some well spacings were driven by factors such as shallow reservoirs with low well costs, multioperator competitive operations, regulatory depth-acreage-based well allowables, oil price controls with two-tier pricing systems, and all vertical wells.

20.4.1 Sandstone Reservoir With Combination Drive Supplemented With Gas and Water Injection. ***Background and Geological Information.*** This field produces from a structure that lies above a deep-seated salt dome (salt has been penetrated at 9,000 ft) and has moderate fault density. A large north/south trending fault divides the field into east and west areas. There is hydraulic communication across the fault. Sands were deposited in aeolian, fluvial, and deltaic environments made up primarily of a meandering, distributary flood plain. Reservoirs are moderate to well sorted; grains are fine to very fine with some interbedded shales. There are 21 mapped producing zones separated by shales within the field but in pressure communication outside the productive limits of the field. The original oil column was 400 ft thick and had an associated gas cap one-third the size of the original oil column. Porosity averages 30%, and permeability varies from 10 to 1500 md.

Program Used. The primary recovery mechanism was a combination of a strong, directional waterdrive in the eastern part of the field, solution-gas drive in the western part of the field, and gas-cap expansion. Initial screened openhole completions resulted in uneven water advance caused by permeability differences between zones. A workover program converted all wells to selective, single-zone completions that allowed better control of aquifer advance. The directional waterdrive resulted in gas-cap tilting to the west despite return of all produced gas to the gas cap and extraneous water production on the east side. Water injection around the downdip periphery of the western part of the field was used to waterflood the western area, equalize pressure between east and west areas, and prevent gas-cap tilting. Because of low dip, recovery by gas-cap expansion was a less efficient recovery mechanism.

When field oil production declined to 4,500 B/D with 98% water cut and an average GOR of 30 Mcf/bbl, gas sales were necessary to maintain acceptable profitability. When gas-cap depressuring was initiated, remaining oil production was thought to be coming from low permeability, thin intervals that would be affected minimally by accelerated water influx accompanying depressuring. Oil production is expected to continue well beyond the time the gas cap is depleted. Gas will be trapped behind advancing water, but reservoir-pressure decline will result in percolation of liberated dissolved gas from the residual oil in the water-invaded zone. This analysis came from a tank-type reservoir simulation model. To date, reservoir pressure has declined from 2,150 to 1,525 psi because the aquifer has been unable to offset the increased withdrawal rates.

Frac packing has provided sand control and near-wellbore stimulation. Conventional gravel packing also has been used when water proximity has been a concern. Plugback workovers and additional well conversions will be used to achieve maximum gas recovery. Original development was on 20-acre spacing. Selective infill drilling has been used to improve the sweep of injected water and to drain isolated parts of various zones.

Recovery Performance. The combined recovery factor for all the recovery mechanisms is 66% of OOIP. A contributing factor to this high recovery is a mixed wettability rock yielding waterdrive residual oil saturation of 12 to 13%, based on single-well chemical-tracer tests.

Field Surveillance and Management. There have been several major reservoir studies of the field during its history to determine and upgrade the depletion strategy. There also has been a sustained surveillance program to monitor reservoir performance. Surveillance has focused on maintaining liquid injection-production balance, monitoring area pressures, and the use of cased-hole logging to monitor gas/oil contacts (GOCs). Several segment and individual well simulation models have been used throughout the life of the field to better understand the waterfront movement, to adjust injection and offtake locations, and to mitigate bypassing of oil. Cased-hole logging programs (neutron or pulsed neutron) are now conducted at 3-month intervals to monitor the uneven water advance. Reservoir pressure is measured monthly. Pulsed-neutron logs are proving very valuable in locating bypassed oil in the shaley sand sections. Pressure maintenance by waterdrive and injection has permitted wells to produce by natural flow/artificial lift. This allowed cost-effective tubing logging and pressure surveillance.

20.4.2 Sandstone Reservoir With Strong Waterdrive and Crestal Gas Injection. ***Background and Geological Information.*** Production from this field is from several Upper Cretaceous sandstone formations. The producing zones are in pressure communication in the gas cap and aquifer but separate in the oil column. The structure is a complexly faulted anticline with a major fault separating the west and east flanks. There is minor communication across the fault. Gross thickness is 384 ft, and net-to-gross thickness is 0.7. Reservoir dip is 6° with somewhat higher dips upstructure. A tar layer exists at the original oil/water contact in the western flank of the field. The tar is 50 ft thick on the north flank and 100 ft thick on the south flank.

Porosity averages 27%, and zone permeability averages range from 1000 to 3000 md. The oil viscosity was 4 cp in the main oil column but grades higher near the tar and oil/water contact.

Program Used. The field was developed and produced competitively by several operators until unitization. The primary producing mechanism was a strong waterdrive that led to some gas-cap shrinkage despite an early gas return program. Performance analysis indicated a recovery efficiency of 65% by water displacement vs. 75% by gas displacement-gravity drainage. Formation of the unit allowed initiation of crestal gas injection. Water is being produced from down-structure wells to aid in moving the GOC downdip. Gas-cap pressure has been increased, water influx has been virtually eliminated, and the waterdrive recovery efficiency has been replaced with better gas-drive gravity-drainage efficiency.

A double displacement process (DDP) is under way to displace water-invaded portions of the original oil column with gas. The objective is to create a gravity-stable gas front that allows the ROS after waterdrive to be remobilized, drain to the base of each zone, and be captured in down-structure producing wells.

Down-structure oil production is from multiple zones. Wells are produced primarily from a single zone to maintain control of producing GOR and WOR. Recompletions to other sands are performed based on well performance, cased-hole logging results, and surveillance maps.

Recovery Performance. The recovery by waterdrive and gas injection is 63% OOIP. The DDP is expected to add another 6% of OOIP for a total of 69%.

Field Surveillance and Management. There have been several major reservoir studies of the field to determine and upgrade depletion strategies. The DDP was studied extensively with regional and detailed fault-block models before starting the field project. A recent sequence stratigraphy study located remaining oil and provided an improved basis for surveillance of the DDP. These studies have been used to guide infill drilling that has resulted in significant additional capture reserves. In addition, an aggressive well recompletion program has been essential in obtaining expected production and recovery performance.

A sustained surveillance program has been in place throughout the life of the field. Data on oil, gas, and water production are collected routinely each month, and annual cased-hole logging and pressure surveys are conducted. Fluid-contact mapping and material-balance models are used to monitor and balance gas-injection rates to offset fluid production and aquifer influx. The annual fieldwide cased-hole logging program to monitor fluid contacts is a very important surveillance tool in judging the effectiveness of the displacement process. Gas coning is one of the biggest operational challenges. Daily surveillance of GORs by field personnel and active involvement of the engineering staff combine to maintain maximum oil production rates. Water has swept heavier oil from deeper depths to shallower depths, reducing well productivity and slowing the gravity drainage process. Frequent measurements of oil gravities are used to understand and predict producing characteristics.

Heavy oil creates oil/water separation problems, which leads to injection of contaminants in saltwater disposal wells. This results in the need for frequent saltwater disposal well cleanout workovers. Keeping the surface-handling facilities clean minimizes these workovers. Techniques include regularly scheduled tank inspections and cleanouts, appropriate internal tank piping/oxygen exclusion, well-managed water chemistry surveillance and chemical programs, and the use and maintenance of appropriate transfer pumps. Periodic backflowing of gas-injection wells has had a significant positive impact in maintaining high levels of injectivity and minimizing gas-injection pressures.

20.4.3 Sandstone Reservoir With Strong Waterdrive and Selective Well Completion Strategy. ***Background and Geological Information.*** This field produces from a series of stacked sandstone reservoirs situated in an angular unconformity trap consisting of an eroded 2 to 6° monocline and an oil/water contact common to all zones. The original oil column was 400 ft thick and contained a 43°API, highly undersaturated crude. Porosity averages 22%, and permeability averages 3 darcy.

Program Used. The primary recovery mechanism is strong waterdrive. All wells are now gas lifted, although this was not necessary in early field life. Wells have typically been drilled through multiple sands, which have been produced sequentially and, in some cases, comingled. Initial development was on 300-acre spacing. Two subsequent infill-drilling programs have been carried out to drain thin zones and areas in the extreme updip portions of the reservoir. The platform is currently constrained by water-handling capacity and, when wells reach high water cut, they are typically cycled on and off so that total production matches facilities limits.

Recovery Performance. The estimated recovery factor is approximately 75% of OOIP. Good rock and fluid properties plus good lateral continuity have contributed to this high recovery through gravity-stable displacement of oil by the invading water. Estimated S_{or} in invaded sections of the reservoir is 10% of pore volume (PV).

Field Surveillance and Management. Cross section and 3D computer models have been used to study reservoir behavior and upgrade the depletion strategy. A sustained surveillance program includes logging to monitor displacement and regular field performance reviews by the geoscience and engineering team members.

20.4.4 Sandstone Field With Waterflood, Gas Injection, and Miscible Projects. ***Background and Geological Information.*** The field is a structural stratigraphic trap that has been divided into several vertical zones. Complex, systematic depositional and diagenetic changes resulted in a dual pore system that was further impacted by structural and hydrocarbon histories, resulting in a highly variable vertical and areal distribution of net pay, porosity, and water saturation. Porosity values range from 10 to 30% with an average of 22%. Average well-zone permeabilities ranged from 100 to 1800 md with a field average of 500 md.

Program Used. The primary depletion mechanisms were gas-cap drive along with a very weak waterdrive (rock properties deteriorate rapidly off structure) and potential solution-gas drive. Waterflooding was planned as part of the initial field development to prevent significant pressure depletion.

Waterflooding was started in conjunction with an infill-drilling program that reduced spacing to 80 acres per well. Inverted seven- and nine-spot injection patterns were applied to areas of the oil column not overlain by the gas cap. A small enriched-hydrocarbon miscible injection project was later expanded to additional areas of the field.

Horizontal wells have been used extensively in areas close to the GOC to capture reserves from relatively thin oil columns that were previously uneconomic because of severe coning problems. A gas-cap cycling project has been expanded three times. A few 40-acre-spaced wells were drilled in areas of poor drainage, especially where locally large shale sections prevent good continuity. However, a more economical approach has been to drill 1,500- to 2,000-ft horizontal sidetracks, which can realize up to 85% of the 40-acre well recovery but at less than one-third the cost.

Recovery Performance. Estimated ultimate recovery is 60% OOIP with a miscible contribution of 10% OOIP in affected areas. The contribution of reservoir management to these high recovery levels is difficult to quantify. Initial estimates of recovery were in the range of 40 to 45% of OOIP. It is clear that a continuing effort to discover improved depletion techniques has added significantly to ultimate recovery.

Field Surveillance and Management. The unit collects surveillance data to manage the reservoir effectively. This includes day-to-day production tests, production/injection profiles for wells, and bottomhole-pressure surveys across the field. Additionally, repeat cased-hole neutron logs are taken to monitor gas movements across the reservoir. On new wells, a comprehensive suite of openhole logs is obtained to provide geologic information and to supplement production data and surveillance. The unit actively collected special reservoir surveillance data from pressure-pulse and pressure-transient well tests, repeat formation tests, and specially obtained cores. Both single-well chemical-tracer tests and log-inject-log tests have been conducted, and specialized core data have been obtained in the waterflood areas to measure and improve the effectiveness of the waterflood/miscible EOR project. Additionally, gas samples are collected routinely from flood area production wells to measure returned miscible injectant to further optimize and improve the miscible EOR project and from gravity-drainage area wells to monitor movement of injected cycle gas. Many different simulation models have been developed and updated over time.

There are frequent discussions, at both the technical and managerial levels, of surveillance techniques, operating strategies, reservoir and facility management programs, and future capital projects. Committees, forums, and teams continue to evolve to facilitate and improve these exchanges. In addition, corporate resources bring technical expertise and a worldwide perspective to the development decisions.

20.4.5 Steeply Dipping Sandstone Reservoir With Gravity Stable Miscible Project. ***Background and Geological Information.*** This field was formed by a piercement salt plug that breached a regional fault system. The reservoir is composed of unconsolidated sands that dip away from the salt dome at 65 to 85°. The reservoir is divided into several fault blocks. Within each block, the sand is relatively homogeneous. Porosity averages 26.5%, and permeability averages 1.3 darcy. Oil gravity is 38° API.

Program Used. This project is operated as a gravity-stable, hydrocarbon-miscible flood. The injection rate corresponds to a velocity of roughly one-half the critical velocity required for gravity-stable operations. A volume, corresponding to 17% PV, of enriched gas (natural gas liquids plus solution gas) was injected. This was followed by injection of solution gas alone. When injection is completed, blowdown of the gas cap is expected to recover approximately 90% of the enriched gas and a significant portion of the injected solution gas, thus reducing the effective cost of the solvent.

3D reservoir simulations were carried out to examine four depletion scenarios: primary depletion by gravity-stable gas-cap expansion, waterflooding at constant pressure, gravity-stable immiscible gas injection at constant pressure, and gravity-stable hydrocarbon miscible flood at constant pressure. Maintaining constant pressure improves recovery by eliminating shrinkage of oil over the course of displacement. Core flood experiments gave recoveries similar to those predicted by the simulations. Miscible residual oil saturation in core floods was 7% of PV. Slimtube experiments were also carried out to determine the level of enrichment required to achieve miscibility at a given pressure level. A minimum miscibility pressure (MMP) was selected that was consistent with the volume of enriching natural gas liquids available for the project.

Recovery Performance. Primary production occurred through gas-cap expansion. Approximately 6.2% of OOIP had been recovered when miscible gas injection began. Estimated ultimate recovery is 50% of OOIP for primary recovery in both sands, 74% for total recovery after miscible flood in one sand, and 86% for total recovery after miscible flood in a second sand. These recovery levels were determined by tracking gas fronts with pulsed-neutron capture logging and material-balance calculations. These recovery levels are consistent with simulations that were carried out during planning of the miscible project. To date, conformance has been excellent, with field recoveries similar to those seen in core floods.

Field Surveillance and Management. Routine pressure measurements, pulsed-neutron capture logs, and pressure-transient testing were used to monitor reservoir performance, contact movements, and identify areas of good and poor communication. Pressure initially was allowed to decline to slightly above the MMP and was then maintained by scheduling injection volumes equal to production. Pressure maintenance became a challenge because of increasing GOR, which resulted in watercut increases and reservoir pressures below the MMP in some areas. Pressures were then increased and maintained by curtailing production from high GOR wells.

Pressure communication between injectors and producers has been good in reservoir A. In reservoir B, pressure communication between injectors and producers has been somewhat hindered by faults and shale barriers that act as baffles.

20.4.6 Low Permeability Sandstone Waterflood. ***Background and Geological Information.*** The reservoir is a series of Cretaceous-age, prograding delta clastic sediments consisting of laminated fine-grained sands and shales that are trapped stratigraphically by overlying shales. Reservoir averages are approximately 50 ft of gross pay, 15% porosity, and 20 md permeability.

Program Used. The primary recovery mechanism was solution-gas drive. The field was converted to waterflood in 1961 with an inverted nine-spot injection pattern. Subsequently, a portion of the field was converted to line-drive water injection for improved sweep efficiency and increased water injection capacity.

Recovery Performance. Under primary recovery, the field produced 5% of OOIP. Incremental recovery to date under waterflood has been an additional 17% of OOIP. An ultimate recovery efficiency of 26% of OOIP is forecast.

Field Surveillance and Management. Exploitation of the reservoir is focused currently on reperforating underperforming wells and infill drilling into portions of the reservoir that are shown to be poorly swept. Continuous rebalancing of water injection is proving effective in displacing oil from uncontacted layers.

20.4.7 Sandstone Reservoir With Solution-Gas Drive and Water- and Gas-Injection Projects. ***Background and Geological Information.*** The highly faulted structure produces from six different productive horizons. The depositional settings range from deepwater marine turbidite fans through near shore and delta san facies to fluvial deposits. Reservoir quality ranges from high permeability (350 md) in highly continuous tidal shoals to fair quality (100 md) in moderate continuity fluvial channels. The reservoir has been developed from two separate platforms and has an extensive subsea development of satellite fields. More than 140 wells have been drilled from two platforms since initial development in 1976 with well lengths varying from 9,800 to 19,000 ft. Six subsea fields have been developed with 28 subsea producing wells. There are 28 reservoirs that are managed separately.

Program Used. The primary recovery mechanism in the majority of these reservoirs was solution-gas drive with weak aquifer support, although some of the satellite fields have strong bottomwater drives. The main reservoir was developed from the two platforms: the Alpha platform began production in 1976, followed by the Bravo platform in 1984. Crestal gas injection and peripheral water injection were used in both areas in the main reservoir to maintain reservoir pressure and displace the oil zone. The secondary reservoir formations were developed with solution-gas drive mechanisms with water and gas injection applied where economically viable. The Alpha main reservoir achieved a 52% oil recovery factor. Gas-cap blowdown started in 1993. The pressure maintenance schemes continue to be used in the Bravo and Satellite fields.

Recovery Performance. The expected recovery in the main formation is expected to exceed 50% of the OOIP on both the Alpha and Bravo platforms, with the average recovery factor for all sands expected to reach 39% of the OOIP. The cumulative recovery to date of 33% of the OOIP compares with the 15 to 20% expected under primary depletion.

Field Surveillance and Management. Reservoir management in this complex field relied on a multidisciplinary team approach. The complex geological and structural natures of the fields are represented in detailed geological models. Improvements in the structural imaging of the fields through the use of ocean-bottom cable seismic and detailed sequence stratagraphic work has improved the understanding of the fields. Black oil reservoir simulation models have been built for all the major producing reservoirs. These models have sufficient detail to represent the geological and structural complexity properly. A surveillance program for each reservoir unit is defined so that cost effective and minimum production deferral interventions can be planned in advance. Asset development and depletion plans are defined for each field/reservoir unit that define the depletion plan and integrates study plans with surveillance and drilling activities. An opportunity list is developed that is reviewed routinely by the multidisciplinary asset team to ensure appropriate prioritization of resources. Reservoir surveillance data is combined with classical reservoir engineering techniques to create surveillance maps for each of the reservoir flow units. Surveillance techniques combined with simulation results are used to ensure quality forecasts and robust infill-drilling targets. Infill wells typically have a full suite of MWD logs and modular formation dynamics testers. More complex horizontal wells tend to have less comprehensive data gathering because of cost and the risk of losing the hole.

20.4.8 Sandstone Reservoir With a Polymer Injection Project. ***Background and Geological Information.*** Production is from three sandstone zones of a Cretaceous-age formation. Productive area of the polymer project was 3,560 acres. Gross thickness was 230 ft, and net-to-gross thickness averaged 0.3. Porosity averages 27%; permeability varies by zone and is less than 100 md in two of them and 170 md in the third, most extensive one. Oil gravity was 20°API, and viscosity varied from 18 to 24 cp. T_R was 150°F.

Program Used. The primary depletion mechanism was solution-gas drive. The field was developed competitively on 40-acre spacing. A unit was formed, and waterflooding began with a staggered line drive. Injection water with a salinity of only 3,500 ppm total dissolved solids (formation water 150,000 ppm) was used without any apparent damage caused by clay swelling or sloughing.

The adverse mobility ratio (see the chapter on foam, polymer, and resin injection in this section of the *Handbook* for a discussion on mobility ratios) was recognized early in the waterflood and polymer injection begun in 1976. Polymer concentrations were targeted to improve the mobility ratio by a factor of 16 (reduced to 0.5). The injection of polymer was targeted to reduce the cycling of water through previously water-swept oil sands and to divert injected-brine polymer into noncontacted zones to displace additional oil.

Recovery Performance. Waterflooding increased oil rates from 300 to 2,000 BOPD. The rate had declined to 1,600 BOPD when polymer injection began. The rate of oil production decline subsequently was reduced in several wells. Primary plus waterflooding recovery factor was 23% of OOIP. Subtle changes in production decline rates and WOR make it difficult to determine the incremental recovery from polymer injection. An incremental increase of 1 to 3% of OOIP has been estimated. Lower injectivity resulted when polymer was added.

Field Surveillance and Management. Surveillance activities focused on maintaining injectivity and pattern balancing. Well stimulations and cleanout were needed. 2D cross-section simulation was used to manage the injection volumes and distribution of polymer and to design the optimum concentrations of the final polymer injection. Rates were adjusted to inject an average of 20% PV of 600 ppm polyacrylamide polymer solution.

20.4.9 Carbonate Field With Waterflood and Miscible Projects. ***Background and Geological Information.*** This field produces primarily from a Jurassic-age limestone-dolomite section that has a simple plunging anticline structure. The updip trap is formed by a combination of facies change from dolomite to dense limestone and a bounding fault. The formation is layered and has been divided into 18 correlative zones.

Program Used. The field was developed competitively by several operators. When production began, the reservoir pressure declined rapidly under a fluid-expansion drive. The field was unitized, waterflooding began, and pressure decline was arrested. Miscible N_2 injection was started eight years later. N_2 was selected rather than methane or CO_2 because of cost and supply considerations.

The field was developed initially with 89 wells on 160-acre spacing. Selective infill drilling in poorer sections of the reservoir (both areally and vertically) improved the sweep of injected water. The need for fieldwide infill drilling to 80-acre spacing was tested with five wells and shown to be uneconomical because little extra recovery was achieved.

The waterflood was implemented with a 3-to-1 line-drive pattern with low salinity water from a water-source well. Produced water is also injected. Peak water-injection rates reached 250 kBOPD. Three air-separation units produced nitrogen, which was reinjected. Peak injection rate was 86 Mcf/D. A water-alternating-gas (WAG) process is used.

There is no downhole corrosion treatment and no internal treatment of flowlines and pipelines. Flowlines and pipelines are protected cathodically and require selective remediation.

Corrosion resistant alloy tubulars and flowlines are used to handle the 2% H_2S (originally 8% but reduced by N_2 contamination) sour gas produced from the field.

Recovery Performance. Primary recovery of 17% of OOIP was expected. Waterflooding increased this to 57%, and the miscible project added another 13% of OOIP.

Field Surveillance and Management. This field has had several periodic reservoir studies to update and refine the depletion strategy and a sustained surveillance program that includes acquisition of data needed to monitor reservoir behavior and injection efficiency. To obtain an accurate geologic description, the entire pay interval in virtually all the wells was cored. The reservoir surveillance program included the monitoring of production data, injection data, bottomhole pressures, flood balancing, WAG ratios, injection-to-withdrawal ratios, and profitability of each pattern. One of the tools used is a history-matched four-component compositional-simulation model that is based on a stochastic geologic model and advanced scale-up technology. The 3D model contains 100,000 gridblocks and 18 layers. Good matches of pressure and oil, water, and nitrogen production at the field and individual well levels were achieved with quality control of key input data and the use of extensive history-match experience.

20.4.10 Carbonate Field With Waterflood and Miscible Projects. ***Background/Geological Information.*** This field is a north/south trending anticline separated into north and south domes by a dense structural saddle running east and west near the center of the field. Deposition was in an intertidal-lagoon-bank sequence. Production comes from formations at depths ranging from 4,200 to 4,800 ft (subsurface). The formation is more than 1,400 ft thick. The upper 200 to 300 ft is productive, and the remaining is a water zone of relatively low permeability. The productive upper portion of the reservoir is divided further into the upper and lower reservoirs. The average porosity is 9%, and the average permeability is 20 md.

Programs Used. *Primary/Waterflood Depletion.* The field was developed initially on 40-acre spacing with 300 wells. The primary producing mechanism was a combination of fluid expansion with a weak waterdrive. In 1963, a unit was created and peripheral water injection began into 36 wells. Production from the unit began declining in 1967 because of insufficient pressure support. A detailed engineering and geologic study identified a flow barrier that was inhibiting pressure support between the upper and lower reservoirs. Implementation of a three-to-one line drive provided the needed pressure support, and production increased. When production began to decline again in 1972, a subsequent reservoir study resulted in a technique to correlate gamma-ray/neutron logs with core data, thus better defining porosity distribution and OOIP. The study resulted in an infill-drilling program to 20-acre spacing, conversion of the injection scheme to an 80-acre inverted nine-spot pattern, and a better reservoir surveillance program.

Infill Drilling/Miscible Process. An oil viscosity of 6 cp makes the waterflood mobility ratio relatively high. From pressure cores and laboratory core floods, waterflood residual oil saturation was estimated to be 34% of PV. Combined, these two factors provided incentive for further infill drilling and evaluation of other recovery methods. A CO_2 miscible project was evaluated with laboratory investigations, a field pilot, and reservoir simulations.

The proposed CO_2 project consisted of 167 patterns on approximately 6,700 acres that encompassed 60% of the productive acres and 82% of the OOIP of the unit. Every attempt was made to use the original 40-acre wells and the 20-acre infill wells. Infill drilling to 10-acre spacing was an integral part of project development. All the WAG injectors and central producers were new 10-acre wells as a part of 40-acre inverted nine-spot patterns.

Within 2 years of project implementation, 205 infill producers and 158 infill injectors had been drilled. Re-evaluation of the project during implementation resulted in changes to project scope. The CO_2 project now consists of 173 patterns on approximately 7,830 acres.

Recovery Performance. Projected recovery from primary and waterflooding methods is 30% of the OOIP. Incremental recovery because of the miscible CO_2 flood is 15% of the OOIP.

Field Surveillance and Management. Throughout the years, a number of reservoir description and engineering studies have been conducted with the goal of developing better reservoir management strategies. A detailed, integrated surveillance and reservoir management program was implemented to achieve areal-flood balancing, vertical-conformance monitoring, production monitoring, injection monitoring, data acquisition and management, pattern-performance monitoring, and optimization. The following are some key objectives of field surveillance and management.

- Integrate all knowledge and data, such as seismic, core, log, laboratory work, and outcrop and field observations, into a fieldwide geologic model and keep it up to date.
- Monitor and understand field performance.
- Increase WAG frequency to minimize problems associated with premature gas breakthrough.
- Maintain a system-operating pressure between the reservoir-parting pressure and the MMP. Falling outside this narrow range would compromise ultimate recovery by fracturing the reservoir or eliminating the miscibility component of the flood.
- Manage the GOR to fit compression limitations for recycled CO_2.
- Update assessments of facility capacity.
- Maintain automated injection, production, and artificial lift monitoring systems to capture data needed to develop programs for maintaining flood-front pressure by balancing the WAG schedule, ratios, and CO_2 slug sizes.
- Implement proactive corrosion- and scale-treatment programs.
- Use new infill wells for injection purposes to minimize downhole mechanical problems.
- Maintain a continuous injection-well profiling program for flood management purposes.

20.4.11 Carbonate Field With Waterflood, Polymer, and Miscible Projects. ***Background and Geological Information.*** The field is a carbonate reef-mound complex of Late Pennsylvanian to Early Permian age with a formation composed of limestone with minor amounts of shale. It reaches a maximum thickness of 918 ft and averages 315 ft. Skeletal and oolitic grainstone shoals form the most significant reservoir facies. More than 90% of the porosity is secondary because of freshwater dissolution of unstable framework grains. Cores indicate the presence of fracturing and enhanced dissolution along fractures. Channeling of water and/or CO_2 is evident between many injector/producer pairs.

Program Used. Primary development was followed by a centerline waterflood that was converted subsequently to a five-spot waterflood with two infill-drilling campaigns. A polymer-augmented waterflood was implemented on the basis of data from the new wells. The possibility of enhanced natural gas liquids production was identified with compositional reservoir simulation studies. A CO_2 miscible WAG injection process was then initiated.

Recovery Performance. CO_2-WAG injection is expected to result in an additional recovery of 12% of the OOIP, bringing the total estimated ultimate recovery to 64% of OOIP. The solvent extraction capability of CO_2 has resulted in an increase of up to 6,000 B/D in natural gas liquids production.

Field Surveillance and Management. Interdisciplinary teams composed of geoscientists; reservoir, production, and facilities engineers; and field operation staff conducted reservoir management. Cores from 30 of the wells were analyzed for stratigraphy and depositional sequences, and these interpretations provided the basis for a reservoir model that has been updated and enhanced throughout the life of the field. 3D geologic and simulation models are integrated into the surveillance process for flood optimization, workover and drill well planning, and WAG management.

20.4.12 Carbonate Waterflood With Strategic Well Placement. ***Background and Geological Information.*** The reservoirs in this field are part of a Devonian atoll reef and carbonate shoal complex consisting of limestone with traces of dolomite, pyrite, and anhydrite. The hydrocarbons are trapped stratigraphically by a calcareous shale seal. The reservoir averages approximately 200 ft of gross pay, 8% porosity, and 50 md permeability. Some stringers have permeabilities of 2000 to 3000 md, which result in high well productivities. Oil characteristics include a gravity of 44°API and a formation volume factor of 1.8 RB/STB. The reservoir depth averages 9,000 ft, and the initial water saturation was 14%.

Program Used. The primary recovery mechanism was solution-gas drive. In 1964, the field was converted to a downdip peripheral waterflood. Originally, the field was developed on 160-acre spacing. Selective infill drilling to 80- and 40-acre spacing has been used in updip portions of the field to develop areas of poor reservoir continuity that contain unswept oil.

Recovery Performance. Under primary recovery, the field produced 1% of OOIP. Recovery to date under waterflood has been an additional 45% of OOIP. An ultimate recovery efficiency of 49% of OOIP is forecast.

Field Surveillance and Management. The waterflood is being balanced in four discrete geologic regions or flow units to replace voidage and improve areal-sweep efficiency. The replacement ratio has averaged 100%.

20.5 Summary

Petroleum reservoir management is a dynamic process that recognizes the uncertainties in reservoir performance resulting from our inability to fully characterize reservoirs and flow processes. It seeks to mitigate the effects of these uncertainties by optimizing reservoir performance through a systematic application of integrated, multidisciplinary technologies. It approaches reservoir operation and control as a system, rather than as a set of disconnected functions. As such, it is a strategy for applying multiple technologies in an optimal way to achieve synergy, rather than being a technology in itself.

The reservoir management concepts in this chapter are keyed to RMT and RMLT teams. The RMT consists of experts in their field of technology who understand the impact of and are able to work with experts in other technologies. The RMLT is responsible for providing resources and determining customer needs. The effectiveness of the process depends on the timely acquisition, appropriate analysis, and use of high-quality data.

The reservoir management process must be tailored to individual fields depending on size, complexity, reservoir and fluid properties, depletion state, regulatory controls, and economics. The field cases discussed give some insight into how reservoir management has been applied and adapted to a variety of reservoir types and recovery processes.

Because reservoir management is a dynamic process tailored to individual fields and recovery processes, it is important to have timely assessments of how it is being applied. This helps ensure that best practices are being implemented, that the process is being improved continually and systematically, and that experience and technology is transferred among teams.

Nomenclature

h = thickness, L, ft
k = permeability, L^2, md
S_{or} = residual oil saturation, % PV
T_R = reservoir temperature, T, °F
μ = viscosity, m/Lt, cp

References

1. Eden, A.L. and Fox, M.L.: "Optimum Plan of Depletion," paper SPE 17458 presented at the 1988 SPE California Regional Meeting, Long Beach, California, 23–25 March.
2. Hickman, T.S.: "A Rationale For Reservoir Management Economics," paper SPE 26411 presented at the 1993 SPE Annual Technical Conference and Exhibition, Houston, 3–6 October.
3. Raza, S.H.: "Data Acquisition and Analysis for Efficient Reservoir Management," *JPT* (April 1992) 466.
4. Satter, A., Varnon, J.E., and Hoang, M.T.: "Reservoir Management: Technical Perspective," paper SPE 22350 presented at the 1992 SPE International Meeting on Petroleum Engineering, Beijing, 24–27 March.
5. Stiles, L.H. and Magruder, J.B.: "Reservoir Management in the Means San Andres Unit," *JPT* (April 1992) 469.
6. Thakur, G.C.: "Implementation of a Reservoir Management Program," paper SPE 20748 presented at the 1990 SPE Annual Technical Conference and Exhibition, New Orleans, 23–26 September.
7. Torvund, T.: "The Oseberg Reservoir Management Planning: A Case History From the Oseberg Field," paper OTC 6140 presented at the 1989 Offshore Technology Conference, Houston, 1–4 May.
8. Wiggins, M.L. and Startzman, R.A.: "An Approach to Reservoir Management," paper SPE 20747 presented at the 1990 SPE Annual Technical Conference and Exhibition, New Orleans, 23–26 September.
9. Holstein, E.D. and Berger, A.R.: "Measuring the Quality of a Reservoir Management Program," *JPT* (January 1997) 52.

General References

"Bibliography," *Reservoir Management,* Reprint Series, SPE, Richardson, Texas (1998) **48,** 193–197.

Abdullah, M.A.Y. and Olsen, B.S.: "Tapis—New Opportunities from a Maturing Field," paper SPE 54339 presented at the 1999 SPE Asia Pacific Oil and Gas Conference and Exhibition, Jakarta, 20–22 April.

Chow, C.V. and Arnondin, M.C.: "Managing Risks Using Integrated Production Models: Process Description," *JPT* (March 2000) 54.

Clayton, C.A.: "Ubit Field Rejuvenation: A Case History of Reservoir Management of a Giant Oilfield, Offshore Nigeria," paper SPE 49165 presented at the 1998 SPE Annual Technical Conference and Exhibition, New Orleans, 27–30 September.

Day, S., Griffin, T., and Martins, P.: "Redevelopment and Management of the Magnus Field for Post-Plateau Production," paper SPE 49130 presented at the 1998 SPE Annual Technical Conference and Exhibition, New Orleans, 27–30 September.

de Carvajal, G.: "Reservoir Management in the B-6/9, SVS-82, Lake Maracaibo, Venezuela–New Exploitation Policy for Maximizing Profit," paper SPE 39052 presented at the 1997 SPE Latin American and Caribbean Petroleum Engineering Conference and Exhibition, Rio de Janeiro, 30 August–3 September.

Gallagher, J.J. *et al.:* "Brent Field Depressurization Management," paper SPE 56973 presented at the 1999 SPE Offshore Europe Conference, Aberdeen, 7–9 September.

Gringarten, A.C.: "Evolution of Reservoir Management Techniques: From Independent Methods to an Integrated Methodology. Impact on Petroleum Engineering Curriculum, Graduate Teaching and Competitive Advantage of Oil Companies," paper SPE 39713 presented at the

1998 SPE Asia Pacific Conference on Integrated Modelling for Asset Management, Kuala Lumpur, 23–24 March.

Hermansen, H.: "Twenty Five Years of Ekofisk Reservoir Management," paper SPE 38927 presented at the 1997 SPE Annual Technical Conference and Exhibition, San Antonio, Texas, 5–8 October.

Khataniar, S.K., Bora, A., and Borah, N.M.: "A Reservoir Management Case Study of the Nahorkatiya Oilfield," paper SPE 39568 presented at the 1998 SPE India Oil and Gas Conference and Exhibition, New Delhi, India, 17–19 February.

Langston, E.P., Shirer, J.A., and Nelson, D.E.: "Innovative Reservoir Management—Key to Highly Successful Jay/LEC Waterflood," *JPT* (May 1981) 783.

Satter, A. and Thakur, G.: *Integrated Petroleum Reservoir Management: A Team Approach*, PennWell, Tulsa (1994).

Schaaf, R.P. and King, G.R.: "Numbi Field: Adapting a Reservoir Management Strategy to Changing Reservoir Conditions," paper SPE 38925 presented at the 1997 SPE Annual Technical Conference and Exhibition, San Antonio, Texas, 5–8 October.

Selemat, S., Goh, S.T., and Lee, K.S.: "Seligi Depletion Management," paper SPE 57251 presented at the 1999 SPE Asia Pacific Improved Oil Recovery Conference, Kuala Lumpur, 25–26 October.

Tewari, R.D. and Raju, A.V.: "Development Strategy and Reservoir Management of a Multilayered Giant Offshore Carbonate Field," paper SPE 64461 presented at the 2000 SPE Asia Pacific Oil and Gas Conference and Exhibition, Brisbane, Australia, 16–18 October.

Thakur, G.C. and Satter, A.: *Integrated Waterflood Asset Management*, PennWell, Tulsa (1998).

SI Metric Conversion Factors

acre	×	4.046 856	E + 03	= m^2
°API		141.5/(131.5 + °API)		= g/cm^3
bbl	×	1.589 873	E − 01	= m^3
cp	×	1.0*	E − 03	= Pa·s
darcy	×	9.869 233	E − 01	= m^2
ft	×	3.048*	E − 01	= m
ft^3	×	2.831 685	E − 02	= m^3
°F		(°F − 32)/1.8		= °C
psi	×	6.894 757	E + 00	= kPa

*Conversion factor is exact.